Lecture Notes in Economics and Mathematical Systems

Managing Editors: M. Beckmann and W. Krelle

346

R. von Randow (Ed.)

Integer Programming and Related Areas

A Classified Bibliography
1984–1987
Compiled at the Institut für Ökonometrie
und Operations Research, University of Bonn

Springer-Verlag
Berlin Heidelberg New York London Paris Tokyo Hong Kong

Managing Editors

Prof. Dr. M. Beckmann
Brown University
Providence, RI 02912, USA

Prof. Dr. W. Krelle
Institut für Gesellschafts- und Wirtschaftswissenschaften
der Universität Bonn
Adenauerallee 24–42, D-5300 Bonn, FRG

Editor

Dr. R. von Randow
Institut für Ökonometrie und Operations Research
Universität Bonn
Nassestr. 2, D-5300 Bonn 1, FRG

ISBN 3-540-52687-0 Springer-Verlag Berlin Heidelberg New York
ISBN 0-387-52687-0 Springer-Verlag New York Berlin Heidelberg

Printing and binding: Druckhaus Beltz, Hemsbach/Bergstr.
2142/3140-543210 – Printed on acid-free paper

TABLE OF CONTENTS

PREFACE

The fields of integer programming and combinatorial optimization continue to be areas of great vitality, with an ever increasing number of publications and journals appearing. A classified bibliography thus continues to be necessary and useful today, even more so than it did when the project, of which this is the fifth volume, was started in 1970 in the Institut für Ökonometrie und Operations Research of the University of Bonn. The pioneering first volume was compiled by Claus Kastning during the years 1970 - 1975 and appeared in 1976 as Volume 128 of the series Lecture Notes in Economics and Mathematical Systems published by the Springer Verlag. Work on the project was continued by Dirk Hausmann, Reinhardt Euler, and Rabe von Randow, and resulted in the publication of the second, third, and fourth volumes in 1978, 1982, and 1985 (Volumes 160, 197, and 243 of the above series).

The present book constitutes the fifth volume of the bibliography and covers the period from autumn 1984 to the end of 1987. It contains 5864 new publications by 4480 authors and was compiled by Rabe von Randow. Its form is practically identical to that of the first four volumes, some additions having been made to the subject list.

With the advent of large on-line systems and easy universal access, a number of new, very up-to-date classified bibliographies have become available. It is thus that we have decided - not without regret - that the integer programming bibliography in its present form should not be continued, and that this will be the last volume.

Finally I would again like to thank the programmers of the Institute for their extensive programming effort.

Bonn, November 1989 Bernhard Korte

INTRODUCTORY REMARKS

This volume continues and concludes the series "Integer Programming and Related Areas: A Classified Bibliography", published as Lecture Notes in Economics and Mathematical Systems Nos. 128, 160, 197, and 243. It contains the publications appearing in the period from autumn 1984 to the end of 1987, as well as any earlier publications not included in the previous volumes.

The following topics are covered:
- theory and methods of general integer programming;
- combinatorial and graph theoretical optimization problems related to integer programming;
- applications of integer programming.

A detailed list of the subjects included is given at the end of this introduction.

PART 1: This part comprises the alphabetical bibliography which lists all publications alphabetically by the first author and chronologically for each author, and includes the full bibliographical data plus a reference code which reflects the author's name and the year of publication. The purpose of the code is to enable the user to identify publications referenced in Parts 2 and 3.

Bibliographical data are given as follows:

a) Textbook:

ref. code	Author(s)
	Title. Where published: Publishers. (Year).

b) Article in a journal:

ref. code	Author(s)
	Title. Journal, vol. no., pages. (Year).

c) Article in a proceedings volume:

ref. code	Author(s) Title. In: Editor(s): title of the proceedings volume, pages. Where published: Publishers. (Year).

d) Paper:

ref. code	Author(s) Title. Classification of the paper. Where published: Publishers, no. of pages. (Year).

Publications in English, French or German appear under their original title. Titles of publications in other languages have been translated into English and the original language indicated in brackets.

PART 2: In this part the publications have been classified according to 50 subjects; within each subject the ordering is again alphabetical by the first author.
The subject list following these remarks contains 50 main subject headings (in capitals), and for 99 further topics the corresponding subject headings have been indicated. The publications appear under all relevant headings; the references are restricted to name(s) of author(s), title, and reference code, with which the full bibliographical data can be retrieved from Part 1.

PART 3: The purpose of this part is to enable the user to locate publications for which he only knows a coauthor's name. It consists of an alphabetical catalogue of all the authors appearing in Part 1, and lists for each author the reference codes of all his publications.

SUBJECT LIST

```
PPPPPPPPPPP    AAAAAAAAAA   RRRRRRRRRRR   TTTTTTTTTTTT                    11
PPPPPPPPPPPP  AAAAAAAAAAAA  RRRRRRRRRRRR  TTTTTTTTTTTT                   111
PP        PP  AA        AA  RR        RR       TT                       1111
PP        PP  AA        AA  RR        RR       TT                         11
PP        PP  AA        AA  RR        RR       TT                         11
PPPPPPPPPPPP  AAAAAAAAAAAA  RRRRRRRRRRRR       TT                         11
PPPPPPPPPPP   AAAAAAAAAAAA  RRRRRRRRRRR        TT                         11
PP            AA        AA  RR    RR           TT                         11
PP            AA        AA  RR     RR          TT                         11
PP            AA        AA  RR      RR         TT                         11
PP            AA        AA  RR       RR        TT                         11
PP            AA        AA  RR        RR       TT                         11
```

ALPHABETICAL BIBLIOGRAPHY

AART 1987-1 Aarts, E. H. L.; Laarhoven, P. J. M. van
A pedestrian review of the theory and application of the simulated annealing algorithm. In: J. L. van Hemmen, I. Morgenstern (Eds.): Heidelberg Colloquium on Glassy Dynamics, 288-306. Lecture Notes in Physics 275, Berlin: Springer. (1987).

ABAI 1987-1 Abaiyldanov, K. N.; Astakhov, N. D.
Problem of determining the optimal order of priority in setting up production centers. Izvestiya Akademii Nauk SSSR, Tekhnicheskaya Kibernetika (Moscow), No. 6, 197-200; translated as Soviet J. Comput. Systems Sci. 25, No. 3, 163-167. (1987).

ABDA 1987-1 Abdallaoui, G.
Maintainability of a grade structure as a transportation problem. Journal of the Operational Research Society 38, 367-368. (1987).

ABDE 1984-1 Abdel-Kader, I.
Path-number of digraphs and graphs. Cahiers du Centre d'Etudes de Recherche Operationelle 26, 263-276. (1984).

ABDE 1986-1 Abdel-Kader, I.
Cheminements remarquables dans les graphes antisymetriques. Bulletin Mathematique de la Societe des Sciences Mathematique de la Republique Socialiste de Roumanie, Nouvelle Serie 30, 223-230. (1986).

ABDEA 1985-1 Abdel-Malek, L. L.
Optimum positioning of moving service facility. Computers & Operations Research 12, 437-444. (1985).

ABDU 1983-1 Abdullaev, A. A.; Rzaev, T. G.
Optimization of operative planning and scheduling in integrated ACS for enterprises (Russian). Elm, Baku. (1983).

ABDUA 1985-1 Abdullaev, A. K.; Kiselev, V. G.
A method for solving the Benders integer subproblem in some problems of plant location (Russian). Reports in Applied Mathematics, 21 p. Acad. Nauk SSSR, Vychil. Tsentr (Moscow). (1985).

ABEL 1986-1 Abelis, E.; Pavlovskii, Y. A.; Spilbergs, A.; Yubase, V.
The use of interactive computers for solving scheduling problems (Russian). Riga: Latv. Gos. Univ. (1986).

ABUO 1986-1 Abuov, K. A.; Cherenin, V. P.
Synthesis of one-product two-stage production-transport networks. U.S.S.R. Computational Mathematics and Mathematical Physics 26, No.4, 129-137; translation from Zhurnal Vychislitelnoi Matematiki i Matematicheskoi Fiziki 26 (1986), No.8, 1160-1174. (1986).

ACHA 1983-1 Achary, K. K.; Tikekar, V. G.
On trade-off solution pairs in a special type of transportation problem. Proceedings, Indian Academy of Sciences, Mathematical Sciences 92, 1-6. (1983).

ACHAA 1983-1 Acharya, B. D.
A characterization of consistent marked graphs. National Academy Science Letters (Allahabad) 6, 433-440. (1983).

ACHAA 1983-2 Acharya, B. D.
On d-sequential graphs. Journal of Mathematical and Physical Sciences 17, 21-35. (1983).

ACHAA 1984-1 Acharya, B. D.
Some further properties of consistent marked graphs. Indian Journal of Pure and Applied Mathematics (New Delhi) 15, 837-842. (1984).

ACHAA 1986-1 Acharya, B. D.
Connectedness and acyclicity of the transposed square of a graph. Indian Journal on Pure and Applied Mathematics 17, 756-759. (1986).

ACHAB 1983-1 Acharya, B. D.; Acharya, B. R.
The contraction polynomial of a graph. National Academy Science Letters 6, 93-95. (1983).

ACHAC 1987-1 Acharya, B. D.; Sampathkumar, E.
Graphoidal covers and graphoidal covering number of a graph. Indian Journal of Pure and Applied Mathematics 18, 882-890. (1987).

ACKE 1984-1 Acketa, D. M.
A construction of non-simple matroids on at most 8 elements. Journal of Combinatorics, Information and System Sciences 9, 121-132. (1984).

ACKE 1984-2 Acketa, D. M.
On the essential chains and squares. In: A. Hajnal; L. Lovasz; V. T. Sos (Eds.): Finite and Infinite Sets, Vol. 1, 25-34. Coll. Math. Soc. J. Bolyai 37, Amsterdam: North-Holland. (1984).

ACKE 1985-1 Acketa, D. M.
Some results on "small" matroids. In: L. Lovasz, A. Recski (Eds.): Matroid Theory, 15-24. Coll. Math. Soc. J. Bolyai Vol. 40. Amsterdam: North-Holland. (1985).

ACKE 1986-1 Acketa, D. M.
A construction of all the non-isomorphic non-simple matroids on 9 elements. Univerzitet u Novom Sadu, Zbornik Radova, Pirodno-Matematichkog Fakulteta, Serija za Matematiku 16, No.2, 269-290. (1986).

ACKE 1986-2 Acketa, D. M.
On the number of graphic representations of graphic matroids. In: R. Tosic; D. Acketa; V. Petrovic (Eds.): Graph theory, Proc. 6th Yugoslav Seminar on Graph Theory, Dubrovnik, Yugoslavia, April 1985, 1-26. University of Novi Sad. (1986).

ADAM 1984-1 Adam, A.
On certain partitions of finite directed graphs and of finite automata. Acta Cybernetica (Szeged) 6, 331-346. (1984).

ADAM 1984-2 Adam, A.
On the laxity of rooted trees with vertices of degree three. Publicationes Mathematicae (Debrecen) 31, 197-208. (1984).

ADAMA 1987-1 Adam, N. R.; Bertrand, J. W. M.; Surkis, J.
Priority assignment procedures in multi-level assembly job shops. IIE Transactions 19, 317-328. (1987).

ADAMB 1986-1 Adams, J.; Balas, E.
The shifting bottleneck procedure for job shop scheduling. Management Science Research Report No. MSRR-525. Pittsburgh: Carnegie-Mellon Univ., Graduate School of Industrial Administration, 30p. (1986).

ADAMC 1985-1 Adams, W. P.; Sherali, H. D.
Mixed-integer bilinear programming problems. Technical Report, 475. Clemson: Clemson Univ., Dept. of Mathematical Sciences, 41 p. (1985).

ADAMC 1986-1 Adams, W. P.; Sherali, H. D.
A tight linearization and an algorithm for zero-one quadratic programming problems. Management Science 32, 1274-1290. (1986).

ADIR 1987-1 Adiri, I.; Aizikowitz, N.; Lev, V.
Reversible-shop scheduling. European Journal of Operational Research 29, 91-97. (1987).

ADIRA 1984-1 Adiri, I.; Amit, N.
Openshop and flowshop scheduling to minimize sum of completion times. Computers & Operations Research 11, 275-284. (1984).

ADIRB 1987-1 Adiri, I.; Yehudai, Z.
Scheduling on machines with variable service rates. Computers and Operations Research 14, 289-298. (1987).

ADRA 1983-1 Adrabinski, A.; Syslo, M. M.
Computational experiments with some approximation algorithms for the travelling salesman problem. Zastosowania Matematyki (Warsaw) 18, 91-95. (1983).

AFEN 1985-1 Afentakis, P.
Simultaneous lot sizing and sequencing for multistage production systems. IIE Transactions 17, 327-331. (1985).

AFEN 1987-1 Afentakis, P.
A parallel heuristic algorithm for lot sizing in multistage production systems. IIE Transactions 19, 34-42. (1987).

AFENA 1986-1 Afentakis, P.; Gavish, B.
Optimal lot-sizing algorithms for complex product structures. Operations Research 34, 237-249. (1986).

AFON 1986-1 Afonin, Y. S.
Blocked branch-and-bound method. Automation and Remote Control 47, 1107-1116; translation from Avtomatika i Telemekhanika (1986), No.8, 98-108. (1986).

AFRA 1985-1 Afrati, F.; Papadimitriou, C. H.; Papageorgiou, G.
The complexity of cubical graphs. Information and Control 66, 53-60. (1985).

AGAR 1985-1 Agarwal, S.; Mittal, A. K.; Sharma, P.
A decomposition algorithm for linear relaxation of the weighted r-covering problem. Mathematical Programming 31, 67-77. (1985).

AGARA 1985-1 Agarwal, Y. K.
Set partitioning approach to vehicle routing. Working Paper. Cleveland:Leaseway Transportation Corp., 21 p. (1985).

AGARA 1985-2 Agarwal, Y. K.
Vehicle routing with limited fleet and common carrier option. Research Paper. Cleveland: Leaseway Transportation Corp. 16 p. (1985).

AGEE 1983-1 Ageev, A. A.
On the complexity of the minimization problems for polynomials of Boolean functions (Russian). Upravlyaemye Sistemy (Novosibirsk) 23, 3-11. (1983).

AGEE 1984-1 Ageev, A. A.
On the minimization of quadratic polynomials of Boolean functions (Russian). Upravlyaemye Sistemy (Novosibirsk) 25, 3-16. (1984).

AGGA 1987-1 Aggarwal, S. P.; Goel, V. P.
Deterministic lot size model with nth power time proportional demand. Indian Journal of Pure and Applied Mathematics 18, 1-5. (1987).

AGGAA 1985-1 Aggarwal, V.
A Lagrangean-relaxation method for the constrained assignment problem. Computers & Operations Research 12, 97-106. (1985).

AGGAB 1985-1 Aggarwal, V.; Suresh, R.
A shortest-route approach for optimum aggregation of data with an application to income data. Computers & Operations Research 12, 227-236. (1985).

AGGAC 1986-1 Aggarwal, V.; Tikekar, V. G.; Hsu, L.-F.
Bottleneck assignment problems under categorization. Computers and Operations Research 13, 11-26. (1986).

AGGAD 1984-1 Aggarwal, V. B.; Bagga, P. C.; Kalra, K. R.
Optimal algorithms for job-blocks in 2xn sequencing problem with transportation times – a computational comparison. Journal of Information and Optimization Sciences (Delhi) 5, 133-139. (1984).

AGRA 1985-1 Agrawal, P. K.
The related activity concept in assembly line balancing. International Journal of Production Research 23, 403-422. (1985).

AGRAA 1984-1 Agrawal, S. C.; Chand, M.
On integer solution to interval linear programming problems by a branch and bound technique. Acta Ciencia Indica, Mathematics (Meerut) 10, 269-274. (1984).

AGRAA 1985-1 Agrawal, S. C.; Chand, M.
An alternative method on integer solutions to complementary programming problems with linear fractional objective function by a branch and bound technique. Acta Ciencia Indica, Mathematics (Meerut) 11, No. 3, 213-221. (1985).

AGRAB 1984-1 Agrawala, A. K.; Coffman, E. G. Jr.; Garey, M. R.; Tripathi, S. K.
A stochastic optimization algorithm minimizing expected flow times on uniform processors. IEEE Transactions on Computers C-33, 351-356. (1984).

AHAR 1985-1 Aharoni, R.
Fractional matchings and covers in infinite hypergraphs. Combinatorica 5, 181-184. (1985).

AHAR 1985-2 Aharoni, R.
Matchings in n-partite n-graphs. Graphs and Combinatorics 1, 303-304. (1985).

AHARA 1985-1 Aharoni, R.; Erdös, P.; Linial, N.
Dual integer linear programs and the relationship between their optima. In: Proc. Seventeenth Annual ACM Symp. on Theory of Computing, Providence 1985, 113-114. ACM. (1985).

AHARB 1985-1 Aharoni, R.; Hartman, I. B.-A.; Hoffman, A. J.
Path partitions and packs of acyclic digraphs. Pacific Journal of Mathematics 118, 249-259. (1985).

AHLF 1985-1 Ahlfield, D. P.; Dembo, R. S.; Mulvey, J. M.; Zenios, S. A.
Nonlinear programming on generalized networks. Report EES-85-7. Princeton: Princton Univ., Dept. of Civil Engineering, 32p. (1985).

AHMA 1986-1 Ahmad, S. H.
GERT models in an educational institution. International Journal of Systems Science 17, 187-192. (1986).

AHME 1982-1 Ahmed, M.
Modeles probabilistes pour choisir un itineraire dans un reseau de trafic. Annales Scientifiques de l'Universite de Clermont-Ferrand II 71, Mathematiques 20, 1-10. (1982).

AHUJ 1986-1 Ahuja, R. K.
Algorithms for the minimax transportation problem. Naval Research Logistics Quarterly 33, 725-740. (1986).

AHUJA 1987-1 Ahuja, R. K.; Murty, V. V. S.
Exact und heuristic algorithms for the optimum communication spanning tree problem. Transportation Science 21, 163-170. (1987).

AHUJB 1987-1 Ahuja, R. K.; Orlin, J. B.
A fast and simple algorithm for the maximum flow problem. Sloan W.P. 1905-87. Cambridge: Mass. Inst. of Technology, 33 p. (1987).

AICA 1986-1 Aicardi, M.; Casalino, G.; Davoli, F.; Minciardi, R.; Zoppoli, R.
A decentralized closed-loop solution to the routing problem in networks. In: Control Applications of Nonlinear Programming and Optimization (Capri 1985), 9-17. Laxenburg: IFAC. (1986).

AICA 1986-2 Aicardi, M.; Casalino, G.; Davoli, F.; Minciardi, R.; Zoppoli, R.
Dynamic routing policies in communication networks: the finite buffer case. In: C. I. Byrnes, A. Lindquist (Ed.): Computational and Combinatorial Methods in Systems Theory, 357-370. Amsterdam: North Holland. (1986).

AIEL 1985-1 Aiello, A.; Burattini, E.; Massarotti, A.; Ventriglia, F.
Heuristic evaluation techniques for bin packing approximation algorithms. Calcolo 22, 319-334. (1985).

AIKE 1985-1 Aikens, C. H.
Facility location models for distribution planning. European Journal of Operational Research 22, 263-279. (1985).

AINO 1985-1 Ainouche, A.; Christofides, N.
Conditions for the existence of Hamiltonian circuits in graphs based on vertex degrees. Journal of the London Mathematical Society, II. Series 32, 385-391. (1985).

AINO 1987-1 Ainouche, A.; Christofides, N.
Semi-independence number of a graph and the existence of hamiltonian circuits. Discrete Applied Mathematics 17, 213-222. (1987).

AIRA 1984-1 Airapetyan, L. R.; Eksuzyan, S. A.
A necessary and sufficient condition for existence of a correct schedule for a two-processor system (Russian). Applied Mathematics (Erevan) No. 3, 25-31. (1984).

AITT 1985-1 Aittoniemi, L.; Oehlandt, K.
A note on the Martello-Toth algorithm for the one-dimensional knapsack problems. European Journal of Operations Research 20, 117. (1985).

AJTA 1984-1 Ajtai, M.; Komlos, J.; Tusnady, G.
On optimal matchings. Combinatorica 4, 259-264. (1984).

AJYA 1984-1 Ajyashi, M.
An optimality criterion in a dynamic problem on minimal cost flow with additional constraints (Russian). Izvestiya Akademii Nauk BSSR, Seriya Fiziko-Matematicheskikh Nauk (Minsk) No. 2, 39-44. (1984).

AKEL 1986-1 Akella, R.; Kumar, P. R.
Optimal scheduling of a flexible manufacturing system: a stochastic control problem for a system with jump Markov disturbances. In: C. I. Byrnes, A. Lindquist (Ed.): Computational and Combinatorial Methods in Systems Theory, 131-134. Amsterdam: North Holland. (1986).

AKIN 1985-1 Akinc, U.
Multi-activity facility design and location problems. Management Science 31, 275-283. (1985).

AKINA 1986-1 Akinc, U.; Roodman, G. M.
A new approach to aggregate production planning. IIE Transactions 18, 88-94. (1986).

AKIY 1984-1 Akiyama, J.; Ando, K.; Harary, F.
A graph and its complement with specified properties. VIII. Interval graphs. Mathematica Japonica (Osaka) 29, 659-670. (1984).

AKIYA 1983-1 Akiyama, J.; Harary, F.
The regulation number of a graph. Publications de l'Institut Mathematique, Nouvelle Serie (Beograd) 34 (48), 3-5. (1983).

AKIYB 1985-1 Akiyama, J.; Kano, M.
Path factors of a graph. In: F. Harary, J. P. Maybee (Eds.): Graphs and Applications, 1-22. New York: Wiley. (1985).

AKIYB 1985-2 Akiyama, J.; Kano, M.
1-factors of triangle graphs. In: Number theory and combinatorics (Japan 1984), 21-35. Singapore: World Science Publishing. (1985).

AKIYC 1985-1 Akiyama, J.; Ruiz, M.-J.
Intersection graphs and their applications. Matimyas Matematika (Quezon City, Phillipinas) 9, No. 1, 8-33. (1985).

AKKA 1985-1 Akka, D. G.
Graph equations for line graphs, entire graphs and semi-entire graphs. Journal of Mathematical and Physical Sciences (Madras) 19, 511-518. (1985).

AKL 1983-1 Akl, S. G.
A note on Euclidean matchings, triangulations and spanning trees. Journal of Combinatorics, Information & System Sciences (Delhi) 8, 169-174. (1983).

AKL 1986-1 Akl, S. G.
An adaptive and cost-optimal parallel algorithm for minimum spanning trees. Computing 36, 271-278. (1986).

AKMA 1987-1 Akman, V.
Unobstructed shortest paths in polyhedral environments. Berlin, Heidelberg: Springer. (1987). (Lecture Notes in Computer Science, 251)

AKOP 1984-1 Akopyan, G. C.
Inclusion of a subset of edges of a hypergraph in some terminal c-covering (Russian). Applied Mathematics (Erevan) No. 3, 32-35. (1984).

AKSE 1984-1 Aksenov, V. A.
The degree of perfection of a graph (Russian). Metody Diskretnogo Analiza (Novosibirsk) No. 41, 3-11. (1984).

AKUL 1983-1 Akulich, F. E.
A problem on maximal flow on bottlenecks (Russian). Izvestiya Akademii Nauk BSSR, Seriya Fisziko-Matematicheskikh Nauk (Minsk), No. 3, 17-20. (1983).

AL-S 1986-1 Al-Salti, B.
Minimization of a concave function on a polymatroid (Russian). Vestnik Belorusskogo Gosudarstvennogo Universiteta im. V.I. Lenina (Minsk), Seriya I, No.2, 53-55. (1986).

ALAM 1983-1 Alameddine, A. F.
Centers of maximal planar graphs with two vertices of degree three. Journal of Combinatorics, Information & System Sciences (Delhi) 8, 90-96. (1983).

ALAM 1987-1 Alameddine, A. F.
On a characterization of a class of outerplanar graphs. Arabian Journal for Science and Engineering 13, No.1, 97-102. (1987).

ALAV 1986-1 Alavi, Y.; Chartrand, G.; Lesniak, L.; Oellermann, O. R.
Bipartite regulation numbers. Discrete Mathematics 62, 113-118. (1986).

ALBE 1987-1 Alberton, I. B.
Polynomial algorithms for problems in the theory of scheduling on "narrow" networks. Izvestiya Akademii Nauk SSSR, Tekhnicheskaya Kibernetika (Moscow), No. 6, 161-171; translated as Soviet J. Comput. Systems Sci. 25, No. 3, 82-92. (1987).

ALBEA 1987-1 Albertson, M. O.
Finding Hamiltonian cycles in ore graphs. Congressus Numerantium 58, 25-28. (1987).

ALBEA 1987-2 Albertson, M. O.
Generalized colorings. In: D. P. Johnson, T. Nishizeki, A. Nozaki, H. P. Wilf (Eds.): Discrete Algorithms and Complexity - Proc. Japan-U.S. Joint Seminar, Kyoto 1986, 35-49. Perspectives in Computing Vol. 15, Boston: Academic Press. (1987).

ALBEB 1983-1 Albertson, M. O.; Berman, D. M.
Cliques, colorings and locally perfect graphs. Congressus Numerantium 39, 69-74. (1983).

ALBEC 1984-1 Albertson, M. O.; Collins, K. L.
Duality and perfection for edges in cliques. Journal of Combinatorial Theory, Series B 36, 298-309. (1984).

ALBR 1984-1 Albrecht, A.; Jung, H.; Bauernöppel, F.; Sammler, O.
Einführende Kapitel der Kompliziertheitstheorie. Seminarbericht 66. Humboldt Univ. (Berlin), Sektion Mathematik. (1984).

ALDR 1984-1 Aldred, R. E. L.
Paths through m vertices in 3-connected cubic graphs. Ars Combinatoria 17A, 85-93. (1984).

ALDRA 1987-1 Aldred, R. E. L.; Holton, D. A.
Cycles through five edges in 3-connected cubic graphs. Graphs and Combinatorics 3, 299-312. (1987).

ALDRB 1985-1 Aldred, R. E. L.; Holton, D. A.; Thomassen, C.
Cycles through four edges in 3-connected cubic graphs. Graphs and Combinatorics 1, 7-12. (1985).

ALDRC 1987-1 Aldred, R. E. L.; Sheng, B.; Holton, D. A.
Primitive graphs. Ars Combinatoria 23B, 183-194. (1987).

ALEK 1980-1 Alekseev, A. E.; Kogan, D. I.
Complexity of determination of the optimal allocation of tasks to persons with two-sided preferences (Russian). in: Combinatorial-Algebraic Methods in Applied Mathematics, 3-11. Gorki: Gor'kovskii Godsudarstvennyi Universitet. (1980).

ALEKA 1987-1 Alekseev, A. G.; Kezhaev, V. A.; Subbotin, A. I.
A special problem of integer programming (Russian). Izvestiya Akademii Nauk SSSR, Tekhnicheskaya Kibernetika (Moscow), No. 4, 16-20; translated as Soviet J. Comput. Systems Sci. 25, No. 1, 14-18. (1987).

ALEKB 1984-1 Alekseev, A. O.
The transportation problem with respect to the time criterion in case of limited amount of transport facilities (Russian). In: Mathematical methods of optimization and control in complex systems, Collect. sci. Works, Kalinin 1984, 60-65. (1984).

ALEKB 1985-1 Alekseev, A. O.
A multivariant travelling-salesman problem. U.S.S.R. Computational Mathematics and Mathematical Physics 25, No.2, 200-201; translation from Zhurnal Vychislitelnoi Matematiki i Matematicheskoi Fiziki 25 (1985), No.4, 631-633. (1985).

ALEKB 1985-2 Alekseev, A. O.
Multivariant transport problem by the time criterion. Soviet Journal of Computer and Systems Sciences 23, No.2, 146-147; translation from Izv. Akad. Nauk SSSR, Tekh. Kibern. (1984), No.6, 188-189. (1985).

ALEKC 1985-1 Alekseev, A. O.; Alekseev, O. G.; Anisimov, V. G.; Anisimov, E. G.
The use of duality to increase the effectiveness of the branch and bound method when solving the knapsack problem. U.S.S.R. Computational Mathematics and Mathematical Physics 24, No.6, 50-54; translation from Zhurnal Vychislitelnoi Matematiki i Matematicheskoi Fiziki 25 (1985), No.11, 1666-1673. (1985).

ALEKD 1987-1 Alekseev, O. G.
Complex application of discrete optimization methods (Russian). Moscow: Nauka. (1987).

ALEKE 1984-1 Alekseev, O. G.; Gregorjev, V. F.
Some algorithms of the set-covering problem and their experimental estimation with computer (Russian). Zhurnal Vychislitel'noi Matematiki i Matematicheskoi Fiziki (Moscow) 24, 1565-1570. (1984).

ALEKF 1982-1 Alekseev, V. E.
The effect of local constraints on the complexity of determination of the graph independence number (Russian). In: A. A. Markov (Ed.): Combinatorial Algebraic Methods in Applied Mathematics, 3-13. Gorkov. Gos. Univ. (Gorki). (1982).

ALEKG 1981-1 Alekseev, V. E.; Markov, A. A.; Noskov, V. V.
Optimization of independent and support sets in hypergraphs (Russian). In: A. A. Markov (Ed.): Combinatorial-algebraic Methods in Applied Mathematics, 3-25. Gorkij: Gorkov. Gosudarst. Univ. (1981).

ALEX 1982-1 Alexander, C.; Hopkins, G.; Staton, W.
Maximal bipartite subgraphs in triangle-free graphs. In: F. Hoffman, K. B. Reid, R. C. Mullin, R. G. Stanton (Eds.): Proc. Thirteenth Southeastern Conf. on Combinatorics, Graph Theory and Computing, Florida 1982, Congressus Numerantium Vol. 35, 33-46. Winnipeg, Canada: Utilitas Mathematica. (1982).

ALI 1984-1 Ali, A. I.
A greedy algorithm for the transshipment along a single road problem. Research Report, 500. Austin: Univ. of Texas, Center for Cybernetic Studies, 11 p. (1984).

ALI 1986-1 Ali, A. I.
An efficient algorithm for the transshipment along a single road problem. Naval Research Logistics Quarterly 33, 673-684. (1986).

ALI A 1986-1 Ali, A. I.; Kennington, J. L.
The asymmetric m-travelling salesmen problem: a duality based branch-and-bound algorithm. Discrete Applied Mathematics 13, 259-276. (1986).

ALI B 1986-1 Ali, A. I.; Kennington, J. L.; Shetty, B.
The linear equal flow problem. Working Paper 85/86-3-5. Austin: Univ. of Texas, Graduate School of Business, Dept. of General Business, 21p. (1986).

ALI C 1986-1 Ali, A. I.; Padman, R.; Thiagarajan, H.
Dual simplex-based reoptimization procedures for network problems. Working Paper 85/ 86-3-4. Austin: Univ. of Texas, Graduate School of Business, 28 p. (1986).

ALI C 1986-2 Ali, A. I.; Padman, R.; Thiagarajan, H.
Dual simplex-based reoptimization procedures for network problems. Working Paper 85/ 86-3-4. Austin: Univ. of Texas, Graduate School of Business, Dept. of General Business, 26p. (1986).

ALI C 1986-3 Ali, A. I.; Padman, R.; Thiagarajan, H.
The dual algorithm for network problems: a computational study. Working Paper 85/86-3-3. Austin: Univ. of Texas, Graduate School of Business, 29 p. (1986).

ALI D 1984-1 Ali, I.; Charnes, A.; Song, T.
Design and implementation of data structures for generalized networks. Research Report CCS 483. Austin: Univ. of Texas, Center for Cybernetic Studies, 31 p. (1984).

ALI D 1985-1 Ali, I.; Charnes, A.; Song, T.
Karmarkar's projective algorithm: a null space variant for multi-commodity generalized networks. Acta Mathematicae Applicatae Sinica, English Series 2, 168-190. (1985).

ALI E 1985-1 Ali, I.; Kennington, J. L.; Shetty, B.
The equal flow problem. Research Report CCS 511. Austin: Univ. of Texas, Center for Cybernetic Studies, 39 p. (1985).

ALIE 1983-1 Aliev, A. A.
A method of solution of the integer linear programming problem (Russian). Mathematical cybernetics and applied mathematics (Baku) No. 6, 3-9. (1983).

ALIEA 1986-1 Aliev, A. A.; Akhmedov, F. B.; Babaev, D. A.
Application package KNAPSACK-1 for approximate solution of the generalized problem of knapsack (Russian). Kibernetika (Kiev), No. 2, 105-108. (1986).

ALLE 1987-1 Allen, J. D.; Helgason, R. V.; Kennington, J. L.
The frequency assignment problem: a solution via nonlinear programming. Naval Research Logistics 34, 133-140. (1987).

ALLEA 1985-1 Allender, E. W.
On the number of cycles possible in digraphs with large girth. Discrete Applied Mathematics 10, 211-226. (1985).

ALLEB 1986-1 Alles, P.
Dimension of amalgamated graphs and trees. Czechoslovak Mathematical Journal 36(111), 393-416. (1986).

ALON 1985-1 Alon, N.
Asynchronous threshold networks. Graphs and Combinatorics 1, 305-310. (1985).

ALON 1986-1 Alon, N.
The number of polytopes, configurations and real matroids. Mathematika 33, 62-71. (1986).

ALON 1987-1 Alon, N.
Monochromatic directed walks in arc-colored directed graphs. Acta Mathematica Hungarica 49, 163-167. (1987).

ALONA 1985-1 Alon, N.; Egawa, Y.
Even edge colorings of a graph. Journal of Combinatorial Theory 38B, 93-94. (1985).

ALONB 1985-1 Alon, N.; Tarsi, M.
Covering multigraphs by simple circuits. SIAM Journal on Algebraic and Discrete Methods 6, 345-350. (1985).

ALSP 1985-1 Alspach, B.; Durnberger, E.; Parsons, T. D.
Hamilton cycles in metacirculant graphs with prime cardinality blocks. In: B. R. Alspach, C. D. Godsil (Eds.): Cycles in Graphs, Annals of Discrete Mathematics Vol. 27, 27-34. Amsterdam: North-Holland. (1985).

ALTI 1985-1 Altinkemer, K.; Gavish, B.
Heuristics with constant error guarantees for the topological design of local access tree networks. Working Paper Series No. QM8540. Rochester: Univ., Graduate School of Management, 14p. (1985).

ALTI 1987-1 Altinkemer, K.; Gavish, B.
Heuristics for unequal weight delivery problems with a fixed error guarantee. Operations Research Letters 6, 149-158. (1987).

ALVA 1985-1 Alvarez-Valdes, R.; Corberan, A.; Tamarit, J. M.
Polyhedral combinatorics and the traveling salesman problem. Application to the case of 103 Spanish cities (Spanish). Questiio 9, No. 3, 199-213. (1985).

ALVAA 1984-1 Alvarez-Valdes, R.; Tamarit, J. M.
Lower bounds for the problem of sequencing with resource restrictions (Spanish). Questiio 8, No. 4, 197-205. (1984).

AMAH 1985-1 Amahashi, A.
On factors with all degrees odd. Graphs and Combinatorics 1, 111-114. (1985).

AMAN 1987-1 Aman, A.; Balakrishnan, A.; Chandru, V.
Maintenance of optimal schedules. CC-87-31. West Lafayette: Purdue Univ. Inst. of Interdisciplinary Engineering Studies, 19 p. (1987).

AMAR 1986-1 Amar, A. D.; Gupta, J. N. D.
Simulated versus real life data in testing the efficiency of scheduling algorithms. IIE Transactions 18, 16-25. (1986).

AMARA 1983-1 Amar, D.
On the connectivity of some telecommunications networks. IEEE Transactions on Computers 32, 512-519. (1983).

AMARA 1986-1 Amar, D.
Partition of a bipartite Hamiltonian graph into two cycles. Discrete Mathematics 58, 1-10. (1986).

AMARB 1985-1 Amar, D.; Fournier, I.; Germa, A.
Some conditions for digraphs to be Hamiltonian. In: M. Rosenfeld, J. Zaks (Eds.): Convexity and Graph Theory, Annals of Discrete Mathematics Vol. 20, 37-42. Amsterdam: North-Holland. (1985).

AMARC 1985-1 Amar, D.; Fournier, I.; Germa, A.; Häggkvist, R.
Covering of vertices of a simple graph with given connectivity and stability number. In: M. Rosenfeld, J. Zaks (Eds.): Convexity and Graph Theory. Annals of Discrete Mathematics Vol. 20, 43-46. Amsterdam: North-Holland. (1985).

ANAS 1985-1 Anastasyan, V. G.
One combinatorial line location problem (Russian). Kibernetika (Kiev) No. 3, 114-115. (1985).

ANCO 1986-1 Ancona, M.; Floriani, L. de; Deogun, J. S.
Path problems in structured graphs. Computer Journal 29, 553-563. (1986).

ANDE 1985-1 Anderson, R.
A parallel algorithm for the maximal path problem. In: Proc. Seventeenth Annual ACM Symp. on Theory of Computing, Providence 1985, 33-37. ACM. (1985).

ANDE 1987-1 Anderson, R.
A parallel algorithm for the maximal path problem. Combinatorica 7, 315-326. (1987).

ANDEA 1987-1 Anderson, R.; Mayr, E.
Parallelism and the maximal path problem. Information Processing Letters 24, 121-126. (1987).

ANDO 1987-1 Ando, K.; Enomoto, H.; Saito, A.
Contractible edges in 3-connected graphs. Journal of Combinatorial Theory B 42, 87-93. (1987).

ANDOA 1985-1 Ando, K.; Mizuno, H.
Biconnectedness of graphs. In: J. Akiyama; Y. Ito; P. Kanemitsu; T. Kano; T. Mitsui; I. Shiokawa (Eds): Number theory and combinatorics, Japan 1984, Proc. Conf. Tokyo, Okayama, Kyoto, 37-41. Singapore: World Scientific Publ. Co. (1985).

ANDOB 1987-1 Ando, K.; Usami, Y.
Critically (k,k)-connected graphs. Discrete Mathematics 66, 15-20. (1987).

ANDOC 1987-1 Andonov, R. A.
Finding facets of the knapsack polytope which cut off a given point. Comptes Rendus de l'Academie Bulgare des Sciences 40, No.8, 41-44. (1987).

ANDR 1983-1 Andrasfai, B.
Graph theory. Network flows. Matrices. (Hungarian). 263p. Budapest: Akademiai Kiado. (1983).

ANDRA 1985-1 Andreae, T.
On an extremal problem concerning the interval number of a graph. Preprint 183. Berlin: Freie Univ. Berlin, 16 p. (1985).

ANDRA 1985-2 Andreae, T.
On superperfect noncomparability graphs. Preprint 186. Berlin: Freie Univ. Berlin, 16 p. (1985).

ANDRA 1985-3 Andreae, T.
On the interval number of a triangulated graph. Preprint 196. Berlin: Freie Univ. Berlin, Fachbereich Mathematik, 13 p. (1985).

ANDRA 1985-4 Andreae, T.
On the unit interval number of a graph. Preprint 199. Berlin: Freie Univ. Berlin, Fachbereich Mathematik, 26 p. (1985).

ANDRB 1982-1 Andreatta, A.; Liburdi, A.; Medagliani, G.
On maximum genus of cubic graphs (Italian). Rendiconti Scienze Matematiche e Applicazioni (Milano) A 116(1985), 351-362. (1982).

ANDRC 1987-1 Andreatta, G.
Shortest path models in stochastic networks. In: G. Andreatta, F. Mason, P. Serafini (Eds.): Advanced School on Stochastics in Combinatorial Optimization (CISM Udine 1986), 178-186. Singapore: World Scientific. (1987).

ANDRD 1986-1 Andreatta, G.; Mason, F.
Review of recent results about the k-centrum of a tree. In: J. P. Osleeb, P. J. Ratick (Eds.): Locational Decisions: Methodology and Applications, 129-136. Annals of Operations Research 6, Basel: Baltzer. (1986).

ANDRE 1985-1 Andreatta, G.; Romanin-Jacur, G.
An integer quadratic allocation problem. Methods of Operations Research 55, 1-12. (1985).

ANDRF 1985-1 Andrews, J. A.; Jacobson, M. S.
On a generalization of chromatic number. Congressus Numerantium 47, 33-48. (1985).

ANDRF 1987-1 Andrews, J. A.; Jacobson, M. S.
On a generalization of chromatic number and two kinds of Ramsey numbers. Ars Combinatoria 23, 97-102. (1987).

ANDRG 1987-1 Androsov, S. M.; Gluzman, B. Y.; Oslon, V. A.
Program realization of shortest path algorithms in transportation MIS. Automation and Remote Control 48, 1246-1250; translation from Avtomatika i Telemekhanika (1987), No. 9, 158-165. (1987).

ANEJ 1985-1 Aneja, Y. P.; Chandrasekaran, R.; Nair, K. P. K.
Classes of linear programs with integral optimal solutions. Mathematical Programming Study 24, 225-237. (1985).

ANEJA 1985-1 Aneja, Y. P.; Nair, K. P. K.
Network flows with age dependent decay rates. European Journal of Operational Research 21, 77-81. (1985).

ANGE 1986-1 Anger, F. D.; Lee, C.-Y.; Martin-Wega, L. A.
Single machine scheduling with tight windows. Research Report, 86-16. Gainesville: Univ. of Florida, Industrial & Systems Engineering Dept., 11 p. (1986).

ANGR 1982-1 Angrisani, M.
Discrete minimax problems: a unified approach using Motzkin's alternative theorem (Italian). Rivista di Matematica per le Scienze Economiche e Sociali 5, 97-113. (1982).

ANGR 1983-1 Angrisani, M.
On a discrete minimax problem (Italian). Rivista di Matematica per le Scienze Economiche e Sociali 6, No. 1-2, 3-17. (1983).

ANIS 1986-1 Anisimov, A. V.
Local algorithm for the one-to-all shortest-path problem (Russian). Kibernetika (Kiev), No. 3, 57-60. (1986).

ANKL 1986-1 Anklesaria, K. P.; Drezner, Z.
A multivariate approach to estimating the completion time for PERT networks. Journal of the Operational Research Society 37, 811-816. (1986).

ANST 1985-1 Anstee, R. P.
An algorithmic proof of Tutte's f-factor theorem. Journal of Algorithms 6, 112-131. (1985).

ANST 1986-1 Anstee, R. P.
Invariant sets of arcs in network flow problems. Discrete Applied Mathematics 13, 1-8. (1986).

ANST 1987-1 Anstee, R. P.
A polynomial algorithm for b-matchings: an alternative approach. Information Processing Letters 24, 153-158. (1987).

ANSTA 1984-1 Anstee, R. P.; Farber, M.
Characterizations of totally balanced matrices. Journal of Algorithms 5, 215-230. (1984).

ANSTB 1985-1 Anstee, R. P.; Murty, U. S. R.
Matrices with forbidden subconfigurations. Discrete Mathematics 54, 113-116. (1985).

ANTA 1983-1 Antamoshkin, A.
On the optimal algorithms for the optimization of functionals with Boolean values (Russian). in: Transactions of the ninth Prague conference on information theory, statistical decision functions, random processes (Prague 1982) A, 137-141. Reidel, Dordrecht-Boston. (1983).

ANTAA 1984-1 Antamoshkin, A. N.; Zajtseva, T. A.; Korsakova, O. N.
Formalization of optimization problems in planning agricultural-industrial combines (English). Soviet Automatic Control 17, No. 3, 75-77; translated from: Avtomatika 1984, No. 3, 77-79. (1984).

ANTH 1987-1 Anthonisse, J. M.; Lenstra, J. K.; Savelsbergh, M. W. P.
Functional description of CAR, an interactive system for computer aided routing. Report OS-R8716. Amsterdam: Centre for Mathematics and Computer Science, 15 p. (1987).

ANTHA 1988-1 Anthonisse, K.; Lageweg, B.
Aircraft-stand allocation at Schiphol airport: a decision support system. In: H. Schellhaas, P. van Beek, H. Isermann, R. Schmidt, M. Zijlstra (Eds.): Operations Research Proceedings 1987 (DGOR/NSOR), 257-258. Berlin: Springer. (1988).

ANTHA 1988-2 Anthonisse, K.; Lageweg, B.
Aircraft-stand allocation at Schiphol airport: an optimization procedure. In: H. Schellhaas, P. van Beek, H. Isermann, R. Schmidt, M. Zijlstra (Eds.): Operations Research Proceedings 1987 (DGOR/NSOR), 258-259. Berlin: Springer. (1988).

ANTO 1983-1 Antonelli, S.; Conte, G.; Delli Guanti, G.
An optimal partitioning algorithm for a new class of graphs (Italian). Calcolo (Pisa) 20, No. 1, 85-100. (1983).

ANTOA 1983-1 Antonucci, S.
Simple and generalized colourings of cliques without Hamiltonian cycles and of regular graphs (Italian). Rivista di Matematica della Universita di Parma, Serie IV, 9, 447-456. (1983).

ANTOA 1984-1 Antonucci, S.
Multicolorings of regular graphs (French). Rivista di Matematica della Universita di Parma, Serie IV 10, 131-136. (1984).

ANTOA 1985-1 Antonucci, S.
The graphs T(m) and L(m) and their chromatic numbers (Italian). Bolletino della Unione Matematica Italiana, VI Serie D, Algebra e Geometria 4, No.1, 137-143. (1985).

ANTOA 1986-1 Antonucci, S.
Observations on the generalized characteristics and the chromatic number of a graph (Italian). Rivista di Matematica della Universita di Parma, Serie IV 12, 101-109. (1986).

APOL 1984-1 Apolloni, B.; Pezzella, F.
Confidence intervals in the solution of stochastic integer linear programming problems. Annals of Operations Research, Vol. 1, 67-78. Basel: Baltzer. (1984).

APOS 1986-1 Apostolico, A.; Hambrusch, S. E.
New cliques and independent set algorithms for circle graphs. CC-86-13. West Lafayette: Purdue Univ. Inst. of Interdisciplinary Engineering Studies, 39 p. (1986).

APOS 1987-1 Apostolico, A.; Hambrusch, S. E.
Finding maximum cliques on circular-arc graphs. Information Processing Letters 26, 209-216. (1987).

ARAI 1986-1 Arai, M.; Yamashita, M.; Hirata, T.; Ibaraki, T.; Honda, N.
Lower bounds on time complexity for some subgraph detection problems (Japanese). Systems Computation Japan 17, No. 6, 95-102. (1986).

ARAN 1985-1 Arani, T.; Karwan, M. H.; Lotfi, V.
A Lagrangian relaxation approach for solving an exam scheduling problem. Working Paper. Buffalo: State Univ. of New York, Dept. of Ind. Eng., 28 p. (1985).

ARANA 1983-1 Arany, I.
Another method for finding pseudo-peripheral nodes. Annales Universitatis Scientiarum Budapestinensis de Rolando Eötvös Nominatae, Sectio Computatorica (Budapest) 4, 39-49. (1983).

ARANA 1985-1 Arany, I.
An efficient algorithm for finding peripheral nodes. In: L. Lovasz, E. Szemeredi (Eds.): Theory of Algorithms, Colloquia Mathematica Societatis Janos Bolyai Vol. 44, 27-36. Amsterdam, Oxford, New York: North-Holland. (1985).

ARANA 1986-1 Arany, I.
An algorithm for getting a minimum cut-set of a graph. In: A. Prekopa, J. Szeleszan, B. Strazicky (Eds.): System Modelling and Optimization - Proc. 12th IFIP Conf., Budapest 1985, 17-24. Lecture Notes in Control and Information Sciences 84, Berlin: Springer. (1986).

ARAO 1984-1 Araoz, J.
Packing problems in semigroup programming. In: R. E. Burkard, R. A. Cuninghame-Green, U. Zimmermann (Eds.): Algebraic and Combinatorial Methods in Operations Research, Annals of Discrete Mathematics Vol. 19, 1-22. Amsterdam: North-Holland. (1984).

ARAOA 1985-1 Araoz, J.; Edmonds, J.
A case of non-convergent dual changes in assignment problems. Discrete Applied Mathematics 11, 95-102. (1985).

ARAOB 1984-1 Araoz, J.; Edmonds, J.; Griffin, V.
Lifting the facets of polyhedra. In: W. R. Pulleyblank (Ed.): Progress in Combinatorial Optimization, 3-12. Toronto: Academic Press. (1984).

ARAU 1984-1 Araujo, E. O.; Franca, P. M.; Soares, S.; Tavares, H. M. F.
Optimal location of telephone exchanges. In: A. Straszak (Ed.): Large Scale Systems: Theory and Applications, Proc.IFAC/IFORS Symp., Warsaw 1983, IFAC Proc. Series, No. 10, 689-694. Oxford: Pergamon Press. (1984).

ARCH 1983-1 Archdeacon, D.
Coupled colorings of planar maps. Congressus Numerantium 39, 89-94. (1983).

ARCH 1986-1 Archdeacon, D.
Coverings of graphs by cycles. Congressus Numerantium 53, 7-14. (1986).

ARCHA 1987-1 Archdeacon, D.; Dinitz, J. H.
Factorizations and orthogonal matchings. Congressus Numerantium 58, 69-74. (1987).

ARCHB 1986-1 Archdeacon, D.; Richter, B.
Circuits in 4-regular plane graphs. Congressus Numerantium 53, 39-48. (1986).

ARDI 1984-1 Arditti, D.
Un nouvel algorithme de recherche d'un ordre induit par des comparaisons par paires. In: E. Diday, M. Jambu, L. Lebart, J. Pages, R. Tomassone (Eds.): Data analysis and informatics, III. Proc. Third Internat. Symp. on Data Analysis and Informatics, Versailles 1983, 323-343. Amsterdam, New York, Oxford: North-Holland. (1984).

ARKI 1985-1 Arkin, E. M.; Papadimitriou, C. H.
On negative cycles in mixed graphs. Operations Research Letters 4, 113-116. (1985).

ARKIA 1987-1 Arkin, E. M.; Silverberg, E. B.
Scheduling jobs with fixed start and end times. Discrete Applied Mathematics 18, 1-8. (1987).

ARNB 1985-1 Arnborg, S.
Efficient algorithms for combinatorial problems on graphs with bounded decomposability - a survey. BIT 25, 2-23. (1985).

ARNBA 1984-1 Arnborg, S.; Corneil, D. G.; Proskurowski, A.
Complexity of finding embeddings in a k-tree. Report TRITA-NA 8407, The Royal Institut of Technology, Stockholm. (1984).

ARNBB 1984-1 Arnborg, S.; Proskurowski, A.
Linear time algorithms for NP-hard problems on graphs embedded in k-trees. Paper No. TRITA-NA-8404, Royal Institute of Technology, Stockholm, 16 p. (1984).

ARNBB 1985-1 Arnborg, S.; Proskurowski, A.
Characterization and recognition of partial k-trees. Congressus Numerantium 47, 69-76. (1985).

ARNBB 1986-1 Arnborg, S.; Proskurowski, A.
Algorithms on graphs with bounded decomposability. Congressus Numerantium 53, 167-170. (1986).

AROC 1984-1 Arocha, J. L.
Properties of the independent polynomial of a graph (Spanish). Ciencia y Matematika (Havana) No. 3, 103-110. (1984).

ARON 1985-1 Aronson, J. E.
A survey of dynamic network flows. Technical Report 85-OR-12. Dallas: Southern Methodist Univ., School of Engineering and Applied Science, Dept. of Operations Research and Engineering Management, 82p. (1985).

ARON 1986-1 Aronson, J. E.
The multiperiod assignment problem: a multicommodity network flow model and specialized branch and bound algorithm. European Journal of Operational Research 23, 367-381. (1986).

ARONA 1985-1 Aronson, J. E.; Barr, R.; Helgason, R. V.; Kennington, J. L.; Loh, A.; Zaki, H.
A projective transformation algorithm by Karmarkar: a computational experiment with assignment problems. Technical Report 85-OR-3. Dallas: Southern Methodist Univ., Dept. of Operations Research, 12 p. (1985).

ARONB 1984-1 Aronson, J. E.; Chen, B. D.
A computational study of empirical decision horizons in infinite horizon, multiperiod network flow problems. Technical Report 84-OR-1. Dallas: Southern Methodist Univ., School of Engineering and Applied Science, Dept. of Operations Research and Engineering Management, 30p. (1984).

ARONB 1985-1 Aronson, J. E.; Chen, B. D.
A forward network simplex algorithm for solving multiperiod network flow problems. Technical Report 83-OR-8. Dallas: Southern Methodist Univ., School of Engineering and Applied Science, Dept. of Operations Research and Engineering Management, 33p. (1985).

ARONB 1985-2 Aronson, J. E.; Chen, B. D.
A primary/secondary memory implementation of a forward network simplex algorithm for multiperiod network flow problems. Technical Report 85-OR-10. Dallas: Southern Methodist Univ., School of Engineering and Applied Science, Dept. of Operations Research and Engineering Management, 41p. (1985).

ARONB 1986-1 Aronson, J. E.; Chen, B. D.
A forward network simplex algorithm for solving multiperiod network flow problems. Naval Research Logistics Quarterly 33, 445-468. (1986).

AROR 1986-1 Arora, S. R.
Multi-index transportation problem with planar constraints. Investigacion Operacional 7, 53-66. (1986).

ARTE 1986-1 Artemenko, V. I.; Chernyakova, A. M.
On the determination of the parameters of algorithms for solving integer linear programming problems (Russian). In: I. V. Sergienko (Ed.): Programming and Algorithms for the Development of Program Packages, 31-34. Kiev: Akad. Nauk Ukrain. SSR, Inst. Kibernet. (1986).

ARTH 1985-1 Arthur, J. L.; Frendewey, J. O.
A computational study of tour construction procedures for the traveling salesman problem. Working Paper. Corvallis: Oregon State Univ., Dept. of Statistics, 19 p. (1985).

ARTH 1985-2 Arthur, J. L.; Frendewey, J. O.
Generating traveling salesman problems with known optimal tours. Working Paper. Corvallis: Oregon State Univ., Dept. of Statistics, 17 p. (1985).

ARTI 1984-1 Artina, S.; Walker, J.
The use of mixed-integer linear programming in designing hydraulic networks (Italian). Atti della Accademia delle Scienze dell'Instituto die Bologna, Classe di Scienze Fisiche, Anno 271, Rendiconti, Serie XIII 10(1982/83), 317-327. (1984).

ARZO 1984-1 Arzola, J.
The task of selection of proposals, formulation, properties, solution method (Spanish). Investigacion Operacional (Cuba) 5, 3-14. (1984).

ASAN 1985-1 Asano, T.
An approach to the subgraph homeomorphism problem. Theoretical Computer Science 38, 249-268. (1985).

ASAN 1987-1 Asano, T.
An application of duality to edge-deletion problems. SIAM Journal on Computing 16, 312-331. (1987).

ASANA 1985-1 Asano, T.; Asano, T.; Guibas, L.; Hershberger, J.; Imai, H.
Visibility-polygon search and euclidean shortest paths. In: 26th Annual Symp. on Foundations of Computer Science, Portland, Oregon 1985, Proceedings, 155-164. IEEE. (1985).

ASANB 1983-1 Asano, T.; Hirata, T.
Edge-contraction problems. Journal of Combinatorics, Information & Systems Sciences (Delhi) 26, 197-208. (1983).

ASANC 1984-1 Asano, T.; Nishizeki, T.; Seymour, P. D.
A note on nongraphic matroids. Journal of Combinatorial Theory, Series B 37, 290-293. (1984).

ASAND 1985-1 Asano, T.; Sato, S.
Long path enumeration algorithms for timing verification on large digital systems. In: Y. Alavi, G. Chartrand, L. Lesniak, D. R. Lick, C. E. Wall (Eds.): Graph Theory with Applications to Algorithms and Computer Science, 25-36. Chichester: Wiley. (1985).

ASH 1983-1 Ash, P.
Two sufficient conditions for the existence of Hamilton cycles in bipartite graphs. Ars Combinatoria 16A, 33-38. (1983).

ASH 1985-1 Ash, P.
The maximum number of diagonals of a cycle in a block. Discrete Mathematics 55, 305-310. (1985).

ASH A 1984-1 Ash, P.; Jackson, B.
Dominating cycles in bipartite graphs. In: J. A. Bondy, U. P. R. Murty (Eds.): Progress in Graph Theory, 81-88. New York: Academic Press. (1984).

ASHI 1985-1 Ashimov, A. A.; Burkov, V. N.; Dzhaparov, B. A.; Kondratev, V. V.
Mechanisms of operative coordinated control of industrial systems. I. Coordinated scheduling of continuous production (Russian). Avtomatika i Telemekhanika (Moscow) No. 5, 112-121. (1985).

ASPV 1984-1 Aspvall, B.; Gilbert, J. R.
Graph coloring using eigenvalue decomposition. SIAM Journal on Algebraic and Discrete Methods 5, 526-538. (1984).

ASRA 1981-1 Asratyan, A. S.
Some sufficient conditions for realizability of a collection of integers in a bipartite multigraph (Russian). In: A. A. Markov (Ed.): Combinatorial-algebraic methods in applied mathematics, 25-34. Gorki: Gor'kov Gruzinskoi Universitatia. (1981).

ASRA 1984-1 Asratyan, A. S.
Investigation of imbedded c-combinations and c-coverings (Russian). Applied Mathematics (Erevan) No. 3, 56-66. (1984).

ASRAA 1985-1 Asratyan, A. S.; Khachatryan, N. K.
Investigation of the Hamiltonian property of a graph using neighborhoods of vertices (Russian). Akademiya Nauk Armyanskoi SSSR, Doklady (Erevan) 81, No. 3, 103-106. (1985).

ASSA 1985-1 Assad, A. A.
Nested optimal policies for set functions with applications to scheduling. Mathematics of Operations Research 10, 82-99. (1985).

ASSAA 1987-1 Assad, A. A.; Pearn, W. L.; Golden, B. L.
The capacitated Chinese postman problem: lower bounds and solvable cases. American Journal of Mathematical and Management Sciences 7, 63-88. (1987).

ASSAB 1985-1 Assad, A. A.; Xu, W.-X.
On lower bounds for a class of quadratic 0, 1 programs. Operations Research Letters 4, 175-180. (1985).

ASSM 1984-1 Assmann, S. F.; Johnson, D. S.; Kleitman, D. J.; Leung, J. Y.-T.
On a dual version of the one-dimensional bin packing problem. Journal of Algorithms 5, 502-525. (1984).

ASTA 1983-1 Astakhov, N. D.
On the dynamical problem of positioning factories with warehouses of production inventory. Engineering Cybernetics 21, No. 3, 13-20; translation from: Izvestiya Akademii Nauk SSSR, Tekhnicheskaya Kibernetika (Moscow) 1983, No. 3, 17-24. (1983).

ASTA 1986-1 Astakhov, N. D.
The production-transportation problem of optimal plant location with startup priorities. Automation and Remote Control 47, 1238-1241; translation from Avtomatika i Telemekhanika (1986), No.9, 83-86. (1986).

ASTAA 1985-1 Astakhov, N. D.; Kaliev, E. Z.
The problem of optimal distribution of points for processing of an agricultural product and its solution by the approximation-combinatorial method (Russian). Izvestiya Akademii Nauk Kazakhskoi SSR, Seriya Fiziko-Matematicheskaya (Alma-Ata), No. 5, 19-22. (1985).

ATAL 1985-1 Atallah, M. J.
A matching problem in the plane. Journal of Computer and System Sciences 31, 63-70. (1985).

ATALA 1985-1 Atallah, M. J.; Hambrusch, S. E.
Efficient solutions to a class of placement problems. In: H. Noltemeier (Ed.): Proc. WG '85: Internat. Workshop on Graphtheoretic Concepts in Computer Science, Castle Schwanberg, Würzburg, 1985, 11-22. Linz: Trauner. (1985).

ATALA 1986-1 Atallah, M. J.; Hambrusch, S. E.
An assignment algorithm with applications to integrated circuit layout. Discrete Applied Mathematics 13, 9-22. (1986).

ATALA 1987-1 Atallah, M. J.; Hambrusch, S. E.
On bipartite matchings of minimum density. Journal of Algorithms 8, 480-502. (1987).

ATALB 1984-1 Atallah, M. J.; Vishkin, U.
Finding Euler tours in parrallel. Journal of Computer and System Sciences 29, 330-337. (1984).

ATKI 1984-1 Atkins, D. R.; Granot, D.; Raghavendra, B. G.
Application of mathematical programming to the plywood design and manufacturing problem. Management Science 30, 1424-1441. (1984).

ATKIA 1987-1 Atkinson, M. D.; Negro, A.; Santoro, N.
Sums of lexicographically ordered sets. SCS-TR-100. Ottawa: Carleton Univ., School of Computer Science, 8 p. (1987).

AUCA 1987-1 Aucamp, D. C.
A lot-sizing policy for production planning with applications in MRP. International Journal of Production Research 25, 1099-1108. (1987).

AUCAA 1986-1 Aucamp, D. C.; Kuzdrall, P. J.
Lot sizes for one-time-only sales. Journal of the Operational Research Society 37, 79-86. (1986).

AUSI 1986-1 Ausiello, G.; D'Atri, A.; Moscarini, M.
Chordality properties on graphs and minimal conceptual connections in semantic data models. Journal of Computer and System Sciences 33, 179-202. (1986).

AUST 1985-1 Austin, L. M.; Ghandforoush, P.
A surrogate cutting plane algorithm for all-integer programming. Computers & Operations Research 12, 241-250. (1985).

AUSTA 1986-1 Austin, L. M.; Ruparel, B. C.
The mixed cutting plane algorithm for all-integer programming. Computers and Operations Research 13, 395-402. (1986).

AVDO 1986-1 Avdonin, L. N.; Kodalaev, V. I.
A combinatorial approach to multicriteria optimization (Russian). Ehkonomika i Matematicheskie Metody 22, 741-743. (1986).

AVEN 1985-1 Aven, O. I.; Lovetskij, S. E.; Moiseenko, G. E.
Optimization of transportation flows (Russian). "Nauka" (Moscow). (1985).

AVIS 1985-1 Avis, D.; Devroye, L.
An analysis of a decomposition heuristic for the assignment problem. Operations Research Letters 3, 279-284. (1985).

AVRA 1981-1 Avram-Nitchi, R.
An application of graph theory to the time-criterion allocation problem (Romanian). Buletinul Stiintific, Institutului Politehnic Cluj-Napoca, Seria Electrotehnica-Energetica-Informatica 24, 11-14. (1981).

AVRA 1984-1 Avram-Nitchi, R.
An algorithm based on graph theory for solving the time-criterion assignment problem. Mathematica - Revue d'Analyse Numerique et de Theorie de l'Approximation (Cluj-Napoca) 13, 99-104. (1984).

AVRA 1984-2 Avram-Nitchi, R.
An operatorial model of transportation type problems. in: Itinerant Seminar on Functional Equations, Approximation and Convexity (Cluj-Napoca), 9-14, Preprint 84-6, Univ. "Babes-Bolyai", Cluj-Napoca. (1984).

AWER 1987-1 Awerbuch, B.
Optimal distributed algorithms for minimum weight spanning tree, counting, leader election, and related problems. In: Proc. Nineteenth Annual ACM Symp. on Theory of Computing, New York City 1987, 230-240. ACM. (1987).

AXSA 1985-1 Axsäter, S.
Performance bounds for lot sizing heuristics. Management Science 31, 634-639. (1985).

AXSA 1986-1 Axsäter, S.
Design and evaluation of a lot sizing heuristic. In: A. Chikan (Ed.): Inventory in Theory and Practice - Proc. 3rd Internat. Symp. on Inventories, Budapest 1984, 431-442. Amsterdam: Elsevier. (1986).

AXSA 1986-2 Axsäter, S.
Evaluation of lot-sizing techniques. International Journal of Production Research 24, 51-58. (1986).

AXSA 1987-1 Axsäter, S.
An extension of the extended basic period approach for economic lot scheduling problems. Journal of Optimization Theory and Applications 52, 179-190. (1987).

AXSAA 1987-1 Axsäter, S.; Nuttle, H. L. W.
Combining items for lot sizing in multi-level assembly systems. International Journal of Production Research 25, 795-808. (1987).

AYEL 1984-1 Ayel, J.; Favaron, O.
Helms are gaceful. In: J. A. Bondy, U. P. R. Murty (Eds.): Progress in Graph Theory, 89-92. New York: Academic Press. (1984).

AYKI 1987-1 Aykin, T.; Babu, A. J. G.
Constrained large-region multifacility location problems. Journal fo the Operational Research Society 38, 241-252. (1987).

AYKI 1987-2 Aykin, T.; Babu, A. J. G.
Multifacility location problems on a sphere. International Journal of Mathematics and Mathematical Sciences 10, 583-596. (1987).

AYUP 1984-1 Ayupov, R. S.
A combinatorial continuous method of solving block problems of discrete programming. Izvestiya Akademii Nauk SSSR, Tekhnicheskaja Kibernetika (Moscow) No. 3, 34-39, 201; translated as Engineering Cybernetics 22, 1-6. (1984).

AZAR 1987-1 Azarenok, A. S.
Packing of graphs of width two (Russian). Vestsi Akademii Navuk BSSR, Seryya Fizika-Matematychnykh Navuk (Minsk), No. 3, 9-16. (1987).

BAAS 1988-1 Baas, S. M.
Berge's strong perfect graph conjecture for 4-chromatic graphs. In: H. Schellhaas, P. van Beek, H. Isermann, R. Schmidt, M. Zijlstra (Eds.): Operations Research Proceedings 1987 (DGOR/NSOR), 594-597. Berlin: Springer. (1988).

BABA 1986-1 Babaev, D. A.; Aliev, A. A.
A method of surrogate constraints for solution of discrete programming problems. Kibernetika (Kiev), No. 2, 112-115. (1986).

BABAA 1986-1 Babaev, D. A.; Babaev, R. D.
Generating 0-1 integer programming test problems. In: M. Arato, I. Katai, L. Varga (Eds.): Topics in the Theoretical Bases and Applications of Computer Science, Proc. 4th Hungarian Computer Science Conf., Györ, Hungary, 1985, 317-329. Budapest: Akademiai Kiado. (1986).

BABAB 1987-1 Babaev, D. A.; Nasibov, E. N.
Dialog approach to the solution of fuzzy integer linear programming problems (Russian). Izvestiya Akademii Nauk Azerbajdzhanskoj SSR, Seriya Fiziko-Tekhnicheskikh i Matematicheskikh Nauk No.1, 113-118. (1987).

BABAC 1983-1 Babaev, R. D.
A problem for the choice of an optimal sequence of construction of objects (Russian). Izvestiya Akademii Nauk Azerbajdzhanskoj SSR, Seriya Fiziko- Tekhnicheskikh i Matematicheskikh Nauk (Baku) No. 4, 120-126. (1983).

BABAC 1985-1 Babaev, R. D.
The construction of test problems in integer-value programming with binary unknowns. USSR Computation of Mathematics and Mathematical Physics 25, No. 1, 98-102; translation from: Zhurnal Vychislitel'noj Matematiki i Matematicheskoj Fiziki (Moscow) 25, No. 1, 146-150. (1985).

BABAC 1987-1 Babaev, R. D.
Generation of test problems on covering and partitioning (Russian). Zhurnal Vychislitelnoi Matematiki i Matematicheskoi Fiziki 27, 1349-1359. (1987).

BABAD 1981-1 Babaev, T. A.
The problem of the shortest tree as a linear programming problem (Russian). In: Mathematical Cybernetics and Applied Mathematics No. 5, 8-15. "Elm", Baku. (1981).

BABAE 1985-1 Babai, L.
On Lovasz' lattice reduction and the nearest lattice point problem. In: K. Mehlhorn (Ed.): STACS 85, Second Annual Symposium on Theoretical Aspects of Computer Science (Saarbrücken, January 1985), Lecture Notes in Computer Science Vol. 182, 13-20. Berlin: Springer-Verlag. (1985).

BABAE 1986-1 Babai, L.
A Las Vegas - NC algorithm for isomorphism of graphs with bounded multiplicity of eigenvalues (draft). Report No. 86416-OR. Bonn: Universität Bonn, Inst. für Ökonometrie und Operations Research, 12p. (1986).

BABAE 1986-2 Babai, L.
On Lovasz' lattice reduction and the nearest lattice point problem. Combinatorica 6, 1-14. (1986).

BABAF 1987-1 Babai, L.; Turan, G.
The complexity of defining a relation on a finite graph. Zeitschrift für Mathematische Logik und Grundlagen der Mathematik 33, 277-288. (1987).

BABAG 1984-1 Babayev, D. A.; Babayev, R. D., Jysifov, M. M.
General multiple-choice knapsack problem. In: K. Lommatzsch (Ed.): Proceedings, 16. Jahrestagung "Mathematische Optimierung", Sellin/Rügen 1984, 1-10. Seminarbericht Nr. 64, Sektion Mathematik, Humboldt-Univ. Berlin (DDR). (1984).

BABAH 1985-1 Babayev, R. D.
On test integer programming problems with binary unknowns. Zhurnal Vychislitel'noj Matematiki i Matematicheskoj Fiziki 25, 146-150. (1985).

BABI 1986-1 Babic, D.; Graovac, A.; Mohar, B.
Recursion for the matching polynomials of fasciagraphs and rotagraphs. In: R. Tosic; D. Acketa; V. Petrovic (Eds.): Graph theory, Proc. 6th Yugoslav Seminar on Graph Theory, Dubrovnik, Yugoslavia, 1985, 27-41. University of Novi Sad. (1986).

BABIA 1986-1 Babic, D.; Graovac, A.; Mohar, B.; Pisanski, T.
Recursion for the matching polynomials of fasciagraphs and rotagraphs. In: R. Tosic, D. Acketa, V. Petrovic (Eds.): Graph Theory - Proc. 6th Yugoslav Seminar, Dubrovnik 1985, 27-41. Novi Sad: Inst. of Math., Univ. of Novi Sad. (1986).

BABIB 1987-1 Babic, O.
Optimization of refuelling truck fleets at an airport. Transportation Research 21B, 479-488. (1987).

BABU 1985-1 Babushkin, A. I.; Belov, I. S.
Optimization of the sequence of an assembly using resources of one kind (Russian). Ehkonomika i Matematicheskie Metody 21, 1128-1132. (1985).

BABU 1986-1 Babushkin, A. I.; Belov, I. S.
Optimal combinations of priority scheduling rules. Automation and Remote Control 47, No.5, 661-668; translation from Avtomatika i Telemekhanika (1986), No.5, 82-90. (1986).

BACH 1988-1 Bach, L.
Regionale Standortplanung für medizinisch-technische Grossgeräte. In: H. Schellhaas, P. van Beek, H. Isermann, R. Schmidt, M. Zijlstra (Eds.): Operations Research Proceedings 1987 (DGOR/NSOR), 394-395. Berlin: Springer. (1988).

BACHA 1984-1 Bachem, A.
Optimization and geometry in discrete structures. In: R. E. Burkard (Ed.): 15. Steiermärkisches mathematisches Symposium, Forschungszentrum Graz, Mathematisch-Statistische Sektion, Bericht No. 232. (1984).

BACHA 1986-1 Bachem, A.
Dualität und Polarität in diskreten Strukturen. In: W. Krelle (Ed.): Ökonomische Prognose-, Entscheidungs- und Gleichgewichtsmodelle, 367-388. Weinheim: VCH-Verlag. (1986).

BACHB 1984-1 Bachem, A.; Euler, R.
Recent trends in combinatorial optimization. Preprints in Optimization 84.2. Köln: Univ. Köln, Mathematisches Institut, 53 p. (1984).

BACHC 1984-1 Bachem, A.; Kannan, R.
Lattices and the basis reduction algorithm. CMU-SC-84-112. Pittsburgh: Carnegie-Mellon Univ., Dept. of Computer Science, 22 p. (1984).

BACHD 1984-1 Bachem, A.; Kern, W.
Adjoints of oriented matroids. Report, 84.1. Köln: Univ. Köln, Mathematisches Inst., 12 p. (1984).

BACHD 1985-1 Bachem, A.; Kern, W.
On sticky matroids. Report 85-18. Köln: Univ. Köln, Mathematisches Institut, 9 p. (1985).

BACHD 1986-1 Bachem, A.; Kern, W.
Adjoints of oriented matroids. Combinatorica 6, 299-308. (1986).

BACHD 1986-2 Bachem, A.; Kern, W.
Extension equivalence of oriented matroids. European Journal of Combinatorics 7, 193-198. (1986).

BACHE 1985-1 Bachem, A.; Wanka, A.
On intersection properties of (oriented) matroids (extended abstract). Preprints in Optimization 85.17. Köln: Univ. Köln, Mathematisches Institut, 3 p. (1985).

BACHE 1985-2 Bachem, A.; Wanka, A.
Separation theorems for oriented matroids. Preprints in Optimization 85.19. Köln: Univ. Köln, Mathematisches Institut, 8 p. (1985).

BACHE 1986-1 Bachem, A.; Wanka, A.
Euclidean intersection properties. Report 88-25. Köln: Univ. Köln, Mathematisches Institut, 11 p. (1986).

BACHE 1986-2 Bachem, A.; Wanka, A.
Matroids without adjoint. Report 86-27. Köln: Univ. Köln, Mathematisches Institut, 5 p. (1986).

BACHF 1984-1 Bachmann, P.
Ein Beitrag zur Maximierung submodularer und isotoner Funktionen über Unabhängigkeitssystemen. Wissenschaftliche Zeitschrift der Technnischen Hochschule Karl-Marx-Stadt 26, 167-172. (1984).

BACHF 1985-1 Bachmann, P.
Ein Beitrag zur optimalen Standortfindung von Teilen im Arbeitsbereich eines Montageroboters. Wissenschaftliche Zeitschrift der Technischen Hochschule Karl-Marx-Stadt 27, 683-687. (1985).

BAGA 1983-1 Bagaev, G. N.
Enumeration of many-colored trees (Russian). Izvestiya Akademii Nauk BSSR, Seriya Fiziko-Matematicheskikh Nauk (Minsk) 1983, No. 6, 45-52. (1983).

BAGAA 1983-1 Bagaev, G. N.; Dmitriev, E. F.
Enumeration of connected two-color graphs (Russian). In: K. A. Rybnikov (Ed.): Combinatorial Analysis, No. 6, 58-64. Moscow: Moskov. Gos. Univ., Mekh.-Mat. Fak. (1983).

BAGAA 1984-1 Bagaev, G. N.; Dmitriev, E. F.
Enumeration of connected labeled bipartite graphs (Russian). Doklady Akademii Nauk BSSR (Minsk) 28, 1061-1063. (1984).

BAGC 1987-1 Bagchi, U.; Chang, Y.-L.; Sullivan, R. S.
Minimizing absolute and squared deviations of completion times with different earliness and tardiness penalties and a common due date. Naval Research Logistics 34, 739-752. (1987).

BAGG 1982-1 Bagga, P. C.
Single machine scheduling with penalty cost as an m-degree monomial function of completion times. Transactions of the Indian Association for Productivity, Quality an Reliability (IAPQR) 7, 75-86. (1982).

BAGGA 1983-1 Bagga, P. C.; Khurana, K.
Minimizing waiting cost of jobs in renting situations. Transactions of the Indian Association for Productivity, Quality an Reliability (IAPQR) 8, 9-13. (1983).

BAGGA 1983-2 Bagga, P. C.; Khurana, K.
The single machine problem with quadratic penalty function. Acta Ciencia Indica Math. (Meerut) 9, 33-36. (1983).

BAGGA 1984-1 Bagga, P. C.; Khurana, K.
Transportation of job-blocks with separated setup times and removal times. Cahiers du Centre d'Etudes de Recherche Operationelle 26, 195-200. (1984).

BAGGA 1985-1 Bagga, P. C.; Khurana, K.
Special cases with set up considerations for n*3 flowshop problem. Journal of Information and Optimization Sciences (Delhi) 6, 75-86. (1985).

BAHL 1984-1 Bahl, H. C.; Ritzman, L. P.
A cyclical scheduling heuristic for lot sizing with capacity constraints. International Journal of Production Research 22, 791-800. (1984).

BAHLA 1987-1 Bahl, H. C.; Ritzman, L. P.; Gupta, J. N. D.
Determining lot sizes and resource requirements: A review. Operations Research 35, 329-345. (1987).

BAHLB 1987-1 Bahl, H. C.; Zionts, S.
Multi-item scheduling by Benders' decomposition. Journal of the Operational Research Society 38, 1141-1148. (1987).

BAKE 1985-1 Bakenrot, V. Y.
The efficiency of planning algorithms in multiprocessor networks. Izvestiya Akademii Nauk SSSR Tekhnicheskaya Kibernetika (Moscow) No. 1, 62-71; translated as: Soviet Journal of Computational Systems Sciences 23, No. 3, 30-39. (1985).

BAKEA 1984-1 Baker, B. M.
A network-flow algorithm for project selection. Journal of the Operational Research Society 35, 847-852. (1984).

BAKEA 1986-1 Baker, B. M.
A partial dual algorithm for the capacitated warehouse location problem. European Journal of Operational Research 23, 48-56. (1986).

BAKEB 1985-1 Baker, B. S.
A new proof for the first-fit decreasing bin-packing algorithm. Journal of Algorithms 6, 49-70. (1985).

BAKEB 1986-1 Baker, B. S.
A provably good algorithm for the two module routing problem. SIAM Journal on Computing 15, 162-188. (1986).

BAKEC 1986-1 Baker, E. K.; Schaffer, J. R.
Solution improvement heuristics for the vehicle routing and scheduling problem with time window constraints. American Journal of Mathematical and Management Sciences 6, 261-300. (1986).

BAKED 1984-1 Baker, K. R.
Sequencing rules and due-date assignments in a job shop. Management Science 30, 1093-1104. (1984).

BAKEE 1984-1 Baker, K. R.; Kanet, J. J.
Improved decision rules in a combined system for minimizing job tardiness. International Journal of Production Research 22, 917-922. (1984).

BAKEF 1981-1 Baker, T. E.
A branch and bound network algorithm for interactive process scheduling. Mathematical Programming Study 15, 43-57. (1981).

BAKH 1981-1 Bakhtin, A. E.
A modification of the algorithm for the solution of location problems (Russian). In: E. L. Berlyand; V. P. Busysin (Eds.): Mathematical analysis of economic interaction models, Novosibirsk 1981, 117-123. Novosibirsk: Izdatel'stvo Nauka Sibisrskoe Otdelenie. (1981).

BAKO 1980-1 Bako, A.
Network realization with flow method. Bulletin of Applied Mathematics 58-65, 105-117. (1980).

BALA 1985-1 Balaban, A. T.; Filip, P. A.; Balaban, T.-S.
Computer program for finding all possible cycles in graphs. Journal of Computational Chemistry 6, 316-329. (1985).

BALA 1986-1 Balaban, A. T.; Filip, P. A.; Balaban, T.-S.
Erratum: Computer program for finding all possible cycles in graphs. Journal of Computational Chemistry 7, 253. (1986).

BALAA 1981-1 Balachandran, V.; Srinivasan, V.; Thompson, G. L.
Applications of the operator theory of parametric programming for the transportation and generalized transportation problems. Mathematical Programming Study 15, 58-85. (1981).

BALAB 1987-1 Balakrishnan, A.; Magnanti, T. L.; Wong, R. T.
A dual-ascent procedure for large-scale uncapacitated network design. CC-87-22. West Lafayette: Purdue Univ. Inst. of Interdisciplinary Engineering Studies, 55 p. (1987).

BALAC 1987-1 Balakrishnan, A.; Ward, J. E.; Wong, R. T.
Integrated facility location and vehicle routing models: recent work and future prospects. American Journal of Mathematical and Management Sciences 7, 35-61. (1987).

BALAD 1987-1 Balakrishnan, N.; Wong, R. T.
A network model for the rotating workforce scheduling problem. CC-87-28. West Lafayette: Purdue Univ. Inst. of Interdisciplinary Engineering Studies, 32 p. (1987).

BALAE 1983-1 Balakrishnan, R.; Paulraja, P.
Existence of graphs with edge-connectivity 3 and 1-Hamiltonian index 2. Journal of Combinatorics, Information & System Sciences (Delhi) 8, 30-32. (1983).

BALAE 1983-2 Balakrishnan, R.; Paulraja, P.
Powers of chordal graphs. Journal of the Australian Mathematical Society, Series A 35, 211-217. (1983).

BALAE 1984-1 Balakrishnan, R.; Paulraja, P.
Counterexample to a conjecture of Häggkvist and Thomassen on circuits containing specified edges. In: K. P. Vijayan, N. M. Singhi (Eds.): Combinatorics and Applications (Proc. Seminar Calcutta 1982), 44-47. Calcutta: Indian Statistical Institute. (1984).

BALAE 1984-2 Balakrishnan, R.; Paulraja, P.
Note on the existence of directed (k+1)-cycles in diconnected complete k-partite digraphs. Journal of Graph Theory 8, 423-426. (1984).

BALAE 1986-1 Balakrishnan, R.; Paulraja, P.
Chordal graphs and some of their derived graphs. Congressus Numerantium 53, 71-74. (1986).

BALAE 1986-2 Balakrishnan, R.; Paulraja, P.
Self-clique graphs and diameters of iterated clique graphs. Utilitas Mathematicae 29, 263-268. (1986).

BALAF 1987-1 Balakrishnan, R.; Paulraja, P.; Basha, A. R.
Packing half-complete graphs with trees. Utilitas Mathematica 31, 131-148. (1987).

BALAG 1985-1 Balas, E.
An O(|V| + |E|) algorithm for finding an edge-maximal subgraph with a tr-formative coloring. Management Science Research Report MSRR-519. Pittsburgh: Carnegie-Mellon Univ., Graduate School of Industrial Administration, 12 p. (1985).

BALAG 1985-2 Balas, E.
Disjunctive programming and a hierarchy of relaxations for discrete optimization problems. SIAM Journal on Algebraic and Discrete Methods 6, 466-486. (1985).

BALAG 1985-3 Balas, E.
On the facial structure of scheduling polyhedra. Mathematical Programming Study 24, 179-218. (1985).

BALAG 1986-1 Balas, E.
A fast algorithm for finding an edge-maximal subgraph with TR-formative coloring. Discrete Applied Mathematics 15, 123-134. (1986).

BALAG 1987-1 Balas, E.
The prize collecting traveling salesman problem. MSRR-539. Pittsburgh: Carnegie Mellon Univ., Graduate School of Industrial Administration, 23 p. (1987).

BALAH 1984-1 Balas, E.; Chang, S. Y.
Finding a maximum clique in an arbitrary graph. Management Science Research Report 515. Pittsburgh: Carnegie-Mellon Univ., Graduate School of Ind. Eng., 27 p. (1984).

BALAI 1987-1 Balas, E.; Chvatal, V.; Nesetril, J.
On the maximum weight clique problem. Mathematics of Operations Research 12, 522-535. (1987).

BALAJ 1987-1 Balas, E.; Nauss, R.; Zemel, E.
Comment on:'Some computational results on real 0-1 knapsack problems'. Operations Research Letters 6, 139-140. (1987).

BALAK 1987-1 Balas, E.; Pulleyblank, W. R.
The perfectly matchable subgraph polytope of an arbitrary graph. Management Sciences Research Report 538. Pittsburgh: Carnegie Mellon Univ., Graduate School of Industrial Engineering, 26 p. (1987).

BALAL 1984-1 Balas, E.; Saltzman, M. J.
Some facets of the three-index assignment polytope. Management Science Research Report, 511. Pittsburgh: Carnegie-Mellon Univ., Graduate School of Industrial Administration, 35 p. (1984).

BALAL 1986-1 Balas, E.; Saltzman, M. J.
Facets of the three-index assignment polytope. Management Science Research Report, MSRR-529. Pittsburgh: Carnegie-Mellon Univ., Graduate School of Industrial Administration, 35 p. (1986).

BALAM 1986-1 Balas, E.; Shu, M.-N.
On the set covering polytope: I. All the facets with coefficients in (0,1,2) (revised version). Management Science Research Report MSRR-522. Pittsburgh: Carnegie-Mellon Univ., Graduate School of Industrial Administration, 28 p. (1986).

BALAN 1986-1 Balas, E.; Tama, J. M.; Tind, J.
Sequential convexification in reverse convex and disjunctive programming. Management Science Research Report, 531. Pittsburgh: Carnegie-Mellon Univ., Graduate School of Industrial Administration, 23 p. (1986).

BALAO 1985-1 Balas, E.; Toth, P.
Branch and bound methods. In: E. L. Lawler, J. K. Lenstra, A. H. G. Rinnooy Kan, D. B. Shmoys (Eds.): The Traveling Salesman Problem, 361-402. New York: Wiley. (1985).

BALAP 1986-1 Balas, E.; Yu, C. S.
Finding a maximum clique in an arbitrary graph. SIAM Journal on Computing 15, 1054-1068. (1986).

BALAP 1986-2 Balas, E.; Yu, C. S.
More on graphs with polynomially solvable maximum-weight clique problem. Management Science Research Report, MSRR-530. Pittsburgh: Carnegie-Mellon Univ., Graduate School of Industrial Administration, 10 p. (1986).

BALI 1984-1 Balinski, M. L.
A good (dual) simplex method for the assignment problem. AD, 275.07.84. Paris: Laboratoire d'Econometrie de l'Ecole Polytechnique, 25 p. (1984).

BALI 1984-2 Balinski, M. L.
The Hirsch conjecture for dual transportation polyhedra. Mathematics of Operations Research 9, 629-633. (1984).

BALI 1985-1 Balinski, M. L.
Signature methods for the assignment problem. Operations Research 33, 527-536. (1985).

BALI 1986-1 Balinski, M. L.
A competitive (dual) simplex method for the assignment problem. Mathematical Programming 34, 125-141. (1986).

BALIA 1986-1 Balinski, M. L.; Demange, G.
Generalizing proportionality; or how to simultaneously apportion seats to regions and to parties. AD 2870186. Paris: Ecole Polytechnique, 36 p. (1986).

BALIB 1984-1 Balinski, M. L.; Russakoff, A.
Faces of dual transportation polyhedra. Mathematical Programming Study 22, 1-8. (1984).

BALL 1987-1 Ball, M. O.; Liu, W. G.; Pulleyblank, W. R.
Two terminal Steiner tree polyhedra. Report 87466-OR. Bonn: Univ. Bonn, Inst. f. Ökonometrie u. Operations Research, 36 p. (1987).

BALLA 1985-1 Ball, M. O.; Roberts, A.
A graph partitioning approach to airline crew scheduling. Transportation Science 19, 107-126. (1985).

BALLB 1986-1 Ball, M. O.; Taverna, R.
Sensitivity analysis for the matching problem and its use in solving matching problems with a single side constraint. In: C. L. Monma (Ed.): Algorithms and Software for Optimization - Part I, 25-56. Annals of Operations Research 4, Basel: Baltzer. (1986).

BAND 1983-1 Bandelt, H. J.
Matchings in squares of odd cycles. Journal of Combinatorics, Information & System Sciences (Delhi) 8, 261-262. (1983).

BAND 1984-1 Bandelt, H. J.
Hereditary modular graphs. Oldenburg: Univ. Oldenburg, 18 p. (1984).

BAND 1985-1 Bandelt, H. J.
Networks with Condorcet solutions. European Journal of Operational Research 20, 314-326. (1985).

BANDA 1986-1 Bandelt, H. J.; Labbe, M.
How bad can a voting locating be. Social Choice and Welfare 3, 125-145. (1986).

BANDB 1985-1 Bandopadhyay, L.
The capacitated solid transportation problem (planar constraints) with upper and lower bounds on rim conditions. Journal of the Korean Operations Research and Management Science Society 10, No.2, 45-53. (1985).

BANG 1987-1 Bange, D. W.; Barkauskas, A. E.; Host, L. H.; Slater, P. J.
Efficient near-domination of grid graphs. Congressus Numerantium 58, 83-92. (1987).

BANK 1987-1 Bank, B.; Mandel, R.
The existence of optimal points for (mixed) integer optimization problems. Preprint 157. Berlin: Humboldt-Univ. Berlin, Sektion Mathematik, 20 p. (1987).

BANK 1987-2 Bank, B.; Mandel, R.
Mixed-integer solutions of quasiconvex polynomial inequalities. Report 108-1987. Graz: T.U. Graz, Inst. f. Mathematik, 19 p. (1987).

BANK 1987-3 Bank, B.; Mandel, R.
Nonlinear parametric integer programming. In: J. Guddat, H. T. Jongen, B. Kummer, F. Nozicka (Eds.): Parametric Optimization and Related Topics, Int. Conf., Plaue/GDR 1985, 16-48. Mathematical Research Vol. 35, Berlin: Akademie-Verlag. (1987).

BANK 1987-4 Bank, B.; Mandel, R.
Quantitative stability of (mixed) integer linear optimization problems. Preprint 158. Berlin: Humboldt-Univ. Berlin, Sektion Mathematik, 14 p. (1987).

BANKA 1982-1 Banks, J.; Mitchem, J.
On the acyclic point-connectivity of the n-cube. International Journal of Mathematics and Mathematical Sciences 5, 737-744. (1982).

BAPT 1984-1 Baptiste, P.; Favrel, J.
Resolution de problemes d'ordonnancement par les treillis de Galois et les graphes d'intervalle. RAIRO Automatique/Systems Analysis and Control 18, 405-416. (1984).

BAR- 1985-1 Bar-Yehuda, R.; Even, S.
A local-ratio theorem for approximating the weighted vertex cover problem. In: G. Ausiello, M. Lucertini (Eds.): Analysis and Design of Algorithms for Combinatorial Problems, Annals of Discrete Mathematics Vol. 25, 27-46. Amsterdam: North-Holland. (1985).

BAR-A 1987-1 Bar-Yehuda, R.; Kutten, S.; Wolfstahl, Y.; Zaks, S.
Making distributed spanning tree algorthms fault-resilient. In: F. J. Brandenburg, G. Vidal-Naquet, M. Wirsing (Eds.): STACS 87, Lecture Notes in Computer Science 247, 432-444. Berlin, New York, Tokyo: Springer. (1987).

BARA 1986-1 Barahona, F.
A solvable case of quadratic 0-1 programming. Discrete Applied Mathematics 13, 23-26. (1986).

BARA 1987-1 Barahona, F.
On via minimization. Report 87491-OR. Bonn: Univ. Bonn, Inst. f. Ökonometrie u. Operations Research, 11 p. (1987).

BARA 1987-2 Barahona, F.
Planar multicommodity flows, max cut and the Chinese postman problem. Report 87454-OR. Bonn: Univ. Bonn, Inst. f. Ökonometrie u. Operations Research, 12 p. (1987).

BARAA 1984-1 Barahona, F.; Cunningham, W. H.
A submodular network simplex method. Mathematical Programming Study 22, 9-31. (1984).

BARAB 1985-1 Barahona, F.; Grötschel, M.
The travelling salesman problem for graphs not contractible to k(5)-e. Preprint 77. Augsburg: Univ. Augsburg, Inst. f. Mathematik, 10 p. (1985).

BARAB 1986-1 Barahona, F.; Grötschel, M.
On the cycle polytope of a binary matroid. Journal of Combinatorial Theory 40B, 40-62. (1986).

BARAC 1986-1 Barahona, F.; Grötschel, M.; Jünger, M.; Reinelt, G.
An application of combinatorial optimization to statistical physics and circuit layout design. Preprint No. 95. Augsburg: Univ., Inst. für Mathematik, 33p. (1986).

BARAD 1985-1 Barahona, F.; Grötschel, M.; Mahjoub, A. R.
Facets of the bipartite subgraph polytope. Mathematics of Operations Research 10, 340-358. (1985).

BARAE 1987-1 Barahona, F.; Jünger, M.; Reinelt, G.
Experiments in quadratic 0-1 programming. Report 87455-OR. Bonn: Univ. Bonn, Inst. f. Ökonometrie u. Operations Research, 11 p. (1987).

BARAF 1985-1 Barahona, F.; Mahjoub, A. R.
Compositions in the acyclic subdigraph polytope. Report No. 85371-OR. Bonn: Univ. Bonn, Inst. f. Ökonometrie u. Operations Research, 40 p. (1985).

BARAF 1987-1 Barahona, F.; Mahjoub, A. R.
Compositions of graphs and polyhedra II: Stable sets. Report 87464-OR. Bonn: Univ. Bonn, Inst. f. Ökonometrie u. Operations Research, 23 p. (1987).

BARAG 1986-1 Barahona, F.; Manjoub, A. R.
On the cut polytope. Mathematical Programming 36, 157-173. (1986).

BARAH 1987-1 Barahona, F.; Pulleyblank, W. R.
Exact arborescences, matchings and cycles. Discrete Applied Mathematics 16, 91-100. (1987).

BARAI 1986-1 Barany, I.; Edmonds, J.; Wolsey, L. A.
Packing and covering a tree by subtrees. Combinatorica 6, 221-234. (1986).

BARAJ 1987-1 Barany, I.; Furedi, Z.
Covering all secants of a square. RUTCOR Research Report 3-86. New Brunswick: Rutgers Center for Operations Research, 10 p. (1987).

BARAK 1984-1 Barany, I.; Vanroy, T. J.; Wolsey, L. A.
Strong formulations for multi-item capacitated lot sizing. Management Science 30, 1255-1261. (1984).

BARAK 1984-2 Barany, I.; Vanroy, T. J.; Wolsey, L. A.
Uncapacitated lot-sizing: the convex hull of solutions. Mathematical Programming Study 22, 32-43. (1984).

BARC 1985-1 Barcelo, J.
Computer experiments with constraint identification processes for some types of integer programs (Spanish). Questiio 9, 121-146. (1985).

BARCA 1985-1 Barcia, P.
The bound improving sequence algorithm. Operations Research Letters 4, 27-30. (1985).

BARCA 1987-1 Barcia, P.
Constructive dual methods for discrete programming. Discrete Applied Mathematics 18, 107-118. (1987).

BARE 1985-1 Barefoot, C. A.
R-partial line graphs. Congressus Numerantium 47, 89-106. (1985).

BARE 1987-1 Barefoot, C. A.
Hamiltonian connectivity of the Halin graphs. Congressus Numerantium 58, 93-102. (1987).

BAREA 1986-1 Barefoot, C. A.; Jones, K. F.; Lundgren, J. R.; Maybee, J. S.
Biclique covers of regular digraphs. Congressus Numerantium 53, 147-154. (1986).

BARK 1985-1 Barker, J. R.; Mac Mahon, G. B.
Scheduling the general job-shop. Management Science 31, 594-598. (1985).

BARN 1985-1 Barnes, E. R.
Partitioning the nodes of a graph. In: Y. Alavi, G. Chartrand, L. Lesniak, D. R. Lick, C. E. Wall (Eds.): Graph Theory with Applications to Algorithms and Computer Science, 57-72. Chichester: Wiley. (1985).

BARNA 1985-1 Barnes, E. R.; Hoffman, A. J.
On transportation problems with upper bounds on leading rectangles. SIAM Journal on Algebraic and Discrete Methods 6, 487-496. (1985).

BARO 1985-1 Baron, G.; Prodinger, H.; Tichy, R. F.; Boesch, F. T.; Wang, J. F.
The number of spanning trees in the square of a cycle. Fibonacci Quarterly 23, 258-264. (1985).
BAROA 1987-1 Baronaigien, D. R. van; Ruskey, F.
A Hamilton path in the rotation lattice of binary trees. Congressus Numerantium 59, 313-318. (1987).
BARR 1981-1 Barr, R. S.; Turner, J.
Microdata file merging through large-scale network technology. Mathematical Programming Study 15, 1-22. (1981).
BARRA 1986-1 Barron, E. N.; Jensen, R.
On the Kleinrock-Nilsson problem of optimal scheduling algorithms for time-shared systems. Journal of Optimization Theory and Applications 50, 49-60. (1986).
BARRB 1986-1 Barros, O.; Weintraub, A.
Spatial market equilibrium problems as network models. Discrete Applied Mathematics 13, 109-130. (1986).
BART 1984-1 Bartholdi, J. J. III; Platzman, L. K.
Heuristics based on spacesfilling curves for combinatorial problems in the plane. PBRC Report Series, 84-08. Atlanta: Georgia Inst. of Technology, 32 p. (1984).
BARTA 1986-1 Bartholdi, J. J. III; Trick, M. A.
Stable matching with preferences derived from a psychological model. Operations Research Letters 5, 165-170. (1986).
BARTB 1987-1 Bartnick, J.
Aggregation of preference relations by a branch and bound algorithm. In: W. Domschke, W. Krabs, J. Lehn, P. Spellucci (Eds.): Proc. XI. Symp. on Operations Research, Darmstadt 1986, 75-84. Methods of Operations Research 57, Frankfurt/Main: Athenäum. (1987).
BARTC 1984-1 Bartolini, G.; Casalino, G.; Davoli, F.; Minciardi, R.; Zoppoli, R.
A team theoretical approach to routing and multiple access in data communication networks. In: Large Scale Systems: Theory and Applications. Proc. IFAC/IFORS Symp., Warsaw 1983. IFAC Proceedings Series 10, 269-276. (1984).
BARTD 1985-1 Barton, D. R.; Zippel, R.
Polynomial decomposition algorithms. Journal of Symbolic Computation 1, 159-168. (1985).
BARTE 1986-1 Bartusch, M.
Optimierung von Netzplänen mit Anordungsbeziehungen bei knappen Betriebsmitteln (Nachdruck der Dissertation, RWTH Aachen, 1983). Passau: Univ. Passau, Fakultät f. Mathematik u. Informatik, 164 p. (1986).
BARTF 1985-1 Bartusch, M.; Möhring, R. H.; Radermacher, F. J.
Some remarks on the m-machine unit time scheduling problem (extended abstract). In: H. Noltemeier (Ed.): Proc. WG '85: Internat. Workshop on Graphtheoretic Concepts in Computer Science, Castle Schwanberg, Würzburg, 1985, 23-26. Linz: Trauner. (1985).
BARTF 1986-1 Bartusch, M.; Möhring, R. H.; Radermacher, F. J.
Scheduling project networks with resource constraints and time windows. Report 86448-OR. Bonn: Univ. Bonn, Inst. f. Ökonometrie u. Operations Research, 45 p. (1986).
BARZ 1986-1 Barzilai, J.; Cook, W. D.; Kress, M.
A generalized network formulation of the pairwise comparison consensus ranking model. Management Science 32, 1007-1014. (1986).
BASA 1983-1 Basangova, E. O.; Erusalimskij, Y. M.
Mixed attainability on partially directed graphs (Russian). In: V. L. Rutman (Ed.): Numerical systems and algorithms, Interuniversity collection of articles, 135-140. Rostov: Izdatel'stvo Rostovskogo Universiteta. (1983).
BASAA 1987-1 Basavayya, D.; Ravindra, G.
Strongly perfect line graphs of (0,1)-graphs. Journal of Mathematical and Physical Sciences (Madras) 21, 391-398. (1987).
BASK 1986-1 Baskaran, G.; Fu, Y.-T.; Anderson, P. W.
On the statistical mechanics of the traveling salesman problem. Journal of Statistical Physics 45, No. 1-2, 1-25. (1986).
BATA 1984-1 Batagelj, V.; Cvetkovic, D.; Gutman, I.
A formula involving the number of 1-factors in a graph. In: D. Cvetcovic et al. (Eds.): Proceedings of the Fourth Yugoslav Seminar on Graph Theory, Novi Sad 1983, 39-42. Novi Sad: Univ. of Novi Sad, Institute of Mathematics. (1984).
BATI 1987-1 Batishchev, D. I.; Kogan, D. I.
Transportation problems with linear estimates of participants. Izvestiya Akademii Nauk SSSR, Tekhnicheskaya Kibernetika (Moscow), No. 1, 45-51; translated as Soviet J. Comput. Systems Sci. 25, No. 3, 93-98. (1987).
BATIA 1983-1 Batishchev, D. I.; Kogan, D. I.; Shakhriev, K.
Transportation problem with dichotomous preferences (Russian). Voprosy Kibernetiki (Tashkent) 122, 17-27. (1983).
BATIA 1985-1 Batishchev, D. I.; Kogan, D. I.; Shakhriev, K.
Solution of a multicriterial transport problem in an interactive regime (Russian). In: Mathematische Optimierung - Theorie und Anwendungen, 30. Int. Wiss. Kolloq., Ilmenau 1985, 4, 7-10. (1985).
BATT 1987-1 Batta, R.
Comment on "The dynamics of plant layout" by M. J. Rosenblatt. Management Science 33, 1065. (1987).
BATTA 1982-1 Batten, L. M.
A problem in transversal theory. In: F. Hoffman, K. B. Reid, R. C. Mullin, R. G. Stanton (Eds.): Proc. Thirteenth Southeastern Conf. on Combinatorics, Graph Theory and Computing, Florida 1982, Congressus Numerantium Vol. 35, 67-76. Winnipeg, Canada: Utilitas Mathematica. (1982).
BATU 1984-1 Batukhtin, V. D.; Maiboroda, L. A.; Tyulyukin, V. A.
Optimization of funcitons of an integer-valued argument (Russian). In: Some Problems on the Optimization of Discontinuous Functions, 17-28. Akademiya Nauk SSSR, Ural. Nauchn. Tsentr (Sverdlovsk). (1984).
BATUA 1983-1 Baturina, L. N.; Lepeshinskij, N. A.
Algorithms for the optimization of parameters ensuring a given vitality level of a network (Russian). Izvestiya Akademii Nauk BSSR, Seriya Fiziko-Matematicheskikh Nauk (Minsk) No. 4, 26-29. (1983).

BATY 1983-1 Batycki, T.; Kasprzak, A.
Flow assignment algorithms for computer communication networks (Polish). Prace Naukowe Instytutu Sterowania i Techniki Systemow, Politechnika Wroclawska, Seria Monografie (Wroclaw) 1, 118p. (1983).

BAUE 1985-1 Bauer, D.
A note on degree conditions for Hamiltonian cycles in line graphs. Congressus Numerantium 49, 11-18. (1985).

BAUEA 1987-1 Bauernöppel, F.
Degree bounds and subhamiltonian cycles in planar graphs. Journal of Information Processing and Cybernetics EIK 23, 529-536. (1987).

BAUEB 1985-1 Bauernöppel, F.; Jung, H.
Fast parallel vertex colouring. In: L. Budach (Ed.): Fundamentals of Computation Theory, FCT '85, Lecture Notes in Computer Science Vol. 199, 28-35. Berlin: Springer-Verlag. (1985).

BAUM 1983-1 Bauman, E. V.; Muchnik, I. B.
Simplification of graph structure with the aid of Boolean matrices. Avtomatika i Telemekhanika (Moscow) No. 11, 90-99, translated as: Automation and Remote Control 44, 1461-1469. (1983).

BAUMA 1985-1 Baumann, U.
On fixed points of partition preserving automorphisms of complete graphs with minimal colourings. 30. Internationales Wissenschaftliches Kolloquium (TH Ilmenau, DDR) Heft 5F, 7-10. (1985).

BAUMB 1985-1 Baumann, U.; Lesch, M.
Klassen von Zerlegungen zu Kantenmenge vollständigen Graphen. Wissenschaftliche Zeitschrift der Pädagogischen Hochschule Karl Friedrich Wilhelm Wander Dresden 19, 65-70. (1985).

BAUMB 1986-1 Baumann, U.; Lesch, M.
Klassen von resistenten Kantenfärbungen. Wissenschaftliche Zeitschrift der Pädagogischen Hochschule "K. F. W. Wander" Dresden 20, 59-62. (1986).

BAUMB 1986-2 Baumann, U.; Lesch, M.
Minimal colourings of complete and regular graphs and partition preserving automorphisms. Wissenschaftliche Zeitschrift der Technischen Hochschule Ilmenau 32, No.3, 29-38. (1986).

BAUMC 1987-1 Baumann, U.; Lesch, M.; Schmeichel, I.
Farb- und zerlegungstreue Automorphismen von Graphen. Mathematische Nachrichten 134, 161-167. (1987).

BAXT 1984-1 Baxter, J.
Depot location: a technique for the avoidance of local optima. European Journal of Operational Research 18, 208-214. (1984).

BAYB 1986-1 Baybars, I.
A survey of exact algorithms for the simple assembly line balancing problem. Management Science 32, 909-932. (1986).

BAYB 1986-2 Baybars, I.
An efficient heuristic method for the simple assembly line balancing problem. International Journal of Production Research 24, 149-166. (1986).

BAYU 1985-1 Bayus, B. L.; Gupta, S. K.
Analyzing floor configurations for casino slot machines. OMEGA 13, 561-567. (1985).

BEAL 1985-1 Beale, E. M. L.
Integer Programming. In: K. Schittkowski (Ed.): Computational Mathematical Programming, 1-24. Berlin: Springer-Verlag. (1985).

BEAN 1984-1 Bean, J. C.
A Langrangian algorithm for the multiple choice integer program. Operations Research 32, 1185-1193. (1984).

BEANA 1985-1 Bean, J. C.; Hall, D. H.
Accuracy of the modified due date rule. Technical Report 85-10. Ann Arbor: Univ. of Michigan, Dept. of Ind. and Operations Eng., 7 p. (1985).

BEANB 1987-1 Bean, J. C.; Smith, R. L.; Yano, C. A.
Forecast horizons for the discounted dynamic lot-size problem allowing speculative motive. Naval Research Logistics 34, 761-774. (1987).

BEAS 1985-1 Beasley, J. E.
A note on solving large p-median problems. European Journal of Operational Research 21, 270-273. (1985).

BEAS 1985-2 Beasley, J. E.
Algorithms for unconstrained two-dimensional guillotine cutting. Journal of the Operational Research Society 36, 297-306. (1985).

BEAS 1985-3 Beasley, J. E.
An algorithm for the two-dimensional assortment problem. European Journal of Operational Research 19, 253-261. (1985).

BEAS 1985-4 Beasley, J. E.
An exact two-dimensional non-guillotine cutting tree search procedure. Operations Research 33, 49-64. (1985).

BEAS 1985-5 Beasley, J. E.
Bounds for two-dimensional cutting. Journal of the Operational Research Society 36, 71-74. (1985).

BEAS 1987-1 Beasley, J. E.
An algorithm for set covering problems. European Journal of Operational Research 31, 85-93. (1987).

BECH 1986-1 Bechara, J. J. B.; Galvao, R. D.
The use of interactive computing for vehicle routing. Recent advances in system modelling and optimization, Proc. IFIP-WG 7/1 Work. Conf., Santiago/Chile 1984, Lect. Notes Control Inf. Sci. 87, 22-32. (1986).

BECHA 1984-1 Bechtold, S. E.; Janaro, R. E.; Sumners, D. L.
Maximization of labor productivity through optimal rest-break schedules. Management Science 30, 1442-1458. (1984).

BECK 1987-1 Becker, B.; Hotz, G.
On the optimal layout of planar graphs with fixed boundary. SIAM Journal on Computing 16, 946-972. (1987).

BECKA 1985-1 Becker, B.; Osthof, H. G.
Layouts with wires of balanced length. In: K. Mehlhorn (Ed.): STACS 85, Second Annual Symposium on Theoretical Aspects of Computer Science (Saarbrücken, January 1985), Lecture Notes in Computer Science Vol. 182, 21-31. Berlin: Springer-Verlag. (1985).

BECKB 1986-1 Becker, M.; Mehlhorn, K.
Algorithms for routing in planar graphs. Acta Informatica 23, 163-176. (1986).

BECKC 1985-1 Becker, R. I.; Perl, Y.
Finding the two-core of a tree. Discrete Applied Mathematics 11, 103-114. (1985).

BECKD 1984-1 Becker, R. I.; Schach, S. R.
A bottom-up algorithm for weight- and height-bounded minimal partition of trees. International Journal of Computer Mathematics 16, 211-228. (1984).

BECKD 1984-2 Becker, R. I.; Schach, S. R.
A bottom-up algorithm for weight- and height-bounded minimal partition of trees. In: B. Courcelle (Ed.): Ninth Colloquium on trees in algebra and programming, Bordeaux 1984, 63-72. Cambridge University Press. (1984).

BECKE 1985-1 Beckert, R.; Girlich, E.
Kombinatorische Typen von Polymatroiden im E(3,+). In: E. Girlich (Ed.): Diskrete Optimierung, 145-155. Jena: Friedrich-Schiller-Universität. (1985).

BECKF 1986-1 Beckmann, M. J.; Thisse, J.-F.
The location of production activities. In: Handbook of Regional and Urban Economics Vol. 1, 21-95. Amsterdam: North-Holland. (1986).

BEEK 1985-1 Beek, P. van; Bremer, A.; Putten, C. van
Design and optimization of multi-echelon assembly networks: savings and potentials. European Journal of Operational Research 19, 57-67. (1985).

BEEKA 1983-1 Beek, P. van; Schuylenburg, P. van
An application of aggregation and decomposition to the N-product, 1-machine problem. Wageningen: Agricultural Univ., Dept. of Mathematics, Technical Note 82-01. (1983).

BEHZ 1987-1 Behzad, M.
Total concepts in graph theory. Ars Combinatoria 23A, 35-40. (1987).

BEIK 1983-1 Beikirch, W.
Kantenfärbungen paarer Graphen, Matrizen und Quasigruppen. Wissenschaftliche Zeitschrift der Pädagogischen Hochschule "Karl Friedrich Wilhelm Wander" Dresden 17, 47-57. (1983).

BEIK 1983-2 Beikirch, W.
Über spezielle Kantenfärbungen und einen Zusammenhang zwischen Kantenfärbungen, Matrizen und Quasigruppen. Wissenschaftliche Zeitschrift der Pädagogischen Hochschule "Karl Friedrich Wilhelm Wander" Dresden 17, 33-45. (1983).

BEIK 1984-1 Beikirch, W.
Equivalence of edge colourings of complete graphs and quasigroups (German). In: Geometrie und Kombinatorik, 2. Kolloq., Karl-Marx-Stadt/DDR 1983, Vortr. 1,2, Teil II, 239-243. (1984).

BEIK 1985-1 Beikirch, W.
Edge colourings of finite simple and undirected graphs and quasigroups (German). In: Graphen und Netzwerke - Theorie und Anwendungen, 30. Int. Wiss. Kolloq., Ilmenau 1985, 5, 11-13. (1985).

BEIN 1986-1 Bein, W. W.; Brucker, P.
Greedy concepts for network flow problems. Discrete Applied Mathematics 15, 135-144. (1986).

BEINA 1985-1 Bein, W. W.; Brucker, P.; Tamir, A.
Minimum cost flow algorithms for series-parallel networks. Discrete Applied Mathematics 10, 117-124. (1985).

BEINB 1985-1 Beineke, L. W.
Packings of complete bipartite graphs. In: F. Harary, J. P. Maybee (Eds.): Graphs and Applications, 41-52. New York: Wiley. (1985).

BEIS 1987-1 Beissinger, J. S.; Peled, U. N.
Enumeration of labelled threshold graphs and a theorem of Frobenius involving Eulerian polynomials. Graphs and Combinatorics 3, 213-220. (1987).

BEKA 1985-1 Beka, J.
On the tripartite conjecture. Mathematica Slovaca (Bratislava) 35, 239-241. (1985).

BELA 1985-1 Belardo, S.; Duchessi, P.; Seagle, J. P.
Microcomputer graphics in support of vehicle fleet routing. Interfaces 15, 84-92. (1985).

BELAA 1984-1 Belardo, S.; Harrald, J.; Wallace, W. A.; Ward, J. E.
A partial covering approach to siting response resources for major maritime oil spills. Management Science 30, 1184-1196. (1984).

BELE 1983-1 Belenkij, A. S.; Ilenkova, S. D.; Levner, E. V.; Rezer, S. M.,; Timkovskii, V. G.; Tsurkan, V. M.
Application of methods of scheduling theory to the optimization of the loading of containers (Russian). In: Dynamics of non-homogeneous systems, Mater. Semin., Moscow 1983, 71-78. (1983).

BELEA 1985-1 Beletskii, S. A.; Levental, V. T.
An effective approximate algorithm for a general problem of scheduling theory (Russian). Izvestiya Akademii Nauk Kazakhskoi SSR, Seriya Fiziko-Matematicheskaza (Alma-Ata), No. 3, 16-19. (1985).

BELI 1985-1 Belikov, S. A.; Zak, Y. A.
Some problems of optimal distribution of limited resources when realizing a network of interrelated works. Kibernetika (Kiev) No. 2, 59-66. (1985).

BELL 1984-1 Belling-Seib, K.
Methods for solving nonlinear network flow problems. In: G. Hammer, D. Pallaschke (Eds.): Methods of Operations Research 51, Vol. 1, 227-238. Königstein/Ts.: Verlag Anton Hain Meisenheim. (1984).

BELL 1985-1 Belling-Seib, K.
A hybrid algorithm for solving convex separable network flow problems. Zeitschrift für Operations Research 29A, 105-124. (1985).

BELL 1987-1 Belling-Seib, K.
Network flow problems with one side constraint: a comparison of three solution methods. In: W. Domschke, W. Krabs, J. Lehn, P. Spellucci (Eds.): Proc. XI. Symp. on Operations Research, Darmstadt 1986, 85-86. Methods of Operations Research 57, Frankfurt/Main: Athenäum. (1987).

BELO 1982-1 Belov, B. I.; Chugunova, G. P.
On a problem concerning the cutting out of rectangles by parallelograms (Russian). In: A. P. Merenkov, V. P. Bulatov (Eds.): Methods of optimization and their applications, Novosibirsk 1982, 52-54. Novosibirsk: Izdatel'stvo Nauka Sibirskoe Otdelenie. (1982).

BELOA 1985-1 Belov, Y. A.; Bondarenko, V. A.
Geometric problems in discrete optimization (Russian). Yaroslavl: Yaroslav. Gos. Univ. (1985).

BELY 1984-1 Belyalov, T. K.; Khabibullin, R. F.
On the problem of choice of optimal sizes of lots (Russian). Issledovaniya po Prikladnoj Matematike 12, 74-84. (1984).

BELYA 1984-1 Belyi, S. B.
Minimal distributions of monochromatic triangles in complete bichromatic graphs (Russian). Vychislitel'naya i Prikladnaya Matematika (Kiev) No. 52, 116-118. (1984).

BENA 1985-1 Benavent, E.; Campos, V.; Corberan, A.; Mota, E.
Analysis of heuristics for the rural postman problem (Spanish). Trabajos de Estadistica y de Investigacion Operativa (Valencia) 36, 27-38. (1985).

BEND 1985-1 Bender, E. A.; Richmond, L. B.; Wormald, N. C.
Almost all chordal graphs split. Journal of the Australian Mathematical Society, Series A 38, 214-221. (1985).

BENDA 1985-1 Bender, E. A.; Wilf, H. S.
A theoretical analysis of backtracking in the graph coloring problem. Journal of Algorithms 6, 275-282. (1985).

BENDB 1985-1 Bender, E. A.; Wormald, N. C.
The number of loopless planar maps. Discrete Mathematics 54, 235-238. (1985).

BENDC 1985-1 Benders, J. F.; Beulens, A. J. M.; Nunen, J. A. E. E. van
On using the linear programming relaxation of assignment type mixed integer problems. In: K. Neumann, D. Pallaschke (Eds.): Contributions to Operations Research, Lecture Notes in Economics and Mathematical Systems Vol. 240, 1-9. Berlin: Springer-Verlag. (1985).

BENH 1986-1 Benhocine, A.
Pancyclism and Meyniel's conditions. Discrete Mathematics 58, 113-120. (1986).

BENHA 1987-1 Benhocine, A.; Fouquet, J. L.
The Chvatal-Erdös condition and pancyclic line-graphs. Discrete Mathematics 66, 21-26. (1987).

BENHB 1987-1 Benhocine, A.; Wojda, A. P.
The Geng-Hua Fan conditions for pancyclic or Hamilton-connected graphs. Journal of Combinatorial Theory 42B, 167-180. (1987).

BENT 1987-1 Bent, S. W.; Manber, U.
On non-intersecting Eulerian circuits. Discrete Applied Mathematics 18, 87-94. (1987).

BENTA 1986-1 Benton, W. C.
Evaluating a modified heuristic for the multiple-vehicle scheduling problem. IIE Transactions 18, 70-78. (1986).

BENV 1986-1 Benveniste, R.
An integrated plant design and scheduling problem in the food industry. Journal of the Operational Research Society 37, 453-462. (1986).

BENZ 1983-1 Benzaken, C.; Boyd, S. C.; Hammer, P. L.; Simeone, B.
Adjoints of pure bidirected graphs. Congressus Numerantium 39, 123-144. (1983).

BENZA 1987-1 Benzaken, C.; Crama, Y.; Hammer, P. L.; Maffray, F.
More characterisations of triangulated and cotriangulated graphs. RUTCOR Research Report 35-87. New Brunswick: Rutgers Univ., Center for Operations Research, 10 p. (1987).

BENZB 1985-1 Benzaken, C.; Hammer, P. L.
Boolean techniques for matroidal decomposition of independence systems and applications to graphs. Discrete Mathematics 56, 7-34. (1985).

BENZC 1985-1 Benzaken, C.; Hammer, P. L.; Werra, D. de
Split graphs of Dilworth number 2. Discrete Mathematics 55, 123-128. (1985).

BERE 1985-1 Berezinski, M.; Krus, L.
Aids for multiobjective decision making on development of transportation network under uncertainty. In: P. Serafini (Ed.): Mathematics of multiobjective optimization (Udine 1984), 421-432. Vienna: Springer. (1985).

BEREA 1983-1 Bereznev, V. A.; Palmova, I. E.; Tretyakov, A. A.
On an optimization problem of the cutting-out of materials (Russian). In: N. P. Bakhvalov, V. G. Karmanov, V. A. Morozov (Eds.): Numerical analysis: methods, algorithms, programs; Collection of articles, 41-50. Moscow: Izdatel'stvo Moskovskogo Universiteta. (1983).

BEREA 1984-1 Bereznev, V. A.; Palmova, I. E.; Tretyakov, A. A.
A method for solving a class of problems of linear integer programming with Boolean variables (Russian). In: Methods and Algorithms in Numerical Analysis, 76-85. Moscow: Moskov. Gos. Univ. (1984).

BERG 1983-1 Berge, C.
Graphes (3. edition). Paris: Gauthier-Villars. (1983).

BERG 1984-1 Berge, C.
A property of k-optimal path-partitions. In: J. A. Bondy, U. P. R. Murty (Eds.): Progress in Graph Theory, 105-108. New York: Academic Press. (1984).

BERG 1984-2 Berge, C.
Diperfect graphs. In: C. Berge, V. Chvatal (Eds.): Topics on perfect graphs, Annals of Discrete Mathematics Vol. 21, 45-56. Amsterdam: North-Holland. (1984).

BERG 1984-3 Berge, C.
Minimax theorems for normal hypergraphs and balanced hypergraphs - a survey. In: C. Berge, V. Chvatal (Eds.): Topics on Perfect Graphs, Annals of Discrete Mathematics Vol. 21, 3-20. Amsterdam: North-Holland. (1984).

BERG 1984-4 Berge, C.
On the minimum number of arcs to reverse to make a graph strongly connected. In: K. M. Koh, H. P. Yap (Eds.): Graph Theory Singapore 1983, Lecture Notes in Mathematics 1073, 31-34. Berlin: Springer-Verlag. (1984).

BERG 1984-5 Berge, C.
Stochastic graphs and strongly perfect graphs: a survey. In: K. P. Vijayan, N. M. Singhi (Eds.): Combinatorics and Applications (Proc. Seminar Calcutta 1982), 60-65. Calcutta: Indian Statistical Institute. (1984).

BERG 1985-1 Berge, C.
Graphs (second revised edition). Amsterdam, New York, Oxford: North-Holland. (1985).

BERG 1985-2 Berge, C.
Path-partitions in directed graphs and posets. In: I. Rival (Ed.): Graphs and Order, 447-466. Dordrecht: Reidel. (1985).
BERG 1986-1 Berge, C.
New classes of perfect graphs. Discrete Applied Mathematics 15, 145-154. (1986).
BERG 1987-1 Berge, C.
A minimax relation for the partial q-colorings of a graph. RUTCOR Research Report 15-87. New Brunswick: Rutgers Univ., Rutgers Center for Operations Research, 12 p. (1987).
BERG 1987-2 Berge, C.
A new color-change to improve the coloring of a graph. RUTCOR Research Report 16-87. New Brunswick: Rutgers Univ., Rutgers Center for Operations Research, 8 p. (1987).
BERG 1987-3 Berge, C.
On two conjectures to generalize Vizing's theorem. RUTCOR Research Report 14-87. New Brunswick: Rutgers Univ., Rutgers Center for Operations Research, 11 p. (1987).
BERGA 1984-1 Berge, C.; Duchet, P.
Strongly perfect graphs. In: C. Berge, V. Chvatal (Eds.): Topics on perfect graphs, Annals of Discrete Mathematics Vol. 21, 57-62. Amsterdam: North-Holland. (1984).
BERGA 1986-1 Berge, C.; Duchet, P.
Solvability of perfect graphs. In: Raspail-Burnside: Workshop on Perfect Graphs, Bellairs Research Institute, McGill University, Barbados, West Indies, 1986, 1-6. (1986).
BERK 1987-1 Berkey, J. O.; Wang, P. Y.
Two-dimensional finite bin-packing algorithms. Journal of the Operational Research Society 38, 423-430. (1987).
BERM 1985-1 Berman, A.
Convexity, graph theory and non-negative matrices. In: M. Rosenfeld, J. Zaks (Eds.): Convexity and Graph Theory. Annals of Discrete Mathematics Vol. 20, 55-60. Amsterdam: North-Holland. (1985).
BERMA 1984-1 Berman, K. A.
A generalization to bases common to r binary matroids of the weighted matrix-tree theorem in the case when the weights are Boolean. European Journal of Combinatorics 5, 197-205. (1984).
BERMA 1986-1 Berman, K. A.
Bicycles and spanning trees. SIAM Journal on Algebraic and Discrete Methods 7, 1-12. (1986).
BERMA 1987-1 Berman, K. A.
Balanced vertex weightings and arborescences. Journal of Combinatorial Theory 42B, 274-301. (1987).
BERMB 1986-1 Berman, K. A.; Paul, J. L.
Large monochromatic components in vertex colored k-trees. Congressus Numerantium 53, 161-166. (1986).
BERMB 1987-1 Berman, K. A.; Paul, J. L.
Long monochromatic paths in vertex 3-colored planar graphs. Congressus Numerantium 58, 115-122. (1987).
BERMC 1987-1 Berman, O.; Larson, R. C.; Parkan, C.
The stochastic queue p-median problem. Transportation Science 21, 207-216. (1987).
BERMD 1984-1 Berman, O.; LeBlanc, B.
Location-relocation of mobile facilities on a stochastic network. Transportation Science 18, 315-330. (1984).
BERME 1986-1 Berman, O.; Mandowsky, R. R.
Location-allocation on congested networks. European Journal of Operational Research 26, 238-250. (1986).
BERMF 1985-1 Berman, O.; Rahnama, M. R.
Optimal location-relocation decisions on stochastic networks. Transportation Science 19, 203-221. (1985).
BERMG 1984-1 Bermond, J. C.; Delorme, C.; Farhi, G.
Large graphs with given degree and diameter. II. Journal of Combinatorial Theory, Series B 36, 32-48. (1984).
BERMH 1984-1 Bermond, J. C.; Fouquet, J. L.; Habib, M.; Peroche, B.
On linear k-arboricity. Discrete Mathematics 52, 123-132. (1984).
BERMI 1985-1 Bermond, J. C.; Paoli, M.
Hamiltonian problems in powers of graphs. Rapport de Recherche 254. Orsay: Univ. de Paris-Sud, Lab. de Recherche en Informatique, 4 p. (1985).
BERN 1986-1 Bern, M. W.
Two probabilistic results on rectilinear Steiner trees. In: Proc. Eighteenth Annual ACM Symp. on Theory of Computing, Berkeley 1986, 433-441. ACM. (1986).
BERNA 1985-1 Bern, M. W.; Lawler, E. L.; Wong, A. L.
Why certain subgraph computations require only linear time. In: 26th Annual Symp. on Foundations of Computer Science, Portland, Oregon 1985, Proceedings, 117-125. IEEE. (1985).
BERNA 1987-1 Bern, M. W.; Lawler, E. L.; Wong, A. L.
Linear-time computation of optimal subgraphs of decomposable graphs. Journal of Algorithms 8, 216-235. (1987).
BERNB 1987-1 Bernardi, C.
On a theorem about vertex colorings of graphs. Discrete Mathematics 64, 95-96. (1987).
BERNC 1984-1 Bernstein, R.
Shortes paths in undirected graphs with negative edges. Congressus Numerantium 43, 117-126. (1984).
BERND 1984-1 Bernussou, J.; LeGall, F.; Garcia, J. M.
Optimisation et acheminement dynamique dans les reseaux telephoniques. In: A.Bensoussan, J. L. Lions (Eds.): Proc. Sixth Internat. Conf. on Analysis and Optimization of Systems, Nice 1984, Part 1, 87-101. Lecture Notes in Control and Inf. Sciences 62, Berlin: Springer. (1984).
BERR 1986-1 Berrada, M.; Stecke, K. E.
A branch and bound approach for machine load balancing in flexible manufacturing systems. Management Science 32, 1316-1335. (1986).
BERT 1986-1 Bertani, L.
On near dimensions in Steiner systems with parallelism and matroids. Journal of Geometry 27, 153-165. (1986).

BERTA 1986-1 Bertolazzi, P.; Sassano, A.
A class of polynomially solvable set-covering problems. R. 167. Roma: Consiglio Nazionale delle Ricerche, 23 p. (1986).

BERTA 1986-2 Bertolazzi, P.; Sassano, A.
Regular clutters. R. 169. Roma: Consiglio Nazionale delle Ricerche, 20 p. (1986).

BERTB 1986-1 Bertossi, A. A.
Total domination in interval graphs. Information Processing Letters 23, 131-134. (1986).

BERTC 1987-1 Bertossi, A. A.; Benuccelli, M. A.
Some parallel algorithms on interval graphs. Discrete Applied Mathematics 16, 101-112. (1987).

BERTD 1985-1 Bertossi, A. A.; Bonuccelli, M. A.
A polynomial feasibility test for preemptive periodic scheduling of unrelated processors. Discrete Applied Mathematics 12, 195-202. (1985).

BERTD 1985-2 Bertossi, A. A.; Bonuccelli, M. A.
Some parallel algorithms on interval graphs. In: P. Bertolazzi, F. Luccio (Eds.): VLSI: Algorithms and Architectures. Proc. Int. Workshop on Parallel Computing and VLSI, Amalfi 1984, 69-78. Amsterdam, New York, Oxford: North-Holland. (1985).

BERTD 1986-1 Bertossi, A. A.; Bonuccelli, M. A.
Hamiltonian circuits in interval graph generalizations. Information Processing Letters 23, 195-200. (1986).

BERTE 1985-1 Bertrand, J. W. M.
Multiproduct optimal batch sizes with in-process inventories and multi work centers. IIE Transactions 17, 157-163. (1985).

BERTF 1982-1 Bertsekas, D. P.
Optimal routing and flow control methods for communication networks. In: A. Bensoussan, J. L. Lions (Eds.): Analysis and Optimization of Systems, Proc. Conf. Versailles 1982, 615-643. Lecture Notes in Control and Information Sciences 44. Berlin, Heidelberg, New York: Springer. (1982).

BERTF 1985-1 Bertsekas, D. P.
A unified framework for primal-dual methods in minimum cost network flow problems. Mathematical Programming 32, 125-145. (1985).

BERTF 1987-1 Bertsekas, D. P.
The auction algorithm: a distributed relaxation method for the assignment problem. Report LIDS - P - 1653. Cambridge: Mass. Inst. of Technology, Laboratory for Information and Decision Systems, 29 p. (1987).

BERTG 1987-1 Bertsekas, D. P.; Eckstein, J.
Distributed asynchronous relaxation methods for linear network flow problems. Cambridge: Mass. Inst. of Technology, 12 p. (1987).

BERTH 1987-1 Bertsekas, D. P.; El-Baz, D.
Distributed asynchronous relaxation methods for convex network flow problems. SIAM Journal on Control and Optimization 25, 74-85. (1987).

BERTI 1987-1 Bertsekas, D. P.; Hossein, P. A.; Tseng, P.
Relaxation methods for network flow problems with convex arc costs. SIAM Journal on Control and Optimization 25, 1219-1243. (1987).

BERTJ 1985-1 Bertsekas, D. P.; Tseng, P.
Relaxation methods for minimum cost network flow problems. LIDS-P-1339. Cambridge: Mass. Inst. of Technology, 62 p. (1985).

BESS 1985-1 Bessonov, Y. E.
On solution of the problem of search for the maximal intersections of graphs on the basis of analysis of projections of subgraphs of the modular product (Russian). Vychslitel'nye Sistemy 112, 3-22. (1985).

BESSA 1981-1 Bessonov, Y. E.; Skorobogatov, V. A.
Recursive analysis of graphs (Russian). Vychislitel'nye Sistemy (Novosibirsk) 85, 3-20. (1981).

BESSA 1982-1 Bessonov, Y. E.; Skorobogatov, V. A.
On a family of schemes of recursive stripping of graphs (Russian). Vychislitel'nye Sistemy (Novosibirsk) 92, 3-49. (1982).

BEUL 1988-1 Beulens, A. J. M.; Kolen, A. W. J.; Niekamp, E. G.
A decision support system for a location-allocation-routing problem. In: H. Schellhaas, P. van Beek, H. Isermann, R. Schmidt, M. Zijlstra (Eds.): Operations Research Proceedings 1987 (DGOR/NSOR), 260-261. Berlin: Springer. (1988).

BEUT 1984-1 Beutelspacher, A.; Hering, P.-R.
Minimal graphs for which the chromatic number equals the maximal degree. Ars Combinatoria 18, 201-216. (1984).

BEYE 1987-1 Beyer, W. A.; Heller, L.
A Steiner tree associated with three quarks. Journal of Symbolic Computation 3, 283-289. (1987).

BHAS 1986-1 Bhaskar, J.; Sahni, S.
Via assignment in single row routing. In: K. V. Nori: Foundations of Software, Technology and Theoretical Computer Science, Lecture Notes in Computer Science 241, 154-176. Berlin, Heidelberg, New York: Springer-Verlag. (1986).

BHAT 1986-1 Bhat-Nayak, V. N.; Gokhale, S. K.
Validity of Hebbare's conjecture. Utilitas Mathematica 29, 49-59. (1986).

BHATA 1987-1 Bhatt, S. N.; Cosmadakis, S. S.
The complexity of minimizing wire lengths in VLSI layouts. Information Processing Letters 25, 263-268. (1987).

BHATB 1984-1 Bhatt, S. N.; Leighton, F. T.
A framework for solving VLSI graph layout problems. Journal of Computer and System Sciences 28, 300-343. (1984).

BHATC 1985-1 Bhattacharya, B. B.; Gupta, B.
Optimal design of hazard-free and easily testable logic networks. International Journal of Systems Sciences 16, 491-502. (1985).

BHATD 1985-1 Bhattacharya, D.
The minimum order of n-connected n-regular graphs with specified diameters. IEEE Transactions on Circuits and Systems 32, 407-409. (1985).

BIAL 1987-1 Bialostocki, A.; Roditty, Y.
A monotone path in an edge-ordered graph. International Journal of Mathematics and Mathematical Sciences 10, 315-320. (1987).

BIALA 1985-1 Bialy, H.
Stabilitätsbetrachtungen zur linearen Transportoptimierung. Wissenschaftliche Zeitschrift der Hochschule für Verkehrswesen Friedrich List Dresden 32, 381-387. (1985).

BIALA 1987-1 Bialy, H.
Duale Transportoptimierung. Optimization 18, 213-224. (1987).

BIAN 1982-1 Bianco, L.; Ricciardelli, S.
On scheduling with ready times to minimize total weighted completion time. in: H. Akashi (Ed.): Control Science and Technology for the Progress of Society. Proc. Eighth Triennial World Congress of the Internat. Fed. of Automatic Control, Koyto, Japan, 1981, Vol. 1, Control Theory, 501-505. Oxford: Pergamon Press. (1982).

BIDA 1985-1 Bidamon, E.; Meyniel, H.
On the Shannon capacity of a directed graph. European Journal of Combinatorics 6, 289-290. (1985).

BIEL 1983-1 Bielak, H.
On graphs with nonisomorphic 2-neighbourhoods. Casopis pro Pestovani Matematiky (Prague) 108, 294-298. (1983).

BIEL 1987-1 Bielak, H.
Some remarks about medians in graphs. In: M. Borowiecki; Z. Skupien (Eds.): Graphs, Hypergraphs and Matroids II, Proc. Sixth Regional Scientific Session of Mathematicians, Zagan, Poland, 1986, 7-15. Section: Graph Theory. Zielona Gora: Higher College of Engineering. (1987).

BIELA 1983-1 Bielak, H.; Soczewinska, E.
Some remarks about digraphs with nonisomorphic 1- or 2-neighbourhoods. Casopis pro Pestovani Matematiky (Prague) 108, 299-304. (1983).

BIELB 1984-1 Bielli, M.; Cini, M.; Calicchio, G.; Magliozzi, L.; Oddone, M.
Network models in planning the new generation of submarine communication systems. In: J. P. Brans (Ed.): Operational Research '84, 1060-2074. Amsterdam: North-Holland. (1984).

BIEN 1983-1 Bienert, K.
Modell- und computergestützte Planung logistischer Systeme. Berlin: Schmidt. (1983).

BIENA 1987-1 Bienia, W.; Cordovil, R.
An axiomatic of non-radon partitions of oriented matroids. European Journal of Combinatorics 8, 1-4. (1987).

BIENB 1986-1 Bienia, W.; Meyniel, H.
Partition of digraphs into paths or circuits. Discrete Mathematics 61, 329-332. (1986).

BIENC 1985-1 Bienstock, D.
An algorithm for reliability analysis of planar graphs. WP 20-85-86. Pittsburgh: Carnegie-Mellon Univ., Graduate School of Industrial Administration, 15 p. (1985).

BIENC 1985-2 Bienstock, D.
Efficient reliability computation of some highly structured networks. Working Paper OR 142-85. Cambridge: Mass. Inst. of Technology, Operations Research Center, 17 p. (1985).

BIENC 1986-1 Bienstock, D.
Some lattice-theoretic tools for network reliability analysis. W.P. 21-85-86. Pittsburgh: Carnegie-Mellon Univ., Graduate School of Admin., 28 p. (1986).

BIENC 1987-1 Bienstock, D.
On embedding graphs in trees. Working Paper. Morristown: Bell Communications Research, 40 p. (1987).

BIEND 1986-1 Bienstock, D.; Brickell, E. F.; Monma, C. L.
Properties of k-connected networks. Research Paper. Morristown: Morris Research and Engineering Center, 20 p. (1986).

BIENE 1987-1 Bienstock, D.; Monma, C. L.
On the complexity of covering vertices by faces in a planar graph. Working Paper. Morristown: Bell Communications Research, 44 p. (1987).

BIENE 1987-2 Bienstock, D.; Monma, C. L.
On the complexity of embedding planar graphs to minimize certain distance measures. Working Paper. Morristown: Bell Communications Research, 20 p. (1987).

BIENE 1987-3 Bienstock, D.; Monma, C. L.
Optimal enclosing regions in planar graphs. Working Paper. Morristown: Bell Communications Research, 20 p. (1987).

BIER 1987-1 Bierbrauer, J.; Gyarfas, A.
On (n,k)-colorings of complete graphs. Congressus Numerantium 58, 123-140. (1987).

BIGG 1985-1 Biggs, J. R.
Priority rules for shop floor control in a material requirements planning system under various levels of capacity. International Journal of Production Research 23, 33-46. (1985).

BIGI 1985-1 Bigildeeva, T. B.
Methods and algorithms of approximate solution of a quadratic assignment problem of large dimensions (a survey) (Russian). In: Optimization and Control, 22-35. Moscow: Moskov. Gos. Univ. (1985).

BILL 1984-1 Billera, L. J.; Munson, B. S.
Oriented matroids and triangulations of convex polytopes. In: W. R. Pulleyblank (Ed.): Progress in Combinatorial Optimization, 27-38. Toronto: Academic Press. (1984).

BILL 1984-2 Billera, L. J.; Munson, B. S.
Polarity and inner products in oriented matroids. European Journal of Combinatorics 5, 293-308. (1984).

BILL 1984-3 Billera, L. J.; Munson, B. S.
Triangulations of oriented matroids and convex polytopes. SIAM Journal on Algebraic and Discrete Methods 5, 515-525. (1984).

BILLA 1986-1 Billington, P. J.
The capacitated multi-item dynamic lot-sizing problem. IIE Transactions 18, 217-219. (1986).

BILLB 1986-1 Billington, P. J.; Mac Clain, J. O.; Thomas, L. J.
Heuristics for multilevel lot-sizing with a bottleneck. Management Science 32, 989-1006. (1986).

BILLC 1986-1 Billionnet, A.
On interval graphs and matrice profiles. RAIRO - Recherche Operationelle 20, 245-256. (1986).

BILLD 1985-1 Billionnet, A.; Minoux, M.
Maximizing a supermodular pseudoboolean function: a polynomial algorithm for supermodular cubic functions. Discrete Applied Mathematics 12, 1-12. (1985).

BIRG 1985-1 Birge, J. R.; Malyshko, V. V.
Methods for a network design problem in solar power systems. Computers & Operations Research 12, 125-137. (1985).

BIRK 1985-1 Birkhoff, I. J. M.
A direct routing algorithm for the bit-reversal permutation on a shuffle-exchange network. Information Processing Letters 21, 259-268. (1985).

BISH 1987-1 Bisht, N. S.; Prakash, S.
A bicriterion machine-assignment problem. Indian Journal of Pure and Applied Mathematics (New Delhi) 18, 779-789. (1987).

BITR 1986-1 Bitran, G. R.; Haas, E. A.; Matsuo, H.
Production planning of style goods with high setup costs and forcast revisions. Operations Research 34, 226-236. (1986).

BITRA 1984-1 Bitran, G. R.; Magnanti, T. L.; Yanasse, H. H.
Approximation methods for the uncapacitated dynamic lot size problem. Management Science 30, 1121-1140. (1984).

BITRB 1986-1 Bitran, G. R.; Matsuo, H.
Approximation formulations for the single-product capacitated lot size problem. Operations Research 34, 63-74. (1986).

BITRB 1986-2 Bitran, G. R.; Matsuo, H.
The multi-item capacitated lot size problem: error bounds of Manne's formulations. Management Science 32, 350-359. (1986).

BIXB 1984-1 Bixby, R. E.
A composition for perfect graphs. In: C. Berge, V. Chvatal (Eds.): Topics on perfect graphs, Annals of Discrete Mathematics Vol. 21, 221-224. Amsterdam: North-Holland. (1984).

BIXB 1984-2 Bixby, R. E.
Recent algorithms for two versions of graph realization and remarks on applications to linear programming. In: W. R. Pulleyblank (Ed.): Progress in Combinatorial Optimization, 39-68. Toronto: Academic Press. (1984).

BIXB 1987-1 Bixby, R. E.
Notes on combinatorial optimization. TR87-21. Houston: Rice Univ., Dept. of Mathematical Sciences, 66 p. (1987).

BIXBA 1986-1 Bixby, R. E.; Coullard, C. R.
On chins of 3-connected matroids. Discrete Applied Mathematics 15, 155-166. (1986).

BIXBA 1987-1 Bixby, R. E.; Coullard, C. R.
Finding a small 3-connected minor maintaining a fixed minor and a fixed element. Combinatorica 7, 231-242. (1987).

BIXBB 1987-1 Bixby, R. E.; Cunningham, W. H.
Short cocircuits in binary matroids. European Journal of Combinatorics 8, 213-226. (1987).

BIXBC 1985-1 Bixby, R. E.; Cunningham, W. H.; Topkis, D. M.
The partial order of a polymatroid extreme point. Mathematics of Operations Research 10, 367-378. (1985).

BIXBD 1986-1 Bixby, R. E.; Fourer, F.
Finding embedded network rows in linear programs I: extraction heuristics. Report No. 86437-OR. Bonn: Universität Bonn, Inst. für Ökonometrie und Operations Research, 63p. (1986).

BIXBE 1986-1 Bixby, R. E.; Rajan, A.
A short proof of a decomposition theorem for max-flow min-cut matroids. Technical Report 86-24. Houston: Rice Univ., Dept. of Math. Sciences, 15 p. (1986).

BIXBE 1987-1 Bixby, R. E.; Rajan, A.
A short proof of the Truemper-Tseng theorem on max-flow min-cut matroids. TR 87-22. Houston: Rice Univ., Dept. of Math. Sciences, 15 p. (1987).

BIXBF 1984-1 Bixby, R. E.; Wagner, D. K.
An almost linear-time algorithm for graph realization. Report No. 84348-OR. Bonn: Univ. Bonn, Inst. für Ökonometrie u. Operations Research, 39 p. (1984).

BIXBF 1986-1 Bixby, R. E.; Wagner, D. K.
A note on detecting simple redundancies in linear systems. CC-86-18. West Lafayette: Purdue Univ. Inst. of Interdisciplinary Engineering Studies, 7 p. (1986).

BJOR 1985-1 Björner, A.
On matroids, groups and exchange matroids. In: L. Lovasz, A. Recski (Eds.): Matroid Theory, 25-60. Coll. Math. Soc. J. Bolyai Vol. 40. Amsterdam: North-Holland. (1985).

BJORA 1985-1 Björner, A.; Korte, B.; Lovasz, L.
Homotopy properties of greedoids. Advances in Applied Mathematics 6, 447-494. (1985).

BJORB 1987-1 Björner, A.; Lovasz, L.
Pseudomodular lattices and continuous matroids. Acta Scientiarum Mathematicarum (Szeged) 51, 295-308. (1987).

BLAC 1985-1 Blackburn, J. D.; Millen, R. A.
An evaluation of heuristic performance in multi-stage lot-sizing systems. International Journal of Production Research 23, 857-866. (1985).

BLAI 1984-1 Blair, C.
Every finite distributive lattice is a set of stable matchings. Journal of Combinatorial Theory, Series A 37, 353-356. (1984).

BLAI 1986-1 Blair, C.
The lattice structure of the set of stable matchings with multiple partners. Research Paper. Urbana: Univ. of Illinois, 19 p. (1986).

BLAIA 1984-1 Blair, C. E.; Jeroslow, R. G.
Constructive characterizations of the value-function of a mixed-integer program, I. Discrete Applied Mathematics 9, 217-234. (1984).

BLAIA 1985-1 Blair, C. E.; Jeroslow, R. G.
Constructive characterizations of the value-function of a mixed-integer program, II. Discrete Applied Mathematics 10, 227-240. (1985).
BLAIA 1986-1 Blair, C. E.; Jeroslow, R. G.
Computational complexity of some problems in parametric discrete programming. I. Mathematics of Operations Research 11, 241-260. (1986).
BLAIB 1984-1 Blair, C. E.; Jeroslow, R. G.; Lowe, J. K.
Some results and experiments in programming techniques for propositional logic. Atlanta: Georgia Inst. of Technology, 33 p. (1984).
BLAIC 1985-1 Blair, E. L.; Miller, S.
An interactive approach to facilities design using microcomputers. Computers and Industrial Engineering 9, 91-102. (1985).
BLAID 1986-1 Blair, J. R. S.; Lloyd, E. L.
Generalized river routing - algorithms and performance bounds. In: F. Makedon, K. Mehlhorn, T. Papatheodorou, P. Spirakis (Eds.): AWOC 86: VLSI Algorithms and Architectures, 317-328. Lecture Notes in Computer Science 227, Berlin: Springer. (1986).
BLAN 1986-1 Bland, R. G.; Dietrich, B. L.
Abstract duality. Technical Report, 713. Ithaca: Cornell Univ., School of Operations Research and Ind. Eng., 11 p. (1986).
BLANA 1984-1 Bland, R. G.; Huang, H. C.; Trotter, L. E. Jr.
Graphical properties related to minimal imperfection. In: C. Berge, V. Chvatal (Eds.): Topics on perfect graphs, Annals of Discrete Mathematics Vol. 21, 181-192. Amsterdam: North-Holland. (1984).
BLANB 1987-1 Bland, R. G.; Shallcross, D. F.
Large traveling salesman problems arising from experiments in x-ray crystallography: a preliminary report on computation. Technical Report, 730. Ithaca: Cornell Univ., School of Operations Research and Industrial Engineering, 11 p. (1987).
BLANC 1982-1 Blankenship, G. L.; Menaldi, J. L.
Optimal stochastic scheduling of power generation systems with scheduling delays and large cost differentials. In: A. Bensoussan, J. L. Lions (Eds.): Analysis and Optimization of Systems, Proc. 5th Internat. Conf. on Analysis and Optimization of Systems (Versailles 1982), 432-456. Berlin, Heidelberg, New York: Springer. (1982).
BLAS 1982-1 Blass, A.; Gurevich, Y.
On the unique satisfiability problem. Information and Control 55, 80-88. (1982).
BLAZ 1984-1 Blazewicz, J.
Scheduling preemptible tasks on parallel processors with information loss. Technique et Science Informatiques 3, 415-420. (1984).
BLAZ 1987-1 Blazewicz, J.
Selected topics in scheduling theory. In: P. Martello, G. Laporte, M. Minoux, C. Ribeiro (Eds.): Surveys in Combinatorial Optimization, Annals of Discrete Mathematics 31, 1-60. Amsterdam, New York, Tokyo: North-Holland. (1987).
BLAZA 1986-1 Blazewicz, J.; Barcelo, J.; Kubiak, W.; Röck, H.
Scheduling tasks on two processors with deadlines and additional resources. European Journal of Operational Research 26, 364-370. (1986).
BLAZB 1984-1 Blazewicz, J.; Bovet, D. P.; Gambosi, G.
Deadlock-resistant flow control procedures for store-and-forward networks. IEEE Transactions on Communications 32, 884-887. (1984).
BLAZC 1984-1 Blazewicz, J.; Brzezinski, J.; Gambosi, G.
Time-stamps approach to store-and-forward deadlock prevention. Report, 88. Roma: C. N. R., Inst. di Analisi dei Sistemi et Informatica, 26 p. (1984).
BLAZD 1986-1 Blazewicz, J.; Cellary, W.; Slowinski, R.; Weglarz, J.
Scheduling under resource constraints - Deterministic models. Basel: Baltzer. (1986). (Annals of Operations Research, 7)
BLAZE 1986-1 Blazewicz, J.; Drabowski, M.; Weglarz, J.
Scheduling multiprocessor tasks to minimize schedule length. IEEE Transactions on Computers 35, 389-393. (1986).
BLAZF 1983-1 Blazewicz, J.; Ecker, K.
A polynomial in time algorithm for task scheduling under fixed resource constraints. Foundations of Control Engineering (Poznan) 8, 3-7. (1983).
BLAZG 1987-1 Blazewicz, J.; Finke, G.
Minimizing mean weighted execution time loss on identical and uniform processors. Information Processing Letters 24, 259-264. (1987).
BLAZH 1984-1 Blazewicz, J.; Kubiak, W.; Röck, H.
Complexity of minimizing mean flow time under resource constraits. Research Report. Poznan: Politechnika Poznanzka, Inst. Automatyki, 19 p. (1984).
BLAZI 1987-1 Blazewicz, J.; Kubiak, W.; Röck, H.; Szwarcfiter, J. L.
Minimizing mean flow-time with parallel processors and resource constraints. Acta Informatica 24, 513-524. (1987).
BLAZJ 1984-1 Blazewicz, J.; Weglarz, J.
The complexity of scheduling under resource constraints (Polish). Archiwum Automatyki i Telemechaniki (Warsaw) 29, 3-33. (1984).
BLID 1986-1 Blidia, M.
A parity digraph has a kernel. Combinatorica 6, 23-28. (1986).
BLOK 1985-1 Blokh, D. A.
The assignment problem for Jacobi matrices (Russian). Sibirskii Matematicheskii Zhurnal (Novosibirsk) 26, No. 5, 190-191. (1985).
BLOO 1982-1 Bloom, G. S.; Hsu, D. F.
On graceful digraphs and a problem in network addressing. In: F. Hoffman, K. B. Reid, R. C. Mullin, R. G. Stanton (Eds.): Proc. Thirteenth Southeastern Conf. on Combinatorics, Graph Theory and Computing, Florida 1982, Congressus Numerantium Vol. 35, 91-104. Winnipeg, Canada: Utilitas Mathematica. (1982).

BLOO 1985-1 Bloom, G. S.; Hsu, D. F.
On graceful directed graphs that are computational models of some algebraic systems. In: Y. Alavi, G. Chartrand, L. Lesniak, D. R. Lick, C. E. Wall (Eds.): Graph Theory with Applications to Algorithms and Computer Science, 89-102. Chichester: Wiley. (1985).

BLOO 1985-2 Bloom, G. S.; Hsu, D. F.
On graceful directed graphs. SIAM Journal on Algebraic and Discrete Methods 6, 519-536. (1985).

BLUM 1985-1 Blum, N.
On the single-operation worst-case time complexity of the disjoint set union problem. In: K. Mehlhorn (Ed.): STACS 85, Second Annual Symposium on Theoretical Aspects of Computer Science (Saarbrücken, January 1985), Lecture Notes in Computer Science Vol. 182, 32-38. Berlin: Springer-Verlag. (1985).

BLYU 1984-1 Blyumin, S. L.; Kozlov, P. A.; Milovidov, S. P.
A dynamic transportation problem with delays (Russian). Avtomatika i Telemekhanika (Moscow) No. 5, 158-161. (1984).

BOCH 1985-1 Bochenek, K.
An algorithm for optimization of packet flow in telecommunication network. In: Real Time Control of Large Scale Systems, Proc. Fust Europ. Workshop, Patras 1983. Lecture Notes in Control and Information Sciences 67, 537-544. Berlin, Heidelberg, New York: Springer-Verlag. (1985).

BOCT 1985-1 Boctor, F. F.
Single machine lot scheduling: a comparison of some solution procedures. RAIRO-APII Automatique Productique Informatique Industrielle 19, 389-402. (1985).

BODE 1980-1 Bodendiek, R.; Schumacher, H.
Zur Numerierung von Raupen. Mitteilungen der Mathematishcen Gesellschaft in Hamburg 10, 701-704. (1980).

BODEA 1984-1 Bodendiek, R.; Schumacher, H.; Wagner, K.
Über 1-optimale Graphen. Mathematische Nachrichten 117, 323-339. (1984).

BOEN 1984-1 Boenchendorf, K.
Identifying redundant facilities in the capacitated facility location problem. In: G. Hammer, D. Pallaschke (Eds.): Methods of Operations Research 51, Vol. 1, 239-246. Königstein/Ts.: Verlag Anton Hain Meisenheim. (1984).

BOEN 1985-1 Boenchendorf, K.
Improving the Kennington/Unger penalty method for the fixed-charge transportation problem. In: P. Brucker, R. Pauly (Eds.): Proc. IX. Symp. on Operations Research, Osnabrück 1984. Methods of Operations Research 49, 139-146. (1985).

BOEN 1986-1 Boenchendorf, K.
Combining penalties for the fixed-charge transportation problem. Methods of Operations Research 53, 231-238. (1986).

BOEN 1987-1 Boenchendorf, K.
An optimality criterion for degenerated transportation problems. In: W. Domschke, W. Krabs, J. Lehn, P. Spellucci (Eds.): Proc. XI. Symp. on Operations Research, Darmstadt 1986, 87-94. Methods of Operations Research 57, Frankfurt/Main: Athenäum. (1987).

BOENA 1987-1 Boender, C. G. E.; Rinnooy Kan, A. H. G.; Vercellis, C.
Stochastic optimization methods. In: G. Andreatta, F. Mason, P. Serafini (Eds.): Advanced School on Stochastics in Combinatorial Optimization (CISM Udine 1986), 94-112. Singapore: World Scientific. (1987).

BOES 1982-1 Boesch, F. T.
Introduction to basic network problems. In: P. A. Burr (Ed.): The Mathematics of Networks, 1-30. Proc. Symp. in Appl. Math. Vol. 26, American Math. Society. (1982).

BOES 1985-1 Boesch, F. T.
Is reliable network synthesis a graph theory problem? Working Paper. Hoboken: Stevens Inst. of Technology, 27 p. (1985).

BOES 1985-2 Boesch, F. T.
The cut frequency vector. In: Y. Alavi, G. Chartrand, L. Lesniak, D. R. Lick, C. E. Wall (Eds.): Graph Theory with Applications to Algorithms and Computer Science, 103-122. Chichester: Wiley. (1985).

BOESA 1984-1 Boesch, F. T.; Suffel, C. L.
An overview of discrete methods in the synthesis of reliable networks. Large Scale Systems 7, No. 2-3, 191-196. (1984).

BOESB 1985-1 Boesch, F. T.; Tindell, R.
Connectivity and symmetry in graphs. In: F. Harary, J. P. Maybee (Eds.): Graphs and Applications, 53-68. New York: Wiley. (1985).

BOESC 1986-1 Boesch, F. T.; Wang, J. F.
Super line-connectivity properties of circulant graphs. SIAM Journal on Algebraic and Discrete Methods 7, 89-98. (1986).

BOHM 1983-1 Böhme, T.; Sachs, H.
Über räumliche Darstellungen endlicher Graphen. Mitteilungen, Mathematische Gesellschaft der Deutschen Demokratischen Republik No. 1, 5-31. (1983).

BOKH 1987-1 Bokhari, S. H.
Assigment Problems in Parallel and Distributed Computing. Boston, Dordrecht, Lancaster: Kluwer. (1987).

BOKO 1985-1 Bokowski, J.; Sturmfels, B.
Coordinatization of oriented matroids. Preprints in Optimization 85.21. Köln: Univ. Köln, Mathematisches Institut, 18 p. (1985).

BOKO 1986-1 Bokowski, J.; Sturmfels, B.
On the coordinatization of oriented matroids. Discrete and Computational Geometry 1, 293-306. (1986).

BOKO 1986-2 Bokowski, J.; Sturmfels, B.
Reell realisierbare orientierte Matroide. In: A. Kerber (Ed.): Diskrete Strukturen, algebraische Methoden und Anwendungen, Sommerschule Bayreuth 1985. Bayreuther Mathematische Schriften, Heft 21, 1-13. (1986).

BOLD 1986-1 Bold, C.
Order-degree sequences. Report, 86.38. Bonn: Univ. Bonn, Sonderforschungsbereich 303, 10 p. (1986).

BOLL 1985-1 Bollobas, B.
Geodesics in oriented graphs. In: M. Rosenfeld, J. Zaks (Eds.): Convexity and Graph Theory. Annals of Discrete Mathematics Vol. 20, 67-74. Amsterdam: North-Holland. (1985).

BOLL 1985-2 Bollobas, B.
Random graphs. London: Academic Press. (1985).

BOLL 1986-1 Bollobas, B.
Extremal graph theory with emphasis on probabilistic methods. Providence: American Mathematical Society. (1986). (AMS Regional Conference Series in Mathematics, 62)

BOLLA 1983-1 Bollobas, B.; Chung, F. R. K.; Graham, R. L.
On complete bipartite subgraphs contained in spanning tree complements. in: P. Erdös (Ed.): Studies in Pure Mathematics. To the Memory of P. Turan, 83-90. Basel, Boston, Stuttgart: Birkhäuser Verlag. Budapest: Akademiai Kiado. (1983).

BOLLB 1985-1 Bollobas, B.; Fenner, T. I.; Frieze, A. M.
An algorithm for finding Hamilton cycles in random graphs. In: Proc. Seventeenth Annual ACM Symp. on Theory of Computing, Providence 1985, 430-439. ACM. (1985).

BOLLB 1987-1 Bollobas, B.; Fenner, T. I.; Frieze, A. M.
An algorithm for finding Hamilton paths and cycles in random graphs. Combinatorica 7, 327-342. (1987).

BOLLC 1983-1 Bollobas, B.; Frieze, A. M.
On matchings and Hamiltonian cycles in random graphs. Cambridge: Univ. of Cambridge, Department of Pure Mathematics and Mathematical Statistics, 35 p. (1983).

BOLLD 1985-1 Bollobas, B.; Harris, A. J.
List-colourings of graphs. Graphs and Combinatorics 1, 115-128. (1985).

BOLLE 1984-1 Bollobas, B.; Hell, P.
Sorting and graphs. Technical Report, TR 84-3. Burnaby: Simon Fraser Univ., Computing Science, 15 p. (1984).

BOLO 1986-1 Bolotashvili, G. G.
Facets of a permutation polyhedron (Russian). Soobshcheniya Akademii Nauk Gruzinskoj SSR (Tbilisi) 121, 281-284. (1986).

BOLO 1986-2 Bolotashvili, G. G.
On the facets of a permutation polyhedron (Russian). Soobshcheniya Akademii Nauk Gruzinskoi SSR 121, 281-284. (1986).

BOND 1982-1 Bondarenko, V. A.
Neighbouring vertices of polyhedra generated by matroids (Russian). Voprosy Teorii Grupp i Gomologicheskoj Algebry, 66-68. (1982).

BOND 1983-1 Bondarenko, V. A.
Nonpolynomial lower bounds for the complexity of the traveling salesman problem in a class of algorithms. Automation and Remote Control 44, 1137-1142; translation from: Avtomatika i Telemekhanika (Moscow) 1983, No. 9, 45-50. (1983).

BOND 1985-1 Bondarenko, V. A.
Nonpolynomial lower bounds for the complexity of optimization problems of clique and vertex covering in a class of direct algorithms (Russian). In: Combinatorial-Algebraic Methods in Applied Mathematics, 59-65. Gorki: Gorkov. Gos. Univ. (1985).

BONDA 1981-1 Bondarenko, V. A.; Kuksa, A. I.
A solution method for the problem of optimally planning independent jobs in time under constraints on stored and unstored resources (Russian). in: V. P. Mikhalevich (Ed.): Investigation methods for extremal problems, Akademiya Nauk Ukrainskoj SSR, Ordena Lenina Institut Kibernetiki (Kiev), 103-108. (1981).

BONDB 1985-1 Bondy, J. A.
Kotzig's conjecture on generalized friendship graphs - a survey. In: B. R. Alspach, C. D. Godsil (Eds.): Cycles in Graphs, Annals of Discrete Mathematics Vol. 27, 351-366. Amsterdam: North-Holland. (1985).

BONDC 1987-1 Bondy, J. A.; Fan, G.
A sufficient condition for dominating cycles. Discrete Mathematics 67, 205-208. (1987).

BONDD 1985-1 Bondy, J. A.; Jackson, B.
Long paths between specified vertices of a block. In: B. R. Alspach, C. D. Godsil (Eds.): Cycles in Graphs, Annals of Discrete Mathematics Vol. 27, 195-200. Amsterdam: North-Holland. (1985).

BONO 1984-1 Bonomi, E.; Lutton, J.-L.
The n-city travelling salesman problem: statistical mechanics and the metropolis algorithm. SIAM Review 26, 551-568. (1984).

BONO 1986-1 Bonomi, E.; Lutton, J.-L.
The asymptotic behaviour of quadratic sum assignment problems: a statistical mechanics approach. European Journal of Operational Research 26, 295-300. (1986).

BONU 1985-1 Bonuccelli, M. A.
Dominating sets and domatic number of circular arc graphs. Discrete Applied Mathematics 12, 203-214. (1985).

BOOK 1985-1 Bookhold, I.
Untersuchungen zum quadratischen Zuordnungsproblem. 30. Internationales Wissenschaftliches Kolloquium (TH Ilmenau, DDR) Heft 4E, 19-20. (1985).

BORA 1984-1 Boranbaev, S. N.
A method for the distribution of resources on networks (Russian). In: Algorithms for Solving Network Problems, 105-113. Moscow: Akad. Nauk SSSR, Tsentral. Ekonom.-Mat. Inst. (1984).

BORC 1985-1 Borchardt, M.
Einige Resultate zur Dualitätstheorie in der diskreten Optimierung. In: Mathematische Optimierung - Theorie und Anwendungen, 30. Int. Wiss. Kolloq., Ilmenau 1985, 4, 21-24. (1985).

BORC 1985-2 Borchardt, M.
Straf- und Multiplikatormethoden für Optimierungsprobleme mit diskreten und stetigen Variablen. I. Wissenschaftliche Zeitschrift der Technischen Hochschule Ilmenau 31, 51-58. (1985).

BORC 1986-1 Borchardt, M.
Penalty and multiplier methods for optimization problems with discrete and continuous variables II (German). Wissenschaftliche Zeitschrift der Technischen Hochschule Ilmenau 32, No.1, 63-72. (1986).

BORC 1987-1 Borchardt, M.
Zur Anwendung eines Dualitätskonzeptes für quadratische Optimierungsprobleme mit stetigen und Booleschen Variablen. I. Wissenschaftliche Zeitschrift der Technischen Hochschule Ilmenau 33, No.3, 25-33. (1987).

BORC 1987-2 Borchardt, M.
Zur Lösung von nichtlinearen Booleschen Optimierungsproblemen. Wissenschaftliche Zeitschrift der Technischen Hochschule Ilmenau 33, No.4, 101-111. (1987).

BORD 1985-1 Bordat, J. P.
Parcours dans les graphes: un outil pour l'algorithmique des ensembles ordonnes. Discrete Applied Mathematics 12, 215-232. (1985).

BORG 1987-1 Borgwardt, K. H.
Probabilistic analysis of optimization algorithms - some aspects from a practical point of view. Acta Applicandae Mathematicae 10, 171-210. (1987).

BORI 1982-1 Borisov, K. Y.
Convex extremal problems on graphs (Russian). Primenenie Matematiki v Ekonomike (Leningrad) No. 13, 26-31. (1982).

BORIA 1985-1 Borisova, L. N.
The telephone center and the plant centroid (Russian). In: Problems in Group Theory and Homological Algebra, 152-153. Yaroslavl: Yaroslav. Gos. Univ. (1985).

BORO 1986-1 Borodich, S. A.; Strusevich, V. A.
Schedules for servicing a partially ordered set of customers for a class of determinate systems (Russian). Vestsi Akademii Navuk BSSR, Seryya Fizika-Matematychnykh Navuk (Minsk), No. 3, 19-22. (1986).

BOROA 1985-1 Borodich, S. A.; Tuzikov, A. V.
On the complexity of constructing time-optimal schedules for some two-stage service systems with non-fixed routes (Russian). V. P. Tanaev (Ed.): Methods, algorithms and programs for the solution of extremal problems, Collect. sci. Works, Minsk 1985, 76-85. (1985).

BOROB 1984-1 Borodin, O. V.
Solution of the Ringel problem on vertex-edge coloring of plane graphs and coloring of 1-planar graphs (Russian). Metody Diskretnogo Analiza (Novosibirsk) No. 41, 12-26. (1984).

BOROB 1987-1 Borodin, O. V.
Simultaneous colorings of graphs on the plane (Russian). Metody Diskretnogo Analiza 45, 21-27. (1987).

BOROC 1983-1 Boronin, A. B.; Pershin, O. Y.
An algorithm for partition of a weighted graph into minimally connected subgraphs (Russian). Problems of control of multiply connected systems, "Nauka", (Moscow), 77-91. (1983).

BOROC 1984-1 Boronin, A. B.; Pershin, O. Y.
Decomposition of a network into minimally connected parts. Automation and Remote Control 45, No. 4, 519-528. Translated from Avtomatika i Telemekhanika (Moscow) No. 4, 128-138. (1984).

BOROC 1984-2 Boronin, A. B.; Pershin, O. Y.
Decomposition of a network into minimally connected parts. Avtomatika i Telemekhanika, No. 4, 128-138; translated as: Automation and Remote Control 45, No. 4, part 2, 519-528. (1984).

BOROD 1986-1 Boros, E.
On the complexity of the surrogate dual of 0-1 programming. Zeitschrift für Operations Research A-30, 145-154. (1986).

BOROE 1986-1 Borosh, I.; Flahive, M.; Treybig, B.
Small solutions of linear Diophantine equations. Discrete Mathematics 58, 215-220. (1986).

BOROF 1984-1 Borowiecki, M.
Point partition numbers and generalized Nordhaus-Gaddum problems. Colloquium Mathematicum (Warsaw) 40, 279-285. (1984).

BOROF 1987-1 Borowiecki, M.
Characterization of graphs having minimaximal kernels cardinality half their order. In: M. Borowiecki; Z. Skupien (Eds.): Graphs, Hypergraphs and Matroids II, Proc. Sixth Regional Scientific Session of Mathematicians, Zagan, Poland, 1986, 27-28. Section: Graph Theory. Zielona Gora: Higher College of Engineering. (1987).

BOROF 1987-2 Borowiecki, M.
Partition numbers, connectivity and Hamiltonicity. Casopis pro Pestovani Matematiky 112, 173-176. (1987).

BOROF 1987-3 Borowiecki, M.
Two extremal problems in the class of uniquely colorable graphs. In: M. Borowiecki; Z. Skupien (Eds.): Graphs, Hypergraphs and Matroids II, Proc. Sixth Regional Scientific Session of Mathematicians, Zagan, Poland, 1986, 17-25. Section: Graph Theory. Zielona Gora: Higher College of Engineering. (1987).

BOROG 1985-1 Borowiecki, M.; Mihok, P.
On (n,P)-partitionable graphs. In: H. Sachs (Ed.): Graphs, Hypergraphs and Applications - Proc. Conf. Graph Theory, Eyba/GDR 1984, 15-18. Leipzig: Teubner. (1985).

BORZ 1987-1 Borzacchini, L.
Graphical partially ordered sets. Discrete Applied Mathematics 18, 247-262. (1987).

BORZA 1984-1 Borzacchini, L.; Pulito, C.
Coverings enumeration for graphical partially ordered sets. In; A. Barlotti, M. Marchi, G. Tallini (Eds.): Proc. Conf. on Combinatorial and Incidence Geometry: Principles and Applications (La Mendola, 1982), 161-168. Milan: Rendiconti del Seminario Matematico die Brescia. (1984).

BORZB 1983-1 Borzenkov, A. V.; Gabasov, R.
An algorithm for solving a special minimax problem (Russian). Vestnik Belorusskogo Gosudarstvennogo Universiteta Imeni V.I. Lenina, Seriya I, No.2, 42-45. (1983).

BOSA 1983-1 Bosak, J.
Enumeration of uniquely edge colourable multigraphs. In: M. Fiedler (Ed.): Graphs and other Combinatorial Topics. Proc. Third Czech. Symp. on Graph Theory, Prague 1982, 17-22. Leipzig: Teubner. (1983).

BOSA 1984-1 Bosak, J.
Induced subgraphs. In: A. Hajnal; L. Lovasz; V. T. Sos (Eds.): Finite and Infinite Sets, Vol. 1, 109-118. Coll. Math. Soc. J. Bolyai 37, Amsterdam: North-Holland. (1984).

BOSA 1984-2 Bosak, J.
Uniquely edge colourable graphs. Mathematica Slovaca (Bratislava) 34, 205-216. (1984).
BOTT 1986-1 Bott, K.; Ballou, R. H.
Research perspectives in vehicle routing and scheduling. Transportation Research 20-A, 239-243. (1986).
BOUC 1987-1 Boucher, T. O.
Choice of assembly line design under task learning. International Journal of Production Research 25, 513-524. (1987).
BOUCA 1985-1 Bouchet, A.
Un algorithme polynomial pour reconnaitre les graphes d'alternance. Comptes Rendus des Sciences de l'Academie des Sciences, Serie I, Mathematique 300, 569-572. (1985).
BOUCA 1986-1 Bouchet, A.
Characterizing and recognizing circle graphs. In: R. Tosic, D. Acketa, V. Petrovic (Eds.): Graph Theory - Proc. 6th Yugoslav Seminar, Dubrovnik 1985, 57-69. Novi Sad: Inst. of Math., Univ. of Novi Sad. (1986).
BOUCA 1987-1 Bouchet, A.
Delta-matroids and matchings. Le Mans: Universite du Maine, 13 p. (1987).
BOUCA 1987-2 Bouchet, A.
Greedy algorithm and symmetric matroids. Mathematical Programming 38, 147-160. (1987).
BOUCA 1987-3 Bouchet, A.
Maps and delta-matroids. Le Mans: Universite du Maine, 18 p. (1987).
BOUCA 1987-4 Bouchet, A.
Reducing prime graphs and recognizing circle graphs. Combinatorica 7, 243-254. (1987).
BOUCA 1987-5 Bouchet, A.
Representability of delta-matroids. Le Mans: Universite du Maine, 18 p. (1987).
BOUCA 1987-6 Bouchet, A.
Unimodularity and circle graphs. Discrete Mathematics 66, 203-208. (1987).
BOUCB 1987-1 Bouchet, A.; Fouquet, J. L.; Jolivet, J. L.; Riviere, M.
On a special face colouring of cubic graphs. Ars Combinatoria 24, 67-76. (1987).
BOUCC 1985-1 Bouchitte, V.
Chordal bipartite graphs and crowns. Order 2, 119-122. (1985).
BOUCD 1986-1 Bouchitte, V.; Habib, M.
Some NP-completeness properties about linear extensions. Rapport de Recherche 11. Brest: Univ. de Bretagne Occidentale, 14 p. (1986).
BOUCE 1985-1 Bouchitte, V.; Habib, M.; Jegou, R.
On the greedy dimension of a partial order. Order 1, 219-224. (1985).
BOUR 1986-1 Bourjolly, J.-M.
Integral and fractional node packings, and pseudo-Boolean programming. Waterloo: Univ. of Waterloo, 269 p. (1986). (Dissertation)
BOURA 1984-1 Bourjolly, J.-M.; Hammer, P. L.; Simeone, B.
Node-weighted graphs having the König-Egervary property. Mathematical Programming Study 22, 44-63. (1984).
BOVE 1986-1 Bovet, J.
Une amelioration de la methode de Dijkstra pour la recherche d'un plus court chemin dans un reseau. Discrete Applied Mathematics 13, 93-96. (1986).
BOXM 1985-1 Boxma, O. J.
A probabilistic analysis of the LPT scheduling rule. In: E. Gelenbe (Ed.): Performance '84 (Paris 1984), 475-490. Amsterdam, New York: North-Holland. (1985).
BOXMA 1986-1 Boxma, O. J.; Forst, F. G.
Minimizing the expected weighted number of tardy jobs in stochastic flow shops. Operations Research Letters 5, 119-126. (1986).
BOYA 1987-1 Boyar, J. F.; Karloff, H. J.
Coloring planar graphs in parallel. Journal of Algorithms 8, 470-479. (1987).
BOYAA 1983-1 Boyarchuk, D. A.
Investigation of the admissible domain of an integer programming problem with a parameter in one of the constraints (Russian). in: Software for program packages, 35-40. Kiev: Akademiya Nauk Ukrainskoj SSR, Inst. Kibernet. (1983).
BOYAA 1985-1 Boyarchuk, D. A.
An algorithm for solving a parametric integer linear programming problem (Russian). In: Discrete optimization methods and efficient organization of program packages, 56-59. Kiev: Akademiya Nauk Ukrainskoi SSSR, Institut Kibernetiki. (1985).
BOYC 1984-1 Boyce, D. E.
Network models in transportation/land use planning. In: M. Florian (Ed.): Transportation Planning Models, 475-498. Amsterdam: North-Holland. (1984).
BOYD 1986-1 Boyd, S. C.
The subtour polytope of the travelling salesman problem. Waterloo: Univ. of Waterloo, 184 p. (1986). (Dissertation)
BOYDA 1984-1 Boyd, S. C.; Pulleyblank, W. R.
Optimizing over the subtour polytope of the travelling salesman problem. Research Report, 84-12. Waterloo: Univ. of Waterloo, Dept. of Combinatorics and Optimization, 29 p. (1984).
BOYDB 1987-1 Boyd, S. C.; Pulleyblank, W. R.; Cornuejols, G.
Travel - an interactive travelling salesman problem package for the IBM-personal computer. Operations Research Letters 6, 141-144. (1987).
BRAC 1984-1 Bracho, R. L.
Le nombre achromatique d'une etoile. Ars Combinatoria 18, 187-194. (1984).
BRAM 1984-1 Brameller, A.
Sparsity in transportation problems. In: D. Evans (Ed.): Sparsity and its applicatins, 229-242. (Meeting Loughborough, England, 1983). Cambridge: Cambridge Univ. Press. (1984).
BRAN 1986-1 Brandeau, M. L.; Chiu, S. S.; Batta, R.
Locating the 2-median of a tree network with continuous link demands. In: J. P. Osleeb, P. J. Ratick (Eds.): Locational Decisions: Methodology and Applications, 223-256. Annals of Operations Research 6, Basel: Baltzer. (1986).

BRANA 1985-1 Brandstädt, A.
A characterization of split permutation graphs. Forschungsergebnisse N/82/21 Friedrich-Schiller-Universität Jena. (1985).

BRANA 1985-2 Brandstädt, A.
Partitions of graphs into one or two independent sets and cliques. In: H. Sachs (Ed.): Graphs, Hypergraphs and Applications, Proc. Conf. on Graph Theory (Eyba 1984), 19-22. Leipzig: Teubner. (1985).

BRANA 1986-1 Brandstädt, A.
Characterization of split permutation graphs. In: J. Machner, G. Schaar (Eds.): Algebra und Graphentheorie, Beiträge der Jahrestagung "Algebra und Grenzgebiete" in Siebenlehn/ DDR 1985, 21-24. Freiberg: Bergakademie Freiberg, Sektion Mathematik. (1986).

BRANA 1987-1 Brandstädt, A.
The computational complexity of feedback vertex set, Hamiltonian circuit, dominating set, Steiner tree, and bandwidth on special perfect graphs. Journal of Information Processing and Cybernetics 23, 471-477. (1987).

BRANB 1985-1 Brandstädt, A.; Kratsch, D.
On the restriction of some NP-complete graph problems to permutation graphs. In: L. Budach (Ed.): Fundamentals of Computation Theory, FCT '85, Lecture Notes in Computer Science Vol. 199, 53-62. Berlin: Springer-Verlag. (1985).

BRANB 1987-1 Brandstädt, A.; Kratsch, D.
On domination problems for permutation and other graphs. Theoretical Computer Science 54, 181-198. (1987).

BRAS 1986-1 Bräsel, H.
Zu einigen Eigenschaften des open shop Problems. Seminarber., Humboldt-Univ. Berlin, Sekt. Math. 85, 11-18. (1986).

BRAS 1987-1 Bräsel, H.
Rank minimal machine scheduling problems (German). Wissenschaftliche Zeitschrift der Technischen Universität Otto von Guericke 31, No.5, 68-72. (1987).

BREB 1985-1 Brebner, G.
Relating routing graphs and 2-dimensional grids. In: P. Bertolazzi, F. Luccio (Eds.): VLSI: Algorithms and Architectures, 221-232. Amsterdam: North-Holland. (1985).

BREC 1985-1 Brecht, T. B.; Colbourn, C. J.
Lower bounds on two-terminal network reliability. Technical Report CCNG E-127. Waterloo: Univ. of Waterloo, Computer Communications Networks Group, 19 p. (1985).

BREC 1986-1 Brecht, T. B.; Colbourn, C. J.
Improving bounds on network reliability: some examples. Congressus Numerantium 51, 93-102. (1986).

BREN 1984-1 Brenner, J. L.; Lyndon, R. C.
Doubly Eulerian trails on rectangular grids. Journal of Graph Theory 8, 379-386. (1984).

BREZ 1986-1 Brezovec, C.; Cornuejols, G.; Glover, F.
Two algorithms for weighted matroid intersection. Mathematical Programming 36, 39-53. (1986).

BREZ 1987-1 Brezovec, C.; Cornuejols, G.; Glover, F.
A matroid algorithm and its application to the efficient solution of two optimization problems on graphs. MSRR 534. Pittsburgh. Carnegie-Mellon Univ., Graduate School of Ind. Admin., 21 p. (1987).

BRIC 1983-1 Brickell, E. F.
Are most low density knapsacks solvable in polynomial time. Congressus Numerantium 39, 145-156. (1983).

BRIC 1984-1 Brickell, E. F.
Solving low density knapsacks. In: Advances in Cryptology (Santa Barbara, Calif., 1983), 25-37. New York, London: Plenum. (1984).

BRID 1987-1 Bridgland, M. F.
Universal traversal sequences for paths and cycles. Journal of Algorithms 8, 395-404. (1987).

BRIG 1984-1 Brigham, R. C.; Chinn, P. Z.; Dutton, R. D.
On the clique graph of the line graph. Congressus Numerantium 43, 175-190. (1984).

BRIGA 1984-1 Brigham, R. C.; Dutton, R. D.
Upper bounds on the edge clique cover number of a graph. Discrete Mathematics 52, 31-38. (1984).

BROD 1984-1 Broder, A.; Dolev, D.; Fischer, M.; Simons, B.
Efficient fault-tolerant routings in networks. IBM Research Report, RJ 4259. Yorktown Heights: IBM Research Center, 18 p. (1984).

BRODA 1982-1 Brodskij, G. M.; Soloveva, N. A.
The partially ordered set of subtrees of a graph (Russian). In: Problems in group theory and homological algebra, 153-154. Yaroslav. Gos. Univ.: Yaroslavl. (1982).

BROE 1986-1 Broere, I.; Burger, M.
Uniquely cocolourable graphs. Graphs and Combinatorics 2, 201-208. (1986).

BROEA 1985-1 Broere, I.; Frick, M.
On the order of color critical graphs. Congressus Numerantium 47, 125-130. (1985).

BROEB 1985-1 Broere, I.; Mynhardt, C. M.
Generalized colorings of outerplanar and planar graphs. In: Y. Alavi, G. Chartrand, L. Lesniak, D. R. Lick, C. E. Wall (Eds.): Graph Theory with Applications to Algorithms and Computer Science, 151-162. Chichester: Wiley. (1985).

BROEC 1987-1 Broersma, H. J.
Subgraph conditions for dominating circuits in graphs and pancyclicity of line graphs. Ars Combinatoria 23, 5-12. (1987).

BROI 1986-1 Broin, M. W.; Lowe, T. J.
A dynamic programming algorithm for covering problems with (greedy) totally balanced constraint matrices. SIAM Journal on Algebraic and Discrete Methods 7, 348-357. (1986).

BRON 1984-1 Bronshtein, M. L.
An algorithm for finding the shortest path for a network with alternate arc lengths (Russian). In: K. V. Kim, B. V. Cherkasskij (Eds.): Algorithms for Solving Network Problems, 42-44. Moscow: Akad. Nauk SSSR, Tsentral. Ekonom.-Mat. Inst. (1984).

BROO 1981-1 Brooks, R. E.
Using generalized networks to forecast natural gas distribution and allocation during periods of shortage. Mathematical Programming Study 15, 23-42. (1981).
BROU 1985-1 Brouwer, A. E.; Mesner, D. M.
The connectivity of strongly regular graphs. European Journal of Combinatorics 6, 215-216. (1985).
BROW 1987-1 Brown, G. G.; Graves, G. W.; Ronen, D.
Scheduling ocean transportation of crude oil. Management Science 33, 335-346. (1987).
BROWA 1984-1 Brown, G. G.; Mac Bride, R. D.
Solving generalized networks. Management Science 30, 1497-1523. (1984).
BROWB 1985-1 Brown, G. G.; Mac Bride, R. D.; Wood, R. K.
Extracting embedded generalized networks from linear programming problems. Mathematical Programming 32, 11-31. (1985).
BROWC 1985-1 Brown, W. G.; Erdös, P.; Simonovits, M.
Algorithmic solution of extremal digraph problems. Transactions of the American Mathematical Society 292, 421-449. (1985).
BRUA 1987-1 Brualdi, R. A.; Katz, M.
An extremal problem concerning matrices of 0's and 1's. Linear and Multilinear Algebra 20, 325-331. (1987).
BRUC 1981-1 Brucker, P.
The Chinese postman problem for mixed graphs. In: H. Noltemeier (Ed.): Graphtheoretic Concepts in Computer Science, Proc. Internat. Workshop WG 80 (Bad Honnef 1980), 354-366. Berlin, Heidelberg, New York: Springer. (1981).
BRUC 1984-1 Brucker, P.
A greedy algorithm for solving network flow problems in trees. In: R. E. Burkard, R. A. Cuninghame-Green, U. Zimmermann (Eds.): Algebraic and Combinatorial Methods in Operations Research, Annals of Discrete Mathematics Vol. 19, 23-34. Amsterdam: North-Holland. (1984).
BRUC 1984-2 Brucker, P.
An o(n) algorithm for quadratic knapsack problems. Operations Research Letters 3, 163-166. (1984).
BRUC 1984-3 Brucker, P.
Über gewisse Prinzipien des Entwurfs effizienter Algorithmen. Bericht Nr. 233 der mathematisch-statistischen Sektion der Forschungsgesellschaft Joanneum, T. U. Graz. (1984).
BRUC 1985-1 Brucker, P.
Parametric programming and circulation problems with one additional linear constraint. In: H. Noltemeier (Ed.): Proc. WG '85: Internat. Workshop on Graphtheoretic Concepts in Computer Science, Castle Schwanberg, Würzburg, 1985, 27-42. Linz: Trauner. (1985).
BRUC 1988-1 Brucker, P.
Scheduling periodically recurring events and processes. In: H. Schellhaas, P. van Beek, H. Isermann, R. Schmidt, M. Zijlstra (Eds.): Operations Research Proceedings 1987 (DGOR/ NSOR), 598-599. Berlin: Springer. (1988).
BRUCA 1987-1 Brucker, P.; Hamacher, H. W.
K-optimal solution sets for some polynomially solvable scheduling problems. Research Report 87-16. Gainesville: Univ. of Florida, Industrial & System Eng. Dept., 27 p. (1987).
BRUCB 1986-1 Brucker, P.; Hurink, J.
A railway scheduling problem. Zeitschrift für Operations Research A-30, 223-228. (1986).
BRUCC 1985-1 Brucker, P.; Papenjohann, W.
An out-of-kilter method for the algebraic circulation problem. Discrete Applied Mathematics 10, 1-18. (1985).
BRUCD 1984-1 Brucker, P.; Papenjohann, W.; Zimmermann, U.
A dual optimality criterion for algebraic linear programs. In: R. E. Burkard, R. A. Cuninghame-Green, U. Zimmermann (Eds.): Algebraic and Combinatorial Methods in Operations Research, Annals of Discrete Mathematics Vol. 19, 35-40. Amsterdam: North-Holland. (1984).
BRUD 1985-1 Brudaru, O.
A parallel algorithm for a variant of the acyclic partition problem. 30. Internationales Wissenschaftliches Kolloquium (TH Ilmenau, DDR) Heft 5F, 29-32. (1985).
BRUD 1987-1 Brudaru, O.
Hamiltonian circuits with generalized cost. Optimization 18, 405-412. (1987).
BRUG 1984-1 Brugha, C. M.
Finding embedded networks in linear programmes. Research Report. Dublin: Dublin Inst. of Technology, College of Marketing and Design, 11 p. (1984).
BRUI 1984-1 Bruin, R. de
Some software for graph theoretical problems. Colloquium Topics in Applied Numerical Analysis 1, 197-214. CWI Syllabi 4, Math. Centrum, Amsterdam. (1984).
BRUN 1983-1 Brunn, B.
Ein Verfahren zur Lösung von ganzzahligen nichtlinearen Optimierungsproblemen. In: 2nd Colloquium for Geometry and Combinatorics (Karl-Marx-Stadt, 1983) Part 1, 2, 55-58. Karl-Marx-Stadt: Technische Hochschule. (1983).
BRUNA 1985-1 Bruno, J. L.
On scheduling tasks with exponential service times and in-tree precedence constraints. Acta Informatica 22, 139-148. (1985).
BRUNB 1985-1 Bruno, J. L.; Downey, P. J.
Probabilistic bounds for dual bin-packing. Acta Informatica 22, 333-345. (1985).
BRUV 1985-1 Bruvold, N. T.; Evans, J. R.
Flexible mixed-integer programming formulations for production scheduling problems. IIE Transactions 17, 2-7. (1985).
BRYA 1984-1 Bryant, V. W.
Undirected strict gammoids. Fundamenta Mathematicae 120, 1-5. (1984).
BRYA 1984-2 Bryant, V. W.
Undirected strict grammoids. Fundamenta Mathematicae (Warsaw) 120, 1-5. (1984).
BRYL 1985-1 Brylawski, T.
Coordinatizing the Dilworth truncation. In: L. Lovasz, A. Recski (Eds.): Matroid Theory, 61-96. Coll. Math. Soc. J. Bolyai Vol. 40. Amsterdam: North-Holland. (1985).

BRYL 1986-1 Brylawski, T.
Blocking sets and the Möbius function. In: Combinatorica, Convegno Roma, 1983, Symposia Mathematica, Vol.28, 231-249. New York: Academic Press. (1986).

BRYL 1986-2 Brylawski, T.
Iterated parallel union of matroids. Rendiconti del Seminario Matematico e Fisico di Milano 53, 229-244. (1986).

BU 1984-1 Bu, Y. H.
Generating trees obtained from a cycle basis (Chinese). Qufu Shiyuan Xuebao (Qufu), No. 3, 53-54. (1984).

BUCK 1985-1 Buckingham, M. A.; Golumbic, M. C.
Recent results on the strong perfect graph conjecture. In: M. Rosenfeld, J. Zaks (Eds.): Convexity and Graph Theory. Annals of Discrete Mathematics Vol. 20, 75-82. Amsterdam: North-Holland. (1985).

BUCKA 1984-1 Buckley, F.
Equalities involving certain graphical distributions. In: K. M. Koh, H. P. Yap (Eds.): Graph Theory Singapore 1983, Lecture Notes in Mathematics 1073, 179-192. Berlin: Springer-Verlag. (1984).

BUCKA 1987-1 Buckley, F.
Facility location problems. College Mathematics Journal 18, No. 1, 24-32. (1987).

BUEL 1983-1 Buell, K. E.; Lesniak, L.; Ringeisen, R. D.
More on mission networks. Congressus Numerantium 39, 157-168. (1983).

BUHL 1988-1 Buhl, H. U.
Zielsetzungen bei der Modellierung von Standortproblemen. In: H. Schellhaas, P. van Beek, H. Isermann, R. Schmidt, M. Zijlstra (Eds.): Operations Research Proceedings 1987 (DGOR/ NSOR), 261-269. Berlin: Springer. (1988).

BUHLA 1986-1 Buhl, H. U.; Siedersleben, J.
Minimizing the transportation costs of a distribution system. In: L. Streitferdt, H. Hauptmann, A. W. Marusev, D. Ohse, U. Pape (Eds.): Operations Research Proceedings 1985, 286-294. Berlin, Heidelberg, New York: Springer-Verlag. (1986).

BUI 1984-1 Bui, T. N.; Chaudhuri, S.; Leighton, F. T.; Sipser, M.
Graph bisection algorithms with good average case behavior. In: 25th Annual Symposium on Foundations of Computer Science, Florida 1984, 181-192. Los Angeles: IEEE Computer Society. (1984).

BUI 1987-1 Bui, T. N.; Chaudhuri, S.; Leighton, F. T.; Sipser, M.
Graph bisection algorithms with good average case behavior. Combinatorica 7, 171-192. (1987).

BUJA 1985-1 Bujan, M. F.
Mathematical model for planning in an industrial branch. Ekonomicko-Matematicky Obzor (Prague) 21, 387-394. (1985).

BULU 1985-1 Bulut, H.
Decomposition properties of the transportation problem (Turkish). Fen Fakültesi Dergisi, Special Issue 2, 302-310. (1985).

BUNN 1985-1 Bunnag, P.; Smith, S. B.
A multifactor priority rule for jobshop scheduling using computer search. IIE Transactions 17, 141-146. (1985).

BURA 1985-1 Burattini, E.; Marra, G.; Sforza, A.
Network design problem: structure of solutions and dominance relations. Acta Cybernetica (Szeged) 7, 225-233. (1985).

BURG 1985-1 Burgin, M. S.
Algorithms and algorithmic problems (Russian). Programmirovanie (Moscow), No. 4, 3-14. (1985).

BURK 1985-1 Burkard, R. E.
Time-slot assignment for TDMA-systems. In: F. Nozicka (Ed.): Optimization - Theory, Methods, Applications, I, International Conference Marianske Lazne 1985, 16-18. Prague: Dum Techniky CSVTS. (1985).

BURK 1985-2 Burkard, R. E.
Time-slot assignment for TDMA-systems. Computing 35, 99-112. (1985).

BURK 1986-1 Burkard, R. E.
Assignment problems: recent solution methods and applications. In: A. Prekopa, J. Szeleszan, B. Strazicky (Eds.): System Modelling and Optimization - Proc. 12th IFIP Conf., Budapest 1985, 153-169. Lecture Notes in Control and Information Sciences 84, Berlin: Springer. (1986).

BURK 1986-2 Burkard, R. E.
Optimal schedules of periodically recurring events. Discrete Applied Mathematics 15, 167-180. (1986).

BURKA 1985-1 Burkard, R. E.; Euler, R.; Grommes, R.
On the facial structure of polytopes related to the planar 3-dimensional assignment problem. In: P. Brucker, R. Pauly (Eds.): Proc. IX. Symp. on Operations Research, Osnabrück 1984. Methods of Operations Research 49, 147-160. (1985).

BURKB 1985-1 Burkard, R. E.; Fincke, U.
Probabilistic asymptotic properties of some combinatorial optimization problems. Discrete Applied Mathematics 12, 21-30. (1985).

BURKC 1985-1 Burkard, R. E.; Hamacher, H. W.; Tind, J.
On general decomposition schemes in mathematical programming. Mathematical Programming Study 24, 238-252. (1985).

BURKD 1984-1 Burkard, R. E.; Krarup, J.; Pruzan, P. M.
Some relationships between multicriteria and parametric discrete optimization problems with bottleneck criteria. Optimization 15, 389-396. (1984).

BURL 1984-1 Burlet, M.; Fonlupt, J.
Polynomial algorithm to recognize a Meyniel graph. In: W. R. Pulleyblank (Ed.): Progress in Combinatorial Optimization, 69-100. Toronto: Academic Press. (1984).

BURL 1984-2 Burlet, M.; Fonlupt, J.
Polynomial algorithm to recognize a Meyniel graph. In: C. Berge, V. Chvatal (Eds.): Topics on perfect graphs, Annals of Discrete Mathematics Vol. 21, 225-252. Amsterdam: North-Holland. (1984).

BURLA 1984-1 Burlet, M.; Uhry, J.-P.
Parity graphs. In: C. Berge, V. Chvatal (Eds.): Topics on perfect graphs, Annals of Discrete Mathematics Vol. 21, 253-278. Amsterdam: North-Holland. (1984).

BURN 1984-1 Burns, D.
Biregular edge-symmetric graphs. Colloquium Mathematicum (Warsaw) 49, 137-140. (1984).

BURN 1984-2 Burns, D.
The n-minimal chromatic multiplicity of a graph. Colloquium Mathematicum (Warsaw) 49, 131-135. (1984).

BURNA 1984-1 Burns, D.; Kapoor, S. F.; Ostrand, P. A.
On line-symmetric graphs. Fundamenta Mathematicae (Warsaw) 122, No. 1, 1-21. (1984).

BURNB 1987-1 Burns, L. D.; Daganzo, C. F.
Assembly line job sequencing principles. International Journal of Production Research 25, 71-100. (1987).

BURNC 1985-1 Burns, R. N.; Carter, M. W.
Work force size and single shift schedules with variable demands. Management Science 31, 599-607. (1985).

BURND 1987-1 Burns, R. N.; Koop, G. J.
A modular approach to optimal multiple-shift manpower scheduling. Operations Research 35, 100-110. (1987).

BURR 1985-1 Burr, S. A.; Falk, J. E.; Karr, A. F.
Integer prim-read solutions to a class of target defense problems. Operations Research 33, 726-745. (1985).

BURRA 1985-1 Burr, S. A.; Nesetril, J.; Rödl, V.
On the use of senders in generalized Ramsey theory for graphs. Discrete Mathematics 54, 1-14. (1985).

BUSN 1984-1 Busnyuk, N. N.
The number of 3-cycles in graphs with given degrees of vertices (Russian). Vestnik Belorusskogo Gosudarstvennogo Universiteta (Minsk), Seriya 1 No. 1, 57-58. (1984).

BUSNA 1985-1 Busnyuk, N. N.; Sarvanov, V. I.
Graphs of partition of a rectangle. I (Russian). Vestsi Akademii Navuk BSSR, Seryya Fizika-Matematychnykh Navuk (Minsk) 1985, No. 6, 26-36, 125. (1985).

BUTK 1984-1 Butkovic, P.
On the uniqueness of the assignment problem solution. In: K.-H. Elster (Ed.): Mathematische Optimierung - Theorie und Anwendungen, Vortragsauszüge, Eisenach 1984, 9-10. Techn. Hochschule Ilmenau. (1984).

BUTT 1985-1 Butterworth, K.
Practical application of linear/integer programming in agriculture. Journal of the Operational Research Society 36, 99-108. (1985).

BUTZ 1984-1 Butz, L.; Hammer, P. L.; Hausmann, D.
Reduction methods for the vertex packing problem. In: M. Iosifescu (Ed.): Probability Theory, Proc. 7th Conf. on Probability Theory (Brasov, Sept. 1982), 73-79. Bucharest: Editura Academiei Republicii Socialiste Romania. (1984).

BUXE 1985-1 Buxey, G.
Designing routes to minimize fleet operating costs. European Journal of Operational Research 21, 57-64. (1985).

BUZA 1984-1 Buzacott, J. A.; Chang, J. S. K.
Cut-set intersections and node partitions. IEEE Transactions on Reliability R-33, 385-389. (1984).

BUZAA 1985-1 Buzacott, J. A.; Shanthikumar, J. G.
On approximate queueing models of dynamic job shops. Management Science 31, 870-887. (1985).

BUZY 1981-1 Buzytskij, P. L.; Freiman, G. A.
Integer programming and number theory. In: J. L. Lebowitz (Ed.): Collective phenomena, 4th Int. Conf., Moscow 1981, Ann. N.Y. Acad. Sci. 373, 191-201. (1981).

BUZY 1983-1 Buzytskij, P. L.; Freiman, G. A.
An effective formula for the number of solutions of a system of 0,1 linear equations. In: J. L. Lebowitz (Ed.): Collective phenomena, 5th Int. Conf., Moscow 1981, Ann. N.Y. Acad. Sci. 410, 83-90. (1983).

BYER 1984-1 Byers, T. H.; Waterman, M. S.
Determining all optimal and near-optimal solutions when solving shortest path problems by dynamic programming. Operations Research 32, 1381-1384. (1984).

BYRD 1987-1 Byrd, R. H.; Goldman, A. J.; Heller, M.
Recognizing unbounded integer programs. Operations Research 35, 140-142. (1987).

BYST 1987-1 Bystrov, V. A.; Shcherbina, O. A.
Solution of block problem of discrete optimization (Russian). Kibernetika (Kiev) No. 3, 96-101. (1987).

CABO 1986-1 Cabot, A. V.; Erenguc, S. S.
A branch and bound algorithm for solving a class of nonlinear integer programming problems. Naval Research Logistics Quarterly 33, 559-568. (1986).

CACC 1987-1 Caccetta, L.; Erdös, P.; Ordman, E. T.; Pullman, N. J.
The difference between the clique numbers of a graph. Ars Combinatoria 19A, 97-106. (1987).

CACCA 1986-1 Caccetta, L.; Vijayan, K.
On minimal graphs with prescribed adjacency property. Ars Combinatoria 21-A, 21-30. (1986).

CACCA 1987-1 Caccetta, L.; Vijayan, K.
Applications of graph theory. Ars Combinatoria 23B, 21-78. (1987).

CAEN 1985-1 Caen, D. de; Gregory, D. A.
Partitions of the edge-set of a multigraph by complete subgraphs. Congressus Numerantium 47, 255-264. (1985).

CAEN 1987-1 Caen, D. de; Gregory, D. A.
On the decomposition of a directed graph into complete bipartite subgraphs. Ars Combinatoria 23B, 139-146. (1987).

CAENA 1985-1 Caen, D. de; Gregory, D. A.; Pullman, N. J.
Clique coverings of complements of paths and cycles. In: B. R. Alspach, C. D. Godsil (Eds.): Cycles in Graphs, Annals of Discrete Mathematics Vol. 27, 257-268. Amsterdam: North-Holland. (1985).

CAHI 1987-1 Cahit, I.
Cordial graphs: a weaker version of graceful and harmonious graphs. Ars Combinatoria 23, 201-208. (1987).

CAI 1987-1 Cai, J.; Meyer, G. E.
On the complexity of graph critical uncolorability. In: T. Ottmann (Ed.): Automata, Languages and Programming, 394-403. Lecture Notes in Computer Science 267, Berlin: Springer. (1987).

CAI A 1983-1 Cai, M.-C.
Edge-disjoint mappings in a bipartite graph (Chinese). Journal of Systems Science and Mathematical Sciences (Beijing) 3, 71-73. (1983).

CAI A 1984-1 Cai, M.-C.
A solution of Chartrand's problem on spanning trees. Acta Mathematicae Applicatae Sinica (English Series) 1, 97-98. (1984).

CAI A 1985-1 Cai, M.-C.
Progress in network flow theory and matroid theory (Chinese). Qufu Shiyuan Xuebao (Qufu) No. 4, 40-50. (1985).

CAI A 1987-1 Cai, M.-C.
The maximal size of a digraph with maximal local connectivity. I. Journal of Systems Science and Mathematical Sciences (Beijing) 7, 113-116. (1987).

CAI B 1985-1 Cai, W.-Q.
The nth minimal spanning trees with degree constraints having two nodes (Chinese). Qufu Shiyuan Xuebao (Qufu), No. 3, 29-35. (1985).

CAI C 1984-1 Cai, X.-T.
On the panconnectivity of Ore graphs. Scientia Sinica, Series A, (Beijing) 27, 684-694. (1984).

CAI C 1985-1 Cai, X.-T.
A long cycle in a 2-connected graph with f>=n-4 (Chinese). Journal of Shanghai Teachers University, Natural Science Edition, No.4, 20-25. (1985).

CALA 1987-1 Calamai, P. H.; Conn, A. R.
A projected Newton method for l(p) norm location problems. Mathematical Programming 38, 75-110. (1987).

CALD 1985-1 Calderbank, A. R.; Coffman, E. G. Jr.; Flatto, L.
Sequencing problems in two-server systems. Mathematics of Operations Research 10, 585-598. (1985).

CALI 1985-1 Calicchio, G.
Introduction to the methodology and techniques for the solution of multicommodity flow problems (Italian). Ricerca Operativa No. 34, 25-42. (1985).

CALV 1983-1 Calvillo, G.; Neumann-Lara, V.
Minimal graphs whose N-path graphs are connected. Anales del Instituto de Matematicas, Universidad Nacional Autonoma de Mexico 23, 1-15. (1983).

CAMA 1985-1 Camarinopoulos, L.; Becker, G.
Zur Anwendung graphentheoretischer Verfahren bei der Modularisierung von Fehlerbäumen. Angewandte Informatik 27, 470-477. (1985).

CAME 1986-1 Camerini, P. M.; Colorni, A.; Maffioli, F.
Some experience in applying a stochastic method to location problems. In: G. Gallo, C. Sandi (Eds.): Netflow at Pisa, Mathematical Programming Study 26, 229-232. (1986).

CAMEA 1985-1 Camerini, P. M.; Galbiati, G.; Maffioli, F.
The complexity of weighted multi-constrained spanning tree problems. In: L. Lovasz, E. Szemeredi (Eds.): Theory of Algorithms, Colloquia Mathematica Societatis Janos Bolyai Vol. 44, 47-52. Amsterdam, Oxford, New York: North-Holland. (1985).

CAMEB 1986-1 Camerini, P. M.; Hamacher, H. W.
Intersections of two matroids: (condensed) border graphs and ranking. Research Report No. 86-11. Gainesville: Univ. of Florida, Industrial and Systems Engineering Dept., 21p. (1986).

CAMEC 1985-1 Camerini, P. M.; Maffioli, F.
Some applications of unlabelled partition systems. In: P. Brucker, R. Pauly (Eds.): Proc. IX. Symp. on Operations Research, Osnabrück 1984. Methods of Operations Research 49, 161-178. (1985).

CAMED 1986-1 Camerini, P. M.; Maffioli, F.; Martello, S.; Toth, P.
Most and least uniform spanning trees. Discrete Applied Mathematics 15, 181-198. (1986).

CAMEE 1984-1 Camerini, P. M.; Vercellis, C.
The matroidal knapsack: a class of (often) well-solvable problems. Operations Research Letters 3, 157-162. (1984).

CAMEF 1985-1 Cameron, J.; Thomas, G.
An approximate graph partitioning algorithm and its computational complexity. Congressus Numerantium 49, 287-294. (1985).

CAMEG 1986-1 Cameron, K.
On k-optimum dipath partitions and partial k-colourings of acyclic digraphs. European Journal of Combinatorics 7, 115-118. (1986).

CAMEH 1986-1 Cameron, K.; Edmonds, J.; Lovasz, L.
A note on perfect graphs. Periodica Mathematica Hungarica 17, 173-175. (1986).

CAMEI 1986-1 Cameron, K.; Sachs, H.
Monotone path systems in simple regions. Report 86429-OR. Bonn: Univ. Bonn, Inst. f. Ökonometrie u. Operations Res., 16 p. (1986).

CAMP 1986-1 Campbell, B. A.; Rardin, R. L.
Polynomial time solution of Steiner tree problems on special planar graphs. Research Memorandum 86-7. West Lafayette: Purdue Univ., School of Industrial Eng., 33 p. (1986).

CAMP 1987-1 Campbell, B. A.; Rardin, R. L.
Steiner tree problems on series-parallel block graphs I: polynomial recognition and solution. CC-87-17. West Lafayette: Purdue Univ. Inst. of Interdisciplinary Engineering Studies, 33 p. (1987).

CAMPA 1985-1 Campello, R. E.
Updating a hybrid algorithm for set partitioning problems. Matematica Aplicada e Computacional 4, 75-90. (1985).

CAMPB 1987-1 Campello, R. E.; Maculan, N. F.
Lagrangean relaxation for a lower bound to a set partitioning problem with side constraints: properties and algorithms. Discrete Applied Mathematics 18, 119-136. (1987).

CAMPC 1987-1 Campers, G.; Henkes, O.; Leclercq, J.-P.
Sur les methodes exactes de coloration de graphes. Cahiers du Centre d'Etudes de Recherche Operationelle 29, 19-30. (1987).

CAO 1985-1 Cao, H.-G.
Storage and search of data with minimal spanning trees (Chinese). Journal Shanxi University (Taiyuan), Natural Science Edition, No. 3, 35-40. (1985).

CAPO 1984-1 Caponi, M.; Pieroni, P.; Saviozzi, G.
A Benders dual decomposition approach to the optimum network orientation. Publicazioni 121. Pisa: Univ. Pisa, Dipt. di Matematica, 22 p. (1984).

CAPT 1985-1 Captivo, M. E.; Branco, I. M.; Coelho, J. D.
An exact algorithm for the problem of the p-median (Portuguese). Portugaliae Mathematica (Lisbon) 42 (1983/84), 151-162. (1985).

CAPT 1986-1 Captivo, M. E.; Branco, I. M.; Coelho, J. D.
Errata: "An exact algorithm for the problem of the p-median" (Portuguese). Portugaliae Mathematica (Lisboa) 42(1983/84), 473. (1986).

CARA 1987-1 Caramanis, M.
Production system design: a discrete event dynamic system and generalized Benders' decomposition approach. International Journal of Production Research 25, 1223-1234. (1987).

CARE 1986-1 Carey, M.
A constraint qualification for a dynamic traffic assignment model. Transportation Science 20, 55-58. (1986).

CARE 1987-1 Carey, M.
Optimal time-varying flows on congested networks. Operations Research 35, 58-69. (1987).

CARL 1987-1 Carlier, J.
The m machines problem. European Journal of Operational Research 29, 298-306. (1987).

CARLA 1985-1 Carlier, J.; Chretienne, P.; Girault, C.
Modelling scheduling problems with timed Petri nets. In: G. Rozenberg (Ed.): Advances in Petri Nets 1984, Lecture Notes in Computer Science Vol. 188, 62-82. Berlin: Springer-Verlag. (1985).

CARO 1986-1 Caro, Y.
Chromatic decomposition of a graph. Ars Combinatoria 21, 29-32. (1986).

CARP 1984-1 Carpaneto, G.
Relaxation techniques and applications to the optimal Hamiltonian tour problem (Italian). In: Methods and algorithms for optimization (Bologna 1982), 53-110. Bologna: Collana Atti Congr., Pitagora. (1984).

CARPA 1984-1 Carpaneto, G.; Gallo, G.
Enumerative methods for solving combinatorial optimization problems (Italian). In: G. Carpaneto, G. Di Pillo (Eds.): Methods and algorithms for optimization, 111-158. Bologna: Pitagora Editrice. (1984).

CARPB 1987-1 Carpaneto, G.; Toth, P.
Primal-dual algorithms for the assignment problem. Discrete Applied Mathematics 18, 137-154. (1987).

CARPC 1985-1 Carpenter, H.; Dowsland, W. B.
Practical considerations of the pallet-loading problem. Journal of the Operational Reserch Society 36, 489-498. (1985).

CARR 1986-1 Carraresi, P.; Sodini, C.
An efficient algorithm for the bipartite matching problem. European Journal of Operational Research 23, 86-93. (1986).

CARRA 1987-1 Carraway, R. L.; Chambers, R. J.; Morin, T. L.; Moskowitz, H.
Scheduling with multiple performance measures: generalized dynamic programming for nonlinear multicriteria cost functions. CC-87-35. West Lafayette: Purdue Univ., Inst. for Interdisciplinary Engineering Studies, 21 p. (1987).

CARRB 1986-1 Carraway, R. L.; Morin, T. L.; Moskowitz, H.
Generalized dynamic programming for multi-criteria optimization. CC-86-3. West Lafayette: Purdue Univ. Inst. of Interdisciplinary Engineering Studies, 39 p. (1986).

CARRB 1987-1 Carraway, R. L.; Morin, T. L.; Moskowitz, H.
The stochastic traveling salesman problem revisited: a generalized dynamic programming approach. CC-86-2. West Lafayette: Purdue Univ. Inst. of Interdisciplinary Engineering Studies, 29 p. (1987).

CARRC 1983-1 Carrera, J.; Szidarovszky, F.
A multiobjective branch and bound algorithm for optimal observation network design. Department of Mathematics, Karl Marx University of Economics (Budapest) 2, 115-135. (1983).

CARRD 1986-1 Carrie, A. S.; Perera, D. T. S.
Work scheduling in FMS under tool availability constraints. International Journal of Production Research 24, 1299-1308. (1986).

CARS 1984-1 Carstensen, P. J.
Parametric cost shortest chain problem. Research Report. Holmdel: Bell Communications Research, 20 p. (1984).

CARSA 1985-1 Carstensen, P. J.; Murty, K. G.; Perin, C.
Intermediate feasibility in 0-1 integer linear systems. Mathematical Programming Study 24, 219-224. (1985).

CART 1984-1 Carter, M.
A practical algorithm for finding the largest clique in a graph. Working Paper, 84-08. Toronto: Univ. of Toronto, Dept. of Industrial Engineering, 23 p. (1984).

CARTA 1986-1 Carter, M. W.
A survey of practical applications of examination timetabling algorithms. Operations Research 34, 193-202. (1986).

CAVA 1986-1 Cavalier, T. M.; Chandra, M. J.
A heuristic algorithm for assigning crews among bases in an airlift operation. Journal of the Operational Research Society 37, 381-386. (1986).

CAVAA 1985-1 Cavalier, T. M.; Sherali, H. D.
Sequential location-allocation problems on chains and trees with probabilistic link demands. Mathematical Programming 32, 249-277. (1985).

CAVAA 1986-1 Cavalier, T. M.; Sherali, H. D.
Euclidean distance location-allocation problems with uniform demands over convex polygons. Transportation Science 20, 107-116. (1986).

CAVAA 1986-2 Cavalier, T. M.; Sherali, H. D.
Network location problems with continuous link demands: p-medians on a chain and 2-medians on a tree. European Journal of Operational Research 23, 246-255. (1986).

CECC 1986-1 Ceccherini, P. V.; Sappa, A.
F-binomial coefficients and related combinatorial topics: perfect matroid designs, posets of full binomial type and F-geodetic graphs. In: A. Barlotti, M. Biliotti, A. Cossu, G. Korchmaros, G. Tallini (Eds.): Combinatorics '84, Annals of Discrete Mathematics 30, 143-158. Amsterdam, New York, Oxford, Tokyo: North-Holland. (1986).

CECH 1986-1 Cechlarova, K.
On some properties of the algebraic assignment problem. Ekonomicko Matematicky Obzor (Prague) 22, 462-478. (1986).

CEDE 1984-1 Ceder, A.; Stern, H. I.
Optimal transit timetables for a fixed vehicle fleet. In: J. Volmuller, R. Hamerslag (Eds.): Proceedings of the ninth international symposium on transportation and traffic theory, 331-355. Utrecht: VNU Science Press. (1984).

CEDEA 1986-1 Ceder, A.; Wilson, N. H. M.
Bus network design. Transportation Research 20-B, 331-344. (1986).

CENE 1985-1 Cenek, P.; Chlebnican, P.; Klima, V.
Computer aided network design. In: F. Nozicka (Ed.): Optimization - Theory, Methods, Applications, I, International Conference Marianske Lazne 1985, 35-42. Prague: Dum Techniky CSVTS. (1985).

CEPU 1984-1 Cepulic, V.
An algorithm for finding minimal intersection sets. Glasnik Matematicki (Zagreb), Seriya III 19 (39), 13-16. (1984).

CERA 1980-1 Cerasoli, M.; Eugeni, F.
On the chromatic polynomial of a graph (Italian). Ricerca: Matematiche Pure ed Applicate (Rome) 31, No. 2, 5-10. (1980).

CERD 1985-1 Cerdeira, J. O.; Luz, C. J.
Covering nodes with a fixed number of trees. Working Paper. Lisboa: Univ.Tecnica de Lisboa, 35 p. (1985).

CERD 1985-2 Cerdeira, J. O.; Luz, C. J.
The 2-components of minimum cost of a graph. In: P. Brucker, R. Pauly (Eds.): Proc. IX. Symp. on Operations Research, Osnabrück 1984. Methods of Operations Research 49, 179-186. (1985).

CERN 1985-1 Cerny, J.; Cerna, A.; Pesko, S.; Vasek, K.
Optimal scheduling of transport. In: F. Nozicka (Ed.): Optimization - Theory, Methods, Applications, I, International Conference Marianske Lazne 1985, 43-55. Prague: Dum Techniky CSVTS. (1985).

CERNA 1987-1 Cerny, J.; Guldan, F.
Location of polygon vertices on circles and its application in transport studies. Aplikace Matematiky 32, 81-95. (1987).

CHAB 1984-1 Chabukiani, T. A.
On a combinatorial geometry of three-dimensional space (Russian). Soobshcheniya Akademii Nauk Gruzinskoj SSR (Tblisi) 113, 489-491. (1984).

CHAK 1986-1 Chakrabarti, B. K.
Directed travelling salesman problem. Journal of Physics A 19, 1273-1275. (1986).

CHAKA 1986-1 Chakraborty, T. K.
A preemptive single sampling attribute plan of given strength. Opsearch 23, 164-174. (1986).

CHAKB 1986-1 Chakravarthy, S.
A single-machine scheduling problem with random processing times. Naval Research Logistics Quarterly 33, 391-398. (1986).

CHAKB 1987-1 Chakravarthy, S.
On two cyclic exponential machines sharing a common tool magazine. American Journal of Mathematical and Management Sciences 7, 205-232. (1987).

CHAKC 1985-1 Chakravarty, A. K.
Multiproduct purchase scheduling with limited budget and/or group discounts. Computers & Operations Research 12, 493-506. (1985).

CHAKD 1987-1 Chakravarty, A. K.; Shtub, A.
Capacity, cost and scheduling analysis for a multiproduct flexible manufacturing cell. International Journal of Production Research 25, 1143-1156. (1987).

CHAKE 1985-1 Chakroun, N.; Manoussakis, Y.
Degrees, diameter and structure of oriented graphs. Rapport de Recherche 247. Orsay: Univ. de Paris-Sud, Lab. de Recherche en Informatique, 19 p. (1985).

CHAL 1985-1 Chalmet, L. G.; Bodt, M. de; Wassenhove, L. N. van
The effect of engineering changes and demand uncertainty on MRP lot sizing: a case study. International Journal of Production Research 23, 233-252. (1985).

CHALA 1986-1 Chalmet, L. G.; Lemonidis, L.; Elzinga, D. J.
An algorithm for the bi-criterion integer programming problem. European Journal of Operational Research 25, 292-300. (1986).

CHAM 1987-1 Chams, M.; Hertz, A.; Werra, D. de
Some experiments with simulated annealing for coloring graphs. European Journal of Operational Research 32, 260-266. (1987).

CHAN 1987-1 Chan, L. K.; Falkenberg, J.; Rosenbloom, E. S.
Implementation problems of a nurse preference mathematical programming approach to scheduling. Congressus Numerantium 56, 251-260. (1987).

CHANA 1987-1 Chan, O.; Chen, C. C.; Slater, P. J.
On a characterization of comparability graphs. Ars Combinatoria 23A, 67-80. (1987).

CHANB 1987-1 Chanas, S.
Fuzzy optimization in networks. In: J. Kacprzyk, P. A. Orlovski (Eds.): Optimization models using fuzzy sets and possibility theory, Theory and Decision Library, Series B 4, 303-327. D. Reidel Publishing Company. (1987).

CHANC 1985-1 Chanas, S.; Kamburowski, J.
The fuzzy shortest route problem. In: J. Albrycht, H. Wisniewski (Eds.): Interval and fuzzy mathematics, Proc. Polish Symp., Poznan 1983, 35-41. Tech. Univ. of Poznan. (1985).

CHAND 1984-1 Chanas, S.; Kolodziejczyk, W.
Real-valued flows in a network with fuzzy arc capacities. Fuzzy Sets and Systems 13, 139-151. (1984).

CHANE 1984-1 Chanas, S.; Kolodziejczyk, W.; Machaj, A.
A fuzzy approach to the transportation problem. Fuzzy Sets and Systems 13, 211-221. (1984).

CHANE 1985-1 Chanas, S.; Kolodziejczyk, W.; Machaj, A.
The fuzzy transportation problem with fuzzy constraints. In: J. Albrycht, H. Wisniewski (Eds.): Interval and fuzzy mathematics, Proc. Polish Symp., Poznan 1983, 43-51. Tech. Univ. of Poznan. (1985).

CHANE 1985-2 Chanas, S.; Kolodziejczyk, W.; Machaj, A.
The min-cost flow problem - a fuzzy approach. In: J. Albrycht, H. Wisniewski (Eds.): Interval and fuzzy mathematics, Proc. Polish Symp., Poznan 1983, 53-61. Tech. Univ. of Poznan. (1985).

CHANF 1986-1 Chand, S.; Morton, T. E.
Minimal forcast horizon procedures for dynamic lot size models. Naval Research Logistics Quarterly 33, 111-122. (1986).

CHANG 1984-1 Chand, S.; Schneeberger, H.
A note on Smith's schedule to minimize the weighted sum of completion time with maximum allowable tardiness, (revised). Research Report. Lafayette: Purdue Univ., Krannert Graduate School of Management, 12 p. (1984).

CHANG 1986-1 Chand, S.; Schneeberger, H.
A note on the single-machine scheduling problem with minimum weighted completion time and maximum allowable tardiness. Naval Research Logistics Quarterly 33, 551-557. (1986).

CHANH 1987-1 Chandra, S.; Saxena, P. K.
Cost/completion-date tradeoffs in quadratic fractional transportation problem. Economic Computation and Economic Cybernetics Studies and Research (Bucharest) 22, No. 3, 67-72. (1987).

CHANI 1987-1 Chandra, S.; Seth, K.; Saxena, P. K.
Time minimizing transportation problem with impurities. Asia-Pacific Journal of Operational Research 4, 19-27. (1987).

CHANJ 1984-1 Chandrasekaran, R.
Integer programming problems for which a simple rounding-type algorithm works. In: W. R. Pulleyblank (Ed.): Progress in Combinatorial Optimization, 101-106. Toronto: Academic Press. (1984).

CHANJ 1985-1 Chandrasekaran, R.
Optimal improvements in bottleneck systems. Opsearch 22, 121-128. (1985).

CHANJ 1986-1 Chandrasekaran, R.
Recognition of Gilmore-Gomory traveling salesman problem. Discrete Applied Mathematics 14, 231-238. (1986).

CHANK 1987-1 Chandrasekaran, R.; Kabadi, S. N.
Pseudomatroids. Working Paper. Richardson: Univ. of Texas at Dallas, 23 p. (1987).

CHANL 1984-1 Chandrasekaran, R.; Shirali, S.
Total weak unimodularity: testing and applications. Discrete Mathematics 51, 137-146. (1984).

CHANM 1984-1 Chandrasekaran, R.; Tamir, A.
On the integrality of an extreme solution to pluperfect graph and balanced systems. Operations Research Letters 3, 215-218. (1984).

CHANN 1985-1 Chandru, V.; Coullard, C. R.; Wagner, D. K.
On the complexity of recognizing a class of generalized networks. Operations Research Letters 4, 75-78. (1985).

CHANO 1987-1 Chandru, V.; Trick, M. A.
On the complexity of Lagrange multiplier search. CC-87-15. West Lafayette: Purdue Univ. Inst. of Interdisciplinary Engineering Studies, 44 p. (1987).

CHANP 1986-1 Chang, F.; Wu, L.
An optimal adaptive routing algorithm. IEEE Transactions on Automatic Control AC-31, 690-700. (1986).

CHANQ 1985-1 Chang, G. J.
Greedoid polyhedra for MPP-greedoids. In: Mathematics Center Reports,Symposium 1985, Taipei/Taiwan, 365-380. (1985).

CHANQ 1986-1 Chang, G. J.
Labeling algorithms for domination problems in sun-free chordal graphs. Report 86442-OR. Bonn: Univ. Bonn, Inst. f. Ökonometrie u. Operations Research, 15 p. (1986).

CHANQ 1986-2 Chang, G. J.
MPP-greedoids. Report 86436-OR. Bonn: Univ. Bonn, Inst. f. Ökonometrie u. Operations Res., 11 p. (1986).

CHANQ 1986-3 Chang, G. J.
Total domination in block graphs. Report 86438-OR. Bonn: Univ. Bonn, Inst. f. Ökonometrie u. Operations Research, 10 p. (1986).

CHANR 1985-1 Chang, G. J.; Edmonds, J.
The poset scheduling problem. Order 2, 113-118. (1985).

CHANS 1987-1 Chang, G. J.; Hao, F. H.
The minimum of the antichains in the factor poset. Order 3, 355-358. (1987).

CHANT 1985-1 Chang, G. J.; Nemhauser, G. L.
Covering, packing and generalized perfection. SIAM Journal on Algebraic and Discrete Methods 6, 109-132. (1985).

CHANU 1986-1 Chang, G. J.; Wu, C.-S.
Graphs whose cycles are of length r modulo k. Report 86439-OR. Bonn: Univ. Bonn, Inst. f. Ökonometrie u. Operations Research, 14 p. (1986).

CHANV 1985-1 Chang, H. C.
A minimax-arc-path problem. Tamkang Journal of Management Science (Taipei) 6, 7-11. (1985).

CHANW 1985-1 Chang, H.-J.
A minimax-arc-path problem. Tamkang Journal of Management Sciences (Taipei) 6, No. 1, 7-11. (1985).
CHANX 1984-1 Chang, L.; Poe, A. T.
Surjections and coverings. International Journal of Computer and Information Sciences 13, 509-520. (1984).
CHANY 1986-1 Chang, M. D.; Chen, C.-H. J.; Engquist, M.
An improved primal simplex variant for pure processing networks. Research Report CCS 552. Austin: Univ. of Texas, Center for Cybernetic Studies, 30 p. (1986).
CHANZ 1986-1 Chang, M. D.; Engquist, M.
On the number of quasi-trees in an optimal generalized network basis. COAL Committee on Algorithms Newsletter No. 14, 5-9. (1986).
CHANAA 1983-1 Chang, R. C.; Lee, R. C. T.
An O(|V|) parallel algorithm to construct minimal spanning trees based upon Voronoi diagrams. in: T. Basar, D. J. Brown, H. V. Poor (Eds.): Twentyfirst Annual Allerton Conference on Communication, Control, and Computing (Monticello, Ill., 1983), 461-470. Urbana-Champaign: Univ. Illinois. (1983).
CHANAA 1984-1 Chang, R. C.; Lee, R. C. T.
The average performance analysis of a closest-pair algorithm. International Journal of Computer Mathematics 16, 125-130. (1984).
CHANAA 1986-1 Chang, R. C.; Lee, R. C. T.
An O(N log N) minimal spanning tree algorithm for N points in the plane. BIT 26, 7-16. (1986).
CHANAB 1987-1 Chang, S. G.
A heuristic for non-bifurcated routing in communication networks. Journal of the Korean Operations Research and Management Science Society 11, No.2, 13-26. (1987).
CHANAC 1985-1 Chang, S. S.
A production planning model for multi-product facilities. Journal of the Operations Research Society of Japan 28, 345-358. (1985).
CHANAD 1987-1 Chang, Y.; Hambrusch, S. E.; Simon, J.
On the computational complexity of continuous routing. CC 87 12. West Lafayette: Purdue Univ., Inst. for Interdisciplinary Engineering Studies, 25 p. (1987).
CHAO 1984-1 Chao, C.-Y.; Zhao, L.-C.
Chromatic polynomials of connected graphs. Archivum Mathematicum (Basel) 43, No. 2, 187-192. (1984).
CHAOA 1986-1 Chao, M.-T.; Franco, J.
Probabilistic analysis of two heuristics for the 3-satisfiability problem. SIAM Journal on Computing 15, 1106-1118. (1986).
CHAP 1985-1 Chapleau, L.; Ferland, J. A.; Rousseau, J.-M.
Clustering for routing in dense area. European Journal of Operational Research 20, 48-57. (1985).
CHAPA 1987-1 Chapman, C. B.; Ward, S. C.
Modular decision analysis - a production-scheduling illustration. Journal of the Operational Research Society 38, 803-814. (1987).
CHAR 1985-1 Charalambous, C.
Acceleration of the HAP approach for the multifacility location problem. Naval Research Logistics Quarterly 32, 373-390. (1985).
CHARA 1984-1 Charnetski, J. R.
Scheduling operating room surgical procedures with early and late completion penalty costs. Journal of Operations Management 5, 91-102. (1984).
CHARB 1983-1 Chartrand, G.; Cooper, J. K. Jr.; Lesniak-Foster, L.; Wall, C. E.
The traceability of a graph. Indian Journal of Mathematics (Allahabad) 25, 17-28. (1983).
CHARC 1985-1 Chartrand, G.; Harary, F.; Oellermann, O. R.
On the regulation number of a multigraph. Graphs and Combinatorics 1, 137-144. (1985).
CHARD 1987-1 Chartrand, G.; Holbert, K. S.; Oellermann, O. R.; Swart, H. C.
F-degrees in graphs. Ars Combinatoria 24, 133-148. (1987).
CHARE 1986-1 Chartrand, G.; Johnson, M.; Oellermann, O. R.
Connected graphs containing a given connected graph as a unique greatest common subgraph. Aequationes Mathematicae 31, 213-222. (1986).
CHARF 1984-1 Chartrand, G.; Kapoor, S. F.; Lesniak, L.; Schuster, S.
Near 1-factors in graphs. Congressus Numerantium 41, 131-148. (1984).
CHARG 1985-1 Chartrand, G.; Kapoor, S. F.; Lick, D. R.; Schuster, S.
The partial complement of graphs. Periodica Mathematica Hungarica (Budapest) 16, 83-95. (1985).
CHARH 1987-1 Chartrand, G.; Kapoor, S. F.; Oellermann, O. R.; Ruiz, S.
On maximum matchings in cubic graphs with a bounded number of bridge-covering paths. Bulletin of the Australian Mathematical Society 36, 441-447. (1987).
CHARI 1986-1 Chartrand, G.; Oellermann, O. R.; Saba, F.; Zou, H.-B.
The uniformity number of a graph. Indian Journal of Mathematics (Allahabad) 28, No. 1, 1-7. (1986).
CHARJ 1986-1 Chartrand, G.; Ruiz, S.; Wall, C. E.
Smallest regular graphs without near 1-factors. Journal of the Australian Mathematical Society, Series A 41, 193-210. (1986).
CHARK 1984-1 Chartrand, G.; Saba, F.; Wormald, N. C.
Bridge and cycle degrees of vertices of graphs. International Journal of Mathematics and Mathematical Sciences 7, 351-360. (1984).
CHARL 1985-1 Chartrand, G.; Saba, F.; Zou, H.-B.
Edge rotations and distance between graphs. Casopis Pro Pestovani Matematiky (Prague) 110, 87-91. (1985).
CHARL 1987-1 Chartrand, G.; Saba, F.; Zou, H.-B.
Greatest common subgraphs of graphs. Casopis pro Pestovani Matematiky 112, 80-88. (1987).
CHARM 1985-1 Chartrand, G.; Wilson, R. J.
The Petersen graph. In: F. Harary, J. P. Maybee (Eds.): Graphs and Applications, 69-100. New York: Wiley. (1985).

CHENO 1986-4 Cheng, T. C. E.
Optimal due-date assignment for a single machine sequencing problem with random processing times. International Journal of Systems Science 17, 1139-1144. (1986).

CHENO 1987-1 Cheng, T. C. E.
An algorithm for the CON due-date determination and sequencing problem. Computers and Operations Research 14, 537-542. (1987).

CHENO 1987-2 Cheng, T. C. E.
Minimizing the average deviation of job completion times about a common due-date: an extension. Mathematical Modelling 9, 13-15. (1987).

CHENO 1987-3 Cheng, T. C. E.
Optimal due-date determination and sequencing with random processing times. Mathematical Modelling 9, 573-576. (1987).

CHENO 1987-4 Cheng, T. C. E.
Optimal TWK-power due-date determination and sequencing. International Journal of Systems Science 18, 1-8. (1987).

CHEP 1986-1 Chepoi, V. D.
Some properties of d-convexity in triangular graphs (Russian). Matematicheskie Issledovaniya No. 87 Model. i Optim. v Zadach. Planirov. i Upravl., 164-177. (1986).

CHER 1987-1 Cherkasskii, B. V.
New algorithms of spanning trees generation (Russian). Kibernetika (Kiev) No. 1, 85-89. (1987).

CHERA 1986-1 Cherkassky, B.
An algorithm for finding shortest paths in a maze. Optimization 17, 237-242. (1986).

CHERB 1984-1 Chernjak, A. A.
Connectivity of graphs with prescribed degree set and order unigraphicity (Russian). Doklady Akademii Nauk BSSR (Minsk) 28, 400-403. (1984).

CHERB 1985-1 Chernjak, A. A.
Switching-complete properties of graphs (Russian). Izvestiya Akademii Nauk BSSR, Seriya Fiziko-Matematicheskikh Nauk (Minsk) No. 1, 29-35. (1985).

CHERC 1987-1 Chernjak, A. A.; Chernjak, Z. A.; Tyshkevich, R. I.
On forcibly hereditary P-graphical seqeuences. Discrete Mathematics 64, 111-128. (1987).

CHERD 1986-1 Chernjak, Z. A.; Chernjak, A. A.
About recognizing (alpha, beta) classes of polar graphs. Discrete Mathematics 62, 133-138. (1986).

CHERE 1985-1 Chernyak, V. V.; Shcherbina, I. N.
On a stochastic model of volume-scheduling planning (Russian). Issledovanie Operatsii i ASU, No. 26, 109-113. (1985).

CHES 1984-1 Cheston, G. A.
Generating a spanning tree with a specified node as a centroid point for biconnected graphs. Congressus Numerantium 43, 211-220. (1984).

CHET 1984-1 Chetwynd, A. G.; Hilton, A. J. W.
The chromatic index of graphs of even order with many edges. Journal of Graph Theory 8, 463-470. (1984).

CHET 1984-2 Chetwynd, A. G.; Hilton, A. J. W.
The chromatic index of graphs with a most four vertices of maximum degree. Congressus Numerantium 43, 221-248. (1984).

CHET 1985-1 Chetwynd, A. G.; Hilton, A. J. W.
Regular graphs of high degree are 1-factorizable. Proceedings of the London Mathematical Society (3) 50, 193-206. (1985).

CHET 1986-1 Chetwynd, A. G.; Hilton, A. J. W.
Critical star multigraphs. Graphs and Combinatorics 2, 209-222. (1986).

CHET 1987-1 Chetwynd, A. G.; Hilton, A. J. W.
The edge-chromatic class of regular graphs of degree 4 and their complements. Discrete Applied Mathematics 16, 125-134. (1987).

CHIA 1985-1 Chia, G.-L.; Lim, C.-K.
On supercompact graphs I: the nucleus. Ars Combinatoria 20, 101-110. (1985).

CHIB 1985-1 Chiba, N.; Nishizeki, T.
Arboricity and subgraph listing algorithms. SIAM Journal on Computing 14, 210-223. (1985).

CHIBA 1985-1 Chiba, N.; Onoguchi, K.; Nishizeki, T.
Drawing plane graphs nicely. Acta Informatica 22, 187-202. (1985).

CHIE 1987-1 Chien, T. W.; Balakrishnan, A.; Wong, R. T.
An integrated inventory allocation and vehicle routing problem. CC-87-27. West Lafayette: Purdue Univ. Inst. of Interdisciplinary Engineering Studies, 35 p. (1987).

CHIN 1985-1 Chin, F.; Ting, H. F.
An almost linear time and O(nlogn + e) messages distributed algorithm for minisum-weight spanning trees. In: 26th Annual Symp. on Foundations of Computer Science, Portland, Oregon 1985, Proceedings, 257-266. IEEE. (1985).

CHINA 1984-1 Chin, F.; Wong, M.
An optimal schedule for printing and binding. International Journal of Computer Mathematics 15, 117-127. (1984).

CHINB 1987-1 Chin, W.-P.; Ntafos, S.
Optimum zoo-keeper routes. Congressus Numerantium 58, 257-266. (1987).

CHINC 1983-1 Chinn, P. Z.; Dewdney, A. K.; Greene, T. M.
Bandwidth and peripheral vertices of trees. Congressus Numerantium 39, 169-178. (1983).

CHIU 1987-1 Chiu, S. S.
The minisum location problem on an undirected network with continuous link demands. Computers and Operations Research 14, 369-384. (1987).

CHIUA 1985-1 Chiu, S. S.; Berman, O.; Larson, R. C.
Locating a mobile server queueing facility on a tree network. Management Science 31, 764-772. (1985).

CHIUB 1985-1 Chiu, S. S.; Larson, R. C.
Locating an n-server facility in a stochastic environment. Computers & Operations Research 12, 509-516. (1985).

CHOG 1984-1 Chogovadze, G. G.; Chachanedze, G. G.; Kachibaya, V. V.; Surguladze, G. G.
A method for determining the minimal structure of functional dependencies (Russian). Analele Stiintifice ale Universitatii Al. I. Cuza din Iasi, Serie Noua, Sectiunea Ia 30, No. 5, 95-103. (1984).

CHOI 1987-1 Choi, H.-A.; Hakimi, S. L.
Scheduling file transfers for trees and odd cycles. SIAM Journal on Computing 16, 162-168. (1987).

CHOIA 1985-1 Choi, W.; Francis, R. L.
Some optimization models of building evacuation problems with flow-dependent arc capacities. Research Report 85-8. Gainesville: Univ. of Florida, Industrial & Systems Eng. Dept., 23 p. (1985).

CHOIB 1984-1 Choi, W.; Francis, R. L.; Hamacher, H. W.; Tufekci, S.
Network models of building evacuation problems with flow-dependent exit capacities. In: J. P. Brans (Ed.): Operational Research '84, 1047-1059. Amsterdam: North-Holland. (1984).

CHOIC 1985-1 Choi, W.; Hamacher, H. W.; Tufekci, S.
Modeling of building evacuation problems by network flows with side constraints. Research Report 85-7. Gainesville: Univ. of Florida, Industrial & Syst. Eng. Dept., 33 p. (1985).

CHOO 1987-1 Choobineh, F.; Khattab, M. M.
Heuristic models for resource-constrained project scheduling. In: M. Ismail (Ed.): 29th Midwest Symposium on Circuits and Systems, 440-443. Amsterdam: North-Holland. (1987).

CHOP 1987-1 Chopra, S.
On ternary problems. Working Paper 87-51. New York: New York Univ., Graduate School of Business Admin., 20 p. (1987).

CHOPA 1987-1 Chopra, S.; Johnson, E. L.
Dual row modules and polyhedra of blocking group problems. Mathematical Programming 38, 229-270. (1987).

CHOU 1984-1 Choudum, S. A.
A necessary condition for disjoint perfect matchings in graphs. In: K. P. Vijayan, N. M. Singhi (Eds.): Combinatorics and Applications (Proc. Seminar Calcutta 1982), 69-72. Calcutta: Indian Statistical Institute. (1984).

CHRE 1984-1 Chretienne, P.
Chemins extremaux d'un graphe doublement value. RAIRO Recherche Operationnelle 18, 221-246. (1984).

CHRI 1985-1 Christofides, N.
Vehicle routing. In: M. O'hEigeartaigh, J. K. Lenstra, A. H. G. Rinnooy Kan (Eds.): Combinatorial Optimization - Annotated Bibliographies, 148-163. Chichester: Wiley. (1985).

CHRI 1985-2 Christofides, N.
Vehicle routing. In: E. L. Lawler, J. K. Lenstra, A. H. G. Rinnooy Kan, D. B. Shmoys (Eds.): The Traveling Salesman Problem, 431-448. New York: Wiley. (1985).

CHRIA 1987-1 Christofides, N.; Alvarez-Valdes, R.; Tamarit, J. M.
Project scheduling with resource contraints: a branch and bound algorithm. European Journal of Operational Research 29, 262-273. (1987).

CHRIB 1984-1 Christofides, N.; Benavent, E.; Campos, V.; Corberan, A.; Mota, E.
An optimal method for the mixed postman problem. In: P. Thoft-Christensen (Ed.): System Modelling and Optimization, 641-649. Lecture Notes in Control and Information Sciences 59, Berlin: Springer. (1984).

CHRIC 1986-1 Christofides, N.; Campos, V.; Corberan, A.; Mota, E.
An algorithm for the rural postman problem on a directed graph. In: G. Gallo, C. Sandi (Eds.): Netflow at Pisa, Mathematical Programming Study 26, 155-166. (1986).

CHRID 1984-1 Christofides, N.; Valls Verdejo, V.
Finding all elementary circuits of zero cost in an optimal incremental graph (Spanish). Estadistica Espanola (Madrid), No. 104, 49-70. (1984).

CHRID 1986-1 Christofides, N.; Valls Verdejo, V.
Finding all optimal solutions to the network flow problem. In: G. Gallo, C. Sandi (Eds.): Netflow at Pisa, Mathematical Programming Study 26, 209-212. (1986).

CHRO 1986-1 Chrobak, M.; Poljak, S.
On common edges in optimal solutions to traveling salesman and other optimization problems. Warsaw: Univ., Inst. of Informatics, 12p. (1986).

CHUN 1986-1 Chung, C.-H.
Recent applications of the maximal covering location planning (M.C.L.P.) model. Journal of the Operational Research Society 37, 735-746. (1986).

CHUNA 1985-1 Chung, F. R. K.
On the cutwidth and the topological bandwidth of a tree. SIAM Journal on Algebraic and Discrete Methods 6, 268-277. (1985).

CHUNB 1983-1 Chung, F. R. K.; Erdös, P.; Spencer, J.
On the decomposition of graphs into complete bipartite subgraphs. In: P. Erdös (Ed.): Studies in Pure Mathematics, 95-101. Basel, Boston: Birkhäuser. (1983).

CHUNC 1984-1 Chung, F. R. K.; Garey, M. R.
Diameter bounds for altered graphs. Journal of Graph Theory 8, 511-534. (1984).

CHUND 1985-1 Chung, F. R. K.; Graham, R. L.
A new bound for Euclidean Steiner minimal trees. In: Discrete Geometry and Convexity (New York, 1982), 328-346. Annals of the New York Academy of Sciences 440. (1985).

CHUNE 1987-1 Chung, F. R. K.; Graham, R. L.; Saks, M.
Dynamic search in graphs. In: D. P. Johnson, T. Nishizeki, A. Nozaki, H. P. Wilf (Eds.): Discrete Algorithms and Complexity - Proc. Japan-U.S. Joint Seminar, Kyoto 1986, 351-387. Perspectives in Computing Vol. 15, Boston: Academic Press. (1987).

CHUNF 1985-1 Chung, F. R. K.; Graham, R. L.; Winkler, P. M.
On the addressing problem for directed graphs. Graphs and Combinatorics 1, 41-50. (1985).

CHUNG 1982-1 Chung, M.-J.; Makedon, F.; Sudborough, I. H.; Turner, J.
Polynomial time algorithms for the MIN CUT problem on degree restricted trees. in: 23rd IEEE Annual Symposium on Foundations of Computer Science (Chicago, Ill.), 262-271. (1982).

CHUNG 1985-1 Chung, M.-J.; Makedon, F.; Sudborough, I. H.; Turner, J.
Polynomial time algorithms for the min-cut problem on degree restricted trees. SIAM Journal on Computing 14, 158-177. (1985).

CHUNH 1986-1 Chung, S.-J.; Murty, K. G.; Chang, S. Y.
On K(delta). Discrete Applied Mathematics 15, 199-212. (1986).

CHUR 1986-1 Church, R. L.; Weaver, J. R.
Theoretical links between median and coverage location problems. In: J. P. Osleeb, P. J. Ratick (Eds.): Locational Decisions: Methodology and Applications, 1-22. Annals of Operations Research 6, Basel: Baltzer. (1986).

CHVA 1984-1 Chvatal, V.
A semi-strong perfect graph conjecture. In: C. Berge, V. Chvatal (Eds.): Topics on perfect graphs, Annals of Discrete Mathematics Vol. 21, 279-280. Amsterdam: North-Holland. (1984).

CHVA 1984-2 Chvatal, V.
An equivalent version of the strong perfect graph conjecture. In: C. Berge, V. Chvatal (Eds.): Topics on perfect graphs, Annals of Discrete Mathematics Vol. 21, 193-196. Amsterdam: North-Holland. (1984).

CHVA 1984-3 Chvatal, V.
Notes on perfect graphs. In: W. R. Pulleyblank (Ed.): Progress in Combinatorial Optimization, 107-116. Toronto: Academic Press. (1984).

CHVA 1984-4 Chvatal, V.
Perfectly ordered graphs. In: C. Berge, V. Chvatal (Eds.): Topics on perfect graphs, Annals of Discrete Mathematics Vol. 21, 63-66. Amsterdam: North-Holland. (1984).

CHVA 1984-5 Chvatal, V.
Probabilistic methods in graph theory. Annals of Operations Research, Vol. 1, 171-182. Basel: Baltzer. (1984).

CHVA 1985-1 Chvatal, V.
Cutting planes in combnatorics. European Journal of Combinatorics 6, 217-226. (1985).

CHVA 1985-2 Chvatal, V.
Hamiltonian cycles. In: E. L. Lawler, J. K. Lenstra, A. H. G. Rinnooy Kan, D. B. Shmoys (Eds.): The Traveling Salesman Problem, 403-430. New York: Wiley. (1985).

CHVA 1985-3 Chvatal, V.
Star-cutsets and perfect graphs. Journal of Combinatorial Theory 39B, 189-199. (1985).

CHVA 1986-1 Chvatal, V.
On the P(4)-structure of perfect graphs III: partner decompositions. In: Raspail-Burnside: Workshop on Perfect Graphs, Bellairs Research Institute, McGill University, Barbados, West Indies, 1986, 7-14. (1986).

CHVA 1987-1 Chvatal, V.
On the P(4)-structure of perfect graphs. III. Partner decompositions. Journal of Combinatorial Theory B-43, 349-353. (1987).

CHVA 1987-2 Chvatal, V.
Perfect graphs. In: C. Whitehead (Ed.): Surveys in Combinatorics 1987, 43-52. London Math. Soc. Lecture Note Series 123, New York: Cambridge University Press. (1987).

CHVA 1987-3 Chvatal, V.
Recent results on perfect graphs. DCS-TR-210. New Brunswick: Rutgers Univ., Dept. of Computer Science, 10 p. (1987).

CHVAA 1984-1 Chvatal, V.; Graham, R. L.; Perold, A. F.; Whitesides, S. H.
Combinatorial designs related to the perfect graph conjecture. In: C. Berge, V. Chvatal (Eds.): Topics on perfect graphs, Annals of Discrete Mathematics Vol. 21, 197-206. Amsterdam: North-Holland. (1984).

CHVAB 1985-1 Chvatal, V.; Hoang, C. T.
On the P(4)-structure of perfect graphs. I. Even decompositions. Journal of Combinatorial Theory 39B, 209-219. (1985).

CHVAC 1986-1 Chvatal, V.; Lenhart, W. J.; Sbihi, N.
2-colourings that decompose perfect graphs. In: Raspail-Burnside: Workshop on Perfect Graphs, Bellairs Research Institute, McGill University, Barbados, West Indies, 1986, 15-26. (1986).

CHVAD 1987-1 Chvatal, V.; Sbihi, N.
Bull-free Berge graphs are perfect. Graphs and Combinatorics 3, 127-140. (1987).

CHVAE 1986-1 Chvatalova, J.; Opatrny, J.
The bandwidth problem and operations on graphs. Discrete Mathematics 61, 141-150. (1986).

CIKA 1987-1 Cikanek, D. G.; Slater, P. J.
The k-processing center of a graph. Congressus Numerantium 60, 199-210. (1987).

CIOB 1984-1 Ciobanu, G.; Ciubotaru, I.
Cybernetic modelling of the manufacturing processes with the help of the co-ordinating graphs. Economic Computation and Economic Cybernetics Studies and Research (Bucharest) 19, 9-20. (1984).

CIOBA 1985-1 Ciobanu, G.; Galupa, A.; Fabian, C.; Hartulari, C.
Reduction of the large size graphs with a view of their scheduling. Economic Computation and Economic Cybernetics Studies and Research 3, 39-42. (1985).

CIUR 1979-1 Ciurea, E.
On dynamic flows (French). Analele Stiinfice ale Universitatii Al, I Cuza din Iasi (Serie Noua), Sectiunea Ia 25, Suppl., 73-77. (1979).

CIUR 1982-1 Ciurea, E.
Two remarks on minimal flows in networks (Romanian). In: Proceedings of the national colloquium on probability theory and operational resaerch (Craiova, 1982), 33-36. Bucharest: Universiteta Bucuresti. (1982).

CIUR 1985-1 Ciurea, E.
Deux extensions sur les flots dynamiques. Buletinul Universitatii din Brasov, Seria C, 25(1983), 13-16. (1985).

CIUR 1985-2 Ciurea, E.
Les flots dynamiques maximaux speciaux. Analele Stiintifice ale Universitatii Al, I Cuza din Iasi (Serie Noua) Sectinuea Ia 31, 297-300. (1985).

CIUR 1985-3 Ciurea, E.
Two classes of maximal dynamic flows. Economic Computation and Economic Cybernetics Studies and Research 20, No. 1, 73-80. (1985).

CIUR 1985-4 Ciurea, E.
Un algorithme pour les flots dynamiques maximals avec stationnements. Buletinul Universitatii di Brasov, Seria C 26, 7-12. (1985).

CIUR 1986-1 Ciurea, E.
Le probleme de l'approvisionnement et de la demande pour les flots dynamiques. Buletinul Universitatii din Brasov, Seria C 27, 13-18. (1986).

CIUR 1987-1 Ciurea, E.
Les flots dynamique avec les temps de parcours nonstationnaires. Bulletin Mathematique de la Societe des Sciences Mathematique de la Republique Socialiste de Roumanie, Nouvelle Serie 31(79), 205-208. (1987).

CIUR 1987-2 Ciurea, E.
Two simple versions of dynamic flow algorithms. Buletinul Universitatii din Brasov, Seria C 29, 19-22. (1987).

CIURA 1983-1 Ciurea, E.; Radescu, N.
Minimal dynamical flows (Romanian). Analele Universitatii din Craiova, Matematica, Fizica-Chimie (Craiova) 11, 47-51. (1983).

CIURA 1984-1 Ciurea, E.; Radescu, N.
Classes of minimal dynamical flows (Romanian). Analele Universitatii din Craiova, Seria Matematica, Fizica-Chimie 12, 62-66. (1984).

CLAR 1985-1 Clark, L.; Entringer, R. C.
Extremal spanning trees of cubic graphs. Congressus Numerantium 47, 205-216. (1985).

CLAU 1986-1 Clausen, J.; Krarup, J.
Combinatorial optimization: challenges and trends. In: L. Streitferdt, H. Hauptmann, A. W. Marusev, D. Ohse, U. Pape (Eds.): Operations Research Proceedings 1985, 24-46. Berlin, Heidelberg, New York: Springer-Verlag. (1986).

CLEG 1987-1 Clegg, W. H.
Optimum machine scheduling. Industrial Engineering 19, 68-69. (1987).

CLEM 1984-1 Clements, G. F.
On uniqueness of maximal antichains of subsets of a multiset. Periodica Mathematica Hungarica (Budapest) 15, 307-313. (1984).

CLEV 1983-1 Cleves, E. M.; Jacobson, M. S.
On mixed Ramsey numbers: total chromatic number versus graphs. Congressus Numerantium 39, 193-202. (1983).

COCA 1985-1 Cocan, A.; Cocan, M.
On a problem concerning multiple cutting. Buletinul Universitatii din Brasov, Seria C 27, 19-22. (1985).

COCAA 1987-1 Cocan, M.
Über Matrizen, die den Multigraphen zugeordnet werden und ihre Anwendung im Studium der Eigenschaften der Multigraphen. Buletinul Universitatii din Brasov, Seria C 28, 29-34. (1987).

COCH 1986-1 Cochand, M.; Werra, D. de
Generalized neighbourhoods and a class of perfectly orderable graphs. Discrete Applied Mathematics 15, 213-220. (1986).

COCK 1986-1 Cockayne, E. J.; Gamble, B.; Shepherd, B.
Domination parameters for the bishops graph. Discrete Mathematics 58, 221-228. (1986).

COCKA 1986-1 Cockayne, E. J.; Hewgill, D. E.
Exact computation of Steiner minimal trees in the plane. Information Processing Letters 22, 151-156. (1986).

COCKB 1983-1 Cockayne, E. J.; Peters, D. B.
MFST82: a program for minimum full Steiner trees in the plane. Congressus Numerantium 39, 203-224. (1983).

COFF 1987-1 Coffman, E. G. Jr.; Flatto, L.; Garey, M. R.; Weber, R. R.
Minimizing expected makespans on uniform processor systems. Advances in Applied Probability 19, 177-201. (1987).

COFFA 1985-1 Coffman, E. G. Jr.; Flatto, L.; Lueker, G. S.
Expected makespans for largest-first multiprocessor scheduling. In: E. Gelenbe (Ed.): Performance '84 (Paris 1984), 491-506. Amsterdam, New York: North-Holland. (1985).

COFFB 1984-1 Coffman, E. G. Jr.; Garey, M. R.; Johnson, D. S.
Approximation algorithms for bin-packing - an updated survey. In: G. Ausiello, M. Lucertini, P. Serafini (Eds.): Algorithm design for computer system design, 49-106. Wien, New York: Springer-Verlag. (1984).

COFFB 1987-1 Coffman, E. G. Jr.; Garey, M. R.; Johnson, D. S.
Bin packing with divisible item sizes. Journal of Complexity 3, 406-428. (1987).

COFFC 1985-1 Coffman, E. G. Jr.; Garey, M. R.; Johnson, D. S.; LaPaugh, A. S.
Scheduling file transfers. SIAM Journal on Computing 14, 743-780. (1985).

COFFD 1984-1 Coffman, E. G. Jr.; Gilbert, E. N.
Dynamic, first-fit packings in two or more dimensions. Information and Control 61, 1-14. (1984).

COFFD 1985-1 Coffman, E. G. Jr.; Gilbert, E. N.
On the expected relative performance of list scheduling. Operations Research 33, 548-561. (1985).

COFFE 1984-1 Coffman, E. G. Jr.; Langston, M. A.
A performance guarantee for the greedy set-partitioning algorithm. Acta Informatica 21, 409-415. (1984).

COFFF 1986-1 Coffman, E. G. Jr.; Lueker, G. S.; Rinnooy Kan, A. H. G.
An introduction to the probabilistic analysis of sequencing and packing heuristics. Report 8602/A. Rotterdam: Erasmus Univ. Rotterdam, Econometric Institute, 71 p. (1986).

COLB 1986-1 Colbourn, C. J.
Exact algorithms for network reliability. Congressus Numerantium 51, 7-58. (1986).

COLB 1986-2 Colbourn, C. J.
The reliability polynomial. Ars Combinatoria 21-A, 31-58. (1986).

COLBA 1985-1 Colbourn, C. J.; Colbourn, M. J.; Rosa, A.
Indecomposable 1-factorizations of the complete multigraph. Journal of the Australian Mathematical Society, Series A39, 334-343. (1985).

COLBB 1985-1 Colbourn, C. J.; Keil, J. M.; Stewart, L. K.
Finding minimum dominating cycles in permutation graphs. Operations Research Letters 4, 13-18. (1985).

COLBC 1985-1 Colbourn, C. J.; Pulleyblank, W. R.
Minimizing setups in ordered sets of fixed width. Order 1, 225-228. (1985).

COLBC 1987-1 Colbourn, C. J.; Pulleyblank, W. R.
Matroid Steiner problems, the Tutte polynomial and network reliability. Report 87458-OR. Bonn: Univ. Bonn, Inst. f. Ökonometrie u. Operations Research, 14 p. (1987).

COLBD 1986-1 Colbourn, C. J.; Pulleyblank, W. R.; Rosa, A.
Hybrid triple systems and cubic feedback sets. Research Report CORR 86-18. Waterloo: Univ. of Waterloo, Dept. of Combinatorics and Optimization, 19 p. (1986).

COLBE 1987-1 Colbourn, C. J.; Slater, P. J.; Stewart, L. K.
Locating dominating sets in series parallel networks. Congressus Numerantium 56, 135-162. (1987).

COLBF 1985-1 Colbourn, C. J.; Stewart, L. K.
Permutation graphs: connected domination and Steiner trees. Research Report CS-85-02. Waterloo: Univ. of Waterloo, Faculty of Mathematics, 11 p. (1985).

COLBF 1987-1 Colbourn, C. J.; Stewart, L. K.
Dominating cycles in series-parallel graphs. Ars Combinatoria 19A, 107-112. (1987).

COLE 1984-1 Cole, R.; Siegel, A.
River routing every which way, but loose. In: 25th Annual Symposium on Foundations of Computer Science, Florida 1984, 65-73. Los Angeles: IEEE Computer Society. (1984).

COLEA 1986-1 Cole, R.; Vishkin, U.
Approximate and exact parallel scheduling with applications to list, tree, and graph problems. In: 27th Annual Symp. on Foundations of Computer Science, Proceedings, 478-491. IEEE. (1986).

COLEB 1986-1 Coleman, T. F.; Cai, J.-Y.
The cyclic coloring problem and estimation of sparse Hessian matrices. SIAM Journal on Algebraic and Discrete Methods 7, 221-235. (1986).

COME 1983-1 Comer, S. D.
Color schemes forbidding monochrome triangles. Congressus Numerantium 39, 231-236. (1983).

CONC 1986-1 Concalves, A. S.
Symmetric duality for disjunctive programming with absolute value functionals. European Journal of Operational Research 26, 301-306. (1986).

CONF 1984-1 Conforti, M.; Corneil, D. G.; Mahjoub, A. R.
K index i - covers 1: complexity and polytopes. Working Paper, 84-30. New York: New York Univ., Graduate School of Business Admin., 40 p. (1984).

CONF 1986-1 Conforti, M.; Corneil, D. G.; Mahjoub, A. R.
K(i)-covers I: complexity and polytopes. Discrete Mathematics 58, 121-142. (1986).

CONFA 1984-1 Conforti, M.; Corneil, D. G.; Majoub, A. R.
K index i - covers 2: K index i - perfect graphs. Working Paper, 84-31. New York: New York Univ., Graduate School of Business Admin., 23 p. (1984).

CONFB 1985-1 Conforti, M.; Laurent, M.
A characterization of matroidal systems of inequalities. Working Paper. Paris: University of Paris VII, 14 p. (1985).

CONFB 1985-2 Conforti, M.; Laurent, M.
On the facial structure of independence systems polyhedra. Working Paper. New York: New York Univ., Dept. of Statistics and Operations Research, 32 p. (1985).

CONFB 1985-3 Conforti, M.; Laurent, M.
On the geometric structure of independence systems. Working Paper. New York: New York Univ., Dept. of Statistics and Operations Research, 39 p. (1985).

CONFB 1986-1 Conforti, M.; Laurent, M.
On the geometric structure of independence systems. New York: Univ., Dept. of Statistics and Operations Research, 39p. (1986).

CONFC 1987-1 Conforti, M.; Rao, M. R.
Some new matroids on graphs: cut sets and the max cut problem. Mathematics of Operations Research 12, 193 204. (1987).

CONFC 1987-2 Conforti, M.; Rao, M. R.
Structural properties and recognition of restricted and strongly unimodular matrices. Mathematical Programming 38, 17-28. (1987).

CONL 1984-1 Conley, W.
Satisfying the customer. International Journal of Mathematical Education in Science and Technology 15, 521-527. (1984).

CONT 1984-1 Conti, A.; Rossetto, S.
Scheduling work loads on one maschine in the presence of an objective function of mixed type (Italian). Ricerca Operativa 30, 21-36. (1984).

COOK 1984-1 Cook, R. J.
Graph factorization and theorems of the Nordhaus-Gaddum class. Periodica Mathematica Hungarica 15, 109-120. (1984).

COOKA 1985-1 Cook, W.
A note on matchings and separability. Discrete Applied Mathematics 10, 203-209. (1985).

COOKA 1986-1 Cook, W.
Cutting-plane proofs in polynomial space. Report No. 86414-OR. Bonn: Universität Bonn, Inst. für Ökonometrie und Operations Research, 9p. (1986).

COOKA 1986-2 Cook, W.
On box totally dual integral polyhedra. Mathematical Programming 34, 48-61. (1986).

COOKB 1987-1 Cook, W.; Coullard, C. R.; Turan, G.
On the complexity of cutting-plane proofs. Discrete Applied Mathematics 18, 25-38. (1987).

COOKC 1986-1 Cook, W.; Fonlupt, J.; Schrijver, A.
An integer analogue of Caratheodory's theorem. Journal of Combinatorial Theory 40B, 63-70. (1986).

COOKD 1986-1 Cook, W.; Gerards, A. M. H.; Schrijver, A.; Tardos, E.
Sensitivity theorems in integer linear programming. Mathematical Programming 34, 251-264. (1986).

COOKE 1986-1 Cook, W.; Kannan, R.; Schrijver, A.
Chvatal closures for mixed integer programming problems. Report 86444-OR. Bonn: Univ. Bonn, Inst. f. Ökonometrie u. Operations Research, 19 p. (1986).

COOKF 1984-1 Cook, W.; Lovasz, L.; Schrijver, A.
A polynomial-time test for total dual integrality in fixed dimension. Mathematical Programming Study 22, 64-69. (1984).

COOKG 1987-1 Cook, W.; Pulleyblank, W. R.
Linear systems for constrained matching problems. Mathematics of Operations Research 12, 97-120. (1987).
COOP 1987-1 Cooper, M. W.
Modelling nonlinearities in multicriteria locational decisions. Mathematical Modelling 8, 207-208. (1987).
COOPA 1984-1 Cooper, M. W.; Farhangian, K.
A dynamic programming algorithm for multiple-choice constraints. Computers & Mathematics with Applications 10, 279-282. (1984).
COOPA 1985-1 Cooper, M. W.; Farhangian, K.
Multicriteria optimization for nonlinear integer-variable problems. Large Scale Systems 9, 73-78. (1985).
COPP 1986-1 Coppersmith, D.; Klawe, M.; Pippenger, N.
Alphabetic minimax trees of degree at most t. SIAM Journal on Computing 15, 189-192. (1986).
COPPA 1987-1 Coppersmith, D.; Raghavan, P.; Tompa, M.
Parallel graph algorithms that are efficient on average. In: 28th Annual IEEE Symposium on Foundations of Computer Science, 260-270. (1987).
COPPB 1985-1 Coppersmith, D.; Vishkin, U.
Solving NP-hard problems in 'almost trees': vertex cover. Discrete Applied Mathematics 10, 27-46. (1985).
CORD 1985-1 Cordovil, R.
On simplicial matroids and Sperner's lemma. In: L. Lovasz, A. Recski (Eds.): Matroid Theory, 97-106. Coll. Math. Soc. J. Bolyai Vol. 40. Amsterdam: North-Holland. (1985).
CORD 1987-1 Cordovil, R.
Polarity and point extensions in oriented matroids. Linear Algebra and Its Applications 90, 15-32. (1987).
CORDA 1986-1 Cordovil, R.; Duchet, P.
Separation par une droite dans les matroides orientes de rang 3. Discrete Mathematics 62, 103-104. (1986).
CORDB 1987-1 Cordovil, R.; Fukuda, K.
Oriented matroids and combinatorial manifolds. Lisboa: C.M.A.F., 16 p. (1987).
CORDC 1987-1 Cordovil, R.; Silva, I. P.
Determining a matroid polytope by non-Radon partitions. Linear Algebra and Its Applications 94, 55-60. (1987).
CORI 1986-1 Cori, R.; Dulucq, S.
Colourings of planar maps and the equality of two languages. In: P. Franchi-Zannettacci (Ed.): CAAP '86, 6-16. Lecture Notes in Computer Science 214, Berlin: Springer. (1986).
CORL 1985-1 Corley, H. W.
Efficient spanning trees. Journal of Optimization Theory and Applications 45, 481-486. (1985).
CORLA 1985-1 Corley, H. W.; Moon, I. D.
Shortest paths in networks with vector weights. Journal of Optimization Theory and Applications 46, 79-86. (1985).
CORN 1985-1 Corneil, D. G.
The complexity of generalized clique packing. Discrete Applied Mathematics 12, 233-240. (1985).
CORNA 1984-1 Corneil, D. G.; Goldberg, M.
A non-factorial algorithm for canonical numbering of a graph. Journal of Algorithms 5, 345-362. (1984).
CORNB 1983-1 Corneil, D. G.; Kirkpatrick, D. G.
Families of recursively defined perfect graphs. Congressus Numerantium 39, 237-246. (1983).
CORNC 1984-1 Corneil, D. G.; Perl, Y.; Stewart, L.
Cographs: recognition, applications and algorithms. Congressus Numerantium 43, 249-258. (1984).
CORND 1985-1 Corneil, D. G.; Perl, Y.; Stewart, L. K.
A linear recognition algorithm for cographs. SIAM Journal on Computing 14, 926-934. (1985).
CORNE 1987-1 Cornuejols, G.
General factors of graphs. Pittsburgh: Carnegie Mellon University, 22 p. (1987).
CORNF 1985-1 Cornuejols, G.; Cunningham, W. H.
Compositions for perfect graphs. Discrete Mathematics 55, 245-254. (1985).
CORNG 1985-1 Cornuejols, G.; Fonlupt, J.; Naddef, D. J.
The traveling salesman problem on a graph and some related integer polyhedra. Mathematical Programming 33, 1-27. (1985).
CORNH 1986-1 Cornuejols, G.; Hartvigsen, D.
An extension of matching theory. Journal of Combinatorial Theory, Series B 40, 285-296. (1986).
CORNI 1985-1 Cornuejols, G.; Naddef, D. J.; Pulleyblank, W. R.
The traveling salesman problem in graphs with 3-edge cutsets. Journal of the ACM 32, 383-410. (1985).
CORNJ 1986-1 Cornuejols, G.; Sassano, A.
On the 0,1 facets of the set covering polytope. R. 153. Roma: Consiglio Nazionale delle Ricerche, 31 p. (1986).
COUL 1986-1 Coullard, C. R.
Counterexamples to conjectures on 4-connected matroids. Combinatorica 6, 315-320. (1986).
COULA 1987-1 Coullard, C. R.; DelGreco, J. G.; Wagner, D. K.
Representations of bicircular matroids. CC-87-21. West Lafayette: Purdue Univ., Inst. for Interdisciplinary Engineering Studies, 20 p. (1987).
COULB 1987-1 Coullard, C. R.; Pulleyblank, W. R.
On cycle cones and polyhedra. CC-87-33. West Lafayette: Purdue Univ. Inst. of Interdisciplinary Engineering Studies, 25 p. (1987).
COUR 1986-1 Courcoubetis, C.; Weber, R. R.
Necessary and sufficient conditions for stability of a bin-packing system. Journal of Applied Probability 23, 989-999. (1986).
COURA 1985-1 Courtois, P.-J.
On time and space decomposition of complex structures. Communications of the ACM 28, 590-604. (1985).

COZZ 1984-1 Cozzens, M. B.; Leibowitz, R.
Threshold dimension of graphs. SIAM Journal on Algebraic and Discrete Methods 5, 579-595. (1984).

COZZA 1982-1 Cozzens, M. B.; Roberts, F. S.
T-colorings of graphs and the channel assigment problem. In: F. Hoffman, K. B. Reid, R. C. Mullin, R. G. Stanton (Eds.): Proc. Thirteenth Southeastern Conf. on Combinatorics, Graph Theory and Computing, Florida 1982, Congressus Numerantium Vol. 35, 191-208. Winnipeg, Canada: Utilitas Mathematica. (1982).

COZZB 1984-1 Cozzens, M. B.; Wang, D. I.
Closed neighborhood containment graphs. Congressus Numerantium 43, 259-272. (1984).

COZZB 1984-2 Cozzens, M. B.; Wang, D. I.
The general channel assignment problem. Congressus Numerantium 41, 115-130. (1984).

CRAI 1986-1 Crainic, T. G.; Rousseau, J.-M.
Multicommodity, multimode freight transportation: a general modeling and algorithmic framework for the service network design problem. Transportation Research 20-B, 225-242. (1986).

CRAI 1987-1 Crainic, T. G.; Rousseau, J.-M.
The column generation principle and the airline crew scheduling problem. INFOR 25, 136-151. (1987).

CRAM 1985-1 Crama, Y.; Hammer, P. L.
The complexity of recognizing partition-quadratic graphs. RUTCOR Research Report 3-85. New Brunswick: Rutgers Univ., Center for Operations Research, 13 p. (1985).

CRAM 1986-1 Crama, Y.; Hammer, P. L.
Some results on the recognition of bimatroidal independence systems. RRR 16-86. New Brunswick: Rutgers Univ., Center for Operations Research, 27 p. (1986).

CRAM 1987-1 Crama, Y.; Hammer, P. L.
Optimization over a bimatroidal independence system. RUTCOR Research Report 3-87. New Brunswick: Rutgers Univ., Rutgers Center for Operations Research, 20 p. (1987).

CRAMA 1986-1 Crama, Y.; Hammer, P. L.; Ibaraki, T.
Strong unimodularity for matrices and hypergraphs. Discrete Applied Mathematics 15, 221-240. (1986).

CRAMA 1987-1 Crama, Y.; Hammer, P. L.; Ibaraki, T.
Packing, covering and partitioning problems with strongly unimodular constraint matrices. RUTCOR Research Report 23-87. New Brunswick: Rutgers Univ., Center for Operations Research, 18 p. (1987).

CRAMB 1986-1 Crama, Y.; Hammer, P. L.; Jaumard, B.; Simeone, B.
Parametric representation of the solutions of a quadratic Boolean equation. RUTCOR Research Report, 15-86. New Brunswick: Rutgers University, 14 p. (1986).

CRAMC 1986-1 Crama, Y.; Hansen, P.; Williams, A. C.
Use of the dual in branch and bound methods for mixed 0-1 programming. Rutcor Research Report No. 1-86. New Brunswick: Rutgers Univ., 9p. (1986).

CRAP 1984-1 Crapo, H.
Selectors: a theory of formal languages, semimodular lattices, and branching and shelling processes. Advances in Mathematics 54, 233-277. (1984).

CRAP 1985-1 Crapo, H.
The combinatorial theory of structures. In: L. Lovasz, A. Recski (Eds.): Matroid Theory, 107-214. Coll. Math. Soc. J. Bolyai Vol. 40. Amsterdam: North-Holland. (1985).

CROI 1986-1 Croitoru, C.; Moscovici, J.
A "greedy" O|V|**3-algorithm for the maximum cardinality matching problem. Analele Stiintifice ale Universitatii Al. I. Cuza din Iasi (Serie Noua), Sectiunea la Matematica 32, No. 3, 93-97. (1986).

CROS 1985-1 Cross, P. A.
Numerical methods in network design. In: E. W. Grafarend, F. Sanso (Eds.): Optimization and design of geodetic networks (Erice, Italy, 1984), 132-168. Berlin: Springer-Verlag. (1985).

CROU 1984-1 Croucher, J. S.
A goal programming model for timetable scheduling. Opsearch 21, 145-152. (1984).

CROW 1985-1 Crowley, J. A.; Scott, P.; Heeney, B.
International truck routing problems with ferry crossing time window and legal constraints in driving hours. Dublin: University College, 16p. (1985).

CRUM 1981-1 Crum, R. L.; Nye, D. J.
A network model of insurance company cash flow management. Mathematical Programming Study 15, 86-101. (1981).

CRYS 1987-1 Crystal Llewellyn, D.
Min algebraic duality. Discrete Applied Mathematics 16, 87-90. (1987).

CRYSA 1985-1 Crystal Llewellyn, D.; Trotter, L. E. Jr.
On abstract integral dependence. Report 85379-OR. Bonn: Univ. Bonn, Inst. für Ökonometrie u. Op. Research, 16 p. (1985).

CSIM 1984-1 Csima, J.; Ralston, T. A.
Crossing-free Hamiltonian paths in Euclidean spaces. Ars Combinatoria 18, 87-98. (1984).

CSIR 1986-1 Csirik, J.
Bin packing as a random walk: a note on Knödel's paper. Operations Research Letters 5, 161-163. (1986).

CSIRA 1987-1 Csirik, J.; Frenk, J. B. G.; Galambos, G.; Rinnooy Kan, A. H. G.
Probabilistic analysis of algorithms for dual bin packing problems. Report 8706/A. Rotterdam: Erasmus Univ., Econometric Institute, 16 p. (1987).

CSIRB 1986-1 Csirik, J.; Galambos, G.
An O(n) bin-packing algorithm for uniformly distributed data. Computing 36, 313-320. (1986).

CSIRB 1986-2 Csirik, J.; Galambos, G.
Nonlinear sequences for bin-packing (Hungarian). Alkalmazott Matematikai Lapok 12, 295-307. (1986).

CSIRC 1986-1 Csirik, J.; Galambos, G.; Frenk, J. B. G.; Frieze, A. M.; Rinnooy Kan, A. H. G.
A probabilistic analysis of the next fit decreasing bin packing heuristic. Operations Research Letters 5, 233-236. (1986).

CSIRD 1986-1 Csirik, J.; Mate, E.
The probabilistic behaviour of the NFD bin packing algorithm. Acta Cybernetica (Beijing) 7, No. 3, 241-245. (1986).

CUNI 1984-1 Cuninghame-Green, R. A.; Borawitz, W. F.
Scheduling by non-commutative algebra. In: R. E. Burkard, R. A. Cuninghame-Green, U. Zimmermann (Eds.): Algebraic and Combinatorial Methods in Operations Research, Annals of Discrete Mathematics Vol. 19, 75-82. Amsterdam: North-Holland. (1984).

CUNIA 1986-1 Cuninghame-Green, R. A.; Burkard, R. E.
Saddlepoints in group and semigroup minimization. Report 86-68. Graz: Univ. Graz, Inst. f. Mathematik, 14 p. (1986).

CUNN 1986-1 Cunningham, P.; Browne, J.
A LISP-based heuristic scheduler for automatic insertion in electronics assembly. International Journal of Production Research 24, 1395-1408. (1986).

CUNNA 1984-1 Cunningham, W. H.
Testing membership in matroid polyhedra. Journal of Combinatorial Theory, Series B 36, 161-188. (1984).

CUNNA 1985-1 Cunningham, W. H.
On submodular function minimization. Combinatorica 5, 185-192. (1985).

CUNNA 1985-2 Cunningham, W. H.
Optimal attack and reinforcement of a network. Journal of the Association for Computing Machinery 32, 549-561. (1985).

CUNNA 1986-1 Cunningham, W. H.
Improved bounds for matroid partition and intersection algorithms. SIAM Journal on Computing 15, 948-957. (1986).

CUNNB 1985-1 Cunningham, W. H.; Frank, A.
A primal-dual algorithm for submodular flows. Mathematics of Operations Research 10, 251-262. (1985).

CUNNC 1986-1 Cunningham, W. H.; Green-Krotki, J.
Dominants and submissives of matching polyhedra. Mathematical Programming 36, 228-237. (1986).

CURR 1985-1 Curran, S. J.; Witte, D.
Hamilton paths in Cartesian products of directed cycles. In: B. R. Alspach, C. D. Godsil (Eds.): Cycles in Graphs, Annals of Discrete Mathematics Vol. 27, 35-74. Amsterdam: North-Holland. (1985).

CURRA 1986-1 Current, J. R.; Min, H.-K.
Multiobjective design of transportation networks: Taxonomy and annotation. European Journal of Operational Research 26, 187-201. (1986).

CURRB 1985-1 Current, J. R.; Revelle, C. S.; Cohon, J. L.
The maximum covering/shortest path problem: a multiobjective network design and routing formulation. European Journal of Operational Research 21, 189-199. (1985).

CURRB 1986-1 Current, J. R.; Revelle, C. S.; Cohon, J. L.
The hierarchical networks design problem. European Journal of Operational Research 27, 57-66. (1986).

CURRB 1987-1 Current, J. R.; Revelle, C. S.; Cohon, J. L.
The median shortest path problem: A multiobjective approach to analyse cost vs. accessibility in the design of transportation networks. Transportation Science 21, 188-197. (1987).

CURRC 1987-1 Current, J. R.; Schilling, D. A.
Elimination of source A and B errors in p-median location problems. Geographical Analysis 19, 95-110. (1987).

CURRD 1986-1 Currin, D. C.
Transportation problems with inadmissible routes. Journal of the Operational Research Society 37, 387-396. (1986).

CURRE 1986-1 Curry, G. L.; Deuermeyer, B. L.
Optimal scheduling of multiple feedstock batch biogas production systems. IIE Transactions 18, 367-373. (1986).

CVET 1983-1 Cvetkovic, D.; Petric, M. V.
Connectedness of the noncomplete extended p-sum of graphs. Univerzitet u Novum Sadu, Zbornik Radova Prirodno-Matematichkog Fakulteta (Novi Sad), Ser. Mat. 13, 345-352. (1983).

CZUC 1986-1 Czuchra, W.
A graphical method to solve a maximin allocation problem. European Journal of Operational Research 26, 259-261. (1986).

DABI 1987-1 Dabija, V.
An algorithm for coloring additive puzzles. Kybernetes 16, No. 1, 19-25. (1987).

DADU 1986-1 Daduna, J. R.
Solving a p-center transportation problem. Methods of Operations Research 53, 239-250. (1986).

DADU 1987-1 Daduna, J. R.
An efficient algorithm for solving bottleneck transportation problems with euclidean distances. In: W. Domschke, W. Krabs, J. Lehn, P. Spellucci (Eds.): Proc. XI. Symp. on Operations Research, Darmstadt 1986, 95-106. Methods of Operations Research 57, Frankfurt/ Main: Athenäum. (1987).

DADU 1988-1 Daduna, J. R.
Verfahren zur Lösung von Vehicle Scheduling Problemen im öffentlichen Personennahverkehr. In: H. Schellhaas, P. van Beek, H. Isermann, R. Schmidt, M. Zijlstra (Eds.): Operations Research Proceedings 1987 (DGOR/NSOR), 270-271. Berlin: Springer. (1988).

DAFE 1984-1 Dafermos, S.; Nagurney, A.
A network formulation of market equilibrium problems and variational inequalities. Providence: Brown Univ., Division of Applied Mathematics, 9 p. (1984).

DAGA 1984-1 Daganzo, C. F.
The distance traveled to visit n points with a maximum of c stops per vehicle: an analytic model and an application. Transportation Science 18, 331-350. (1984).

DAHB 1984-1 Dahbura, A. T.
An efficient minimum vertex cover set algorithm for a broad class of undirected graphs. Congressus Numerantium 43, 273-278. (1984).

DAHL 1987-1 Dahlhaus, E.; Duchet, P.
On strongly chordal graphs. Ars Combinatoria 24B, 23-30. (1987).

DAHLA 1986-1 Dahlhaus, E.; Karpinski, M.
Perfect matching for regular graphs is AC**0-hard for the general matching problem. Research Report 858-CS. Bonn: Univ. Bonn, Inst. f. Informatik, 4 p. (1986).

DAHLA 1986-2 Dahlhaus, E.; Karpinski, M.
The matching problem for strongly chordal graphs is in NC. Research Report 855-CS. Bonn: Univ. Bonn, Inst. f. Informatik, 6 p. (1986).

DAHLA 1987-1 Dahlhaus, E.; Karpinski, M.
A fast parallel algorithm for computing all maximal cliques in a graph and the related problems. Research Report 8516-CS. Bonn: Univ. Bonn, Institut für Informatik, 7 p. (1987).

DAHLA 1987-2 Dahlhaus, E.; Karpinski, M.
Fast parallel computation of perfect and strongly perfect elimination schemes. Research Report 8513-CS. Bonn: Univ. Bonn, Institut für Informatik, 21 p. (1987).

DAHLA 1987-3 Dahlhaus, E.; Karpinski, M.
On the parallel complexity of matching for chordal and path graphs. Research Report 8512-CS. Bonn: Univ. Bonn, Inst. F. Informatik, 13 p. (1987).

DAHLB 1986-1 Dählmann, A; Schütte, H.
Optimale Standorte und Abstimmungsparadoxa auf Graphen. In: L. Streitferdt, H. Hauptmann, A. W. Marusev, D. Ohse, U. Pape (Eds.): Operations Research Proceedings 1985, 59-65. Berlin, Heidelberg, New York: Springer-Verlag. (1986).

DAI 1986-1 Dai, H. T.
A graceful labelling of the complete binary trees (Chinese). Qufu Shifan Daxue Xuebao. Ziran Kexue Ban 12, No. 3, 70-72. (1986).

DAI A 1985-1 Dai, X.-S.; Xue, Y.-F.
Some Hamiltonian properties of connected graph (Chinese). Journal of Kunming Institute of Technology, No.2., 74-80. (1985).

DAI A 1985-2 Dai, X.-S.; Xue, Y.-F.
Some properties of nth order minimally 2-connected graphs (Chinese). Journal of Kunming Institute of Technology, No.4, 67-74. (1985).

DALA 1986-1 Dalang, R. C.; Trotter, L. E. Jr.; Werra, D. de
On randomized stopping points and perfect graphs. O.R.W.P. 86/1. Lausanne: Ecole Polytechnique Federale de Lausanne, 34 p. (1986).

DALL 1984-1 Dallery, Y.
Optimisation de la repartition des palettes dans un atelier flexible. In: Analysis and Optimization of Systems (Nice 1984), Part 2, 633-644. Lecture Notes in Control and Information Science 63, Berlin: Springer. (1984).

DAMA 1985-1 Damaschke, P.; Stern, M.
A characterization of generalized matroid lattices. Studia Scientiarum Mathematicarum Hungarica 20, 425-431. (1985).

DANH 1983-1 Danhof, K. J.; Varol, Y. L.
Generalized set covering problem. Congressus Numerantium 39, 263-272. (1983).

DANI 1985-1 Danilchenko, A. F.
An approach to the solution of the sorting problem (Russian). Matematicheskie Issledovaniya (Kishinev) 81, 39-41. (1985).

DANIA 1985-1 Danilchenko, A. M.; Levchenko, S. N.; Panishev, A. V.
An approximate algorithm for the three-machine problem. Automation and Remote Control 46, 896-902; translation from Avtomatika i Telemekhanika (1985), No. 7, 133-139. (1985).

DANU 1984-1 Danut, M.
Note on the flow and tension in a network. Mathematica Revue d'Analyse Numerque et de Theorie de l'Approximation, Mathematica (Cluj-Napoca) 26(49), 125-128. (1984).

DARB 1982-1 Darbinyan, S. K.
Cycles of arbitrary length in digraphs with large semi degrees (Russian). Doklady, Akademiya Nauk Armyanskoj SSR 75, 147-152. (1982).

DARB 1985-1 Darbinyan, S. K.
Pancyclicity of digraphs with large semidegrees (Russian). Akademiya Nauk Armyanskoi SSR, Doklady 80, No. 2, 51-54. (1985).

DARB 1985-2 Darbinyan, S. K.
Pancyclicity of digraphs with the Meyniel condition (Russian). Studia Scientiarum Mathematicarum Hungarica 20, 95-117. (1985).

DARB 1986-1 Darbinyan, S. K.
A sufficient condition for the Hamilton property of digraphs with large semidegrees (Russian). Doklady, Akademiya Nauk Armyanskoj SSR 82, No. 1, 6-8. (1986).

DARB 1986-2 Darbinyan, S. K.
On the pancyclicity of digraphs with large semidegrees (Russian). Doklady, Akademiya Nauk Armyanskoj SSR 83, No. 3, 99-101. (1986).

DARBA 1985-1 Darby-Dowman, K.; Mitra, G.
An extension of set partitioning with application to scheduling problems. European Journal of Operational Research 21, 200-205. (1985).

DARZ 1987-1 Darzentas, J.
On fuzzy location models. In: J. Kacprzyk, P. A. Orlovski (Eds.): Optimization Models Using Fuzzy Sets and Possibility Theory, 328-341. Dordrecht: Reidel. (1987).

DARZA 1986-1 Darzentas, J.; Bairaktaris, P.
On the set partitioning type formulation for the discrete location problem. Computers and Operations Research 13, 671-680. (1986).

DAS 1987-1 Das, C.; Patel, G.
A variant of the solid transportation problem with mixed constraints. Ricerca Operativa 17, No.42, 103-130. (1987).

DAS A 1985-1 Das, C.; Verma, A.
A heuristic method of finding the optimal location and size of facilities. The Logistics and Transportation Review 21, 115-132. (1985).

DAS B 1983-1 Das, P.; Gangopadhyay, T.
Some results on alpha-valuations and average numberings. Utilitas Mathematica 23, 209-222. (1983).

DAS C 1984-1 Das, P.; Rao, S. B.
Center graphs of chordal graphs. In: K. P. Vijayan, N. M. Singhi (Eds.): Combinatorics and Applications (Proc. Seminar Calcutta 1982), 81-94. Calcutta: Indian Statistical Institute. (1984).

DAS D 1986-1 Das, S. K.; Deo, N.
Stirling graphs and their properties. Congressus Numerantium 54, 5-20. (1986).

DAUL 1987-1 Dauler, J. C.; Nuttle, H. L. W.
Addendum to "A comparison of heuristics for a school bus scheduling problem". Transportation Research 21B, 311-316. (1987).

DAUS 1983-1 Dauscha, W.; Modrow, H. D.; Neumann, A.
An approach for periodic networks. Zentralinstitut für Kybernetik und Informationsprozesse (ZKI) Informationen, Akademie der Wissenschaften der DDR 2/83, 4-22. (1983).

DAUS 1985-1 Dauscha, W.; Modrow, H. D.; Neumann, A.
On cyclic sequence types for constructing cyclic schedules. Zeitschrift für Operations Research 29, Series A, 1-30. (1985).

DAVI 1985-1 Davitt, R. M.; Fink, J. F.; Jacobson, M. S.
On isopart parameters of complete bipartite graphs and n-cubes. Mathematica Slovaca (Bratislava) 35, 409-415. (1985).

DAVY 1986-1 Davydov, E. G.; Bolshakov, S. Y.
Group-theoretic methods for aggregation in network problems (Russian). Reports in Applied Mathematics, Akad. Nauk SSSR, Vychisl. Tsentr, Moscow. (1986).

DAVYA 1985-1 Davydov, G. V.; Davydova, I. M.
A duality criterion for optimality in allocation problems. Soviet Mathematics 29, No. 1, 73-77. Translation from: Izvestiya Vysshykh Uchebnykh Zavedenij, Matematika (Kazan) 1985, No. 1, 57-59. (1985).

DAVYA 1987-1 Davydov, G. V.; Davydova, I. M.
Dual algorithms in discrete optimization (Russian). Voprosy Kibernetiki (Moscow) 131, 90-117. (1987).

DAVYB 1985-1 Davydova, I. M.
Enumeration schemes in location problems (Russian). Leningrad: Leningradskij Gosudarstvennyi Universitet im. A. A. Zhdanova. (1985).

DAWE 1983-1 Dawes, R. W.
Constructions of minimally k-connected graphs for k=1,2 and 3. Congressus Numerantium 39, 273-290. (1983).

DAWE 1986-1 Dawes, R. W.
Minimally 3-connected graphs. Journal of Combinatorial Theory, Series B, 40, 159-168. (1986).

DAWS 1983-1 Dawson, J. E.
Balanced sets in an independence structure induced by a submodular function. Journal of Mathematical Analysis and Applications 95, 214-222. (1983).

DAWS 1984-1 Dawson, J. E.
A collection of sets related to the Tutte polynomial of a matroid. In: K. M. Koh, H. P. Yap (Eds.): Graph Theory Singapore 1983, Lecture Notes in Mathematics 1073, 193-204. Berlin: Springer-Verlag. (1984).

DAWS 1984-2 Dawson, J. E.
Solubility of the matroid sum equations. Ars Combinatoria 17A, 103-116. (1984).

DAWS 1985-1 Dawson, J. E.
Algorithms to find directed packings. In: C. J. Colbourn, M. J. Colbourn (Eds.): Algorithms in Combinatorial Design Theory, Annals of Discrete Mathematics Vol. 26, 137-142. Amsterdam: North-Holland. (1985).

DAWS 1985-2 Dawson, J. E.
Decomposition of binary matroids. Combinatorica 5, 1-10. (1985).

DAX 1986-1 Dax, A.
A note on optimality conditions for the euclidean multifacility location problem. Mathematical Programming 36, 72-80. (1986).

DAX 1986-2 Dax, A.
An efficient algorithm for solving the rectilinear multifacility location problem. IMA Journal of Numerical Analysis 6, 343-356. (1986).

DAYA 1986-1 Dayan, I.; Harel, D.
Fair termination with cruel schedulers. Fundamenta Informaticae (Warsaw) 9, No. 1, 1-11. (1986).

DEAK 1983-1 Deak, I.; Hoffer, J.; Mayer, J.; Nemeth, A.; Potecz, B.; Prekopa, A.; Strazicky, B.
A large scale, mixed variable mathematical programming model for the short-range optimal schedule of electric power systems with thermal power plants taking network constraints into account (Hungarian). Alkalmazott Matematikai Lapok (Budapest) 9, 221-337. (1983).

DEAK 1983-2 Deak, I.; Hoffer, J.; Mayer, J.; Nemeth, A.; Potecz, B.; Prekopa, A.; Strazicky, B.
Optimal daily scheduling of electricity production in Hungary. Tanulmanyok - MTA Szamitastechnikai es Automatizalasi Kutato Intezet Budapest No. 152, 43-67. (1983).

DEAK 1984-1 Deak, I.; Hoffer, J.; Mayer, J.; Nemeth, A.; Potecz, B.; Prekopa, A.; Strazicky, B.
Optimal, short-distance plant scheduling for electrical energy systems based on thermal plants, taking conditions of network locations into consideration (Hungarian). Tanulmanyok - MTA Szamitastechnikai es Automatizalasi Kutato Intezete Budapest No. 155. (1984).

DEAR 1985-1 Dearing, P. M.
Location problems. Operations Research Letters 4, 95-98. (1985).

DEBA 1984-1 Debanne, J. G.
Tailoring decomposition algorithms to efficiently solve large scale energy planning problems. In: J. P. Brans (Ed.): Operational Research '84, 840-854. Amsterdam: North-Holland. (1984).

DECA 1985-1 Decaestecker, C.
On stable colourings of exclusive boolean graphs (French). Bulletin de la Classe des Sciences, 5e Serie, Academie Royale de Belgique 71, 297-300. (1985).

DEIN 1985-1 Deineko, V. G.
A polynomial algorithm for the search for the optimum in exponential neighborhoods for traveling salesman problems (Russian). In: Methods for Solving Nonlinear Problems and for Data Processing, 82-86. Dnepropetrovsk: Dnepropetrovsk. Gos. Univ. (1985).

DEL 1985-1 Del Ojo Mesa, J.; Fernandez Garcia, F. R.; Pelegrin, B.
A global otpimization algorithm for integer programming (Spanish). In: Proceedings of the tenth Spanish-Portuguese conference of mathematics, VI (Murcia 1985), 73-81. Murcia: Univ. (1985).
DELF 1983-1 Delfrate, M. G.
Oriented 2-graphs and their generalization to dimension n (Italian). Rivista di Matematica della Universita di Parma, Serie IV 9, 153-160. (1983).
DELF 1983-2 Delfrate, M. G.
The oriented 2-graphs and a generalization to the n-dimensional case (Italian). Rivista di Matematica della Universita di Parma, Serie IV/9, 153-160. (1983).
DELG 1987-1 Delgado, M.; Verdegay, J. L.; Vila, M. A.
Fuzzy transportation problems: a general analysis. In: J. Kacprzyk, P. A. Orlovski (Eds.): Optimization Models Using Fuzzy Sets and Possibility Theory, 342-358. Dordrecht: Reidel. (1987).
DELL 1986-1 Dellaert, N.; Wessels, J.
Production scheduling with uncertain demand and no stockkeeping. In: L. Streitferdt, H. Hauptmann, A. W. Marusev, D. Ohse, U. Pape (Eds.): Operations Research Proceedings 1985, 247-252. Berlin, Heidelberg, New York: Springer-Verlag. (1986).
DELO 1985-1 Delorme, C.
Grands graphes de degre et diametre donnes. European Journal of Combinatorics 6, 291-302. (1985).
DELOA 1984-1 Delorme, C.; Farhi, G.
Large graphs with given degree and diameter. I. IEEE Transactions on Computers 33, 857-860. (1984).
DEMA 1987-1 Demange, G.; Gale, D.; Sotomayor, M.
A further note on the stable matching problem. Discrete Applied Mathematics 16, 217-222. (1987).
DEMAA 1979-1 DeMaria, F.; Gionfriddo, M.
Some parameters associated with graphs and hypergraphs (Italian). Atti della Socieata Peloritana die Scienze Fisiche, Matematiche e Naturali (Messina) 25, 227-239. (1979).
DEMAA 1980-1 DeMaria, F.; Gionfriddo, M.
A theorem on the s-chromatic number of a graph (Italian). Atti della Societa Peloritana di Scienze Fisiche, Matematiche e Naturali (Messina) 26, 315-322. (1980).
DEMB 1986-1 Dembo, R. S.
The performance of NLPNET, a large-scale nonlinear network optimizer. In: G. Gallo, C. Sandi (Eds.): Netflow at Pisa, Mathematical Programming Study 26, 245-248. (1986).
DEMB 1987-1 Dembo, R. S.
A primal truncated Newton algorithm with application to large-scale nonlinear network optimization. Mathematical Programming Study 31, 43-72. (1987).
DEMBA 1981-1 Dembo, R. S.; Klincewicz, J. G.
An approximate second-order algorithm for network flow problems with convex separable costs. Mathematical Programming Study 15, 125-147. (1981).
DEMP 1983-1 Dempe, S.
Zwei greedy-Algorithmen für ein gemischt-ganzzahliges Optimierungsproblem. In: 2nd Colloquium for Geometry and Combinatorics (Karl-Marx-Stadt, 1983) Part 1, 2, 59-64. Karl-Marx-Stadt: Technische Hochschule. (1983).
DEMP 1984-1 Dempe, S.
Gradient algorithms for solving a generalized knapsack problem (Russian). Vestnik Belorusskogo Gosudarstvennogo Universiteta (Minsk), Serya I No. 3, 48-50. (1984).
DEMP 1985-1 Dempe, S.
A simple but NP-hard problem of mixed-discrete programming and its solution by approximate algorithms. Optimization 16, 705-714. (1985).
DEOR 1985-1 Deorditsa, Y. S.; Morozova, T. Y.
Method of block decomposition for engineer networks (Russian). Izvestiya Akademii Nauk Turkmenskoj SSR, Seriya Fiziko-Tekhnicheskikh, Khimicheskikh i Geologicheskikh Nauk (Ashkhabad) No. 1, 3-7. (1985).
DEPA 1987-1 DePalma, A.; Hansen, P.; Labbe, M.
Commuters' paths with penalties for early or late arrival time. CORE Discussion Paper 8712. Louvain-la-Neuve: Center for Operations Research & Econometrics, 24 p. (1987).
DERI 1981-1 Derigs, U.
F-factors, perfect matchings and related concepts. In: H. Noltemeier (Ed.): Graphtheoretic Concepts in Computer Science, Proc. Internat. Workshop WG 80 (Bad Honnef 1980), 388-403. Berlin, Heidelberg, New York: Springer. (1981).
DERI 1984-1 Derigs, U.
Alternate strategies for solving bottleneck assignment problems - analysis and computational results. Computing 33, 95-106. (1984).
DERI 1984-2 Derigs, U.
Exchange properties and K-best strategies in combinatorial optimization. In: J. P. Brans (Ed.): Operational Research '84, 393-406. Amsterdam: North-Holland. (1984).
DERI 1984-3 Derigs, U.
Programming in networks and graphs - on the combinatorial background and near-equivalence of network flow and matching algorithms. Bonn: Univ. Bonn, Inst. f. Operations Research, 323 p. (1984). (Habilitationsschrift)
DERI 1984-4 Derigs, U.
The cardinality matching problem - a state of the art tutorial. In: U. Pape (Ed.): Proc. WG '84: Internat. Workshop on Graphtheoretic Concepts in Computer Science, Berlin 1984, 58-78. Linz: Trauner. (1984).
DERI 1985-1 Derigs, U.
An efficient Dijkstra-like labeling method for computing shortest odd/even paths. Information Processing Letters 21, 253-258. (1985).
DERI 1985-2 Derigs, U.
Postoptimal analysis for matching problems. In: P. Brucker, R. Pauly (Eds.): Proc. IX. Symp. on Operations Research, Osnabrück 1984. Methods of Operations Research 49, 215-222. (1985).
DERI 1985-3 Derigs, U.
Some basic exchange properties in combinatorial optimization and their application to constructing the K-best solutions. Discrete Applied Mathematics 11, 129-142. (1985).

DERI 1985-4 Derigs, U.
Using confidence limits for the global optimum in combinatorial optimization. Operation Research 33, 1024-1049. (1985).

DERI 1986-1 Derigs, U.
A short note on matching algorithms. In: G. Gallo, C. Sandi (Eds.): Netflow at Pisa, Mathematical Programming Study 26, 200-204. (1986).

DERI 1986-2 Derigs, U.
The shortest augmenting path method for solving assignment problems - motivation and computational experience. In: C. L. Monma (Ed.): Algorithms and Software for Optimization - Part I, 57-102. Annals of Operations Research 4, Basel: Baltzer. (1986).

DERIA 1984-1 Derigs, U.; Goecke, O.; Schrader, R.
Bisimplicial edges, Gaussian elimination and matchings in bipartite graphs. In: U. Pape (Ed.): Proc. WG '84: Internat. Workshop on Graphtheoretic Concepts in Computer Science, Berlin 1984, 79-87. Linz: Trauner. (1984).

DERIA 1986-1 Derigs, U.; Goecke, O.; Schrader, R.
Monge sequences and a simple assignment algorithm. Discrete Applied Mathematics 15, 241-248. (1986).

DERIB 1985-1 Derigs, U.; Hagberg, B.
Ein iterativer Lösungsansatz für das Rostering Problem mittels optimaler Zuordnungen. In: D. Ohse, A. C. Esprester, H.-U. Küpper, P. Stähly, H.Steckhan (Eds.): Operations Research Proceedings 1984, 417-420. Berlin: Springer-Verlag. (1985).

DERIC 1987-1 Derigs, U.; Hilbrand, C.
Optimal matchings with GUP-constraints. In: W. Domschke, W. Krabs, J. Lehn, P. Spellucci (Eds.): Proc. XI. Symp. on Operations Research, Darmstadt 1986, 107-108. Methods of Operations Research 57, Frankfurt/Main: Athenäum. (1987).

DERID 1985-1 Derigs, U.; Metz, A.
An efficient labeling technique for solving sparse assignment problems. Report No. 85367-OR. Bonn: Univ. Bonn, Inst. f. Ökonometrie u. Operations Research, 11 p. (1985).

DERID 1986-1 Derigs, U.; Metz, A.
An efficient labeling technique for solving sparse assignment problems. Computing 36, 301-312. (1986).

DERID 1986-2 Derigs, U.; Metz, A.
An in-core/out-of-core method for solving large scale assignment problems. Zeitschrift für Operations Research A-30, 181-196. (1986).

DERID 1986-3 Derigs, U.; Metz, A.
On the use of optimal fractional matchings for solving the (integer) matching problem. Computing 36, 263-270. (1986).

DESI 1986-1 DeSimone, C.; Lucertini, M.; Pallottino, S.; Simeone, B.
Fair dissections of spiders, worms and caterpillars. RUTCOR Research Report 17-86. New Brunswick: Rutgers Univ., Center for Operations Research, 23 p. (1986).

DESR 1986-1 Desrochers, M.
An algorithm for the shortest path problem with resource constraints. Publication No. 421A. Montreal: Univ., Centre de recherche sur les transports, 31p. (1986).

DESR 1987-1 Desrochers, M.
A note on the partitioning shortest path algorithm. Operations Research Letters 6, 183-188. (1987).

DESRA 1988-1 Desrochers, M.; Kolen, A. W. J.; Lucena, A.
A 2-commodity flow formulation for the vehicle routing problem. In: H. Schellhaas, P. van Beek, H. Isermann, R. Schmidt, M. Zijlstra (Eds.): Operations Research Proceedings 1987 (DGOR/NSOR), 599-600. Berlin: Springer. (1988).

DESRB 1987-1 Desrochers, M.; Lenstra, J. K.; Savelsbergh, M. W. P.
A classification scheme for vehicle routing and scheduling problems. Report OS-R8721. Amsterdam: Centre for Mathematics and Computer Science, 12 p. (1987).

DESRC 1987-1 Desrochers, M.; Lenstra, J. K.; Savelsbergh, M. W. P.; Soumis, F.
Vehicle routing with time windows: optimization and approximation. Report OS-R8715. Amsterdam: Centre for Mathematics and Computer Science, 15 p. (1987).

DESRD 1986-1 Desrosiers, J.; Dumas, Y.; Soumis, F.
A dynamic programming solution of the large-scale single-vehicle dial-a-ride problem with time windows. American Journal of Mathematical and Management Sciences 6, 301-325. (1986).

DESRE 1984-1 Desrosiers, J.; Soumis, F.; Desrochers, M.
Routing with time windows by column generation. Networks 14, 545-566. (1984).

DESRF 1986-1 Desrosiers, J.; Soumis, F.; Desrochers, M.; Sauve, M.
Methods for routing with time windows. European Journal of Operational Research 23, 236-245. (1986).

DESRF 1986-2 Desrosiers, J.; Soumis, F.; Desrochers, M.; Sauve, M.
Vehicle routing and scheduling with time windows. In: G. Gallo, C. Sandi (Eds.): Netflow at Pisa, Mathematical Programming Study 26, 249-251. (1986).

DESRG 1984-1 Desrosiers, J.; Soumis, F.; Sauve, M.
Lagrangian relaxation for routing with time windows. In: J. P. Brans (Ed.): Operational Research '84, 673-684. Amsterdam: North-Holland. (1984).

DEWD 1985-1 Dewdney, A. K.; Walsh, T. R.
A methodology for average-time testing of exact and approximate graph algorithms. Congressus Numerantium 47, 273-284. (1985).

DEWE 1984-1 Dewess, G.
Zur Anwendung der Filtermethode auf das kombinatorische Fundamentalproblem der Ablaufplanung. In: K. Lommatzsch (Ed.): Proceedings, 16. Jahrestagung "Mathematische Optimierung", Sellin/Rügen 1984, 18-26. Seminarbericht Nr. 64, Sektion Mathematik, Humboldt-Univ. Berlin (DDR). (1984).

DEZA 1984-1 Deza, M.
Perfect matroid designs. Report No. 8, Department of Mathematics, University of Stockholm. (1984).

DEZAA 1986-1 Deza, M.; Fukuda, K.
On bouquets of matroids and orientations. Research Report on Information Sciences Series B, 176. Tokyo: Tokyo Inst. of Technology, Dept. of Information Sciences, 20 p. (1986).

DEZAB 1987-1 Deza, M.; Laurent, M.
Bouquets of matroids, d-injection geometries and diagrams. Journal of Geometry 29, 12-35. (1987).

DHAR 1987-1 Dhar, D.; Barma, M.; Chakrabarti, B. K.; Taraphder, A.
The travelling salesman problem on a randomly diluted lattice. Journal of Physics A 20, 5289-5298. (1987).

DHARA 1986-1 Dhar, U. R.
Spherical location under restricted distance. Defence Science Journal 36, 353-365. (1986).

DHARB 1981-1 Dhar, U. R.; Rao, J. R.
Large region location problems under distance constraints. Ganit (Lucknow) 1, 99-105. (1981).

DIAK 1987-1 Diakonu, M. M.
An algorithm for solving a bicriteria problem of integer linear programming (Russian). Mat. Issled. No. 96 Optim. i Obrabot. Dannykh, 29-35. (1987).

DICA 1986-1 Dicasali, R. L.
Job shops can use repetitive manufacturing methods to facilitate just-in-time production. Industrial Engineering 18, 48-52. (1986).

DICK 1984-1 Dick, W. E.
Discrete characterization of planarity. I: The classical viewpoint. Linear and Multilinear Algebra 16, 39-65. (1984).

DIEH 1985-1 Diehr, G.
Evaluation of a branch and bound algorithm for clustering. SIAM Journal on Scientific and Statistical Computing 6, 268-284. (1985).

DIET 1987-1 Dietrich, B. L.
A circuit set characterization of antimatroids. Journal of Combinatorial Theory B-43, 314-321. (1987).

DILL 1987-1 Dillencourt, M. B.
Traveling salesman cycles are not always subgraphs of Delaunay triangulations or of minimum weight triangulations. Information Processing Letters 24, 339-342. (1987).

DILLA 1988-1 Dillmann, R.
Hierarchisierung von Restriktionen und der Balas-Algorithmus. In: H. Schellhaas, P. van Beek, H. Isermann, R. Schmidt, M. Zijlstra (Eds.): Operations Research Proceedings 1987 (DGOR/NSOR), 600-601. Berlin: Springer. (1988).

DIMO 1985-1 Dimow, E. T.; Kovalev, M. M.; Milanov, P. B.
Qualitätsabschätzungen von Greedy-Algorithmen (Russian). In: E. Girlich (Ed.): Diskrete Optimierung, 54-72. Jena: Friedrich-Schiller-Universität. (1985).

DINE 1984-1 Dinescu, C.; Savulescu, B.
On transportation of a maximum flow under some conditions on arcs. Analele Universitatii Bucuresti, Matematica (Bucarest) 33, 25-29. (1984).

DING 1987-1 Ding, G.-L.
On the maximum 2-1 matching. Acta Mathematicae Applicatae Sinica, English Series 3, 305-312. (1987).

DINGA 1984-1 Ding, J. Y.; Song, G. D.; Pan, Z. T.
The Steiner problem for triangle-trapezoid zigzag lines (Chinese). Heilongjiang Daxue Ziran Kexue Xuebao No. 4, 38-50. (1984).

DINGB 1987-1 Ding, L.-I.; Yue, M.-Y.
On a generalization of the Rado-Hall theorem to greedoids. Asia-Pacific Journal of Operational Research 4, 28-38. (1987).

DINGC 1985-1 Ding, S. K.
The structure of a maximum critical 2-edge-connected graph (Chinese). Chinese Annals in Mathematics, Series A, 447-453. (1985).

DINY 1985-1 Diny, D. A.; Harary, F.; Mac Morris, F. R.
A survey of results on a graph and its complement with specified hereditary properties. Ars Combinatoria 20-B, 193-206. (1985).

DIVO 1987-1 Divoky, J. J.
Improvements for the Thresh X2 shortest path algorithm. Operations Research Letters 6, 227-232. (1987).

DIXO 1987-1 Dixon, P. S.
Comments on 'lot sizing for material requirements planning systems'. International Journal of Production Research 25, 1239-1240. (1987).

DJID 1984-1 Djidjev, H. N.
On some properties of nonplanar graphs. Comptes Rendus de l'Academie Bulgare des Sciences (Sofia) 37, 1183-1184. (1984).

DJID 1985-1 Djidjev, H. N.
A linear algorithm for partitioning graphs of fixed genus. Serdica 11, No. 4, 369-387. (1985).

DJID 1985-2 Djidjev, H. N.
A separator theorem for graphs of fixed genus. Serdica 11, No. 4, 319-329. (1985).

DJID 1985-3 Djidjev, H. N.
Efficient parallel algorithms of graphs. Comptes Rendus de l'Academie Bulgare des Sciences (Sofia) 38, 687-689. (1985).

DJID 1985-4 Djidjev, H. N.
Solving graph problems on VLSI models. Comptes Rendus de l'Academie Bulgare des Sciences (Sofia) 38, 1623-1625. (1985).

DMIT 1982-1 Dmitriev, I. G.
Minimal degree of uniquely colored graphs with minimal number of edges (Russian). Vychislitel'nye Sistemy (Novosibirsk) 92, 50-55. (1982).

DO 1987-1 Do, Z. C.
Transformations of the noninteger maximum of a concave function into the integer maximum (Russian). Vestnik Belorusskogo Gosudarstvennogo Universiteta im. V.I. Lenina (Minsk), Seriya I, No. 1, 62-64. (1987).

DOBS 1987-1 Dobson, G.
The economic lot-scheduling problem: Achieving feasibility using time-varying lot sizes. Operations Research 35, 764-771. (1987).

CHAU 1985-1 Chaudhry, S. S.; Mac Cormick, S. T.; Moon, I. D.
Locating independent facilities with maximum weight: greedy heuristics. Working Paper. La Crosse: Univ. of Wisconsin, Dept. of Management, 19 p. (1985).

CHAUA 1987-1 Chaudhry, S. S.; Moon, I. D.; Mac Cormick, S. T.
Conditional covering: greedy heuristics and computational results. Computers and Operations Research 14, 11-18. (1987).

CHAUB 1987-1 Chaudhuri, P.
Algorithms for some graph problems on a distributed computational model. Information Sciences 43, 205-228. (1987).

CHAUB 1987-2 Chaudhuri, P.
Efficient algorithms for some graph problems on a tree-structured parallel computer. International Journal of Electronics 62, 599-615. (1987).

CHAUC 1986-1 Chaudhuri, P.; Ghosh, R. K.
Parallel algorithms for analyzing activity networks. BIT 26, 418-429. (1986).

CHAUD 1987-1 Chauny, F.; Loulou, R.; Sadones, S.; Soumis, F.
A two-phase heuristic for strip packing: algorithm and probabilistic analysis. Operations Research Letters 6, 25-34. (1987).

CHAZ 1983-1 Chazelle, B.
The bottom-left bin-packing heuristic: an efficient implementation. IEEE Transactions on Computers C-32, 697-707. (1983).

CHEB 1986-1 Cheban, Y. I.
Investigation of a generalized graph coloring problem (Russian). In: Investigation of Methods for Solving Extremal Problems, 77-81. Kiev: Akad. Nauk Ukrain. SSR, Inst. Kibernet. (1986).

CHEBA 1984-1 Chebotar, K. S.
A problem of optimization on a set of permutations (Russian). Izvestiya Akademii Nauk Moldavskoj SSR, Seriya Fiziko-Tekhnicheskik i Matematicheskikh Nauk (Kishinev) No. 1, 51-53. (1984).

CHEG 1987-1 Chegireddy, C. R.; Hamacher, H. W.
Algorithms for finding K-best perfect matchings. Discrete Applied Mathematics 18, 155-166. (1987).

CHEN 1987-1 Chen, B.-L.; Lih, K.-W.
Hamiltonian uniform subset graphs. Journal of Combinatorial Theory 42B, 257-263. (1987).

CHENA 1986-1 Chen, C.-H. J.; Engquist, M.
A primal simplex approach to pure processing networks. Management Science 32, 1582-1598. (1986).

CHENB 1986-1 Chen, J. H.
Spanning trees with the maximum number of end vertices (Chinese). Acta Scientiarun Naturalium Universitatis Amoiensis (Xiamen) 25, 508-512. (1986).

CHENC 1984-1 Chen, J. Y.
The satisfiability problem for simple Boolean expressions belongs to P (Chinese). Beijing Daxue Xuebao No. 1, 14-21. (1984).

CHEND 1987-1 Chen, M. S.; Meditch, J. S.
A distributed adaptive routing algorithm. Computers & Electrical Engineering 13, 105-115. (1987).

CHENE 1983-1 Chen, Q.-H.
Finding r-dimensional minimum spanning trees with a fixed incident number at an assigned simplex D(S) (Chinese). Journal of Shandong University (Natural Science Edition), No.3, 1-8. (1983).

CHENF 1984-1 Chen, R.
Location problems with costs being sums of powers of Euclidean distances. Computers & Operations Research 11, 285-294. (1984).

CHENG 1987-1 Chen, R.; Handler, G. Y.
Relaxation method for the solution of the minimax location-allocation problem in Euclidean space. Naval Research Logistics 34, 775-788. (1987).

CHENH 1984-1 Chen, R. S.
An algorithm for finding generalized systems of distinct representatives (Chinese). Journal of the Xinjiang University of Natural Sciences No. 4, 41-52. (1984).

CHENI 1985-1 Chen, R. Y.
Long cycles in regular 2-connected bipartite graphs (Chinese). Journal of Mathematics, Wuhan University, 5, No. 3, 209-216. (1985).

CHENJ 1986-1 Chen, Z. B.
A simple proof of a theorem about strongly connected digraphs (Chinese). Journal Xinjiang University of Natural Science (Urumqi) 3, No. 2, 105-106. (1986).

CHENK 1982-1 Chen, Z. M.
Some results about the graphs with unique 1-factor (Chinese). Journal of Shanghai Teachers University, Natural Science Edition, No.3, 29-34. (1982).

CHENL 1986-1 Chen, Z. Z.
A generalization of the Bodendiek conjecture about graceful graphs (Chinese). Qufu Shifan Daxue Xuebao. Ziran Kexue Ban 12, No. 4, 45-46. (1986).

CHENM 1985-1 Cheng, D.-Q.
The solution to the Chinese postman problem by branch and bound (Chinese). Chinese Journal of Operations Research 4, 60-62. (1985).

CHENN 1987-1 Cheng, S.-C.; Stankovic, J. A.; Ramaritham, K.
Dynamic scheduling of groups of tasks with precedence constraints in distributed hard real-time systems. In: Real-time Systems Symposium (New Orleans 1986), 166-174. Washington, D.C.: IEEE Comput. Soc. (1987).

CHENO 1985-1 Cheng, T. C. E.
Analysis of job flow-time in a job shop. Journal of the Operational Research Society 36, 225-230. (1985).

CHENO 1986-1 Cheng, T. C. E.
A note on minimizing the average deviation of job completion times about a common due-date. Industrial Mathematics 36, 87-89. (1986).

CHENO 1986-2 Cheng, T. C. E.
A note on the common due-date assignment problem. Journal of the Operational Research Society 37, 1089-1092. (1986).

CHENO 1986-3 Cheng, T. C. E.
On optimal common due-date determination. IMA Journal of Mathematics in Management 1, 39-43. (1986).

DOBSA 1986-1 Dobson, G.; Karmarkar, U. S.
Large scale shop scheduling: formulations and decomposition. Working Paper QM 86-31. Rochester: Univ. of Rochester, Graduate School of Management, 37 p. (1986).
DOBSA 1986-2 Dobson, G.; Karmarkar, U. S.
Simultaneous resource scheduling to minimize flow times. Working Paper QM 8623. Rochester: Univ. of Rochester, Graduate School of Management, 20 p. (1986).
DOBSA 1987-1 Dobson, G.; Karmarkar, U. S.
Competitive location on a network. Operations Research 35, 565-574. (1987).
DOBSB 1985-1 Dobson, G.; Karmarkar, U. S.; Rummel, J. L.
Single machine sequencing with lot sizing. Working Paper Series QM8419. Rochester: Univ. of Rochester, Graduate School of Management, 34 p. (1985).
DOBSB 1985-2 Dobson, G.; Karmarkar, U. S.; Rummel, J. L.
Single-product multi-machine flow minimization with batching. Worling Paper. Rochester: Univ. of Rochester, Graduate School of Management, 17 p. (1985).
DOBSB 1987-1 Dobson, G.; Karmarkar, U. S.; Rummel, J. L.
Batching to minimize flow times on one machine. Management Science 33, 784-799. (1987).
DODI 1984-1 Dodin, B.
Determining the k most critical paths in PERT networks. Operations Research 32, 859-877. (1984).
DODI 1984-2 Dodin, B.
Scheduling N products on a single facility with backordering and fixed terminal inventory. INFOR 22, 317-342. (1984).
DODI 1985-1 Dodin, B.
Bounding the project completion time distribution in PERT networks. Operations Research 33, 862-881. (1985).
DODI 1985-2 Dodin, B.
Scheduling N products on a single facility with allowed backordering. International Journal of Production Research 23, 329-344. (1985).
DODIA 1985-1 Dodin, B.; Elmaghraby, S. E.
Approximating the criticality indices of the activities in PERT networks. Managemnet Science 31, 207-223. (1985).
DODIB 1985-1 Dodin, B.; Sirvanci, M.
The extreme value distribution and stochastic activity networks. Working Paper. Riverside: Univ. of California, Graduate School of Management, 29 p. (1985).
DOKE 1985-1 Dokev, N. B.
An algorithm for solving a class of scheduling tasks (Bulgarian). Problemi na Tekhnicheskata Kibernetika i Robotikata 21, 56-63. (1985).
DOLE 1987-1 Dolev, D.; Halpern, J. Y.; Simons, B.; Strong, H. R.
A new look at fault-tolerant network routing. Information and Computation 72, No. 3, 180-196. (1987).
DOLEA 1985-1 Dolev, D.; Upfal, E.; Warmuth, M. K.
Scheduling trees in parallel. In: P. Bertolazzi, F. Luccio (Eds.): VLSI: Algorithms and Architectures, 91-102. Amsterdam: North-Holland. (1985).
DOLEB 1984-1 Dolev, D.; Warmuth, M. K.
Scheduling precedence graphs of bounded height. Journal of Algorithms 5, 48-59. (1984).
DOLEB 1985-1 Dolev, D.; Warmuth, M. K.
Profile scheduling of opposing forests and level orders. SIAM Journal on Algebraic and Discrete Methods 6, 665-687. (1985).
DOLEB 1985-2 Dolev, D.; Warmuth, M. K.
Scheduling flat graphs. SIAM Journal on Computing 14, 638-657. (1985).
DOMK 1986-1 Domke, G. S.; Laskar, R.; Hedetniemi, S. T.; Peters, K.
The partite-chromatic number of a graph. Congressus Numerantium 53, 235-246. (1986).
DOMS 1981-1 Domschke, W.
Logistik: Transport. Grundlagen, lineare Transport- und Umladeprobleme. München, Wien: Oldenbourg. (1981).
DOMSA 1985-1 Domschke, W.; Drexl, A.
ADD- and DROP-procedures for capacitated plant location models with different capacities. European Journal of Operational Research 21, 47-53. (1985).
DOMSA 1985-2 Domschke, W.; Drexl, A.
Location and layout planning. An international bibliography. Berlin, Heidelberg: Springer. (1985). (Lecture Notes in Economics and Mathematical Systems, 238)
DONA 1987-1 Donald, J.; Elwin, J.; Hager, R.; Salamon, P.
Handle bases and bounds on the number of subgraphs. Journal of Combinatorial Theory B 42, 1-13. (1987).
DONAA 1986-1 Donath, G.; Hofstedt, K.
Über eine spezielle lineare 0-1-Optimierungsaufgabe. Optimization 17, 785-792. (1986).
DONG 1984-1 Dong, C. F.; Liu, W. G.
The shortest path problem on a kind of variable weighted networks (Chinese). Chinese Journal of Operations Research 3 No. 1, 61-62. (1984).
DONGA 1985-1 Dong, J. Q.
Graceful labelings of a special class of lobster trees (Chinese). Neimenggu Daxue Xuebao (Hohhot) 16, 337-341. (1985).
DONGB 1986-1 Dong, Z. F.
An investigation of degenerate cases of the transportation problem (Chinese). Heilongjiang Daxue Ziran Kexue Xuebao (Harbin), No. 2, 13-16. (1986).
DOOB 1986-1 Doob, M.
Pseudocyclic graphs. In: R. Tosic, D. Acketa, V. Petrovic (Eds.): Graph Theory - Proc. 6th Yugoslav Seminar, Dubrovnik 1985, 107-114. Novi Sad: Inst. of Math., Univ. of Novi Sad. (1986).
DORH 1988-1 Dorhout, B.
Solution of a tinned iron purchasing problem using Lagrangean relaxation. In: H. Schellhaas, P. van Beek, H. Isermann, R. Schmidt, M. Zijlstra (Eds.): Operations Research Proceedings 1987 (DGOR/NSOR), 601-602. Berlin: Springer. (1988).

DORI 1985-1 Döring, S.
Optimization of node-disjoint paths in networks. In: A. Iwainsky (ed.): Optimization of connection structures in graphs, 202-210. Berlin: Central Institute of Cybernetics and Infomations Processes, Academy of Sciences of the German Democratic Republic. (1985).

DORIA 1985-1 Döring, S.; Iwainsky, A.
Graph reduction for the optimization of node-disjoint paths in networks. In: A. Iwainsky (Ed.): Optimization of connection structures in graphs, 177-201. Berlin: Central Institute of Cybernetics and Information Processes, Academy of Sciences of the German Democratic Republic. (1985).

DORO 1984-1 Dorofeeva, M. P.; Ovsyankin, B. P.; Smirnov, A. V.
Multiprocessor schedules with due dates (Russian). Akademiya Nauk Armyanskoi SSR Doklady (Erevan) 78, 147-150. (1984).

DOSS 1984-1 Doss, H.
Bayesian estimation in the symmetric location problem. Zeitschrift für Wahrscheinlichkeitstheorie und verwandte Gebiete 68, 127-147. (1984).

DOWL 1986-1 Dowling, P. D.; Love, R. F.
Bounding methods for facilities location algorithms. Naval Research Logistics Quarterly 33, 775-788. (1986).

DOWL 1987-1 Dowling, P. D.; Love, R. F.
An evaluation of the dual as a lower bound in facilities location problems. IIE Transactions 19, 160-166. (1987).

DOWS 1984-1 Dowsland, K. A.
The three-dimensional pallet chart: an analysis of the factors affecting the set of feasible layouts for a class of two-dimensional packing problems. Journal of the Operational Research Society 35, 895-906. (1984).

DOWS 1985-1 Dowsland, K. A.
A graph-theoretic approach to a pallet loading problem. New Zealand Operational Research 13, 77-86. (1985).

DOWS 1985-2 Dowsland, K. A.
Determining an upper bound for a class of rectangular packing problems. Computers & Operations Research 12, 201-206. (1985).

DOWS 1987-1 Dowsland, K. A.
A combined data-base and algorithmic approach to the pallet-loading problem. Journal of the Operational Research Society 38, 341-346. (1987).

DOWS 1987-2 Dowsland, K. A.
An exact algorithm for the pallet loading problem. European Journal of Operational Research 31, 78-84. (1987).

DOWSA 1985-1 Dowsland, W. B.
Two and three dimensional packing problems and solution methods. New Zealand Operational Research 13, 1-18. (1985).

DOYA 1981-1 Doyar, A. J.; Peltek, Y. A.
An algorithm to solve scheduling problems in project tasks (Russian). In: A. N. Antamoshkin (Ed.): Applications of random search to the solution of applied problems, 38-45. Kemerovo: Kemerovskij Gosud. Univ. (1981).

DRES 1986-1 Dress, A.
Chirotopes and oriented matroids. In: A. Kerber (Ed.): Diskrete Strukturen, algebraische Methoden und Anwendungen, Sommerschule Bayreuth 1985. Bayreuther Mathematische Schriften, Heft 21, 14-68. (1986).

DRES 1986-2 Dress, A.
Duality theory for finite and infinite matroids with coefficients. Advances in Mathematics 59, 97-123. (1986)

DRESA 1986-1 Dress, A.; Havel, T. F.
Some combinatorial properties of discriminants in metric vector spaces. Advances in Mathematics 62, 285-312. (1986).

DRESB 1987-1 Dress, A.; Krüger, M.
Parsimonious phylogenetic trees in metric spaces and simulated annealing. Advances in Applied Mathematics 8, 8-37. (1987).

DRESC 1987-1 Dress, A.; Lovasz, L.
On some combinatorial properties of algebraic matroids. Combinatorica 7, 39-48. (1987).

DRESD 1987-1 Dress, A.; Wenzel, W.
Endliche Matroide mit Koeffizienten. Bielefeld: Univ. Bielefeld, 62 p. (1987).

DREX 1988-1 Drexl, A.
A stochastic assignment approach to resource-constrained multi-project scheduling. In: H. Schellhaas, P. van Beek, H. Isermann, R. Schmidt, M. Zijlstra (Eds.): Operations Research Proceedings 1987 (DGOR/NSOR), 603-610. Berlin: Springer. (1988).

DREZ 1984-1 Drezner, Z.
The planar two-center and two-median problems. Transportation Science 18, 351-361. (1984).

DREZ 1985-1 Drezner, Z.
A solution to the Weber location problem on the sphere. Journal of the Operational Research Society 36, 333-334. (1985).

DREZ 1985-2 Drezner, Z.
O (N log N) algorithm for the rectilinear round-trip location problem. Transportation Science 19, 91-93. (1985).

DREZ 1985-3 Drezner, Z.
Sensitivity analysis of the optimal location of a facility. Naval Research Logistics Quaterly 32, 209-224. (1985).

DREZ 1986-1 Drezner, Z.
Location of regional facilities. Naval Research Logistics Quarterly 33, 523-530. (1986).

DREZ 1986-2 Drezner, Z.
The p-cover problem. European Journal of Operational Research 26, 312-313. (1986).

DREZ 1987-1 Drezner, Z.
A heuristic procedure for the layout of a large number of facilities. Management Science 33, 907-915. (1987).

DREZ 1987-2 Drezner, Z.
Heuristic solution methods for two location problems with unreliable facilities. Journal of the Operational Research Society 38, 509-514. (1987).

DREZ 1987-3 Drezner, Z.
On the rectangular p-center problem. Naval Research Logistics 34, 229-234. (1987).

DREZA 1986-1 Drezner, Z.; Barak, A.
Efficient algorithms for routing information in a multicomputer system. In: E. Gafni, N. Santoro (Eds.): Distributed Algorithms on Graphs, 41-48. Ottawa: Carleton Univ. Press. (1986).

DREZB 1985-1 Drezner, Z.; Gavish, B.
Epsilon-approximations for multi-dimensional weighted location problems. Operations Research 33, 772-783. (1985).

DREZC 1984-1 Drezner, Z.; Nof, S. Y.
On optimizing bin picking and insertion plans for assembly robots. IIE Transactions 16, 262-270. (1984).

DREZD 1987-1 Drezner, Z.; Shelah, S.
On the complexity of the Elzinga-Hearn algorithm for the 1-center problem. Mathematics of Operations Researchs 12, 255-261. (1987).

DREZE 1985-1 Drezner, Z.; Steiner, G.; Wesolowsky, G. O.
One-facility location with rectilinear tour distances. Naval Research Logistics Quarterly 32, 391-406. (1985).

DREZF 1985-1 Drezner, Z.; Wesolowsky, G. O.
Layout of facilities with some fixed points. Computers & Operations Research 12, 603-610. (1985).

DREZF 1985-2 Drezner, Z.; Wesolowsky, G. O.
Location of multiple obnoxious facilities. Transportation Science 19, 193-202. (1985).

DRID 1986-1 Dridi, N.; Portmann, M. C.; Proth, J. M.
A decomposition method for production scheduling problems (French). In: A. Bensoussan, J. L. Lions (Eds.): Analysis and optimization of systems, Proc. 7th. Int. Conf., Antibes/ France 1986, Lecture Notes in Control and Information Sciences 83, 247-254. Berlin: Springer Verlag. (1986).

DRIDA 1984-1 Dridi, N.; Proth, J. M.; Sedillot, S.
Ordonnancement de taches constraintes dans le cas d'un serveur. Institut National de Recherche en Informatique et Automatique (INRIA) no. 325, 21 p. (1984).

DROR 1987-1 Dror, M.; Ball, M. O.
Inventory/routing: reduction from an annual to a short-period problem. Naval Research Logistics 34, 891-906. (1987).

DRORA 1986-1 Dror, M.; Ball, M. O.; Golden, B. L.
A computational comparison of algorithms for the inventory routing problem. In: C. L. Monma (Ed.): Algorithms and Software for Optimization - Part I, 3-24. Annals of Operations Research 4, Basel: Baltzer. (1986).

DRORB 1986-1 Dror, M.; Levy, L.
A vehicle routing improvement algorithm comparison of a greedy and a matching implementation for inventory routing. Computers and Operations Research 13, 33-46. (1986).

DRORC 1987-1 Dror, M.; Stern, H. I.; Lenstra, J. K.
Parallel machine scheduling: Processing rates dependent on number of jobs in operation. Management Science 33, 1001-1009. (1987).

DRORD 1986-1 Dror, M.; Trudeau, P.
Stochastic vehicle routing with modified savings algorithm. European Journal of Operational Research 23, 228-235. (1986).

DRUD 1986-1 Drud, A.; Zenios, S. A.; Mulvey, J. M.
Balancing large social accounting matrices with nonlinear network programming. Nat. Science Foundation, Grant No. DCR-84/01098. Washington: World Bank, Development Research Dept., 11p. (1986).

DU 1986-1 Du, D. Z.
An optimization problem on graphs. Discrete Applied Mathematics 14, 101-104. (1986).

DU A 1987-1 Du, D. Z.; Hwang, F. K.
Steiner minimal trees for bar waves. Acta Mathematicae Applicatae Sinica (English Series) 3, 246-256. (1987).

DU B 1985-1 Du, D. Z.; Hwang, F. K.; Chao, S. C.
Steiner minimal tree for points on a circle. Proceedings of the Amercian Mathematical Society 95, 613-618. (1985).

DU C 1987-1 Du, D. Z.; Hwang, F. K.; Song, G. D.; Ting, G. Y.
Steiner minimal trees on sets of four points. Discrete and Computational Geometry 2, 401-414. (1987).

DU D 1987-1 Du, D. Z.; Hwang, F. K.; Weng, J. F.
Steiner minimal trees for regular polygons. Discrete and Computational Geometry 2, 65-84. (1987).

DUBO 1983-1 Dubois, D.
A fuzzy, heuristic, interactive approach to the optimal network problem. in: Advances in fuzzy sets, possibility theory, and applications, 253-276. Plenum (New York). (1983).

DUBU 1986-1 Dubuc, S.; Tanguay, M.
Deplacement de materiel continu unidimensionnel a moindre cout. RAIRO Recherche Operationelle 20, 139-162. (1986).

DUCH 1984-1 Duchet, P.
Classical perfect graphs. In: C. Berge, V. Chvatal (Eds.): Topics on perfect graphs, Annals of Discrete Mathematics Vol. 21, 67-96. Amsterdam: North-Holland. (1984).

DUCHA 1987-1 Duchet, P.; Hamidoune, Y. O.; Meyniel, H.
Sur les quasi-noyaux d'un graphe. Discrete Mathematics 65, 231-236. (1987).

DUCHB 1986-1 Duchet, P.; Olariu, S.
Graphes parfaitement ordonnables generalises. In: Raspail-Burnside: Workshop on Perfect Graphs, Bellairs Research Institute, McGill University, Barbados, West Indies, 1986, 27-30. (1986).

DUCHC 1986-1 Duchet, P.; Vergnas, M. Las; Meyniel, H.
Connected cutsets of a graph and triangle bases of the cycle space. Discrete Mathematics 62, 145-154. (1986).

DUCR 1985-1 Ducrocq, J.-C.
Optimal energy management in a complex system with mixed-integer programming. RAIRO-APII Automatique Productique Informatique Industrielle 19, 403-414. (1985).

DUDZ 1984-1 Dudzinski, K.
A dynamic programming approach to solving the multiple choice knapsack problem. Bulletin de l'Academie Polonaise des Sciences, Serie des Sciences Techniques (Warsaw) 32, 325-332. (1984).

DUDZA 1984-1 Dudzinski, K.; Walukiewicz, S.
A fast algorithm for the linear multiple-choice knapsack problem. Operations Research Letters 3, 205-210. (1984).

DUDZA 1984-2 Dudzinski, K.; Walukiewicz, S.
An algorithm for multiple choice knapsack problem. In: P. Thoft-Christensen (Ed.): System Modelling and Optimization, 175-182. Lecture Notes in Control and Information Sciences 59, Berlin: Springer. (1984).

DUDZA 1985-1 Dudzinski, K.; Walukiewicz, S.
On the multiperiod binary knapsack problem. Bulletin of the Polish Academy of Sciences, Technical Sciences (Warsaw) 33, 85-91. (1985).

DUDZA 1985-2 Dudzinski, K.; Walukiewicz, S.
On the multiperiod binary knapsack problem. In: P. Brucker, R. Pauly (Eds.): Proc. IX. Symp. on Operations Research, Osnabrück 1984. Methods of Operations Research 49, 223-232. (1985).

DUDZA 1986-1 Dudzinski, K.; Walukiewicz, S.
A multiperiod binary knapsack problem (Polish). Archiwum Automatyki i Telemechaniki 31, 5-20. (1986).

DUKE 1982-1 Duke, R. A.; Erdös, P.
Subgraphs in which each pair of edges lies in a short common cycle. In: F. Hoffman, K. B. Reid, R. C. Mullin, R. G. Stanton (Eds.): Proc. Thirteenth Southeastern Conf. on Combinatorics, Graph Theory and Computing, Florida 1982, Congressus Numerantium Vol. 35, 253-260. Winnipeg, Canada: Utilitas Mathematica. (1982).

DUNN 1987-1 Dunne, P. E.
A result on k-valent graphs and its application to a graph embedding problem. Acta Informatica 24, 447-460. (1987).

DUPA 1984-1 Dupac, V.; Herkenrath, U.
On integer stochastic approximation. Aplikace Matematiky (Prague) 29, 372-383. (1984).

DURA 1987-1 Duran, F.
A large mixed integer production and distribution program. European Journal of Operational Research 28, 207-217. (1987).

DURAA 1986-1 Duran, M. A.; Grossmann, I. E.
An outer-approximation algorithm for class of mixed-integer nonlinear programs. Mathematical Programming 36, 307-339. (1986).

DURI 1985-1 Durier, R.; Michelot, C.
Geometrical properties of the Fermat-Weber problem. European Journal of Operational Research 20, 332-343. (1985).

DURR 1984-1 Durrett, R.
Oriented percolation in two dimensions. Annals of Probability 12, 999-1040. (1984).

DUTA 1983-1 Duta, L. D.; Fabian, C.
Mathematical methods in optimal cutting (Romanian). Editura Tehnica (Bucharest). (1983).

DUTA 1984-1 Duta, L. D.; Fabian, C.
Solving cutting-stock problems through the Monte-Carlo method. Economic Computation and Economic Cybernetics Studies and Research (Bucharest) 19, 35-54. (1984).

DYER 1986-1 Dyer, M. E.
On a multidimensional search technique and its application to the Euclidean one-centre problem. SIAM Journal on Computing 15, 725-738. (1986).

DYERA 1985-1 Dyer, M. E.; Foulds, L. R.; Frieze, A. M.
Analysis of heuristics for finding a maximum weight planar subgraph. European Journal of Operations Research 20, 102-114. (1985).

DYERB 1985-1 Dyer, M. E.; Frieze, A. M.
A simple heuristic for the p-center problem. Operations Research Letters 3, 285-288. (1985).

DYERB 1985-2 Dyer, M. E.; Frieze, A. M.
On the complexity of partitioning graphs into connected subgraphs. Discrete Applied Mathematics 10, 139-154. (1985).

DYERC 1984-1 Dyer, M. E.; Frieze, A. M.; Mac Diarmid, C.
Partitioning heuristics for two geometric maximization problems. Operations Research Letters 3, 267-270. (1984).

DYERC 1986-1 Dyer, M. E.; Frieze, A. M.; Mac Diarmid, C.
On linear programs with random costs. Mathematical Programming 35, 3-16. (1986).

DYERD 1984-1 Dyer, M. E.; Kayal, N.; Walker, J.
A branch and bound algorithm for solving the multiple-choice knapsack problem. Journal of Computational and Applied Mathematics 11, 231-250. (1984).

DYERE 1987-1 Dyer, M. E.; Walker, J.
An algorithm for a separable integer programming problem with cumulatively bounded variables. Discrete Applied Mathematics 16, 135-150. (1987).

DZHI 1985-1 Dzhidzhev, K. N.
A linear algorithm for partitioning graphs of fixed genus. Serdica (Sofia) 11, 369-387. (1985).

EADE 1986-1 Eades, P.; Kelly, D.
Heuristics for drawing two-layered networks. Ars Combinatoria 21-A, 89-98. (1986).

EADEA 1984-1 Eades, P.; Pullman, N. J.; Robinson, P. J.
On clique partition numbers of regular graphs. In: J. A. Bondy, U. P. R. Murty (Eds.): Progress in Graph Theory, 191-202. New York: Academic Press. (1984).

EASA 1985-1 Easa, S. M.
Shortest route algorithm with movement prohibitions. Transportation Research 19B, 197-208. (1985).

EAST 1986-1 Easton, F.; Faaland, B.; Klastorin, T. D.; Schmitt, T.
An improved network based algorithm for the assembly line balancing problem. Working Paper, 6-86. Seattle: Univ. of Washington, Dept. of Management Science, 23 p. (1986).

EBEN 1984-1 Ebenegger, C.; Hammer, P. L.; Werra, D. de
Pseudo-boolean functions and stability of graphs. In: R. E. Burkard, R. A. Cuninghame-Green, U. Zimmermann (Eds.): Algebraic and Combinatorial Methods in Operations Research, Annals of Discrete Mathematics Vol. 19, 83-98. Amsterdam: North-Holland. (1984).

EBER 1988-1 Eberhard, U.; Vaterrodt, H.-J.
Mehr-Depot-Tourenplanung. In: H. Schellhaas, P. van Beek, H. Isermann, R. Schmidt, M. Zijlstra (Eds.): Operations Research Proceedings 1987 (DGOR/NSOR), 271-272. Berlin: Springer. (1988).

EBERA 1987-1 Ebert, J.
A versatile data structure for edge-oriented graph algorithms. Communications of the ACM 30, 513-519. (1987).

EDAH 1984-1 Edahiro, M.; Kokubo, I.; Asano, T.
A new-point location algorithm and its practical efficiency - comparison with existing algorithms. ACM Transactions on Graphics 3, 86-109. (1984).

EDEL 1984-1 Edelman, P. H.
The acyclic sets of an oriented matroid. Journal of Combinatorial Theory, Series B, 36 26-31. (1984).

EDELA 1987-1 Edelsbrunner, H.; Rote, G.; Welzl, E.
Testing the necklace condition for shortest tours and optimal factors in the plane. In: T. Ottmann (Ed.): Automata, Languages and Programming, 364-375. Lecture Notes in Computer Science 267, Berlin: Springer. (1987).

EDEN 1986-1 Edenbrandt, A.
Quotient tree partitioning of undirected graphs. BIT 26, 148-155. (1986).

EDEN 1987-1 Edenbrandt, A.
Chordal graph recognition is in NC. Information Processing Letters 24, 239-242. (1987).

EDMO 1984-1 Edmonds, J.; Giles, R.
Total dual integrality of linear inequality systems. In: W. R. Pulleyblank (Ed.): Progress in Combinatorial Optimization, 117-130. Toronto: Academic Press. (1984).

EDMOA 1985-1 Edmonds, J.; Lubiw, A.
Bipartition systems and how to partition polygons. Report No. 85406-OR. Bonn: Universität Bonn, Inst. für Ökonometrie und Operations Research, 10p. (1985).

EDWA 1986-1 Edwards, K.
The complexity of colouring problems on dense graphs. Theoretical Computer Science 43, 337-344. (1986).

EGAW 1987-1 Egawa, Y.
Cycles in k-connected graphs whose deletion results in a (k-2)-connected graph. Journal of Combinatorial Theory 42B, 371-377. (1987).

EGAWA 1986-1 Egawa, Y.; Enomoto, H.; Saito, A.
Contractible edges in triangle-free graphs. Combinatorica 6, 269-274. (1986).

EGAWA 1986-2 Egawa, Y.; Enomoto, H.; Saito, A.
On component factors. Graphs and Combinatorics 2, 223-226. (1986).

EGAWB 1986-1 Egawa, Y.; Fukuda, T.; Nagoya, S.; Urabe, M.
A decomposition of complete bipartite graphs into edge-disjoint subgraph with star components. Discrete Mathematics 58, 93-96. (1986).

EGGA 1986-1 Eggan, L.; Plantholt, M.
The chromatic index of nearly bipartite multigraphs. Journal of Combinatorial Theory 40B, 71-80. (1986).

EGGL 1984-1 Eggleton, R. B.; Skilton, D. K.
Chain decompositions of graphs, 1: Abstract graphs. In: K. M. Koh, H. P. Yap (Eds.): Graph Theory Singapore 1983, Lecture Notes in Mathematics 1073, 294-306. Berlin: Springer-Verlag. (1984).

EGGL 1984-2 Eggleton, R. B.; Skilton, D. K.
Chain decompositions of graphs, 2: Surface embeddings. In: K. M. Koh, H. P. Yap (Eds.): Graph Theory Singapore 1983, Lecture Notes in Mathematics 1073, 307-323. Berlin: Springer-Verlag. (1984).

EGGL 1984-3 Eggleton, R. B.; Skilton, D. K.
Double tracings of graphs. Ars Combinatoria 17A, 307-324. (1984).

EGGL 1984-4 Eggleton, R. B.; Skilton, D. K.
Graphs with Eulerian chains. Bulletin of the Australian Mathematical Society 29, 389-399. (1984).

EGIA 1983-1 Egiazaryan, E. V.
The external stability number of graphs (Russian). Applied Mathematics (Erevan) No. 2, 67-74. (1983).

EGLE 1987-1 Eglese, R. W.; Rand, G. K.
Conference seminar timetabling. Journal of the Operational Research Society 38, 591-598. (1987).

EGYH 1985-1 Egyhazy, C. J.
An algorithm for generating the minimum length fundamental cycle set in a graph. Congressus Numerantium 50, 219-230. (1985).

EHNE 1984-1 Ehnert, G.; Harant, J.; Walther, H.
Ein exakter in Polynomialzeit arbeitender Algorithmus zur Fahrzeugminimierung in einem Kraftverkehrsbetrieb. Wissenschaftliche Zeitschrift der Technischen Hochschule Ilmenau 30, No. 6, 29-32. (1984).

EHNE 1985-1 Ehnert, G.; Harant, J.; Walther, H.
Ein in Polynomialzeit arbeitender Algorithmus zur Minimierung einer skalarisierten Zielfunktion in einem Kraftverkehrsbetrieb mit einem Depot, einem Fahrzeugtyp und Aufträgen mit unterschiedlichen Anforderungen an die zu transportierenden Mengen. In: H. Sachs (Ed.): Graphs, Hypergraphs and Applications, Proc. Conf. on Graph Theory (Eyba 1984), 31-35. Leipzig: Teubner. (1985).

EHRE 1984-1 Ehrenfeucht, A.; Faber, V.; Kierstead, H. A.
A new method of proving theorems on chromatic index. Discrete Mathematics 52, 159-164. (1984).

EIGE 1985-1 Eiger, A.; Mirchandani, P. B.; Soroush, H.
Path preferences and optimal paths in probabilistic networks. Tansportation Science 19, 75-83. (1985).

EISE 1985-1 Eiselt, F.
Unabhängige bunte Mengen in k-chromatischen Graphen der Radien 4 und 5. Wissenschaftliche Zeitschrift der Pädagogischen Hochschule Karl Friedrich Wilhelm Wander Dresden 19, 111-116. (1985).

EISEA 1985-1 Eiselt, H. A.
A location problem with minimax objective. In: H. Noltemeier (Ed.): Proc. WG '85: Internat. Workshop on Graphtheoretic Concepts in Computer Science, Castle Schwanberg, Würzburg, 1985, 85-100. Linz: Trauner. (1985).

EISEA 1986-1 Eiselt, H. A.
Continuous maximin knapsack problems with GLB constraints. Mathematical Programming 36, 114-121. (1986).

EISEB 1986-1 Eiselt, H. A.; Charlesworth, G.
A note on p-center problems in the plane. Transportation Science 20, 130-132. (1986).

EISEC 1984-1 Eiselt, H. A.; Gerchak, Y.
Solution structures and sensitivity of special assignment problems. Computers & Operations Research 11, 397-400.B (1984).

EISED 1987-1 Eiselt, H. A.; Laporte, G.
Combinatorial optimization problems with soft and hard requirements. Journal of the Operational Research Society 38, 785-796. (1987).

EL-B 1985-1 El-Bakry, M.; Nassar, S.
The solution of the 0-1 programming problem by parallel processing. In: P. Brucker, R. Pauly (Eds.): Proc. IX. Symp. on Operations Research, Osnabrück 1984. Methods of Operations Research 49, 361-370. (1985).

EL-BA 1985-1 El-Bakry, M.; Wahsh, S.; Nassar, S.
A new pulse width modulation technique based on 0-1 programming. Methods of Operations Research 55, 57-64. (1985).

EL-D 1986-1 El-Dessouki, O.; Huen, W.; Evens, M.
Optimal partitioning of sequential jobs on a network computer. Journal of VLSI and Computer Systems 1, 377-397. (1986).

EL-G 1986-1 El-Gamal, A.; Hemachandra, L. A.; Shperling, I.; Wei, V. K.
Using simulated annealing to design good codes. Working Paper. Stanford: Stanford Univ., Information Systems Laboratory, 33 p. (1986).

EL-M 1984-1 El-Mallah, E. S.; Colbourn, C. J.
Optimum communication spanning trees in chordal graphs. Congressus Numerantium 43, 301-316. (1984).

EL-M 1985-1 El-Mallah, E. S.; Colbourn, C. J.
Optimum communication spanning trees in series-parallel networks. SIAM Journal on Computing 14, 915-925. (1985).

EL-M 1985-2 El-Mallah, E. S.; Colbourn, C. J.
Reliability of delta-Y reducible networks. Congressus Numerantium 48, 49-54. (1985).

EL-M 1987-1 El-Mallah, E. S.; Colbourn, C. J.
On two dual classes of planar graphs. Research Report CORR 87-5. Waterloo: Univ. of Waterloo, Dept. of Combinatorics and Optimization, 22 p. (1987).

EL-Z 1985-1 El-Zahar, M.; Rival, I.
Examples of jump-critical ordered sets. SIAM Journal on Algebraic and Discrete Methods 6, 713-720. (1985).

EL-ZA 1985-1 El-Zahar, M.; Sauer, N. W.
The chromatic number of the product of two 4-chromatic graphs is 4. Combinatorica 5, 121-126. (1985).

ELBA 1986-1 Elbas, J.
Substitution and atomic extension on greedy posets. Order 3, 235-244. (1986).

ELIM 1985-1 Elimam, A. A.; Elmaghraby, S. E.
On the reduction method for integer linear programs, II. Discrete Applied Mathematics 12, 241-260. (1985).

ELIZ 1984-1 Elizarov, Y. D.; Stolov, E. L.
An analogue of methods of a deformable polyhedron in problems of scheduling theory (Russian). Issledovaniya po Prikladnoi Matematike (Kazan) No. 11, part 1, 66-72. (1984).

ELLI 1984-1 Ellingham, M. N.
Peterson subdivisions in some regular graphs. Congressus Numerantium 44, 33-40. (1984).

ELLIA 1986-1 Ellingham, M. N.; Caro, Y.
Uniqueness of impartial edge-colorings. Ars Combinatoria 22, 59-72. (1986).

ELLIB 1984-1 Ellingham, M. N.; Holton, D. A.; Little, C. H. C.
Cycles through ten vertices in 3-connected cubic graphs. Combinatorica 4, 265-274. (1984).

ELSA 1986-1 Elsayed, E. A.; Nasr, N. Z.
Heuristics for resource constrained scheduling. International Journal of Production Research 24, 299-310. (1986).

ELST 1984-1 Elster, K.-H.; Harant, J.; Hutschenreuther, H.
Ein Partitionierungsalgorithmus zur Distanzberechnung auf Verkehrs- netzen. Optimization 15, 429-438. (1984).

ELVE 1985-1 Elvers, D. A.; Treleven, M. D.
Job-shop vs. hybrid flow-shop routing in a dual resource constrained system. Decision Sciences 16, 213-222. (1985).

EMEL 1981-1 Emelichev, V. A.; Kravtsov, M. K.
Transport polytopes (Russian). Algebraic Systems, Ivanov. Gos. Univ. (Ivanovo), 134-152. (1981).

EMEL 1981-2 Emelichev, V. A.; Kravtsov, M. K.
Transportation polyhedra (Russian). in: D. I. Moldavanskij (Ed.): Algebraic Systems. Interuniv. Collection of scientific works (Ivanovo), 134-152. (1981).

EMEL 1983-1 Emelichev, V. A.; Kravtsov, M. K.
Combinatorial problems on transport polyhedra - a survey (Russian). Ehkonomika i Matematicheskie Metody (Moscow) 19, 301-320. (1983).

EMEL 1984-1 Emelichev, V. A.; Kravtsov, M. K.
Transportpolyeder mit maximaler Anzahl von Seiten aller Dimensionen. In: K.-H. Elster (Ed.): Mathematische Optimierung - Theorie und Anwendungen, Vortragsauszüge, Eisenach 1984, 35-36. Techn. Hochschule Ilmenau. (1984).

EMELA 1985-1 Emelichev, V. A.; Kravtsov, M. K.; Krachkovskii, A. P.
Kombinatorische und Extremalprobleme auf Transportpolyedern. In: E. Girlich (Ed.): Diskrete Optimierung, 31-35. Wissensch. Beitr., Jena: Friedrich-Schiller-Univ. (1985).

EMELA 1985-2 Emelichev, V. A.; Kravtsov, M. K.; Krachkovskii, A. P.
Transportation polyhedra with a maximum number of k-faces (Russian). Doklady Akademii Nauk SSSR (Moscow) 282, 784-788. (1985).

EMELB 1985-1 Emelichev, V. A.; Ovchinnikov, V. G.
Greedy algorithms for solution some problems of discrete optimization. In: F. Nozicka (Ed.): Optimization - Theory, Methods, Applications, I, International Conference Marianske Lazne 1985, 78-80. Prague: Dum Techniky CSVTS. (1985).

EMELB 1985-2 Emelichev, V. A.; Ovchinnikov, V. G.
Strafmethoden in der diskreten Optimierung. 30. Internationales Wissenschaftliches Kolloquium (TH Ilmenau, DDR) Heft 4E, 45-46. (1985).

EMELC 1985-1 Emelichev, V. A.; Perepelitsa, V. A.
On multicriterial assignment problems. In: F. Nozicka (Ed.): Optimization - Theory, Methods, Applications, I, International Conference Marianske Lazne 1985, 81-83. Prague: Dum Techniky CSVTS. (1985).

EMELC 1986-1 Emelichev, V. A.; Perepelitsa, V. A.
On the estimation of the complexity of multicriterial transportation problems (Russian). Doklady Akademii Nauk BSSR (Minsk) 30, 593-596. (1986).

EMELD 1987-1 Emelichev, V. A.; Perepelitsa, V. A.; Shungarov, K. D.
Asymptotic approach to a multicriteria problem of the covering of a graph by stars (Russian). Doklady Akademii Nauk BSSR (Minsk) 31, 430-433. (1987).

EMELE 1985-1 Emelicheva, E. V.; Kovalev, M. M.
The method of partial orderings on matroid structures (Russian). Soobshcheniya Akademii Nauk Gruzinskoj SSR 120, 269-272. (1985).

EMELE 1986-1 Emelicheva, E. V.; Kovalev, M. M.
Treelike polymatroids and extremal problems on permutations (Russian). Soobshcheniya Akademii Nauk Gruzinskoj SSR (Tbilisi) 121, 277-280. (1986).

EMELF 1985-1 Emelicheva, E. V.; Kovalev, M. M.; Rogov, S. F.
Methode der Halbordnung zur Lösung von Optimierungsproblemen über Permutationen (Russian). In: E. Girlich (Ed.): Diskrete Optimierung, 83-92. Jena: Friedrich-Schiller-Universität. (1985).

EMMO 1985-1 Emmons, H.
Work-force scheduling with cyclic requirements and constraints on days off, weekends off, and work stretch. IIE Transactions 17, 8-16. (1985).

EMMO 1987-1 Emmons, H.
Scheduling to a common due date on parallel uniform processors. Naval Research Logistics 34, 803-810. (1987).

ENGQ 1985-1 Engquist, M.; Chang, M. D.
New labeling procedures for the basis graph in generalized networks. Operations Research Letters 4, 151-156. (1985).

ENGQA 1986-1 Engquist, M.; Chen, C.-H. J.
Computational comparison of two solution procedures for allocation/processing networks. In: G. Gallo, C. Sandi (Eds.): Netflow at Pisa, Mathematical Programming Study 26, 218-220. (1986).

ENNS 1984-1 Enns, T. C.
An efficient algorithm determining when a fabric hangs together. Geometriae Dedicata 15, 259-260. (1984).

ENOM 1986-1 Enomoto, H.
Toughness and the existence of k-factors.II. Graphs and Combinatorics 2, 37-42. (1986).

ENOMA 1984-1 Enomoto, H.; Saito, A.
Disjoint shortest paths in graphs. Combinatorica 4, 275-280. (1984).

ENTR 1985-1 Entringer, R. C.
A short proof of Rubin's block theorem. In: B. R. Alspach, C. D. Godsil (Eds.): Cycles in Graphs, Annals of Discrete Mathematics Vol. 27, 367-368. Amsterdam: North-Holland. (1985).

ENTRA 1983-1 Entringer, R. C.; Jackson, D. E.
Degree sums of paths in extremal trees. Congressus Numerantium 39, 353-366. (1983).

ENTRB 1982-1 Entringer, R. C.; Mac Kendrick, S.
Longest paths in locally Hamiltonian graphs. In: F. Hoffman, K. B. Reid, R. C. Mullin, R. G. Stanton (Eds.): Proc. Thirteenth Southeastern Conf. on Combinatorics, Graph Theory and Computing, Florida 1982, Congressus Numerantium Vol. 35, 275-282. Winnipeg, Canada: Utilitas Mathematica. (1982).

EPPE 1985-1 Eppen, G. D.; Martin, R. K.
Solving multi-item capacitated lot sizing problems using variable redefinition. Research Paper. Chicago: Univ. of Chicago, Graduate School of Business, 42 p. (1985).

EPPEA 1986-1 Eppen, G. D.; Martin, R. K.; Schrage, L.
A mixed-integer programming model for evaluating distribution configurations in the steel industry. Computers and Operations Research 13, 575-586. (1986).

ERA 1984-1 Era, H.
A note on semiregular factorizations of regular graphs. Proceedings of the Faculty of Science of Tokai University (Tokyo) 19, 7-13. (1984).

ERA 1985-1 Era, H.
Semiregular factorizations of regular graphs. In: F. Harary, J. P. Maybee (Eds.): Graphs and Applications, 101-116. New York: Wiley. (1985).

ERDO 1984-1 Erdös, P.
Some new and old problems on chromatic graphs. In: K. P. Vijayan, N. M. Singhi (Eds.): Combinatorics and Applications (Proc. Seminar Calcutta 1982), 118-126. Calcutta: Indian Statistical Institute. (1984).

ERDOA 1986-1 Erdös, P.; Füredi, Z.; Hajnal, A.; Komjath, P.; Rödl, V.; Seress, A.
Coloring graphs with locally few colors. Discrete Mathematics 59, 21-34. (1986).

ERDOB 1985-1 Erdös, P.; Hajnal, A.
Chromatic number of finite and infinite graphs and hypergraphs. Discrete Mathematics 53, 281-286. (1985).

ERDOC 1985-1 Erdös, P.; West, D. B.
A note on the interval number of a graph. Discrete Mathematics 55, 129-134. (1985).

ERDOD 1985-1 Erdös, P. L.; Györi, E.
Any four independent edges of a 4-connected graph are contained in a circuit. Acta Mathematica Hungarica (Budapest) 46, 311-313. (1985).

EREM 1987-1 Eremin, G. S.
Partition of an arbitrary set of vertices of an acyclic digraph into convex subsets (Russian). Avtomatika i Telemekhanika, No.8, 137-143. (1987).

EREN 1984-1 Erenguc, S. S.; Tufekci, S.
A linear order algorithm for a transportation type production scheduling problem. Research Report 84-28. Gainesville: Univ. of Florida, Industrial & Systems Engineering Dept., 7 p. (1984).

EREN 1984-2 Erenguc, S. S.; Tufekci, S.
A transportation type aggregate production model with bounds on inventory and backordering. Research Report 84-29. Gainesville: Univ. of Florida, Ind. & Systems Eng. Dept., 26 p. (1984).

EREN 1987-1 Erenguc, S. S.; Tufekci, S.
A branch and bound algorithm for a single-item multi-source dynamic lot sizing problem with capacity constraints. IIE Transactions 19, 73-80. (1987).

ERFL 1986-1 Erfle, W.
Ein Branch-and-Bound-Verfahren zur optimalen Allokation von hierarchisch geordneten Prüfstellen in den Fertigungsablauf. In: L. Streitferdt, H. Hauptmann, A. W. Marusev, D. Ohse, U. Pape (Eds.): Operations Research Proceedings 1985, 274-281. Berlin, Heidelberg, New York: Springer-Verlag. (1986).

ERIC 1987-1 Erickson, R. E.; Monma, C. L.; Veinott, A. F. Jr.
Send-and-split method for minimum-concave-cost network flows. Mathematics of Operations Research 12, 634-664. (1987).

ERKU 1985-1 Erkut, E.; Francis, R. L.; Lowe, T. J.
Locating two facilities on a tree subject to a distance constraint. Research Report 85-24. Gainesville: Univ. of Florida, Ind. & Systems Eng. Dept., 22 p. (1985).

ERKU 1987-1 Erkut, E.; Francis, R. L.; Lowe, T. J.
A multimedian problem with interdistance constraints. Research Report 87-1. Gainesville: Univ. of Florida, Industrial & Systems Eng. Dept., 17 p. (1987).

ERLE 1985-1 Erlenkotter, D.; Leonardi, G.
Facility location with spatially interactive travel behavior. Sistemi Urbani 1 (1985) 29-41 (1985).

ERNV 1984-1 Ernvall, J.
On the construction of spanning paths by Gray-code in compression of files. Technique et Science Informatiques 3, 411-414. (1984).

ERNVA 1985-1 Ernvall, J.; Katajainen, J.; Penttonen, M.
NP-completeness of the Hamming salesman problem. BIT, 289-292. (1985).

ERSC 1985-1 Erschler, J.; Fontan, G.; Merce, C.
Un nouveau concept de dominance pour l'ordonnancement de travaux sur une machine. RAIRO Recherche Operationelle 19, 15-26. (1985).

ERZI 1983-1 Erzin, A. I.; Mordvinova, T. B.
A problem of the construction of an optimal tree (Russian). Upravlyaemye Sistemy (Novosibirsk) 23, 44-54. (1983).

ESCU 1986-1 Escudero, L. F.
A motivation for using the truncated Newton approach in a very large scale nonlinar network problem. In: G. Gallo, C. Sandi (Eds.): Netflow at Pisa, Mathematical Programming Study 26, 240-244. (1986).

ESFA 1985-1 Esfahanian, A.-H.; Hakimi, S. L.
Fault-tolerant routing in de Bruijn communication networks. IEEE Transactions on Computers 34, 777-788. (1985).

ETSC 1985-1 Etschmaier, M. M.; Mathaisel, D. F. X.
Airline scheduling: an overview. Transportation Science 19, 127-138. (1985).

EULE 1984-1 Euler, R.
Independence systems and perfect k-matroid-intersections. In: R. E. Burkard, R. A. Cuninghame-Green, U. Zimmermann (Eds.): Algebraic and Combinatorial Methods in Operations Research, Annals of Discrete Mathematics Vol. 19, 99-114. Amsterdam: North-Holland. (1984).

EULEA 1986-1 Euler, R.; Burkard, R. E.; Grommes, R.
On latin squares and the facial structure of related polytopes. Discrete Mathematics 62, 155-182. (1986).

EULEB 1987-1 Euler, R.; Jünger, M.; Reinelt, G.
Generalizations of cliques, odd cycles and anticycles and their relation to independence system polyhedra. Mathematics of Operations Research 12, 451-462. (1987).

EULEC 1985-1 Euler, R.; Mahjoub, A. R.
On a composition of independence systems by circiut-identification. Research Report. Grenoble: Univ. de Grenoble, Laboratoire Artemis (IMAG), 46 p. (1985).

EVAN 1984-1 Evans, J. R.
The factored transportation problem. Management Science 30, 1021-1024. (1984).

EVAN 1985-1 Evans, J. R.
On equivalent transportation models for production planning problems. IIE TRansactions 17, 102-107. (1985).

EVAN 1987-1 Evans, J. R.
Structural analysis of local search heuristics in combinatorial optimization. Computers and Operations Research 14, 465-478. (1987).

EVANA 1984-1 Evans, J. R.; Hall, R. A.
Probabilistic analysis of assignment ranking: the traveling salesman problem. American Journal of Mathematical and Management Sciences 4, 71-87. (1984).

EVANB 1985-1 Evans, S. R.; Norback, J. P.
The impact of a decision-support system for vehicle routing in a foodservice supply situation. Journal of the Operational Research Society 36, 467-472. (1985).

EVEN 1985-1 Even, S.; Gazit, H.
Updating distances in dynamic graphs. In: P. Brucker, R. Pauly (Eds.): Proc. IX. Symp. on Operations Research, Osnabrück 1984. Methods of Operations Research 49, 371-388. (1985).

EVER 1986-1 Everett, H.
Finding Hamiltonian paths in non-rectangular grid graphs. Congressus Numerantium 53, 185-192. (1986).
EVERA 1985-1 Everett, M. G.; Seidman, S. B.
The hull number of a graph. Discrete Mathematics 57, 217-224. (1985).
EVST 1979-1 Evstigneev, V. A.
A local algorithm for isolating blocks in a graph (Russian). in: Combinatorial-Algebraic Methods in Applied Mathematics, 35-42. Gorki: Gor'kovskii Gosidarstvennyi Universitet. (1979).
EVST 1981-1 Evstigneev, V. A.
Local numerical algorithms and determination of vertices with the greatest degree in a non-directed graph (Russian). in: A. A. Markov (Ed.), Combinatorial-Algebraic Methods in Applied Mathematics, Interuniv. Collect. (Gor'kij), 60-66. (1981).
EVST 1983-1 Evstigneev, V. A.
On properties of local algorithms on graphs (Russian). In: A. A. Markov (Ed.): Combinatorial-algebraic methods in applied mathematics, 72-105. Gor'kij: Gor'kovskij Gosudars. Univ. im. N. I. Labachevskogo. (1983).
EVST 1985-1 Evstigneev, V. A.
Application of graph theory to computer programming (Russian). Moscow: Nauka. (1985).
EXOO 1984-1 Exoo, G.; Harary, F.
Smallest cubic graphs with prescribed properties. Southeast Asian Bulletin of Mathematics (Singapore) 8, 1-4. (1984).
FAAL 1987-1 Faaland, B.; Schmitt, T.
Scheduling tasks with due dates in a fabrication/assembly process. Operations Research 35, 378-388. (1987).
FABE 1987-1 Faber, Z.; Vannelli, A.
Partitioning a graph with weighted nodes and edges. In: M. Ismail (Ed.): 29th Midwest Symposium on Circuits and Systems, 435-439. Amsterdam: North-Holland. (1987).
FABI 1985-1 Fabian, C.
Economic pre-optimizing algorithms for integer programming. Economic Computation and Economic Cybernetics Studies and Research 20, No. 1, 33-42. (1985).
FABI 1985-2 Fabian, C.
Greedy algorithm for a type of integer programming problems. In: I. Maruscíac, W. W. Breckner (Eds.): Approximation and Optimization, Proc. Colloq., Cluj-Napoca 1984, 245-254. University of Cluj-Napoca. (1985).
FABIA 1984-1 Fabian, C.; Stoica, M.
Fuzzy integer programming. In: H.-J. Zimmermann, L. A. Zadeh, B. R. Gaines (Eds.): Fuzzy Sets and Decision Analysis, 123-132. Amsterdam: North-Holland. (1984).
FAHR 1987-1 Fahrion, R.
A parallelization approach for sequencing problems. In: W. Domschke, W. Krabs, J. Lehn, P. Spellucci (Eds.): Proc. XI. Symp. on Operations Research, Darmstadt 1986, 607-618. Methods of Operations Research 57, Frankfurt/Main: Athenäum. (1987).
FAIG 1977-1 Faigle, U.
Geometries on partially ordered sets. Chapel Hill: Univ. of North Carolina, Dept. of Mathematics, 80 p. (1977). (Dissertation)
FAIG 1984-1 Faigle, U.
Matroids on ordered sets and the greedy algorithm. In: R. E. Burkard, R. A. Cuninghame-Green, U. Zimmermann (Eds.): Algebraic and Combinatorial Methods in Operations Research, Annals of Discrete Mathematics Vol. 19, 115-128. Amsterdam: North-Holland. (1984).
FAIG 1984-2 Faigle, U.
A vector exchange property of submodular systems. Discrete Applied Mathematics 9, 209-212. (1984).
FAIG 1985-1 Faigle, U.
On ordered languages and the optimization of linear functions by greedy algorithms. Journal of the Association for Computing Machinery 32, 861-870. (1985).
FAIG 1985-2 Faigle, U.
Orthogonal sets, matroids, and theorems of the alternative. Bollettino della Unione Matematica Italiana, IV, Serie B 4, 139-153. (1985).
FAIG 1986-1 Faigle, U.
Exchange properties of combinatorial closure spaces. Discrete Applied Mathematics 15, 249-260. (1986).
FAIG 1987-1 Faigle, U.
The clique polytope of a graph. Report 87474-OR. Bonn: Univ. Bonn, Inst. f. Ökonometrie u. Operations Research, 9 p. (1987).
FAIGA 1987-1 Faigle, U.; Gierz, G.
The bandwidth of planar distributive lattices. In: G. Tinhofer, G. Schmidt (Eds.): Graph-Theoretic Concepts in Computer Science, 94-105. Lecture Notes in Computer Science 246, Berlin: Springer. (1987).
FAIGB 1985-1 Faigle, U.; Gierz, G.; Schrader, R.
Algorithmic approaches to setup minimization. SIAM Journal on Computing 14, 954-965. (1985).
FAIGC 1984-1 Faigle, U.; Goecke, O.; Schrader, R.
On simplicial decomposition in combinatorial structures. Report No. 84349-OR. Bonn: Univ. Bonn, Inst. für Ökonometrie und Operations Research, 15 p. (1984).
FAIGC 1986-1 Faigle, U.; Goecke, O.; Schrader, R.
Church-Rosser decomposition in combinatorial structures. Report No. 86426-OR. Bonn: Universität Bonn, Inst. für Ökonometrie und Operations Research, 16p. (1986).
FAIGD 1985-1 Faigle, U.; Lovasz, L.; Schrader, R.; Turan, G.
Search problems in ordered sets. In: D. Ohse, A. C. Esprester, H.-U. Küpper, P. Stähly, H.Steckhan (Eds.): Operations Research Proceedings 1984, 411-415. Berlin: Springer-Verlag. (1985).
FAIGD 1986-1 Faigle, U.; Lovasz, L.; Schrader, R.; Turan, G.
Searching in trees, series-parallel and interval orders. SIAM Journal on Computing 15, 1075-1084. (1986).
FAIGE 1986-1 Faigle, U.; Sands, B.
A size-width inequality for distributive lattices. Combinatorica 6, 29-34. (1986).

FAIGF 1984-1 Faigle, U.; Schrader, R.
Comparability graphs and order invariants. In: U. Pape (Ed.): Proc. WG '84: Internat. Workshop on Graphtheoretic Concepts in Computer Science, Berlin 1984, 136-145. Linz: Trauner. (1984).

FAIGF 1985-1 Faigle, U.; Schrader, R.
A setup heuristic for interval orders. Operations Research Letters 4, 185-188. (1985).

FAIGF 1985-2 Faigle, U.; Schrader, R.
Linear arrangements of interval orders. In: D. Ohse, A. C. Esprester, H.-U. Küpper, P. Stähly, H.Steckhan (Eds.): Operations Research Proceedings 1984, 416. Berlin: Springer-Verlag. (1985).

FAIGF 1985-3 Faigle, U.; Schrader, R.
Setup minimization techniques for comparability graphs. In: H. Noltemeier (Ed.): Proc. WG '85: Internat. Workshop on Graphtheoretic Concepts in Computer Science, Castle Schwanberg, Würzburg, 1985, 101-118. Linz: Trauner. (1985).

FAIGF 1986-1 Faigle, U.; Schrader, R.
A combinatorial bijection between linear extensions of equivalent orders. Discrete Mathematics 58, 295-302. (1986).

FAIGF 1986-2 Faigle, U.; Schrader, R.
Greene-Kleitman extensions of ordered sets. Methods of Operations Research 53, 251-256. (1986).

FAIGF 1986-3 Faigle, U.; Schrader, R.
On the computational complexity of the order polynomial. Discrete Applied Mathematics 15, 261-270. (1986).

FAIGF 1986-4 Faigle, U.; Schrader, R.
On the convergence of stationary distributions in simulated annealing algorithms. Report No. 86428-OR. Bonn: Universität Bonn, Inst. für Ökonometrie und Operations Research, 7p. (1986).

FAIGF 1987-1 Faigle, U.; Schrader, R.
Interval orders without odd crowns are defect optimal. Computing 38, 59-70. (1987).

FAIGF 1987-2 Faigle, U.; Schrader, R.
Setup optimization problems with matroid structure. Order 4, 43-54. (1987).

FAIGG 1987-1 Faigle, U.; Schrader, R.; Suletzki, R.
A cutting-plane algorithm for optimal graph partitioning. In: W. Domschke, W. Krabs, J. Lehn, P. Spellucci (Eds.): Proc. XI. Symp. on Operations Research, Darmstadt 1986, 109-116. Methods of Operations Research 57, Frankfurt/Main: Athenäum. (1987).

FAIGH 1985-1 Faigle, U.; Turan, G.
Sorting and recognition problems for ordered sets. In: K. Mehlhorn (Ed.): STACS 85, Second Annual Symposium on Theoretical Aspects of Computer Science (Saarbrücken, January 1985), Lecture Notes in Computer Science Vol. 182, 109-118. Berlin: Springer-Verlag. (1985).

FAIGH 1987-1 Faigle, U.; Turan, G.
On the complexity of interval orders and semiorders. Discrete Mathematics 63, 131-142. (1987).

FALC 1987-1 Falconer, K. J.
Cut-set sums and tree processes. Proceedings of the AMS 101, 337-346. (1987).

FALK 1987-1 Falkner, J. C.; Ryan, D. M.
A bus crew scheduling system using a set partitioning model. Asia-Pacific Journal of Operational Research 4, 39-56. (1987).

FAN 1984-1 Fan, G.-H.
New sufficient conditions for cycles in graphs. Journal of Combinatorial Theory, Series B 37, 221-227. (1984).

FAN 1985-1 Fan, G.-H.
Longest cycles in regular graphs. Journal of Combinatorial Theory 39B, 325-345. (1985).

FAND 1985-1 Fandel, G.; Reese, J.
Organisation und Planung des Fahrzeugeinsatzes in einem Chemieunternehmen. OR Spektrum 7, 115-124. (1985).

FANE 1985-1 Fanelli, S.; Lombari, T.
An application of the technique of Q-GERT networks to the evaluation of logistic parameters of electronic systems (Italian). Ricerca Operativa No. 34, 111-138. (1985).

FANG 1985-1 Fang, S. C.; Peterson, E. L.
An economic equilibrium model on a multicommodity network. International Journal of Systems Science 16, 479-490. (1985).

FANGA 1984-1 Fang, W.-W.
An algorithm for the directed T-immersion of a graph. In: K. M. Koh, H. P. Yap (Eds.): Graph Theory Singapore 1983, Lecture Notes in Mathematics 1073, 254-267. Berlin: Springer-Verlag. (1984).

FARB 1986-1 Farber, M.; Hahn, G.; Hell, P.; Miller, D.
Concerning the achromatic number of graphs. Journal of Combinatorial Theory 40B, 21-39. (1986).

FARBA 1986-1 Farber, M.; Jamison, R. E.
Convexity in graphs and hypergraphs. SIAM Journal on Algebraic and Discrete Methods 7, 433-444. (1986).

FARBB 1985-1 Farber, M.; Keil, J. M.
Domination in permutation graphs. Journal of Algorithms 6, 309-321. (1985).

FARD 1986-1 Fardanesh, B.; Villaseca, F. E.
Two-step optimal thermal generation scheduling. Automatica 22, 361-366. (1986).

FARL 1984-1 Farley, A. M.; Proskurowski, A.
Computing the maximum order of an open irredundant set in a tree. Congressus Numerantium 41, 219-228. (1984).

FARL 1984-2 Farley, A. M.; Proskurowski, A.
Extremal graphs with no disconnecting matching. Congressus Numerantium 41, 153-166. (1984).

FARR 1985-1 Farr, G.; Mac Diarmid, C.
The complexity of counting homeomorphs. Theoretical Computer Science 36, 345-348. (1985).

FARRA 1983-1 Farrell, E. J.
Decomposition of complete graphs and complete bipartite graphs into node disjoint paths. Pure and Applied Mathematika Sciences (Saharanpur) 17, 7-14. (1983).

FARRA 1983-2 Farrell, E. J.
On the number of 2-factors in the short ladder. Journal of Combinatorics, Information & System Sciences (Delhi) 8, 175-180. (1983).

FARRA 1983-3 Farrell, E. J.
The subgraph polynomial and its relation to some other graph polynomials. Caribbean Journal of Mathematics 2, No. 3, 39-53. (1983).
FARRA 1984-1 Farrell, E. J.
A note on cycle decompositions of complete bipartite graphs. Caribbean Journal of Mathematics 3, 9-15. (1984).
FARRA 1984-2 Farrell, E. J.
Decompositions of graphs with cyclomatic number 2 into node disjoint paths. Indian Journal of Mathematics 26, 1-13. (1984).
FARRA 1984-3 Farrell, E. J.
On coefficients of path polynomials. International Journal of Mathematics and Mathematical Sciences 7, 65-73. (1984).
FARRA 1985-1 Farrell, E. J.
On coefficients of circuit polynomials and characteristic polynomials. International Journal of Mathematics and Mathematical Sciences 8, 697-705. (1985).
FARRA 1986-1 Farrell, E. J.
A survey of the unifying effects of F-polynomials in combinatorics and graph theory. In: D. Cvetkovic, I. Gutman, T. Pisanski, R. Tosic (Ed.): Graph Theory, Proc. Fourth Yugoslav Seminar on Graph Theory, Dubrovnik 1983, 137-150. Novi Sad: University of Novi Sad. (1986).
FARRA 1986-2 Farrell, E. J.
On the occurrences of Fibonacci sequences in the counting of matchings in linear polygonal chains. Fibonacci Quarterly 24, 238-246. (1986).
FARRA 1987-1 Farrell, E. J.
Counting matchings in graphs. Journal of the Franklin Institute 324, No. 3, 331-339. (1987).
FARRA 1987-2 Farrell, E. J.
Matchings in hexagonal cacti. International Journal of Mathematics and Mathematical Sciences 10, 321-338. (1987).
FARRB 1983-1 Farrell, E. J.; Grell, J. C.
Some analytical properties of the circuit polynomial of a graph. Caribbean Journal of Mathematics 2, 69-76. (1983).
FARRB 1983-2 Farrell, E. J.; Grell, J. C.
The circuit polynomial and its relation to other graph polynomials. Caribbean Journal of Mathematics 2, 15-24. (1983).
FARRB 1985-1 Farrell, E. J.; Grell, J. C.
Some further constructions of cocircuit and cospectral graphs. Caribbean Journal of Mathematics 4, 17-28. (1985).
FARRB 1986-1 Farrell, E. J.; Grell, J. C.
An application of circuit polynomials to the counting of spanning trees in graphs. Publications de l'Institut Mathematique (Belgrade), Nouvelle Serie 39(53), 63-67. (1986).
FARRC 1987-1 Farrell, E. J.; Guo, J. M.
Circuit characterizations of unions of graphs. Journal of Combinatorial Mathematics and Combinatorial Computing 2, 161-177. (1987).
FARRD 1985-1 Farrell, E. J.; Gutman, I.
A note on the circuit polynomial and its relation to the mu-polynomial. Match No. 18, 55-61. (1985).
FARRE 1984-1 Farrell, E. J.; Wahid, S. A.
Matchings in heptagonal and octagonal chains. Caribbean Journal of Mathematics 3, 45-57. (1984).
FARRE 1984-2 Farrell, E. J.; Wahid, S. A.
Matchings in nonagonal and decagonal chains. Caribbean Journal of Mathematics (Barbados) 3, No. 2, 73-84. (1984).
FARRE 1985-1 Farrell, E. J.; Wahid, S. A.
Matchings in long benzene chains. Match 18, 37-48. (1985).
FARRE 1986-1 Farrell, E. J.; Wahid, S. A.
Matching polynomials: a matrix approach and its applications. Journal of the Franklin Institute 322, No. 1, 13-21. (1986).
FARRE 1987-1 Farrell, E. J.; Wahid, S. A.
On the reconstruction of the matching polynomial and the reconstruction conjecture. International Journal of Mathematics and Mathematical Sciences 10, 155-162. (1987).
FARRE 1987-2 Farrell, E. J.; Wahid, S. A.
Some general classes of comatching graphs. International Journal of Mathematics and Mathematical Sciences 10, 519-524. (1987).
FAUD 1987-1 Faudree, R. J.; Gould, R. J.; Schelp, R. H.
Neighborhood conditions and edge disjoint Hamiltonian cycles. Congressus Numerantium 59, 55-68. (1987).
FAUDA 1987-1 Faudree, R. J.; Ordman, E. T.; Schelp, R. H.; Jacobson, M. S.; Tuza, Z.
Menger's theorem and short paths. Journal of Combinatorial Mathematics and Combinatorial Computing 2, 235-253. (1987).
FAUDB 1985-1 Faudree, R. J.; Rousseau, C. C.; Schelp, R. H.
Edge-disjoint Hamiltonian cycles. In: Y. Alavi, G. Chartrand, L. Lesniak, D. R. Lick, C. E. Wall (Eds.): Graph Theory with Applications to Algorithms and Computer Science, 231-250. Chichester: Wiley. (1985).
FAVA 1985-1 Favaron, O.
On a conjecture of Fink and Jacobson concerning k-domination and k-dependence. Journal of Combinatorial Theory 39B, 101-102. (1985).
FAVAA 1985-1 Favaron, O.; Kouider, M.
Path partitions and cycle partitions of Eulerian graphs of maximum degree 4. Rapport de Recherche 202. Orsay: Univ. de Paris-Sud, Lab. de Recherche en Informatique, 10 p. (1985).
FAVAB 1985-1 Favaron, O.; Kouider, M.; Maheo, M.
Edge-vulnerability and mean distance. Rapport de Recherche 248. Orsay: Univ. de Paris-Sud, Lab. de Recherche en Informatique, 15 p. (1985).
FAVAC 1985-1 Favaron, O.; Lonc, Z.; Truszczynski, M.
Decompositions of graphs into graphs with three edges. Ars Combinatoria 20, 125-146. (1985).

FAVAD 1986-1 Favaron, O.; Ordaz, O.
A sufficient condition for oriented graphs to be Hamiltonian. Discrete Mathematics 58, 243-252. (1986).

FEDE 1986-1 Federgruen, A.; Groenevelt, H.
Optimal flows in networks with multiple sources and sinks, with applications to oil and gas lease investment programs. Operations Research 34, 218-225. (1986).

FEDE 1986-2 Federgruen, A.; Groenevelt, H.
Preemptive scheduling of uniform machines by ordinary network flow techniques. Management Science 32, 341-349. (1986).

FEDE 1986-3 Federgruen, A.; Groenevelt, H.
The greedy procedure for resource allocation problems: necessary and sufficient conditions for optimality. Operations Research 34, 909-918. (1986).

FEDEA 1984-1 Federgruen, A.; Zipkin, P.
A combined vehicle routing and inventory allocation problem. Operations Research 32, 1019-1037. (1984).

FEDO 1985-1 Fedorova, N. P.
On homogeneous Hamiltonian graphs contractible to complete ones (Russian). In: Y. A. Abramov (Ed.): Mathematical methods of optimization and control in systems, Collect. sci. Works, Kalinin 1985, 32-36. (1985).

FEIR 1986-1 Feiring, B. R.
A parametric programming procedure for solving pure integer programming problems. Asia-Pacific Journal of Operational Research 3, 37-42. (1986).

FEIR 1987-1 Feiring, B. R.
A knapsack procedure for solving integer programming problems. Asia-Pacific Journal of Operational Research 4, 123-130. (1987).

FEIR 1987-2 Feiring, B. R.
Some properties of the solution space of the N-city traveling-salesman problem. Applied Mathematics and Computation 23, 291-340. (1987).

FEJG 1983-1 Fejgin, L. I.
Scheduling problems with fuzzy activity durations. Soviet Physics, Doklady 28, 833-835. Translated from Doklady Akademii Nauk SSSR (Moscow) 272, 812-815. (1983).

FELD 1985-1 Feldman, P.
Fault tolerance of minimal path routings in a network. In: Proc. Seventeenth Annual ACM Symp. on Theory of Computing, Providence 1985, 327-334. ACM. (1985).

FELDA 1986-1 Feldman, P.; Friedman, J.; Pippenger, N.
Non-blocking networks. In: Proc. Eighteenth Annual ACM Symp. on Theory of Computing, Berkeley 1986, 247-254. ACM. (1986).

FELL 1987-1 Fellows, M.; Hickling, F.; Syslo, M. M.
A topological parametrization and hard graph problems. Congressus Numerantium 59, 69-78. (1987).

FENG 1983-1 Feng, C. J.
A necessary and sufficient condition for a class of graceful graphs (Chinese). Qufu Shiyuan Xuebao (Qufu) No. 1, 23-29. (1983).

FENGA 1985-1 Feng, T.; Shi, R.-H.
Smallest regular graphs with girth pair (4,2t+1). Graphs and Combinatorics 1, 201-202. (1985).

FENN 1984-1 Fenner, T. I.; Frieze, A. M.
Hamiltonian cycles in random regular graphs. Journal of Combinatorial Theory, Series B 37, 103-112. (1984).

FENN 1984-2 Fenner, T. I.; Frieze, A. M.
On the complexity of finding sets of edges in graphs and digraphs. In: J. A. Bondy, U. P. R. Murty (Eds.): Progress in Graph Theory, 219-232. New York: Academic Press. (1984).

FENN 1985-1 Fenner, T. I.; Frieze, A. M.
An algorithm for finding a matroid basis which maximizes the products of the weights of the elements. BIT 25, 434-438. (1985).

FENT 1984-1 Fenton, N. E.
Matroid representations - an algebraic treatment. Quarterly Journal of Mathematics, Oxford Second Series 35, 263-280. (1984).

FENT 1985-1 Fenton, N. E.
The structural complexity of flow graphs. In: Y. Alavi, G. Chartrand, L. Lesniak, D. R. Lick, C. E. Wall (Eds.): Graph Theory with Applications to Algorithms and Computer Science, 273-282. Chichester: Wiley. (1985).

FENTA 1987-1 Fenton, W. E.
Completeness in oriented matroids. Discrete Mathematics 66, 79-90. (1987).

FEO 1986-1 Feo, T. A.; Hochbaum, D. S.
Lagrangian relaxation for testing infeasibility in VLSI routing. Operations Research 34, 819-831. (1986).

FEOF 1987-1 Feofiloff, P.; Younger, D. H.
Directed cut transversal packing for source-sink connected graphs. Combinatorica 7, 255-264. (1987).

FERL 1986-1 Ferland, J. A.
The single depot vehicle scheduling problem. In: Recent Advances in System Modelling and Optimization (Santiago 1984), 99-112. Lecture Notes in Control and Information Sciences 87, Berlin: Springer. (1986).

FERLA 1985-1 Ferland, J. A.; Roy, S.
Timetabling problem for university as assignment of activities to resources. Computers & Operations Research 12, 207-218. (1985).

FERN 1982-1 Fernandez Garcia, F. R.; Pino Mejias, J. L.
Study of multiple objectives in sequencing problems on one machine (Spanish). In: Proceedings of the ninth Spanish-Portuguese Conference on Mathematics, 2 (Salamanca, 1982), 615-618. Acta Salmanticensia, Ciencias 46, Univ. Salamanca. (1982).

FERNA 1987-1 Fernandez, E. B.; Ngo, T. A.
Scheduling of task graphs into transputer networks. Congressus Numerantium 59, 79-88. (1987).

FIAM 1983-1 Fiamcik, J.
Acyclic chromatic index of a subdivided graph. In: M. Fiedler (Ed.): Graphs and Other Combinatorial Topics. Proc. Third Czechoslovak Symposium on Graph Theory, Prague 1982, 64-69. Leibzig: Teubner. (1983).

FIAM 1984-1 Fiamcik, J.
Acyclic chromatic index of a subdivided graph. Archivum Mathematicum (Brno) 20, 69-82. (1984).

FIED 1985-1 Fiedler, M.
Signed bigraphs of monotone matrices. In: H. Sachs (Ed.): Graphs, Hypergraphs and Applications, Proc. Conf. on Graph Theory (Eyba 1984), 36-40. Leipzig: Teubner. (1985).

FILI 1983-1 Filipiak, J.
Joint optimization of routing and congestion control in data communication networks. Archiwum Automatyki i Telemechaniki (Warsaw) 28, 57-67. (1983).

FILI 1984-1 Filipiak, J.
A dynamic flow assignment problem for a telephone trunk network. in: A. Straszak (Ed.): Large Scale Systems: Theory and Applications, Proc. IFAC/IFORS Symp., Warsaw 1983, IFAC Proc. Series, No. 10, 681-687. Oxford: Pergamon Press. (1984).

FINC 1987-1 Fincke, U.; Hamacher, H. W.
On the expected value of stochastic linear programs and (dynamic) network flow problems. European Journal of Operational Research 29, 307-316. (1987).

FINK 1983-1 Fink, J. F.
Randomly bitraceable graphs. Congressus Numerantium 39, 383-398. (1983).

FINKA 1985-1 Fink, J. F.; Jacobson, M. S.
N-domination in graphs. In: Y. Alavi, G. Chartrand, L. Lesniak, D. R. Lick, C. E. Wall (Eds.): Graph Theory with Applications to Algorithms and Computer Science, 283-300. Chichester: Wiley. (1985).

FINKA 1985-2 Fink, J. F.; Jacobson, M. S.
On n-domination, n-dependence and forbidden subgraphs. In: Y. Alavi, G. Chartrand, L. Lesniak, D. R. Lick, C. E. Wall (Eds.): Graph Theory with Applications to Algorithms and Computer Science, 301-312. Chichester: Wiley. (1985).

FINKB 1985-1 Fink, J. F.; Jacobson, M. S.; Kinch, L. F.; Roberts, J. A.
On graphs having domination number half their order. Periodica Mathematica Hungarica (Budapest) 16, 287-293. (1985).

FINKC 1984-1 Finke, G.
Bulk shipment formulations for the traveling salesman problem. In: G. Hammer, D. Pallaschke (Eds.): Methods of Operations Research 51, Vol. 1, 247-256. Königstein/Ts.: Verlag Anton Hain Meisenheim. (1984).

FINKC 1987-1 Finke, G.
Network flow based branch and bound method for asymmetric traveling salesman problems. In: W. Domschke, W. Krabs, J. Lehn, P. Spellucci (Eds.): Proc. XI. Symp. on Operations Research, Darmstadt 1986, 117-120. Methods of Operations Research 57, Frankfurt/Main: Athenäum. (1987).

FINKD 1987-1 Finke, G.; Ahmed, S.
Network flow methods for the asymmetric traveling salesman problem. Congressus Numerantium 57, 223-234. (1987).

FINKE 1987-1 Finke, G.; Burkard, R. E.; Rendl, F.
Quadratic assignment problems. In: P. Martello, G. Laporte, M. Minoux, C. Ribeiro (Eds.): Surveys in Combinatorial Optimization, Annals of Discrete Mathematics 31, 61-82. Amsterdam, New York, Tokyo: North-Holland. (1987).

FINKF 1984-1 Finke, G.; Claus, A.; Gunn, E. A.
A two-commodity network flow approach to the traveling salesman problem. Congressus Numerantium 41, 167-178. (1984).

FINKG 1984-1 Finke, U.; Hamacher, H. W.
Some bounds for stochastic network flow problems. In: U. Pape (Ed.): Proc. WG '84: Internat. Workshop on Graphtheoretic Concepts in Computer Science, Berlin 1984, 146-160. Linz: Trauner. (1984).

FINKH 1984-1 Finkelstein, Y. Y.; Shmelev, A. A.
Über die Abschätzung der Koeffizienten von Seiten gewisser ganzzahliger Polyeder (Russisch). In: K.-H. Elster (Ed.): Mathematische Optimierung - Theorie und Anwendungen, Vortragsauszüge, Eisenach 1984, 45-48. Techn. Hochschule Ilmenau. (1984).

FINKH 1986-1 Finkelstein, Y. Y.; Shmelev, A. A.
On an estimate of coefficients of the faces of some integer polyhedra (Russian). Ehkonomika i Matematicheskie Metody 22, 744-747. (1986).

FIOL 1985-1 Fiol, M. A.; Alegre, I.; Yebra, J. L. A.; Fabrega, J.
Digraphs with walks of equal length between vertices. In: Y. Alavi, G. Chartrand, L. Lesniak, D. R. Lick, C. E. Wall (Eds.): Graph Theory with Applications to Algorithms and Computer Science, 313-322. Chichester: Wiley. (1985).

FIOLA 1983-1 Fiol, M. A.; Yebra, J. L. A.; Fabrega, J.
Sequence graphs and interconnection networks. Ars Combinatoria 16A, 7-14. (1983).

FIOLA 1983-2 Fiol, M. A.; Yebra, J. L. A.; Fabrega, J.
The vertex-edge duality in the colouring of graphs and hypergraphs. Ars Combinatoria 16B, 245-256. (1983).

FISC 1986-1 Fischetti, M.
An improved bound for the asymmetric traveling salesman problem. Ricerca Operativa 16, No. 37, 71-86. (1986).

FISC 1987-1 Fischetti, M.
A new linear storage, polynomial time approximation scheme for the subset-sum problem. Ricerca Operativa 17, No.42, 7-38. (1987).

FISCA 1987-1 Fischetti, M.; Toth, P.
Bounding and dominating procedure for combinatorial optimization problems (Italian). Ricerca Operativa 17, No.42, 61-102. (1987).

FISH 1985-1 Fishburn, P. C.
Interval graphs and interval orders. Discrete Mathematics 55, 135-150. (1985).

FISH 1985-2 Fishburn, P. C.
Interval orders and interval graphs. A study of partially ordered sets. New York, Chichester: Wiley. (1985).

FISH 1987-1 Fishburn, P. C.
Decomposing weighted digraphs into sums of chains. Discrete Applied Mathematics 16, 223-238. (1987).

FISHA 1985-1 Fishburn, P. C.; Kochenberger, G. A.
Aggregating assignment constraints. Naval Research Logistics Quarterly 32, 653-664. (1985).

FISHB 1985-1 Fisher, D. L.; Saisi, D.; Goldstein, W. M.
Stochastic PERT networks: OP diagrams, critical paths and the project completion time. Computers & Operations Research 12, 471-482. (1985).

FISHC 1983-1 Fisher, J. G.; Shier, D. R.
A heuristic procedure for large-scale examination scheduling problems. Congressus Numerantium 39, 399-410. (1983).

FISHD 1985-1 Fisher, M. L.
An applications oriented guide to Lagrangian relaxation. Interfaces 15, 10-21. (1985).

FISHD 1986-1 Fisher, M. L.
A dual algorithm for large sale set partitioning. Paper No. 894. West Lafayette: Purdue Univ., Krannert Graduate School of Management, 14p. (1986).

FISHE 1986-1 Fisher, M. L.; Huang, J.-G.; Tang, B.-X.
Scheduling bulk-pickup-delivery vehicles in Shanghai. Interfaces 16, No. 2, 18-23. (1986).

FISHF 1984-1 Fisher, M. L.; Jaikumar, R.
The local delivery problem: algorithms and applications. In: M. Florian (Ed.): Transportation Planning Models, 419-438. Amsterdam: North-Holland. (1984).

FISHG 1986-1 Fisher, M. L.; Jaikumar, R.; Wassenhove, L. N. van
A multiplier adjustment method for the generalized assignment problem. Management Science 32, 1095-1103. (1986).

FISHH 1981-1 Fisher, M. L.; Krieger, A. M.
Analysis of a linearization heuristic for single-machine scheduling to maximize profit. 81-06-01. Philadelphia: Univ. of Pennsylvania, The Wharton School, 39 p. (1981).

FISHI 1981-1 Fisher, M. L.; Wolsey, L. A.
On the greedy heuristic for covering and packing problems. 81-07-04. Philadelphia: Univ. of Pennsylvania, The Wharton School, 15 p. (1981).

FISHJ 1987-1 Fishman, G. S.
Maximum flow and critical cutset as descriptors of multi-stage systems with randomly capacitated components. Computers and Operations Research 14, 507-520. (1987).

FLAC 1987-1 Flach, P.; Volkmann, L.
Abschätzungen gesättigter Matchings nach unten. Analele Universitatii Bucuresti Mathematica 36, 25-30. (1987).

FLAM 1985-1 Flament, C.
Comparability graphs with constraint, partial semi-orders and interval orders. Discrete Mathematics 53, 79-90. (1985).

FLAN 1985-1 Flandrin, E.
Cycles hamiltoniens et dominants dans le carre d'un graphe connexe. Rapport de Recherche 206. Orsay: Univ. de Paris-Sud, Lab. de Recherche en Informatique, 7 p. (1985).

FLEI 1985-1 Fleischmann, B.
A cutting plane procedure for the travelling salesman problem on road networks. European Journal of Operational Research 21, 307-317. (1985).

FLEI 1988-1 Fleischmann, B.
Cutting planes for the symmetric travelling salesman problem. In: H. Schellhaas, P. van Beek, H. Isermann, R. Schmidt, M. Zijlstra (Eds.): Operations Research Proceedings 1987 (DGOR/NSOR), 611-612. Berlin: Springer. (1988).

FLEIA 1983-1 Fleischner, H.
A way of enumerating Eulerian trails in a connected Eulerian graph. Ars Combinatoria 16B, 177-196. (1983).

FLEIA 1984-1 Fleischner, H.
Cycle decompositions, 2-coverings, removable cycles and the four-color-disease. In: J. A. Bondy, U. P. R. Murty (Eds.): Progress in Graph Theory, 233-246. New York: Academic Press. (1984).

FLEIA 1986-1 Fleischner, H.
Proof of the strong 2-cover conjecture for planar graphs. Journal of Combinatorial Theory, Series B, 40, 229-230. (1986).

FLEIB 1985-1 Fleischner, H.; Jackson, B.
Removable cycles in planar graphs. Journal of the London Mathematical Society, Second Series 31, 193-199. (1985).

FLEIB 1987-1 Fleischner, H.; Jackson, B.
Compatible path-cycle-decompositions on plane graphs. Journal of Combinatorial Theory B 42, 94-121. (1987).

FLEJ 1985-1 Flejshman, S. B.
A pseudo-polynomial algorithm for partitioning graphs of special form. Soviet Journal of Automation and Information Sciences 18, No.3, 18-25; translation from Avtomatika (1985), No.3, 18-25. (1985).

FLIK 1985-1 Flikop, Z.
Routing optimization in packet switching communication networks. European Journal of Operational Research 19, 262-267. (1985).

FLOR 1986-1 Florescu, G.
Aspects of modelling transportation through the dynamic transportation problem. Economic Computation and Economic Cybernetics Studies and Research 21, 55-58. (1986).

FLORA 1984-1 Florian, M.
An introduction to network models used in transportation planning. In: M. Florian (Ed.): Transportation Planning Models, 137-152. Amsterdam: North-Holland. (1984).

FLORA 1986-1 Florian, M.
Nonlinear cost network models in transportation analysis. In: G. Gallo, C. Sandi (Eds.): Netflow at Pisa, Mathematical Programming Study 26, 167-196. (1986).

FLOY 1986-1 Floyd, S.; Karp, R. M.
FFD bin-packing for item sizes with distributions on (0, 1/2). In: 27th Annual Symp. on Foundations of Computer Science, Proceedings, 322-330. IEEE. (1986).

FOLE 1984-1 Foley, R. D.; Suresh, S.
Minimizimg the expected flowtime in stochastic flow shops. IIE Transactions 16, 391-395. (1984).
FOLE 1984-2 Foley, R. D.; Suresh, S.
Stochastically minimizing the makespan in flow shops. Naval Research Logistics Quarterly 31, 551-558. (1984).
FOLE 1986-1 Foley, R. D.; Suresh, S.
Scheduling n overlapping jobs and two stochastic jobs in a flow shop. Naval Research Logistics Quarterly 33, 123-128. (1986).
FONL 1985-1 Fonlupt, J.; Naddef, D. J.
The traveling salesman problem in graphs with some excluded minors. Rapport de Recherche 557. Grenoble: Univ. Scientifique et Medicale de Grenoble, 31 p. (1985).
FONLA 1984-1 Fonlupt, J.; Raco, M.
Orientation of matrices. Mathematical Programming Study 22, 86-98. (1984).
FONLB 1986-1 Fonlupt, J.; Zemirline, A.
Polynomial recognition algorithm of perfect 3-chromatic K(4)/(e)-free graphs. Rapport Technique 13. Grenoble: IMAG, Artemis, 37 p. (1986).
FORL 1986-1 Forler, C.
Modified transport optimization under restrictions. Methods of Operations Research 56, 33-54. (1986).
FORS 1984-1 Forst, F. G.
A review of the static, stochastic job sequencing literature. Opsearch 21, 127-144. (1984).
FORT 1985-1 Fortenberry, J. C.; Cox, J. F.
Multiple criteria approach to the facilities layout problem. International Journal of Production Research 23, 773-782. (1985).
FORTA 1986-1 Fortenberry, J. C.; Mitra, A.
A multiple criteria approach to the location-allocation problem. Computers and Industrial Engineering 10, 77-87. (1986).
FORTB 1985-1 Fortenberry, J. C.; Snyder, C. A.; Cox, J. F.
A multiple criteria decision support model for facilities layout. In: P. Brucker, R. Pauly (Eds.): Proc. IX. Symp. on Operations Research, Osnabrück 1984. Methods of Operations Research 50, 559-564. (1985).
FOTI 1985-1 Foti, G.; Tadei, R.
New transportation models for urban planning: fundamental theory and use (Italian). Ricerca Operativa No. 34, 169-192. (1985).
FOUL 1984-1 Foulds, L. R.
Combinatorial optimization for undergraduates. New York, Berlin: Springer. (1984).
FOUL 1985-1 Foulds, L. R.
Enumeration of graph theoretic solutions for facilities layout. Congressus Numerantium 48, 87-100. (1985).
FOUL 1985-2 Foulds, L. R.
One-way street system design. In: P. Brucker, R. Pauly (Eds.): Proc. IX. Symp. on Operations Research, Osnabrück 1984. Methods of Operations Research 50, 565-572. (1985).
FOUL 1985-3 Foulds, L. R.
Traffic network arc elimination by branch and bound enumeration. Arabian Journal for Science and Engineering 10, 149-157. (1985).
FOUL 1986-1 Foulds, L. R.
Braess's paradox: which arcs should be eliminated in order to minimize congestion? Methods of Operations Research 54, 497-506. (1986).
FOUL 1986-2 Foulds, L. R.
Testing the theory of evolution: a novel application of combinatorial optimization. Discrete Applied Mathematics 15, 271-282. (1986).
FOULA 1985-1 Foulds, L. R.; Gibbons, P. B.; Giffin, J. W.
Facilities layout adjacency determination: An experimental comparison of three graph theoretic heuristics. Operations Research 33, 1091-1106. (1985).
FOULB 1985-1 Foulds, L. R.; Giffin, J. W.
A graph-theoretic heuristic for minimizing total transport cost in facilities layout. International Journal of Production Research 23, 1247-1258. (1985).
FOULB 1985-2 Foulds, L. R.; Giffin, J. W.
Facilities layout via new graph theoretic methods. New Zealand Operational Research 13, 107-114. (1985).
FOULC 1986-1 Foulds, L. R.; Giffin, J. W.; Evans, J. D.
Towards a new decision support system for facilities layout planners. Research Report 85-11. Gainesville: Univ. of Florida, Industrial & Systems Eng. Dept., 54 p. (1986).
FOULD 1984-1 Foulds, L. R.; Robinson, R. W.
Enumeration of phylogenetic trees without points of degree two. Research Report, 84-15. Gainesville: Univ. of Florida, Ind. & Systems Eng. Dept., 17 p. (1984).
FOUQ 1983-1 Fouquet, J. L.; Jolivet, J. L.
Strong edge-colorings of graphs and applications to multi k-gons. Ars Combinatoria 16A, 141-150. (1983).
FOUQ 1984-1 Fouquet, J. L.; Jolivet, J. L.
Strong edge-coloring of cubic planar graphs. In: J. A. Bondy, U. P. R. Murty (Eds.): Progress in Graph Theory, 247-264. New York: Academic Press. (1984).
FOUQA 1985-1 Fouquet, J. L.; Thuillier, H.
Cycles through given vertices in planar 3-connected cubic graphs. Ars Combinatoria 20-B, 75-106. (1985).
FOUR 1985-1 Fournier, I.
Longest cycles in 2-connected graphs of independence number alpha. In: B. R. Alspach, C. D. Godsil (Eds.): Cycles in Graphs, Annals of Discrete Mathematics Vol. 27, 201-204. Amsterdam: North-Holland. (1985).
FOURA 1985-1 Fournier, I.; Fraisse, P.
On a conjecture of Bondy. Journal of Combinatorial Theory 39B, 17-26. (1985).

FOURB 1984-1 Fournier, J.-C.; Vergnas, M. Las
A class of bichromatic hypergraphs. In: C. Berge, V. Chvatal (Eds.): Topics on perfect graphs, Annals of Discrete Mathematics Vol. 21, 21-28. Amsterdam: North-Holland. (1984).
FOX 1985-1 Fox, G. E.; Scudder, G. D.
A heuristic with tie breaking for certain 0-1 integer programming models. Naval Research Logistics Quarterly 32, 613-624. (1985).
FOX 1986-1 Fox, G. E.; Scudder, G. D.
A simple strategy for solving a class of 0-1 integer programming models. Computers and Operations Research 13, 707-712. (1986).
FRAE 1985-1 Fraenkel, A. S.; Hammer, P. L.
Pseudo-Boolean functions and their graphs. In: M. Rosenfeld, J. Zaks (Eds.): Convexity and Graph Theory. Annals of Discrete Mathematics Vol. 20, 137-146. Amsterdam: North-Holland. (1985).
FRAI 1984-1 Frair, L.
Airport noise modelling and aircraft scheduling so as to minimize community annoyance. Applied Mathematical Modelling 8, 271-280. (1984).
FRAIA 1985-1 Fraisse, P.
A new sufficient condition for Hamiltonian graphs. Rapport de Recherche 241. Orsay: Univ. de Paris-Sud, Lab. de Recherche en Informatique, 4 p. (1985).
FRAIA 1985-2 Fraisse, P.
Cycle covering in bridgeless graphs. Journal of Combinatorial Theory 39B, 146-152. (1985).
FRAIB 1986-1 Fraisse, P.; Hell, P.; Kirkpatrick, D. G.
A note on f-factors in directed and undirected multigraphs. Graphs and Combinatorics 2, 61-66. (1986).
FRAIC 1985-1 Fraisse, P.; Ordaz, O.
Chvatal-Erdos condition for (tau, tau)-factors in digraphs using given arcs. Rapport de Recherche 212. Orsay: Univ. de Paris-Sud, Lab. de Recherche en Informatique, 9 p. (1985).
FRAID 1987-1 Fraisse, P.; Thomassen, C.
Hamiltonian dicycles avoiding prescribed arcs in tournaments. Graphs and Combinatorics 3, 239-250. (1987).
FRAN 1985-1 Francis, N. D.; Fleming, D. I.
Optimum allocation of places to students in a national university system. BIT 25, 307-317. (1985).
FRANA 1984-1 Francis, R. L.
A negative exponential solution to an evacuation problem. Research Report 84-36. Gainesville: Univ. of Florida, Ind. & Systems Eng. Dept., 13 p. (1984).
FRANB 1986-1 Franco, J.
On the probabilistic performance of algorithms for the satisfiability problem. Information Processing Letters 23, 103-106. (1986).
FRANC 1984-1 Frank, A.
Finding feasible vectors of Edmonds-Giles polyhedra. Journal of Combinatorial Theory, Series B 36, 221-239. (1984).
FRANC 1984-2 Frank, A.
Generalized polymatroids. In: A. Hajnal; L. Lovasz; V. T. Sos (Eds.): Finite and Infinite Sets, Vol. 1, 285-294. Coll. Math. Soc. J. Bolyai 37, Amsterdam: North-Holland. (1984).
FRANC 1984-3 Frank, A.
Submodular flows. In: W. R. Pulleyblank (Ed.): Progress in Combinatorial Optimization, 147-166. Toronto: Academic Press. (1984).
FRANC 1985-1 Frank, A.
Edge-disjoint paths in planar graphs. Journal of Combinatorial Theory 39B, 164-178. (1985).
FRANC 1986-1 Frank, A.
On connectivity properties of Eulerian digraphs. Report No. 86419-OR. Bonn: Universität Bonn, Inst. für Ökonometrie und Operations Research, 23p. (1986).
FRANC 1987-1 Frank, A.
Graph connectivity and network flows. Report 87473-OR. Bonn: Univ. Bonn, Inst. f. Ökonometrie u. Operations Research, 72 p. (1987).
FRAND 1984-1 Frank, A.; Sebö, A.; Tardos, E.
Covering directed and odd cuts. Mathematical Programming Study 22, 99-112. (1984).
FRANE 1984-1 Frank, A.; Tardos, E.
An algorithm for the unbounded matroid intersection polyhedron. In: R. E. Burkard, R. A. Cuninghame-Green, U. Zimmermann (Eds.): Algebraic and Combinatorial Methods in Operations Research, Annals of Discrete Mathematics Vol. 19, 129-134. Amsterdam: North-Holland. (1984).
FRANE 1984-2 Frank, A.; Tardos, E.
Matroids from crossing families. In: A. Hajnal; L. Lovasz; V. T. Sos (Eds.): Finite and Infinite Sets, Vol. 1, 295-304. Coll. Math. Soc. J. Bolyai 37, Amsterdam: North-Holland. (1984).
FRANE 1985-1 Frank, A.; Tardos, E.
An application of simultaneous approximation in combinatorial optimization. In: 26th Annual Symp. on Foundations of Computer Science, Portland, Oregon 1985, Proceedings, 459-463. IEEE. (1985).
FRANE 1985-2 Frank, A.; Tardos, E.
Generalized polymatroids and submodular flows. Report 85389-OR. Bonn: Univ. Bonn, Inst. f. Ökonometrie u. Operations Research, 91 p. (1985).
FRANE 1987-1 Frank, A.; Tardos, E.
An application of simultaneous Diophantine approximation in combinatorial optimization. Combinatorica 7, 49-66. (1987).
FRANF 1986-1 Frank, S.; Shier, D. R.
The chromatic polynomial revisited. Congressus Numerantium 55, 57-68. (1986).
FRANG 1983-1 Franke, H.-G.
Analyse der mittleren Rechenzeit für den transitiven Abschluß azyklischer Graphen. Dissertation im Fachbereich Mathematik der Johann Wolfgang Goethe- Universität zu Frankfurt am Main. (1983).
FRAY 1984-1 Fraysseix, H. de
A characterization of circle graphs. European Journal of Combinatorics 5, 223-238. (1984).
FRAYA 1985-1 Fraysseix, H. de; Rosenstiehl, P.
A characterization of planar graphs by Tremaux orders. Combinatorica 5, 127-136. (1985).

FRED 1984-1 Frederickson, G. N.
On linear-time algorithms for five-coloring planar graphs. Information Processing Letters 19, 219-224. (1984).
FRED 1985-1 Frederickson, G. N.
A single source shortest path algorithm for a planar distributed network. In: K. Mehlhorn (Ed.): STACS 85, Second Annual Symposium on Theoretical Aspects of Computer Science (Saarbrücken, January 1985), Lecture Notes in Computer Science Vol. 182, 143-150. Berlin: Springer-Verlag. (1985).
FRED 1985-2 Frederickson, G. N.
Data structures for on-line updating of minimum spanning trees, with applications. SIAM Journal on Computing 14, 781-798. (1985).
FRED 1987-1 Frederickson, G. N.
A new approach to all pairs shortest paths in planar graphs. In: Proc. Nineteenth Annual ACM Symp. on Theory of Computing, New York City 1987, 19-28. ACM. (1987).
FRED 1987-2 Frederickson, G. N.
Fast algorithms for shortest paths in planar graphs, with applications. SIAM Journal on Computing 16, 1004-1022. (1987).
FREDA 1987-1 Frederickson, G. N.; Hambrusch, S. E.
Planar linear arrangements of outerplanar graphs. CC-87-7. West Lafayette: Purdue Univ. Inst. of Interdisciplinary Engineering Studies, 27 p. (1987).
FREDB 1986-1 Frederickson, G. N.; Janardan, R.
Space-efficient message routing in c-decomposable networks. CC-86-7. West Lafayette: Purdue Univ., Inst. for Interdisciplinary Engineering Studies, 25 p. (1986).
FREDB 1987-1 Frederickson, G. N.; Janardan, R.
Designing networks with compact routing tables. CC-86-6. West Lafayette: Purdue Univ. Inst. of Interdisciplinary Engineering Studies, 29 p. (1987).
FREDB 1987-2 Frederickson, G. N.; Janardan, R.
Efficient message routing in planar networks. CC-87-5. West Lafayette: Purdue Univ. Inst. of Interdisciplinary Engineering Studies, 22 p. (1987).
FREDC 1987-1 Frederickson, G. N.; Srinivas, M. A.
Algorithms and data structures for an expanded family of matroid intersection problems. CC-87-26. West Lafayette: Purdue Univ. Inst. of Interdisciplinary Engineering Studies, 50 p. (1987).
FREDC 1987-2 Frederickson, G. N.; Srinivas, M. A.
On-line updating of solutions to a class of matroid intersection problems. Information and Computation 74, No. 2, 113-139. (1987).
FREDD 1984-1 Fredman, M. L.; Tarjan, R. E.
Fibonacci heaps and their uses in improved network optimization algorithms. In: 25th Annual Symposium on Foundations of Computer Science, Florida 1984, 338-346. Los Angeles: IEEE Computer Society. (1984).
FREE 1986-1 Freeman, D. S.
Design of automated assembly systems: issues and procedures. Research Report No. 86-7. Gainesville: Univ. of Florida, Industrial and Systems Engineering Dept., 17p. (1986).
FREEA 1986-1 Freeman, D. S.; Swain, R. W.
A heuristic technique for balancing automated assembly lines. Research Report No. 86-6. Gainesville: Univ. of Florida, Industrial and Systems Engineering Dept., 21p. (1986).
FREN 1986-1 French, S.; Hartley, R.; Thomas, L. C.; White, D. J.
Operational Research Techniques. London: Arnold. (1986).
FRENA 1987-1 Frenk, J. B. G.; Galambos, G.
Hybrid next-fit algorithm for the two-dimensional rectangle bin-packing problem. Computing 39, 201-218. (1987).
FRENB 1985-1 Frenk, J. B. G.; Houweninge, M. van; Rinnooy Kan, A. H. G.
Asymptotic properties of the quadratic assignment problem. Mathematics of Operations Research 10, 100-116. (1985).
FRENB 1986-1 Frenk, J. B. G.; Houweninge, M. van; Rinnooy Kan, A. H. G.
Order statistics and the linear assignment problem. Report 8609/A. Rotterdam: Erasmus Univ. Rotterdam, Econometric Institute, 17 p. (1986).
FRENC 1987-1 Frenk, J. B. G.; Rinnooy Kan, A. H. G.
The asymptotic optimality of the LPT rule. Mathematics of Operations Researchs 12, 241-254. (1987).
FREND 1984-1 Frenk, J. B. G.; Rinnooy Kan, A. H. G.; Stougie, L.
A hierarchical scheduling problem with a well-solvable second stage. Annals of Operations Research 1, 43-58. (1984).
FRENE 1986-1 Frenkel, K. A.
Complexity and parallel processing: an interview with Richard M. Karp. Communications of the ACM 29, 112-117. (1986).
FREV 1986-1 Freville, A.; Plateau, G.
Heuristics and reduction methods for multiple constraints 0-1 linear programming problems. European Journal of Operational Research 24, 206-215. (1986).
FREY 1986-1 Frey, J.-J.; Yehia-Alcoutlabi, A.
Comparaisons par paires: une interpretation et une generalisation de la methode des scores. RAIRO - Recherche Operationelle 20, 213-228. (1986).
FRIC 1985-1 Frick, H.
On the solution of the facility location problem when the dominance method fails. Zeitschrift für Operations Research 29B, 101-106. (1985).
FRICA 1987-1 Fricker, J.; Nafakh, J.
Finding optimal routes in doubly weighted graphs using the multiplane technique. Working Paper. West Lafayette: Purdue Univ., School of Civil Engineering, 19 p. (1987).
FRID 1983-1 Fridman, V. V.
A transportation problem with respect to time for networks of a special type (Russian). In: Numerical Methods and Optimization Problems, 110-114. Tomsk. Gos. Univ. (1983).
FRIE 1987-1 Friesen, D. K.
Tighter bounds for LPT scheduling on uniform processors. SIAM Journal on Computing 16, 554-560. (1987).

FRIEA 1983-1 Friesen, D. K.; Langston, M. A.
Variable sized bin packing. in: T. Basar, D. J. Brown, H. V. Poor (Eds.): Twentyfirst Annual Allerton Conference on Communication, Control, and Computing (Monticello, Ill., 1983), 61-69. Urbana-Champaign: Univ. Illinois. (1983).

FRIEA 1986-1 Friesen, D. K.; Langston, M. A.
Variable sized bin packing. SIAM Journal on Computing 15, 222-230. (1986).

FRIEA 1987-1 Friesen, D. K.; Langston, M. A.
Bin packing: on optimizing the number of pieces packed. BIT 27, 148-156. (1987).

FRIEB 1985-1 Friesz, T. L.
Transportation network equilibrium, design and aggregation: Key developments and research opportunities. Transportation Research 19A, 413-428. (1985).

FRIEC 1984-1 Frieze, A. M.
Algebraic flows. In: R. E. Burkard, R. A. Cuninghame-Green, U. Zimmermann (Eds.): Algebraic and Combinatorial Methods in Operations Research, Annals of Discrete Mathematics Vol. 19, 135-146. Amsterdam: North-Holland. (1984).

FRIEC 1984-2 Frieze, A. M.
Limit distribution for the existence of Hamiltonian cycles in random bibartite graphs. Research Paper. London: Queen Mary College, 16 p. (1984).

FRIEC 1986-1 Frieze, A. M.
Maximum matchings in a class of random graphs. Journal of Combinatorial Theory, Series B, 40, 196-212. (1986).

FRIEC 1986-2 Frieze, A. M.
On large matchings and cycles in sparse random graphs. Discrete Mathematics 59, 243-256. (1986).

FRIEC 1987-1 Frieze, A. M.
On the exact solution of random travelling salesman problems with medium size integer coefficients. SIAM Journal on Computing 16, 1052-1072. (1987).

FRIEC 1987-2 Frieze, A. M.
Parallel algorithms for finding hamilton cycles in random graphs. Information Processing Letters 25, 111-118. (1987).

FRIED 1986-1 Frieze, A. M.; Baybars, I.
Expected behaviour of line-balancing heuristics. IMA Journal of Mathematics in Management 1, 1-11. (1986).

FRIEE 1985-1 Frieze, A. M.; Grimmett, G. R.
The shortest-path problem for graphs with random arc-lengths. Discrete Applied Mathematics 10, 57-78. (1985).

FRIS 1983-1 Fris, P.; Volgenant, A.
Solving the traveling salesman problem in Basic. Kwantitatieve Methoden 10/4, 139-158. (1983).

FROL 1987-1 Froleyks, B.; Korte, B.; Prömel, H. J.
Routing in VLSI-layout. Report 87494-OR. Bonn: Univ. Bonn, Forschungsinst. f. Diskrete Mathematik und Inst. f. Ökonometrie und Operations Research, 20 p. (1987).

FROS 1985-1 Frostig, E.; Adiri, I.
Stochastic flowshop no-wait scheduling. Journal of Applied Probability 22, 240-246. (1985).

FROS 1985-2 Frostig, E.; Adiri, I.
Three-machine flowshop stochastic scheduling to minimize distribution of schedule length. Naval Research Logistics Quarterly 32, 165-178. (1985).

FRUC 1985-1 Frucht, R. W.
A new method of computing chromatic polynomials of graphs. In: Chuaqui, R. (Ed.): Analysis, geometry, and probability, First Chilean Symp. on Mathematics, Valparaiso 1981, Lecture Notes in Pure and Applied Mathematics 96, 69-77. Basel: Marcel Dekker. (1985).

FRUCA 1985-1 Frucht, R. W.; Salinas, L. C.
Graceful numbering of snakes with constraints on the first label. Ars Combinatoria 20-B, 143-158. (1985).

FRUH 1987-1 Fruhwirth, B.
Untersuchung über genetisch motivierte Heuristiken für das Euklidische Rundreiseproblem. Techn. Bericht 87-103. Graz: TU Graz, Inst. f. Mathematik, 31 p. (1987).

FRY 1987-1 Fry, T. D.; Armstrong, R. D.; Blackstone, J. H.
Minimizing weighted absolute deviation in single machine scheduling. IIE Transactions 19, 445-450. (1987).

FRY A 1987-1 Fry, T. D.; Leong, G. K.
A bi-criterion approach to minimizing inventory costs on a single machine when early shipments are forbidden. Computers and Operations Research 14, 363-368. (1987).

FU 1986-1 Fu, Y.-T.; Anderson, P. W.
Application of statistical mechanics to NP-complete problems in combinatorial optimization. Journal of Physics A 19, 1605-1620. (1986).

FUCH 1985-1 Fuchs, A.
1-N Verfahren zur Bestimmung von k-kürzesten Wegen. In: D. Ohse, A. C. Esprester, H.-U. Küpper, P. Stähly, H.Steckhan (Eds.): Operations Research Proceedings 1984, 421-428. Berlin: Springer-Verlag. (1985).

FUCHA 1985-1 Fuchs, K.; Kafura, D.
Memory-constrained task scheduling on a network of dual processors. Journal of the ACM 32, 102-129. (1985).

FUJI 1984-1 Fujii, T.; Wakabayashi, S.; Kikuno, T.
Partition of graph into forest in LSI layout design (Japanese). In: Basic Theory of Computer Science, Proceedings of a Symposium held at the Research Institute for Mathematical Sciences (Kyoto), 105-119. (1984).

FUJIA 1984-1 Fujishige, S.
A system of linear inequalities with a submodular function on $\{0,+-1\}$vectors. Linear Algebra and its Applications 63, 253-266. (1984).

FUJIA 1984-2 Fujishige, S.
Combinatorial optimization problems described by submodular functions. In: J. P. Brans (Ed.): Operational Research '84, 379-392. Amsterdam: North-Holland. (1984).

FUJIA 1984-3 Fujishige, S.
Submodular systems and related topics. Mathematical Programming Study 22, 113-131. (1984).

FUJIA 1985-1 Fujishige, S.
A decomposition of distributive lattices. Discrete Mathematics 55, 35-56. (1985).
FUJIA 1986-1 Fujishige, S.
A capacity-rounding algorithm for the minimum-cost circulation problem: a dual framework of the Tardos algorithm. Mathematical Programming 35, 298-308. (1986).
FUJIA 1987-1 Fujishige, S.
An out-of-kilter method for submodular flows. Discrete Applied Mathematics 17, 3-16. (1987).
FUJIA 1987-2 Fujishige, S.
Optimization over the polyhedron determined by a submodular function on a co-intersection family. Discussion Paper, 330. Sakura: Univ. of Tsukuba, Institute of Socio-Economic Planning, 19 p. (1987).
FUJIB 1986-1 Fujishige, S.; Katoh, N.; Ichimori, T.
The fair resource allocation problem with submodular constraints and its solution algorithm. Discussion Paper Series, 291. Ibaraki: Univ. of Tsukuba, Inst. of Socio-Economic Planning, 15 p. (1986).
FUJIC 1986-1 Fujishige, S.; Nakayama, A.; Cui, W.-T.
On the equivalence of the maximum balanced flow problem and the weighted minimax flow problem. Operations Research Letters 5, 206-210. (1986).
FUJID 1986-1 Fujishige, S.; Röck, H.; Zimmermann, U.
A strongly polynomial algorithm for minimum cost submodular flow problems. Preprint 102. Kaiserslautern: Univ. Kaiserslautern, Fachbereich Mathematik, 14 p. (1986).
FUKU 1987-1 Fukuda, K.; Tamura, A.
Local deformation and orientation transformation in oriented matroids. Research Report on Information Science. Tokyo: Tokyo Inst. of Technology, Dept. of Information Sciences, 16 p. (1987).
FUKUA 1985-1 Fukui, M.; Kamae, T.
Information polynomials of graphs. In: J. Akiyama; Y. Ito; P. Kanemitsu; T. Kano; T. Mitsui; I. Shiokawa (Eds.): Number theory and combinatorics, Japan, Proc. Conf. Tokyo, Okayama, Kyoto 1984, 97-104. Singapore: World Scientific Publ. Co. (1985).
FUKUB 1987-1 Fukumori, K.; Sano, H.; Hasegawa, T.; Sakai, T.
Fundamental algorithm for train scheduling based on artificial intelligence. Systems Comput. Japan 18, No. 3, 52-64. (1987).
FULO 1987-1 Fülöp, J.
On the facial disjunctive programming problem (Hungarian). Alkalmazott Matematikai Lapok 13, 9-15. (1987).
FUNG 1985-1 Fung, K. T.
An entropy approach to solving some network reliability problems. Computer Journal 28, 353-356. (1985).
FURU 1984-1 Furukawa, M.
The 1-3-dimensional knapsack problem in production planning: formulation and heuristic methods (Japanese). Journal of the Asahikawa Technical College (Asahikawa) 21, 23-32. (1984).
GABB 1986-1 Gabbani, D.; Magazine, M.
An interactive heuristic approach for multiobjective integer-programming problems. Journal of the Operational Research Society 37, 285-292. (1986).
GABO 1985-1 Gabor, C. P.; Hsu, W.-L.; Supowit, K. J.
Recognizing circle graphs in polynomial time. In: 26th Annual Symp. on Foundations of Computer Science, Portland, Oregon 1985, Proceedings, 106-116. IEEE. (1985).
GABOA 1985-1 Gabow, H. N.
A scaling algorithm for weighted matching on general graphs. In: 26th Annual Symp. on Foundations of Computer Science, Portland, Oregon 1985, Proceedings, 90-100. IEEE. (1985).
GABOA 1985-2 Gabow, H. N.
Scaling algorithms for network problems. Journal of Computer and System Sciences 31, 148-168. (1985).
GABOB 1984-1 Gabow, H. N.; Galil, Z.; Spencer, T.
Efficient implementation of graph algorithms using contraction. In: 25th Annual Symposium on Foundations of Computer Science, Florida 1984, 347-357. Los Angeles: IEEE Computer Society. (1984).
GABOC 1986-1 Gabow, H. N.; Galil, Z.; Spencer, T.; Tarjan, R. E.
Efficient algorithms for finding minimum spanning trees in undirected and directed graphs. Combinatorica 6, 109-122. (1986).
GABOD 1985-1 Gabow, H. N.; Stallmann, M.
Efficient algorithms for graphic matroid intersection and parity. In: W. Brauer (Ed.): Automata, Languages and Programming, 210-220. Lecture Notes in Computer Science 194, Berlin: Springer. (1985).
GABOD 1986-1 Gabow, H. N.; Stallmann, M.
An augmenting path algorithm for linear matroid parity. Combinatorica 6, 123-150. (1986).
GABOE 1984-1 Gabow, H. N.; Tarjan, R. E.
Efficient algorithms for a family of matroid intersection problems. Journal of Algorithms 5, 80-131. (1984).
GABOE 1985-1 Gabow, H. N.; Tarjan, R. E.
A linear-time algorithm for a special case of disjoint set union. Journal of Computer and System Sciences 30, 209-221. (1985).
GAFN 1987-1 Gafni, E. M.; Bertsekas, D. P.
Asymptotic optimality of shortest path routing algorithms. IEEE Transactions on Information Theory 33, 83-90. (1987).
GAIM 1987-1 Gaimon, C.; Mohan, S.; Thompson, G. L.
A network approach to cohort personnel planning using cross-sectional data. Computers and Operations Research 14, 183-196. (1987).
GAIN 1986-1 Gainanov, D. N.
A graph-theoretic algorithm for the construction of the committee for inconsistent system of linear inequalities. Zhurnal Vychislitelnoj Matematiki i Matematicheskoj Fiziki (Moscow) 26, 1431-1432. (1986).

GAIN 1987-1 Gainanov, D. N.
On the connectedness of the graphs of some classes of independence systems (Russian). In: Y. A. Shashkin, E. G. Pytkeev (Eds.): Investigations in the Theory of Convex Sets and Graphs, 16-23. Collection of Scientific Works, Sverdlovsk: Akad. Nauk SSSR, Ural. Nauchn. Tsentr. (1987).

GAINA 1987-1 Gainanov, D. N.; Gusak, I. Y.
On the number of bases of a vector matroid that is generated by a positive basis (Russian). In: Y. A. Shashkin, E. G. Pytkeev (Eds.): Investigations in the Theory of Convex Sets and Graphs, 24-32. Collection of Scientific Works, Sverdlovsk: Akad. Nauk SSSR, Ural. Nauchn. Tsentr. (1987).

GALA 1985-1 Galambos, G.
A new heuristic for the classical bin-packing problem. Preprint 82. Augsburg: Univ. Augsburg, Inst. f. Mathematik, 8 p. (1985).

GALA 1985-2 Galambos, G.
Algorithms for different bin packing problems. In: Graphen und Netzwerke - Theorie und Anwendungen, 30. Int. Wiss. Kolloq., Ilmenau 1985, 5, 33-36. (1985).

GALA 1986-1 Galambos, G.
Parametric lower bound for on-line bin-packing. SIAM Journal on Algebraic and Discrete Methods 7, 362-367. (1986).

GALAA 1985-1 Galambos, G.; Csirik, J.
Next-fit type algorithms for solving different bin-packing problems. Preprint 81. Augsburg: Univ. Augsburg, Inst. f. Mathematik, 13 p. (1985).

GALAB 1984-1 Galambos, G.; Imreh, B.
Solution of one-dimensional cutting stock problems by column-generation (Hungarian). Alkalmazott Matematikai Lapok (Budapest) 10, No. 1-2, 73-85. (1984).

GALE 1985-1 Gale, D.; Sotomayor, M.
Some remarks on the stable matching problem. Discrete Applied Mathematics 11, 223-232. (1985).

GALEA 1986-1 Galeana-Sanchez, H.
A theorem about a conjecture of H. Meyniel on kernel-perfect graphs. Discrete Mathematics 59, 35-42. (1986).

GALEB 1986-1 Galeana-Sanchez, H.; Neumann-Lara, V.
On kernel-perfect critical digraphs. Discrete Mathematics 59, 257-266. (1986).

GALI 1986-1 Galil, Z.
Efficient algorithms for finding maximum matching in graphs. Computing Surveys (ACM) 18, 23-38. (1986).

GALI 1986-2 Galil, Z.
Sequential and parallel algorithms for finding maximum matchings in graphs. In: Annual Review of Computer Science Vol. 1, 197-224. Palo Alto, CA: Annual Reviews. (1986).

GALIA 1987-1 Galil, Z.; Hoffmann, C. M.; Luks, E. M.; Schnorr, C. P.; Weber, A.
An O(n**3 log n) deterministic and an O(n**3) Las Vegas isomorphism test for trivalent graphs. Journal of the Association for Computing Machinery 34, 513-531. (1987).

GALIB 1982-1 Galil, Z.; Micali, S.; Gabow, H.
Priority queues with variable priority and an O(EVlogV) algorithm for finding a maximal weighted matching in general graphs. In: 23rd Annual Symp. on Foundations of Computer Science, Chicago, Illinois 1982, Proceedings, 255-261. IEEE. (1982).

GALIB 1986-1 Galil, Z.; Micali, S.; Gabow, H.
An 0(EVlogV) algorithm for finding a maximal weighted matching in general graphs. SIAM Journal on Computing 15, 120-130. (1986).

GALIC 1986-1 Galil, Z.; Tardos, E.
An O(n**2(m+nlogn)logn) min-cost flow algorithm. In: 27th Annual Symp. on Foundations of Computer Science, Proceedings, 1-9. IEEE. (1986).

GALL 1984-1 Gallo, G.; Pallottino, S.
Shortest path methods in transportation models. In: M. Florian (Ed.): Transportation Planning Models, 227-256. Amsterdam: North-Holland. (1984).

GALL 1986-1 Gallo, G.; Pallottino, S.
Shortest path methods: A unifying approach. In: G. Gallo, C. Sandi (Eds.): Netflow at Pisa, Mathematical Programming Study 26, 38-64. (1986).

GALLA 1987-1 Gallo, G.; Simeone, B.
On the supermodular knapsack problem. RRR 10-87. New Brunswick: Rutgers Univ., Center for Operations Research, 13 p. (1987).

GALV 1985-1 Galvin, B.
Machine scheduling models with application to the loading of bulk carriers in tidal ports. Bulletin of the Institute of Mathematics and its Applications (Essex) 21, 21-25. (1985).

GAMB 1987-1 Gambarov, L. A.
On a method of decomposition for a multiproduct transportation problem with intermediate nodes (Russian). Ehkonomika i Matematicheskie Metody 23, 165-168. (1987).

GAMBA 1986-1 Gamble, A. B.; Conn, A. R.; Pulleyblank, W. R.
A network penalty method. Report 86445-OR. Bonn: Univ. Bonn, Inst. f. Ökonometrie u. Operations Research, 28 p. (1986).

GAMBB 1987-1 Gamble, A. B.; Pulleyblank, W. R.
Forest covers and a polyhedral intersection theorem. Report 87467-OR. Bonn: Univ. Bonn, Inst. f. Ökonometrie u. Operations Research, 11 p. (1987).

GANI 1987-1 Gani, A. Z.
The application of the interaction theory for solving the traveling salesman problem. Working Paper. Bandung: Institut Teknologi Bandung, Dept. of Industrial Engineering, 25 p. (1987).

GANS 1987-1 Gansner, E. R.; Vo, K.-P.
The chromatic generating function. Linear Mulitlinear Algebra 22, No.1, 87-93. (1987).

GAO 1987-1 Gao, J.-Z.
Quasi-1-Hamilton connectedness of tree graphs (Chinese). Journal of Mathematical Research & Exposition 7, 498-504. (1987).

GAO A 1985-1 Gao, S.-D.
An algorithm for two-layer channel routing. In: K. Mehlhorn (Ed.): STACS 85, Second Annual Symposium on Theoretical Aspects of Computer Science (Saarbrücken, January 1985), Lecture Notes in Computer Science Vol. 182, 151-160. Berlin: Springer-Verlag. (1985).

GAO B 1986-1 Gao, S.-D.; Hambrusch, S. E.
Two-layer channel routing with vertical unit-length overlap. Algorithmica 1, 223-232. (1986).
GARA 1987-1 Garashchuk, M. M.
On the testing of paths in schemes of functional elements (Russian). Matematicheskie Zametki (Moscow) 41, 854-861. (1987).
GARC 1987-1 Garcia-Aviles, A.
Writing cut-set and loop equations by inspection. In: M. Ismail (Ed.): 29th Midwest Symposium on Circuits and Systems, 448-451. Amsterdam: North-Holland. (1987).
GARCA 1985-1 Garcia, R.
An optimization method for multiplexer locations. OR Spektrum 7, 111-114. (1985).
GARD 1985-1 Gardner, M. L.
An algorithm to aid in the design of large scale networks. Large Scale Systems (Amsterdam) 8, 147-156. (1985).
GARF 1985-1 Garfinkel, R. S.
Motivation and modeling. In: E. L. Lawler, J. K. Lenstra, A. H. G. Rinnooy Kan, D. B. Shmoys (Eds.): The Traveling Salesman Problem, 17-36. New York: Wiley. (1985).
GARFA 1984-1 Garfinkel, R. S.; Sundararaghavan, P. S.
Hamiltonian cycles in striped graphs: the two-stripe problem. SIAM Journal on Algebraic and Discrete Methods 5, 463-466. (1984).
GARG 1985-1 Garg, R.; Prakash, S.
Time minimizing transshipment problem. Indian Journal of Pure and Applied Mathematics (New Delhi) 16, 449-460. (1985).
GARGA 1985-1 Gargano, M.; Quintas, L. V.
A digraph generalization of balanced signed graphs. Congressus Numerantium 48, 133-144. (1985).
GARR 1986-1 Garrett, S. E.
A microcomputer implementation of a facility location model. Methods of Operations Research 54, 507-518. (1986).
GAST 1986-1 Gastou, G.; Johnson, E. L.
Binary group and Chinese postman polyhedra. Mathematical Programming 34, 1-33. (1986).
GAUL 1984-1 Gaul, W.
Optimal routes in compound transportation systems. In: H. Hauptmann, W. Krelle, K. C. Mosler (Eds.): Operations Research and Economic Theory, 309-320. Berlin: Springer-Verlag. (1984).
GAVI 1985-1 Gavish, B.; Pirkul, H.
Efficient algorithms for solving multiconstraint zero-one knapsack problems to optimality. Mathematical Programming 31, 78-105. (1985).
GAVI 1985-2 Gavish, B.; Pirkul, H.
Zero-one integer programs with few constraints - lower bounding theory. European Journal of Operational Research 21, 213-224. (1985).
GAVI 1985-3 Gavish, B.; Pirkul, H.
Zero-one integer programs with few constraints - efficient branch and bound algorithms. European Journal of Operational Research 22, 35-43. (1985).
GAVIA 1986-1 Gavish, B.; Srikanth, K.
An optimal solution method for large-scale multiple traveling salesmen problems. Operations Research 34, 698-717. (1986).
GAVR 1987-1 Gavril, F.
Generating the maximum spanning trees of a weighted graph. Journal of Algorithms 8, 592-597. (1987).
GAZD 1983-1 Gazdik, I.
Fuzzy-network planning - FNET. IEEE Transactions on Reliability R-32, 304-313. (1983).
GAZI 1986-1 Gazit, H.
An optimal randomized parallel algorithm for finding connected components in a graph. In: 27th Annual Symp. on Foundations of Computer Science, Proceedings, 492-501. IEEE. (1986).
GAZIA 1987-1 Gazit, H.; Miller, G. L.
A parallel algorithm for finding a separator in planar graphs. In: 28th Annual IEEE Symposium on Foundations of Computer Science, 238-248. (1987).
GAZM 1985-1 Gazmuri, P. G.
Probabilistic analysis of a machine scheduling problem. Mathematics of Operations Research 10, 328-339. (1985).
GAZM 1986-1 Gazmuri, P. G.
Probabilistic analysis of algorithms for some combinatorial optimization problems. Recent advances in system modelling and optimization, Proc. IFIP-WG 7/1 Work. Conf., Santiago/ Chile 1984, Lect. Notes Control Inf. Sci. 87, 113-126. (1986).
GDAN 1982-1 Gdanskij, N. I.
Investigation of some kinds of edge colroings of graphs (Russian). In: V. V. Fedorov, Yu. I. Mokin (Eds.): Some Questions of Applied Mathematics and Software of Computers, 44-48. Moscow: Izdat. Moskovs. Univ. (1982).
GE 1984-1 Ge, X. K.; Zhou, G. R.; Jiang, J. R.; Guo, R.
Heuristic algorithms for vehicle routing of a fleet (Chinese). Chinese Journal of Operations Research 3 No. 2, 53-58. (1984).
GEHR 1987-1 Gehrlein, W. V.; Fishburn, P. C.
A comparison of algorithms for minimizing bumps in linear extensions of partial orders. Operations Research Letters 6, 115-120. (1987).
GELD 1987-1 Gelders, L. F.; Pintelon, L.; Wassenhove, L. N. van
A location-allocation problem in a large Belgian brewery. European Journal of Operational Research 28, 196-206. (1987).
GELF 1987-1 Gelfand, I. M.; Goresky, R. M.; Mac Pherson, R. D.; Serganova, V. V.
Combinatorial geometries, convex polyhedra, and Schubert cells. Advances in Mathematics 63, 301-316. (1987).
GELFA 1987-1 Gelfand, I. M.; Serganova, V. V.
On the general definition of a matroid and a greedoid. Soviet Mathematics, Doklady 35, 6-10; translation from Doklady Akademii Nauk SSSR 292 (1987), 15-20. (1987).

GELFB 1987-1 Gelfand, S. B.; Mitter, S. K.
Simulated annealing. In: G. Andreatta, F. Mason, P. Serafini (Eds.): Advanced School on Stochastics in Combinatorial Optimization (CISM Udine 1986), 1-51. Singapore: World Scientific. (1987).

GEOR 1987-1 Georgakopoulos, G.; Papadimitriou, C. H.
The 1-Steiner tree problem. Journal of Algorithms 8, 122-130. (1987).

GERA 1985-1 Gerards, A. M. H.
Testing the odd bicycle wheel inequalities for the bipartite subgraph polytope. Mathematics of Operations Research 10, 359-360. (1985).

GERAA 1987-1 Gerards, A. M. H.; Kolen, A. W. J.
Polyhedral combinatorics in combinatorial optimization. Statistica Neerlandica 41, 1-26. (1987).

GERAB 1986-1 Gerards, A. M. H.; Schrijver, A.
Matrices with the Edmonds-Johnson property. Combinatorica 6, 365-380. (1986).

GERAB 1986-2 Gerards, A. M. H.; Schrijver, A.
Signed graphs - regular matroids - grafts. Research Memorandum 237, Dept. of Econ., Cath. Univ. Brabant. (1986).

GERAC 1987-1 Gerards, A. M. H.; Sebö, A.
Total dual integrality implies local strong unimodularity. Mathematical Programming 38, 69-74. (1987).

GERAD 1983-1 Gerasimov, V. A.
On a modification of the traveling salesman problem (Russian). Avtomatika i Telemekhanika (Moscow) 1983, No. 9, 161-164. (1983).

GERN 1984-1 Gernert, D.
Recent results on chromatic polynomials. In: G. Hammer, D. Pallaschke (Eds.): Methods of Operations Research 51, Vol. 1, 307-314. Königstein/Ts.: Verlag Anton Hain Meisenheim. (1984).

GERN 1985-1 Gernert, D.
A survey of partial proofs for Read's conjecture and some recent results. In: P. Brucker, R. Pauly (Eds.): Proc. IX. Symp. on Operations Research, Osnabrück 1984. Methods of Operations Research 49, 233-238. (1985).

GERN 1986-1 Gernert, D.
Equations and inequalities related to the diameter of a graph. Methods of Operations Research 53, 257-266. (1986).

GERN 1987-1 Gernert, D.
The cycle-length sequence of a graph and some of its applications. In: W. Domschke, W. Krabs, J. Lehn, P. Spellucci (Eds.): Proc. XI. Symp. on Operations Research, Darmstadt 1986, 121-130. Methods of Operations Research 57, Frankfurt/Main: Athenäum. (1987).

GERS 1985-1 Gershwin, S. B.; Akella, R.; Choong, Y. F.
Short-term production scheduling of an automated manufacturing facility. IBM Journal of Research and Development 29, 392-400. (1985).

GERSA 1984-1 Gershwin, S. B.; Akella, R.; Choong, Y. F.; Mitter, S. K.
Short term production scheduling of an automated manufacturing facility. In: A. Bensoussan, J. L. Lions (Eds.): Proc. Sixth Internat. Conf. on Analysis and Optimization of Systems, Nice 1984, Part 2, 615-632. Lecture Notes in Control and Inf. Sciences 63, Berlin: Springer-Verlag. (1984).

GERSB 1984-1 Gershwin, S. B.; Orlicki, D. M.; Platzman, L. K.
An equilibrium model for traffic assignment in urban transportation networks. Large Scale Systems 7, 115-121. (1984).

GERT 1986-1 Gerth, C.
Duality properties of the vector-valued location problem (German). Seminarber., Humboldt-Univ. Berlin, Sekt. Math. 85, 25-38. (1986).

GEY 1985-1 Gey, H.; Lehne, D.
Netzplantechnik: Grundlagen, Methoden, praxisnahe Einführung, Anwendungsbeispiele, Netzplanprogramm in MBASIC. Gensingen: Luther. (1985).

GHAN 1985-1 Ghandforoush, P.; Austin, L. M.
Solution approaches for highly primal- and dual-degenerate all-integer programming problems. Computers & Operations Research 12, 91-96. (1985).

GHANA 1986-1 Ghandforoush, P.; Greber, B. J.
Solving allocation and scheduling problems inherent in forest resource management using mixed-integer programming. Computers and Operations Research 13, 551-562. (1986).

GHEY 1984-1 Gheysens, F.; Golden, B. L.; Assad, A. A.
A comparison of techniques for solving the fleet size and mix vehicle routing problem. OR Spektrum 6, 207-216. (1984).

GHEY 1986-1 Gheysens, F.; Golden, B. L.; Assad, A. A.
A new heuristic for determining fleet size and composition. In: G. Gallo, C. Sandi (Eds.): Netflow at Pisa, Mathematical Programming Study 26, 233-236. (1986).

GHIU 1985-1 Ghius, V.
The types of flows in fan-shaped graphs (French). Buletinul Institutului Politehnic din Iasi, Sectia I, suppl., 219-223. (1985).

GHOS 1984-1 Ghosh, R. K.; Bhattacharjee, G. P.
Parallel breadth-first search algorithms for trees and graphs. International Journal of Computer Mathematics 15, 255-268. (1984).

GHOS 1986-1 Ghosh, R. K.; Bhattacharjee, G. P.
Parallel algorithm for shortest paths. IEE Proceedings E: Computers and Digital Techniques 133, No. 2, 87-93. (1986).

GIBB 1985-1 Gibbons, A.
Algorithmic graph theory. Cambridge: Cambridge Univ. Press. (1985).

GIFF 1984-1 Giffin, J. W.; Foulds, L. R.
Efficient graph planarity updating tests. Ars Combinatoria 17A, 185-202. (1984).

GIFF 1987-1 Giffin, J. W.; Foulds, L. R.
Facilities layout generalized model solved by n-boundary shortest path heuristics. European Journal of Operational Research 28, 382-391. (1987).

GILB 1984-1 Gilbert, J. R.; Hutchinson, J. P.; Tarjan, R. E.
A separator theorem for graphs of bounded genus. Journal of Algorithms 5, 391-407. (1984).

GILBA 1987-1 Gilbert, K. C.; Hofstra, R. B.
An algorithm for a class of three-dimensional assignment problems arising in scheduling applications. IIE Transactions 19, 29-33. (1987).
GILBB 1985-1 Gilbert, K. C.; Holmes, D. D.; Rosenthal, R. E.
A multiobjective discrete optimization model for land allocation. Management Science 31, 1509-1522. (1985).
GILE 1984-1 Giles, R.; Trotter, L. E. Jr.; Tucker, A.
The strong perfect graph theorem for a class of partitionable graphs. In: C. Berge, V. Chvatal (Eds.): Topics on perfect graphs, Annals of Discrete Mathematics Vol. 21, 161-168. Amsterdam: North-Holland. (1984).
GILM 1985-1 Gilmore, P. C.; Lawler, E. L.; Shmoys, D. B.
Well-solved special cases. In: E. L. Lawler, J. K. Lenstra, A. H. G. Rinnooy Kan, D. B. Shmoys (Eds.): The Traveling Salesman Problem, 87-144. New York: Wiley. (1985).
GIMA 1983-1 Gimadi, E. K.
An effective algorithm for the solution of the location problem with service domains connected with respect to an acyclic network (Russian). Upravlyaemye Sistemy (Novosibirsk) 23, 12-23. (1983).
GIMAA 1984-1 Gimadi, E. K.; Zalyubovskij, V. V.
An asymptotically exact approach to the solution of one-dimensional problem of packing in containers (Russian). Upravylaemye Sistemy 25, 48-57. (1984).
GIMB 1986-1 Gimbel, J.
Three extremal problems in cochromatic theory. Rostocker Mathematisches Kolloquium 30, 73-78. (1986).
GIMBA 1987-1 Gimbel, J.; Straight, H. J.
Some topics in cochromatic theory. Graphs and Combinatorics 3, 255-266. (1987).
GION 1987-1 Gionfriddo, M.
A short survey on some generalized colourings of graphs. Ars Combinatoria 24B, 155-164. (1987).
GIONA 1985-1 Gionfriddo, M.; Vanstone, S. A.
On L(2)-colourings of a graph. Journal of Information & Optimization Sciences (Delhi) 6, 243-246. (1985).
GIRA 1985-1 Girard, A.
Nonhierarchical sequential routing: An approximate solution to the optimal routing problem. Large Scale Systems 8, 257-265. (1985).
GIRAA 1984-1 Girard, A.; Cote, Y.
Sequential routing optimization for circuit switched networks. IEEE Transactions on Communications COM-32, 1234-1242. (1984).
GIRL 1985-1 Girlich, E.; Kovalev, M. M.
Klassifizierung von Polymatroiden. In: E. Girlich (Ed.): Diskrete Optimierung, 6-21. Jena: Friedrich-Schiller-Universität. (1985).
GIRLA 1985-1 Girlich, E.; Schlechtweg, M.
Struktur von Polymatroiden mit submodularen Funktionen des Problems der optimalen Standardisierung. In: E. Girlich (Ed.): Diskrete Optimierung, 22-30. Jena: Friedrich-Schiller-Universität. (1985).
GIUD 1985-1 Giudici, R. E.
Some new families of chromatically unique graphs. In: Chuaqui, R. (Ed.): Analysis, geometry, and probability, First Chilean Symp. on Mathematics, Valparaiso 1981, Lecture Notes in Pure and Applied Mathematics 96, 147-158. Basel: Marcel Dekker. (1985).
GIUDA 1986-1 Giudici, R. E.; Lopez, M. A.; Salzberg, P. M.
Chromatic uniqueness for some bipartite graphs (Spanish). Acta Cientifica Venezolana 37, 484-494. (1986).
GLAN 1984-1 Glanert, W.
A timetable approach to the assignment of pilots to rotations. Working Paper. Strasbourg (XXIV. Annual AGIFORS Symposium, September 1984) 22 p. (1984).
GLAS 1986-1 Glassey, C. R.; Mizrach, M.
A decision support system for assigning classes to rooms. Interfaces 16, 92-100. (1986).
GLAZ 1985-1 Glazebrook, K. D.
Semi-Markov models for single-machine stochastic scheduling problems. International Journal of Systems Science 16, 573-587. (1985).
GLAZ 1987-1 Glazebrook, K. D.
Evaluating the effects of machine breakdowns in stochastic scheduling problems. Naval Research Logistics 34, 319-336. (1987).
GLAZ 1987-2 Glazebrook, K. D.
Sensitivity analysis for stochastic scheduling problems. Mathematics of Operations Research 12, 205-223. (1987).
GLAZA 1987-1 Glazebrook, K. D.; Fay, N. A.
On the scheduling of alternative stochastic jobs on a single machine. Advances in Applied Probability 19, 955-973. (1987).
GLEB 1983-1 Glebov, N. I.
Minimization of a convex separable function on an intersection of polymatroids (Russian). Upravlyaemye Sistemy (Novosibirsk) No. 23, 33-43. (1983).
GLEB 1983-2 Glebov, N. I.
On the problem of minimizing a convex separable function on the intersection of polymatroids. Upravlyaemye Sistemy (Novosibirsk) 23, 33-43. (1983).
GLIC 1985-1 Glickman, T. S.; Sherali, H. D.
Large-scale network distribution of pooled empty freight cars over time, with limited substitution and equitable benefits. Transportation Research 19B, 85-94. (1985).
GLOV 1986-1 Glover, F.
Future paths for integer programming and links to artificial intelligence. Computers and Operations Research 13, 533-550. (1986).
GLOV 1987-1 Glover, F.
Tabu search. Working Paper. Boulder: Univ. of Colorado, Graduate School of Business, 43 p. (1987).

GLOVA 1986-1 Glover, F.; Glover, R.; Klingman, D.
Threshold assignment algorithm. In: G. Gallo, C. Sandi (Eds.): Netflow at Pisa, Mathematical Programming Study 26, 12-37. (1986).

GLOVB 1987-1 Glover, F.; Harary, F.
A simple criterion for a graph to have a perfect matching. Journal of Information & Optimization Sciences (Delhi) 8, 271-273. (1987).

GLOVC 1981-1 Glover, F.; Klingman, D.
The simplex SON algorithm for LP/embedded network problems. Mathematical Programming Study 15, 148-176. (1981).

GLOVC 1985-1 Glover, F.; Klingman, D.
Basis exchange characterization for the simplex SON algorithm for LP/embedded networks. Mathematical Programming Study 24, 141-157. (1985).

GLOVC 1985-2 Glover, F.; Klingman, D.
Layering strategies for creating exploitable structure in linear and integer programs. CBDA 119. Austin: Univ. of Texas, Center for Business Decision Analysis, 15 p. (1985).

GLOVD 1985-1 Glover, F.; Klingman, D.; Phillips, N.
A new polynomially bounded shortest path algorithm. Operations Research 33, 65-73. (1985).

GLOVE 1986-1 Glover, F.; Klingman, D.; Phillips, N.; Ross, G. T.
Integrating modeling, algorithm design, and computational implementation to solve a large-scale nonlinear mixed integer programming problem. In: C. L. Monma (Ed.): Algorithms and Software for Optimization - Part II, 395-412. Annals of Operations Research 5, Basel: Baltzer. (1986).

GLOVF 1985-1 Glover, F.; Klingman, D.; Phillips, N. V.; Schneider, R. F.
New polynomial shortest path algorithms and their computational attributes. Management Science 31, 1106-1128. (1985).

GLOVG 1986-1 Glover, F.; Mac Millan, C.
The general employee scheduling problem: an integration of MS and AI. Computers and Operations Research 13, 563-574. (1986).

GLOVH 1986-1 Glover, F.; Novick, B.
The 2-quasi-greedy algorithm for cardinality constrained matroid bases. Discrete Applied Mathematics 13, 277-286. (1986).

GLUS 1985-1 Glushkova, V. V.
An algorithm for solving the traveling salesman problem with moving objects (Russian). In: Discrete optimization methods and efficient organization of program packages, 35-39. Kiev: Akademiya Nauk Ukrainskoi SSR, Institut Kibernetika. (1985).

GLYA 1983-1 Glyakov, P. V.
Algorithm for a partitioning problem and its investigation on an electronic computer (Russian). Vestnik Belorusskogo Gosudarstvennogo Universiteta Imeni V.I. Lenina, Seriya I, No.1, 40-44. (1983).

GLYAA 1987-1 Glyakov, P. V.; Kovalev, M. M.; Kotov, V. M.
A series of gradient algorithms for a problem of the partitioning of a tree (Russian). Vestnik Belorusskogo Gosudarstvennogo Universiteta im. V.I. Lenina (Minsk), Seriya I, No. 1, 36-39. (1987).

GNDO 1984-1 Gndoyan, A. K.; Igitkhanyan, S. A.
Plane arrangement of vertices of a tree on a line with minimization of the length of the largest edge (Russian). Applied Mathematics (Erevan) No. 3, 87-92. (1984).

GODD 1986-1 Goddard, W.; Winter, P. A.
A characterization of stable graphs on a maximum (minimum) number of edges. Quaestiones Mathematicae 10, No. 2, 175-178. (1986).

GODDA 1985-1 Goddyn, L.
A girth requirement for the double cycle cover conjecture. In: B. R. Alspach, C. D. Godsil (Eds.): Cycles in Graphs, Annals of Discrete Mathematics Vol. 27, 13-26. Amsterdam: North-Holland. (1985).

GODS 1984-1 Godsil, C. D.
Real graph polynomials. In: J. A. Bondy, U. P. R. Murty (Eds.): Progress in Graph Theory, 281-294. New York: Academic Press. (1984).

GODSA 1987-1 Godsil, C. D.; Nowakowski, R.; Nesetril, J.
The chromatic connectivity of graphs. KAM Series 87-56. Praha: Charles Univ., Dept. of Applied Mathematics, 4 p. (1987).

GOEC 1986-1 Goecke, O.
A greedy algorithm for hereditary set systems and a generalization of the Rado-Edmonds characterization of matroids. Report No. 86423-OR. Bonn: Universität Bonn, Inst. für Ökonometrie und Operations Research, 13p. (1986).

GOEC 1986-2 Goecke, O.
Eliminationsprozesse in der Kombinatorischen Optimierung - ein Beitrag zur Greedoidtheorie. Bonn: Univ. Bonn, Rechts- und Staatswissenschaftliche Fakultät, 158 p. (1986). (Dissertation)

GOECA 1987-1 Goecke, O.; Korte, B.; Lovasz, L.
Examples and algorithmic properties of greedoids. Report 87456-OR. Bonn: Univ. Bonn, Inst. f. Ökonometrie u. Operations Research, 63 p. (1987).

GOECB 1986-1 Goecke, O.; Schrader, R.
Minor characterization of undirected branching greedoids - a short proof. Report No. 86435-OR. Bonn: Universität Bonn, Inst. für Ökonometrie und Operations Research, 6p. (1986).

GOET 1986-1 Goetschel, R. H. Jr.
Linear objective functions on certain classes of greedoids. Discrete Applied Mathematics 14, 11-16. (1986).

GOETA 1985-1 Goetz, J.; Makuchowski, M.
Heuristic methods for the general problem of resources allocation and operation scheduling (Polish). Postepy Cybernetyki 8, No.3, 115-131. (1985).

GOETB 1985-1 Goetze, B.; Staedler, G.
Algorithms on graphs with costly adjacency-information. In: H. Sachs (Ed.): Graphs, Hypergraphs and Applications, Proc. Conf. on Graph Theory (Eyba 1984), 41-47. Leipzig: Teubner. (1985).

GOIC 1985-1 Goicoechea, A.; Soland, R. M.; Sharp, F. M.
A transportation model for inland waterways navigation via mathematical programming. Large Scale Systems 8, 105-118. (1985).

GOLB 1987-1 Golberg, A. I.; Gurvich, V. A.
On the maximal number of edges for a graph with n vertices in which a given subgraph with k vertices has no more than l edges (Russian). Doklady Akademii Nauk SSSR (Moscow) 293, 27-32. (1987).

GOLD 1987-1 Goldberg, A. V.; Plotkin, S. A.
Parallel(delta + 1)-coloring of constant-degree graphs. Information Processing Letters 25, 241-246. (1987).

GOLDA 1986-1 Goldberg, A. V.; Tarjan, R. E.
A new approach to the maximum flow problem. In: Proc. Eighteenth Annual ACM Symp. on Theory of Computing, Berkeley 1986, 136-146. ACM. (1986).

GOLDA 1987-1 Goldberg, A. V.; Tarjan, R. E.
Solving minimum-cost flow problems by successive approximation. In: Proc. Nineteenth Annual ACM Symp. on Theory of Computing, New York City 1987, 7-18. ACM. (1987).

GOLDB 1984-1 Goldberg, M. K.
An approximate algorithm for the edge-coloring problem. Congressus Numerantium 43, 317-319. (1984).

GOLDB 1986-1 Goldberg, M. K.
Parallel algorithms for three graph problems. Congressus Numerantium 54, 111-122. (1986).

GOLDC 1984-1 Goldberg, M. K.; Gardner, R.
On the minimal cut problem. In: J. A. Bondy, U. P. R. Murty (Eds.): Progress in Graph Theory, 295-306. New York: Academic Press. (1984).

GOLDD 1987-1 Goldberg, M. K.; Spencer, T.
A new parallel algorithm for the maximal independent set problem. In: 28th Annual IEEE Symposium on Foundations of Computer Science, 161-165. (1987).

GOLDE 1984-1 Golden, B. L.
Introduction to and recent advances in vehicle routing methods. In: M. Florian (Ed.): Transportation Planning Models, 383-418. Amsterdam: North-Holland. (1984).

GOLDF 1984-1 Golden, B. L.; Assad, A. A.
A decision-theoretic framework for comparing heuristics. European Journal of Operational Research 18, 167-171. (1984).

GOLDF 1986-1 Golden, B. L.; Assad, A. A.
Perspectives on vehicle routing: exciting new developments. Operations Research 34, 803-810. (1986).

GOLDG 1984-1 Golden, B. L.; Assad, A. A.; Dahl, R.
Analysis of a large scale vehicle routing problem with an inventory component. Large Scale Systems (Amsterdam) 7, 181-190. (1984).

GOLDH 1986-1 Golden, B. L.; Bodin, L. D.; Goodwin, T.
Microcomputer-based vehicle routing and scheduling software. Computers and Operations Research 13, 277-286. (1986).

GOLDI 1985-1 Golden, B. L.; Levy, L.; Vohra, R. V.
Some heuristics for the generalized traveling salesman problem. Working Paper Series MS/S 85-036. College Park: Univ. of Maryland, College of Business and Management, 3 p. (1985).

GOLDI 1987-1 Golden, B. L.; Levy, L.; Vohra, R. V.
The orienteering problem. Naval Research Logistics 34, 307-318. (1987).

GOLDJ 1986-1 Golden, B. L.; Skiscim, C. C.
Using simulated annealing to solve routing and location problems. Naval Research Logistics Quarterly 33, 261-280. (1986).

GOLDK 1985-1 Golden, B. L.; Stewart, W. R. Jr.
Empirical analysis of heuristics. In: E. L. Lawler, J. K. Lenstra, A. H. G. Rinnooy Kan, D. B. Shmoys (Eds.): The Traveling Salesman Problem, 207-250. New York: Wiley. (1985).

GOLDL 1987-1 Golden, B. L.; Wasil, E. A.
Computerized vehicle routing in the soft drink industry. Operations Research 35, 6-17. (1987).

GOLDM 1982-1 Goldengorin, B. I.
Stability of solutions in problems with one component of connection of local minima (Russian). In: Models and Methods for Solving Problems of the Interaction of Economic Systems, 149-160. Mat. Anal. Ekonom. Modelei, "Nauka" Sibirsk. Otdel., Novosibirsk. (1982).

GOLDM 1984-1 Goldengorin, B. I.
A correcting algorithm for solving allocation type problems. Automation and Remote Control 45, 590-598. Translated from Avtomatika i Telemekhanika, No. 5, 53-62. (1984).

GOLDM 1986-1 Goldengorin, B. I.
Decomposition of the location problem. Automation and Remote Control 47, No.5, 668-677; translation from Avtomatika i Telemekhanika (1986), No.5, 91-101. (1986).

GOLDN 1985-1 Goldfarb, D.
Efficient dual simplex algorithms for the assignment problem. Mathematical Programming 33, 187-203. (1985).

GOLDO 1986-1 Goldman, A. J.; Tiwari, P.
Allowable processing orders in the accelerated cascade algorithm. Discrete Applied Mathematics 13, 213-222. (1986).

GOLE 1985-1 Golenko, D. I.
Approximated statistical methods in optimization problems of scheduling theory. Sixth Int. Conf. on Collective Phenomena, (Moscow 1983-85, Stockholm 1983, Tel Aviv 1984), 23-30. Ann. New York Acad. Sci. 452. (1985).

GOLE 1985-2 Golenko, D. I.
Statistical methods of optimizing varied-route problems in the scheduling theory. In: I. Fischer-Hjalmars, J. L. Lebowitz (Eds.): 6th Int. Conf. on Collective Phenomena, Stockholm 1983 and 7th Conf., Tel Aviv 1984, Annals of the New York Academy of Sciences 452, 1-10. (1985).

GOLE 1985-3 Golenko, D. I.
Using metric spaces in problems of optimal calendar planning. In: I. Fischer-Hjalmars, J. L. Lebowitz (Eds.): 6th Int. Conf. on Collective Phenomena, Stockholm 1983 and 7th Conf., Tel Aviv 1984, Annals of the New York Academy of Sciences 452, 11-22. (1985).

GOLI 1987-1 Golitschek, M. von
The cost-to-time ratio problem for large or infinite graphs. Discrete Applied Mathematics 16, 1-10. (1987).

GOLU 1984-1 Golumbic, M. C.
Algorithmic aspects of perfect graphs. In: C. Berge, V. Chvatal (Eds.): Topics on perfect graphs, Annals of Discrete Mathematics Vol. 21, 301-324. Amsterdam: North-Holland. (1984).

GOLU 1985-1 Golumbic, M. C.
Interval graphs and related topics. Discrete Mathematics 55, 113-122. (1985).

GOLU 1987-1 Golumbic, M. C.
A general method for avoiding cycling in a network. Information Processing Letters 24, 251-254. (1987).

GOLUA 1985-1 Golumbic, M. C.; Jamison, R. E.
Edge and vertex intersection of paths in a tree. Discrete Mathematics 55, 151-160. (1985).

GOLUA 1985-2 Golumbic, M. C.; Jamison, R. E.
The edge intersection graphs of paths in a tree. Journal of Combinatorial Theory 38B, 8-22. (1985).

GOLUB 1982-1 Golumbic, M. C.; Monma, C. L.
A generalization of interval graphs with tolerances. In: F. Hoffman, K. B. Reid, R. C. Mullin, R. G. Stanton (Eds.): Proc. Thirteenth Southeastern Conf. on Combinatorics, Graph Theory and Computing, Florida 1982, Congressus Numerantium Vol. 35, 321-332. Winnipeg, Canada: Utilitas Mathematica. (1982).

GOLUB 1982-2 Golumbic, M. C.; Monma, C. L.
A generalization of interval graphs with tolerances. Congressus Numerantium 35, 321-332. (1982).

GOLUC 1984-1 Golumbic, M. C.; Monma, C. L.; Trotter, W. T. Jr.
Tolerance graphs. Discrete Applied Mathematics 9, 157-170. (1984).

GOME 1987-1 Gomes, L. F. A. M.
On the problem of assigning traffic flows to a multicriteria, multicommodity, free-flow network. Foundations of Control Engineering 12, No. 2, 85-92. (1987).

GOMM 1984-1 Gommers, J. H. M.; Volgenant, A.
A Langrangean appraoch to the degree-constrained minimum spanning tree problem. A&E Reports 4/84. Amsterdam: Univ. of Amsterdam, Institute of Actuarial Sciences and Econometrics. (1984).

GONC 1985-1 Gonczarowski, J.; Warmuth, M. K.
Applications of scheduling theory to formal language theory. Theoretical Computer Science 37, 217-243. (1985).

GONS 1987-1 Gonsalvez, D. J.; Hall, N. G.; Rhee, W. T.; Siferd, S. P.
Heuristic solutions and confidence intervals for the multicovering problem. European Journal of Operational Research 31, 94-101. (1987).

GONZ 1985-1 Gonzalez, J. J.; Reeves, G. R.; Franz, L. S.
An interactive procedure for solving multiple-objective integer linear programming problems. In: Y. Y. Haimes, V. Chankong (Eds.): Decision Making with Multiple Objectives, Lecture Notes in Economics and Mathematical Systems Vol. 242, 250-260. Berlin: Springer-Verlag. (1985).

GONZA 1986-1 Gonzalez, R. L.; Rofman, E.
Computational methods in scheduling optimization. In: J.-B. Hiriart-Urruty (Ed.): FERMAT Days 85: Mathematics for Optimization, 135-155. Amsterdam: North-Holland. (1986).

GONZB 1982-1 Gonzalez, T. F.
On the computational complexity of clustering and related problems. In: System Modelling and Optimization (New York, 1981), 174-182. Lecture Notes in Control and Information Sciences 38. Berlin, New York: Springer. (1982).

GONZC 1987-1 Gonzalez, T. F.; Lee, S.-L.
A 1.6 approximation algorithm for routing multiterminal nets. SIAM Journal on Computing 16, 669-704. (1987).

GOOD 1985-1 Goodman, A. W.
Triangles in a complete chromatic graph with three colors. Discrete Mathematics 57, 225-236. (1985)

GOOD 1985-2 Goodman, A. W.
Triangles in a complete chromatic graph. Journal of the Australian Mathematical Society, Series A, 39, 86-93. (1985).

GOODA 1986-1 Goodwin, J. S.; Weeks, J. K.
Evaluating scheduling policies in a multi-level assembly system. International Journal of Production Research 24, 247-258. (1986).

GORD 1981-1 Gordeev, E. N.
New estimates in the covering problem (Russian). In: Modeling and Optimization of Complex Control Systems, 116-122. "Nauka", Moscow. (1981).

GORDA 1984-1 Gordon, G.
Matroids over F(p) which are rational excluded minors. Discrete Mathematics 52, 51-66. (1984).

GORDB 1985-1 Gordon, V. S.; Yanova, O. V.
Uniform parallel machine scheduling under precedence constraints and due-dates (Russian). In: V. P. Tanaev (Ed.): Methods, Algorithms and Programs for the Solution of Extremal Problems, 68-75. Minsk: Inst. Tekh. Kibernet. AN BSSR. (1985).

GORG 1982-1 Gorgos, I. M.
Internally stable sets and the method of alternating sequences (Russian). In: B. A. Shcherbakov (Ed.): Operations research and programming, 37-44. Matematicheskie Nauki. Kishinev: Shtiintsa. (1982).

GORK 1981-1 Gorkovoi, V. F.
Coloring of graphs (Russian). In: Mathematical Methods of Optimization and Control in Complex Systems, 151-157. Kalinin. Gos. Univ. (1981).

GORK 1983-1 Gorkovoi, V. F.
The topology of 3-colored graphs (Russian). In: Y. A. Abramov (Ed.): Mathematical methods of optimization and control in complex systems, Collection of scientific works, 3-8. Kalinin: Kalininskij Gosudarstvennzj Universitet. (1983).

GORK 1984-1 Gorkovoi, V. F.
The minimal coloring of graphs (Russian). In: Control of Dynamical Systems, 147-149. Voprosy Mekh. Protsess. Upravl. 7, Leningrad Univ. (1984).

GORKA 1986-1 Gorkovoi, V. F.; Egorov, M. Y.
Decomposition of graphs (Russian). In: Mathematical Methods in the Analysis of Controllable Processes, 70-76. Voprosy Mekh. Protsess. Upravl. 8, Leningrad University. (1986).

GORS 1985-1 Gorshkov, A. F.
A method for finding extremal subgraphs (Russian). Sibirskij Matematicheskij Zhurnal (Novosibirsk) 26, No. 1, 44-48. (1985).
GORS 1987-1 Gorshkov, A. F.
A method for finding extremal subgraphs in bipartite graphs. Izvestiya Akademii Nauk SSSR, Tekhnicheskaya Kibernetika (Moscow), No. 4, 136-142; translated as Soviet J. Comput. Systems Sci. 25, No. 1, 85-91. (1987).
GORU 1985-1 Gorunovich, S. A.
An algorithm for finding p-medians of a graph (Russian). Kibernetika No. 5, 67-70. (1985).
GOTT 1986-1 Gottlieb, E. S.; Rao, M. R.
The generalized assignment problem I: valid inequalities and facets. New York: City Univ. of New York, 34 p. (1986).
GOTT 1986-2 Gottlieb, E. S.; Rao, M. R.
The generalized assignment problem II: three classes of facets. New York: City Univ. of New York, 49 p. (1986).
GOUL 1983-1 Gould, R. J.
Traceability in the square of a tree. Journal of Combinatorics, Information & System Sciences (Delhi) 8, 253-260. (1983).
GOULA 1983-1 Gould, R. J.; Jacobson, M. S.
A note on mixed Ramsey numbers: total chromatic number versus complete graphs. Journal of Combinatorics, Information & System Sciences (Delhi) 8, 181-184. (1983).
GOULB 1984-1 Gould, R. J.; Lick, D. R.
Degree sets and graph factorizations. Colloquium Mathematicum (Warsaw) 48, 269-277. (1984).
GOYA 1984-1 Goyal, S. K.
Improving VAM for unbalanced transportation problems. Journal of the Operational Research Society 35, 1113-1114. (1984).
GOYAA 1985-1 Goyal, S. K.; Demirbag-Büyükkurt, M.
A note on "Determination of coordinated maintenance scheduling frequencies for a group of machines". IIE Transactions 17, 400-402. (1985).
GOYAB 1986-1 Goyal, S. K.; Szendrovits, A. Z.
A constant lot size model with equal and unequal sized batch shipments between production stages. Engineering Costs and Production Economics 10, 203-210. (1986).
GRAB 1984-1 Grabowski, J.; Janiak, A.
On job-shop scheduling with resource constraints. In: P. Thoft-Christensen (Ed.): System Modelling and Optimization, 193-200. Lecture Notes in Control and Information Sciences 59, Berlin: Springer. (1984).
GRAB 1987-1 Grabowski, J.; Janiak, A.
Job-shop scheduling with resource-time models of operations. European Journal of Operational Research 28, 58-73. (1987).
GRABA 1986-1 Grabowski, J.; Nowicki, E.; Zdrzalka, S.
A block approach for single-machine scheduling with release dates and due dates. European Journal of Operational Research 26, 278-285. (1986).
GRABB 1985-1 Grabowski, J.; Skubalska, E.
Optimization of the transportation network structure with respect to the flow costs (Polish). Archiwum Automatyki i Telemechaniki (Warsaw) 30, 3-21. (1985).
GRABC 1983-1 Grabowski, J.; Smutnicki, C.
The flow-shop problems. Properties, algorithms and applications. I. (Polish). Przeglad Statystyczny (Warsaw) 30, 275-288. (1983).
GRABC 1984-1 Grabowski, J.; Smutnicki, C.
Minimizing maximum cost in flow-shop problems. I: basic properties and algorithms (Polish). Archiwum Automatyki i Telemechaniki (Warsaw) 29, 35-55. (1984).
GRABC 1984-2 Grabowski, J.; Smutnicki, C.
Minimizing maximum cost in flow-shop problems. II: Lower bounds, computational results, applications (Polish). Archiwum Automatyki i Telemechaniki (Warsaw) 29, 57-74. (1984).
GRABC 1984-3 Grabowski, J.; Smutnicki, C.
The flow-shop problems, Properties, algorithms and applications. II (Polish). Przeglad Statystyczny (Warsaw) 31, 31-45. (1984).
GRABC 1986-1 Grabowski, J.; Smutnicki, C.
Scheduling problems with minimax goal function (Polish). Archiwum Automatyki i Telemechaniki 31, 21-37. (1986).
GRABD 1985-1 Grabowski, J.; Zdrzalka, S.
Problems on sequencing jobs on an assembly line (Polish). Prace Naukowe, Instytut Cybernetyki Techicznej, Politechnicynej Wroclaw, Seria Konferencje 68, No. 27, 143-152. (1985).
GRAC 1982-1 Grace, T.
New proof techniques for gracefully labelling graphs. In: F. Hoffman, K. B. Reid, R. C. Mullin, R. G. Stanton (Eds.): Proc. Thirteenth Southeastern Conf. on Combinatorics, Graph Theory and Computing, Florida 1982, Congressus Numerantium Vol. 36, 403-408. Winnipeg, Canada: Utilitas Mathematica. (1982).
GRAC 1983-1 Grace, T.
More graceful trees. Congressus Numerantium 39, 433-440. (1983).
GRACA 1985-1 Gracia, Z.; Salzberg, P. M.
Chromatic classification of K(p)-G(6). Ars Combinatoria 20-B, 107-112. (1985).
GRAH 1986-1 Graham, D.; Nuttle, H. L. W.
A comparison of heuristics for a school bus scheduling problem. Transportation Research 20-B, 175-182. (1986).
GRAHA 1984-1 Graham, R. L.; Foulds, L. R.
It is unlikely that minimal phylogenies for a realistic biological study can be constructed in reasonable computational time. Research Report, 84-18. Gainesville: Univ. of Florida, Ind. & Systems Eng. Dept., 15 p. (1984).
GRAHB 1985-1 Graham, R. L.; Hell, P.
On the history of the minimum spanning tree problem. Annals of the History of Computing 7, 43-57. (1985).

GRAN 1987-1 Granot, D.
The role of cost allocation in locational models. Operations Research 35, 234-248. (1987).

GRANA 1984-1 Granot, D.; Granot, F.
Minimal covers, minimal sets and canonical facets of the posynomial knapsack polytope. Discrete Applied Mathematics 9, 171-186. (1984).

GRANB 1986-1 Granot, F.; Hassin, R.
Multi-terminal maximum flows in node-capacitated networks. Discrete Applied Mathematics 13, 157-164. (1986).

GRANC 1985-1 Granot, F.; Veinott, A. F. Jr.
Substitutes, complements and ripples in network flows. Mathematics of Operations Research 10, 471-497. (1985).

GRAND 1987-1 Grant, D. D.; Mac Avaney, K. L.
Antidirected cycles in planar graphs. Ars Combinatoria 19A, 143-148. (1987).

GRANE 1986-1 Grant, T. J.
Lessons for O. R. from A. I.: a scheduling case study. Journal of the Operational Research Society 37, 41-58. (1986).

GRAO 1985-1 Graovac, A.; Babic, D.
Comment on "Matchings in long benzene chains" by E. J. Farrell and S. A. Wahid. Match 18, 49-53. (1985).

GRAOA 1984-1 Graovac, A.; Jashari, G.; Strunje, M.
On the distance spectrum of a cycle. Aplikace Matematiky (Prague) 30, 286-290. (1984).

GRAV 1986-1 Graves, S. C.
A tactical planning model for a job shop. Operations Research 34, 522-533. (1986).

GREE 1985-1 Green, C. D.; Suraweera, F.
A path entropy function for rooted acyclic digraphs. Journal of the University of Kuwait (Science) 12, 15-21. (1985).

GREEA 1985-1 Greenberg, H.
An algorithm for the periodic solutions in the knapsack problem. Journal of Mathematical Analysis and Applications 111, 327-331. (1985).

GREEA 1986-1 Greenberg, H.
On equivalent knapsack problems. Discrete Applied Mathematics 14, 263-268. (1986).

GREEB 1986-1 Greenberg, H. J.
Diagnosing infeasibility in min-cost network flow problems. I: Dual infeasibility. IMA Journal of Mathematics in Management 1, 99-109. (1986).

GREEC 1983-1 Greenberg, H. J.; Lundgren, J. R.; Maybee, J. S.
The inversion of 2-step graphs. Journal of Combinatorics, Information & System Sciences (Delhi) 8, 33-43. (1983).

GREED 1983-1 Greene, C.
On the interpretation of Whitney numbers through arrangements of hyperplanes, zonotopes, non-Radon partitions, and orientations of graphs. Transactions of the American Mathematical Society 280, 97-126. (1983).

GREEE 1986-1 Greene, T. J.; Sadowski, R. P.
A mixed integer program for loading and scheduling multiple flexible manufacturing cells. European Journal of Operational Research 24, 379-386. (1986).

GREG 1985-1 Grega, W.
A decomposition algorithm for the determination of optimal bus frequencies. In: System modelling and optimization, Proc. 12th IFIP Conf., Budapest/Hungary 1985, Lect. Notes Control Inf. Sci. 84, 292-297. (1985).

GREGA 1986-1 Gregory, D. A.; Mac Guinness, S.; Wallis, W. D.
Clique partitions of the cocktail party graphs. Discrete Mathematics 59, 267-274. (1986).

GRES 1984-1 Greshnev, S. N.
The polyhedron of a problem on an m-vertex subgraph of a complete graph. USSR Computational Mathematics and Mathematical Physics 24, 109-111. (1984).

GRIB 1985-1 Gribik, P. R.; Kortanek, K. O.
Extremal methods of operations research. New York, Basel: Dekker. (1985).

GRIE 1984-1 Griesshaber, R.
Rotation system for sales representatives and station managers. Working Paper. Strasbourg (XXIV. Annual AGIFORS Symposium, September 1984) 33 p. (1984).

GRIG 1986-1 Grigoriadis, M. D.
An efficient implementation of the network simplex method. In: G. Gallo, C. Sandi (Eds.): Netflow at Pisa, Mathematical Programming Study 26, 83-111. (1986).

GRIGA 1984-1 Grigoriadis, M. D.; Hsu, T.
Numerical methods for basic solutions of generalized flow networks. Mathematical Programming Study 22, 147-166. (1984).

GRIGB 1986-1 Grigoriadis, M. D.; Kalantari, B.
A lower bound to the complexity of Euclidean and rectilinear matching algorithms. Information Processing Letters 22, 73-76. (1986).

GRIGC 1986-1 Grigoriadis, M. D.; Kalantari, B.; Lai, C. Y.
On the existence of weak greedy matching heuristics. Operations Research Letters 5, 201-206. (1986).

GRIGC 1987-1 Grigoriadis, M. D.; Kalantari, B.; Lai, C. Y.
Weak greedy heuristics for perfect matching. In: Trans. of the 4th Army Conf. on Applied Mathematics and Computing (Ithaca NY 1986), 909-915. ARO-Report 87-1, US Army Res. Office, Research Triangle Park, NC. (1987).

GRIGD 1987-1 Grigoriev, D. Y.; Karpinski, M.
The matching problem for bipartite graphs with polynomially bounded permanents is in NC. In: 28th Annual IEEE Symposium on Foundations of Computer Science, 166-172. (1987).

GRIM 1985-1 Grimaldi, R. P.
Properties of integer-interval graphs. Congressus Numerantium 48, 145-152. (1985).

GRIMA 1985-1 Grimmett, G. R.
Large deviations in subadditive processes and first-passage percolation. In: Particle systems, random media and large deviations, 175-194. Contemporam Math. Vol. 41. Providence, R. I.: American Mathematical Society. (1985).

GRIMA 1985-2 Grimmett, G. R.
Random flows: network flows and electrical flows through random media. In: I Anderson (Ed.): Surveys in Combinatorics 1985, London Mathematical Society Lecture Notes Series Vol. 103, 59-96. Cambridge: University Press. (1985).
GRIMA 1986-1 Grimmett, G. R.
An exact threshold theorem for random graphs and the node-packing problem. Journal of Combinatorial Theory, Series B, 40, 187-195. (1986).
GRIMB 1985-1 Grimmett, G. R.; Pulleyblank, W. R.
Random near-regular graphs and the node packing problem. Operations Research Letters 4, 169-174. (1985).
GRIN 1984-1 Grinstead, C.
The perfect graph conjecture for toroidal graphs. In: C. Berge, V. Chvatal (Eds.): Topics on perfect graphs, Annals of Discrete Mathematics Vol. 21, 97-102. Amsterdam: North-Holland. (1984).
GRIS 1983-1 Grishukhin, V. P.
Submodular problems and algorithms for solving them. Engineering Cybernetics 21, No. 1, 147-152. Translation from: Izvestiya Akademii Nauk SSSR (Moscow), (Tekh. Kibern.) 1983, No. 1, 188-193. (1983).
GROB 1987-1 Grobelny, J.
On one possible 'fuzzy' approach to facilities layout problems. International Journal of Production Research 25, 1123-1142. (1987).
GROB 1987-2 Grobelny, J.
The fuzzy approach to facilities layout problems. Fuzzy Sets and Systems 23, 175-190. (1987).
GROF 1984-1 Gröflin, H.
Path-closed sets. Combinatorica 4, 281-290. (1984).
GROF 1987-1 Gröflin, H.
On switching path polyhedra. Combinatorica 7, 193-204. (1987).
GROG 1985-1 Gröger, M.
Location problems in graphs. Annales Universitatis Scientiarum Budapestinensis, Sectio Computatorica 5, 21-28. (1985).
GROP 1984-1 Groppen, V. O.
Models and algorithms of combinatorial programming (Russian). Izdatel'stvo Rostovskogo Univ. (Rostov). (1984).
GROP 1985-1 Groppen, V. O.
Optimization of dynamic programming methods in the solution of extremal combinatorial problems (Russian). Avtomatika i Telemekhanika (Moscow) No. 12, 79-84. (1985).
GROP 1985-2 Groppen, V. O.
Optimization of dynamic programming methods when solving extremal combinatorial problems. Automation and Remote Control 46, 1561-1566; translation from Avtomatika i Telemekhanika (1985), No.12, 79-84. (1985).
GROP 1987-1 Groppen, V. O.
Optimizing the realization of dynamic programming for a class of combinatorial problems. Wissenschaftliche Zeitschrift der Technischen Hochschule Leipzig 11, 25-29. (1987).
GROS 1985-1 Grosser, E.
Über eine Methode zur Lösung des Zuordnungsproblems. Wissenschaftliche Zeitschrift der Technischen Hochschule Karl-Marx-Stadt 27, 39-47. (1985).
GROSA 1982-1 Grossman, J. W.
On edge-partitioned graphs with degree restrictions. In: F. Hoffman, K. B. Reid, R. C. Mullin, R. G. Stanton (Eds.): Proc. Thirteenth Southeastern Conf. on Combinatorics, Graph Theory and Computing, Florida 1982, Congressus Numerantium Vol. 35, 349-360. Winnipeg, Canada: Utilitas Mathematica. (1982).
GROT 1984-1 Grötschel, M.
Developments in combinatorial optimization. In: W. Jäger, J. Moser, R. Remmert (Eds.): Perspectives in Mathematics, Anniversary of Oberwolfach 1984, 249-294. Basel, Boston, Stuttgart: Birkhäuser Verlag. (1984).
GROT 1985-1 Grötschel, M.
Polyhedral Combinatorics. In: M. O'hEigeartaigh, J. K. Lenstra, A. H. G. Rinnooy Kan (Eds.): Combinatorial Optimization - Annotated Bibliographies, 1-10. Chichester: Wiley. (1985).
GROT 1986-1 Grötschel, M.
Fortschritte in der polyedrischen Kombinatorik. In: W. Krelle (Ed.): Ökonomische Prognose-, Entscheidungs- und Gleichgewichtsmodelle, 341-366. Weinheim: VCH-Verlag. (1986).
GROT 1986-2 Grötschel, M.
Polyedrische Kombinatorik und Schnittebenenverfahren. In: A. Kerber (Ed.): Diskrete Strukturen, algebraische Methoden und Anwendungen, Sommerschule Bayreuth 1985. Bayreuther Mathematische Schriften, Heft 21, 69-100. (1986).
GROTA 1985-1 Grötschel, M.; Holland, O.
Solving matching problems with linear programming. Mathematical Programming 33, 243-259. (1985).
GROTA 1987-1 Grötschel, M.; Holland, O.
A cutting plane algorithm for minimum perfect 2-matchings. Computing 39, 327-344. (1987).
GROTB 1984-1 Grötschel, M.; Jünger, M.; Reinelt, G.
A cutting plane algorithm for the linear ordering problem. Operations Research 32, 1195-1220. (1984).
GROTB 1984-2 Grötschel, M.; Jünger, M.; Reinelt, G.
Optimal triangulation of large real world input-output matrices. Statistische Hefte 25, 261-296. (1984).
GROTB 1985-1 Grötschel, M.; Jünger, M.; Reinelt, G.
Acyclic subdigraphs and linear orderings: polytopes, facets, and a cutting plane algorithm. In: I. Rival (Ed.): Graphs and Order, 217-266. Dordrecht: Reidel. (1985).
GROTB 1985-2 Grötschel, M.; Jünger, M.; Reinelt, G.
Facets of the linear ordering polytope. Mathematical Programming 33, 43-60. (1985).
GROTB 1985-3 Grötschel, M.; Jünger, M.; Reinelt, G.
On the acyclic subgraph polytope. Mathematical Programming 33, 28-42. (1985).

GROTB 1987-1 Grötschel, M.; Jünger, M.; Reinelt, G.
Calculating exact ground states of spin glasses: a polyhedral approach. In: J. L. van Hemmen, I. Morgenstern (Eds.): Heidelberg Colloquium on Glassy Dynamics, 325-353. Lecture Notes in Physics 275, Berlin: Springer. (1987).

GROTB 1987-2 Grötschel, M.; Jünger, M.; Reinelt, G.
Via minimization with pin preassignments and layer preference. Report 19. Augsburg: Univ. Augsburg, Inst. f. Mathematik, 13 p. (1987).

GROTC 1984-1 Grötschel, M.; Lovasz, L.; Schrijver, A.
Corrigendum to our paper "The ellipsoid method and its consequences in combinatorial optimization". Combinatorica 4, 291-296. (1984).

GROTC 1984-2 Grötschel, M.; Lovasz, L.; Schrijver, A.
Geometric methods in combinatorial optimization. In: W. R. Pulleyblank (Ed.): Progress in Combinatorial Optimization, 167-184. Toronto: Academic Press. (1984).

GROTC 1984-3 Grötschel, M.; Lovasz, L.; Schrijver, A.
Polynomial algorithms for perfect graphs. In: C. Berge, V. Chvatal (Eds.): Topics on perfect graphs, Annals of Discrete Mathematics Vol. 21, 325-356. Amsterdam: North-Holland. (1984).

GROTC 1986-1 Grötschel, M.; Lovasz, L.; Schrijver, A.
Relaxations of vertex packing. Journal of Combinatorial Theory, Series B 40, 330-343. (1986).

GROTD 1985-1 Grötschel, M.; Padberg, M. W.
Polyhedral theory. In: E. L. Lawler, J. K. Lenstra, A. H. G. Rinnooy Kan, D. B. Shmoys (Eds.): The Traveling Salesman Problem, 251-306. New York: Wiley. (1985).

GROTE 1986-1 Grötschel, M.; Pulleyblank, W. R.
Clique tree inequalities and the symmetric travelling salesman problem. Mathematics of Operations Research 11, 537-569. (1986).

GROTE 1987-1 Grötschel, M.; Pulleyblank, W. R.
Corrigenda: clique tree inequalities and the symmetric travelling salesman problem. Mathematics of Operations Research 12, 182-184. (1987).

GROTF 1986-1 Grötschel, M.; Truemper, K.
Decomposition and optimization over cycles in binary matroids. Preprint No. 108. Augsburg: Univ., Inst. für Mathematik, 27p. (1986).

GROTF 1986-2 Grötschel, M.; Truemper, K.
Master polytopes for cycles of binary matroids. Preprint, 111. Augsburg: Univ. Augsburg, Institut für Mathematik, 15 p. (1986).

GROTG 1987-1 Grötschel, M.; Wakabayashi, Y.
A cutting plane algorithm for a clustering problem. Report 9. Augsburg: Univ. Augsburg, Institut f. Mathematik, 52 p. (1987).

GROTG 1987-2 Grötschel, M.; Wakabayashi, Y.
Composition of facets of the clique partitioning polytope. Report 18. Augsburg: Univ. Augsburg, Institut f. Mathematik, 14 p. (1987).

GROTG 1987-3 Grötschel, M.; Wakabayashi, Y.
Facets of the clique partitioning polytope. Report 6. Augsburg: Univ. Augsburg, Institut f. Mathematik, 21 p. (1987).

GUAN 1983-1 Guan, M.-G.
Another improvement in the assignment method on graphs. Qufu Shiyuan Xuebao (Qufu) No. 2, 1-4. (1983).

GUAN 1984-1 Guan, M.-G.
The maximum weighted cycle-packing problem and its relation to the Chinese postman problem. In: J. A. Bondy, U. P. R. Murty (Eds.): Progress in Graph Theory, 323-326. New York: Academic Press. (1984).

GUANA 1985-1 Guan, M.-G.; Chen, Q.-H.
A simple algorithm for finding the second order-constrained base of matroid. Kexue Tongbao, Foreign Language Edition 30, 862-865. (1985).

GUANA 1985-2 Guan, M.-G.; Chen, Q.-H.
Order of minimal bases of restricted degree in a matroid (Chinese). Kexue Tongbao 30, 488-490. (1985).

GUANB 1985-1 Guan, M.-G.; Fleischner, H.
On the minimum weighted cycle covering problem for planar graphs. Ars Combinatoria 20, 61-68. (1985).

GUANC 1985-1 Guan, M.-G.; Pulleyblank, W. R.
Eulerian orientations and circulations. SIAM Journal on Algebraic and Discrete Methods 6, 657-664. (1985).

GUGL 1986-1 Guglielmann Grieco, M.; Zuchetti, R.; Maria, B.
Digraphs and bigraphs: relationship between strong connectedness and strong alpha-connectedness, alpha-hamiltonian digraphs. Istituto Lombardo, Accademia di Scienze e Lettere (Milan), Rendiconti A 117, 253-263. (1986).

GUIB 1987-1 Guibas, L.; Hershberger, J.; Leven, D.; Sharir, M.; Tarjan, R. E.
Linear-time algorithms for visibility and shortest path problems inside triangulated simple polygons. Algorithmica 2, 209-234. (1987).

GUID 1985-1 Guidici, R. E.
Some new families of chromatically unique graphs. In: Analysis, Geometry, and Probability (Valparaiso, 1981), 147-158. Lecture Notes in Pure and Applied Mathematics 96, New York: Dekker. (1985).

GUIG 1983-1 Guignard, M.
A Lagrangean dual ascent method based on (separable) relaxation strengthened by valid inequalities (illustrated for the matching problem). Technical Report, 55. Philadelphia: Univ. of Pennsylvania, Dept. of Statistics, 19 p. (1983).

GUIG 1984-1 Guignard, M.
A Lagrangean dual ascent algorithm for simple plant location problems. Technical Report, 59. Philadelphia: Univ. of Pennsylvania, Dept. of Statistics, 18 p. (1984).

GUIG 1984-2 Guignard, M.
Dual and primal heuristics for network location problems. In: G. Hammer, D. Pallaschke (Eds.): Methods of Operations Research 51, Vol. 1, 257-270. Königstein/Ts.: Verlag Anton Hain Meisenheim. (1984).

GUIGA 1983-1 Guignard, M.; Kim, S.
A strong Lagrangean relaxation for capacitated plant location problems. Technical Report, 56. Philadelphia: Univ. of Pennsylvania, Dept. of Statistics, 15 p. (1983).

GUIGA 1986-1 Guignard, M.; Kim, S.
Lagrangean decomposition: a model yielding better bounds (French). Comptes Rendus de l'Academie des Sciences, Paris, Serie I 303, 667-670. (1986).

GUIGA 1987-1 Guignard, M.; Kim, S.
Lagrangean decomposition for integer programming: theory and applications. RAIRO Recherche Operationelle 21, 307-324. (1987).

GUIGB 1984-1 Guignard, M.; Opaswongkarn, K.
Langrangean dual ascent algorithms for the capacitated plant location problem. Technical Report, 57. Philadelphia: Univ. of Pennsylvania, Dept. of Statistics, 17 p. (1984).

GUIGC 1985-1 Guignard, M.; Spielberg, K.
Separable-Lagrangean relaxation for weakly linked mixed integer programming problems. In: P. Brucker, R. Pauly (Eds.): Proc. IX. Symp. on Operations Research, Osnabrück 1984. Methods of Operations Research 49, 239-254. (1985).

GUIGC 1987-1 Guignard, M.; Spielberg, K.
On combining Lagrangean relaxation, Benders inequalities and Lagrangean dual ascent. In: W. Domschke, W. Krabs, J. Lehn, P. Spellucci (Eds.): Proc. XI. Symp. on Operations Research, Darmstadt 1986, 131-142. Methods of Operations Research 57, Frankfurt/Main: Athenäum. (1987).

GUIN 1984-1 Guinet, A.
Transports industriels routiers, un probleme d'affectation avec reemploi sous contraintes. RAIRO Recherche Operationelle 18, 353-380. (1984).

GULD 1987-1 Guldan, F.
On a problem of linear arboricity. Casopis pro Pestovani Matematiky 112, 395-400. (1987).

GULY 1983-1 Gulyanitskii, L. F.; Khodzinskii, A. N.
Two algorithms for solution of optimization problems on permutations (Russian). In: Efficient Organization of Computations and Numerical Methods, 9-17. Akademiya Nauk Ukrainskoi SSR (Kiev), Inst. Kibernet. (1983).

GULYA 1986-1 Gulyanitskii, L. F.; Sapko, A. V.
On an approach to the solution of nonsymmetric travelling salesman problems (Russian). In: I. V. Sergienko (Ed.): Programming and Algorithms for the Development of Program Packages, 10-16. Kiev: Akad. Nauk Ukrain. SSR, Inst. Kibernet. (1986).

GUMI 1986-1 Guminskij, A. P.; Martynov, V. V.
Construction and implementation of a scheduling algorithm in a calculus based on universal semantic code. Soviet Journal of Computer and Systems Sciences 24, No.3, 48-53; translation from Izvestiya Akademii Nauk SSSR, Tekhnicheskaya Kibernetika (1985), No.5, 76-82. (1986).

GUNN 1986-1 Gunn, E. A.; Kusiak, A.
The procurement problem: an integer programming problem well suited to a solution using duality. Naval Research Logistics Quarterly 33, 613-620. (1986).

GUNT 1985-1 Gunther, G.
Neighbour-connectivity in regular graphs. Discrete Applied Mathematics 11, 233-244. (1985).

GUNTA 1987-1 Gunther, H. O.
Planning lot sizes and capacity requirements in a single stage production system. European Journal of Operational Research 31, 223-231. (1987).

GUO 1986-1 Guo, X. F.; Zhu, B. W.
A construction of critically 2-connected graphs (Chinese). Journal Xinjiang University of Natural Science (Urumqi) 3, No. 2, 1-9. (1986).

GUO A 1985-1 Guo, Z.-Y.
A theorem on cohesiveness of points of a graph (Chinese). Journal of the Changsha Railway Institute, No.4, 21-24. (1985).

GUO A 1985-2 Guo, Z.-Y.
On complete k-edge colorable subgraphs (Chinese). Journal of Huazhong (Central China) University of Science and Technology 13, No. 6, 41-45. (1985).

GUO A 1986-1 Guo, Z.-Y.
Minimally and critically 2-connected graph (Chinese). Journal of Huazhong (Central China) University of Science and Technology 14, 281-283. (1986).

GUPT 1986-1 Gupta, A. K.; Hambrusch, S. E.
On optimal three-dimensional layouts of complete binary trees. CC-86-14. West Lafayette: Purdue Univ. Inst. of Interdisciplinary Engineering Studies, 12 p. (1986).

GUPT 1987-1 Gupta, A. K.; Hambrusch, S. E.
Simulating tree algorithms with limited parallelism. CC-87-23. West Lafayette: Purdue Univ. Inst. of Interdisciplinary Engineering Studies, 25 p. (1987).

GUPTA 1987-1 Gupta, A. K.; Sharma, J. K.
An iterative technique for time-cost trade-off in solid transportation problems. Journal of Mathemetical and Physical Sciences 21, 133-142. (1987).

GUPTB 1984-1 Gupta, J. N. D.
Optimal schedules for single facility with two job classes. Computers & Operations Research 11, 409-414. (1984).

GUPTB 1986-1 Gupta, J. N. D.
Flowshop schedules with sequence dependent setup times. Journal of the Operations Research Society of Japan 29, 206-219. (1986).

GUPTC 1986-1 Gupta, J. N. D.; Darrow, W. P.
The two-machine sequence dependent flowshop scheduling problem. European Journal of Operational Research 24, 439-446. (1986).

GUPTD 1987-1 Gupta, J. N. D.; Shanthikumar, J. G.; Szwarc, W.
Generating improved dominance conditions for the flowshop problem. Computers and Operations Research 14, 41-46. (1987).

GUPTE 1985-1 Gupta, O. K.; Ravindran, A.
Branch and bound experiments in convex nonlinear integer programming. Management Science 31, 1533-1546. (1985).

GUPTF 1984-1 Gupta, R. G.; Srivastava, V. S. P.
On synthesis of scheduling algorithms. Information Processing Letters 19, 147-150. (1984).

GUPTG 1984-1 Gupta, S.; Sen, T.
Minimizing the range of lateness on a single machine. Journal of the Operational Research Society 35, 853-858. (1984).

GUPTH 1985-1 Gupta, S. K.
Linear programming and network models. New Delhi, Madras, Hyderabad: Affiliated East-West Press Pvt. Ltd., 232p. (1985).

GURA 1984-1 Gurari, E. M.; Sudborough, I. H.
Improved dynamic programming algorithms for bandwidth minimization and the min-cut linear arrangement problem. Journal of Algorithms 5, 531-546. (1984).

GURE 1987-1 Gurevich, Y.; Shelah, S.
Expected computation time for Hamiltonian path problem. SIAM Journal on Computing 16, 486-502. (1987).

GURG 1986-1 Gurgel, M. A.; Wakabayashi, Y.
On k-leaf-connected graphs. Journal of Combinatorial Theory, Series B 41, 1-16. (1986).

GURV 1984-1 Gurvich, V. A.
Kirchhoff's second law for optimal transportation flows. Soviet Physics, Doklady 29, 431-433; translation from Doklady Akademii Nauk SSSR 276 (1984), 812-817. (1984).

GURV 1984-2 Gurvich, V. A.
Some properties and applications of complete edge-chromatic graphs and hypergraphs. Soviet Mathematics, Doklady 30, 803-807; translation from Doklady Akademii Nauk SSSR 279 (1984), 1306-1310. (1984).

GUSE 1983-1 Guseinov, B. A.; Stanevichyus, A.-I. A.
Method of solution of the crossing problem as an integer programming problem (Russian). In: D. E. Allakhverdiev (Ed.): Mathematical cybernetics and applied mathematics, No. 6, 29-40. Baku: Elm. (1983).

GUSF 1984-1 Gusfield, D.
Bounds for naive multiple machine scheduling with release times and deadlines. Journal of Algorithms 5, 1-6. (1984).

GUSF 1987-1 Gusfield, D.
Optimal mixed graph augmentation. SIAM Journal on Computing 16, 599-612. (1987).

GUSF 1987-2 Gusfield, D.
Three fast algorithms for four problems in stable marriage. SIAM Journal on Computing 16, 111-128. (1987).

GUSF 1987-3 Gusfield, D.
Very simple methods for all pairs network flow analysis. Working Paper. Davis: Univ. of California, Computer Science Division, 22 p. (1987).

GUSFA 1987-1 Gusfield, D.; Martel, C.; Fernandez-Baca, D.
Fast algorithms for bipartite network flow. SIAM Journal on Computing 16, 237-251. (1987).

GUSFB 1986-1 Gusfield, D.; Pitt, L.
Equivalent approximation algorithms for node cover. Information Processing Letters 22, 291-294. (1986).

GUSH 1987-1 Gushchinskii, N. N.; Levin, G. M.
Two-level minimization of a composite function and its application to a problem of the optimization of a path in a graph (Russian). Vestsi Akademii Navuk BSSR, Seryya Fizika-Matematychnykh Navuk (Minsk), No. 2, 28-32. (1987).

GUST 1985-1 Gustafson, K.; Harary, F.
The curl of graphs and networks. Mathematical Modelling 6, 145-155. (1985).

GUSTA 1985-1 Gustafson, K.; Hartman, R.
Graph theory and fluid dynamics. SIAM Journal on Algebraic and Discrete Methods 6, 643-656. (1985).

GUTI 1983-1 Gutierrez Diez, J. M.
A variant of the Edmonds Algorithm for maximum matching (Spanish). Trabajos de Estadistica y de Investigacion Operativa (Valencia) 34, No. 2, 52-66. (1983).

GUTIA 1984-1 Gutin, G. M.
Cycles in strong n-partite tournaments (Russian). Vestsi Akademii Navuk BSSR, Seryya Fizika-Matematychnykh Navuk (Minsk), No. 5, 105-106. (1984).

GUTIA 1987-1 Gutin, G. M.
Finding of the largest contour in a complete bipartite graph (Russian). Kibernetika (Kiev) No. 2, 117-118. (1987).

GUTM 1983-1 Gutman, I.
Graphs with maximum and minimum independence numbers. Publications de l'Institut Mathematique, Nouvelle Serie (Beograd) 34(48), 73-79. (1983).

GUTM 1985-1 Gutman, I.
On some graphic polynomials whose zeros are real. Publications de l'Institut Mathematique (Beograd), Nouvelle Serie 37(51), 29-32. (1985).

GUTM 1986-1 Gutman, I.
Some relations for graphic polynomials. Publications de l'Institut Mathematique (Belgrade), Nouvelle Serie 39(53), 55-62. (1986).

GUTMA 1987-1 Gutman, I.; Cioslowski, J.
Bounds for the number of perfect matchings in hexagonal systems. Publications de l'Institut Mathematique (Belgrade), Nouvelle Serie 42(56), 21-27. (1987).

GUTMB 1984-1 Gutman, I.; Cvetkovic, D.
Finding tricyclic graphs with a maximal number of matchings - another example of computer aided research in graph theory. Publications de l'Institut Mahtematique. Nouvelle Serie. (Beograd) 35(49), 33-40. (1984).

GUTMC 1983-1 Gutman, I.; Farrell, E. J.; Wahid, S. A.
On the matching polynomials of graphs containing benzenoid chains. Journal of Combinatorics, Information and System Sciences 8, 159-168. (1983).

GUTMD 1986-1 Gutman, I.; Zhang, F. J.
On a quasi-ordering of bipartite graphs. Publications de l'Institut Mathematique (Belgrade), Nouvelle Serie 40(54), 11-15. (1986).

GVIS 1984-1 Gvishiani, A. D.; Gurvich, V. A.
Balanced flow in a multipole network. Soviet Physics, Doklady 29, 268-270; translation from: Doklady Akademii Nauk SSSR (Moscow) 275, 1327-1331 (1984). (1984).

GYAR 1985-1 Gyarfas, A.
On the chromatic number of multiple interval graphs and overlap graphs. Discrete Mathematics 55, 161-166. (1985).

GYAR 1985-2 Gyarfas, A.
Problems from the world surrounding perfect graphs. Studies 177/1985. Budapest: Hungarian Academy of Sciences, Computer and Automation Institute, 53 p. (1985).

GYARA 1985-1 Gyarfas, A.; Lehel, J.
Covering and coloring problems for relatives of intervals. Discrete Mathematics 55, 167-180. (1985).

GYARA 1986-1 Gyarfas, A.; Lehel, J.
On-line and first fit colorings of graphs. Working Paper. Budapest: Hungarian Academy of Sciences, Computer and Automation Institute, 25 p. (1986).

GYARB 1987-1 Gyarfas, A.; Lehel, J.; Nesetril, J.; Rödl, V.; Schelp, R. H.; Tuza, Z.
Local k-colorings of graphs and hypergraphs. Journal of Combinatorial Theory 43B, 127-139. (1987).

GYARC 1987-1 Gyarfas, A.; Lehel, J.; Schelp, R. H.; Tuza, Z.
Ramsey numbers for local colorings. Graphs and Combinatorics 3, 267-278. (1987).

GYOR 1984-1 Györi, E.
A minimax theorem on intervals. Journal of Combinatorial Theory, Series B 37, 1-9. (1984).

GYORA 1985-1 Györi, E.; Rothschild, B. L.; Rucinski, A.
Every graph is contained in a sparsest possible balanced graph. Mathematical Proceedings of the Cambridge Philosophical Society 98, 397-401. (1985).

GYUL 1987-1 Gyulumyan, S. M.
Colour-critical graphs with an independent line cut (Russian). Kibernetika (Kiev) No. 2, 6-9. (1987).

HABE 1984-1 Habenicht, W.
Interaktive Lösungsverfahren für diskrete Vektoroptimierungsprobleme unter besonderer Berücksichtigung von Wegeproblemen in Graphen. Mathematical Systems in Economics, 90. Verlagsgruppe Athenäum/Hain Hanstein, Königstein/Ts., 219 p. (1984).

HABI 1984-1 Habib, M.
Comparability invariants. In: M. Pouzet, D. Richard (Eds.): Orders: Description and Roles, Annals of Discrete Mathematics Vol. 23, 371-386. Amsterdam: Norht-Holland. (1984).

HABIA 1985-1 Habib, M.; Jegou, R.
N-free posets as generalizations of series-parallel posets. Discrete Applied Mathematics 12, 279-292. (1985).

HABIB 1985-1 Habib, M.; Maurer, M. C.
1-intersecting families. Discrete Mathematics 53, 91-102. (1985).

HABIC 1987-1 Habib, M.; Möhring, R. H.
On some complexity properties of N-free posets and posets with bounded decomposition diameter. Discrete Mathematics 63, 157-183. (1987).

HABID 1987-1 Habib, M.; Möhring, R. H.; Steiner, G.
Computing the bump number of ordered sets is easy. Report 87475-OR. Bonn: Univ. Bonn, Inst. f. Ökonometrie u. Operations Research, 28 p. (1987).

HACK 1984-1 Hackbarth, K.-D.
Implementation of algorithms for telecommunication network planning. In: U. Pape (Ed.): Proc. WG '84: Internat. Workshop on Graphtheoretic Concepts in Computer Science, Berlin 1984, 174-188. Linz: Trauner. (1984).

HACK 1986-1 Hackbarth, K.-D.
A heuristic procedure for calculating telecommunication transmission networks in consideration of network reliability. In: G. Gallo, C. Sandi (Eds.): Netflow at Pisa, Mathematical Programming Study 26, 213-214. (1986).

HAES 1985-1 Haessler, R. W.
Production planning and scheduling for an integrated container company. Automatica 21, 445-452. (1985).

HAGE 1985-1 Hager, M.
Pendant tree-connectivity. Journal of Combinatorial Theory 38B, 179-189. (1985).

HAGE 1986-1 Hager, M.
Path-connectivity in graphs. Discrete Mathematics 59, 53-60. (1986).

HAGEA 1987-1 Hagerup, T.; Chrobak, M.; Diks, K.
Parallel 5-colouring of planar graphs. In: T. Ottmann (Ed.): Automata, Languages and Programming, 304-313. Lecture Notes in Computer Science 267, Berlin: Springer. (1987).

HAGG 1980-1 Haggard, G.
Excursions in graph theory. Orono: Univ. of Maine at Orono. (1980).

HAGGA 1985-1 Häggkvist, R.
A lemma on cycle decompositions. In: B. R. Alspach, C. D. Godsil (Eds.): Cycles in Graphs, Annals of Discrete Mathematics Vol. 27, 227-232. Amsterdam: North-Holland. (1985).

HAGGA 1985-2 Häggkvist, R.
A note on Hamilton cycles. In: B. R. Alspach, C. D. Godsil (Eds.): Cycles in Graphs, Annals of Discrete Mathematics Vol. 27, 233-234. Amsterdam: Norht-Holland. (1985).

HAGGA 1985-3 Häggkvist, R.
Equicardinal disjoint cycles in sparse graphs. In: B. R. Alspach, C. D. Godsil (Eds.): Cycles in Graphs, Annals of Discrete Mathematics Vol. 27, 269-274. Amsterdam: North-Holland. (1985).

HAGGB 1985-1 Häggkvist, R.; Jackson, B.
A note on maximal cycles in 2-connected graphs. In: B. R. Alspach, C. D. Godsil (Eds.): Cycles in Graphs, Annals of Discrete Mathematics Vol. 27, 205-208. Amsterdam: North-Holland. (1985).

HAGH 1987-1 Haghani, A. E.; Daskin, M. S.
A combined model of train routing, makeup, and empty car distribution. The Logistics and Transportation Review 23, 173-188. (1987).

HAGS 1984-1 Hagstrom, J. N.
Note on independence of arcs in antiparallel for network flow problems. Networks 14, 567-570. (1984).

HAIM 1985-1 Haimovich, M.; Rinnooy Kan, A. H. G.
Bounds and heuristics for capacitated routing problems. Mathematics of Operations Research 10, 527-542. (1985).

HAIN 1985-1 Haines, C. L.
An algorithm for carrier routing in a flexible material-handling system. IBM Journal of Research and Development 29, 356-362. (1985).

HAJE 1984-1 Hajek, B.; Ogier, R. G.
Optimal dynamic routing in communication networks with continuous traffic. Networks 14, 457-487. (1984).

HAKI 1985-1 Hakimi, S. L.
Further results on a generalization of edge-coloring. In: Y. Alavi, G. Chartrand, L. Lesniak, D. R. Lick, C. E. Wall (Eds.): Graph Theory with Applications to Algorithms and Computer Science, 371-390. Chichester: Wiley. (1985).

HAKI 1986-1 Hakimi, S. L.
P-median theorems for competitive locations. In: J. P. Osleeb, P. J. Ratick (Eds.): Locational Decisions: Methodology and Applications, 77-98. Annals of Operations Research 6, Basel: Baltzer. (1986).

HALI 1981-1 Halin, R.
Graphentheorie 2. Darmstadt: Wissenschaftliche Buchgesellschaft. (1981).

HALL 1985-1 Hall, J. I.; Shult, E. E.
Locally cotriangular graphs. Geometriae Dedicata (Dordrecht) 18, 113-159. (1985).

HALLA 1986-1 Hall, N. G.
Production problems with deadline penalties. International Journal of Production Research 24, 1383-1394. (1986).

HALLA 1986-2 Hall, N. G.
Scheduling problems with generalized due dates. IIE Transactions 18, 220-222. (1986).

HALLB 1986-1 Hall, N. G.; Hochbaum, D. S.
A fast approximation algorithm for the multicovering problem. Discrete Applied Mathematics 15, 35-40. (1986).

HALLC 1986-1 Hall, N. G.; Rhee, W. T.
Average and worst-case analysis of heuristics for the maximum tardiness problem. European Journal of Operational Research 26, 272-277. (1986).

HALLD 1986-1 Hall, N. G.; Vohra, R. V.
Absolute bounds on optimal costs for a class of set covering problems. Working Paper Series No. 86-22. Columbus: Ohio State Univ., College of Administrative Science, 11p. (1986).

HALLD 1987-1 Hall, N. G.; Vohra, R. V.
An on-line assignment problem with random effectiveness and costly information. Operations Research Letters 6, 163-168. (1987).

HALLE 1985-1 Hall, R. W.
Heuristics for selecting facility locations. The Logistics and Transportation Review 21, 353-374. (1985).

HALLE 1985-2 Hall, R. W.
Vehicle scheduling at a transportation terminal with random delay en route. Transportation Science 19, 308-320. (1985).

HALLE 1986-1 Hall, R. W.
The fastest path through a network with random time-dependent travel times. Transportation Science 20, 182-188. (1986).

HALLE 1987-1 Hall, R. W.
Comparison of strategies for routing shipments through transportation terminals. Transportation Research 21A, 421-430. (1987).

HALLE 1987-2 Hall, R. W.
Direct versus terminal freight routing on a network with concave costs. Transportation Research 21B, 287-298. (1987).

HAMA 1981-1 Hamacher, H. W.
Optimal (s,t)-cuts. In: H. Noltemeier (Ed.): Graphtheoretic Concepts in Computer Science, Proc. Internat. Workshop WG 80 (Bad Honnef 1980), 383-387. Berlin, Heidelberg, New York: Springer. (1981).

HAMA 1985-1 Hamacher, H. W.
A time expanded matroid algorithm for finding optimal dynamic matroid intersections. Zeitschrift für Operations Research 29A, 203-216. (1985).

HAMA 1985-2 Hamacher, H. W.
Min cost tensions. Journal of Information & Optimization Science (Delhi) 6, 285-304. (1985).

HAMA 1986-1 Hamacher, H. W.
Maximal dynamic polymatroid flows and applications. Discrete Applied Mathematics 15, 41-54. (1986).

HAMAA 1986-1 Hamacher, H. W.; Foulds, L. R.
Algorithms for flows with parametric capacities. Research Report No. 86-10. Gainesville: Univ. of Florida, Industrial and Systems Engineering Dept., 26p. (1986).

HAMAB 1984-1 Hamacher, H. W.; Picard, J.-C.; Queyranne, M.
Ranking the cuts and cut-sets of a network. In: R. E. Burkard, R. A. Cuninghame-Green, U. Zimmermann (Eds.): Algebraic and Combinatorial Methods in Operations Research, Annals of Discrete Mathematics Vol. 19, 183-200. Amsterdam: North-Holland. (1984).

HAMAC 1986-1 Hamacher, H. W.; Queyranne, M.
K best solutions to combinatorial optimization problems. In: C. L. Monma (Ed.): Algorithms and Software for Optimization - Part I, 123-144. Annals of Operations Research 4, Basel: Baltzer. (1986).

HAMAD 1984-1 Hamacher, H. W.; Tufekci, S.
Algebraic flows and time-cost tradeoff problems. In: R. E. Burkard, R. A. Cuninghame-Green, U. Zimmermann (Eds.): Algebraic and Combinatorial Methods in Operations Research, Annals of Discrete Mathematics Vol. 19, 165-182. Amsterdam: North-Holland. (1984).

HAMAD 1987-1 Hamacher, H. W.; Tufekci, S.
On the use of lexiographic min cost flows in evacuation modeling. Naval Research Logistics 34, 487-504. (1987).

HAMI	1984-1	Hamidoune, Y. O.; Vergnas, M. Las Directed switching games on graphs and matroids (French). Comptes Rendus des Seances de l'Academie des Sciences, Serie I, Mathematique 298, 497-499. (1984).
HAMI	1986-1	Hamidoune, Y. O.; Vergnas, M. Las Switching games on graphs and matroids. Journal of Combinatorial Theory, Series B 40, 237-269. (1986).
HAMM	1984-1	Hammer, G. Greedy solutions for general set covering problems. In: H. Hauptmann, W. Krelle, K. C. Mosler (Eds.): Operations Research and Economic Theory, 321-326. Berlin: Springer-Verlag. (1984).
HAMMA	1986-1	Hammer, P. L.; Maffray, F. Completely separable graphs. RUTCOR Research Report 29-86. New Brunswick: Rutgers Center for Operations Research, 20 p. (1986).
HAMMB	1985-1	Hammer, P. L.; Mahadev, N. V. R. Bithreshold graphs. SIAM Journal on Algebraic and Discrete Methods 6, 497-506. (1985).
HAMMB	1985-2	Hammer, P. L.; Mahadev, N. V. R. Intershold graphs. Utilitas Mathematica 27, 207-215. (1985).
HAMMC	1985-1	Hammer, P. L.; Mahadev, N. V. R.; Werra, D. de Stability in CAN-free graphs. Journal of Combinatorial Theory 38B, 23-30. (1985).
HAMMC	1985-2	Hammer, P. L.; Mahadev, N. V. R.; Werra, D. de The struction of a graph: application to CN-free graphs. Combinatorica 5, 141-148. (1985).
HAMMD	1987-1	Hammer, P. L.; Simeone, B. Order relations of variables in 0-1 programming. In: P. Martello, G. Laporte, M. Minoux, C. Ribeiro (Eds.): Surveys in Combinatorial Optimization, Annals of Discrete Mathematics 31, 83-112. Amsterdam, New York, Tokyo: North-Holland. (1987).
HAMMD	1987-2	Hammer, P. L.; Simeone, B. Quadratic functions of binary variables. RUTCOR Research Report 20-87. New Brunswick: Rutgers Univ., Center for Operations Research, 92 p. (1987).
HAMME	1984-1	Hammer, P. L.; Simeone, B.; Liebling, T. M.; Werra, D. de From linear separability to unimodality: a hierarchy of pseudo-Boolean functions. O.R. Working Paper 84/3. Lausanne: Ecole Polytechnique Federale de Lausanne, Dept. de Mathematiques, 34 p. (1984).
HAMMF	1985-1	Hammouche, A.; Webster, D. B. Evaluation of an application of graph theory to the layout problem. International Journal of Production Research 23, 987-1000. (1985).
HAMO	1984-1	Hamon, J. P.; Weissert, F. ROSSINI: a general model for optimizing aircraft assignment to parking positions at an airport. In: J. P. Brans (Ed.): Operational Research '84, 748-760. Amsterdam: North-Holland. (1984).
HAN	1982-1	Han, F. An integer programming algorithm for the plan of a freight car classification system (Chinese). Mathematics in Practice and Theory, No. 2, 1-13. (1982).
HAN A	1987-1	Han, J.-Y.; Xu, B.-S. Advances in the theory of sequencing and scheduling (Chinese). Qufu Shifan Daxue Xuebao. Ziran Kexue Ban 13, No. 2, 19-28. (1987).
HAND	1984-1	Handler, G. Y. The absolute m-center problem on a directed network. Transportation Science 18, 300-302. (1984).
HAND	1985-1	Handler, G. Y. Medi-centers of a tree. Transportation SCience 19, 246-260. (1985).
HANJ	1985-1	Hanjoul, P.; Peeters, D. Comparison of two dual-based procedures for solving the p-median problem. European Journal of Operational Research 20, 387-396. (1985).
HANL	1985-1	Hanlon, P. The chromatic polynomial of an unlabeled graph. Journal of Combinatorial Theory 38B, 226-239. (1985).
HANN	1985-1	Hanna, M. E.; Austin, L. M. An advanced start algorithm for all-integer programming. Computers & Operations Research 12, 301-310. (1985).
HANS	1984-1	Hanscom, M. A.; Nguyen, V. H.; Strodiot, J.-J. A reduced subgradient algorithm for network flow problems with convex nondifferentiable costs. In: V. F. Demyanov, D. Pallaschke (Eds.): Nondifferentiable optimization: motivations and applications, 318-322. Berlin, New York: Springer. (1984).
HANSA	1986-1	Hansen, J. V.; Giauque, W. C. Task allocation in distributed processing systems. Operations Research Letters 5, 137-144. (1986).
HANSB	1984-1	Hansen, P. Shortest paths in signed graphs. In: R. E. Burkard, R. A. Cuninghame-Green, U. Zimmermann (Eds.): Algebraic and Combinatorial Methods in Operations Research, Annals of Discrete Mathematics Vol. 19, 201-214. Amsterdam: North-Holland. (1984).
HANSC	1987-1	Hansen, P.; Huge, J. Implicit treatment of zero or range constraints in a model for minimun cost foundry alloys. RUTCOR Research Report 19-87. New Brunswick: Rutgers Univ., Center for Operations Research, 7 p. (1987).
HANSD	1987-1	Hansen, P.; Jaumard, B. Algorithms for the maximum satisfiability problem. RUTCOR Research Report 43-87. New Brunswick: Rutgers Center for Operations Research, 44 p. (1987).
HANSD	1987-2	Hansen, P.; Jaumard, B. Minimum sum of diameters clustering. RUTCOR Research Report 31-87. New Brunswick: Rutgers Center for Operations Research, 14 p. (1987).
HANSE	1987-1	Hansen, P.; Labbe, M.; Peeters, D.; Thisse, J.-F. Single facility location on networks. In: P. Martello, G. Laporte, M. Minoux, C. Ribeiro (Eds.): Surveys in Combinatorial Optimization, Annals of Discrete Mathematics 31, 113-146. Amsterdam, New York, Tokyo: North-Holland. (1987).
HANSF	1987-1	Hansen, P.; Lu, S. H. On the equivalence of paved-duality and standard linearization in 0-1 optimization. RUTCOR Research Report 30-87. New Brunswick: Rutgers Center for Operations Research, 7 p. (1987).

HANSG 1987-1 Hansen, P.; Minoux, M.; Labbe, M.
Extension de la programmation lineaire generalisee au cas des programmes mixtes. Comptes Rendus de l'Academie des Sciences, Paris, Serie I, Mathematique 305, 569-572. (1987).

HANSH 1985-1 Hansen, P.; Peeters, D.; Richard, D.; Thisse, J.-F.
The minisum and minimax locations problems revisited. Operations Research 33, 1251-1265. (1985).

HANSI 1985-1 Hansen, P.; Peeters, D.; Thisse, J.-F.
Facility location analysis. RUTCOR Research Report 4-85. New Brunswick: Rutgers Center for Operations Research, 56 p. (1985).

HANSJ 1986-1 Hansen, P.; Simeone, B.
Unimodular functions. Discrete Applied Mathematics 14, 269-282. (1986).

HANSK 1985-1 Hansen, P.; Thisse, J.-F.; Wendell, R. E.
Location by competitive and voting processes. RUTCOR Research Report 6-85. New Brunswick: Rutgers Center for Operations Research, 27 p. (1985).

HANSK 1986-1 Hansen, P.; Thisse, J.-F.; Wendell, R. E.
Equivalence of solutions to network location problems. Mathematics of Operations Research 11, 672-678. (1986).

HANSL 1986-1 Hanson, D.; Robinson, G. C.; Toft, B.
Remarks on the graph colour theorem of Hajos. Congressus Numerantium 55, 69-76. (1986).

HANSM 1985-1 Hanson, D.; Seyffarth, K.
Point saturation of graphs. Ars Combinatoria 20, 147-158. (1985).

HARA 1985-1 Harant, J.
Neue Ergebnisse in der Dienstplanoptimierung. 30. Internationales Wissenschaftliches Kolloquium (TH Ilmenau, DDR) Heft 4E, 57-58. (1985).

HARAA 1984-1 Harao, M.; Lomtong, P.
Set-up and routing algorithms for a cellular interconnection network. In: Basic Theory of Computer Science, Proceedings of a Symposium held at the Research Institute for Mathematical Sciences (Kyoto), 120-130. (1984).

HARAB 1984-1 Harary, F.
General connectivity. In: K. M. Koh, H. P. Yap (Eds.): Graph Theory Singapore 1983, Lecture Notes in Mathematics 1073, 83-92. Berlin: Springer-Verlag. (1984).

HARAB 1985-1 Harary, F.
Conditional colorability in graphs. In: F. Harary, J. P. Maybee (Eds.): Graphs and Applications, 127-136. New York: Wiley. (1985).

HARAB 1985-2 Harary, F.
Graph theoretic approaches to finite mathematical structures. In: Y. Alavi, G. Chartrand, L. Lesniak, D. R. Lick, C. E. Wall (Eds.): Graph Theory with Applications to Algorithms and Computer Science, 401-412. Chichester: Wiley. (1985).

HARAB 1985-3 Harary, F.
Petersen's graph and its unique properties (Danish). Normat (Oslo) 33, 113-118. (1985).

HARAC 1982-1 Harary, F.; Hoede, C.; Kadlecek, D.
Graph-valued functions related to step graphs. Journal of Combinatorics, Information & System Sciences (Delhi) 7, 231-245. (1982).

HARAD 1985-1 Harary, F.; Iri, M.
Generalized Ramsey theory for graphs. XVI. Reduced Ramsey numbers. Bolletino della Unione Matematica Italiana A (6) 4, 13-18. (1985).

HARAE 1985-1 Harary, F.; Jones, K. F.
Conditional colorability II: bipartite variations. Congressus Numerantium 50, 205-218. (1985).

HARAE 1986-1 Harary, F.; Jones, K. F.
Degree conditional bipartition numbers in graphs. Congressus Numerantium 55, 39-50. (1986).

HARAF 1984-1 Harary, F.; Karabed, R.
Regular digraphs containing a given digraph. Canadian Mathematical Bulletin 27, No. 2, 129-133. (1984).

HARAG 1985-1 Harary, F.; Kolasinska, E.; Syslo, M. M.
Cycle basis interpolation theorems. In: B. R. Alspach, C. D. Godsil (Eds.): Cycles in Graphs, Annals of Discrete Mathematics Vol. 27, 369-380. Amsterdam: North-Holland. (1985).

HARAH 1987-1 Harary, F.; Peled, U. N.
Hamiltonian threshold graphs. Discrete Applied Mathematics 16, 11-16. (1987).

HARAI 1983-1 Harary, F.; Plantholt, M.
The derived signed graph of a digraph. Expositiones Mathematicae 1, 343-347. (1983).

HARAI 1985-1 Harary, F.; Plantholt, M.
The point-distinguishing chromatic index. In: F. Harary, J. P. Maybee (Eds.): Graphs and Applications, 147-162. New York: Wiley. (1985).

HARAJ 1985-1 Harary, F.; Robinson, R. W.
The diameter of a graph and its complement. American Mathematical Monthly 92, 211-212. (1985).

HARAK 1984-1 Harary, F.; Schmidt, R.
Realization of graph regulation numbers. In: D. Cvetkovic, I. Gutman, T. Pisanski, R. Tosic (Eds.): Graph Theory (Novi Sad, 1983), 161-166. Univ. Novi Sad. (1984).

HARAL 1987-1 Harary, F.; Schuster, S.; Vestergaard, P. D.
Interpolation theorems for the invariants of spanning trees of a given graph: edge-covering. Congressus Numerantium 59, 107-114. (1987).

HARAM 1986-1 Harary, F.; Slater, P. J.
A linear algorithm for the cutting center of a tree. Information Processing Letters 23, 317-320. (1986).

HARAN 1985-1 Harary, F.; Tindell, R.
The minimal blocks of diameter two and three. In: F. Harary, J. P. Maybee (Eds.): Graphs and Applications, 163-182. New York: Wiley. (1985).

HARB 1985-1 Harborth, H.
Darstellungen von Graphen in der Ebene. In: R. Bodendiek, H. Schumacher, G. Walther (Eds.): Graphen in Forschung und Unterricht, Festschrift Klaus Wagner, 54-64. Salzdetfurth: Franzbecker. (1985).

HARB 1985-2 Harborth, H.
Drawings of graphs and multiple crossings. In: Y. Alavi, G. Chartrand, L. Lesniak, D. R. Lick, C. E. Wall (Eds.): Graph Theory with Applications to Algorithms and Computer Science, 413-422. Chichester: Wiley. (1985).

HARE 1985-1 Hare, E. O.; Hare, W. R.; Hedetniemi, S. T.
Another upper bound for the bandwidth of trees. Congressus Numerantium 50, 77-84. (1985).

HAREA 1986-1 Hare, E. O.; Hedetniemi, S. T.; Hare, W. R.
Algorithms for computing the domination number of KxN complete grid graphs. Congressus Numerantium 55, 81-92. (1986).

HAREB 1987-1 Hare, E. O.; Hedetniemi, S. T.; Laskar, R.; Peters, K.; Wimer, T. V.
Linear-time computability of combinatorial problems on generalized series-parallel graphs. In: D. P. Johnson, A. Nozaki, T. Nishizeki, H. P. Wilf (Eds.): Discrete Algorithms and Complexity, 437-458. Boston: Academic Press. (1987).

HAREC 1985-1 Hare, W. R.; Hare, E. O.; Hedetniemi, S. T.
Bandwidth of grid graphs. Congressus Numerantium 50, 67-76. (1985).

HARED 1985-1 Hare, W. R.; Hedetniemi, S. T.; Laskar, R.; Pfaff, J.
Complete coloring parameters of graphs. Congressus Numerantium 48, 171-178. (1985).

HAREE 1984-1 Hare, W. R.; Laskar, R.; Peters, K.
Max-min parameters and disconnected colorings. Congressus Numerantium 44, 139-150. (1984).

HAREF 1985-1 Harel, D.
A linear algorithm for finding dominators in flow graphs and related problems. In: Proc. Seventeenth Annual ACM Symp. on Theory of Computing, Providence 1985, 185-194. ACM. (1985).

HARK 1982-1 Harker, P. T.; Friesz, T. L.
A simultaneous freight network equilibrium model. In: F. Hoffman, K. B. Reid, R. C. Mullin, R. G. Stanton (Eds.): Proc. Thirteenth Southeastern Conf. on Combinatorics, Graph Theory and Computing, Florida 1982, Congressus Numerantium Vol. 36, 365-402. Winnipeg, Canada: Utilitas Mathematica. (1982).

HARN 1984-1 Harnett, R. M.
A single-branch implicit enumeration algorithm for zero-one programs with geometrical constraints. Computers & Operations Research 11, 321-332. (1984).

HARR 1986-1 Harris, B.
A continuous dual method for the Hitchcock problem. Transportation Research 20B, 71-75. (1986).

HARRA 1984-1 Harrison, P.
The two dimensional cutting problem of the furniture manufacturer. In: J. P. Brans (Ed.): Operational Research '84, 625-640. Amsterdam: North-Holland. (1984).

HART 1987-1 Hart, J. P.; Shogan, A. W.
Semi-greedy heuristics: an empirical study. Operations Research Letters 6, 107-114. (1987).

HARTA 1985-1 Hartley, R.
Vector optimal routing by dynamic programming. In: P. Serafini (Ed.): Mathematics of Multi Objective Optimization, 215-226. ICMS No. 289. Wien, New York: Springer. (1985).

HARTB 1985-1 Hartman, A.; Rosa, A.
Cyclic one-factorization of the complete graph. European Journal of Combinatorics 6, 45-48. (1985).

HARTC 1983-1 Hartman, J.; Katchalski, M.
On k-cycled refinements of certain graphs. Canadian Mathematical Bulletin 26, 152-156. (1983).

HARTD 1984-1 Hartulari, C.
The systemic approach on the problem of production operative programming and scheduling. Economic Computation and Economic Cybernetics Studies and Research (Bucharest) No. 2, 63-68. (1984).

HARTE 1987-1 Hartvigsen, D.; Zemel, E.
Is every cycle basis fundamental? Discussion Paper. Evanston: Northwestern Univ., Dept. of Managerial Economics and Decision Sciences, 35 p. (1987).

HARTE 1987-2 Hartvigsen, D.; Zemel, E.
On the complexity of lifted inequalities for the knapsack problem. Discussion Paper 740. Evanston: Northwestern Univ., Dept. of Managerial Economics and Decision Sciences, 19 p. (1987).

HARTF 1985-1 Hartwig, A.
Recursive branch and bound. Optimization 16, 219-228. (1985).

HASH 1987-1 Hashizume, S.; Fukushima, M.; Katoh, N.; Ibaraki, T.
Approximation algorithms for combinatorial fractional programming problems. Mathematical Programming 37, 255-268. (1987).

HASS 1985-1 Hassan, M. M. D.; Hogg, G. L.; Smith, D. R.
A construction algorithm for the selection and assignment of materials handling equipment. International Journal of Production Research 23, 381-392. (1985).

HASSA 1984-1 Hässelbarth, W.
Die Verzweigtheit von Graphen. Match No. 16, 3-17. (1984).

HASSB 1987-1 Hassin, R.
A computing scheme for network problems with random edge lengths. In: G. Andreatta, F. Mason, P. Serafini (Eds.): Advanced School on Stochastics in Combinatorial Optimization (CISM Udine 1986), 228-232. Singapore: World Scientific. (1987).

HASSB 1987-2 Hassin, R.
Solution bases of multiterminal cut problems. Working Paper. Tel-Aviv: Tel-Aviv Univ., Statistics Dept., 12 p. (1987).

HASSC 1985-1 Hassin, R.; Johnson, D. B.
An 0(n(log n)**2) algorithm for maximum flow in undirected planar networks. SIAM Journal on Computing 14, 612-624. (1985).

HASSD 1986-1 Hassin, R.; Tamir, A.
Efficient algorithms for optimization and selection on series-parallel graphs. SIAM Journal on Algebraic and Discrete Methods 7, 379-389. (1986).

HASSD 1987-1 Hassin, R.; Tamir, A.
Maximizing classes of two parameter objectives over matroids. Working Paper. Tel-Aviv: Tel-Aviv Univ., Statistics Dept., 22 p. (1987).

HASSE 1985-1 Hassin, R.; Zemel, E.
On shortest paths in graphs with random weights. Mathematics of Operations Research 10, 557-564. (1985).

HAST 1985-1 Hastad, J.; Helfrich, B.; Lagarias, J. C.; Schnorr, C. P.
Polynomial time algorithms for finding integer relations among real numbers. Working Paper. Cambridge: Mass. Inst. of Techology, Dept. of Computer Science, 14 p. (1985).

HASTA 1986-1 Hastad, J.; Just, B.; Lagarias, J. C.; Schnorr, C. P.
Polynomial time algorithms for finding integer relations among real numbers. Cambridge: MIT Laboratory for Computer Science, 14p. (1986).

HAUR 1986-1 Haurie, A.; Marcotte, P.
A game-theoretic approach to network equilibrium. In: G. Gallo, C. Sand (Eds.): Netflow at Pisa, Mathematical Programming Study 26, 252-255. (1986).

HAVE 1984-1 Havel, I.
On Hamiltonian circuits and spanning trees of hypercubes. Casopis Pro Pestovani Matematiky (Prague) 109, No. 2, 135-152. (1984).

HAVEA 1984-1 Haverly, C. A.
The number of simplex iterations for a general LP code on network problems. COAL - Committee on Algorithms, Newsletter 11, 16-19. (1984).

HAYA 1985-1 Hayashi, Y.
Constuction of the F-, P- and K-trees of a knapsack problem and their computational experiments. Journal of the Operations Research Society of Japan 28, 241. (1985).

HAYA 1986-1 Hayashi, Y.
A solution method for a knapsack problem and its variant. Japan Journal of Applied Mathematics 3, 73-91. (1986).

HAYM 1984-1 Haymond, R. E.; Jarvis, J. P.; Shier, D. R.
Algorithm 613. Minimum spanning tree for moderate integer weights. ACM Transactions on Mathematical Software 10, 108-111. (1984).

HAYW 1985-1 Hayward, R. B.
Weakly triangulated graphs. Journal of Combinatorial Theory 39B, 200-208. (1985).

HAYW 1986-1 Hayward, R. B.
Murky graphs. In: Raspail-Burnside: Workshop on Perfect Graphs, Bellairs Research Institute, McGill University, Barbados, West Indies, 1986, 31-99. (1986).

HAYW 1987-1 Hayward, R. B.
A lower bound for the optimal crossing-free Hamiltonian cycle problem. Discrete and Computational Geometry 2, 327-344. (1987).

HE 1985-1 He, W.-J.
On Jackson's conjecture of 2-connected and "almost regular" graphs (Chinese). Chinese Journal of Operations Research 4, 67-68. (1985).

HEAR 1984-1 Hearn, D. W.
Practical and theoretical aspects of aggregation problems in transportation planning models. In: M. Florian (Ed.): Transportation planning models, Proc. Course Internat. Center Transport Stud., Amalfi/Italy 1982, 257-287. North-Holland. (1984).

HEARA 1984-1 Hearn, D. W.; Lawphongpanich, S.; Ventura, J. A.
Restricted simplicial decomposition: computation and extensions. Research Report 84-38. Gainesville: Univ. of Florida, Ind. & Systems Eng. Dept., 39 p. (1984).

HEARA 1987-1 Hearn, D. W.; Lawphongpanich, S.; Ventura, J. A.
Optimization algorithms for congested network models. Research Report, 87-3. Gainesville: Univ. of Florida, Industrial & Systems Eng. Dept., 14 p. (1987).

HEAT 1987-1 Heath, L. S.
Embedding outerplanar graphs in small books. SIAM Journal on Algebraic and Discrete Methods 8, 198-218. (1987).

HEDE 1985-1 Hedetniemi, S.; Hedetniemi, S.; Laskar, R.
Domination in trees: models and algorithms. In: Y. Alavi, G. Chartrand, L. Lesniak, D. R. Lick, C. E. Wall (Eds.): Graph Theory with Applications to Algorithms and Computer Science, 423-442. Chichester: Wiley. (1985).

HEDEA 1986-1 Hedetniemi, S.; Laskar, R.
A bipartite theory of graphs: I. Congressus Numerantium 55, 5-14. (1986).

HEDEB 1982-1 Hedetniemi, S. M.; Hedetniemi, S.; Beyer, T.
A linear algorithm for the Grundy (coloring) number of a tree. In: F. Hoffman, K. B. Reid, R. C. Mullin, R. G. Stanton (Eds.): Proc. Thirteenth Southeastern Conf. on Combinatorics, Graph Theory and Computing, Florida 1982, Congressus Numerantium Vol. 36, 351-364. Winnipeg, Canada: Utilitas Mathematica. (1982).

HEDEC 1985-1 Hedetniemi, S. M.; Proskurowski, A.; Syslo, M. M.
Interior graphs of maximal outerplane graphs. Journal of Combinatorial Theory 38B, 156-167. (1985).

HEDED 1983-1 Hedetniemi, S. T.
A max-min relationship between matchings and domination in graphs. Congressus Numerantium 40, 23-34. (1983).

HEDEE 1985-1 Hedetniemi, S. T.; Laskar, R.; Pfaff, J.
Irredundance in graphs: a survey. Congressus Numerantium 48, 183-194. (1985).

HEDEE 1986-1 Hedetniemi, S. T.; Laskar, R.; Pfaff, J.
A linear algorithm for finding a minimum dominating set in a cactus. Discrete Applied Mathematics 13, 287-294. (1986).

HEDG 1983-1 Hedge, M.; Sridharan, M. R.
Partitions of some classes of digraphs. Journal of Combinatorics, Information & System Sciences (Delhi) 8, 213-222. (1983).

HEDM 1985-1 Hedman, B.
The maximum number of cliques in dense graphs. Discrete Mathematics 54, 161-166. (1985).

HEDM 1986-1 Hedman, B.
A polynomial algorithm for constructing the clique graph of a line graph. Discrete Applied Mathematics 15, 61-66. (1986).

HEDM 1987-1 Hedman, B.
Another extremal problem for Turan graphs. Discrete Mathematics 65, 173-176. (1987).

HEES 1985-1 Heesch, H.
E-Reduktion. In: R. Bodendiek, H. Schumacher, G. Walther (Eds.): Graphen in Forschung und Unterricht, Festschrift Klaus Wagner, 78-84. Salzdetfurth: Franzbecker. (1985).

HEFN 1987-1 Hefner, K. A. S.; Lundgren, J. R.
Biclique covers and partitions of acyclic digraphs. Congressus Numerantium 59, 135-146. (1987).

HEIN 1984-1 Heinrich, H.
Ein heuristisches Lösungsverfahren zur Stundenplanung und Raumbelegung an Hochschulen. Angewandte Informatik 9, 379-388. (1984).

HEINA 1987-1 Heinrich, K.; Hell, P.
On the problem of bandsize. Graphs and Combinatorics 3, 279-284. (1987).

HEINB 1985-1 Heinrich, K.; Nonay, G. M.
Path and cycle decompositions of complete multigraphs. In: B. R. Alspach, C. D. Godsil (Eds.): Cycles in Graphs, Annals of Discrete Mathematics Vol. 27, 275-286. Amsterdam: North-Holland. (1985).

HEINC 1985-1 Heinssen, D. K.; Powers, D. L.
Generalized floorplans from assigned room adjacencies. International Journal of Systems Science 16, 503-508. (1985).

HELD 1984-1 Held, M.; Hoffman, A. J.; Johnson, E. L.; Wolfe, P.
Aspects of the traveling salesman problem. IBM Journal of Research and Development 28, 476-486. (1984).

HELL 1986-1 Hell, P.; Kirkpatrick, D. G.
Packings by complete bipartite graphs. SIAM Journal on Algebraic and Discrete Methods 7, 199-209. (1986).

HELLA 1986-1 Hell, P.; Li, W.-X.; Schmerl, J. H.
Jump number and width. Order 3, 227-234. (1986).

HELLB 1985-1 Hell, P.; Speer, E. R.
Matroids with weighted bases and Feynman integrals. In: M. Rosenfeld, J. Zaks (Eds.): Convexity and Graph Theory. Annals of Discrete Mathematics Vol. 20, 165-176. Amsterdam: North-Holland. (1985).

HELLC 1984-1 Hellmann, A.
Theorie und Praxis von Routing-Problemen: Lösungsverfahren zur Tourenplanung im Eindepot- und Mehrdepot-Fall. Veröffentlichung der SNV Studiengesellschaft Nahverkehr Nordrhein-Westfalen. Schwarzenbek: Verlag Dr. Rüdiger Martienss. (1984).

HELM 1986-1 Helmbold, D.; Mayr, E.
Two processor scheduling is in NC. In: F. Makedon, K. Mehlhorn, T. Papatheodorou, P. Spirakis (Eds.): AWOC 86: VLSI Algorithms and Architectures, 12-25. Lecture Notes in Computer Science 227, Berlin: Springer. (1986).

HELM 1987-1 Helmbold, D.; Mayr, E.
Applications of parallel scheduling to perfect graphs. In: G. Tinhofer, G. Schmidt (Eds.): Graph-Theoretic Concepts in Computer Science, 188-203. Lecture Notes in Computer Science 246, Berlin: Springer. (1987).

HELM 1987-2 Helmbold, D.; Mayr, E.
Two processor scheduling is in NC . SIAM Journal on Computing 16, 747-759. (1987).

HEMM 1985-1 Hemminger, R. L.; Wilson, E. L.
Gamma-regularity: multigraphs and digraphs. Congressus Numerantium 49, 339-349. (1985).

HEMP 1985-1 Hempel, L.; Herrmann, U.; Hölzer, W.
Einige spezielle Probleme im Zusammenhang mit Multiströmen. In: H. Sachs (Ed.): Graphs, Hypergraphs and Applications, Proc. Conf. on Graph Theory (Eyba 1984), 62-65. Leipzig: Teubner. (1985).

HEND 1986-1 Hendry, G. R. T.
The size of graphs uniquely Hamiltonian-connected from a vertex. Discrete Mathematics 61, 57-60. (1986).

HENDA 1987-1 Hendy, M. D.; Penny, D.
Edge lengths of trees from sequence data. Mathematical Biosciences 83, 157-166. (1987).

HENI 1986-1 Henig, M. I.
Extensions of the dynamic programming method in the deterministic and stochastic assembly-line balancing problems. Computers and Operations Research 13, 443-449. (1986).

HENI 1986-2 Henig, M. I.
The shortest path problem with two objective functions. European Journal of Operational Research 25, 281-291. (1986).

HENR 1987-1 Henriques, N.; Safayeni, F.; Fuller, D.
Human heuristics for the travelling salesman problem. Working Paper. Waterloo: Univ. of Waterloo, Dept. of Management Sciences, 34 p. (1987).

HERN 1985-1 Hernandez, M. G.; Alvarez, A.; Ruiz J.; Blanco, R.
Synthesis of a transportation graph with a minimum number of edges, given an upper bound for the flow (Spanish). Investigacion Operational (Havanna) 6, No. 2-3, 3-9. (1985).

HERNA 1985-1 Hernandez, M. G.; Canizares, L. C.; Gonzalez, I. Y.; Cedeno, R. V.
Synthesis of a transportation graph with a minimum number of intermediate nodes (Spanish). Investigacion Operational (Havanna) 6, No. 2-3, 10-17. (1985).

HERNB 1985-1 Hernandez, M. G.; Cedeno, R. V.; Gonzalez, I. Y.
Synthesis of a transportation network with linear restrictions on the flow capacities (Spanish). Investigacion Operacional (Habana) 6, No. 1, 3-10. (1985).

HERNC 1986-1 Hernandez, M. G.; Cornell, A. M.; Gamboa, D.
Synthesis of transport-type networks with restrictions in the output-edges (Spanish). Investigacion Operacional 7, 3-12. (1986).

HERND 1985-1 Hernandez, M. G.; Gonzalez, I. Y.; Canizares, L. C.; Cedeno, R. V.
Analysis of the existence of a linear flow in a transportation graph (Spanish). Investigacion Operational (Havanna) 6, No. 2-3, 32-36. (1985).

HERR 1985-1 Herroelen, W.; Gils, A. van
On the use of flow dominance in complexity measures for facility layout problems. International Journal of Production Research 23, 97-108. (1985).

HERT 1987-1 Hertz, A.; Werra, D. de
Using tabu search techniques for graph coloring. Computing 39, 345-352. (1987).

HEXE 1985-1 Hexel, E.
On a class of greedoids (German). In: Graphen und Netzwerke - Theorie und Anwendungen, 30. Int. Wiss. Kolloq., Ilmenau 1985, 5, 37-39. (1985).

HEXE 1985-2 Hexel, E.
On vertex sets in graphs with neighbourhood constraints - the component greedoid of a graph. In: H. Sachs (Ed.): Graphs, Hypergraphs and Applications, Proc. Conf. on Graph Theory (Eyba 1984), 66-71. Leipzig: Teubner. (1985).

HEYD 1985-1 Heydemann, M. C.
Minimum number of circuits covering the vertices of a strong digraph. In: B. R. Alspach, C. D. Godsil (Eds.): Cycles in Graphs, Annals of Discrete Mathematics Vol. 27, 287-296. Amsterdam: North-Holland. (1985).

HEYDA 1984-1 Heydemann, M. C.; Sotteau, D.
Number of arcs and cycles in digraphs. Discrete Mathematics 52, 199-208. (1984).

HEYDA 1985-1 Heydemann, M. C.; Sotteau, D.
About some cyclic properties in digraphs. Journal of Combinatorial Theory 38B, 261-278. (1985).

HIC 1984-1 Hic, P.
On partially directed p-graphs. Mathematica Slovaca 34, 26-31. (1984).

HIC 1985-1 Hic, P.
A construction of geodetic blocks. Mathematica Slovaca (Bratislava) 35, 251-261. (1985).

HIC 1986-1 Hic, P.
Criticality concepts in geodetic graphs. Mathematica Slovaca 36, 329-333. (1986).

HIGA 1984-1 Higashiyama, Y.; Ariyoshi, H.
An algorithm for generation of s-t acyclic graphs. Electronics and Communication Engineers of Japan 67, 10-18. (1984).

HIGG 1984-1 Higgs, D.
Topological spaces which are also matroidal. Revue Roumaine de Mathematiques Pures et Appliquees 29, 763-768. (1984).

HILB 1986-1 Hilbig, F.
Edge structures in nonhamiltonian graphs (German). Fachbereich Mathematik der Technischen Universität Berlin. (1986).

HILT 1984-1 Hilton, A. J. W.
Hamiltonian decompositions of complete graphs. Journal of Combinatorial Theory, Series B 36, 125-134. (1984).

HILT 1987-1 Hilton, A. J. W.
Recent progress on edge-coloring graphs. Discrete Mathematics 64, 303-308. (1987).

HILTA 1987-1 Hilton, A. J. W.; Johnson, P. D.
Graphs which are vertex-critical with respect to the edge-chromatic number. Mathematical Proceedings of the Cambridge Philosophical Society 102, 211-221. (1987).

HILTB 1986-1 Hilton, A. J. W.; Rodger, C. A.
Hamiltonian decompositions of complete regular s-partite graphs. Discrete Mathematics 58, 63-78. (1986).

HIRA 1984-1 Hirata, T.; Kubota, K.; Saito, O.
A sufficient condition for a graph to be weakly k-linked. Journal of Combinatorial Theory, Series B 36, 85-94. (1984).

HIRC 1985-1 Hirche, J.
Optimalität der Nordwesteckenlösung beim Transportproblem mit monoton kreuzdifferenten Kosten. Wissenschaftliche Zeitschrift der Martin-Luther-Universität Halle-Wittenberg, Mathematisch-Naturwissenschaftliche Reihe, 34, No. 4, 59-62. (1985).

HIRS 1985-1 Hirschberg, D. S.; Larmore, L. L.
The least weight subsequence problem. In: 26th Annual Symp. on Foundations of Computer Science, Portland, Oregon 1985, Proceedings, 137-143. IEEE. (1985).

HIRSA 1984-1 Hirshfeld, D. S.
An overview of mathematical programming for strategic planning. In: T. H. Naylor, C. Thomas (Eds.): Optimization Models for Strategic Planning, 121-138. Amsterdam: North-Holland. (1984).

HOAN 1985-1 Hoang, C. T.
On the P(4)-structure of perfect graphs. II. Odd decompositions. Journal of Combinatorial Theory 39B, 220-232. (1985).

HOAN 1987-1 Hoang, C. T.
Alternating orientation and alternating colouration of perfect graphs. Journal of Combinatorial Theory 42B, 264-273. (1987).

HOAN 1987-2 Hoang, C. T.
On a conjecture of Meyniel. Journal of Combinatorial Theory 42B, 302-312. (1987).

HOANA 1986-1 Hoang, C. T.; Maffray, F.
Weakly triangulated graphs are strict quasi-parity graphs. In: Raspail-Burnside: Workshop on Perfect Graphs, Bellairs Research Institute, McGill University, Barbados, West Indies, 1986, 100-105. (1986).

HOANB 1987-1 Hoang, C. T.; Reed, B.
A note on short cycles in digraphs. Discrete Mathematics 66, 103-108. (1987).

HOBB 1984-1 Hobbs, A. M.
Packing sequences. Journal of Combinatorial Theory, Series A 36, 174-182. (1984).

HOBBA 1987-1 Hobbs, A. M.; Bourgeois, B. A.; Kasiraj, J.
Packing trees in complete graphs. Discrete Mathematics 67, 27-42. (1987).

HOCH 1984-1 Hochbaum, D. S.
When are NP-hard location problems easy? Annals of Operations Research, Vol. 1, 201-214. Basel: Baltzer. (1984).

HOCH 1985-1 Hochbaum, D. S.
Easy solutions for the K-center problem or the dominating set problem on random graphs. In: G. Ausiello, M. Lucertini (Eds.): Analysis and Design of Algorithms for Combinatorial Problems, Annals of Discrete Mathematics Vol. 25, 189-210. Amsterdam: North-Holland. (1985).

HOCHA 1983-1 Hochbaum, D. S.; Maass, W.
Fast approximation algorithms for the robot placement problem. Research Paper. Berkeley: Univ. of California, 37 p. (1983).

HOCHA 1985-1 Hochbaum, D. S.; Maass, W.
Approximation schemes for covering and packing problems in image processing and VLSI. Journal of the ACM 32, 130-136. (1985).

HOCHA 1987-1 Hochbaum, D. S.; Maass, W.
Fast approximation algorithms for a nonconvex covering problem. Journal of Algorithms 8, 305-323. (1987).

HOCHB 1985-1 Hochbaum, D. S.; Shmoys, D. B.
A best possible heuristic for the k-center problem. Mathematics of Operations Research 10, 180-184. (1985).

HOCHB 1985-2 Hochbaum, D. S.; Shmoys, D. B.
An 0(||V||**2) algorithm for the planar 3-cut problem. SIAM Journal on Algebraic and Discrete Methods 6, 707-712. (1985).

HOCHB 1985-3 Hochbaum, D. S.; Shmoys, D. B.
Using dual approximation algorithms for scheduling problems: theoretical and practical results. In: 26th Annual Symp. on Foundations of Computer Science, Portland, Oregon 1985, Proceedings, 79-89. IEEE. (1985).

HOCHB 1986-1 Hochbaum, D. S.; Shmoys, D. B.
A packing problem you can almost solve by sitting on your suitcase. SIAM Journal on Algebraic and Discrete Methods 7, 247-257. (1986).

HOCHB 1986-2 Hochbaum, D. S.; Shmoys, D. B.
A polynomial approximation scheme for machine scheduling on uniform processors: using the dual approximation approach. In: K. V. Nori: Foundations of Software, Technology and Theoretical Computer Science, Lecture Notes in Computer Science 241, 382-393. Berlin, Heidelberg, New York: Springer-Verlag. (1986).

HOCHB 1986-3 Hochbaum, D. S.; Shmoys, D. B.
A unified approach to approximation algorithms for bottleneck problems. Journal of the Association for Computing Machinery 33, 533-550. (1986).

HOCHB 1986-4 Hochbaum, D. S.; Shmoys, D. B.
Best possible heuristics for the bottleneck wandering salesperson and bottleneck vehicle routing problems. European Journal of Operational Research 26, 380-384. (1986).

HOCHB 1987-1 Hochbaum, D. S.; Shmoys, D. B.
Using dual approximation algorithms for scheduling problems: theoretical and practical results. Journal of the Association for Computing Machinery 34, 144-162. (1987).

HOCHC 1986-1 Hochbaum, D. S.; Wigderson, A.
The Linzertorte problem, or a unified approach to painting, baking and weaving. Discrete Applied Mathematics 14, 17-32. (1986).

HODD 1984-1 Hodder, J. E.
Financial market approaches to facility location under uncertainty. Operations Research 32, 1374-1380. (1984).

HODDA 1986-1 Hodder, J. E.; Dincer, M. C.
A multifactor model for international plant location and financing under uncertainty. Computers and Operations Research 13, 601-610. (1986).

HODDB 1985-1 Hodder, J. E.; Jucker, J. V.
A simple plant-location model for quantity-setting firms subject to price uncertainty. European Journal of Operational Research 21, 39-46. (1985).

HODG 1986-1 Hodgson, M. J.
A hierarchical location-allocation model with allocations based on facility size. In: J. P. Osleeb, P. J. Ratick (Eds.): Locational Decisions: Methodology and Applications, 273-292. Annals of Operations Research 6, Basel: Baltzer. (1986).

HODGA 1987-1 Hodgson, M. J.; Wong, R. T.; Honsaker, J.
The P-centroid problem on an inclined plane. Operations Research 35, 221-233. (1987).

HODGB 1986-1 Hodgson, T. J.; Nuttle, H. L. W.
A note on linear programming and the single machine lot size scheduling problem. International Journal of Production Research 24, 939-944. (1986).

HODI 1987-1 Höding, M.
Algebraic structures of special discrete optimization problems (German). Wissenschaftliche Zeitschrift der Technischen Universität Otto von Guericke 31, No.5, 63-67. (1987).

HODS 1985-1 Hodson, A.; Muhlemann, A. P.; Price, D. H. R.
A microcomputer based solution to a practical scheduling problem. Journal of the Operational Research Society 36, 903-914. (1985).

HOEK 1983-1 Hoek, G. van der; Meerendonk, H. W. van den; Wijmenga, R. T.
The elimination of convexly dominated cutting patterns in trimloss problems. Rotterdam: Erasmus Univ., Econometric Inst., Report 8333/0. (1983).

HOES 1986-1 Hoesel, C. van; Schrijver, A.
Edge-disjoint homotopic paths in a planar graph with one hole. Report OS-R8608. Amsterdam: Centre for Mathematics and Computer Science, 14 p. (1986).

HOFF 1985-1 Hoffman, A. J.
On greedy algorithms that succeed. In: I Anderson (Ed.): Surveys in Combinatorics 1985, London Mathematical Society Lecture Notes Series Vol. 103, 97-112. Cambridge: University Press. (1985).

HOFFA 1985-1 Hoffman, A. J.; Kolen, A. W. J.; Sakarovitch, M.
Totally-balanced and greedy matrices. SIAM Journal on Algebraic and Discrete Methods 6, 721-730. (1985).

HOFFB 1986-1 Hoffman, A. J.; Lee, C. W.
On the cone of nonnegative circuits. Discrete & Computational Geometry 1, 229-239. (1986).

HOFFC 1985-1 Hoffman, A. J.; Wolfe, P.
History. In: E. L. Lawler, J. K. Lenstra, A. H. G. Rinnooy Kan, D. B. Shmoys (Eds.): The Traveling Salesman Problem, 1-16. New York: Wiley. (1985).

HOFFC 1985-2 Hoffman, A. J.; Wolfe, P.
Minimizing a unimodal function of two integer variables. In: R. W. Cottle (Ed.): Mathematical Programming Essays in Honor of G. B. Dantzig, Mathematical Programming Study 25, 76-87. (1985).

HOFFD 1985-1 Hoffman, K.; Padberg, M. W.
LP-based combinatorial problem solving. In: K. Schittkowski (Ed.): Computational Mathematical Programming, 55-124. Berlin: Springer-Verlag. (1985).

HOFFD 1986-1 Hoffman, K.; Padberg, M. W.
LP-based combinatorial problem solving. In: C. L. Monma (Ed.): Algorithms and Software for Optimization - Part I, 145-194. Annals of Operations Research 4, Basel: Baltzer. (1986).

HOFFE 1985-1 Hoffmann, T. R.; Scudder, G. D.
An evaluation of value-based dispatching rules in a flow shop. European Journal of Operational Research 22, 310-318. (1985).

HOFR 1984-1 Hofri, M.
A probabilistic analysis of the next-fit bin packing algorithm. Journal of Algorithms 5, 547-556. (1984).

HOFS 1985-1 Hofstedt, K.
Zur Kopplung eines exakten Verfahrens mit einem heuristischen Verfahren für die Lösung ganzzahliger linearer Optimierungsprobleme. Optimization 16, 391-400. (1985).

HOHM 1986-1 Hohmann, C.; Kern, W.
Optimization and optimality test for the max-cut problem. Report 86/35. Köln: Univ. Köln, Mathematisches Institut, 13 p. (1986).

HOLL 1983-1 Holland, O.
Optimierung kapazitierter und nicht kapazitierter b-Matchings unter Verwendung der linearen Programmierung. Bonn: Univ. Bonn, 105 p. (1983). (Diplomarbeit)

HOLM 1987-1 Holm, S.
Adding activities to the dual instead of cuts to the primal problem. Discrete Applied Mathematics 18, 167-174. (1987).

HOLR 1985-1 Holroyd, F. C.; Wingate, W. J. G.
Cycles in the complement of a tree or other graph. Discrete Mathematics 55, 267-282. (1985).

HOLS 1984-1 Holst, O.; Sorensen, B.
Combined scheduling and maitainance planning for an aircraft fleet. In: J. P. Brans (Ed.): Operational Research '84, 735-747. Amsterdam: North-Holland. (1984).

HOLT 1982-1 Holton, D. A.
Cycles in 3-connected cubic graphs. In: F. Hoffman, K. B. Reid, R. C. Mullin, R. G. Stanton (Eds.): Proc. Thirteenth Southeastern Conf. on Combinatorics, Graph Theory and Computing, Florida 1982, Congressus Numerantium Vol. 35, 373-378. Winnipeg, Canada: Utilitas Mathematica. (1982).

HOLT 1985-1 Holton, D. A.
Cycles in 3-connected cubic planar graphs. In: B. R. Alspach, C. D. Godsil (Eds.): Cycles in Graphs, Annals of Discrete Mathematics Vol. 27, 219-226. Amsterdam: North-Holland. (1985).

HOLTA 1986-1 Holton, D. A.; Mac Kay, B. D.
Cycles in 3-connected cubic planar graphs. II. Ars Combinatoria 21-A, 107-114. (1986).

HOLTB 1985-1 Holton, D. A.; Manvel, B.; Mac Kay, B. D.
Hamiltonian cycles in cubic 3-connected bipartite planar graphs. Journal of Combinatorial Theory 38B, 279-297. (1985).

HOLTC 1987-1 Holton, D. A.; Plummer, M. D.
On the cyclability of k-connected (k+1)-regular graphs. Ars Combinatoria 23, 37-56. (1987).

HOLU 1985-1 Holub, P.
A remark on covering graphs. Order 2, 321-322. (1985).

HOLZ 1986-1 Holzman, R.
An axiomatic approach to location on networks. RUTCOR Research Report 27-86. New Brunswick: Rutgers Univ., Rutgers Center for Operations Research, 19 p. (1986).

HONG 1987-1 Hong, D. W.
An efficient algorithm for determining the intersection of all minimal coverings in a bipartite graph, and a related algorithm for finding all minimal coverings (Chinese). Journal of the East China Normal University, Natural Science Edition (Shanghai), No. 3, 21-26. (1987).

HONGA 1984-1 Hong, Y.
A note on coefficients of chromatic polynomials (Chinese). Journal of East China Normal University, Natural Science Edition (Shanghai) No. 4, 33- 35. (1984).

HOOK 1986-1 Hooker, J.
Solving nonlinear single-facility network location problems. Operations Research 34, 732-743. (1986).

HOPK 1985-1 Hopkins, G.; Staton, W.
Graphs with unique maximum independent sets. Discrete Mathematics 57, 245-252. (1985).

HOPP 1987-1 Hoppe, V.
Polymatroide im E(3,+). Wissenschaftliche Zeitschrift der Technischen Hochschule Ilmenau 33, No.3, 35-46. (1987).

HORA 1982-1 Horak, P.; Niepel, L.
A short proof of a linear arboricity theorem for cubic graphs. Acta Mathematica Universitatis Comenianae (Bratislava) 40 (41), 275-277. (1982).

HORN 1986-1 Hornung, R.
Methods for solving the discrete capacity assignment problem in data networks: numerical comparison. Methods of Operations Research 54, 519-522. (1986).

HORO 1987-1 Horowitz, A. J.; Pithavadian, A. R.
Toward a data structure for computer-aided design of transportation networks. Transportation Research 21B, 317-322. (1987).

HORT 1987-1 Horton, J. D.
A polynomial-time algorithm to find the shortest cycle basis of a graph. SIAM Journal on Computing 16, 358-366. (1987).

HORTA 1985-1 Horton, J. D.; Roy, B. K.; Schellenberg, P. J.; Stinson, D. R.
On decomposing graphs into isomorphic uniform 2-factors. In: B. R. Alspach, C. D. Godsil (Eds.): Cycles in Graphs, Annals of Discrete Mathematics Vol. 27, 297-320. Amsterdam: North-Holland. (1985).

HOSA 1986-1 Hosage, C. M.; Goodchild, M. F.
Discrete space location-allocation solutions from genetic algorithms. In: J. P. Osleeb, P. J. Ratick (Eds.): Locational Decisions: Methodology and Applications, 35-48. Annals of Operations Research 6, Basel: Baltzer. (1986).

HOSK 1984-1 Hoskins, W. D.; Lambert, T. D.; Stanton, R. G.
An improved algorithm for determining optimal sums of disjoint sets. Ars Combinatoria 17, 5-14. (1984).

HOU 1987-1 Hou, M.
Enumeration of 2-matchings of complete graphs (Chinese). Journal Xinjiang University of Natural Science (Urumqi) 4, No. 1, 6-10. (1987).

HOUS 1984-1 Houska, R. V.
Using stochastic optimal control to find shortest paths in networks. in: Mathematical Modelling in Science and Technology, 4th internat. Conference (Zürich 1983), 295-298. New York: Pergamon Press. (1984).

HOUSA 1982-1 Houska, R. V.; Brockus, C. G.
Stochastic optimal control applied to deterministic network problems. In: Modeling and Simulation, Vol 13, Part 1, 429-435. Research Triangle Park, N. C.: Instrument Soc. of America. (1982).

HSIE 1984-1 Hsieh, W. N.
Routing strategies in computer networks. Computer 17, 46-56. (1984).

HSU 1985-1 Hsü, L.-H.
On a strongly multiplicative graph function conjecture (Chinese). Chinese Journal of Mathematics (Taipei) 13, 103-108. (1985).

HSU A 1984-1 Hsu, W.-L.
Efficient algorithms for the maximum weight clique problem on circular-arc graphs and circle graphs. In: J. A. Bondy, U. P. R. Murty (Eds.): Progress in Graph Theory, 335-346. New York: Academic Press. (1984).

HSU A 1984-2 Hsu, W.-L.
Finding maximum independent sets in planar perfect graphs. Research Paper. Evanston: Northwestern Univ., Dept. of Industrial Engineering and Management Sciences, 27 p. (1984).

HSU A 1984-3 Hsu, W.-L.
On the existence of induced odd cycles in planar graphs (revised). Research Paper. Evanston: Northwestern Univ., Dept. of Industrial Engineering and Management Sciences, 52 p. (1984).

HSU A 1984-4 Hsu, W.-L.
The perfect graph conjecture on special graphs - a survey. In: C. Berge, V. Chvatal (Eds.): Topics on perfect graphs, Annals of Discrete Mathematics Vol. 21, 103-114. Amsterdam: North-Holland. (1984).

HSU A 1985-1 Hsu, W.-L.
Maximum weight clique algorithms for circular-arc graphs and circle graphs. SIAM Journal on Computing 14, 224-231. (1985).

HSU A 1986-1 Hsu, W.-L.
Coloring planar perfect graphs by decomposition. Combinatorica 6, 381-386. (1986).

HSU A 1987-1 Hsu, W.-L.
Decomposition of perfect graphs. Journal of Combinatorial Theory 43B, 70-94. (1987).

HSU A 1987-2 Hsu, W.-L.
Recognizing planar perfect graphs. Journal of the Association for Computing Machinery 34, 255-288. (1987).

HSU B 1984-1 Hsu, W.-L.; Nemhauser, G. L.
Algorithms for maximum weight cliques, minimum weighted clique covers and minimum colorings of claw-free perfect graphs. In: C. Berge, V. Chvatal (Eds.): Topics on perfect graphs, Annals of Discrete Mathematics Vol. 21, 357-369. Amsterdam: North-Holland. (1984).

HU 1987-1 Hu, T. C.; Kuo, Y. S.; Ruskey, F.
Some optimum algorithms for scheduling problems with changeover costs. Operations Research 35, 94-99. (1987).

HU A 1985-1 Hu, T. C.; Shing, M. T.
A decomposition algorithm for circuit routing. Mathematical Programming Study 24, 87-103. (1985).

HU B 1987-1 Hu, Z.-Q.
On the maximal path in connected graphs (Chinese). Journal of Systems Science and Mathematical Sciences (Beijing) 7, 35-39. (1987).

HUAN 1987-1 Huang, B. F.; Yin, Q.
A note on edge chromatic numbers of compositions of graphs (Chinese). Acta Mathematicae Applicatae Sinica (Beijing) 10, 254-256. (1987).

HUANA 1984-1 Huang, H. C.; Ong, H. L.
Expected performance of some lot-sizing heuristics in a rolling-schedule environment. IIE Transactions 16, 248-256. (1984).

HUANB 1984-1 Huang, J.-H.
Uniquely 3-edge-colourable graphs (Chinese). Acta Mathematicae Applicatae Sinica (Peking) 7, 285-292. (1984).

HUANC 1984-1 Huang, J.-H.; Cai, C.-T.
Numbers of Hamilton cycles on generalized Petersen graph P(n,2) (Chinese). Journal of Shanghai Teachers University, Natural Science Edition, No.2, 17-20. (1984).

HUAND 1984-1 Huang, S.-H. S.; Wong, C. K.
Optimal binary split trees. Journal of Algorithms 5, 69-79. (1984).

HUANE 1984-1 Huang, S.-M.
A note on graphs with equal edge-connectivity and minimum degree (Chinese). Chinese Journal of Operations Research 3 No. 1, 67-68. (1984).

HUANF 1986-1 Huang, W.-Q.; Li, Q.-H.; Yu, X.-D.
A quasi-physical method for solving the three-dimensional packing problem. Acta Mathematicae Applicatae Sinica 9, 443-453. (1986).

HULM 1984-1 Hulme, B. L.; Shiver, A. W.; Slater, P. J.
A boolean algebraic analysis of fire protection. In: R. E. Burkard, R. A. Cuninghame-Green, U. Zimmermann (Eds.): Algebraic and Combinatorial Methods in Operations Research, Annals of Discrete Mathematics Vol. 19, 215-228. Amsterdam: North-Holland. (1984).

HUMM 1985-1 Hummeltenberg, W.
Capital budgeting with Benders decomposition. European Journal of Operational Research 21, 318-329. (1985).

HUNG 1986-1 Hung, M. S.; Rom, W. O.; Waren, A. D.
Degeneracy in transportation problems. Discrete Applied Mathematics 13, 223-238. (1986).

HURK 1986-1 Hurkens, C. A. J.; Schrijver, A.; Tardos, E.
On fractional multicommodity flows and distance functions. Report OS-R8611, CWI, Amsterdam. (1986).

HUTS 1984-1 Hutschenreuther, H.
Die Backwarenversorgung in einem abgeschlossenen Gebiet - ein typisches Beispiel für die Produktions- Transport- Standort- Optimierung. In: K.-H. Elster (Ed.): Mathematische Optimierung - Theorie und Anwendungen, Vortragsauszüge, Eisenach 1984, 57-63. Techn. Hochschule Ilmenau. (1984).

HWAN 1985-1 Hwang, C.-L.
Manufacturing plant location analysis by multiple attribute decision making: part II - multi-plant strategy and plant relocation. International Journal of Production Research 23, 361-370. (1985).

HWANA 1984-1 Hwang, C.-L.; Lee, H. B.; Tillmann, F. A.; Lie, C. H.
Nonlinear integer goal programming applied to optimal system reliability. IEEE Transactions on Reliability R-33, 431-438. (1984).

HWANB 1986-1 Hwang, F. K.
A linear time algorithm for full Steiner trees. Operations Research Letters 4, 235-238. (1986).

HWANB 1987-1 Hwang, F. K.
The Hamiltonian property of linear functions. Operations Research Letters 6, 125-128. (1987).

HWANC 1986-1 Hwang, F. K.; Weng, J. F.
Hexagonal coordinate systems and Steiner minimal trees. Discrete Mathematics 62, 49-58. (1986).

IBAR 1987-1 Ibaraki, T.
Enumerative Approaches to Combinatorial Optimization - Part 1. Basel: Baltzer. (1987). (Annals of Operations Research, 10)

IBAR 1987-2 Ibaraki, T.
Enumerative Approaches to Combinatorial Optimization - Part 2. Basel: Baltzer. (1987). (Annals of Operations Research, 11)

ICHI 1985-1 Ichimori, T.
A shortest path approach to a multifacility minimax location problem with rectilinear distances. Journal of the Operations Research Society of Japan 28, 269-284. (1985).

ICHIA 1985-1 Ichimori, T.; Nishida, T.
Note on a rectilinear distance round-trip location problem. Transportation Science 19, 84-90. (1985).

ICHIB 1984-1 Ichino, M.; Sklansky, J.
Optimum feature selection by zero-one integer programming. IEEE Transactions on Systems, Man and Cybernetics 14, 737-746. (1984).

IDZI 1987-1 Idzik, A.; Komar, J.; Malawski, M.
Edge-coloured complete graphs: connectedness of some subgraphs. Discrete Mathematics 66, 119-126. (1987).

IGNA 1984-1 Ignatev, V. M.
Some algorithms for inactive schedules (Russian). In: New Directions in the Development of Automatic Control of Production, 32-36. Akademiya Nauk Ukrainskoj SSR, Inst. Kibernet. (Kiev). (1984).

IGNI 1984-1 Ignizio, J. P.
A generalized goal programming approach to the minimal interference, multicriteria N*1 scheduling problem. IIE Transactions 16, 316-322. (1984).

IGNI 1985-1 Ignizio, J. P.
Integer goal programming via goal aggregation. Large Scale Systems (Amsterdam) 8, 81-86. (1985).

IGNIA 1984-1 Ignizio, J. P.; Thomas, L. C.
An enhanced conversion scheme for lexicographic, multiobjective integer programs. European Journal of Operational Research 18, 57-61. (1984).

IHM 1982-1 Ihm, H. S.; Ntafos, S.
On finding legal paths. In: F. Hoffman, K. B. Reid, R. C. Mullin, R. G. Stanton (Eds.): Proc. Thirteenth Southeastern Conf. on Combinatorics, Graph Theory and Computing, Florida 1982, Congressus Numerantium Vol. 36, 311-324. Winnipeg, Canada: Utilitas Mathematica. (1982).

IKRA 1986-1 Ikram, M.
A note on minimization of lateness cost function and determination of optimal due-date in a 2-machine scheduling problem. Pure and Applied Mathematika Sciences 23, No.1-2, 89-93. (1986).

IKUR 1986-1 Ikura, Y.; Gimple, M.
Efficient scheduling algorithms for a single batch processing machine. Operations Research Letters 5, 61-66. (1986).

IKURA 1986-1 Ikura, Y.; Gross, G.; Hall, G. S.
PGandE's state-of-the-art scheduling tool for hydro systems. Interfaces 16, 65-82. (1986).

IKURB 1983-1 Ikura, Y.; Nemhauser, G. L.
A polynomial-time dual simplex algorithm for the transportation problem: condensed version. Technical Report 602a. Ithaca: Cornell Univ., School of Operations Research and Industrial Engineering, 36 p. (1983).

IKURB 1985-1 Ikura, Y.; Nemhauser, G. L.
Simplex pivots on the set packing polytope. Mathematical Programmimg 33, 123-138. (1985).

IKURB 1986-1 Ikura, Y.; Nemhauser, G. L.
Computational experience with a polynomial-time dual simplex algorithm for the transportation problem. Discrete Applied Mathematics 13, 239-248. (1986).

ILEV 1984-1 Ilev, V. P.
Basis graphs of polymatroids (Russian). Metody Diskretnogo Analiza (Novosibirsk) 41, 35-48, 109. (1984).

ILEV 1984-2 Ilev, V. P.
On the basis graphs of polymatroids (Russian). Metody Diskretnogo Analiza (Novosibirsk) 41, 35-48. (1984).

ILEV 1985-1 Ilev, V. P.
Structure of intersection of the partitioning matroids (Russian). Metody Diskretnogo Analiza 42, 17-29. (1985).

ILEVA 1983-1 Ilev, V. P.; Fridman, G. S.
On a problem of approximation by three-component graphs (Russian). In: Numerical Methods and Optimization Problems, 80-95. Tomsk. Gos. Univ. (1983).

ILIC 1981-1 Ilichev, A. P.; Shevchenko, V. N.
Extreme points of polyhedra of multi-index transportation problems (Russian). In: A. A. Markov (Ed.): Combinatorial-algebraic methods in applied mathematics, 66-72. Gorki: Gor'kov Gos. Univ. (1981).

ILUM 1987-1 Ilumoka, A. A.
A tolerance analysis approach to network scheduling for engineering project management. International Journal of Production Research 25, 531-548. (1987).

IMAI 1985-1 Imai, H.
On combinatorial structures of line drawings of polyhedra. Discrete Applied Mathematics 10, 79-92. (1985).

IMAI 1986-1 Imai, H.
Worst-case analysis for planar matching and tour heuristics with bucketing techniques and spacefilling curves. Journal of the Operations Research Society of Japan 29, 43-68. (1986).

IMAIA 1986-1 Imai, H.; Asano, T.
Efficient algorithms for geometric graph search problems. SIAM Journal on Computing 15, 478-494. (1986).

IMAS 1985-1 Imase, M.; Soneoka, T.; Okada, K.
Connectivity of regular directed graphs with small diameters. IEEE Transactions on Computers C-34, 267-273. (1985).

INOU 1984-1 Inoue, M.; Onozawa, A.
On quasimatroids (Japanese). In: Graph theory and applications, Proc. Symp. Kyoto University 1984, 173-184. Kyoto University, Research Inst. for Mathematical Sciences. (1984).

INTR 1985-1 Intrator, J.
A note on "The assignment problem under categorized jobs". European Journal of Operations Research 20, 115-116. (1985).

INTR 1985-2 Intrator, J.
Experimental comparisons of codes for long transportation problems. Computers & Operations Research 12, 391-394. (1985).

INTRA 1984-1 Intrator, J.; Ehrlich, G.
Reusing augmenting paths in long transportation problems. NZOR 12, 99-109. (1984).

INTRA 1984-2 Intrator, J.; Ehrlich, G.
Tree modification in the transportation problem: a new method. NZOR 12, 135-140. (1984).

INTRB 1987-1 Intrator, J.; Peleg, D.
An efficient algorithm for the long transportation problem. Asia-Pacific Journal of Operational Research 4, 57-68. (1987).

INTRC 1986-1 Intrator, J.; Weiss, J.
Improving techniques for transportation problems. OR Spektrum 8, 191-196. (1986).

INTRC 1987-1 Intrator, J.; Weiss, J.
Exhaustive iterations - an improving approach for long transportation problems. Economic Computation and Economic Cybernetics Studies and Research (Bucharest) 22, No. 3, 73-87. (1987).

IOI 1986-1 Ioi, S.
A special "tree" problem. Bulletin of the System Science Institute, Waseda University No. 17, 17-26. (1986).

IORD 1986-1 Iordache, M.; Dumitriu, L.; Mierlea, C.; Georgescu, M.
Generation without duplication of all trees in a connected graph. Buletinul Institutuli Politehnic Bucuresti, Seria Electrotehnica 48, 57-64. (1986).

IORDA 1982-1 Iordanskii, M. A.
Problems of graph arrangement. II. (Russian). In: Combinatorial Algebraic Methods in Applied Mathematics, 26-65. Gorkov. Gos. Univ. (Gorki). (1982).

IPEK 1983-1 Ipek, M.; Dökmeci, V.
Optimum design of transportation networks. Bulletin of the Technical Univ. of Istanbul 36, 461-474. (1983).

IRI 1984-1 Iri, M.
Structural theory for the combinatorial systems characterized by submodular flows. In: W. R. Pulleyblank (Ed.): Progress in Combinatorial Optimization, 197-220. Toronto: Academic Press. (1984).

IRI 1987-1 Iri, M.
A very personal reminiscence on the problem of computational complexity. Discrete Applied Mathematics 17, 17-28. (1987).

IRI A 1984-1 Iri, M.; Murota, K.; Ohya, T.
A fast Voronoi-diagram algorithm with applications to geographical optimization problems. In: P. Thoft-Christensen (Ed.): System modelling and optimization (Copenhagen 1983), 273-288, Lecture Notes in Control and Information Science 59. Berlin, New-York: Springer. (1984).

IRI B 1975-1 Iri, M.; Tomizawa, N.
A unifying approach to fundamental problems in network theory by means of matroids. Electronics and Communications in Japan, Vol. 58-A, p. 28-35 (1975).

IRVI 1985-1 Irving, R. W.
An efficient algorithm for the stable roommates problem. Journal of Algorithms 6, 577-595. (1985).

IRVIA 1986-1 Irving, R. W.; Leather, P.
The complexity of counting stable marriages. SIAM Journal on Computing 15, 655-667. (1986).

IRVIB 1987-1 Irving, R. W.; Leather, P.; Gusfield, D.
An efficient algorithm for the "optimal" stable marriage. Journal of the Association for Computing Machinery 34, 532-543. (1987).

ISAC 1984-1 Isachenko, A. N.
Polynomial reducibility of matroid oracles (Russian). Izvestiya Akademii Nauk BSSR, Seriya Fiziko-Matematicheskikh Nauk (Minsk) No. 6, 33-36. (1984).

ISER 1984-1 Isermann, H.
Linear bottleneck transportation problems. Asia-Pacific Journal of Operational Research 1, 38-52. (1984).

ISER 1987-1 Isermann, H.
Ein Planungssystem zur Optimierung der Palettenbeladung mit kongruenten rechteckigen Versandgebinden. OR Spektrum 9, 235-250. (1987).

ISHI 1987-1 Ishii, H.; Masuda, T.; Nishida, T.
Two machine mixed shop scheduling problem with controllable machine speeds. Discrete Applied Mathematics 17, 29-38. (1987).

ISHIA 1984-1 Ishii, H.; Nishida, T.
Stochastic linear knapsack problem : probability maximization model. Mathematica Japonica (Osaka) 29, 273-281. (1984).

ISHIA 1986-1 Ishii, H.; Nishida, T.
Two machine open shop scheduling problem with controllable machine speeds. Journal of the Operations Research Society of Japan 29, 123-132. (1986).

ISRA 1986-1 Israeli, A.; Itai, A.
A fast and simple randomized parallel algorithm for maximal matching. Information Processing Letters 22, 77-80. (1986).

ISRAA 1986-1 Israeli, A.; Shiloach, Y.
An improved parallel algorithm for maximal matching. Information Processing Letters 22, 57-60. (1986).

ISTO 1984-1 Istomin, L. A.; Golovnya, S. G.
An approach to the solution of large-scale transportation problems (Russian). In; K. N. Gizatullin, L. A. Istomin (Eds.): Optimization of planning in regional social-economic systems, 68-76. Sverdlovsk: Uralskij Nauchnyj Tsentr AN SSSR, 68-76. (1984).

ITAI 1985-1 Itai, A.; Rodeh, M.
Scheduling transmissions in a network. Journal of Algorithms 6, 409-429. (1985).

ITAK 1984-1 Itakura, H.; Nishikawa, Y.
Fuzzy network technique for technological forecasting. Fuzzy Sets and Systems 14, 99-113. (1984).

IVAN 1984-1 Ivanchev, D. T.
Minimization of the processing time for metal cutting machines considered as a shortest path problem in a network (Bulgarian). Godishnik na Visshite Uchebni Zavedniya, Prilozhna Matematika (Sofia) 19, No.4, 155-162. (1984).

IVAN 1985-1 Ivanchev, D. T.
Finding the foremost arcs in a network with parametric arc capacities. Optimization 16, 909-920. (1985).

IVANA 1984-1 Ivanchev, D. T.; Ruhe, G.
Flows in generalized networks with parametric arc capacities. Comptes Rendus de l'Academie Bulgare des Sciences (Sofia) 37, No. 4, 477-480. (1984).

IVANA 1986-1 Ivanchev, D. T.; Ruhe, G.
Finding the k bicriterial foremost edges in planar flow networks. Comptes Rendus de l'Academie Bulgare des Sciences 39, No.9, 35-38. (1986).

IVANB 1984-1 Ivanchev, M. D.; Janev, N. I.; Mitev, J. G.
A knapsack approach to a given class of distribution problems. In: K. Lommatzsch (Ed.): Proceedings, 16. Jahrestagung "Mathematische Optimierung", Sellin/Rügen 1984, 41-44. Seminarbericht Nr. 64, Sektion Mathematik, Humboldt-Univ. Berlin (DDR). (1984).

IVANC 1984-1 Ivani, A. M.
Estimation of efficiency of algorithms for packing into containers (Russian). Problemy Kibernetiki (Moscow) 41, 253-256. (1984).

IVAND 1983-1 Ivanov, A. A.
Bounding the diameter of a distance-regular graph (Russian). Doklady Akademii Nauk SSSR (Moscow) 271, 789-792. (1983).

IVANE 1984-1 Ivanov, A. A.; Ivanov, A. V.; Faradzhev, I. A.
Distance-transitive graphs of valency 5, 6 and 7 (Russian). Zhurnal Vychislitelnoj Matematiki i Matematicheskoj Fiziki (Moscow) 24, 1704-1718. (1984).

IVANF 1984-1 Ivanyi, A.
Estimation of the efficiency of bin-packing algorithms (Russian). Problemy Kibernetiki (Moscow) No. 41, 253-256. (1984).

IVANF 1984-2 Ivanyi, A.
Performance bounds for simple bin packing algorithms. Annales Universitatis Scientarium Budapestinensis de Rolando Eötvös Nominatae, Sectio Mathematica 5, 77-82. (1984).

IVANF 1985-1 Ivanyi, A.
Tight worst-case bounds for bin packing algorithms. In: L. Lovasz, E. Szemeredi (Eds.): Theory of Algorithms, Colloquia Mathematica Societatis Janos Bolyai Vol. 44, 233-240. Amsterdam, Oxford, New York: North-Holland. (1985).

IWAH 1983-1 Iwahori, N.
On some algorithm of the number of the branch points and their degrees starting from the distance matrix of a given tree. Scientific Papers of the College of Arts and Sciences (Tokyo) 33, No. 2, 63-81. (1983).

IWAI 1985-1 Iwainsky, A.
Layout optimization for a hierarchy of star networks. In: A. Iwainsky (Ed.): Optimization of connection structures in graphs, 38-54. Berlin: Central Institute of Cybernetics and Information Processes. Academy of Sciences of the German Democratic Republic. (1985).

IWAI 1985-2 Iwainsky, A.
Optimal trees - a short overview on problem formulations. In: A. Iwainsky (Ed.): Optimization of connection structures in graphs, 121-137. Berlin: Central Institute of Cybernetics and Information Processes. Academy of Sciences of the German Democratic Republic. (1985).

IWAI 1985-3 Iwainsky, A.
Some notes on the Steiner tree problem in graphs. In: A.Iwainsky (Ed.): Optimization of connection structures in graphs, 57-73. Berlin: Central institute of Cybernetics and Information Processes. Academy of Sciences of the German Democratic Republic. (1985).

IWAM 1987-1 Iwamura, K.
Lexicographically optimal base of a submodular system with respect to a weight vector. Report 87482-OR. Bonn: Univ. Bonn, Inst. f. Ökonometrie u. Operations Research, 11 p. (1987).

IYER 1987-1 Iyer, A. V.; Ratliff, H. D.
Location issues in guaranteed time distribution systems. Working Paper. Chicago: Univ. of Chicago, Graduate School of Business, 18 p. (1987).

JABL 1984-1 Jablonsky, J.
Integer vector optimization problems and their solutions (Czechoslovakian). Ekonomicko-Matematicky Obzor (Prague) 20, 293-304. (1984).

JACK 1985-1 Jackson, B.
Long cycles in bipartite graphs. Journal of Combinatorial Theory 38B, 118-131. (1985).

JACK 1986-1 Jackson, B.
Longest cycles in 3-connected cubic graphs. Journal of Combinatorial Theory, Series B 41, 17-26. (1986).
JACK 1987-1 Jackson, B.
A Chvatal-Erdös condition for Hamilton Cycles in Digraphs. Journal of Combinatorial Theory B-43, 245-252. (1987).
JACK 1987-2 Jackson, B.
Compatible Euler tours for transition systems in Eulerian graphs. Discrete Mathematics 66, 127-132. (1987).
JACK 1987-3 Jackson, B.
Small r-regular r-connected nonhamiltonian graphs. Ars Combinatoria 24, 77-84. (1987).
JACKA 1985-1 Jackson, B.; Ordaz, O.
A Chvatal-Erdös condition for (1,1)-factors on digraphs. Discrete Mathematics 57, 199-202. (1985).
JACKB 1984-1 Jackson, B.; Ringel, G.
Coloring island maps. Bulletin of the Australian Mathematical Society 29, 151-165. (1984).
JACKB 1984-2 Jackson, B.; Ringel, G.
The splitting number of complete bipartite graphs. Archiv der Mathematik (Basel) 42, 178-184. (1984).
JACKC 1982-1 Jackson, D. E.; Entringer, R. C.
Minimum k-geodetically connected graphs. In: F. Hoffman, K. B. Reid, R. C. Mullin, R. G. Stanton (Eds.): Proc. Thirteenth Southeastern Conf. on Combinatorics, Graph Theory and Computing, Florida 1982, Congressus Numerantium Vol. 36, 303-310. Winnipeg, Canada: Utilitas Mathematica. (1982).
JACO 1985-1 Jacobs, K.
A measure theoretical max-flow-problem (Italian). In: Proc. Conf. commemorating the 1st Centennial of the Circolo Matematico di Palermo, Rendiconti del Circolo Matematico di Palermo, Serie II, Suppl. No. 8, 359-360. (1985).
JAEG 1984-1 Jaeger, F.
On circular flows in graphs. In: A. Hajnal; L. Lovasz; V. T. Sos (Eds.): Finite and Infinite Sets, Vol. 1, 391-402. Coll. Math. Soc. J. Bolyai 37, Amsterdam: North-Holland. (1984).
JAEG 1984-2 Jaeger, F.
On some algebraic properties of graphs. In: J. A. Bondy, U. P. R. Murty (Eds.): Progress in Graph Theory, 347-366. New York: Academic Press. (1984).
JAEG 1985-1 Jaeger, F.
A survey of the cycle double cover conjecture. In: B. R. Alspach, C. D. Godsil (Eds.): Cycles in Graphs, Annals of Discrete Mathematics Vol. 27, 1-12. Amsterdam: North-Holland. (1985).
JAEG 1985-2 Jaeger, F.
Graphic description of binary spaces. In: L. Lovasz, A. Recski (Eds.): Matroid Theory, 215-232. Coll. Math. Soc. J. Bolyai Vol. 40. Amsterdam: North-Holland. (1985).
JAEG 1985-3 Jaeger, F.
On five-edge colorings of cubic graphs and nowhere-zero flow. Ars Combinatoria 20-B, 229-244. (1985).
JAEGA 1985-1 Jaeger, F.; Khelladi, A.; Mollard, M.
On shortest cocycle covers of graphs. Journal of Combinatorial Theory 39B, 153-163. (1985).
JAFF 1985-1 Jaffe, J. M.
Distributed multi-destination routing: the constraints of local information. SIAM Journal on Computing 14, 875-888. (1985).
JAIK 1987-1 Jaikumar, R.; Solomon, M. M.
The tug fleet size problem for barge line operations: a polynomial algorithm. Transportation Science 21, 264-272. (1987).
JAIL 1985-1 Jaillet, P.
Analysis of the probabilistic traveling salesman problem in the plane. Ricerca Operativa 36, 57-68. (1985).
JAIL 1987-1 Jaillet, P.
On some probabilistic combinatorial optimization problems defined on graphs. In: Flow Control of Congested Networks (Capri 1986), 255-267. NATO Adv. Sci. Inst. Ser. F: Comput. Systems Sci. 38, Berlin: Springer. (1987).
JAIL 1987-2 Jaillet, P.
Stochastic routing problems. In: G. Andreatta, F. Mason, P. Serafini (Eds.): Advanced School on Stochastics in Combinatorial Optimization (CISM Udine 1986), 197-213. Singapore: World Scientific. (1987).
JAIN 1987-1 Jain, A.
Probabilistic analysis of LP relaxation bounds for the Steiner problem in networks. Philadelphia: Univ. of Pennsylvania, 12 p. (1987).
JAINA 1986-1 Jain, M.; Sharma, G. C.
A seminumerical iterative method for solving a machine interference problem. Indian Journal of Pure and Applied Masthematics (New Delhi) 17, 1-7. (1986).
JAMI 1984-1 Jamison, R. E.
Whitney sums and matchings in trees (part I). Research Report. Clemson: Clemson Univ., Dept. of Mathematical Science, 20 p. (1984).
JAMI 1987-1 Jamison, R. E.
Alternating Whitney sums and matchings in trees, Part 1. Discrete Mathematics 67, 177-190. (1987).
JAMIA 1984-1 Jamison, R. E.; Laskar, R.
Elimination orderings of chordal graphs. In: K. P. Vijayan, N. M. Singhi (Eds.): Combinatorics and Applications (Proc. Seminar Calcutta 1982), 192-200. Calcutta: Indian Statistical Institute. (1984).
JAMIB 1984-1 Jamison, R. E.; White, M. M.
Coronation numbers and circular systems in block graphs. Congressus Numerantium 44, 175-184. (1984).
JAMR 1985-1 Jamroz, L.
Suboptimal algorithm for job-shop problem with intermediate constraints (Polish). Zeszyty Naukowe Akademii Gorniczo-Hutniczej Im. Stanislawa Staszica 1064, Automatyka 39, 139-147. (1985).

JAMS 1987-1 Jamshy, U.; Raspaud, A.; Tarsi, M.
Short circuit covers for regular matroids with a nowhere zero 5-flow. Journal of Combinatorial Theory B-43, 354-357. (1987).

JANI 1986-1 Janiak, A.
On a single machine sequencing to minimize the maximum job cost subject to resource and precedence constraints. Archiwum Automatyki i Telemechaniki 31, 415-417. (1986).

JANI 1986-2 Janiak, A.
One-machine scheduling problems with resource constraints. In: A. Prekopa, J. Szeleszan, B. Strazicky (Eds.): System Modelling and Optimization - Proc. 12th IFIP Conf., Budapest 1985, 358-364. Lecture Notes in Control and Information Sciences 84, Berlin: Springer. (1986).

JANI 1986-3 Janiak, A.
Time-optimal control in a single machine problem with resource constraints. Automatica 22, 745-747. (1986).

JANI 1987-1 Janiak, A.
One-machine scheduling with allocation of continuously-divisible resource and with no precedence constraints. Kybernetika (Prague) 23, 289-293. (1987).

JANIA 1984-1 Janiak, A.; Grabowski, J.
Job-shop problem with resource constraints. In: A. Straszak (Ed.): Large Scale Systems: Theory and Applications 1983 (Warsaw 1983), 475-480. IFAC Proc. Series 10, Laxenburg, Pergamon Press. (1984).

JANIB 1985-1 Janiga, L.; Koubek, V.
A note on finding minimum cuts in directed planar networks by parallel computations. Information Processing Letters 21, 75-78. (1985).

JANK 1985-1 Jankowska-Zorychta, Z.
An assignment problem of specific structure (Polish). Przeglad Statystyczny 32, 255-262. (1985).

JASI 1984-1 Jasinska, E.; Woytych, E.
Location of depots in a sugar-beet distribution system (case study). European Journal of Operational Research 18, 396-402. (1984).

JASIA 1984-1 Jasinski, L. J.
An exact method for solving the vehicle routing problem (Polish). Przeglad Statystyczny (Warsaw) 31, 319-332. (1984).

JASM 1985-1 Jasmon, G. B.
Cutset analysis of networks using basic minimal paths and network decomposition. IEEE Transactions on Reliability R-34, 303-307. (1985).

JASMA 1987-1 Jasmon, G. B.; Foong, K. W.
A method for evaluating all the minimal cuts of a graph. IEEE Transactions on Reliability R-36, 539-545. (1987).

JAUM 1986-1 Jaumard, B.; Minoux, M.
An efficient algorithm for the transitive closure and a linear worst-case complexity result for a class of sparse graphs. Information Processing Letters 22, 163-170. (1986).

JAW 1986-1 Jaw, J.-J.; Odoni, A. R.; Psaraftis, H. N.; Wilson, N. H. M.
A heuristic algorithm for the multi-vehicle advance request dial-a-ride problem with time windows. Transportation Research 20-B, 243-258. (1986).

JAYA 1984-1 Jayakumar, R.; Thulasiraman, K.; Swamy, M. N. S.
Complexity of computation of a spanning tree enumeration algorithm. IEEE Transactions on Circuits and Systems CAS-31, 853-860. (1984).

JAYAA 1987-1 Jayaram, S. R.
Line domination in graphs. Graphs and Combinatorics 3, 357-364. (1987).

JAYAA 1987-2 Jayaram, S. R.
The nomatic number of a graph. National Academy Science Letters (Allahabad) 10, No. 1, 23-25. (1987).

JENK 1987-1 Jenkins, L.
Parametric-objective integer programming using knapsack facets and Gomory cutting planes. European Journal of Operational Research 31, 102-109. (1987).

JENK 1987-2 Jenkins, L.
Using parametric integer programming to plan the mix of an air transport fleet. INFOR 25, 117-135. (1987).

JENKA 1987-1 Jenkins, L.; Peters, D.
A computational comparison of Gomory and knapsack cuts. Computers and Operations Research 14, 449-456. (1987).

JENKB 1986-1 Jenkyns, T. A.
The greedy algorithm is a shady lady. Congressus Numerantium 51, 209-216. (1986).

JERO 1984-1 Jeromin, B.; Körner, F.
A hybrid method for solving integer linear programming problems. Wissenschaftliche Informationen 44, Technische Hochschule Karl-Marx-Stadt, Sektion Mathematik. 11 p. (1984).

JERO 1985-1 Jeromin, B.; Körner, F.
On the refinement of bounds of heuristic algorithms for the traveling salesman problem. Mathematical Programming 32, 114-117. (1985).

JERO 1986-1 Jeromin, B.; Körner, F.
A hybrid method for solving integer linear programming problems. Ekonomicko Matematicky Obzor (Prague) 22, 400-407. (1986).

JERO 1986-2 Jeromin, B.; Körner, F.
Sharp bounds for Karp's "patching"-algorithm for the approximate solution of the traveling salesman problem. Optimization 17, 85-92. (1986).

JEROA 1984-1 Jeroslow, R. G.
On monotone chaining procedures of the CF type. Atlanta: Georgia Inst. of Technology, 10 p. (1984).

JEROA 1985-1 Jeroslow, R. G.
A simplification for some disjunctive formulations. Working Paper. Atlanta: Georgia Inst. of Technology, 19 p. (1985).

JEROA 1985-2 Jeroslow, R. G.
Alternative formulations of mixed integer programs. Working Paper. Atlanta: Georgia Inst. of Technology, 53 p. (1985).

JEROA 1985-3 Jeroslow, R. G.
An extension of mixed-integer programming models and techniques to some database and artificial intelligence settings. Research Paper. Atlanta: Georgia Inst. of Technology, 93 p. (1985).

JEROA 1985-4 Jeroslow, R. G.
Computation-oriented reductions of predicate to propositional logic. Working Paper. Atlanta: Georgia Inst. of Technology, 53 p. (1985).

JEROA 1986-1 Jeroslow, R. G.
Ten lectures on mixed integer model formulation for logic-based decision support. Atlanta: Georgia Inst. of Technology, 257 p. (1986).

JEROA 1987-1 Jeroslow, R. G.
Representability in mixed integer programming, I: characterization results. Discrete Applied Mathematics 17, 223-244. (1987).

JEROB 1984-1 Jeroslow, R. G.; Lowe, J. K.
Modelling with integer variables. Mathematical Programming Study 22, 167-184. (1984).

JEROB 1985-1 Jeroslow, R. G.; Lowe, J. K.
Experimental results on the new techniques for integer programming formulations. Journal of the Operational Research Society 36, 393-404. (1985).

JEROC 1987-1 Jeroslow, R. G.; Wang, J.
Dynamic programming, integral polyhedra and Horn clause knowledge bases. Atlanta: Georgia Inst. of Technology, 41 p. (1987).

JEUR 1985-1 Jeurissen, R. H.
The Petersen graph as a box of Pandora. Nieuw Archief voor Wiskunde (Amsterdam) (4) 3, 219-133. (1985).

JEWO 1987-1 Jeworrek, T.; Schulz, G.
Zur Lösung des Briefträgerproblems auf gemischten Graphen. Wissenschaftliche Zeitschrift der Technischen Universität Otto von Guericke 31, No.5, 58-62. (1987).

JEZN 1983-1 Jezny, S.; Trenkler, M.
Characterization of magic graphs. Czechoslovak Mathematical Journal (Prague) 33, 435-438. (1983).

JHA 1986-1 Jha, N. K.
Optimizing the number of tools and cutting parameters in multi-tool turning for multiple objectives through geometric programming. Applied Mathematical Modelling 10, 162-169. (1986).

JIAN 1987-1 Jiang, H. G.
Optimality conditions for a nonsmooth optimization problem (Chinese). Journal of Shanghai Jiaotong University, No. 5, 116-120. (1987).

JIANA 1986-1 Jiang, J.-W.; Tang, P.-S.
An algorithm for rectilinear Steiner trees (Chinese). Journal of Fudan University (Natural Science Edition) 25, 343-349. (1986).

JIN 1986-1 Jin, M. Y.; Diao, Z. Y.; Zhang, S. L.
An optimal-depth ordering algorithm on large scale graphs (Chinese). Numerical Mathematics, A Journal of Chinese Universities 8, No. 4, 289-297. (1986).

JIRA 1987-1 Jirasek, J.
On a certain class of multidigraphs, for which reversal of no arc decreases the number of their cycles. Commentationes Mathematicae Universitatis Carolinae 28, 185-189. (1987).

JOHA 1986-1 Johansson, B.; Leonardi, G.
Public facility location: a multiregional and multiauthority decision context. In: Handbook of Regional and Urban Economics Vol. 1, 133-170. Amsterdam: North-Holland. (1986).

JOHN 1985-1 John, C. G.; Kochenberger, G. A.
Using surrogate constraints in a Lagrangian relaxation approach to set covering problems. Working Paper. University Park: Pennsylvania State Univ., Dept. of Management Science, 8 p. (1985).

JOHNA 1985-1 John, P.; Rempel, J.
Counting perfect matchings in hexagonal systems. In: H. Sachs (Ed.): Graphs, Hypergraphs and Applications, Proc. Conf. on Graph Theory (Eyba 1984), 72-79. Leipzig: Teubner. (1985).

JOHNB 1985-1 John, P.; Sachs, H.
Calculating the number of perfect matchings and Pauling's bond orders in hexagonal systems whose inner dual is a tree. In: H. Sachs (Ed.): Graphs, Hypergraphs and Applications, Proc. Conf. on Graph Theory (Eyba 1984), 80-91. Leipzig: Teubner. (1985).

JOHNC 1984-1 John, T. C.; Sadowski, R. P.
Compromise solutions in single stage production scheduling for minimizing total flow time and maximum penalty. Research Report. Tucson: Univ. of Arizona, Systems & Industrial Engineering, 38 p. (1984).

JOHND 1987-1 Johnson, D. B.
Parallel algorithms for minimum cuts and maximum flows in planar networks. Journal of the Association for Computing Machinery 34, 950-967. (1987).

JOHNE 1982-1 Johnson, D. B.; Venkatesan, S. M.
Parallel algorithms for minimum cuts and maximum flows in planar networks (preliminary version). In: 23rd Annual Symp. on Foundations of Computer Science, Chicago, Illinois 1982, Proceedings, 244-254. IEEE. (1982).

JOHNF 1985-1 Johnson, D. S.; Garey, M. R.
A 71/60 theorem for bin packing. Journal of Complexity 1, 65-106. (1985).

JOHNG 1985-1 Johnson, D. S.; Papadimitriou, C. H.
Computational complexity. In: E. L. Lawler, J. K. Lenstra, A. H. G. Rinnooy Kan, D. B. Shmoys (Eds.): The Traveling Salesman Problem, 37-86. New York: Wiley. (1985).

JOHNG 1985-2 Johnson, D. S.; Papadimitriou, C. H.
Performance guarantees for heuristics. In: E. L. Lawler, J. K. Lenstra, A. H. G. Rinnooy Kan, D. B. Shmoys (Eds.): The Traveling Salesman Problem, 145-180. New York: Wiley. (1985).

JOHNH 1985-1 Johnson, D. S.; Papadimitriou, C. H.; Yannakakis, M.
How easy is local search? In: 26th Annual Symp. on Foundations of Computer Science, Portland, Oregon 1985, Proceedings, 39-42. IEEE. (1985).

JOHNI 1985-1 Johnson, E. J.; Kostreva, M. M.; Suhl, U. H.
Solving 0-1 integer programming problems arising from large scale planning models. Operations Research 33, 803-819. (1985).

JOHNJ 1987-1 Johnson, E. L.; Mosterts, S.
On four problems in graph theory. SIAM Journal on Algebraic and Discrete Methods 8, 163-185. (1987).

JOHNK 1986-1 Johnson, E. L.; Pieroni, P.
A linear programming approach to the optimum network orientation problem. In: G. Gallo, C. Sandi (Eds.): Netflow at Pisa, Mathematical Programming Study 26, 215-217. (1986).

JOHNL 1984-1 Johnson, R.
Network reliability and acyclic orientations. Networks 14, 489-506. (1984).

JOHNM 1986-1 Johnston, R. E.
Rounding algorithms for cutting stock problems. Asia-Pacific Journal of Operational Research 3, 166-171. (1986).

JONE 1985-1 Jones, K. F.
Maximum bipartite subgraphs and independence. Congressus Numerantium 48, 219-224. (1985).

JONEA 1985-1 Jones, K. F.; Lundgren, J. R.; Maybee, J. S.
Clique covers of digraphs II. Congressus Numerantium 48, 211-218. (1985).

JONG 1985-1 Jongens, K.; Volgenant, T.
The symmetric clustered traveling salesman problem. European Journal of Operational Research 19, 68-75. (1985).

JONK 1986-1 Jonker, R.
Traveling salesman and assignment algorithms: design and implementation. Doctoral dissertation, University of Amsterdam. (1986).

JONKA 1984-1 Jonker, R.; Volgenant, T.
Nonoptimal edges for the symmetric traveling salesman problem. Operations Research 32, 837-846. (1984).

JONKA 1986-1 Jonker, R.; Volgenant, T.
Improving the Hungarian assignment algorithm. Operations Research Letters 5, 171-176. (1986).

JONKA 1986-2 Jonker, R.; Volgenant, T.
Transforming asymmetric into symmetric traveling salesman problems: Erratum. Operations Research Letters 5, 215-216. (1986).

JONKA 1987-1 Jonker, R.; Volgenant, T.
A shortest augmenting path algorithm for dense and sparse linear assignment problems. Computing 38, 325-340. (1987).

JONKA 1988-1 Jonker, R.; Volgenant, T.
A shortest augmenting path algorithm for dense and sparse linear assignment problems. In: H. Schellhaas, P. van Beek, H. Isermann, R. Schmidt, M. Zijlstra (Eds.): Operations Research Proceedings 1987 (DGOR/NSOR), 622-623. Berlin: Springer. (1988).

JORD 1982-1 Jordan, S. A.; Shier, D. R.
Algorithms for determining fundamental cycles in a graph. In: F. Hoffman, K. B. Reid, R. C. Mullin, R. G. Stanton (Eds.): Proc. Thirteenth Southeastern Conf. on Combinatorics, Graph Theory and Computing, Florida 1982, Congressus Numerantium Vol. 36, 279-294. Winnipeg, Canada: Utilitas Mathematica. (1982).

JORG 1987-1 Jorgensen, C.; Powell, S.
Solving 0-1 minimax problems. Journal of the Operational Research Society 38, 515-522. (1987).

JORN 1986-1 Jörnsten, K. O.; Näsberg, M.
A new Lagrangian relaxation approach to the genralized assignment problem. European Journal of Operational Research 27, 313-323. (1986).

JORNA 1986-1 Jörnsten, K. O.; Värbrand, P.
An augmented Lagrangian approach to multi-commodity flow problems with conversions. Opsearch 23, 96-106. (1986).

JUEL 1985-1 Juel, H.; Love, R. F.
The facility location problem for hyper-rectilinear distances. IIE Transactions 17, 94-98. (1985).

JUEL 1986-1 Juel, H.; Love, R. F.
A geometrical interpretation of the existing facility solution condition for the Weber problem. Journal of the Operational Research Society 37, 1129-1132. (1986).

JUEL 1987-1 Juel, H.; Love, R. F.
Duality in constrained location problems. Operations Research Letters 6, 281-284. (1987).

JUNG 1984-1 Jung, H. A.
Longest circuits in 3-connected graphs. In: A. Hajnal; L. Lovasz; V. T. Sos (Eds.): Finite and Infinite Sets, Vol. 1, 403-442. Coll. Math. Soc. J. Bolyai 37, Amsterdam: North-Holland. (1984).

JUNG 1986-1 Jung, H. A.
Longest paths joining given vertices in a graph. Abhandlungen aus dem Mathematischen Seminar der Universität Hamburg 56, 127-137. (1986).

JUNGA 1985-1 Jünger, M.
Polyhedral combinatorics and the acyclic subdigraph problem. Berlin: Heldermann. (1985).

JUNGB 1985-1 Junginger, W.
Stundenpläne aus dem Computer - Ein Bericht zur momentanen Lage in Deutschland. In: D. Ohse, A. C. Esprester, H.-U. Küpper, P. Stähly, H. Steckhan (Eds.): Operations Research Proceedings 1984, 36. Berlin: Springer-Verlag. (1985).

JUNGB 1985-2 Junginger, W.
Stundenpläne aus dem Computer - ein Bericht zur momentanen Lage in Deutschland. Angewandte Informatik, 168-174. (1985).

JUNGB 1986-1 Junginger, W.
Timetabling in Germany - a survey. Interfaces 16, No. 4, 66-74. (1986).

JUNGC 1986-1 Jungnickel, D.
Transversaltheorie: ein Überblick. In: A. Kerber (Ed.): Diskrete Strukturen, algebraische Methoden und Anwendungen, Sommerschule Bayreuth 1985. Bayreuther Mathematische Schriften, Heft 21, 122-155. (1986).

JUNGC 1987-1 Jungnickel, D.
Graphen, Netzwerke und Algorithmen. Mannheim, Wien, Zürich: Bibliographisches Institut. (1987).

JURK 1984-1 Jurkowska, T.; Komorowska, E.; Mazbic-Kulma, B.; Partyka, A.; Pogorzelec, A.
Computer method of bus routes determining for the inter-urban transportation. Systems Science (Wroclaw) 10, No. 2, 83-91. (1984).

KABA 1985-1 Kabadi, S. N.; Chandrasekaran, R.
On totally dual integral systems. Richardson: Univ. of Texas at Dallas, School of Management, 41 p. (1985).
KABE 1985-1 Kabe, D. G.
On discrete linear and quadratic programming problems. Utilitas Mathematica 27, 275-286. (1985).
KABE 1986-1 Kabe, D. G.
On a linear costs standard transportation problem optimality test. Industrial Mathematics 36, 197-209. (1986).
KABE 1986-2 Kabe, D. G.
On solving linear and quadratic integer programming problems. Industrial Mathematics 36, 51-66. (1986).
KABE 1986-3 Kabe, D. G.
On solving two-variables programming problems. Industrial Mathematics 36, 35-49. (1986).
KABU 1986-1 Kabulov, A. V.; Bajzhumanov, A. A.
Local methods for solving systems of Boolean equations (Russian). Doklady Akademii Nauk UzSSR, No.3, 3-5. (1986).
KAHN 1984-1 Kahn, J.
A family of perfect graphs associated with directed graphs. Journal of Combinatorial Theory, Series B 37, 279-282. (1984).
KAHN 1984-2 Kahn, J.
A geometric approach to forbidden minors for GF(3). Journal of Combinatorial Theory, Series A 37, 1-12. (1984).
KAHN 1985-1 Kahn, J.
A problem of P. Seymour on nonbinary matroids. Combinatorica 5, 319-324. (1985).
KAHNA 1987-1 Kahn, J.; Seymour, P. D.
On forbidden minors for GF(3). RUTCOR Research Report 1-87. New Brunswick: Rutgers Univ., Rutgers Center for Operations Research, 7 p. (1987).
KAJI 1981-1 Kajitani, Y.; Ueno, S.
Redundant elements of a fundamental tie matrix of a graph. Electronics and Communications in Japan 62 (1979), 1-8. (1981).
KAJI 1984-1 Kajitani, Y.; Ueno, S.
A matroid generalization of theorems of Lewin and Gallai. Discrete Applied Mathematics 9, 213-216. (1984).
KAKH 1985-1 Kakhichko, A. A.
On the problem of the existence of perfect matchings of a graph (Russian). In: Methods for Solving Nonlinear Problems and for Data Processing, 105-110. Dnepropetrovsk: Dnepropetrovsk. Gos. Univ. (1985).
KAKU 1986-1 Kaku, B. K.; Thompson, G. L.
Comment on "A perturbation scheme to improve Hillier's solution to the facilities layout problem". Management Science 32, 251-252. (1986).
KALA 1985-1 Kalai, G.
Hyperconnectivity of graphs. Graphs and Combinatorics 1, 65-80. (1985).
KALAA 1985-1 Kalantari, B.; Hill, A. V.; Arora, S. R.
An algorithm for the traveling salesman problem with pickup and delivery customers. European Journal of Operational Research 22, 377-386. (1985).
KALAB 1987-1 Kalantari, B.; Rosen, J. B.
Penalty formulation for zero-one nonlinear programmming. Discrete Applied Mathematics 16, 179-182. (1987).
KALB 1984-1 Kalb, G.; Röck, H.; Schmidt, G.
Netzflussmodelle zur präferenzorientierten dynamischen Einsatz- und Ablaufplanung. Zeitschrift für Operations Research, Series B28, 283-296. (1984).
KALI 1983-1 Kalinin, V. B.
On graphs of intersections (Russian). In: Algorithmic Constructions and Their Efficiency, 72-76. Yaroslavl: Yaroslav. Gos. Univ. (1983).
KALIA 1987-1 Kalinin, V. N.; Sokolov, B. V.
Dynamic model and optimal scheduling algorithm for a set of jobs with preemption constraints. Automation and Remote Control 48, No.1, 88-94; translation from Avtomatika i Telemekhanika (1987), No.1, 106-114. (1987).
KALM 1984-1 Kalmar, J.; Lengyel, I.
The consideration of stock within the Gilmore-Gomory method (Hungarian). Szigma 17, 167-183. (1984).
KALR 1982-1 Kalra, K. R.; Bagga, P. C.
Single machine sequencing problem with polynomial production cost functions: modifications in the branch-and-bound technique of Townsend. Acta Ciencia Indica (Meerut) 8, 119-126. (1982).
KALRA 1985-1 Kalra, K. R.; Bagga, P. C.; Singh, S. P.
Some remarks on the paper "Minimizing the average deviation of job completion times about a common due-date" by John J. Kanet. Acta Ciencia Indica 11, 202-207. (1985).
KALT 1987-1 Kaltinska, R. N.
A method for solving a class of quadratic transportation problems (Russian). Comptes Rendus de l'Academie Bulgare des Sciences 40, No.7, 35-38. (1987).
KALU 1983-1 Kaluza, T.
Existenz von Graphen jeden Grades mit Hamilton-Färbung. Abhandlungen der Braunschweigischen Wissenschaftlichen Gesellschaft (Göttingen) 35, 7-13. (1983).
KAMB 1984-1 Kamburowski, J.
Criticality indices in the PERT-type network (Polish). Przeglad Statystyczny (Warsaw) 31, 333-341. (1984).
KAMB 1984-2 Kamburowski, J.
Flow distributions through a stochastic acyclic network. Systems Science 10, No. 2, 25-35. (1984).
KAMB 1985-1 Kamburowski, J.
A note on the stochastic shortest route problem. Operations Research 33, 696-698. (1985).
KAMB 1985-2 Kamburowski, J.
An upper approximation of the expected completion time of a project described by the PERT-type network (Polish). Przeglad Statystyczny (Warsaw) 32, 63-68. (1985).

KAMB 1985-3 Kamburowski, J.
An upper bound on the expected completion time of PERT networks. European Journal of Operational Research 21, 206-212. (1985).
KAMB 1985-4 Kamburowski, J.
Two-point approximations for activity times in PERT network. RAIRO Recherche Operationelle 19, 301-313. (1985).
KAMB 1986-1 Kamburowski, J.
On the computational complexity of the shortest route and maximum flow problems in stochastic networks. Foundations of Control Engineering (Poznan) 11, 168-175. (1986).
KAMB 1986-2 Kamburowski, J.
Upper bounds on the expected project duration under incomplete probabilistic information (Polish). Przeglad Statystyczny 33, 41-47. (1986).
KAMB 1987-1 Kamburowski, J.
An overview of the computational complexity of the PERT shortest route and maximum flow problems in stochastic networks. In: G. Andreatta, F. Mason, P. Serafini (Eds.): Advanced School on Stochastics in Combinatorial Optimization (CISM Udine 1986), 187-196. Singapore: World Scientific. (1987).
KAME 1986-1 Kamenarov, S. N.
Two special cases for the traveling salesman problem with premiums (Bulgarian). Godishnik na Visshite Uchebni Zavedniya, Prilozhna Matematika (Sofia) 21, No.1, 193-200. (1986).
KAMP 1987-1 Kämpke, T.
Necessary optimality conditions for priority policies in stochastic weighted flowtime scheduling problems. Advances in Applied Probability 19, 749-750. (1987).
KAMP 1987-2 Kämpke, T.
On the optimality of static priority policies in stochastic scheduling on parallel machines. Journal of Applied Probability 24, 430-448. (1987).
KANA 1984-1 Kanaeva, N. N.; Shcherbina, O. A.
On the efficiency of a modified local algorithm for solving discrete optimization problems. USSR Computation of Mathematics and Mathematical Physics 24, No. 5, 145-152; translation from: Zhurnal Vychislitel'noj Matematiki i Matematicheskoj Fiziki (Moscow) 24, 1520-1530. (1984).
KANA 1986-1 Kanaeva, N. N.; Shcherbina, O. A.
On a modified local algorithm for problems of discrete programming. U.S.S.R. Computational Mathematics and Mathematical Physics 26, No.1, 162-170; translation from Zhurnal Vychislitelnoi Matematiki i Matematicheskoi Fiziki 26 (1986), No.2, 263-275. (1986).
KANE 1986-1 Kaneko, H.; Fushimi, M.
A polymatroid associated with convex games. Discrete Applied Mathematics 14, 33-46. (1986).
KANEA 1986-1 Kanet, J. J.
Tactically delayed versus non-delay scheduling: An experimental investigation. European Journal of Operational Research 24, 99-105. (1986).
KANEB 1987-1 Kanet, J. J.; Adelsberger, H. H.
Expert systems in production scheduling. European Journal of Operational Research 29, 51-59. (1987).
KANEC 1984-1 Kanetkar, S. V.; Rao, P. R.
Connected, locally 2-connected, K(1,3)-free graphs are panconnected. Journal of Graph Theory 8, 347-354. (1984).
KANED 1987-1 Kanevsky, A.; Ramachandran, V.
Improved algorithms for graph four-connectivity. In: 28th Annual IEEE Symposium on Foundations of Computer Science, 252-259. (1987).
KANG 1984-1 Kang, T.; Lin, K. R.
Necessary and sufficient conditions for a planar graph G with chi(G)=3 (Chinese). Fuzhou Daxue Xuebao (Fuzhou), No. 3, 10-17. (1984).
KANN 1985-1 Kannan, R.
Unraveling k-page graphs. Information and Control 66, 1-5. (1985).
KANN 1987-1 Kannan, R.
Algorithmic geometry of numbers. Report 87453-OR. Bonn: Univ. Bonn, Inst. f. Ökonometrie u. Operations Research, 48 p. (1987).
KANN 1987-2 Kannan, R.
Minkowski's convex body theorem and integer programming. Mathematics of Operations Research 12, 415-440. (1987).
KANNA 1986-1 Kannan, R.; Lovasz, L.
Covering minima and lattice point free convex bodies. In: K. V. Nori: Foundations of Software, Technology and Theoretical Computer Science, Lecture Notes in Computer Science 241, 193-213. Berlin, Heidelberg, New York: Springer-Verlag. (1986).
KANNA 1987-1 Kannan, R.; Lovasz, L.
Covering minima and lattice point free convex bodies. CS-TR-114-87. Princeton: Princeton Univ., Dept. of Computer Science, 36 p. (1987).
KANO 1984-1 Kano, M.
Graph factors with given properties. In: K. M. Koh, H. P. Yap (Eds.): Graph Theory Singapore 1983, Lecture Notes in Mathematics 1073, 161-168. Berlin: Springer-Verlag. (1984).
KANO 1985-1 Kano, M.
(a,b)-factorizations of nearly bipartite graphs. In: Y. Alavi, G. Chartrand, L. Lesniak, D. R. Lick, C. E. Wall (Eds.): Graph Theory with Applications to Algorithms and Computer Science, 471-474. Chichester: Wiley. (1985).
KANO 1986-1 Kano, M.
Factors of regular graphs. Journal of Combinatorial Theory, Series B 41, 27-36. (1986).
KANO 1987-1 Kano, M.
Maximum and k-th maximal spanning trees of a weighted graph. Combinatorica 7, 205-214. (1987).
KANO 1987-2 Kano, M.
Ranking the vertices of an r-partite paired comparison digraph. Discrete Applied Mathematics 17, 245-254. (1987).
KANOA 1985-1 Kano, M.; Sakamoto, A.
Ranking the vertices of a paired comparison digraph. SIAM Journal on Algebraic and Discrete Methods 6, 79-92. (1985).

KANOB 1983-1 Kanopa, R.
On the analysis of the heuristic solution for a problem concerning the composition of a time-table of works in network planning (Russian). Avtomatizatsiya Protsessov Planirovaniya i Upravleniya (Vil'nyus) 10, 48-57. (1983).

KANT 1985-1 Kant, K.
Finding interferences between rectangular paths. IEEE Transactions on Computers C-34, 1045-1049. (1985).

KANTA 1987-1 Kanter, I.; Sompolinsky, H.
Graph optimization problems and the Potts glass. Journal of Physics A 20, L673-L679. (1987).

KANTB 1983-1 Kantorovich, E. G.
Investigation of a model of the location of service in an urban system (Russian). Dynamics of non-homogeneous systems, Mater., Semin, Moskva 1983, 164-173. (1983).

KANTC 1983-1 Kantsedal, S. A.
A decomposition approach to the solution of large-scale scheduling problems. Avtomatika i Telemekhanika (Moscow) No. 10, 144-151, translated as Automation and Remote Control 44, 1360-1366. (1983).

KAPO 1986-1 Kapoor, S. F.; Vaidya, P. M.
Fast algorithms for convex quadratic programming and multicommodity flows. In: Proc. Eighteenth Annual ACM Symp. on Theory of Computing, Berkeley 1986, 147-159. ACM. (1986).

KAPU 1985-1 Kapur, D.; Krishnamoorthy, M. S.
Worst-case choice for the stable marriage problem. Information Processing Letters 21, 27-30. (1985).

KARA 1985-1 Karapetyan, I. A.
Chord graphs (Russian). Matematicheskie Voprosy Kibernetiki i Vychislitelnoi Tekhniki (Erevan), No. 14, 6-10. (1985).

KARG 1987-1 Karger, P. B.
Untersuchungen zur Matching-Relaxation für Set-Partitioning-Probleme. Bonn: Univ. Bonn, 132 p. (1987). (Diplomarbeit)

KARK 1986-1 Karkazis, J.; Karagiorgis, P.
A method to locate the maximum circle(s) inscribed in a polygon. Belgian Journal of Operations Research, Statistics and Computer Sience 26, No. 3, 3-36. (1986).

KARL 1983-1 Karla, K. R.; Khurana, K.
Single machine scheduling to minimize waiting cost with secondary criterion. Journal of Mathematical Sciences (Delhi) 16, 9-15. (1983).

KARLA 1986-1 Karloff, H. J.
A Las Vegas RNC algorithm for maximum matching. Combinatorica 6, 387-392. (1986).

KARLB 1987-1 Karloff, H. J.; Shmoys, D. B.
Efficient parallel algorithms for edge coloring problems. Journal of Algorithms 8, 39-52. (1987).

KARM 1984-1 Karmakar, S. B.
An algorithm for finding a circuit of even length in a directed graph. International Journal of Systems Science 15, 1197-1202. (1984).

KARM 1986-1 Karmakar, S. B.
Determination of a circuit in a directed graph. International Journal of Systems Science 17, 717-724. (1986).

KARM 1987-1 Karmakar, S. B.
A heuristic method for the determination of a Hamiltonian circuit in a graph. Journal of the Australian Mathematical Society, Series B 28, 328-339. (1987).

KARMA 1982-1 Karmarkar, N.
Probabilistic analysis of some bin-packing problems. In: 23rd Annual Symp. on Foundations of Computer Science, Chicago, Illinois 1982, Proceedings, 107-111. IEEE. (1982).

KARMB 1982-1 Karmarkar, N.; Karp, R. M.
An efficient approximation scheme for the one-dimensional bin-packing problem. in: 23rd IEEE Annual Symposium on Foundations of Computer Science (Chicago, Ill.), 312-320. (1982).

KARMC 1987-1 Karmarkar, U. S.
Lot-sizing and sequencing delays. Management Science 33, 419-423. (1987).

KARMD 1985-1 Karmarkar, U. S.; Kekre, S.; Kekre, S.
Lotsizing in multi-item multi-machine job shops. IIE Transactions 17, 290-298. (1985).

KARMD 1987-1 Karmarkar, U. S.; Kekre, S.; Kekre, S.
The dynamic lot-sizing problem with startup and reservation costs. Operations Research 35, 389-398. (1987).

KARME 1985-1 Karmarkar, U. S.; Kekre, S.; Kekre, S.; Freeman, S.
Lot-sizing and lead-time performance in a manufacturing cell. Interfaces 15, 1-9. (1985).

KARN 1986-1 Karni, R.
A study of lot sizing algorithms. IIE Transactions 18, 356-366. (1986).

KARNA 1984-1 Karnin, E. D.
A parallel algorithm for the knapsack problem. IEEE Transactions on Computers C-33, 404-408. (1984).

KARP 1982-1 Karp, R. M.
The computational complexity of network problems. In: P. A. Burr (Ed.): The Mathematics of Networks, 45-62. Proc. Symp. in Appl. Math. Vol. 26, American Math. Society. (1982).

KARP 1986-1 Karp, R. M.
Combinatorics, complexity and randomness. Communications of the ACM 29, 98-109. (1986).

KARP 1987-1 Karp, R. M.
An upper bound on the expected cost of an optimal assignment. In: D. P. Johnson, T. Nishizeki, A. Nozaki, H. P. Wilf (Eds.): Discrete Algorithms and Complexity - Proc. Japan-U.S. Joint Seminar, Kyoto 1986, 1-4. Boston: Academic Press. (1987).

KARPA 1987-1 Karp, R. M.; Leighton, F. T.; Rivest, R. L.; Thompson, C. D.; Vazirani, U. V.; Vazirani, V. V.
Global wire routing in two-dimensional arrays. Algorithmica 2, 113-130. (1987).

KARPB 1985-1 Karp, R. M.; Lenstra, J. K.; Mac Diarmid, C.; Rinnooy Kan, A. H. G.
Probabilistic analysis. In: M. O'hEigeartaigh, J. K. Lenstra, A. H. G. Rinnooy Kan (Eds.): Combinatorial Optimization - Annotated Bibliographies, 52-88. Chichester: Wiley. (1985).

KARPC 1980-1 Karp, R. M.; Papadimitriou, C. H.
On linear characterizations of combinatorial optimization problems. In: 21st Annual Symp. on Foundations of Computer Science, Syracuse, N.Y. 1980, Proceedings, 1-9. IEEE. (1980).

KARPD 1983-1 Karp, R. M.; Pearl, J.
Searching for an optimal path in a tree with random costs. Artificial Intelligence 21, 99-116. (1983).

KARPE 1986-1 Karp, R. M.; Saks, M.; Wigderson, A.
On a search problem related to branch-and-bound procedures. In: 27th Annual Symp. on Foundations of Computer Science, Proceedings, 19-28. IEEE. (1986).

KARPF 1985-1 Karp, R. M.; Steele, J. M.
Probabilistic analysis of heurictics. In: E. L. Lawler, J. K. Lenstra, A. H. G. Rinnooy Kan, D. B. Shmoys (Eds.): The Traveling Salesman Problem, 181-206. New York: Wiley. (1985).

KARPG 1985-1 Karp, R. M.; Upfal, E.; Wigderson, A.
Constructing a perfect matching is in random NC. In: Proc. Seventeenth Annual ACM Symp. on Theory of Computing, Providence 1985, 22-32. ACM. (1985).

KARPG 1985-2 Karp, R. M.; Upfal, E.; Wigderson, A.
The complexity of parallel computation on matroids. In: 26th Annual Symp. on Foundations of Computer Science, Portland, Oregon 1985, Proceedings, 541-550. IEEE. (1985).

KARPG 1986-1 Karp, R. M.; Upfal, E.; Wigderson, A.
Constructing a perfect matching is in random NC. Combinatorica 6, 35-48. (1986).

KARPH 1985-1 Karp, R. M.; Wigderson, A.
A fast parallel algorithm for the maximal independent set problem. Journal of the Association for Computing Machinery 32, 762-773. (1985).

KARPI 1987-1 Karpinski, M.; Lingas, A.
Subtree isomorphism and bipartite perfect matchings are mutually nc-reducible. MSRI 02318-87. Berkeley: Mathematical Sciences Research Institute, 9 p. (1987).

KARW 1984-1 Karwan, M. H.; Rardin, R. L.; Sarin, S.
A new surrogate dual multiplier search procedure. Technical Report NCATSU-IE-84-S2. Greensboro: North Carolina A & T State Univ., Dept. of Industrial Eng., 50 p. (1984).

KARW 1985-1 Karwan, M. H.; Rardin, R. L.; Sarin, S.
Surrogate duality in a branch-and-bound procedure for integer programming. Technical Report, NCATSU-IE-85-S1. Greensboro: North Carolina A & T State Univ., Dept. of Industrial Engineering, 31 p. (1985).

KARWA 1985-1 Karwan, M. H.; Zionts, S.; Villarreal, B.; Ramesh, R.
An improved interactive multicriteria integer programming algorithm. In: Y. Y. Haimes, V. Chankong (Eds.): Decision Making with Multiple Objectives, Lecture Notes in Economics and Mathematical Systems Vol. 242, 261-271. Berlin: Springer-Verlag. (1985).

KARZ 1984-1 Karzanov, A. V.
A generalized MFMC-property and multicommodity cut problems. In: A. Hajnal; L. Lovasz; V. T. Sos (Eds.): Finite and Infinite Sets, Vol. 2, 443-486. Coll. Math. Soc. J. Bolyai 37, Amsterdam: North-Holland. (1984).

KARZ 1984-2 Karzanov, A. V.
On closed sets of a directed graph. U.S.S.R. Computational Mathematics and Mathematical Physics 24, No.6, 197-200; translation from Zhurnal Vychislitelnoi Matematiki i Matematicheskoi Fiziki 24 (1984), No.12, 1903-1906. (1984).

KARZ 1985-1 Karzanov, A. V.
Cuts that are minimal with respect to mean weight, and cycles of a digraph (Russian). In: Qualitative and Approximate Methods for Investigating Operator Equations, 72-83. Yaroslavl: Yaroslav. Gos. Univ. (1985).

KARZ 1985-2 Karzanov, A. V.
Families of cuts with the MFMC-property. Combinatorica 5, 325-336. (1985).

KARZ 1985-3 Karzanov, A. V.
Metrics and undirected cuts. Mathematical Programming 32, 183-198. (1985).

KARZ 1985-4 Karzanov, A. V.
Multiproduct flow problems with integer optimal solutions (Russian). Doklady Akademii Nauk SSSR (Moscow) 280, 789-792. (1985).

KARZ 1985-5 Karzanov, A. V.
On multicommodity flow problems with integer-valued optimal solutions. Soviet Mathematics, Doklady 31, 151-154. Translation from: Doklady Akademii Nauk SSSR (Moscow) 280, 789-792. (1985).

KARZ 1986-1 Karzanov, A. V.
An algorithm for the maximal packing of odd-pole cuts and its applications (Russian). In: Investigations in Applied Graph Theory, 126-140. Novosibirsk: "Nauka" Sibirsk. Otdel. (1986).

KARZ 1987-1 Karzanov, A. V.
Half-integral five-terminus flows. Discrete Applied Mathematics 18, 263-278. (1987).

KARZ 1987-2 Karzanov, A. V.
On maximal matchings of a prescribed weight in complete and complete bipartite graphs. Kibernetika (Kiev) No. 1, 7-11. (1987).

KARZA 1986-1 Karzanov, A. V.; Timofeev, E. A.
An effective algorithm for finding all minimal branch sections of a non-oriented graph (Russian). Kibernetika (Kiev), No. 2, 8-12. (1986).

KASA 1984-1 Kasai, T.; Yamada, S.; Ohno, T.
An approximation method for planar assignment problem. Bulletin of Univ. of Osaka Prefecture, Series A33, No.1, 47-64. (1984).

KASC 1985-1 Kaschube, P. A.
Determining when a graphic matroid is transversal in linear time. Linear and Multilinear Algebra 17, 41-83. (1985).

KASP 1985-1 Kaspshitskaya, M. F.
Stability of the solution of a certain combinatorial optimization problem (Russian). Doklady Akademii Nauk Ukrainskoj SSR (Kiev), Seriya A, No.10, 60-64. (1985).

KASPA 1986-1 Kaspshitskaya, M. F.; Glushkova, V. V.
On the stability to solve travelling salesman problem with moving objects (Russian). Kibernetika (Kiev), No. 3, 121-122. (1986).

KASPB 1985-1 Kaspshitskaya, M. F.; Sergienko, I. V.; Stiranka, A. I.
A special class of problems of geometric optimization (Russian). Zhurnal Vychislitel'noi Matematiki i Matematicheskoi Fiziki (Moscow) 25, 1656-1665. (1985).

KASZ 1985-1 Kaszonyi, L.
On homogeneous forms related to four-colourings. In: L. Lovasz, A. Recski (Eds.): Matroid Theory, 233-238. Coll. Math. Soc. J. Bolyai Vol. 40. Amsterdam: North-Holland. (1985).

KATA 1987-1 Katajainen, J.; Nevalainen, O.; Teuhola, J.
A linear expected-time algorithm for computing planar relative neigbourhood graphs. Information Processing Letters 25, 77-86. (1987).

KATAA 1985-1 Kataranchuk, S. G.
D-convex simple planar graphs (Russian). In: Investigations in numerical methods and theoretical cybernetics, 68-75. Kishinev: "Shtiintsa". (1985).

KATAA 1986-1 Kataranchuk, S. G.
D-convex simple bipartite graphs (Russian). In: Investigations in General Algebra, Geometry, and their Applications, 92-96. Kishinev: "Shtiintsa". (1986).

KATAB 1986-1 Katayama, N.; Kasugai, H.
An algorithm for the discrete road network design problem using branch and bound method (Japanese). Bulletin of Science and Engineering Research Laboratory, Waseda University, No. 116, 1-6. (1986).

KATE 1985-1 Katehakis, M. N.; Levine, A.
A dynamic routing problem - numerical procedures for light traffic conditions. Applied Mathematics and Computation 17, 267-276. (1985).

KATEA 1983-1 Katerinis, P.
Some results on the existence of 2n-factors in terms of vertex-deleted subgraphs. Ars Combinatoria 16B, 271-278. (1983).

KATEA 1985-1 Katerinis, P.
A Chvatal-Erdös condition for an r-factor in a graph. Ars Combinatoria 20-B, 185-192. (1985).

KATEA 1985-2 Katerinis, P.
Minimum degree of a graph and the existence of k-factors. Proceedings of the Indian Academy of Sciences, Mathematical Sciences (Bangalore) 94, 123-127. (1985).

KATEB 1987-1 Katerinis, P.; Woodall, D. R.
Binding numbers of graphs and the existence of k-factors. Quarterly Journal of Mathematics, Oxford II Series 38, 221-228. (1987).

KATH 1987-1 Kathawala, Y.; Gholamnezhad, H.
New approach to facility locations decisions. International Journal of Systems Science 18, 389-402. (1987).

KATO 1985-1 Katoh, N.; Ibaraki, T.; Mine, H.
An algorithm for the equipollent resource allocation problem. Mathematics of Operations Research 10, 44-53. (1985).

KAUF 1985-1 Kaufman, L.; Tubeeckx, S.; Plastria, F.
The zero-one knapsack problem with equality constraint. European Journal of Operational Research 19, 384-389. (1985).

KAUFA 1984-1 Kaufmann, M.
Nearly tight bounds for the longest edge of a tree in a VLSI layout. in: Ninth Colloquium on Trees in Algebra and Programming (Bordeaux), 169-179. Cambridge-New York: Cambridge Univ. Press. (1984).

KAUFB 1985-1 Kaufmann, M.; Mehlhorn, K.
Routing through a generalized switchbox. In: W. Brauer (Ed.): Automata, Languages and Programming, 328-337. Lecture Notes in Computer Science 194, Berlin: Springer. (1985).

KAUFB 1986-1 Kaufmann, M.; Mehlhorn, K.
On local routing of two-terminal nets. Working Paper 03/1986. Saarbrücken. Univ. des Saarlandes, Fachbereich 10, 37 p. (1986).

KAWA 1986-1 Kawaguchi, T.; Kyan, S.
Worst case bound of an LRF schedule for the mean weighted flow-time problem. SIAM Journal on Computing 15, 1119-1129. (1986).

KAWAA 1981-1 Kawamoto, T.; Kajitani, Y.; Shinoda, S.
Eccentricity of tree set. in: Circuits and Systems, Proc. IEEE Int. Symp., Chicago 1981, Vol. 3, 786-789. (1981).

KAWAB 1984-1 Kawanago, T.; Yamashiro, M.
A study on the single machine production-line system with multiple arrival sources. II: On the development of simulation programs (Japanese). Research Reports, Ashikaga Institute of Technology 10, 129-134. (1984).

KEAT 1985-1 Keating, K.
Multiple-ply Hamiltonian graphs and digraphs. In: B. R. Alspach, C. D. Godsil (Eds.): Cycles in Graphs, Annals of Discrete Mathematics Vol. 27, 81-88. Amsterdam: North-Holland. (1985).

KEIL 1985-1 Keil, J. M.
Finding Hamiltonian circuits in interval graphs. Information Processing Letters 20, 201-206. (1985).

KEIL 1986-1 Keil, J. M.
Total domination in interval graphs. Information Processing Letters 22, 171-174. (1986).

KELL 1985-1 Kelly, D.
Comparability graphs. In: I. Rival (Ed.): Graphs and Order, 3-40. Dordrecht: Reidel. (1985).

KELL 1986-1 Kelly, D.
Invariants of finite comparability graphs. Order 3, 155-158. (1986).

KELLA 1984-1 Kelly, D. G.; Oxley, J. G.
On random representable matroids. Studies in Applied Mathematics 71, 181-205. (1984).

KELLB 1985-1 Kelly, F. P.
Some probabilistic aspects of network flow. In: Computer Communications (Denver, 1983), 15-43. Proceedings of Symposia in Applied Mathematics 31, American Mathematical Society. (1985).

KELM 1983-1 Kelmans, A. K.
A criterion for the existence of two crossing chain-chords of given cycle (Russian). In: Algorithmic Constructions and Their Efficiency, 50-60. Yaroslavl: Yaroslav. Gos. Univ. (1983).

KELM 1984-1 Kelmans, A. K.
A strengthening of the Kuratowski planarity criterion for 3-connected graphs. Discrete Mathematics 51, 215-220. (1984).

KELM 1986-1 Kelmans, A. K.
On 3-connected graphs without essential 3-cuts or triangles. Soviet Mathematics, Doklady 33, 698-703; translation from Doklady Akademii Nauk SSSR 288 (1986), 531-535. (1986).

KELMA 1986-1 Kelmans, A. K.; Polesskii, V. P.
Extremal sets and covering and packing problems in matroids (Russian). In: Investigations in Applied Graph Theory, 140-168. Novosibirsk: "Nauka" Sibirsk. Otdel. (1986).

KENN 1985-1 Kennedy, J. W.; Quintas, L. V.; Syslo, M. M.
The theorem on planar graphs. Historia Mathematica 12, 356-368. (1985).

KERN 1985-1 Kern, W.
An efficient algorithm for solving a special class of LP's. Preprints in Optimization 85.16. Köln: Univ. Köln, Mathematisches Institut, 6 p. (1985).

KERN 1985-2 Kern, W.
Lattice theoretic duality in combinatorial geometries and oriented matroids (German). Mathematisch-Naturwissenschaftliche Fakultät der Universität zu Köln. (1985).

KERN 1985-3 Kern, W.
On finite locally projective planar spaces. Preprints in Optimization 85.23. Köln: Univ. Köln, Mathematisches Institut, 7 p. (1985).

KERN 1985-4 Kern, W.
On modular embeddings of geometric lattices (extended abstract). Preprints in Optimization 85.20. Köln: Univ. Köln, Mathematisches Institut, 5 p. (1985).

KERN 1985-5 Kern, W.
Verbandstheoretische Dualität in kombinatorischen Geometrien und orientierten Matroiden. Preprints in Optimization 85.15. Köln: Univ. Köln, Mathematisches Institut, 57 p. (1985).

KERN 1986-1 Kern, W.
A probabilistic analysis of the switching algorithm for the euclidean TSP. Report 86.28. Köln: Univ. Köln, Mathematisches Institut, 8 p. (1986).

KERN 1986-2 Kern, W.
On the depth of combinatorial optimization problems. Report 86/33. Köln: Univ. Köln, Mathematisches Institut, 12 p. (1986).

KESE 1987-1 Keselman, D. Y.
Cycle coverage of the graph ribs (Russian). Kibernetika (Kiev) No. 3, 16-22. (1987).

KESEA 1984-1 Keselman, V. O.
The statistical approach to the solution of a class of transportation problems with fuzzy constraints (Russian). Doklady, Akademiya Nauk Ukrainskoi SSR, Seriya A No. 4, 11-14. (1984).

KHAB 1984-1 Khabibullin, R. F.; Khalimov, F. R.
The problem of optimal operation-wise scheduling (Russian). Issledovaniya po Prikladnoj Matematike (Kazan), No.10, 121-127. (1984).

KHAB 1984-2 Khabibullin, R. F.; Khalimov, F. R.
The problem of optimization of a manager's workload in the execution of a set of of independent jobs (Russian). Issledovaniya po Prikladnoj Matematike (Kazan), No.10, 114-120. (1984).

KHAC 1986-1 Khachatryan, N. K.
The Hamilton property and systems of different representatives in families of subsets of vertices of a graph (Russian). Doklady, Akademiya Nauk Armyanskoj SSR 82, No. 5, 198-201. (1986).

KHACA 1986-1 Khachaturov, V. R.
Models and methods for solving multiextremal allocation problems by the use of properties of supermodular functions defined on Boolean lattices (Russian). In: Applied Program Packages, 63-98. Moscow: Algoritmy Algoritm. Yazyki, "Nauka". (1986).

KHACB 1985-1 Khachaturov, V. R.; Zlotov, A. V.; Kityan, S. P.
Optimization of a structure of two networks taking account of the agglomeration effect (Russian). Soobshcheniya po Prikladnoj Matematike. Moscow: Akademii Nauk SSSR, Vychislitel'noj Tsentra. (1985).

KHACC 1983-1 Khachiyan, L. G.
Polynomial solvability of convex diophantine programming problems with a fixed number of variables. Engineering Cybernetics 21, No. 1, 135-139. Translation from: Izvestiya Akademii Nauk SSSR (Moscow), (Tekh. Kibern.) 1983, No. 1, 177-180. (1983).

KHAD 1986-1 Khadzhiivanov, N. G.
Edge colourings of the graph K(n)-e (Russian). Comptes Rendus de l'Academie Bulgare des Sciences 39, No.4, 23-26. (1986).

KHADA 1985-1 Khadzhiivanov, N. G.; Nenov, N. D.
Relations and graphs (Bulgarian). Fiziko-Matematichesko Spisanie, Bulgarska Akademiya na Naukite (Sofia) 27(60), No. 4, 314-318. (1985).

KHADA 1986-1 Khadzhiivanov, N. G.; Nenov, N. D.
Edge colorings of complete graphs in which monochromatic triangles do not cover the set of all vertices (Russian). Comptes Rendus de l'Academie Bulgare des Sciences 39, No.6, 31-34. (1986).

KHADA 1986-2 Khadzhiivanov, N. G.; Nenov, N. D.
On two-colorings of the edges of complete graphs with a minimal number of monochromatic triangles (Russian). Godishnik na Visshite Uchebni Zavedniya, Prilozhna Matematika (Sofia) 21, No.4, 181-188. (1986).

KHAN 1986-1 Khan, A. M.
Solid-waste disposal with intermediate transfer stations: an application of the fixed-charge location problem. Journal of the Operational Research Society 38, 31-38. (1986).

KHANA 1987-1 Khan, M. R.; Lewis, D. A.
A network model for nursing staff scheduling. Zeitschrift für Operations Research B 31, 161-172. (1987).

KHANB 1984-1 Khan, Z. A.; Ahsan, M. J.
On integer solutions to the concave quadratic programs using branch and bound technique. Soochow Journal of Mathematics (Taipei) 10, 49-54. (1984).

KHANC 1983-1 Khan, Z. A.; Ahsan, M. J.; Khan, E. A.
On mixed integer concave quadratic programs. Soochow Journal of Mathematics (Taipei) 9, 105-110. (1983).

KHANC 1986-1 Khan, Z. A.; Ahsan, M. J.; Khan, E. A.
A branch and bound method for integer solutions to the concave quadratic programs. Pure and Applied Mathematika Sciences 24, No.1-2, 47-52. (1986).

KHAND 1983-1 Khanna, S.
Multi-index time transportation problem. Ricerca Operativa, No. 28, 27-36. (1983).

KHAR 1984-1 Kharazishvili, A. B.
Independent families and universal colored graphs (Russian). Seminar, Institut Prikladnoi Matematiki, Doklady No. 18, 42-46. (1984).

KHAR 1986-1 Kharazishvili, A. B.
Coloured quasihomogeneous graphs (Russian). Soobshcheniya Akademii Nauk Gruzinskoj SSR 120, No. 2, 249-252. (1986).

KHEL 1984-1 Khelladi, A.; Payan, C.
Generalized n-tuple colorings of a graph: a counterexample to a conjecture of Brigham and Dutton. Journal of Combinatorial Theory, Series B 37, 283-289. (1984).

KHO 1985-1 Kho, P.-L.
Tree growing. Matimyas Matematika (Quezon City, Phillipinas) 9, No. 2, 20-33. (1985).

KHOD 1985-1 Khodzinskii, A. N.
Sequential algorithm for the solution of problems of combinatorial optimization on permutations. Cybernetics 21, 788-795; translation from Kibernetika (1985), No.6, 56-60. (1985).

KHOD 1985-2 Khodzinskii, A. N.
Solution of the quadratic assignment problem by the mean value method (Russian). In: Discrete optimization methods and efficient organization of program packages, 31-35. Kiev: Akademiya Nauk Ukrainskoi SSSR, Institut Kibernetiki. (1985).

KHOK 1982-1 Khokhlyuk, V. I.
Algorithms of integer optimization (Russian). Lecture Notes, Novosibirsk. (1982).

KHOK 1983-1 Khokhlyuk, V. I.
Cutting plane algorithms (Russian). Lecture Notes, Novosibirsk. (1983).

KHOK 1987-1 Khokhlyuk, V. I.
Parallel algorithms for integer optimization (Russian). Moscow: "Radio i Svyaz". (1987).

KHRI 1985-1 Khristov, G.; Karamiteva, Z.; Stoyanov, T.
Numerical method and program for the solution of a linear fractional integer optimization problem (Bulgarian). In: Mathematics and education in mathematics, Proc. 14th Spring Conf. of the Union of Bulgarian Mathemticians (Sunny Beach, Apr. 1980), 566-570. Sofia: Izdatelstvo na B'lgarskata Akademiya na Naukite. (1985).

KHUR 1984-1 Khurana, K.; Bagga, P. C.
Minimizing the makespan in a 2-machine flowshop with time lags and setup conditions. Zeitschrift für Operations Research, Series A 28, 163-174. (1984).

KHUR 1985-1 Khurana, K.; Bagga, P. C.
Scheduling of job-block with deadline in n*2 flowshop problem with separated set up times. Indian Journal of Pure and Applied Mathematics 16, 213-224. (1985).

KIDW 1987-1 Kidwell, M. E.; Richter, R. B.
Trees and Euler tours in a planar graph and its relatives. American Mathematical Monthly 94, 618-630. (1987).

KIE 1986-1 Kie, L.-T.
On the legitimate edge deck problem. Shandong: Inst. of Mathematics, Shandong Univ., 16p. (1986).

KIER 1984-1 Kierstead, H. A.
On the chromatic index of multigraphs without large triangles. Journal of Combinatorial Theory, Series B 36, 156-160. (1984).

KIER 1986-1 Kierstead, H. A.
NP-completeness results concerning greedy and super greedy linear extensions. Order 3, 123-134. (1986).

KIERA 1986-1 Kierstead, H. A.; Schmerl, J. H.
The chromatic number of graphs which induce neither K(1,3) nor K(5)-e. Discrete Mathematics 58, 253-262. (1986).

KIERB 1984-1 Kierstead, H. A.; Szemeredi, E.; Trotter, W. T. Jr.
On coloring graphs with locally small chromatic number. Combinatorica 4, 183-186. (1984).

KIERC 1987-1 Kierstead, H. A.; Trotter, W. T. Jr.; Zhou, B.
Representing an ordered set as the intersection of super greedy linear extensions. Order 4, 293-312. (1987).

KIKU 1985-1 Kikuchi, S.
Empty freight car dispatching model under freight car pool concept. Transportation Research 19B, 169-186. (1985).

KIKUA 1986-1 Kikusts, P.
Another algorithm determining the independence number of a graph. Elektronische Informationsverarbeitung und Kybernetik 22, 157-166. (1986).

KIM 1985-1 Kim, J.-G.
Lagrangean relaxation approach for interactive multiobjective integer programming problem. UOU Report (Ulsan Univ., Korea) 16, 185-190. (1985).

KIM A 1982-1 Kim, K. H.
Boolean matrix theory and applications. New York, Basel: Dekker. (1982).

KIM B 1984-1 Kim, K. V.; Cherkasskii, B. V.
A program for solving a multicommodity transportation problem (Russian). In: Algorithms for Solving Network Problems, 6-24. Moscow: Akad. Nauk SSSR, Tsentral. Ekonom.-Mat. Inst. (1984).

KIM C 1985-1 Kim, S.-H.
Sufficient conditions for zero duality gap of Lagrangean relaxation. Journal of the Korean Operations Research and Management Science Society 10, No.2, 24-27. (1985).

KIM D 1985-1 Kim, S.-I.; Park, Y. S.
Models maximizing covering reliability. Journal of the Korean Operations Research and Management Science Society 10, No.1, 1-8. (1985).

KIM E 1987-1 Kim, T.; Chwa, K.-Y.
An O(nlognloglogn) parallel maximum matching algorithm for bipartite graphs. Information Processing Letters 24, 15-18. (1987).

KIM F 1986-1 Kim, T. U.; Lowe, T. J.; Ward, J. E.; Francis, R. L.
A minimal length covering subtree of a tree. Research Report 86-1. Gainesville: Univ. of Florida, Ind. & Systems Eng. Dept., 22 p. (1986).

KIM G 1986-1 Kim, Y.
Computational study on the simple plant location problem: variations of the Benders decomposition method. Journal of the Korean Operations Research and Management Science Society 11, No.1, 24-35. (1986).

KIME 1985-1 Kimemia, J.; Gershwin, S. B.
Flow optimization in flexible manufacturing systems. International Journal of Production Research 23, 81-96. (1985).

KIMO 1984-1 Kimoto, T.; Tsukiyama, S.; Shirakawa, I.
On the computational complexity of assignment problems with restrictions (Japanese). In: Graph theory and applications, Proc. Symp. Kyoto University 1984, 286-297. Kyoto University, Research Inst. for Mathematical Sciences. (1984).

KINC 1987-1 Kincaid, R. K.
Finding two-centers of graphs that are almost trees. Williamsburg: College of William and Mary, Dept. of Mathematics, 21 p. (1987).

KINCA 1987-1 Kincaid, R. K.; Lowe, T. J.; Morin, T. L.
The location of central structures in trees. CC-87-3. West Lafayette: Purdue Univ., Inst. for Interdisciplinary Engineering Studies, 18 p. (1987).

KINCB 1987-1 Kincaid, R. K.; Maimon, O.
The variance measure for location problems on 3-cacti. Williamsburg: College of William and Mary, Dept. of Mathematics, 15 p. (1987).

KIND 1985-1 Kindervater, G. A. P.; Lenstra, J. K.
An introduction to parallelism in combinatorial optimization. In: J. van Leeuwen, J. K. Lenstra (Eds.): Parallel computers and computations, 163-184. Amsterdam: Mathematisch Centrum. (1985).

KIND 1985-2 Kindervater, G. A. P.; Lenstra, J. K.
Parallel algorithms. In: M. O'hEigeartaigh, J. K. Lenstra, A. H. G. Rinnooy Kan (Eds.): Combinatorial Optimization - Annotated Bibliographies, 106-128. Chichester: Wiley. (1985).

KIND 1986-1 Kindervater, G. A. P.; Lenstra, J. K.
An introduction to parallelism in combinatorial optimization. Discrete Applied Mathematics 14, 135-156. (1986).

KIND 1986-2 Kindervater, G. A. P.; Lenstra, J. K.
Parallel computing in combinatorial optimization. Report OS-R8614, CWI, Amsterdam. (1986).

KIND 1986-3 Kindervater, G. A. P.; Lenstra, J. K.
The parallel complexity of TSP heuristics. Report OS-R8609, CWI, Amsterdam. (1986).

KINDA 1985-1 Kindervater, G. A. P.; Volgenant, A.; Leve, G. de; Gijlswijk, V. van
On dual solutions of the linear assignment problem. European Journal of Operational Research 19, 76-81. (1985).

KINDB 1986-1 Kindler, J.
A Mazur-Orlicz type theorem for submodular set functions. Journal of Mathematical Analysis and Applications 120, 533-546. (1986).

KIRA 1984-1 Kiran, A. S.
On ordered flow shop-mean flowtime scheduling problem. Working Paper. Los Angeles: Univ. of Southern Calif., Dept. of Ind. and Systems Eng., 8 p. (1984).

KIRA 1986-1 Kiran, A. S.
Complexity of FMS loading and scheduling problems. Working Paper 86-02. Los Angeles: Univ. of Southern Calif., Industrial & Systems Eng., 18 p. (1986).

KIRK 1985-1 Kirkpatrick, D. G.; Klawe, M.; Pippenger, N.
Some graph-colouring theorems with applications to generalized connection networks. SIAM Journal on Algebraic and Discrete Methods 6, 576-582. (1985).

KIRKA 1985-1 Kirkpatrick, S.; Toulouse, G.
Configuration space analysis of travelling salesman problems. Journal de Physique (Orsay) 46, 1277-1292. (1985).

KIRO 1985-1 Kirousis, L. M.; Papadimitriou, C. H.
Interval graphs and searching. Discrete Mathematics 55, 181-184. (1985).

KISE 1985-1 Kiseleva, E. M.; Shor, N. Z.
Calculation algorithm for multiproduct problem of optimal division with constraints (Russian). Kibernetika (Kiev) No. 1, 76-81, 85. (1985).

KIVI 1984-1 Kivistik, L.
A generalized rule for selection of a generating row for Gomory's all-integer algorithm (Russian). Eesti NSV Teaduste Akadeemia Toimetised, Füüsika-Matemaatika (Tallin) 33, 197-200. (1984).

KIVI 1985-1 Kivistik, L.
Acceleration of Gomory's all-integer algorithm (Russian). Eesti NSV Teaduste Akadeemia Toimetised Füüsika-Matemaatika (Tallin) 34, 11-19. (1985).

KIVI 1985-2 Kivistik, L.
New sufficient finiteness conditions for dual cutoff algorithms (Russian). Izvestiya Akademii Nauk Ehstonskoj SSR, Fizika, Matematika (Tallin) 34, 429-431. (1985).

KLAF 1985-1 Klafszky, E.; Terlaky, T.
Remarks on the feasibility problem of oriented matroids. Working Paper. Budapest: Eötvös Lorand Unvi., Dept. of Operations Res., 8 p. (1985).

KLAF 1985-2 Klafszky, E.; Terlaky, T.
Variants of the "Hungarian Method" for solving linear programming problems. Working Paper. Budapest: Eötvös Lorand Univ., Dept. of Operations Res., 26 p. (1985).

KLAF 1986-1 Klafszky, E.; Terlaky, T.
A new approach to the feasibility problem of oriented matroids (Hungarian). Alkalmazott Matematikai Lapok 12, 279-282. (1986).

KLAS 1985-1 Klastorin, T. D.
The p-median problem for cluster analysis: a comparative test using the mixture model approach. Management Science 31, 84-95. (1985).

KLAU 1983-1 Klauz, S. A.; Kulkova, L. I.
On a method of routing in a network of common signaling channels. Engineering Cybernetics 21, No. 3, 28-34; translation from: Izvestiya Akademii Nauk SSSR, Tekhnicheskaya Kibernetika (Moscow) 1983, No. 3, 34-39. (1983).

KLEB 1987-1 Klebanovich, D. M.; Malyshko, V. V.
An approximate algorithm for the tree layout on a rectangular lattice with area minimization (Russian). Izvestiya Akademii Nauk BSSR, Seriya Fiziko-Matematicheskikh Nauk, No.5, 25-30. (1987).

KLEE 1983-1 Klee, V.
Combinatorial optimization: what is the state of the art? (Chinese). Applied Mathematics & Mathematics of Computation (Chongqing) No. 6, 49-65. (1983).

KLEEA 1984-1 Klee, V.; Kleinschmidt, P.
An upper bound on graph diameters. Bericht, 28. Bochum: Ruhr-Univ. Bochum, 10 p. (1984).

KLEI 1986-1 Klein Haneveld, W. K.
Robustness against dependence in PERT: an application of duality and distributions with known marginalS. In: A. Prekopa, R. J.-B. Wets (Eds.): Stochastic Programming 84, Mathematical Programming Study 27, 153-182. (1986).

KLEIA 1984-1 Klein Haneveld, W. K.; Schweigman, C.
Robustness against dependence in PERT: an application of duality and distributions with known marginals. EIG-Report 84-02-OR. Groningen: Univ. of Groningen, Econometric Institute. (1984).

KLEIB 1986-1 Klein, P. N.; Reif, J. H.
An efficient parallel algorithm for planarity. In: 27th Annual Symp. on Foundations of Computer Science, Proceedings, 465-477. IEEE. (1986).

KLEIC 1987-1 Kleinschmidt, P.; Lee, C. W.; Schannath, H.
Transportation problems which can be solved by the use of Hirsch-paths for the dual problems. Mathematical Programming 37, 153-168. (1987).

KLEID 1982-1 Kleitman, D. J.
Some practical network problems. In: P. A. Burr (Ed.): The Mathematics of Networks, 85-100. Proc. Symp. in Appl. Math. Vol. 26, American Math. Society. (1982).

KLIM 1984-1 Klimov, A. N.
Structural functions of labeled graphs (Russian). Avtomatika i Telemekhanika (Moscow) No. 6, 165-168. (1984).

KLIN 1986-1 Klincewicz, J. G.; Luss, H.
A Lagrangian relaxation heuristic for capacitated facility location with single-source constraints. Journal of the Operational Research Society 37, 495-500. (1986).

KLIN 1987-1 Klincewicz, J. G.; Luss, H.
A dual-based algorithm for multiproduct uncapacitated facility location. Transportation Science 21, 198-206. (1987).

KLINA 1986-1 Klincewicz, J. G.; Luss, H.; Rosenberg, E.
Optimal and heuristic algorithms for multiproduct uncapacitated facility location. European Journal of Operational Research 26, 251-258. (1986).

KLINB 1985-1 Klingman, D.; Armstrong, R. D.; Partow-Navid, P.
Implementation and analysis of alternative algorithms for generalized shortest path problems. Computers & Operations Research 12, 39-50. (1985).

KLINC 1984-1 Klingman, D.; Mead, M.; Phillips, N. V.
Network optimization models for military manpower planning. In: J. P. Brans (Ed.): Operational Research '84, 786-800. Amsterdam: North-Holland. (1984).

KLIND 1984-1 Klingman, D.; Phillips, N. V.
Topological and computational aspects of preemptive multicriteria military personnel assignment problems. Management Science 30, 1362-1375. (1984).

KLINE 1987-1 Klingman, D.; Phillips, N. V.; Steiger, D.; Wirth, R.; Padman, R.; Krishnan, R.
An optimization based integrated short-term refined petroleum product planning system. Management Science 33, 813-830. (1987).

KLINF 1984-1 Klingman, D.; Ross, G. T.; Schneider, R. F.
Optimization-based facility location planning: an application and some insights. In: T. H. Naylor, C. Thomas (Eds.): Optimization Models for Strategic Planning, 79-90. Amsterdam: North-Holland. (1984).

KLING 1986-1 Klingman, D.; Schneider, R. F.
Microcomputer-based algorithms for large scale shortest path problems. Discrete Applied Mathematics 13, 183-206. (1986).

KNOP 1984-1 Knop, J. V.; Szymanski, K.; Trinajstic, N.; Krivka, P.
Computer generation of all 1-factors for a class of graphs with all vertices of degree two or three. Computers & Mathematics with Applications 10, 369-382. (1984).

KOBA 1984-1 Kobayashi, T.
The lexico-bounded flow algorithm for solving the minimum cost project scheduling problem with an additional linear constraint. Journal of the Operations Research Society of Japan 27, 230-248. (1984).

KOCH 1985-1 Kochkarov, A. M.
Asymptotically exact algorithms for solving a multicriterial problem of covering a graph by paths (Russian). Izvestiya Akademii Nauk BSSR, Seriza Fiyiko-Matematicheskikh Nauk (Minsk), No. 4, 39-43. (1985).

KOCHA 1985-1 Kochkarov, A. M.; Perepelitsa, V. A.
A multicriterial covering problem of a graph by paths of large and small length (Russian). Izvestiya Akademii Nauk BSSR, Seriya Fiziko-Matematicheskikh Nauk (Minsk), No.5, 39-44. (1985).

KOEB 1985-1 Koebe, M.
Graphentheoretische Modelle zur Behandlung des Kanaltrassierungsproblems. 30. Internationales Wissenschaftliches Kolloquium (TH Ilmenau, DDR) Heft 5F, 45-48. (1985).

KOEB 1986-1 Koebe, M.
Über die Anzahl der Spuren für die Lösung des Kanaltrassierungsproblems. Elektronische Informationsverarbeitung und Kybernetik 22, 371-384. (1986).

KOEL 1984-1 Koelling, C. P.; Bailey, J. E.
A multiple criteria decision aid for personnel scheduling. IIE Transactions 16, 299-307. (1984).

KOEN 1983-1 Koene, J.
Minimal cost flow in processing networks, a primal approach. Amsterdam: Mathematisch Centrum, CWI Tract 4, 157 p. (1983).

KOES 1985-1 Koester, G.
Coloring problems on a class of 4-regular planar graphs. In: H. Sachs (Ed.): Graphs, Hypergraphs and Applications, Proc. Conf. on Graph Theory (Eyba 1984), 102-105. Leipzig: Teubner. (1985).

KOGA 1987-1 Koganov, L. M.
An oriented graph connected with an ordered pair of ordered partitions (Russian). Matematicheskie Zametki (Moscow) 41, 101-103. (1987).

KOH 1985-1 Koh, K. M.; Poh, K. S.
On a construction of critical graphs which are not sensitive. Graphs and Combinatorics 1, 265-270. (1985).

KOH A 1985-1 Koh, K. M.; Yap, K. Y.
Graceful numberings of cycles with a P(3)-chord. Bulletin of the Institute of Mathematics, Academia Sinica (Taipei) 13, 41-48. (1985).

KOHL 1983-1 Kohlberg, E.
Equilibrium store locations when consumers minimize travel time plus waiting time. Economics Letters 11, 211-216. (1983).

KOHLA 1984-1 Köhler, J.
A method for the solution of a special nonlinear integer programming problem. In: K.-H. Elster (Ed.): Mathematische Optimierung - Theorie und Anwendungen, Vortragsauszüge, Eisenach 1984, 81-82. Techn. Hochschule Ilmenau. (1984).

KOHLA 1984-2 Köhler, J.
A note on the solution of integer linear programming problems using duality properties. In: Computational Mathematics (Warsaw, 1980), 785-792. Banach Center Publ., 13. Warsaw: PWN. (1984).

KOIT 1984-1 Koita, Y.
The dynamic form of the generalized problem of minimal cost flow (Russian). Vestsi Akademii Navuk BSSR Seriya Fizika-Matematychnykh Navuk (Minsk) No. 2, 33-39. (1984).

KOLA 1987-1 Kolaitis, P. G.; Prömel, H. J.; Rothschild, B. L.
K(l+1)-free graphs: asymptotic structure and a 0-1 law. Transactions of the AMS 303, 637-671. (1987).

KOLB 1986-1 Kolbin, V. V.
Optimization problems in semiordered spaces (Russian). Izvestiya Akademii Nauk Azerbajdzhanskoj SSR, Seriya Fiziko-Tekhnicheskikh i Matematicheskikh Nauk No.6, 118-123. (1986).

KOLE 1985-1 Kolen, A. W. J.
The round-trip p-center and covering problem on a tree. Transportation Science 19, 222-234. (1985).

KOLE 1986-1 Kolen, A. W. J.
Tree network and planar rectilinear location theory. CWI Tract 25, Stichting Mathematisch Centrum, 85p. Amsterdam: Centrum voor Wiskunde en Informatica. (1986).

KOLEA 1987-1 Kolen, A. W. J.; Rinnooy Kan, A. H. G.; Trienekens, H. W. J. M.
Vehicle routing with time windows. Operations Research 35, 266-273. (1987).

KOLEB 1984-1 Kolen, A. W. J.; Tamir, A.
Covering problems. Report 8426/O. Rotterdam: Erasmus Univ., Econometric Institute. (1984).

KOLEC 1988-1 Kolenbrander, W. G.
Tourenplanung mit einem Personal Computer. In: H. Schellhaas, P. van Beek, H. Isermann, R. Schmidt, M. Zijlstra (Eds.): Operations Research Proceedings 1987 (DGOR/NSOR), 274-275. Berlin: Springer. (1988).

KOLI 1987-1 Kolinek, M.
A polynomial-time linear decision tree for the traveling salesman problem and other NP-complete problems. Discrete and Computational Geometry 2, 37-48. (1987).

KOLO 1983-1 Kolokolov, A. A.
Lower bound for the number of iterations for a class of truncation algorithms (Russian). Upravlyaemye Sistemy (Novosibirsk) No. 23, 64-69. (1983).

KOLOA 1984-1 Kolokolov, A. A.; Tsepkova, E. V.
An upper estimate of the number of regular cutting planes for a family of problems on a cone (Russian). Upravlyaemye Sistemy (Novosibirsk) 25, 75-79. (1984).

KOMA 1984-1 Komarek, P.
Some new good characterizations for directed graphs. Casopis pro Pestovane Matematiky (Prague) 109, 348-354. (1984).

KOMA 1986-1 Komarek, P.
On good characterizations for digraphs. Casopis pro Pestovani Matematiky 111, 424-430. (1986).

KOMAA 1986-1 Komarovskii, I. V.
Alpha-weighted t-ary trees (Russian). Vestnik Belorusskogo Gosudarstvennogo Universiteta im. V.I. Lenina (Minsk), Seriya I, No.2, 55-58. (1986).

KOMJ 1987-1 Komjath, P.
The colouring number. Proceedings of the London Mathematical Society (3) 54, No. 1, 1-14. (1987).

KOML 1984-1 Komlos, J.
Linear verification for spanning trees. In: 25th Annual Symposium on Foundations of Computer Science, Florida 1984, 201-206. Los Angeles: IEEE Computer Society. (1984).

KOML 1985-1 Komlos, J.
Linear verification for spanning trees. Combinatorica 5, 57-66. (1985).

KOOP 1986-1 Koop, G. J.
Cyclic scheduling of offweekends. Operations Research Letters 4, 259-264. (1986).

KOPY 1984-1 Kopylov, G. N.; Levin, A. Y.
On the realizability of a stochastic multiproduct flow. Soviet Mathematics, Doklady 29, 631-633; translation from Doklady Akademii Nauk SSSR 276 (1984), 1053-1055. (1984).

KORA 1982-1 Korach, E.
Packing of t-cuts, and other aspects of dual integrality. Waterloo: Univ. of Waterloo, Dept. of Combinatorics and Optimization, 292 p. (1982). (Dissertation)

KORAA 1987-1 Korach, E.; Moran, S.; Zaks, S.
The optimality of distributive constructions of minimum weight and degree restricted spanning trees in a complete network of processors. SIAM Journal on Computing 16, 231-236. (1987).

KORB 1983-1 Korbut, A. A.; Finkelstein, Y. Y.
Approximate methods of discrete programming. Engineering Cybernetics 21, No. 1, 124-134. Translation from: Izvestiya Akademii Nauk SSSR (Moscow), (Tekh. Kibern.) 1983, No. 1, 165-176. (1983).

KORH 1984-1 Korhonen, P.; Wallenius, J.; Zionts, S.
Solving the discrete multiple criteria problem using convex cones. Management Science 30, 1336-1345. (1984).

KORK 1987-1 Körkel, M.
On Karkazis' bounding procedure for the simple plant location problem. In: W. Domschke, W. Krabs, J. Lehn, P. Spellucci (Eds.): Proc. XI. Symp. on Operations Research, Darmstadt 1986, 143-144. Methods of Operations Research 57, Frankfurt/Main: Athenäum. (1987).

KORK 1987-2 Körkel, M.
On the exact solution of large-scale simple plant location problems. Technischer Bereich, Deutsche Bundespost, Forschungsinstitut FTZ, No.22e. (1987).

KORN 1985-1 Körner, F.
Ein effektiver Branch and Bound-Algorithmus für Boolesche quadratische Optimierungsprobleme. Zeitschrift für Angewandte Mathematik und Mechanik 65, 392-394. (1985).

KORN 1985-2 Körner, F.
Integer quadratic optimization. European Journal of Operational Research 19, 268-273. (1985).

KORN 1985-3 Körner, F.
Linear transformations associated with the traveling salesman problem. In: H. Sachs (Ed.): Graphs, Hypergraphs and Applications, Proc. Conf. on Graph Theory (Eyba 1984), 106-109. Leipzig: Teubner. (1985).

KORN 1985-4 Körner, F.
Nichtlineare, ganzzahlige Optimierung. 30. Internationales Wissenschaftliches Kolloquium (TH Ilmenau, DDR) Heft 4E, 71-74. (1985).

KORN 1986-1 Körner, F.
A new bound for the quadratic knap sack problem and its use in a branch and bound algorithm. Optimization 17, 643-648. (1986).

KORN 1986-2 Körner, F.
On the relationship of approximation algorithms for the minimum and the maximum travelling salesman problem. European Journal of Operational Research 26, 262-265. (1986).

KORN 1987-1 Körner, F.
On the degree of asymmetry in the travelling salesman problem. Zastosowania Matematyki 19, 117-123. (1987).

KORNA 1984-1 Kornienko, N. M.
Combinatorial algorithms on a class of graphs (Russian). Isvestiya Akademii Nauk BSSR, Seriya Fiziko-Matematicheskikh Nauk (Minsk), No. 3, 109-111. (1984).

KORNB 1986-1 Kornienko, N. M.; Lepin, V. V.
Optimal packings of a weighted cycle in a line (Russian). Vestsi Akademii Navuk BSSR, Seryya Fizika-Matematychnykh Navuk (Minsk), No. 4, 8-12. (1986).

KORNB 1986-2 Kornienko, N. M.; Lepin, V. V.
Optimal packings of a weighted cycle into a line (Russian). Izvestiya Akademii Nauk BSSR, Seriya Fiziko-Matematicheskikh Nauk, No.4, 8-12. (1986).

KORO 1987-1 Korolev, A. G.
A cubic algorithm for the construction of matching with minimal weight using linear memory (Russian). Ehkonomika i Matematicheskie Metody 23, 168-170. (1987).

KOROA 1983-1 Korotkin, A. A.
The accuracy of a greedy algorithm in a problem of the choice of an optimal acyclic subgraph (Russian). In: Algorithmic Constructions and Their Efficiency, 77-82. Yaroslavl: Yaroslav. Gos. Univ. (1983).

KORS 1985-1 Korshunov, A. D.
Discrete extremal problems on covering. In: L. Budach (Ed.): Fundamentals of Computation Theory, FCT '85, Lecture Notes in Computer Science Vol. 199, 191-207. Berlin: Springer-Verlag. (1985).

KORSA 1987-1 Korsnikov, A. D.
Planar three-index transportation problems with dominating index. Report 87-91. Graz: Univ. Graz, Institut für Mathematik, 6 p. (1987).

KORSB 1987-1 Korsnikov, A. D.; Burkard, R. E.
On the dimension of polytopes of planar three-index transportation problems. Report 87-91. Graz: Univ. Graz, Institut für Mathematik, 20 p. (1987).

KORT 1985-1 Kortanek, K. O.
Network design and dynamic routing under queueing demand. Zeitschrift für Operations Research 29A, 141-160. (1985).

KORTA 1984-1 Korte, B.
Kombinatorische Optimierung und algorithmische Prinzipien. In: R. E. Burkard (Ed.): 15. Steiermärkisches mathematisches Symposium, Forschungszentrum Graz, Mathematisch-Statistische Sektion, Bericht No. 235. (1984).

KORTA 1985-1 Korte, B.
Combinatorial optimization (Part I) (Chinese). Chinese Journal of Operations Research 4, 1-3. (1985).

KORTA 1985-2 Korte, B.
Was ist kombinatorische Optimierung? Report No. 85370-OR. Bonn: Univ. Bonn, Inst. f. Ökonometrie u. Operations Research, 27 p. (1985).

KORTA 1986-1 Korte, B.
Kombinatorische Optimierung und algorithmische Prinzipien. In: W. Krelle (Ed.): Ökonomische Prognose-, Entscheidungs- und Gleichgewichtsmodelle, 286-340. Weinheim: VCH-Verlag. (1986).

KORTB 1984-1 Korte, B.; Lovasz, L.
Greedoids - a structural framework for the greedy algorithm. In: W. R. Pulleyblank (Ed.): Progress in Combinatorial Optimization, 221-244. Toronto: Academic Press. (1984).

KORTB 1985-1 Korte, B.; Lovasz, L.
A note on selectors and greedoids. European Journal of Combinatorics 6, 59-68. (1985).

KORTB 1985-2 Korte, B.; Lovasz, L.
Basis graphs of greedoids and two-connectivity. Mathematical Programming Study 24, 158-165. (1985).

KORTB 1985-3 Korte, B.; Lovasz, L.
On submodularity in greedoids and a counterexample. Report 85384-OR. Bonn: Univ. Bonn, Inst. für Ökonometrie u. Op. Research, 14 p. (1985).

KORTB 1985-4 Korte, B.; Lovasz, L.
Polyhedral results for antimatroids. Report 85390-OR. Bonn: Univ. Bonn, Inst. für Ökonometrie u. Op. Research, 16 p. (1985).

KORTB 1985-5 Korte, B.; Lovasz, L.
Polymatroid greedoids. Journal of Combinatorial Theory 38B, 41-72. (1985).

KORTB 1985-6 Korte, B.; Lovasz, L.
Posets, matroids and greedoids. In: L. Lovasz, A. Recski (Eds.): Matroid Theory, 239-266. Coll. Math. Soc. J. Bolyai Vol. 40. Amsterdam: North-Holland. (1985).

KORTB 1985-7 Korte, B.; Lovasz, L.
Relations between subclasses of greedoids. Zeitschrift für Operations Research 29 A, 249-267. (1985).

KORTB 1986-1 Korte, B.; Lovasz, L.
Homomorphisms and Ramsey properties of antimatroids. Discrete Applied Mathematics 15, 283-290. (1986).

KORTB 1986-2 Korte, B.; Lovasz, L.
Non-interval greedoids and the transposition property. Discrete Mathematics 59, 297-314. (1986).

KORTB 1986-3 Korte, B.; Lovasz, L.
The intersection of matroids and antimatroids. RUTCOR Research Report 30-86. New Brunswick: Rutgers Center for Operations Research, 18 p. (1986).

KORTC 1986-1 Korte, B.; Möhring, R. H.
Zur Bedeutung der diskreten Mathematik für die Konstruktion hochintegrierter Schaltkreise. Report 86441-OR. Bonn: Univ. Bonn, Inst. f. Ökonometrie u. Operations Res., 27 p. (1986).

KORTD 1987-1 Korte, N.
Intervallgraphen und MPQ-Bäume: Algorithmen und Datenstrukturen zur Manipulation von Intervallgraphen und der Lösung von Seriationsproblem. Aachen: RWTH Aachen, Math. Nat. Fakultät, 125 p. (1987). (Dissertation)

KORTE 1985-1 Korte, N.; Möhring, R. H.
Transitive orientation of graphs with side constraints. In: H. Noltemeier (Ed.): Proc. WG '85: Internat. Workshop on Graphtheoretic Concepts in Computer Science, Castle Schwanberg, Würzburg, 1985, 143-160. Linz: Trauner. (1985).

KORTE 1987-1 Korte, N.; Möhring, R. H.
A simple linear-time algorithm to recognize interval graphs. In: G. Tinhofer, G. Schmidt (Eds.): Graph-Theoretic Concepts in Computer Science, 1-16. Lecture Notes in Computer Science 246, Berlin: Springer. (1987).

KOSK 1983-1 Koska, A.
On partition graphs. Discussiones Mathematicae (Warszawa) 6, 59-64. (1983).

KOST 1985-1 Kostochka, A. V.
Criterion for existence of perfect matchings in hexagonal systems (Russian). In: Graphen und Netzwerke - Theorie und Anwendungen, 30. Int. Wiss. Kolloq., Ilmenau 1985, 5, 49-52. (1985).

KOST 1985-2 Kostochka, A. V.
Maximum set of edges no two covered by a clique. Combinatorica 5, 229-236. (1985).

KOST 1985-3 Kostochka, A. V.
On the number of graph edges not contained in common cliques. In: H. Sachs (Ed.): Graphs, Hypergraphs and Applications, 110-114. Leipzig: Teubner. (1985).

KOTI 1984-1 Kotin, L.
Connectivity and the design of optimally survivable systems. Congressus Numerantium 41, 179-196. (1984).

KOTIA 1985-1 Kotin, L.; Stark, W. R.
A parallel minimum spanning tree algorithm in LISP. Congressus Numerantium 48, 243-252. (1985).

KOTO 1985-1 Kotov, V. M.; Cherenkova, E. P.
A parametric algorithm for the solution of the travelling salesman problem in the plane (Russian). In: V. P. Tanaev (Ed.): Methods, algorithms and programs for the solution of extremal problems, Collect. sci. Works, Minsk 1985, 94-97. (1985).

KOTZ 1983-1 Kotzig, A.
Regularly k-path connected graphs. Congressus Numerantium 40, 137-142. (1983).

KOTZ 1984-1 Kotzig, A.
Recent results and open problems in graceful graphs. Congressus Numerantium 44, 197-220. (1984).

KOU 1984-1 Kou, L. T.
Fixed length sources and sinks of a graph. Mathematica Japonica (Osaka) 29, 411-417. (1984).

KOU A 1987-1 Kou, L. T.; Makki, K.
An even faster approximation algorithm for the Steiner tree problem in graphs. Congressus Numerantium 59, 147-154. (1987).

KOUB 1985-1 Koubek, V.; Krsnakova, J.
Parallel algorithms for connected components in a graph. In: L. Budach (Ed.): Fundamentals of Computation Theory, FCT '85, Lecture Notes in Computer Science Vol. 199, 208-217. Berlin: Springer-Verlag. (1985).

KOUR 1984-1 Kourtz, P.
A network approach to least-cost daily transfers of forest fire control resources. INFOR 22, 283-290. (1984).

KOVA 1984-1 Kovacs, L. B.; Boros, E.; Inotay, F.
A two-stage mathematical model and interactive program system for planning networks of sewer systems and waste water treatment plants - with application to the Lake Balaton area (Hungarian). Alkalmazott Matematikai Lapok (Budapest) 10, 87-102. (1984).

KOVA 1986-1 Kovacs, L. B.; Boros, E.; Inotay, F.
A two-stage approach for large-scale sewer systems design with application to the Lake Balaton resort area. European Journal of Operational Research 23, 169-178. (1986).

KOVAA 1985-1 Kovacs, P.
The minimal trivalent graphs with given smallest odd cycle. Discrete Mathematics 54, 295-300. (1985).

KOVAB 1985-1 Kovacs, W. J.; Goodin, D. T.
A statistical approach to the traveling salesman problem. Transportation Research 19B, 239-252. (1985).

KOVAC 1987-1 Kovalenko, A. G.
Algorithms of intervals and their application to solve problems of discrete optimization of multistage processes (Russian). Kibernetika (Kiev) No. 3, 102-105. (1987).

KOVAD 1985-1 Kovalev, M. M.
Cones of symmetric and flow submodular functions (Russian). Kibernetika No. 5, 122-123. (1985).

KOVAD 1985-2 Kovalev, M. M.
Gradient methods of maximization of convex functions on discrete structures. Cybernetics 21, 819-830; translation from Kibernetika (1985), No.6, 77-85. (1985).

KOVAD 1987-1 Kovalev, M. M.
Matroids in discrete optimization (Russian). Minsk: "Universitet-skoe". (1987).

KOVAE 1984-1 Kovalev, M. M.; Kotov, V. M.
Analysis of algorithms for constructing a Hamiltonian cycle of maximal weight (Russian). Izvestiya Akademii Nauk BSSR, Seriya Fiziko-Matematicheskikh Nauk (Minsk) No. 4, 45-50. (1984).

KOVAE 1985-1 Kovalev, M. M.; Kotov, V. M.
Serien von Greedy-Algorithmen (Russian). In: E. Girlich (Ed.): Diskrete Optimierung, 93-103. Jena: Friedrich-Schiller-Universität. (1985).

KOVAF 1984-1 Kovalev, M. M.; Milanov, P. B.
Monotonic functions of many-valued logic and supermatroids. U.S.S.R. Computation of Mathematics and Mathematical Physics 24, 106-108. Translation from: Zhurnal Vychislitel'noj Matematiki i Matematicheskoj Fiziki (Moscow) 24, 786-790. (1984).

KOVAG 1984-1 Kovalev, M. M.; Pisaruk, N. N.
Generalized supermatroids (Russian). Doklady Akademii Nauk BSSR (Minsk) 28, 972-975. (1984).

KOVAG 1984-2 Kovalev, M. M.; Pisaruk, N. N.
Independent flows and polymatroids (Russian). Vestnik Belorusskogo Gosudarstvennogo Univ., Seriya I, Fizika, Matematika, Makhnanika (Minsk) No. 2, 41-43. (1984).

KOVAG 1985-1 Kovalev, M. M.; Pisaruk, N. N.
Gradient methods in convex integer programming. Soviet Physics, Doklady 30, No.10, 830-832; translation from Doklady Akademii Nauk SSSR 284, No.6, (1985), 1322-1326. (1985).

KOVAG 1985-2 Kovalev, M. M.; Pisaruk, N. N.
Independent flows with a discrete-concave expenditure function. U.S.S.R. Computational Mathematics and Mathematical Physics 25, No.3, 72-81; translation from Zhurnal Vychislitelnoi Matematiki i Matematicheskoi Fiziki 25 (1985), No.5, 757-771. (1985).

KOVAG 1985-3 Kovalev, M. M.; Pisaruk, N. N.
Ordnungskonvexität (Russian). In: E. Girlich (Ed.): Diskrete Optimierung, 31-35. Jena: Friedrich-Schiller-Universität. (1985).

KOVAG 1986-1 Kovalev, M. M.; Pisaruk, N. N.
Generalized flows with discrete convex cost function (Russian). Izvestiya Akademii Nauk BSSR. Seriya Fiziko-Matematicheskikh Nauk, No.1, 28-33. (1986).

KOVAH 1980-1 Kovalev, S. M.
On the determination of minimal coverings in hypergraphs (Russian). In: N. G. Malyshev (Ed.): Methods for the construction of algorithmic models of complex systems, Work Collection No. 5, 39-46. Taganrog: Radiotekh. Inst. im. V. D. Kalmykov. (1980).

KOZE 1985-1 Kozen, D.; Vazirani, U. V.; Vazirani, V. V.
NC algorithms for comparability graphs, interval graphs, and testing for unique perfect matching. In: P. N. Maheshwari (Ed.): Foundations of Software Technology and Theoretical Computer Science, 496-503. Lecture Notes in Computer Science 206, Berlin: Springer. (1985).

KOZEA 1986-1 Kozeratskaya, L. N.
Stability regions for some integer programming problems (Russian). Doklady Akademii Nauk Ukrainskoj SSR, Seriya A (Kiev) No. 2, 57-60. (1986).

KOZY 1985-1 Kozyrev, V. P.; Yushmanov, S. V.
Graph theory (algorithmic, algebraic and metric problems) (Russian). Probability Theory, Mathematical Statistics, Theoretical Cybernetics Vol. 23, 68-117. Igoti Nauki i Tekhniki, Akademiya Nauk SSSR, Vsesoyuz. Inst. Nauchn. i Tekhn. Informatsii (Moscow). (1985).

KRAH 1986-1 Krahl, R.
Surrogate constraints und Schnittverfahren in der linearen 0-1-Optimierung. In: 17. Jahrestagung "Mathematische Optimierung" (Sellin 1984), 54-64. Seminarbericht 80, Humboldt Univ., Berlin. (1986).

KRAM 1986-1 Kramer, F.; Kramer, H.
On the generalized chromatic number. In: A. Barlotti, M. Biliotti, A. Cossu, G. Korchmaros, G. Tallini (Eds.): Combinatorics '84, Annals of Discrete Mathematics 30, 275-284. Amsterdam, New York, Oxford, Tokyo: North-Holland. (1986).

KRAMA 1985-1 Krämer, M.
A processor-sharing model with several types of jobs and preemptive priorities. In: E. Gelenbe (Ed.): Performance '84 (Paris 1984), 337-344. Amsterdam, New York: North-Holland. (1985).

KRAMB 1982-1 Kramer, M. R.; Leeuwen, J. van
Wire-routing is NP-complete. RUU-CS-82-4. Utrecht: Rijksuniversiteit Utrecht, Vakgroep Informatica, 11 p. (1982).

KRAR 1986-1 Krarup, J.; Pruzan, P. M.
Assessment of approximate algorithms: the error measure's crucial role. BIT 26, 284-294. (1986).

KRAT 1986-1 Kratochvil, J.; Goljan, M.; Kucera, P.
String graphs. Rozpravy Ceskoslovenske Akademie Ved Rada Matematickych a Prirodnich Ved (Prague) 96, No. 3. (1986).

KRATA 1986-1 Kratochvil, J.; Zeps, D.
On the minimum number of Hamiltonian cycles in triangulations. KAM Series, 15. Praha: Charles Univ., Dept. of Applied Mathematics, 8 p. (1986).

KRATB 1987-1 Kratsch, D.
Finding the minimum bandwidth of an interval graph. Information and Computation 74, No. 2, 140-158. (1987).

KRATB 1987-2 Kratsch, D.
On the partition of graphs into cliques or independent sets. Forschungsergebnisse, Friedrich-Schiller-Universität Jena N/87/25. (1987).

KRAU 1985-1 Krause-Solberg, H.
Über die Konstruktion regulärer Pläne gegebener Weite. In: R. Bodendiek, H. Schumacher, G. Walther (Eds.): Graphen in Forschung und Unterricht, Festschrift Klaus Wagner, 112-122. Salzdetfurth: Franzbecker. (1985).

KRAUA 1987-1 Krause, K.; Larmore, L. L.; Volper, D. J.
Packing items from a triangular distribution. Information Processing Letters 25, 351-362. (1987).

KRAV 1984-1 Kravtsov, M. K.
On an upper bound for the radius of a transportation polyhedron (Russian). Vestnik Belorusskogo Gosudarstvennogo Universiteta Imeni V.I. Lenina, Seriya I, No.2, 49-53. (1984).

KRAV 1985-1 Kravtsov, M. K.
A proof of the maximal diameter conjecture for the transportation polyhedron. Cybernetics 21, 514-519; translation from Kibernetika (1985), No.4, 79-82. (1985).

KRAV 1986-1 Kravtsov, M. K.
On the structure of the set of admissible solutions of a transportation problem with limited traffic capacities (Russian). Ehkonomika i Matematicheskie Metody 22, 524-531. (1986).

KRAVA 1984-1 Kravtsov, M. K.; Emelichev, V. A.
Facets, diameter and radius of a transportation polytope (Russian). Doklady Akademii Nauk BSSR (Minsk) 28, 965-967. (1984).

KRAVA 1984-2 Kravtsov, M. K.; Emelichev, V. A.
On transportation polytopes with a maximum number of facets of all dimensions (Russian). Doklady Akademii Nauk BSSR (Minsk) 28, 698-701. (1984).

KRAVB 1987-1 Kravtsov, M. K.; Galaktionova, E. S.
The number of lattice points in integer transportation polyhedra (Russian). Vestsi Akademii Navuk BSSR, Seryya Fizika-Matematychnykh Navuk (Minsk), No. 4, 39-45. (1987).

KREJ 1983-1 Krejnes, M. G.; Novikova, N. M.
An algorithm for solving a class of discrete multicriterion problems. USSR Journal on Computational Mathematics and Mathematical Physics 23, No. 3, 39-47. Translation from: Zhurnal Vychislitel'noj Matematiki i Matematicheskoj Fiziki (Moscow) 23, No. 3, 576-589. (1983).

KRES 1984-1 Kress, M.
The chance constrained critical path with location-scale distributions. European Journal of Operational Research 18, 359-363. (1984).

KREU 1985-1 Kreuter, P.
Kreuztrassierung - ein algebraischer Ansatz zur Trassierungstheorie mikroelektronischer Schaltkreise. 30. Internationales Wissenschaftliches Kolloquium (TH Ilmenau, DDR) Heft 5F, 57-60. (1985).

KRIA 1981-1 Kriauciukas, V.
A tree-matching algorithm (Russian). Matematicheskaya Logika i ee Primeneniya (Vilnius) No. 1, 21-32. (1981).

KRIA 1985-1 Kriauciukas, V.
Free decomposition of graph search problems (Russian). Matematicheskaya Logika i ee Primenenjya (Vilnius) No. 4, 73-88. (1985).

KRIS 1986-1 Krishnamoorthy, M. S.; Murthy, K.
On the total dominating set problem. Congressus Numerantium 54, 265-278. (1986).

KRIZ 1986-1 Kriz, I.
A hypergraph-less construction of highly chromatic graphs without short cycles. KAM Series, ? Praha: Charles Univ., Dept. of Applied Mathematics, 4 p. (1986).

KRUS 1986-1 Kruse, H.-J.
Degeneracy graphs and the neighbourhood problem. Berlin, Heidelberg, New York, Tokyo: Springer. (1986). (Lecture Notes in Economics and Mathematical Systems, 260)

KRUSA 1985-1 Kruskal, C. P.; West, D. B.
Compatible matchings in bipartite graphs. In: Y. Alavi, G. Chartrand, L. Lesniak, D. R. Lick, C. E. Wall (Eds.): Graph Theory with Applications to Algorithms and Computer Science, 489-500. Chichester: Wiley. (1985).

KU 1986-1 Ku, P.-S.; Niu, S.-C.
On Johnson's two-machine flow shop with random processing times. Operations Research 34, 130-136. (1986).

KUBA 1984-1 Kubale, M.
Ordered edge colorings of weighted multigraphs. In: U. Pape (Ed.): Proc. WG '84: Internat. Workshop on Graphtheoretic Concepts in Computer Science, Berlin 1984, 201-212. Linz: Trauner. (1984).

KUBA 1985-1 Kubale, M.
On the ochromatic index of a graph. 30. Internationales Wissenschaftliches Kolloquium (TH Ilmenau, DDR) Heft 5F, 61-64. (1985).

KUBA 1986-1 Kubale, M.
Application of graph edge colouring to the control of rearrangeable Clos networks (Polish). Archiwum Automatyki i Telemechaniki 31, 39-50. (1986).

KUBA 1987-1 Kubale, M.
The complexity of scheduling independent two-processor tasks on dedicated processors. Information Processing Letters 24, 141-148. (1987).

KUBAA 1985-1 Kubale, M.; Jackowski, B.
A generalized implicit enumeration algorithm for graph coloring. Communications of the ACM 28, 412-418. (1985).

KUBO 1986-1 Kubo, M.; Kasugai, H.
Traveling purchaser problem (Japanese). Bulletin of Science and Engineering Research Laboratory, Waseda University, No. 116, 7-12. (1986).

KUBO 1987-1 Kubo, M.; Kasugai, H.
Errata: "Traveling purchaser problem" (Bull. Sci. Engrg. Res. Lab. Waseda Univ., No. 116 (1986), 7-12). Bulletin of Science and Engineering Research Laboratory, Waseda University, No. 117, 92. (1987).

KUBY 1987-1 Kuby, M. J.
Programming models for facility dispersion: the p-dispersion and maxisum-dispersion problems. Geographical Analysis 19, 315-329. (1987).

KUCE 1986-1 Kucera, L.
Greedy coloring is a bad probabilistic algorithm. KAM Series 10. Praha: Charles Univ., Dept. of Applied Mathematics, 7 p. (1986).

KULI 1983-1 Kulikowski, J. L.
On some formal properties of tri-partite graphs. Prace Instytut Podstaw Informatyki Polskiej Akademii Nauk (Warsaw) 531. (1983).

KULK 1985-1 Kulkarni, R. V.; Bhave, P. R.
Integer programming formulations of vehicle routing problems. European Journal of Operations Research 20, 58-67. (1985).

KULKA 1984-1 Kulkarni, V. G.
A compact hash function for paths in PERT networks. Operations Research Letters 3, 137-140. (1984).

KULKB 1985-1 Kulkarni, V. G.; Adlakha, V. G.
Maximum flow in planar networks with exponentially distributed arc capacities. Communications in Statistics, Stochastic Models 1, 263-289. (1985).

KULKB 1986-1 Kulkarni, V. G.; Adlakha, V. G.
Markov and Markov-regenerative PERT networks. Operations Research 34, 769-781. (1986).

KULS 1987-1 Kulshrestha, D. K.
A mini-max location problem with demand points arbitrarily distributed in a compact connected space. Journal of the Operational Research Society 38, 447-452. (1987).

KUMA 1986-1 Kumar, A.; Sharma, S.
On additional job assignment problem. Journal of the Indian Society of Statistics and Operations Research 7, 15-21. (1986).

KUMAA 1984-1 Kumar, K.
Efficient algorithms for grouping component-processor families. Research Paper. Champaign: Univ. of Illinois, 25 p. (1984).

KUMAB 1986-1 Kumar, K. R.; Kusiak, A.; Vannelli, A.
Grouping of parts and components in flexible manufacturing systems. European Journal of Operational Research 24, 387-397. (1986).

KUMAC 1985-1 Kumar, P.; Miller, F. G.
Shortest route network heuristic for the dynamic vehicle routing problem. Working Paper. Naperville: AT&T Bell Laboratories, 16 p. (1985).

KUMAD 1987-1 Kumara, S. R. T.; Kashyap, R. L.; Moodie, C. L.
Expert system for industrial facilities layout planning and analysis. Computers and Industrial Engineering 12, 143-152. (1987).

KUND 1985-1 Kunde, M.; Steppat, H.
First fit decreasing scheduling on uniform multiprocessors. Discrete Applied Mathematics 10, 165-178. (1985).

KUND 1987-1 Kunde, M.; Steppat, H.
On the worst-case ratio of a compound multiprocessor scheduling algorithm. Information Processing Letters 25, 389-396. (1987).

KUNG 1986-1 Kung, J. P. S.
A source book in matroid theory. Boston, Basel, Stuttgart: Birkhäuser. (1986).

KUNG 1986-2 Kung, J. P. S.
Growth rates and critical exponents of classes of binary combinatorial geometries. Transactions of the American Mathematical Society 293, 837-859. (1986).

KUNG 1987-1 Kung, J. P. S.
Matchings and Radon transforms in lattices. II. Concordant sets. Mathematical Proceedings of the Cambridge Philosophical Society 101, 221-231. (1987).

KURT 1985-1 Kurtulus, I.
Multiproject scheduling: Analysis of scheduling strategies under unequal delay penalties. Journal of Operations Management 5, 291-307. (1985).

KURTA 1985-1 Kurtulus, I.; Narula, S. C.
Multi-project scheduling: analysis of project performance. IIE Transactions 17, 58-66. (1985).

KUSI 1986-1 Kusiak, A.
Efficient implementation of Johnson's scheduling algorithm. IIE Transactions 18, 215-216. (1986).

KUSI 1987-1 Kusiak, A.
Designing expert systems for scheduling automated manufacturing. Industrial Engineering 19, No. 7, 42-46. (1987).

KUSIA 1987-1 Kusiak, A.; Finke, G.
Hierarchical approach to the process planning problem. Discrete Applied Mathematics 18, 175-184. (1987).

KUSIB 1987-1 Kusiak, A.; Heragu, S. S.
The facility layout problem. European Journal of Operational Research 29, 229-251. (1987).

KUZM 1985-1 Kuzmenko, V. N.
Solution of a discrete programming problem of large dimensions (Russian). In: V. M. Mikhalevich (Ed.): Methods for the Solution of Complex Mathematical Programming Problems, 70-76. Kiev: Akad. Nauk Ukrain. SSR, Inst. Kibernet. (1985).

KUZMA 1984-1 Kuzmin, S. L.
A problem of maximum flow in networks with relaying. Izvestiya Akademii Nauk SSSR, Tekhnicheskaya Kibernetika (Moscow) No. 3, 133-135; translated as: Engineering Cybernetics 22, 117-120 (1985). (1984).

KUZMA 1985-1 Kuzmin, S. L.
A problem on a flow with limitations on the number of transits. Soviet Journal of Computer and Systems Sciences 23, No.3, 110-116; translation from Izv. Akad. Nauk SSSR, Tekh. Kibern. (1985), No.2, 77-83. (1985).

KUZMA	1985-2	Kuzmin, S. L. On the complexity of flow problems with path length constraints. Soviet Journal of Computer and Systems Sciences 23, No.6, 57-64; translation from Izvestiya Akademii Nauk SSSR, Tekhnicheskaya Kibernetika (1985), No.4, 91-97. (1985).
KUZY	1982-1	Kuzyurin, N. N. On the problem of the alpha-depth of (0,1)-matrices (Russian). Voprosy Kibernetiki (Moscow) No. 86, 131-139. (1982).
KUZY	1984-1	Kuzyurin, N. N. The complexity of approximate algorithms for the problem of integer programming. USSR Computational Mathematics and Mathematical Physics 24, No. 1, 100-103. (1984).
KWAK	1985-1	Kwak, N. K.; Schniederjans, M. J. A goal programming model as an aid in facility location analysis. Computers & Operations Research 12, 151-162. (1985).
KWAK	1985-2	Kwak, N. K.; Schniederjans, M. J. A goal programming model for selecting a facility location site. RAIRO Recherche Operationelle 19, 1-14. (1985).
KWAK	1985-3	Kwak, N. K.; Schniederjans, M. J. Goal programming solutions to transportation problems with variable supply and demand requirements. Socio-Economic Planning Sciences 19, No. 2, 95-100. (1985).
KWAK	1987-1	Kwak, N. K.; Schniederjans, M. J. Introduction to Mathematical Programming. Malabar: Krieger. (1987).
KWAS	1985-1	Kwasnik, M. Primitivity of products of directed graphs. In: H. Sachs (Ed.): Graphs, Hypergraphs and Applications, Proc. Conf. on Graph Theory (Eyba 1984), 115-118. Leipzig: Teubner. (1985).
KYS	1982-1	Kys, P. A note on characterization of diameter stable graphs. Bulletin Mathematique de la Societe des Sciences Mathematiques de la Republique Socialiste de Roumaine (Bucharest), Nouvelle Serie 26 (74), 165-168. (1982).
KYS	1985-1	Kys, P. Remarks on domatic number. Acta Mathematica Universitatis Comenianae 46/47, 237-241. (1985).
KYS A	1983-1	Kys, P.; Wawruch, A. Point-maximal subgraphs with a given diameter. Acta Mathematica Universitatis Comenianae (Bratislava) 42-43, 285-289. (1983).
KYS A	1985-1	Kys, P.; Wawruch, A. Maximal sizes of planar digraphs without transititve edges. Acta Mathematica Universitatis Comenianae 46/47, 243-249. (1985).
LAAR	1987-1	Laarhoven, P. J. M. van; Aarts, E. H. L. Simulated annealing: theory and applications. Dordrecht: Reidel. (1987).
LABA	1986-1	Labahn, R.; Kanthack, H. A colouring problem. Rostocker Mathematisches Kolloquium No. 30, 67-72. (1986).
LABB	1985-1	Labbe, M. Essays in network location theory. Cahiers du Centre d'Etudes de Recherche Operationelle 27, 5-130. (1985).
LABB	1985-2	Labbe, M. Outcomes of voting and planning in single facility location problems. European Journal of Operational Research 20, 299-313. (1985).
LABBA	1986-1	Labbe, M.; Laporte, G. Maximizing user convenience and postal service efficiency in post box location. Belgian Journal of Operations Research, Statistics and Computer Science 26, No. 2, 21-36. (1986).
LABE	1984-1	Labetoulle, J.; Lawler, E. L.; Lenstra, J. K.; Rinnooy Kan, A. H. G. Preemptive scheduling of uniform machines subject to release dates. In: W. R. Pulleyblank (Ed.): Progress in Combinatorial Optimization, 245-262. Toronto: Academic Press. (1984).
LACH	1982-1	Lachlan, A. H. Finite homogeneous simple digraphs. in: Proceedings of the Herbrand Symposium (Marseilles, 1981), 189-208. North-Holland (Amsterdam, New York). (1982).
LAFO	1985-1	Laforge, R. L. MRP lot sizing with multiple purchase discounts. Computers & Operations Research 12, 579-588. (1985).
LAGA	1982-1	Lagarias, J. C. The computational complexity of simultaneous diophantine approximation problems. In: 23rd Annual Symp. on Foundations of Computer Science, Chicago, Illinois 1982, Proceedings, 32-39. IEEE. (1982).
LAGE	1985-1	Lageweg, B.; Lenstra, J. K.; Stougie, L.; Rinnooy Kan, A. H. G. Stochastic integer programming by dynamic programming. Statistica Neerlandica 39, 97-114. (1985).
LAI	1987-1	Lai, P.-Y.; Goldschmidt, Y. Y. Application of statistical mechanics to combinatorial optimization problems: the chromatic number problem and q-partitioning of a graph. Journal of Statistical Physics 48, 513-529. (1987).
LAI A	1983-1	Lai, T.-H.; Sahni, S. Preemptive scheduling of uniform processors with memory. in: T. Basar, D. J. Brown, H. V. Poor (Eds.): Twentyfirst Annual Allerton Conference on Communication, Control, and Computing (Monticello, Ill., 1983), 886-895. Urbana-Champaign: Univ. Illinois. (1983).
LAI B	1985-1	Lai, T.-H.; Sprague, A. On the complexity of a family of generalized matching problems. Discrete Applied Mathematics 12, 303-318. (1985).
LAI B	1985-2	Lai, T.-H.; Sprague, A. Performance of parallel branch-and-bound algorithms. IEEE Transactions on Computers C-34, 962-964. (1985).
LAI B	1986-1	Lai, T.-H.; Sprague, A. A note on anomalies in parallel branch-and-bound algorithms with one-to-one bounding functions. Information Processing Letters 23, 119-122. (1986).
LAKH	1984-1	Lakhani, G. D. An improved distribution algorithm for shortest paths problem. IEEE Transactions on Computers C-33, 855-857. (1984).

LAKS 1986-1 Lakshmipathy, N.; Winklmann, K.
"Global" graph problems tend to be intractable. Journal of Computer and System Sciences 32, 407-428. (1986).

LALO 1986-1 Lalov, P. I.; Vasileva, T. V.
Extremal problems of the triangular metric (Bulgarian). Godishnik na Visshite Uchebni Zavedniya, Prilozhna Matematika 21, No.1, 223-229. (1986).

LAMA 1988-1 Lamatsch, A.
Wagenumlaufplanung im ÖPNV bei begrenzten Betriebshofkapazitäten - ein ganzzahliges Mehrgüterflussmodell mit Lösungsansätzen. In: H. Schellhaas, P. van Beek, H. Isermann, R. Schmidt, M. Zijlstra (Eds.): Operations Research Proceedings 1987 (DGOR/ NSOR), 275-282. Berlin: Springer. (1988).

LAN 1985-1 Lan, J.-L.; Chen, W.-K.
On f-factors of a graph. Journal of the Franklin Institute 320, 55-62. (1985).

LANG 1985-1 Lang, Z.
Untersuchung spezieller Fälle des Layout-Problems mit Hilfe von Graphenmethoden. 30. Internationales Wissenschaftliches Kolloquium (TH Ilmenau, DDR) Heft 5F, 69-72. (1985).

LANGA 1986-1 Lange, E. G.; Batyrkanov, M. S.
Sufficient conditions for the existence of a nonstoppage schedule for the work of a transportation-loading-unloading complex (Russian). In: Application of Mathematics in Economics, 170-180. Frunze: "Ilim". (1986).

LANGB 1986-1 Langen, D.; Jandl, H.
T O U R - Transport Optimization Under Restrictions. Methods of Operations Research 53, 273-284. (1986).

LANGC 1987-1 Langston, M. A.
A study of composite heuristic algorithms. Journal of the Operational Research Society 38, 539-544. (1987).

LANGC 1987-2 Langston, M. A.
Interstage transportation planning in the deterministic flow-shop environment. Operations Research 35, 556-564. (1987).

LAPA 1980-1 LaPaugh, A. S.
A polynomial time algorithm for optimal routing around a rectangle. In: 21st Annual Symp. on Foundations of Computer Science, Syracuse, N.Y. 1980, Proceedings, 282-293. IEEE. (1980).

LAPAA 1984-1 LaPaugh, A. S.; Papadimitriou, C. H.
The even-path problem for graphs and digraphs. Networks 14, 507-514. (1984).

LAPO 1986-1 Laporte, G.
Generalized subtour elimination constraints and connectivity constraints. Journal of the Operational Research Society 37, 509-514. (1986).

LAPO 1987-1 Laporte, G.
Solving a family of permutation problems on 0-1 matrices. RAIRO Recherche Operationelle 20, 65-86. (1987).

LAPOA 1984-1 Laporte, G.; Desroches, S.
Examination timetabling by computer. Computers & Operations Research 11, 351-360. (1984).

LAPOA 1986-1 Laporte, G.; Desroches, S.
The problem of assigning students to course sections in a large engineering school. Computers and Operations Research 13, 387-394. (1986).

LAPOB 1987-1 Laporte, G.; Martello, S.
The selective travelling salesman problem. G-87-15. Montreal: Ecole des Hautes Etudes Commerciales, 17 p. (1987).

LAPOC 1985-1 Laporte, G.; Mercure, H.; Nobert, Y.
Finding the shortest Hamiltonian circuit through n-clusters: a Lagrangean relaxation approach. Congressus Numerantium 48, 277-290. (1985).

LAPOC 1987-1 Laporte, G.; Mercure, H.; Nobert, Y.
Cutting planes based on bin packing solutions for the capacitated vehicle routing problem. Congressus Numerantium 59, 165-178. (1987).

LAPOC 1987-2 Laporte, G.; Mercure, H.; Nobert, Y.
Generalized travelling salesman problem through n sets of nodes: the asymmetrical case. Discrete Applied Mathematics 18, 185-198. (1987).

LAPOD 1983-1 Laporte, G.; Nobert, Y.
Finding the shortest cycle through k specified nodes. Congressus Numerantium 40, 155-168. (1983).

LAPOD 1984-1 Laporte, G.; Nobert, Y.
Comb inequalities for the vehicle routing problem. In: G. Hammer, D. Pallaschke (Eds.): Methods of Operations Research 51, Vol. 1, 271-278. Königstein/Ts.: Verlag Anton Hain Meisenheim. (1984).

LAPOD 1987-1 Laporte, G.; Nobert, Y.
Exact algorithms for the vehicle routing problem. In: P. Martello, G. Laporte, M. Minoux, C. Ribeiro (Eds.): Surveys in Combinatorial Optimization, Annals of Discrete Mathematics 31, 147-184. Amsterdam, New York, Tokyo: North-Holland. (1987).

LAPOE 1984-1 Laporte, G.; Nobert, Y.; Arpin, D.
Optimal solutions to capacitated multidepot vehicle routing problems. Congressus Numerantium 44, 283-292. (1984).

LAPOE 1986-1 Laporte, G.; Nobert, Y.; Arpin, D.
An exact algorithm for solving a capacitated location-routing problem. In: J. P. Osleeb, P. J. Ratick (Eds.): Locational Decisions: Methodology and Applications, 293-312. Annals of Operations Research 6, Basel: Baltzer. (1986).

LAPOF 1985-1 Laporte, G.; Nobert, Y.; Desrochers, M.
Optimal routing under capacity and distance restrictions. Operations Research 33, 1050-1073. (1985).

LAPOG 1987-1 Laporte, G.; Nobert, Y.; Nga, T. K.
An optimal flow circulation algorithm for the asymmetrical multiple travelling salesman problem. Congressus Numerantium 57, 235-249. (1987).

LAPOH 1987-1 Laporte, G.; Nobert, Y.; Taillefer, S.
A branch-and-bound algorithm for the asymmetrical distance-constrained vehicle routing problem. Mathematical Modelling 9, 857-868. (1987).

LARS 1985-1 Larson, R. E.; Dessouky, M. I.; Devor, R. E.
A forward-backward procedure for the single machine problem to minimize maximum lateness. IIE Transactions 17, 252-260. (1985).

LASK 1985-1 Laskar, R.; Peters, K.
Vertex and edge domination parameters. Congressus Numerantium 48, 291-306. (1985).

LASS 1985-1 Lasserre, J. B.; Bes, C.; Roubellat, F.
The stochastic discrete dynamic lot size problem: an open-loop solution. Operations Research 33, 684-689. (1985).

LAU 1986-1 Lau, H. T.
A note on the shortest path through a given link in a network. IEEE Transactions on Circuits and Systems 33, 109-110. (1986).

LAUM 1985-1 Laumond, J.-P.
Enumeration of articulation pairs of a planar graph. Information Processing Letters 21, 173-180. (1985).

LAUN 1985-1 Laundy, R. S.
A tree search algorithm for the multi-commodity location problem. European Journal of Operational Research 20, 344-351. (1985).

LAUR 1985-1 Laurent, M.
Upper bounds for the cardinality of s-distance codes. Working Paper. Paris: University of Paris VII, 29 p. (1985).

LAUR 1986-1 Laurent, M.
A generalization of antiwebs to independence systems and their relation to set covering polytopes. Working Paper. Issy les Moulineaux: CNET, 15 p. (1986).

LAURA 1984-1 Lauritzen, S. L.; Speed, T. P.; Vijayan, K.
Decomposable graphs and hypergraphs. Journal of the Australian Mathematical Society, Series A 36, 12-29. (1984).

LAUT 1985-1 Lautemann, C.; Meyer auf der Heide, F.
Lower time bounds for integer programming with two variables. Information Processing Letters 21, 101-106. (1985).

LAVA 1984-1 Lavallee, I.
An efficient parallel algorithm for computing a minimum spanning tree. In: M. Feilmeier, G. Joubert, U. Schendel (Eds.): Parallel Computing 83, Proceedings of the International Conference (Berlin, 1983), 259-262. Amsterdam, New York: North-Holland. (1984).

LAVA 1985-1 Lavallee, I.
Un algorithme parallele efficace pour construire un arbre de poids minimal dans un graph. RAIRO Recherche Operationelle 19, 57-70. (1985).

LAVAA 1986-1 Lavallee, I.; Roucairol, C.
A fully distributed (minimal) spanning tree algorithm. Information Processing Letters 23, 55-62. (1986).

LAWL 1985-1 Lawler, E. L.
Submodular functions and polymatroid optimization. In: M. O'hEigeartaigh, J. K. Lenstra, A. H. G. Rinnooy Kan (Eds.): Combinatorial Optimization - Annotated Bibliographies, 32-38. Chichester: Wiley. (1985).

LAWLA 1981-1 Lawler, E. L.; Lenstra, J. K.
Machine scheduling with precedence constraints. BW, 148/81. Amsterdam: Mathematisch Centrum, 18 p. (1981).

LAWLB 1981-1 Lawler, E. L.; Lenstra, J. K.; Rinnooy Kan, A. H. G.
Recent developments in deterministic sequencing and scheduling: a survey. BW, 146/81. Amsterdam: Mathematisch Centrum, 39 p. (1981).

LAWLC 1986-1 Lawler, E. L.; Martel, C. U.
Polymatroidal flows with lower bounds. Discrete Applied Mathematics 15, 291-314. (1986).

LAWLD 1985-1 Lawler, E. L.; Slater, P. J.
A linear time algorithm for finding an optimal dominating subforest of a tree. In: Y. Alavi, G. Chartrand, L. Lesniak, D. R. Lick, C. E. Wall (Eds.): Graph Theory with Applications to Algorithms and Computer Science, 501-506. Chichester: Wiley. (1985).

LAWR 1984-1 Lawrence, K. D.; Reeves, G. R.; Lawrence, S. M.
A multiple objective shift allocation model. IIE Transactions 16, 323-328. (1984).

LAZA 1984-1 Lazarev, A. A.
Algorithms in scheduling theory that are based on necessary conditions for optimality (Russian). Issledovaniya po Prikladnoj Matematike (Kazan), No.10, 102-110. (1984).

LAZA 1984-2 Lazarev, A. A.
Investigation of problems of scheduling theory by means of transformations (Russian). Issledovaniya po Prikladnoi Matematike (Kazan) No. 12, 63-74. (1984).

LAZI 1985-1 Lazimy, R.
Improved algorithm for mixed-integer quadratic programs and a computational study. Mathematical Programming 32, 100-113. (1985).

LAZI 1986-1 Lazimy, R.
Interactive relaxation method for a broad class of integer and continuous nonlinear multiple criteria problems. Journal of Mathematical Analysis and Applications 116, 553-573. (1986).

LE 1984-1 Le, W.-L.; Liu, Z. Y.; Wang, Z.-Y.
The algorithm for mixed programming with upper bounded variables and its application to energy planning (Chinese). Journal of Xi'an Jiaotong University 18, No. 4, 113-122. (1984).

LEAH 1985-1 Leahu, V.
Über p-konforme Unabhängigkeitssysteme. Analele Universitatii din Galati, Metalurgie, 3(8), No. 2, 29-32. (1985).

LEBE 1986-1 Lebedev, S. S.; Medvedovskaya, I. T.
A generalized method of marks for a problem of integer linear programming. Soviet Journal of Computer and Systems Sciences 24, No.5, 90-97; translation from Izvestiya Akademii Nauk SSSR, Tekhnicheskaya Kibernetika (1986), No.3, 16-22. (1986).

LEBEA 1983-1 Lebedev, S. S.; Shejnman, O. K.
Dual approach to integer programming. Engineering Cybernetics 21, No. 1, 140-146. Translation from: Izvestiya Akademii Nauk SSSR (Moscow), (Tekh. Kibern.) 1983, No. 1, 181-187. (1983).

LEBEB 1984-1 Lebedeva, T. T.; Sergienko, I. V.; Soltan, V. P.
Concerning conditions of local and global extrema coincidence in the problems of discrete optimization (Russian). Kibernetika (Kiev), No. 5, 58-65. (1984).

LEBEC 1985-1 Lebedinskaya, N. B.
An algorithm for constructing an optimal structural schedule. Journal of Soviet Mathematics 29, 1649-1659. (1985).

LEBED 1987-1 Lebedinskaya, N. B.; Safonova, T. E.
The stability of an optimal structural schedule (Russian). Zapiski Nauchnykh Seminarov Leningradskogo Otdeleniya Matematicheskogo Instituta imeni V. A. Steklova (LOMI) 159, Chisl. Metody i Voprosy Organiz. Vychisl. 8, 56-68. (1987).

LEBL 1986-1 LeBlanc, L. J.; Boyce, D. E.
A bilevel programming algorithm for exact solution of the network design problem with user-optimal flows. Transportation Research 20-B, 259-266. (1986).

LECL 1987-1 Leclerc, B.
Arbres minimums communs et compatibilite de donnees de types varies. Mathematiques et Sciences Humaines, No. 98, 41-67. (1987).

LECLA 1986-1 Leclerc, M.
A graph-theoretical approach to a problem in circuit design. Mitteilungen aus dem Mathematischen Seminar Giessen No. 175, 19-26. (1986).

LECLA 1986-2 Leclerc, M.
Slices of the matching polytopes. Report, 86.32. Bonn: Univ. Bonn, Sonderforschungsbereich 303, 5 p. (1986).

LECLA 1987-1 Leclerc, M.
Eine Min-Max Beziehung für das Exakte Matroid Problem. Archives of Mathematics (Basel) 49, 103-105. (1987).

LEDE 1980-1 Ledermann, W.; Vajda, S.
Handbook of applicable mathematics. Vol. 1: Algebra. Chichester, New York: Wiley. (1980).

LEDE 1985-1 Ledermann, W.; Vajda, S.
Handbook of applicable mathematics. Vol. 5 : Combinatorics and geometry. Part A. Chichester, New York: Wiley. (1985).

LEE 1985-1 Lee, C. C.; Lee, D. T.
A simple on-line packing algorithm. Journal of the Association for Computing Machinery 32, 562-572. (1985).

LEE A 1985-1 Lee, C.-Y.
Worst case performance of heuristics for the joint replenishment problem. Research Report 85-12. Gainesville: Univ. of Florida, Ind. & Syst. Eng. Dept., 22 p. (1985).

LEE A 1986-1 Lee, C.-Y.
The dynamic lot size model with quantity discount. Research Report No. 86-8. Gainesville: Univ. of Florida, Industrial and Engineering Dept., 21p. (1986).

LEE B 1987-1 Lee, C.-Y.; Danusaputro, S.
Economic lot scheduling problem: a resolution of two-product problem. Research Report 87-13. Gainesville: Univ. of Florida, Industrial & Systems Eng. Dept., 29 p. (1987).

LEE C 1985-1 Lee, C.-Y.; Denardo, E. V.
Economic lot scheduling problem: test of feasibility and heuristic method. Research Report 85-5. Gainesville: Univ. of Florida, Ind. & Systems Eng. Dept., 35 p. (1985).

LEE C 1986-1 Lee, C.-Y.; Denardo, E. V.
Rolling planning horizons: error bounds for the dynamic lot size model. Mathematics of Operations Research 11, 423-432. (1986).

LEE D 1987-1 Lee, C.-Y.; Hwang, J.-J.; Chow, Y.-C.; Anger, F. D.
Multiprocessor scheduling with interprocessor communication delays. Research Report 87-12. Gainesville: Univ. of Florida, Industrial & Systems Engineering Dept., 17 p. (1987).

LEE E 1986-1 Lee, C.-Y.; Massey, J. D.
Multiprocessor scheduling: an extension of multifit algorithm. Research Report, 86-9. Gainesville: Univ. of Florida, Industrial & Systems Engineering Department, 14 p. (1986).

LEE E 1986-2 Lee, C.-Y.; Massey, J. D.
Multiprocessor scheduling: combining LPT and MULTIFIT. Research Report 86-14. Gainesville: Univ. of Florida, Industrial & Systems Eng. Dept., 18 p. (1986).

LEE F 1984-1 Lee, D. T.; Leung, J. Y.-T.
On the 2-dimensional channel assignment problem. IEEE Transactions on Computers C-33, 2-6. (1984).

LEE G 1984-1 Lee, D. T.; Preparata, F. P.
Euclidean shortest paths in the presence of rectilinear barriers. Networks 14, 393-410. (1984).

LEE H 1986-1 Lee, D. T.; Wu, Y. F.
Geometric complexity of some location problems. Algorithmica 1, 193-212. (1986).

LEE I 1986-1 Lee, I.-S.
On sensitivity analysis for shortest-path dynamic programming models. Working Paper No. 337. Los Angeles: Univ. of California, Western Management Science Inst., 33p. (1986).

LEE J 1986-1 Lee, J.
The incidence structure of subspaces with well-scaled frames. Working Paper, B 95. New Haven: Yale School of Organization and Management, 30 p. (1986).

LEE J 1987-1 Lee, J.
Turan's triangle theorem and binary matroids. Working Paper Series B 97. New Haven: Yale School of Organization and Management, 14 p. (1987).

LEE K 1985-1 Lee, K. T.
Knapsack problems solved by dynamic programming. Menemui Matematik (Kuala Lumpur) 7, 68-77. (1985).

LEE L 1985-1 Lee, S.-M.
K-graceful labelling of Mongolian tents and related graphs. Congressus Numerantium 50, 85-96. (1985).

LEE M 1987-1 Lee, S.-M.; Luebbe, R. L.
A zero-one goal-programming algorithm using partitioning and constraint aggregation. Journal of the Operational Research Society 38, 633-642. (1987).

LEE N 1986-1 Lee, Y.-D.
Integer programming applications and their relationship (Korean). UOU Report 17, 237-247. (1986).

LEEU 1985-1 Leeuwen, J. van
Parallel computers and algorithms. In: J. van Leeuwen, J. K. Lenstra (Eds.): Parallel computers and computations, 1-32. Amsterdam: Matematisch Centrum. (1985).

LEFE 1985-1 Lefevre, C.; Laloyaux, P.
Note sur les problemes d'ordonnancement Pert aleatoires. RAIRO Recherche Operationelle 19, 27-34. (1985).

LEFK 1985-1 Lefkovitch, L. P.
Entropy and set covering. Information Sciences 36, 283-294. (1985).

LEHE 1986-1 Lehel, J.; Tuza, Z.
Neighborhood perfect graphs. Discrete Mathematics 61, 93-102. (1986).

LEHM 1985-1 Lehmer, D. H.
The chromatic polynomial of a graph. Pacific Journal of Mathematics 118, 463-469. (1985).

LEIS 1985-1 Leiserson, C. E.; Maley, F. M.
Algorithms for routing and testing routability of planar VLSI layouts. In: Proc. Seventeenth Annual ACM Symp. on Theory of Computing, Providence 1985, 69-78. ACM. (1985).

LEISA 1983-1 Leiserson, C. E.; Saxe, J. B.
A mixed-integer linear programming problem which is efficiently solvable. in: T. Basar, D. J. Brown, H. V. Poor (Eds.): Twentyfirst Annual Allerton Conference on Communication, Control, and Computing (Monticello, Ill., 1983), 204-213. Urbana-Champaign: Univ. Illinois. (1983).

LEMO 1985-1 Lemos, M.
On Seymour's question about packing and covering with matroid circuits. Ars Combinatoria 20-B, 27-34. (1985).

LENG 1983-1 Lengauer, T.
Efficient algorithms for the constraint generation for integrated circuit layout compaction. Report 09/1983. Saarbrücken: Univ. des Saarlandes, Fachbereich 10, 22 p. (1983).

LENG 1985-1 Lengauer, T.
Efficient solution of biconnectivity problems on hierarchically defined graphs. In: H. Noltemeier (Ed.): Proc. WG '85: Internat. Workshop on Graphtheoretic Concepts in Computer Science, Castle Schwanberg, Würzburg, 1985, 201-216. Linz: Trauner. (1985).

LENG 1986-1 Lengauer, T.
Efficient algorithms for finding minimum spanning forests of hierarchically defined graphs. In: B. Monien, G. Vidal-Naquet (Eds.): STACS 86, 153-170. Lecture Notes in Computer Science 210, Berlin: Springer. (1986).

LENG 1987-1 Lengauer, T.
Efficient algorithms for finding minimum spanning forests of hierarchically defined graphs. Journal of Algorithms 8, 260-284. (1987).

LENGA 1986-1 Lengyel, I.
A practical generalization of the knapsack problem for cutting stock (Hungarian). Szigma, Matematikai-Közgazdasagi Folyoirat (Budapest) 19, No. 3, 193-208. (1986).

LENS 1984-1 Lenstra, H. W.; Schnorr, C. P.
On the successive minima of a pair of polar lattices. Research Report. Amsterdam: Univ. of Amsterdam, Dept. of Mathematics, 14 p. (1984).

LENSA 1984-1 Lenstra, J. K.; Rinnooy Kan, A. H. G.
Two open problems in precedence constrained scheduling. In: M. Pouzet, D. Richard (Eds.): Orders: Description and Roles, Annals of Discrete Mathematics Vol. 23, 509-522. Amsterdam: Norht-Holland. (1984).

LENSA 1985-1 Lenstra, J. K.; Rinnooy Kan, A. H. G.
Sequencing and scheduling. In: M. O'hEigeartaigh, J. K. Lenstra, A. H. G. Rinnooy Kan (Eds.): Combinatorial Optimization - Annotated Bibliographies, 164-189. Chichester: Wiley. (1985).

LENSB 1982-1 Lenstra, J. K.; Rinnooy Kan, A. H. G.; Emde Boas, P. van
An appraisal of computational complexity for operations research. BW, 159/82. Amsterdam: Mathematisch Centrum, 19 p. (1982).

LENSC 1985-1 Lenstra, J. K.; Rinnooy Kan, A. H. G.; Guan, M.-G.
On travelling salesman problem (Chinese). Chinese Journal of Operations Research 4, 22-32. (1985).

LENSD 1987-1 Lenstra, J. K.; Shmoys, D. B.; Tardos, E.
Approximation algorithms for scheduling unrelated parallel machines. In: 28th Annual IEEE Symposium on Foundations of Computer Science, 217-224. (1987).

LENSD 1988-1 Lenstra, J. K.; Shmoys, D. B.; Tardos, E.
Approximation algorithms for scheduling unrelated parallel machines. In: H. Schellhaas, P. van Beek, H. Isermann, R. Schmidt, M. Zijlstra (Eds.): Operations Research Proceedings 1987 (DGOR/NSOR), 623-624. Berlin: Springer. (1988).

LENT 1985-1 Lentsevichyus, R.
Clustering of the graph with vertex labels using a rooted tree (Russian). Teoriya Optimalnykh Reshenij 11, 61-68. (1985).

LEON 1985-1 Leonas, V.; Simniskis, S.
Automatic partitioning of large graphs and systems of linear algebraic equations of network systems into blocks (Russian). Trudy Akademii Nauk Litovskoi SSR, Seriya B (Vilnius) 1985, No. 6(151), 72-84. (1985).

LEPI 1985-1 Lepin, V. V.
Minimax numberings of weighted trees (Russian). Izvestiya Akademii Nauk BSSR, Seriya Fiziko-Matematicheskikh Nauk, No. 2, 16-21. (1985).

LEPS 1986-1 Lepschy, A.; Romanin-Jacur, G.; Viaro, U.
An extension to 'An algorithm for finding a circuit of even length in a directed graph'. International Journal of Systems Science 17, 829-831. (1986).

LEPS 1986-2 Lepschy, A.; Romanin-Jacur, G.; Viaro, U.
An extension to: "An algorithm for finding a circuit of even length in a directed graph" by S. Karmakar. International Journal of Systems Science 17, 829-831. (1986).

LESC 1983-1 Lesch, M.
Kantenfärbungen schlichter Graphen. In: Geometrie und Kombinatorik, 2. Kolloq., Karl-Marx-Stadt/DDR 1983, Teil I, 113-116. Karl-Marx-Stadt: Technische Hochschule. (1983).

LESC 1985-1 Lesch, M.
A note on resistent edge-colourings. In: H. Sachs (Ed.): Graphs, Hypergraphs and Applications, Proc. Conf. on Graph Theory (Eyba 1984), 119-121. Leipzig: Teubner. (1985).
LESCA 1983-1 Lesch, M.; Schmeichel, I.
Über farbtreue Automorphismen schlichter Graphen mit Kantenfärbungen. Wissenschaftliche Zeitschrift der Pädagogischen Hochschule "Karl Friedrich Wilhelm Wander" Dresden 17, 7-12. (1983).
LESH 1984-1 Leshchenko, L. L.
On a stochastic transportation problem (Russian). Issledovanie Operatsij i ASU (Kiew) 23, 43-50. (1984).
LESK 1984-1 Lesk, M.
Diametre de graphes et qualite de service d'un reseau de donnees. RAIRO Recherche Operationnelle 18, 247-262. (1984).
LEUN 1986-1 Leung, J. M. Y.; Magnanti, T. L.
Valid inequalities and facets of the capacitated plant location problem. Working Paper OR 149-86. Boston: Massachusetts Inst. of Technology, Operations Research Center, 59p. (1986).
LEUNA 1984-1 Leung, J. Y.-T.
Fast algorithms for generating all maximal independent sets of interval, circular-arc and chordal graphs. Journal of Algorithms 5, 22-35. (1984).
LEV 1984-1 Lev, V.; Adiri, I.
V-shop scheduling. European Journal of Operational Research 18, 51-56. (1984).
LEVA 1984-1 Levary, R. R.
An experimental sequential solution procedure to stochastic linear programming problems with 0-1 variables. International Journal of Systems Science 15, 1073-1086. (1984).
LEVAA 1985-1 Levary, R. R.; Kalchik, S.
Facilities layout - a survey of solution procedures. Computers and Industrial Engineering 9, 141-148. (1985).
LEVC 1985-1 Levchenko, V. I.
A generalization of the notion of homogeneity of a graph (Russian). Matematicheskie Issledovaniya (Kishinev), No. 82, Matematicheskie Modeli Met. Optim., 100-104. (1985).
LEVCA 1985-1 Levcopoulos, C.
On covering regions with minimum number of rectangles. In: P. Bertolazzi, F. Luccio (Eds.): VLSI: Algorithms and Architectures, 243-254. Amsterdam: North-Holland. (1985).
LEVI 1983-1 Levin, A. A.
Gluing of Boolean functions and its application in estimates of the comparative complexity of DNF (Russian). Metody Diskretnogo Analiza (Novosibirsk) No. 40, 54-71,101. (1983).
LEVIA 1987-1 Levin, V. I.
The m-machine problem with parts-sequencing constraints. Automation and Remote Control 48, No.3, 361-367; translation from Avtomatika i Telemekhanika (1987), No.3, 107-116. (1987).
LEWI 1986-1 Lewinter, M.
Spanning trees of product graphs. Congressus Numerantium 54, 157-162. (1986).
LEWI 1987-1 Lewinter, M.
Interpolation theorem for the number of degree-preserving vertices of spanning trees. IEEE Transactions on Circuits and Systems CAS-34, 205. (1987).
LEWIA 1986-1 Lewis, A.
Extreme points of infinite transportation problems. Methods of Operations Research 53, 115-126. (1986).
LEZI 1984-1 Lezin, V. I.
Operations control system for a multiproduct plant. Automation Remote Control 45, 1341-1346; translation from: Avtomatika Telemekhanika 1984, No. 10, 114-120. (1984).
LI 1985-1 Li, C.-K.
On the extremal solutions for capacitated network problems. Linear and Multilinear Algebra 18, 117-125. (1985).
LI A 1986-1 Li, C.-L.; Mac Cormick, S. T.; David, S.-L.
The complexity of finding two disjoint paths with min-max objective function. New York: Columbia Univ., Dept. of Industrial Eng. and Operations Research, 20 p. (1986).
LI B 1984-1 Li, H.; Lai, Z. K.
Some theorems on self-centered graphs (Chinese). Journal of the Xinjiang University of Natural Sciences No. 4, 17-20. (1984).
LI B 1985-1 Li, H.; Lai, Z. K.
Self-centered tree graphs (Chinese). Journal of the Xinjiang University of Natural Sciences No. 1, 62-65. (1985).
LI C 1981-1 Li, Q.; Tao, M. Q.; Shen, Y. Q.
The bandwidth of the discrete tori C(m)*C(n) (Chinese). Journal of China University of Science and Technology (Hefei) 11, 1-16. (1981).
LI D 1984-1 Li, W.-X.
Graphs decomposable into paths of length at least three. Ars Combinatoria 17, 27-32. (1984).
LI D 1984-2 Li, W.-X.
The cycle-chromatic number of a hypergraph and an inequality of Lovasz. In: J. A. Bondy, U. P. R. Murty (Eds.): Progress in Graph Theory, 381-388. New York: Academic Press. (1984).
LI E 1986-1 Li, W.-X.; Schmerl, J. H.
Ordered sets with small width and large jump number. Order 3, 1-2. (1986).
LI F 1978-1 Li, W.-X.; Tian, F.
Some problems concerning the chromatic polynomials of graphs (Chinese). Acta Mathematica Sinica (Beijing) 21, 223-230. (1978).
LI G 1984-1 Li, W.-Z.
On lower bounds of Ramsey numbers (Chinese). J. Central China Teachers College No. 2, 1-4. (1984).
LI H 1984-1 Li, X.-M.; Slyke, R. van
Primal-dual heuristics for multiple set covering. Congressus Numerantium 41, 63-74. (1984).
LI I 1985-1 Li, Y.-S.; Li, S.-H.; Liu, J.-J.
A new algorithm for determining the minimum spanning tree. BUSEFAL - Bulletin pour les Sous Ensembles Flous et leurs Application (Toulouse) 22, 80-84. (1985).

LI J 1987-1 Li, Z.-J.
A matrix representation for the overall planning network and relative calculations (Chinese). Acta Mathematicae Applicatae Sinica 10, 117-120. (1987).

LIBU 1984-1 Libura, M.
Integer programming duality and its application in sensitivity analysis and approximate algorithms (Polish). Archiwum Automatyki i Telemechaniki (Warsaw) 29, 75-92. (1984).

LICA 1986-1 Licata, R.; Johnston, R. E.
A scheduling procedure to minimize wastages in cable manufacturing processes. Ricerca Operativa 16, Nuova Serie, No. 40, 87-116. (1986).

LIEB 1986-1 Liebl, F.
Der simulative Test verschiedener Routing-Algorithmen in Datennetzen. In: L. Streitferdt, H. Hauptmann, A. W. Marusev, D. Ohse, U. Pape (Eds.): Operations Research Proceedings 1985, 66-76. Berlin, Heidelberg, New York: Springer-Verlag. (1986).

LIEB 1987-1 Liebl, F.
Routing Algorithmen in Datennetzen - eine vergleichende Simulationsstudie. OR Spektrum 9, 167-186. (1987).

LIEP 1983-1 Lieponite-Yukhimenko, B. I.
Effectiveness of preference functions in integer linear programming algorithms (Russian). Litovskij Matematicheskij Sbornik 23, No.1, 134-140. (1983).

LIEP 1986-1 Lieponite-Yukhimenko, B. I.
Sequential design algorithm for solving linear programming problems with Boolean variables (Russian). Litovskij Matematicheskij Sbornik 26, No.2, 272-282. (1986).

LIES 1986-1 Liestman, A. L.; Campbell, R. H.
A fault-tolerant scheduling problem. IEEE Transactions on Software Engineering SE-12, 1089-1095. (1986).

LIGG 1985-1 Liggett, R. S.
Optimal spatial arrangement as a quadratic assignment problem. In: J. P. Gero (Ed.): Design optimization, Notes and Reports in Mathematics in Science and Engineering, Vol. 1, 1-40. Orlando, Fla.: Academic Press. (1985).

LIH 1981-1 Lih, K.-W.
Connected planar graphs are Sperner posets. Tamkang Journal of Mathematics (Tamsui, Taipei, Taiwan) 12, No. 2, 209-213. (1981).

LIKH 1985-1 Likholat, N. A.
On an algebraic algorithm for solving a problem of the covering of a graph by a growing directed tree (Russian). In: Methods for Solving Nonlinear Problems and for Data Processing, 136-138. Dnepropetrovsk: Dnepropetrovsk. Gos. Univ. (1985).

LIMN 1986-1 Limnios, N.; Ziani, R.
An algorithm for reducing cut sets in fault-tree analysis. IEEE Transactions on Reliability R-35, 559-562. (1986).

LIN 1986-1 Lin, G. D.
A note on aggregating constraints in integer programming. Zeitschrift für Operations Research 30 A, 79-82. (1986).

LIN A 1986-1 Lin, G. N.; Li, X. L.
On strongly harmonious labelings of graphs (Chinese). Journal Xinjiang University of Natural Science (Urumqi) 3, No. 4, 17-23. (1986).

LIN B 1986-1 Lin, T. F.
A system of linear equations related to the transportation problem with application to probability theory. Discrete Applied Mathematics 14, 47-56. (1986).

LIN C 1986-1 Lin, Y. C.
The adjunction algorithm for planarity of graphs (Chinese). Chinese Journal of Computers (Beijing) 9, No. 1, 44-50. (1986).

LIN D 1984-1 Lin, Y. X.
The discrete location problems (Chinese). Chinese Journal of Operations Research 3 No. 1, 7-13. (1984).

LIN D 1985-1 Lin, Y. X.
A simpler proof of interpolation theorem for spanning trees. Kexue Tongbao (Engl. Edition) (Beijing) 30, 134. (1985).

LIN D 1985-2 Lin, Y. X.
All minimum-weight spanning trees in a graph (Chinese). Mathematics in Practice and Theory No. 2, 42-47. (1985).

LIN D 1986-1 Lin, Y. X.
Dynamic programming method for m x n flow-shop scheduling problems (Chinese). Advances in Mathematics (Beijing) 15, 337-346. (1986).

LIN E 1985-1 Lin, Y. X.; Zhang, F. J.
Hamilton circuits in minimum tree graphs and generation of all minimum trees (Chinese). Chinese Annals in Mathematics, Series A 6, 715-718. (1985).

LIND 1984-1 Lindberg, P. O.; Olafsson, S.
On the length of simplex paths: the assignment case. Mathematical Programming 30, 243-260. (1984).

LINDA 1983-1 Lindström, B.
Some examples of non-regular simplicial matroids. Ars Combinatoria 16A, 199-204. (1983).

LINDA 1983-2 Lindström, B.
The non-Pappus matroid is algebraic. Ars Combinatoria 16B, 95-96. (1983).

LINDA 1984-1 Lindström, B.
A simple nonalgebraic matroid of rank three. Utilitas Mathematica 25, 95-97. (1984).

LINDA 1984-2 Lindström, B.
On algebraic representations of matroids. Reports No. 9, Department of Mathematics, University of Stockholm. (1984).

LINDA 1985-1 Lindström, B.
A Desarguesian theorem for algebraic combinatorial geometries. Combinatorica 5, 237-240. (1985).

LINDA 1985-2 Lindström, B.
On the algebraic characteristic set for a class of matroids. Proceedings of the American Mathematical Society 95, 147-151. (1985).

LINDA 1986-1 Lindström, B.
A non-linear algebraic matroid with infinite characteristic set. Discrete Mathematics 59, 319-320. (1986).
LINDA 1986-2 Lindström, B.
On harmonic conjugates in full algebraic combinatorial geometries. European Journal of Combinatorics 7, 259-262. (1986).
LINDA 1986-3 Lindström, B.
The non-Pappus matroid is algebraic over any finite field. Utilitas Mathematica 30, 53-55. (1986).
LINDA 1987-1 Lindström, B.
A class of nonalgebraic matroids of rank 3. Geometriae Dedicata 23, 255-258. (1987).
LINDA 1987-2 Lindström, B.
A reduction of algebraic representations of matroids. Proceedings of the AMS 100, 388-389. (1987).
LINDA 1987-3 Lindström, B.
An elementary proof in matroid theory using Tutte's coordinatization theorem. Utilitas Mathematica 31, 189-190. (1987).
LINI 1984-1 Linial, N.
Graph decompositions without isolates. Journal of Combinatorial Theory, Series B, 36 16-25. (1984).
LINI 1986-1 Linial, N.
Graph coloring and monotone functions on posets. Discrete Mathematics 58, 97-98. (1986).
LINI 1986-2 Linial, N.
Hard enumeration problems in geometry and combinatorics. SIAM Journal on Algebraic and Discrete Methods 7, 331-335. (1986).
LINI 1986-3 Linial, N.
Legal coloring of graphs. Combinatorica 6, 49-54. (1986).
LINIA 1986-1 Linial, N.; Lovasz, L.; Wigderson, A.
A physical interpretation of graph connectivity and its algorithmic applications. In: 27th Annual Symp. on Foundations of Computer Science, Proceedings, 39-48. IEEE. (1986).
LINIB 1985-1 Linial, N.; Tarsi, M.
Deciding hypergraph 2-colourability by H-resolution. Theoretical Computer Science 38, 343-348. (1985).
LINS 1985-1 Lins, S.
An uncrossing algorithm fo the multicommodity flow problem in plane graphs. Notas e Communicacoes de Matematica No. 133. Recife: Universidade Federal de Pernambuco, Centro de Ciencias Exatas e da Natureza, Departamento de Matematica, 17p. (1985).
LIPA 1986-1 Lipatov, E. P.
Graph theory and its applications (Russian). In: Novoe v Bieni, Nauke, Teknike: Serie "Matematika, Kibernetika". Moscow: Znanie. (1986).
LIPM 1985-1 Lipman, M. J.
Hamiltonian cycles and paths in vertex-transitive graphs with abelian and nilpotent groups. Discrete Mathematics 54, 15-22. (1985).
LIPO 1985-1 Lipovetskii, A. I.
Change in the structure of a finite set in discrete optimization (Russian). Ekonomika i Matematicheskie Metody (Moscow) 21, 937-939. (1985).
LIPS 1983-1 Lipski, W. Jr.
The consecutive retrieval property, interval graphs and related topics - a survey. In: Data Base File Organization (Warsaw, 1981), 17-54. Notes and Reports in Computer Science and Applied Mathematics 6. New York, London: Academic Press. (1983).
LIPS 1984-1 Lipski, W. Jr.
An O(n log n) Manhattan path algorithm. Information Processing Letters 19, 99-102. (1984).
LIRO 1987-1 Lirov, Y.
A locally optimal fast obstacle-avoiding path-planning algorithm. Mathematical Modelling 9, 63-68. (1987).
LITK 1984-1 Litke, J. D.
An improved solution to the traveling salesman problem with thousands of nodes. Communications of the ACM 27, 1227-1236. (1984).
LITT 1985-1 Little, C. H. C.; Holton, D. A.
No graph has a maximal 3-ring of bonds. Journal of Combinatorial Theory 38B, 139-142. (1985).
LIU 1986-1 Liu, B. L.
Errata to: "Several classes of self-centered graphs with cyclomatic number 3" (Chinese). Journal Xinjiang University of Natural Science (Urumqi) 3, No. 2, 88. (1986).
LIU A 1986-1 Liu, C. C.
An improved lower bound on the max flow of postoutage network configurations. IEEE Transactions on Circuits and Systems CAS-33, 1153-1154. (1986).
LIU B 1985-1 Liu, C. Y.; Bulfin, R. L.
On the complexity of preemptive open-shop scheduling problems. Operations Research Letters 4, 71-74. (1985).
LIU B 1987-1 Liu, C. Y.; Bulfin, R. L.
Scheduling ordered open shops. Computers and Operations Research 14, 257-264. (1987).
LIU C 1982-1 Liu, G.-Z.
Pseudo-matchings in graphs (Chinese). Journal of Shandong University (Natural Science Edition), No.4, 6-9. (1982).
LIU C 1984-1 Liu, G.-Z.
A lower bound on solutions of Chartrand's problem. Acta Mathematicae Applicatae Sinica, English Series (Beijing) 1, 93-96. (1984).
LIU C 1984-2 Liu, G.-Z.
A result on Welsh's conjecture (Chinese). Journal of Shandong University (Natural Science Edition), No.4, 6-10. (1984).
LIU C 1984-3 Liu, G.-Z.
A theorem on the 1-factors of hypergraphs without cycles (Chinese). Journal of Mathematics (Wuhan) 4, 109-111. (1984).
LIU C 1984-4 Liu, G.-Z.
A theorem on the 1-factors of 2-trees (Chinese). Chinese Journal of Operations Research 3 No. 2, 67-68. (1984).

LIU C 1984-5 Liu, G.-Z.
On independent circuits of 3-circuit graphs (Chinese). Journal of Shandong University (Natural Science Edition), Suppl., 1-4. (1984).

LIU C 1984-6 Liu, G.-Z.
The connectivity of matroid base graphs (Chinese). Chinese Journal of Operations Research 3 No. 1, 59-60. (1984).

LIU C 1985-1 Liu, G.-Z.
A dual algorithm for optimum matching. Journal of Mathematical Research and Exposition 5, 101-104. (1985).

LIU C 1985-2 Liu, G.-Z.
A property of minimally 2-connected graphs (Chinese). Hunan Annals of Mathematics 5, No.1, 20-23. (1985).

LIU C 1985-3 Liu, G.-Z.
A theorem on the 1-factors of (m,n)-trees. Acta Mathematica Scientia (English Edition) (Wuhan) 5, No. 3, 267-270. (1985).

LIU C 1985-4 Liu, G.-Z.
An algorithm for lexicographically generating ordered rooted trees (Chinese). Journal of Systems Science and Mathematical Sciences (Beijing) 5, 15-19. (1985).

LIU C 1985-5 Liu, G.-Z.
Matroid complexes - geometrical representations of matroids. Acta Mathematica Scientia, English Edition (Wuhan) 5, No. 1, 35-42. (1985).

LIU C 1985-6 Liu, G.-Z.
The perfect matching in quasi-forest complexes (Chinese). Journal of Shandong University, Natural Science Edition, No.3, 11-17. (1985).

LIU C 1985-7 Liu, G.-Z.
Verification of the Welsh conjecture for simple binary matroids (Chinese). Kexue Tongbao (Bejiing) 30, 270-272. (1985).

LIU C 1985-8 Liu, G.-Z.
Welsh's conjecture is true for simple binary matroids. Kexue Tongbao, Foreign Language Edition 30, 1147-1149. (1985).

LIU C 1987-1 Liu, G.-Z.
The connectivities of adjacent tree graphs. Acta Mathematicae Applicatae Sinica, English Series 3, 313-317. (1987).

LIU D 1984-1 Liu, J.; Wang, X.-M.
A kind of efficient algorithm on the optimal k-partitioning of the graphs. Journal of the Huazhong (Central China) University of Science and Technology (Wuhan), English Edition, 6, 1-12. (1984).

LIU E 1986-1 Liu, J.-H.
Sequences of path lengths and their applications. Shandong: Inst. of Mathematics, Shandong Univ., 16p. (1986).

LIU F 1984-1 Liu, J.-Q.; Tian, S.-L.
The binding number of the cartesian product of n circuits (Chinese). Journal of Shandong College of Oceanology (Qingdao) 14, No. 3, 97-101. (1984).

LIU G 1983-1 Liu, J.-Z.
Lexicographic generation of rooted trees and trees. Kexue Tongbao, Foreign Lang. Ed. 28, 448-451. (1983).

LIU H 1984-1 Liu, Y.-P.
Enumeration of rooted separable planar maps. Utilitas Mathematica 25, 77-94. (1984).

LIU H 1985-1 Liu, Y.-P.
Enumeration equation for inseparable planar map node decompositions (Chinese). Kexue Tongbao 30, 646-649. (1985).

LIU H 1985-2 Liu, Y.-P.
Some enumeration problems for node decompositions (Chinese). Kexue Tongbao (Beijing) 30, 1521-1524. (1985).

LIU H 1986-1 Liu, Y.-P.
An equation for the enumeration of the node decompositions of nonseparable Eulerian planar maps (Chinese). Kexue Tongbao (Chinese) (Beijing) 31, No. 2, 81-84. (1986).

LIU H 1986-2 Liu, Y.-P.
Enumeration of node decompositions of nonseparable outerplanar maps (Chinese). Kexue Tongbao (Chinese) (Beijing) 31, 1681-1684. (1986).

LIU H 1986-3 Liu, Y.-P.
On chromatic sum equations for rooted 3-regular planar maps (Chinese). Kexue Tongbao (Chinese) (Beijing) 31, No. 17, 1285-1289. (1986).

LIU H 1986-4 Liu, Y.-P.
On the linearity of testing planarity of graphs. Chinese Annals of Mathematics (Shanghai), Series B 7, 425-434. (1986).

LIU H 1987-1 Liu, Y.-P.
Boolean planarity characterization of graphs. RUTCOR Research Report 38-87. New Brunswick: Rutgers Univ., Center for Operations Research, 30 p. (1987).

LIU H 1987-2 Liu, Y.-P.
Chromatic sum equations for rooted cubic planar maps. Acta Mathematicae Applicatae Sinica (English Series) 3, 136-167. (1987).

LIU H 1987-3 Liu, Y.-P.
Enumeration of rooted outerplanar maps with vertex partition. Kexue Tongbao (English Edition) 32, No. 5, 295-299. (1987).

LIU I 1983-1 Liu, Z. H.
Approximate algorithms for combinatorial optimization problems. I (Chinese). Mathematics in Practice and Theory (Beijing) No. 3, 66-74. (1983).

LIU I 1983-2 Liu, Z. H.
Approximate algorithms for combinatorial optimization problems. II (Chinese). Mathematics in Practice and Theory (Beijing) No. 4, 65-75. (1983).

LIU I 1983-3 Liu, Z. H.
NP-complete problems and some approximate algorithms II (Chinese). Qufu Shiyuan Xuebao (Qufu), No. 3, 5-15. (1983).

LIU I 1983-4 Liu, Z. H.
NP-complete problems and some approximate algorithms I (Chinese). Qufu Shiyuan Xuebao (Qufu), No. 2, 1-12. (1983).

LIU I 1987-1 Liu, Z. H.
A new sufficient condition for the existence of a Hamiltonian cycle (Chinese). Qufu Shifan Daxue Xuebao. Ziran Kexue Ban 13, No. 4, 1-6. (1987).

LIU I 1987-2 Liu, Z. H.
Matroid theory and its applications (Chinese). Qufu Shifan Daxue Xuebao. Ziran Kexue Ban 13, No. 2, 1-18. (1987).

LIU J 1984-1 Liu, Z. S.
Two generalizations of Rosa's theorem on graceful graphs (Chinese). Journal of the Xinjiang University of Natural Sciences, No. 3, 91-93. (1984).

LIU J 1985-1 Liu, Z. S.
A class of graceful graphs and two classes of graceful graphs derived from it (Chinese). Neimenggu Daxue Xuebao (Hohhot) 16, 183-186. (1985).

LIU K 1985-1 Liu, Z. S.; Li, H.
A problem on uniquely k-colorable graphs (Chinese). Journal of the Xinjiang University of Natural Sciences No. 2, 123-124. (1985).

LIZZ 1983-1 Lizzio, A.; Milici, S.
Determination of V(3,1), V(3,2), V(4,1) in a graph and sufficient conditions for delta(3) = 0 (Italian). Rivista di Matematica della Universita di Parma, Serie IV/9, 379-386. (1983).

LLOY 1986-1 Lloyd, E. L.
Two processor scheduling with limited preemption. Performance Evaluation 6, 307-312. (1986).

LLOY 1987-1 Lloyd, E. L.
Scheduling with semaphore constraints. Operations Research Letters 6, 121-124. (1987).

LLOYA 1984-1 Lloyd, E. L.; Ravi, S. S.
One-layer routing without component constraints. Journal of Computer and System Sciences 28, 420-438. (1984).

LO 1985-1 Lo, S.-P.
On edge-graceful labellings of graphs. Congressus Numerantium 50, 231-242. (1985).

LOCK 1985-1 Locke, S. C.
A basis for the cycle space of a 3-connected graph. In: B. R. Alspach, C. D. Godsil (Eds.): Cycles in Graphs, Annals of Discrete Mathematics Vol. 27, 381-398. Amsterdam: North-Holland. (1985).

LOCK 1985-2 Locke, S. C.
A generalization of Dirac's theorem. Combinatorica 5, 149-160. (1985).

LODI 1986-1 Lodi, E.; Pagli, L.
Some new results on a restricted channel routing problem. In: F. Makedon, K. Mehlhorn, T. Papatheodorou, P. Spirakis (Eds.): AWOC 86: VLSI Algorithms and Architectures, 133-143. Lecture Notes in Computer Science 227, Berlin: Springer. (1986).

LOEB 1986-1 Loebl, M.; Poljak, S.
On matroids induced by packing subgraphs. KAM Series, 4. Praha: Charles Univ., Dept. of Applied Mathematics, 10 p. (1986).

LOEB 1986-2 Loebl, M.; Poljak, S.
Remark on restricted and strongly unimodular matrices and a class between them. KAM Series, 20. Praha: Charles Univ., Dept. of Applied Mathematics, 7 p. (1986).

LOGA 1987-1 Loganathan, G. V.; Sherali, H. D.
A convergent interactive cutting-plane algorithm for multiobjective optimization. Operations Research 35, 365-377. (1987).

LOIZ 1984-1 Loizou, G.
Certain NP-complete matching problems. Information Sciences 32, 117-121. (1984).

LOMO 1985-1 Lomonosov, M. V.
A multiflow problem without the max-flow min-cut property (Abstract). In: M. Rosenfeld, J. Zaks (Eds.): Convexity and Graph Theory. Annals of Discrete Mathematics Vol. 20, 329-330. Amsterdam: North-Holland. (1985).

LOMO 1985-2 Lomonosov, M. V.
Combinatorial approaches to multiflow problems. Discrete Applied Mathematics 11, 3-93. (1985).

LONC 1985-1 Lonc, Z.; Truszczynski, M.
Decomposition of graphs into graphs with bounded maximum degrees. Demonstratio Mathematica (Warsaw) 18, 295-302. (1985).

LONG 1982-1 Long, T. D.
Application of graph theory in the study of the reliability of large systems. in: Analysis and Optimization of Systems, Proceedings of the 5th internat. Conference (Versailles 1982), 685-698. Lecture Notes in Control and Information Sciences 44. Berlin: Springer-Verlag. (1982).

LORI 1984-1 Loridan, P.
A dual approach to the generalized Weber problem under locational uncertainty. Cahiers du Centre d'Etudes de Recherche Operationelle 26, 241-254. (1984).

LOTF 1986-1 Lotfi, V.; Sarin, S.
A graph coloring algorithm for large scale scheduling problems. Computers and Operations Research 13, 27-32. (1986).

LOUK 1983-1 Loukakis, E.
Two algorithms for generating and ranking the family of solutions of a finite irreflexive relation. Mathematical Social Sciences 6, 75-86. (1983).

LOUK 1984-1 Loukakis, E.
Two algorithms for determining a minimum independent dominating set. International Journal of Computer Mathematics 15, 213-229. (1984).

LOUL 1984-1 Loulou, R.
Probabilistic behaviour of optimal bin-packing solutions. Operations Research Letters 3, 129-136. (1984).

LOUV 1986-1 Louveaux, F.
Discrete stochastic location models. In: J. P. Osleeb, P. J. Ratick (Eds.): Locational Decisions: Methodology and Applications, 23-34. Annals of Operations Research 6, Basel: Baltzer. (1986).

LOUVA 1985-1 Louveaux, F.; Thisse, J.-F.
Production and location on a network under demand uncertainty. Operations Research Letters 4, 145-150. (1985).

LOVA 1983-1 Lovasz, L.
Self-dual polytopes and the chromatic number of distance graphs on the sphere. Acta Scientiarum Mathematicarum (Szeged) 45, 317-323. (1983).

LOVA 1984-1 Lovasz, L.
Normal hypergraphs and the weak perfect graph conjecture. In: C. Berge, V. Chvatal (Eds.): Topics on perfect graphs, Annals of Discrete Mathematics Vol. 21, 29-42. Amsterdam: North-Holland. (1984).

LOVA 1985-1 Lovasz, L.
An algorithmic theory of numbers, graphs, and convexity. Report No. 85368-OR. Bonn: Univ. Bonn, Inst. f. Ökonometrie u. Operations Research, 81 p. (1985).

LOVA 1985-2 Lovasz, L.
Computing ears and branchings in parallel. In: 26th Annual Symp. on Foundations of Computer Science, Portland, Oregon 1985, Proceedings, 464-467. IEEE. (1985).

LOVA 1985-3 Lovasz, L.
Some algorithmic problems on lattices. In: L. Lovasz, E. Szemeredi (Eds.): Theory of Algorithms, Colloquia Mathematica Societatis Janos Bolyai Vol. 44, 323-338. Amsterdam, Oxford, New York: North-Holland. (1985).

LOVA 1985-4 Lovasz, L.
Vertex packing algorithms. In: W. Brauer (Ed.): Automata, Languages and Programming, 1-14. Lecture Notes in Computer Science 194, Berlin: Springer. (1985).

LOVA 1986-1 Lovasz, L.
An algorithmic theory of numbers, graphs and convexity. CBMS-NSF Regional Conf. Series in Appl. Math, Vol. 50. Philadelphia: SIAM. (1986).

LOVA 1986-2 Lovasz, L.
Connectivity algorithms using rubber-bands. In: K. V. Nori: Foundations of Software, Technology and Theoretical Computer Science, Lecture Notes in Computer Science 241, 394-411. Berlin, Heidelberg, New York: Springer-Verlag. (1986).

LOVA 1986-3 Lovasz, L.
Theory of numbers, graphs and convexity. Philadelphia: Society for Industrial and Appl. Mathematics. (1986). (CBMS-NSF Regional Conference Series in Applied Mathematics, 50)

LOVA 1987-1 Lovasz, L.
Matching structure and the matching lattice. Journal of Combinatorial Theory 43B, 187-222. (1987).

LOVAA 1986-1 Lovasz, L.; Plummer, M. D.
Matching theory. Amsterdam, New York, Oxford, Tokyo: North-Holland. (1986). (Annals of Discrete Mathematics, 29)

LOVAB 1987-1 Lovasz, L.; Saks, M.; Schrijver, A.
Orthogonal representations and connectivity of graphs. RUTCOR Research Report 45-87. New Brunswick: Rutgers Center for Operations Research, 19 p. (1987).

LOVAC 1983-1 Lovasz, L.; Simonovits, M.
On the number of complete subgraphs of a graph. II. In: P. Erdös (Ed.): Studies in Pure Mathematics, 459-495. Basel, Boston: Birkhäuser. (1983).

LOVAD 1986-1 Lovasz, L.; Spencer, J.; Vesztergombi, K.
Discrepancy of set-systems and matrices. European Journal of Combinatorics 7, 151-160. (1986).

LOVE 1985-1 Love, R. F.; Dowling, P. D.
Optimal weighted l(p) norm parameters for facilities layout distance characterizations. Management Science 31, 200-206. (1985).

LOVEA 1985-1 Love, R. F.; Truscott, W. G.; Walker, J.
Terminal location problem: a case study supporting the status quo. Journal of the Operational Research Society 36, 131-136. (1985).

LOVEB 1981-1 Love, R. R. Jr.
Traffic scheduling via Benders decomposition. Mathematical Programming Study 15, 102-124. (1981).

LOVEC 1987-1 Lovetskij, S. E.; Melamed, I. I.
Static flows in networks. Automation and Remote Control 48, No.10, 1269-1291; translation from Avtomatika i Telemekhanika (1987), No.10, 3-29. (1987).

LOVI 1985-1 Lovinescu, C.
Conditions for the existence of flux in a matroid. Buletinul Institutului Politehnic din Iasi, Sectia I, Suppl. Cercet. Mat. 201-204. (1985).

LOVI 1986-1 Lovinescu, C.
Flow in matroids with parallel elements. Buletinul Institutului Politehnic din Iasi, Sectia I, 32(36), No. 1-4, 43-46. (1986).

LOZO 1984-1 Lozovanu, D. D.
A generalized problem of network synthesis and some of its properties (Russian). Izvestiya Akademii Nauk Moldavskoj SSR, Seriya Fiziko-Tekhnicheskik i Matematicheskikh Nauk (Kishinev) 1984, No. 2, 13-18. (1984).

LOZO 1984-2 Lozovanu, D. D.
Some properties of a network nonlinear transportation problem and the problem of synthesis of a network in the form of an outgoing tree with a source (Russian). Mathematical Modelling of Processes in Economics, Matematicheskie Issledovaniya (Kishinev) No. 78, 79-88. (1984).

LU 1987-1 Lu, S. H.; Simeone, B.
On the equivalence of roof-duality and paved-duality in quadratic 0-1 optimization. RUTCOR Research Report 22-87. New Brunswick: Rutgers Univ., Center for Operations Research, 5 p. (1987).

LU A 1985-1 Lu, S. H.; Williams, A. C.
Roof duality for nonlinear 0-1 optimization. Rutcor Research Report No. 2-85. New Brunswick: Rutgers Univ., 9p. (1985).

LU A 1987-1 Lu, S. H.; Williams, A. C.
Roof duality and linear relaxation for quadratic and polynomial 0-1 optimization. RRR 8-87. New Brunswick: Rutgers Univ., Center for Operations Research, 9 p. (1987).

LU A 1987-2 Lu, S. H.; Williams, A. C.
Roof duality for polynomial 0-1 optimization. Mathematical Programming 37, 357-360. (1987).

LU B 1980-1 Lu, S. X.
Necessary and sufficient conditions for planar graphs to be Hamiltonian (Chinese). Kexue Tongbao 25, Special Issue, 74-75. (1980).

LU C 1987-1 Lü, W. J.
A new convergent algorithm for location problems of the type min(sum||A(i)x-b(i)||) (Chinese). Qufu Shifan Daxue Xuebao. Ziran Kexue Ban 13, No. 4, 130-132. (1987).

LU D 1985-1 Lu, Z. D.
A job scheduling algorithm for duplex processor systems (Chinese). Journal of Huazhong (Central China) University of Science and Technology 13, No. 6, 75-80. (1985).

LUBI 1985-1 Lubiw, A.
Orderings and some combinatorial optimization problems with geometric applications. Toronto: Univ. of Toronto, 177 p. (1985). (Dissertation)

LUBY 1985-1 Luby, M.
A simple parallel algorithm for the maximal independent set problem. In: Proc. Seventeenth Annual ACM Symp. on Theory of Computing, Providence 1985, 1-10. ACM. (1985).

LUBY 1986-1 Luby, M.
A simple parallel algorithm for the maximal independent set problem. SIAM Journal on Computing 15, 1036-1053. (1986).

LUBYA 1985-1 Luby, M.; Ragde, P.
A bidirectional shortest-path algorithm with good average-case behavior. In: W. Brauer (Ed.): Automata, Languages and Programming, 394-403. Lecture Notes in Computer Science 194, Berlin: Springer. (1985).

LUCE 1984-1 Lucertini, M.
Lagrangian and group-theoretic relaxation in integer programming (Italian). In: G. Carpaneto, G. Di Pillo (Eds.): Methods and algorithms for optimization, 253-303. Bologna: Pitagora Editrice. (1984).

LUCE 1984-2 Lucertini, M.
Models of the task assignment problem in distributed systems. in: Algorithm Design for Computer System Design, CISM Courses and Lectures Vol. 284, 31-48. Vienna: Springer-Verlag. (1984).

LUCEA 1984-1 Lucertini, M.; Paletta, G.
A decomposition method for network design with multicommodity non-simultaneous flows. In: U. Pape (Ed.): Proc. WG '84: Internat. Workshop on Graphtheoretic Concepts in Computer Science, Berlin 1984, 249-258. Linz: Trauner. (1984).

LUCEA 1984-2 Lucertini, M.; Paletta, G.
Network design with non-simultaneous flows. In: G. Ausiello, M. Lucertini, P. Serafini (Eds.): Algorithm Design for Computer System Design. CISM Courses and Lectures 284, 107-123. Wien, New York: Springer-Verlag. (1984).

LUCEA 1985-1 Lucertini, M.; Paletta, G.
Network design with multiple demand: a new approach. In: G. Ausiello, M. Lucertini (Eds.): Analysis and Design of Algorithms for Combinatorial Problems, Annals of Discrete Mathematics Vol. 25, 211-238. Amsterdam: North-Holland. (1985).

LUCEA 1986-1 Lucertini, M.; Paletta, G.
A class of network design problems with multiple demand: Model formulation and an algorithmic approach. In: G. Gallo, C. Sandi (Eds.): Netflow at Pisa, Mathematical Programming Study 26, 225-228. (1986).

LUH 1985-1 Luh, J. Y. S.; Lin, C. S.
Scheduling parallel operations in automation for minimum execution time based on PERT. Computers and Industrial Engineering 9, 149-164. (1985).

LUNA 1987-1 Luna, H. P. L.; Ziviani, N.; Cabral, R. H. B.
The telephonic switching centre network problem: formalization and computational experience. Discrete Applied Mathematics 18, 199-210. (1987).

LUND 1983-1 Lundgren, J. R.; Maybee, J. S.
A characterization of upper bound graphs. Congressus Numerantium 40, 189-194. (1983).

LUND 1985-1 Lundgren, J. R.; Maybee, J. S.
Block graphs and their matrix duals. In: F. Harary, J. P. Maybee (Eds.): Graphs and Applications, 231-244. New York: Wiley. (1985).

LUND 1985-2 Lundgren, J. R.; Maybee, J. S.
Food webs with interval competition graph. In: F. Harary, J. P. Maybee (Eds.): Graphs and Applications, 245-256. New York: Wiley. (1985).

LUO 1984-1 Luo, Y.-X.
The binding number of a product graph - proof of a conjecture of V. C. Kane et al. (Chinese). Fuzhou Daxue Xuebao 1984, No. 2, 19-24. (1984).

LUO A 1985-1 Luo, Z.-H.
Fundamental characteristics of distributions of random flow in PERT (Chinese). Journal of Xiamen University, Natural Science 24, 173-181. (1985).

LUPS 1984-1 Lupsa, L.
Duality theorems for integer programming problems. in: Itinerant Seminar on Functional Equations, Approximation and Convexity (Cluj-Napoca), 99-102, Preprint 84-6, Univ. "Babes-Bolyai", Cluj-Napoca. (1984).

LUPS 1985-1 Lupsa, L.
On quasiconvex integer programming problems. In: I. Maruscіac, W. W. Breckner (Eds.): Approximation and Optimization, Proc. Colloq., Cluj-Napoca 1984, 261-268. University of Cluj-Napoca. (1985).

LUPS 1986-1 Lupsa, L.
An integer theorem of Farkas lemma type. Preprint, "Babes-Bolyai" University, Faculty of Mathematics and Physics, Research Seminaries 7, 147-150. (1986).

LUQU 1984-1 Luque, E.; Ripoll, A.
Integer linear programming for microprograms register allocation. Information Processing Letters 19, 81-86. (1984).

LUTH 1985-1 Lüthi, H.-J.; Polymeris, A.
Scheduling to minimize maximum workload. Management Science 31, 1409-1415. (1985).

LUTT 1986-1 Lutton, J.-L.; Bonomi, E.
Simulated annealing algorithm for the minimum weighted perfect euclidean matching problem. RAIRO - Recherche Operationelle 20, 177-198. (1986).

LYSK	1985-1	Lyska, W. Ein kombinatorischer Algorithmus zur Lösung der linearen 0-1 Vektoroptimierungsaufgabe. In: K.-H. Elster, M. Peschel (Eds.): Vorträge der Arbeitstagung "Vektoroptimierung", Plaue (DDR) 1984, 101-110. Wissenschaftliche Zeitschrift 31 (1985), Heft 2, Techn. Hochschule Ilmenau. (1985).
LYSK	1985-2	Lyska, W. Über ein Ersatzproblem zur ganzzahligen linearen Vektoroptimierung. In: E. Girlich (Ed.): Diskrete Optimierung, 117-124. Jena: Friedrich-Schiller-Universität. (1985).
LYUL	1986-1	Lyulyakov, A. V. Algorithms for the construction of dense schedules for pipeline computers (Russian). Preprint 86-650, Akad. Nauk SSSR Sibirsk. Otdel., Vychisl. Tsentr, Novosibirsk. (1986).
MA	1984-1	Ma, H.-K. A counterexample to the conjecture of Aksionov and Mel'nikov on non-3-colorable planar graphs. Journal of Combinatorial Theory, Series B 36, 218-219. (1984).
MA A	1986-1	Ma, J.-L. An optimum solution for a knapsack problem (Chinese). Journal of Lanzhou Railway College 5, No.3, 72-76. (1986).
MA A	1987-1	Ma, J.-L. A class of scheduling problems for multiprocessors (Chinese). Journal of Lanzhou Railway College 6, No.2, 40-45. (1987).
MA B	1984-1	Ma, K. J.; Feng, C. J. About the Bodendiek's conjecture of graceful graph (Chinese). Journal of Mathematical Research and Exposition 4, 15-18. (1984).
MA B	1984-2	Ma, K. J.; Feng, C. J. On gracefulness of gear graphs (Chinese). Mathematics in Practice and Theory (Beijing) No. 4, 72-73. (1984).
MA C	1985-1	Ma, S.-H. Solvability of a class of knapsack problems (Chinese). Journal of Shandong University, Natural Science Edition, No.4, 37-42. (1985).
MA C	1985-2	Ma, S.-H. The Steiner tree problem on graphs and a heuristic algorithm for it (Chinese). Chinese Journal of Computers (Beijing) 8, No. 3, 237-239. (1985).
MA D	1986-1	Ma, S.-H.; Wallis, W. D. Clique numbers of threshold graphs. Caribbean Journal of Mathematics 5, 17-28. (1986).
MA D	1986-2	Ma, S.-H.; Wallis, W. D. Clique partitions of triangulated graphs. Congressus Numerantium 55, 93-102. (1986).
MA E	1986-1	Ma, S.-H.; Wang, Y. Interval colouring of graphs with cost and an algorithm for colouring of graphs (Chinese). Journal of Shandong University (Natural Science Edition) 21, No.4, 23-28. (1986).
MA F	1986-1	Ma, X. D. Some classes of graceful graphs (Chinese). Journal Xinjiang University of Natural Science (Urumqi) 3, No. 2, 106-107. (1986).
MAAM	1985-1	Maamoun, M. Decompositions of digraphs into paths and cycles. Journal of Combinatorial Theory 38B, 97-101. (1985).
MAAMA	1984-1	Maamoun, M.; Meyniel, H. On a problem of G. Hahn about coloured hamiltonian paths in K(2n). Discrete Mathematics 51, 213-214. (1984).
MAAS	1985-1	Maas, C. Ein Zuordnungsproblem im Zusammenhang mit der Einteilung von Übungsgruppen. Zeitschrift für Angewandte Mathematik und Mechanik 65, 313-314. (1985).
MAC	1984-1	Mac Avaney, K. L. Composite graphs with stability index one. Bulletin of the Australian Mathematical Society 29, 423-424. (1984).
MAC A	1985-1	Mac Bride, R. D. Solving embedded generalized network problems. European Journal of Operational Research 21, 82-92. (1985).
MAC B	1985-1	Mac Call, J. On the basis number of some complete bipartite graphs. Bulletin of the Australian Mathematical Society 31, 433-437. (1985).
MAC C	1983-1	Mac Carthy, P. J. Pseudographs with degrees in prescribed intervals. Malaysian Mathematical Society, Bulletin, Second Series (2) 6, 23-29. (1983).
MAC D	1985-1	Mac Clain, J. O.; Trigeiro, W. W. Cyclic assembly schedules. IIE Transactions 17, 346-353. (1985).
MAC E	1985-1	Mac Cuaig, W. D.; Rosenfeld, M. Parity of cycles containing specified edges. In: B. R. Alspach, C. D. Godsil (Eds.): Cycles in Graphs, Annals of Discrete Mathematics Vol. 27, 419-432. Amsterdam: Norht-Holland. (1985).
MAC F	1986-1	Mac Curley, R.; Schneider, F. B. Derivation of a distributed algorithm for finding paths in directed networks. Science of Computer Programming 6, 1-9. (1986).
MAC G	1984-1	Mac Diarmid, C. Colouring random graphs. Annals of Operations Research 1, 183-200. (1984).
MAC G	1985-1	Mac Diarmid, C. On some conditioning results in the probabilistic analysis of algorithms. Discrete Applied Mathematics 10, 197-202. (1985).
MAC G	1986-1	Mac Diarmid, C. On the greedy algorithm with random costs. Mathematical Programming 36, 245-255. (1986).
MAC H	1984-1	Mac Hugh, J. A. M. Hu's precedence tree scheduling algorithm: a simple proof. Naval Research Logistics Quarterly 31, 409-412. (1984).
MAC I	1983-1	Mac Kay, B. D.; Tollis, I. G. A resource location problem on graphs. Congressus Numerantium 40, 223-234. (1983).

MAC J 1983-1 Mac Kee, T. A.
Validity up to complementation in graph theory. Fundamenta Mathematicae (Warsaw) 116, 93-98. (1983).
MAC J 1984-1 Mac Kee, T. A.
Balance and duality in signed graphs. Congressus Numerantium 44, 11-18. (1984).
MAC J 1984-2 Mac Kee, T. A.
Recharacterizing Eulerian: intimations of new duality. Discrete Mathematics 51, 237-242. (1984).
MAC J 1984-3 Mac Kee, T. A.
Relative expressiveness of the edge/adjacency language for graph theory. Fundamenta Mathematicae 123, 163-167. (1984).
MAC J 1985-1 Mac Kee, T. A.
The logic of graph-theoretic duality. American Mathematical Monthly 92, 457-464. (1985).
MAC J 1987-1 Mac Kee, T. A.
A local analogy between directed and signed graphs. Utilitas Mathematica 32, 175-180. (1987).
MAC J 1987-2 Mac Kee, T. A.
Dual properties within graph theory. Fundamenta Mathematicae (Warsaw) 128, No. 2, 91-97. (1987).
MAC J 1987-3 Mac Kee, T. A.
Generalized complementation. Journal of Combinatorial Theory 42B, 378-383. (1987).
MAC J 1987-4 Mac Kee, T. A.
Multiterminal duality and three-terminal series-parallelness. Discrete Applied Mathematics 18, 47-54. (1987).
MAC K 1983-1 Mac Morris, F. R.; Shier, D. R.
Representing chordal graphs on K(1,n). Commentationes Mathematicae Universitatis Carolinae 24, 489-494. (1983).
MACH 1987-1 Machida, H.
A remark on the complexity of the knapsack problem. In: D. P. Johnson, T. Nishizeki, A. Nozaki, H. P. Wilf (Eds.): Discrete Algorithms and Complexity - Proc. Japan-U.S. Joint Seminar, Kyoto 1986, 107-118. Boston: Academic Press. (1987).
MACU 1987-1 Maculan, N. F.
The Steiner problem in graphs. In: P. Martello, G. Laporte, M. Minoux, C. Ribeiro (Eds.): Surveys in Combinatorial Optimization, Annals of Discrete Mathematics 31, 185-212. Amsterdam, New York, Tokyo: North-Holland. (1987).
MADE 1984-1 Mader, W.
On k-critically n-connected graphs. In: J. A. Bondy, U. P. R. Murty (Eds.): Progress in Graph Theory, 389-398. New York: Academic Press. (1984).
MADE 1985-1 Mader, W.
Degree and local connectivity in digraphs. Combinatorica 5, 161-166. (1985).
MADE 1985-2 Mader, W.
Disjunkte Fragmente in kritisch n-fach zusammenhängenden Graphen. European Journal of Combinatorics 6, 353-360. (1985).
MADE 1985-3 Mader, W.
Minimal n-fach zusammenhängende Digraphen. Journal of Combinatorial Theory 38B, 102-117. (1985).
MADE 1985-4 Mader, W.
Paths in graphs, reducing the edge connectivity only by two. Graphs and Combinatorics 1, 81-90. (1985).
MADE 1986-1 Mader, W.
Kritisch n-fach kantenzusammenhängende Graphen. Journal of Combinatorial Theory, Series B, 40, 121-151. (1986).
MAES 1986-1 Maes, J.; Wassenhove, L. N. van
A simple heuristic for the multi-item single level capacitated lot sizing problem. Operations Research Letters 4, 265-274. (1986).
MAES 1986-2 Maes, J.; Wassenhove, L. N. van
Multi item single level capacitated dynamic lotsizing heuristics: A computational comparison (Part I: Static case). IIE Transactions 18, 114-123. (1986).
MAES 1986-3 Maes, J.; Wassenhove, L. N. van
Multi item single level capacitated dynamic lotsizing heuristics: A computational comparison (Part II: Rolling horizon). IIE Transactions 18, 124-129. (1986).
MAFF 1986-1 Maffioli, F.
Randomized algorithms in combinatorial optimization: a survey. Discrete Applied Mathematics 14, 157-170. (1986).
MAFFA 1985-1 Maffioli, F.; Speranza, M. G.; Vercellis, C.
Randomized algorithms. In: M. O'hEigeartaigh, J. K. Lenstra, A. H. G. Rinnooy Kan (Eds.): Combinatorial Optimization - Annotated Bibliographies, 89-105. Chichester: Wiley. (1985).
MAFFB 1986-1 Maffray, F.
On kernels in i-triangulated graphs. Discrete Mathematics 61, 247-252. (1986).
MAGA 1986-1 Maga, F.; Vizvari, B.
The relaxation of a special polynomial zero-one programming problem to a set covering problem (Hungarian). Alkalmazott Matematikai Lapok 12, 41-49. (1986).
MAGG 1981-1 Maggu, P. L.
The two-machine maximum flow time problem with break-down intervals of machines. Journal of the Indian Society of Statistics and Operations Reserch 2, 33-38. (1981).
MAGGA 1983-1 Maggu, P. L.; Mohd, H. N.; Prakash, O.
On an extension of n*3 flow-shop scheduling problem involving 'job-weight' and 'job-transportation' time. Journal of the Indian Society of Statistics and Operations Research 4, 47-51. (1983).
MAGGA 1983-2 Maggu, P. L.; Mohd, H. N.; Prakash, O.
Weighted-job flow-shop scheduling problem involving transportation time for 'n-job, 3-machine' problem. Journal of the Indian Society of Statistics and Operations Research 4, 41-45. (1983).
MAGGB 1984-1 Maggu, P. L.; Prakash, O.
Principle of dominance in flow-shop scheduling models. Journal of the Indian Society of Statistics and Operations Research (Bangalore) 5, 11-16. (1984).

MAGGC 1984-1 Maggu, P. L.; Prakash, O.; Singh, T.
A note on transportation time in dual three-stage production scheduling problem with set up time included. Journal of the Indian Society of Statistics and Operations Research (Saharanpur) 5, 51-52. (1984).

MAGGC 1984-2 Maggu, P. L.; Prakash, O.; Singh, T.
On dual optimal three-stage production scheduling problem with set-up times included. Pure and Applied Mathematika Sciences (Saharanpur) 19, 85-91. (1984).

MAGGD 1985-1 Maggu, P. L.; Sharma, B. K.
'n-job, 2-machine' sequencing problem with stochastic service time involving equivalent-jobs for job-blocks. Pure and Applied Mathematika Sciences (Saharanpur) 21, 1-5. (1985).

MAGGD 1986-1 Maggu, P. L.; Sharma, B. K.
On equivalent-job for job-block in a 'n-job, 2-machine' flow-shop scheduling with random processing times. Journal of the Indian Society of Statistics and Operations Research 7, 1-4. (1986).

MAGGD 1986-2 Maggu, P. L.; Sharma, B. K.
On 2-machine flow shop with additional times and equivalent jobs. Pure and Applied Mathematika Sciences 24, No.1-2, 59-64. (1986).

MAGGE 1984-1 Maggu, P. L.; Singh, T.; Sharma, V. K.
A short note on first generalized dual scheduling problem. Pure and Applied Mathematika Sciences (Saharanpur) 20, 159-160. (1984).

MAGGE 1984-2 Maggu, P. L.; Singh, T.; Sharma, V. K.
A short note on second generalised dual scheduling problem. Pure and Applied Mathematika Sciences (Saharanpur) 20, 161-162. (1984).

MAGGE 1984-3 Maggu, P. L.; Singh, T.; Sharma, V. K.
A short note on third generalised dual scheduling problem. Pure and Applied Mathematika Sciences (Saharanpur) 20, 163-164. (1984).

MAGGF 1984-1 Maggu, P. L.; Yadav, S. K.; Singh, T. P.; Dev, A.
Flow-shop scheduling problem involving job-weight and transportation time. Pure and Applied Mathematika Sciences (Saharanpur) 20, 153-158. (1984).

MAGI 1986-1 Magirou, V. F.
The efficient drilling of printed circuit boards. Interfaces 16, No. 4, 13-23. (1986).

MAGN 1986-1 Magnanti, T. L.; Mireault, P.; Wong, R. T.
Tailoring Benders decomposition for uncapacitated network design. In: G. Gallo, C. Sandi (Eds.): Netflow at Pisa, Mathematical Programming Study 26, 112-154. (1986).

MAHA 1986-1 Mahadev, N. V. R.; Peled, U. N.
Strict 2-threshold graphs. Working Paper. Chicago: Univ. of Illinois, Dept. of Mathematics, Statistics, and Computer Science, 24 p. (1986).

MAHE 1985-1 Mahey, P.
Subgradient techniques and combinatorial optimization. Report 85397-OR. Bonn: Univ. Bonn, Inst. für Ökonometrie u. Op. Research, 34 p. (1985).

MAHO 1985-1 Mahoney, C.
On the unimodality of the independent set numbers of a class of matroids. Journal of Combinatorial Theory 39B, 77-85. (1985).

MAHR 1981-1 Mahr, B.
A bird's-eye view to path problems. In: H. Noltemeier (Ed.): Graphtheoretic Concepts in Computer Science, Proc. Internat. Workshop WG 80 (Bad Honnef 1980), 335-353. Berlin, Heidelberg, New York: Springer. (1981).

MAI 1984-1 Mai, C.-C.; Shieh, Y.-N.
Transport rates structure, optimum location, and theory of production: reexamination. Journal of Urban Economics 16, 225-231. (1984).

MAI A 1984-1 Mai, J.-H.
The bandwidth of the product of two gamma-type condensed graphs. Kexue Tongbao, Foreign Language Edition (Beijing) 29, 1575-1578. (1984).

MAI B 1984-1 Mai, J.-H.; Luo, H.-P.
Some theorems for the bandwidth of a graph (Chinese). Acta Mathematicae Applicatae Sinica (Peking) 7, 86-95. (1984).

MAI B 1984-2 Mai, J.-H.; Luo, H.-P.
Some theorems on the bandwidth of a graph (Chinese). Acta Mathematicae Applicatae Sinica (Peking) 7, 86-95. (1984).

MAI C 1984-1 Mai, J.-H.; Wang, S. H.; Luo, H.-P.
The number of edges of graphs with girth exceeding n+1 (Chinese). Journal of the China University of Science and Technics 14, 467-474. (1984).

MAI D 1984-1 Mai, S.-W.; Evans, D. J.
A parallel algorithm for the enumeration of the spanning trees of a graph. Parallel Computing (Amsterdam) 1, 275-286. (1984).

MAIE 1986-1 Maier, G.
Postoptimal analysis of minimal cost network flow problems. Methods of Operations Research 53, 291-302. (1986).

MAIE 1987-1 Maier, G.
Determining all nondegenerate bounds in sensitivity analysis of minimal cost network flow problems. In: W. Domschke, W. Krabs, J. Lehn, P. Spellucci (Eds.): Proc. XI. Symp. on Operations Research, Darmstadt 1986, 145-154. Methods of Operations Research 57, Frankfurt/Main: Athenäum. (1987).

MAIM 1986-1 Maimon, O.
The variance equity measure in locational decision theory. In: J. P. Osleeb, P. J. Ratick (Eds.): Locational Decisions: Methodology and Applications, 147-162. Annals of Operations Research 6, Basel: Baltzer. (1986).

MAJC 1984-1 Majcher, Z.
On some regularities of graphs. II. Casopis pro Pestovane Matematiky (Prague) 109, 380-388. (1984).

MAJC 1985-1 Majcher, Z.
Matrices representable by directed graphs. Archivum Mathematicum, Brno 21, 205-218. (1985).

MAJS 1983-1 Majstrenko, V. E.
A method for the determination of the smallest dominating set of vertices for a graph of a certain form (Russian). In: Theoretical Cybernetics - 2. Collection of articles, 49-53. Tbilisi: Metsniereba. (1983).

MAJS 1983-2 Majstrenko, V. E.
The graph in problems of the description of recognition objects (Russian). In: Theoretical Cybernetics - 2. Collection of articles, 42-48. Tbilisi: Metsniereba. (1983).

MAKR 1985-1 Makris, A.; Tziafetas, G.; Koutsopoulos, K.
Simultaneous location allocation of facilities and servers. Methods of Operations Research 55, 167-174. (1985).

MALE 1983-1 Malenchenko, S. I.
An algorithm for enumeration of cliques of a graph (Russian). Problems of Bionics (Kharkov) No. 31, 20-24. (1983).

MALH 1984-1 Malhotra, R.; Bhatia, H. L.
Variants of the time minimization assignment problem. Trabajos de Estadistica y de Ivestigacion Operativa (Valencia) 35, 331-338. (1984).

MALHA 1985-1 Malhotra, R.; Bhatia, H. L.; Puri, M. C.
The three dimensional bottleneck assignment problem and its variants. Optimization 16, 245-256. (1985).

MALO 1983-1 Malova, S.; Milashin, S.
A problem of scheduling theory for two identical parallel devices (Russian). in: Numerical methods: theoretical bases and algorithms, 57-65. Ivanovo: Ivanov. Gos. Univ. (1983).

MALS 1984-1 Malstrom, E. M.; Moraniec, D.
Complex lot sizing analyses with the pocket calculator. Production and Inventory Management Review 4, No. 7, 40-44. (1984).

MALY 1985-1 Malyshko, V. V.
An effective approach to some practical capacitated tree problems. In: Y. Alavi, G. Chartrand, L. Lesniak, D. R. Lick, C. E. Wall (Eds.): Graph Theory with Applications to Algorithms and Computer Science, 531-542. Chichester: Wiley. (1985).

MALYA 1986-1 Malyshko, V. V.; Harary, F.
A problem of the minimum blocking flow in a network (Russian). Kibernetika (Kiev), No. 2, 44-48. (1986).

MAME 1986-1 Mamedov, K. S.
On a method for calculating estimates in Boolean programming problems (Russian). In: Mathematical Cybernetics and Applied Mathematics, No. VII, 70-77. Baku: "Elm". (1986).

MAMEA 1986-1 Mamedov, K. S.; Shikhaliev, N. S.
Application of the subgradient descent method for approximation solution of a Boolean programming problem with an estimate of the deviation from the optimum (Russian). Izvestiya Akademii Nauk Azerbajdzhanskoj SSR, Seriya Fiziko-Tekhnicheskikh i Matematicheskikh Nauk No.2, 128-133. (1986).

MANA 1986-1 Manabe, R.
Scheduling buses by the network flow method on a personal computer. Asia-Pacific Journal of Operational Research 3, 158-165. (1986).

MANAA 1986-1 Manaster, A. B.; Remmel, J. B.; Schmerl, J. H.
Planarity and minimal path algorithms. Australian Mathematical Society, Journal, Series A 40, 131-142. (1986).

MAND 1985-1 Mandrescu, E.
On the alpha-partitionability of X-join of graphs. Analele Universitatii din Galati, Metalurgie, Fascicula II 3(8), 21-24. (1985).

MANG 1985-1 Mangoubi, R. S.; Mathaisel, D. F. X.
Optimizing gate assignments at airport terminals. Transportation Science 19, 173-188. (1985).

MANN 1985-1 Manning, J. R.
How not to schedule a stitching room. Journal of the Operational Reserch Society 36, 697-704. (1985).

MANV 1985-1 Manvel, B.
Extremely greedy coloring algorithms. In: F. Harary, J. P. Maybee (Eds.): Graphs and Applications, 257-270. New York: Wiley. (1985).

MAO 1986-1 Mao, J. Z.
An extremal problem for minimally k-edge-connected graphs (Chinese). Journal of Mathematics (Wuhan) 6, No. 1, 17-20. (1986).

MAON 1986-1 Maon, Y.; Schieber, B.; Vishkin, U.
Parallel ear decomposition search (EDS) and st-numbering in graphs. Theoretical Computer Science 47, 277-298. (1986).

MAON 1986-2 Maon, Y.; Schieber, B.; Vishkin, U.
Parallel ear decomposition search (EDS) and st-numbering in graphs. In: F. Makedon, K. Mehlhorn, T. Papatheodorou, P. Spirakis (Eds.): AWOC 86: VLSI Algorithms and Architectures, 34-45. Lecture Notes in Computer Science 227, Berlin: Springer. (1986).

MARC 1987-1 Marchetti-Spaccamela, A.; Pelaggi, A.
Worst case analysis of two heuristics for the set partitioning problem. RAIRO - Theoretical Informatics and Applications 21, 11-24. (1987).

MARCA 1984-1 Marchetti-Spaccamela, A.; Rinnooy Kan, A. H. G.; Stougie, L.
Hierarchical vehicle routing problems. Networks 14, 571-586. (1984).

MARCB 1985-1 Marchetti-Spaccamela, A.; Romano, G.
On different approximation criteria for subset product problems. Information Processing Letters 21, 213-218. (1985).

MARCC 1987-1 Marchetti-Spaccamela, A.; Vercellis, C.
Efficient on-line algorithms for the knapsack problem. In: T. Ottmann (Ed.): Automata, Languages and Programming, 445-456. Lecture Notes in Computer Science 267, Berlin: Springer. (1987).

MARCD 1985-1 Marchioro, P.; Morgana, A.; Petreschi, R.; Simeone, B.
Adamant digraphs. Rutcor Research Report No. 7-85. New Brunswick: Rutgers Univ., 23p. (1985).

MARCE 1987-1 Marcotorchino, F.
Block seriation problems: a unified approach. Reply to the problem of H. Garcia and J. M. Proth. Applied Stochastic Models and Data Analysis 3, 73-91. (1987).

MARCF 1985-1 Marcotte, O.
The cutting stock problem and integer rounding. Mathematical Programming 33, 82-92. (1985).

MARCF 1986-1 Marcotte, O.
An instance of the cutting stock problem for which the rounding property does not hold. Operations Research Letters 4, 239-244. (1986).

MARCF 1986-2 Marcotte, O.
On the chromatic index of multigraphs and a conjecture of Seymour. I. Journal of Combinatorial Theory B-41, 306-331. (1986).

MARCG 1986-1 Marcotte, O.; Soland, R. M.
An interactive branch-and-bound algorithm for multiple criteria optimization. Management Science 32, 61-75. (1986).

MARCH 1984-1 Marcotte, O.; Trotter, L. E. Jr.
An application of matroid polyhedral theory to unit-execution time, tree-precedence job scheduling. In: W. R. Pulleyblank (Ed.): Progress in Combinatorial Optimization, 263-272. Toronto: Academic Press. (1984).

MARCI 1986-1 Marcotte, P.
Network design problem with congestion effects: a case of bilevel programming. Mathematical Programming 34, 142-162. (1986).

MARCJ 1983-1 Marcu, D.
Note on the standard matrix representation of a matroid. Zbornik Radova Prirodno - Review of Research (Novi Sad) 13, 339-344. (1983).

MARCJ 1983-2 Marcu, D.
On some vector spaces associated to a digraph. Bulletin Mathematique de la Societe des Sciences Mathematiques de la Republique Socialiste de Roumanie (Bucharest), Nouvelle Serie 27 (75), 335-341. (1983).

MARCJ 1983-3 Marcu, D.
Sur les bases d'un digraphe. Scripta Facultatis Scientiarum Naturalium Universitatis Purkynianae Brunensis (Brno) 13, No. 8, 291-295. (1983).

MARCJ 1984-1 Marcu, D.
Note on the circuits of a perfect matroid design. Publications de l'Institut Mathematique (Beograd), Nouvelle Serie 36 (50), 9-11. (1984).

MARCJ 1984-2 Marcu, D.
Note on the flow and tension in a network. Mathematica (Cluj) 26 (49), 125-128. (1984).

MARCJ 1984-3 Marcu, D.
Note on the spanning trees and cotrees of a connected graph. Demonstratio Mathematica (Warsaw) 17, 809-812. (1984).

MARCJ 1984-4 Marcu, D.
On the strictly quasistrong connected digraphs without bridges. Bulletin Mathematique de la Societe des Sciences Mathematiques de la Republique Socialiste de Roumaine (Bucharest), Nouvelle Serie 28 (76), 37-38. (1984).

MARCJ 1985-1 Marcu, D.
A new upper bound for the domination number of a graph. Quarterly Journal of Mathematics, Oxford Second Series 36, 221-223. (1985).

MARCJ 1985-2 Marcu, D.
K-graphs in regular graphs. Buletinul Institutului Politehnic Bucuresti, Seria Electrotehnica 46(47), 15-18. (1985).

MARCJ 1985-3 Marcu, D.
Note on the regular digraphs. Rendiconti del Seminario Matematico della Universita di Padova 73, 95-98. (1985).

MARCJ 1985-4 Marcu, D.
Note on the spanning trees of a connected graph. Rivista di Matematica della Universita di Parma, Serie IV 11, 131-134. (1985).

MARCJ 1985-5 Marcu, D.
Note on the 2-connectivity of a graph. Scripta Facultatis Scientiarum Naturalium Universitatis Purkynianae Brunensis (Brno) 15, No. 4, 213-215. (1985).

MARCJ 1985-6 Marcu, D.
On finding the elementary paths of a digraph. Mathematica - Revue d'Analyse Numerique et de Theorie de l'Approximation (Cluj-Napoca) 14, 117-121. (1985).

MARCJ 1985-7 Marcu, D.
On finding the Eulerian paths and circuits of a digraph. Buletinul Institutului Politehnic Bucuresti, Seria Automatica-Calculatoare 46/47, 143-147. (1985).

MARCJ 1985-8 Marcu, D.
On the paths of a digraph without circuits. Gazeta Matematica, Perfectionare Metodica si Metodologica in Matematica si Informatica 6, 93-94. (1985).

MARCJ 1985-9 Marcu, D.
Some results on the bipartite graphs. Buletinul Universitatii di Brasov, Seria C 26, 16-20. (1985).

MARCJ 1986-1 Marcu, D.
A note on Bermond's conjecture. Publications de l'Institut Mathematique (Belgrade), Nouvelle Serie 39(53), 33-34. (1986).

MARCJ 1986-2 Marcu, D.
A note on the independence number of an identically self-dual perfect matroid design. Publications de l'Institut Mathematique (Belgrade), Nouvelle Serie 39(53), 29-31. (1986).

MARCJ 1986-3 Marcu, D.
An upper bound on the domination number of a graph. Mathematica Scandinavica 59, No. 1, 41-44. (1986).

MARCJ 1986-4 Marcu, D.
Another upper bound for the domination number of a graph. Revista Colombiana de Matematicas (Bogota) 20, 51-55. (1986).

MARCJ 1986-5 Marcu, D.
Note on the number of Hamiltonian paths in tournaments. Studia Universitatas Babes-Bolyai Mathematica 31, No. 1, 3-5. (1986).

MARCJ 1986-6 Marcu, D.
On the closed sets of a matroid (Romanian). Buletinul Institutului Politehnic Bucuresti, Seria Electrotehnica 48, 25-30. (1986).

MARCJ 1986-7 Marcu, D.
Paths, connectivity and matchings in graphs (French). Analele Universitatii Bucuresti Mathematica 35, 37-43. (1986).

MARCJ 1987-1 Marcu, D.
Note on the domination number of a graph. Gazeta Matematica, Perfectionare Metodica si Metodologica in Matematica si Informatica 8, 95-96. (1987).

MARCJ 1987-2 Marcu, D.
Note on the square of a graph. Annales Societatis Mathematicae Polonae, Seria I, Commentationes Mathematicae 27, 141-142. (1987).

MARCK 1984-1 Marcus, R.
An application of artificial intelligence to operations research. Communications of the ACM 27, 1044-1047. (1984).

MARCL 1985-1 Marczyk, A.; Skupien, Z.
General approach to line graphs of graphs. Demonstratio Mathematica (Warsaw) 18, 447-465. (1985).

MARE 1985-1 Marecki, F.
Asymmetrical assembly line balancing with the aid of a multi-stage programming method (Polish). Podstawy Sterowania (Warsaw) 15, 183-200. (1985).

MARE 1985-2 Marecki, F.
Generalized assembly line balancing problem. In: F. Nozicka (Ed.): Optimization - Theory, Methods, Applications, I, International Conference Marianske Lazne 1985, 134-141. Prague: Dum Techniky CSVTS. (1985).

MARE 1985-3 Marecki, F.
Multiobjective resources allocation of assembly line (Polish). Postepy Cybernetyki 8, No.3, 21-35. (1985).

MAREA 1984-1 Marello, S.; Pulleyblank, W. R.; Toth, P.; Werra, D. de
Balanced optimization problems. Operations Research Letters 3, 275-278. (1984).

MARI 1979-1 Maria, F. de; Gionfriddo, M.
On some parameters associated with graphs and hypergraphs (Italian). Atti della Societa Peloritana di Scienze Fisiche, Matematiche e Naturali (Messina) 25, 227-239. (1979).

MARI 1980-1 Maria, F. de; Gionfriddo, M.
A theorem on the s-chromatic number of a graph (Italian). Atti della Societa Peloritana di Scienze Fisiche, Matematiche e Naturali (Messina) 26, 315-322. (1980).

MARIA 1985-1 Marinescu, D.
An integer programming model for two-dimensional cutting stock-problems. In: I. Marusciac, W. W. Breckner (Eds.): Approximation and optimization, Proc. Coll. on Approximation and Optimization, Cluj-Napoca, Romania 1984, 269-276. Cluj-Napoca: University, Department of Mathematics. (1985).

MARIB 1982-1 Marino, M. C.
Some bounds for the parameter triangle in a graph (Italian). Matematiche (Catania) 37, 28-35. (1982).

MARIC 1983-1 Marino, M. C.; Puccio, L.
On the parameter delta(s,G) of a planar graph (Italian). Rivista di Matematica della Universita di Parma, Serie IV/9, 9-13. (1983).

MARK 1986-1 Markland, R. E.; Vickery, S. K.
The efficient computer implementation of a large-scale integer goal programming model. European Journal of Operational Research 26, 341-354. (1986).

MARKA 1985-1 Markosyan, A. S.
On imperfect critical graphs with incomplete critical component (Russian). Uchenye Zapiski Erevanskogo Gosudarsvenngo Universiteta, Estestvennye Nauki 158, No.1, 33-37. (1985).

MARKB 1984-1 Markosyan, S. E.; Karapetyan, I. A.
Critically imperfect graphs (Russian). Applied Mathematics (Erevan) No. 3, 131-138. (1984).

MART 1983-1 Martel, C. U.
Scheduling multiple processors with memory constraits. in: Parallel and large-scale computers: performance, architecture, applications (Montreal, Que., 1982), 67-75. IMACS (New Brunswick, N. Y.). (1983).

MART 1985-1 Martel, C. U.
A linear time bin-packing algorithm. Operations Research Letters 4, 189-192. (1985).

MART 1985-2 Martel, C. U.
Preemptive scheduling to minimize maximum completion time on uniform processors with memory constraints. Operations Research 33, 1360-1380. (1985).

MART 1986-1 Martel, C. U.
Lower bounds on parallel algorithms for finding the first maximal independent set. Information Processing Letters 22, 81-86. (1986).

MARTA 1984-1 Martello, S.; Pulleyblank, W. R.; Toth, P.; Werra, D. de
Balanced optimization problems. Research Paper. Bologna: Univ. of Bologna, 9 p. (1984).

MARTB 1984-1 Martello, S.; Toth, P.
Lower bounds and reduction procedures for the bin packing problem. Research Paper. San Francisco (TIMS/ORSA Meeting, May 1984), 14 p. (1984).

MARTB 1985-1 Martello, S.; Toth, P.
Algorithm 632. A program for the 0-1 multiple knapsack problem. ACM Transactions on Mathematical Software 11, 135-140. (1985).

MARTB 1985-2 Martello, S.; Toth, P.
Approximation schemes for the subset-sum problem: survey and experimental analysis. European Journal of Operational Research 22, 56-69. (1985).

MARTB 1986-1 Martello, S.; Toth, P.
A heuristic approach to the bus driver scheduling problem. European Journal of Operational Research 24, 106-117. (1986).

MARTB 1987-1 Martello, S.; Toth, P.
Algorithms for knapsack problems. In: P. Martello, G. Laporte, M. Minoux, C. Ribeiro (Eds.): Surveys in Combinatorial Optimization, Annals of Discrete Mathematics 31, 213-258. Amsterdam, New York, Tokyo: North-Holland. (1987).

MARTB 1987-2 Martello, S.; Toth, P.
Linear assignment problems. In: P. Martello, G. Laporte, M. Minoux, C. Ribeiro (Eds.): Surveys in Combinatorial Optimization, Annals of Discrete Mathematics 31, 259-282. Amsterdam, New York, Tokyo: North-Holland. (1987).

MARTC 1985-1 Martin, C. H.; Lubin, S. L.
Optimization modeling for business planning at Trumbull Asphalt. Interfaces 15, 66-72. (1985).

MARTD 1983-1 Martin, E.
Lösung von Transportproblemen mit dem Programmsystem OPSI. Wissenschaftliche Beiträge, Martin-Luther-Universität Halle Wittenberg 1983/44(M30), 89-98. (1983).

MARTE 1985-1 Martin, R. K.
Generating alternative mixed - 0/1 linear programming models using variable redefintion (revised version). Research Paper. Chicago: Univ. of Chicago, Graduate School of Business, 32 p. (1985).

MARTF 1987-1 Martin, R. K.; Rardin, R. L.; Campbell, B. A.
Polyhedral characterization of discrete dynamic programming. VV-87-24. West Lafayette: Purdue Univ. Inst. of Interdisciplinary Engineering Studies, 30 p. (1987).

MARTG 1985-1 Martin, R. K.; Schrage, L.
Subset coefficient reduction cuts for 0/1 mixed-integer programming. Operations Research 33, 505-526. (1985).

MARTH 1985-1 Martin, R. K.; Sweeney, D. J.; Doherty, M. E.
The reduced cost branch and bound algorithm for mixed integer programming. Computers & Operations Research 12, 139-150. (1985).

MARTI 1985-1 Martinec, I.
Methods for a class of discrete location problems. Optimization 16, 581-596. (1985).

MARTI 1987-1 Martinec, I.
An algorithm for the epsilon-optimal solution of a discrete location problem (Czech). Ekonomicko-Matematicky Obzor 23, 332-348. (1987).

MARTJ 1983-1 Martinov, M. K.
Some decompositions of a complete graph into wheels (Bulgarian). Godishnik na Visshite Uchebni Zavedeniya, Prilozhna Matematika (Sofiya) 18 (1982), No. 3, 141-152. (1983).

MARTK 1980-1 Martinov, N.
Uncontractable 4-connected graphs (Bulgarian). Annuaire l'Universite de Sofia, Faculte de Mathematiques et Mechanique 74, 61-65. (1980).

MARTK 1986-1 Martinov, N.
Locally tree-like graphs and their generalizations (Russian). Comptes Rendus de l'Academie Bulgare des Sciences 39, No.7, 45-48. (1986).

MARTL 1983-1 Martinova, M. K.
A planimetric problem with a combinatorial significance (Bulgarian). Godishnik na Visshite Uchebni Zavedeniya, Prilozhna Matematika (Sofia) 19, No. 4, 177-182. (1983).

MARTL 1983-2 Martinova, M. K.
On the decomposition of a complete graph into wheels (Bulgarian). Godishnik na Visshite Uchebni Zavedeniya, Prilozhna Matematika (Sofiya) 18 (1982), No. 2, 155-166. (1983).

MARTL 1984-1 Martinova, M. K.
Cyclic and acyclic invariants for a class of outerplanar graphs (Bulgarian). Godishnik na Visshite Uchebni Yavedeniya, Prilozhna Matematika (Sofia) 19(1983), 153-162. (1984).

MARTL 1984-2 Martinova, M. K.
On the cyclicity of a complete n-partite graph (Bulgarian). Godishnik na Visshite Uchebni Zavedeniya, Prilozhna Matematika (Sofia) 19(1983), 145-152. (1984).

MARTM 1984-1 Martins, E. Q. V.
An algorithm for ranking paths that may contain cycles. European Journal of Operational Research 18, 123-130. (1984).

MARTM 1987-1 Martins, E. Q. V.
On a particular quadratic network problem. European Journal of Operational Research 29, 317-327. (1987).

MARTN 1985-1 Martins, E. Q. V.; Rosa, M. S.
An algorithm for the maximal multicommodity funnel-node flow in an undirected network. Operations Research 33, 537-547. (1985).

MARU 1987-1 Marucheck, A. S.; Sulek, J. M.
An algorithm for technology choice in local area network design. Management Science 33, 58-72. (1987).

MARUA 1987-1 Marusic, D.
Hamiltonian cycles in vertex symmetric graphs of order 2p**2. Discrete Mathematics 66, 169-174. (1987).

MASH 1983-1 Mashchenko, S. O.
Sequential algorithm to solve problems of mixed linear programming (Russian). Issledovanie Operatsij i ASU (Kiew) 21, 33-40. (1983).

MASO 1985-1 Mason, L. G.
Equilibrium flows, routing patterns and algorithms for store-and-forward networks. Large Scale Systems 8, 187-209. (1985).

MASU 1987-1 Masuda, S.; Nakajima, K.
An optimal algorithm for generating all maximal independent sets of an interval graph. In: M. Ismail (Ed.): 29th Midwest Symposium on Circuits and Systems, 444-447. Amsterdam: North-Holland. (1987).

MASU 1987-2 Masuda, S.; Nakajima, K.
Optimal algorithms for interval containment graphs. In: M. Ismail (Ed.): 29th Midwest Symposium on Circuits and Systems, 431-434. Amsterdam: North-Holland. (1987).

MASUA 1985-1 Masuda, T.; Ishii, H.
Two-machine scheduling problem for jobs with generalized due dates. Mathematica Japonica 30, 117-125. (1985).

MASUB 1985-1 Masuda, T.; Ishii, H.; Nishida, T.
The mixed shop scheduling problem. Discrete Applied Mathematics 11, 175-186. (1985).
MASUB 1985-2 Masuda, T.; Ishii, H.; Nishida, T.
Two-machine scheduling problem for jobs with generalized due dates. Mathematica Japonica 30, 117-125. (1985).
MASUC 1987-1 Masuzawa, T.; Hagihara, K.; Tokura, N.
An optimal time algorithm for the k-vertex-connectivity unweighted augmentation problem for rooted directed trees. Discrete Applied Mathematics 17, 67-106. (1987).
MASUD 1986-1 Masuzawa, T.; Hagihara, K.; Tokura, N.; Wada, K.
Optimal-time algorithm for the k-node-connectivity augmentation problem for ternary trees. Systems and Computers, Japan (Kyoto) 17, No. 3, 56-65. (1986).
MASUE 1984-1 Masuzawa, T.; Hagihara, K.; Wada, K.; Tokura, N.
The k-node connectivity augmentation problem for directed binary trees. Systems-Computer-Controls 15, 75-83. (1984).
MATH 1986-1 Mathews, B. P.; Waters, C. D. J.
Computerized routeing for community nurses - a pilot study. Journal of the Operational Research Society 37, 677-684. (1986).
MATHA 1986-1 Mathur, K.; Salkin, H. M.; Bohanty, B. B.
A note on a general non-linear knapsack problem. Operations Research Letters 5, 79-82. (1986).
MATHB 1986-1 Mathur, K.; Salkin, H. M.; Morito, S.
An efficient algorithm for the general multiple-choice knapsack problem (GMKP). In: C. L. Monma (Ed.): Algorithms and Software for Optimization - Part I, 253-284. Annals of Operations Research 4, Basel: Baltzer. (1986).
MATHC 1985-1 Mathur, K.; Salkin, H. M.; Nishimura, K.; Morito, S.
Applications of integer programming in radio frequency management. Management Science 31, 829-839. (1985).
MATI 1987-1 Matitskas, I.-K. L.; Palubetskis, G. S.
Graph cutting and 0-1 quadratic programming (Russian). Izvestiya Akademii Nauk SSSR, Tekhnicheskaya Kibernetika (Moscow), No. 4, 152-159; translated as Soviet J. Comput. Systems Sci. 25, No. 1, 100-107. (1987).
MATO 1986-1 Matousek, J.
Few colored cuts or cycles in edge colored graphs. KAM Series, 7. Praha: Charles Univ., Dept. of Applied Mathematics, 13 p. (1986).
MATS 1983-1 Matsui, K.; Takada, J.
SLP method for a nonlinear multicommodity flow assignment problem. Memoirs of the National Defense Academy (Yokosuka) 23, 133-144. (1983).
MATSA 1984-1 Matsumoto, K.; Nishizeki, T.; Saito, N.
Algorithms for plane multicommodity flows. Electronics and Communications in Japan 67, 9-16. (1984).
MATSA 1985-1 Matsumoto, K.; Nishizeki, T.; Saito, N.
An efficient algorithm for finding multicommodity flows in planar networks. SIAM Journal on Computing 14, 289-302. (1985).
MATSA 1986-1 Matsumoto, K.; Nishizeki, T.; Saito, N.
Planar multicommodity flows, maximum matchings and negative cycles. SIAM Journal on Computing 15, 495-510. (1986).
MATSB 1985-1 Matsutomi, T.; Ishii, H.; Nishida, T.
Stochastic location problem of emergency service facility. Technology Reports of the Osaka University 35, No. 1800-1818, 157-166. (1985).
MATSB 1987-1 Matsutomi, T.; Ishii, H.; Nishida, T.
Chance constrained round trip location problem. Mathematica Japonica 32, 609-620. (1987).
MATT 1984-1 Matta, K. F.
Job sequencing: economic approach provides best order. Industrial Engineering 16, 22-28. (1984).
MATTA 1986-1 Mattsson, L.-G.
Residential location and school planning in a tightening urban economy. In: J. P. Osleeb, P. J. Ratick (Eds.): Locational Decisions: Methodology and Applications, 181-202. Annals of Operations Research 6, Basel: Baltzer. (1986).
MATU 1985-1 Matula, D. W.
Concurrent flow and concurrent connectivity in graphs. In: Y. Alavi, G. Chartrand, L. Lesniak, D. R. Lick, C. E. Wall (Eds.): Graph Theory with Applications to Algorithms and Computer Science, 543-560. Chichester: Wiley. (1985).
MATU 1987-1 Matula, D. W.
Determining edge connectivity in O(nm). In: 28th Annual IEEE Symposium on Foundations of Computer Science, 249-251. (1987).
MATUA 1987-1 Matula, D. W.; Shahrokhi, F.
Sparsest cuts and bottlenecks in graphs. CSR 181. Socorro: New Mexico Inst. of Mining and Technology, Computer Science Dept., 16 p. (1987).
MATV 1986-1 Matveenko, V. D.
A combinatorial approach to the problem of solvability of the three-index transportation problem. Mathematical Notes 40, 638-643; translation from Matematicheskie Zametki 40 (1986), No.2, 243-251. (1986).
MATVA 1984-1 Matveev, V. V.; Shcherbina, O. A.
On one class of local algorithm for solving quasi-unit problems of integer linear programming (Russian). Kibernetika (Kiev) No. 6, 116-118. (1984).
MAUR 1981-1 Maurer, H.
The post-office problem and related questions. In: H. Noltemeier (Ed.): Graphtheoretic Concepts in Computer Science, Proc. Internat. Workshop WG 80 (Bad Honnef 1980), 1-19. Berlin, Heidelberg, New York: Springer. (1981).
MAWE 1986-1 Mawengkang, H.; Murtagh, B. A.
Solving nonlinear integer programs with large-scale optimization software. In: C. L. Monma (Ed.): Algorithms and Software for Optimization - Part II, 425-438. Annals of Operations Research 5, Basel: Baltzer. (1986).

MAXE 1987-1 Maxemchuk, N. F.
Routing in the Manhattan street network. IEEE Transactions on Communications 35, 503-512. (1987).

MAXW 1986-1 Maxwell, W. L.; Singh, H.
Scheduling cyclic production on several identical machines. Operations Research 34, 460-463. (1986).

MAY 1985-1 May, M.
On longest paths in digraphs (German). In: Graphen und Netzwerke - Theorie und Anwendungen, 30. Int. Wiss. Kolloq., Ilmenau 1985, 5, 77-80. (1985).

MAY 1985-2 May, M.
Über längste Wege in Digraphen. 30. Internationales Wissenschaftliches Kolloquium (TH Ilmenau, DDR) Heft 5F, 77-80. (1985).

MAYB 1983-1 Maybee, J. S.; Maybee, S. J.
An algorithm for identifying Morishima and anti-Morishima matrices and balanced digraphs. Mathematical Social Sciences 6, 99-103. (1983).

MAZE 1985-1 Maze, T. H.; Khasnabis, S.
Bus garage location planning with dynamic vehicle assignments: a methodology. Transportation Research 19B, 1-14. (1985).

MAZU 1984-1 Mazurov, V. D.
Committee solution of an unbalanced transportation problem (Russian). In: V. M. Kislyak, V. P. Kazantsev (Eds.): Mathematical methods for planning industrial production, 88-90. Sverdlovsk: Akademiya Nauk SSSR. (1984).

MAZZ 1983-1 Mazzocca, F.
On a characterization of Dilworth truncations of combinatorial geometries. Journal of Geometry 20, 63-73. (1983).

MAZZ 1984-1 Mazzocca, F.
Extensions of combinatorial geometries by the addition of a unique line. Journal of Combinatorial Theory, Series A 37, 32-45. (1984).

MAZZA 1987-1 Mazzola, J. B.; Mac Coy, W. F.; Wagner, H. M.
Algorithms and heuristics for variable-yield lot sizing. Naval Research Logistics 34, 67-86. (1987).

MAZZB 1986-1 Mazzola, J. B.; Neebe, A. W.
Resource-constrained assignment scheduling. Operations Research 34, 560-572. (1986).

MEAN 1984-1 Meanti, M.; Rinnooy Kan, A. H. G.; Stougie, L.; Vercellis, C.
A probabilistic analysis of the multi-knapsack value function. Report 8435/O. Rotterdam: Erasmus Univ., Econometric Institute. (1984).

MEAN 1987-1 Meanti, M.; Rinnooy Kan, A. H. G.; Stougie, L.; Vercellis, C.
A probabilistic analysis of the multi-knapsack value function. Report 8709/A. Rotterdam: Erasmus Univ., Econometric Institute, 10 p. (1987).

MEDV 1985-1 Medvedev, Y. T.; Pushchinskii, E. V.
On an integer minimax problem arising in the print industry. Izvestya Akademii Nauk SSSR Tekhnicheskaya Kibernetika (Moscow) No. 6, 15-24. Translated as: Soviet Journal on Computational Systems Sciences 23, No. 1, 110-118. (1985).

MEGI 1986-1 Megiddo, N.
Dynamic location problems. In: J. P. Osleeb, P. J. Ratick (Eds.): Locational Decisions: Methodology and Applications, 313-320. Annals of Operations Research 6, Basel: Baltzer. (1986).

MEGIA 1986-1 Megiddo, N.; Zemel, E.
An O(n log n) randomizing algorithm for the weighted Euclidean 1-center problem. Journal of Algorithms 7, 358-368. (1986).

MEHL 1984-1 Mehlhorn, K.
Data structures and algorithms 2: Graph algorithms and NP-completeness. Berlin, Heidelberg, New York, Tokyo: Springer. (1984).

MEHL 1986-1 Mehlhorn, K.
Über Verdrahtungsalgorithmen. Informatik-Spektrum 9, 227-234. (1986).

MEHLA 1987-1 Mehlhorn, K.; Kaufmann, M.
On local routing of 2-terminal nets. In: F. J. Brandenburg, G. Vidal-Naquet, M. Wirsing (Eds.): STACS 87, Lecture Notes in Computer Science 247, 40-52. Berlin, New York, Tokyo: Springer. (1987).

MEHLB 1986-1 Mehlhorn, K.; Preparata, F. P.
Routing through a rectangle. Journal of the Association for Computing Machinery 33, 60-85. (1986).

MEHLC 1986-1 Mehlhorn, K.; Preparata, F. P.; Sarrafzadeh, M.
Channel routing in knock-knee mode: simplified algorithms and proofs. Algorithmica 1, 213-222. (1986).

MEHLD 1986-1 Mehlhorn, K.; Schmidt, B. H.
On BF-orderable sets. Discrete Applied Mathematics 15, 315-328. (1986).

MEHR 1985-1 Mehrez, A.
A resource allocation problem with the aid of an integer zero-one programming model in the plastic industry. Journal of Operations Management 5, 455-460. (1985).

MEHRA 1985-1 Mehrez, A.; Sinuany-Stern, Z.; Stulman, A.
A single facility location problem with a weighted maximin-minimax rectilinear distance. Computers & Operations Research 12, 51-60. (1985).

MEHRA 1986-1 Mehrez, A.; Sinuany-Stern, Z.; Stulman, A.
An enhancement of the Drezner-Wesolowsky algorithm for single-facility location with maximin of rectilinear distance. Journal of the Operational Research Society 37, 971-978. (1986).

MEHRB 1985-1 Mehrez, A.; Stern, H. I.
Optimal refueling strategies for a mixed-vehicle fleet. Naval Research Logistics Quaterly 32, 315-328. (1985).

MEHRC 1984-1 Mehrez, A.; Stulman, A.
The facility location problem when the underlying distribution is either dominated or dominating. Zeitschrift für Operations Research, Series B 28, 157-162. (1984).

MEIN 1984-1 Meinel, C.
Effiziente Algorithem. Seminarbericht, Humboldt-Universität Berlin, Sektion Mathematik 61, 175 P. (1984).

MELA 1985-1 Melachrinoudis, E.
Determining an optimum location for an undesirable facility in a workroom environment. Applied Mathematical Modelling 9, 365-369. (1985).
MELAA 1982-1 Melachrinoudis, E.; Cullinane, T. P.
A maximin approach to the location of an undesirable facility in a nonconvex region. In: Modeling and Simulation, Vol 13, Part 2, 533-538. Research Triangle Park, N. C.: Instrument Soc. of America. (1982).
MELAA 1985-1 Melachrinoudis, E.; Cullinane, T. P.
A heuristic approach to the single facility maximin location problem. International Journal of Production Research 23, 523-532. (1985).
MELAA 1986-1 Melachrinoudis, E.; Cullinane, T. P.
Locating an obnoxious facility within a polygonal region. In: J. P. Osleeb, P. J. Ratick (Eds.): Locational Decisions: Methodology and Applications, 137-146. Annals of Operations Research 6, Basel: Baltzer. (1986).
MELAA 1986-2 Melachrinoudis, E.; Cullinane, T. P.
Locating an undesirable facility with a minimax criterion. European Journal of Operational Research 24, 239-246. (1986).
MELAB 1986-1 Melachrinoudis, E.; Rumpf, D.; Venegas, R.
Spray operation planning aided by a mixed integer programming model. Boston: Dept. of Industrial Engineering and Information Systems, 23p. (1986).
MELC 1986-1 Melching, C. S.; Liebman, J. S.
Solving binary knapsack problems with precedence constraints. ORL 86-2. Urbana: Univ. of Illinois, Operations Research Laboratory, 31 p. (1986).
MELN 1984-1 Melnikov, L. S.; Ponomarev, V. V.
Solution of a problem of the Turan type (Russian). Metody Diskretnogo Analiza (Novosibirsk) 41, 62-71. (1984).
MELNA 1984-1 Melnikov, O. I.
On Bellman-Johnson's (2*n)-problem for which all schedules are optimal (Russian). Izvestiya Akademii Nauk BSSR, Seriya Fisziko-Matematicheskikh Nauk (Minsk), No. 6, 31-33. (1984).
MELNB 1985-1 Melnikov, O. I.; Kozhich, P. P.
Algorithms for recognizing the polarity of a graph with bounded parameters (Russian). Vestsi Akademii Navuk BSSR, Seryya Fizika-Matematychnykh Navuk (Minsk) 1985, No. 6, 50-54. (1985).
MELNC 1985-1 Melnyk, S. A.; Piper, C. J.
Leadtime errors in MRP: the lot-sizing effect. International Journal of Production Research 23, 253-264. (1985).
MENG 1984-1 Meng, J. X.
Some classes of hypomatchable graphs (Chinese). Journal of the Xinjiang University of Natural Sciences No. 1, 14-17. (1984).
MERE 1986-1 Merenzon, M. N.
A dynamic programming algorithm for optimal scheduling of a nonhomogeneous multiprocessor system (English). Soviet Journal of Computer and Systems Sciences 24, No.4, 13-16; translation from Izvestiya Akademii Nauk SSSR, Tekhnicheskaya Kibernetika (1986), No.1, 17-20. (1986).
MESA 1985-1 Mesa Lopez-Colmenar, J. A.
Stochastic dynamic location with the minimax criterion (Spanish). In: Proceedings of the tenth Spanish-Portuguese conference of mathematics, VI (Murcia 1985), 55-62. Murcia: Univ. (1985).
METE 1984-1 Metelskij, N. N.
Methods of local optimization for the problem of permutating bipartite graphs. U.S.S.R. Computational Mathematics and Mathematical Physics 24, No.5, 96-98; translation from Zhurnal Vychislitelnoi Matematiki i Matematicheskoi Fiziki 24 (1984), No.9, 1428-1432. (1984).
MEYE 1985-1 Meyer, R. R.
Network optimization. In: K. Schittkowski (Ed.): Computational Mathematical Programming, 125-140. Berlin: Springer-Verlag. (1985).
MEYN 1984-1 Meyniel, H.
The graphs whose odd cycles have at least two chords. In: C. Berge, V. Chvatal (Eds.): Topics on perfect graphs, Annals of Discrete Mathematics Vol. 21, 115-120. Amsterdam: North-Holland. (1984).
MEYN 1985-1 Meyniel, H.
A note about a theorem by Maamoun on decompositions of digraphs into elementary directed paths or cycles. Journal of Combinatorial Theory 39B, 368-370. (1985).
MEYN 1986-1 Meyniel, H.
A new property of critical imperfect graphs and some consequences. In: Raspail-Burnside: Workshop on Perfect Graphs, Bellairs Research Institute, McGill University, Barbados, West Indies, 1986, 106-119. (1986).
MEYN 1987-1 Meyniel, H.
A new property of critical imperfect graphs and some consequences. European Journal of Combinatorics 8, 313-316. (1987).
MEZA 1987-1 Mezard, M.
Spin glasses and optimization. In: J. L. van Hemmen, I. Morgenstern (Eds.): Heidelberg Colloquium on Glassy Dynamics, 354-372. Lecture Notes in Physics 275, Berlin: Springer. (1987).
MICA 1980-1 Micali, S.; Vazirani, V. V.
An O(|E|.root|v|) algorithm for finding maximum matching in general graphs. In: 21st Annual Symp. on Foundations of Computer Science, Syracuse, N.Y. 1980, Proceedings, 17-27. IEEE. (1980).
MICH 1983-1 Michalak, D.
On line and middle graphs with point-coarseness number one. Discussiones Mathematicae (Poznan) 6, 73-78. (1983).
MICH 1983-2 Michalak, D.
On line graphs with coarseness number 1. Discussiones Mathematicae (Poznan) 6, 65-72. (1983).
MICH 1985-1 Michalak, D.
On the binding number of a graph. In: M. Borowiecki; Z. Skupien (Eds.): Graphs, hypergraphs and matroids, Proc. Fifth Regional Scientific Session of Mathematicians, Zagan, Poland, 1985, 45-48. Section: Graph theory. Zielona Gora: Higher College of Engineering. (1985).

MICHA 1983-1 Michalski, M.
Spectra and direct derivations of graphs. Journal of Combinatorics, Information & System Sciences (Delhi) 8, 116-122. (1983).

MICHB 1987-1 Michelot, C.
Localization in multifacility location theory. European Journal of Operational Research 31, 177-184. (1987).

MICHB 1987-2 Michelot, C.
Properties of a class of location-allocation problems. Cahiers du Centre d'Etudes de Recherche Operationelle 29, 105-114. (1987).

MIDD 1987-1 Middendorf, M.
Symmetric matroids and connectivity properties of graphs. Report 87485-OR. Bonn: Univ. Bonn, Inst. f. Ökonometrie u. Operations Research, 9 p. (1987).

MIHA 1987-1 Mihalyi, M.; Raischi, C.
An algorithm for optimal ordering of the arcs of a network. Economic Computation and Economic Cybernetics Studies and Research 22, No.2, 29-32. (1987).

MIHO 1985-1 Mihok, P.
Additive hereditary properties and uniquely partitionable graphs. In: M. Borowiecki; Z. Skupien (Eds.): Graphs, hypergraphs and matroids, Proc. Fifth Regional Scientific Session of Mathematicians, Zagan, Poland, 1985, Section: Graph theory, 49-58. Zielona Gora: Higher College of Engineering. (1985).

MIHO 1987-1 Mihok, P.
On graphs matroidal with respect to additive hereditary properties. In: M. Borowiecki; Z. Skupien (Eds.): Graphs, hypergraphs and matroids II, Proc. Sixth Regional Scientific Session of Mathematicians, Zagan, Poland, 1986, 53-64. Section: Graph theory. Zielona Gora: Higher College of Engineering. (1987).

MIKA 1984-1 Mikami, K.; Machida, H.
On approximate solutions of the rectilinear travelling salesman problem (Japanese). In: Basic Theory of Computer Science, Proceedings of a Symposium held at the Research Institute for Mathematical Sciences (Kyoto), 90-104. (1984).

MIKH 1986-1 Mikhailyuk, V. A.
Certain aspects of postoptimal analysis of the minimization problem for the Boolean functions (Russian). Kibernetika, No. 4, 109-111. (1986).

MIKHA 1986-1 Mikhalevich, M. V.
A stochastic algorithm for solving nonlinear mixed integer problems (Russian). In: Mathematical Methods for the Analysis and Optimization of Complex Systems that Function under Conditions of Uncertainty, 20-27. Kiev: Akad. Nauk Ukrain. SSR, Inst. Kibernet. (1986).

MIKHB 1987-1 Mikhalevich, M. V.; Sapko, A. V.
The efficiency of an algorithm for solving the traveling salesman problem (Russian). Issledovanie Operatsii i ASU (Kiev) 29, 37-42. (1987).

MIKHC 1986-1 Mikhalevich, V. S.; Trubin, V. A.; Shor, N. Z.
Optimization problems of production-transportation planning. Models, Methods, Algorithms (Russian). Moscow: "Nauka". (1986).

MILA 1984-1 Milanov, P. B.
A polynomial time algorithm for a large subclass of the integer knapsack problem. Comptes Rendus de l'Academie Bulgare des Sciences (Sofia) 37, 309-312. (1984).

MILA 1985-1 Milanov, P. B.
The linear diophantine problem of Frobenius and discrete optimization. In: E. Girlich (Ed.): Diskrete Optimierung, 73-82. Jena: Friedrich-Schiller-Universität. (1985).

MILA 1986-1 Milanov, P. B.
On the worst behaviour of the greedy procedure. Comptes Rendus de l'Academie Bulgare des Sciences 39, No.2, 43-45. (1986).

MILA 1987-1 Milanov, P. B.
On the treshold independence systems. Report 87487-OR. Bonn: Univ. Bonn, Forschungsinst. f. Diskrete Mathematik und Inst. f. Ökonometrie und Operations Research, 14 p. (1987).

MILAA 1987-1 Milanov, P. B.; Brimkov, V. E.
On the knapsack problem. Comptes Rendus de l'Academie Bulgare des Sciences (Sofia) 40, No.3, 25-27. (1987).

MILAB 1983-1 Milanov, P. B.; Girlich, E.
Zum Kegel der monoton wachsenden submodularen Funktionen. In: 15th Annual Conference on Mathematical Optimization, Part 2 (Sellin), 241-247. Seminarberichte 50, Humboldt Univ., Berlin. (1983).

MILAC 1982-1 Milazzo, F.
A generalization of the s-chromatic number of a directed graph (Italian). Matematiche (Catania) 37, 233-238. (1982).

MILAC 1982-2 Milazzo, F.
Automorphism groups of T(s)-colorings and relations on the s-chromatic numbers of a graph (Italian). Atti del Seminario Matematico e Fisico dell'Universita di Modena 31, 78-85. (1982).

MILAC 1982-3 Milazzo, F.
Groups of T(s)-colored automorphisms and relations between the s-chromatic numbers of a graph (Italian). Atti del Seminario Matematico e Fisiko dell'Universita di Modena 31, 78-85. (1982).

MILAD 1984-1 Milazzo, F.; Vacirca, V.
Total colorings of higher kind in a graph (Italian). Rivista di Matematica della Universita di Parma, Serie IV 10, 137-142. (1984).

MILAD 1987-1 Milazzo, F.; Vacirca, V.
On the achromatic number G x K(m). Ars Combinatoria 24B, 173-178. (1987).

MILAD 1987-2 Milazzo, F.; Vacirca, V.
On the achromatic number of permutation graphs. Ars Combinatoria 24B, 71-76. (1987).

MILI 1983-1 Militaru, V.
A procedure of gradual generating trees in a linear graph. Studia Universitatis Babes-Bolyai (Cluj-Napoca), Physika 28, 29-32. (1983).

MILIA 1984-1 Militaru, V.; Nicula, A.
Base transformations through the technique of the structural numbers. Studia Universitatis Babes-Bolyai, Physica (Cluj-Napoca) 29, 16-20. (1984).

MILK 1984-1 Milkiewicz, F.
Decomposition of production scheduling in production systems with switched over continuous technological processes. In: A. Straszak (Ed.): Large Scale Systems: Theory and Applications 1983 (Warsaw 1983), 447-454. IFAC Proc. Series 10, Laxenburg, Pergamon Press. (1984).

MILL 1987-1 Miller, D. A.
Oriented matroids from smooth manifolds. Journal of Combinatorial Theory 43B, 173-186. (1987).

MILLA 1986-1 Miller, D. M.
A decomposition technique for the design of internally-unate combinational networks. Congressus Numerantium 51, 217-232. (1986).

MILLA 1987-1 Miller, D. M.
An interactive, computer-aided ship scheduling system. European Journal of Operational Research 32, 363-379. (1987).

MILLB 1987-1 Miller, G.; Ramachandran, V.
A new graph triconnectivity algorithm and its parallelization. In: Proc. Nineteenth Annual ACM Symp. on Theory of Computing, New York City 1987, 335-344. ACM. (1987).

MILLC 1986-1 Miller, G. L.
Finding small simple cycle separators for 2-connected planar graphs. Journal of Computer and System Sciences 32, 265-279. (1986).

MILN 1984-1 Milner, E. C.
Lectures on the marriage theorem of Aharoni, Nash-Williams and Shelah. In: K. M. Koh, H. P. Yap (Eds.): Graph Theory Singapore 1983, Lecture Notes in Mathematics 1073, 55-79. Berlin: Springer-Verlag. (1984).

MINA 1985-1 Minaev, Y. N.
On a formally heuristic algorithm for solving integer optimization problems (linear and quadratic models) (Russian). Ehkonomika i Matematicheskie Metody 21, 357-360. (1985).

MINI 1987-1 Minieka, E.
The delivery man problem on a tree network. Chicago: Univ. of Illinois, Dept. of Quantitative Methods, 9 p. (1987).

MINK 1980-1 Minker, J.; Minker, R. G.
Optimization of Boolean expressions - historical developments. Annals of the History of Computing 2, 227-238. (1980).

MINO 1984-1 Minoux, M.
A polynomial algorithm for minimum quadratic cost flow problems. European Journal of Operational Research 18, 377-387. (1984).

MINO 1986-1 Minoux, M.
Mathematical Programming. Theory and Algorithms. Chichester, New York: Wiley. (1986).

MINO 1986-2 Minoux, M.
Optimal traffic assignment in a SS/TDMA frame: a new approach by set covering and column generation. RAIRO Recherche Operationnelle 20, 273-286. (1986).

MINO 1986-3 Minoux, M.
Solving integer minimum cost flows with separable convex cost objective polynomially. In: G. Gallo, C. Sandi (Eds.): Netflow at Pisa, Mathematical Programming Study 26, 237-239. (1986).

MINO 1987-1 Minoux, M.
A class of combinatorial problems with polynomially solvable large scale set covering/ partitioning relaxations. RAIRO Recherche Operationnelle 20, 105-136. (1987).

MINO 1987-2 Minoux, M.
Network synthesis and dynamic network optimization. In: P. Martello, G. Laporte, M. Minoux, C. Ribeiro (Eds.): Surveys in Combinatorial Optimization, Annals of Discrete Mathematics 31, 283-324. Amsterdam, New York, Tokyo: North-Holland. (1987).

MINOA 1986-1 Minoux, M.; Bartnik, G.
Graphes, algorithmes, logiciels. Paris: Dunod. (1986).

MINOB 1987-1 Minoux, M.; Pinson, E.
Lower bounds to the graph partitioning problem through generalized linear programming and network flows. RAIRO Recherche Operationelle 21, 349-364. (1987).

MINOC 1984-1 Minoux, M.; Ribeiro, C. C.
A transformation of hard (equality constrained) knapsack problems into constrained shortest path problems. Operations Research Letters 3, 211-214. (1984).

MINOC 1985-1 Minoux, M.; Ribeiro, C. C.
Une nouvelle approche du probleme d'affectation quadratique MIN-MAX. Note Technique, 15-3. Issy: C. N. E. T., Division Telematique, Informatique et Mathematique Appliquees (TIM), 13 p. (1985).

MIRC 1987-1 Mirchandani, P. B.
Generalized hierarchical facility locations. Transportation Science 21, 123-125. (1987).

MIRCA 1985-1 Mirchandani, P. B.; Oudjit, A.; Wong, R. T.
Multidimensional extensions and a nested dual approach for the m-median problem. European Journal of Operational Research 21, 121-137. (1985).

MIRCB 1986-1 Mirchandani, P. B.; Reilly, J. M.
"Spatial nodes" in discrete location problems. In: J. P. Osleeb, P. J. Ratick (Eds.): Locational Decisions: Methodology and Applications, 203-222. Annals of Operations Research 6, Basel: Baltzer. (1986).

MIRCC 1985-1 Mirchandani, P. B.; Soroush, H.
Optimal paths in probabilistic networks: a case with temporary preferences. Computers & Operations Research 12, 365-382. (1985).

MIRCC 1987-1 Mirchandani, P. B.; Soroush, H.
Routes and flows in stochastic networks. In: G. Andreatta, F. Mason, P. Serafini (Eds.): Advanced School on Stochastics in Combinatorial Optimization (CISM Udine 1986), 128-177. Singapore: World Scientific. (1987).

MIRCD 1986-1 Mirchandani, P. B.; Veatch, M. H.
"Hot job" routing through a stochastic job-shop network. Large Scale Systems 11, 131-148. (1986).

MIRZ 1987-1 Mirzaian, A.
River routing in VLSI. Journal of Computer and System Sciences 34, 43-54. (1987).

MISH 1983-1 Misha, M. K.; Das, C.
Maximin problem and symmetric duality for mixed-integer quadratic programs. Ricerca Operativa, No. 28, 37-50. (1983).

MISHA 1986-1 Mishin, Y. S.
A method to describe graph on the plane. Kibernetika (Kiev) No. 1, 46-50. (1986).

MISHB 1984-1 Mishra, B.
An efficient algorithm to find all bidirectional edges of an undirected graph. In: 25th Annual Symposium on Foundations of Computer Science, Florida 1984, 207-216. Los Angeles: IEEE Computer Society. (1984).

MISHC 1984-1 Mishra, B. K.; Das, C.
Minimax and symmetric duality for nonlinear mixed-integer programming problem. Utilitas Mathematica 26, 215-234. (1984).

MISHC 1985-1 Mishra, B. K.; Das, C.
Maximin problem and a duality theorem for mixed-integer quadratic programming. Zeitschrift für Angewandte Mathematik and Mechanik 65, 310-311. (1985).

MISHD 1985-1 Mishra, M. S.; Nanda, S.
On a pair of nonlinear mixed integer programming problems. European Journal of Operational Research 19, 98-103. (1985).

MITC 1987-1 Mitchell, J. S. B.
98%-effective lot-sizing for one-warehouse, multi-retailer inventory systems with backlogging. Operations Research 35, 399-404. (1987).

MITCA 1985-1 Mitchem, J.; Schmeichel, E.
Pancyclic and bipancyclic graphs - a survey. In: F Harary, J. P. Maybee (Eds.): Graphs and Applications, 271-278. New York: Wiley. (1985).

MITCB 1986-1 Mitchison, G.; Durbin, R.
Optimal numberings of an N x N array. SIAM Journal on Algebraic and Discrete Methods 7, 571-582. (1986).

MITE 1984-1 Mitev, J. G.
Numerical method for combinatorial optimization problem (Bulgarian). Mathematics and Education in Mathematics, Proceedings of the 13th Spring Conference (Sunny Beach/Bulg.), 375-378. Sofia: Bulg. Acad. (1984).

MIYA 1987-1 Miyazaki, S.
A scheduling method based on local optimum for minimizing total tardiness. Pure and Applied Mathematika Sciences 26, 19-34. (1987).

MIYAA 1984-1 Miyazaki, S.; Nishiyama, N.
Scheduling to minimize total penalties due to earliness and tardiness. Bulletin of University of Osaka Prefecture, Series A 33, 127-143. (1984).

MIZU 1984-1 Mizukami, K.; Sikorski, J.
Three algorithms for calculating surrogate constraints in integer programming problems. Control and Cybernetics (Warsaw) 13, 375-397. (1984).

MLYN 1985-1 Mlynarovic, V.
Benders' decomposition and some problems of the production-transport type (Czechoslovakian). Economicko-Matematicky Obzor 21, 280-290. (1985).

MOFF 1985-1 Moffat, A.; Takaoka, T.
An all pairs shortest path algorithm with expected running time O(n**2.logn). In: 26th Annual Symp. on Foundations of Computer Science, Portland, Oregon 1985, Proceedings, 101-105. IEEE. (1985).

MOFF 1987-1 Moffat, A.; Takaoka, T.
An all pairs shortest path algorithm with expected time O(n**2 log n). SIAM Journal on Computing 16, 1023-1031. (1987).

MOHA 1986-1 Mohanty, H.; Bhattacharjee, G. P.
A distributed algorithm for edge-disjoint path problem. In: K. V. Nori: Foundations of Software, Technology and Theoretical Computer Science, Lecture Notes in Computer Science 241, 344-361 Berlin, Heidelberg, New York: Springer-Verlag. (1986).

MOHAA 1984-1 Mohar, B.
On edge-colorability products of graphs. Publications de l'Institut Mathematique, Nouvelle Serie (Beograd) 36(50), 13-16. (1984).

MOHAA 1984-2 Mohar, B.
Search for minimal non-Hamiltonian simple 3-polytopes. In: D. Cvetkovic, I. Gutman, T. Pisanski, R. Tosic (Eds.): Graph Theory (Novi Sad, 1983), 191-208. Univ. Novi Sad. (1984).

MOHAA 1984-3 Mohar, B.
Simplicial schemes and some combinatorial applications. In: D. Cvetkovic, I. Gutman, T. Pisanski, R. Tosic (Eds.): Graph Theory (Novi Sad 1983), 209-229. Novi Sad: Universiati Novi Sad. (1984).

MOHAA 1986-1 Mohar, B.
Colouring of graphs (Slovenian). Obzornik za Matematiko in Fiziko (Ljubljana) 33, 75-80. (1986).

MOHAA 1986-2 Mohar, B.
On the cycle space of graphs. Discrete Mathematics 58, 79-80. (1986).

MOHAA 1987-1 Mohar, B.
Generalization of Maclane's characterization of planar graphs. Congressus Numerantium 56, 231-237. (1987).

MOHAB 1984-1 Mohar, B.; Pisanski, T.; Shawe-Taylor, J.
Edge-coloring of composite regular graphs. In: A. Hajnal; L. Lovasz; V. T. Sos (Eds.): Finite and Infinite Sets, Vol. 2, 591-600. Coll. Math. Soc. J. Bolyai 37, Amsterdam: North-Holland. (1984).

MOHR 1985-1 Möhring, R. H.
Algorithmic aspects of comparability graphs and interval graphs. In: I. Rival (Ed.): Graphs and Order, 41-102. Dordrecht: Reidel. (1985).

MOHR 1986-1 Möhring, R. H.
Algorithmic aspects of the substitution decomposition in optimization over relations, set systems and Boolean functions. In: C. L. Monma (Ed.): Algorithms and Software for Optimization - Part I, 195-226. Annals of Operations Research 4, Basel: Baltzer. (1986).

MOHR 1987-1 Möhring, R. H.
Computationally tractable classes of ordered sets. Report 87468-OR. Bonn: Univ. Bonn, Inst. f. Ökonometrie u. Operations Research, 90 p. (1987).

MOHRA 1984-1 Möhring, R. H.; Radermacher, F. J.
Substitution decomposition for discrete structures and connections with combinatorial optimization. In: R. E. Burkard, R. A. Cuninghame-Green, U. Zimmermann (Eds.): Algebraic and Combinatorial Methods in Operations Research, Annals of Discrete Mathematics Vol. 19, 257-356. Amsterdam: North-Holland. (1984).

MOHRA 1985-1 Möhring, R. H.; Radermacher, F. J.
Generalized results on the polynomiality of certain weighted sum scheduling problems. In: P. Brucker, R. Pauly (Eds.): Proc. IX. Symp. on Operations Research, Osnabrück 1984. Methods of Operations Research 49, 405-418. (1985).

MOHRA 1985-2 Möhring, R. H.; Radermacher, F. J.
Introduction to stochastic scheduling problems. In: K. Neumann, D. Pallaschke (Eds.): Contributions to Operations Research, Lecture Notes in Economics and Mathematical Systems Vol. 240, 72-130. Berlin: Springer-Verlag. (1985).

MOHRB 1984-1 Möhring, R. H.; Radermacher, F. J.; Weiss, G.
Stochastic scheduling problems I - general strategies. Zeitschrift für Operations Research, Series A28, 193-260. (1984).

MOHRB 1985-1 Möhring, R. H.; Radermacher, F. J.; Weiss, G.
Stochastic scheduling problems II - set strategies. Zeitschrift für Operations Research 29A, 65-104. (1985).

MOIL 1986-1 Moily, J. P.
Optimal and heuristic procedures for component lot-splitting in multi-stage manufacturing systems. Management Science 32, 113-125. (1986).

MOIS 1984-1 Moiseenko, G. E.
A multicommodity network flow problem with commodity conversion and dependent flows. Automation and Remote Control 45, 1464-1470; translated from Avtomatika i Telemekhanika 1984, No. 11, 93-100. (1984).

MOIT 1986-1 Moitra, A.
Scheduling of hard real-time systems. In: K. V. Nori: Foundations of Software, Technology and Theoretical Computer Science, Lecture Notes in Computer Science 241, 362-381. Berlin, Heidelberg, New York: Springer-Verlag. (1986).

MOITA 1985-1 Moitra, A.; Iyengar, S. S.
A maximally parallel balancing algorithm for obtaining complete balanced binary trees. IEEE Transactions on Computers 34, 563-565. (1985).

MOLE 1983-1 Mole, R. H.; Johnson, D. G.; Wells, K.
Combinatorial analysis for route first-cluster second vehicle routing. OMEGA, International Journal of Management Science 11, 507-512. (1983).

MONA 1983-1 Monakhov, O. G.
One method of decentralized job allocation in computer systems with a programmable structure. Automatic Control and Computer Sciences 17, 5-11; translated from Avtomatika i Vychislitel'naya Tekhnika (Riga) No. 1, 7-13. (1983).

MONH 1987-1 Monhor, D.
An approach to PERT: application of Dirichlet distribution. Optimization 18, 113-118. (1987).

MONI 1985-1 Monien, B.
How to find long paths efficiently. In: G. Ausiello, M. Lucertini (Eds.): Analysis and Design of Algorithms for Combinatorial Problems, Annals of Discrete Mathematics Vol. 25, 239-254. Amsterdam: North-Holland. (1985).

MONI 1985-2 Monien, B.
The complexity of embedding graphs into binary trees. In: L. Budach (Ed.): Fundamentals of Computation Theory, FCT '85, Lecture Notes in Computer Science Vol. 199, 300-309. Berlin: Springer-Verlag. (1985).

MONI 1986-1 Monien, B.
The bandwidth minimization problem for caterpillars with hair length 3 is NP-complete. SIAM Journal on Algebraic and Discrete Methods 7, 505-512. (1986).

MONIA 1984-1 Monien, B.; Speckenmeyer, E.
Ramsey numbers and an approximation algorithm for the vertex cover problem. in: G. Wechsung (Ed.): Proc. Second International Frege Conference (Schwerin 1984), 345-353. Berlin: Akademie-Verlag. (1984).

MONIA 1985-1 Monien, B.; Speckenmeyer, E.
Ramsey numbers and an approximation algorithm for the vertex cover problem. Acta Informatica 22, 115-123. (1985).

MONIB 1985-1 Monien, B.; Sudborough, I. H.
Bandwidth constrained NP-complete problems. Theoretical Computer Science 41, 141-168. (1985).

MONM 1986-1 Monma, C. L.; Erickson, R. E.; Veinott, A. F. Jr.
Send-and-split method for minimum-concave-cost network flows. Research Paper. Morristown: Morris Research and Engineering Center, 36 p. (1986).

MONMA 1986-1 Monma, C. L.; Munson, B. S.; Pulleyblank, W. R.
Minimum-weight two-connected spanning networks. Research Paper. Morristown: Bell communications Research,.40 p. (1986).

MONMB 1987-1 Monma, C. L.; Potts, C. N.
On the complexity of scheduling with batch set-up times. Morristown: Bell Communications Research, 16 p. (1987).

MONMC 1986-1 Monma, C. L.; Reed, B.; Trotter, W. T. Jr.
A generalization of threshold graphs with tolerances. Congressus Numerantium 55, 187-198. (1986).

MONMC 1987-1 Monma, C. L.; Reed, B.; Trotter, W. T. Jr.
Threshold tolerance graphs. Morristown: Bell Communications Research, 36 p. (1987).

MONMD 1986-1 Monma, C. L.; Schrijver, A.; Todd, M. J.; Wei, V. K.
A class of convex programs for resource allocation problems on directed acyclic graphs. Morristown: Bell Communications Research, 30 p. (1986).

MONME 1986-1 Monma, C. L.; Shallcross, D. F.
Methods for designing survivable communication networks. Research Paper. Morristown: Morris Research and Engineering Center, 19 p. (1986).

MONMF 1986-1 Monma, C. L.; Sheng, D. D.
Backbone network design and performance analysis: a methodology for packet switching networks. Morristown: Bell Communications Research, 108 p. (1986).

MONMG 1987-1 Monma, C. L.; Sidney, J. B.
Optimal sequencing via modular decomposition: characterization of sequencing functions. Mathematics of Operations Research 12, 22-31. (1987).

MONMH 1985-1 Monma, C. L.; Wei, V. K.
Intersection graphs of paths in a tree. Congressus Numerantium 49, 57-74. (1985).

MONMH 1986-1 Monma, C. L.; Wei, V. K.
Intersection graphs of paths in a tree. Journal of Combinatorial Theory, Series B 41, 141-181. (1986).

MONT 1984-1 Montejano, L.; Neumann-Lara, V.
A variation of Menger's theorem for long paths. Journal of Combinatorial Theory, Series B 36, 213-217. (1984).

MONTA 1985-1 Monterosso, C. D. B.; Wright, C. L.; Lacerda, M. C. S.; Ofugi, N.
Grain storage in developing areas: Location and size of facilities. American Journal of Agricultural Economics 67, 101-111. (1985).

MONTB 1987-1 Montreuil, B.; Ratliff, H. D.; Goetschalckx, M.
Matching based interactive facility layout. IIE Transactions 19, 271-279. (1987).

MORA 1985-1 Moran, G.
On planar Eulerian graphs and permutations. Transactions of the American Mathematical Society 287, 323-341. (1985).

MORA 1987-1 Moran, G.
A remark on embedded bipartite graphs. Journal of Combinatorial Theory 43B, 241-243. (1987).

MORAA 1984-1 Moran, S.
On the length of optimal TSP circuits in sets of bounded diameter. Journal of Combinatorial Theory, Series B 37, 113-141. (1984).

MORE 1985-1 Moreno, J. A.
A correction to the definition of local center. European Journal of Operational Research 20, 382-386. (1985).

MORG 1987-1 Morgenstern, I.
Spin glasses, optimization and neural networks. In: J. L. van Hemmen, I. Morgenstern (Eds.): Heidelberg Colloquium on Glassy Dynamics, 399-429. Lecture Notes in Physics 275, Berlin: Springer. (1987).

MORI 1985-1 Morikawa, T.
Chemical matroids, I: independence and dependence in chemical reaction systems. Match 17, 203-217. (1985).

MORO 1987-1 Morozov, V. V.
On an application of discrete optimization methods (Russian). In: Systems Programming and Problems of Optimization, 120-133. Moscow: Moskov. Gos. Univ. (1987).

MORR 1986-1 Morris, W. D.
Oriented matroids and the linear complementarity problem. Technical Report 717. Ithaca: Cornell Univ., School of Operations Research and Ind. Eng., 111 p. (1986).

MORT 1985-1 Morton, G.; Randow, R. von; Ringwald, R.
A greedy algorithm for solving a class of convex programming problems and its connection with polymatroid theory. Mathematical Programming 32, 238-241. (1985).

MOSE 1986-1 Mosevich, J.
Balancing hydraulic turbine runners - a discrete combinatorial optimization problem. European Journal of Operational Research 26, 202-204. (1986).

MOSH 1986-1 Moshchenskii, A. V.
Semimatroids and Gale diagrams (Russian). Vestnik Belorusskogo Gosudarstvennogo Universiteta im V.I. Lenina (Minsk), Seriya I, No.3, 70-71. (1986).

MOSI 1984-1 Mosier, C. T.; Elvers, D. A.; Kelly, D.
Analysis of group technology scheduling heuristics. International Journal of Production Research 22, 857-876. (1984).

MOSK 1985-1 Moskalyuk, L. V.
The discrete maximum principle in the solution of scheduling problems (Russian). Preprint 85-21, Akademii Nauk Ukrainskoj SSR, Inst. Kibernet., Kiev. (1985).

MOSL 1987-1 Mosler, K. C.
Continuous location of transportation networks. Berlin: Springer-Verlag. (1987).

MOTE 1986-1 Mote, J.
The family constrained network problem. Discrete Applied Mathematics 13, 249-258. (1986).

MOTS 1985-1 Motskus, J.; Teshis, V.; Shaltyanis, V.; Yushka, F.
On a problem allocation of resources (Russian). Teoriya Optimal'nykh Reshenij 11, 22-28. (1985).

MOZH 1983-1 Mozhan, N. N.
The chromatic number of graphs with a density not exceeding two-thirds of the maximal degree (Russian). Metody Diskretnogo Analiza (Novosibirsk) No. 39, 52-65. (1983).

MROZ 1984-1 Mrozek, M.
A note on minimum-dummy-activities Pert networks. RAIRO Recherche Operationelle 18, 415-424. (1984).

MUCH 1987-1 Muchnik, I. B.; Shvartser, L. V.
Submodular set functions and monotone systems in aggregation problems. I (Russian). Avtomatika i Telemekhanika, No.5, 135-148. (1987).

MUCH 1987-2 Muchnik, I. B.; Shvartser, L. V.
Submodular set functions and monotone systems in aggregation problems. II (Russian). Avtomatika i Telemekhanika, No.6, 138-147. (1987).

MUHL 1984-1 Mühlbacher, J. R.; Steinparz, F. X.; Tinhofer, G.
On certain classes of fractional matchings. Discrete Applied Mathematics 9, 235-244. (1984).

MUKH 1983-1 Mukhacheva, E. A.
Search for the rational solution in the two-criterial problem of guillotine cutting-out (Russian). Optimizatsiya (Novosibirsk) 33(50), 56-62. (1983).

MUKHA 1984-1 Mukhacheva, E. A.; Ibatullina, S. M.
Optimization of cutting material according to a given assortment ratio (Russian). Optimizatsiya (Novosibirsk) 34(51), 122-127. (1984).

MUKHB 1986-1 Mukhamadiev, E. M.; Aralbaev, T. Z.; Nazimov, A. B.
An algorithm for calculating routes in a problem with cyclic servicing of the subscribers (Russian). Izvestiya Akademii Nauk Tadzhikskoi SSR, Otdelenie Fiziko-Matematicheskikh, Khimicheskikh i Geologicheskikh Nauk (Dushanbe) No. 3(101), 33-36. (1986).

MUKHC 1985-1 Mukherjee, P.; Mukherjee, T.
Studies on interrelations of matrices associated with a graph. Journal of Mathematical and Physical Sciences (Madras) 19, 547-554. (1985).

MUKHC 1986-1 Mukherjee, P.; Mukherjee, T.
An algorithmic approach for determination of iterated line graphs. Journal of Mathematical and Physical Sciences 20, 27-31. (1986).

MULL 1983-1 Müller-Merbach, H.
Head-and-shoulder-branching in integer programming. SIGMAP Bulletin No. 32, 17-23. (1983).

MULLA 1987-1 Müller, H.; Brandstädt, A.
The NP-completeness of STEINER TREE and DOMINATING SET for chordal bipartite graphs. Theoretical Computer Science 53, 257-266. (1987).

MULLB 1987-1 Müller, P.
Untersuchungen zu speziellen linearen gemischt-ganzzahligen 0-1-Optimierungsaufgaben. Optimization 18, 83-96. (1987).

MULM 1987-1 Mulmuley, K.; Vazirani, U. V.; Vazirani, V. V.
Matching is as easy as matrix inversion. Combinatorica 7, 105-114. (1987).

MULV 1984-1 Mulvey, J. M.; Beck, M. P.
Solving capacitated clustering problems. European Journal of Operational Research 18, 339-348. (1984).

MULVA 1985-1 Mulvey, J. M.; Zenios, S. A.
Integrated risk/cost planning models for the U.S. air traffic system. Report EES-85-9. Princeton: Princton Univ., Dept. of Civil Engineering, 39p. (1985).

MULVA 1985-2 Mulvey, J. M.; Zenios, S. A.
Solving large scale generalized networks. Journal of Information and Optimization Sciences (Delhi) 6, 95-112. (1985).

MUMM 1986-1 Mummy, M. S.
Full transversal matroids, strict gammoids, and the matroid components problem. Linear and Multilinear Algebra 19, 97-116. (1986).

MUMM 1986-2 Mummy, M. S.
Strict matching matroids and matroid algorithms. Linear and Multilinear Algebra 19, 221-247. (1986).

MUMM 1986-3 Mummy, M. S.
The generalized column incidence graph and a matroid base-listing algorithm. Linear and Multilinear Algebra 19, 33-49. (1986).

MUNO 1984-1 Munoz Perez, J.
Minimax location when destinations are positioned randomly (Spanish). Trabajos de Estadistica y de Investigacion Operativa (Valencia) 35, 83-91. (1984).

MURA 1986-1 Muradyan, D. O.
A polynomial algorithm for finding the bandwidth of interval graphs (Russian). Doklady, Akademiya Nauk Armyanskoj SSR 82, No. 2, 64-66. (1986).

MURD 1985-1 Murdeshwar, T. M.; Sathe, Y. S.
Some aspects of lot size models with two levels of storage. Opsearch 22, 255-262. (1985).

MURG 1987-1 Murgolo, F. D.
An efficient approximation scheme for variable-sized bin packing. SIAM Journal on Computing 16, 149-161. (1987).

MURO 1985-1 Murota, K.
Combinatorial canonical form of layered mixed matrices and block-triangularization of large-scale systems of linear/nonlinear equations. Discussion Paper 257. Ibaraki: Univ. of Tsukuba, Inst. of Socio-Economic Planning, 49 p. (1985).

MURO 1986-1 Murota, K.
Bimatroid network and its application to the structural solvability of systems of equations. Discussion Paper, 319. Sakura: Univ. of Tsukuba, Inst. of Socio-Economic Planning, 55 p. (1986).

MURO 1986-2 Murota, K.
Combinatorial dynamical system theory. Research Memorandum RMI 86-02. Tokyo: Univ. of Tokyo, Dept. of Math. Eng. and Instrumentation Physics, 49 p. (1986).

MURO 1987-1 Murota, K.
Homotopy base of an acyclic graph - a combinatorial analysis of commutative diagrams by means of preordered matroid. Discrete Applied Mathematics 17, 135-156. (1987).

MURO 1987-2 Murota, K.
Refined study on structural controllability of descriptor systems by means of matroids. SIAM Journal on Control and Optimization 25, 967-989. (1987).

MURO 1987-3 Murota, K.
Systems analysis by graphs and matroids. Structural solvability and controllability. Algorithms and Combinatorics Vol. 3, Berlin: Springer. (1987).

MUROA 1987-1 Murota, K.; Fujishige, S.
Finding a homotopy base for directed paths in an acyclic graph. Discrete Applied Mathematics 17, 157-162. (1987).

MUROB 1985-1 Murota, K.; Iri, M.
Structural solvability of systems of equations - a mathematical formulation for distinguishing accurate and inaccurate numbers in structural analysis of systems. Japan Journal of Applied Mathematics 2, 247-271. (1985).

MURP 1986-1 Murphy, R. A.
Some computational results on real 0-1 knapsack problems. Operations Research Letters 5, 67-72. (1986).

MURPA 1987-1 Murphy, R. A.; Mueller, T. E.
Routing full truckload movements. Working Paper. Cincinnati: Optimal Decision Systems Inc., 14 p. (1987).

MURT 1985-1 Murtagh, F.
A survey of algorithms for contiguity-constrained clustering and relating problems. Computer Journal 28, 82-88. (1985).

MURTA 1985-1 Murty, K. G.; Kabadi, S. N.
Some NP-complete problems in quadratic and nonlinear programming. Technical Report 85-23. Ann Arbor: Univ. of Michigan, College of Engineering, 15 p. (1985).

MUSA 1984-1 Musa, A. A.; Saxena, U.
Scheduling nurses using goal-programming techniques. IIE Transactions 16, 216-221. (1984).

MUST 1985-1 Mustafi, C. K.; Xavier, M. J.
Mixed-integer linear-programming formulation of a multi-attribute threshold model of choice. Journal of the Operational Research Society 36, 935-942. (1985).

MYNH 1985-1 Mynhardt, C. M.; Broere, I.
Generalized colorings of graphs. In: Y. Alavi, G. Chartrand, L. Lesniak, D. R. Lick, C. E. Wall (Eds.): Graph Theory with Applications to Algorithms and Computer Science, 583-594. Chichester: Wiley. (1985).

MYNHA 1987-1 Mynhardt, C. M.; Cockayne, E. J.
K-minimal domination numbers of cycles. Ars Combinatoria 23A, 195-206. (1987).

MYUN 1987-1 Myung, Y.-S.; Tcha, D.-W.
The stochastic uncapacitated facility location problem: a new formulation and extensions to the capacitated problem. IIE Transactions 19, 460-462. (1987).

NABE 1982-1 Nabeshima, I.
Notes on the analytical results in flow-shop scheduling. XVII. Special single criterion sequences.I. Reports of the University of Electro-Communications 33, 17-25. (1982).

NABE 1982-2 Nabeshima, I.
Notes on the analytical results in flow-shop scheduling. XVIII. Special single criterion sequences. II. Reports of the University of Electro-Communications 33, 27-33. (1982).

NABE 1983-1 Nabeshima, I.
Notes on the analytical results in flow-shop scheduling. XIX. Generalization of job-base results in part 13. Reports of the University of Electro-Communications 33, 211-221. (1983).

NABE 1983-2 Nabeshima, I.
Notes on the analytical results in flow-shop scheduling. XX. Another generalization of job-base results in part 13. Reports of the University of Electro-Communications 33, 223-233. (1983).

NABE 1983-3 Nabeshima, I.
Notes on the analytical results in flow-shop scheduling. XXI. Machine-base counterpart of job-base results for r=2 in part 20. Reports of the University of Electro-Communications 34, 21-28. (1983).

NABE 1984-1 Nabeshima, I.
Notes on the analytical results in flow shop scheduling. XXIII. Weaker sufficient optimality conditions for w(2) by machine-base transformation method. Reports of the Univ. of Electro-Communications (Tokyo) 35, 23-31. (1984).

NABE 1984-2 Nabeshima, I.
Notes on the analytical results in flow shop scheduling. XXIV. Job-base counterpart of the results in part XXII. Reports of the Univ. of Electro-Communications (Tokyo) 35, 33-42. (1984).

NABE 1984-3 Nabeshima, I.
Notes on the analytical results in flow-shop scheduling. XXII. Weaker sufficient optimality conditions for a specified sequence w(2) in parts 2 and 8. Reports of the University of Electro-Communications 34, 199-212. (1984).

NABE 1984-4 Nabeshima, I.
On sufficient optimality conditions with convex parameter in three-machine flow-shop makespan problem: a supplement to: "Notes on the analytical results in flow shop scheduling. XI". Reports of the University of Electro-Communications 34, 191-197. (1984).

NABE 1985-1 Nabeshima, I.
General formulation and applications of a machine-base transformation method for the NP-complete permutation scheduling problems. Reports of the Univ. of Electro-Communications (Tokyo) 35, 139-148. (1985).

NABE 1985-2 Nabeshima, I.
Notes on the analytical results in flow shop scheduling. XXV. Results by job-base transformation method for conditions in part XXIV. Reports of the Univ. of Electro-Communications (Tokyo) 35, 131-138. (1985).

NABE 1985-3 Nabeshima, I.
Notes on the analytical results in flow shop scheduling. XXVI. Generalization of job-base results in part XXIV. Reports of the University of Electro-Communications (Tokyo) 36, No. 1, 29-38. (1985).

NABE 1985-4 Nabeshima, I.
Solvable cases in the m-machine flow-shop problem with separated sequence-independent setup times. Reports of the University of Electro-Communications (Tokyo) 36, No. 1, 39-48. (1985).

NABE 1986-1 Nabeshima, I.
Another solvable case in the three-machine flow-shop makespan scheduling problem with separated sequence-independent setup times. Reports of the University of Electro-Communications (Tokyo) 37, No. 1, 35-44. (1986).

NABE 1986-2 Nabeshima, I.
Job-base conditions in the three-machine flow-shop makespan scheduling problem with separated sequence-independent setup times. Reports of the University of Electro-Communications (Tokyo) 37, No. 1, 25-33. (1986).

NABE 1986-3 Nabeshima, I.
Solvable cases extended by transformation method in the m-machine flow-shop problem with separated sequence-independent setup times. I, II. Reports of the University of Electro-Communications (Tokyo) 36, No. 2, 213-220, 221-228. (1986).

NABE 1987-1 Nabeshima, I.
Solvable cases in the m-machine flow shop mean completion time scheduling problem with separated sequence-independent setup times. Reports of the University of Electro-Communications (Tokyo) 37, No. 2, 143-152. (1987).

NABEA 1984-1 Nabeshima, I.; Maruyama, S.
Two- and three-machine flow-shop makespan scheduling problems with additional times separated from processing times. Journal of the Operations Research Society of Japan 27, 348-366. (1984).

NABEA 1985-1 Nabeshima, I.; Maruyama, S.
Correction to: Two- and three-machine flow-shop makespan scheduling problems with additional times separated from processing times. Journal of the Operations Research Society of Japan 28, 85. (1985).

NACH 1987-1 Nachtigall, K.
Residuated lattice spaces. In: W. Domschke, W. Krabs, J. Lehn, P. Spellucci (Eds.): Proc. XI. Symp. on Operations Research, Darmstadt 1986, 155-160. Methods of Operations Research 57, Frankfurt/Main: Athenäum. (1987).

NADD 1984-1 Naddef, D. J.
Pancyclic properties of the graph of some 0-1-polyhedra. Journal of Combinatorial Theory, Series B 37, 10-26. (1984).

NADD 1987-1 Naddef, D. J.
The symmetric traveling salesman polytope I: valid inequalities derived from the graphical relaxation. Rapport de Recherche 660-M. Grenoble: Univ. Scientifique, Technologique et Medicale de Grenoble, 42 p. (1987).

NADDA 1984-1 Naddef, D. J.; Pulleyblank, W. R.
Hamiltonicity in (0-1)-polyhedra. Journal of Combinatorial Theory, Series B 37, 41-52. (1984).

NAGA 1985-1 Nagasawa, H.; Nishiyama, N.; Hitomi, K.
Decision analysis for determining the optimum planning horizon in aggregate production planning. Part 2: Difference between planning horizons in weekly and in monthly schedulings. International Journal of Production Research 23, 423-436. (1985).

NAGE 1986-1 Nageswara, S. V.; Viswanadham, N.
Fault diagnosis in dynamical systems: a graph theoretic approach. In: C. I. Byrnes, A. Lindquist (Ed.): Computational and Combinatorial Methods in Systems Theory, 149-158. Amsterdam: North Holland. (1986).

NAGL 1985-1 Nägler, G.
Lokal optimale Lösungen der kombinierten Strom- und Spannungsoptimierung auf Netzwerken im Anwendungsfall der Ablaufplanung. Optimization 16, 229-244. (1985).

NAGL 1986-1 Nägler, G.
Coupling between network flow and network tension and its application in project planning. Godishnik na Visshite Uchebni Zavedniya, Prilozhna Matematika (Sofia) 21, No.4, 213-224. (1986).

NAID 1986-1 Naidu, M. M.; Singh, N.
Lot sizing for material requirements planning systems - an incremental cost approach. International Journal of Production Research 24, 223-240. (1986).

NAID 1987-1 Naidu, M. M.; Singh, N.
Further investigations on the performance of incremental cost approach for lot sizing for material requirements planning systems. International Journal of Production Research 25, 1241-1246. (1987).

NAJI 1985-1 Naji, W.
Reconnaissance des graphes de cordes. Discrete Mathematics 54, 329-338. (1985).

NAKA 1985-1 Nakai, T.
Optimal assignment for a random sequence with an unknown number of jobs. Journal of the Operations Research Society of Japan 28, 179-193. (1985).

NAKAA 1984-1 Nakamori, M.; Mizutani, A.; Kakegawa, M.; Shimizu, K.
Solution of the minimum flow problem in selective networks and application to the design of arrangements and distributions of integrated circuits (Japanese). In: Graph theory and applications, Proc. Symp. Kyoto University 1984, 234-246. Kyoto University, Research Inst. for Mathematical Sciences. (1984).

NAKAB 1986-1 Nakamura, M.
A decomposition of poly-linking systems. Working Paper. Tokyo: Univ. of Tokyo, Dept. of Natural and Artificial Systems, 13 p. (1986).

NAKAB 1986-2 Nakamura, M.
Structural theorems for submodular functions, polymatroids and polymatroid intersections. Working Paper. Tokyo: Univ. of Tokyo, Dept. of Natural and Artificial Systems, 47 p. (1986).

NAKAC 1981-1 Nakamura, M.; Iri, M.
A structural theory for submodular functions, polymatroids and polymatroid intersections. Research Memorandum RMI 81-06. Tokyo: Univ. of Tokyo, Dept. of Math. Eng. and Instrumentation Physics, 70 p. (1981).

NAKAD 1984-1 Nakano, H.; Nakanishi, Y.
Simulated annealing method for combinatorial optimization problems (Japanese). In: Graph theory and applications, Proc. Symp. Kyoto University 1984, 274-285. Kyoto University, Research Inst. for Mathematical Sciences. (1984).

NAKAE 1986-1 Nakayama, A.
A polynomial algorithm for the maximum balanced flow problem with a constant balancing rate function. Journal of the Operations Research Society of Japan 29, 400-410. (1986).

NAKAE 1987-1 Nakayama, A.
A polynomial-time dual simplex algorithm for the minimum cost flow problem. Journal of the Operations Research Society of Japan 30, 265-290. (1987).

NAKAF 1985-1 Nakayama, H.; Nishizeki, T.; Saito, N.
Lower bounds for combinatorial problems on graphs. Journal of Algorithms 6, 393-399. (1985).

NAKE 1985-1 Nakechbandi, M.; Trehel, M.
Systolic array for integer and linear programming. in: Dynamical systems and cellular automata (Luminy 1983), 339-346. London, New York: Academic Press. (1985).

NAND 1987-1 Nanda, S.; Mishra, M. S.; Acharya, D.
A note on a pair of nonlinear mixed integer programming problems. European Journal of Operational Research 28, 89-92. (1987).

NAOR 1987-1 Naor, J.; Naor, M.; Schaffer, A.
Fast parallel algorithms for chordal graphs. In: Proc. Nineteenth Annual ACM Symp. on Theory of Computing, New York City 1987, 355-364. ACM. (1987).

NARA 1984-1 Nara, C.
The maximal number of edges of a self-placeable graph. Revue Roumaine de Mathematiques Pures et Appliquees (Bucharest) 29, 323-328. (1984).

NARAA 1984-1 Narasimhan, R.; Melnyk, S. A.
Assessing the transient impact of lot sizing rules following MRP implementation. International Journal of Production Research 22, 759-772. (1984).

NARE 1985-1 Narendra, K. S.; Wheeler, R. M. Jr.
Routing in communication networks - a case study of learning in large scale systems. Large Scale Systems (Amsterdam) 8, 211-222. (1985).

NARU 1986-1 Narula, S. C.
Minisum hierarchical location-allocation problems on a network: a survey. In: J. P. Osleeb, P. J. Ratick (Eds.): Locational Decisions: Methodology and Applications, 257-272. Annals of Operations Research 6, Basel: Baltzer. (1986).

NARUA 1985-1 Narula, S. C.; Ogbu, U. I.
Lagrangean relaxation and decomposition in a uncapacitated 2-hierarchal location-allocation problem. Computers & Operations Research 12, 169-180. (1985).

NARUB 1983-1 Narushima, H.; Kusaka, Y.
A move problem of weighted digraphs. Proceedings of the Faculty of Sciences of Tokai University 18, 23-27. (1983).

NASH 1985-1 Nash-Williams, C. St. J. A.
Connected detachments of graphs and generalized Euler trails. Journal of the London Mathematical Society, Second Series 31, 17-29. (1985).

NASH 1985-2 Nash-Williams, C. St. J. A.
Detachments of graphs and generalised Euler trails. In: I Anderson (Ed.): Surveys in Combinatorics 1985, London Mathematical Society Lecture Notes Series Vol. 103, 137-152. Cambridge: University Press. (1985).

NASH 1987-1 Nash-Williams, C. St. J. A.
Amalgamations of almost regular edge-colourings of simple graphs. Journal of Combinatorial Theory B-43, 322-342. (1987).

NAUM 1986-1 Naumovich, N. A.
Dichotomous realization of gradient algorithms for solving integer programming problems (Russian). Vestnik Belorusskogo Gosudarstvennogo Universiteta Imeni V.I. Lenina, Seriya I, No.1, 39-43. (1986).

NEBE 1984-1 Nebesky, L.
Edge-disjoint 1-factors in powers of connected graphs. Czechoslovak Mathematical Journal 34(109), 499-505. (1984).

NEBE 1984-2 Nebesky, L.
On quasistars in n-cubes. Casopis Pro Pestovani Matematiky (Prague) 109, No. 2, 153-156. (1984).

NEBE 1984-3 Nebesky, L.
Upper embeddable factorizations of graphs. Czechoslovak Mathematical Journal (Prague) 34(109), 423-438. (1984).

NEBE 1985-1 Nebesky, L.
On locally quasiconnected graphs and their upper embeddability. Czechloslovak Mathematical Journal (Prague) 35 (110), 162-166. (1985).

NEBEA 1985-1 Nebesky, L.; Wisztova, E.
Two edge-disjoint Hamiltonian cycles of powers of a graph. Casopis pro Pestovani Matematiky (Prague) 110, 294-301. (1985).

NEEB 1986-1 Neebe, A. W.; Rao, M. R.
Sequencing capacity expansion projects in continuous time. Management Science 32, 1467-1479. (1986).

NEGA 1987-1 Negami, S.
Polynomial invariants of graphs. Transactions of the AMS 299, 601-622. (1987).

NEGAA 1986-1 Negami, S.; Xu, G.-H.
Locally geodesic cycles in 2-self-centered graphs. Discrete Mathematics 58, 263-268. (1986).

NEHS 1985-1 Nehse, R.
Diskrete Vektoroptimierung - Möglichkeiten der numerischen Bearbeitung. 30. Internationales Wissenschaftliches Kolloquium (TH Ilmenau, DDR) Heft 4E, 95-98. (1985).

NEKR 1983-1 Nekrasov, V. P.
Planar graphs having an external access (Russian). In: Combinatorial Properties of Convex Sets and Graphs, 34-44. Akademiya Nauk SSSR, Ural. Nauchn. Tsentr (Sverdlovsk). (1983).

NELS 1985-1 Nelson, M. D.; Nygard, K. E.; Griffin, J. H.; Shreve, W. E.
Implementation techniques for the vehicle routing problem. Computers & Operations Research 12, 273-284. (1985).

NELSA 1986-1 Nelson, R. T.; Sarin, R. K.; Daniels, R. L.
Scheduling with multiple performance measures: The one-machine case. Management Science 32, 464-479. (1986).

NEMH 1985-1 Nemhauser, G. L.
Duality for integer optimization. In: M. O'hEigeartaigh, J. K. Lenstra, A. H. G. Rinnooy Kan (Eds.): Combinatorial Optimization - Annotated Bibliographies, 11-20. Chichester: Wiley. (1985).

NEMHA 1984-1 Nemhauser, G. L.; Wolsey, L. A.
A recursive procedure for generating all mixed integer cuts. CORE Discussion Paper 8439. Louvain-la-Neuve: Center of Operations Research & Econometrics, 19 p. (1984).

NENO 1982-1 Nenov, N. D.
On proper 3-coloring of the edges of complete graphs (Bulgarian). in: Mathematics and education in mathematics, Proc. 11th Spring Conf., Sunny Beach/Bulg., 242-247. (1982).

NENO 1984-1 Nenov, N. D.
Lower estimates for some constants related to Ramsey graphs (Bulgarian). Godishnik na Sofijskiya Universitet, Fakultet po Matematika i Mekhanika (Sofiya) 75 (1981), 27-38. (1984).

NENO 1984-2 Nenov, N. D.
The chromatic number of any 10-vertex graph without 4-cliques is at most 4 (Russian). Comptes Rendus de l'Academie Bulgare des Sciences (Sofia) 37, 301-304. (1984).

NENO 1986-1 Nenov, N. D.
On the existence of a minimal t-graph with nine vertices (Bulgarian). Godishnik na Sofijskiya Universitet, Fakultet po Matematika i Mekhanika 72, 271-276. (1986).

NENO 1986-2 Nenov, N. D.
Up to isomorphism there exists only one minimal t-graph with nine vertices (Bulgarian). Godishnik na Sofijskiya Universitet, Fakultet po Matematika i Mekhanika 73, 169-184. (1986).

NENOA 1986-1 Nenov, N. D.; Khadzhiivanov, N. G.
Some colorings in two colors of the edges of a complete graph with nine vertices (Russian). Annuaire l'Universite de Sofia, Fac. Phys. 71, No. 2, 95-115. (1986).
NERA 1984-1 Neralic, L.; Sirovitsa, L.; Skok, S.; Khunet, D.
Solution of a production-transportation problem by Benders' decomposition method (Russian). Ehkonomika i Matematicheskie Metody (Moscow) 20, 1140-1143. (1984).
NESE 1986-1 Nesetril, J.
For graphs there are only 4 types of hereditary Ramsey classes. KAM Series, 16. Praha: Charles Univ., Dept. of Applied Mathematics, 10 p. (1986).
NESEA 1985-1 Nesetril, J.; Poljak, S.
On the complexity of the subgraph problem. Commentationes Mathematicae Universitatis Carolinae (Prague) 26, 415-419. (1985).
NESEA 1986-1 Nesetril, J.; Poljak, S.
A remark on max-cut problem with an application to digital-analogue convertors. Operations Research Letters 4, 289-292. (1986).
NESEB 1985-1 Nesetril, J.; Poljak, S.; Turzik, D.
Special amalgams and Ramsey matroids. In: L. Lovasz, A. Recski (Eds.): Matroid Theory, 267-298. Coll. Math. Soc. J. Bolyai Vol. 40. Amsterdam: North-Holland. (1985).
NESEC 1987-1 Nesetril, J.; Rödl, V.
Complexity of diagrams. Order 3, 321-330. (1987).
NEUF 1985-1 Neufeld, E. M.
Optimal expansion of existing networks. Congressus Numerantium 49, 79-86. (1985).
NEUM 1985-1 Neumaier, A.
Characterization of a class of distance regular graphs. Journal für die Reine und Angewandte Mathematik 357, 182-192. (1985).
NEUMA 1984-1 Neumann-Lara, V.
The generalized dichromatic number of a digraph. In: A. Hajnal; L. Lovasz; V. T. Sos (Eds.): Finite and Infinite Sets, Vol. 2, 601-606. Coll. Math. Soc. J. Bolyai 37, Amsterdam: North-Holland. (1984).
NEUMB 1986-1 Neumann-Lara, V.; Santoro, N.; Urrutia, J.
Uniquely colourable m-dichromatic oriented graphs. Discrete Mathematics 62, 65-70. (1986).
NEUMC 1984-1 Neumann, K.
Recent developments in stochastic activity networks. INFOR 22, 219-248. (1984).
NEUMC 1985-1 Neumann, K.
EOR project networks. Computing 34, 1-16. (1985).
NEUMC 1985-2 Neumann, K.
GERT networks with tree stucture: properties, temporal analysis, cost minimization and scheduling. In: K. Neumann, D. Pallaschke (Eds.): Contributions to Operations Research, Lecture Notes in Economics and Mathematical Systems Vol. 240, 142-158. Berlin: Springer-Verlag. (1985).
NEUMC 1985-3 Neumann, K.
Stochastic single-machine scheduling with GERT-like precedence constraints. Research Report 85-3. Gainesville: Univ. of Florida, Ind. & Systems Eng. Dept., 24 p. (1985).
NEUMC 1987-1 Neumann, K.
The postman problem in mixed graphs (German). In: M. J. Beckmann; K.-W. Gaede; K. Ritter (Eds.): Beiträge zur angewandten Mathematik und Statistik, Prof. Heinhold zum 75. Geburtstag, 195-218. München, Wien: Carl Hanser Verlag. (1987).
NEVI 1985-1 Nevison, C. H.
A cost adjustment heuristic for dynamic lot-sizing with uncertain demand timing. Operations Research 33, 1342-1352. (1985).
NEWE 1986-1 Newell, G. F.
Design of multiple-vehicle delivery tours - III. Valuable goods. Transportation Research 20B, 377-390. (1986).
NEWEA 1986-1 Newell, G. F.; Daganzo, C. F.
Design of multiple-vehicle delivery tours - I. A ring-radial network. Transportation Research 20B, 345-364. (1986).
NEWEA 1986-2 Newell, G. F.; Daganzo, C. F.
Design of multiple-vehicle delivery tours - II. Other metrics. Transportation Research 20B, 365-376. (1986).
NEWF 1984-1 Newfeld, E. M.; Colbourn, C. J.
The construction of reliable series - parallel networks. Congressus Numerantium 41, 21-26. (1984).
NGUE 1986-1 Nguen Ngoc Tu
Algorithm for maximizing a linear function on the set of integral points of a convex polyhedron. Cybernetics 22, 622-627; translation from Kibernetika (1986), No.5, 71-74. (1986).
NGUY 1983-1 Nguyen Van Thieu
On the problem of production location in an interbranch formulation (Vietnamese). Tap Chi Toan Hoc (Hanoi) 11, No. 3, 17-26. (1983).
NGUYA 1984-1 Nguyen, S.
Estimating origin-destination matrices from observed flows. In: M. Florian (Ed.): Transportation planning models, Proc. Course Internat. Center Transport Stud., Amalfi/ Italy 1982, 363-380. North-Holland. (1984).
NGUYB 1986-1 Nguyen, S.; Pallottino, S.
Assignment of passengers to an urban transport system: Determination of minimal hyperpaths (Italian). Ricerca Operativa (Milano) 17, No. 38, 29-74. (1986).
NGUYB 1986-2 Nguyen, S.; Pallottino, S.
Estimating origin destination flows for transit networks. Ricerca Operativa (Milano) 17, No. 38, 9-28. (1986).
NI 1985-1 Ni, L. M.; Hwang, K.
Optimal load balancing in a multiple processor system with many job classes. IEEE Transactions on Software Engineering SE-11, 491-496. (1985).
NI A 1986-1 Ni, Z. Z.
Bond graph construction from planar circuits (Chinese). Sichuan Daxue Xuebao, No. 4, 59-62. (1986).

NIEM 1983-1 Nieminen, J.
On linear and semilinear pointsets in graphs. Bulletin Mathematique de la Societe des Sciences Mathematiques de la Republique Scoaliste de Roumanie, Nouvelle Serie (Bucharest) 27(75), 343-351. (1983).

NIEM 1984-1 Nieminen, J.
A characterization of median graphs. Indian Journal of Pure and Applied Mathematics 15, 727-730. (1984).

NIEM 1984-2 Nieminen, J.
Centrality, convexity and intersections in graphs. Bulletin Mathematique de la Societe des Sciences Mathematiques de la Republique Socialiste de Roumaine (Bucharest), Nouvelle Serie 28 (76), 337-344. (1984).

NIEM 1984-3 Nieminen, J.
New binary operations on graphs. Bulletin Mathematique de la Societe des Sciences Mathematiques de la Republique Socialiste de Roumanie (Bucarest), Nouvelle Serie 28 (76), 55-62. (1984).

NIEM 1984-4 Nieminen, J.
On convexies and related congruences on graphs. Serdica (Sofia) 10, 106-113. (1984).

NIEM 1987-1 Nieminen, J.
Distance center and centroid of a median graph. Journal of the Franklin Institute 323, 89-94. (1987).

NIEP 1983-1 Niepel, L.; Safarikova, D.
On a generalization of Menger's theorem. Acta Mathematica Universitatis Comenianae 42-43, 275-284. (1983).

NIGM 1985-1 Nigmatullin, R. G.; Shtivelman, V. K.
Unique and multiple coverings of graphs by edges (Russian). In: Probabilistic methods and cybernetics, No. 21, 86-94. Kazan: Gosnioj Universiteta. (1985).

NIKI 1983-1 Nikitin, A. I.; Nuriev, U. G.; Krasilovets, L. V.
Software for a multi-user computer center. Some discrete optimization algorithms (Russian). Kiev: Akademiya Nauk Ukrainskoi SSR, Inst. Kibernet., Preprint 83-6, 32 p. (1983).

NIKO 1982-1 Nikogosyan, Z. G.
N-spanning and hypo-n-spanning graphs (Russian). Tanulmanyok - Szamitastechnikai es Automatizalasi Kutato Intezete, Magyar Tudomanyos Akademia (Budapest) No. 135, 153-167. (1982).

NIKO 1983-1 Nikogosyan, Z. G.
The maximum cycle of a graph (Russian). Studia Scientiarum Mathematicarum Hungarica (Budapest) 17, 251-282. (1983).

NIKO 1984-1 Nikogosyan, Z. G.
A sufficient condition for a graph to be Hamiltonian (Russian). Akademiya Nauk Armyanskoi SSR Doklady (Erevan) 78, No. 1, 12-16. (1984).

NIKO 1985-1 Nikogosyan, Z. G.
A sufficient condition for the Hamilton property of a graph (Russian). Matematicheskie Voprosy Kibernetiki i Vychislitelnoi Tekhniki (Erevan), No. 14, 34-54. (1985).

NIKO 1985-2 Nikogosyan, Z. G.
On longest cycles in graphs (Russian). Doklady, Akademiya Nauk Armyanskoj SSR 81, 166-170. (1985).

NINC 1982-1 Ninchak, J.
Chromatical number for a class of graphs (Russian). Bulletin for Applied Mathematics (Budapest) 122-136, 19-30. (1982).

NINC 1984-1 Ninchak, J.
On a problem of R. Häggkvist concerning the edge colourings of bipartite graphs. Bulletins for Applied Mathematics (Budapest) 34, No. 255, 117-126. (1984).

NISH 1985-1 Nishizeki, T., Kashiwagi, K.
An upper bound on the chromatic index of multigraphs. In: Y. Alavi, G. Chartrand, L. Lesniak, D. R. Lick, C. E. Wall (Eds.): Graph Theory with Applications to Algorithms and Computer Science, 595-604. Chichester: Wiley. (1985).

NISHA 1984-1 Nishizeki, T.; Sato, M.
An algorithm for edge-coloring of multigraphs. Systems-Computers-Controls 15, 58-63. (1984).

NKAN 1986-1 Nkansah, P. T.; David, H. T.
Network median problems with continuously distributed demand. Transportation Science 20, 213-219. (1986).

NODA 1985-1 Noda, R.
Partitioning strongly regular graphs. Osaka Journal of Mathematics 22, 379-389. (1985).

NOGO 1985-1 Nogova, E. G.
Construction of an optimal system of isolated routes for passenger transport (Russian). In: Theoretical and Applied Problems of Optimization, 35-40. "Nauka", Moscow. (1985).

NOMU 1987-1 Nomura, K.
An inequality between intersection numbers of a distance-regular graph. Journal of Combinatorial Theory B-43, 358-359. (1987).

NORB 1986-1 Norbis, M. I.; Smith, J. M.
Two level heuristic for the resource constrained scheduling problem. International Journal of Production Research 24, 1203-1220. (1986).

NOSH 1985-1 Noshita, K.
A theorem on the expected complexity of Dijkstra's shortest path algorithm. Journal of Algorithms 6, 400-408. (1985).

NOSO 1985-1 Nosov, V. A.; Sachkov, V. N.; Tarakanov, V. E.
Combinatorial analysis (nonnegative matrices, algorithmic problems). Journal of Soviet Mathematics 29, 1051-1099; translation from: Itogi Nauki i Tekhniki, Seriya Teoriya Veroyatnostej, Matematicheskaya Statistika, Teoreticheskaya Kibernetika 21, 120-178. (1985).

NOUH 1987-1 Nouh, A.
A sequential aggregation algorithm for the set partitioning problem. Computer Methods in Applied Mechanics and Engineering 63, No. 3, 225-232. (1987).

NOVI 1983-1 Novikov, N. A.
An algorithm for solving an all-integer problem of linear programming (Russian). In: Algorithmic Constructions and Their Efficiency, 23-26. Yaroslavl: Yaroslav. Gos. Univ. (1983).

NOVIA 1982-1 Novitskaya, I. I.; Safronenko, V. A.
Determination of the shortest way satisfying the given requirements on a sequential or-graph (Russian). In: V. P. Tanaev (Ed.): Algorithms and programs for the solution of extremal problems and related questions, 54-61. Minsk: Inst. Tekh. Kibernetiki AN BSSR. (1982).

NOVO 1986-1 Novozhilova, L. M.
Ordered sorting, with a successive decrease in the size of the problem, of enumeration problems on graphs (Russian). In: Mathematical Models of Complex Systems. Reliability and Processing of Information, 79-81, 177-178. Voprosy Mekh. Protsess. Upravl. 9, Leningrad Univ. (1986).

NOWI 1987-1 Nowicki, E.; Smutnicki, C.
On lower bounds on the minimum maximum lateness on one machine subject to release date. Opsearch 24, 106-110. (1987).

NOWIA 1985-1 Nowicki, E.; Zdrzalka, S.
Two-machine flow-shop scheduling problem with controllable job processing times (Polish). Zeszyty Naukowe Akademii Gorniczo-Hutniczej Im. Stanislawa Staszica 1064, Automatyka 39, 161-169. (1985).

NOWIA 1986-1 Nowicki, E.; Zdrzalka, S.
A note on minimizing maximum lateness in a one-machine sequencing problem with release dates. European Journal of Operational Research 23, 266-267. (1986).

NOWIB 1984-1 Nowicki, T.; Stanczak, W.
An algorithmic approach to partitioning a set of elements due to their similarity. in: Data Analysis and Informatics, III (Versailles, 1983), 375-383. Amsterdam-New York: North-Holland. (1984).

NTAF 1984-1 Ntafos, S.; Gonzalez, T.
On the computational complexity of path cover problems. Journal of Computer and System Sciences 29, 225-242. (1984).

NUDE 1983-1 Nudelman, G. A.
An algorithm for solution of a class of location problems (Russian). In: Numerical Methods and Optimization Problems, 115-124. Tomsk. Gos. Univ. (1983).

NUNE 1983-1 Nunen, J. A. E. E. van; Beulens, A. J. M.; Benders, J. F.
On using the linear programming relaxation of assignment type mixed integer problems. Delft: Graduate School of Management, Working Paper. (1983).

NUNE 1984-1 Nunen, J. A. E. E. van; Beulens, A. J. M.; Benders, J. F.
On solving assignment type mixed integer linear programming problems within decision support systems. Wissenschaftliche Zeitschrift, Technische Hochschule Leipzig 8, 89-95. (1984).

NURI 1983-1 Nuriev, U. G.
A hybrid method for solving the multidimensional knapsack problem (Russian). Preprint 83-43, Akademiya Nauk Ukrainskoi SSR, Inst. Kibernet. (Kiev). (1983).

NURI 1986-1 Nuriev, U. G.
On the solution of the knapsack problem with guaranteed estimates (Russian). In: Mathematical Cybernetics and Applied Mathematics, No. VII, 65-69. Baku: "Elm". (1986).

NYGR 1987-1 Nygreen, B.
A possible way to reduce degeneracy in integer programming computations. Operations Research Letters 6, 47-52. (1987).

NYKO 1986-1 Nykowski, I.
Bicriterion linear transportation problem (Polish). Przeglad Statystyczny 33, 9-25. (1986).

NYU 1982-1 Nyu, V.
Covering of a set of words by circuits (Russian). Metody Diskretnogo Analiza, Akademkya Nauk SSSR Sibirsk (Novosibirsk) No. 38, 59-73. (1982).

NYU A 1983-1 Nyu, V.; Evdokimov, A. A.
Coverings of graphs by paths (Russian). Metody Diskretnogo Analiza (Novosibirsk) No. 40, 72-86, 101. (1983).

O'GR 1985-1 O'Grady, P. J.; Byrne, M. D.
A combined switching algorithm and linear decision rule approach to production planning. International Journal of Production Research 23, 285-296. (1985).

O'GRA 1985-1 O'Grady, P. J.; Harrison, C.
A general search sequencing rule for job shop sequencing. International Journal of Production Research 23, 961-974. (1985).

O'KE 1986-1 O'Kelly, M. E.
The location of interacting hub facilities. Transportation Science 20, 92-106. (1986).

O'KE 1987-1 O'Kelly, M. E.
A quadratic integer program for the location of interacting hub facilities. European Journal of Operational Research 32, 393-404. (1987).

O'NE 1984-1 O'Neill, R. P.; Heinkel, J.; Kurator, W. G.; Mariner-Volpe, B.; Stokes, R.; Tawshunsky, A.; Trapmann, W.
The gas analysis modeling system (GAMS). Operations Research Letters 3, 227-236. (1984).

O'RO 1985-1 O'Rourke, J.; Suri, S.; Booth, H.
Shortest paths on polyhedral surfaces. In: K. Mehlhorn (Ed.): STACS 85, Second Annual Symposium on Theoretical Aspects of Computer Science (Saarbrücken, January 1985), Lecture Notes in Computer Science Vol. 182, 243-254. Berlin: Springer-Verlag. (1985).

ODLY 1987-1 Odlyzko, A. M.; Wilf, H. S.
Bandwidths and profiles of trees. Journal of Combinatorial Theory 42B, 348-370. (1987).

ODON 1985-1 Odoni, A. R.
The "third generation" of vehicle routing and scheduling problems: two illustrative examples. Ricerca Operativa 36, 7-56. (1985).

ODON 1987-1 Odoni, A. R.
Stochastic facility location problems. In: G. Andreatta, F. Mason, P. Serafini (Eds.): Advanced School on Stochastics in Combinatorial Optimization (CISM Udine 1986), 113-127. Singapore: World Scientific. (1987).

OELL 1987-1 Oellermann, O. R.
On the l-connectivity of a graph. Graphs and Combinatorics 3, 285-292. (1987).

OELLA 1984-1 Oellermann, O. R.; Swart, H. C.
Graphs for which all strong orientations are Eulerian. Expositiones Mathematicae 2, 183-184. (1984).

OFIT 1986-1 Ofitserov, D. V.
A multiextremal discrete optimization algorithm (Russian). Avtomatika i Vychislitelnaya Tekhnika (Minsk), No. 15, 58-60. (1986).

OGRY 1984-1 Ogryczak, W.
Branch-and-bound procedures for multicriteria integer programming (Polish). Sprawozdania Instytutu Informatyki Universytetu Warszawskiego 135, 36p. (1984).

OGUN 1987-1 Ogunyode, O. A.
Parallel algorithms for approximate edge coloring of simple graphs. In: K. V. Nori (Ed.): Foundations of Software Technology and Theoretical Computer Science, 207-225. Lecture Notes in Computer Science 287, Berlin, Heidelberg, New York: Springer. (1987).

OGUZ 1985-1 Oguz, O.
An integral transformation for integer programming problems. Operations Research Letters 4, 117-120. (1985).

OHSA 1987-1 Ohsawa, Y.; Suzuki, A.
Location-allocation problem of multi-person facility. Journal of the Operations Research Society of Japan 30, 368-395. (1987).

OHTA 1985-1 Ohtake, Y.; Nishida, N.
A branch-and-bound algorithm for 0-1 parametric mixed integer programming. Operations Research Letters 4, 41-45. (1985).

OHUC 1984-1 Ohuchi, A.; Kaji, I.
Lagrangian dual coordinatewise maximization algorithm for network transportation problems with quadratic costs. Networks 14, 515-530. (1984).

OKAM 1984-1 Okamura, H.
Multicommodity flows in graphs, II. Japanese Journal of Mathematics, New Series (Tokyo) 10, 99-116. (1984).

OKAM 1984-2 Okamura, H.
Path and edge-connectivity in graphs. Journal of Combinatorial Theory, Series B 37, 151-172. (1984).

OKAM 1984-3 Okamura, H.
Paths and edge-connectivity in graphs. In: Problems in Combinatorics, Proceedings of a Symposium held at the Research Institute for Mathematical Sciences, Kyoto Univ., 1-23. (1984).

OKAM 1985-1 Okamura, H.
Paths and edge-connectivity in graphs. II. In: J. Akiyama; Y. Ito; P. Kanemitsu; T. Kano; T. Mitsui; I. Shiokawa (Eds.): Number theory and combinatorics, Japan 1984, Proc. Conf. Tokyo, Okayama, Kyoto, 337-352. Singapore: World Scientific Publ. Co. (1985).

OKAM 1987-1 Okamura, H.
Paths and edge-connectivity in graphs III. Six-terminal k paths. Graphs and Combinatorics 3, 159-190. (1987).

OLAH 1981-1 Olah, G.
Path-matrices of trees (Hungarian). Matematikai Lapok (Budapest) No. 1-3, 9-12. (1981).

OLAR 1986-1 Olariu, S.
No antitwins in minimal imperfect graphs. In: Raspail-Burnside: Workshop on Perfect Graphs, Bellairs Research Institute, McGill University, Barbados, West Indies, 1986, 120-124. (1986).

OLARA 1983-1 Olaru, E.; Antohe, S.
Decompositions with respect to the graph-join operation. Analele Universitatii din Galati, Metalurgie, Fascicula II 1(6), 15-26. (1983).

OLARA 1984-1 Olaru, E.; Antohe, S.
Algebraic aspects of S-partitions of graphs (German). Analele Universitatii din Galati, Metalurgie, Fascicula II 2(7), 21-26. (1984).

OLARA 1985-1 Olaru, E.; Antohe, S.
Some results concerning graph decomposition. Analele Universitatii din Galati, Metalurgie, Fascicula II 3(8), 15-20. (1985).

OLARB 1984-1 Olaru, E.; Sachs, H.
Contributions to a characterization of the structure of perfect graphs. In: C. Berge, V. Chvatal (Eds.): Topics on perfect graphs, Annals of Discrete Mathematics Vol. 21, 121-144. Amsterdam: North-Holland. (1984).

OLEJ 1984-1 Olejnik, F.
On total matching numbers and total covering numbers for k-uniform hypergraphs. Mathematica Slovaca (Bratislava) 34, No.3, 319-328. (1984).

OLEY 1983-1 Oley, L. A.; Sjoquist, R. J.
Automatic reformulation of mixed and pure integer models to reduce solution time in APEX IV. SIGMAP Bulletin No. 32, 39-51. (1983).

OLHA 1984-1 Olhager, J.; Rapp, B.
Lot sizing and turnover rate. In: J. P. Brans (Ed.): Operational Research '84, 613-624. Amsterdam: North-Holland. (1984).

OLHA 1985-1 Olhager, J.; Rapp, B.
Balancing capacity and lot sizes. European Journal of Operational Research 19, 337-344. (1985).

OLIF 1985-1 Oliff, M. D.; Burch, E. E.
Multiproduct production scheduling at Owens-Corning Fiberglas. Interfaces 15, 25-34. (1985).

OLIN 1986-1 Olinger, W.
Interactive algorithm for scheduling under renewable resource constraints (Polish). Archiwum Automatyki i Telemechaniki 31, 201-207. (1986).

OLIV 1983-1 Olivo, P.
A mixed algorithm for the multidimensional assignment problem (Italian). Rivista di Matematica per le Scienze Economiche e Sociali (Milan) 6, 67-78. (1983).

ONAG 1982-1 Onaga, K.
Time-optimal transmission of time-variant particle flow streams through time-invariant networks. Electronics and Communications in Japan 65, No. 12, 20-29. (1982).

ONDA 1984-1 Ondas, J.
Algorithms for scheduling inhomogeneous multiprocessor computers. in: J. Miklosko, V. Kotov (Eds.): Algorithms, Software and Hardware of Parallel Computers, 259-271. Berlin: Springer-Verlag. (1984).

ONG 1984-1 Ong, H. L.
Probabilistic analysis of some travelling salesman heuristics. In: P. Thoft-Christensen (Ed.): System modelling and optimization (Copenhagen 1983), 631-640, Lecture Notes in Control and Information Science 59. Berlin, New-York: Springer. (1984).

ONG A 1984-1 Ong, H. L.; Magazine, M.; Wee, T. S.
Probabilistic analysis of bin packing heuristics. Operations Research 32, 983-998. (1984).

ONOZ 1984-1 Onozawa, A.; Inoue, M.
Some problems on principal partition of 2-matroids (Japanese). In: Graph theory and applications, Proc. Symp. Kyoto University 1984, 161-172. Kyoto University, Research Inst. for Mathematical Sciences. (1984).

ONUR 1987-1 Onur, L.; Fabrycky, W. J.
An input/output control system for the dynamic job shop. IIE Transactions 19, 88-97. (1987).

OPAS 1984-1 Opaswongkarn, K.; Guignard, M.
A multi-commodity location problem. Research Report. San Francisco (TIMS/ORSA Meeting, May 1984), 36 P. (1984).

ORDA 1986-1 Orda, A.; Rom, R.
Duplicate routing in distributed networks. In: E. Gafni, N. Santoro (Eds.): Distributed Algorithms on Graphs, 103-114. Ottawa: Carleton Univ. Press. (1986).

ORDI 1987-1 Ordin, V. V.
The method of Markov iterations for finding paths in a network of general form (Russian). Avtomatika i Telemekhanika, No.1, 183-185. (1987).

ORDM 1987-1 Ordman, E. T.
Dining philosophers and graph covering problems. Journal of Combinatorial Mathematics and Combinatorial Computing 1, 181-190. (1987).

ORLI 1980-1 Orlin, J. B.
Fast approximation schemes for combinatorial optimization problems (extended abstract). Working Paper. Cambridge: Mass. Inst. of Technology, Sloan School of Management, 10 p. (1980).

ORLI 1985-1 Orlin, J. B.
A finitely converging cutting plane technique. Operations Research Letters 4, 1-4. (1985).

ORLI 1985-2 Orlin, J. B.
Genuinely polynomial simplex and nonsimplex algorithms for the minimum cost flow problem. CWI Reports OS, R8504. Stichting Mathematisch Centrum, Centrum voor Wiskunde en Informatica (Amsterdam). (1985).

ORLI 1985-3 Orlin, J. B.
On the simplex algorithm for networks and generalized networks. Mathematical Programming Study 24, 166-178. (1985).

ORLI 1985-4 Orlin, J. B.
Some very easy knapsack/partition problems. Operations Research 33, 1154-1160. (1985).

ORLI 1985-5 Orlin, J. B.
The complexity of dynamic/periodic languages and optimization problems. Report OS-R8505. Amsterdam: Centre for Mathematics and Computer Science, 19 p. (1985).

ORLI 1986-1 Orlin, J. B.
A dual version of Tardos's algorithm for linear programming. Operations Research Letters 5, 221-226. (1986).

ORLIA 1985-1 Orlin, J. B.; Rothblum, U. G.
Computing optimal scalings by parametric network algorithms. Mathematical Programming 32, 1-10. (1985).

ORLIB 1987-1 Orlin, J. B.; Vandevate, J. H.
Solving the linear matroid parity problem as a sequence of matroid intersection problems. Working Paper. Cambridge: Mass. Inst. of Technology, 47 p. (1987).

ORZE 1984-1 Orzech, G.
Elementary properties of vector space graphs. Bulletin of the Australian Mathematical Society 30, 411-420. (1984).

OSMA 1986-1 Osman, M. S. A.; Ellaimony, E. E. M.
An algorithm for solving large scale transhipment problems. Indian Journal of Technology 24, 296-302. (1986).

OTIM 1985-1 Otiman, P.
Optimizing the location of crop fields. Economic Computation and Economic Cybernetics Studies and Research (Bucharest) 4, 79-82. (1985).

OUBI 1984-1 Oubina, L.; Zucchello, R.
A generalization of outerplanar graphs. Discrete Mathematics 51, 243-250. (1984).

OUDH 1985-1 Oudheusden, D. L. van; Khan, M. L. R.
Note on ILP flow formulation for subgraph selection. Belgian Journal of Operations Research, Statistics and Computer Science 25, 31-38. (1985).

OUDH 1987-1 Oudheusden, D. L. van; Khan, M. L. R.
Planning and development of rural road networks in developing countries. European Journal of Operational Research 32, 353-362. (1987).

OW 1985-1 Ow, P. S.
Focused scheduling in proportionate flowshops. Management Science 31, 852-869. (1985).

OW A 1985-1 Ow, P. S.; Morton, T. E.
An investigation of beam search for scheduling. Research Paper. Pittsburgh: Carnegie-Mellon Univ., Graduate School of Industrial Administration, 39 p. (1985).

OXLE 1984-1 Oxley, J. G.
On singleton 1-rounded sets of matroids. Journal of Combinatorial Theory, Series B 37, 189-197. (1984).

OXLE 1984-2 Oxley, J. G.
On the intersections of circuits and cocircuits in matroids. Combinatorica 4, 187-196. (1984).

OXLE 1985-1 Oxley, J. G.
Proof of a conjecture of Kahn for non-binary matroids. Combinatorica 5, 343-346. (1985).

OXLE 1986-1 Oxley, J. G.
On matroids with no 5-point-line-minor. Baton Rouge: Louisiana State Univ., Mathematics Dept., 10p. (1986).

OXLE 1986-2 Oxley, J. G.
On ternary transversal matroids. Discrete Mathematics 62, 71-84. (1986).

OXLE 1986-3 Oxley, J. G.
On the matroids representable over GF(4). Journal of Combinatorial Theory, Series B 41, 250-252. (1986).

OXLE 1987-1 Oxley, J. G.
A characterization of the ternary matroids with no M(K(4))-minor. Journal of Combinatorial Theory 42B, 212-249. (1987).

OXLE 1987-2 Oxley, J. G.
On nonbinary 3-connected matroids. Transactions of the AMS 300, 663-679. (1987).

OXLE 1987-3 Oxley, J. G.
The binary matroids with no 4-wheel minor. Transactions of the AMS 301, 63-75. (1987).

OXLEA 1987-1 Oxley, J. G.; Reid, T. J.
The smallest rounded sets of binary matroids. Working Paper. Baton Rouge: Louisiana State Univ., Dept. of Mathematics, 17 p. (1987).

OYAM 1987-1 Oyama, T.; Han-I, S.-J.
Application of discrete optimization techniques to train scheduling problems. Asia-Pacific Journal of Operational Research 4, 158-186. (1987).

OZAW 1987-1 Ozawa, T.; Kajitani, Y.; Ueno, S.
The principal partition of vertex-weighted graphs and its applications. In: D. P. Johnson, T. Nishizeki, A. Nozaki, H. P. Wilf (Eds.): Discrete Algorithms and Complexity - Proc. Japan-U.S. Joint Seminar, Kyoto 1986, 5-33. Perspectives in Computing Vol. 15, Boston: Academic Press. (1987).

PADB 1984-1 Padberg, M. W.
A characterization of perfect matrices. In: C. Berge, V. Chvatal (Eds.): Topics on perfect graphs, Annals of Discrete Mathematics Vol. 21, 169-178. Amsterdam: North-Holland. (1984).

PADBA 1985-1 Padberg, M. W.; Grötschel, M.
Polyhedral computations. In: E. L. Lawler, J. K. Lenstra, A. H. G. Rinnooy Kan, D. B. Shmoys (Eds.): The Traveling Salesman Problem, 307-360. New York: Wiley. (1985).

PADBB 1987-1 Padberg, M. W.; Rinaldi, G.
Optimization of a 532-city symmetric traveling salesman problem by branch and cut. Operations Research Letters 6, 1-8. (1987).

PADBC 1985-1 Padberg, M. W.; Vanroy, T. J.; Wolsey, L. A.
Valid linear inequalities for fixed charge problems. Operations Research 33, 842-861. (1985).

PAES 1988-1 Paessens, H.
Fahrzeugeinsatzprobleme in Städtereinigungsunternehmen. In: H. Schellhaas, P. van Beek, H. Isermann, R. Schmidt, M. Zijlstra (Eds.): Operations Research Proceedings 1987 (DGOR/ NSOR), 290-291. Berlin: Springer. (1988).

PAIB 1986-1 Paiboon, C.-P.; Pojana, P.-P.; Rosenthal, R. E.
Optimal ship routing and personnel assignment for naval recruitment in Thailand. Interfaces 16, No. 4, 47-52. (1986).

PAIX 1985-1 Paixao, J. P.; Almeida, M. T.; Coelho, J. D.
Problems in graph theory and their relation with set covering and partitioning problems (Portuguese). Portugaliae Mathematica (Lisboa) 39, 299-329 (1980). (1985).

PAL 1984-1 Pal, B. B.; Basu, M.
Optimal solution of a transportation problem by goal programming technique. Bulletin of the Clacutta Mathematical Society 76, 342-348. (1984).

PALE 1987-1 Palekar, U. S.; Karwan, M. H.; Zionts, S.
A branch-and-bound method for the fixed charge transportation problem. Working Paper. Urbana: Univ. of Illinois, Dept. of Mechanical and Industrial Engineering, 35 p. (1987).

PALE 1987-2 Palekar, U. S.; Karwan, M. H.; Zionts, S.
Duality-based bounds for the fixed charge transportation problem. Working Paper. Urbana: Univ. of Illinois, Dept. of Mechanical and Industrial Engineering, 27 p. (1987).

PALEA 1984-1 Palem, K. V.; Fussel, D. S.
The complexity of scheduling systems of non-identical processors. Congressus Numerantium 45, 13-18. (1984).

PALEB 1987-1 Paletta, G.
A heuristic algorithm for a pseudo-cyclic delivery problem under window constraints. European Journal of Operational Research 31, 71-77. (1987).

PALK 1983-1 Palka, Z.
A new algorithm calculating the distance between binary arborescences. Zastosowania Matematyki (Warsaw) 18, 151-157. (1983).

PALU 1985-1 Palubetskis, G. S.
Subgraph polytopes (Russian). Litovskij Matematicheskij Sbornik (Vil'nyus) 25, No. 3, 147-162. (1985).

PAN 1985-1 Pan, Z. T.; Zhao, S. C.
Graph-theoretic proof of an inequality (Chinese). Heilongjiang Daxue Ziran Kexue Xuebao, 31-34. (1985).

PAND 1983-1 Pandit, S. N. N.; Reddy, B. K.
A maximum flow problem with intermediate node requirements. Proceedings of the Indian Academy of Sciences, Mathematical Sciences, 92, 171-190. (1983).

PANI 1984-1 Panishev, A. V.
An algorithm for optimization of treelike structures (Russian). In: Problems of bionics, No. 32, 96-99. Kharkov: Vishcha Shkola. (1984).

PANI 1987-1 Panishev, A. V.
The dichotomy method in the three-machine problem. Izvestiya Akademii Nauk SSSR, Tekhnicheskaya Kibernetika (Moscow), No. 4, 181-186; translated as Soviet J. Comput. Systems Sci. 25, No. 1, 181-187. (1987).

PANY 1985-1 Panyukov, A. V.; Shtraus, V. A.
Nonatomic problems of location of extended objects (Russian). Avtomatika i Telemekhanika (Moscow) 1985, No. 11, 54-61. (1985).

PAOL 1984-1 Paoli, M.
Powers of connected graphs and hamiltonicity. Discrete Mathematics 52, 91-100. (1984).

PAOL 1986-1 Paoli, M.
Some Hamiltonian properties of the powers of a graph. Congressus Numerantium 54, 173-180. (1986).

PAOLA 1984-1 Paoli, M.; Trotter, W. T. Jr.
The maximum number of edges in a strongly connected oriented graph without long directed paths. Congressus Numerantium 45, 19-26. (1984).

PAPA 1985-1 Papadimitriou, C. H.
An algorithm for shortest-path motion in three dimensions. Information Processing Letters 20, 259-264. (1985).

PAPA 1985-2 Papadimitriou, C. H.
Computational complexity. In: M. O'hEigeartaigh, J. K. Lenstra, A. H. G. Rinnooy Kan (Eds.): Combinatorial Optimization - Annotated Bibliographies, 39-51. Chichester: Wiley. (1985).

PAPA 1986-1 Papadimitriou, C. H.
Shortest-path motion. In: K. V. Nori: Foundations of Software, Technology and Theoretical Computer Science, Lecture Notes in Computer Science 241, 144-153. Berlin, Heidelberg, New York: Springer-Verlag. (1986).

PAPAA 1987-1 Papadimitriou, C. H.; Tsitsiklis, J. N.
On stochastic scheduling with in-tree precedence constraints. SIAM Journal on Computing 16, 1-6. (1987).

PAPAB 1984-1 Papadimitriou, C. H.; Vazirani, U. V.
On two geometric problems related to the traveling salesman problem. Journal of Algorithms 5, 231-246. (1984).

PAPAC 1985-1 Papadimitriou, C. H.; Wolfe, D.
The complexity of facets resolved. In: 26th Annual Symp. on Foundations of Computer Science, Portland, Oregon 1985, Proceedings, 74-78. IEEE. (1985).

PAPAD 1986-1 Papadimitriou, C. H.; Yannakakis, M.
A note on succinct representations of graphs. Information and Control 71, No. 3, 181-185. (1986).

PAPAE 1986-1 Papakostantinou, N.
The complexity of the travelling repairman problem. Informatique Theorique et Applications 20, 79-88. (1986).

PAPAF 1986-1 Papathanassiu, B.; Tsouros, C.
Determining the optimal place and capacity of spare parts warehouses in the construction industry. European Journal of Operational Research 27, 91-94. (1986).

PARA 1984-1 Paramonov, A. V.; Popov, Y. V.; Stepanenko, I. D.; Frenkin, B. R.
The use of a distributive problem for determining the optimal schedules for processing stocks of sugar beets (Russian). In: Algorithms for Solving Network Problems, 96-104. Moscow: Akad. Nauk SSSR, Tsentral. Ekonom.-Mat. Inst. (1984).

PARD 1987-1 Pardalos, P. M.; Rosen, J. B.
Constrained global optimization: Algorithms and applications. Berlin Heidelberg New York: Springer. (1987). (Lecture Notes in Computer Science, 268)

PARK 1984-1 Park, K. S.; Yun, D. K.
A stepwise partial enumeration algorithm for the economic lot scheduling problem. IIE Transactions 16, 363-370. (1984).

PARK 1985-1 Park, K. S.; Yun, D. K.
Optimal scheduling of periodic activities. Operations Research 33, 690-695. (1985).

PARKA 1986-1 Parkhomenko, N. V.; Korneiko, V. G.
An attempt at solving, in an automatic subject-indexed control system, an allocation problem in a discrete formulation (Russian). Ehkonomika i Matematicheskie Metody 22, 1105-1109. (1986).

PASC 1987-1 Pasche, C.
Flot a cout convexe lineaire par morceaux. RAIRO Recherche Operationelle 21, 205-218. (1987).

PASCA 1986-1 Pascoletti, K.-H.
Diophantische Systeme und Lösungsmethoden zur Bestimmung aller Invarianten in Petri-Netzen. München, Wien: Oldenbourg. (1986).

PATI 1984-1 Patil, H. P.
A note on the edge-arboricity of maximal k-degenerate graphs. Malaysian Mathematical Society, Bulletin, Second Series (Kuala Lumpur) (2) 7, 57-59. (1984).

PATK 1985-1 Patkar, V.; Stancu-Minasian, I. M.
Duality in disjunctive linear fractional programming. European Journal of Operational Research 21, 101-105. (1985).

PAUL 1987-1 Paulraja, P.
Counterexample to a conjecture on Hamilton cycles. Discrete Mathematics 64, 99-100. (1987).

PAUL 1987-2 Paulraja, P.
On graphs admitting spanning Eulerian subgraphs. Ars Combinatoria 24, 57-66. (1987).

PAUM 1986-1 Paumelle, I.
Un indicateur informatique d'itineraire pour les reseaux d'autobus. RAIRO - Recherche Operationelle 20, 199-212. (1986).

PAVL 1985-1 Pavlov, A. A.
Analysis of complexity estimates of exact algorithms for solving the general linear integer programming problem. Soviet Journal of Automation and Information Sciences 18, No.5, 40-46; translation from Avtomatika (1985), No.5, 42-48. (1985).

PAVLA 1985-1 Pavlov, A. A.; Gershgorin, A. E.
Reducing a general linear integer programming problem to a knapsack problem. Soviet Journal of Automation and Information Sciences 18, No. 1, 53-57; translation from: Avtomatika 1985, No. 1, 52-56. (1985).

PAWA 1987-1 Pawagi, S.
Maximum weight independent set in trees. BIT 27, 170-180. (1987).

PAWAA 1986-1 Pawagi, S.; Ramakrishnan, I. V.
An O(log n) algorithm for parallel update of minimum spanning trees. Information Processing Letters 22, 223-230. (1986).

PAZ 1986-1 Paz, A.
A duality property for the set of all feasible solutions to an integer program. Discrete Applied Mathematics 14, 57-66. (1986).

PAZZ 1984-1 Pazzoli, L.
Hyperplanes and circuits of matroids on partially ordered sets (Italian). in: A. Barlotti, M. Marchi, G. Tallini (Eds.): Proceedings of the conference on combinatorial and incidence geometry: principles and applications (La Mendola, 1982), 539-551. Milan: Rendiconti del Seminario Matematico di Brescia 7. (1984).

PEAR 1987-1 Pearn, W. L.; Assad, A. A.; Golden, B. L.
Transforming arc routing into node routing problems. Computers and Operations Research 14, 285-288. (1987).

PEDE 1987-1 Pederzoli, G.; Sancho, N. G. F.
A shortest path routing problem with resource allocation. Journal of Mathematical Analysis and Applications 124, 33-42. (1987).

PEEL 1985-1 Peele, R.
Finite partition sublattice representations for some geometric lattices. In: L. Lovasz, A. Recski (Eds.): Matroid Theory, 299-310. Coll. Math. Soc. J. Bolyai Vol. 40. Amsterdam: North-Holland. (1985).

PEEM 1984-1 Peemöller, J.
Hamiltonsche Graphen - notwendige Bedingungen und Algorithmen. Fachbereich Wirtschafts- und Organisationswissenschaften der Hochschule der Bundeswehr Hamburg. (1984).

PEEM 1985-1 Peemöller, J.
Necessary conditions for Hamiltonian split graphs. Discrete Mathematics 54, 39-48. (1985).

PEEM 1986-1 Peemöller, J.
Numerical experiences with graph coloring algorithms. European Journal of Operational Research 24, 146-151. (1986).

PELE 1987-1 Peled, U. N.
The threshold measure of a graph. RUTCOR Research Report 9-87. New Brunswick: Rutgers Univ., Rutgers Center for Operations Research, 22 p. (1987).

PELEA 1985-1 Peled, U. N.; Simeone, B.
Polynominal-time algorithms for regular set-covering and threshold synthesis. Discrete Applied Mathematics 12, 57-70. (1985).

PELEB 1987-1 Peled, U. N.; Srinivasan, M. K.
The polytope of degree sequences. Discussion Paper 751. Evanston: Northwestern Univ., Center for Math. Studies in Economics and Management Science, 50 p. (1987).

PELEC 1987-1 Peleg, D.; Simons, B.
On fault tolerant routings in general networks. Information and Computation 74, No. 1, 33-49. (1987).

PELED 1987-1 Peleg, D.; Upfal, E.
Constructing disjoint paths on expander graphs. In: Proc. Nineteenth Annual ACM Symp. on Theory of Computing, New York City 1987, 264-273. ACM. (1987).

PELED 1987-2 Peleg, D.; Upfal, E.
The generalized packet routing problem. Theoretical Computer Science 53, 281-294. (1987).

PELEE 1983-1 Pelegrin, B.
The expected value criterion in an industrial location model (Spanish). Trabajos de Estadistica y de Investigacion Operativa (Valencia) 34, 61-67. (1983).

PELEE 1985-1 Pelegrin, B.
Minimax location: the problem of a center (Spanish). In: Proceedings of the tenth Spanish-Portuguese conference of mathematics, VI (Murcia 1985), 45-54. Murcia: Univ. (1985).

PELEF 1985-1 Pelegrin, B.; Michelot, C.; Plastria, F.
On the uniqueness of optimal solutions in continuous location theory. European Journal of Operational Research 20, 327-331. (1985).

PELI 1984-1 Pelikan, J.
Container transportation problem (Czechoslovakian). Review of Econometrics 20, 131-142. (1984).

PELI 1985-1 Pelikan, J.
Container transportation problem. In: F. Nozicka (Ed.): Optimization - Theory, Methods, Applications, I, International Conference Marianske Lazne 1985, 199-207. Prauge: Dum Techniky CSVTS. (1985).

PELT 1987-1 Peltsverger, B. V.; Khavronin, O. V.
A decompositional approach to the problem of synthesizing a transport network with a fuzzy use period (Russian). Izvestiya Akademii Nauk SSSR, Tekhnicheskaya Kibernetika (Moscow), No. 4, 32-36; translated as Soviet J. Comput. Systems Sci. 25, No. 1, 29-34. (1987).

PERE 1985-1 Perez Prados, A.
Obtaining extremal paths in an associated graph (Spanish). In: Proceedings of the tenth Spanish-Portuguese conference on mathematics, VI (Murcia 1985), 35-44. Murcia: Univ. (1985).

PERF 1986-1 Perfilova, V. N.
On the number of transversal matroids. Moscow University Mathematics Bulletin 41, No.1, 16-19; translation from Vestnik Moskovskogo Universiteta, Seriya I (1986), No.1, 13-15. (1986).

PERK 1984-1 Perko, A.
Storing paths in kth shortest path problems. In: U. Pape (Ed.): Proc. WG '84: Internat. Workshop on Graphtheoretic Concepts in Computer Science, Berlin 1984, 301-306. Linz: Trauner. (1984).

PERL 1987-1 Perl, J.
The multidepot routing allocation problem. American Journal of Mathematical and Management Sciences 7, 7-34. (1987).

PERLA 1985-1 Perl, J.; Daskin, M. S.
A warehouse location-routing problem. Transportation Research 19B, 381-396. (1985).

PERLB 1984-1 Perl, Y.
Optimum split trees. Journal of Algorithms 5, 367-374. (1984).

PERM 1984-1 Perminov, E. A.
Imbeddings of graphs in rigid graphs (Russian). Ural'skii Gosudarstvennyi Universitet Matematicheskie Zapiski 13, No. 4, 104-115. (1984).

PERO 1984-1 Peroche, B.
Products and path numbers I: the cartesian sum. Ars Combinatoria 17, 147-166. (1984).

PERO 1985-1 Peroche, B.
Complexite de l'arboricite lineaire d'une graph II. RAIRO Recherche Operationelle 19, 293-300. (1985).

PEROA 1986-1 Peroche, B.; Sadi, B.
Recouvrement et partition en chaines des aretes d'un graphe cubique. RAIRO Recherche Operationnelle 20, 163-170. (1986).

PERS 1982-1 Pershin, O. Y.
General problem of the optimal design of multi-component networks (Russian). In: B. N. Petrov, M. V. Meerov (Ed.): Investigations in the theory of multiply connected systems (Russian), 100-105. Moscow: Izdatel'stvo Nauka. (1982).

PERS 1985-1 Pershin, O. Y.
One class of problems for synthesis of the optimal networks (Russian). Kibernetika No. 5, 76-80. (1985).

PERS 1986-1 Pershin, O. Y.
A class of extremal combinatorial problems for multicomponent network design. Automation and Remote Control 47, 262-269; translation from Avtomatika i Telemekhanika (1986), No.2, 127-135. (1986).

PERU 1986-1 Perusch, M.
Simulated annealing applied to a single machine scheduling problem with sequence dependent set up times and due dates. Report 86-84. Graz: Univ. Graz, Institut für Mathematik, 14 p. (1986).

PESA 1987-1 Pesamosca, G.
On the solution of the generalized Steiner problem by the subgradient method. Publicazioni 142. Pisa: Univ. di Pisa, Dipartimento di Matematica, 23 p. (1987).

PETE 1984-1 Peteanu, V.
Algebraical structures in the network theory. L'Analyse Numerique et la Theory de L'Approximation (Budapest) 13, No. 1, 57-63. (1984).

PETEA 1984-1 Peteanu, V.; Tigan, S.
On some discrete fractional max-min problems. Application to max-min problems in graphs. Mathematica - Revue d'Analyse Numerique et de Theorie de l'Approximation (Bucarest) 13, 167-173. (1984).

PETEB 1987-1 Peter-Orth, C.
On solving computationally difficult set covering problems. Ars Combinatoria 24A, 71-96. (1987).

PETEC 1987-1 Peters, J. G.; Rudolph, L.
Parallel approximation schemes for subset sum and knapsack problems. Acta Informatica 24, 417-432. (1987).

PETED 1986-1 Peters, K.; Laskar, R.; Hedetniemi, S.
Maximal/minimal connectivity in graphs. Ars Combinatoria 21, 59-70. (1986).

PETEE 1986-1 Peters, K.; Laskar, R.; Hedetniemi, S. T.
On the strength of a graph. Congressus Numerantium 54, 247-262. (1986).

PETK 1986-1 Petkov, M. G.
The union of two transportation flows - additional investigations (Bulgarian). Godishnik na Visshite Uchebni Zavedniya, Prilozhna Matematika (Sofia) 21, No.1, 201-208. (1986).

PETK 1986-2 Petkov, M. G.
Transportation flow passing through a bottleneck (Bulgarian). Godishnik na Visshite Uchebni Zavedniya, Prilozhna Matematika (Sofia) 21, No.4, 225-234. (1986).

PETKA 1984-1 Petkovsek, M.
Comparability line graphs. In: D. Cvetkovic, I. Gutman, T. Pisanski, R. Tosic (Eds.): Graph Theory (Novi Sad, 1983), 231-244. Univ. Novi Sad. (1984).

PETR 1986-1 Petrov, K.
On strongly regular graphs (Bulgarian). Annuaire l'Universite de Sofia, Fac. Phys. 71, No. 2, 75-78. (1986).

PETRA 1985-1 Petrova, G. L.
Perfect matchings in n-dimensional parallelepipeds (Russian). In: Graphen und Netzwerke - Theorie und Anwendungen, 30. Int. Wiss. Kolloq., Ilmenau 1985, 5, 83-84. (1985).

PETRB 1986-1 Petrova, V. A.
Application of combinatorial methods for estimating the reliability of flow systems (Russian). In: Mathematical Models of Complex Systems. Reliability and Processing of Information, 51-57, 175-176. Voprosy Mekh. Protsess. Upravl. 9, Leningrad Univ. (1986).

PETRC 1984-1 Petrovic, V.
Antidirected Hamiltonian circuits in tournaments. In: D. Cvetkovic, I. Gutman, T. Pisanski, R. Tosic (Eds.): Graph Theory, Proc. Fourth Yugoslav Seminar on Graph Theory, Novi Sad 1983, 259-269. Novi Sad: University of Novi Sad. (1984).

PEZZ 1984-1 Pezzoli, L.
Hyperplanes and circuits of matroids on partially ordered sets (Italian). Rendiconti del Seminario Matematico die Brescia (Milano) 7, 539-551. (1984).

PFAF 1984-1 Pfaff, J.; Laskar, R.; Hedetniemi, S. T.
Linear algorithms for independent domination and total domination in series-parallel graphs. Congressus Numerantium 45, 71-82. (1984).

PHAN 1986-1 Phan Dinh Dieu; Le Cong Thanh; Le Tuan Hoa
Average polynomial time complexity of some NP-complete problems. Theoretical Computer Science 46, 219-238. (1986).

PHIL 1986-1 Phillips, N. V.
A new variant of the partitioning shortest path algorithm. Discrete Applied Mathematics 13, 207-212. (1986).

PHIL 1987-1 Phillips, N. V.
A weigthing function for pre-emptive multicriteria assignment problems. Journal of the Operational Research Society 38, 797-802. (1987).

PIAZ 1984-1 Piazza, B. L.
On Hamiltonian cycles in cycle permutation graphs. Congressus Numerantium 45, 83-90. (1984).

PIAZA 1985-1 Piazza, B. L.; Stueckle, S.
On the cut frequency vector of permutation graphs. Congressus Numerantium 49, 143-146. (1985).

PICA 1983-1 Pica, G.
Generalized chromatic numbers of some graphs. Rivista di Matematica della Univ. di Parma, Serie IV/9, 259-264. (1983).

PICAA 1985-1 Pica, G.; Pisanski, T.; Shawe-Taylor, J.
Generalized chromatic numbers of some graphs. II. Rivista di Matematica della Universita di Parma, Serie IV 11, 375-379. (1985).

PICAB 1982-1 Pica, G.; Pisanski, T.; Shawe-Taylor, J.; Ventre, A. G. S.
The permutation girth of a graph. Atti del Seminario Matematico e Fisico dell'Universita di Modena 31 297-301. (1982).

PICO 1984-1 Picone, C. J.; Wilhelm, W. E.
A perturbation scheme to improve Hillier's solution to the facilities layout problem. Management Science 30, 1238-1249. (1984).

PICO 1986-1 Picone, C. J.; Wilhelm, W. E.
"A perturbation scheme to improve Hillier's solution to the facilities layout problem": a clarification. Management Science 32, 253-254. (1986).

PIEH 1985-1 Piehler, J.
Über die irreduziblen Punkte des Eckenpolyeders. Optimization 16, 367-372. (1985).

PIEH 1987-1 Piehler, J.
Parallelschnitte in der ganzzahligen Optimierung. Optimization 18, 387-392. (1987).

PILC 1987-1 Pilcher, M. G.; Rardin, R. L.
A random cut generator for symmetric traveling salesman problems with known optimal solutions. CC-87-4. West Lafayette: Purdue Univ., Inst. for Interdisciplinary Engineering Studies, 28 p. (1987).

PILI 1984-1 Pilipchuk, L. A.
An algorithm for solving a two-commodity transport problem with supplementary matrix of constraints by a direct method (Russian). Vestnik Belorusskogo Gosudarstvennogo Universiteta Imeni V.I. Lenina, Seriya I, No.1, 42-46. (1984).

PILI 1984-2 Pilipchuk, L. A.
On dual methods for solving a two-commodity transportation network problem (Russian). Vestnik Belorusskogo Gosudarstvennogo Universiteta Imeni V.I. Lenina (Minsk), Seriya I, No.3, 52-54. (1984).

PILI 1986-1 Pilipchuk, L. A.
Algorithms for solving an inhomogeneous distributive network problem (Russian). Izvestiya Akademii Nauk BSSR, Seriya Fiziko-Matematicheskikh Nauk, No.3, 37-43. (1986).

PINE 1984-1 Pinedo, M.
A note on the flow time and the number of tardy jobs in stochastic open shops. European Journal of Operational Research 18, 81-85. (1984).

PINE 1984-2 Pinedo, M.
Optimal policies in stochastic shop scheduling. Annals of Operations Research, Vol. 1, 305-330. Basel: Baltzer. (1984).

PINE 1985-1 Pinedo, M.
A note on stochastic shop models in which jobs have the same processing requirements on each machine. Management Science 31, 840-846. (1985).

PINEA 1985-1 Pinedo, M.; Schechner, Z.
Inequalities and bounds for the scheduling of stochastic jobs on parallel machines. Journal of Applied Probability 22, 739-744. (1985).

PINEB 1985-1 Pinedo, M.; Weiss, G.
Scheduling jobs with exponentially distributed processing times and intree precedence constraints on two parallel machines. Operations Research 33, 1381-1388. (1985).

PINEC 1986-1 Pinedo, M.; Wie, S.-H.
Inequalities for stochastic flow shops and job shops. Applied Stochastic Models and Data Analysis 2, 61-69. (1986).

PINO 1982-1 Pino Mejias, J. L.
Solving scheduling problems by means of dynamic programming (Spanish). In: Proceedings of the ninth Spanish-Portuguese Conference on Mathematics, 2 (Salamanca, 1982), 707-710. Acta Salmanticensia, Ciencias 46, Univ. Salamanca. (1982).

PIPP 1982-1 Pippenger, N.
Telephone switching networks. In: P. A. Burr (Ed.): The Mathematics of Networks, 101-134. Proc. Symp. in Appl. Math. Vol. 26, American Math. Society. (1982).

PIRK 1986-1 Pirkul, H.
An integer programming model for the allocation of databases in a distributed computer system. European Journal of Operational Research 26, 401-411. (1986).

PIRK 1987-1 Pirkul, H.
A heuristic solution procedure for the multiconstraint zero-one knapsack problem. Naval Research Logistics 34, 161-172. (1987).

PIRK 1987-2 Pirkul, H.
Efficient algorithms for the capacitated concentrator location problem. Computers and Operations Research 14, 197-208. (1987).

PIRKA 1986-1 Pirkul, H.; Narasimhan, S.
Efficient algorithms for the multiconstraint general knapsack problem. IIE Transactions 18, 195-203. (1986).

PISA 1983-1 Pisanski, T.; Shawe-Taylor, J.; Mohar, B.
1-factorization of the composition of regular graphs. Institut Mathematique, Publications (Beograd), Nouvelle Serie 33 (47), 193-196. (1983).

PLAI 1984-1 Plaisted, D. A.
Heuristic matching for graphs satisfying the triangle inequality. Journal of Algorithms 5, 163-179. (1984).

PLAN 1987-1 Plante, R. D.; Lowe, T. J.; Chandrasekaran, R.
The product matrix traveling salesman problem: An application and solution heuristic. Operations Research 35, 772-783. (1987).

PLAS 1985-1 Plasil, J.
Optimal organizing of flows in transportation network. In: F. Nozicka (Ed.): Optimization - Theory, Methods, Applications, I, International Conference Marianske Lazne 1985, 208-220. Prauge: Dum Techniky CSVTS. (1985).

PLASA 1984-1 Plastria, F.
Localization in single facility location. European Journal of Operational Research 18, 215-219. (1984).

PLASA 1987-1 Plastria, F.
Solving general continuous single facility location problems by cutting planes. European Journal of Operational Research 29, 98-110. (1987).

PLAT 1985-1 Plateau, G.; Elkihel, M.
A hybrid method for the 0-1 knapsack problem. In: P. Brucker, R. Pauly (Eds.): Proc. IX. Symp. on Operations Research, Osnabrück 1984. Methods of Operations Research 49, 277-294. (1985).

PLATA 1984-1 Plateau, G.; Guerch, M. T.
Aggregation of equalities in integer programming. In: P. Thoft-Christensen (Ed.): System Modelling and Optimization, 183-192. Lecture Notes in Control and Information Sciences 59, Berlin: Springer. (1984).

PLATB 1984-1 Platzman, L. K.; Bartholdi, J. J. III
Spacefilling curves and the planar travelling salesman problem (revised). Research Report. Atlanta: Georgia Inst. of Technology, School of Industrial and Systems Engineering, 19 p. (1984).

PLES 1985-1 Plesnik, J.
Remarks on diameters of orientations of graphs. Acta Mathematica Universitatis Comenianae 46/47, 225-236. (1985).

PLES 1986-1 Plesnik, J.
Bad examples of the metric traveling salesman problem for the 2-change heuristic. Acta Mathematica Universitatis Comenianae 48/49, 203-208. (1986).

PLES 1986-2 Plesnik, J.
On minimal graphs of diameter 2 with every edge in a 3-cycle. Mathematica Slovaca 36, 145-149. (1986).

PLES 1987-1 Plesnik, J.
A heuristic for the p-center problem in graphs. Discrete Applied Mathematics 17, 263-268. (1987).

PLES 1987-2 Plesnik, J.
On the interchange heuristic for locating centers and medians in a graph. Mathematica Slovaca 37, 209-216. (1987).

PLON 1984-1 Plonka, J.
Graphs of parital partitions. Demonstratio Mathematica (Warsaw) 17, 219-232. (1984).

PLUM 1985-1 Plummer, M. D.
Some recent results on cycle traversability in graphs. In: M. Rosenfeld, J. Zaks (Eds.): Convexity and Graph Theory. Annals of Discrete Mathematics Vol. 20, 255-262. Amsterdam: North-Holland. (1985).

PLUM 1986-1 Plummer, M. D.
Matching extension in bipartite graphs. Congressus Numerantium 54, 245-258. (1986).

POCH 1986-1 Pochet, Y.
Valid inequalities and separation for capacitated economic lot sizing. CORE Discussion Paper, 8649. Louvain-la-Neuve: Center for Operations Research & Econometrics, 13 p. (1986).

POCHA 1986-1 Pochet, Y.; Wolsey, L. A.
Lot-size models with backlogging: strong reformulations and cutting planes. CORE Discussion Paper, 8618. Louvain-la-Neuve: Center for Operations Research & Econometrics, 22 p. (1986).

POGU 1986-1 Pögün, G.; Satir, A.
Alternative bus scheduling policies for an exclusive bus lane. Transportation Research 20-A, 437-446. (1986).

POGUA 1986-1 Poguntke, W.
Order-theoretic aspects of scheduling. In: Combinatorics and Ordered Sets (Arcata, CA, 1985), 1-32. Contemporary Mathematics 57, Amer. Math. Soc. (1986).

POHL 1986-1 Pöhler, K.
An application of duality for the solution of special vector optimization problems (German). Seminarber., Humboldt-Univ. Berlin, Sekt. Math. 85, 79-90. (1986).

POIA 1986-1 Poiaga, L.
Resolution by reduction of a network, by means of its assimilation to a semi-Markov process (Italian). Ricerca Operativa 16, No. 37, 7-28. (1986).

POLA 1985-1 Polansky, O. E.
Die Entwicklung des mü-Polynoms eines schlichten Graphen, entsprechend seiner Zerlegung an einem Knoten. Match 18, 71-81. (1985).

POLAA 1983-1 Polat, N.
Couplages dans les systemes combinatoires flous. I. Bulletin pour les Sous Ensembles Flous et leurs Application BUSEFAL (Univ. Toulouse) 12, 14-23. (1983).

POLAA 1983-2 Polat, N.
Couplages dans les systemes combinatoires flous. II. Bulletin pour les Sous Ensembles Flous et leurs Application BUSEFAL (Univ. Toulouse) 12, 4-14. (1983).

POLI 1986-1 Politof, T.; Satyanarayana, A.
Network reliability and inner-four-cycle-free graphs. Mathematics of Operations Research 11, 484-505. (1986).

POLJ 1986-1 Poljak, S.; Chrobak, M.
On common edges in optimal solutions to traveling salesman and other optimization problems. KAM Series, 11. Praha: Charles Univ., Dept. of Applied Mathematics, 14 p. (1986).

POLJA 1984-1 Poljak, S.; Turzik, D.
Amalgamation of matroids along hypergraphs. In: A. Hajnal; L. Lovasz; V. T. Sos (Eds.): Finite and Infinite Sets, Vol. 2, 607-620. Coll. Math. Soc. J. Bolyai 37, Amsterdam: North-Holland. (1984).

POLJA 1984-2 Poljak, S.; Turzik, D.
Amalgamation over uniform matroids. Czechoslovak Mathematical Journal 34(109), 239-246. (1984).

POLJA 1986-1 Poljak, S.; Turzik, D.
A polynomial time heuristic for certain subgraph optimization problems with guaranteed worst case bound. Discrete Mathematics 58, 99-104. (1986).

POLJA 1987-1 Poljak, S.; Turzik, D.
On a facet of the balanced subgraph polytope. Casopis pro Pestovani Matematiky 112, 373-380. (1987).

POLJB 1987-1 Poljak, S.; Tuza, Z.
Maximum bipartite subgraphs of Kneser graphs. Graphs and Combinatorics 3, 191-200. (1987).

PONO 1982-1 Ponomarenko, T. Y.
The telephone center of a tree (Russian). In: Problems in group theory and homological algebra, 142-143. Yaroslav. Gos. Univ.: Yaroslavl'. (1982).

POPE 1983-1 Popescu, E.
Algorithme matriciel pour la determination du graphe minimal qui conserve la relation d'ordre donne d'un graphe d'ordre sans circuits. Buletinul Universitatii din Brasov, Seria C 25, 29-32. (1983).

POPEA 1983-1 Popescu, M. F.
Efficiency of implicit enumeration algorithms and rejection power of the tests. Revue Roumaine de Mathematiques Pures et Appliquees (Bucarest) 28, 875-892. (1983).

POPP 1985-1 Popp, W.
Zur Analyse strategischer Aspekte mit gemischt-ganzzahliger Programmierung. In: D. Ohse, A. C. Esprester, H.-U. Küpper, P. Stähly, H.Steckhan (Eds.): Operations Research Proceedings 1984, 53. Berlin: Springer-Verlag. (1985).

PORT 1986-1 Porteus, E. L.
Optimal lot sizing, process quality improvement and setup cost reduction. Operations Research 34, 137-144. (1986).

POSH 1985-1 Poshtaru, A. I.
Medians of planar graphs (Russian). In: Investigations in numerical methods and theoretical cybernetics, 89-92. Kishinev: Shtiintsa. (1985).

POSN 1985-1 Posner, M. E.
A sequencing problem with release dates and clustered jobs. Research Paper. New York: New York Univ., Graduate School of Business Admin., 16 p. (1985).

POSN 1985-2 Posner, M. E.
Minimizing weighted completion times with deadlines. Operations Research 33, 562-574. (1985).

POSNA 1985-1 Posner, M. E.; Szwarc, W.
Transshipment along a single road. Naval Research Logistics Quarterly 32, 363-372. (1985).

POTA 1984-1 Potapova, L. I.
On a procedure for solving a transportation problem with respect to a time criterion (Russian). In: K. V. Kim; B. V Cherkasskij (Eds.): Algorithms for Solving Network Problems, Collect. Artic., 33-41. Moskva: Tsentr. Ehkon.-Mat. Inst. (1984).

POTE 1980-1 Potemkina, A. V.
Solution of a problem of integer convex programming (Russian). in: Combinatorial-Algebraic Methods in Applied Mathematics, 139-160. Gorki: Gor'kovskii Gosudarstvennyi Universitet. (1980).

POTT 1985-1 Potts, C. N.
A Lagrangean based branch and bound algorithm for single machine sequencing with precedence constraints to minimize total weighted completion time. Management Science 31, 1300-1311. (1985).

POTT 1985-2 Potts, C. N.
Analysis of a linear programming heuristic for scheduling unrelated parallel machines. Discrete Applied Mathematics 10, 155-164. (1985).

POTT 1985-3 Potts, C. N.
Analysis of heuristics for two-machine flow-shop sequencing subject to release dates. Mathematics of Operations Research 10, 576-584. (1985).

POTTA 1985-1 Potts, C. N.; Wassenhove, L. N. van
A branch and bound algorithm for the total weighted tardiness problem. Operations Research 33, 363-377. (1985).

POTTA 1987-1 Potts, C. N.; Wassenhove, L. N. van
Dynamic programming and decomposition approaches for the single machine total tardiness problem. European Journal of Operational Research 32, 405-414. (1987).

POTY 1983-1 Potyagailo, Z. D.
Construction of a matroid of states of a dynamical system and algorithms for its investigation (Russian). in: Systems-theoretic Methods and their Application in Automated Systems, 27-35. Kiev: Akademiya Nauk Ukrainskoi SSR, Institut Kibernet. (1983).

POTY 1984-1 Potyagailo, Z. D.
Matroid theory methods in the study of information systems dynamics (Russian). In: Local and distributed data processing systems, 47-50. Kiev: Akademija Nauk. (1984).

POWE 1985-1 Powell, S.
Software. In: M. O'hEigeartaigh, J. K. Lenstra, A. H. G. Rinnooy Kan (Eds.): Combinatorial Optimization - Annotated Bibliographies, 190-194. Chichester: Wiley. (1985).

PRAB 1984-1 Prabhu, G. M.; Deo, N.
The ellipsoid algorithm and the graph isomorphism problem. T.S.I.-Technique et Science Informatiques 3, 327-334. (1984).

PRAK 1984-1 Prakash, J.; Sinha, S. B.
A goal programming model for express bus-stop location. Indian Journal of Technology (New Delhi) 22, 369-372. (1984).

PRAKA 1983-1 Prakash, S.
On the machine-assignment problem with objectives to minimize duration of completing work and processing cost. Proceedings of the National Academy of Sciences, India. Section A. (Allahabad) 53, 364-368. (1983).

PRAS 1985-1 Prasad, V. R.
Scheduling on a single processor subject to interruptions in processing. Opsearch 22, 49-52. (1985).

PREI 1985-1 Preissmann, M.
A class of strongly perfect graphs. Discrete Mathematics 54, 117-120. (1985).

PREIA 1985-1 Preissmann, M.; Werra, D. de
A note on strong perfectness of graphs. Mathematical Programming 31, 321-326. (1985).

PREIB 1986-1 Preissmann, M.; Werra, D. de; Mahadev, N. V. R.
A note on superbrittle graphs. Discrete Mathematics 61, 259-268. (1986).

PREP 1985-1 Preparata, F. P.; Sarrafzadeh, M.
Channel routing of nets bounded degree. In: P. Bertolazzi, F. Luccio (Eds.): VLSI: Algorithms and Architectures, 189-204. Amsterdam: North-Holland. (1985).

PRET 1985-1 Pretzel, O.
On graphs that can be oriented as diagrams of ordered sets. Order 2, 25-40. (1985).

PRET 1986-1 Pretzel, O.
On reorienting graphs by pushing down maximal vertices. Order 3, 135-154. (1986).

PRET 1986-2 Pretzel, O.
Orientations and reorientations of graphs. In: I. Rival (Ed.): Combinatorics and Ordered Sets, Proc. AMS-IMS-SIAM Joint Summer Research Conf., August 1985, 103-126. Providence: AMS. (1986).

PRIC 1986-1 Price, W. L.; Turcotte, M.
Locating a blood bank. Interfaces 16, 17-26. (1986).

PRIS 1987-1 Prisakar, K. F.; Prisakar, A. V.
On the minimal covering of the vertices of a graph by d-convex subsets (Russian). Mat. Issled. No. 96 Optim. i Obrabot. Dannykh, 114-118. (1987).

PROC 1983-1 Procesi Ciampi, R.
A criterion for acyclic orientations. Bollettino della Unione Matematica Italiana, Serie VI, D, Algebra e Geometria (Bologna) 2, No. 1, 1-7. (1983).

PROC 1983-2 Procesi Ciampi, R.
A minimality property for acyclic orientations. Rendiconti die Matematica e delle sue Applicazione, Seria VII. (Rome) 3, 681-685. (1983).

PROC 1984-1 Procesi Ciampi, R.
Acyclic orientations and domination of a graph. Coxeter Festschrift II, Mitteilungen aus dem Mathematischen Seminar Giessen 164, 51-57. (1984).

PROC 1985-1 Procesi Ciampi, R.
Spanning trees and acyclic orientations. Rendiconti di Matematica e delle sue Applicazioni (Rome) VII, Ser. 5, 93-100. (1985).

PROD 1985-1 Prodinger, H.
Einige Bemerkungen zu einer Arbeit von W. Knödel über das mittlere Verhalten von on-line-Packungsalgorithmen. Elektronische Informationsverarbeitung und Kybernetik 21, 3-8. (1985).

PRODA 1984-1 Prodon, A.
A note on the separation problem for the matching matroid. Discrete Mathematics 52, 307-312. (1984).

PRODB 1985-1 Prodon, A.; Liebling, T. M.; Gröflin, H.
Steiner's problem on two-trees. RO 850315. Lausanne: Ecole Polytechnique Federale de Lausanne, Dept. de Mathematiques, 24 p. (1985).

PROM 1985-1 Prömel, H. J.
Some remarks on natural orders for combinatorial cubes. Report 85398-OR. Bonn: Univ. Bonn, Inst. für Ökonometrie u. Op. Research, 13 p. (1985).

PROM 1986-1 Prömel, H. J.
Almost bipartite-making graphs. Report 86425-OR. Bonn: Univ. Bonn, Inst. f. Ökonometrie u. Operations Research, 8 p. (1986).

PROM 1986-2 Prömel, H. J.
Asymptotic enumeration of l-colorable graphs. Report No. 85405-OR. Bonn: Universität Bonn, Inst. für Ökonometrie und Operations Research, 11p. (1986).

PROM 1986-3 Prömel, H. J.
Counting unlabeled structures. Report No. 85409-OR. Bonn: Universität Bonn, Inst. für Ökonometrie und Operations Research, 12p. (1986).

PROMA 1986-1 Prömel, H. J.; Voigt, B.
Ramsey theorems for finite graphs. Report 86447-OR. Bonn: Univ. Bonn, Inst. f. Ökonometrie u. Operations Research, 27 p. (1986).

PROS 1986-1 Proskurowski, A.; Syslo, M. M.
Efficient vertex- and edge-coloring of outerplanar graphs. SIAM Journal on Algebraic and Discrete Methods 7, 131-136. (1986).

PROV 1986-1 Provan, J. S.
The complexity of reliability computations in planar and acyclic graphs. SIAM Journal on Computing 15, 694-702. (1986).

PROV 1987-1 Provan, J. S.
An approximation scheme for finding Steiner trees with obstacles. Technical Report UNC/OR/ TR-87/9. Chapel Hill: Univ. of North Carolina, Dept. of Operations Research, 27 p. (1987).

PRUS 1984-1 Pruss, V.
A characterization of matroidal families of multigraphs. Discrete Mathematics 52, 101-106. (1984).

PSAR 1984-1 Psaraftis, H. N.
On the practical importance of asymptotic optimality in certain heuristic algorithms. Networks 14, 587-596. (1984).

PSAR 1986-1 Psaraftis, H. N.
Scheduling large-scale advance-request dial-a-ride systems. American Journal of Mathematical and Management Sciences 6, 327-367. (1986).

PUCC 1988-1 Puccio, L.
An algorithm for L(s)-colourations. In: A. Barlotti, M. Marchi, G. Tallini (Eds.): Combinatorics '86, 385-390. Annals of Discrete Mathematics 37, Amsterdam: North-Holland. (1988).

PUEB 1987-1 Puebla, J. G.
Spatial structures of network flows: A graph theoretical approach. Transportation Research 21B, 489-502. (1987).

PUEL 1985-1 Puel, L.
Unavoidable sets of trees and an application to generalizing Kruskal's theorem (French). Comptes Rendus des Seances de l'Academie des Sciences, Serie I, Mathematique 301, No. 18, 849-851. (1985).

PUJA 1983-1 Pujari, A. K.; Das, P. K.
Monotonicity of polymatroids. Indian Academy of Sciences, Proceedings, Mathematical Sciences (Bangalore) 92, 103-107. (1983).

PULA 1984-1 Pulat, P. S.; Elmaghraby, S. E.
On maximizing flow in generalized flow networks. OR Report, 202. Raleigh: North Carolina State Univ. at Raleigh, 41 p. (1984).

PULL 1987-1 Pulleyblank, W. R.
Fractional matchings and the Edmonds-Gallai theorem. Discrete Applied Mathematics 16, 51-58. (1987).

PULLA 1987-1 Pullman, N. J.; Stanford, M.
The biclique numbers of regular graphs. Congressus Numerantium 56, 237-250. (1987).

PUNN 1987-1 Punnim, N.; Pabhapote, N.
On graceful graphs: cycles with a P*(k)-chord, k>=4. Ars Combinatoria 23A, 225-228. (1987).

PURD 1985-1 Purdom, P. W.; Brown, C. A.
The analysis of algorithms. New York, London: Holt, Rinehart and Winston. (1985).

PURU 1984-1 Purushotam, S.; Prakash, S.
A transportation problem with minimization of duration and total cost of transportation as high and low priority objectives respectively. Bulletin of the Technical University of Istanbul 37, 1-11. (1984).

PUTU 1983-1 Puturidze, Z. S.
A modified algorithm for constructing a sequence of schemes for the solution of discrete optimization problems (Russian). Soobshcheniya Akademii Nauk Gruzinskoj SSR (Tbilisi) 111, 249-252. (1983).

PUTU 1984-1 Puturidze, Z. S.
Problem on the wiring of the electrical net in technical products (Russian). Soobshcheniya Akademii Nauk Gruzinskoj SSR (Tbilisi) 115, 509-512. (1984).

PUTU 1986-1 Puturidze, Z. S.
Computational aspects of a modified algorithm for constructing a sequence of plans for the solution of discrete optimization problems (Russian). Soobshcheniya Akademii Nauk Gruzinskoj SSR (Tbilisi) 122, 505-508. (1986).

PUTU 1986-2 Puturidze, Z. S.
Optimization of some implementation of a modified algorithm for the method of constructing a sequence of plans (Russian). Soobshcheniya Akademii Nauk Gruzinskoj SSR 124, 65-68. (1986).

PYBE 1986-1 Pyber, L.
Clique covering of graphs. Combinatorica 6, 393-398. (1986).

PYBE 1987-1 Pyber, L.
On one-factorizations of the complete graph. Discrete Mathematics 67, 209-210. (1987).

QI 1985-1 Qi, L.-Q.
Forest iteration method for stochastic transportation problem. In: R. W. Cottle (Ed.): Mathematical Programming Essays in Honor of G. B. Dantzig, Mathematical Programming Study 25, 142-163. (1985).

QI 1986-1 Qi, L.-Q.
Directed submodular functions, ditroids and total dual integrality. Science Report 86009. Beijing: Tsinghua Univ., Dept. of Applied Mathematics, 20 p. (1986).

QI 1987-1 Qi, L.-Q.
A rank characterization of 2-matroid intersections. IE Tech. Report 87-13. Madison: Univ. of Wisconsin, Dept. of Industrial Eng., 16 p. (1987).

QI 1987-2 Qi, L.-Q.
Odd submodular functions, Dilworth functions and discrete convex functions. Working Paper. Beijing: Tsinghua Univ., Dept. of Applied Mathematics, 16 p. (1987).

QI 1987-3 Qi, L.-Q.
The A-forest iteration method for the stochastic generalized transportation problem. Mathematics of Operations Research 12, 1-21. (1987).

QI A 1987-1 Qi, N.
On separation and adjacency problems for perfectly matchable subgraph polytopes of a graph. Operations Research Letters 6, 239-242. (1987).

QIN 1985-1 Qin, Y.
On generalized Floyd algorithm. Working Paper. Wuhan: Wuhan Institute of Iron and Steel Technology, 19 p. (1985).

QIN 1985-2 Qin, Y.
On Jar-metric principle - a unified approach to solve optimum paths problems on multistage directed graphs. Working Paper. Wuhan: Wuhan Iron and Steel University, 40 p. (1985).

QUAD 1987-1 Quaddus, M. A.
A generalized model of optimal due-date assignment by linear programming. Journal of the Operational Research Society 38, 353-360. (1987).

QUEY 1985-1 Queyranne, M.
Bounds for assembly line balancing heuristics. Operations Research 33, 1353-1359. (1985).

QUEY 1986-1 Queyranne, M.
Performance ratio of polynomial heuristics for triangle inequality quadratic assignment problems. Operations Research Letters 4, 231-234. (1986).

QUIN 1985-1 Quinn, M. J.
A note on two parallel algorithms to solve the stable marriage problem. BIT 25, 473-476. (1985).

QUINA 1984-1 Quinn, M. J.; Deo, N.
Parallel graph algorithms. ACM Computing Surveys 16, 319-348. (1984).

QUINA 1986-1 Quinn, M. J.; Deo, N.
An upper bound for the speedup of parallel best-bound branch-and-bound algorithms. BIT 26, 35-43. (1986).

RABI 1985-1 Rabinowitz, J. H.; Proulx, V. K.
An asymptotic approach to the channel assignment problem. SIAM Journal on Algebraic and Discrete Methods 6, 507-518. (1985).

RACH 1987-1 Rachamadugu, R. M. V.
A note on the weighted tardiness problem. Operations Research 35, 450-452. (1987).

RADA 1986-1 Radaramanan, R.; Choi, L. I.
A branch and bound algorithm for the travelling salesman and the transportation routing problems. Computers and Industrial Engineering 11, No. 1-4, 236-40. (1986).

RADE 1986-1 Radermacher, F. J.
Schedule-induced posets. Discrete Applied Mathematics 14, 67-92. (1986).

RADE 1986-2 Radermacher, F. J.
Scheduling of project networks. In: C. L. Monma (Ed.): Algorithms and Software for Optimization - Part I, 227-252. Annals of Operations Research 4, Basel: Baltzer. (1986).

RADEA 1982-1 Radescu, N.
An algorithm for determining maximum flow through a transportation network (Romanian). In: Proceedings of the national colloquium on probability theory and operational research (Craiova, 1982), 160-163. Bucharest: Universiteta Bucuresti. (1982).

RADEA 1982-2 Radescu, N.
Determination of a flow compatible with capacity intervals by means of epsilon-minimal paths (Romanian). In: Proceedings of the national colloquium on probability theory and operational research (Craiova, 1982), 155-159. Bucharest: Universiteta Bucuresti. (1982).

RADEA 1982-3 Radescu, N.
Determination of minimal flow through a network with accumulated capacities (Romanian). In: Proceedings of the national colloquium on probability theory and operational research (Craiova, 1982), 150-154. Bucharest: Universiteta Bucuresti. (1982).

RADEB 1985-1 Radev, E. I.
A heuristic algorithm for the coloring of planar graphs (Russian). Preprint 85-25, Akademii Nauk Ukrainskoj SSR, Inst. Kibernet., Kiev. (1985).

RADO 1983-1 Radosavljevic, Z.; Simic, S.; Syslo, M. M.; Topp, J.
A note on generalized line graphs. Publications de l'Institute Mathematique (Nouvelle Serie) (Beograd) 34 (48), 193-198. (1983).

RAGH 1986-1 Raghavachari, M.
A V-shape property of optimal schedule of jobs about a common due date. European Journal of Operational Research 23, 401-402. (1986).

RAGHA 1986-1 Raghavan, P.
Probabilistic construction of deterministic algorithms: approximating packing integer programs. In: 27th Annual Symp. on Foundations of Computer Science, Proceedings, 10-18. IEEE. (1986).

RAGHB 1985-1 Raghavan, P.; Thompson, C. D.
Provably good routing in graphs: regular arrays. In: Proc. Seventeenth Annual ACM Symp. on Theory of Computing, Providence 1985, 79-87. ACM. (1985).

RAIS 1987-1 Raiszadeh, F. M. E.; Dileepan, P.; Sen, T.
A single machine bicriterion scheduling problem and an optimizing branch-and-bound procedure. Journal of Information & Optimization Sciences (Delhi) 8, 311-321. (1987).

RAJA 1987-1 Rajasekaran, S.; Tsantilas, T.
An optimal randomized routing algorithm for the mesh and a class of efficient mesh-like routing networks. In: K. V. Nori (Ed.): Foundations of Software Technology and Theoretical Computer Science, 226-241. Lecture Notes in Computer Science 287, Berlin, Heidelberg, New York: Springer. (1987).

RALL 1984-1 Rall, D. F.; Slater, P. J.
On location-domination numbers for certain classes of graphs. Congressus Numerantium 45, 97-106. (1984).

RAMA 1986-1 Ramamurthy, K. G.; Parthasarathy, T.
An algorithm to find the smallest committee containing a given set. Opsearch 23, 1-6. (1986).

RAMA 1986-2 Ramamurthy, K. G.; Parthasarathy, T.
Matroidal games. International Journal of Game Theory 15, 21-30. (1986).

RAMAA 1984-1 Ramanan, P.; Deogun, J. S.; Liu, C. L.
A personnel assignment problem. Journal of Algorithms 5, 132-144. (1984).

RAMAB 1987-1 Ramanathan, A.; Colbourn, C. J.
Bounds for all-terminal reliability by arc-packing. Ars Combinatoria 23A, 229-236. (1987).

RAME 1986-1 Ramesh, R.; Karwan, M. H.; Zionts, S.
A class of practical interactive branch and bound algorithms for multicriteria integer programming. European Journal of Operational Research 26, 161-172. (1986).

RAMO 1986-1 Ramos, R. E.
On 1-factorization of line graphs. Matimyas Matematika (Quezon City) 10, No. 3, 18-19. (1986).

RAMR 1983-1 Ramras, M.
Maximal independent sets in graphs with vertex m-covers. Congressus Numerantium 39, 463-472. (1983).

RANC 1983-1 Ranchich, M. V.
Determination of shortest paths of a graph (Russian). in: Systems-theoretic Methods and their Application in Automated Systems, 53-55. Kiev: Akademiya Nauk Ukrainskoi SSR, Institut Kibernet. (1983).

RAND 1986-1 Randelman, R. E.; Grest, G. S.
N-city travelling salesman problem: optimization by simulated annealings. Journal of Statistical Physics 45, 885-890. (1986).

RANDA 1985-1 Randhawa, S. U.; Mac Dowell, E. D.; Faruqui, S.-D.
An integer programming application to solve sequencer mix problems in printed circuit board production. International Journal of Production Research 23, 543-552. (1985).

RANDB 1986-1 Randow, R. von
On an optimal location problem in a triangle. The Mathematical Scientist 11, No. 2, 107-110. (1986).

RAO 1985-1 Rao, I. H. N.; Narasimham, M. S.
A note on graceful and supergraceful numbering of a class of trees. Pure Mathematics Manuscript (Calcutta) 4, 157-168. (1985).

RAO A 1983-1 Rao, J. R.; Rao, S. N.
Facility location in stochastic environment. Ricerca Operativa, No. 28, 51-62. (1983).

RAO B 1985-1 Rao, J. R.; Varma, N. H.
Stochastic multi-facility minisum location problem involving Euclidean distances. Opsearch 22, 232-240. (1985).

RAO C 1987-1 Rao, S.
Finding near optimal separators in planar graphs. In: 28th Annual IEEE Symposium on Foundations of Computer Science, 225-237. (1987).

RAO D 1985-1 Rao, S. B.
On regular and strongly-regular self-complementary graphs. Discrete Mathematics 54, 73-82. (1985).

RASM 1986-1 Rasmussen, L. M.
Zero-one programming with multiple criteria. European Journal of Operational Research 26, 83-95. (1986).

RAVI 1984-1 Ravindra, G.
Meyniel's graphs are strongly perfect. In: C. Berge, V. Chvatal (Eds.): Topics on perfect graphs, Annals of Discrete Mathematics Vol. 21, 145-148. Amsterdam: North-Holland. (1984).

RAVI 1984-2 Ravindra, G.
Strongly perfect line graphs and total graphs. In: A. Hajnal; L. Lovasz; V. T. Sos (Eds.): Finite and Infinite Sets, Vol. 2, 621-634. Coll. Math. Soc. J. Bolyai 37, Amsterdam: North-Holland. (1984).

RAVIA 1986-1 Ravindran, A.; Shin, W. S.; Arthur, J. L.; Moskowitz, H.
Nonlinear integer goal programming models for acceptance sampling. Computers and Operations Research 13, 611-622. (1986).

RAVIB 1985-1 Ravindranath, K.; Vrat, P.; Singh, N.
Bicriteria single facility rectilinear location problems in the presence of a single forbidden region. Opsearch 22, 1-16. (1985).

RAYC 1985-1 Raychaudhuri, A.; Roberts, F. S.
Generalized competition graphs and their applications. In: P. Brucker, R. Pauly (Eds.): Proc. IX. Symp. on Operations Research, Osnabrück 1984. Methods of Operations Research 49, 295-312. (1985).

RAYW 1987-1 Rayward-Smith, V. J.
The complexity of preemptive scheduling given interprocessor communication delays. Information Processing Letters 25, 123-126. (1987).

RAYW 1987-2 Rayward-Smith, V. J.
UET scheduling with unit interprocessor communication delays. Discrete Applied Mathematics 18, 55-72. (1987).

RAZ 1987-1 Raz, T.; Bricker, D.
Sequencing of imperfect inspection operations subject to constraints on the quality of accepted and rejected units. International Journal of Production Research 25, 809-822. (1987).

READ 1987-1 Read, R. C.
A new method for drawing a planar graph given the cyclic order of the edges at each vertex. Congressus Numerantium 56, 31-44. (1987).

READ 1987-2 Read, R. C.
Connectivity and chromatic uniqueness. Ars Combinatoria 23, 209-218. (1987).

READ 1987-3 Read, R. C.
On the chromatic properties of graphs up to 10 vertices. Congressus Numerantium 59, 243-256. (1987).

RECS 1984-1 Recski, A.
Some problems of self-dual matroids. In: A. Hajnal; L. Lovasz; V. T. Sos (Eds.): Finite and Infinite Sets, Vol. 2, 635-648. Coll. Math. Soc. J. Bolyai 37, Amsterdam: North-Holland. (1984).

RECS 1984-2 Recski, A.
Statics and electric network theory: a unifying role of matroids. In: W. R. Pulleyblank (Ed.): Progress in Combinatorial Optimization, 307-314. Toronto: Academic Press. (1984).

RECS 1985-1 Recski, A.
Some open problems of matroid theory, suggested by its applications. In: L. Lovasz, A. Recski (Eds.): Matroid Theory, 311-326. Coll. Math. Soc. J. Bolyai Vol. 40. Amsterdam: North-Holland. (1985).

RECS 1987-1 Recski, A.
Applications of combinatorics to electrical engineering and rigidity. Report 87483-OR. Bonn: Univ. Bonn, Inst. f. Ökonometrie u. Operations Research, 13 p. (1987).

RECS 1987-2 Recski, A.
Elementary strong maps of graphic matroids. Graphs and Combinatorics 3, 379-382. (1987).

RECS 1987-3 Recski, A.
One-story buildings as tensegrity frameworks, II. Report 87492-OR. Bonn: Univ. Bonn, Forschungsinst. f. Diskrete Mathematik und Inst. f. Ökonometrie und Operations Research, 6 p. (1987).

RECS 1987-4 Recski, A.
Some algorithmic problems of linear network solvability. Report 87496-OR. Bonn: Univ. Bonn, Inst. f. Ökonometrie u. Operations Research, 4 p. (1987).

REDD 1984-1 Reddy, K. P.; Mittal, A. K.; Gupta, S. K.
Bivalent quadratic programming problem - a computational study. Opsearch 21, 153-166. (1984).

REED 1985-1 Reed, B.
A note on the semi-strong perfect graph conjecture. Discrete Mathematics 54, 111-112. (1985).

REED 1987-1 Reed, B.
A note on even pairs. Discrete Mathematics 65, 317-318. (1987).

REED 1987-2 Reed, B.
A semi-strong perfect graph theorem. Journal of Combinatorial Theory 43B, 223-240. (1987).

REES 1987-1 Rees, S.; Ball, R. C.
Criteria for an optimum simulated annealing schedule for problems of the traveling salesman type. Journal of Physics A 20, 1239-1249. (1987).

REEV 1985-1 Reeves, C. R.
An improved heuristic for the quadratic assignment problem. Journal of the Operational Research Society 36, 163-168. (1985).

REHF 1985-1 Rehfeldt, M.
Eine netzwerktheoretische Methode zur Untersuchung der Disjunktheit bestimmter endlicher Mengen großer Mächtigkeit. Zeitschrift für Operations Research 29 B, 167-176. (1985).

REIC 1984-1 Reichmeider, P. F.
The equivalence of some combinatorial matching theorems. Washington: Polygonal Publ. House. (1984).

REID 1985-1 Reid, K. B.
Two complementary circuits in two-connected tournaments. In: B. R. Alspach, C. D. Godsil (Eds.): Cycles in Graphs, Annals of Discrete Mathematics Vol. 27, 321-334. Amsterdam: North-Holland. (1985).

REIDA 1983-1 Reid, K. B.; Wormald, N. C.
Embedding oriented n-trees in tournaments. Studia Scientiarum Mathematicarum Hungaria (Budapest) 18, 377-387. (1983).

REIF 1987-1 Reif, J. H.
A topological approach to dynamic graph connectivity. Information Processing Letters 25, 65-70. (1987).

REIN 1985-1 Reinelt, G.
The linear ordering problem: algorithms and application. Berlin: Helderman. (1985).

REIT 1984-1 Reiter, S.
How a network of processors can schedule its work. Discussion Paper, 605. Evanston: Northwestern Univ., Center for Math. Studies in Economics and Management Science, 28 p. (1984).

REND 1984-1 Rendl, F.
On the complexity of decomposing matrices arising in satellite communication. Bericht, 84-47. Graz: Techn. Univ. Graz, Inst. für Mathematik, 11 p. (1984).

REND 1985-1 Rendl, F.
Ranking scalar products to improve bounds for the quadratic assignment problem. European Journal of Operational Research 20, 363-372. (1985).

REND 1986-1 Rendl, F.
On the Euclidean assignment problem. Research Report CORR 86-28. Waterloo: Univ. of Waterloo, Dept. of Combinatorics and Optimization, 12 p. (1986).

REND 1986-2 Rendl, F.
Quadratic assignment problems on series-parallel digraphs. Zeitschrift für Operations Research A-30, 161-173. (1986).

REND 1988-1 Rendl, F.
Combinatorial improvements of the 1-tree bound for the traveling salesman problem. In: H. Schellhaas, P. van Beek, H. Isermann, R. Schmidt, M. Zijlstra (Eds.): Operations Research Proceedings 1987 (DGOR/NSOR), 624-625. Berlin: Springer. (1988).

REUT 1986-1 Reutenauer, C.
Free sets of paths in a graph (French). Bulletin de la Societe Mathematique de France 114, No. 2, 135-152. (1986).

RHEE 1985-1 Rhee, W.-S. T.
Convergence of optimal stochastic bin packing. Operations Research Letters 4, 121-124. (1985).

RHEE 1987-1 Rhee, W.-S. T.
Probabilistic analysis of the next fit decreasing algorithm for bin-packing. Operations Research Letters 6, 189-192. (1987).

RIBE 1983-1 Ribeiro, C. C.
Algorithmes de recherche de plus courts chemins avec contraintes: Etude theorique, implementation et parallelisation. Paris: Ecole Nationale Superieure des Telecommunications, 237 p. (1983). (Dissertation)

RIBE 1987-1 Ribeiro, C. C.
Parallel computer models and combinatorial algorithms. In: P. Martello, G. Laporte, M. Minoux, C. Ribeiro (Eds.): Surveys in Combinatorial Optimization, Annals of Discrete Mathematics 31, 325-364. Amsterdam, New York, Tokyo: North-Holland. (1987).

RIBEA 1985-1 Ribeiro, C. C.; Minoux, M.
A heuristic approach to hard constrained shortest path problems. Discrete Applied Mathematics 10, 125-138. (1985).

RIBEA 1986-1 Ribeiro, C. C.; Minoux, M.
Solving hard constrained shortest path problems by Lagrangean relaxation and branch-and-bound algorithms. Methods of Operations Research 53, 303-316. (1986).

RICH 1984-1 Richards, D.
Complexity of single-layer routing. IEEE Transactions on Computers C-33, 286-288. (1984).

RICH 1986-1 Richards, D.
Finding short cycles in planar graphs using separators. Journal of Algorithms 7, 382-394. (1986).

RICHA 1985-1 Richards, D.; Liestman, A. L.
Finding cycles of a given length. In: B. R. Alspach, C. D. Godsil (Eds.): Cycles in Graphs, Annals of Discrete Mathematics Vol. 27, 249-256. Amsterdam: North-Holland. (1985).

RICHB 1985-1 Richey, M. B.; Parker, R. G.
On multiple Steiner subgraph problems. Research Paper. Atlanta: Georgia Inst. of Technology, School of Ind. and Systems Eng., 27 p. (1985).

RICHB 1987-1 Richey, M. B.; Parker, R. G.
On computational issues related to series-parallel solvability. Working Paper. Fairfax: George Mason Univ., Dept. of Operations Research and Applied Statistics, 29 p. (1987).

RICHC 1985-1 Richey, M. B.; Parker, R. G.; Rardin, R. L.
On finding spanning Eulerian subgraphs. Naval Research Logistics Quarterly 32, 443-456. (1985).

RICHD 1985-1 Richmond, L. B.; Robinson, R. W.; Wormald, N. C.
On Hamilton cycles in 3-connected cubic maps. In: B. R. Alspach, C. D. Godsil (Eds.): Cycles in Graphs, Annals of Discrete Mathematics Vol. 27, 141-150. Amsterdam: North-Holland. (1985).

RICHE 1984-1 Richter, B.; Shank, H.
The cycle space of an embedded graph. Journal of Graph Theory 8, 365-370. (1984).

RICHF 1985-1 Richter, K.
Complexity of the robot sequencing problem. Seminarberichte, Humboldt-Universität zu Berlin, Sektion Mathematik 71, 94-99. (1985).

RICHF 1985-2 Richter, K.
Stability of the constant cost dynamic lot size model. In: F. Nozicka (Ed.): Optimization - Theory, Methods, Applications, I, International Conference Marianske Lazne 1985, 234-241. Prague: Dum Techniky CSVTS. (1985).

RICHF 1985-3 Richter, K.
The robot sequencing problem: polynomial algorithm and complexity. Optimization 16, 597-606. (1985).

RICHF 1986-1 Richter, K.
The two-criterial dynamic lot size problem. Systems Analysis, Modelling Simulation 3, 99-105. (1986).

RICHF 1987-1 Richter, K.
Stability of the constant cost dynamic lot size model. European Journal of Operational Research 31, 61-65. (1987).

RIED 1986-1 Rieder, J.
On k-tuple-colorings of graphs. Report, 86.29. Bonn: Univ. Bonn, Sonderforschungsbereich 303, 16 p. (1986).

RIHA 1983-1 Riha, S.; Sekanina, M.
Squares of triangular cacti. Archiv der Mathematik (Brno) 19, 153-160. (1983).

RIIV 1984-1 Riives, K.
A classification problem for polymatroids (Russian). Izvestiya Akademij Nauk Ehstonskoj SSR. Fizika. Matematika. (Tallin) 33, 408-423. (1984).

RIIV 1986-1 Riives, K.
A classification of three-dimensional polymatroids (Russian). Izvestiya Akademii Nauk Estonskoi SSR, Seriya Fizika-Matematika (Tallinn) 35, No. 2, 127-131. (1986).

RIJC 1986-1 Rijckaert, M. J.
Heuristics and the role of microcomputers for the one-dimensional cutting stock problem. Methods of Operations Research 56, 85-92. (1986).

RING 1985-1 Ringel, G.
250 Jahre Graphentheorie. In: R. Bodendiek, H. Schumacher, G. Walther (Eds.): Graphen in Forschung und Unterricht, Festschrift Klaus Wagner, 136-152. Salzdetfurth: Franzbecker. (1985).

RINGA 1987-1 Ringuest, J. L.; Rinks, D. B.
Interactive solutions for the linear multiobjective transportation problem. European Journal of Operational Research 32, 96-106. (1987).

RINN 1984-1 Rinnooy Kan, A. H. G.
An introduction to approximation algorithms. Report 8420/O. Rotterdam: Erasmus Univ., Econometric Institute. (1984).

RINN 1986-1 Rinnooy Kan, A. H. G.
An introduction to the analysis of approximation algorithms. Discrete Applied Mathematics 14, 171-186. (1986).

RINN 1986-2 Rinnooy Kan, A. H. G.
Stochastic integer programming: the distribution problem. In: F. Archetti, G. Di Pillo, M. Lucertini (Eds.): Stochastic Programming (Gargnano 1983), 140-150. Lecture Notes in Control and Information Sciences 76, Berlin: Springer. (1986).

RINN 1987-1 Rinnooy Kan, A. H. G.
Probabilistic analysis of algorithms. In: P. Martello, G. Laporte, M. Minoux, C. Ribeiro (Eds.): Surveys in Combinatorial Optimization, Annals of Discrete Mathematics 31, 365-384. Amsterdam, New York, Tokyo: North-Holland. (1987).

RINNA 1986-1 Rinnooy Kan, A. H. G.; Frenk, J. B. G.
The rate of convergence to optimality of the LPT rule. Discrete Applied Mathematics 14, 187-198. (1986).

RINNB 1987-1 Rinnooy Kan, A. H. G.; Wit, J. R. de; Wijmenga, R. T.
Nonorthogonal two-dimensional cutting patterns. Management Science 33, 670-684. (1987).

RINS 1987-1 Rinsma, I.
Embedding a tree in a maximal outerplanar graph. Ars Combinatoria 23B, 147-182. (1987).

RIST 1984-1 Ristroph, J. H.; Ashley, C. E.
Measurement of the effectiveness of alternative priority rules in network scheduling. Congressus Numerantium 45, 27-44. (1984).

RITC 1986-1 Ritchie, E.
Lot-sizing for non-stationary demand. European Journal of Operational Research 27, 267-273. (1986).

RIVA 1984-1 Rival, I.
Linear extensions of finite ordered sets. In: M. Pouzet, D. Richard (Eds.): Orders: Description and Roles, Annals of Discrete Mathematics Vol. 23, 355-370. Amsterdam: Norht-Holland. (1984).

RIVA 1985-1 Rival, I.
Some order theoretical ideas about scheduling. In: P. Brucker, R. Pauly (Eds.): Proc. IX. Symp. on Operations Research, Osnabrück 1984. Methods of Operations Research 49, 419-430. (1985).

RIVA 1986-1 Rival, I.
Stories about order and the letter N. In: I. Rival (Ed.): Combinatorics and Ordered Sets, Proc. AMS-IMS-SIAM Joint Summer Research Conf., August 1985, 263-285. Providence: AMS. (1986).

RIVAA 1986-1 Rival, I.; Zaguia, N.
Constructing greedy linear extensions by interchanging chains. Order 3,107-122. (1986).

RIVAA 1986-2 Rival, I.; Zaguia, N.
Constructing N-free, jump-critical ordered sets. Congressus Numerantium 55, 199-204. (1986).

RIVAA 1987-1 Rival, I.; Zaguia, N.
Greedy linear extensions with constraints. Discrete Mathematics 63, 249-260. (1987).

RIVE 1986-1 Rivera Campo, E.
Interchanging of edges in graceful graphs (Spanish). In: M. Clapp, J. A. Seade (Eds.): Research Program of the 18th Nat. Congr. of the Mexican Math. Soc., Merida 1984, 349-358. Aportaciones Matematicas, Communicaciones, 1. Mexico: Inst. de Matematicas de la UNAM. (1986).

RO 1984-1 Ro, H.-B.; Tcha, D.-W.
A branch and bound algorithm for the two-level uncapacitated facility location problem with some side constraints. European Journal of Operational Research 18, 349-358. (1984).

ROBE 1985-1 Robert, P.
Etude de la version temps reel d'un algorithme de bin-packing asymptotiquement optimal. Rapport de Recherche 444. Le Chesnay: Inst. National de Recherche en Informatique et en Automatique, 14 p. (1985).

ROBEA 1985-1 Robert, Y.; Tchuente, M.
Connection-graph and iteration-graph of monotone boolean functions. Discrete Applied Mathematics 11, 245-254. (1985).

ROBEB 1986-1 Robert, Y.; Trystram, D.
An orthogonal systolic array for the algebraic shortest-path problem (French). Comptes Rendus des Seances de l'Academie des Sciences, Serie I Mathematique (Paris) 302, 241-244. (1986).

ROBEC 1985-1 Roberts, F. S.
Applications of edge coverings by cliques. Discrete Applied Mathematics 10, 93-110. (1985).

ROBED 1984-1 Robertson, N.
Minimal cyclic-4-connected graphs. Transcations of the American Mathematical Society 284, 665-687. (1984).

ROBEE 1984-1 Robertson, N.; Seymour, P. D.
Graph minors. III. Planar tree-width. Journal of Combinatorial Theory, Series B 36, 49-64. (1984).

ROBEE 1984-2 Robertson, N.; Seymour, P. D.
Graph width and well-quasi-ordering: a survey. In: J. A. Bondy, U. P. R. Murty (Eds.): Progress in Graph Theory, 399-406. New York: Academic Press. (1984).

ROBEE 1985-1 Robertson, N.; Seymour, P. D.
Disjoint paths - a survey. SIAM Journal on Algebraic and Discrete Methods 6, 300-305. (1985).

ROBEE 1985-2 Robertson, N.; Seymour, P. D.
Graph minors - a survey. In: I Anderson (Ed.): Surveys in Combinatorics 1985, London Mathematical Society Lecture Notes Series Vol. 103, 153-171. Cambridge: University Press. (1985).

ROBEE 1985-3 Robertson, N.; Seymour, P. D.
Graph minors IV. Tree-width and well-quasi-ordering. Working Paper. Columbus: Ohio State Univ., Dept. of Mathematics, 61 p. (1985).

ROBEE 1985-4 Robertson, N.; Seymour, P. D.
Graph minors VII: Disjoint paths on a surface. Working Paper. Morristown: Bell Communications Research, 56 p. (1985).

ROBEE 1985-5 Robertson, N.; Seymour, P. D.
Graph minors VIII. A Kuratowski theorem for general surfaces. Working Paper. Columbus: Ohio State Univ., Dept. of Mathematics, 37 p. (1985).

ROBEE 1986-1 Robertson, N.; Seymour, P. D.
Graph minors. II. Algorithmic aspects of tree-width. Journal of Algorithms 7, 309-322. (1986).

ROBEE 1986-2 Robertson, N.; Seymour, P. D.
Graph minors. V. Excluding a planar graph. Journal of Combinatorial Theory, Series B 41, 92-114. (1986).

ROBEE 1986-3 Robertson, N.; Seymour, P. D.
Graph minors. VI. Disjoint paths across a disc. Journal of Combinatorial Theory, Series B 41, 115-138. (1986).

ROBI 1985-1 Robinson, D. F.; Janjic, I.
The constructability of floorplans with given outerplanar adjacency graph and room areas. Ars Combinatoria 20-B, 133-142. (1985).

ROBS 1986-1 Robson, J. M.
Algorithms for maximum independent sets. Journal of Algorithms 7, 425-440. (1986).

ROCK 1984-1 Rockafellar, R. T.
Network flows and monotropic optimization. New York, Chichester: Wiley. (1984).

RODI 1986-1 Roditty, Y.
The packing and covering of the complete graph. I: The forests of order five. International Journal of Mathematics and Mathematical Sciences 9, 277-282. (1986).

RODL 1985-1 Rödl, V.; Tuza, Z.
On color critical graphs. Journal of Combinatorial Theory 38B, 204-213. (1985).

RODR 1984-1 Rodriguez, R.; Kakes, A.; Casas, O.
Suboptimal partial trees. Investigacion Operacional 5, 3-16. (1984).

ROHN 1985-1 Rohnert, H.
A dynamization of the all pairs least cost path problem. In: K. Mehlhorn (Ed.): STACS 85, Second Annual Symposium on Theoretical Aspects of Computer Science (Saarbrücken, January 1985), Lecture Notes in Computer Science Vol. 182, 279-286. Berlin: Springer-Verlag. (1985).

ROHN 1986-1 Rohnert, H.
Shortest paths in the plane with convex polygonal obstacles. Information Processing Letters 23, 71-76. (1986).

RONE 1986-1 Ronen, D.
Short term scheduling of vessels for shipping bulk or semi-bulk commodities originating in a single area. Operations Research 34, 164-173. (1986).

RONEA 1984-1 Ronen, D.; Perl, Y.
Heuristics for finding a maximum number of disjoint bounded paths. Networks 14, 531-544. (1984).

ROOD 1986-1 Roodman, G. M.
Near-optimal solutions to one-dimensional cutting stock problems. Computers and Operations Research 13, 713-720. (1986).

ROOS 1984-1 Roos, C.
Network programming with umbrellas. Report 84-32. Delft: Univ. of Technology, Department of Mathematics and Computer Science. (1984).

ROSB 1985-1 Rosberg, Z.
Process scheduling in a computer system. IEEE Transactions on Computers 34, 633-645. (1985).

ROSBA 1985-1 Rosberg, Z.; Towsley, D.
Customer routing to parallel servers with different rates. IEEE Transactions on Automatic Control 30, 1140-1143. (1985).

ROSE 1987-1 Rosenberg, E.
A nonlinear programming heuristic for computing optimal link capacities in a multi-hour alternate routing communications network. Operations Research 35, 354-364. (1987).

ROSEA 1987-1 Rosenberg, I. G.
Characterization of diagrams by 0-1 inequalities. Discrete Mathematics 63, 261-270. (1987).

ROSEB 1956-1 Rosenblatt, D.
On aggregation and consolidation in linear systems. Technical Report C to the U.S. Office of Naval Research. Department of Statistics, American University, Washington, D.C. (1956).

ROSEC 1986-1 Rosenblatt, M. J.
The dynamics of plant layout. Management Science 32, 76-86. (1986).

ROSED 1987-1 Rosenblatt, M. J.; Lee, H. L.
A robustness approach to facilities design. International Journal of Production Research 25, 479-486. (1987).

ROSEE 1987-1 Rosenbloom, E. S.; Goertzen, N. F.
Cyclic nurse scheduling. European Journal of Operational Research 31, 19-23. (1987).

ROSEF 1984-1 Rosenstiehl, P.; Tarjan, R. E.
Gauss codes, planar hamiltonian graphs, and stack-sortable permutations. Journal of Algorithms 5, 375-390. (1984).

ROSEF 1986-1 Rosenstiehl, P.; Tarjan, R. E.
Rectilinear planar layouts and bipolar orientations of planar graphs. Discrete and Computational Geometry 1, 343-354. (1986).

ROSEG 1986-1 Rosenthal, R. E.
Representing inverses in pure network flow optimization. European Journal of Operational Research 23, 356-366. (1986).

ROSH 1986-1 Roshchin, V. A.
An algorithm for solving a covering problem by means of parallel computations (Russian). In: I. V. Sergienko (Ed.): Programming and Algorithms for the Development of Program Packages, 21-26. Kiev: Akad. Nauk Ukrain. SSR, Inst. Kibernet. (1986).

ROSHA 1984-1 Roshchin, V. A.; Sergienko, I. V.
An approach to the solution of the set-covering problem. Cybernetics 20, 849-855; translation from: Kibernetika 1984, No. 6, 65-69. (1984).

ROSHA 1986-1 Roshchin, V. A.; Sergienko, I. V.
An algorithm for the exact solution of the set partitioning problem (Russian). Doklady Akademiya Nauk Ukrainskoi SSR, Seriya A, No.1, 63-66. (1986).

ROSK 1985-1 Roskind, J.; Tarjan, R. E.
A note on finding minimum-cost edge-disjoint spanning trees. Mathematics of Operations Research 10, 701-708. (1985).

ROSS 1987-1 Rossi, F. A.; Gavioli, I.
Aspects of heuristic methods in the "Probabilistic Travelling Salesman Problem". In: G. Andreatta, F. Mason, P. Serafini (Eds.): Advanced School on Stochastics in Combinatorial Optimization (CISM Udine 1986), 214-227. Singapore: World Scientific. (1987).

ROSSA 1986-1 Rossier, Y.; Troyon, M.; Liebling, T. M.
Probabilistic exchange algorithms and Euclidean traveling salesman problems. Operations Research Spektrum 8, 151-164. (1986).

ROTE 1985-1 Rote, G.
A systolic array algorithm for the algebraic path problem (shortest paths; matrix inversion). Computing 34, 191-220. (1985).

ROTEA 1986-1 Rote, G.; Rendl, F.
Minimizing the density of terminal assignments in layout design. Operations Research Letters 5, 111-118. (1986).

ROTH 1985-1 Roth, A. E.
The college admissions problem is not equivalent to the marriage problem. Journal of Economic Theory 36, 277-288. (1985).

ROTHA 1985-1 Rothe, G.; Rendl, F.
Minimizing the density of terminal assignments in layout design. Bericht 85-64. Graz: T.U. Graz, Inst. f. Mathematik, 13 p. (1985).

ROTK 1986-1 Rotko, V. F.
Efficient algorithms for synthesis of graphs with a prescribed set of fundamental sections or cycles. Kibernetika (Kiev) No. 1, 39-45. (1986).

ROUB 1985-1 Roubens, M.
Families of nested interval graphs (French). Belgian Journal of Operations Research, Statistics and Computer Science 25, 63-76. (1985).

ROUBA 1983-1 Roubens, M.; Vincke, P.
Linear fuzzy graphs. Fuzzy Sets and Systems 10, 79-86. (1983).

ROUBA 1985-1 Roubens, M.; Vincke, P.
Preference modelling. Berlin, Heidelberg: Springer. (1985). (Lecture Notes in Economics and Mathematical Systems, 250)

ROUC 1987-1 Roucairol, C.
A parallel branch and bound algorithm for the quadratic assignment problem. Discrete Applied Mathematics 18, 211-226. (1987).

ROUCA 1987-1 Roucairol, C.; Hansen, P.
Probleme de la bipartition minimale d'un graphe. RAIRO Recherche Operationelle 21, 325-348. (1987).

ROUH 1985-1 Rouhani, R.; Lasdon, L.; Lebow, W.; Warren, A. D.
A generalized Benders decomposition approach to reactive source planning in power systems. In: R. W. Cottle (Ed.): Mathematical Programming Essays in Honor of G. B. Dantzig, Mathematical Programming Study 25, 62-75. (1985).

ROUN 1987-1 Rounault, A.
Superconcentrateurs. Publications Mathematique d'Orsay 87-01. (1987).

ROUNA 1985-1 Roundy, R.
98%-effective integer-ratio lot-sizing for one-warehouse multi-retailer systems. Management Science 31, 1416-1430. (1985).

ROUNA 1986-1 Roundy, R.
A 98% effective lot sizing rule for a multi-product, multi-stage production/inventory system. Mathematics of Operations Research 11, 699-727. (1986).

ROUS 1985-1 Rousseau, C. K.; Rousseau, C. C.; Liu, A. L.; Klamkin, M. S.
Greedy algorithm and coinage systems. Rocky Mountain Journal of Mathematics 15, 571. (1985).

ROUSA 1984-1 Rousseau, J.-M.
Crew scheduling methods in the operation and planning of transportation systems. In: M. Florian (Ed.): Transportation Planning Models, 439-474. Amsterdam: North-Holland. (1984).

ROWI 1986-1 Rowicki, A.
Preemptive scheduling for two-processor systems. Prace Naukowo-Badawcze Instytutu Maszyn Matematycznych (Warsaw) 28, No. 1, 25-38. (1986).

RUBI 1985-1 Rubin, D. S.
Polynomial algorithms for m*(m+1) integer programs and m*(m+k) diophantine systems. Operations Research Letters 3, 289-292. (1985).

RUBIA 1986-1 Rubinshtein, M. I.
Algorithm for the minimax assignment problem with a sparse rectangular matrix. Automation and Remote Control 47, 73-80; translation from Avtomatika i Telemekhanika (1986), No.1, 81-89. (1986).

RUBL 1983-1 Rublev, V. S.
A generalized assignment problem and the scheduling of a transportation route (Russian). In: Algorithmic Constructions and Their Efficiency, 26-33. Yaroslavl: Yaroslav. Gos. Univ. (1983).

RUCI 1982-1 Rucinski, A.
Matchings and k-factors in a random graph. Studia Scientiarum Mathematicarum Hungaria (Budapest) 17, 335-340. (1982).

RUDN 1984-1 Rudnev, Y. P.
A method for solution of a calendar planning problem with discrete time (Russian). Kibernetika (Kiev), No. 5, 89-93. (1984).

RUDNA 1980-1 Rudneva, T. L.
A modification of Balas' algorithm for a class of integer programming problems. (Russian). in: V. A. Bereznev, V. G. Karmanov (Eds.): Questions of optimiztion and control, Moscow 1980, 12-15. Moscow: Izdatel'stvo Moskovskogo Universiteta. (1980).

RUDNB 1981-1 Rudneva, T. L.; Yaranov, M. G.
Program realization of the completely integer Gomory algorithm (Russian). In: V. A. Beteynev, V. G. Katmanov, B. M. Pavlov (Eds.): Mathematical Questions of Optimization and Control Problems. Collection of Articles. 3-14. Moscow: Iydat. Moskov. Univ. (1981).

RUDO 1984-1 Rudowski, M. H.
Hypothesis on the length of minimal Steiner tree in multidimensional spaces with a rectilinear metric. Archiwum Automatyki i Telemechaniki (Warsaw) 29, 353-358. (1984).

RUGE 1984-1 Rüger, W.
Berechnung der Zusammenhangswahrscheinlichkeit von Netzgraphen. Wissenschaftliche Zeitschrift, Hochschule für Verkehrswesen "Friedrich List" Dresden 31, 961-966. (1984).

RUGE 1984-2 Rüger, W.
Ermittlung der maximalen Anzahl kantendisjunkter Wege in einem Netz. Wissenschaftliche Zeitschrift, Hochschule für Verkehrswesen Friedrich List Dresden 31, 957-960. (1984).

RUHE 1984-1 Ruhe, G.
Ein Verfahren zum Auffinden eingebetteter Netzwerkstrukturen in linearen Programmen. In: K. Lommatzsch (Ed.): Proceedings, 16. Jahrestagung "Mathematische Optimierung", Sellin/ Rügen 1984, 80-92. Seminarbericht Nr. 64, Sektion Mathematik, Humboldt-Univ. Berlin (DDR). (1984).

RUHE 1985-1 Ruhe, G.
Characterization of all optimal solutions and parametric maximal flows in networks. Optimization 16, 51-62. (1985).

RUHEA 1985-1 Ruhe, G.; Ivanchev, D. T.
The solution of generalized network flow problems with parametric capacities and stochastic demands. Foundations of Control Engineering (Poznan) 9 (1984), 123-139. (1985).

RUHEB 1986-1 Ruhe, G.; Sieber, N.
Zur Komplexität zweier vektorieller Netzwerkflussprobleme. Seminarber., Humboldt-Univ. Berlin, Sekt. Math. 85, 91-97. (1986).

RULL 1986-1 Rülling, W.
Constructing legal layouts from stick diagrams is an NP-complete problem. Report 09/1986. Saarbrücken: Univ. des Saarlandes, Fachbereich 10, 17 p. (1986).

RUMC 1984-1 Rumchev, V. G.; Mateev, L. N.
Solution of a dynamic transportation problem in network formulation by the method of a time expanded network (Russian). Godishnik na Visshite Uchebni Zavedeniya, Prilozhna Matematika (Sofia) 19, No. 3, 149-161. (1984).

RUSC 1981-1 Rüschendorf, L.
On the multidimensional assignment problem. Working Paper. Freiburg: Univ. Freiburg, 10 p. (1981).

RUSS 1986-1 Russell, R. A.
A comparison of heuristics for scheduling projects with cash flows and resource restrictions. Management Science 32, 1291-1300. (1986).

RUSSA 1986-1 Russell, R. A.; Morrel, R. B.
Routing special-education school buses. Interfaces 16, 56-64. (1986).

RUSSB 1985-1 Russell, R. S.; Taylor, B. W. III
An evaluation of scheduling policies in a dual resource constrained assembly shop. IIE Transactions 17, 219-232. (1985).

RUSSB 1985-2 Russell, R. S.; Taylor, B. W. III
An evaluation of sequencing rules for an assembly shop. Decision Sciences 16, 196-212. (1985).

RYAN 1986-1 Ryang, S. Y.
A solution method of the fractional n-acyclic transportation problem (Korean). Cho-son In-min Kong-hwa-kuk Kwa-hak-won Tong-bo, No. 4, 1-4. (1986).

RYAN 1987-1 Ryang, S. Y.
A solution method of the fractional n-index transportation problem with pass capacity (Korean). Su-hak (Pyongyang), No. 1, 21-28. (1987).

RYJA 1987-1 Ryjacek, Z.; Zelinka, B.
Locally disconnected graphs with large numbers of edges. Mathematica Slovaca 37, 195-198. (1987).

RYU 1983-1 Ryu, H. D.
On invariants of graphs. IV. (Korean). Cho-son In-min Kong-hwa-kuk Kwa-hak-won T'ong-bo No. 6, 8-10. (1983).

RYU 1984-1 Ryu, H. D.
On invariants of graphs. V. (Korean). Cho-son In-min Kong-hwa-kuk Kwa-hak-won T'ong-bo, No. 4, 16-18. (1984).

RYU 1985-1 Ryu, H. D.
On invariants of graphs, VI (Korean). Cho-son In-min Kong-hwa-kuk Kwa-hak-won Tong-bo No. 3, 16-17. (1985).

RYU A 1985-1 Ryu, H. D.; Bak, T. B.
On dual problem of Hadwiger's conjecture (Korean). Mathematics and Physics (Pyongyang) No. 3, 30-33. (1985).

RYU B 1982-1 Ryu, H. D.; Chong, C. J.
On a graph invariant and its application (Korean). Su-hak kwa Mul-li No. 3, 33-38. (1982).

RYU C 1984-1 Ryu, H. D.; Li, C. G.
Criteria for the efficiency of algorithms in graph theory and its application (Korean). Mathematics and Physics (Pyongyang) No. 3, 30-33. (1984).

RYU D 1984-1 Ryu, H. D.; Ri, C. K.
Criteria for the efficiency of algorithms in graph theory and its application (Korean). Su-hak kwa Mul-li No. 3, 30-34. (1984).

RZEM 1986-1 Rzemykowski, Z.
On the method of maximal sets of activities (Polish). Przeglad Statystyczny 33, 415-428. (1986).

SAAT 1987-1 Saatcioglu, Ö.
A multi-attribute assignment goal-programming model with incentives. Journal of the Operational Research Society 38, 361-366. (1987).

SABI 1986-1 Sabin, G. C. W.; Winter, G. K.
The impact of automated timetabling on universities - a case study. Journal of the Operational Research Society 37, 689-694. (1986).

SACH 1983-1 Sachs, H.
On a spatial analogue of Kuratowski's theorem on planar graphs - an open problem. Graph Theory (Lagow, 1981), 230-241. Lecture Notes in Mathematics 1018. Berlin, New York: Springer-Verlag. (1983).

SACH 1985-1 Sachs, H.
On colour critical graphs. In: L. Budach (Ed.): Fundamentals of Computation Theory, FCT '85, Lecture Notes in Computer Science Vol. 199, 390-401. Berlin: Springer-Verlag. (1985).

SACH 1986-1 Sachs, H.
On the number of perfect matchings in bipartite plane graphs. In: J. Machner, G. Schaar (Eds.): Algebra und Graphentheorie, Beiträge der Jahrestagung "Algebra und Grenzgebiete" in Siebenlehn/DDR 1985, 65-70. Freiberg: Bergakademie Freiberg, Sektion Mathematik. (1986).

SACHA 1982-1 Sachs, H.; Stiebitz, M.
Simple eigenvalues of transitive graphs. Studia Scientiarum Mathematicarum Hungaria (Budapest) 17, 77-90. (1982).

SAFO 1985-1 Safonova, T. E.
Stability of compatible schedules. Journal of Soviet Mathematics 29, 1660-1670. (1985).

SAFO 1987-1 Safonova, T. E.
Reconstruction of a consistent schedule in connection with change of the volume of a job. Journal of Soviet Mathematics 36, 173-185; translation from Zapiski Nauchnykh Seminarov Leningradskogo Otdeleniya Ordena Lenina Matematicheskogo Instituta Steklova 139, 5-21. (1987).

SAIT 1985-1 Saito, A.; Tian, S.-L.
The binding number of line graphs and total graphs. Graphs and Combinatorics 1, 351-356. (1985).

SAKA 1984-1 Sakaguchi, M.
A sequential stochastic assignment problem associated with a non-homogeneous Markov process. Mathematica Japonica (Osaka) 29, 13-22. (1984).

SAKA 1984-2 Sakaguchi, M.
A sequential stochastic assignment problem with an unknown number of jobs. Mathematica Japonica (Osaka) 29, 141-152. (1984).

SAKAA 1984-1 Sakarovitch, M.
Optimisation combinatoire - methodes mathematiques et algorithmiques. Vol. 1: Programmation discrete. Paris: Hermann. (1984).

SAKAA 1984-2 Sakarovitch, M.
Optimisation combinatoire - methodes mathematiques et algorithmiques. Vol. 2: Graphes et programmation lineaire. Paris: Hermann. (1984).

SAKS 1986-1 Saks, M.
Some sequences associated with combintorial structures. Discrete Mathematics 59, 135-166. (1986).

SAKSA 1985-1 Saksena, J. P.
Applications of decision networks. European Journal of Operational Research 19, 41-44. (1985).

SAKSB 1986-1 Saksena, J. P.; Ramachandran, A.
An optimal fleet utilisation. European Journal of Operational Research 25, 406-420. (1986).

SALH 1987-1 Salhi, S.; Rand, G. K.
Improvements to vehicle routing heuristics. Journal fo the Operational Research Society 38, 293-296. (1987).

SALT 1987-1 Saltzman, M. J.; Baybars, I.
A two-process implicit enumeration algorithm for the simple assembly line balancing problem. European Journal of Operational Research 32, 118-129. (1987).

SAMO 1984-1 Samodivkin, V. D.
Hamiltonian line graphs (Russian). Godishnik na Visshite Uchebni Zavedeniya, Prilozhna Matematika (Sofia) 19(1983), 163-170. (1984).

SAMP 1983-1 Sampathkumar, E.
A generalization of Menger's theorem for trees. Journal of Combinatorics, Information and System Sciences (Delhi) 8, 79-80. (1983).

SAMP 1984-1 Sampathkumar, E.
Connectivity of a graph - a generalization. Journal of Combinatorics, Information and System Sciences (Delhi) 9, 71-78. (1984).

SAMP 1984-2 Sampathkumar, E.
Convex sets in a graph. Indian Journal of Pure and Applied Mathematics (New Delhi) 15, 1065-1071. (1984).

SAMP 1984-3 Sampathkumar, E.
Point signed and line signed graphs. National Academy Science Letters (Allahabad) 7, No. 3, 91-93. (1984).

SAMPA 1982-1 Sampathkumar, E.; Bhave, V. N.
Arbitrarily traceable graphs. The Mathematics Student 43, 428-430. (1982).

SAMPB 1985-1 Sampathkumar, E.; Neeralagi, P. S.
The neighbourhood number of a graph. Indian Journal of Pure and Applied Mathematics (New Delhi) 16, 126-132. (1985).

SAMPB 1986-1 Sampathkumar, E.; Neeralagi, P. S.
The line neighbourhood number of a graph. Indian Journal of Pure and Applied Mathematics 17, 142-149. (1986).

SAMPC 1982-1 Sampathkumar, E.; Walikar, H. B.
On a conjecture of Cockayne and Hedetniemi. Journal of Combinatorics, Information & Systems Sciences (Delhi) 7, 212-215. (1982).

SANC 1987-1 Sanchez-Flores, A.
A counterexample to a generalization of Richardson's theorem. Discrete Mathematics 65, 319-320. (1987).

SANCA 1985-1 Sancho, N. G. F.
Maximum reliability through a network with resource constraints. Journal of the Operational Reserch Society 36, 537-540. (1985).

SANCA 1986-1 Sancho, N. G. F.
Optimal and near-optimal solutions to the travelling salesman problem. Montreal: McGill Univ., Dept. of Mathematics and Statistics, 10p. (1986).

SAND 1986-1 Sandi, C.
On a nonbasic dual method for the transportation problem. In: G. Gallo, C. Sandi (Eds.): Netflow at Pisa, Mathematical Programming Study 26, 65-82. (1986).

SANK 1985-1 Sankaran, V.; Krishnamoorthy, V.
On the diameters of spanning trees. IEEE Transactions on Circuits and Systems 32, 1060-1062. (1985).

SANN 1985-1 Sannikova, A. K.; Shafranskij, Y. M.
NP-completeness of some optimal sequencing problems with a given grouping of elements. Cybernetics 21, 628-633; translation from Kibernetika (1985), No.5, 55-58. (1985).

SANNA 1985-1 Sannomiya, N.; Okamoto, K.
A method for decomposing mixed-integer linear programs with staircase structure. International Journal of Systems Science 16, 99-112. (1985).

SANT 1985-1 Santoro, N.; Khatib, R.
Labelling and implicit routing in networks. Computer Journal 28, 5-8. (1985).

SANTA 1984-1 Santoro, N.; Urrutia, J.
An improved algorithm for Boolean matrix multiplication. SCS-TR-38. Ottawa: Carleton Univ., School of Computer Science, 16 p. (1984).

SANTB 1985-1 Santos, C.; Magazine, M.
Batching in single operation manufacturing systems. Operations Research Letters 4, 99-104. (1985).

SARR 1985-1 Sarrafzadeh, M.; Preparata, F. P.
Compact channel routing of multiterminal nets. In: G. Ausiello, M. Lucertini (Eds.): Analysis and Design of Algorithms for Combinatorial Problems, Annals of Discrete Mathematics Vol. 25, 255-280. Amsterdam: North-Holland. (1985).

SARV 1984-1 Sarvanov, V. I.
Sufficient conditions for traceability of planar 3-connected graphs (Russian). Izvestiya Akademii Nauk BSSR, Seriya Fiziko-Matematicheskikh Nauk (Minsk) 1984, No. 3, 11-14. (1984).

SARW 1983-1 Sarwate, D. G.
A note on strongly regular Hamiltonian equivalence of K(2n) and K(2n-1,2n-1) and its generalization. Journal of the Indian Academy of Mathematics (Shankarbag, Indore) 5, 65-67. (1983).

SASS 1985-1 Sassano, A.
On the facial structure of the set covering polytope. Report 139. Roma: Consiglio Nazionale delle Ricerche, 47 p. (1985).

SAST 1987-1 Sastry, P. S.; Jayakumar, N.; Veni Madhavan, C. E.
Efficient algorithms for domination and Hamilton circuit problems on permutation graphs. In: K. V. Nori (Ed.): Foundations of Software Technology and Theoretical Computer Science, 65-78. Lecture Notes in Computer Science 287, Berlin, Heidelberg, New York: Springer. (1987).

SATY 1987-1 Satyanarayana, A.; Tindell, R.
Chromatic polynomials and network reliability. Discrete Mathematics 67, 57-80. (1987).

SAVA 1984-1 Savage, C.
A systolic design for connectivity problems. IEEE Transactions on Computers C-33, 99-104. (1984).

SAVE 1985-1 Savelsbergh, M. W. P.
Vehicle routing and computer graphics. CWI Newsletter (Centre for Mathematics and Computer Science Amsterdam) 7, 15-24. (1985).

SAVE 1986-1 Savelsbergh, M. W. P.
Local search in routing problems with time windows. In: C. L. Monma (Ed.): Algorithms and Software for Optimization - Part I, 285-306. Annals of Operations Research 4, Basel: Baltzer. (1986).

SAVE 1988-1 Savelsbergh, M. W. P.
Local search for constrained routing problems. In: H. Schellhaas, P. van Beek, H. Isermann, R. Schmidt, M. Zijlstra (Eds.): Operations Research Proceedings 1987 (DGOR/ NSOR), 625-626. Berlin: Springer. (1988).

SAVEA 1985-1 Savelsbergh, M. W. P.; Volgenant, T.
Edge exchanges in the degree-constrained minimum spanning tree problem. Computers & Operations Research 12, 341-348. (1985).

SAVI 1986-1 Saviozzi, G.
Advanced start for the multicommodity network flow problem. In: G. Gallo, C. Sandi (Eds.): Netflow at Pisa, Mathematical Programming Study 26, 221-224. (1986).

SAWI 1985-1 Sawik, T. J.
Scheduling lots of dependent unit-time operations on identical machines to minimize schedule length. European Journal of Operational Research 19, 331-336. (1985).

SAWI 1987-1 Sawik, T. J.
A critical path algorithm for scheduling a batch-type production on nonidentical machines. Foundations of Control Engineering 12, No. 1, 31-44. (1987).

SAWI 1987-2 Sawik, T. J.
Multilevel scheduling of multistage production with limited in-process inventory. Journal of the Operational Research Society 38, 651-664. (1987).

SAXE 1985-1 Saxena, P. K.
Shipping cost/completion-date tradeoffs in fractional transportation problems with penalty cost. Ricerca Operativa 35, 3-16. (1985).

SBIH 1984-1 Sbihi, N.; Uhry, J.-P.
A class of h-perfect graphs. Discrete Mathematics 51, 191-206. (1984).

SCAR 1985-1 Scarf, H. E.
Integral polyhedra in three-space. Mathematics of Operations Research 10, 403-438. (1985).

SCHA 1986-1 Schaar, G.
Remarks on k-path traceable graphs. In: J. Machner, G. Schaar (Eds.): Algebra und Graphentheorie, Beiträge der Jahrestagung "Algebra und Grenzgebiete" in Siebenlehn/DDR Oktober 1985, 75-78. Freiberg: Bergakademie Freiberg, Sektion Mathematik. (1986).

SCHAA 1985-1 Schaffrin, B.
Aspects of network design. In: E. W. Grafarend, F. Sanso (Eds.): Optimization and design of geodetic networks (Erice, Italy, 1984), 548-597. Berlin: Springer-Verlag. (1985).

SCHE 1985-1 Scheinerman, E. R.
Characterization and computational complexity questions for representation classes of graphs. Congressus Numerantium 49, 195-204. (1985).

SCHE 1985-2 Scheinerman, E. R.
Characterizing intersection classes of graphs. Discrete Mathematics 55, 185-194. (1985).

SCHE 1985-3 Scheinerman, E. R.
Irrepresentability by multiple intersection, or why the interval number is unbounded. Discrete Mathematics 55, 195-212. (1985).

SCHEA 1985-1 Scheithauer, G.
On recursive computation of minimum spanning trees for special partial and subgraphs. In: H. Sachs (Ed.): Graphs, Hypergraphs and Applications, Proc. Conf. on Graph Theory (Eyba 1984), 141-144. Leipzig: Teubner. (1985).

SCHEA 1985-2 Scheithauer, G.
On recursive computation of minimum spanning trees for special partial graphs. Optimization 16, 41-50. (1985).

SCHEA 1985-3 Scheithauer, G.
Rekursive Berechnung von Minimalgerüsten für spezielle Subgraphen. Elektronische Informationsverarbeitung und Kybernetik 21, 613-624. (1985).

SCHEA 1987-1 Scheithauer, G.
On the determination of a restricted shortest path (German). Seminarber., Humboldt-Univ. Berlin, Sekt. Math. 90, 107-114. (1987).

SCHEB 1985-1 Scheithauer, G.; Terno, J.
Lösung von Guillotine Zuschnittproblemen mittels revidierter Simplexmethode und Spaltengenerierung. 30. Internationales Wissenschaftliches Kolloquium (TH Ilmenau, DDR) Heft 4E, 135-138. (1985).

SCHEB 1986-1 Scheithauer, G.; Terno, J.
Effektive Lösung von Guillotine-Zuschnittproblemen. Wissenschaftliche Zeitschrift der Technischen Universität Dresden 35, No. 4, 35-40. (1986).

SCHI 1985-1 Schiemangk, C.
Convergence of the thermodynamically motivated simulation and experiments for solving the Steiner-Tree-Problem. In: Mathematische Optimierung - Theorie und Anwendungen, 30. Int. Wiss. Kolloq., Ilmenau 1985, 4, 139-142. (1985).

SCHL 1983-1 Schlechtweg, M.; Hoppe, V.
Polymatroide mit baumartigen Facettensystemen. I. Jena: Friedrich-Schiller Univ. Jena, Research Report N/83/23, 8 p. (1983).

SCHL 1983-2 Schlechtweg, M.; Hoppe, V.
Polymatroide mit baumartigen Facettensystemen. II. Jena: Friedrich-Schiller Univ. Jena, Research Report N/83/26, 8 p. (1983).

SCHL 1983-3 Schlechtweg, M.; Hoppe, V.
Polymatroide mit baumartigen Facettensystemen. III. Jena: Friedrich-Schiller Univ. Jena, Research Report N/83/27, 8 p. (1983).

SCHM 1982-1 Schmerl, J. H.
The effective version of Brooks' theorem. Canadian Journal of Mathematics 34, 1036-1046. (1982).

SCHMA 1984-1 Schmidt Stoup, M.
Hamiltonsche und Hypohamiltonsche unendliche Graphen. Dissertation, Abteilung Mathematik der Universität Dortmund. (1984).

SCHMB 1984-1 Schmidt, G.
Scheduling on semi-identical processors. Zeitschrift für Operations Research, Series A 28, 153-162. (1984).

SCHMB 1986-1 Schmidt, G.
A fertilizer transportation problem. Zeitschrift für Operations Research, Series B 30, 135-146. (1986).

SCHMC 1985-1 Schmidt, W.
A characterization of undirected branching greedoids. Report 85387-OR. Bonn: Univ. Bonn, Inst. für Ökonometrie u. Op. Research, 35 p. (1985).

SCHMC 1985-2 Schmidt, W.
A min-max theorem for greedoids. Report 85396-OR. Bonn: Univ. Bonn, Inst. für Ökonometrie u. Op. Research, 12 p. (1985).

SCHMC 1985-3 Schmidt, W.
Greedoids and searches in directed graphs. Report 85388-OR. Bonn: Univ. Bonn, Inst. für Ökonometrie u. Op. Research, 18 p. (1985).

SCHN 1986-1 Schneider, M. H.
A complementary pivoting algorithm for linear network problems. In: C. L. Monma (Ed.): Algorithms and Software for Optimization - Part II, 439-462. Annals of Operations Research 5, Basel: Baltzer. (1986).

SCHNA 1987-1 Schniederjans, M. J.; Kim, G. C.
A goal programming model to optimize departmental preference in course assignments. Computers and Operations Research 14, 87-96. (1987).

SCHNB 1987-1 Schnorr, C. P.
Geometry of numbers and integer programming (summary). Technical Report 87-019. Chicago: Univ. of Chicago, Dept. of Computer Science, 9 p. (1987).

SCHO 1982-1 Schoch, M.
Zur Lösung einer speziellen linearen diskreten Optimierungsaufgabe mit Zusatzforderungen. In: Proceedings of the 13th Annual Conference on Mathematical Optimization (Berlin, 1981), 185-196. Seminarberichte 39, Humboldt Univ., Berlin. (1982).

SCHO 1985-1 Schoch, M.
Optimale Fertigungsplanung für einen zweistufigen Produktionsprozess. In: E. Girlich (Ed.): Diskrete Optimierung, 104-116. Jena: Friedrich-Schiller-Universität. (1985).

SCHOA 1985-1 Schöning, U.
Complexity and structure. Berlin, Heidelberg: Springer. (1985). (Lecture Notes in Computer Science, 211)

SCHOB 1986-1 Schoone, A. A.; Bodlaender, H. L.; Leeuwen, J. van
Improved diameter bounds for altered graphs. In: Graph-theoretic concepts in computer science, Proc. Int. Workshop, Bernried/FRG 1986, Lecture Notes in Computer Science 246, 227-236. (1986).

SCHR 1985-1 Schrage, L.; Wolsey, L. A.
Sensitivity analysis for branch and bound integer programming. Operations Research 33, 1008-1023. (1985).

SCHRA 1986-1 Schreier, H.
A variant for checking a solution criterion for the balanced multicommodity transportation problem (German). Seminarber., Humboldt-Univ. Berlin, Sekt. Math. 85, 108-117. (1986).

SCHRB 1985-1 Schreier, H.; Tzscharschuch, D.
Ein Verfahren zur Lösung eines mehrstufigen Transportproblems mit nach oben beschränkten Variablen und rechentechnische Realisierung. In: 8th Conf. on Mathematcial Methods in Economics (Sellin 1984), Seminarberichte 71, 100-109. Berlin: Humboldt-Universität. (1985).

SCHRC 1983-1 Schreuder, J. A. M.
The challenge of program-timetables: a new concept. Enschede: Univ. of Technology, Department of Applied Mathematics, Memorandum 425. (1983).

SCHRC 1988-1 Schreuder, J. A. M.
Location of Rotterdam fire stations. In: H. Schellhaas, P. van Beek, H. Isermann, R. Schmidt, M. Zijlstra (Eds.): Operations Research Proceedings 1987 (DGOR/NSOR), 299-305. Berlin: Springer. (1988).

SCHRD 1984-1 Schreuder, J. A. M.; Velde, J. A. van der
Timetables in Dutch high schools. In: J. P. Brans (Ed.): Operational Research '84, 601-612. Amsterdam: North-Holland. (1984).

SCHRE 1983-1 Schrijver, A.
Packing and covering of crossing families of cuts. Report AE 1/83. Amsterdam: Univ. of Amsterdam, Faculty of Actuarial Science & Econometrics, 29 p. (1983).

SCHRE 1984-1 Schrijver, A.
Total dual integrality from directed graphs, crossing families, and sub- and supermodular functions. In: W. R. Pulleyblank (Ed.): Progress in Combinatorial Optimization, 315-362. Toronto: Academic Press. (1984).

SCHRE 1985-1 Schrijver, A.
Supermodular colourings. In: L. Lovasz, A. Recski (Eds.): Matroid Theory, 327-344. Coll. Math. Soc. J. Bolyai Vol. 40. Amsterdam: North-Holland. (1985).

SCHRE 1986-1 Schrijver, A.
Distances and cuts in planar graphs. Report OS-R8610, CWI, Amsterdam. (1986).

SCHRE 1986-2 Schrijver, A.
Geometric methods in discrete optimization. In: M. Hazewinkel, J. K. Lenstra, L. G. L. T. Meertens (Eds.): Mathematics and Computer Science II, 111-138. CWI Monograph 4, Amsterdam: North-Holland. (1986).

SCHRE 1986-3 Schrijver, A.
Polyhedral combinatorics - some recent developments and results. Report OS-R8607. Amsterdam: Centre for Mathematics and Computer Science, 15 p. (1986).

SCHRE 1986-4 Schrijver, A.
Polyhedral proof methods in combinatorial optimization. Discrete Applied Mathematics 14, 111-134. (1986).

SCHRE 1986-5 Schrijver, A.
Theory of linear and integer programming. Chichester, New York, Brisbane: Wiley. (1986).

SCHRE 1987-1 Schrijver, A.
Disjoint circuits of prescribed homotopies in a graph on a surface. RUTCOR Research Report 40-87. New Brunswick: Rutgers Univ., Center for Operations Research, 32 p. (1987).

SCHRE 1987-2 Schrijver, A.
Polyhedral combinatorics. Report OS-R8701. Amsterdam: Centre for Mathematics and Computer Science, 67 p. (1987).

SCHRF 1984-1 Schröter, N.
Average-case analysis of an approximative algorithm for the knapsack problem. In: G. Hammer, D. Pallaschke (Eds.): Methods of Operations Research 51, Vol. 2, 455-468. Königstein/Ts.: Verlag Anton Hain Meisenheim. (1984).

SCHU 1986-1 Schubert, I. S.; Zimmermann, U.
One-parametric bottleneck transportation problems. In: C. L. Monma (Ed.): Algorithms and Software for Optimization - Part I, 343-372. Annals of Operations Research 4, Basel: Baltzer. (1986).

SCHUA 1987-1 Schulz, G.
Ein duales Verfahren zur Lösung des Transportproblems mit zusätzlichen linearen Restriktionen. Wissenschaftliche Zeitschrift der Technischen Universität Otto von Guericke 31, No.5, 55-57. (1987).

SCHUB 1985-1 Schulz, R.
Über das Feedback-Vertex-Set-Problem. Dissertation, Univ.-Gesamthochsch.-Paderborn, Fachber. Math.-Informatik. (1985).

SCHUC 1985-1 Schumacher, H.; Wagner, K.
Zur Finitisierung von Färbungsproblemen. Mitteilungen der Mathematischen Gesellschaft Hamburg 11, 335-340. (1985).

SCHW 1983-1 Schwarze, J.
Netzplantheorie. Herne, Berlin: Verlag Neue Wirtschafts-Briefe. (1983).

SCOB 1984-1 Scobey, P.; Kabe, D. G.
Direct solutions to some constrained transportation problems. Industrial Mathematics 34, 109-121. (1984).

SCOT 1985-1 Scott, C. H.; Murtagh, B. A.; Sirri, E.
Solution of constrained minmax location with Euclidean distances via conjugate duality. New Zealand Operational Research 13, 61-68. (1985).

SCOTA 1987-1 Scott, D. D.
The competition-common enemy graph of a digraph. Discrete Applied Mathematics 17, 269-280. (1987).

SCOTB 1985-1 Scott, J. L.; Vignaux, G. A.
A Benders' decomposition algorithm to help airline schedules planning. New Zealand Operational Research 13, 87-96. (1985).
SCRI 1985-1 Scriabin, M.; Vergin, R. C.
A cluster-analytic approach to facility layout. Management Science 31, 33-49. (1985).
SCUD 1985-1 Scudder, G. D.; Hoffmann, T. R.
Composite cost-based rules for priority scheduling in a randomly routed job shop. International Journal of Production Research 23, 1185-1196. (1985).
SCUL 1987-1 Sculli, D.; Mok, K. C.; Cheung, S. H.
Scheduling vehicles for refuse collection. Journal of the Operational Research Society 38, 233-240. (1987).
SCUT 1985-1 Scutella, M. G.; Scevola, G.
An algorithm for the maximum cardinality matching problem in convex bipartite graphs. Nota Interna S-15-85. Pisa: Univ. di Pisa, Dipt. di Informatica, 27 p. (1985).
SEBA 1986-1 Sebastian, H.-J.
Optimization and optimal control of a general transportation, exchange and inventory process II: routing problems with time interval restrictions (German). Wissenschaftliche Zeitschrift der Technischen Hochschule Leipzig 10, 193-202. (1986).
SEBO 1985-1 Sebö, A.
On the Chinese postman problem: algorithms, structure and applications. Working Paper MO/ 61. Budapest: Eötvös Lorand Univ., 39 p. (1985).
SEBO 1985-2 Sebö, A.
The Schrijver system of odd join polyhedra. Report 85394-OR. Bonn: Univ. Bonn, Inst. für Ökonometrie u. Op. Research, 17 p. (1985).
SEBO 1986-1 Sebö, A.
Application of network flow methods in the scheduling of terminal sessions (Hungarian). Alkalmazott Matematikai Lapok 12, 15-39. (1986).
SEBO 1986-2 Sebö, A.
Finding the t-join structure of graphs. Mathematical Programming 36, 123-134. (1986).
SEBO 1987-1 Sebö, A.
A particular timetable problem: terminal scheduling. Optimization 18, 97-112. (1987).
SEBO 1987-2 Sebö, A.
A quick proof of Seymour's theorem on t-joins. Discrete Mathematics 64, 101-104. (1987).
SEDG 1983-1 Sedgewick, R.
Mathematical analysis of combinatorial algorithms. in: G. Louchard, G. Latouche (Eds.): Probability Theory and Computer Science, 123-205. Academic Press (London, New York). (1983).
SEDGA 1984-1 Sedgewick, R.; Vitter, J. S.
Shortest paths in Euclidean graphs. In: 25th Annual Symposium on Foundations of Computer Science, Florida 1984, 417-424. Los Angeles: IEEE Computer Society. (1984).
SEDGA 1986-1 Sedgewick, R.; Vitter, J. S.
Shortest paths in Euclidean graphs. Algorithmica 1, 31-48. (1986).
SEDL 1985-1 Sedlacek, J.
Finite graphs with distinct neighbourhoods. In: H. Sachs (Ed.): Graphs, Hypergraphs and Applications, Proc. Conf. on Graph Theory (Eyba 1984), 152-156. Leipzig: Teubner. (1985).
SEDL 1985-2 Sedlacek, J.
Über eine spezielle Klasse von asymmetrischen Graphen. In: R. Bodendiek, H. Schumacher, G. Walther (Eds.): Graphen in Forschung und Unterricht, Festschrift Klaus Wagner, 167-177. Salzdetfurth: Franzbecker. (1985).
SEDU 1985-1 Sedukhin, S. G.
Computational structures of algorithms for finding paths in a graph (Russian). Izvestiya Akademii Nauk SSSR, Tekhnicheskaya Kibernetika (Moscow), No. 1, 29-36; translated from: Soviet Journal of Computational Systems Systems Sciences 23, No. 3, 1-7. (1985).
SEEL 1984-1 Seeländer, J.
Einige Bemerkungen zur Lösung von rein ganzzahlingen linearen Optimierungsproblemen. Wissenschaftliche Zeitschrift der Technischen Hochschule Leuna-Merseburg 26, No. 2, 318-331. (1984).
SEEL 1984-2 Seeländer, J.
Über ein Kriterium zur Lösbarkeit reinganzzahliger linearer Optimierungsprobleme. Optimization 15, 397-404. (1984).
SEEL 1986-1 Seeländer, J.
Comments on special scalar auxiliary problems in discrete optimization (German). In: Seminarber., Humboldt-Univ. Berlin, Sekt. Math. 80, 98-109. (1986).
SEEL 1987-1 Seeländer, J.
On two duality concepts of discrete vector optimization (German). Seminarber., Humboldt-Univ. Berlin, Sekt. Math. 90, 115-131. (1987).
SEES 1985-1 Seese, D.
Baum-partite Graphen und die Komplexität von Algorithmen. In: Graphen und Netzwerke - Theorie und Anwendungen, 30. Int. Wiss. Kolloq., Ilmenau 1985, 5, 101-103. (1985).
SEES 1985-2 Seese, D.
Tree-partite graphs and the complexity of algorithms. In: L. Budach (Ed.): Fundamentals of Computation Theory, FCT '85, Lecture Notes in Computer Science Vol. 199, 412-421. Berlin: Springer-Verlag. (1985).
SEGE 1985-1 Segev, A.
The node-weighted Steiner tree problem. Working Paper. Berkeley: Univ. of California, School of Business Admin., 28 p. (1985).
SEID 1983-1 Seidman, S. B.
Internal cohesion of LS sets in graphs. Social Networks 5, 97-107. (1983).
SEID 1983-2 Seidman, S. B.
LS sets as cohesive subsets of graphs and hypergraphs. Mathematical Social Sciences 6, 87-91. (1983).
SEID 1983-3 Seidman, S. B.
Network structure and minimum degree. Social Networks 5, 269-287. (1983).

SEIDA 1987-1 Seidmann, A.
On-line scheduling of a robotic manufacturing cell with stochastic sequence-dependent processing rates. International Journal of Production Research 25, 907-924. (1987).
SEKA 1985-1 Sekanina, M.
Factors in powers of graphs. In: H. Sachs (Ed.): Graphs, Hypergraphs and Applications, Proc. Conf. on Graph Theory (Eyba 1984), 157-160. Leipzig: Teubner. (1985).
SEKI 1987-1 Sekiguchi, Y.
A decomposition theory based on a dominance relation and composite jobs. Discrete Applied Mathematics 17, 187-211. (1987).
SELE 1986-1 Selen, W. J.; Hott, D. D.
A mixed-integer goal-programming formulation of the standard flow-shop scheduling problem. Journal of the Operational Research Society 37, 1121-1128. (1986).
SELE 1986-2 Selen, W. J.; Hott, D. D.
A new formulation and solution of the flowshop scheduling problem with no in-process waiting. Applied Mathematical Modelling 10, 246-248. (1986).
SEME 1985-1 Semenkin, E.
Optimization of monotone pseudo-Boolean functions (Russian). Teoriya Optimalnykh Reshenii (Vilnius), No. 11, 40-50. (1985).
SEMEA 1984-1 Semenov, A. S.; Semenov, A. V.
On the Steiner-Weber problem in a rectangular metric (Russian). In: Y. A. Abramov (Ed.): Mathematical methods of optimization and control in complex systems, Collect. sci. Works, Kalinin 1984, 73-78. (1984).
SEMEB 1984-1 Semenova, N. V.
Solution of one generalized integer programming problem(Russian). Kibernetika (Kiev), No. 5, 25-31. (1984).
SEN 1983-1 Sen, M.; Roy, A. B.
A Boolean approach to graph labelling. II. Ganit (Dacca) 3, 99-106. (1983).
SEN A 1985-1 Sen, S.; Sherali, H. D.
A branch and bound algorithm for extreme point mathematical programming problems. Discrete Applied Mathematics 11, 265-280. (1985).
SEN A 1986-1 Sen, S.; Sherali, H. D.
Facet inequalities from simple disjunctions in cutting plane theory. Mahtematical Programming 34, 72-83. (1986).
SENG 1987-1 Sengoku, M.; Shinoda, S.; Chen, W.-K.
A fault diagnosis in a nonlinear location network. In: M. Ismail (Ed.): 29th Midwest Symposium on Circuits and Systems, 456-459. Amsterdam: North-Holland. (1987).
SENG 1987-2 Sengoku, M.; Shinoda, S.; Chen, W.-K.
A fault diagnosis in a nonlinear location network. Journal of the Franklin Institute 323, No. 3, 395-405. (1987).
SEPE 1986-1 Sepehri, M.; Silver, E. A.; New, C.
A heuristic for multiple lot sizing for an order under variable yield. IIE Transactions 18, 63-69. (1986).
SERA 1984-1 Serafini, P.
Dual relaxation and branch-and-bound techniques for multiobjective optimization. in: Interactive Decision Analysis, Proc. Int. Workshop, Luxemburg 1983. Lecture Notes in Economics and Mathematical Systems 229, 84-90. Berlin, Heidelberg, New York: Springer-Verlag. (1984).
SERAA 1985-1 Serafini, P.; Ukovich, W.
An approach towards solving periodic scheduling problems. Ricerca Operativa 35, 17-40. (1985).
SERAB 1984-1 Serazzi, G.
A bottleneck-driven scheduler. In: F. Baccelli, G. Fayolle (Eds.): Modelling and performance evaluation methodology. Proc. Int. Seminar, Paris 1983 (INRIA). Lecture Notes in Control and Information Sciences 60, 335-350. Berlin: Springer-Verlag. (1984).
SERD 1984-1 Serdyukov, A. I.
An algorithm with an estimate for the travelling salesman problem of the maximum (Russian). Upravlyaemye Sistemy (Novosibirsk) No. 25, 80-86. (1984).
SERD 1986-1 Serdyukov, A. I.
A polynomial algorithm with an estimate of accuracy of (6+root2)/8 for a problem on the maximal Hamiltonian cycle on the Euclidean plane (Russian). Metody Diskretnogo Analiza, No. 43, 87-96. (1986).
SERG 1987-1 Sergeev, S. I.
A new lower bound for the quadratic assignment problem (Russian). Zhurnal Vychislitelnoi Matematiki i Matematicheskoi Fiziki 27, 1802-1811. (1987).
SERGA 1982-1 Sergienko, I. V.
Some questions of the development and use of approximation methods of solution of discrete optimization problems (Russian). In: Cybernetic Methods in Planning, Design and Control, 45-62. Akademiya Nauk Ukrainskoi SSR, Inst. Kibernet. (Kiev). (1982).
SERGA 1985-1 Sergienko, I. V.
Mathematical models and methods for solving problems of discrete optimization. Kiev: Naukova Dumka. (1985).
SERGB 1984-1 Sergienko, I. V.; Artemenko, V. I.
Search for approximate solutions of integer linear problems by the methods of linear and parametric programming (Russian). Doklady Akademii Nauk Ukrainskoj SSR, Seriya A (Kiev) No. 7, 74-77. (1984).
SERGC 1984-1 Sergienko, I. V.; Chervak, Y. Y.; Grendzhi, Y. I.
Paralleling of discrete optimization in lexicographic algorithms (Russian). Kibernetika (Kiev), No. 5, 82-85. (1984).
SERGD 1983-1 Sergienko, I. V.; Filonenko, N. V.
Investigation of the stability of the vector of the right-hand sides of a system of constraints for an integer linear programming problem (Russian). In: I. V. Sergienko (Ed.): Software for program packages, 3-6. Kiev: Akademiya Nauk Ukrainskoi SSR, Institut Kibernetiki. (1983).
SERGE 1983-1 Sergienko, I. V.; Kaspshitskaya, M. F.
Stability of algorithms of the method of descent vector in a class of combinatorial optimization problems (Russian). Doklady Akademii Nauk Ukrainskoj SSR, Seriya A No. 10, 64-66. (1983).

SERGE 1984-1 Sergienko, I. V.; Kaspshitskaya, M. F.
Stability of a discrete allocation problem (Russian). Doklady Akademii Nauk Ukrainkoj SSR, Seriya A, No. 11, 72-75. (1984).

SERGF 1985-1 Sergienko, I. V.; Lebedeva, T. T.; Konzeratskaya, L. N.
Some parametric problems of integer convex programming and their approximate solution (Russian). Kibernetika (Kiev) No. 4, 1-5. (1985).

SERGG 1980-1 Sergienko, I. V.; Lebedeva, T. T.; Roshchin, V. A.
Approximate methods for solving discrete problems of optimization. "Naukova Dumka" (Kiev). (1980).

SERGH 1987-1 Sergienko, I. V.; Shilo, V. P.
Polynomial asymptotically-optimal random search algorithm for a multidimensional problem of Boolean linear programming (Russian). Doklady Akademii Nauk Ukrainskoj SSR, Seriya A, No.3, 70-72. (1987).

SERGH 1987-2 Sergienko, I. V.; Shilo, V. P.
Probabilistic approach to evaluate heuristics for problems of the Boolean linear programming (Russian). Kibernetika (Kiev) No. 1, 1-6. (1987).

SERGI 1986-1 Sergienko, I. V.; Sokirko, S. V.
Solution of a nonlinear integer programming problem (Russian). Issledovanie Operatsii i ASU (Kiev), No. 27, 36-44. (1986).

SERGJ 1985-1 Sergienko, T. I.
Method for solving a special integer programming problem. Cybernetics 21, 364-368; translation from Kibernetika (1985), No.3, 72-76. (1985).

SERV 1983-1 Servakh, V. V.
On the Akers-Friedman problem (Russian). Upravlyaemye Sistemy (Novosibirsk) 23, 70-81. (1983).

SETH 1987-1 Seth, A.
Wastage reduction in wood cutting. Opsearch 24, 94-105. (1987).

SETHA 1986-1 Seth, A.; Prasad, V. R.; Ramamurthy, K. G.
A heuristic approach to one-dimensional cutting stock problem. Opsearch 23, 235-243. (1986).

SEVA 1982-1 Sevastyanov, S. V.
Algorithms with estimates for the Johnson and Akers-Fridman problems in the case of three benches (Russian). Upravlyaemye Sistemy (Novosibirsk) 22, 51-57. (1982).

SEVA 1984-1 Sevastyanov, S. V.
Efficient construction of near-optimal schedules for the cases of arbitrary and alternative routes of parts. Soviet Mathematics, Doklady 29, 447-450. Translation from: Doklady Akademii Nauk SSSR (Moscow) 276, 46-48. (1984).

SEXT 1985-1 Sexton, T. R.; Bodin, L. D.
Optimizing single vehicle many-to-many operations with desired delivery times: I. Scheduling. Transportation Science 19, 378-410. (1985).

SEXT 1985-2 Sexton, T. R.; Bodin, L. D.
Optimizing single vehicle many-to-many operations with desired delivery times: II. Routing. Transportation Science 19, 411-435. (1985).

SEYM 1985-1 Seymour, P. D.
Applications of the regular matroid decomposition. In: L. Lovasz, A. Recski (Eds.): Matroid Theory, 345-358. Coll. Math. Soc. J. Bolyai Vol. 40. Amsterdam: North-Holland. (1985).

SEYM 1985-2 Seymour, P. D.
Minors of 3-connected matroids. European Journal of Combinatorics 6, 375-382. (1985).

SEYM 1986-1 Seymour, P. D.
Adjacency in binary matroids. European Journal of Combinatorics 7, 171-176. (1986).

SEYM 1986-2 Seymour, P. D.
Triples in matroid circuits. European Journal of Combinatorics 7, 177-186. (1986).

SEYMA 1984-1 Seymour, P. D.; Robertson, N.
Some new results on the well-quasi ordering of graphs. In: M. Pouzet, D. Richard (Eds.): Orders: Description and Roles, Annals of Discrete Mathematics Vol. 23, 343-354. Amsterdam: North-Holland. (1984).

SEYMB 1987-1 Seymour, P. D.; Thomassen, C.
Characterization of even directed graphs. Journal of Combinatorial Theory B 42, 36-45. (1987).

SFOR 1986-1 Sforza, A.; Sepe, A.
An algorithm for the problem of the absolute centre of a network (Italian). Ricerca Operativa 39, 71-106. (1986).

SGUR 1983-1 Sgurev, V. S.
A flow on a net with linear capacity constraints (Bulgarian). Problemi na Tekhnicheskata Kibernetika i Robotikata 17, 3-10. (1983).

SGUR 1983-2 Sgurev, V. S.
A min cost network flow with linear capacity constraints (Bulgarian). Avtomatika i Izchislitelna Tekhnika (Sofiya) 17, No. 2, 29-37. (1983).

SGUR 1983-3 Sgurev, V. S.
Capacity and maximum flow in a net with linear capacity constraints (Bulgarian). Problemi na Tekhnicheskata Kibernetika i Robotikata 18, 3-10. (1983).

SGUR 1985-1 Sgurev, V. S.
Flows on graphs with multiplicative capacity constraints. Problemi na Tekhnicheskata Kibernetika i Robotikata (Sofia) 21, 3-9. (1985).

SGURA 1986-1 Sgurev, V. S.; Dzhelatova, M. N.
Maximal linear flow and capacity of cutting set. Problemi na Tekhnicheskata Kibernetika i Robotikata 25, 3-10. (1986).

SGURB 1985-1 Sgurev, V. S.; Nikolova, M.
A linear network flow (Russian). Problemi Tekhnicheskata Kibernetika i Robotika (Sofia) 22, 3-10. (1985).

SGURC 1981-1 Sgurev, V. S.; Vasilev, V. S.
Approximate algorithm for integer programming (Russian). Problemi na Tekhnicheskata Kibernetika i Robotikata 12, 48-53. (1981).

SGURC 1985-1 Sgurev, V. S.; Vasilev, V. S.
An approximate algorithm solving mixed integer programming problems (Bulgarian). Problemi na Tekhnicheskata Kibernetika i Robotikata 14, 8-14. (1985).

SGURD 1984-1 Sgurev, V. S.; Vasilev, V. S.; Dokev, N. B.; Genova, K. B.
An algorithm for the solution of operative scheduling problems in an averaging mode (Bulgarian). Problemi na Tekhnicheskata Kibernetika i Robotikata (Sofia) 20, 28-36. (1984).

SGURE 1982-1 Sgurev, V. S.; Vasilev, V. S.; Dzhelatova, M. N.
An experimental investigation of algorithms solving linear problems with binary variables (Bulgarian). Problemi na Tekhnicheskata Kibernetika i Robotikata 14, 3-7. (1982).

SGURF 1982-1 Sgurev, V. S.; Vasilev, V. S.; Ivanov, S. S.; Dzhelatova, M. N.
Algorithmic maintenance of a packet of application programmes solving optimization problems with binary variables (Bulgarian). Problemi na Tekhnicheskata Kibernetika i Robotikata 15, 8-13. (1982).

SHAH 1987-1 Shahrokhi, F.
A primal dual approximation algorithm for a multicommodity flow problem. CSR 182. Socorro: New Mexico Inst. of Mining and Technology, 14 p. (1987).

SHAHA 1987-1 Shahrokhi, F.; Matula, D. W.
The maximum concurrent flow problem. CSR 179. Socorro: New Mexico Inst. of Mining and Technology, Computer Science Dept., 19 p. (1987).

SHAK 1985-1 Shakhbazyan, K. V.
Algorithm of shifts for optimal structural schedules. Journal of Soviet Mathematics 29, 1699-1713. (1985).

SHAK 1985-2 Shakhbazyan, K. V.
Solution of two sequencing problems. Journal of Soviet Mathematics 29, 1677-1699. (1985).

SHAK 1985-3 Shakhbazyan, K. V.
Structured solutions in scheduling problems. Journal of Soviet Mathematics 28, 440-447. (1985).

SHAKA 1984-1 Shakhbazyan, K. V.; Tushkina, T. A.
On the maximum number of flows in a pipeline mode (Russian). Zapiski Nauchnykh Seminarov Leningradskogo Otdeleniya Matematicheskogo Instituta imeni V. A. Steklova Academii Nauk SSSR (LOMI) (Leningrad) 139, 22-40. (1984).

SHAKB 1987-1 Shakhbazyan, K. V.; Tushkina, T. A.; Lebedinskii, M. M.
A spanning tree algorithm for the construction of flows in unconditional pipeline computers (Russian). Zapiski Nauchnykh Seminarov Leningradskogo Otdeleniya Matematicheskogo Instituta imeni V. A. Steklova (LOMI) 159, Chisl. Metody i Voprosy Organiz. Vychisl. 8, 40-46. (1987).

SHAM 1985-1 Shameeva, O. V.
Algebraic representability of matroids (Russian). Vestnik Moskovskogo Universiteta, Seriya I, Matematika, Mekhanika (Moscow) No. 4, 29-32. (1985).

SHAM 1985-2 Shameeva, O. V.
On the algebraic representability of matroids (Russian). Vestnik Moskovskogo Universiteta, Seriya I, No. 4, 29-32. (1985).

SHAMA 1984-1 Shamir, E.; Upfal, E.
Sequential and distributed graph coloring algorithms with performance analysis in random graph spaces. Journal of Algorithms 5, 488-501. (1984).

SHAMA 1985-1 Shamir, E.; Upfal, E.
A fast parallel construction of disjoint paths in networks. In: M. Karpinski, J. van Leeuwen (Eds.): Topics in the Theory of Computation, Annals of Discrete Mathematics Vol. 24, 141-154. Amsterdam: North-Holland. (1985).

SHAN 1983-1 Shan, K. Y.; Zhuang, S. Q.
On graceful labelings of some particular lobsters (Chinese). Qufu Shiyuan Xuebao (Qufu) No. 1, 17-22. (1983).

SHANA 1984-1 Shanker, K.; Kumar, P.
Coordinate repair scheduling for a group of goods-producting equipment. IIE Transactions 16, 240-247. (1984).

SHANB 1985-1 Shanthikumar, J. G.; Wu, Y. B.
Decomposition approaches in permutation scheduling problems with application to the M-machine flow shop scheduling problems. European Journal of Operational Research 19, 125-141. (1985).

SHAO 1985-1 Shao, J.-Y.
The connectivity of the interchange graph of class U(R,S) of (0,1) matrices. Acta Mathematicae Applicatae Sinica 2, 304-308. (1985).

SHAR 1986-1 Sharir, M.; Schorr, A.
On shortest paths in polyhedral spaces. SIAM Journal on Computing 15, 193-215. (1986).

SHARA 1984-1 Sharma, R. K.
A note on three machines flowshop sequencing problem (in special case). Acta Ciencia Indica, Mathematics (Meerut) 10, 238-240. (1984).

SHARB 1987-1 Sharma, S.; Rani, P.
On weighted-jobs scheduling problem involving machine break-down time interval. Pure and Applied Mathematika Sciences 25, No.1/2, 135-137. (1987).

SHAT 1983-1 Shatalina, L. S.
The compiling of a guaranteed spare-parts supply schedule formulated as a mixed linear programming problem. Automation and Remote Control 44, 536-541; translated from Avtomatika i Telemekhanika (Moscow) No. 4, 157-164. (1983).

SHAW 1982-1 Shawe-Taylor, J.; Pisanski, T.
Cycle permutation graphs with large girth. Glasnik Matematicki (Zagreb), Serija III 17 (37), 233-236. (1982).

SHCH 1985-1 Shcherbina, O. A.
Local algorithms for block-tree-like problems of discrete programming. U.S.S.R. Computational Mathematics and Mathematical Physics 25, No.4, 114-121; translation from Zhurnal Vychislitel'noj Matematiki i Matematicheskoj Fiziki (Moscow) 25, 1143-1154. (1985).

SHCH 1986-1 Shcherbina, O. A.
A modified local algorithm for solving block problems of discrete programming. U.S.S.R. Computational Mathematics and Mathematical Physics 26, No.5, 40-48; translation from Zhurnal Vychislitelnoi Matematiki i Matematicheskoi Fiziki 26 (1986), No.9, 1339-1349. (1986).

SHCH 1986-2 Shcherbina, O. A.
A unimodular problem of integer programming. U.S.S.R. Computational Mathematics and Mathematical Physics 26, No.4, 88-90; translation from Zhurnal Vychislitelnoi Matematiki i Matematicheskoi Fiziki 26 (1986), No.7, 1096-1099. (1986).

SHCH 1986-3 Shcherbina, O. A.
On local algorithms with variable neighbourhoods and indicator information (Russian). Zhurnal Vychislitelnoi Matematiki i Matematicheskoi Fiziki 26, No.10, 1588-1591. (1986).

SHCH 1987-1 Shcherbina, O. A.
On quasiblock problems of discrete programming with additional restrictions (Russian). Zhurnal Vychislitelnoi Matematiki i Matematicheskoi Fiziki 27, 676-687. (1987).

SHCHA 1983-1 Shcherbina, O. A.; Matveev, V. V.
On the efficiency of local algorithms for the solution of quasi-block problems of integer linear programming (Russian). Dinamicheskie Sistemy (Kiev) 2, 94-97. (1983).

SHEE 1987-1 Shee, S.-C.
On harmonious and related graphs. Ars Combinatoria 23A, 237-248. (1987).

SHEN 1982-1 Shen, L.
Effective use of heuristic algorithms in network design. In: P. A. Burr (Ed.): The Mathematics of Networks, 63-84. Proc. Symp. in Appl. Math. Vol. 26, American Math. Society. (1982).

SHER 1981-1 Sher, S. Y.
On a problem of composing a schedule (Russian). Issledovaniya po Priklasnoj Matematike (Kazan) 9, 133-141. (1981).

SHERA 1985-1 Sherali, H. D.
A disjunctive cutting plane algorithm for the extreme point mathematical programming problem. Opsearch 22, 83-94. (1985).

SHERB 1984-1 Sherali, H. D.; Adams, W. P.
A decomposition algorithm for a discrete location-allocation problem. Operations Research 32, 878-900. (1984).

SHERC 1984-1 Sherali, H. D.; Baines, S. G.
Advanced basis or block pivoting methods for network structured linear programs. The Aligarh Journal of Statistics 3-4, 17-35. (1984).

SHERD 1986-1 Sherali, H. D.; Myers, D. C.
The design of branch and bound algorithms for a class of nonlinear integer programs. In: C. L. Monma (Ed.): Algorithms and Software for Optimization - Part II, 463-484. Annals of Operations Research 5, Basel: Baltzer. (1986).

SHERE 1985-1 Sherali, H. D.; Nordai, F. L.
A capacitated, balanced, 2-median location-allocation problem on a tree network with a continuum of link demands. Research Paper. Blacksburg: Virginia Polytechnic Inst. and State Univ., 28 p. (1985).

SHERE 1985-2 Sherali, H. D.; Nordai, F. L.
Capacitated, balanced location-allocation problems on a chain graph with a continuum of link demands. Research Paper. Blacksburg: Virginia Polytechnic Inst. and State Univ., 39 p. (1985).

SHERF 1986-1 Sherali, H. D.; Rajgopal, P.
A flexible, polynomial-time, construction and improvement heuristic for the quadratic assignment problem. Computers and Operations Research 13, 587-600. (1986).

SHERG 1985-1 Sherali, H. D.; Sen, S.
On generating cutting planes from combinatorial disjunctions. Operations Research 33, 928-933. (1985).

SHERH 1987-1 Sherali, H. D.; Staschus, K; Huacuz, J. M.
An integer programming approach and implementation for an electric utility capacity planning problem with renewable energy sources. Management Science 33, 831-847. (1987).

SHERI 1985-1 Sherlekar, D. D.; Pawagi, S.; Ramakrishnan, I. V.
O(1) parallel time incremental graph algorithms. In: P. N. Maheshwari (Ed.): Foundations of Software Technology and Theoretical Computer Science, 477-495. Lecture Notes in Computer Science 206, Berlin: Springer. (1985).

SHET 1986-1 Shetty, B.
The linear equal flow problem. College Station: Texas A&M Univ., College of Business Administration, Dept. of Business Analysis and Research, 21p. (1986).

SHEV 1984-1 Shevchenko, A. N.
Application of the dynamic programming method and duality theory to a multi-index transportation problem (Russian). In: Algorithms for Solving Network Problems, 45-48. Moscow: Akad. Nauk SSSR, Tsentral. Ekonom.-Mat. Inst. (1984).

SHEVA 1979-1 Shevchenko, V. N.
Convex polyhedral cones, comparison systems and regular truncations in integer programming (Russian). in: Combinatorial-Algebraic Methods in Applied Mathematics, 109-119. Gorki: Gor'kovskii Gosudarstvennyi Universitet. (1979).

SHEVA 1985-1 Shevchenko, V. N.
On some functions of many-valued logic connected with integer programming (Russian). Metody Diskretnogo Analiza, No. 42, 99-108. (1985).

SHEVB 1986-1 Shevchik, V. I.
A problem of multicommodity flow (Russian). In: Dynamic Systems No. 5, 95-97. Kiev: "Vishcha Shkola". (1986).

SHI 1985-1 Shi, R.-H.
Graph decomposition under a minimum degree restriction (Chinese). Kexue Tongbao (Chinese) 30, 650-652. (1985).

SHI 1985-2 Shi, R.-H.
Graph decomposition with constraint on the minimum degree. Kexue Tongbao (English Ed.) (Beijing) 30, 1143-1146. (1985).

SHI 1985-3 Shi, R.-H.
Smallest regular graphs with girth pair (4,5). Journal of Systems Science and Mathematical Sciences (Beijing) 5, 34-42. (1985).

SHI 1985-4 Shi, R.-H.
The binding number of a graph and its circuits. Acta Mathematicae Applicatae Sinica (Beijing) 2, 154-160. (1985).

SHI 1985-5 Shi, R.-H.
The binding number of a graph and its triangle. Acta Mathematicae Applicatae Sinica, English Series, 2, 79-86. (1985).

SHI 1987-1 Shi, R.-H.
The binding number of a graph and its pancyclism. Acta Mathematicae Applicatae Sinica (English Series) 3, 257-269. (1987).
SHI A 1986-1 Shi, R.-H.; Sun, R.-G.; Nian, K.
Hamiltonian property in the square of a connected graph. Journal of Systems Science and Mathematical Sciences (Beijing) 6, 281-285. (1986).
SHI B 1984-1 Shi, W.-G.; Chen, Y.-N.
A graph approach to the function testing of sequential circuits (Chinese). Chinese Journal of Computers 7, No. 3, 228-236. (1984).
SHI C 1981-1 Shi, Y.-B.
On USP graphs (Chinese). Journal of Shanghai Teachers University, Natural Science Edition, No.3, 9-16. (1981).
SHI C 1985-1 Shi, Y.-B.
On UPC graphs (Chinese). Kexue Tongbao (Chinese) 30, 252-254. (1985).
SHI C 1986-1 Shi, Y.-B.
A counting theorem for Hamilton cyclic number of a graph (Chinese). Journal of Mathematical Research and Exposition 6, No. 1, 173-178. (1986).
SHI C 1986-2 Shi, Y.-B.
Further results on uniquely pancyclic graphs, II (Chinese). Journal of Shanghai Teachers University, Natural Science Edition, No.4, 1-7. (1986).
SHI C 1986-3 Shi, Y.-B.
Further results on uniquely pancyclic graphs, I (Chinese). Journal of Shanghai Teachers University, Natural Science Edition, No.2, 20-27. (1986).
SHIE 1986-1 Shier, D. R.
Algebraic aspects of computing network reliability. Technical Report No. 517. Clemson: Clemson Univ., Dept. of Mathematical Sciences, 22p. (1986).
SHIEA 1987-1 Shier, D. R.; Shandrasekharan, N.
Algorithms for computing the chromatic polynomial. Williamsburg: College of William and Mary, 9 p. (1987).
SHIEB 1985-1 Shier, D. R.; Whited, D. E.
Algorithms for generating minimal cutsets by inversion. IEEE Transactions on Reliability R-34, 314-319. (1985).
SHIH 1984-1 Shih, C.-H.
Cographic subspaces of cocycle spaces for line graphs of complete graphs. In: J. A. Bondy, U. P. R. Murty (Eds.): Progress in Graph Theory, 407-418. New York: Academic Press. (1984).
SHIM 1984-1 Shimizu, M.; Nakamura, G.
Expressions for the number of 2-3 trees with n leaves. Systems-Computers-Controls 15, 84-91. (1984).
SHIN 1987-1 Shin, K. G.
Performance analysis of distributed routing strategies free of ping-pong-type looping. IEEE Transactions of Computers C-36, 129-137. (1987).
SHINA 1986-1 Shing, M. T.; Hu, T. C.
A decomposition algorithm for multi-terminal network flows. Discrete Applied Mathematics 13, 165-182. (1986).
SHINB 1985-1 Shinoda, S.; Tsukiyama, S.; Shirakawa, I.; Omura, K.
On a second shortest k-tuple of edge-disjoint paths. In: V. Zima, J. Kvasil (Eds.): Circuit theory and design 85. Proc. 1985 European Conf. on Circuit Theory and Design, Prague, Czechoslovakia, 1985, 29-32. Prague: Academia, and Amsterdam, New York, Oxford: North Holland. (1985).
SHIO 1985-1 Shiode, S.; Ishii, H.; Nishida, T.
A chance constrained minimax facility location problem. Mathematica Japonica (Osaka) 30, 783-803. (1985).
SHIO 1987-1 Shiode, S.; Ishii, H.; Nishida, T.
A stochastic linear knapsack problem. Naval Research Logistics 34, 753-759. (1987).
SHIR 1987-1 Shirakawa, I.; Shinoda, S.; Sengoku, M.; Chen, W.-K.
A problem of diagnosis of mixed networks. In: M. Ismail (Ed.): 29th Midwest Symposium on Circuits and Systems, 452-455. Amsterdam: North-Holland. (1987).
SHLO 1987-1 Shlosman, S. B.
Graph colouring: a way to variety of new correlation inequalities. In: Proc. 8th Int. Congr. on Mathematical Physics (Marseille 1986), 839-847. Singapore: World Sci. Publ. (1987).
SHLY 1986-1 Shlyk, V. A.
Polyhedral paths connecting two vertices of a graph (Russian). Vestsi Akademii Navuk BSSR Serie Fiziko-Matematiko Navuk No. 1, 23-28. (1986).
SHOR 1985-1 Shor, N. Z.; Davydov, A. S.
On the method of obtaining estimates in quadratic extremal problems with Boolean variables. Kibernetika (Kiev) No. 2, 48-50. (1985).
SHORA 1984-1 Shor, P. W.
The average-case analysis of some on-line algorithms for bin packing. In: 25th Annual Symposium on Foundations of Computer Science, Florida 1984, 193-200. Los Angeles: IEEE Computer Society. (1984).
SHORA 1986-1 Shor, P. W.
The average-case analysis of some on-line algorithms for bin packing. Combinatorica 6, 179-200. (1986).
SHTI 1985-1 Shtilman, B. M.
A formal linguistic model for solving discrete optimization problems. I. Optimization tools. Language of trajectories. Isvestiza Akademii Nauk SSSR, Tekhnicheskaya Kibernetika (Moscow), No. 3, 110-122. Translated as: Soviet Journal of Computational Systems Science 23, No. 5, 53-64. (1985).
SHTU 1983-1 Shtujkis, V.; Sabalyauskas, G.
An algorithm for the solution of a problem of colouring edges of a tree (Russian). Teoriya Optimal'nykh Reshenij (Vil'nius) 9, 83-91. (1983).
SHU 1986-1 Shu, J.-J.
The minimum degree of a critically n-line-connected graph (Chinese). Journal of Mathematical Research and Exposition 6, No. 1, 169-172. (1986).

SIDN 1986-1 Sidney, J. B.; Steiner, G.
Optimal sequencing by modular decomposition: polynomial algorithms. Operations Research 34, 606-612. (1986).

SIDO 1980-1 Sidorenko, A. F.
B(r) properties and chromatic numbers of hypergraphs (Russian). in: K. A. Rybnikov (Ed.): Combinatorial analysis, Moskovskii Gosudarstvennyi Universitet (Moscow) No. 5, 66-69. (1980).

SIEB 1984-1 Sieber, N.
Einige Ergebnisse der Forschungsrichtung "Optimierung auf Netzwerken und Netzplantechnik". In: K. Lommatzsch (Ed.): Proceedings, 16. Jahrestagung "Mathematische Optimierung", Sellin/Rügen 1984, 93-105. Seminarbericht Nr. 64, Sektion Mathematik, Humboldt-Univ. Berlin (DDR). (1984).

SIEB 1985-1 Sieber, N.
Entwicklungsrichtungen in der Netzwerkoptimierung. 30. Internationales Wissenschaftliches Kolloquium (TH Ilmenau, DDR) Heft 5F, 109-110. (1985).

SIED 1982-1 Siedersleben, J.
Optimierung von deterministischen und stochastischen Entscheidungsnetz- plänen (Dissertation). Fakultät für Wirtschaftswissenschaften der Univ. Fridericiana, Karlsruhe. (1982).

SIED 1986-1 Siedersleben, J.
Common aspects of several network flow algorithms. In: G. Gallo, C. Sandi (Eds.): Netflow at Pisa, Mathematical Programming Study 26, 205-208. (1986).

SIEM 1984-1 Siemons, J.
Regularity in a graph. Rendiconti del Seminario Matematico di Brescia 7, 603-607. (1984).

SIGA 1985-1 Sigal, I. K.
Composite algorithms for solving the traveling salesman problem (Russian). Akademiya Nauk SSSR, Vychislitel'nyi Tsentr, Soobshcheniya po Prikladnoj Matematike, Moskva No.7. (1985).

SIGA 1986-1 Sigal, I. K.
Computer implementation of a combined branch and bound algorithm for the travelling salesman problem. Zhurnal Vychislitel'noi Matematiki i Matematicheskoi Fiziki (Moscow) 26, 664-672. Translated as: U.S.S.R. Computational Mathematics and Mathematical Physics. (1986).

SIGA 1987-1 Sigal, I. K.
An approximate algorithm for the large-scale travelling salesman problem and its computer implementation (Russian). Zhurnal Vychislitelnoi Matematiki i Matematicheskoi Fiziki 27, 1145-1153. (1987).

SILL 1986-1 Sillaber, A. H.
A continuous linear optimization model for the exact solution of travelling-salesman-problems in connexion with expansion planning of ring networks. Zeitschrift für Operations Research, Series B, 30, 77-90. (1986).

SILV 1985-1 Silver, E. A.; Miltenburg, J.
Errata: two modifications of the Silver-Meal lot sizing heuristic. INFOR 23, 205-206. (1985).

SILVA 1986-1 Silverman, F. N.; Carter, J. C.
A cost-based methodology for stochastic line balancing with intermittent line stoppages. Management Science 32, 455-463. (1986).

SIMC 1987-1 Simchi-Levy, D.; Berman, O.
A heuristic algorithm for the travelling salesman location problem on networks. New York: Columbia Univ., Dept. of Industrial Eng. and Operations Research, 19 p. (1987).

SIMC 1987-2 Simchi-Levy, D.; Berman, O.
Heuristics and bounds for the travelling salesman location problem on the plane. Operations Research Letters 6, 243-248. (1987).

SIME 1985-1 Simeone, B.
Consistency of quadratic boolean equations and the König-Egervary property for graphs. In: G. Ausiello, M. Lucertini (Eds.): Analysis and Design of Algorithms for Combinatorial Problems, Annals of Discrete Mathematics Vol. 25, 281-290. Amsterdam: North-Holland. (1985).

SIME 1986-1 Simeone, B.
An asymptotically exact polynomial algorithm for equipartition problems. Discrete Applied Mathematics 14, 283-294. (1986).

SIMM 1983-1 Simmons, G. J.
On the ochromatic number of a graph. Congressus Numerantium 40, 339-366. (1983).

SIMM 1985-1 Simmons, G. J.
The ochromatic number of planar graphs. In: F Harary, J. P. Maybee (Eds.): Graphs and Applications, 295-316. New York: Wiley. (1985).

SIMO 1981-1 Simon, I.
Probabilistic algorithms and complexity (Portuguese). In: Proceedings of the twelfth Brazilian Mathematical Colloquium (Pocos de Caldas, 1979), 37-49. Cons. Nac. Desenvolvimento Ci. Tec., Inst. Mat. Pura Apl., Rio de Janeiro. (1981).

SIMOA 1980-1 Simons, B.
A fast algorithm for multiprocessor scheduling. In: 21st Annual Symp. on Foundations of Computer Science, Syracuse, N.Y. 1980, Proceedings, 50-53. IEEE. (1980).

SIMOB 1987-1 Simonson, S.
A variation on the min cut linear arrangement problem. Mathematical Systems Theory 20, 235-252. (1987).

SINC 1986-1 Sinclair, M.
An exact penalty function approach for nonlinear integer programming problems. European Journal of Operational Research 27, 50-56. (1986).

SINC 1986-2 Sinclair, M.
The assignment of workers to tasks - an example from an academic department. Orion 1, No. 2, 46-69. (1986).

SINCA 1987-1 Sinclair, M.; Dyk, E. van
Combined routeing and scheduling for the transportation of containerized cargo. Journal of the Operational Research Society 38, 487-498. (1987).

SING 1985-1 Singer, D.; Elek, J.
Algorithm for determining the near optimal centrum locations in large graph structures. Acta Technica Academiae Scientarium Hungaricae 98, 387-397. (1985).

SINGA 1987-1 Singer, D.; Moritz, P.
Expert system for scheduling cleaning operations for municipal canal systems. International Journal of Systems Science 18, 1217-1226. (1987).

SINGB 1986-1 Singer, I.
A general duality approach to min-max results in combinatorial optimization. I. Preprint Series in Mathematics No. 45/1986. Bucarest: Inst. National Pentru Creatie Stiintifica si Tehnica, Inst. de Matematica, 42p. (1986).

SINGC 1986-1 Singh, C.; Asgarpoor, S.
Reliability evaluation of flow networks using delta-star transformations. IEEE Transactions on Reliability R-35, 472-477. (1986).

SINGD 1987-1 Singh, H.; Foster, J. B.
Production scheduling with sequence dependent setup costs. IIE Transactions, 19, 43-49. (1987).

SINGE 1987-1 Singh, H. G.; Pargas, R. P.
A parallel implementation for the domination number of a grid graph. Congressus Numerantium 59, 297-312. (1987).

SINGF 1984-1 Singh, N.; Vrat, P.
A location-allocation model for a two-echelon repair-inventory system. IIE Transactions 16, 222-228. (1984).

SINGG 1986-1 Singh, S.
Improved methods for storing and updating information in the out-of-kilter algorithm. Journal of the Association for Computing Machinery 33, 551-567. (1986).

SINGH 1985-1 Singh, T. P.
On 2*n flow-shop problem involving job-block, transportation-time, arbitrary time and break-down machine times. Pure Applied Mathematika Sciences (Saharanpur) 21, 59-63. (1985).

SINH 1983-1 Sinha, S.
An algorithm for network decompositions. Bulletin of the Calcutta Mathematical Society 75, No. 3, 185-190. (1983).

SINU 1986-1 Sinuany-Stern, Z.; Teomi, Y.
Multi-objective scheduling plans for security guards. Journal of the Operational Research Society 37, 67-78. (1986).

SIRA 1983-1 Siran, J.
Edges and Kuratowski subgraphs of non-planar graphs. Mathematische Nachrichten 113, 187-190. (1983).

SIRA 1984-1 Siran, J.
Additivity of the crossing number of graphs with connectivity 2. Periodica Mathematica Hungarica (Budapest) 15, 301-315. (1984).

SITN 1984-1 Sitnikova, O. D.
Calculation of integer linear programming problems by the method of consecutive investigation of the level planes using filter-limitation (Russian). Kibernetika (Kiev) No. 6, 119-121. (1984).

SIVA 1987-1 Sivazlian, B. D.
Optimum scheduling of a new maintenance program under stochastic degradation. Research Report 87-21. Gainesville: Univ. of Florida, Industrial & Systems Eng. Dept., 21 p. (1987).

SKAL 1983-1 Skaletskii, V. V.
An algorithm for solution of a problem of integer programming (Russian). In: Differential and Operator Equations in Function Spaces, 66-69. Akademiya Nauk SSSR, Dalnevostochn. Nauchn. Tsentr (Vladivostok). (1983).

SKIL 1986-1 Skillicorn, D. B.
Graph problems in network design. Ars Combinatoria 21-A, 169-188. (1986).

SKILA 1984-1 Skilton, D. K.
Eulerian chains and segment reversals. In: K. M. Koh, H. P. Yap (Eds.): Graph Theory Singapore 1983, Lecture Notes in Mathematics 1073, 228-235. Berlin: Springer-Verlag. (1984).

SKILA 1985-1 Skilton, D. K.
Chain decompositions of graphs. Bulletin of the Australian Mathematical Society 31, 471-473. (1985).

SKILA 1986-1 Skilton, D. K.
Digraphs with Eulerian chains. Journal of the Australian Mathematical Society, Series A 41, 298-303. (1986).

SKIT 1985-1 Skitt, R. A.; Levary, R. R.
Vehicle routing via column generation. European Journal of Operational Research 21, 65-76. (1985).

SKLA 1984-1 Sklar, M. G.; Armstrong, R. D.; Davis, K. R.
A routing problem for a plant nursery. Mathematical Modelling 5, 63-67. (1984).

SKOR 1987-1 Skorin-Kapov, J.; Granot, F.
Non-linear integer programming: sensitivity analysis for branch and bound. Operations Research Letters 6, 269-274. (1987).

SKORA 1981-1 Skorobogatov, V. A.; Khvorostov, P. V.
Analysis of the metric properties of graphs (Russian). Vychislitelnye Sistemy (Novosibirsk) No. 91, 3-20. (1981).

SKORA 1984-1 Skorobogatov, V. A.; Khvorostov, P. V.
Methods and algorithms for the analysis of symmetries of graphs (Russian). Vychislitelnye Sistemy (Novosibirsk), No. 103, 6-25. (1984).

SKOW 1983-1 Skowronska, M.
Characterizations of middle graphs and properties of Halin graphs (Polish). Prepr., Inst. Mat. Uniw. Mikolaja Kopernika 1. (1983).

SKOW 1983-2 Skowronska, M.
Hamiltonian properties of Halin-like graphs. Ars Combinatoria 16B, 97-110. (1983).

SKOW 1985-1 Skowronska, M.
The pancyclicity of Halin graphs and their exterior contractions. In: B. R. Alspach, C. D. Godsil (Eds.): Cycles in Graphs, Annals of Discrete Mathematics Vol. 27, 179-194. Amsterdam: North-Holland. (1985).

SKRI 1984-1 Skrien, D.
Interval line graphs. Discrete Mathematics 51, 317-318. (1984).

SKRIA 1985-1 Skrien, D.; Gimbel, J.
Homogeneously representable interval graphs. Discrete Mathematics 55, 213-216. (1985).

SKUB 1984-1 Skubalska, E.
Application of the branch-and-bound method to the optimal vehicle routing (Czechoslovakian). Przeglad Statystyczny (Warsaw) 31, 65-81. (1984).

SKUB 1984-2 Skubalska, E.
Vehicle routing problems, mathematical models (Polish). Archiwum Automatyki i Telemechaniki (Warsaw) 29, 483-499. (1984).

SKUP 1984-1 Skupien, Z.
Homogeneously traceable and Hamiltonian connected graphs. Demonstratio Mathematica (Warsaw) 17, 1051-1067. (1984).

SKUP 1985-1 Skupien, Z.
An improvement of Jung's condition for hamiltonicity. In: Graphen und Netzwerke - Theorie und Anwendungen, 30. Int. Wiss. Kolloq., TH Ilmenau DDR 1985, 5, 111-113. (1985).

SKUP 1985-2 Skupien, Z.
N-closures of large nonhamiltonian n-vertex graphs. In: M. Borowiecki, Z. Skupien (Eds.): Graphs, hypergraphs and matroids, Proc. 5th Reg. Sci. Sess. Math., Zagan/Pol. 1985, 59-66. Zielona Gora: Higher Coll. of Engineering. (1985).

SKUP 1986-1 Skupien, Z.
On counting maximum path-factors of a tree. In: J. Machner, G. Schaar (Eds.): Algebra und Graphentheorie, Beiträge der Jahrestagung "Algebra und Grenzgebiete" in Siebenlehn/DDR Oktober 1985, 91-94. Freiberg: Bergakademie Freiberg, Sektion Mathematik. (1986).

SLAT 1982-1 Slater, P. J.
On k-graceful graphs. In: F. Hoffman, K. B. Reid, R. C. Mullin, R. G. Stanton (Eds.): Proc. Thirteenth Southeastern Conf. on Combinatorics, Graph Theory and Computing, Florida 1982, Congressus Numerantium Vol. 36, 53-58. Winnipeg, Canada: Utilitas Mathematica. (1982).

SLAT 1984-1 Slater, P. J.
Some definitions of central structures. In: K. M. Koh, H. P. Yap (Eds.): Graph Theory Singapore 1983, Lecture Notes in Mathematics 1073, 169-178. Berlin: Springer-Verlag. (1984).

SLAT 1985-1 Slater, P. J.
Minimum average distance partitions of the vertices on a line network. Congressus Numerantium 49, 215-222. (1985).

SLOM 1986-1 Slominski, L.
On existence of assignments in zero-one matrices. In: A. Prekopa, J. Szeleszan, B. Strazicky (Eds.): System Modelling and Optimization - Proc. 12th IFIP Conf., Budapest 1985, 838-844. Lecture Notes in Control and Information Sciences 84, Berlin: Springer. (1986).

SLOW 1984-1 Slowinski, R.
Modelling and solving multicriteria project scheduling problems. Foundations of Control Engineering 9, 43-52. (1984).

SLOW 1984-2 Slowinski, R.
Modelling and solving multicriteria project scheduling problems. In: A. Straszak (Ed.): Large Scale Systems: Theory and Applications 1983 (Warsaw 1983), 469-474. IFAC Proc. Series 10, Laxenburg, Pergamon Press. (1984).

SLUS 1985-1 Slusarek, M.
A note on the dynamic storage allocation problem. Information Processing Letters 20, 223-228. (1985).

SMAD 1986-1 Smadici, C.
A note on interval graphs. Analele Stiintifice ale Universitatii Al. I. Cuza din Iasi (Serie Noua), Sectiunea Ia Matematica 32, No. 3, 85-92. (1986).

SMAR 1984-1 Smarandache, F. G.
General solution properties in whole numbers for linear equations. Bulletin of the University of Brasov, Serie C, 24(1982), 37-39. (1984).

SMET 1987-1 Smetanin, Y. G.; Khachiyan, L. G.
Use of pseudopolynomial algorithms for some problems of combinatorial optimization with constraints. Izvestiya Akademii Nauk SSSR, Tekhnicheskaya Kibernetika (Moscow), No. 6, 139-144; translated as Soviet J. Comput. Systems Sci. 25, No. 2, 161-165. (1987).

SMIR 1985-1 Smirnov, A. N.
Mathematical model of three-dimensional scheduling of technologically linked activities. Automation and Remote Control 46, 1297-1303; translation from Avtomatika i Telemekhanika (1985), No.10, 108-115. (1985).

SMIT 1987-1 Smith-Daniels, D. E.; Smith-Daniels, V. L.
Optimal project scheduling with materials ordering. IIE Transactions 19, 122-129. (1987).

SMITA 1986-1 Smith-Daniels, V. L.; Smith-Daniels, D. E.
A mixed integer programming model for lot sizing and sequencing packaging lines in the process industries. IIE Transactions 18, 278-285. (1986).

SMITB 1983-1 Smith, D. H.
Reliable networks and (k,2k-2)-connected graphs. Ars Combinatoria 16B, 171-176. (1983).

SMITB 1985-1 Smith, D. H.
Connectivity preserving transformations of graphs and the reliability of mobile communication networks. Ars Combinatoria 20-A, 135-140. (1985).

SMITC 1985-1 Smith, J. M.
Generalized Steiner network problems in engineering design. In: J. P. Gero (Ed.): Design optimization, Notes and Reports in Mathematics in Science and Engineering 1, 119-161. Orlando, Fla.: Academic Press. (1985).

SMOR 1983-1 Smorodinskii, S. S.; Tanaevskii, V. G.
A means of scalarization of vector estimates for the solution of discrete problems of multicriteria choice (Russian). Avtomatika i Vychislitelnaya Tekhnika (Minsk) No. 13, 62-64. (1983).

SMYT 1987-1 Smyth, W. F.
Sharp bounds on the diameter of a graph. Canadian Mathematical Bulletin 30, No. 1, 72-74. (1987).

SOBI 1981-1 Sobik, F.; Sommerfeld, E.
Untersuchungen zur Klassifikation strukturierter Objekte auf der Grundlage des Enthaltenseins bestimmter Untergraphen. Rostocker Mathematisches Kolloquium No. 18, 113-122. (1981).

SOCZ 1985-1 Soczewinska, E.; Szamkolowicz, L.
Chordal characterization of binary matroids. In: M. Borowiecki; Z. Skupien (Eds.): Graphs, hypergraphs and matroids, Proc. Fifth Regional Scientific Session of Mathematicians, Zagan, Poland, June 1985. Section: Graph theory, 67-73. Zielona Gora: Higher College of Engineering. (1985).
SOKO 1983-1 Sokolov, A. V.
A heuristic algorithm for the distribution of multicommodity flow (Russian). In: Algorithmic Constructions and Their Efficiency, 107-111. Yaroslavl: Yaroslav. Gos. Univ. (1983).
SOKOA 1987-1 Sokolova, M.
The chromatic number of extended odd graphs is four. Casopis pro Pestovani Matematiky 112, 308-311. (1987).
SOKOB 1983-1 Sokolova, T. E.
Determining the group of shortest paths in a graph (Russian). In: Algorithmic Constructions and Their Efficiency, 68-72. Yaroslavl: Yaroslav. Gos. Univ. (1983).
SOKOC 1982-1 Sokolovskij, V. Z.; Zadorozhnyj, M. V.
Experiences with random search in problems of the traveling salesman (Russian). In: A. N. Antamoshkin (Ed.): Applications of random search to the solution of applied problems, 3-8. Kemerovo: Kemerovskij Gosud. Univ. (1982).
SOLI 1984-1 Solis, R.
Evaluation of heuristic algorithms for the TSP: a new statistical approach. American Journal of Mathematical and Management Sciences 4, 89-108. (1984).
SOLO 1983-1 Sologashvili, V. G.; Dularidze, N. G.
An algorithm for solution of a problem of integer linear programming (Russian). In: D. Z. Avazashvili (Ed.): Mathematical analysis, 126-129. Tbilisi: Trudy Gruzinskoi Politekhnicheskie Institut. (1983).
SOLOA 1984-1 Solomon, D. I.
Generalization of the linear and fractional-linear transportation problem (Russian). Izvestiya Akademii Nauk Moldavskoj SSR, Seriya Fiziko-Tekhnicheskik i Matematicheskikh Nauk (Kishinev) No. 1, 13-18. (1984).
SOLOB 1986-1 Solomon, M. M.
The minimum spanning tree problem with time window constraints. American Journal of Mathematical and Management Sciences 6, 399-421. (1986).
SOLOB 1987-1 Solomon, M. M.
Algorithms for the vehicle routing and scheduling problem with time window constraints. Operations Research 35, 254-265. (1987).
SOLT 1983-1 Soltan, V. P.
D-convexity in graphs. Soviet Mathematics, Doklady 28, 419-421. Translated from Doklady Akademii Nauk SSSR (Moscow) 272, No. 3, 535-537. (1983).
SOLTA 1984-1 Soltan, V. P.; Chepoi, V. D.
D-convex sets in triangulated graphs (Russian). Matematicheskie Issledovaniya (Kishinev) 78, 105-124. (1984).
SOLTA 1984-2 Soltan, V. P.; Chepoi, V. D.
The number of d-convex sets in a graph (Russian). Izvestiya Akademii Nauk Moldavskoj SSR, Seriya Fiziko-Tekhnicheskikh i Matematicheskikh Nauk (Kishinev) No. 2, 19-23. (1984).
SOLTA 1985-1 Soltan, V. P.; Chepoi, V. D.
D-convexity and Steiner functions on a graph (Russian). Doklady Akademii Nauk BSSR (Minsk) 29, 407-408. (1985).
SOLTB 1986-1 Soltes, L.
Orientations of graphs minimizing the radius or the diameter. Mathematica Slovaca 36, 289-296. (1986).
SOMM 1985-1 Sommerfeld, E.
Zur Stabilität der Korona zweier Graphen. In: H. Sachs (Ed.): Graphs, Hypergraphs and Applications, Proc. Conf. on Graph Theory (Eyba 1984), 165-168. Leipzig: Teubner. (1985).
SONE 1987-1 Soneoka, T.; Nakada, H.; Imase, M.; Peyrat, C.
Sufficient conditions for maximally connected dense graphs. Discrete Mathematics 63, 53-66. (1987).
SONG 1986-1 Song, G. D.; Ding, J. Y.
A note on the four-point complete Steiner tree theorem (Chinese). Heilongjiang Daxue Ziran Kexue Xuebao, No. 1, 11-16. (1986).
SONGA 1987-1 Song, S. H.; Park, S. D.
An exact algorithm for the asymmetrical vehicle routing problem (Korean). Journal of the Korean Operations Research and Management Science Society 12, No.1, 34-44. (1987).
SONGB 1985-1 Song, Z.-M.
The arc constrained optimum arborescence problem (Chinese). Journal of Nanjing Institute of Technology, No.1, 110-112. (1985).
SONGB 1986-1 Song, Z.-M.
Long paths and cycles in oriented graphs (Chinese). Journal of Nanjing Institute of Technology 16, No.5, 102-108. (1986).
SONGB 1986-2 Song, Z.-M.
Longest paths and cycles in bipartite oriented graphs (Chinese). Journal of Nanjing Institute of Technology 16, No.2, 40-44. (1986).
SONGB 1987-1 Song, Z.-M.
Number of arcs and cycles in digraphs (Chinese). Journal of Nanjing Institute of Technology 17, No.2, 121-125. (1987).
SONN 1985-1 Sonntag, M.
Some Hamiltonian properties of products of digraphs. Elektronische Informationsverarbeitung und Kybernetik 21, 275-282. (1985).
SONN 1986-1 Sonntag, M.
Pancyclicity and strong path-connectedness of the Cartesian sum of digraphs. In: J. Machner, G. Schaar (Eds.): Algebra und Graphentheorie, Beiträge der Jahrestagung "Algebra und Grenzgebiete" in Siebenlehn/DDR Oktober 1985, 95-98. Freiberg: Bergakademie Freiberg, Sektion Mathematik. (1986).
SONN 1987-1 Sonntag, M.
Some results on Hamiltonian properties of products of digraphs I. t-Hamiltonian properties and strong path-connectedness. Journal of Information Processing and Cybernetics EIK 23, 181-194. (1987).

SONN 1987-2 Sonntag, M.
Some results on Hamiltonian properties of products of digraphs II. Pancyclicity properties. Journal of Information Processing and Cybernetics EIK 23, 323-330. (1987).
SOTS 1985-1 Sotskov, Y. N.
The network approach for the solution of problems of scheduling theory (Russian). In: V. P. Tanaev (Ed.): Methods, Algorithms and Programs for the Solution of Extremal Problems, 52-62. Minsk: Inst. Tekh. Kibernet. AN BSSR. (1985).
SOTSA 1982-1 Sotskov, Y. N.; Chernjak, Z. A.
Determination of parameters of a network graph (Russian). In: V. P. Tanae (Ed.): Algorithms and programs for the solution of extremal problems and related questions (Russian), 42-53. Minsk: Institut Tekhnicheskoj Kibernetiki AN BSSR. (1982).
SOTT 1986-1 Sotteau, D.; Wojda, A. P.
Digraphs without directed path of length two or three. Discrete Mathematics 58, 105-108. (1986).
SOUM 1984-1 Soumis, F.; Desrosiers, J.; Desrochers, M.
Optimal urban bus routing with scheduling flexibilities. In: P. Thoft-Christensen (Ed.): System Modelling and Optimization, 155-165. Lecture Notes in Control and Information Sciences 59, Berlin: Springer. (1984).
SPAT 1985-1 Späth, H.
Heuristically determining cliques of given cardinality and with minimal cost within weighted complete graphs. Zeitschrift für Operations Research 29A, 125-131. (1985).
SPAT 1986-1 Späth, H.
Heuristic solution methods for the m-mean median location problem. In: L. Streitferdt, H. Hauptmann, A. W. Marusev, D. Ohse, U. Pape (Eds.): Operations Research Proceedings 1985, 410-417. Berlin, Heidelberg, New York: Springer-Verlag. (1986).
SPEC 1983-1 Speckenmeyer, E.
Untersuchungen zum Feedback Vertex Problem in ungerichteten Graphen (Dissertation). Fachbereich Mathematik/Informatik der Univ.-Gesamthochschule- Paderborn. (1983).
SPEN 1985-1 Spencer, J.
Probabilistic methods. Graphs and Combinatorics 1, 357-382. (1985).
SPER 1985-1 Speranza, M. G.
A remark on the Karp algorithm for the set covering problem (Italian). Istituto Lombardo, Accademia di Scienze e Lettere, Rendiconti, A (Milan) 116(1982), 55-64. (1985).
SPES 1986-1 Spesivtsev, A. V.
The error of certain approximate list scheduling algorithms (English). Soviet Journal of Computer and Systems Sciences 24, No.1, 96-101; translation from Izvestiya Akademii Nauk SSSR, Tekhnicheskaya Kibernetika (1985), No.6, 22-28. (1986).
SPIL 1988-1 Spilker, I.
Aircraft-stand allocation at Schiphol airport: problem description. In: H. Schellhaas, P. van Beek, H. Isermann, R. Schmidt, M. Zijlstra (Eds.): Operations Research Proceedings 1987 (DGOR/NSOR), 306-307. Berlin: Springer. (1988).
SPIN 1982-1 Spinrad, J.
Two dimensional partial orders. Princeton: Princeton Univ.; Ann Arbor: University Microfilms. (1982). (Dissertation)
SPIN 1985-1 Spinrad, J.
On comparability and permutation graphs. SIAM Journal on Computing 14, 658-670. (1985).
SPIN 1985-2 Spinrad, J.
Worst-case analysis of a scheduling algorithm. Operations Research Letters 4, 9-12. (1985).
SPINA 1987-1 Spinrad, J.; Brandstädt, A.; Stewart, L.
Bipartite permutation graphs. Discrete Applied Mathematics 18, 279-292. (1987).
SPINB 1985-1 Spinrad, J.; Vijayan, G.
Worst case analysis of a graph coloring algorithm. Discrete Applied Mathematics 12, 89-92. (1985).
SPRI 1986-1 Springer, J.
Algorithm 641. Exact solution of general integer systems of linear equations. ACM Transactions on Mathematical Software 12, 149. (1986).
SRID 1983-1 Sridharan, M. R.; George, O. T.
On indecomposable graphs. Journal of Mathematical and Physical Sciences (Madras) 17, 95-102. (1983).
SRIDA 1986-1 Sridharan, R.
A heuristic Lagrangean algorithm for the capacitated plant location problem - a comment. European Journal of Operational Research 23, 264-165. (1986).
SRIDB 1987-1 Sridharan, S.
Biregular subgraphs of biregular graphs. Discrete Mathematics 65, 213-214. (1987).
SRIDC 1987-1 Sridharan, V.; Berry, W. L.; Udayabhanu, V.
Freezing the master production schedule under rolling planning horizons. Management Science 33, 1137-1149. (1987).
SRIK 1986-1 Srikar, B. N.; Ghosh, S.
A MILP model for the n-job, M-stage flowshop with sequence dependent set-up times. International Journal of Production Research 24, 1459-1474. (1986).
SRIN 1983-1 Srinivasan, N.
Block-cut-point incidence pattern of separable geodetic graphs of diameter three and four. Journal of Mathematical and Physical Sciences 18 (1983/84), Special Issue, S109-S113. (1983).
SRINA 1985-1 Srinivasan, V.; Kim, Y. H.
A network optimization approach to efficient cash management in India. Asia-Pacific Journal of Operational Research 2, 54-65. (1985).
SRIS 1986-1 Sriskandarajah, C.; Ladet, P.
Some no-wait shops scheduling problems: complexity aspects. European Journal of Operational Research 24, 424-438. (1986).
STAH 1985-1 Stahl, S.
Homology as a tool in integer programming. SIAM Journal on Algebraic and Discrete Methods 6, 641-642. (1985).

STAL 1984-1 Stallmann, M.; Gabow, H. N.
An augmenting path algorithm for the parity problem on linear matroids. In: 25th Annual Symposium on Foundations of Computer Science, Florida 1984, 217-228. Los Angeles: IEEE Computer Society. (1984).

STAN 1985-1 Stancu-Minasian, I. M.; Tigan, S.
The minimum-risk approach to the bottleneck transportation problem. In: Itinerant seminar on functional equations, approximation and convexity (Cluj-Napoca 1985), 203-208, Preprint 85-6. Cluj-Napoca: University Babes-Bolyai. (1985).

STANA 1984-1 Stanczak, W.
An efficient algorithm for partitioning a network into minimally interconnected subnetworks. Control and Cybernetics (Warsaw) 13, No. 1-2, 97-112. (1984).

STANB 1985-1 Stanczak, W.; Kacprzyk, J.
On an extension of Weber problem with Euclidean distance. Optimization 16, 849-862. (1985).

STANC 1983-1 Standley, H. M.
Minimal area graph embedding into a directed grid is NP-hard. in: T. Basar, D. J. Brown, H. V. Poor (Eds.): Twentyfirst Annual Allerton Conference on Communication, Control, and Computing (Monticello, Ill., 1983), 234-235. Urbana-Champaign: Univ. Illinois. (1983).

STAND 1985-1 Stanfel, L. E.
Partitioning problems in cellular manufacturing. Rairo APII 19, 523-536. (1985).

STAND 1986-1 Stanfel, L. E.
A recursive Lagrangian method for clustering problems. European Journal of Operational Research 27, 332-342. (1986).

STANE 1985-1 Stanley, J. D.
Bottleneck structures and planning horizons. Working Paper. Eau Claire: Univ. of Wisconsin, Dept. of Business Administratin, 28 p. (1985).

STANF 1983-1 Stanton, R. G.; Stinson, D. R.
Perfect pair-coverings with block sizes two, three, and four. Journal of Combinatorics, Information & System Sciences (Delhi) 8, 21-25. (1983).

STAS 1987-1 Stassinopoulos, G. I.
Optimal dynamic routing in multidestination networks. IEEE Transactions on Communications COM-35, 472-475. (1987).

STASA 1985-1 Stassinopoulos, G. I.; Konstantopoulos, P.
Optimal congestion control in single destination networks. IEEE Transactions on Communications COM-33, 792-800. (1985).

STAV 1982-1 Stavre, P.
Applications of mathematics in engineering and to the controversial lot-size problem (Romanian). In: Proceedings of the national colloquium on probability theory and operational research (Craiova, 1982), 171-178. Bucharest: Universiteta Bucuresti. (1982).

STEE 1986-1 Steele, J. M.
Probabilistic algorithm for the directed traveling salesman problem. Mathematics of Operations Research 11, 343-350. (1986).

STEE 1987-1 Steele, J. M.
On Frieze's zeta(3) limit for lengths of minimal spanning trees. Discrete Applied Mathematics 18, 99-103. (1987).

STEF 1985-1 Steffens, K.
Maximal tight sets and the Edmonds-Gallai decomposition for matchings. Combinatorica 5, 359-366. (1985).

STEI 1985-1 Steiner, G.
A compact labeling scheme for series-parallel graphs. Discrete Applied Mathematics 11, 281-298. (1985).

STEI 1985-2 Steiner, G.
An algorithm for minimizing setups in precedence constrained scheduling. Journal of Computational and Applied Mathematics 12 & 13, 579-590. (1985).

STEI 1985-3 Steiner, G.
On finding the jump number of a partial order by substitution decomposition. Order 2, 9-24. (1985).

STEI 1987-1 Steiner, G.
On computing the information theoretic bound for sorting: counting the linear extensions of posets. Report 87459-OR. Bonn: Univ. Bonn, Inst. f. Ökonometrie u. Operations Research, 16 p. (1987).

STEI 1987-2 Steiner, G.
On computing the information theoretic bound for sorting: balancing the linear extensions of posets. Report 87460-OR. Bonn: Univ. Bonn, Inst. f. Ökonometrie u. Operations Research, 21 p. (1987).

STEIA 1987-1 Steiner, G.; Stewart, L. K.
A linear time algorithm to find the jump number of 2-dimensional bipartite partial orders. Order 3, 359-368. (1987).

STEIB 1983-1 Steinmüller, J.
Graphentheoretische und kombinatorische Probleme in der technologischen Fertigungsvorbereitung. In: 2nd Colloquium for Geometry and Combinatorics (Karl-Marx-Stadt, 1983) Part 1, 2, 185-190. Karl-Marx-Stadt: Technische Hochschule. (1983).

STEIC 1982-1 Steinparz, F. X.
On the existence of F-factors. In: J. R. Mühlbacher (Ed.): Graphtheoretic Concepts in Computer Science, Proc. 7th Conf. on Graphtheoretic Concepts in Computer Science, Linz 1981, 61-73. München-Wien: Carl Hanser Verlag. (1982).

STEN 1984-1 Stenzel, L.
On automorphisms and edge colourings of quasigroups (German). In: Geometrie und Kombinatorik, 2. Kolloq., Karl-Marx-Stadt/DDR 1983, Vortr. 1,2, Teil II, 250-254. (1984).

STEP 1983-1 Stepanov, G. D.
An approach to the solution of a problem of the multimedian on a tree (Russian). In: Algorithmic Constructions and Their Efficiency, 43-50. Yaroslavl: Yaroslav. Gos. Univ. (1983).

STET 1985-1 Stetsenko, S. I.
Numerical experiments in the solution of quadratic problems with Boolean variables (Russian). In: V. M. Mikhalevich (Ed.): Methods for the Solution of Complex Mathematical Programming Problems, 7-11. Kiev: Akad. Nauk Ukrain. SSR, Inst. Kibernet. (1985).

STEU 1986-1 Steuer, R. E.; Wood, E. F.
On the 0-1 implementation of the Tchebycheff solution approach: a water quality illustration. Large Scale Systems 10, 243-255. (1986).

STEW 1987-1 Stewart, I. A.
An algorithm for colouring perfect planar graphs. In: K. V. Nori (Ed.): Foundations of Software Technology and Theoretical Computer Science, 58-64. Lecture Notes in Computer Science 287, Berlin, Heidelberg, New York: Springer. (1987).

STEWA 1985-1 Stewart, W. R. Jr.
An accelerated branch exchange heuristic for the traveling salesman problem. Working Paper, 85-004. Williamsburg: College of William & Mary, School of Buisiness Administration, 22 p. (1985).

STEWA 1985-2 Stewart, W. R. Jr.
An improved assignment lower bound for the Euclidean traveling salesman problem. Operations Research Letters 4, 55-60. (1985).

STIE 1985-1 Stiebitz, M.
Colour-critical graphs with complete major-vertex subgraph. In: H. Sachs (Ed.): Graphs, Hypergraphs and Applications, Proc. Conf. on Graph Theory (Eyba 1984), 169-181. Leipzig: Teubner. (1985).

STIE 1985-2 Stiebitz, M.
Subgraphs of colour-critical graphs. In: Graphen und Netzwerke - Theorie und Anwendungen, 30. Int. Wiss. Kolloq., TH Ilmenau DDR 1985, 5, 115-117. (1985).

STIE 1986-1 Stiebitz, M.
Colour-critical graphs and linear algebra. In: J. Machner, G. Schaar (Eds.): Algebra und Graphentheorie, Beiträge der Jahrestagung "Algebra und Grenzgebiete" in Siebenlehn/DDR Oktober 1985, 103-106. Freiberg: Bergakademie Freiberg, Sektion Mathematik. (1986).

STIE 1987-1 Stiebitz, M.
K*(5) is the only double-critical 5-chromatic graph. Discrete Mathematics 64, 91-94. (1987).

STIE 1987-2 Stiebitz, M.
Subgraphs of colour-critical graphs. Combinatorica 7, 303-312. (1987).

STIN 1983-1 Stinson, J. P.; Khumawala, B. M.; Steinberg, E.
An improved algorithm for the multi-activity, multi-facility location problem. Journal of the Indian Society of Statistics and Operations Research 4, 1-8. (1983).

STIR 1984-1 Stiranka, A. I.
A problem of distribution of productive capacities (Russian). Doklady Akademiya Nauk Ukrainskoi SSR Seriya A No. 5, 80-83. (1984).

STOI 1984-1 Stoica, M.; Angheluta, G.
An application of the problem of three-dimensional allocation. Economic Computation and Economic Cybernetics Studies and Research 19, No. 4, 61-66. (1984).

STOIA 1986-1 Stoica, M.; Hartulari, C.
Modalities and principles to design heuristic algorithms. Economic Computation and Economic Cybernetics Studies and Research 21, 31-36. (1986).

STOK 1988-1 Stokx, C. F. M.
A strategic model for the solution of the location-allocation problem of a major oil company. In: H. Schellhaas, P. van Beek, H. Isermann, R. Schmidt, M. Zijlstra (Eds.): Operations Research Proceedings 1987 (DGOR/NSOR), 307-308. Berlin: Springer. (1988).

STOU 1987-1 Stougie, L.
Design and analysis of algorithms for stochastic integer programming. CWI Tract No. 37. Amsterdam: Centrum voor Wiskunde en Informatica, Stichting Math. Centrum. (1987).

STOU 1988-1 Stougie, L.
The stochastic knapsack problem. In: H. Schellhaas, P. van Beek, H. Isermann, R. Schmidt, M. Zijlstra (Eds.): Operations Research Proceedings 1987 (DGOR/NSOR), 626-627. Berlin: Springer. (1988)

STOW 1985-1 Stowe, J. D.
An integer programming solution for the optimal credit investigation credit granting sequence. Financial Management 14, 66-76. (1985).

STRA 1982-1 Straight, H. J.
Extremal problems concerning the cochromatic number of a graph. Journal of the Indian Mathematical Society (Delhi) (New Series) 44 (1980), 137-142. (1982).

STRAA 1985-1 Strassner, K.
Decomposing graphs and hypergraphs (German). In: H. Sachs (Ed.): Graphs, hypergraphs and applications, Proc. Conf. Graph Theory, Eyba/GDR 1984, 182-185. Teubner-Verlag. (1985).

STRU 1985-1 Strusevich, V. A.
The three-machine problem: sufficient condition for reducibility in a convex form (Russian). Vestnik Belorusskogo Gosudarstvennogo Universiteta (Minsk), Serya I No. 1, 36-39, 79. (1985).

STRU 1986-1 Strusevich, V. A.
On the possiblity for constructing time-optimal schedules for a multi-stage system with non-fixed routes of stage passage (Russian). Izvestiya Akademii Nauk BSSR, Seriya Fiziko-Matematicheskikh Nauk (Minsk), No.6, 43-48. (1986).

STRZ 1985-1 Strzelec, K. R.; Gadecki, H.
Exact or approximate algorithms? A combinatorial approach to integer programming. In: F. Nozicka (Ed.): Optimization - Theory, Methods, Applications, I, International Conference Marianske Lazne 1985, 260-269. Prague: Dum Techniky CSVTS. (1985).

STUE 1984-1 Stueckle, S.; Ringeisen, R. D.
Generalized Petersen graphs which are cycle permutation graphs. Journal of Combinatorial Theory, Series B 37, 142-150. (1984).

STUR 1986-1 Sturmfels, B.
Oriented matroids and combinatorial convex geometry. Darmstadt: TH Darmstadt, 95 p. (1986). (Dissertation)

STURA 1985-1 Sturmfels, B.; Bokowski, J.
Programmsystem zur Realisierung orientierter Matroide. Preprints in Optimization 85.22. Köln: Univ. Köln, Mathematisches Institut, 33 p. (1985).

SU 1984-1 Su, J. J.
Minimally k-edge connected graphs of low order with maximum number of edges (Chinese). Journal of the Xinjiang University of Natural Sciences No. 3, 72-77. (1984).

SU 1985-1 Su, J. J.
Upper bounds for the maximum degree and for the number of vertices having degree greater than four in a critically 2-edge-connected graph (Chinese). Journal of the Xinjiang University of Natural Sciences No. 1, 56-61. (1985).
SU 1986-1 Su, J. J.
The number of vertices with low degree in critically h-connected graphs (Chinese). Acta Mathematicae Applicatae Sinica 9, 479-486. (1986).
SU A 1986-1 Su, X. Y.
A note on clique graphs (Chinese). Fujian Shifan Daxue Xuebao Ziran Kexue Ban 2, No. 2, 1-2. (1986).
SUCI 1985-1 Suciu, D.
Labelled graphs and their associated values (Romanian). Buletinul Institutului Politehnic Bucuresti, Seria Transport. Aeronave 46/47, 52-54. (1985).
SUHL 1985-1 Suhl, U. H.
Solving large scale mixed-integer programs with fixed charge variables. Mathematical Programming 32, 165-182. (1985).
SUMI 1986-1 Sumichrast, R. T.
A new cutting-stock heuristic for scheduling production. Computers and Operations Research 13, 403-410. (1986).
SUMI 1987-1 Sumichrast, R. T.
Scheduling parallel processors to minimize setup time. Computers and Operations Research 14, 305-314. (1987).
SUN 1984-1 Sun, H.-C.
Some theorems on edge chromatic number of graphs (Chinese). Acta Mathematicae Applagatae Sinica (Beijing) 7, 510-512. (1984).
SUN 1985-1 Sun, H.-C.
On the cycle decomposition of complete graphs (Chinese). Nanjing Daxue Xuebao Ziran Kexue Ban 21, 421-426. (1985).
SUN 1985-2 Sun, H.-C.
Two sufficient conditions for simple graphs to be of class one (Chinese). Journal of Nanjing University, Mathematical Biquarterly 2, 203-207. (1985).
SUN 1987-1 Sun, H.-C.
A note to: "A theorem on colouring edges" by F. Z. Zhong, X. Z. Jian and F. W. Jian. Nanjing Daxue Xuebao Shuxue Bannian Kan 4, No. 1, 75-76. (1987).
SUN A 1986-1 Sun, H.-Q.
The connectivity of line graphs (Chinese). Chinese Annals of Mathematics (Shanghai), Series A 7, 602-605. (1986).
SUNA 1985-1 Sunaga, T.; Hayakawa, A. J.; Maruki, Y.
A computational method for all-integer interval linear programming. Journal of the Operations Research Society of Japan 28, 87-111. (1985).
SUND 1984-1 Sundararaghavan, P. S.
The unit capacity pickup and delivery problem on a one-way loop. Opsearch 21, 209-226. (1984).
SUNG 1985-1 Sung, C. S.
A multiproduct multifacility production planning model with capacity constraints. Journal of the Korean Operations Research and Management Science Society 10, No.2, 15-23. (1985).
SUNG 1986-1 Sung, C. S.
A single-product parallel-facilities production-planning model. International Journal of Systems Science 17, 983-990. (1986).
SUNG 1987-1 Sung, C. S.
Multifacility in-series production-planning model with capacity constraints. International Journal of Systems Science 18, 623-632. (1987).
SUNGA 1986-1 Sung, C. S.; Chang, S. H.
A capacity-constrained single-facility multi-product production planning model. Journal of the Operations Research Society of Japan 29, 232-245. (1986).
SUNGB 1986-1 Sung, C. S.; Park, M. H.
An approach for the optimal production rate of a single product system with dynamic demands. Journal of the Operations Research Society of Japan 29, 177-190. (1986).
SURE 1985-1 Suresh, S.; Foley, R. D.; Dickey, S. E.
On Pinedo's conjecture for scheduling in a stochastic flow shop. Operations Research 33, 1146-1153. (1985).
SURO 1986-1 Surovtsov, L. K.; Vilkov, V. B.
An approach to the solution of linear problems of integer programming (Russian). Primenenie Matematiki v Ekonomike (Leningrad), No. 14, 25-33. (1986).
SUSH 1985-1 Sushkov, Y. A.
Matroids and hypergraphs. Vestnik Leningrad University, Mathematics 18, No. 4, 51-56. Translation from: Vestnik Leningradskogo Universiteta, Matematika, Mekhanika, Astronomiza (Leningrad) 1985, No. 4, 42-46. (1985).
SUSS 1984-1 Sussams, J.
The future for computerized vehicle-load planning systems. Journal of the Operational Research Society 35, 963-966. (1984).
SUZU 1985-1 Suzuki, H.; Nishizeki, T.; Saito, N.
Multicommodity flows in planar undirected graphs and shortest paths. In: Proc. Seventeenth Annual ACM Symp. on Theory of Computing, Providence 1985, 195-204. ACM. (1985).
SUZU 1987-1 Suzuki, H.; Nishizeki, T.; Saito, N.
Algorithms for routing around a rectangle (extended abstract). In: D. P. Johnson, A. Nozaki, T. Nishizeki, H. P. Wilf (Eds.): Discrete Algorithms and Complexity, 99-106. Boston: Academic Press. (1987).
SWEE 1984-1 Sweeney, D. J.
A practical evaluation of the potential of mixed integer programming as a strategic planning tool. In: T. H. Naylor, C. Thomas (Eds.): Optimization Models for Strategic Planning, 111-120. Amsterdam: North-Holland. (1984).

SYSL 1984-1 Syslo, M. M.
Series-parallel graphs and depth-first search trees. IEEE Transactions on Circuits and Systems CAS-31, 1029-1033. (1984).

SYSL 1985-1 Syslo, M. M.
A graph-theoretic approach to the jump-number problem. In: I. Rival (Ed.): Graphs and Order, 185-216. Dordrecht: Reidel. (1985).

SYSL 1985-2 Syslo, M. M.
A reduction lemma for the jump-number problem. In: H. Sachs (Ed.): Graphs, Hypergraphs and Applications, Proc. Conf. on Graph Theory (Eyba 1984), 186-189. Leipzig: Teubner. (1985).

SYSL 1985-3 Syslo, M. M.
Remarks on Dilworth partially ordered sets. In: H. Noltemeier (Ed.): Proc. WG '85: Internat. Workshop on Graphtheoretic Concepts in Computer Science, Castle Schwanberg, Würzburg, 1985, 355-362. Linz: Trauner. (1985).

SYSL 1985-4 Syslo, M. M.
Triangulated edge intersection graphs of paths in a tree. Discrete Mathematics 55, 217-220. (1985).

SYSL 1986-1 Syslo, M. M.
An algorithm for solving the jump number problem. Report 158. Wroclaw: Wroclaw Univ., Institute of Computer Science, 15 p. (1986).

SYSL 1986-2 Syslo, M. M.
On some new types of greedy chains and greedy linear extensions of partially ordered sets. Report 157. Wroclaw: Wroclaw Univ., Institute of Computer Science, 15 p. (1986).

SYSL 1987-1 Syslo, M. M.
A solvable case of the set-partitioning problem. Zastosowania Matematyki 19, 265-277. (1987).

SYSL 1987-2 Syslo, M. M.
Minimizing the jump number for partially-ordered sets: a graph-theoretic approach, II. Discrete Mathematics 63, 279-296. (1987).

SYSL 1987-3 Syslo, M. M.
On some generalizations of outerplanar graphs: results and open problems. In: G. Tinhofer, G. Schmidt (Eds.): Graph-Theoretic Concepts in Computer Science, 146-164. Lecture Notes in Computer Science 246, Berlin: Springer. (1987).

SYSLA 1983-1 Syslo, M. M.; Topp, J.
A characterization of generalized line graphs whose complements are also generalized line graphs. Report N-129. Wroclaw: Wroclaw Univ., Inst. of Computer Science, 25 p. (1983).

SYSLA 1985-1 Syslo, M. M.; Topp, J.
On the planarity and outerplanarity of generalized line, middle and total graphs. Journal of Combinatorics, Information and System Sciences 10, 69-78. (1985).

SZAM 1983-1 Szamkolowicz, L.
On some classes of three-chromatic graphs. Discussiones Mathematicae (Poznan) 6, 79-80. (1983).

SZID 1984-1 Szidarovszky, F.; Carrera, J.; Usunoff, E.
Optimum location of measuring points in the quality analysis of groundwater (Hungarian). Szigma (Budapest) 17, 157-166. (1984).

SZIL 1983-1 Szilagyine-Szemkeö, J.
A multiple objective scheduling problem (Hungarian). Szigma (Budapest) 16, 85-91. (1983).

SZKA 1987-1 Szkatula, K.; Libura, M.
On probabilistic properties of greedy-like algorithms for the binary knapsack problem. In: G. Andreatta, F. Mason, P. Serafini (Eds.): Advanced School on Stochastics in Combinatorial Optimization (CISM Udine 1986), 233-254. Singapore: World Scientific. (1987).

SZWA 1986-1 Szwarc, W.
The flow shop problem with time lags and separated setup times. Zeitschrift für Operations Research 30 B, 15 22. (1986).

SZWAA 1987-1 Szwarc, W.; Gupta, J. N. D.
A flow-shop problem with sequence-dependent additive setup times. Naval Research Logistics 34, 619-628. (1987).

SZWAB 1985-1 Szwarc, W.; Posner, M. E.
The caterer problem. Operations Research 33, 1215-1224. (1985).

SZWAC 1987-1 Szwarcfiter, J. L.
A note on the computation of the k-closure of a graph. Information Processing Letters 24, 279-280. (1987).

SZWAC 1987-2 Szwarcfiter, J. L.
Job shop scheduling with unit time operations under resource constraints and release dates. Discrete Applied Mathematics 18, 227-234. (1987).

SZWAD 1985-1 Szwarcfiter, J. L.; Persiano, R. C. M.; Oliveira, A. A. F.
Orientations with single source and sink. Discrete Applied Mathematics 10, 313-322. (1985).

SZYM 1986-1 Szymanowski, W.
COLOS: a collection optimization system as an application of the vehicle scheduling problem. European Journal of Operational Research 27, 82-90. (1986).

TADA 1986-1 Tada, M.; Ishii, H.; Nishida, T.
A generalized version of 1-machine mean flow time minimization problem. Technology Reports of the Osaka University 36, No. 1819-1843, 1-4. (1986).

TADE 1983-1 Tadeusz, U.
Schnelle Routing-Algorithmen in Rechnernetzen. Univ. Dortmund, Abteilung Informatik, Report 149, 17 p. (1983).

TAKA 1984-1 Takaoka, T.; Ko, K.
An 0((N**2)LOGN) expected time algorithm for the all pairs shortes path problem over a certain class of graphs. Congressus Numerantium 45, 225-234. (1984).

TALA 1985-1 Talamo, M.; Marchetti-Spaccamela, A.; Protasi, M.
On some relationships between combinatorics and probabilistic analysis. In: G. Ausiello, M. Lucertini (Eds.): Analysis and Design of Algorithms for Combinatorial Problems, Annals of Discrete Mathematics Vol. 25, 291-310. Amsterdam: North-Holland. (1985).

TALAA 1979-1 Talanov, V. A.; Shevchenko, V. N.
A generalization of the assignment problem (Russian). in: Combinatorial-Algebraic Methods in Applied Mathematics, 101-103. Gorki: Gor'kovskii Gosudarstvennyi Universitet. (1979).

TALB 1986-1 Talbot, F. B.; Patterson, J. H.; Gehrlein, W. V.
A comparative evaluation of heuristic line balancing techniques. Management Science 32, 430-454. (1986).

TAMA 1984-1 Tamaki, M.
The secretary problem with optimal assignment. Operations Research 32, 847-858. (1984).

TAMAA 1987-1 Tamassia, R.
On embedding a graph in the grid with the minimum number of bends. SIAM Journal on Computing 16, 421-444. (1987).

TAMI 1985-1 Tamir, A.
A finite algorithm for the continuous p-center location problem on a graph. Mathematical Programming 31, 298-306. (1985).

TAMI 1987-1 Tamir, A.
On the solution value of the continuous p-center location problem on a graph. Mathematics of Operations Researchs 12, 340-349. (1987).

TAMI 1987-2 Tamir, A.
Totally balanced and totally unimodular matrices defined by center location problems. Discrete Applied Mathematics 16, 245-264. (1987).

TAN 1986-1 Tan, E. L.
Some classes of cycle graphs. Research Report No. 238. Singapore: National Univ., Dept. of Mathematics, 20p. (1986).

TAN A 1985-1 Tan, Z.-G.; Zheng, L.-N.; Li, L.-N.
A mathematical model for the traffic assignment problem (Chinese). Mathematics in Practice and Theory (Beijing), No. 1, 76-80. (1985).

TAN A 1985-2 Tan, Z.-G.; Zheng, L.-N.; Li, L.-N.
Introduction to mathematical models and algorithms for the traffic assignment problem (Chinese). Mathematics in Practice and Theory (Beijing) No. 1, 76-80. (1985).

TANA 1984-1 Tanaev, V. S.; Gordon, V. S.; Shafranskij, Y. M.
Scheduling theory, one-machine systems (Russian). In: Library of Mathematical Economics 18, 383 pp. Moscow: Nauka. (1984).

TANAA 1984-1 Tanaka, K.; Maruyama, Y.
The multiobjective optimization problem of set function. Journal of Information & Optimization Sciences 5, 293-306. (1984).

TANE 1985-1 Taneja, H. R.
On a new simple heuristic algorithm for scheduling n weighted jobs in a flow shop of two machines. Journal of the Indian Society of Statistics and Operations Research (Saharanpur) 6, 13-17. (1985).

TANG 1986-1 Tang, C.; Denardo, E. V.
Models arising from a flexible manufacturing machine. Part 1: minimization of the number of tool switches. Working Paper, 341. Los Angeles: Western Management Science Institute, 31 p. (1986).

TANG 1986-2 Tang, C.; Denardo, E. V.
Models arising from a flexible manufacturing machine. Part 2: minimization of the number of switching instants. Working Paper, 342. Los Angeles: Western Management Science Institute, 18 p. (1986).

TANGA 1984-1 Tang, C. Y.; Lee, R. C. T.
Optimal speeding up of parallel algorithms based upon the divide-and-conquer strategy. Information Sciences 32, 173-186. (1984).

TANGB 1987-1 Tang, H.; Ni, M.
An algorithm for solving surrogate dual problems in integer linear programming (Chinese). Journal of Dalian Institute of Technology 26, No.4, 1-6. (1987).

TANGC 1986-1 Tang, H.-W.; Sun, Z.-X.; Gao, G.-Q.
An algorithm for a class of nonlinear mixed integer programming problems, and its convergence (Chinese). Journal of Dalian Institute of Technology 25, Special Issue on Mathematics, Supplement, 74-78. (1986).

TANGD 1986-1 Tang, Q. F.
An algorithm for constructing minimal trees with designated roots (Chinese). Heilongjiang Daxue Ziran Kexue Xuebao, No. 4, 27-31. (1986).

TAO 1987-1 Tao, G.-Y.; Zhou, P.-D.; Wang, Q.
The heuristic shortest-path algorithm for transportation (Chinese). Journal of Nanjing Institute of Technology 17, No.6, 72-77. (1987).

TAO A 1982-1 Tao, K. M.; Hsia, P.
On the structure of directed graphs with applications. I. A rapprochement with systems theory. International Journal of General Systems 8, 147-159. (1982).

TAO B 1983-1 Tao, M. Q.; Shen, Y. Q.
A conjecture of Erdös - the planar case (Chinese). Journal of China University of Science and Technology (Hefei) 13, No. 3, Math. Issue, 279-287. (1983).

TARA 1985-1 Tarakanov, V. E.
Combinatorial problems and (0,1)-matrices (Russian). Problemy Nauki i Tekhnicheskogo Progressa. Moscow: Nauka. (1985).

TARAA 1985-1 Tarasov, O.; Richter, P.
Efficient algorithms for the determination of optimal paths in non-negatively weighted digraphs. In: A. Iwainsky (Ed.): Optimization of connection structures in grahps, 13-28. Berlin: Central Insitute of Cybernetics and Information Processes, Academy of Sciences of the German Democratic Republic. (1985).

TARAA 1985-2 Tarasov, O.; Richter, P.
Placement and routing for the optimization of star networks in graphs. In: A. Iwainsky (Ed.): Optimization of connection structures in grahps, 29-37. Berlin: Central Insitute of Cybernetics and Information Processes, Academy of Sciences of the German Democratic Republic. (1985).

TARAB 1986-1 Tarasova, L. G.
Clutters of the set of paths between two vertices of a graph (Russian). In: Mathematical Methods for the Analysis and Optimization of Complex Systems that Function under Conditions of Uncertainty, 17-20. Kiev: Akad. Nauk Ukrain. SSR, Inst. Kibernet. (1986).

TARD 1985-1 Tardos, E.
A strongly polynomial minimum cost circulation algorithm. Combinatorica 5, 247-256. (1985).

TARD 1985-2 Tardos, E.
Generalized matroids and supermodular colourings. In: L. Lovasz, A. Recski (Eds.): Matroid Theory, 359-382. Coll. Math. Soc. J. Bolyai Vol. 40. Amsterdam: North-Holland. (1985).

TARD 1986-1 Tardos, E.
A strongly polynomial algorithm to solve combinatorial linear programs. Operations Research 34, 250-256. (1986).

TARDA 1986-1 Tardos, E.; Tovey, C. A.; Trick, M. A.
Layered augmenting path algorithms. Mathematics of Operations Research 11, 362-370. (1986).

TARJ 1983-1 Tarjan, R. E.
Data structures and network algorithms. CBMS-NSF Regional Conference Series in Applied Mathematics (Philadelphia) 44, 131p. (1983).

TARJ 1985-1 Tarjan, R. E.
Decomposition by clique separators. Discrete Mathematics 55, 221-232. (1985).

TARJ 1985-2 Tarjan, R. E.
Shortest path algorithms. In: Y. Alavi, G. Chartrand, L. Lesniak, D. R. Lick, C. E. Wall (Eds.): Graph Theory with Applications to Algorithms and Computer Science, 753-760. Chichester: Wiley. (1985).

TARJ 1986-1 Tarjan, R. E.
Algorithms for maximum network flow. In: G. Gallo, C. Sandi (Eds.): Netflow at Pisa, Mathematical Programming Study 26, 1-11. (1986).

TARJ 1986-2 Tarjan, R. E.
Corrigendum to: Sensitivity analysis of minimum spanning trees and shortest path trees. Information Processing Letters 23, 219. (1986).

TARJ 1987-1 Tarjan, R. E.
Algorithm design. Communications of the ACM 30, 204-213. (1987).

TARJA 1985-1 Tarjan, R. E.; Vishkin, U.
An efficient parallel biconnectivity algorithm. SIAM Journal on Computing 14, 862-874. (1985).

TARJB 1985-1 Tarjan, R. E.; Yannakakis, M.
Simple linear-time algorithms to test chordality of graphs, test acyclicity of hypergraphs, and selectively reduce acyclic hypergraphs (Addendum). SIAM Journal on Computing 14, 254-255. (1985).

TARS 1985-1 Tarsi, M.
Nowhere zero flow and circuit covering in regular matroids. Journal of Combinatorial Theory 39B, 346-352. (1985).

TASH 1985-1 Tashchuk, D. A.
Properties of special type subgraphs in an oriented graph (Russian). Kibernetika No. 5, 15-20. (1985).

TASHA 1984-1 Tashkinov, V. A.
Three-regular parts of four-regular graphs. Mathematical Notes of the Academy of Sciences of the USSR 36, 612-623. Translated from Matematicheskie Zametki 36, No. 2, 239-259. (1984).

TAVA 1986-1 Tavares, L. V.
Multicriteria scheduling of a railway renewal program. European Journal of Operational Research 25, 395-405. (1986).

TAVA 1987-1 Tavares, L. V.
Optimal resource profiles for program scheduling. European Journal of Operational Research 29, 83-90. (1987).

TAY 1984-1 Tay, T.-S.
Rigidity of multi-graphs II. In: K. M. Koh, H. P. Yap (Eds.): Graph Theory Singapore 1983, Lecture Notes in Mathematics 1073, 129-134. Berlin: Springer-Verlag. (1984).

TAYI 1985-1 Tayi, G. K.
Sensitivity analysis of mixed integer programs: an application to environmental policy making. European Journal of Operational Research 22, 224-233. (1985).

TAYI 1986-1 Tayi, G. K.
Bicriteria transportation problem: an alternate approach. Socio-Economic Planning Sciences 20, 127-130. (1986).

TAYL 1985-1 Taylor, B. W. III; Moore, L. J.; Clayton, E. R.; Davis, K. R.; Rakes, T. R.
An integer nonlinear goal programming model for the deployment of state highway patrol units. Management Science 31, 1335-1347. (1985).

TAZA 1984-1 Tazawa, S.
A short proof of Turan theorem. Journal of Science and Technology, Kinki University 19, 31-32. (1984).

TAZA 1985-1 Tazawa, S.
Decomposition of a complete multi-partite graph into isomorphic claws. SIAM Journal on Algebraic and Discrete Mathods 6, 413-417. (1985).

TAZZ 1983-1 Tazzi, G. C.; Aragno Marauta, E.
The maximum genus of certain graphs (Italian). Bolletino, Unione Matematica Italiana, Serie B (6) 2, 857-866. (1983).

TCHA 1985-1 Tcha, D.-W.
On the selection of primary paths for a communication network. Computer Networks and ISDN Systems (Amsterdam) 9, 257-265. (1985).

TCHAA 1987-1 Tcha, D.-W.; Jung, S.-H.; Myung, Y.-S.
A comparative study on optimization algorithms for capacitated facility location (Korean). Journal of the Korean Operations Research and Management Science Society 11, No.2, 1-6. (1987).

TCHAB 1984-1 Tcha, D.-W.; Lee, B.-I.
A branch-and-bound algorithm for the multi-level uncapacitated facility location problem. European Journal of Operational Research 18, 35-43. (1984).

TCHAC 1985-1 Tcha, D.-W.; Yoon, M.-G.
A dual-based heuristic for the simple facility location problem with stochastic demand. IIE Transactions 17, 364-369. (1985).

TEGH 1986-1 Teghem, J. Jr.; Kunsch, P. L.
A survey of techniques for finding efficient solutions to multi-objective integer linear programming. Asia-Pacific Journal of Operational Research 3, 95-108. (1986).

TEGH 1986-2 Teghem, J. Jr.; Kunsch, P. L.
Interactive methods for multiobjective integer linear programming. In: Large-Scale Modelling and Interactive Decision Analysis (Eisenach 1985), 75-87. Lecture Notes in Economics and Mathematical Systems 273, Berlin: Springer. (1986).

TEGZ 1984-1 Tegze, M.; Vlach, M.
Parallel flow shop problems. In: K.-H. Elster (Ed.): Mathematische Optimierung - Theorie und Anwendungen, Vortragsauszüge, Eisenach 1984, 127-128. Techn. Hochschule Ilmenau. (1984).

TEGZ 1985-1 Tegze, M.; Vlach, M.
Matrix given scheduling problems. In: 8th Conf. on Mathematical Methods in Economics (Sellin 1984), Seminarberichte 71, 127-132. Berlin: Humboldt-Universität. (1985).

TEGZ 1987-1 Tegze, M.; Vlach, M.
Improved bounds for the range of lateness on a single machine. Working Paper. Prague: Charles Univ., Dept. of Cybernetics and Informatics, 20 p. (1987).

TEIC 1983-1 Teichert, H.-M.
Hamiltonian properties of the lexicographic product of undirected graphs. Elektronische Informationsverarbeitung und Kybernetik 19, 67-77. (1983).

TEIC 1984-1 Teichert, H.-M.
Zusammenhangs- und Durchlaufbarkeitseigenschaften der Disjunktion ungerichteter Graphen. In: Geometrie und Kombinatorik, 2. Kolloq., Karl-Marx-Stadt/DDR 1983, Vortr. 1,2, Teil II, 255-259. (1984).

TELE 1986-1 Telemtaev, M. M.
An algorithm for the travelling salesman problem (Russian). Izvestiya Akademii Nauk Kazakhskoj SSR, Seriya Fiziko-Matmaticheskaya (Alma-Ata) No. 1, 40-43. (1986).

TELL 1987-1 Telley, H.; Liebling, T. M.; Mocellin, A.
Reconstruction of polycrystalline structures: a new application of combinatorial optimization. Computing 38, 1-12. (1987).

TEMP 1985-1 Templeton, W. D.; Daellenbach, H. G.
Interactive simulation for production scheduling. New Zealand Operational Research 13, 27-38. (1985).

TENG 1985-1 Teng, A.; Tucker, A.
An 0(qn) algorithm to q-color a proper family of circular arcs. Discrete Mathematics 55, 233-242. (1985).

TENGA 1985-1 Teng, C.-L.
The stepping-stone method for solving assignment problems (Chinese). Journal of Lanzhou Railway College 4, No.3, 59-66. (1985).

TEOD 1985-1 Teodorovic, D.
Model for designing meteorologically most reliable airline schedules. European Journal of Operational Research 21, 156-164. (1985).

TERL 1985-1 Terlaky, T.
A finite criss-cross method for oriented matroids (Hungarian). Alkalmazott Matematikai Lapok 11, 385-398. (1985).

TERL 1986-1 Terlaky, T.
A finite criss-cross method and applications (Hungarian). Tanulmanyok - MTA Szamitastech. Automat. Kutato Int. Budapest No. 179. (1986).

TERL 1987-1 Terlaky, T.
A finite crisscross method for oriented matroids. Journal of Combinatorial Theory 42B, 319-327. (1987).

TERS 1986-1 Tersine, R. J.; Fisher, W. W.; Morris, J. S.
Varying lot sizes as an alternative to undertime and layoffs in aggregate scheduling. International Journal of Production Research 24, 97-106. (1986).

TERW 1985-1 Terwilliger, P.
Distance-regular graphs with Girth 3 or 4: I. Journal of Combinatorial Theory 39B, 265-281. (1985).

TERZ 1985-1 Terzi, D. G.
Generators of vectors in problems of discrete optimization (Russian). Matematicheskie Issledovaniya (Kishinev), No. 82, Matematicheskie Modeli Met Optim., 135-141. (1985).

THIS 1984-1 Thisse, J.-F.; Ward, J. E.; Wendell, R. E.
Some properties of location problems with block and round norms. Operations Research 32, 1309-1327. (1984).

THIZ 1985-1 Thizy, J. M.; Wassenhove, L. N. van
Lagrangean relaxation for the multi-item capacitated lot-sizing problem: a heuristic implementation. IIE Transactions 17, 308-313. (1985).

THOM 1984-1 Thomas, R.
Series-parallel graphs and well- and better-quasi-orderings. Commentationes Mathematicae Universitatis Carolinae (Prague) 25, 719-723. (1984).

THOMA 1984-1 Thomassen, C.
A refinement of Kuratowski's theorem. Journal of Combinatorial Theory, Series B 37, 245-253. (1984).

THOMA 1984-2 Thomassen, C.
Graph theory - an introduction (Danish). Normat (Oslo) 32, 111-124. (1984).

THOMA 1984-3 Thomassen, C.
Plane representations of graphs. In: J. A. Bondy, U. P. R. Murty (Eds.): Progress in Graph Theory, 43-70. New York: Academic Press. (1984).

THOMA 1985-1 Thomassen, C.
Even cycles in directed graphs. European Journal of Combinatorics 6, 85-90. (1985).

THOMA 1985-2 Thomassen, C.
Hamilton circuits in regular tournaments. In: B. R. Alspach, C. D. Godsil (Eds.): Cycles in Graphs, Annals of Discrete Mathematics Vol. 27, 159-162. Amsterdam: North-Holland. (1985).

THOMA 1985-3 Thomassen, C.
The 2-linkage problem for acyclic digraphs. Discrete Mathematics 55, 73-88. (1985).

THOMA 1986-1 Thomassen, C.
Decomposition of graphs into chains. Bulletin of the London Mathematical Society 18, 248-254. (1986).

THOMA 1986-2 Thomassen, C.
Interval representations of planar graphs. Journal of Combinatorial Theory 40B, 9-20. (1986).

THOMA 1986-3 Thomassen, C.
Sign-nonnegative matrices and even cycles in directed graphs. Linear Algebra and Its Appplications 75, 27-42. (1986).

THOMA 1987-1 Thomassen, C.
A characterization of locally finite vertex-transitive graphs. Journal of Combinatorial Theory 43B, 116-119. (1987).

THOMA 1987-2 Thomassen, C.
Counterexamples to Adam's conjecture on arc reversals in directed graphs. Journal of Combinatorial Theory B 42, 128-130. (1987).

THOMA 1987-3 Thomassen, C.
On digraphs with no two disjoint directed cycles. Combinatorica 7, 145-150. (1987).

THOMB 1986-1 Thompson, G. L.; Kaku, B. K.
An exact algorithm for the general quadratic assignment problem. European Journal of Operational Research 23, 382-390. (1986).

THOMC 1986-1 Thompson, G. L.; Sethi, A. P.
Solution of constrained generalized transportation problems using the pivot and probe algorithm. Computers and Operations Research 13, 1-10. (1986).

THOMD 1985-1 Thompson, G. L.; Singhal, S.
A successful algorithm for the undirected Hamiltonian path problem. Discrete Applied Mathematics 10, 179-196. (1985).

THOME 1986-1 Thompson, G. L.; Zawack, D. J.
A problem expanding parametric programming method for solving the job shop scheduling problem. In: C. L. Monma (Ed.): Algorithms and Software for Optimization - Part I, 327-342. Annals of Operations Research 4, Basel: Baltzer. (1986).

THUL 1987-1 Thulasiraman, K.; Comeau, M. A.
Maximum-weight markings in marked graphs: algorithms and interpretations based on the simplex method. IEEE Transactions on Circuits and Systems CAS-34, 1535-1545. (1987).

THUL 1987-2 Thulasiraman, K.; Comeau, M. A.
Structure of the reachability problem for (0,1)-capacitated marked graphs. IEEE Transactions on Circuits and Systems CAS-34, 317-319. (1987).

TIAN 1986-1 Tian, F.
The progress on some problems in graph theory (Chinese). Qufu Shifan Daxue Xuebao. Ziran Kexue Ban, No. 2, 30-36. (1986).

TIANA 1986-1 Tian, F.; Shi, R.-H.
A new class of pancyclic graphs (Chinese). Journal of Systems Science and Mathematical Sciences (Beijing) 6, 258-262. (1986).

TIANB 1985-1 Tian, S.-L.; Wang, J. F.; Liu, J.-Q.
The binding number of two product graphs. Journal of Mathematical Research and Exposition (Dalian) 5, No.1, 154. (1985).

TIEN 1983-1 Tien, J. M.; El-Tell, K.; Simons, G. R.
Improved formulations to the hierarchical health facility location-allocation problem. IEEE Transactions on Systems, Man, and Cybernetics 13, 1128-1132. (1983).

TIES 1987-1 Tiesis, V.; Saltenis, V.; Juska, F.
A problem of the allocation of a discrete resource (Russian). Teoriya Optimalnykh Reshenii (Vilnius), No. 12, 100-113. (1987).

TIGA 1984-1 Tigan, S.
On some nonlinear knapsack problems. in: Itinerant Seminar on Functional Equations, Approximation and Convexity (Cluj Napoca), 203-208, Preprint 84-6, Univ. "Babes-Bolyai", Cluj-Napoca. (1984).

TIGAA 1985-1 Tigan, S.; Stancu-Minasian, I. M.
The stochastic bottleneck transportation problem. Mathematica - Revue d'Analyse Numerique et de Theorie de l'Approximation (Cluj-Napoca) 14, 153-158. (1985).

TILA 1988-1 Tilanus, C. B.
Transportation planning: recent developments in the Netherlands. In: H. Schellhaas, P. van Beek, H. Isermann, R. Schmidt, M. Zijlstra (Eds.): Operations Research Proceedings 1987 (DGOR/NSOR), 308-313. Berlin: Springer. (1988).

TIMK 1985-1 Timkovskii, V. G.
On the complexity of scheduling an arbitrary system. Izvestiya Akademii Nauk SSSR Teckhnicheskaya Kibernetika (Moscow) No. 3, 102-109; translated as: Soviet Journal of Computational Systems Science 23, No. 5, 46-52. (1985).

TIMK 1986-1 Timkovskii, V. G.
Approximate solution of the scheduling problem for a cyclic system (Russian). Ekononomika i Matematicheskie Metody (Moscow) 22, 171-174. (1986).

TIMM 1988-1 Timmer, G. T.
Scheduling the construction of Dutch roads. In: H. Schellhaas, P. van Beek, H. Isermann, R. Schmidt, M. Zijlstra (Eds.): Operations Research Proceedings 1987 (DGOR/NSOR), 314-315. Berlin: Springer. (1988).

TING 1985-1 Ting, S.-K.
The structure of a maximum critical 2-edge-connected graph (Chinese). Chinese Annals of Mathematics, Series A (Shanghai) 6, 447-453. (1985).

TINH 1984-1 Tinhofer, G.
A probabilistic analysis of some greedy cardinality matching algorithms. Annals of Operations Research, Vol. 1, 239-254. Basel: Baltzer. (1984).

TINH 1984-2 Tinhofer, G.
Naive greedy matching - a probabilistic analysis. In: U. Pape (Ed.): Proc. WG '84: Internat. Workshop on Graphtheoretic Concepts in Computer Science, Berlin 1984, 354-368. Linz: Trauner. (1984).

TINH 1984-3 Tinhofer, G.
Probabilistische Ansätze in der diskreten Optimierung. In: R. E. Burkard (Ed.): 15. Steiermärkisches mathematisches Symposium, Forschungszentrum Graz, Mathematisch-Statistische Sektion, Bericht No. 230. (1984).

TINHA	1984-1	Tinhofer, G.; Schreck, H. Linear time tree codes. Computing 33, 211-226. (1984).
TINHA	1986-1	Tinhofer, G.; Schreck, H. The bounded subset sum problem is almost everywhere randomly decidable in O(n). Information Processing Letters 23, 11-18. (1986).
TOBI	1985-1	Tobin, R. L.; Friesz, T. L. A new look at spacially competitive facility location models. In: Spatial Price Equilibrium: Advances in Theory, Computation and Application (Denver 1984), 1-19. Lecture Notes in Econ. and Math. Systems 249, Berlin: Springer. (1985).
TOBI	1986-1	Tobin, R. L.; Friesz, T. L. Spatial competition facility location models: definition, formulation, and solution approach. In: J. P. Osleeb, P. J. Ratick (Eds.): Locational Decisions: Methodology and Applications, 49-76. Annals of Operations Research 6, Basel: Baltzer. (1986).
TOCZ	1984-1	Toczylowski, E. A perfect matching algorithm for sparse bipartite graphs. Discrete Applied Mathematics 9, 263-268. (1984).
TOCZ	1986-1	Toczylowski, E. On aggregation in a two-dimensional cutting stock scheduling problem. Large Scale Systems 10, 165-174. (1986).
TOCZ	1986-2	Toczylowski, E. On aggregation of items in the single-stage lot size scheduling problem. Large Scale Systems 10, 157-164. (1986).
TOCZ	1987-1	Toczylowski, E. Two-phase algorithm for lot size scheduling in single-stage production systems with parallel facilities. Bulletin of the Polish Academy of Sciences, Technical Sciences 35, 77-81. (1987).
TOCZA	1986-1	Toczylowski, E.; Nowosad, K.; Jagdev, H.; Hindi, K. S. Aggregate lot-size scheduling for a class of two-stage batch manufacturing systems. Large Scale Systems 11, 165-175. (1986).
TODD	1984-1	Todd, M. J. Complementarity in oriented matroids. SIAM Journal on Algebraic and Discrete Methods 5, 467-485. (1984).
TODD	1985-1	Todd, M. J. Linear and quadratic programming in oriented matroids. Journal of Combinatorial Theory 39B, 105-133. (1985).
TODO	1984-1	Todorov, D. T. A method of constructing coverings. Mathematical Notes of the Academy of Sciences of the USSR 35, 456-460; translated from: Matematicheskie Zametki 35, No. 6, 869-876. (1984).
TOFT	1985-1	Toft, B. Some problems and results related to subgraphs of colour critical graphs. In: R. Bodendiek, H. Schumacher, G. Walther (Eds.): Graphen in Forschung und Unterricht, Festschrift Klaus Wagner, 178-191. Salzdetfurth: Franzbecker. (1985).
TOID	1985-1	Toida, S. A decomposition of a graph into dense subgraphs. IEEE Transactions on Circuits and Systems CAS-32, 583-589. (1985).
TOKA	1986-1	Tokarev, D. D.; Dishliev, A. B. On a problem of scheduling theory (Bulgarian). Godishnik na Visshite Uchebni Zavedniya, Prilozhna Matematika (Sofia) 21, No.4, 235-245. (1986).
TOKU	1985-1	Tokuyama, H.; Ueno, N. The cutting stock problem for large sections in the iron and steel industries. European Journal of Operational Research 22, 280-292. (1985).
TOME	1984-1	Tomescu, I. A Hamiltonian connectivity property of regular graphs with forbidden subgraphs. Quarterly Journal of Mathematics, Oxford Series (2) 35, 507-512. (1984).
TOME	1984-2	Tomescu, I. Colorings and irreducible coverings by cliques of graphs and hypergraphs. Analele Universitatii din Galati, Metalurgie 2 (7), No. 2, 15-20. (1984).
TOME	1985-1	Tomescu, I. Problems in combinatorics and graph theory. New York, Chichester: Wiley. (1985).
TOME	1986-1	Tomescu, I. On hypergraph colourings. Quarterly Journal of Mathematics, Oxford, Series (2), 37, 239-243. (1986).
TOML	1985-1	Tomlin, J. A.; Welch, J. S. Integration of a primal simplex network algorithm with a large-scale mathematical programming system. ACM Transactions on Mathematical Software 11, 1-11. (1985).
TOMO	1984-1	Tomova, E. Decomposition of complete bipartite graphs into factors with given diameters and radii. Mathematica Slovaca (Bratislava) 34, 249-253. (1984).
TONG	1984-1	Tong, P.; Lawler, E. L.; Vazirani, V. V. Solving the weighted parity problem for gammoids by reduction to graphic matching. In: W. R. Pulleyblank (Ed.): Progress in Combinatorial Optimization, 363-374. Toronto: Academic Press. (1984).
TOPA	1984-1	Topala, O. I. Star property on graphs (Russian). Matematicheskie Issledovaniya (Kishinev) 78, 130-133. (1984).
TOPC	1985-1	Topchishvili, A. L. On the problem of choosing a maximum solvable system from a system of linear pseudo-Boolean equations (Russian). Soobshcheniza Akademii Nauk Gruzinskoj SSR (Tbilisi) 117, 49-52. (1985).
TOPK	1984-1	Topka, V. V. An exact algorithm for the uniform distribution of resources in a network (Russian). In: Y.Z. Tyspkin (Ed.): Deterministic and stochastic control systems, Work Collect., 58-64. Moscow: "Nauka" 1984. (1984).

TOPK 1985-1 Topka, V. V.
Decomposition in a problem of resource allocation on a network (Russian). In: Y.Z. Tyspkin (Ed.): Theoretical and applied problems of optimization, Work Collect., 48-54. Moscow: "Nauka" 1985. (1985).

TOPKA 1984-1 Topkis, D. M.
Adjacency on polymatroids. Mathematical Programming 30, 229-237. (1984).

TOSI 1986-1 Tosic, R.
Vertex-to-vertex search in a graph. In: R. Tosic; D. Acketa; V. Petrovic (Eds.): Graph theory, Proc. 6th Yugoslav Seminar on Graph Theory, Dubrovnik, Yugoslavia, April 1985, 233-237. University of Novi Sad. (1986).

TOVE 1981-1 Tovey, C. A.
Polynomial local improvement algorithms in combinatorial optimization. Technical Report SOL 81-21. Stanford: Stanford Univ., Dept. of Operations Research, 63 p. (1981).

TOVE 1986-1 Tovey, C. A.
Rescheduling to minimize makespan on a changing number of identical processors. Naval Research Logistics Quarterly 33, 717-724. (1986).

TOVEA 1985-1 Tovey, C. A.; West, D. B.
Networks and chain coverings in partial orders and their products. Order 2, 49-60. (1985).

TOWN 1986-1 Townsend, W.
An application of the assignment model to bus crew rostering. IMA Journal of Mathematics in Management 1, 45-52. (1986).

TRAU 1987-1 Trauwaert, E.; Decrock, A.
A heuristic algorithm for allocation of components in manufacturing processes. European Journal of Operational Research 31, 232-239. (1987).

TRIC 1987-1 Trick, M. A.
Matching problems and generalized networks. Chapter IV of: Networks with additional structured constraints. Atlanta: Georgia Inst. of Technology, pp. 28-72 (1987).

TRIE 1988-1 Trienekens, H. W. J. M.
A simulation tool for the performance evaluation of parallel branch and bound algorithms. In: H. Schellhaas, P. van Beek, H. Isermann, R. Schmidt, M. Zijlstra (Eds.): Operations Research Proceedings 1987 (DGOR/NSOR), 628-629. Berlin: Springer. (1988).

TRIEA 1984-1 Trietsch, D.
An improved algorithm for Steiner trees. Discussion Paper, 635. Evanston: Northwestern Univ., 22 p. (1984).

TRIEA 1985-1 Trietsch, D.
Augmenting Euclidean networks - the Steiner case. SIAM Journal on Applied Mathematics 45, 855-860. (1985).

TRIEA 1985-2 Trietsch, D.
Connecting Euclidean networks–the Steiner case. Discussion Paper 640. Evanston: Northwestern Univ., Center for Math. Studies in Economics and Management Science, 24 p. (1985).

TRIG 1987-1 Trigeiro, W. W.
A dual-cost heuristic for the capacitated lot sizing problem. IIE Transactions 19, 67-72. (1987).

TRIP 1984-1 Tripathy, A.
School timetabling - a case in large binary integer linear programming. Management Science 30, 1473-1489. (1984).

TROT 1985-1 Trotter, L. E. Jr.
Discrete packing and covering. In: M. O'hEigeartaigh, J. K. Lenstra, A. H. G. Rinnooy Kan (Eds.): Combinatorial Optimization - Annotated Bibliographies, 21-31. Chichester: Wiley. (1985).

TROTA 1987-1 Trotter, L. E. Jr.; West, D. B.
Two easy duality theorems for product partial orders. Discrete Applied Mathematics 16, 283-286. (1987).

TROTB 1986-1 Trotter, W. T. Jr.
Problems and conjectures in the combinatorial theory of ordered sets. Research Paper. Columbia: Univ. of South Carolina, Dept. of Mathematics, 31 p. (1986).

TRUB 1985-1 Trubin, V. A.
On one subclass of the Steiner problems on the plane with rectangular metric (Russian). Kibernetika (Kiev) No. 3, 37-40. (1985).

TRUBA 1986-1 Trubin, V. A.; Sharifov, F. A.
An efficient method for solving a class of allocation problems (Russian). Akademiya Nauk Azerbaidzhanskoi SSR Doklady (Baku) 42, No. 11, 7-11. (1986).

TRUE 1984-1 Truemper, K.
Elements of a decomposition theory for matroids. In: J. A. Bondy, U. P. R. Murty (Eds.): Progress in Graph Theory, 439-476. New York: Academic Press. (1984).

TRUE 1984-2 Truemper, K.
Partial matroid representations. European Journal of Combinatorics 5, 377-394. (1984).

TRUE 1985-1 Truemper, K.
A decomposition theory for matroids. I. General results. Journal of Combinatorial Theory 39B, 43-76. (1985).

TRUE 1985-2 Truemper, K.
A decomposition theory for matroids. II. Minimal violation matroids. Journal of Combinatorial Theory 39B, 282-297. (1985).

TRUE 1985-3 Truemper, K.
A flexible decomposition approach for finding submatrices with certain inherited properties. Technical Report. Dallas: Univ. of Texas, 61 p. (1985).

TRUE 1986-1 Truemper, K.
A decomposition theory for matroids. III. Decomposition conditions. Journal of Combinatorial Theory B-41, 275-305. (1986).

TRUE 1987-1 Truemper, K.
A decomposition theory for matroids V. Testing of matrix total unimodularity. Research Report. Richardson: Univ. of Texas, Computer Science Program, 56 p. (1987).

TRUE 1987-2 Truemper, K.
Max-flow min-cut matroids: polynomial testing and polynomial algorithms for maximum flow and shortest routes. Mathematics of Operations Research 12, 72-96. (1987).

TRUS 1985-1 Truscott, W. G.
Scheduling production activities in multi-stage batch manufactoring systems. International Journal of Production Research 23, 315-328. (1985).
TRUSA 1987-1 Trushkina, V. I.
The three-color method of coloring graphs (Russian). Sibirskii Matematicheskii Zhurnal (Novosibirsk) 28, No. 2, 186-200. (1987).
TRUSB 1983-1 Truszczynski, M.
Decompositions of complete multipartite graphs into isomorphic stars. In: Graphs and other Combinatorial Topics (Prague, 1982), 333-343. Leipzig: Teubner. (1983).
TRUSB 1983-2 Truszczynski, M.
On a conjecture of Mohar and Pisanski. Demonstratio Mathematica (Warsaw) 16, 755-759. (1983).
TRUSB 1984-1 Truszczynski, M.
Graceful unicyclic graphs. Demonstratio Mathematica (Warsaw) 17, 377-387. (1984).
TRUSB 1984-2 Truszczynski, M.
Some results on uniquely colourable graphs. In: A. Hajnal; L. Lovasz; V. T. Sos (Eds.): Finite and Infinite Sets, Vol. 2, 733-748. Coll. Math. Soc. J. Bolyai 37, Amsterdam: North-Holland. (1984).
TRUSB 1985-1 Truszczynski, M.
Centers and centroids of unicyclic graphs. Mathematica Slovaca (Bratislava) 35, 223-228. (1985).
TRUSB 1985-2 Truszczynski, M.
Jump number problem: the role of matroids. Order 2, 1-8. (1985).
TRUSB 1985-3 Truszczynski, M.
Note on decomposition of LambdaK(m,n) (LambdaK(m,n)*) into paths. Discrete Mathematics 55, 89-96. (1985).
TRUSB 1986-1 Truszczynski, M.
Decomposing graphs into forests of stars. Congressus Numerantium 54, 73-86. (1986).
TRUSB 1986-2 Truszczynski, M.
Decompositions of graphs into forests of equal size with bounded maximum degrees. Ars Combinatoria 21, 201-216. (1986).
TRYF 1986-1 Tryfos, P.
An integer programming approach to the apparel sizing problem. Journal of the Operational Research Society 37, 1001-1006. (1986).
TRZA 1984-1 Trzaskalik, T.
Optimal multiperiod planning for manufacturing plant - general problems. In: A. Straszak (Ed.): Large Scale Systems: Theory and Applications 1983 (Warsaw 1983), 481-486. IFAC Proc. Series 10, Laxenburg, Pergamon Press. (1984).
TSAR 1985-1 Tsarevskij, N. A.
An approximational decomposition algorithm for a combinatorial problem of developing production capacities of an industry with block production constraints II. Soviet Journal of Computer and Systems Sciences 23, No.4, 115-122; translation from Izvestiya Akademii Nauk SSSR, Tekhnicheskaya Kibernetika (1985), No.3, 94-101. (1985).
TSEN 1986-1 Tseng, F. T.; Truemper, K.
A decomposition of the matroids with the max-flow min-cut property. Discrete Applied Mathematics 15, 329-364. (1986).
TSENA 1984-1 Tseng, S. S.; Lee, R. C. T.
A parallel algorithm to solve the stable marriage problem. BIT 24, 308-316. (1984).
TSIL 1984-1 Tsiligirides, T.
Heuristic methods applied to orienteering. Journal of the Operational Research Society 35, 797-810. (1984).
TSIT 1986-1 Tsitsiklis, J. N.; Bertsekas, D. P.
Distributed asynchronous optimal routing in data networks. IEEE Transactions on Automatic Control AC-31, 325-332. (1986).
TSO 1986-1 Tso, M.
Network flow models in image processing. Journal of the Operational Research Society 37, 31-34. (1986).
TSUC 1983-1 Tsuchiya, M.
Another proof of Bela Bollobas' theorem on edge disjoint cycles. Proceedings of the Faculty of Science of Tokai University (Tokyo) 18, 29-40. (1983).
TSYG 1985-1 Tsygankov, I. V.
Solving the transportation problem by using the method of invariants (Russian). In: I. I. Eremin, L. D. Popov (Eds.): Parametric optimization and approximation methods for improper problems of mathematical programming, 114-119. Sverdlovsk: Ural'skij Nauchnyj Tsentr AN SSSR. (1985).
TUCK 1984-1 Tucker, A.
Applied Combinatorics (second edition). New York, Chichester: Wiley. (1984).
TUCK 1984-2 Tucker, A.
The validity of the perfect graph conjecture for K(4)-free graphs. In: C. Berge, V. Chvatal (Eds.): Topics on perfect graphs, Annals of Discrete Mathematics Vol. 21, 149-158. Amsterdam: North-Holland. (1984).
TUCK 1987-1 Tucker, A.
A reduction procedure for coloring perfect K(4)-free graphs. Journal of Combinatorial Theory 43B, 151-172. (1987).
TUCK 1987-2 Tucker, A.
Coloring Perfect (K(4)-e)-free graphs. Journal of Combinatorial Theory 42B, 313-318. (1987).
TUCKA 1984-1 Tucker, A.; Wilson, D.
An O(N**2) algorithm for coloring perfect planar graphs. Journal of Algorithms 5, 60-68. (1984).
TUMA 1985-1 Tuma, J.
Dilworth truncations and modular cuts. In: L. Lovasz, A. Recski (Eds.): Matroid Theory, 383-400. Coll. Math. Soc. J. Bolyai Vol. 40. Amsterdam: North-Holland. (1985).
TUNG 1986-1 Tung, P. C. T.
A bicriterion Pareto path algorithm. Research Report No. 251. Singapore: National Univ., Dept. of Mathematics, 10p. (1986).
TUNG 1986-2 Tung, P. C. T.
Finding a path with minimal cost-to-time ratio in a network. Asia-Pacific Journal of Operational Research 3, 43-52. (1986).

TUNGA 1987-1 Tung, P. C. T.; Chan, G. H.; Chew, K. L.
Finding a minimal-ratio elementary path in a network. Asia-Pacific Journal of Operational Research 4, 151-157. (1987).

TUNGB 1987-1 Tung, S. P.
Computational complexities of diophantine equations with parameters. Journal of Algorithms 8, 324-336. (1987).

TURA 1985-1 Turan, G.
On the greedy algorithm for an edge-partitioning problem. In: L. Lovasz, E. Szemeredi (Eds.): Theory of Algorithms, Colloquia Mathematica Societatis Janos Bolyai Vol. 44, 405-424. Amsterdam, Oxford, New York: North-Holland. (1985).

TUSH 1985-1 Tushkina, T. A.
Computation of a lower bound of the lengths of schedules. Journal of Soviet Mathematics 29, 1671-1675. (1985).

TUTT 1985-1 Tutte, W. T.
Graph theory as I have known it (Part III) (Chinese). Chinese Journal of Operations Research 4, 33-53. (1985).

TUTZ 1985-1 Tutzauer, F.
Toward a theory of disintegration in communication networks. Social Networks 7, 263-285. (1985).

TUYT 1986-1 Tuyttens, D.; Teghem, J. Jr.
Theorie des matroides et optimisation combinatoire. Belgian Journal of Operations Research, Statistics and Computer Science 26, No. 1, 27-62. (1986).

TUZI 1984-1 Tuzikov, A. V.
A bicriterion problem of the scheduling theory taking the variation of servicing time into account. USSR Journal on Computational Mathematics and Mathematical Physics 24, No. 5, 191-194. Translation from: 1984,Zhurnal Vychislitel'noj Matematiki i Matematicheskoj Fiziki (Moscow) 24, No. 10, 1585-1590. (1984).

TUZI 1984-2 Tuzikov, A. V.
A two-criterion scheduling problem allowing for variation in job execution time (Russian). Zhurnal Vzchislitel'noi Matematiki i Matematicheskoi Fiziki (Moscow) 24, 1585-1590. (1984).

TUZI 1984-3 Tuzikov, A. V.
On a bicriterion problem of scheduling theory with regard to the change of service length (Russian). Zhurnal Vychislitel'noj Matematiki i Matematicheskoj Fiziki (Moscow) 24, 1585-1590. (1984).

TUZI 1984-4 Tuzikov, A. V.
On bicriterion scheduling problem with variable executing time of jobs (Russian). Zhurnal Vychislitel'noi Matematiki i Matematicheskoi Fiziki (Moscow) 24, 1581-1584. (1984).

TYSH 1985-1 Tyshkevich, R. I.
Polar graphs and the decomposition. In: H. Sachs (Ed.): Graphs, Hypergraphs and Applications, Proc. Conf. on Graph Theory (Eyba 1984), 190-197. Leipzig: Teubner. (1985).

TYSHA 1985-1 Tyshkevich, R. I.; Chernjak, A. A.
Algorithms of a canonical decomposition of a graph and polarity recognition (Russian). Izvestiya Akademii Nauk BSSR, Seriya Fiziko-Matematicheskikh Nauk, No.6, 16-23. (1985).

TYSHA 1985-2 Tyshkevich, R. I.; Chernjak, A. A.
Box-threshold graphs: The structure and the enumeration. In: Graphen und Netzwerke - Theorie und Anwendungen, 30. Int. Wiss. Kolloq., Ilmenau 1985, 5, 119-121. (1985).

TYSHA 1985-3 Tyshkevich, R. I.; Chernjak, A. A.
Decomposition of graphs. Kibernetika (Kiev) No. 2, 67-74. (1985).

TYUL 1986-1 Tyulyukin, V. A.
On a numerical method for the minimization of functions of an integer argument (Russian). In: Methods for Nonsmooth Optimization and Control Problems, 106-117. Sverdlovsk: Akad. Nauk SSSR, Ural. Nauchn. Tsentr. (1986).

ULUS 1985-1 Ulusoy, G.
The fleet size and mix problem for capacitated arc routing. European Journal of Operational Research 22, 329-337. (1985).

URRU 1984-1 Urrutia, J.
Intersection graphs, (B*1)-orientable graphs and proper circular arc graphs. SCS-TR-41. Ottawa: Carleton Univ., School of Computer Science, 13 p. (1984).

URSI 1983-1 Ursic, S.
A linear characterization of NP-complete problems. in: T. Basar, D. J. Brown, H. V. Poor (Eds.): Twentyfirst Annual Allerton Conference on Communication, Control, and Computing (Monticello, Ill., 1983), 100-109. Urbana-Champaign: Univ. Illinois. (1983).

USAM 1984-1 Usami, Y.
Minimum number of edges in P(4)-connected graphs. Utilitas Mathematica 26, 106-169. (1984).

USAN 1979-1 Usanov, V. V.
Representation and recognition of triangulated graphs (Russian). in: Combinatorial-Algebraic Methods in Applied Mathematics, 103-109. Gorki: Gor'kovskii Gosudarstvennyi Universitet. (1979).

USHI 1985-1 Ushio, K.
Factorization of complete bipartite graphs into Hamilton paths and Hamilton cycles. Journal of the Faculty of Science and Technology, Kinki University 21, 411-413. (1985).

USHI 1986-1 Ushio, K.
K(2,2)-factorization of complete multipartite graphs. Journal of the Faculty of Science and Technology, Kinki University 22, 379-380. (1986).

UVAR 1985-1 Uvarov, S. I.
Estimate of the quality of a method for all construction of a schedule for carrying out a set of operations in a sequence specified by a graph (Russian). Avtomatika i Telemekhanika (Moscow) No. 12, 161-165. (1985).

UVAR 1985-2 Uvarov, S. I.
Investigation of a method for constructing a schedule of jobs with graph-specified precedence (Russian). Automatika i Telemekhanika (Moscow), No. 7, 172-176. (1985).

UYTT 1988-1 Uyttenhove, H. J. J.
Transportation planning - as easy as 1-2-3. In: H. Schellhaas, P. van Beek, H. Isermann, R. Schmidt, M. Zijlstra (Eds.): Operations Research Proceedings 1987 (DGOR/NSOR), 315-316. Berlin: Springer. (1988).

VACH 1983-1 Vachajitpan, P.
A computer program for solving job sequencing problems. Computers and Industrial Engineering 7, 173-185. (1983).

VACI 1982-1 Vacirca, V.
T(r,s)-colorings of a graph and their associated automorphism groups (Italian). Atti del Seminario Matematico e Fisico dell'Universita di Modena 31, 86-94. (1982).

VACI 1986-1 Vacirca, V.
On some parameters associated with the modulus of a totally coloured digraph (Italian). Matematiche 37, 189-204. (1986).

VADE 1986-1 Vaderlind, P.
Clutters and atomistic lattices. European Journal of Combinatorics 7, 389-396. (1986).

VADE 1986-2 Vaderlind, P.
Clutters and semimatroids. European Journal of Combinatorics 7, 271-282. (1986).

VAGI 1983-1 Vagin, V. N.; Georgieva, N. T.; Pospelov, D. A.
On segmentation of graphs. Izvestiya Akademii Nauk SSSR, Tekhnicheskaya Kibernetika (Moscow) No. 4, 95-103. Translated as: Engineering Cybernetics 21, No. 6, 9-17 (1984). (1983).

VAID 1984-1 Vaidya, P. M.
A fast approximation algorithm for minimum spanning trees in k-dimensional space. In: 25th Annual Symposium on Foundations of Computer Science, Florida 1984, 403-407. Los Angeles: IEEE Computer Society. (1984).

VAIN 1982-1 Vainer, V. G.
Estimates of the accuracy of algorithms for construction of connecting networks (Russian). Akademiya Nauk Ukrainskoi SSR (Kiev), Numerical methods of cybernetics, 46-55. (1982).

VAINA 1984-1 Vainshtein, A. D.
The problem of packing of rectangles into a strip - a survey (Russian). Upravlyaemye Sistemy (Novosibirsk) No. 25, 17-37, 87. (1984).

VAINA 1985-1 Vainshtein, A. D.
A vector problem of the shortest path in a uniform norm (Russian). Ekonomika i Matematicheskie Metody (Moscow) 21, 1132-1137. (1985).

VAINB 1986-1 Vainshtein, Y. B.
An algorithm for complete planarization of graphs of schemes (Russian). Soobshcheniya Akademii Nauk Gruzinskoj SSR (Tbilisi) 122, 53-55. (1986).

VALK 1985-1 Valk, M.
Generalized flows and stresses on graphs (Russian). Vestnik Belorusskogo Gosudarstvennogo Universiteta , Seriya I, Fizika, Matematika, Mekhanika No. 2, 51-55. (1985).

VAND 1987-1 Vandevate, J. H.
Structural properties of matroid matchings. Industrial and Systems Engineering Report J-87-5. Atlanta: Georgia Inst. of Technology, 29 p. (1987).

VANI 1987-1 Vani, V.; Raghavachari, M.
Deterministic and random single machine sequencing with variance minimization. Operations Research 35, 111-120. (1987).

VANN 1984-1 Vannelli, A.
Approximating a class of graph decomposition problems by linear transportation problems. IBM Research Report. Yorktown Heights: IBM Research Center, 28 p. (1984).

VANNA 1986-1 Vannelli, A.; Kumar, K. R.
A method for finding minimal bottle-neck cells for grouping part-machine families. International Journal of Production Research 24, 387-400. (1986).

VANNB 1984-1 Vannelli, A.; Vidyasagar, M.
Three approximate solution techniques for the optimal k-decomposition problem. IBM Research Report. Yorktown Heights: IBM Research Center, 24 p. (1984).

VANR 1986-1 Vanroy, T. J.
A cross decomposition algorithm for capacitated facility location. Operations Research 34, 145-163. (1986).

VANRA 1985-1 Vanroy, T. J.; Wolsey, L. A.
Valid inequalities and separation for uncapacitated fixed charge networks. Operations Research Letters 4, 105-112. (1985).

VANRA 1986-1 Vanroy, T. J.; Wolsey, L. A.
Valid inequalities for mixed 0-1 programs. Discrete Applied Mathematics 14, 199-213. (1986).

VANRA 1987-1 Vanroy, T. J.; Wolsey, L. A.
Solving mixed integer programming problems using automatic reformulation. Operations Research 35, 45-57. (1987).

VARA 1985-1 Varakin, A. S.; Danilchenko, A. M.; Panishev, A. V.
Optimization of schedules in a multiprocessor system. Izvestiya Akademii Nauk SSSR Tekhnicheskaya Kibernetika (Moscow) No. 2, 138-141; translated as: Soviet Journal of Computational Systems Sciences 23, No. 3, 133-136. (1985).

VARM 1986-1 Varma, B. N.
On pancyclic line graphs. Congressus Numerantium 54, 203-208. (1986).

VASH 1982-1 Vashchenko, V. P.
An algorithm for graph coloring on the basis of fixed homomorphisms (Russian). Trudy Moskovskogo Ordena Lenina Energeticheskogo Instituta (Moscow) No. 587, 130-134. (1982).

VASI 1985-1 Vasilev, N. S.
On the properties of a flow polyhedron and their applications in the numerical solution of extremal problems on graphs. U.S.S.R. Computational Mathematics and Mathematical Physics 25, No.3, 87-89; translation from Zhurnal Vychislitelnoi Matematiki i Matematicheskoi Fiziki 25 (1985), No.5, 780-783. (1985).

VASI 1986-1 Vasilev, N. S.
A model of a communication network development. Soviet Journal of Computer and Systems Sciences 24, No.2, 20-27; translation from Izvestiya Akademii Nauk SSSR, Tekhnicheskaya Kibernetika (1985), No.6, 15-21. (1986).

VASIA 1983-1 Vasilev, V. S.
An approximate algorithm, solving a class of nonlinear discrete programming problems (Bulgarian). Problemi na Tekhnicheskata Kibernetika i Robotikata 17, 36-41. (1983).

VASIA 1985-1 Vasilev, V. S.
Problems of discrete programming and algorithms for their solution (Bulgarian). Problemi na Tekhnicheskata Kibernetika i Robotikata 22, 28-36. (1985).

VASK 1986-1 Vasko, F. J.; Wilson, G. R.
Hybrid heuristics for minimum cardinality set covering problems. Naval Research Logistics Quarterly 33, 241-250. (1986).

VASKA 1987-1 Vasko, F. J.; Wolf, F. E.; Stott, K. L.
Optimal selection of ingot sizes via set covering. Operations Research 35, 346-353. (1987).

VASS 1985-1 Vassilacopoulos, G.
Allocating doctors to shifts in an accident and emergency department. Journal of the Operational Reserch Society 36, 517-524. (1985).

VASY 1983-1 Vasyanin, V. A.
On a problem of discrete optimization in processes of control over transportations in multiflow transportation networks (Russian). Kibernetika i Vychislitel'naya Tekhnika (Kiev) 60, 82-87. (1983).

VASYA 1982-1 Vasyanin, V. A.; Savenkov, A. I.
A suboptimal algorithm for solving a transportation routing problem. Soviet Automatic Control 15, No. 6, 1-5; translated from Avtomatika (Kiev) No. 6, 5-9. (1982).

VAZI 1985-1 Vazirani, U. V.; Vazirani, V. V.
The 2-processor scheduling problem is in R-NC. In: Proc. Seventeenth Annual ACM Symp. on Theory of Computing, Providence 1985, 11-21. ACM. (1985).

VELD 1988-1 Velde, S. L. van de
Minimizing makespan on unrelated parallel machines. In: H. Schellhaas, P. van Beek, H. Isermann, R. Schmidt, M. Zijlstra (Eds.): Operations Research Proceedings 1987 (DGOR/ NSOR), 629-630. Berlin: Springer. (1988).

VELO 1985-1 Veloso, P. A. S.
How to "divide-and-conquer": a logical basis. In: P. Brucker, R. Pauly (Eds.): Proc. IX. Symp. on Operations Research, Osnabrück 1984. Methods of Operations Research 49, 431-442. (1985).

VEN 1985-1 Ven, V. L.
Partial aggregation in problems of integer linear programming with variable parameters (Russian). Ekonomika i Matematicheskie Metody (Moscow) 21, 501-510. (1985).

VENE 1985-1 Venetsanopoulos, A. N.; Waung, W.
Adaptive bifurcation routing algorithms for computer-communication networks. Kybernetika 21, 178-196. (1985).

VENI 1984-1 Veni Madhavan, C. E.
Approximation algorithm for maximum independent set in planar triangle-free graphs. In: M. Joseph; R. Shyamasundar (Eds.): Foundations of Software Technology and Theoretical Computer Science, Lecture Notes in Computer Science Vol. 181, 381-392. Berlin: Springer-Verlag. (1984).

VENIA 1984-1 Veniaminov, S. S.
Additional cutoff rule for the branch-and-bound method. Izvestiya Akademii Nauk SSSR, Tekhnicheskaya Kibernetika (Moscow) No. 3, 59-63; translated as: Engineering Cybernetics 22, 75-80 (1985). (1984).

VENIA 1986-1 Veniaminov, S. S.
The effectiveness of the hybrid method for solving discrete optimization problems. Soviet Journal of Computer and Systems Sciences 24, No.4, 17-23; translation from Izvestiya Akademii Nauk SSSR, Tekhnicheskaya Kibernetika (1986), No.1, 21-26. (1986).

VENK 1985-1 Venkatesan, S. M.
Approximation algorithms for planar matching. In: P. N. Maheshwari (Ed.): Foundations of Software Technology and Theoretical Computer Science, 162-175. Lecture Notes in Computer Science 206, Berlin: Springer. (1985).

VENK 1987-1 Venkatesan, S. M.
Approximation algorithms for weighted matching. Theoretical Computer Science 54, 129-137. (1987).

VENT 1985-1 Venta, E. R.; Hurter, A. P. Jr.
Production-location problems with demand considerations. Naval Research Logistics Quarterly 32, 625-630. (1985).

VENTA 1987-1 Ventura, J. A.; Hearn, D. W.
A Lagrangian dual approach for networks with quadratic costs. Research Report 87-8. Gainesville: Univ. of Florida, Industrial & Systems Engineering Dept., 16 p. (1987).

VEPS 1987-1 Vepsalainen, A. P. J.; Morton, T. E.
Priority rules for job shops with weighted tardiness costs. Management Science 33, 1035-1047. (1987).

VERC 1984-1 Vercellis, C.
A probabilistic analysis of the set covering problem. Annals of Operations Research, Vol. 1, 255-272. Basel: Baltzer. (1984).

VERC 1986-1 Vercellis, C.
A probabilistic analysis of the set packing problem. In: F. Archetti, G. Di Pillo, M. Lucertini (Eds.): Stochastic Programming, 272-285. Lecture Notes in Control and Information Sciences 76, Berlin: Springer. (1986).

VERG 1984-1 Vergnas, M. Las
Oriented matroids as signed geometries real in corank 2. In: A. Hajnal; L. Lovasz; V. T. Sos (Eds.): Finite and Infinite Sets, Vol. 2, 555-566. Coll. Math. Soc. J. Bolyai 37, Amsterdam: North-Holland. (1984).

VERG 1984-2 Vergnas, M. Las
The Tutte polynomial of a morphism of matroids II. Activities of orientations. In: J. A. Bondy, U. P. R. Murty (Eds.): Progress in Graph Theory, 367-380. New York: Academic Press. (1984).

VERG 1986-1 Vergnas, M. Las
Hamilton paths in tournaments and a problem of McMullen on projective transformations in R**d. Bulletin of the London Mathematical Society 18, 571-572. (1986).

VESE 1985-1 Veselov, S. I.
Aggregation of linear integer equations. Cybernetics 21, 485-489; translation from Kibernetika (1985), No.4, 58-60. (1985).

VESEA 1981-1 Veselov, S. I.; Shevchenko, V. N.
Faces and extreme points of discrete programming problems (Russian). In: A. A. Markov (Ed.): Combinatorial-algebraic methods in applied mathematics, 39-49. Gorki: Gor'kov Gos. Univ. (1981).

VEST 1987-1 Vestergaard, P. D.
On graphs with prescribed spanning subgraphs. Ars Combinatoria 24A, 47-58. (1987).
VESZ 1985-1 Vesztergombi, K.
Bounds on the number of small distances in a finite planar set. Report No. 85365-OR. Bonn: Univ. Bonn, Inst. f. Ökonometrie u. Operations Research, 8 p. (1985).
VESZ 1985-2 Vesztergombi, K.
On the distribution of distances in finite sets in the plane. Report, 84359-OR. Bonn: Univ. Bonn, Inst. für Ökonometrie und Operations Research, 21 p. (1985).
VESZ 1987-1 Vesztergombi, K.
On large distances in planar sets. Discrete Mathematics 67, 191-198. (1987).
VICH 1984-1 Viches, S. A.
An asymptotically optimal algorithm for enumerating intersections of edges of a bipartite graph (Russian). Avtomatika i Telemekhanika (Moscow) No. 12, 133-137. (1984).
VICK 1985-1 Vickery, S. K.; Markland, R. E.
Integer goal programming for multistage lot sizing: experimentation and implementation. Journal of Operations Management 5, 169-182. (1985).
VIDA 1984-1 Vidal, R. V. V.
Large-scale production planning. In: A. Straszak (Ed.): Large Scale Systems: Theory and Applications 1983 (Warsaw 1983), 461-467. IFAC Proc. Series 10, Laxenburg, Pergamon Press. (1984).
VIEG 1985-1 Viegas, J.; Hansen, P.
Finding shortest paths in the plane in the presence of barriers to travel (for any l(p)-norm). European Journal of Operational Research 20, 373-381. (1985).
VIJA 1985-1 Vijay, J.
An algorithm for the p-center problem in the plane. Transportation Science 19, 235-245. (1985).
VILA 1984-1 Vilar, J.
Minimal cuts up to third order in a planar graph. IEEE Transactions on Reliability R-33, 250-256. (1984).
VINC 1983-1 Vince, A.
Generalization of the cycle matroid of a graph. Congressus Numerantium 40, 399-408. (1983).
VINI 1983-1 Vinichenko, S. M.
Calculation of Hamiltonian cycles and paths in complete n-colored graphs (Russian). In: K. A. Rybnikov (Ed.): Combinatorial Analysis, No. 6, 65-67. Moscow: Moskov. Gos. Univ., Mekh.-Mat. Fak. (1983).
VINN 1985-1 Vinnichenko, N. G.
On graphs with the given nondensity and length of a maximal cycle (Russian). Kibernetika (Kiev) No.3, 115-117. (1985).
VINO 1985-1 Vinokur, A. B.
One method for storage and retrieval of information on the shortest paths of graphs with distinguished characteristics (Russian). Kibernetika (Kiev) No. 4, 52-57. (1985).
VISH 1984-1 Vishkin, U.
An optimal parallel connectivity algorithm. Discrete Applied Mathematics 9, 197-208. (1984).
VIZI 1985-1 Vizing, V. G.
Schedules ensuring the most uniform machine loading. Soviet Mathematics, Doklady 31, 95-98; translation from Doklady Akademii Nauk SSSR 280 (1985), 525-528. (1985).
VIZV 1983-1 Vizvari, B.
The heuristic methods of discrete programming. I. The method of neighbourhood. Tanulmanyok - MTA Szamitastechnikai es Automatizalasi Kutato Intezet Budapest No. 152, 109-138. (1983).
VIZV 1984-1 Vizvari, B.
On the connection of the Frobenius problem and the knapsack problem. In: A. Hajnal; L. Lovasz; V. T. Sos (Eds.): Finite and Infinite Sets, Vol. 2, 799-820. Coll. Math. Soc. J. Bolyai 37, Amsterdam: North-Holland. (1984).
VIZV 1986-1 Vizvari, B.
Some recent Hungarian results in discrete programming (Hungarian). Szigma, Matematikai-Közgazdasagi Folyoirat (Budapest) 19, 93-123. (1986).
VIZV 1987-1 Vizvari, B.
On the greedy solution in integer linear programming. Zeitschrift für Operations Research 31A, 55-68. (1987).
VLAC 1986-1 Vlach, M.
Conditions for the existence of solutions of the three-dimensional planar transportation problem. Discrete Applied Mathematics 13, 61-78. (1986).
VO 1987-1 Vo, K.-P.
Graph colorings and acyclic orientations. Linear & Multilinear Algebra 22, 161-170. (1987).
VOHR 1985-1 Vohra, R. V.
On the greedy heuristic for the vertex packing problem. Working Paper Series 85-98. Columbus: Ohio State Univ., College of Administrative Science, 11 p. (1985).
VOHRA 1987-1 Vohra, R. V.; Golden, B. L.
Finding the most vital arcs in a network. Working Paper 87-78. Columbus: Ohio State Univ., College of Business, 10 p. (1987).
VOHRB 1986-1 Vohra, R. V.; Hall, N. G.
Bounds and heuristics for the maximal covering location problem: a preliminary report. WPS 86-68. Columbus: Ohio State Univ., College of Administrative Science, 10 p. (1986).
VOHRC 1984-1 Vohra, R. V.; Washington, L.
Counting spanning trees in the graphs of Kleitman and Golden and a generalization. Journal of the Franklin Institute 318, 349-355. (1984).
VOIG 1987-1 Voigt, J.-U.
Solving cutting stock problems of the furniture industry via man-machine communication. Wissenschaftliche Zeitschrift der Technischen Hochschule Ilmenau 33, No.6, 113-118. (1987).
VOIT 1984-1 Voitishin, Y. V.
Application of network fronts for solution of certain problems on graphs (Russian). In: Investigation and design of "man-machine" systems, 66-70. Kiev: Institut Kibernetiki, Akad. Nauk. Ukrain. SSR. (1984).

VOIT 1986-1 Voitishin, Y. V.
A network approach to the solution of problems of set partitioning (Russian). Preprint 86-13, Akademii Nauk Ukrainskoj SSR, Inst. Kibernet., Kiev. (1986).

VOLG 1986-1 Volgenant, A.; Duin, C. W.
Reduction tests for the Steiner problem in graphs. Report AE 16/86, IAWE, Univ. of Amsterdam. (1986).

VOLGA 1985-1 Volgenant, A.; Jonker, R.
Improving Christofides' lower bound for the traveling salesman problem. Optimization 16, 691-704. (1985).

VOLGA 1986-1 Volgenant, A.; Jonker, R.
A transformation of a generalized traveling salesman problem. Report AE 5/86, IAWE, Univ. of Amsterdam. (1986).

VOLGA 1986-2 Volgenant, A.; Jonker, R.
An improved transformation for the symmetric multiple traveling salesman problem. Report AE 1/86, IAWE, Univ. of Amsterdam. (1986).

VOLGB 1986-1 Volgenant, A.; Jonker, R.; Kindervater, G. A. P.
A note on finding a shortest complete cycle in an undirected graph. European Journal of Operational Research 23, 82-85. (1986).

VOLGC 1987-1 Volgenant, A.; Sluis, H. J. van der; Jonker, R.
Better assignment lower bounds for the Euclidean traveling salesman problem. Optimization 18, 393-404. (1987).

VOLGD 1986-1 Volgenant, A.; Zoon, J. A.
An improved heuristic for multidimensional 0-1 knapsack problems (Dutch). Report AE N3/86, IAWE, Univ. of Amsterdam. (1986).

VOLGE 1987-1 Volgenant, T.; Jonker, R.
On some generalizations of the travelling-salesman problem. Journal of the Operational Research Society 38, 1073-1080. (1987).

VOLK 1986-1 Volkovich, O. V.
Determination of an admissible solution for the problem of integer linear programming (Russian). Vestnik Kievskogo Universiteta, Modelirovanie i Optimizatsiya Slozhnykh Sistem (Kiev) 5, 52-54. (1986).

VOLK 1987-1 Volkovich, O. V.
An algorithm for finding a feasible solution of the integral linear programming (Russian). Kibernetika (Kiev) No. 2, 65-69. (1987).

VOLK 1987-2 Volkovich, O. V.
Approximation of the admissible domain in a problem of integer programming (Russian). Doklady Akademii Nauk Ukrainskoj SSR, Seriya A, No.2, 76-78. (1987).

VOLKA 1987-1 Volkovich, O. V.; Roshchin, V. A.; Sergienko, I. V.
Models and methods for solution of integer quadratic programming problems (Russian). Kibernetika (Kiev) No. 3, 1-15. (1987).

VOLKB 1986-1 Volkovich, S. V.
A parallel algorithm for the sequential analysis of variants for a problem of the block-diagonal structure of mixed linear programming (Russian). In: Models and Systems of Information Processing, No. 5, 31-35. Kiev: "Vishcha Shkola". (1986).

VOLKC 1987-1 Volkovich, S. V.; Voloshin, A. F.
A successive approach to the solution of problems of linear partially integer programming of special structure (Russian). Doklady Akademii Nauk Ukrainskoj SSR (Kiev), Seriya A, No.8, 63-67. (1987).

VOLO 1984-1 Volodina, V. V.; Cherkasskii, B. V.
On the realization of an algorithm for recognizing the planarity and packing of a graph in O(v) operations (Russian). In: Algorithms for Solving Network Problems, 59-91. Moscow: Akad. Nauk SSSR, Tsentral. Ekonom.-Mat. Inst. (1984).

VOLOA 1987-1 Voloshin, A. F.
A method for localization of the optimum region in problems of mathematical programming (Russian). Doklady Akademii Nauk SSSR (Moscow) 293, 549-553. (1987).

VOLOB 1986-1 Voloshin, A. F.; Mashchenko, S. O.
A subexponential algorithm for solving the knapsack problem (Russian). Doklady Akademii Nauk Ukrainskoj SSR (Kiev), Seriya A, No.10, 55-58. (1986).

VOLOC 1982-1 Voloshin, V. I.
Properties of triangulated graphs (Russian). In: B. A. Shcherbakov (Ed.): Operations research and programming, 24-32. Matematicheskie Nauki. Kishinev: Shtiintsa. (1982).

VOLOD 1984-1 Voloshina, A. A.; Feinberg, V. Z.
Planarity of hypergraphs (Russian). Doklady Akademii Nauk BSSR (Minsk) 28, 309-311. (1984).

VOLY 1982-1 Volynskij, E. I.; Nilova, L. I.; Sejts, A. I.; Filonenko, V. L.
Application of random search to solve a problem of time scheduling (Russian). In: A. N. Antamoshkin (Ed.): Applications of random search to the solution of applied problems, 40-47. Kemerovo: Kemerovskij Gosud. Univ. (1982).

VORN 1981-1 Vornberger, O.
Alternating cycle covers and paths. In: H. Noltemeier (Ed.): Graphtheoretic Concepts in Computer Science, Proc. Internat. Workshop WG 80 (Bad Honnef 1980), 367-378. Berlin, Heidelberg, New York: Springer. (1981).

VORO 1983-1 Vorobev, V. A.
Relative addressing of elements of a cyclic graph (Russian). Vychislitel'nye Sistemy (Novosibirsk) No. 97, 87-103, 155. (1983).

VOSS 1983-1 Voss, H.-J.
Unabhängige Mengen in k-chromatischen Graphen. In: 2nd Colloquium for Geometry and Combinatorics (Karl-Marx-Stadt, 1983) Part 1, 2, 199-206. Karl-Marx-Stadt: Technische Hochschule. (1983).

VOSS 1984-1 Voss, H.-J.
Färbungsprobleme auf Graphen. Wissenschaftliche Zeitschrift der Pädagogischen Hochschule Karl Friedrich Wilhelm Wander Dresden 18, 9-27. (1984).

VOSS 1985-1 Voss, H.-J.
Bridges of circuits through specified edges. In: H. Sachs (Ed.): Graphs, Hypergraphs and Applications, Proc. Conf. on Graph Theory (Eyba 1984), 198-202. Leipzig: Teubner. (1985).

VOSS 1985-2 Voss, H.-J.
Chords of circuits in graphs with given minimum degree. Elektronische Informationsverarbeitung und Kybernetik 21, 201-208. (1985).
VOSS 1985-3 Voss, H.-J.
Independent sets in k-chromatic graphs. Combinatorica 5, 261-269. (1985).
VOSS 1985-4 Voss, H.-J.
Symmetric subgraphs and the reconstruction conjecture. 30. Internationales Wissenschaftliches Kolloquium (TH Ilmenau, DDR) Heft 5F, 123-126. (1985).
VOSSA 1987-1 Voss, S.; Mehr, K. D.
On a generalized Steiner problem with 2-edge connectivity. In: W. Domschke, W. Krabs, J. Lehn, P. Spellucci (Eds.): Proc. XI. Symp. on Operations Research, Darmstadt 1986, 161-172. Methods of Operations Research 57, Frankfurt/Main: Athenäum. (1987).
VU 1984-1 Vu Dinh Hoa
Hamiltonkreise in ungerichteten und schlichten Graphen. In: Geometrie und Kombinatorik, 2. Kolloq., Karl-Marx-Stadt/DDR 1983, Vortr. 1,2, Teil II, 264-268. Karl-Marx-Stadt: Technische Hochschule. (1984).
VU 1986-1 Vu Dinh Hoa
Ein Struktursatz für 2-fach zusammenhängende Graphen mit grosser Minimalvalenz. Mathematische Nachrichten 128, 151-160. (1986).
VULI 1985-1 Vulis, V. I.; Sherman, Y. S.
Decomposition method for solving large-scale problems of transportation type (Russian). Ekonomika i Matematicheskie Metody (Moscow) 21, 304-314. (1985).
WADA 1987-1 Wada, K.; Kawaguchi, K.; Luo, Y.-P.
Fault-tolerant network routings for (k+1)-node connected and (k+1)-edge connected graphs. Systems Comput. Japan 18, No. 11, 50-60. (1987).
WAGN 1985-1 Wagner, D. K.
Connectivity in bicircular matroids. Journal of Combinatorial Theory 39B, 308-324. (1985).
WAGN 1985-2 Wagner, D. K.
On theorems of Whitney and Tutte. Discrete Mathematics 57, 147-154. (1985).
WAGN 1986-1 Wagner, D. K.
Equivalent factor matroids of graphs. CC-86-17. West Lafayette: Purdue Univ. Inst. of Interdisciplinary Engineering Studies, 12 p. (1986).
WAGNA 1984-1 Wagner, K.
The complexity of combinatorial problems with compactly described instances. Forschungsergebnisse, Friedrich-Schiller-Univ. Jena N\84\23. (1984).
WAGNA 1984-2 Wagner, K.
The complexity of problems concerning graphs with regularities. Forschungsergebnisse, Friedrich-Schiller-Univ. Jena N\84\52. (1984).
WAH 1984-1 Wah, B. W.; Chen, K. L.
A partitioning approach to the design of selection networks. IEEE Transactions on Computers 33, 261-268. (1984).
WAJS 1984-1 Wajs, W.
Scheduling problem for metallurgical processes (Polish). Archiwum Automatyki i Telemechaniki (Warsaw) 29, 389-403. (1984).
WAJS 1986-1 Wajs, W.
Polynomial algorithm for dynamic sequencing problem (Polish). Archiwum Automatyki i Telemechaniki 31, 209-213. (1986).
WAKA 1986-1 Wakabayashi, Y.
Aggregation of binary relations: algorithmic and polyhedral investigations. Augsburg: Univ. Augsburg, 191 p. (1986). (Dissertation)
WALA 1984-1 Walas, R. A.; Askin, R. G.
An algorithm for NC turret punch press tool location and hit sequencing. IIE Transactions 16, 280-287. (1984).
WALD 1982-1 Wald, J. A.; Colbourn, C. J.
Steiner trees in outerplanar graphs. In: F. Hoffman, K. B. Reid, R. C. Mullin, R. G. Stanton (Eds.): Proc. Thirteenth Southeastern Conf. on Combinatorics, Graph Theory and Computing, Florida 1982, Congressus Numerantium Vol. 36, 15-22. Winnipeg, Canada: Utilitas Mathematica. (1982).
WALDA 1984-1 Wald, J. A.; Sorenson, P. G.
Resolving the query inference problem using Steiner trees. ACM Transactions on Database Systems 9, 348-368. (1984).
WALK 1985-1 Walk, M.
Generalized flows and tensions on graphs (Russian). Vestnik Belorusskogo Gosudarstvennogo Universiteta Imeni V.I. Lenina, Seriya I, No.2, 51-55. (1985).
WALKA 1987-1 Walker, K.
Fully-coloured Hamiltonian cycles in edge-colorings of K(n) (n odd) when there are no fully-coloured C(4)'s. Ars Combinatoria 24A, 97-106. (1987).
WALL 1986-1 Wallace, S.
Decomposing the requirement space of a transportation problem into polyhedral cones. In: A. Prekopa, R. J.-B. Wets (Eds.): Stochastic Programming 84, Mathematical Programming Study 28, 29-47. (1986).
WALLA 1987-1 Wallace, S. W.
A piecewise linear upper bound on the network recourse problem. Mathematical Programming 38, 133-146. (1987).
WALLB 1985-1 Wallis, W. D.
The clique partition number of the complement of a cycle. In: B. R. Alspach, C. D. Godsil (Eds.): Cycles in Graphs, Annals of Discrete Mathematics Vol. 27, 335-344. Amsterdam: North-Holland. (1985).
WALLC 1985-1 Wallis, W. D.; Wang, Z.-J.
On one-factorizations of Cartesian products. Congressus Numerantium 49, 237-246. (1985).
WALS 1984-1 Walsh, T. R.
How evenly should one divide to conquer quickly? Information Processing Letters 19, 203-208. (1984).

WALT 1984-1 Walther, H.
Ten applications of graph theory. Dordrecht: Reidel. (1984).
WALT 1987-1 Walther, H.
On a conjecture of W. Klotz concerning clique-decomposition of graphs. In: M. Borowiecki; Z. Skupien (Eds.): Graphs, hypergraphs and matroids II, Proc. Sixth Regional Scientific Session of Mathematicians, Zagan, Poland, 1986, 83-88. Section: Graph theory. Zielona Gora: Higher College of Engineering. (1987).
WALTA 1987-1 Walther, H.; Nägler, G.
Graphen, Algorithmen, Programme. Wien, New York: Springer. (1987).
WALU 1987-1 Walukiewicz, S.; Dudzinski, K.
Exact methods for the knapsack problem and its generalizations. European Journal of Operational Research 28, 3-21. (1987).
WALUA 1987-1 Walukiewicz, S.; Lucertini, M.
Constraint generation for the graph partitioning problem. In: Systems Theory in Economics, Managemnent and Technology (Zakopane-Warsaw-Krakow 1985), Control and Cybernetics (Warsaw) 15, 447-459. (1987).
WAN 1985-1 Wan, W. X.
A special integer programming problem in management science (Chinese). Mathematics in Practice and Theory No. 2, 35-42. (1985).
WANG 1985-1 Wang, C.-C.; Lloyd, E. L.; Soffa, M. L.
Feedback vertex sets and cyclically reducible graphs. Journal of the ACM 32, 296-313. (1985).
WANGA 1985-1 Wang, C.-D.
On the sum of two parameters concerning independence and irredundance in graphs (Chinese). Journal of Mathematics, Wuhan University 5, 201-204. (1985).
WANGB 1985-1 Wang, H.
A combinatorial optimal algorithm with least partition times (Chinese). Journal of Dalian Institute of Technology 24, No. 1, 93-96. (1985).
WANGB 1986-1 Wang, H.
The maximum number of edges of a critical 3-edge connected graph on n vertices (Chinese). Journal of Mathematics (Wuhan) 6, No. 4, 381-384. (1986).
WANGC 1987-1 Wang, J. F.; Tian, S.-L.; Liu, J.-Q.
On binding numbers of graphs (Chinese). Acta Mathematica Sinica 30, 653-660. (1987).
WANGD 1984-1 Wang, J. F.; Yang, C. S.
On the number of spanning trees of circulant graphs. International Journal of Computer Mathematics 16, 229-241. (1984).
WANGE 1984-1 Wang, J. S.
A branch and bound algorithm for a logic partitioning problem (Chinese). Chinese Journal of Computers (Beijing) 7, 458-472. (1984).
WANGE 1986-1 Wang, J. S.
A general algorithm for enumerating some subgraphs of a simple graph (Chinese). Chinese Journal of Computers (Beijing) 9, No. 1, 37-43. (1986).
WANGF 1982-1 Wang, J.-Z.
A necessary and sufficient condition for a Hamiltonian tournament (Chinese). Qufu Shiyuan Xuebao (Qufu), No. 1, 14. (1982).
WANGG 1985-1 Wang, K.-Y.; Zhan, J.; Wang, Z.-T.
Models and algorithms for integrated energy transportation system planning (Chinese). Journal of Dalian Institute of Technology 24, No.3, 103-108. (1985).
WANGH 1986-1 Wang, N. S.
On the specific properties of the characteristic polynomial of a tree (Chinese). Journal of Lanzhou Railway College 5, No.4, 89-94. (1986).
WANGI 1986-1 Wang, S. H.; Tao, M. Q.
The structure of k-connected digraphs and mixed (p,q)-connected graphs (Chinese). Journal of China University of Science and Technology (Hefei) 16, No. 4, 385-390. (1986).
WANGJ 1985-1 Wang, S.-M.
A multiple source algorithm for suboptimum Steiner trees in graphs. In: H. Noltemeier (Ed.): Proc. WG '85: Internat. Workshop on Graphtheoretic Concepts in Computer Science, Castle Schwanberg, Würzburg, 1985, 387-396. Linz: Trauner. (1985).
WANGK 1986-1 Wang, S.-M.; Sun, Q.-H.; Wang, X.-F.
A study of flows in complete stochastic networks - theoretical study of a new method for evaluating the reliability of interconnected electrical networks (Chinese). Xian Jiaotong Daxue Xuebao 20, No. 3, 11-24. (1986).
WANGL 1985-1 Wang, Y. C.
An improvement on an algorithm for a transport problem with a feasibility constraint (Chinese). Journal on Numerical Methods and Computer Applications (Beijing) 6, 42-48. (1985).
WANGM 1984-1 Wang, Z.-M.
A proof of constructive type for a theorem by Welsh (Chinese). Fudan Journal (Fudan Xuebao) (Shanghai), Natural Science 23, 211-214. (1984).
WANGM 1987-1 Wang, Z.-M.
A finite conformal-elimination free algorithm over oriented matroid programming. Chinese Annals of Mathematics (Shanghai), Series B 8, 120-125. (1987).
WANGM 1987-2 Wang, Z.-M.
An extended version of the cardinality matroid intersection algorithm (Chinese). Chinese Annals of Mathematics (Shanghai), Series A 8, 504-507. (1987).
WANGN 1983-1 Wang, Z.-Y.
Additive enumeration problems over trees. Acta Mathematica Scientica (Wuhan) 3, 445-453. (1983).
WANGN 1983-2 Wang, Z.-Y.
On the endpoint number of free trees. Journal of Mathematics, Wuhan University 3, 169-173. (1983).
WANGN 1983-3 Wang, Z.-Y.
On the problem of the external path length of trees (Chinese). Kexue Tongbao (Beijing) 28, 842-844. (1983).

WANGN 1983-4 Wang, Z.-Y.
Some combinatorial problems on ordered trees. Chinese Annals of Mathematics (Shanghai), Series 4B, 501-504. (1983).

WANGN 1984-1 Wang, Z.-Y.
Some combinatorial problems on t-ary trees. Kexue Tongbao, Foreign Language Edition (Beijing) 29, 1-5. (1984).

WANGN 1984-2 Wang, Z.-Y.
Some combinatorial problems on t-ary trees. Kexue Tongbao, English Edition (Beijing) 29, 1-5. (1984).

WANGN 1985-1 Wang, Z.-Y.
On the degree path length of trees. Acta Mathematica Scientia (English Edition) (Wuhan) 5, No. 3, 289-293. (1985).

WANGN 1986-1 Wang, Z.-Y.
An enumeration problem on ordered trees. Journal of Mathematics (Wuhan) 6, No. 2, 201-208. (1986).

WANK 1986-1 Wanka, A.
Matroiderweiterungen zur Existenz endlicher LP-Algorithmen, von Hahn-Banach Sätzen und Polarität in orientierten Matroiden. Köln: Univ. Köln, 79 p. (1986). (Dissertation)

WARB 1985-1 Warburton, A.
Worst case analysis of greedy and related heuristics for some min-max combinatorial optimization problems. Mathematical Programming 33, 234-241. (1985).

WARB 1987-1 Warburton, A.
Approximation of Pareto optima in multiple-objective, shortest path problems. Operations Research 35, 70-79. (1987).

WARD 1985-1 Ward, J. E.; Wendell, R. E.
Using block norms for location modeling. Operations Research 33, 1074-1090. (1985).

WARDA 1987-1 Ward, J. E.; Wong, R. T.; Lemke, P.; Oudjit, A.
Properties of the tree k-median linear programming relaxation. CC-87-29. West Lafayette: Purdue Univ. Inst. of Interdisciplinary Engineering Studies, 40 p. (1987).

WASC 1985-1 Wäscher, G.; Carow, P.; Müller, H.
Ein flexibles Lösungsverfahren für Zuschneideprobleme in einem Kaltwalzwerk. In: D. Ohse, A. C. Esprester, H.-U. Küpper, P. Stähly, H.Steckhan (Eds.): Operations Research Proceedings 1984, 150. Berlin: Springer-Verlag. (1985).

WASC 1985-2 Wäscher, G.; Carow, P.; Müller, H.
Entwicklung eines flexiblen Verfahrens für Zuschneideprobleme in einem Kaltwalzwerk. Zeitschrift für Operations Research 29 B, 209-230. (1985).

WASCA 1987-1 Wäscher, G.; Chamoni, P.
MICROLAY: an interactive computer program for factory layout planning on microcomputers. European Journal of Operational Research 31, 185-193. (1987).

WATA 1987-1 Watanabe, T.; Nakamura, A.
Edge-connectivity augmentation problems. Journal of Computer and System Sciences 35, 96-144. (1987).

WATAA 1984-1 Watanabe, T.; Nakamura, A.; Takahashi, M.
Problems of graph augmentation by the addition of sides (Japanese). In: Basic Theory of Computer Science, Proceedings of a Symposium held at the Research Institute for Mathematical Sciences (Kyoto), 16-25. (1984).

WATE 1984-1 Waters, C. D. J.
Interactive vehicle routeing. Journal of the Operational Research Society 35, 821-826. (1984).

WATE 1984-2 Waters, C. D. J.
Interactive vehicle routing. An approach to complex problems. In: J. P. Brans (Ed.): Operational Research '84, 685-695. Amsterdam: North-Holland. (1984).

WATE 1987-1 Waters, C. D. J.
A solution procedure for the vehicle-scheduling problem based on iterative route improvement. Journal of the Operational Research Society 38, 833-840. (1987).

WATS 1984-1 Watson-Gandy, C. D. T.
The multi-facility min-max Weber problem. European Journal of Operational Research 18, 44-50. (1984).

WATS 1985-1 Watson-Gandy, C. D. T.
The solution of distance constrained mini-sum location problems. Operations Research 33, 784-802. (1985).

WEAV 1985-1 Weaver, J. R.; Church, R. L.
A median location model with nonclosest facility service. Transportation Science 19, 58-74. (1985).

WEAV 1986-1 Weaver, J. R.; Church, R. L.
A location model based on multiple metrics and multiple facility assignment. Transportation Research 20-B, 283-296. (1986).

WEAV 1987-1 Weaver, J. R.; Church, R. L.
The formal and computational relationship of the supporting median problem to the p-median problem. Transportation Research 21B, 323-330. (1987).

WEBE 1986-1 Weber, M.; Liebling, T. M.
Euclidean matching problems and the Metropolis algorithm. Zeitschrift für Operations Research A-30, 85-110. (1986).

WEBEA 1986-1 Weber, R. R.; Varaiya, P.; Walrand, J.
Scheduling jobs with stochastically ordered processing times on parallel machines to minimize expected flowtime. Journal of Applied Probability 23, 841-847. (1986).

WEGE 1985-1 Wegener, I.
The critical complexity of all (monotone) Boolean functions and monotone graph properties. Information and Control 67, 212-222. (1985).

WEGE 1985-2 Wegener, I.
The critical complexity of all (monotone) Boolean functions and monotone graph properties. In: L. Budach (Ed.): Fundamentals of Computation Theory, FCT '85, Lecture Notes in Computer Science Vol. 199, 494-502. Berlin: Springer-Verlag. (1985).

WEGL 1985-1 Weglarz, J.
Scheduling independent activities to meet deadlines with continuous constrained resources. In: P. Brucker, R. Pauly (Eds.): Proc. IX. Symp. on Operations Research, Osnabrück 1984. Methods of Operations Research 49, 443-452. (1985).

WEI 1987-1 Wei, S.-B.
Some classes of hypomatchable graphs (Chinese). Journal of Jiangxi Normal University, Natural Science Edition, No.1, 38-42. (1987).

WEI A 1986-1 Wei, S.-B.; Ye, D.-B.
Some classes of secondary randomly matchable graphs (Chinese). Journal of Jiangxi Normal University, Natural Science Edition, No.2, 37-41. (1986).

WEI B 1986-1 Wei, W. D.; Cai, Y. Z.; Liu, C. L.; Odlyzko, A. M.
Balloting labelling and personnel assignment. SIAM Journal on Algebraic and Discrete Methods 7, 150-158. (1986).

WEID 1986-1 Weidner, H. G.
Ein Periodizitätssatz der ganzzahligen Optimierung. Arbeitspapier. Erlangen: Univ. Erlangen-Nürnberg, Math. Institut, 27 p. (1986).

WEIS 1987-1 Weiss, E. N.
An optimization based heuristic for scheduling parallel project networks with constrained renewable resources. Ithaca: Cornell University, 21 p. (1987).

WEISA 1984-1 Weiss, G.
Scheduling spares with exponential lifetimes in a two-component parallel system. Naval Research Logistics Quarterly 31, 431-446. (1984).

WEISA 1984-2 Weiss, G.
Stochastic bounds on distributions of optimal value functions with applications to PERT, network flows and reliability. Annals of Operations Research, Vol. 1, 59-66. Basel: Baltzer. (1984).

WEISA 1986-1 Weiss, G.
Stochastic bounds on distributions of optimal value functions with applications to PERT. Operations Research 34, 595-605. (1986).

WENG 1985-1 Weng, J. F.
Generalized Steiner problem and hexagonal coordinate system (Chinese). Acta Mathematicae Applicatae Sinica (Beijing) 8, 383-397. (1985).

WENG 1985-2 Weng, J. F.
Steiner minimal trees on vertices of regular polygons (Chinese). Acta Mathematicae Applicatae Sinica (Beijing) 8, 129-141. (1985).

WENT 1987-1 Wentzlau, M.
On classes of preuniquely k-colourable graphs. In: M. Borowiecki; Z. Skupien (Eds.): Graphs, hypergraphs and matroids II, Proc. Sixth Regional Scientific Session of Mathematicians, Zagan, Poland, 1986, 89-96. Section: Graph theory. Zielona Gora: Higher College of Engineering. (1987).

WENT 1987-2 Wentzlau, M.
On uniquely and preuniquely k-colourable graphs. Journal of Information Processing and Cybernetics EIK 23, 331-338. (1987).

WENZ 1986-1 Wenzel, W.
Über die Struktur der Tutte-Gruppe von Matroiden. Bielefeld: Univ. Bielefeld, Fakultät für Mathematik, 108 p. (1986). (Dissertation)

WERN 1986-1 Werner, F.
On the method of interchanging networks for permutation problems. Optimization 17, 629-642. (1986).

WERN 1986-2 Werner, F.
On the structure of special 2-machine problems. Seminarber., Humboldt-Univ. Berlin, Sekt. Math. 85, 148-155. (1986).

WERN 1987-1 Werner, F.
On some neighborhood graphs for the development of suitable iteration methods to approximately solve a special permutation problem (German). Wissenschaftliche Zeitschrift der Technischen Universität Otto von Guericke 31, No.5, 48-54. (1987).

WERN 1987-2 Werner, F.
On the quasi-convexity of a special job shop scheduling problem. Wissenschaftliche Zeitschrift der Technischen Universität Otto von Guericke 31, No.5, 43-47. (1987).

WERNA 1986-1 Werner, R.
Kreise im Strukturgraphen eines Netzplanes. Elektronische Informationsverarbeitung und Kybernetik 22, 615-622. (1986).

WERR 1984-1 Werra, D. de
A decomposition property of polyhedra. Mathematical Programming 30, 261-266. (1984).

WERR 1984-2 Werra, D. de
Variation on a theorem of König. Discrete Mathematics 51, 319-322. (1984).

WERR 1985-1 Werra, D. de
An introduction to timetabling. European Journal of Operational Research 19, 151-162. (1985).

WERR 1985-2 Werra, D. de
Graphs, hypergraphs and timetabling. In: P. Brucker, R. Pauly (Eds.): Proc. IX. Symp. on Operations Research, Osnabrück 1984. Methods of Operations Research 49, 201-214. (1985).

WERR 1985-3 Werra, D. de
Some uses of hypergraphs in timetabling. Asia-Pacific Journal of Operational Research 2, 2-12. (1985).

WERR 1986-1 Werra, D. de
Some models of graphs for scheduling sports competitions. Lausanne-Ecublens: Ecole Polytechnique federale, Dept. de Mathematiques, 39p. (1986).

WERR 1986-2 Werra, D. de
Variations on the integral decomposition property. In: G. Gallo, C. Sandi (Eds.): Netflow at Pisa, Mathematical Programming Study 26, 197-199. (1986).

WERR 1987-1 Werra, D. de
Design and operation of flexible manufacturing systems: the kingdom of heuristic methods. RAIRO Recherche Operationelle 21, 365-382. (1987).

WERR 1987-2 Werra, D. de
Partitions into odd chains. Mathematical Programming 37, 41-50. (1987).

WERRA 1986-1 Werra, D. de; Pasche, C.; Petter, A.
Time-tabling problems: should they be canonical? INFOR 24, 304-308. (1986).
WESS 1985-1 Wessel, W.
Kreissehnengraphen - ein neues Hilfsmittel zur Darstellung von Graphen. 30. Internationales Wissenschaftliches Kolloquium (TH Ilmenau, DDR) Heft 5F, 131-134. (1985).
WEST 1985-1 West, D. B.
Parameters of partial orders and graphs: packing, covering and representation. In: I. Rival (Ed.): Graphs and Order, 267-350. Dordrecht: Reidel. (1985).
WESTA 1985-1 West, D. B.; Banerjee, P.
Partial matching in degree-restricted bipartite graphs. Congressus Numerantium 49, 259-266. (1985).
WETZ 1985-1 Wetzel, R.
Ein Branch and Bound Algorithmus zur Lösung eines Problems der Fahrplangestaltung. Wissenschaftliche Zeitschrift der Technischen Hochschule Leipzig 9, 279-285. (1985).
WETZ 1986-1 Wetzel, R.
A branch and bound concept for solving a vehicle schedule designing problem. In: 17. Jahrestagung "Mathematische Optimierung" (Sellin 1984), 121-128. Seminarbericht 80, Humboldt Univ., Berlin. (1986).
WHIT 1985-1 Whitaker, R. A.
Some add-drop and drop-add interchange heuristics for non-linear warehouse location. Journal of the Operational Research Society 36, 61-70. (1985).
WHITA 1985-1 White, C.
Production scheduling with applications in the printing industry. European Journal of Operational Research 22, 304-309. (1985).
WHITB 1984-1 White, K.; Faber, M.; Pulleyblank, W. R.
Steiner trees, connected domination and strongly chordal graphs. Waterloo: Univ. of Waterloo, Dept. of Combinatorics & Optimization, 29 p. (1984).
WHITC 1986-1 White, N.
Theory of matroids. Cambridge: Cambridge Univ. Press. (1986).
WHITC 1987-1 White, N.
Combinatorial Geometries. Cambridge: Cambridge Univ. Press. (1987).
WHITD 1986-1 White, N. L.
A pruning theorem for linear count matroids. Congressus Numerantium 54, 259-264. (1986).
WHITE 1985-1 White, P. M.; Caplin, A. S.; Heyden, L. van der
Scarf's procedure for integer programming and a dual simplex algorithm. Mathematics of Operations Research 10, 439-449. (1985).
WHITF 1984-1 Whitehead, E. G. Jr.; Zhao, L.-C.
Cutpoints and the chromatic polynomial. Journal of Graph Theory 8, 371-378. (1984).
WHITG 1984-1 Whitesides, S. H.
A classification of certain graphs with minimal imperfection properties. In: C. Berge, V. Chvatal (Eds.): Topics on perfect graphs, Annals of Discrete Mathematics Vol. 21, 207-218. Amsterdam: North-Holland. (1984).
WHITG 1984-2 Whitesides, S. H.
A method for solving certain graph recognition and optimization problems, with applications to perfect graphs. In: C. Berge, V. Chvatal (Eds.): Topics on perfect graphs, Annals of Discrete Mathematics Vol. 21, 281-298. Amsterdam: North-Holland. (1984).
WHITH 1984-1 Whittle, G. P.
On the critical exponent of transversal matroids. Journal of Combinatorial Theory, Series B 37, 94-95. (1984).
WHITH 1985-1 Whittle, G. P.
An elementary proof that every matroid is an intersection of principal transversal matroids. Discrete Mathematics 54, 239. (1985).
WHITH 1987-1 Whittle, G. P.
Modularity in tangential k-blocks. Journal of Combinatorial Theory B 42, 24-35. (1987).
WHITH 1987-2 Whittle, G. P.
Some aspects of the critical problem for matroids. Bulletin of the Australian Mathematical Society (St. Lucia) 35, 153-154. (1987).
WHITI 1984-1 Whittle, P.
Optimal routing in Jackson networks. Asia-Pacific Journal of Operational Research 1, 32-37. (1984).
WHITI 1985-1 Whittle, P.
Scheduling and characterization problems for stochastic networks. With discussion. Journal of the Royal Statistical Society, Series B 47, 407-428. (1985).
WHITJ 1985-1 Whitty, R. W.
Spanning trees in program flowgraphs. In: Y. Alavi, G. Chartrand, L. Lesniak, D. R. Lick, C. E. Wall (Eds.): Graph Theory with Applications to Algorithms and Computer Science, 777-788. Chichester: Wiley. (1985).
WIDM 1987-1 Widmayer, P.
On approximation algorithms for Steiner's problem in graphs. In: G. Tinhofer, G. Schmidt (Eds.): Graph-Theoretic Concepts in Computer Science, 17-28. Lecture Notes in Computer Science 246, Berlin: Springer. (1987).
WIDMA 1985-1 Widmayer, P.; Wu, Y. F.; Schlag, M. D. F.; Wong, C. K.
Rectilinear shortest paths and minimum spanning trees in the presence of rectilinear obstacles. In: H. Noltemeier (Ed.): Proc. WG '85: Internat. Workshop on Graphtheoretic Concepts in Computer Science, Castle Schwanberg, Würzburg, 1985, 409-420. Linz: Trauner. (1985).
WIE 1986-1 Wie, S.-H.; Pinedo, M.
On minimizing the expected makespan and flow time in stochastic flow shops with blocking. Mathematics of Operations Research 11, 336-342. (1986).
WIEG 1987-1 Wiegers, M.
Recognizing outerplanar graphs in linear time. In: G. Tinhofer, G. Schmidt (Eds.): Graph-Theoretic Concepts in Computer Science, 165-176. Lecture Notes in Computer Science 246, Berlin: Springer. (1987).

WIJN 1988-1 Wijngaard, P.
A heuristic for scheduling problems especially for scheduling farm operations. In: H. Schellhaas, P. van Beek, H. Isermann, R. Schmidt, M. Zijlstra (Eds.): Operations Research Proceedings 1987 (DGOR/NSOR), 630-631. Berlin: Springer. (1988).

WILF 1986-1 Wilf, H. S.
Algorithms and complexity. Englewood Cliffs: Prentice-Hall. (1986).

WILF 1986-2 Wilf, H. S.
The number of maximal independent sets in a tree. SIAM Journal on Algebraic and Discrete Methods 7, 125-130. (1986).

WILFA 1987-1 Wilf, H. S.; Yoshimura, N. A.
Ranking rooted trees, and a graceful application. In: D. P. Johnson, T. Nishizeki, A. Nozaki, H. P. Wilf (Eds.): Discrete Algorithms and Complexity - Proc. Japan-U.S. Joint Seminar, Kyoto 1986, 341-349. Perspectives in Computing Vol. 15, Boston: Academic Press. (1987).

WILH 1985-1 Wilhelm, M. R.; Evans, G. W.
The quadratic assignment problem: applications and solution techniques. IE News-Operations Research 20, No. 2, 1-4. (1985).

WILHA 1987-1 Wilhelm, M. R.; Ward, T. L.
Solving quadratic assignment problems by "simulated annealing". IIE Transactions 19, 107-119. (1987).

WILHB 1986-1 Wilhelm, M. R.; Ward, T. L.; Lichtefeld, R. A.
Computer-aided facility location. Computers and Industrial Engineering 10, 215-225. (1986).

WILHC 1985-1 Wilhelm, W. E.; Sarin, S.
A structure for sequencing robot activities in machine tending applications. International Journal of Production Research 23, 47-64. (1985).

WILHD 1985-1 Wilhelm, W. E.; Shin, H.-M.
Effectiveness of alternate operations in a flexible manufacturing system. International Journal of Production Research 23, 65-80. (1985).

WILL 1985-1 Williams, A. C.
Quadratic 0-1 programming using the proof dual - with computational results. Rutcor Research Report No. 8-85. New Brunswick: Rutgers Univ., 25p. (1985).

WILL 1986-1 Williams, A. C.
Marginal values in mixed integer linear programming. RUTCOR Research Report 25-86. New Brunswick: Rutgers Center for Operations Research, 8 p. (1986).

WILLA 1985-1 Williams, H. P.
Model building in linear and integer programming. In: K. Schittkowski (Ed.): Computational Mathematical Programming, 25-54. Berlin: Springer-Verlag. (1985).

WILLB 1985-1 Williams, M. H.
A linear algorithm for colouring planar graphs with five colours. Computer Journal 28, 78-81. (1985).

WILLC 1984-1 Williamson, S. G.
Depth-first search and Kuratowski subgraphs. Journal of the ACM 31, 681-693. (1984).

WILLD 1985-1 Willis, R. J.
An algorithm for constructing project network diagrams on a ordinary line printer. Computers & Operations Research 12, 163-168. (1985).

WILLD 1985-2 Willis, R. J.
Critical path analysis and resource constrained project scheduling - theory and practice. European Journal of Operational Research 21, 149-155. (1985).

WILS 1984-1 Wilson, H. G.; Ezzat, M. O.
Scheduling of material handling vehicles in a jobshop environment. Logistics and Transportation Review 20, 517-526. (1984).

WILSA 1984-1 Wilson, J. M.
A 2-phase method for the minimisation of Boolean polynomials. International Journal of Computer Mathematics 16, 131-135. (1984).

WILSA 1985-1 Wilson, J. M.
A method of enumerating all integer extreme solutions of an integer programming problem. Methods of Operations Research 55, 387-396. (1985).

WILSA 1986-1 Wilson, J. M.
Formulation of a problem involving assembly lines with multiple manning of work stations. International Journal of Production Research 24, 59-64. (1986).

WILSB 1985-1 Wilson, R. J.
Introduction to graph theory (third edition). New York: Longman. (1985).

WIME 1987-1 Wimer, T. V.
Linear algorithms for the dominating cycle problems in series-parallel graphs, partial k-trees, and Halin graphs. Congressus Numerantium 57, 289-298. (1987).

WIMEA 1985-1 Wimer, T. V.; Hedetniemi, S. T.; Laskar, R.
A methodology for constructing linear graph algorithms. Congressus Numerantium 50, 43-60. (1985).

WIMM 1984-1 Wimmer, W.
Using barrier graphs for deadlock prevention in communication networks. IEEE Transactions on Communications 32, 897-901. (1984).

WINT 1985-1 Winter, P.
Generalized Steiner problems in outerplanar networks. BIT 25, 485-496. (1985).

WINT 1986-1 Winter, P.
An algorithm for the enumeration of spanning trees. BIT 26, 44-62. (1986).

WINT 1987-1 Winter, P.
Steiner problem in Halin networks. Discrete Applied Mathematics 17, 281-294. (1987).

WINTA 1984-1 Winter, P. A.; Swart, H. C.
On alpha, beta destructible graphs. Quaestiones Mathematicae (Pretoria) 7, 161-178. (1984).

WINTA 1984-2 Winter, P. A.; Swart, H. C.
On stable graphs. Quaestiones Mathematicae 7, 397-405. (1984).

WISZ 1985-1 Wisztova, E.
A Hamiltonian cycle and a 1-factor in the fourth power of a graph. Casopis Pro Pestovani Matematiky (Prague) 110, 403-412. (1985).

WITT 1985-1 Wittrock, R. J.
Scheduling algorithms for flexible flow lines. IBM Journal of Research and Development 29, 401-412. (1985).

WOER 1988-1 Woerlee, A. P.
Scheduling techniques with limited intermediate storage. In: H. Schellhaas, P. van Beek, H. Isermann, R. Schmidt, M. Zijlstra (Eds.): Operations Research Proceedings 1987 (DGOR/ NSOR), 179-180. Berlin: Springer. (1988).

WOJC 1984-1 Wojciechowski, J.
Generation of trees of a graph with the use of decomposition. Journal of the Franklin Institute 318, 215-231. (1984).

WOJC 1985-1 Wojciechowski, J.
An approximate formula for counting trees in a graph. IEEE Transactions on Circuits and Systems 32, 382-385. (1985).

WOLL 1985-1 Wollmer, R. D.
Critical path planning under uncertainty. In: R. W. Cottle (Ed.): Mathematical Programming Essays in Honor of G. B. Dantzig, Mathematical Programming Study 25, 164-171. (1985).

WOLS 1981-1 Wolsey, L. A.
An analysis of the greedy algorithm for the submodular (set) covering problem. CORE Discussion Paper 8125. Louvain-la-Neuve: Center for Operations Research & Econometrics, 13 p. (1981).

WOLT 1987-1 Wolter, H.
Lösungsverfahren für eine Klasse diskreter Optimierungsaufgaben spezieller Struktur. Optimization 18, 561-572. (1987).

WONG 1984-1 Wong, K. F.; Beasley, J. E.
Vehicle routing using fixed delivery areas. OMEGA International Journal of Management Science 12, 591-600. (1984).

WONGA 1984-1 Wong, R. T.
Introduction and recent advances in network design: models and algorithms. In: M. Florian (Ed.): Transportation Planning Models, 187-226. Amsterdam: North-Holland. (1984).

WONGA 1985-1 Wong, R. T.
Location and network design. In: M. O'hEigeartaigh, J. K. Lenstra, A. H. G. Rinnooy Kan (Eds.): Combinatorial Optimization - Annotated Bibliographies, 129-147. Chichester: Wiley. (1985).

WONGA 1986-1 Wong, R. T.
Recent advances in location and network design models for transportation planning. Ricerca Operativa 39, 37-70. (1986).

WOOL 1986-1 Woollam, C. R.
Flowshop with no idle machine time allowed. Computers and Industrial Engineering 10, 69-76. (1986).

WORM 1985-1 Wormald, N. C.
Counting labelled chordal graphs. Graphs and Combinatorics 1, 193-200. (1985).

WORT 1984-1 Wortmann, J. C.; Monhemius, W.
KANBAN: its use as final assembly scheduling tool within MRP II. In: J. P. Brans (Ed.): Operational Research '84, 641-655. Amsterdam: North-Holland. (1984).

WRIG 1986-1 Wright, D. D.; Haehling von Lanzenauer, C.
COAL: A new heuristic approach for solving the fixed charge problem - computational results. Working Paper. Wolfville: Acadia Univ., School of Business Administration, 27 p. (1986).

WRIG 1986-2 Wright, D. D.; Haehling von Lanzenauer, C.
Solving the fixed charge problem with Lagrangian relaxation and cost allocation heuristics. Working Paper. Wolfville: Acadia Univ., School of Business Administration, 23 p. (1986).

WU 1984-1 Wu, S.-Z.; Wu, X.-M.
Generalized principle of optimality in pansystems network analysis. Kybernetes 13, 231-235. (1984).

WU A 1985-1 Wu, W.-J.
On the planar imbedding of linear graphs. I. Journal of Systems Science and Mathematical Sciences (Beijing) 5, 290-302. (1985).

WU A 1986-1 Wu, W.-J.
On the planar imbedding of linear graphs (continued). Journal of Systems Science and Mathematical Sciences (Beijing), 6, 23-35. (1986).

WU B 1983-1 Wu, W.-M.
Robbins' theorem for fuzzy mixed multigraphs (Chinese). Fuzzy Mathematics (Wuhan) 3, 1-6. (1983).

WU C 1984-1 Wu, X.-M.
Recent developments on graph theory (Chinese). Advances in Mathematics, Beijing 13, 241-253. (1984).

WU D 1986-1 Wu, Y. F.; Widmayer, P.; Wong, C. K.
A faster approximation algorithm for the Steiner problem in graphs. Acta Informatica 23, 223-229. (1986).

XIA 1985-1 Xia, D.-L.
A new method of integer linear programming - branch direction search method. Applied Mathematics and Mechanics, English Edition (Hongkong) 6, 283-290. (1985).

XIA A 1985-1 Xia, W.-M.; Huang, B.; Cheng, S.-Z.
An optimization design of vehicle routes in urban public transportation systems (Chinese). Chinese Journal of Operations Research 4, 54-59. (1985).

XIAO 1987-1 Xiao, D.
Optimal design of pipe network systems (Chinese). Kexue Tongbao (Chinese) (Beijing) 32, 332-334. (1987).

XIE 1982-1 Xie, L.-T.
On a kind of integer programming problems on n-complexes (Chinese). Journal of Shandong University (Natural Science Edition), No.4, 1-5. (1982).

XIE 1983-1 Xie, L.-T.
Methods of algebraic topology applied to the theory of graphs and matroids (Chinese). Advances in Mathematics (Beijing) 12, No. 3, 179-192. (1983).

XIE 1984-1 Xie, L.-T.
Connected partition problems on graphs and n-dimensional complexes (Chinese). Journal of Shandong University (Natural Science Edition), No.2, 1-9. (1984).

XIE 1987-1 Xie, L.-T.
A supplement to the paper: "Method of algebraic topology applied to the theory of graphs and matroids" (Chinese). Advances in Mathematics, Beijing 16, 251-257. (1987).

XIE A 1984-1 Xie, L.-T.; Liu, G.-Z.
A survey of the problem of graceful trees (Chinese). Qufu Shiyuan Xuebao (Qufu), No. 1, 8-15. (1984).

XIE B 1982-1 Xie, L.-T.; Zheng, H.-D.; Liu, J.-Z.
On the valence sequence of an ordered tree. Kexue Tongbao (English Edition) (Beijing) 27, 807-810. (1982).

XU 1984-1 Xu, B.-S.
A family of sequencing problems with earliness and tardiness penalties. Journal, Huazhong University of Science and Technology (Central China, Wuhan) (Engl. Ed.) 6, No. 2, 19-23. (1984).

XU A 1982-1 Xu, C.-D.
Diameter-stable graphs and line-persistence (Chinese). Journal of Shanghai Teachers University, Natural Science Edition, No.3, 17-28. (1982).

XU B 1984-1 Xu, J.-M.
On the proof of Simon and Murty conjecture (Chinese). Journal of Mathematical Research and Exposition 4, 85-86. (1984).

XU B 1984-2 Xu, J.-M.
Proof of the mixed form of Menger's theorem in graph theory (Chinese). Journal, Huazhong Univ. of Science and Technology (Wuhan) 12, No. 2, 17-18. (1984).

XU B 1986-1 Xu, J.-M.
A note on the algebraic connectivity of graphs (Chinese). Journal Xinjiang University of Natural Science (Urumqi) 3, No. 3, 113-114. (1986).

XU C 1986-1 Xu, J.-M.; Tian, F.
On the lower bound of the maximum size of a critically h-edge-connected and a critically h-connected graph (Chinese). Journal of Huazhong (Central China) University of Science and Technology 14, 147-151. (1986).

XU D 1987-1 Xu, M. L.
A simple method for finding shortest paths (Chinese). Qufu Shifan Daxue Xuebao. Ziran Kexue Ban 13, No. 2, 70-73. (1987).

XU E 1986-1 Xu, O.
A class of graceful graphs (Chinese). Journal of the Changsha Railway Institute, No.3, 69-72. (1986).

XU F 1983-1 Xu, S.-J.
A further discussion on bichromaticity (Chinese). Journal of Shanghai Teachers University, Natural Science Edition, No.1, 7-12. (1983).

XU F 1983-2 Xu, S.-J.
Complete-graph-basis and graphs with more edges (Chinese). Journal of Shanghai Teachers University, Natural Science Edition, No.3, 18-23. (1983).

XU F 1984-1 Xu, S.-J.
The chromaticity of regular graphs and bigraphs with more edges (Chinese). Journal of Shanghai Teachers University, Natural Science Edition, No.1, 10-18. (1984).

XU F 1987-1 Xu, S.-J.
Some parameters of graph and its complement. Discrete Mathematics 65, 197-208. (1987).

XU G 1984-1 Xu, S.-J.; Li, N.-Z.
The chromaticity of wheels. Discrete Mathematics 51, 207-212. (1984).

XU H 1987-1 Xu, W.-X.
A new bounding technique for the quadratic assignment problem (Chinese). Acta Mathematicae Applicatae Sinica 10, 333-339. (1987).

XUE 1987-1 Xue, G. L.
A quickly converging algorithm for optimal location problems of the form min(sum c(i)||x-a(i)||) on a closed convex set (Chinese). Qufu Shifan Daxue Xuebao. Ziran Kexue Ban 13, No. 3, 15-20. (1987).

XUE A 1986-1 Xue, Y.-F.; Dai, X.-S.
On maximum degree and the number of vertices of degree >4 in minimally 2-(edge) connected graphs (Chinese). Journal of Kunming Institute of Technology, No.3, 107-113. (1986).

XUE B 1987-1 Xue, Y.-F.; Li, Y.
A minimum spanning tree algorithm for solving the transportation problem (Chinese). Journal of Dalian Institute of Technology 26, No.1, 109-113. (1987).

YAKO 1981-1 Yakovlev, S. V.
Solution of scheduling theory problems by directed random search (Russian). In: A. N. Antamoshkin (Ed.): Applications of random search to the solution of applied problems, 46-49. Kemerovo: Kemerovskij Gosud. Univ. (1981).

YAKO 1987-1 Yakovlev, S. V.
Probabilistic methods for optimization on discrete sets (Russian). Acta Cybernetica (Szeged) 8, No. 2, 219-226. (1987).

YAKU 1983-1 Yaku, T.
Partitions of graphs into chains of the minimum block number. Proceedings of the Faculty of Science of Tokai University (Tokyo) 18, 41-44. (1983).

YAMA 1985-1 Yamamoto, M.; Nof, S. Y.
Scheduling/rescheduling in the manufacturing operating system environment. International Journal of Production Research 23, 705-722. (1985).

YAMAA 1984-1 Yamashiro, M.; Kawanago, T.
A study on the single machine production-line system with multiple arrival sources. I: On the analysis of the system with two different types of arrival sources (Japanese). Research Reports, Ashikaga Institute of Technology 10, 121-128. (1984).

YAN 1985-1 Yan, J.-Y.
Greedoid polyhedra. Acta Mathematicae Applicatae Sinica, English Series 2, 263-268. (1985).

YAN A 1982-1 Yan, T. Y.; Tan, H. H.
Quadratic integer programming for large scale banned matrices. In: Modeling and simulation, Vol. 13, Part 1, 99-103. Research Triangle Park, N. C.: Instrument Soc. of America. (1982).

YANA 1987-1 Yanasse, H. H.; Soma, N. Y.
A new enumeration scheme for the knapsack problem. Discrete Applied Mathematics 18, 235-245. (1987).

YANE 1985-1 Yanev, N. I.
Algorithms for three polynomially solvable cases of the lot-size problem. In: Mathematics and Mathematical Education (Sunny Beach, 1985), 585-589. Bulgar. Akad. Nauk (Sofia). (1985).

YANG 1987-1 Yang, C. E.
A class of scheduling problems and their solution (Chinese). Acta Mathematicae Applicatae Sinica (Beijing) 10, 235-246. (1987).

YANGA 1985-1 Yang, C. E.; Lin, Y. X.
Board welding problem (Chinese). Journal of the Changsha Railway Institute, No.4, 9-20. (1985).

YANGB 1982-1 Yang, S. H.
Isomorphic factorizations of complete tripartite graphs (Chinese). Qufu Shizuan Xuebao (Qufu), No. 2, 20-21. (1982).

YANGC 1986-1 Yang, S. M.; Zhang, Z. F.
The enumeration of the circles of a lattice plane graph (Chinese). Journal of Lanzhou Railway College 5, No.2, 26-32. (1986).

YANGD 1987-1 Yang, T.-Y.; Ignizio, J. P.
An algorithm for the scheduling of army battalion training exercises. Computers and Operations Research 14, 479-492. (1987).

YANGE 1985-1 Yang, W.-L.
On graphs which contain all trees with n edges I. A lower bound of S(T(n)) (Chinese). Acta Mathematicae Applicatae Sinica 8, 284-292. (1985).

YANGE 1985-2 Yang, W.-L.
On graphs which contain all trees with n edges II - the exact values of s(T(n)) for n<or=16 (Chinese). Acta Mathematicae Applicatae Sinica 8, 398-411. (1985).

YANGF 1987-1 Yang, Y. M.
A comment on the discussion of the PERT time distribution (Chinese). Qufu Shifan Daxue Xuebao. Ziran Kexue Ban 13, No. 4, 133-137. (1987).

YANN 1980-1 Yannakakis, M.
On a class of totally unimodular matrices. In: 21st Annual Symp. on Foundations of Computer Science, Syracuse, N.Y. 1980, Proceedings, 10-16. IEEE. (1980).

YANN 1985-1 Yannakakis, M.
A polynomial algorithm for the min-cut linear arrangement of trees. Journal of the Association for Computing Machinery 32, 950-989. (1985).

YANN 1985-2 Yannakakis, M.
On a class of totally unimodular matrices. Mathematics of Operations Research 10, 280-304. (1985).

YANN 1986-1 Yannakakis, M.
Four pages are necessary and sufficient for planar graphs. In: Proc. Eighteenth Annual ACM Symp. on Theory of Computing, Berkeley 1986, 104-108. ACM. (1986).

YANNA 1987-1 Yannakakis, M.; Gavril, F.
The maximum k-colorable subgraph problem for chordal graphs. Information Processing Letters 24, 133-138. (1987).

YANO 1985-1 Yano, C. A.; Carlson, R. C.
An analysis of scheduling policies in multiechelon production systems. IIE Transactions 17, 370-377. (1985).

YANOA 1987-1 Yano, C. A.; Chan, T. J.; Richter, L. K.; Cutler, T.; Murty, K. G.; Mac Gettigari, D.
Vehicle routing at quality stores. Interfaces 17, 52-63. (1987).

YAO 1987-1 Yao, D. D.; Kim, S. C.
Reducing the congestion in a class of job shops. Management Science 33, 1165-1172. (1987).

YAO A 1982-1 Yao, F.
Maximum flows in networks. In: P. A. Burr (Ed.): The Mathematics of Networks, 31-44. Proc. Symp. in Appl. Math. Vol. 26, American Math. Society. (1982).

YAP 1983-1 Yap, H. P.
A class of chromatic index critical graphs. Bulletin of the Malaysian Mathematical Society (2) 6, 45-46. (1983).

YAP 1984-1 Yap, H. P.
Computational complexity of graph properties. In: K. M. Koh, H. P. Yap (Eds.): Graph Theory Singapore 1983, Lecture Notes in Mathematics 1073, 35-54. Berlin: Springer-Verlag. (1984).

YARL 1983-1 Yarlagadda, R.
Unit integer quadratic binary programming. Journal of Mathematical Analysis and Applications 94, 246-267. (1983).

YASU 1984-1 Yasuda, H.; Tokuyama, H.; Tarui, K.; Tanimoto, Y.; Nagano, M.
Two-stage algorithm for production scheduling of hot strip mill. In: J. P. Brans (Ed.): Operational Research '84, 588-600. Amsterdam: North-Holland. (1984).

YE 1983-1 Ye, M.-D.
On the passing through a rod problem and multi-machine scheduling. II (Chinese). Journal of Zhejiang Univ. No. 3, 100-106. (1983).

YEH 1986-1 Yeh, D. Y.
A dynamic programming approach to the complete set partitioning problem. BIT 26, 467-474. (1986).

YEH 1986-2 Yeh, D. Y.
Finding fundamental cycles and bridges on a tree-structured parallel computer. Information Sciences 40, 67-73. (1986).

YEH A 1984-1 Yeh, D. Y.; Lee, D. T.
Graph algorithms on a tree-structured parallel computer. BIT 24, 333-340. (1984).

YON 1986-1 Yon, R. S.
A solution method of the fractional n-index transportation problem. Suhak, No.3, 1-8. (1986).

YON 1987-1 Yon, R. S.
A solution method of the fractional n-index transportation problem with pass capacity (Korean). Suhak, No.1, 21-28. (1987).

YOON 1985-1 Yoon, K.; Hwang, C.-L.
Manufacturing plant location analysis by multiple attribute decision making: part I - single-plant strategy. International Journal of Production Research 23, 345-360. (1985).

YOSH 1984-1 Yoshimura, M.; Hori, E.; Sakamoto, C.; Kashimoto, M.
Active scheduling system for an integrated manufacuting system. In: J. P. Brans (Ed.): Operational Research '84, 561-572. Amsterdam: North-Holland. (1984).

YOU 1985-1 You, Z. Y.; Zhu, S. Q.
Convex combined iteration and non-tip point iteration for solving constrained optimal location problems (Chinese). Journal of Xian Jiaotong University 19, No.4, 119-122. (1985).

YOUN 1985-1 Young, M. H.; Muroga, S.
Symmetric minimal covering problem and minimal PLA's with symmetric variables. IEEE Transactions on Computers 34, 523-541. (1985).

YU 1984-1 Yu, J.-L.
Remarks on the critically n-connected graphs (Chinese). Journal, Huazhong University of Science and Technology (Wuhan) 12, No. 1, 13-14. (1984).

YU A 1984-1 Yu, J.-L.; Ma, Y.
A remark on critically n-connected graphs (Chinese). Journal of Huazhong University of Science and Technology (Central China) 12, No. 1, 13-14. (1984).

YU B 1984-1 Yu, Z.-G.
A generalization of Bondy-Chvatal's theorem on the k-closure of a bipartite graph (Chinese). Journal of the Tsinghua University 24, 101-109. (1984).

YUE 1985-1 Yue, M.-Y.; Han, J.-Y.; Chen, Y.-M.
Elimination conditions and lower bounds for the permutation flow-shop sequencing problem. Acta Mathematicae Applicatae Sinica, English Series 2, 321-331. (1985).

YUNU 1984-1 Yunussa, K.
Dynamic form of generalized problem on minimal cost flow (Russian). Izvestiya Akademii Nauk BSSR, Seriya Fiziko-Matematicheskikh Nauk (Minsk) No. 2, 33-39. (1984).

ZABL 1984-1 Zablotskaya, O. A.
On the comparisons of completely regular cutting-off algorithms. Upravlyaemye Sistemy 25, 68-74. (1984).

ZABLA 1983-1 Zablotskaya, O. A.; Kolokolov, A. A.
Completely regular truncations in Boolean programming (Russian). Upravlyaemye Sistemy (Novosibirsk) No. 23, 55-63. (1983).

ZACH 1984-1 Zachary, W. W.
Modeling social network processes using constrained flow representations. Social Networks 6, 259-292. (1984).

ZAGA 1983-1 Zagaglia Salvi, N.
On the chromatic index of permutation graphs (Italian). Rivista die Matematica del Universita di Parma, Serie IV, 9, 333-337. (1983).

ZAGU 1987-1 Zaguia, N.
Greedy posets for the bump-minimizing problem. Order 4, 257-268. (1987).

ZAIT 1985-1 Zaitman, A. A.
Factor methods in a discrete minimax problem (Russian). In: V. M. Mikhalevich (Ed.): Methods for the Solution of Complex Mathematical Programming Problems, 59-64. Kiev: Akad. Nauk Ukrain. SSR, Inst. Kibernet. (1985).

ZAITA 1986-1 Zaitsev, I. D.; Vainer, V. G.
An optimal scheduling algorithm taking account of the injury factor (Russian). Doklady Akademii Nauk Ukrainskoj SSR, Seriya A, No.6, 68-71. (1986).

ZAITB 1983-1 Zaitsev, R. V.; Sibirko, A. N.; Kostenko, E. I.
Application of sequential optimization algorithms for the solution of a problem of transportation type (Russian). in: Software for program packages, 87-91. Kiev: Akademiya Nauk Ukrainskoj SSR, Inst. Kibernet. (1983).

ZAKS 1985-1 Zaks, J.
How does a complete graph split into bipartite graphs and how are neighborly cubes arranged? American Mathematical Monthly 92, 568-571. (1985).

ZALO 1987-1 Zaloom, V. A.; Chu, H.-W.; Li, D.
An algorithm for sequencing jobs to minimize total tardiness. Beaumont: Lamar Univ., Dept. of Industrial Engineering, 11 p. (1987).

ZAMA 1986-1 Zamanskij, L. Y.
Properties of the number of spanning trees and of partial connected graphs (Russian). Doklady Akademii Nauk Ukrainskoj SSR (Kiev), Seriya A, No.7, 72-77. (1986).

ZAMAA 1984-1 Zamanskij, L. Y.; Cherkasskii, V. L.
Formula for computation of the number of integer points under straight line and its application (Russian). Ehkonomika i Matematicheskie Metody (Moscow) 20, 1132-1137. (1984).

ZAMB 1985-1 Zambitskij, D. K.
Study of the stability of the optimal solution of Weber's problem in a metric space (Russian). Matematicheskie Issledovaniya (Kishinev), No. 82, Matematicheskie Modeli Met. Optim., 49-57. (1985).

ZAMBA 1985-1 Zambitskij, D. K.; Sibirskij, V. K.
Nonlinear Weber problem on graphs (Russian). In: Graphen und Netzwerke - Theorie und Anwendungen, 30. Int. Wiss. Kolloq., Ilmenau 1985, 5, 97-99. (1985).

ZAMF 1986-1 Zamfirescu, T.; Zucco, A.
On the least Hamiltonian traceable graphs (French). Mathematica Japonica 31, 493-502. (1986).

ZASL 1984-1 Zaslavsky, T.
Generalized matchings and generalized Hermite polynomials. In: A. Hajnal; L. Lovasz; V. T. Sos (Eds.): Finite and Infinite Sets, Vol. 2, 851-866. Coll. Math. Soc. J. Bolyai 37, Amsterdam: North-Holland. (1984).

ZASL 1984-2 Zaslavsky, T.
How colorful the signed graph? Discrete Mathematics 52, 279-284. (1984).

ZASL 1987-1 Zaslavsky, T.
Balanced decompositions of a signed graph. Journal of Combinatorial Theory 43B, 1-13. (1987).

ZASL 1987-2 Zaslavsky, T.
The biased graphs whose matroids are binary. Journal of Combinatorial Theory 42B, 337-347. (1987).

ZAW 1986-1 Zaw Win
On the windy postman problem on Eulerian graphs. Preprint 127. Augsburg: Univ. Augsburg, Institut für Mathematik, 17 p. (1986).

ZAW 1987-1 Zaw Win
On the windy postman problem. In: W. Domschke, W. Krabs, J. Lehn, P. Spellucci (Eds.): Proc. XI. Symp. on Operations Research, Darmstadt 1986, 173-176. Methods of Operations Research 57, Frankfurt/Main: Athenäum. (1987).

ZAWA 1987-1 Zawack, D. J.; Thompson, G. L.
A dynamic space-time network flow model for city traffic congestion. Transportation Science 21, 153-162. (1987).

ZDRZ 1985-1 Zdrzalka, S.
Single machine scheduling on an assembly line (Polish). Zeszyty Naukowe Akademii Gorniczo-Hutniczej Im. Stanislawa Staszica, Automatyka 39, 185-196. (1985).

ZELI 1983-1 Zelinka, B.
Double covers and logics of graphs II. Mathematica Slovaca (Bratislava) 33, No. 4, 329-334. (1983).

ZELI 1983-2 Zelinka, B.
Double covers and logics of graphs. Czechoslovak Mathematical Journal (Prague) 33(108), No. 3, 354-360. (1983).

ZELI 1984-1 Zelinka, B.
On k-ply domatic numbers of graphs. Mathematica Slovaca (Bratislava) 34, 313-318. (1984).

ZELI 1984-2 Zelinka, B.
Semidomatic numbers of directed graphs. Mathematica Slovaca 34, 371-374. (1984).

ZELI 1985-1 Zelinka, B.
Bichromaticity and domatic number of a bipartite graph. Casopis Pro Pestovani Matematiky (Prague) 110, 113-115. (1985).

ZELI 1985-2 Zelinka, B.
Comparison of various distances between isomorphism classes of graphs. Casopis pro Pestovani Matematiky (Prague) 110, 289-293. (1985).

ZELI 1985-3 Zelinka, B.
Cyclicity of a complete 3-equipartite graph. Archivum Mathematicum (Brno) 21, 1-3. (1985).

ZELI 1985-4 Zelinka, B.
Domatic numbers of uniform hypergraphs. Archivum Mathematicum (Brno) 21, No. 3, 129-134. (1985).

ZELI 1985-5 Zelinka, B.
Graphs with prescribed neighbourhood graphs. Mathematica Slovaca (Bratislava) 35, 195-197. (1985).

ZELI 1985-6 Zelinka, B.
Odd graphs. Archivum Mathematicum (Brno) 21, No. 4, 181-187. (1985).

ZELI 1986-1 Zelinka, B.
Connected domatic number of a graph. Mathematica Slovaca 36, 387-392. (1986).

ZELI 1986-2 Zelinka, B.
Disconnected neighbourhood graphs. Mathematica Slovaca 36, 109-110. (1986).

ZELI 1986-3 Zelinka, B.
Domatic number and linear arboricity of cacti. Mathematica Slovaca (Bratislava) 36, No. 1, 49-54. (1986).

ZELI 1986-4 Zelinka, B.
Edge neighbourhood graphs. Czechoslovak Mathematical Journal (Prague) 36(111), 44-47. (1986).

ZELI 1986-5 Zelinka, B.
Neighbourhood tournaments. Mathematica Slovaca 36, 241-244. (1986).

ZELI 1987-1 Zelinka, B.
Distances between directed graphs. Casopis pro Pestovani Matematiky 112, 359-367. (1987).

ZELI 1987-2 Zelinka, B.
Natural bijections between directed and undirected self-complementary graphs. Mathematica Slovaca 37, 363-366. (1987).

ZELI 1987-3 Zelinka, B.
On a problem of P. Vestergaard concerning circuits in graphs. Czechoslovak Mathematical Journal 37(112), 318-319. (1987).

ZELI 1987-4 Zelinka, B.
The distance between a graph and its complement. Czechoslovak Mathematical Journal 37(112), 120-123. (1987).

ZEME 1984-1 Zemel, E.
Probabilistic analysis of geometric location problems. Annals of Operations Research, Vol. 1, 215-238. Basel: Baltzer. (1984).

ZEME 1985-1 Zemel, E.
Probabilistic analysis of geometric location problems. SIAM Journal on Algebraic and Discrete Methods 6, 189-200. (1985).

ZEME 1986-1 Zemel, E.
On the computational complexity of facets of the knapsack problem. Working Paper, 713. Evanston: Northwestern University, 21 p. (1986).

ZENG 1983-1 Zeng, F. K.
Finding the tree-set of linear graphs by the DIDORTG-method (Chinese). Journal of Huazhong (Cent. China) Univ. Sci. Technol. 11, No. 4, 17-22. (1983).

ZENI 1985-1 Zenios, S. A.; Mulvey, J. M.
A distributed algorithm for convex network optimization problems. Report EES-85-10. Princeton: Princeton Univ., Civil Engineering Dept., 23 p. (1985).

ZENI 1986-1 Zenios, S. A.; Mulvey, J. M.
Nonlinear network programming on vector supercomputers. Report EES-85-13. Princeton: Princeton Univ., Civil Engineering Dept., 36 p. (1986).

ZENI 1986-2 Zenios, S. A.; Mulvey, J. M.
Relaxation techniques for strictly convex network problems. In: C. L. Monma (Ed.): Algorithms and Software for Optimization - Part II, 517-540. Annals of Operations Research 5, Basel: Baltzer. (1986).

ZHAN 1986-1 Zhan, S.-M.
Hamiltonian connectedness of line graphs. Ars Combinatoria 22, 89-96. (1986).

ZHANA 1987-1 Zhang, B. K.
On a class of integer bottleneck problems (Chinese). Journal on Numerical Methods and Computer Applications (Beijing) 8, No. 3, 178-182. (1987).

ZHANB 1985-1 Zhang, C.-Q.
Circuits and Hamilton circuits in domination orientable tournaments. In: B. R. Alspach, C. D. Godsil (Eds.): Cycles in Graphs, Annals of Discrete Mathematics Vol. 27, 163-172. Amsterdam: North-Holland. (1985).

ZHANC 1985-1 Zhang, F. J.; Chen, R. S.
On two kinds of 2-edge-connected graphs (Chinese). Acta Mathematicae Applicatae Sinica 8, No. 2, 227-234. (1985).

ZHAND 1985-1 Zhang, F. J.; Chen, R. S.; Guo, X. F.
Perfect matchings in hexagonal systems. Graphs and Combinatorics 1, 383-386. (1985).

ZHANE 1986-1 Zhang, F. J.; Chen, Z. B.
Connectivity of (adjacency) tree graphs. Journal Xinjiang University of Natural Science (Urumqi) 3, No. 4, 1-5. (1986).

ZHANE 1987-1 Zhang, F. J.; Chen, Z. B.
Lower bounds for the maximum number of end-vertices of spanning trees (Chinese). Kexue Tongbao (Chinese) (Beijing) 32, No. 7, 481-483. (1987).

ZHANF 1983-1 Zhang, F. J.; Guo, X. F.
H-transformations of Hamiltonian cycles (Chinese). Journal of Mathematics (Wuhan) 3, 307-314. (1983).

ZHANF 1985-1 Zhang, F. J.; Guo, X. F.
Connectedness of generalized directed tree graphs and an interpolation theorem for the numbers of terminal vertices of directed trees (Chinese). Journal Xinjiang University of Natural Science (Urumqi), No. 4, 5-7. (1985).

ZHANF 1985-2 Zhang, F. J.; Guo, X. F.
Some properties of minimally strongly connected digraphs (Chinese). Journal Xinjiang University of Natural Science (Urumqi), No. 3, 1-6. (1985).

ZHANF 1986-1 Zhang, F. J.; Guo, X. F.
Hamilton cycles in Euler tour graph. Journal of Combinatorial Theory 40B, 1-8. (1986).

ZHANF 1986-2 Zhang, F. J.; Guo, X. F.
Hamiltonian cycles in perfect matching graphs (Chinese). Journal Xinjiang University of Natural Science (Urumqi) 3, No. 4, 10-16. (1986).

ZHANF 1987-1 Zhang, F. J.; Guo, X. F.
Hamilton cycles in directed Euler tour graphs. Discrete Mathematics 64, 289-298. (1987).

ZHANG 1983-1 Zhang, F. J.; Lin, G. N.
A graphical method for generating all Euler tours of an Eulerian digraph (Chinese). Lanzhou Daxue Xuebao 19, No. 2, 1-9. (1983).

ZHANH 1987-1 Zhang, F. J.; Lin, G. N.; Zheng, M. L.
An algorithm for generating all perfect matchings of a graph (Chinese). Journal Xinjiang University of Natural Science (Urumqi) 4, No. 3, 1-8. (1987).

ZHANI 1985-1 Zhang, F. J.; Lin, Y. X.
Two classes of general polynomials for graphs (Chinese). Acta Mathematica Sinica (Beijing) 28, 122-130. (1985).

ZHANJ 1985-1 Zhang, F. J.; Liu, Y.-T.; Guo, X. F.
The enumeration of perfect matchings of treelike polyhexes (Chinese). Journal Xinjiang University of Natural Science (Urumqi), No. 3, 7-11. (1985).

ZHANK 1986-1 Zhang, F. J.; Zheng, M. L.
Some notes on an algorithm for minimal spanning trees of undirected graphs, and their uniqueness (Chinese). Journal Xinjiang University of Natural Science (Urumqi), 3, No. 2, 10-13. (1986).

ZHANL 1987-1 Zhang, F. J.; Zhou, M. K.
The characteristic and matching polynomials of a class of graphs (Chinese). Journal Xinjiang University of Natural Science (Urumqi) 4, No. 1, 1-5. (1987).

ZHANM 1983-1 Zhang, G. Q.
NP-completeness and restricted partitions (Chinese). Beijing Daxue Xuebao, No. 6, 1-6. (1983).

ZHANM 1984-1 Zhang, G. Q.
"NP=P?" and restricted partitions. Information Sciences 32, 231-242. (1984).

ZHANN 1986-1 Zhang, L. A.; Geng, S. Y.
The complexity of the 0/1 multi-knapsack problem. Journal of Computer Science and Technology 1, No.1, 46-50. (1986).

ZHANN 1987-1 Zhang, L. A.; Geng, S. Y.
Computation of multiknapsack problems (Chinese). Beijing Daxue Xuebao, No. 1, 65-76. (1987).

ZHANO 1982-1 Zhang, S. K.
Generalization of sequencing problems of the optimum service for model P-J. Kexue Tongbao, Foreign Language Edition (Beijing) 27, 594-597. (1982).

ZHANP 1985-1 Zhang, T.-C.
A brief survey of plane location problems (Chinese). Chinese Journal of Operations Research 4, 4-11. (1985).

ZHANQ 1984-1 Zhang, Y.; Zhang, Y. C.
A limit theorem for N(0n)/n in first-passage percolation. Annals of Probability 12, 1068-1076. (1984).

ZHANR 1986-1 Zhang, Z. F.; Chen, Y.
Operations on graphs and the chromatic number of a graph (Chinese). Journal of Lanzhou Railway College 5, No.1, 24-30. (1986).

ZHANS 1983-1 Zhang, Z. F.; Yao, B.; Xie, J. G.
On the bandwidth of graphs (Chinese). Journal of Mathematics (Wuhan) 3, 323-330. (1983).

ZHANT 1985-1 Zhang, Z. F.; Zhang, J. X.
On some sufficient conditions for graphs to be of first kind (Chinese). Journal of Mathematics, Wuhan University 5, 161-165. (1985).

ZHANU 1984-1 Zhang, Z. F.; Zhang, J. X.; Wang, J. F.
A theorem on colouring edges. Kexue Tongbao (Engl. Edition) 29, 1137-1139. (1984).

ZHANU 1984-2 Zhang, Z. F.; Zhang, J. X.; Wang, J. F.
An edge coloring theorem (Chinese). Kexue Tongbao (Beijing) 29, 139-140. (1984).
ZHANV 1987-1 Zhang, Z. Z.; Masuyama, S.; Ibaraki, T.; Mine, H.
Graph packing over a rooted tree. Memoirs of the Faculty of Engineering, Kyoto University 49, 206-215. (1987).
ZHAO 1985-1 Zhao, C. K.
A conjecture on matroids (Chinese). Neimenggu Daxue Xuebao (Hohhot) 16, 312-326. (1985).
ZHAO 1985-2 Zhao, C. K.
A note on bipartite graphs (Chinese). Neimenggu Daxue Xuebao (Hohhot) 16, 343-345. (1985).
ZHAOA 1982-1 Zhao, F.-Z.
The potential method for solving the multicommodity capacity transportation problem (Chinese). Mathematica Numerica Sinica (Beijing) 4, 427-435. (1982).
ZHAOB 1987-1 Zhao, G.-F.
Graphs isomorphic to subgraphs of their iterated line graphs. Acta Mathematicae Applicatae Sinica (English Series) 3, 111-121. (1987).
ZHAOC 1986-1 Zhao, L.-C.; Liu, C. F.; Wang, H.
A note on Hamiltonian line graphs (Chinese). Acta Mathematicae Applicatae Sinica (Peking) 9, 17-20. (1986).
ZHAOD 1984-1 Zhao, S. L.
The theory of pleasant stringing trees and some kinds of pleasant trees (Chinese). Acta Mathematicae Applicatae Sinica (Peking) 7, 370-372. (1984).
ZHAOD 1986-1 Zhao, S. L.
Double isomorphic graceful trees are graceful stringing trees (Chinese). Qufu Shifan Daxue Xuebao. Ziran Kexue Ban, No. 2, 54-55. (1986).
ZHEN 1986-1 Zheng, M. L.
The Prim algorithm and the uniqueness of minimal trees of graphs (Chinese). Journal Xinjiang University of Natural Science (Urumqi) 3, No. 3, 15-17. (1986).
ZHENA 1984-1 Zheng, Q.
Global optimization of integer and mixed programming (Chinese). Numerical Mathematics, A Journal of Chinese Universities (Nanjing) 6, No. 1, 91-94. (1984).
ZHOU 1984-1 Zhou, B.
A note on the Erdös-Sos conjecture. Acta Mathematica Scientia (Wuhan) 4, 287-289. (1984).
ZHOUA 1985-1 Zhou, M. K.
Some properties about distance degree regular graphs. Journal of the Xinjiang University of Natural Sciences No. 1, 15-22. (1985).
ZHOUA 1986-1 Zhou, M. K.
Matching polynomials of polyphenylenes. Journal Xinjiang University of Natural Science (Urumqi) 3, No. 3, 44-48. (1986).
ZHOUA 1986-2 Zhou, M. K.
Numbers of perfect matchings in three classes of graphs. Journal Xinjiang University of Natural Science (Urumqi), 3, No. 2, 17-26. (1986).
ZHOUB 1984-1 Zhou, Z. H.
Minimal and maximal line digraphs (Chinese). Journal of Mathematics (Wuhan) 4, No. 3, 297-302. (1984).
ZHU 1985-1 Zhu, A. X.
Non-coloring-contradiction graphs and a graph coloring algorithm (Chinese). Chinese Journal of Computers (Beijing) 8, No. 3, 189-197. (1985).
ZHU A 1985-1 Zhu, B. W.
A partition of the set of vertices of a connected graph (Chinese). Journal Xinjiang University of Natural Science (Urumqi), No. 4, 1-4. (1985).
ZHU A 1985-2 Zhu, B. W.
The circumference of minimal 2-edge-connected graphs (Chinese). Journal of the Xinjiang University of Natural Sciences No. 2, 118-122. (1985).
ZHU B 1985-1 Zhu, B. W.; Liu, Z. S.; Chen, Z. Q.
A theorem on the edge-cohesiveness of a vertex of a graph (Chinese). Neimenggu Daxue Xuebao (Hohhot) 16, 319-320. (1985).
ZHU C 1987-1 Zhu, B. W.; Su, J. J.
Some problems in the research on the connectedness of graphs (Chinese). Advances in Mathematics (Beijing) 16, 113-119. (1987).
ZHU D 1985-1 Zhu, D. X.; Cai, G. Q.
On an exhaustive circuit matrix of an oriented graph (Chinese). Shanghai Keji Daxue Xuebao, No. 2, 53-59. (1985).
ZHU E 1985-1 Zhu, F.
Notes on the elimination method in the m x n flow-shop scheduling problem (Chinese). Acta Mathematicae Applicatae Sinica (Beijing) 8, 305-310. (1985).
ZHU F 1983-1 Zhu, R. Y.
A remark on Jackson's theorem on regular graphs (Chinese). Journal of Mathematics (Wuhan) 3, 301-306. (1983).
ZHU F 1983-2 Zhu, R. Y.
On maximal circuits in 2-çonnected graphs (Chinese). Qufu Shiyuan Xuebao (Qufu), No. 4, 8-9. (1983).
ZHU G 1986-1 Zhu, S. Y.
A strong graceful path (Chinese). Journal of Shanghai Jiaotong University, No. 4, 36-43. (1986).
ZHU H 1985-1 Zhu, W. S.
Proof of Bodendiek's conjecture on graceful graphs (Chinese). Kexue Tongbao (Beijing) 30, 1052-1055. (1985).
ZHU I 1983-1 Zhu, X. D.
Pansystems graph theory and (l,d)-stable graphs. Science Exploration (Changsha) 3, No. 2, 77-86. (1983).
ZHU J 1986-1 Zhu, X. F.
On arc-Hamiltonian tournaments (Chinese). Advances in Mathematics (Beijing) 15, 321-324. (1986).

ZHU K 1983-1 Zhu, Y.-J.
A survey of the research on the Hamilton problem in graph theory (Chinese). Qufu Shizuan Xuebao (Qufu), No. 4, 1-7. (1983).

ZHU K 1986-1 Zhu, Y.-J.
Constructive characteristics appearing in some problems of graph theory and algorithms (Chinese). Qufu Shifan Daxue Xuebao. Ziran Kexue Ban 12, No. 4, 1-12. (1986).

ZHU L 1985-1 Zhu, Y.-J.; Li, H.
Progress on the Hamilton problem in graph theory (Chinese). Qufu Shizuan Xuebao (Qufu) No. 2, 12-22. (1985).

ZHU M 1985-1 Zhu, Y.-J.; Liu, Z. H.; Yu, Z.-G.
An improvement of Jackson's result on Hamilton cycles in 2-connected regular graphs. In: B. R. Alspach, C. D. Godsil (Eds.): Cycles in Graphs, Annals of Discrete Mathematics Vol. 27, 237-248. Amsterdam: North-Holland. (1985).

ZHU M 1986-1 Zhu, Y.-J.; Liu, Z. H.; Yu, Z.-G.
2-connected k-regular graphs (k greater equal 6) on at most 3k+3 vertices are Hamiltonian. Journal of Systems Science and Mathematical Sciences (Beijing) 6, 36-49. (1986).

ZHU M 1986-2 Zhu, Y.-J.; Liu, Z. H.; Yu, Z.-G.
2-connected k-regular graphs (k greater equal 6) on at most 3k+3 vertices are Hamiltonian. II. Journal of Systems Science and Mathematical Sciences (Beijing) 6, 136-145. (1986).

ZHU N 1984-1 Zhu, Y.-J.; Tian, T.; Chen, C.-P.; Zhang, C.-Q.
Arc-pancyclic property of tournaments under some degree conditions. Journal of Information and Optimization Sciences (Delhi) 5, 1-16. (1984).

ZHUR 1984-1 Zhuravlev, Y. I.; Platonenko, I. M.
On an economical product of Boolean equations (Russian). Zhurnal Vychislitelnoj Matematiki i Matematicheskoj Fiziki (Moscow) 24, 164-166. (1984).

ZIKA 1985-1 Zikan, K.; Schmeichel, E.
A path counting problem in digraphs. Fibonacci Quarterly 23, 3-6. (1985).

ZIMM 1987-1 Zimmermann, H.-J.
Methoden und Modelle des Operations Research. Für Ingenieure, Ökonomen und Informatiker. Braunschweig, Wiesbaden: Vieweg. (1987).

ZIMMA 1984-1 Zimmermann, H.-J.; Pollatschek, M. A.
Fuzzy 0-1 linear programs. In: H.-J. Zimmermann, L. A. Zadeh, B. R. Gaines (Eds.): Fuzzy Sets and Decision Analysis, 133-146. Amsterdam: North-Holland. (1984).

ZIMMB 1984-1 Zimmermann, K.
On max-separable optimization problems. In: R. E. Burkard, R. A. Cuninghame-Green, U. Zimmermann (Eds.): Algebraic and Combinatorial Methods in Operations Research, Annals of Discrete Mathematics Vol. 19, 357-362. Amsterdam: North-Holland. (1984).

ZIMMC 1984-1 Zimmermann, U.
Minimization of combined objective functions on integral submodular flows. In: R. E. Burkard, R. A. Cuninghame-Green, U. Zimmermann (Eds.): Algebraic and Combinatorial Methods in Operations Research, Annals of Discrete Mathematics Vol. 19, 363-381. Amsterdam: North-Holland. (1984).

ZIMMC 1984-2 Zimmermann, U.
Submodulare Flüsse: Verfahren zur Minimierung linearer Zielfunktionen. In: R. E. Burkard (Ed.): 15. Steiermärkisches mathematisches Symposium, Forschungszentrum Graz, Mathematisch-Statistische Sektion, Bericht No. 234. (1984).

ZIMMC 1985-1 Zimmermann, U.
Augmenting circuit methods for submodular flow problems. In: L. Lovasz, A. Recski (Eds.): Matroid Theory, 401-439. Coll. Math. Soc. J. Bolyai Vol. 40. Amsterdam: North-Holland. (1985).

ZIMMC 1986-1 Zimmermann, U.
Duality for balanced submodular flows. Discrete Applied Mathematics 15, 365-376. (1986).

ZIMMC 1986-2 Zimmermann, U.
Linear and combinatorial sharing problems. Discrete Applied Mathematics 15, 85-104. (1986).

ZIMMC 1986-3 Zimmermann, U.
Sharing problems. Optimization 17, 31-48. (1986).

ZINC 1984-1 Zinchenko, A. B.
Properties and conditions for solvability of a special problem of optimal geometric allocation (Russian). Izvestiya Severo-Kavkazskogo Nauchnogo Tsentra Vysshej Shkoly, Estestvennye Nauki No. 3(47), 17-20. (1984).

ZIND 1986-1 Zinder, Y. A.
An efficient algorithm for a deterministic problem of the scheduling theory with parallel devices (Russian). Kibernetika (Kiev), No. 2, 120-122. (1986).

ZINDA 1985-1 Zinder, Y. A.; Shkurba, V. V.
Effective iterative algorithms of scheduling theory (Russian). Kibernetika (Kiev) No. 1, 72-75. (1985).

ZISS 1985-1 Zissimopoulos, V.
Heuristic methods for solving (un)constrained 2-dimensional cutting stock problems. In: P. Brucker, R. Pauly (Eds.): Proc. IX. Symp. on Operations Research, Osnabrück 1984. Methods of Operations Research 49, 345-360. (1985).

ZIYA 1984-1 Ziyatdinov, R. K.
Estimate of the accuracy of certain heuristic algorithms for the solution of the travelling salesman problem (Russian). Probabilistic Methods and Cybernetics No. 20, 56-61. Kazan: Gosudarstvennyi Universitet. (1984).

ZNOI 1983-1 Znoiko, D. V.
Vertex-transitive graphs with a given degree and girth (Russian). Matematicheski Zametki (Moscow) 33, 780-783. (1983).

ZOLL 1987-1 Zoller, K.; Robrade, A.
Dynamische Bestellmengen- und Losgrössenplanung. Verfahrensübersicht und Vergleich. OR Spektrum 9, 219-234. (1987).

ZORK 1983-1 Zorkin, V. N.; Stepanov, G. D.
Optimal arrangement of additional centers on a graph of simple structure (Russian). In: Algorithmic Constructions and Their Efficiency, 90-97. Yaroslavl: Yaroslav. Gos. Univ. (1983).

ZORN 1985-1 Zörnig, P.
Strukturuntersuchungen an 2xn-Entartungsgraphen. Diskussionsbeitrag 87. Hagen: Fernuniv. Hagen, Fachbereich Wirtschaftwissensch., 21 p. (1985).

ZUCC 1984-1 Zucco, A.
On hyperhomogeneously traceable oriented graphs. Atti della Accademia delle Scienze di Torino, Classe di Scienze Fisiche, Matematiche e Naturali 116, 231-236. (1984).

ZYKA 1987-1 Zyka, O.
Nowhere-zero 30-flow on bidirected graphs. KAM Series, 87-26. Praha: Charles Univ., Dept. of Applied Mathematics, 11 p. (1987).

ZYKO 1982-1 Zykov, A. A.
The generalized assignment problem (Russian). In: B. A. Shcherbakov (Ed.): Operations research and programming, 59-77. Kishinev: Shtiintsa. (1982).

```
PPPPPPPPPPP   AAAAAAAAAA   RRRRRRRRRRR  TTTTTTTTTTTT           2222222222
PPPPPPPPPPPP AAAAAAAAAAAA RRRRRRRRRRRR TTTTTTTTTTTT          222222222222
PP        PP AA        AA RR        RR     TT                22        22
PP        PP AA        AA RR        RR     TT                          22
PP        PP AA        AA RR        RR     TT                          22
PPPPPPPPPPPP AAAAAAAAAAAA RRRRRRRRRRR      TT                         22
PPPPPPPPPPP  AAAAAAAAAAAA RRRRRRRRRR       TT                       22
PP           AA        AA RR    RR         TT                     22
PP           AA        AA RR     RR        TT                   22
PP           AA        AA RR      RR       TT                 22
PP           AA        AA RR       RR      TT                222222222222
PP           AA        AA RR        RR     TT                222222222222
```

SUBJECT CLASSIFICATION

BRUCC 1985-1 Brucker, P.; Papenjohann, W.
An out-of-kilter method for the algebraic circulation problem.
BRUCD 1984-1 Brucker, P.; Papenjohann, W.; Zimmermann, U.
A dual optimality criterion for algebraic linear programs.
BURKC 1985-1 Burkard, R. E.; Hamacher, H. W.; Tind, J.
On general decomposition schemes in mathematical programming.
CECH 1986-1 Cechlarova, K.
On some properties of the algebraic assignment problem.
HAMAD 1984-1 Hamacher, H. W.; Tufekci, S.
Algebraic flows and time-cost tradeoff problems.
HODI 1987-1 Höding, M.
Algebraic structures of special discrete optimization problems (German).
QIN 1985-1 Qin, Y.
On generalized Floyd algorithm.
QIN 1985-2 Qin, Y.
On Jar-metric principle - a unified approach to solve optimum paths problems on multistage directed graphs.
ZIMMB 1984-1 Zimmermann, K.
On max-separable optimization problems.
ZIMMC 1986-2 Zimmermann, U.
Linear and combinatorial sharing problems.
ZIMMC 1986-3 Zimmermann, U.
Sharing problems.

ABDA 1987-1 Abdallaoui, G.
Maintainability of a grade structure as a transportation problem.

ABDU 1983-1 Abdullaev, A. A.; Rzaev, T. G.
Optimization of operative planning and scheduling in integrated ACS for enterprises (Russian).

AGGAB 1985-1 Aggarwal, V.; Suresh, R.
A shortest-route approach for optimum aggregation of data with an application to income data.

AHME 1982-1 Ahmed, M.
Modeles probabilistes pour choisir un itineraire dans un reseau de trafic.

AKINA 1986-1 Akinc, U.; Roodman, G. M.
A new approach to aggregate production planning.

ANTAA 1984-1 Antamoshkin, A. N.; Zajtseva, T. A.; Korsakova, O. N.
Formalization of optimization problems in planning agricultural-industrial combines (English).

ARTI 1984-1 Artina, S.; Walker, J.
The use of mixed-integer linear programming in designing hydraulic networks (Italian).

ARZO 1984-1 Arzola, J.
The task of selection of proposals, formulation, properties, solution method (Spanish).

ASHI 1985-1 Ashimov, A. A.; Burkov, V. N.; Dzhaparov, B. A.; Kondratev, V. V.
Mechanisms of operative coordinated control of industrial systems. I. Coordinated scheduling of continuous production (Russian).

ASTA 1986-1 Astakhov, N. D.
The production-transportation problem of optimal plant location with startup priorities.

ATALA 1986-1 Atallah, M. J.; Hambrusch, S. E.
An assignment algorithm with applications to integrated circuit layout.

ATKI 1984-1 Atkins, D. R.; Granot, D.; Raghavendra, B. G.
Application of mathematical programming to the plywood design and manufacturing problem.

AVRA 1981-1 Avram-Nitchi, R.
An application of graph theory to the time-criterion allocation problem (Romanian).

BABAC 1983-1 Babaev, R. D.
A problem for the choice of an optimal sequence of construction of objects (Russian).

BAKEA 1984-1 Baker, B. M.
A network-flow algorithm for project selection.

BARA 1987-1 Barahona, F.
On via minimization.

BARAC 1986-1 Barahona, F.; Grötschel, M.; Jünger, M.; Reinelt, G.
An application of combinatorial optimization to statistical physics and circuit layout design.

BARR 1981-1 Barr, R. S.; Turner, J.
Microdata file merging through large-scale network technology.

BARRB 1986-1 Barros, O.; Weintraub, A.
Spatial market equilibrium problems as network models.

BARZ 1986-1 Barzilai, J.; Cook, W. D.; Kress, M.
A generalized network formulation of the pairwise comparison consensus ranking model.

BAYU 1985-1 Bayus, B. L.; Gupta, S. K.
Analyzing floor configurations for casino slot machines.

BEIS 1987-1 Beissinger, J. S.; Peled, U. N.
Enumeration of labelled threshold graphs and a theorem of Frobenius involving Eulerian polynomials.

BELAA 1984-1 Belardo, S.; Harrald, J.; Wallace, W. A.; Ward, J. E.
A partial covering approach to siting response resources for major maritime oil spills.

BELI 1985-1 Belikov, S. A.; Zak, Y. A.
Some problems of optimal distribution of limited resources when realizing a network of interrelated works.

BERND 1984-1 Bernussou, J.; LeGall, F.; Garcia, J. M.
Optimisation et acheminement dynamique dans les reseaux telephoniques.

BERTF 1982-1 Bertsekas, D. P.
Optimal routing and flow control methods for communication networks.

BEYE 1987-1 Beyer, W. A.; Heller, L.
A Steiner tree associated with three quarks.

BIELB 1984-1 Bielli, M.; Cini, M.; Calicchio, G.; Magliozzi, L.; Oddone, M.
Network models in planning the new generation of submarine communication systems.

BIGG 1985-1 Biggs, J. R.
Priority rules for shop floor control in a material requirements planning system under various levels of capacity.

BITR 1986-1 Bitran, G. R.; Haas, E. A.; Matsuo, H.
Production planning of style goods with high setup costs and forcast revisions.

BIXB 1984-2 Bixby, R. E.
Recent algorithms for two versions of graph realization and remarks on applications to linear programming.

BLOO 1982-1 Bloom, G. S.; Hsu, D. F.
On graceful digraphs and a problem in network addressing.

BOESA 1984-1 Boesch, F. T.; Suffel, C. L.
An overview of discrete methods in the synthesis of reliable networks.

BOYC 1984-1 Boyce, D. E.
Network models in transportation/land use planning.

BREC 1985-1 Brecht, T. B.; Colbourn, C. J.
Lower bounds on two-terminal network reliability.

BROO 1981-1 Brooks, R. E.
Using generalized networks to forecast natural gas distribution and allocation during periods of shortage.

BRUV 1985-1 Bruvold, N. T.; Evans, J. R.
Flexible mixed-integer programming formulations for production scheduling problems.

BUHLA 1986-1 Buhl, H. U.; Siedersleben, J.
Minimizing the transportation costs of a distribution system.

BUJA 1985-1 Bujan, M. F.
Mathematical model for planning in an industrial branch.

BURR	1985-1	Burr, S. A.; Falk, J. E.; Karr, A. F. Integer prim-read solutions to a class of target defense problems.
BUTT	1985-1	Butterworth, K. Practical application of linear/integer programming in agriculture.
CACCA	1987-1	Caccetta, L.; Vijayan, K. Applications of graph theory.
CAMA	1985-1	Camarinopoulos, L.; Becker, G. Zur Anwendung graphentheoretischer Verfahren bei der Modularisierung von Fehlerbäumen.
CAO	1985-1	Cao, H.-G. Storage and search of data with minimal spanning trees (Chinese).
CARA	1987-1	Caramanis, M. Production system design: a discrete event dynamic system and generalized Benders' decomposition approach.
CARE	1986-1	Carey, M. A constraint qualification for a dynamic traffic assignment model.
CARRD	1986-1	Carrie, A. S.; Perera, D. T. S. Work scheduling in FMS under tool availability constraints.
CHAKA	1986-1	Chakraborty, T. K. A preemptive single sampling attribute plan of given strength.
CHAKC	1985-1	Chakravarty, A. K. Multiproduct purchase scheduling with limited budget and/or group discounts.
CHANAC	1985-1	Chang, S. S. A production planning model for multi-product facilities.
CHAPA	1987-1	Chapman, C. B.; Ward, S. C. Modular decision analysis - a production-scheduling illustration.
CHAUA	1987-1	Chaudhry, S. S.; Moon, I. D.; Mac Cormick, S. T. Conditional covering: greedy heuristics and computational results.
CHAUB	1987-1	Chaudhuri, P. Algorithms for some graph problems on a distributed computational model.
CHINA	1984-1	Chin, F.; Wong, M. An optimal schedule for printing and binding.
CHOIA	1985-1	Choi, W.; Francis, R. L. Some optimization models of building evacuation problems with flow-dependent arc capacities.
CHOIB	1984-1	Choi, W.; Francis, R. L.; Hamacher, H. W.; Tufekci, S. Network models of building evacuation problems with flow-dependent exit capacities.
CHOIC	1985-1	Choi, W.; Hamacher, H. W.; Tufekci, S. Modeling of building evacuation problems by network flows with side constraints.
CIOB	1984-1	Ciobanu, G.; Ciubotaru, I. Cybernetic modelling of the manufacturing processes with the help of the co-ordinating graphs.
COLEB	1986-1	Coleman, T. F.; Cai, J.-Y. The cyclic coloring problem and estimation of sparse Hessian matrices.
CONL	1984-1	Conley, W. Satisfying the customer.
COZZA	1982-1	Cozzens, M. B.; Roberts, F. S. T-colorings of graphs and the channel assigment problem.
COZZB	1984-2	Cozzens, M. B.; Wang, D. I. The general channel assignment problem.
CRUM	1981-1	Crum, R. L.; Nye, D. J. A network model of insurance company cash flow management.
DAFE	1984-1	Dafermos, S.; Nagurney, A. A network formulation of market equilibrium problems and variational inequalities.
DALA	1986-1	Dalang, R. C.; Trotter, L. E. Jr.; Werra, D. de On randomized stopping points and perfect graphs.
DAUS	1983-1	Dauscha, W.; Modrow, H. D.; Neumann, A. An approach for periodic networks.
DEAK	1983-1	Deak, I.; Hoffer, J.; Mayer, J.; Nemeth, A.; Potecz, B.; Prekopa, A.; Strazicky, B. A large scale, mixed variable mathematical programming model for the short-range optimal schedule of electric power systems with thermal power plants taking network constraints into account (Hungarian).
DEAK	1983-2	Deak, I.; Hoffer, J.; Mayer, J.; Nemeth, A.; Potecz, B.; Prekopa, A.; Strazicky, B. Optimal daily scheduling of electricity production in Hungary.
DEBA	1984-1	Debanne, J. G. Tailoring decomposition algorithms to efficiently solve large scale energy planning problems.
DELL	1986-1	Dellaert, N.; Wessels, J. Production scheduling with uncertain demand and no stockkeeping.
DERIB	1985-1	Derigs, U.; Hagberg, B. Ein iterativer Lösungsansatz für das Rostering Problem mittels optimaler Zuordnungen.
DJID	1985-4	Djidjev, H. N. Solving graph problems on VLSI models.
DROR	1987-1	Dror, M.; Ball, M. O. Inventory/routing: reduction from an annual to a short-period problem.
DRUD	1986-1	Drud, A.; Zenios, S. A.; Mulvey, J. M. Balancing large social accounting matrices with nonlinear network programming.
DUCR	1985-1	Ducrocq, J.-C. Optimal energy management in a complex system with mixed-integer programming.
DURA	1987-1	Duran, F. A large mixed integer production and distribution program.
EL-BA	1985-1	El-Bakry, M.; Wahsh, S.; Nassar, S. A new pulse width modulation technique based on 0-1 programming.
EL-G	1986-1	El-Gamal, A.; Hemachandra, L. A.; Shperling, I.; Wei, V. K. Using simulated annealing to design good codes.
EPPEA	1986-1	Eppen, G. D.; Martin, R. K.; Schrage, L. A mixed-integer programming model for evaluating distribution configurations in the steel industry.

EREN 1984-1 Erenguc, S. S.; Tufekci, S.
A linear order algorithm for a transportation type production scheduling problem.

EREN 1984-2 Erenguc, S. S.; Tufekci, S.
A transportation type aggregate production model with bounds on inventory and backordering.

ERFL 1986-1 Erfle, W.
Ein Branch-and-Bound-Verfahren zur optimalen Allokation von hierarchisch geordneten Prüfstellen in den Fertigungsablauf.

ERNV 1984-1 Ernvall, J.
On the construction of spanning paths by Gray-code in compression of files.

ESCU 1986-1 Escudero, L. F.
A motivation for using the truncated Newton approach in a very large scale nonlinar network problem.

EVAN 1985-1 Evans, J. R.
On equivalent transportation models for production planning problems.

EVST 1985-1 Evstigneev, V. A.
Application of graph theory to computer programming (Russian).

FAND 1985-1 Fandel, G.; Reese, J.
Organisation und Planung des Fahrzeugeinsatzes in einem Chemieunternehmen.

FANE 1985-1 Fanelli, S.; Lombari, T.
An application of the technique of Q-GERT networks to the evaluation of logistic parameters of electronic systems (Italian).

FEDE 1986-1 Federgruen, A.; Groenevelt, H.
Optimal flows in networks with multiple sources and sinks, with applications to oil and gas lease investment programs.

FELD 1985-1 Feldman, P.
Fault tolerance of minimal path routings in a network.

FILI 1984-1 Filipiak, J.
A dynamic flow assignment problem for a telephone trunk network.

FISHE 1986-1 Fisher, M. L.; Huang, J.-G.; Tang, B.-X.
Scheduling bulk-pickup-delivery vehicles in Shanghai.

FLORA 1984-1 Florian, M.
An introduction to network models used in transportation planning.

FOUL 1985-2 Foulds, L. R.
One-way street system design.

FOUL 1986-1 Foulds, L. R.
Braess's paradox: which arcs should be eliminated in order to minimize congestion?

FOUL 1986-2 Foulds, L. R.
Testing the theory of evolution: a novel application of combinatorial optimization.

FRAI 1984-1 Frair, L.
Airport noise modelling and aircraft scheduling so as to minimize community annoyance.

FRAN 1985-1 Francis, N. D.; Fleming, D. I.
Optimum allocation of places to students in a national university system.

FRANA 1984-1 Francis, R. L.
A negative exponential solution to an evacuation problem.

FURU 1984-1 Furukawa, M.
The 1-3-dimensional knapsack problem in production planning: formulation and heuristic methods (Japanese).

GAIM 1987-1 Gaimon, C.; Mohan, S.; Thompson, G. L.
A network approach to cohort personnel planning using cross-sectional data.

GAIN 1986-1 Gainanov, D. N.
A graph-theoretic algorithm for the construction of the committee for inconsistent system of linear inequalities.

GALL 1984-1 Gallo, G.; Pallottino, S.
Shortest path methods in transportation models.

GELD 1987-1 Gelders, L. F.; Pintelon, L.; Wassenhove, L. N. van
A location-allocation problem in a large Belgian brewery.

GERSA 1984-1 Gershwin, S. B.; Akella, R.; Choong, Y. F.; Mitter, S. K.
Short term production scheduling of an automated manufacturing facility.

GERSB 1984-1 Gershwin, S. B.; Orlicki, D. M.; Platzman, L. K.
An equilibrium model for traffic assignment in urban transportation networks.

GILBB 1985-1 Gilbert, K. C.; Holmes, D. D.; Rosenthal, R. E.
A multiobjective discrete optimization model for land allocation.

GLIC 1985-1 Glickman, T. S.; Sherali, H. D.
Large-scale network distribution of pooled empty freight cars over time, with limited substitution and equitable benefits.

GLOVC 1985-1 Glover, F.; Klingman, D.
Basis exchange characterization for the simplex SON algorithm for LP/embedded networks.

GOETA 1985-1 Goetz, J.; Makuchowski, M.
Heuristic methods for the general problem of resources allocation and operation scheduling (Polish).

GOIC 1985-1 Goicoechea, A.; Soland, R. M.; Sharp, F. M.
A transportation model for inland waterways navigation via mathematical programming.

GOLDL 1987-1 Golden, B. L.; Wasil, E. A.
Computerized vehicle routing in the soft drink industry.

GOME 1987-1 Gomes, L. F. A. M.
On the problem of assigning traffic flows to a multicriteria, multicommodity, free-flow network.

GONC 1985-1 Gonczarowski, J.; Warmuth, M. K.
Applications of scheduling theory to formal language theory.

GRABB 1985-1 Grabowski, J.; Skubalska, E.
Optimization of the transportation network structure with respect to the flow costs (Polish).

GRANE 1986-1 Grant, T. J.
Lessons for O. R. from A. I.: a scheduling case study.

GRIE 1984-1 Griesshaber, R.
Rotation system for sales representatives and station managers.

GROTB 1984-1 Grötschel, M.; Jünger, M.; Reinelt, G.
A cutting plane algorithm for the linear ordering problem.

GROTB 1984-2 Grötschel, M.; Jünger, M.; Reinelt, G.
Optimal triangulation of large real world input-output matrices.

GROTB 1987-1 Grötschel, M.; Jünger, M.; Reinelt, G.
Calculating exact ground states of spin glasses: a polyhedral approach.

GROTB 1987-2 Grötschel, M.; Jünger, M.; Reinelt, G.
Via minimization with pin preassignments and layer preference.

GROTG 1987-1 Grötschel, M.; Wakabayashi, Y.
A cutting plane algorithm for a clustering problem.

GUNN 1986-1 Gunn, E. A.; Kusiak, A.
The procurement problem: an integer programming problem well suited to a solution using duality.

GUSTA 1985-1 Gustafson, K.; Hartman, R.
Graph theory and fluid dynamics.

HAES 1985-1 Haessler, R. W.
Production planning and scheduling for an integrated container company.

HAGH 1987-1 Haghani, A. E.; Daskin, M. S.
A combined model of train routing, makeup, and empty car distribution.

HAMO 1984-1 Hamon, J. P.; Weissert, F.
ROSSINI: a general model for optimizing aircraft assignment to parking positions at an airport.

HAN 1982-1 Han, F.
An integer programming algorithm for the plan of a freight car classification system (Chinese).

HANSC 1987-1 Hansen, P.; Huge, J.
Implicit treatment of zero or range constraints in a model for minimun cost foundry alloys.

HARAB 1985-2 Harary, F.
Graph theoretic approaches to finite mathematical structures.

HARK 1982-1 Harker, P. T.; Friesz, T. L.
A simultaneous freight network equilibrium model.

HARTD 1984-1 Hartulari, C.
The systemic approach on the problem of production operative programming and scheduling.

HASS 1985-1 Hassan, M. M. D.; Hogg, G. L.; Smith, D. R.
A construction algorithm for the selection and assignment of materials handling equipment.

HEAR 1984-1 Hearn, D. W.
Practical and theoretical aspects of aggregation problems in transportation planning models.

HEARA 1984-1 Hearn, D. W.; Lawphongpanich, S.; Ventura, J. A.
Restricted simplicial decomposition: computation and extensions.

HEINC 1985-1 Heinssen, D. K.; Powers, D. L.
Generalized floorplans from assigned room adjacencies.

HIRSA 1984-1 Hirshfeld, D. S.
An overview of mathematical programming for strategic planning.

HOSK 1984-1 Hoskins, W. D.; Lambert, T. D.; Stanton, R. G.
An improved algorithm for determining optimal sums of disjoint sets.

HULM 1984-1 Hulme, B. L.; Shiver, A. W.; Slater, P. J.
A boolean algebraic analysis of fire protection.

HUMM 1985-1 Hummeltenberg, W.
Capital budgeting with Benders decomposition.

HUTS 1984-1 Hutschenreuther, H.
Die Backwarenversorgung in einem abgeschlossenen Gebiet - ein typisches Beispiel für die Produktions- Transport- Standort- Optimierung.

HWANA 1984-1 Hwang, C.-L.; Lee, H. B.; Tillmann, F. A.; Lie, C. H.
Nonlinear integer goal programming applied to optimal system reliability.

ICHIB 1984-1 Ichino, M.; Sklansky, J.
Optimum feature selection by zero-one integer programming.

IRI B 1975-1 Iri, M.; Tomizawa, N.
A unifying approach to fundamental problems in network theory by means of matroids.

ITAI 1985-1 Itai, A.; Rodeh, M.
Scheduling transmissions in a network.

IVANB 1984-1 Ivanchev, M. D.; Janev, N. I.; Mitev, J. G.
A knapsack approach to a given class of distribution problems.

IWAI 1985-1 Iwainsky, A.
Layout optimization for a hierarchy of star networks.

JASI 1984-1 Jasinska, E.; Woytych, E.
Location of depots in a sugar-beet distribution system (case study).

JEROA 1986-1 Jeroslow, R. G.
Ten lectures on mixed integer model formulation for logic-based decision support.

JOHNA 1985-1 John, P.; Rempel, J.
Counting perfect matchings in hexagonal systems.

JOHNB 1985-1 John, P.; Sachs, H.
Calculating the number of perfect matchings and Pauling's bond orders in hexagonal systems whose inner dual is a tree.

KANEB 1987-1 Kanet, J. J.; Adelsberger, H. H.
Expert systems in production scheduling.

KANTA 1987-1 Kanter, I.; Sompolinsky, H.
Graph optimization problems and the Potts glass.

KHACA 1986-1 Khachaturov, V. R.
Models and methods for solving multiextremal allocation problems by the use of properties of supermodular functions defined on Boolean lattices (Russian).

KHAN 1986-1 Khan, A. M.
Solid-waste disposal with intermediate transfer stations: an application of the fixed-charge location problem.

KLAF 1985-2 Klafszky, E.; Terlaky, T.
Variants of the "Hungarian Method" for solving linear programming problems.

KLEID 1982-1 Kleitman, D. J.
Some practical network problems.

KLINC 1984-1 Klingman, D.; Mead, M.; Phillips, N. V.
Network optimization models for military manpower planning.

KLINE 1987-1 Klingman, D.; Phillips, N. V.; Steiger, D.; Wirth, R.; Padman, R.; Krishnan, R.
An optimization based integrated short-term refined petroleum product planning system.

KORTC 1986-1 Korte, B.; Möhring, R. H.
Zur Bedeutung der diskreten Mathematik für die Konstruktion hochintegrierter Schaltkreise.

KOTI 1984-1 Kotin, L.
Connectivity and the design of optimally survivable systems.

KOUR 1984-1 Kourtz, P.
A network approach to least-cost daily transfers of forest fire control resources.

KOVA 1984-1 Kovacs, L. B.; Boros, E.; Inotay, F.
A two-stage mathematical model and interactive program system for planning networks of sewer systems and waste water treatment plants - with application to the Lake Balaton area (Hungarian).

KOVA 1986-1 Kovacs, L. B.; Boros, E.; Inotay, F.
A two-stage approach for large-scale sewer systems design with application to the Lake Balaton resort area.

KRUS 1986-1 Kruse, H.-J.
Degeneracy graphs and the neighbourhood problem.

KUBA 1986-1 Kubale, M.
Application of graph edge colouring to the control of rearrangeable Clos networks (Polish).

KUSIA 1987-1 Kusiak, A.; Finke, G.
Hierarchical approach to the process planning problem.

KUZM 1985-1 Kuzmenko, V. N.
Solution of a discrete programming problem of large dimensions (Russian).

LAPO 1987-1 Laporte, G.
Solving a family of permutation problems on 0-1 matrices.

LAWR 1984-1 Lawrence, K. D.; Reeves, G. R.; Lawrence, S. M.
A multiple objective shift allocation model.

LE 1984-1 Le, W.-L.; Liu, Z. Y.; Wang, Z.-Y.
The algorithm for mixed programming with upper bounded variables and its application to energy planning (Chinese).

LECL 1987-1 Leclerc, B.
Arbres minimums communs et compatibilite de donnees de types varies.

LECLA 1986-1 Leclerc, M.
A graph-theoretical approach to a problem in circuit design.

LEE N 1986-1 Lee, Y.-D.
Integer programming applications and their relationship (Korean).

LESK 1984-1 Lesk, M.
Diametre de graphes et qualite de service d'un reseau de donnees.

LEZI 1984-1 Lezin, V. I.
Operations control system for a multiproduct plant.

LONG 1982-1 Long, T. D.
Application of graph theory in the study of the reliability of large systems.

LOVEA 1985-1 Love, R. F.; Truscott, W. G.; Walker, J.
Terminal location problem: a case study supporting the status quo.

LUCE 1984-2 Lucertini, M.
Models of the task assignment problem in distributed systems.

LUNA 1987-1 Luna, H. P. L.; Ziviani, N.; Cabral, R. H. B.
The telephonic switching centre network problem: formalization and computational experience.

LUQU 1984-1 Luque, E.; Ripoll, A.
Integer linear programming for microprograms register allocation.

MAGI 1986-1 Magirou, V. F.
The efficient drilling of printed circuit boards.

MAJS 1983-2 Majstrenko, V. E.
The graph in problems of the description of recognition objects (Russian).

MANG 1985-1 Mangoubi, R. S.; Mathaisel, D. F. X.
Optimizing gate assignments at airport terminals.

MANN 1985-1 Manning, J. R.
How not to schedule a stitching room.

MARE 1985-3 Marecki, F.
Multiobjective resources allocation of assembly line (Polish).

MARTC 1985-1 Martin, C. H.; Lubin, S. L.
Optimization modeling for business planning at Trumbull Asphalt.

MATHC 1985-1 Mathur, K.; Salkin, H. M.; Nishimura, K.; Morito, S.
Applications of integer programming in radio frequency management.

MATTA 1986-1 Mattsson, L.-G.
Residential location and school planning in a tightening urban economy.

MEDV 1985-1 Medvedev, Y. T.; Pushchinskii, E. V.
On an integer minimax problem arising in the print industry.

MEHR 1985-1 Mehrez, A.
A resource allocation problem with the aid of an integer zero-one programming model in the plastic industry.

MEHRB 1985-1 Mehrez, A.; Stern, H. I.
Optimal refueling strategies for a mixed-vehicle fleet.

MELAB 1986-1 Melachrinoudis, E.; Rumpf, D.; Venegas, R.
Spray operation planning aided by a mixed integer programming model.

MEZA 1987-1 Mezard, M.
Spin glasses and optimization.

MIKHC 1986-1 Mikhalevich, V. S.; Trubin, V. A.; Shor, N. Z.
Optimization problems of production-transportation planning. Models, Methods, Algorithms (Russian).

MILK 1984-1 Milkiewicz, F.
Decomposition of production scheduling in production systems with switched over continuous technological processes.

MINO 1986-2 Minoux, M.
Optimal traffic assignment in a SS/TDMA frame: a new approach by set covering and column generation.

MLYN 1985-1 Mlynarovic, V.
Benders' decomposition and some problems of the production-transport type (Czechoslovakian).

MORG 1987-1 Morgenstern, I.
Spin glasses, optimization and neural networks.

MORO 1987-1 Morozov, V. V.
On an application of discrete optimization methods (Russian).

MOSE 1986-1 Mosevich, J.
Balancing hydraulic turbine runners - a discrete combinatorial optimization problem.

MULVA 1985-1 Mulvey, J. M.; Zenios, S. A.
Integrated risk/cost planning models for the U.S. air traffic system.

MURO 1987-3 Murota, K.
Systems analysis by graphs and matroids. Structural solvability and controllability.

MUST 1985-1 Mustafi, C. K.; Xavier, M. J.
Mixed-integer linear-programming formulation of a multi-attribute threshold model of choice.

NAGA 1985-1 Nagasawa, H.; Nishiyama, N.; Hitomi, K.
Decision analysis for determining the optimum planning horizon in aggregate production planning. Part 2: Difference between planning horizons in weekly and in monthly schedulings.

NAGE 1986-1 Nageswara, S. V.; Viswanadham, N.
Fault diagnosis in dynamical systems: a graph theoretic approach.

NERA 1984-1 Neralic, L.; Sirovitsa, L.; Skok, S.; Khunet, D.
Solution of a production-transportation problem by Benders' decomposition method (Russian).

NGUYA 1984-1 Nguyen, S.
Estimating origin-destination matrices from observed flows.

NGUYB 1986-2 Nguyen, S.; Pallottino, S.
Estimating origin destination flows for transit networks.

O'GR 1985-1 O'Grady, P. J.; Byrne, M. D.
A combined switching algorithm and linear decision rule approach to production planning.

O'NE 1984-1 O'Neill, R. P.; Heinkel, J.; Kurator, W. G.; Mariner-Volpe, B.; Stokes, R.; Tawshunsky, A.; Trapmann, W.
The gas analysis modeling system (GAMS).

OLIF 1985-1 Oliff, M. D.; Burch, E. E.
Multiproduct production scheduling at Owens-Corning Fiberglas.

ONUR 1987-1 Onur, L.; Fabrycky, W. J.
An input/output control system for the dynamic job shop.

ORLI 1986-1 Orlin, J. B.
A dual version of Tardos's algorithm for linear programming.

ORLIA 1985-1 Orlin, J. B.; Rothblum, U. G.
Computing optimal scalings by parametric network algorithms.

OTIM 1985-1 Otiman, P.
Optimizing the location of crop fields.

PAN 1985-1 Pan, Z. T.; Zhao, S. C.
Graph-theoretic proof of an inequality (Chinese).

PETRB 1986-1 Petrova, V. A.
Application of combinatorial methods for estimating the reliability of flow systems (Russian).

PIPP 1982-1 Pippenger, N.
Telephone switching networks.

PIRK 1986-1 Pirkul, H.
An integer programming model for the allocation of databases in a distributed computer system.

POPP 1985-1 Popp, W.
Zur Analyse strategischer Aspekte mit gemischt-ganzzahliger Programmierung.

PUTU 1986-2 Puturidze, Z. S.
Optimization of some implementation of a modified algorithm for the method of constructing a sequence of plans (Russian).

RANDA 1985-1 Randhawa, S. U.; Mac Dowell, E. D.; Faruqui, S.-D.
An integer programming application to solve sequencer mix problems in printed circuit board production.

RECS 1987-1 Recski, A.
Applications of combinatorics to electrical engineering and rigidity.

REHF 1985-1 Rehfeldt, M.
Eine netzwerktheoretische Methode zur Untersuchung der Disjunktheit bestimmter endlicher Mengen großer Mächtigkeit.

REIT 1984-1 Reiter, S.
How a network of processors can schedule its work.

REND 1984-1 Rendl, F.
On the complexity of decomposing matrices arising in satellite communication.

RICHF 1985-1 Richter, K.
Complexity of the robot sequencing problem.

ROBI 1985-1 Robinson, D. F.; Janjic, I.
The constructability of floorplans with given outerplanar adjacency graph and room areas.

ROSE 1987-1 Rosenberg, E.
A nonlinear programming heuristic for computing optimal link capacities in a multi-hour alternate routing communications network.

ROUH 1985-1 Rouhani, R.; Lasdon, L.; Lebow, W.; Warren, A. D.
A generalized Benders decomposition approach to reactive source planning in power systems.

SAKSB 1986-1 Saksena, J. P.; Ramachandran, A.
An optimal fleet utilisation.

SCHNA 1987-1 Schniederjans, M. J.; Kim, G. C.
A goal programming model to optimize departmental preference in course assignments.

SCHO 1985-1 Schoch, M.
Optimale Fertigungsplanung für einen zweistufigen Produktionsprozess.

SHAT 1983-1 Shatalina, L. S.
The compiling of a guaranteed spare-parts supply schedule formulated as a mixed linear programming problem.

SHCH 1986-2 Shcherbina, O. A.
A unimodular problem of integer programming.

SHERH 1987-1 Sherali, H. D.; Staschus, K; Huacuz, J. M.
An integer programming approach and implementation for an electric utility capacity planning problem with renewable energy sources.

SHI B 1984-1 Shi, W.-G.; Chen, Y.-N.
A graph approach to the function testing of sequential circuits (Chinese).

SHLO 1987-1 Shlosman, S. B.
Graph colouring: a way to variety of new correlation inequalities.

SIED 1982-1 Siedersleben, J.
Optimierung von deterministischen und stochastischen Entscheidungsnetz- plänen (Dissertation).

SILL 1986-1 Sillaber, A. H.
A continuous linear optimization model for the exact solution of travelling-salesman-problems in connexion with expansion planning of ring networks.

SINC 1986-2 Sinclair, M.
The assignment of workers to tasks - an example from an academic department.

SRINA 1985-1 Srinivasan, V.; Kim, Y. H.
A network optimization approach to efficient cash management in India.

STAND 1985-1 Stanfel, L. E.
Partitioning problems in cellular manufacturing.

STASA 1985-1 Stassinopoulos, G. I.; Konstantopoulos, P.
Optimal congestion control in single destination networks.

STEIB 1983-1 Steinmüller, J.
Graphentheoretische und kombinatorische Probleme in der technologischen Fertigungsvorbereitung.

STOW 1985-1 Stowe, J. D.
An integer programming solution for the optimal credit investigation credit granting sequence.

SUND 1984-1 Sundararaghavan, P. S.
The unit capacity pickup and delivery problem on a one-way loop.

SUNG 1985-1 Sung, C. S.
A multiproduct multifacility production planning model with capacity constraints.

SUNG 1986-1 Sung, C. S.
A single-product parallel-facilities production-planning model.

SUNG 1987-1 Sung, C. S.
Multifacility in-series production-planning model with capacity constraints.

SWEE 1984-1 Sweeney, D. J.
A practical evaluation of the potential of mixed integer programming as a strategic planning tool.

SZID 1984-1 Szidarovszky, F.; Carrera, J.; Usunoff, E.
Optimum location of measuring points in the quality analysis of groundwater (Hungarian).

SZWAB 1985-1 Szwarc, W.; Posner, M. E.
The caterer problem.

SZYM 1986-1 Szymanowski, W.
COLOS: a collection optimization system as an application of the vehicle scheduling problem.

TAN A 1985-1 Tan, Z.-G.; Zheng, L.-N.; Li, L.-N.
A mathematical model for the traffic assignment problem (Chinese).

TAN A 1985-2 Tan, Z.-G.; Zheng, L.-N.; Li, L.-N.
Introduction to mathematical models and algorithms for the traffic assignment problem (Chinese).

TAO 1987-1 Tao, G.-Y.; Zhou, P.-D.; Wang, Q.
The heuristic shortest-path algorithm for transportation (Chinese).

TAO A 1982-1 Tao, K. M.; Hsia, P.
On the structure of directed graphs with applications. I. A rapprochement with systems theory.

TAYI 1985-1 Tayi, G. K.
Sensitivity analysis of mixed integer programs: an application to environmental policy making.

TAYL 1985-1 Taylor, B. W. III; Moore, L. J.; Clayton, E. R.; Davis, K. R.; Rakes, T. R.
An integer nonlinear goal programming model for the deployment of state highway patrol units.

TELL 1987-1 Telley, H.; Liebling, T. M.; Mocellin, A.
Reconstruction of polycrystalline structures: a new application of combinatorial optimization.

TEMP 1985-1 Templeton, W. D.; Daellenbach, H. G.
Interactive simulation for production scheduling.

THOMA 1986-3 Thomassen, C.
Sign-nonnegative matrices and even cycles in directed graphs.

TRAU 1987-1 Trauwaert, E.; Decrock, A.
A heuristic algorithm for allocation of components in manufacturing processes.

TRUS 1985-1 Truscott, W. G.
Scheduling production activities in multi-stage batch manufactoring systems.

TRYF 1986-1 Tryfos, P.
An integer programming approach to the apparel sizing problem.

TSIL 1984-1 Tsiligirides, T.
Heuristic methods applied to orienteering.

TSO 1986-1 Tso, M.
Network flow models in image processing.

VAIN 1982-1 Vainer, V. G.
Estimates of the accuracy of algorithms for construction of connecting networks (Russian).

VANNA 1986-1 Vannelli, A.; Kumar, K. R.
A method for finding minimal bottle-neck cells for grouping part-machine families.

VASI 1986-1 Vasilev, N. S.
A model of a communication network development.

VASKA 1987-1 Vasko, F. J.; Wolf, F. E.; Stott, K. L.
Optimal selection of ingot sizes via set covering.

VASS 1985-1 Vassilacopoulos, G.
Allocating doctors to shifts in an accident and emergency department.

VIDA 1984-1 Vidal, R. V. V.
Large-scale production planning.

WALA 1984-1 Walas, R. A.; Askin, R. G.
An algorithm for NC turret punch press tool location and hit sequencing.

WALDA 1984-1 Wald, J. A.; Sorenson, P. G.
Resolving the query inference problem using Steiner trees.

WAN 1985-1 Wan, W. X.
A special integer programming problem in management science (Chinese).

WANGG 1985-1 Wang, K.-Y.; Zhan, J.; Wang, Z.-T.
Models and algorithms for integrated energy transportation system planning (Chinese).

WEBE 1986-1 Weber, M.; Liebling, T. M.
Euclidean matching problems and the Metropolis algorithm.

WERR 1987-1 Werra, D. de
Design and operation of flexible manufacturing systems: the kingdom of heuristic methods.

WHITA 1985-1 White, C.
Production scheduling with applications in the printing industry.

WILHD 1985-1 Wilhelm, W. E.; Shin, H.-M.
Effectiveness of alternate operations in a flexible manufacturing system.

WILSA 1986-1 Wilson, J. M.
Formulation of a problem involving assembly lines with multiple manning of work stations.

WIMM 1984-1 Wimmer, W.
Using barrier graphs for deadlock prevention in communication networks.

WONGA 1984-1 Wong, R. T.
Introduction and recent advances in network design: models and algorithms.

WORT 1984-1 Wortmann, J. C.; Monhemius, W.
KANBAN: its use as final assembly scheduling tool within MRP II.

XIAO 1987-1 Xiao, D.
Optimal design of pipe network systems (Chinese).

YASU 1984-1 Yasuda, H.; Tokuyama, H.; Tarui, K.; Tanimoto, Y.; Nagano, M.
Two-stage algorithm for production scheduling of hot strip mill.

YEH 1986-2 Yeh, D. Y.
Finding fundamental cycles and bridges on a tree-structured parallel computer.

YOSH 1984-1 Yoshimura, M.; Hori, E.; Sakamoto, C.; Kashimoto, M.
Active scheduling system for an integrated manufacuting system.

ZACH 1984-1 Zachary, W. W.
Modeling social network processes using constrained flow representations.

ZAIT 1985-1 Zaitman, A. A.
Factor methods in a discrete minimax problem (Russian).

ZAMAA 1984-1 Zamanskij, L. Y.; Cherkasskii, V. L.
Formula for computation of the number of integer points under straight line and its application (Russian).

ZAWA 1987-1 Zawack, D. J.; Thompson, G. L.
A dynamic space-time network flow model for city traffic congestion.

ZHOUA 1986-1 Zhou, M. K.
Matching polynomials of polyphenylenes.

AGGAA 1985-1 Aggarwal, V.
A Lagrangean-relaxation method for the constrained assignment problem.

AGGAC 1986-1 Aggarwal, V.; Tikekar, V. G.; Hsu, L.-F.
Bottleneck assignment problems under categorization.

ALEK 1980-1 Alekseev, A. E.; Kogan, D. I.
Complexity of determination of the optimal allocation of tasks to persons with two-sided preferences (Russian).

ALLE 1987-1 Allen, J. D.; Helgason, R. V.; Kennington, J. L.
The frequency assignment problem: a solution via nonlinear programming.

ARAOA 1985-1 Araoz, J.; Edmonds, J.
A case of non-convergent dual changes in assignment problems.

ARON 1986-1 Aronson, J. E.
The multiperiod assignment problem: a multicommodity network flow model and specialized branch and bound algorithm.

ARONA 1985-1 Aronson, J. E.; Barr, R.; Helgason, R. V.; Kennington, J. L.; Loh, A.; Zaki, H.
A projective transformation algorithm by Karmarkar: a computational experiment with assignment problems.

ATALA 1986-1 Atallah, M. J.; Hambrusch, S. E.
An assignment algorithm with applications to integrated circuit layout.

AVIS 1985-1 Avis, D.; Devroye, L.
An analysis of a decomposition heuristic for the assignment problem.

AVRA 1984-1 Avram-Nitchi, R.
An algorithm based on graph theory for solving the time-criterion assignment problem.

BALAL 1984-1 Balas, E.; Saltzman, M. J.
Some facets of the three-index assignment polytope.

BALAL 1986-1 Balas, E.; Saltzman, M. J.
Facets of the three-index assignment polytope.

BALI 1984-1 Balinski, M. L.
A good (dual) simplex method for the assignment problem.

BALI 1985-1 Balinski, M. L.
Signature methods for the assignment problem.

BALI 1986-1 Balinski, M. L.
A competitive (dual) simplex method for the assignment problem.

BENDC 1985-1 Benders, J. F.; Beulens, A. J. M.; Nunen, J. A. E. E. van
On using the linear programming relaxation of assignment type mixed integer problems.

BERTF 1987-1 Bertsekas, D. P.
The auction algorithm: a distributed relaxation method for the assignment problem.

BIGI 1985-1 Bigildeeva, T. B.
Methods and algorithms of approximate solution of a quadratic assignment problem of large dimensions (a survey) (Russian).

BISH 1987-1 Bisht, N. S.; Prakash, S.
A bicriterion machine-assignment problem.

BLAI 1984-1 Blair, C.
Every finite distributive lattice is a set of stable matchings.

BLOK 1985-1 Blokh, D. A.
The assignment problem for Jacobi matrices (Russian).

BOKH 1987-1 Bokhari, S. H.
Assigment Problems in Parallel and Distributed Computing.

BONO 1986-1 Bonomi, E.; Lutton, J.-L.
The asymptotic behaviour of quadratic sum assignment problems: a statistical mechanics approach.

BOOK 1985-1 Bookhold, I.
Untersuchungen zum quadratischen Zuordnungsproblem.

BURK 1985-1 Burkard, R. E.
Time-slot assignment for TDMA-systems.

BURK 1985-2 Burkard, R. E.
Time-slot assignment for TDMA-systems.

BURK 1986-1 Burkard, R. E.
Assignment problems: recent solution methods and applications.

BURKA 1985-1 Burkard, R. E.; Euler, R.; Grommes, R.
On the facial structure of polytopes related to the planar 3-dimensional assignment problem.

BUTK 1984-1 Butkovic, P.
On the uniqueness of the assignment problem solution.

CARPB 1987-1 Carpaneto, G.; Toth, P.
Primal-dual algorithms for the assignment problem.

CECH 1986-1 Cechlarova, K.
On some properties of the algebraic assignment problem.

CHANW 1985-1 Chang, H.-J.
A minimax-arc-path problem.

DERI 1984-1 Derigs, U.
Alternate strategies for solving bottleneck assignment problems - analysis and computational results.

DERI 1986-2 Derigs, U.
The shortest augmenting path method for solving assignment problems - motivation and computational experience.

DERIA 1986-1 Derigs, U.; Goecke, O.; Schrader, R.
Monge sequences and a simple assignment algorithm.

DERIB 1985-1 Derigs, U.; Hagberg, B.
Ein iterativer Lösungsansatz für das Rostering Problem mittels optimaler Zuordnungen.

DERID 1985-1 Derigs, U.; Metz, A.
An efficient labeling technique for solving sparse assignment problems.

DERID 1986-1 Derigs, U.; Metz, A.
An efficient labeling technique for solving sparse assignment problems.

DERID 1986-2 Derigs, U.; Metz, A.
An in-core/out-of-core method for solving large scale assignment problems.

DYERC 1986-1 Dyer, M. E.; Frieze, A. M.; Mac Diarmid, C.
On linear programs with random costs.
EISEC 1984-1 Eiselt, H. A.; Gerchak, Y.
Solution structures and sensitivity of special assignment problems.
EISED 1987-1 Eiselt, H. A.; Laporte, G.
Combinatorial optimization problems with soft and hard requirements.
EMELC 1985-1 Emelichev, V. A.; Perepelitsa, V. A.
On multicriterial assignment problems.
EULEA 1986-1 Euler, R.; Burkard, R. E.; Grommes, R.
On latin squares and the facial structure of related polytopes.
FINKE 1987-1 Finke, G.; Burkard, R. E.; Rendl, F.
Quadratic assignment problems.
FISHA 1985-1 Fishburn, P. C.; Kochenberger, G. A.
Aggregating assignment constraints.
FISHG 1986-1 Fisher, M. L.; Jaikumar, R.; Wassenhove, L. N. van
A multiplier adjustment method for the generalized assignment problem.
FRAN 1985-1 Francis, N. D.; Fleming, D. I.
Optimum allocation of places to students in a national university system.
FRENB 1985-1 Frenk, J. B. G.; Houweninge, M. van; Rinnooy Kan, A. H. G.
Asymptotic properties of the quadratic assignment problem.
FRENB 1986-1 Frenk, J. B. G.; Houweninge, M. van; Rinnooy Kan, A. H. G.
Order statistics and the linear assignment problem.
FREY 1986-1 Frey, J.-J.; Yehia-Alcoutlabi, A.
Comparaisons par paires: une interpretation et une generalisation de la methode des scores.
GILBA 1987-1 Gilbert, K. C.; Hofstra, R. B.
An algorithm for a class of three-dimensional assignment problems arising in scheduling applications.
GLOVA 1986-1 Glover, F.; Glover, R.; Klingman, D.
Threshold assignment algorithm.
GOLDN 1985-1 Goldfarb, D.
Efficient dual simplex algorithms for the assignment problem.
GOTT 1986-1 Gottlieb, E. S.; Rao, M. R.
The generalized assignment problem I: valid inequalities and facets.
GOTT 1986-2 Gottlieb, E. S.; Rao, M. R.
The generalized assignment problem II: three classes of facets.
GREG 1985-1 Grega, W.
A decomposition algorithm for the determination of optimal bus frequencies.
GROS 1985-1 Grosser, E.
Über eine Methode zur Lösung des Zuordnungsproblems.
GUIN 1984-1 Guinet, A.
Transports industriels routiers, un probleme d'affectation avec reemploi sous contraintes.
GULY 1983-1 Gulyanitskii, L. F.; Khodzinskii, A. N.
Two algorithms for solution of optimization problems on permutations (Russian).
GUSF 1987-2 Gusfield, D.
Three fast algorithms for four problems in stable marriage.
HALLD 1987-1 Hall, N. G.; Vohra, R. V.
An on-line assignment problem with random effectiveness and costly information.
HANSA 1986-1 Hansen, J. V.; Giauque, W. C.
Task allocation in distributed processing systems.
HORN 1986-1 Hornung, R.
Methods for solving the discrete capacity assignment problem in data networks: numerical comparison.
IRVI 1985-1 Irving, R. W.
An efficient algorithm for the stable roommates problem.
IRVIA 1986-1 Irving, R. W.; Leather, P.
The complexity of counting stable marriages.
IRVIB 1987-1 Irving, R. W.; Leather, P.; Gusfield, D.
An efficient algorithm for the "optimal" stable marriage.
JANK 1985-1 Jankowska-Zorychta, Z.
An assignment problem of specific structure (Polish).
JONK 1986-1 Jonker, R.
Traveling salesman and assignment algorithms: design and implementation.
JONKA 1986-1 Jonker, R.; Volgenant, T.
Improving the Hungarian assignment algorithm.
JONKA 1987-1 Jonker, R.; Volgenant, T.
A shortest augmenting path algorithm for dense and sparse linear assignment problems.
JONKA 1988-1 Jonker, R.; Volgenant, T.
A shortest augmenting path algorithm for dense and sparse linear assignment problems.
JORN 1986-1 Jörnsten, K. O.; Näsberg, M.
A new Lagrangian relaxation approach to the genralized assignment problem.
KAPU 1985-1 Kapur, D.; Krishnamoorthy, M. S.
Worst-case choice for the stable marriage problem.
KARP 1987-1 Karp, R. M.
An upper bound on the expected cost of an optimal assignment.
KASA 1984-1 Kasai, T.; Yamada, S.; Ohno, T.
An approximation method for planar assignment problem.
KHOD 1985-1 Khodzinskii, A. N.
Sequential algorithm for the solution of problems of combinatorial optimization on permutations.
KHOD 1985-2 Khodzinskii, A. N.
Solution of the quadratic assignment problem by the mean value method (Russian).
KIMO 1984-1 Kimoto, T.; Tsukiyama, S.; Shirakawa, I.
On the computational complexity of assignment problems with restrictions (Japanese).
KINDA 1985-1 Kindervater, G. A. P.; Volgenant, A.; Leve, G. de; Gijlswijk, V. van
On dual solutions of the linear assignment problem.

KLIND 1984-1 Klingman, D.; Phillips, N. V.
Topological and computational aspects of preemptive multicriteria military personnel assignment problems.
KUMA 1986-1 Kumar, A.; Sharma, S.
On additional job assignment problem.
LAPOA 1986-1 Laporte, G.; Desroches, S.
The problem of assigning students to course sections in a large engineering school.
LEE F 1984-1 Lee, D. T.; Leung, J. Y.-T.
On the 2-dimensional channel assignment problem.
LEE N 1986-1 Lee, Y.-D.
Integer programming applications and their relationship (Korean).
LIGG 1985-1 Liggett, R. S.
Optimal spatial arrangement as a quadratic assignment problem.
LIND 1984-1 Lindberg, P. O.; Olafsson, S.
On the length of simplex paths: the assignment case.
MAAS 1985-1 Maas, C.
Ein Zuordnungsproblem im Zusammenhang mit der Einteilung von Übungsgruppen.
MALH 1984-1 Malhotra, R.; Bhatia, H. L.
Variants of the time minimization assignment problem.
MALHA 1985-1 Malhotra, R.; Bhatia, H. L.; Puri, M. C.
The three dimensional bottleneck assignment problem and its variants.
MANG 1985-1 Mangoubi, R. S.; Mathaisel, D. F. X.
Optimizing gate assignments at airport terminals.
MAREA 1984-1 Marello, S.; Pulleyblank, W. R.; Toth, P.; Werra, D. de
Balanced optimization problems.
MARTA 1984-1 Martello, S.; Pulleyblank, W. R.; Toth, P.; Werra, D. de
Balanced optimization problems.
MARTB 1987-2 Martello, S.; Toth, P.
Linear assignment problems.
MATS 1983-1 Matsui, K.; Takada, J.
SLP method for a nonlinear multicommodity flow assignment problem.
MAZZB 1986-1 Mazzola, J. B.; Neebe, A. W.
Resource-constrained assignment scheduling.
MINOC 1985-1 Minoux, M.; Ribeiro, C. C.
Une nouvelle approche du probleme d'affectation quadratique MIN-MAX.
MONA 1983-1 Monakhov, O. G.
One method of decentralized job allocation in computer systems with a programmable structure.
NAKA 1985-1 Nakai, T.
Optimal assignment for a random sequence with an unknown number of jobs.
NUNE 1983-1 Nunen, J. A. E. E. van; Beulens, A. J. M.; Benders, J. F.
On using the linear programming relaxation of assignment type mixed integer problems.
NUNE 1984-1 Nunen, J. A. E. E. van; Beulens, A. J. M.; Benders, J. F.
On solving assignment type mixed integer linear programming problems within decision support systems.
OLIV 1983-1 Olivo, P.
A mixed algorithm for the multidimensional assignment problem (Italian).
OZAW 1987-1 Ozawa, T.; Kajitani, Y.; Ueno, S.
The principal partition of vertex-weighted graphs and its applications.
PAIB 1986-1 Paiboon, C.-P.; Pojana, P.-P.; Rosenthal, R. E.
Optimal ship routing and personnel assignment for naval recruitment in Thailand.
PHIL 1987-1 Phillips, N. V.
A weigthing function for pre-emptive multicriteria assignment problems.
PICO 1984-1 Picone, C. J.; Wilhelm, W. E.
A perturbation scheme to improve Hillier's solution to the facilities layout problem.
PICO 1986-1 Picone, C. J.; Wilhelm, W. E.
"A perturbation scheme to improve Hillier's solution to the facilities layout problem": a clarification.
QUEY 1986-1 Queyranne, M.
Performance ratio of polynomial heuristics for triangle inequality quadratic assignment problems.
QUIN 1985-1 Quinn, M. J.
A note on two parallel algorithms to solve the stable marriage problem.
RAMAA 1984-1 Ramanan, P.; Deogun, J. S.; Liu, C. L.
A personnel assignment problem.
REEV 1985-1 Reeves, C. R.
An improved heuristic for the quadratic assignment problem.
REND 1985-1 Rendl, F.
Ranking scalar products to improve bounds for the quadratic assignment problem.
REND 1986-2 Rendl, F.
Quadratic assignment problems on series-parallel digraphs.
ROCK 1984-1 Rockafellar, R. T.
Network flows and monotropic optimization.
ROUC 1987-1 Roucairol, C.
A parallel branch and bound algorithm for the quadratic assignment problem.
RUBIA 1986-1 Rubinshtein, M. I.
Algorithm for the minimax assignment problem with a sparse rectangular matrix.
RUBL 1983-1 Rublev, V. S.
A generalized assignment problem and the scheduling of a transportation route (Russian).
RUSC 1981-1 Rüschendorf, L.
On the multidimensional assignment problem.
SAAT 1987-1 Saatcioglu, Ö.
A multi-attribute assignment goal-programming model with incentives.
SAKA 1984-1 Sakaguchi, M.
A sequential stochastic assignment problem associated with a non-homogeneous Markov process.

SAKA 1984-2 Sakaguchi, M.
A sequential stochastic assignment problem with an unknown number of jobs.
SERG 1987-1 Sergeev, S. I.
A new lower bound for the quadratic assignment problem (Russian).
SHERC 1984-1 Sherali, H. D.; Baines, S. G.
Advanced basis or block pivoting methods for network structured linear programs.
SHERF 1986-1 Sherali, H. D.; Rajgopal, P.
A flexible, polynomial-time, construction and improvement heuristic for the quadratic assignment problem.
SIME 1986-1 Simeone, B.
An asymptotically exact polynomial algorithm for equipartition problems.
SLOM 1986-1 Slominski, L.
On existence of assignments in zero-one matrices.
STOI 1984-1 Stoica, M.; Angheluta, G.
An application of the problem of three-dimensional allocation.
TALAA 1979-1 Talanov, V. A.; Shevchenko, V. N.
A generalization of the assignment problem (Russian).
TENGA 1985-1 Teng, C.-L.
The stepping-stone method for solving assignment problems (Chinese).
THOMB 1986-1 Thompson, G. L.; Kaku, B. K.
An exact algorithm for the general quadratic assignment problem.
TOWN 1986-1 Townsend, W.
An application of the assignment model to bus crew rostering.
TSENA 1984-1 Tseng, S. S.; Lee, R. C. T.
A parallel algorithm to solve the stable marriage problem.
WALA 1984-1 Walas, R. A.; Askin, R. G.
An algorithm for NC turret punch press tool location and hit sequencing.
WALT 1984-1 Walther, H.
Ten applications of graph theory.
WILH 1985-1 Wilhelm, M. R.; Evans, G. W.
The quadratic assignment problem: applications and solution techniques.
WILHA 1987-1 Wilhelm, M. R.; Ward, T. L.
Solving quadratic assignment problems by "simulated annealing".
XU H 1987-1 Xu, W.-X.
A new bounding technique for the quadratic assignment problem (Chinese).
ZHUR 1984-1 Zhuravlev, Y. I.; Platonenko, I. M.
On an economical product of Boolean equations (Russian).
ZYKO 1982-1 Zykov, A. A.
The generalized assignment problem (Russian).

CHANL 1984-1 Chandrasekaran, R.; Shirali, S.
Total weak unimodularity: testing and applications.
MARCF 1985-1 Marcotte, O.
The cutting stock problem and integer rounding.
SCHRE 1987-2 Schrijver, A.
Polyhedral combinatorics.

AGEE 1983-1 Ageev, A. A.
On the complexity of the minimization problems for polynomials of Boolean functions (Russian).
ANTA 1983-1 Antamoshkin, A.
On the optimal algorithms for the optimization of functionals with Boolean values (Russian).
ANTAA 1984-1 Antamoshkin, A. N.; Zajtseva, T. A.; Korsakova, O. N.
Formalization of optimization problems in planning agricultural-industrial combines (English).
BAUM 1983-1 Bauman, E. V.; Muchnik, I. B.
Simplification of graph structure with the aid of Boolean matrices.
BENZ 1983-1 Benzaken, C.; Boyd, S. C.; Hammer, P. L.; Simeone, B.
Adjoints of pure bidirected graphs.
BENZB 1985-1 Benzaken, C.; Hammer, P. L.
Boolean techniques for matroidal decomposition of independence systems and applications to graphs.
BEREA 1984-1 Bereznev, V. A.; Palmova, I. E.; Tretyakov, A. A.
A method for solving a class of problems of linear integer programming with Boolean variables (Russian).
BILLD 1985-1 Billionnet, A.; Minoux, M.
Maximizing a supermodular pseudoboolean function: a polynomial algorithm for supermodular cubic functions.
BLAS 1982-1 Blass, A.; Gurevich, Y.
On the unique satisfiability problem.
BORC 1987-2 Borchardt, M.
Zur Lösung von nichtlinearen Booleschen Optimierungsproblemen.
BOUR 1986-1 Bourjolly, J.-M.
Integral and fractional node packings, and pseudo-Boolean programming.
CRAMB 1986-1 Crama, Y.; Hammer, P. L.; Jaumard, B.; Simeone, B.
Parametric representation of the solutions of a quadratic Boolean equation.
EBEN 1984-1 Ebenegger, C.; Hammer, P. L.; Werra, D. de
Pseudo-boolean functions and stability of graphs.
FRAE 1985-1 Fraenkel, A. S.; Hammer, P. L.
Pseudo-Boolean functions and their graphs.
FRANB 1986-1 Franco, J.
On the probabilistic performance of algorithms for the satisfiability problem.
FULO 1987-1 Fülöp, J.
On the facial disjunctive programming problem (Hungarian).
GOLDM 1984-1 Goldengorin, B. I.
A correcting algorithm for solving allocation type problems.
HAMME 1984-1 Hammer, P. L.; Simeone, B.; Liebling, T. M.; Werra, D. de
From linear separability to unimodality: a hierarchy of pseudo-Boolean functions.
HANSJ 1986-1 Hansen, P.; Simeone, B.
Unimodular functions.
HULM 1984-1 Hulme, B. L.; Shiver, A. W.; Slater, P. J.
A boolean algebraic analysis of fire protection.
KABU 1986-1 Kabulov, A. V.; Bajzhumanov, A. A.
Local methods for solving systems of Boolean equations (Russian).
KHACA 1986-1 Khachaturov, V. R.
Models and methods for solving multiextremal allocation problems by the use of properties of supermodular functions defined on Boolean lattices (Russian).
KIM A 1982-1 Kim, K. H.
Boolean matrix theory and applications.
KORN 1985-1 Körner, F.
Ein effektiver Branch and Bound-Algorithmus für Boolesche quadratische Optimierungsprobleme.
LEVI 1983-1 Levin, A. A.
Gluing of Boolean functions and its application in estimates of the comparative complexity of DNF (Russian).
LIU H 1987-1 Liu, Y.-P.
Boolean planarity characterization of graphs.
LUBI 1985-1 Lubiw, A.
Orderings and some combinatorial optimization problems with geometric applications.
MAME 1986-1 Mamedov, K. S.
On a method for calculating estimates in Boolean programming problems (Russian).
MIKH 1986-1 Mikhailyuk, V. A.
Certain aspects of postoptimal analysis of the minimization problem for the Boolean functions (Russian).
MINK 1980-1 Minker, J.; Minker, R. G.
Optimization of Boolean expressions - historical developments.
MOHR 1986-1 Möhring, R. H.
Algorithmic aspects of the substitution decomposition in optimization over relations, set systems and Boolean functions.
MOHRA 1984-1 Möhring, R. H.; Radermacher, F. J.
Substitution decomposition for discrete structures and connections with combinatorial optimization.
PELEA 1985-1 Peled, U. N.; Simeone, B.
Polynominal-time algorithms for regular set-covering and threshold synthesis.
ROBEA 1985-1 Robert, Y.; Tchuente, M.
Connection-graph and iteration-graph of monotone boolean functions.
ROSEB 1956-1 Rosenblatt, D.
On aggregation and consolidation in linear systems.
SANTA 1984-1 Santoro, N.; Urrutia, J.
An improved algorithm for Boolean matrix multiplication.
SEME 1985-1 Semenkin, E.
Optimization of monotone pseudo-Boolean functions (Russian).
SHEVA 1985-1 Shevchenko, V. N.
On some functions of many-valued logic connected with integer programming (Russian).

TOPC 1985-1 Topchishvili, A. L.
On the problem of choosing a maximum solvable system from a system of linear pseudo-Boolean equations (Russian).
WEGE 1985-1 Wegener, I.
The critical complexity of all (monotone) Boolean functions and monotone graph properties.
WEGE 1985-2 Wegener, I.
The critical complexity of all (monotone) Boolean functions and monotone graph properties.
WILLA 1985-1 Williams, H. P.
Model building in linear and integer programming.
WILSA 1984-1 Wilson, J. M.
A 2-phase method for the minimisation of Boolean polynomials.
ZHUR 1984-1 Zhuravlev, Y. I.; Platonenko, I. M.
On an economical product of Boolean equations (Russian).

ADAMB 1986-1 Adams, J.; Balas, E.
The shifting bottleneck procedure for job shop scheduling.

AGGAC 1986-1 Aggarwal, V.; Tikekar, V. G.; Hsu, L.-F.
Bottleneck assignment problems under categorization.

AKUL 1983-1 Akulich, F. E.
A problem on maximal flow on bottlenecks (Russian).

BILLB 1986-1 Billington, P. J.; Mac Clain, J. O.; Thomas, L. J.
Heuristics for multilevel lot-sizing with a bottleneck.

BOKH 1987-1 Bokhari, S. H.
Assigment Problems in Parallel and Distributed Computing.

BURKB 1985-1 Burkard, R. E.; Fincke, U.
Probabilistic asymptotic properties of some combinatorial optimization problems.

BURKD 1984-1 Burkard, R. E.; Krarup, J.; Pruzan, P. M.
Some relationships between multicriteria and parametric discrete optimization problems with bottleneck criteria.

CHANJ 1985-1 Chandrasekaran, R.
Optimal improvements in bottleneck systems.

DADU 1987-1 Daduna, J. R.
An efficient algorithm for solving bottleneck transportation problems with euclidean distances.

DELL 1986-1 Dellaert, N.; Wessels, J.
Production scheduling with uncertain demand and no stockkeeping.

DERI 1984-1 Derigs, U.
Alternate strategies for solving bottleneck assignment problems - analysis and computational results.

EISEC 1984-1 Eiselt, H. A.; Gerchak, Y.
Solution structures and sensitivity of special assignment problems.

HOCHB 1986-3 Hochbaum, D. S.; Shmoys, D. B.
A unified approach to approximation algorithms for bottleneck problems.

ISER 1984-1 Isermann, H.
Linear bottleneck transportation problems.

MALHA 1985-1 Malhotra, R.; Bhatia, H. L.; Puri, M. C.
The three dimensional bottleneck assignment problem and its variants.

MINOC 1985-1 Minoux, M.; Ribeiro, C. C.
Une nouvelle approche du probleme d'affectation quadratique MIN-MAX.

PETK 1986-2 Petkov, M. G.
Transportation flow passing through a bottleneck (Bulgarian).

SCHU 1986-1 Schubert, I. S.; Zimmermann, U.
One-parametric bottleneck transportation problems.

SERAB 1984-1 Serazzi, G.
A bottleneck-driven scheduler.

STAN 1985-1 Stancu-Minasian, I. M.; Tigan, S.
The minimum-risk approach to the bottleneck transportation problem.

STANE 1985-1 Stanley, J. D.
Bottleneck structures and planning horizons.

TIGAA 1985-1 Tigan, S.; Stancu-Minasian, I. M.
The stochastic bottleneck transportation problem.

VANNA 1986-1 Vannelli, A.; Kumar, K. R.
A method for finding minimal bottle-neck cells for grouping part-machine families.

ZHANA 1987-1 Zhang, B. K.
On a class of integer bottleneck problems (Chinese).

ZIMMC 1986-3 Zimmermann, U.
Sharing problems.

AFON	1986-1	Afonin, Y. S. Blocked branch-and-bound method.
AGRAA	1984-1	Agrawal, S. C.; Chand, M. On integer solution to interval linear programming problems by a branch and bound technique.
AGRAA	1985-1	Agrawal, S. C.; Chand, M. An alternative method on integer solutions to complementary programming problems with linear fractional objective function by a branch and bound technique.
ALEKC	1985-1	Alekseev, A. O.; Alekseev, O. G.; Anisimov, V. G.; Anisimov, E. G. The use of duality to increase the effectiveness of the branch and bound method when solving the knapsack problem.
ALEKD	1987-1	Alekseev, O. G. Complex application of discrete optimization methods (Russian).
ALI A	1986-1	Ali, A. I.; Kennington, J. L. The asymmetric m-travelling salesmen problem: a duality based branch-and-bound algorithm.
ARON	1986-1	Aronson, J. E. The multiperiod assignment problem: a multicommodity network flow model and specialized branch and bound algorithm.
BABIB	1987-1	Babic, O. Optimization of refuelling truck fleets at an airport.
BAKEF	1981-1	Baker, T. E. A branch and bound network algorithm for interactive process scheduling.
BALAO	1985-1	Balas, E.; Toth, P. Branch and bound methods.
BARTB	1987-1	Bartnick, J. Aggregation of preference relations by a branch and bound algorithm.
BATY	1983-1	Batycki, T.; Kasprzak, A. Flow assignment algorithms for computer communication networks (Polish).
BEAL	1985-1	Beale, E. M. L. Integer Programming.
BERR	1986-1	Berrada, M.; Stecke, K. E. A branch and bound approach for machine load balancing in flexible manufacturing systems.
BOEN	1986-1	Boenchendorf, K. Combining penalties for the fixed-charge transportation problem.
BONDA	1981-1	Bondarenko, V. A.; Kuksa, A. I. A solution method for the problem of optimally planning independent jobs in time under constraints on stored and unstored resources (Russian).
BORC	1985-2	Borchardt, M. Straf- und Multiplikatormethoden für Optimierungsprobleme mit diskreten und stetigen Variablen. I.
BORC	1986-1	Borchardt, M. Penalty and multiplier methods for optimization problems with discrete and continuous variables II (German).
CABO	1986-1	Cabot, A. V.; Erenguc, S. S. A branch and bound algorithm for solving a class of nonlinear integer programming problems.
CARPA	1984-1	Carpaneto, G.; Gallo, G. Enumerative methods for solving combinatorial optimization problems (Italian).
CARRC	1983-1	Carrera, J.; Szidarovszky, F. A multiobjective branch and bound algorithm for optimal observation network design.
CHENM	1985-1	Cheng, D.-Q. The solution to the Chinese postman problem by branch and bound (Chinese).
CHRIA	1987-1	Christofides, N.; Alvarez-Valdes, R.; Tamarit, J. M. Project scheduling with resource contraints: a branch and bound algorithm.
CRAMC	1986-1	Crama, Y.; Hansen, P.; Williams, A. C. Use of the dual in branch and bound methods for mixed 0-1 programming.
DIEH	1985-1	Diehr, G. Evaluation of a branch and bound algorithm for clustering.
DYERC	1986-1	Dyer, M. E.; Frieze, A. M.; Mac Diarmid, C. On linear programs with random costs.
DYERD	1984-1	Dyer, M. E.; Kayal, N.; Walker, J. A branch and bound algorithm for solving the multiple-choice knapsack problem.
EMELB	1985-2	Emelichev, V. A.; Ovchinnikov, V. G. Strafmethoden in der diskreten Optimierung.
EREN	1987-1	Erenguc, S. S.; Tufekci, S. A branch and bound algorithm for a single-item multi-source dynamic lot sizing problem with capacity constraints.
ERFL	1986-1	Erfle, W. Ein Branch-and-Bound-Verfahren zur optimalen Allokation von hierarchisch geordneten Prüfstellen in den Fertigungsablauf.
FINKC	1987-1	Finke, G. Network flow based branch and bound method for asymmetric traveling salesman problems.
FISCA	1987-1	Fischetti, M.; Toth, P. Bounding and dominating procedure for combinatorial optimization problems (Italian).
FISHD	1986-1	Fisher, M. L. A dual algorithm for large sale set partitioning.
FISHG	1986-1	Fisher, M. L.; Jaikumar, R.; Wassenhove, L. N. van A multiplier adjustment method for the generalized assignment problem.
FOUL	1985-3	Foulds, L. R. Traffic network arc elimination by branch and bound enumeration.
FOUL	1986-1	Foulds, L. R. Braess's paradox: which arcs should be eliminated in order to minimize congestion?
GAMBA	1986-1	Gamble, A. B.; Conn, A. R.; Pulleyblank, W. R. A network penalty method.
GAVI	1985-3	Gavish, B.; Pirkul, H. Zero-one integer programs with few constraints - efficient branch and bound algorithms.

GOMM	1984-1	Gommers, J. H. M.; Volgenant, A. A Langrangean appraoch to the degree-constrained minimum spanning tree problem.
GULY	1983-1	Gulyanitskii, L. F.; Khodzinskii, A. N. Two algorithms for solution of optimization problems on permutations (Russian).
GUPTE	1985-1	Gupta, O. K.; Ravindran, A. Branch and bound experiments in convex nonlinear integer programming.
HAN	1982-1	Han, F. An integer programming algorithm for the plan of a freight car classification system (Chinese).
HARN	1984-1	Harnett, R. M. A single-branch implicit enumeration algorithm for zero-one programs with geometrical constraints.
HARTF	1985-1	Hartwig, A. Recursive branch and bound.
IBAR	1987-1	Ibaraki, T. Enumerative Approaches to Combinatorial Optimization - Part 1.
IBAR	1987-2	Ibaraki, T. Enumerative Approaches to Combinatorial Optimization - Part 2.
JERO	1984-1	Jeromin, B.; Körner, F. A hybrid method for solving integer linear programming problems.
JERO	1986-1	Jeromin, B.; Körner, F. A hybrid method for solving integer linear programming problems.
JEROA	1985-4	Jeroslow, R. G. Computation-oriented reductions of predicate to propositional logic.
KALAB	1987-1	Kalantari, B.; Rosen, J. B. Penalty formulation for zero-one nonlinear programmming.
KALR	1982-1	Kalra, K. R.; Bagga, P. C. Single machine sequencing problem with polynomial production cost functions: modifications in the branch-and-bound technique of Townsend.
KARPE	1986-1	Karp, R. M.; Saks, M.; Wigderson, A. On a search problem related to branch-and-bound procedures.
KARW	1984-1	Karwan, M. H.; Rardin, R. L.; Sarin, S. A new surrogate dual multiplier search procedure.
KARW	1985-1	Karwan, M. H.; Rardin, R. L.; Sarin, S. Surrogate duality in a branch-and-bound procedure for integer programming.
KATAB	1986-1	Katayama, N.; Kasugai, H. An algorithm for the discrete road network design problem using branch and bound method (Japanese).
KHANB	1984-1	Khan, Z. A.; Ahsan, M. J. On integer solutions to the concave quadratic programs using branch and bound technique.
KHANC	1986-1	Khan, Z. A.; Ahsan, M. J.; Khan, E. A. A branch and bound method for integer solutions to the concave quadratic programs.
KHRI	1985-1	Khristov, G.; Karamiteva, Z.; Stoyanov, T. Numerical method and program for the solution of a linear fractional integer optimization problem (Bulgarian).
KORN	1985-1	Körner, F. Ein effektiver Branch and Bound-Algorithmus für Boolesche quadratische Optimierungsprobleme.
KORN	1986-1	Körner, F. A new bound for the quadratic knap sack problem and its use in a branch and bound algorithm.
KOVAH	1980-1	Kovalev, S. M. On the determination of minimal coverings in hypergraphs (Russian).
KUBAA	1985-1	Kubale, M.; Jackowski, B. A generalized implicit enumeration algorithm for graph coloring.
LAI B	1985-2	Lai, T.-H.; Sprague, A. Performance of parallel branch-and-bound algorithms.
LAI B	1986-1	Lai, T.-H.; Sprague, A. A note on anomalies in parallel branch-and-bound algorithms with one-to-one bounding functions.
LAPOH	1987-1	Laporte, G.; Nobert, Y.; Taillefer, S. A branch-and-bound algorithm for the asymmetrical distance-constrained vehicle routing problem.
LUH	1985-1	Luh, J. Y. S.; Lin, C. S. Scheduling parallel operations in automation for minimum execution time based on PERT.
MARCG	1986-1	Marcotte, O.; Soland, R. M. An interactive branch-and-bound algorithm for multiple criteria optimization.
MARE	1985-3	Marecki, F. Multiobjective resources allocation of assembly line (Polish).
MARTH	1985-1	Martin, R. K.; Sweeney, D. J.; Doherty, M. E. The reduced cost branch and bound algorithm for mixed integer programming.
MARTI	1985-1	Martinec, I. Methods for a class of discrete location problems.
MINO	1986-1	Minoux, M. Mathematical Programming. Theory and Algorithms.
MIZU	1984-1	Mizukami, K.; Sikorski, J. Three algorithms for calculating surrogate constraints in integer programming problems.
MULL	1983-1	Müller-Merbach, H. Head-and-shoulder-branching in integer programming.
NEEB	1986-1	Neebe, A. W.; Rao, M. R. Sequencing capacity expansion projects in continuous time.
NGUY	1983-1	Nguyen Van Thieu On the problem of production location in an interbranch formulation (Vietnamese).
OGRY	1984-1	Ogryczak, W. Branch-and-bound procedures for multicriteria integer programming (Polish).
OLEY	1983-1	Oley, L. A.; Sjoquist, R. J. Automatic reformulation of mixed and pure integer models to reduce solution time in APEX IV.
OLIV	1983-1	Olivo, P. A mixed algorithm for the multidimensional assignment problem (Italian).

PALE 1987-1 Palekar, U. S.; Karwan, M. H.; Zionts, S.
A branch-and-bound method for the fixed charge transportation problem.

PELI 1984-1 Pelikan, J.
Container transportation problem (Czechoslovakian).

PELI 1985-1 Pelikan, J.
Container transportation problem.

POPEA 1983-1 Popescu, M. F.
Efficiency of implicit enumeration algorithms and rejection power of the tests.

POTT 1985-1 Potts, C. N.
A Lagrangean based branch and bound algorithm for single machine sequencing with precedence constraints to minimize total weighted completion time.

POTTA 1985-1 Potts, C. N.; Wassenhove, L. N. van
A branch and bound algorithm for the total weighted tardiness problem.

QUINA 1986-1 Quinn, M. J.; Deo, N.
An upper bound for the speedup of parallel best-bound branch-and-bound algorithms.

RADA 1986-1 Radaramanan, R.; Choi, L. I.
A branch and bound algorithm for the travelling salesman and the transportation routing problems.

RAME 1986-1 Ramesh, R.; Karwan, M. H.; Zionts, S.
A class of practical interactive branch and bound algorithms for multicriteria integer programming.

REIN 1985-1 Reinelt, G.
The linear ordering problem: algorithms and application.

RIBEA 1986-1 Ribeiro, C. C.; Minoux, M.
Solving hard constrained shortest path problems by Lagrangean relaxation and branch-and-bound algorithms.

RO 1984-1 Ro, H.-B.; Tcha, D.-W.
A branch and bound algorithm for the two-level uncapacitated facility location problem with some side constraints.

ROUC 1987-1 Roucairol, C.
A parallel branch and bound algorithm for the quadratic assignment problem.

SAKAA 1984-1 Sakarovitch, M.
Optimisation combinatoire - methodes mathematiques et algorithmiques. Vol. 1: Programmation discrete.

SALT 1987-1 Saltzman, M. J.; Baybars, I.
A two-process implicit enumeration algorithm for the simple assembly line balancing problem.

SAXE 1985-1 Saxena, P. K.
Shipping cost/completion-date tradeoffs in fractional transportation problems with penalty cost.

SCHR 1985-1 Schrage, L.; Wolsey, L. A.
Sensitivity analysis for branch and bound integer programming.

SEN A 1985-1 Sen, S.; Sherali, H. D.
A branch and bound algorithm for extreme point mathematical programming problems.

SERA 1984-1 Serafini, P.
Dual relaxation and branch-and-bound techniques for multiobjective optimization.

SERGG 1980-1 Sergienko, I. V.; Lebedeva, T. T.; Roshchin, V. A.
Approximate methods for solving discrete problems of optimization.

SHERD 1986-1 Sherali, H. D.; Myers, D. C.
The design of branch and bound algorithms for a class of nonlinear integer programs.

SHERH 1987-1 Sherali, H. D.; Staschus, K; Huacuz, J. M.
An integer programming approach and implementation for an electric utility capacity planning problem with renewable energy sources.

SIED 1982-1 Siedersleben, J.
Optimierung von deterministischen und stochastischen Entscheidungsnetz- plänen (Dissertation).

SIGA 1986-1 Sigal, I. K.
Computer implementation of a combined branch and bound algorithm for the travelling salesman problem.

SINC 1986-1 Sinclair, M.
An exact penalty function approach for nonlinear integer programming problems.

SKOR 1987-1 Skorin-Kapov, J.; Granot, F.
Non-linear integer programming: sensitivity analysis for branch and bound.

SKUB 1984-1 Skubalska, E.
Application of the branch-and-bound method to the optimal vehicle routing (Czechoslovakian).

TCHAB 1984-1 Tcha, D.-W.; Lee, B.-I.
A branch-and-bound algorithm for the multi-level uncapacitated facility location problem.

TRIE 1988-1 Trienekens, H. W. J. M.
A simulation tool for the performance evaluation of parallel branch and bound algorithms.

TRIEA 1984-1 Trietsch, D.
An improved algorithm for Steiner trees.

VASYA 1982-1 Vasyanin, V. A.; Savenkov, A. I.
A suboptimal algorithm for solving a transportation routing problem.

VENIA 1984-1 Veniaminov, S. S.
Additional cutoff rule for the branch-and-bound method.

VENIA 1986-1 Veniaminov, S. S.
The effectiveness of the hybrid method for solving discrete optimization problems.

WANGE 1984-1 Wang, J. S.
A branch and bound algorithm for a logic partitioning problem (Chinese).

WETZ 1985-1 Wetzel, R.
Ein Branch and Bound Algorithmus zur Lösung eines Problems der Fahrplangestaltung.

WETZ 1986-1 Wetzel, R.
A branch and bound concept for solving a vehicle schedule designing problem.

WILSA 1985-1 Wilson, J. M.
A method of enumerating all integer extreme solutions of an integer programming problem.

WOLT 1987-1 Wolter, H.
Lösungsverfahren für eine Klasse diskreter Optimierungsaufgaben spezieller Struktur.

YANA 1987-1 Yanasse, H. H.; Soma, N. Y.
A new enumeration scheme for the knapsack problem.

YANGA 1985-1 Yang, C. E.; Lin, Y. X.
Board welding problem (Chinese).

AHARB 1985-1 Aharoni, R.; Hartman, I. B.-A.; Hoffman, A. J.
Path partitions and packs of acyclic digraphs.
ALBEA 1987-2 Albertson, M. O.
Generalized colorings.
ALBEB 1983-1 Albertson, M. O.; Berman, D. M.
Cliques, colorings and locally perfect graphs.
ALON 1987-1 Alon, N.
Monochromatic directed walks in arc-colored directed graphs.
ALONA 1985-1 Alon, N.; Egawa, Y.
Even edge colorings of a graph.
ANDRF 1985-1 Andrews, J. A.; Jacobson, M. S.
On a generalization of chromatic number.
ANTOA 1983-1 Antonucci, S.
Simple and generalized colourings of cliques without Hamiltonian cycles and of regular graphs (Italian).
ANTOA 1984-1 Antonucci, S.
Multicolorings of regular graphs (French).
ANTOA 1985-1 Antonucci, S.
The graphs T(m) and L(m) and their chromatic numbers (Italian).
ANTOA 1986-1 Antonucci, S.
Observations on the generalized characteristics and the chromatic number of a graph (Italian).
ARCH 1983-1 Archdeacon, D.
Coupled colorings of planar maps.
ASPV 1984-1 Aspvall, B.; Gilbert, J. R.
Graph coloring using eigenvalue decomposition.
ASRA 1981-1 Asratyan, A. S.
Some sufficient conditions for realizability of a collection of integers in a bipartite multigraph (Russian).
BAGAA 1983-1 Bagaev, G. N.; Dmitriev, E. F.
Enumeration of connected two-color graphs (Russian).
BALAG 1985-1 Balas, E.
An $O(|V| + |E|)$ algorithm for finding an edge-maximal subgraph with a tr-formative coloring.
BALAG 1986-1 Balas, E.
A fast algorithm for finding an edge-maximal subgraph with TR-formative coloring.
BAUEB 1985-1 Bauernöppel, F.; Jung, H.
Fast parallel vertex colouring.
BAUMA 1985-1 Baumann, U.
On fixed points of partition preserving automorphisms of complete graphs with minimal colourings.
BAUMB 1986-1 Baumann, U.; Lesch, M.
Klassen von resistenten Kantenfärbungen.
BAUMB 1986-2 Baumann, U.; Lesch, M.
Minimal colourings of complete and regular graphs and partition preserving automorphisms.
BEIK 1983-1 Beikirch, W.
Kantenfärbungen paarer Graphen, Matrizen und Quasigruppen.
BEIK 1983-2 Beikirch, W.
Über spezielle Kantenfärbungen und einen Zusammenhang zwischen Kantenfärbungen, Matrizen und Quasigruppen.
BEIK 1984-1 Beikirch, W.
Equivalence of edge colourings of complete graphs and quasigroups (German).
BEIK 1985-1 Beikirch, W.
Edge colourings of finite simple and undirected graphs and quasigroups (German).
BELYA 1984-1 Belyi, S. B.
Minimal distributions of monochromatic triangles in complete bichromatic graphs (Russian).
BENDA 1985-1 Bender, E. A.; Wilf, H. S.
A theoretical analysis of backtracking in the graph coloring problem.
BERG 1983-1 Berge, C.
Graphes (3. edition).
BERG 1985-1 Berge, C.
Graphs (second revised edition).
BERG 1987-1 Berge, C.
A minimax relation for the partial q-colorings of a graph.
BERG 1987-2 Berge, C.
A new color-change to improve the coloring of a graph.
BERMB 1986-1 Berman, K. A.; Paul, J. L.
Large monochromatic components in vertex colored k-trees.
BERMB 1987-1 Berman, K. A.; Paul, J. L.
Long monochromatic paths in vertex 3-colored planar graphs.
BERNB 1987-1 Bernardi, C.
On a theorem about vertex colorings of graphs.
BEUT 1984-1 Beutelspacher, A.; Hering, P.-R.
Minimal graphs for which the chromatic number equals the maximal degree.
BIER 1987-1 Bierbrauer, J.; Gyarfas, A.
On (n,k)-colorings of complete graphs.
BOLLD 1985-1 Bollobas, B.; Harris, A. J.
List-colourings of graphs.
BOROB 1984-1 Borodin, O. V.
Solution of the Ringel problem on vertex-edge coloring of plane graphs and coloring of 1-planar graphs (Russian).
BOROB 1987-1 Borodin, O. V.
Simultaneous colorings of graphs on the plane (Russian).
BOROF 1984-1 Borowiecki, M.
Point partition numbers and generalized Nordhaus-Gaddum problems.

BOROF	1987-3	Borowiecki, M. Two extremal problems in the class of uniquely colorable graphs.
BOSA	1983-1	Bosak, J. Enumeration of uniquely edge colourable multigraphs.
BOSA	1984-2	Bosak, J. Uniquely edge colourable graphs.
BOUCB	1987-1	Bouchet, A.; Fouquet, J. L.; Jolivet, J. L.; Riviere, M. On a special face colouring of cubic graphs.
BOYA	1987-1	Boyar, J. F.; Karloff, H. J. Coloring planar graphs in parallel.
BRAC	1984-1	Bracho, R. L. Le nombre achromatique d'une etoile.
BROE	1986-1	Broere, I.; Burger, M. Uniquely cocolourable graphs.
BROEA	1985-1	Broere, I.; Frick, M. On the order of color critical graphs.
BROEB	1985-1	Broere, I.; Mynhardt, C. M. Generalized colorings of outerplanar and planar graphs.
BURN	1984-2	Burns, D. The n-minimal chromatic multiplicity of a graph.
CAI	1987-1	Cai, J.; Meyer, G. E. On the complexity of graph critical uncolorability.
CAMEG	1986-1	Cameron, K. On k-optimum dipath partitions and partial k-colourings of acyclic digraphs.
CAMPC	1987-1	Campers, G.; Henkes, O.; Leclercq, J.-P. Sur les methodes exactes de coloration de graphes.
CARO	1986-1	Caro, Y. Chromatic decomposition of a graph.
CERA	1980-1	Cerasoli, M.; Eugeni, F. On the chromatic polynomial of a graph (Italian).
CHAM	1987-1	Chams, M.; Hertz, A.; Werra, D. de Some experiments with simulated annealing for coloring graphs.
CHAO	1984-1	Chao, C.-Y.; Zhao, L.-C. Chromatic polynomials of connected graphs.
CHEB	1986-1	Cheban, Y. I. Investigation of a generalized graph coloring problem (Russian).
CHET	1984-1	Chetwynd, A. G.; Hilton, A. J. W. The chromatic index of graphs of even order with many edges.
CHET	1984-2	Chetwynd, A. G.; Hilton, A. J. W. The chromatic index of graphs with a most four vertices of maximum degree.
CHET	1987-1	Chetwynd, A. G.; Hilton, A. J. W. The edge-chromatic class of regular graphs of degree 4 and their complements.
CHVAC	1986-1	Chvatal, V.; Lenhart, W. J.; Sbihi, N. 2-colourings that decompose perfect graphs.
CLEV	1983-1	Cleves, E. M.; Jacobson, M. S. On mixed Ramsey numbers: total chromatic number versus graphs.
COLEB	1986-1	Coleman, T. F.; Cai, J.-Y. The cyclic coloring problem and estimation of sparse Hessian matrices.
COME	1983-1	Comer, S. D. Color schemes forbidding monochrome triangles.
COOK	1984-1	Cook, R. J. Graph factorization and theorems of the Nordhaus-Gaddum class.
CORI	1986-1	Cori, R.; Dulucq, S. Colourings of planar maps and the equality of two languages.
COZZA	1982-1	Cozzens, M. B.; Roberts, F. S. T-colorings of graphs and the channel assigment problem.
COZZB	1984-2	Cozzens, M. B.; Wang, D. I. The general channel assignment problem.
DABI	1987-1	Dabija, V. An algorithm for coloring additive puzzles.
DECA	1985-1	Decaestecker, C. On stable colourings of exclusive boolean graphs (French).
DEMAA	1980-1	DeMaria, F.; Gionfriddo, M. A theorem on the s-chromatic number of a graph (Italian).
DMIT	1982-1	Dmitriev, I. G. Minimal degree of uniquely colored graphs with minimal number of edges (Russian).
DOMK	1986-1	Domke, G. S.; Laskar, R.; Hedetniemi, S. T.; Peters, K. The partite-chromatic number of a graph.
EDWA	1986-1	Edwards, K. The complexity of colouring problems on dense graphs.
EGGA	1986-1	Eggan, L.; Plantholt, M. The chromatic index of nearly bipartite multigraphs.
EHRE	1984-1	Ehrenfeucht, A.; Faber, V.; Kierstead, H. A. A new method of proving theorems on chromatic index.
EISE	1985-1	Eiselt, F. Unabhängige bunte Mengen in k-chromatischen Graphen der Radien 4 und 5.
EL-ZA	1985-1	El-Zahar, M.; Sauer, N. W. The chromatic number of the product of two 4-chromatic graphs is 4.
ELLI	1984-1	Ellingham, M. N. Peterson subdivisions in some regular graphs.
ELLIA	1986-1	Ellingham, M. N.; Caro, Y. Uniqueness of impartial edge-colorings.

ERDO 1984-1 Erdös, P.
Some new and old problems on chromatic graphs.
ERDOA 1986-1 Erdös, P.; Füredi, Z.; Hajnal, A.; Komjath, P.; Rödl, V.; Seress, A.
Coloring graphs with locally few colors.
ERDOB 1985-1 Erdös, P.; Hajnal, A.
Chromatic number of finite and infinite graphs and hypergraphs.
FARB 1986-1 Farber, M.; Hahn, G.; Hell, P.; Miller, D.
Concerning the achromatic number of graphs.
FENN 1984-2 Fenner, T. I.; Frieze, A. M.
On the complexity of finding sets of edges in graphs and digraphs.
FIAM 1983-1 Fiamcik, J.
Acyclic chromatic index of a subdivided graph.
FIAM 1984-1 Fiamcik, J.
Acyclic chromatic index of a subdivided graph.
FIOLA 1983-2 Fiol, M. A.; Yebra, J. L. A.; Fabrega, J.
The vertex-edge duality in the colouring of graphs and hypergraphs.
FOUQ 1983-1 Fouquet, J. L.; Jolivet, J. L.
Strong edge-colorings of graphs and applications to multi k-gons.
FOUQ 1984-1 Fouquet, J. L.; Jolivet, J. L.
Strong edge-coloring of cubic planar graphs.
FRANF 1986-1 Frank, S.; Shier, D. R.
The chromatic polynomial revisited.
FRED 1984-1 Frederickson, G. N.
On linear-time algorithms for five-coloring planar graphs.
FRUC 1985-1 Frucht, R. W.
A new method of computing chromatic polynomials of graphs.
GANS 1987-1 Gansner, E. R.; Vo, K.-P.
The chromatic generating function.
GAZM 1986-1 Gazmuri, P. G.
Probabilistic analysis of algorithms for some combinatorial optimization problems.
GDAN 1982-1 Gdanskij, N. I.
Investigation of some kinds of edge colroings of graphs (Russian).
GERN 1984-1 Gernert, D.
Recent results on chromatic polynomials.
GERN 1985-1 Gernert, D.
A survey of partial proofs for Read's conjecture and some recent results.
GIBB 1985-1 Gibbons, A.
Algorithmic graph theory.
GIMB 1986-1 Gimbel, J.
Three extremal problems in cochromatic theory.
GIMBA 1987-1 Gimbel, J.; Straight, H. J.
Some topics in cochromatic theory.
GION 1987-1 Gionfriddo, M.
A short survey on some generalized colourings of graphs.
GIONA 1985-1 Gionfriddo, M.; Vanstone, S. A.
On L(2)-colourings of a graph.
GIUD 1985-1 Giudici, R. E.
Some new families of chromatically unique graphs.
GIUDA 1986-1 Giudici, R. E.; Lopez, M. A.; Salzberg, P. M.
Chromatic uniqueness for some bipartite graphs (Spanish).
GODSA 1987-1 Godsil, C. D.; Nowakowski, R.; Nesetril, J.
The chromatic connectivity of graphs.
GOLD 1987-1 Goldberg, A. V.; Plotkin, S. A.
Parallel(delta + 1)-coloring of constant-degree graphs.
GOLDB 1984-1 Goldberg, M. K.
An approximate algorithm for the edge-coloring problem.
GOOD 1985-1 Goodman, A. W.
Triangles in a complete chromatic graph with three colors.
GOOD 1985-2 Goodman, A. W.
Triangles in a complete chromatic graph.
GORK 1981-1 Gorkovoi, V. F.
Coloring of graphs (Russian).
GORK 1983-1 Gorkovoi, V. F.
The topology of 3-colored graphs (Russian).
GORK 1984-1 Gorkovoi, V. F.
The minimal coloring of graphs (Russian).
GOULA 1983-1 Gould, R. J.; Jacobson, M. S.
A note on mixed Ramsey numbers: total chromatic number versus complete graphs.
GRACA 1985-1 Gracia, Z.; Salzberg, P. M.
Chromatic classification of K(p)-G(6).
GROSA 1982-1 Grossman, J. W.
On edge-partitioned graphs with degree restrictions.
GUID 1985-1 Guidici, R. E.
Some new families of chromatically unique graphs.
GUO A 1985-2 Guo, Z.-Y.
On complete k-edge colorable subgraphs (Chinese).
GURV 1984-2 Gurvich, V. A.
Some properties and applications of complete edge-chromatic graphs and hypergraphs.
GYAR 1985-1 Gyarfas, A.
On the chromatic number of multiple interval graphs and overlap graphs.
GYARA 1985-1 Gyarfas, A.; Lehel, J.
Covering and coloring problems for relatives of intervals.

GYARA 1986-1 Gyarfas, A.; Lehel, J.
On-line and first fit colorings of graphs.
GYARB 1987-1 Gyarfas, A.; Lehel, J.; Nesetril, J.; Rödl, V.; Schelp, R. H.; Tuza, Z.
Local k-colorings of graphs and hypergraphs.
GYARC 1987-1 Gyarfas, A.; Lehel, J.; Schelp, R. H.; Tuza, Z.
Ramsey numbers for local colorings.
GYUL 1987-1 Gyulumyan, S. M.
Colour-critical graphs with an independent line cut (Russian).
HAGEA 1987-1 Hagerup, T.; Chrobak, M.; Diks, K.
Parallel 5-colouring of planar graphs.
HAKI 1985-1 Hakimi, S. L.
Further results on a generalization of edge-coloring.
HANL 1985-1 Hanlon, P.
The chromatic polynomial of an unlabeled graph.
HANSL 1986-1 Hanson, D.; Robinson, G. C.; Toft, B.
Remarks on the graph colour theorem of Hajos.
HARAB 1985-1 Harary, F.
Conditional colorability in graphs.
HARAE 1985-1 Harary, F.; Jones, K. F.
Conditional colorability II: bipartite variations.
HARAI 1985-1 Harary, F.; Plantholt, M.
The point-distinguishing chromatic index.
HARED 1985-1 Hare, W. R.; Hedetniemi, S. T.; Laskar, R.; Pfaff, J.
Complete coloring parameters of graphs.
HEDEB 1982-1 Hedetniemi, S. M.; Hedetniemi, S.; Beyer, T.
A linear algorithm for the Grundy (coloring) number of a tree.
HEES 1985-1 Heesch, H.
E-Reduktion.
HERT 1987-1 Hertz, A.; Werra, D. de
Using tabu search techniques for graph coloring.
HILT 1987-1 Hilton, A. J. W.
Recent progress on edge-coloring graphs.
HILTA 1987-1 Hilton, A. J. W.; Johnson, P. D.
Graphs which are vertex-critical with respect to the edge-chromatic number.
HOAN 1987-1 Hoang, C. T.
Alternating orientation and alternating colouration of perfect graphs.
HOCHC 1986-1 Hochbaum, D. S.; Wigderson, A.
The Linzertorte problem, or a unified approach to painting, baking and weaving.
HONGA 1984-1 Hong, Y.
A note on coefficients of chromatic polynomials (Chinese).
HSU A 1986-1 Hsu, W.-L.
Coloring planar perfect graphs by decomposition.
HUAN 1987-1 Huang, B. F.; Yin, Q.
A note on edge chromatic numbers of compositions of graphs (Chinese).
HUANB 1984-1 Huang, J.-H.
Uniquely 3-edge-colourable graphs (Chinese).
IDZI 1987-1 Idzik, A.; Komar, J.; Malawski, M.
Edge-coloured complete graphs: connectedness of some subgraphs.
JACKB 1984-1 Jackson, B.; Ringel, G.
Coloring island maps.
JAEG 1985-3 Jaeger, F.
On five-edge colorings of cubic graphs and nowhere-zero flow.
KALU 1983-1 Kaluza, T.
Existenz von Graphen jeden Grades mit Hamilton-Färbung.
KARLB 1987-1 Karloff, H. J.; Shmoys, D. B.
Efficient parallel algorithms for edge coloring problems.
KASZ 1985-1 Kaszonyi, L.
On homogeneous forms related to four-colourings.
KHAD 1986-1 Khadzhiivanov, N. G.
Edge colourings of the graph K(n)-e (Russian).
KHADA 1986-1 Khadzhiivanov, N. G.; Nenov, N. D.
Edge colorings of complete graphs in which monochromatic triangles do not cover the set of all vertices (Russian).
KHADA 1986-2 Khadzhiivanov, N. G.; Nenov, N. D.
On two-colorings of the edges of complete graphs with a minimal number of monochromatic triangles (Russian).
KHAR 1984-1 Kharazishvili, A. B.
Independent families and universal colored graphs (Russian).
KHAR 1986-1 Kharazishvili, A. B.
Coloured quasihomogeneous graphs (Russian).
KHEL 1984-1 Khelladi, A.; Payan, C.
Generalized n-tuple colorings of a graph: a counterexample to a conjecture of Brigham and Dutton.
KIER 1984-1 Kierstead, H. A.
On the chromatic index of multigraphs without large triangles.
KIERA 1986-1 Kierstead, H. A.; Schmerl, J. H.
The chromatic number of graphs which induce neither K(1,3) nor K(5)-e.
KIERB 1984-1 Kierstead, H. A.; Szemeredi, E.; Trotter, W. T. Jr.
On coloring graphs with locally small chromatic number.
KIRK 1985-1 Kirkpatrick, D. G.; Klawe, M.; Pippenger, N.
Some graph-colouring theorems with applications to generalized connection networks.
KOES 1985-1 Koester, G.
Coloring problems on a class of 4-regular planar graphs.

KOMJ 1987-1 Komjath, P.
The colouring number.
KRAM 1986-1 Kramer, F.; Kramer, H.
On the generalized chromatic number.
KRIZ 1986-1 Kriz, I.
A hypergraph-less construction of highly chromatic graphs without short cycles.
KUBA 1984-1 Kubale, M.
Ordered edge colorings of weighted multigraphs.
KUBA 1985-1 Kubale, M.
On the ochromatic index of a graph.
KUBA 1986-1 Kubale, M.
Application of graph edge colouring to the control of rearrangeable Clos networks (Polish).
KUBAA 1985-1 Kubale, M.; Jackowski, B.
A generalized implicit enumeration algorithm for graph coloring.
KUCE 1986-1 Kucera, L.
Greedy coloring is a bad probabilistic algorithm.
KULI 1983-1 Kulikowski, J. L.
On some formal properties of tri-partite graphs.
LABA 1986-1 Labahn, R.; Kanthack, H.
A colouring problem.
LESC 1983-1 Lesch, M.
Kantenfärbungen schlichter Graphen.
LESC 1985-1 Lesch, M.
A note on resistent edge-colourings.
LESCA 1983-1 Lesch, M.; Schmeichel, I.
Über farbtreue Automorphismen schlichter Graphen mit Kantenfärbungen.
LI D 1984-2 Li, W.-X.
The cycle-chromatic number of a hypergraph and an inequality of Lovasz.
LI F 1978-1 Li, W.-X.; Tian, F.
Some problems concerning the chromatic polynomials of graphs (Chinese).
LINI 1986-1 Linial, N.
Graph coloring and monotone functions on posets.
LINI 1986-3 Linial, N.
Legal coloring of graphs.
LINIB 1985-1 Linial, N.; Tarsi, M.
Deciding hypergraph 2-colourability by H-resolution.
LIU H 1986-3 Liu, Y.-P.
On chromatic sum equations for rooted 3-regular planar maps (Chinese).
LIU H 1987-2 Liu, Y.-P.
Chromatic sum equations for rooted cubic planar maps.
LIU K 1985-1 Liu, Z. S.; Li, H.
A problem on uniquely k-colorable graphs (Chinese).
LIZZ 1983-1 Lizzio, A.; Milici, S.
Determination of V(3,1), V(3,2), V(4,1) in a graph and sufficient conditions for delta(3) = 0 (Italian).
LOTF 1986-1 Lotfi, V.; Sarin, S.
A graph coloring algorithm for large scale scheduling problems.
LOVA 1983-1 Lovasz, L.
Self-dual polytopes and the chromatic number of distance graphs on the sphere.
LOVAD 1986-1 Lovasz, L.; Spencer, J.; Vesztergombi, K.
Discrepancy of set-systems and matrices.
MA E 1986-1 Ma, S.-H.; Wang, Y.
Interval colouring of graphs with cost and an algorithm for colouring of graphs (Chinese).
MAAMA 1984-1 Maamoun, M.; Meyniel, H.
On a problem of G. Hahn about coloured hamiltonian paths in K(2n).
MAC G 1984-1 Mac Diarmid, C.
Colouring random graphs.
MAC J 1987-3 Mac Kee, T. A.
Generalized complementation.
MARCF 1986-2 Marcotte, O.
On the chromatic index of multigraphs and a conjecture of Seymour. I.
MARI 1980-1 Maria, F. de; Gionfriddo, M.
A theorem on the s-chromatic number of a graph (Italian).
MARIC 1983-1 Marino, M. C.; Puccio, L.
On the parameter delta(s,G) of a planar graph (Italian).
MATO 1986-1 Matousek, J.
Few colored cuts or cycles in edge colored graphs.
MILAC 1982-1 Milazzo, F.
A generalization of the s-chromatic number of a directed graph (Italian).
MILAC 1982-2 Milazzo, F.
Automorphism groups of T(s)-colorings and relations on the s-chromatic numbers of a graph (Italian).
MILAC 1982-3 Milazzo, F.
Groups of T(s)-colored automorphisms and relations between the s-chromatic numbers of a graph (Italian).
MILAD 1984-1 Milazzo, F.; Vacirca, V.
Total colorings of higher kind in a graph (Italian).
MILAD 1987-1 Milazzo, F.; Vacirca, V.
On the achromatic number G x K(m).
MILAD 1987-2 Milazzo, F.; Vacirca, V.
On the achromatic number of permutation graphs.
MOHAA 1984-1 Mohar, B.
On edge-colorability products of graphs.
MOHAA 1984-3 Mohar, B.
Simplicial schemes and some combinatorial applications.

MOHAA 1986-1 Mohar, B.
Colouring of graphs (Slovenian).

MOHAB 1984-1 Mohar, B.; Pisanski, T.; Shawe-Taylor, J.
Edge-coloring of composite regular graphs.

MOZH 1983-1 Mozhan, N. N.
The chromatic number of graphs with a density not exceeding two-thirds of the maximal degree (Russian).

MYNH 1985-1 Mynhardt, C. M.; Broere, I.
Generalized colorings of graphs.

NASH 1987-1 Nash-Williams, C. St. J. A.
Amalgamations of almost regular edge-colourings of simple graphs.

NENO 1982-1 Nenov, N. D.
On proper 3-coloring of the edges of complete graphs (Bulgarian).

NENO 1984-1 Nenov, N. D.
Lower estimates for some constants related to Ramsey graphs (Bulgarian).

NENO 1984-2 Nenov, N. D.
The chromatic number of any 10-vertex graph without 4-cliques is at most 4 (Russian).

NENOA 1986-1 Nenov, N. D.; Khadzhiivanov, N. G.
Some colorings in two colors of the edges of a complete graph with nine vertices (Russian).

NEUMA 1984-1 Neumann-Lara, V.
The generalized dichromatic number of a digraph.

NEUMB 1986-1 Neumann-Lara, V.; Santoro, N.; Urrutia, J.
Uniquely colourable m-dichromatic oriented graphs.

NINC 1982-1 Ninchak, J.
Chromatical number for a class of graphs (Russian).

NINC 1984-1 Ninchak, J.
On a problem of R. Häggkvist concerning the edge colourings of bipartite graphs.

NISH 1985-1 Nishizeki, T.; Kashiwagi, K.
An upper bound on the chromatic index of multigraphs.

NISHA 1984-1 Nishizeki, T.; Sato, M.
An algorithm for edge-coloring of multigraphs.

OGUN 1987-1 Ogunyode, O. A.
Parallel algorithms for approximate edge coloring of simple graphs.

PEEM 1986-1 Peemöller, J.
Numerical experiences with graph coloring algorithms.

PICA 1983-1 Pica, G.
Generalized chromatic numbers of some graphs.

PICAA 1985-1 Pica, G.; Pisanski, T.; Shawe-Taylor, J.
Generalized chromatic numbers of some graphs. II.

PRET 1985-1 Pretzel, O.
On graphs that can be oriented as diagrams of ordered sets.

PROM 1986-2 Prömel, H. J.
Asymptotic enumeration of l-colorable graphs.

PROS 1986-1 Proskurowski, A.; Syslo, M. M.
Efficient vertex- and edge-coloring of outerplanar graphs.

PUCC 1988-1 Puccio, L.
An algorithm for L(s)-colourations.

RABI 1985-1 Rabinowitz, J. H.; Proulx, V. K.
An asymptotic approach to the channel assignment problem.

RADEB 1985-1 Radev, E. I.
A heuristic algorithm for the coloring of planar graphs (Russian).

READ 1987-3 Read, R. C.
On the chromatic properties of graphs up to 10 vertices.

RIED 1986-1 Rieder, J.
On k-tuple-colorings of graphs.

RODL 1985-1 Rödl, V.; Tuza, Z.
On color critical graphs.

RYU A 1985-1 Ryu, H. D.; Bak, T. B.
On dual problem of Hadwiger's conjecture (Korean).

SACH 1985-1 Sachs, H.
On colour critical graphs.

SCHM 1982-1 Schmerl, J. H.
The effective version of Brooks' theorem.

SCHRE 1985-1 Schrijver, A.
Supermodular colourings.

SCHUC 1985-1 Schumacher, H.; Wagner, K.
Zur Finitisierung von Färbungsproblemen.

SHAMA 1984-1 Shamir, E.; Upfal, E.
Sequential and distributed graph coloring algorithms with performance analysis in random graph spaces.

SHIEA 1987-1 Shier, D. R.; Shandrasekharan, N.
Algorithms for computing the chromatic polynomial.

SHLO 1987-1 Shlosman, S. B.
Graph colouring: a way to variety of new correlation inequalities.

SHTU 1983-1 Shtujkis, V.; Sabalyauskas, G.
An algorithm for the solution of a problem of colouring edges of a tree (Russian).

SIDO 1980-1 Sidorenko, A. F.
B(r) properties and chromatic numbers of hypergraphs (Russian).

SIEM 1984-1 Siemons, J.
Regularity in a graph.

SIMM 1983-1 Simmons, G. J.
On the ochromatic number of a graph.

SIMM 1985-1 Simmons, G. J.
The ochromatic number of planar graphs.

SOKOA 1987-1 Sokolova, M.
The chromatic number of extended odd graphs is four.

SPINB 1985-1 Spinrad, J.; Vijayan, G.
Worst case analysis of a graph coloring algorithm.

STAH 1985-1 Stahl, S.
Homology as a tool in integer programming.

STEN 1984-1 Stenzel, L.
On automorphisms and edge colourings of quasigroups (German).

STEW 1987-1 Stewart, I. A.
An algorithm for colouring perfect planar graphs.

STIE 1985-1 Stiebitz, M.
Colour-critical graphs with complete major-vertex subgraph.

STIE 1985-2 Stiebitz, M.
Subgraphs of colour-critical graphs.

STIE 1986-1 Stiebitz, M.
Colour-critical graphs and linear algebra.

STIE 1987-2 Stiebitz, M.
Subgraphs of colour-critical graphs.

SUN 1984-1 Sun, H.-C.
Some theorems on edge chromatic number of graphs (Chinese).

SUN 1987-1 Sun, H.-C.
A note to: "A theorem on colouring edges" by F. Z. Zhong, X. Z. Jian and F. W. Jian.

SZAM 1983-1 Szamkolowicz, L.
On some classes of three-chromatic graphs.

TARD 1985-2 Tardos, E.
Generalized matroids and supermodular colourings.

TENG 1985-1 Teng, A.; Tucker, A.
An 0(qn) algorithm to q-color a proper family of circular arcs.

TOFT 1985-1 Toft, B.
Some problems and results related to subgraphs of colour critical graphs.

TOME 1984-2 Tomescu, I.
Colorings and irreducible coverings by cliques of graphs and hypergraphs.

TOME 1985-1 Tomescu, I.
Problems in combinatorics and graph theory.

TOME 1986-1 Tomescu, I.
On hypergraph colourings.

TRUSA 1987-1 Trushkina, V. I.
The three-color method of coloring graphs (Russian).

TRUSB 1984-2 Truszczynski, M.
Some results on uniquely colourable graphs.

TUCK 1987-1 Tucker, A.
A reduction procedure for coloring perfect K(4)-free graphs.

TUCK 1987-2 Tucker, A.
Coloring Perfect (K(4)-e)-free graphs.

TUCKA 1984-1 Tucker, A.; Wilson, D.
An O(N**2) algorithm for coloring perfect planar graphs.

VACI 1982-1 Vacirca, V.
T(r,s)-colorings of a graph and their associated automorphism groups (Italian).

VACI 1986-1 Vacirca, V.
On some parameters associated with the modulus of a totally coloured digraph (Italian).

VASH 1982-1 Vashchenko, V. P.
An algorithm for graph coloring on the basis of fixed homomorphisms (Russian).

VO 1987-1 Vo, K.-P.
Graph colorings and acyclic orientations.

VOSS 1984-1 Voss, H.-J.
Färbungsprobleme auf Graphen.

VOSS 1985-3 Voss, H.-J.
Independent sets in k-chromatic graphs.

WALKA 1987-1 Walker, K.
Fully-coloured Hamiltonian cycles in edge-colorings of K(n) (n odd) when there are no fully-coloured C(4)'s.

WALLC 1985-1 Wallis, W. D.; Wang, Z.-J.
On one-factorizations of Cartesian products.

WENT 1987-1 Wentzlau, M.
On classes of preuniquely k-colourable graphs.

WENT 1987-2 Wentzlau, M.
On uniquely and preuniquely k-colourable graphs.

WERR 1985-2 Werra, D. de
Graphs, hypergraphs and timetabling.

WHITF 1984-1 Whitehead, E. G. Jr.; Zhao, L.-C.
Cutpoints and the chromatic polynomial.

WILLB 1985-1 Williams, M. H.
A linear algorithm for colouring planar graphs with five colours.

XU F 1983-1 Xu, S.-J.
A further discussion on bichromaticity (Chinese).

XU F 1984-1 Xu, S.-J.
The chromaticity of regular graphs and bigraphs with more edges (Chinese).

XU G 1984-1 Xu, S.-J.; Li, N.-Z.
The chromaticity of wheels.

YANNA 1987-1 Yannakakis, M.; Gavril, F.
The maximum k-colorable subgraph problem for chordal graphs.

YAP 1983-1 Yap, H. P.
A class of chromatic index critical graphs.

ZAGA 1983-1 Zagaglia Salvi, N.
On the chromatic index of permutation graphs (Italian).

ZASL 1984-2 Zaslavsky, T.
How colorful the signed graph?

ZHANR 1986-1 Zhang, Z. F.; Chen, Y.
Operations on graphs and the chromatic number of a graph (Chinese).

ZHANU 1984-1 Zhang, Z. F.; Zhang, J. X.; Wang, J. F.
A theorem on colouring edges.

ZHANU 1984-2 Zhang, Z. F.; Zhang, J. X.; Wang, J. F.
An edge coloring theorem (Chinese).

ZHU 1985-1 Zhu, A. X.
Non-coloring-contradiction graphs and a graph coloring algorithm (Chinese).

ABUO	1986-1	Abuov, K. A.; Cherenin, V. P. Synthesis of one-product two-stage production-transport networks.
AGRAA	1984-1	Agrawal, S. C.; Chand, M. On integer solution to interval linear programming problems by a branch and bound technique.
AHUJA	1987-1	Ahuja, R. K.; Murty, V. V. S. Exact und heuristic algorithms for the optimum communication spanning tree problem.
AKIN	1985-1	Akinc, U. Multi-activity facility design and location problems.
ALEKA	1987-1	Alekseev, A. G.; Kezhaev, V. A.; Subbotin, A. I. A special problem of integer programming (Russian).
ALTI	1987-1	Altinkemer, K.; Gavish, B. Heuristics for unequal weight delivery problems with a fixed error guarantee.
ANAS	1985-1	Anastasyan, V. G. One combinatorial line location problem (Russian).
ANGR	1982-1	Angrisani, M. Discrete minimax problems: a unified approach using Motzkin's alternative theorem (Italian).
ANGR	1983-1	Angrisani, M. On a discrete minimax problem (Italian).
ANSTB	1985-1	Anstee, R. P.; Murty, U. S. R. Matrices with forbidden subconfigurations.
ARDI	1984-1	Arditti, D. Un nouvel algorithme de recherche d'un ordre induit par des comparaisons par paires.
ASANA	1985-1	Asano, T.; Asano, T.; Guibas, L.; Hershberger, J.; Imai, H. Visibility-polygon search and euclidean shortest paths.
ATAL	1985-1	Atallah, M. J. A matching problem in the plane.
ATALA	1985-1	Atallah, M. J.; Hambrusch, S. E. Efficient solutions to a class of placement problems.
ATKIA	1987-1	Atkinson, M. D.; Negro, A.; Santoro, N. Sums of lexicographically ordered sets.
AYUP	1984-1	Ayupov, R. S. A combinatorial continuous method of solving block problems of discrete programming.
BABAE	1985-1	Babai, L. On Lovasz' lattice reduction and the nearest lattice point problem.
BABAE	1986-2	Babai, L. On Lovasz' lattice reduction and the nearest lattice point problem.
BACHC	1984-1	Bachem, A.; Kannan, R. Lattices and the basis reduction algorithm.
BALAB	1987-1	Balakrishnan, A.; Magnanti, T. L.; Wong, R. T. A dual-ascent procedure for large-scale uncapacitated network design.
BALIA	1986-1	Balinski, M. L.; Demange, G. Generalizing proportionality; or how to simultaneously apportion seats to regions and to parties.
BARTB	1987-1	Bartnick, J. Aggregation of preference relations by a branch and bound algorithm.
BEAS	1985-2	Beasley, J. E. Algorithms for unconstrained two-dimensional guillotine cutting.
BEAS	1985-3	Beasley, J. E. An algorithm for the two-dimensional assortment problem.
BEAS	1985-4	Beasley, J. E. An exact two-dimensional non-guillotine cutting tree search procedure.
BEAS	1985-5	Beasley, J. E. Bounds for two-dimensional cutting.
BECK	1987-1	Becker, B.; Hotz, G. On the optimal layout of planar graphs with fixed boundary.
BECKA	1985-1	Becker, B.; Osthof, H. G. Layouts with wires of balanced length.
BELO	1982-1	Belov, B. I.; Chugunova, G. P. On a problem concerning the cutting out of rectangles by parallelograms (Russian).
BELOA	1985-1	Belov, Y. A.; Bondarenko, V. A. Geometric problems in discrete optimization (Russian).
BENV	1986-1	Benveniste, R. An integrated plant design and scheduling problem in the food industry.
BEREA	1983-1	Bereznev, V. A.; Palmova, I. E.; Tretyakov, A. A. On an optimization problem of the cutting-out of materials (Russian).
BERTA	1986-2	Bertolazzi, P.; Sassano, A. Regular clutters.
BHATA	1987-1	Bhatt, S. N.; Cosmadakis, S. S. The complexity of minimizing wire lengths in VLSI layouts.
BHATB	1984-1	Bhatt, S. N.; Leighton, F. T. A framework for solving VLSI graph layout problems.
BHATC	1985-1	Bhattacharya, B. B.; Gupta, B. Optimal design of hazard-free and easily testable logic networks.
BIRG	1985-1	Birge, J. R.; Malyshko, V. V. Methods for a network design problem in solar power systems.
BLAIB	1984-1	Blair, C. E.; Jeroslow, R. G.; Lowe, J. K. Some results and experiments in programming techniques for propositional logic.
BLAIC	1985-1	Blair, E. L.; Miller, S. An interactive approach to facilities design using microcomputers.
BLAZC	1984-1	Blazewicz, J.; Brzezinski, J.; Gambosi, G. Time-stamps approach to store-and-forward deadlock prevention.
BLUM	1985-1	Blum, N. On the single-operation worst-case time complexity of the disjoint set union problem.

BOCH	1985-1	Bochenek, K. An algorithm for optimization of packet flow in telecommunication network.
BOES	1985-1	Boesch, F. T. Is reliable network synthesis a graph theory problem?
BOLLE	1984-1	Bollobas, B.; Hell, P. Sorting and graphs.
BORA	1984-1	Boranbaev, S. N. A method for the distribution of resources on networks (Russian).
BORD	1985-1	Bordat, J. P. Parcours dans les graphes: un outil pour l'algorithmique des ensembles ordonnes.
BORZ	1987-1	Borzacchini, L. Graphical partially ordered sets.
BORZA	1984-1	Borzacchini, L.; Pulito, C. Coverings enumeration for graphical partially ordered sets.
BOUCD	1986-1	Bouchitte, V.; Habib, M. Some NP-completeness properties about linear extensions.
BOUCE	1985-1	Bouchitte, V.; Habib, M.; Jegou, R. On the greedy dimension of a partial order.
BRODA	1982-1	Brodskij, G. M.; Soloveva, N. A. The partially ordered set of subtrees of a graph (Russian).
BRUC	1984-3	Brucker, P. Über gewisse Prinzipien des Entwurfs effizienter Algorithmen.
BURA	1985-1	Burattini, E.; Marra, G.; Sforza, A. Network design problem: structure of solutions and dominance relations.
BURG	1985-1	Burgin, M. S. Algorithms and algorithmic problems (Russian).
BURKB	1985-1	Burkard, R. E.; Fincke, U. Probabilistic asymptotic properties of some combinatorial optimization problems.
CAMEI	1986-1	Cameron, K.; Sachs, H. Monotone path systems in simple regions.
CARRC	1983-1	Carrera, J.; Szidarovszky, F. A multiobjective branch and bound algorithm for optimal observation network design.
CEDEA	1986-1	Ceder, A.; Wilson, N. H. M. Bus network design.
CENE	1985-1	Cenek, P.; Chlebnican, P.; Klima, V. Computer aided network design.
CEPU	1984-1	Cepulic, V. An algorithm for finding minimal intersection sets.
CHANAA	1984-1	Chang, R. C.; Lee, R. C. T. The average performance analysis of a closest-pair algorithm.
CHANS	1987-1	Chang, G. J.; Hao, F. H. The minimum of the antichains in the factor poset.
CHAUD	1987-1	Chauny, F.; Loulou, R.; Sadones, S.; Soumis, F. A two-phase heuristic for strip packing: algorithm and probabilistic analysis.
CHEBA	1984-1	Chebotar, K. S. A problem of optimization on a set of permutations (Russian).
CHENH	1984-1	Chen, R. S. An algorithm for finding generalized systems of distinct representatives (Chinese).
CHUNH	1986-1	Chung, S.-J.; Murty, K. G.; Chang, S. Y. On K(delta).
CHVA	1985-1	Chvatal, V. Cutting planes in combnatorics.
CIUR	1986-1	Ciurea, E. Le probleme de l'approvisionnement et de la demande pour les flots dynamiques.
CLAU	1986-1	Clausen, J.; Krarup, J. Combinatorial optimization: challenges and trends.
CLEM	1984-1	Clements, G. F. On uniqueness of maximal antichains of subsets of a multiset.
COCA	1985-1	Cocan, A.; Cocan, M. On a problem concerning multiple cutting.
COLB	1986-2	Colbourn, C. J. The reliability polynomial.
COOKC	1986-1	Cook, W.; Fonlupt, J.; Schrijver, A. An integer analogue of Caratheodory's theorem.
CRAI	1986-1	Crainic, T. G.; Rousseau, J.-M. Multicommodity, multimode freight transportation: a general modeling and algorithmic framework for the service network design problem.
CROS	1985-1	Cross, P. A. Numerical methods in network design.
CRYS	1987-1	Crystal Llewellyn, D. Min algebraic duality.
CRYSA	1985-1	Crystal Llewellyn, D.; Trotter, L. E. Jr. On abstract integral dependence.
CURRB	1985-1	Current, J. R.; Revelle, C. S.; Cohon, J. L. The maximum covering/shortest path problem: a multiobjective network design and routing formulation.
CURRB	1986-1	Current, J. R.; Revelle, C. S.; Cohon, J. L. The hierarchical networks design problem.
DALL	1984-1	Dallery, Y. Optimisation de la repartition des palettes dans un atelier flexible.
DANI	1985-1	Danilchenko, A. F. An approach to the solution of the sorting problem (Russian).

DAVYA 1987-1 Davydov, G. V.; Davydova, I. M.
Dual algorithms in discrete optimization (Russian).
DERI 1984-2 Derigs, U.
Exchange properties and K-best strategies in combinatorial optimization.
DERI 1985-3 Derigs, U.
Some basic exchange properties in combinatorial optimization and their application to constructing the K-best solutions.
DOLE 1987-1 Dolev, D.; Halpern, J. Y.; Simons, B.; Strong, H. R.
A new look at fault-tolerant network routing.
DOMSA 1985-2 Domschke, W.; Drexl, A.
Location and layout planning. An international bibliography.
DUTA 1983-1 Duta, L. D.; Fabian, C.
Mathematical methods in optimal cutting (Romanian).
DUTA 1984-1 Duta, L. D.; Fabian, C.
Solving cutting-stock problems through the Monte-Carlo method.
DYERC 1984-1 Dyer, M. E.; Frieze, A. M.; Mac Diarmid, C.
Partitioning heuristics for two geometric maximization problems.
EADE 1986-1 Eades, P.; Kelly, D.
Heuristics for drawing two-layered networks.
EDMOA 1985-1 Edmonds, J.; Lubiw, A.
Bipartition systems and how to partition polygons.
EL-Z 1985-1 El-Zahar, M.; Rival, I.
Examples of jump-critical ordered sets.
ELBA 1986-1 Elbas, J.
Substitution and atomic extension on greedy posets.
EMELA 1985-1 Emelichev, V. A.; Kravtsov, M. K.; Krachkovskii, A. P.
Kombinatorische und Extremalprobleme auf Transportpolyedern.
EMELE 1985-1 Emelicheva, E. V.; Kovalev, M. M.
The method of partial orderings on matroid structures (Russian).
EMELF 1985-1 Emelicheva, E. V.; Kovalev, M. M.; Rogov, S. F.
Methode der Halbordnung zur Lösung von Optimierungsproblemen über Permutationen (Russian).
ENTRA 1983-1 Entringer, R. C.; Jackson, D. E.
Degree sums of paths in extremal trees.
FAIG 1977-1 Faigle, U.
Geometries on partially ordered sets.
FAIG 1985-1 Faigle, U.
On ordered languages and the optimization of linear functions by greedy algorithms.
FAIGA 1987-1 Faigle, U.; Gierz, G.
The bandwidth of planar distributive lattices.
FAIGC 1984-1 Faigle, U.; Goecke, O.; Schrader, R.
On simplicial decomposition in combinatorial structures.
FAIGC 1986-1 Faigle, U.; Goecke, O.; Schrader, R.
Church-Rosser decomposition in combinatorial structures.
FAIGD 1985-1 Faigle, U.; Lovasz, L.; Schrader, R.; Turan, G.
Search problems in ordered sets.
FAIGD 1986-1 Faigle, U.; Lovasz, L.; Schrader, R.; Turan, G.
Searching in trees, series-parallel and interval orders.
FAIGE 1986-1 Faigle, U.; Sands, B.
A size-width inequality for distributive lattices.
FAIGF 1984-1 Faigle, U.; Schrader, R.
Comparability graphs and order invariants.
FAIGF 1985-1 Faigle, U.; Schrader, R.
A setup heuristic for interval orders.
FAIGF 1985-2 Faigle, U.; Schrader, R.
Linear arrangements of interval orders.
FAIGF 1985-3 Faigle, U.; Schrader, R.
Setup minimization techniques for comparability graphs.
FAIGF 1986-1 Faigle, U.; Schrader, R.
A combinatorial bijection between linear extensions of equivalent orders.
FAIGF 1986-2 Faigle, U.; Schrader, R.
Greene-Kleitman extensions of ordered sets.
FAIGF 1986-3 Faigle, U.; Schrader, R.
On the computational complexity of the order polynomial.
FAIGF 1987-1 Faigle, U.; Schrader, R.
Interval orders without odd crowns are defect optimal.
FAIGF 1987-2 Faigle, U.; Schrader, R.
Setup optimization problems with matroid structure.
FAIGH 1985-1 Faigle, U.; Turan, G.
Sorting and recognition problems for ordered sets.
FAIGH 1987-1 Faigle, U.; Turan, G.
On the complexity of interval orders and semiorders.
FARRA 1986-1 Farrell, E. J.
A survey of the unifying effects of F-polynomials in combinatorics and graph theory.
FELDA 1986-1 Feldman, P.; Friedman, J.; Pippenger, N.
Non-blocking networks.
FISC 1987-1 Fischetti, M.
A new linear storage, polynomial time approximation scheme for the subset-sum problem.
FORT 1985-1 Fortenberry, J. C.; Cox, J. F.
Multiple criteria approach to the facilities layout problem.
FOUL 1984-1 Foulds, L. R.
Combinatorial optimization for undergraduates.
FOUL 1985-1 Foulds, L. R.
Enumeration of graph theoretic solutions for facilities layout.

FOUL	1985-3	Foulds, L. R. Traffic network arc elimination by branch and bound enumeration.
FOUL	1986-1	Foulds, L. R. Braess's paradox: which arcs should be eliminated in order to minimize congestion?
FOULA	1985-1	Foulds, L. R.; Gibbons, P. B.; Giffin, J. W. Facilities layout adjacency determination: An experimental comparison of three graph theoretic heuristics.
FOULB	1985-1	Foulds, L. R.; Giffin, J. W. A graph-theoretic heuristic for minimizing total transport cost in facilities layout.
FOULB	1985-2	Foulds, L. R.; Giffin, J. W. Facilities layout via new graph theoretic methods.
FOULC	1986-1	Foulds, L. R.; Giffin, J. W.; Evans, J. D. Towards a new decision support system for facilities layout planners.
FRANE	1985-1	Frank, A.; Tardos, E. An application of simultaneous approximation in combinatorial optimization.
FRANE	1987-1	Frank, A.; Tardos, E. An application of simultaneous Diophantine approximation in combinatorial optimization.
FREDB	1987-1	Frederickson, G. N.; Janardan, R. Designing networks with compact routing tables.
FREY	1986-1	Frey, J.-J.; Yehia-Alcoutlabi, A. Comparaisons par paires: une interpretation et une generalisation de la methode des scores.
FRIEB	1985-1	Friesz, T. L. Transportation network equilibrium, design and aggregation: Key developments and research opportunities.
FUJIA	1985-1	Fujishige, S. A decomposition of distributive lattices.
GABOE	1985-1	Gabow, H. N.; Tarjan, R. E. A linear-time algorithm for a special case of disjoint set union.
GALAB	1984-1	Galambos, G.; Imreh, B. Solution of one-dimensional cutting stock problems by column-generation (Hungarian).
GARD	1985-1	Gardner, M. L. An algorithm to aid in the design of large scale networks.
GARF	1985-1	Garfinkel, R. S. Motivation and modeling.
GEHR	1987-1	Gehrlein, W. V.; Fishburn, P. C. A comparison of algorithms for minimizing bumps in linear extensions of partial orders.
GERAA	1987-1	Gerards, A. M. H.; Kolen, A. W. J. Polyhedral combinatorics in combinatorial optimization.
GERAB	1986-1	Gerards, A. M. H.; Schrijver, A. Matrices with the Edmonds-Johnson property.
GLOV	1987-1	Glover, F. Tabu search.
GNDO	1984-1	Gndoyan, A. K.; Igitkhanyan, S. A. Plane arrangement of vertices of a tree on a line with minimization of the length of the largest edge (Russian).
GOLDD	1987-1	Goldberg, M. K.; Spencer, T. A new parallel algorithm for the maximal independent set problem.
GOLDI	1987-1	Golden, B. L.; Levy, L.; Vohra, R. V. The orienteering problem.
GOLDM	1982-1	Goldengorin, B. I. Stability of solutions in problems with one component of connection of local minima (Russian).
GROP	1984-1	Groppen, V. O. Models and algorithms of combinatorial programming (Russian).
GROP	1985-2	Groppen, V. O. Optimization of dynamic programming methods when solving extremal combinatorial problems.
GROT	1984-1	Grötschel, M. Developments in combinatorial optimization.
GROT	1986-1	Grötschel, M. Fortschritte in der polyedrischen Kombinatorik.
GROTB	1984-1	Grötschel, M.; Jünger, M.; Reinelt, G. A cutting plane algorithm for the linear ordering problem.
GROTB	1985-1	Grötschel, M.; Jünger, M.; Reinelt, G. Acyclic subdigraphs and linear orderings: polytopes, facets, and a cutting plane algorithm.
GROTB	1985-2	Grötschel, M.; Jünger, M.; Reinelt, G. Facets of the linear ordering polytope.
GROTC	1984-2	Grötschel, M.; Lovasz, L.; Schrijver, A. Geometric methods in combinatorial optimization.
GULY	1983-1	Gulyanitskii, L. F.; Khodzinskii, A. N. Two algorithms for solution of optimization problems on permutations (Russian).
GUPT	1986-1	Gupta, A. K.; Hambrusch, S. E. On optimal three-dimensional layouts of complete binary trees.
GUPT	1987-1	Gupta, A. K.; Hambrusch, S. E. Simulating tree algorithms with limited parallelism.
GURA	1984-1	Gurari, E. M.; Sudborough, I. H. Improved dynamic programming algorithms for bandwidth minimization and the min-cut linear arrangement problem.
GUSE	1983-1	Guseinov, B. A.; Stanevichyus, A.-I. A. Method of solution of the crossing problem as an integer programming problem (Russian).
GYOR	1984-1	Györi, E. A minimax theorem on intervals.
HABIA	1985-1	Habib, M.; Jegou, R. N-free posets as generalizations of series-parallel posets.

HABIB 1985-1 Habib, M.; Maurer, M. C.
1-intersecting families.
HABIC 1987-1 Habib, M.; Möhring, R. H.
On some complexity properties of N-free posets and posets with bounded decomposition diameter.
HABID 1987-1 Habib, M.; Möhring, R. H.; Steiner, G.
Computing the bump number of ordered sets is easy.
HACK 1984-1 Hackbarth, K.-D.
Implementation of algorithms for telecommunication network planning.
HAMAC 1986-1 Hamacher, H. W.; Queyranne, M.
K best solutions to combinatorial optimization problems.
HAMMD 1987-1 Hammer, P. L.; Simeone, B.
Order relations of variables in 0-1 programming.
HAMMF 1985-1 Hammouche, A.; Webster, D. B.
Evaluation of an application of graph theory to the layout problem.
HANSD 1987-1 Hansen, P.; Jaumard, B.
Algorithms for the maximum satisfiability problem.
HARRA 1984-1 Harrison, P.
The two dimensional cutting problem of the furniture manufacturer.
HASH 1987-1 Hashizume, S.; Fukushima, M.; Katoh, N.; Ibaraki, T.
Approximation algorithms for combinatorial fractional programming problems.
HAUR 1986-1 Haurie, A.; Marcotte, P.
A game-theoretic approach to network equilibrium.
HELLA 1986-1 Hell, P.; Li, W.-X.; Schmerl, J. H.
Jump number and width.
HERR 1985-1 Herroelen, W.; Gils, A. van
On the use of flow dominance in complexity measures for facility layout problems.
HIRS 1985-1 Hirschberg, D. S.; Larmore, L. L.
The least weight subsequence problem.
HOEK 1983-1 Hoek, G. van der; Meerendonk, H. W. van den; Wijmenga, R. T.
The elimination of convexly dominated cutting patterns in trimloss problems.
HORO 1987-1 Horowitz, A. J.; Pithavadian, A. R.
Toward a data structure for computer-aided design of transportation networks.
INTRC 1987-1 Intrator, J.; Weiss, J.
Exhaustive iterations - an improving approach for long transportation problems.
IVAN 1984-1 Ivanchev, D. T.
Minimization of the processing time for metal cutting machines considered as a shortest path problem in a network (Bulgarian).
JAIL 1987-1 Jaillet, P.
On some probabilistic combinatorial optimization problems defined on graphs.
JAMI 1984-1 Jamison, R. E.
Whitney sums and matchings in trees (part I).
JIAN 1987-1 Jiang, H. G.
Optimality conditions for a nonsmooth optimization problem (Chinese).
JOHNH 1985-1 Johnson, D. S.; Papadimitriou, C. H.; Yannakakis, M.
How easy is local search?
JOHNM 1986-1 Johnston, R. E.
Rounding algorithms for cutting stock problems.
KAKU 1986-1 Kaku, B. K.; Thompson, G. L.
Comment on "A perturbation scheme to improve Hillier's solution to the facilities layout problem".
KALM 1984-1 Kalmar, J.; Lengyel, I.
The consideration of stock within the Gilmore-Gomory method (Hungarian).
KANA 1984-1 Kanaeva, N. N.; Shcherbina, O. A.
On the efficiency of a modified local algorithm for solving discrete optimization problems.
KANA 1986-1 Kanaeva, N. N.; Shcherbina, O. A.
On a modified local algorithm for problems of discrete programming.
KANN 1987-1 Kannan, R.
Algorithmic geometry of numbers.
KANN 1987-2 Kannan, R.
Minkowski's convex body theorem and integer programming.
KANNA 1986-1 Kannan, R.; Lovasz, L.
Covering minima and lattice point free convex bodies.
KANNA 1987-1 Kannan, R.; Lovasz, L.
Covering minima and lattice point free convex bodies.
KANT 1985-1 Kant, K.
Finding interferences between rectangular paths.
KASPB 1985-1 Kaspshitskaya, M. F.; Sergienko, I. V.; Stiranka, A. I.
A special class of problems of geometric optimization (Russian).
KATAB 1986-1 Katayama, N.; Kasugai, H.
An algorithm for the discrete road network design problem using branch and bound method (Japanese).
KERN 1985-1 Kern, W.
An efficient algorithm for solving a special class of LP's.
KERN 1986-2 Kern, W.
On the depth of combinatorial optimization problems.
KHACB 1985-1 Khachaturov, V. R.; Zlotov, A. V.; Kityan, S. P.
Optimization of a structure of two networks taking account of the agglomeration effect (Russian).
KHOK 1982-1 Khokhlyuk, V. I.
Algorithms of integer optimization (Russian).
KHOK 1987-1 Khokhlyuk, V. I.
Parallel algorithms for integer optimization (Russian).
KIER 1986-1 Kierstead, H. A.
NP-completeness results concerning greedy and super greedy linear extensions.

KIERC 1987-1 Kierstead, H. A.; Trotter, W. T. Jr.; Zhou, B.
Representing an ordered set as the intersection of super greedy linear extensions.

KIND 1985-1 Kindervater, G. A. P.; Lenstra, J. K.
An introduction to parallelism in combinatorial optimization.

KIND 1986-1 Kindervater, G. A. P.; Lenstra, J. K.
An introduction to parallelism in combinatorial optimization.

KLEB 1987-1 Klebanovich, D. M.; Malyshko, V. V.
An approximate algorithm for the tree layout on a rectangular lattice with area minimization (Russian).

KLEE 1983-1 Klee, V.
Combinatorial optimization: what is the state of the art? (Chinese).

KOGA 1987-1 Koganov, L. M.
An oriented graph connected with an ordered pair of ordered partitions (Russian).

KOLB 1986-1 Kolbin, V. V.
Optimization problems in semiordered spaces (Russian).

KORNA 1984-1 Kornienko, N. M.
Combinatorial algorithms on a class of graphs (Russian).

KORT 1985-1 Kortanek, K. O.
Network design and dynamic routing under queueing demand.

KORTA 1984-1 Korte, B.
Kombinatorische Optimierung und algorithmische Prinzipien.

KORTA 1985-1 Korte, B.
Combinatorial optimization (Part I) (Chinese).

KORTA 1985-2 Korte, B.
Was ist kombinatorische Optimierung?

KORTA 1986-1 Korte, B.
Kombinatorische Optimierung und algorithmische Prinzipien.

KOVAC 1987-1 Kovalenko, A. G.
Algorithms of intervals and their application to solve problems of discrete optimization of multistage processes (Russian).

KUBO 1986-1 Kubo, M.; Kasugai, H.
Traveling purchaser problem (Japanese).

KUZY 1982-1 Kuzyurin, N. N.
On the problem of the alpha-depth of (0,1)-matrices (Russian).

LANG 1985-1 Lang, Z.
Untersuchung spezieller Fälle des Layout-Problems mit Hilfe von Graphenmethoden.

LAUT 1985-1 Lautemann, C.; Meyer auf der Heide, F.
Lower time bounds for integer programming with two variables.

LEBEB 1984-1 Lebedeva, T. T.; Sergienko, I. V.; Soltan, V. P.
Concerning conditions of local and global extrema coincidence in the problems of discrete optimization (Russian).

LEBL 1986-1 LeBlanc, L. J.; Boyce, D. E.
A bilevel programming algorithm for exact solution of the network design problem with user-optimal flows.

LEE J 1986-1 Lee, J.
The incidence structure of subspaces with well-scaled frames.

LEEU 1985-1 Leeuwen, J. van
Parallel computers and algorithms.

LENG 1983-1 Lengauer, T.
Efficient algorithms for the constraint generation for integrated circuit layout compaction.

LENGA 1986-1 Lengyel, I.
A practical generalization of the knapsack problem for cutting stock (Hungarian).

LENS 1984-1 Lenstra, H. W.; Schnorr, C. P.
On the successive minima of a pair of polar lattices.

LEVAA 1985-1 Levary, R. R.; Kalchik, S.
Facilities layout - a survey of solution procedures.

LI E 1986-1 Li, W.-X.; Schmerl, J. H.
Ordered sets with small width and large jump number.

LINI 1986-1 Linial, N.
Graph coloring and monotone functions on posets.

LINI 1986-2 Linial, N.
Hard enumeration problems in geometry and combinatorics.

LIPO 1985-1 Lipovetskii, A. I.
Change in the structure of a finite set in discrete optimization (Russian).

LIRO 1987-1 Lirov, Y.
A locally optimal fast obstacle-avoiding path-planning algorithm.

LIU I 1983-1 Liu, Z. H.
Approximate algorithms for combinatorial optimization problems. I (Chinese).

LIU I 1983-2 Liu, Z. H.
Approximate algorithms for combinatorial optimization problems. II (Chinese).

LOVA 1985-1 Lovasz, L.
An algorithmic theory of numbers, graphs, and convexity.

LOVA 1985-3 Lovasz, L.
Some algorithmic problems on lattices.

LOVA 1986-3 Lovasz, L.
Theory of numbers, graphs and convexity.

LOVAD 1986-1 Lovasz, L.; Spencer, J.; Vesztergombi, K.
Discrepancy of set-systems and matrices.

LOVE 1985-1 Love, R. F.; Dowling, P. D.
Optimal weighted l(p) norm parameters for facilities layout distance characterizations.

LOZO 1984-2 Lozovanu, D. D.
Some properties of a network nonlinear transportation problem and the problem of synthesis of a network in the form of an outgoing tree with a source (Russian).

LUBI 1985-1 Lubiw, A.
Orderings and some combinatorial optimization problems with geometric applications.

LUCEA 1984-1 Lucertini, M.; Paletta, G.
A decomposition method for network design with multicommodity non-simultaneous flows.

LUCEA 1984-2 Lucertini, M.; Paletta, G.
Network design with non-simultaneous flows.

LUCEA 1985-1 Lucertini, M.; Paletta, G.
Network design with multiple demand: a new approach.

LUCEA 1986-1 Lucertini, M.; Paletta, G.
A class of network design problems with multiple demand: Model formulation and an algorithmic approach.

MAGN 1986-1 Magnanti, T. L.; Mireault, P.; Wong, R. T.
Tailoring Benders decomposition for uncapacitated network design.

MARCE 1987-1 Marcotorchino, F.
Block seriation problems: a unified approach. Reply to the problem of H. Garcia and J. M. Proth.

MARCF 1985-1 Marcotte, O.
The cutting stock problem and integer rounding.

MARCF 1986-1 Marcotte, O.
An instance of the cutting stock problem for which the rounding property does not hold.

MARCI 1986-1 Marcotte, P.
Network design problem with congestion effects: a case of bilevel programming.

MAREA 1984-1 Marello, S.; Pulleyblank, W. R.; Toth, P.; Werra, D. de
Balanced optimization problems.

MARIA 1985-1 Marinescu, D.
An integer programming model for two-dimensional cutting stock-problems.

MARTA 1984-1 Martello, S.; Pulleyblank, W. R.; Toth, P.; Werra, D. de
Balanced optimization problems.

MARTB 1985-2 Martello, S.; Toth, P.
Approximation schemes for the subset-sum problem: survey and experimental analysis.

MARTF 1987-1 Martin, R. K.; Rardin, R. L.; Campbell, B. A.
Polyhedral characterization of discrete dynamic programming.

MARTL 1983-1 Martinova, M. K.
A planimetric problem with a combinatorial significance (Bulgarian).

MARU 1987-1 Marucheck, A. S.; Sulek, J. M.
An algorithm for technology choice in local area network design.

MIHA 1987-1 Mihalyi, M.; Raischi, C.
An algorithm for optimal ordering of the arcs of a network.

MILLA 1986-1 Miller, D. M.
A decomposition technique for the design of internally-unate combinational networks.

MINI 1987-1 Minieka, E.
The delivery man problem on a tree network.

MINO 1987-2 Minoux, M.
Network synthesis and dynamic network optimization.

MIRZ 1987-1 Mirzaian, A.
River routing in VLSI.

MOHR 1987-1 Möhring, R. H.
Computationally tractable classes of ordered sets.

MOHRA 1984-1 Möhring, R. H.; Radermacher, F. J.
Substitution decomposition for discrete structures and connections with combinatorial optimization.

MONIA 1984-1 Monien, B.; Speckenmeyer, E.
Ramsey numbers and an approximation algorithm for the vertex cover problem.

MONIA 1985-1 Monien, B.; Speckenmeyer, E.
Ramsey numbers and an approximation algorithm for the vertex cover problem.

MONME 1986-1 Monma, C. L.; Shallcross, D. F.
Methods for designing survivable communication networks.

MONMF 1986-1 Monma, C. L.; Sheng, D. D.
Backbone network design and performance analysis: a methodology for packet switching networks.

MONTB 1987-1 Montreuil, B.; Ratliff, H. D.; Goetschalckx, M.
Matching based interactive facility layout.

MOSL 1987-1 Mosler, K. C.
Continuous location of transportation networks.

MUKH 1983-1 Mukhacheva, E. A.
Search for the rational solution in the two-criterial problem of guillotine cutting-out (Russian).

MUKHA 1984-1 Mukhacheva, E. A.; Ibatullina, S. M.
Optimization of cutting material according to a given assortment ratio (Russian).

MURT 1985-1 Murtagh, F.
A survey of algorithms for contiguity-constrained clustering and relating problems.

NACH 1987-1 Nachtigall, K.
Residuated lattice spaces.

NESEA 1986-1 Nesetril, J.; Poljak, S.
A remark on max-cut problem with an application to digital-analogue convertors.

NESEC 1987-1 Nesetril, J.; Rödl, V.
Complexity of diagrams.

NEWF 1984-1 Newfeld, E. M.; Colbourn, C. J.
The construction of reliable series - parallel networks.

NIKI 1983-1 Nikitin, A. I.; Nuriev, U. G.; Krasilovets, L. V.
Software for a multi-user computer center. Some discrete optimization algorithms (Russian).

OFIT 1986-1 Ofitserov, D. V.
A multiextremal discrete optimization algorithm (Russian).

ORLI 1980-1 Orlin, J. B.
Fast approximation schemes for combinatorial optimization problems (extended abstract).

OUDH 1987-1 Oudheusden, D. L. van; Khan, M. L. R.
Planning and development of rural road networks in developing countries.

PANI	1984-1	Panishev, A. V. An algorithm for optimization of treelike structures (Russian).
PAPAC	1985-1	Papadimitriou, C. H.; Wolfe, D. The complexity of facets resolved.
PAPAE	1986-1	Papakostantinou, N. The complexity of the travelling repairman problem.
PARD	1987-1	Pardalos, P. M.; Rosen, J. B. Constrained global optimization: Algorithms and applications.
PEEL	1985-1	Peele, R. Finite partition sublattice representations for some geometric lattices.
PELT	1987-1	Peltsverger, B. V.; Khavronin, O. V. A decompositional approach to the problem of synthesizing a transport network with a fuzzy use period (Russian).
PERS	1982-1	Pershin, O. Y. General problem of the optimal design of multi-component networks (Russian).
PERS	1986-1	Pershin, O. Y. A class of extremal combinatorial problems for multicomponent network design.
PETEC	1987-1	Peters, J. G.; Rudolph, L. Parallel approximation schemes for subset sum and knapsack problems.
PICO	1984-1	Picone, C. J.; Wilhelm, W. E. A perturbation scheme to improve Hillier's solution to the facilities layout problem.
PICO	1986-1	Picone, C. J.; Wilhelm, W. E. "A perturbation scheme to improve Hillier's solution to the facilities layout problem": a clarification.
PROM	1985-1	Prömel, H. J. Some remarks on natural orders for combinatorial cubes.
PROM	1986-3	Prömel, H. J. Counting unlabeled structures.
PUTU	1986-1	Puturidze, Z. S. Computational aspects of a modified algorithm for constructing a sequence of plans for the solution of discrete optimization problems (Russian).
RADE	1986-1	Radermacher, F. J. Schedule-induced posets.
RAMA	1986-1	Ramamurthy, K. G.; Parthasarathy, T. An algorithm to find the smallest committee containing a given set.
RECS	1987-3	Recski, A. One-story buildings as tensegrity frameworks, II.
RECS	1987-4	Recski, A. Some algorithmic problems of linear network solvability.
REND	1984-1	Rendl, F. On the complexity of decomposing matrices arising in satellite communication.
RIJC	1986-1	Rijckaert, M. J. Heuristics and the role of microcomputers for the one-dimensional cutting stock problem.
RINNB	1987-1	Rinnooy Kan, A. H. G.; Wit, J. R. de; Wijmenga, R. T. Nonorthogonal two-dimensional cutting patterns.
RIVA	1984-1	Rival, I. Linear extensions of finite ordered sets.
RIVA	1985-1	Rival, I. Some order theoretical ideas about scheduling.
RIVA	1986-1	Rival, I. Stories about order and the letter N.
RIVAA	1986-1	Rival, I.; Zaguia, N. Constructing greedy linear extensions by interchanging chains.
RIVAA	1986-2	Rival, I.; Zaguia, N. Constructing N-free, jump-critical ordered sets.
RIVAA	1987-1	Rival, I.; Zaguia, N. Greedy linear extensions with constraints.
ROOD	1986-1	Roodman, G. M. Near-optimal solutions to one-dimensional cutting stock problems.
ROSEC	1986-1	Rosenblatt, M. J. The dynamics of plant layout.
ROSEF	1984-1	Rosenstiehl, P.; Tarjan, R. E. Gauss codes, planar hamiltonian graphs, and stack-sortable permutations.
ROTEA	1986-1	Rote, G.; Rendl, F. Minimizing the density of terminal assignments in layout design.
ROTH	1985-1	Roth, A. E. The college admissions problem is not equivalent to the marriage problem.
ROTHA	1985-1	Rothe, G.; Rendl, F. Minimizing the density of terminal assignments in layout design.
ROUBA	1985-1	Roubens, M.; Vincke, P. Preference modelling.
RULL	1986-1	Rülling, W. Constructing legal layouts from stick diagrams is an NP-complete problem.
SAKAA	1984-2	Sakarovitch, M. Optimisation combinatoire - methodes mathematiques et algorithmiques. Vol. 2: Graphes et programmation lineaire.
SAKS	1986-1	Saks, M. Some sequences associated with combintorial structures.
SCHAA	1985-1	Schaffrin, B. Aspects of network design.
SCHEB	1985-1	Scheithauer, G.; Terno, J. Lösung von Guillotine-Zuschnittproblemen mittels revidierter Simplexmethode und Spaltengenerierung.
SCHEB	1986-1	Scheithauer, G.; Terno, J. Effektive Lösung von Guillotine-Zuschnittproblemen.

SCHRE 1986-2 Schrijver, A.
Geometric methods in discrete optimization.
SCHRE 1986-3 Schrijver, A.
Polyhedral combinatorics - some recent developments and results.
SCHRE 1986-4 Schrijver, A.
Polyhedral proof methods in combinatorial optimization.
SCHRE 1986-5 Schrijver, A.
Theory of linear and integer programming.
SCRI 1985-1 Scriabin, M.; Vergin, R. C.
A cluster-analytic approach to facility layout.
SEDG 1983-1 Sedgewick, R.
Mathematical analysis of combinatorial algorithms.
SERGA 1982-1 Sergienko, I. V.
Some questions of the development and use of approximation methods of solution of discrete optimization problems (Russian).
SERGA 1985-1 Sergienko, I. V.
Mathematical models and methods for solving problems of discrete optimization.
SERGC 1984-1 Sergienko, I. V.; Chervak, Y. Y.; Grendzhi, Y. I.
Paralleling of discrete optimization in lexicographic algorithms (Russian).
SERGE 1983-1 Sergienko, I. V.; Kaspshitskaya, M. F.
Stability of algorithms of the method of descent vector in a class of combinatorial optimization problems (Russian).
SETH 1987-1 Seth, A.
Wastage reduction in wood cutting.
SETHA 1986-1 Seth, A.; Prasad, V. R.; Ramamurthy, K. G.
A heuristic approach to one-dimensional cutting stock problem.
SHCH 1986-1 Shcherbina, O. A.
A modified local algorithm for solving block problems of discrete programming.
SHCH 1987-1 Shcherbina, O. A.
On quasiblock problems of discrete programming with additional restrictions (Russian).
SHEN 1982-1 Shen, L.
Effective use of heuristic algorithms in network design.
SHEVA 1979-1 Shevchenko, V. N.
Convex polyhedral cones, comparison systems and regular truncations in integer programming (Russian).
SHTI 1985-1 Shtilman, B. M.
A formal linguistic model for solving discrete optimization problems. I. Optimization tools. Language of trajectories.
SINGB 1986-1 Singer, I.
A general duality approach to min-max results in combinatorial optimization. I.
SKIL 1986-1 Skillicorn, D. B.
Graph problems in network design.
SMET 1987-1 Smetanin, Y. G.; Khachiyan, L. G.
Use of pseudopolynomial algorithms for some problems of combinatorial optimization with constraints.
SMITB 1983-1 Smith, D. H.
Reliable networks and (k,2k-2)-connected graphs.
SPIN 1982-1 Spinrad, J.
Two dimensional partial orders.
STEI 1985-3 Steiner, G.
On finding the jump number of a partial order by substitution decomposition.
STEI 1987-1 Steiner, G.
On computing the information theoretic bound for sorting: counting the linear extensions of posets.
STEI 1987-2 Steiner, G.
On computing the information theoretic bound for sorting: balancing the linear extensions of posets.
STOI 1984-1 Stoica, M.; Angheluta, G.
An application of the problem of three-dimensional allocation.
SUMI 1986-1 Sumichrast, R. T.
A new cutting-stock heuristic for scheduling production.
SYSL 1985-1 Syslo, M. M.
A graph-theoretic approach to the jump-number problem.
SYSL 1985-2 Syslo, M. M.
A reduction lemma for the jump-number problem.
SYSL 1985-3 Syslo, M. M.
Remarks on Dilworth partially ordered sets.
SYSL 1986-1 Syslo, M. M.
An algorithm for solving the jump number problem.
SYSL 1986-2 Syslo, M. M.
On some new types of greedy chains and greedy linear extensions of partially ordered sets.
SYSL 1987-2 Syslo, M. M.
Minimizing the jump number for partially-ordered sets: a graph-theoretic approach, II.
TALA 1985-1 Talamo, M.; Marchetti-Spaccamela, A.; Protasi, M.
On some relationships between combinatorics and probabilistic analysis.
TAMA 1984-1 Tamaki, M.
The secretary problem with optimal assignment.
TARA 1985-1 Tarakanov, V. E.
Combinatorial problems and (0,1)-matrices (Russian).
TARD 1986-1 Tardos, E.
A strongly polynomial algorithm to solve combinatorial linear programs.
TELL 1987-1 Telley, H.; Liebling, T. M.; Mocellin, A.
Reconstruction of polycrystalline structures: a new application of combinatorial optimization.
TERL 1986-1 Terlaky, T.
A finite criss-cross method and applications (Hungarian).

THOM 1984-1 Thomas, R.
Series-parallel graphs and well- and better-quasi-orderings.
THUL 1987-2 Thulasiraman, K.; Comeau, M. A.
Structure of the reachability problem for (0,1)-capacitated marked graphs.
TINH 1984-3 Tinhofer, G.
Probabilistische Ansätze in der diskreten Optimierung.
TOCZ 1986-1 Toczylowski, E.
On aggregation in a two-dimensional cutting stock scheduling problem.
TOKU 1985-1 Tokuyama, H.; Ueno, N.
The cutting stock problem for large sections in the iron and steel industries.
TOVE 1981-1 Tovey, C. A.
Polynomial local improvement algorithms in combinatorial optimization.
TOVEA 1985-1 Tovey, C. A.; West, D. B.
Networks and chain coverings in partial orders and their products.
TRIEA 1985-1 Trietsch, D.
Augmenting Euclidean networks - the Steiner case.
TRIEA 1985-2 Trietsch, D.
Connecting Euclidean networks–the Steiner case.
TROTA 1987-1 Trotter, L. E. Jr.; West, D. B.
Two easy duality theorems for product partial orders.
TROTB 1986-1 Trotter, W. T. Jr.
Problems and conjectures in the combinatorial theory of ordered sets.
TRUSB 1985-2 Truszczynski, M.
Jump number problem: the role of matroids.
TSENA 1984-1 Tseng, S. S.; Lee, R. C. T.
A parallel algorithm to solve the stable marriage problem.
TUTZ 1985-1 Tutzauer, F.
Toward a theory of disintegration in communication networks.
VESZ 1985-1 Vesztergombi, K.
Bounds on the number of small distances in a finite planar set.
VESZ 1985-2 Vesztergombi, K.
On the distribution of distances in finite sets in the plane.
VESZ 1987-1 Vesztergombi, K.
On large distances in planar sets.
VIZV 1986-1 Vizvari, B.
Some recent Hungarian results in discrete programming (Hungarian).
VOIG 1987-1 Voigt, J.-U.
Solving cutting stock problems of the furniture industry via man-machine communication.
WAGNA 1984-1 Wagner, K.
The complexity of combinatorial problems with compactly described instances.
WAH 1984-1 Wah, B. W.; Chen, K. L.
A partitioning approach to the design of selection networks.
WANGB 1985-1 Wang, H.
A combinatorial optimal algorithm with least partition times (Chinese).
WANGE 1984-1 Wang, J. S.
A branch and bound algorithm for a logic partitioning problem (Chinese).
WANGN 1984-1 Wang, Z.-Y.
Some combinatorial problems on t-ary trees.
WANGN 1984-2 Wang, Z.-Y.
Some combinatorial problems on t-ary trees.
WASC 1985-1 Wäscher, G.; Carow, P.; Müller, H.
Ein flexibles Lösungsverfahren für Zuschneideprobleme in einem Kaltwalzwerk.
WASC 1985-2 Wäscher, G.; Carow, P.; Müller, H.
Entwicklung eines flexiblen Verfahrens für Zuschneideprobleme in einem Kaltwalzwerk.
WASCA 1987-1 Wäscher, G.; Chamoni, P.
MICROLAY: an interactive computer program for factory layout planning on microcomputers.
WEI B 1986-1 Wei, W. D.; Cai, Y. Z.; Liu, C. L.; Odlyzko, A. M.
Balloting labelling and personnel assignment.
WERN 1986-1 Werner, F.
On the method of interchanging networks for permutation problems.
WEST 1985-1 West, D. B.
Parameters of partial orders and graphs: packing, covering and representation.
WONGA 1984-1 Wong, R. T.
Introduction and recent advances in network design: models and algorithms.
WONGA 1985-1 Wong, R. T.
Location and network design.
WONGA 1986-1 Wong, R. T.
Recent advances in location and network design models for transportation planning.
YANGA 1985-1 Yang, C. E.; Lin, Y. X.
Board welding problem (Chinese).
ZAGU 1987-1 Zaguia, N.
Greedy posets for the bump-minimizing problem.
ZHANM 1983-1 Zhang, G. Q.
NP-completeness and restricted partitions (Chinese).
ZIMMC 1986-2 Zimmermann, U.
Linear and combinatorial sharing problems.
ZINC 1984-1 Zinchenko, A. B.
Properties and conditions for solvability of a special problem of optimal geometric allocation (Russian).
ZISS 1985-1 Zissimopoulos, V.
Heuristic methods for solving (un)constrained 2-dimensional cutting stock problems.

AFRA 1985-1 Afrati, F.; Papadimitriou, C. H.; Papageorgiou, G.
The complexity of cubical graphs.
AGEE 1983-1 Ageev, A. A.
On the complexity of the minimization problems for polynomials of Boolean functions (Russian).
ALBR 1984-1 Albrecht, A.; Jung, H.; Bauernöppel, F.; Sammler, O.
Einführende Kapitel der Kompliziertheitstheorie.
ALEK 1980-1 Alekseev, A. E.; Kogan, D. I.
Complexity of determination of the optimal allocation of tasks to persons with two-sided preferences (Russian).
ALEKF 1982-1 Alekseev, V. E.
The effect of local constraints on the complexity of determination of the graph independence number (Russian).
ARAI 1986-1 Arai, M.; Yamashita, M.; Hirata, T.; Ibaraki, T.; Honda, N.
Lower bounds on time complexity for some subgraph detection problems (Japanese).
ARNBA 1984-1 Arnborg, S.; Corneil, D. G.; Proskurowski, A.
Complexity of finding embeddings in a k-tree.
ARNBB 1984-1 Arnborg, S.; Proskurowski, A.
Linear time algorithms for NP-hard problems on graphs embedded in k-trees.
ASAN 1987-1 Asano, T.
An application of duality to edge-deletion problems.
ASANB 1983-1 Asano, T.; Hirata, T.
Edge-contraction problems.
BABAF 1987-1 Babai, L.; Turan, G.
The complexity of defining a relation on a finite graph.
BACHA 1984-1 Bachem, A.
Optimization and geometry in discrete structures.
BACHB 1984-1 Bachem, A.; Euler, R.
Recent trends in combinatorial optimization.
BENDA 1985-1 Bender, E. A.; Wilf, H. S.
A theoretical analysis of backtracking in the graph coloring problem.
BENT 1987-1 Bent, S. W.; Manber, U.
On non-intersecting Eulerian circuits.
BERNA 1985-1 Bern, M. W.; Lawler, E. L.; Wong, A. L.
Why certain subgraph computations require only linear time.
BESSA 1981-1 Bessonov, Y. E.; Skorobogatov, V. A.
Recursive analysis of graphs (Russian).
BESSA 1982-1 Bessonov, Y. E.; Skorobogatov, V. A.
On a family of schemes of recursive stripping of graphs (Russian).
BHATA 1987-1 Bhatt, S. N.; Cosmadakis, S. S.
The complexity of minimizing wire lengths in VLSI layouts.
BIENE 1987-1 Bienstock, D.; Monma, C. L.
On the complexity of covering vertices by faces in a planar graph.
BIENE 1987-2 Bienstock, D.; Monma, C. L.
On the complexity of embedding planar graphs to minimize certain distance measures.
BLAIA 1986-1 Blair, C. E.; Jeroslow, R. G.
Computational complexity of some problems in parametric discrete programming. I.
BLAS 1982-1 Blass, A.; Gurevich, Y.
On the unique satisfiability problem.
BLAZH 1984-1 Blazewicz, J.; Kubiak, W.; Röck, H.
Complexity of minimizing mean flow time under resource constraits.
BLAZJ 1984-1 Blazewicz, J.; Weglarz, J.
The complexity of scheduling under resource constraints (Polish).
BLUM 1985-1 Blum, N.
On the single-operation worst-case time complexity of the disjoint set union problem.
BOND 1985-1 Bondarenko, V. A.
Nonpolynomial lower bounds for the complexity of optimization problems of clique and vertex covering in a class of direct algorithms (Russian).
BORG 1987-1 Borgwardt, K. H.
Probabilistic analysis of optimization algorithms - some aspects from a practical point of view.
BOROD 1986-1 Boros, E.
On the complexity of the surrogate dual of 0-1 programming.
BOUCD 1986-1 Bouchitte, V.; Habib, M.
Some NP-completeness properties about linear extensions.
BRANA 1987-1 Brandstädt, A.
The computational complexity of feedback vertex set, Hamiltonian circuit, dominating set, Steiner tree, and bandwidth on special perfect graphs.
BRANB 1985-1 Brandstädt, A.; Kratsch, D.
On the restriction of some NP-complete graph problems to permutation graphs.
BRUNB 1985-1 Bruno, J. L.; Downey, P. J.
Probabilistic bounds for dual bin-packing.
BURKB 1985-1 Burkard, R. E.; Fincke, U.
Probabilistic asymptotic properties of some combinatorial optimization problems.
CAI 1987-1 Cai, J.; Meyer, G. E.
On the complexity of graph critical uncolorability.
CAMEA 1985-1 Camerini, P. M.; Galbiati, G.; Maffioli, F.
The complexity of weighted multi-constrained spanning tree problems.
CAMEF 1985-1 Cameron, J.; Thomas, G.
An approximate graph partitioning algorithm and its computational complexity.
CARSA 1985-1 Carstensen, P. J.; Murty, K. G.; Perin, C.
Intermediate feasibility in 0-1 integer linear systems.
CHANAD 1987-1 Chang, Y.; Hambrusch, S. E.; Simon, J.
On the computational complexity of continuous routing.

CHANN	1985-1	Chandru, V.; Coullard, C. R.; Wagner, D. K. On the complexity of recognizing a class of generalized networks.
CHANO	1987-1	Chandru, V.; Trick, M. A. On the complexity of Lagrange multiplier search.
CHAOA	1986-1	Chao, M.-T.; Franco, J. Probabilistic analysis of two heuristics for the 3-satisfiability problem.
CHENC	1984-1	Chen, J. Y. The satisfiability problem for simple Boolean expressions belongs to P (Chinese).
CHERD	1986-1	Chernjak, Z. A.; Chernjak, A. A. About recognizing (alpha, beta) classes of polar graphs.
COFFF	1986-1	Coffman, E. G. Jr.; Lueker, G. S.; Rinnooy Kan, A. H. G. An introduction to the probabilistic analysis of sequencing and packing heuristics.
CONF	1984-1	Conforti, M.; Corneil, D. G.; Mahjoub, A. R. K index i - covers 1: complexity and polytopes.
CONF	1986-1	Conforti, M.; Corneil, D. G.; Mahjoub, A. R. K(i)-covers I: complexity and polytopes.
COOKB	1987-1	Cook, W.; Coullard, C. R.; Turan, G. On the complexity of cutting-plane proofs.
COOKF	1984-1	Cook, W.; Lovasz, L.; Schrijver, A. A polynomial-time test for total dual integrality in fixed dimension.
CORN	1985-1	Corneil, D. G. The complexity of generalized clique packing.
CRAM	1985-1	Crama, Y.; Hammer, P. L. The complexity of recognizing partition-quadratic graphs.
CSIRC	1986-1	Csirik, J.; Galambos, G.; Frenk, J. B. G.; Frieze, A. M.; Rinnooy Kan, A. H. G. A probabilistic analysis of the next fit decreasing bin packing heuristic.
CUNNA	1985-2	Cunningham, W. H. Optimal attack and reinforcement of a network.
DAHLA	1986-1	Dahlhaus, E.; Karpinski, M. Perfect matching for regular graphs is AC**0-hard for the general matching problem.
DAHLA	1986-2	Dahlhaus, E.; Karpinski, M. The matching problem for strongly chordal graphs is in NC.
DAHLA	1987-3	Dahlhaus, E.; Karpinski, M. On the parallel complexity of matching for chordal and path graphs.
DEPA	1987-1	DePalma, A.; Hansen, P.; Labbe, M. Commuters' paths with penalties for early or late arrival time.
DERI	1985-4	Derigs, U. Using confidence limits for the global optimum in combinatorial optimization.
DESI	1986-1	DeSimone, C.; Lucertini, M.; Pallottino, S.; Simeone, B. Fair dissections of spiders, worms and caterpillars.
DEWD	1985-1	Dewdney, A. K.; Walsh, T. R. A methodology for average-time testing of exact and approximate graph algorithms.
DYERB	1985-2	Dyer, M. E.; Frieze, A. M. On the complexity of partitioning graphs into connected subgraphs.
EDEN	1987-1	Edenbrandt, A. Chordal graph recognition is in NC.
EDWA	1986-1	Edwards, K. The complexity of colouring problems on dense graphs.
EMELC	1986-1	Emelichev, V. A.; Perepelitsa, V. A. On the estimation of the complexity of multicriterial transportation problems (Russian).
FAIGF	1986-3	Faigle, U.; Schrader, R. On the computational complexity of the order polynomial.
FAIGH	1987-1	Faigle, U.; Turan, G. On the complexity of interval orders and semiorders.
FARR	1985-1	Farr, G.; Mac Diarmid, C. The complexity of counting homeomorphs.
FENN	1984-2	Fenner, T. I.; Frieze, A. M. On the complexity of finding sets of edges in graphs and digraphs.
FISHI	1981-1	Fisher, M. L.; Wolsey, L. A. On the greedy heuristic for covering and packing problems.
FRANE	1985-1	Frank, A.; Tardos, E. An application of simultaneous approximation in combinatorial optimization.
FRANE	1987-1	Frank, A.; Tardos, E. An application of simultaneous Diophantine approximation in combinatorial optimization.
FRENE	1986-1	Frenkel, K. A. Complexity and parallel processing: an interview with Richard M. Karp.
FU	1986-1	Fu, Y.-T.; Anderson, P. W. Application of statistical mechanics to NP-complete problems in combinatorial optimization.
GAZM	1986-1	Gazmuri, P. G. Probabilistic analysis of algorithms for some combinatorial optimization problems.
GIBB	1985-1	Gibbons, A. Algorithmic graph theory.
GOETB	1985-1	Goetze, B.; Staedler, G. Algorithms on graphs with costly adjacency-information.
GONZB	1982-1	Gonzalez, T. F. On the computational complexity of clustering and related problems.
GRAHA	1984-1	Graham, R. L.; Foulds, L. R. It is unlikely that minimal phylogenies for a realistic biological study can be constructed in reasonable computational time.
GRIGD	1987-1	Grigoriev, D. Y.; Karpinski, M. The matching problem for bipartite graphs with polynomially bounded permanents is in NC.
GROT	1984-1	Grötschel, M. Developments in combinatorial optimization.

HABIC	1987-1	Habib, M.; Möhring, R. H. On some complexity properties of N-free posets and posets with bounded decomposition diameter.
HALLC	1986-1	Hall, N. G.; Rhee, W. T. Average and worst-case analysis of heuristics for the maximum tardiness problem.
HAREB	1987-1	Hare, E. O.; Hedetniemi, S. T.; Laskar, R.; Peters, K.; Wimer, T. V. Linear-time computability of combinatorial problems on generalized series-parallel graphs.
HARTE	1987-2	Hartvigsen, D.; Zemel, E. On the complexity of lifted inequalities for the knapsack problem.
HELM	1986-1	Helmbold, D.; Mayr, E. Two processor scheduling is in NC.
HELM	1987-2	Helmbold, D.; Mayr, E. Two processor scheduling is in NC .
HOCH	1984-1	Hochbaum, D. S. When are NP-hard location problems easy?
HOCH	1985-1	Hochbaum, D. S. Easy solutions for the K-center problem or the dominating set problem on random graphs.
HOCHB	1985-1	Hochbaum, D. S.; Shmoys, D. B. A best possible heuristic for the k-center problem.
HOCHC	1986-1	Hochbaum, D. S.; Wigderson, A. The Linzertorte problem, or a unified approach to painting, baking and weaving.
IBAR	1987-1	Ibaraki, T. Enumerative Approaches to Combinatorial Optimization - Part 1.
IRI	1987-1	Iri, M. A very personal reminiscence on the problem of computational complexity.
IRVIA	1986-1	Irving, R. W.; Leather, P. The complexity of counting stable marriages.
IVANF	1984-1	Ivanyi, A. Estimation of the efficiency of bin-packing algorithms (Russian).
JAUM	1986-1	Jaumard, B.; Minoux, M. An efficient algorithm for the transitive closure and a linear worst-case complexity result for a class of sparse graphs.
JAYA	1984-1	Jayakumar, R.; Thulasiraman, K.; Swamy, M. N. S. Complexity of computation of a spanning tree enumeration algorithm.
JOHNG	1985-1	Johnson, D. S.; Papadimitriou, C. H. Computational complexity.
KAMB	1986-1	Kamburowski, J. On the computational complexity of the shortest route and maximum flow problems in stochastic networks.
KAMB	1987-1	Kamburowski, J. An overview of the computational complexity of the PERT shortest route and maximum flow problems in stochastic networks.
KANA	1984-1	Kanaeva, N. N.; Shcherbina, O. A. On the efficiency of a modified local algorithm for solving discrete optimization problems.
KARMA	1982-1	Karmarkar, N. Probabilistic analysis of some bin-packing problems.
KARP	1982-1	Karp, R. M. The computational complexity of network problems.
KARP	1986-1	Karp, R. M. Combinatorics, complexity and randomness.
KARPB	1985-1	Karp, R. M.; Lenstra, J. K.; Mac Diarmid, C.; Rinnooy Kan, A. H. G. Probabilistic analysis.
KARPC	1980-1	Karp, R. M.; Papadimitriou, C. H. On linear characterizations of combinatorial optimization problems.
KARPF	1985-1	Karp, R. M.; Steele, J. M. Probabilistic analysis of heurictics.
KARPG	1985-1	Karp, R. M.; Upfal, E.; Wigderson, A. Constructing a perfect matching is in random NC.
KARPG	1985-2	Karp, R. M.; Upfal, E.; Wigderson, A. The complexity of parallel computation on matroids.
KARPI	1987-1	Karpinski, M.; Lingas, A. Subtree isomorphism and bipartite perfect matchings are mutually nc-reducible.
KERN	1986-1	Kern, W. A probabilistic analysis of the switching algorithm for the euclidean TSP.
KIER	1986-1	Kierstead, H. A. NP-completeness results concerning greedy and super greedy linear extensions.
KIMO	1984-1	Kimoto, T.; Tsukiyama, S.; Shirakawa, I. On the computational complexity of assignment problems with restrictions (Japanese).
KIND	1986-3	Kindervater, G. A. P.; Lenstra, J. K. The parallel complexity of TSP heuristics.
KIRA	1986-1	Kiran, A. S. Complexity of FMS loading and scheduling problems.
KOLI	1987-1	Kolinek, M. A polynomial-time linear decision tree for the traveling salesman problem and other NP-complete problems.
KORNB	1986-2	Kornienko, N. M.; Lepin, V. V. Optimal packings of a weighted cycle into a line (Russian).
KRAMB	1982-1	Kramer, M. R.; Leeuwen, J. van Wire-routing is NP-complete.
KUBA	1987-1	Kubale, M. The complexity of scheduling independent two-processor tasks on dedicated processors.
KUND	1987-1	Kunde, M.; Steppat, H. On the worst-case ratio of a compound multiprocessor scheduling algorithm.

KUZMA 1985-2 Kuzmin, S. L.
On the complexity of flow problems with path length constraints.

KUZY 1984-1 Kuzyurin, N. N.
The complexity of approximate algorithms for the problem of integer programming.

LAGA 1982-1 Lagarias, J. C.
The computational complexity of simultaneous diophantine approximation problems.

LAI B 1985-1 Lai, T.-H.; Sprague, A.
On the complexity of a family of generalized matching problems.

LAKS 1986-1 Lakshmipathy, N.; Winklmann, K.
"Global" graph problems tend to be intractable.

LEE H 1986-1 Lee, D. T.; Wu, Y. F.
Geometric complexity of some location problems.

LENSB 1982-1 Lenstra, J. K.; Rinnooy Kan, A. H. G.; Emde Boas, P. van
An appraisal of computational complexity for operations research.

LEVI 1983-1 Levin, A. A.
Gluing of Boolean functions and its application in estimates of the comparative complexity of DNF (Russian).

LI A 1986-1 Li, C.-L.; Mac Cormick, S. T.; David, S.-L.
The complexity of finding two disjoint paths with min-max objective function.

LINI 1986-2 Linial, N.
Hard enumeration problems in geometry and combinatorics.

LINIB 1985-1 Linial, N.; Tarsi, M.
Deciding hypergraph 2-colourability by H-resolution.

LIU B 1985-1 Liu, C. Y.; Bulfin, R. L.
On the complexity of preemptive open-shop scheduling problems.

LIU I 1983-3 Liu, Z. H.
NP-complete problems and some approximate algorithms II (Chinese).

LIU I 1983-4 Liu, Z. H.
NP-complete problems and some approximate algorithms I (Chinese).

LOIZ 1984-1 Loizou, G.
Certain NP-complete matching problems.

LOVA 1985-1 Lovasz, L.
An algorithmic theory of numbers, graphs, and convexity.

LOVA 1985-2 Lovasz, L.
Computing ears and branchings in parallel.

LOVA 1986-3 Lovasz, L.
Theory of numbers, graphs and convexity.

MAC G 1985-1 Mac Diarmid, C.
On some conditioning results in the probabilistic analysis of algorithms.

MACH 1987-1 Machida, H.
A remark on the complexity of the knapsack problem.

MARC 1987-1 Marchetti-Spaccamela, A.; Pelaggi, A.
Worst case analysis of two heuristics for the set partitioning problem.

MEHL 1984-1 Mehlhorn, K.
Data structures and algorithms 2: Graph algorithms and NP-completeness.

MEIN 1984-1 Meinel, C.
Effiziente Algorithem.

MIKHB 1987-1 Mikhalevich, M. V.; Sapko, A. V.
The efficiency of an algorithm for solving the traveling salesman problem (Russian).

MILA 1986-1 Milanov, P. B.
On the worst behaviour of the greedy procedure.

MONI 1985-1 Monien, B.
How to find long paths efficiently.

MONI 1985-2 Monien, B.
The complexity of embedding graphs into binary trees.

MONI 1986-1 Monien, B.
The bandwidth minimization problem for caterpillars with hair length 3 is NP-complete.

MONIB 1985-1 Monien, B.; Sudborough, I. H.
Bandwidth constrained NP-complete problems.

MULLA 1987-1 Müller, H.; Brandstädt, A.
The NP-completeness of STEINER TREE and DOMINATING SET for chordal bipartite graphs.

MURTA 1985-1 Murty, K. G.; Kabadi, S. N.
Some NP-complete problems in quadratic and nonlinear programming.

NESEA 1985-1 Nesetril, J.; Poljak, S.
On the complexity of the subgraph problem.

NESEC 1987-1 Nesetril, J.; Rödl, V.
Complexity of diagrams.

NOSH 1985-1 Noshita, K.
A theorem on the expected complexity of Dijkstra's shortest path algorithm.

NOSO 1985-1 Nosov, V. A.; Sachkov, V. N.; Tarakanov, V. E.
Combinatorial analysis (nonnegative matrices, algorithmic problems).

NTAF 1984-1 Ntafos, S.; Gonzalez, T.
On the computational complexity of path cover problems.

ONG 1984-1 Ong, H. L.
Probabilistic analysis of some travelling salesman heuristics.

ORLI 1985-5 Orlin, J. B.
The complexity of dynamic/periodic languages and optimization problems.

PAPA 1985-2 Papadimitriou, C. H.
Computational complexity.

PAPAC 1985-1 Papadimitriou, C. H.; Wolfe, D.
The complexity of facets resolved.

PAPAE 1986-1 Papakostantinou, N.
The complexity of the travelling repairman problem.

PAVL	1985-1	Pavlov, A. A. Analysis of complexity estimates of exact algorithms for solving the general linear integer programming problem.
PERO	1985-1	Peroche, B. Complexite de l'arboricite lineaire d'une graph II.
PERS	1985-1	Pershin, O. Y. One class of problems for synthesis of the optimal networks (Russian).
PHAN	1986-1	Phan Dinh Dieu; Le Cong Thanh; Le Tuan Hoa Average polynomial time complexity of some NP-complete problems.
POLJA	1986-1	Poljak, S.; Turzik, D. A polynomial time heuristic for certain subgraph optimization problems with guaranteed worst case bound.
PROV	1986-1	Provan, J. S. The complexity of reliability computations in planar and acyclic graphs.
PURD	1985-1	Purdom, P. W.; Brown, C. A. The analysis of algorithms.
QUEY	1986-1	Queyranne, M. Performance ratio of polynomial heuristics for triangle inequality quadratic assignment problems.
REND	1984-1	Rendl, F. On the complexity of decomposing matrices arising in satellite communication.
RHEE	1987-1	Rhee, W.-S. T. Probabilistic analysis of the next fit decreasing algorithm for bin-packing.
RICHB	1987-1	Richey, M. B.; Parker, R. G. On computational issues related to series-parallel solvability.
RICHF	1985-1	Richter, K. Complexity of the robot sequencing problem.
RICHF	1985-3	Richter, K. The robot sequencing problem: polynomial algorithm and complexity.
RINN	1987-1	Rinnooy Kan, A. H. G. Probabilistic analysis of algorithms.
RUHEB	1986-1	Ruhe, G.; Sieber, N. Zur Komplexität zweier vektorieller Netzwerkflussprobleme.
RULL	1986-1	Rülling, W. Constructing legal layouts from stick diagrams is an NP-complete problem.
SAKAA	1984-1	Sakarovitch, M. Optimisation combinatoire - methodes mathematiques et algorithmiques. Vol. 1: Programmation discrete.
SCHE	1985-1	Scheinerman, E. R. Characterization and computational complexity questions for representation classes of graphs.
SCHOA	1985-1	Schöning, U. Complexity and structure.
SCHRE	1986-5	Schrijver, A. Theory of linear and integer programming.
SCHRE	1987-2	Schrijver, A. Polyhedral combinatorics.
SCHRF	1984-1	Schröter, N. Average-case analysis of an approximative algorithm for the knapsack problem.
SEES	1985-1	Seese, D. Baum-partite Graphen und die Komplexität von Algorithmen.
SEES	1985-2	Seese, D. Tree-partite graphs and the complexity of algorithms.
SHORA	1984-1	Shor, P. W. The average-case analysis of some on-line algorithms for bin packing.
SHORA	1986-1	Shor, P. W. The average-case analysis of some on-line algorithms for bin packing.
SIMO	1981-1	Simon, I. Probabilistic algorithms and complexity (Portuguese).
SPIN	1985-2	Spinrad, J. Worst-case analysis of a scheduling algorithm.
SPINA	1987-1	Spinrad, J.; Brandstädt, A.; Stewart, L. Bipartite permutation graphs.
SPINB	1985-1	Spinrad, J.; Vijayan, G. Worst case analysis of a graph coloring algorithm.
SRIS	1986-1	Sriskandarajah, C.; Ladet, P. Some no-wait shops scheduling problems: complexity aspects.
STANC	1983-1	Standley, H. M. Minimal area graph embedding into a directed grid is NP-hard.
TALA	1985-1	Talamo, M.; Marchetti-Spaccamela, A.; Protasi, M. On some relationships between combinatorics and probabilistic analysis.
TARJ	1987-1	Tarjan, R. E. Algorithm design.
TIMK	1985-1	Timkovskii, V. G. On the complexity of scheduling an arbitrary system.
TINH	1984-1	Tinhofer, G. A probabilistic analysis of some greedy cardinality matching algorithms.
TINH	1984-2	Tinhofer, G. Naive greedy matching - a probabilistic analysis.
TUNGB	1987-1	Tung, S. P. Computational complexities of diophantine equations with parameters.
URSI	1983-1	Ursic, S. A linear characterization of NP-complete problems.
VAIN	1982-1	Vainer, V. G. Estimates of the accuracy of algorithms for construction of connecting networks (Russian).

VAZI 1985-1 Vazirani, U. V.; Vazirani, V. V.
The 2-processor scheduling problem is in R-NC.
VERC 1984-1 Vercellis, C.
A probabilistic analysis of the set covering problem.
VERC 1986-1 Vercellis, C.
A probabilistic analysis of the set packing problem.
VORN 1981-1 Vornberger, O.
Alternating cycle covers and paths.
WAGNA 1984-1 Wagner, K.
The complexity of combinatorial problems with compactly described instances.
WAGNA 1984-2 Wagner, K.
The complexity of problems concerning graphs with regularities.
WANG 1985-1 Wang, C.-C.; Lloyd, E. L.; Soffa, M. L.
Feedback vertex sets and cyclically reducible graphs.
WEGE 1985-1 Wegener, I.
The critical complexity of all (monotone) Boolean functions and monotone graph properties.
WEGE 1985-2 Wegener, I.
The critical complexity of all (monotone) Boolean functions and monotone graph properties.
WILF 1986-1 Wilf, H. S.
Algorithms and complexity.
YAP 1984-1 Yap, H. P.
Computational complexity of graph properties.
ZEME 1984-1 Zemel, E.
Probabilistic analysis of geometric location problems.
ZEME 1985-1 Zemel, E.
Probabilistic analysis of geometric location problems.
ZEME 1986-1 Zemel, E.
On the computational complexity of facets of the knapsack problem.
ZHANM 1983-1 Zhang, G. Q.
NP-completeness and restricted partitions (Chinese).
ZHANM 1984-1 Zhang, G. Q.
"NP=P?" and restricted partitions.
ZHANN 1986-1 Zhang, L. A.; Geng, S. Y.
The complexity of the 0/1 multi-knapsack problem.

ADRA 1983-1 Adrabinski, A.; Syslo, M. M.
Computational experiments with some approximation algorithms for the travelling salesman problem.

ALEKE 1984-1 Alekseev, O. G.; Gregorjev, V. F.
Some algorithms of the set-covering problem and their experimental estimation with computer (Russian).

ALI C 1986-3 Ali, A. I.; Padman, R.; Thiagarajan, H.
The dual algorithm for network problems: a computational study.

ARONA 1985-1 Aronson, J. E.; Barr, R.; Helgason, R. V.; Kennington, J. L.; Loh, A.; Zaki, H.
A projective transformation algorithm by Karmarkar: a computational experiment with assignment problems.

BALAJ 1987-1 Balas, E.; Nauss, R.; Zemel, E.
Comment on:'Some computational results on real 0-1 knapsack problems'.

BARAE 1987-1 Barahona, F.; Jünger, M.; Reinelt, G.
Experiments in quadratic 0-1 programming.

BARC 1985-1 Barcelo, J.
Computer experiments with constraint identification processes for some types of integer programs (Spanish).

DEMB 1987-1 Dembo, R. S.
A primal truncated Newton algorithm with application to large-scale nonlinear network optimization.

DERI 1984-1 Derigs, U.
Alternate strategies for solving bottleneck assignment problems - analysis and computational results.

DERI 1986-2 Derigs, U.
The shortest augmenting path method for solving assignment problems - motivation and computational experience.

DERID 1985-1 Derigs, U.; Metz, A.
An efficient labeling technique for solving sparse assignment problems.

DERID 1986-1 Derigs, U.; Metz, A.
An efficient labeling technique for solving sparse assignment problems.

DRORB 1986-1 Dror, M.; Levy, L.
A vehicle routing improvement algorithm comparison of a greedy and a matching implementation for inventory routing.

ENGQA 1986-1 Engquist, M.; Chen, C.-H. J.
Computational comparison of two solution procedures for allocation/processing networks.

FRANG 1983-1 Franke, H.-G.
Analyse der mittleren Rechenzeit für den transitiven Abschluß azyklischer Graphen.

GARR 1986-1 Garrett, S. E.
A microcomputer implementation of a facility location model.

GLYA 1983-1 Glyakov, P. V.
Algorithm for a partitioning problem and its investigation on an electronic computer (Russian).

GOLDK 1985-1 Golden, B. L.; Stewart, W. R. Jr.
Empirical analysis of heuristics.

GRAH 1986-1 Graham, D.; Nuttle, H. L. W.
A comparison of heuristics for a school bus scheduling problem.

GROTB 1984-1 Grötschel, M.; Jünger, M.; Reinelt, G.
A cutting plane algorithm for the linear ordering problem.

HOCHB 1985-3 Hochbaum, D. S.; Shmoys, D. B.
Using dual approximation algorithms for scheduling problems: theoretical and practical results.

HOCHB 1987-1 Hochbaum, D. S.; Shmoys, D. B.
Using dual approximation algorithms for scheduling problems: theoretical and practical results.

HORN 1986-1 Hornung, R.
Methods for solving the discrete capacity assignment problem in data networks: numerical comparison.

IKURB 1986-1 Ikura, Y.; Nemhauser, G. L.
Computational experience with a polynomial-time dual simplex algorithm for the transportation problem.

JENKA 1987-1 Jenkins, L.; Peters, D.
A computational comparison of Gomory and knapsack cuts.

KLINB 1985-1 Klingman, D.; Armstrong, R. D.; Partow-Navid, P.
Implementation and analysis of alternative algorithms for generalized shortest path problems.

LAZI 1985-1 Lazimy, R.
Improved algorithm for mixed-integer quadratic programs and a computational study.

LUNA 1987-1 Luna, H. P. L.; Ziviani, N.; Cabral, R. H. B.
The telephonic switching centre network problem: formalization and computational experience.

MOTE 1986-1 Mote, J.
The family constrained network problem.

MURP 1986-1 Murphy, R. A.
Some computational results on real 0-1 knapsack problems.

PEEM 1986-1 Peemöller, J.
Numerical experiences with graph coloring algorithms.

RUSSB 1985-2 Russell, R. S.; Taylor, B. W. III
An evaluation of sequencing rules for an assembly shop.

TALB 1986-1 Talbot, F. B.; Patterson, J. H.; Gehrlein, W. V.
A comparative evaluation of heuristic line balancing techniques.

WAKA 1986-1 Wakabayashi, Y.
Aggregation of binary relations: algorithmic and polyhedral investigations.

WALL 1986-1 Wallace, S.
Decomposing the requirement space of a transportation problem into polyhedral cones.

WONGA 1984-1 Wong, R. T.
Introduction and recent advances in network design: models and algorithms.

WRIG 1986-1 Wright, D. D.; Haehling von Lanzenauer, C.
COAL: A new heuristic approach for solving the fixed charge problem - computational results.

AMAR 1986-1 Amar, A. D.; Gupta, J. N. D.
Simulated versus real life data in testing the efficiency of scheduling algorithms.

BALA 1985-1 Balaban, A. T.; Filip, P. A.; Balaban, T.-S.
Computer program for finding all possible cycles in graphs.

BALA 1986-1 Balaban, A. T.; Filip, P. A.; Balaban, T.-S.
Erratum: Computer program for finding all possible cycles in graphs.

BOYDB 1987-1 Boyd, S. C.; Pulleyblank, W. R.; Cornuejols, G.
Travel - an interactive travelling salesman problem package for the IBM-personal computer.

BRUI 1984-1 Bruin, R. de
Some software for graph theoretical problems.

CHAUB 1987-2 Chaudhuri, P.
Efficient algorithms for some graph problems on a tree-structured parallel computer.

CHIBA 1985-1 Chiba, N.; Onoguchi, K.; Nishizeki, T.
Drawing plane graphs nicely.

EBERA 1987-1 Ebert, J.
A versatile data structure for edge-oriented graph algorithms.

EL-B 1985-1 El-Bakry, M.; Nassar, S.
The solution of the 0-1 programming problem by parallel processing.

FRED 1985-2 Frederickson, G. N.
Data structures for on-line updating of minimum spanning trees, with applications.

FRIS 1983-1 Fris, P.; Volgenant, A.
Solving the traveling salesman problem in Basic.

GARR 1986-1 Garrett, S. E.
A microcomputer implementation of a facility location model.

GOLDH 1986-1 Golden, B. L.; Bodin, L. D.; Goodwin, T.
Microcomputer-based vehicle routing and scheduling software.

GUTMB 1984-1 Gutman, I.; Cvetkovic, D.
Finding tricyclic graphs with a maximal number of matchings - another example of computer aided research in graph theory.

JENK 1987-2 Jenkins, L.
Using parametric integer programming to plan the mix of an air transport fleet.

KHRI 1985-1 Khristov, G.; Karamiteva, Z.; Stoyanov, T.
Numerical method and program for the solution of a linear fractional integer optimization problem (Bulgarian).

KIND 1985-1 Kindervater, G. A. P.; Lenstra, J. K.
An introduction to parallelism in combinatorial optimization.

KIND 1985-2 Kindervater, G. A. P.; Lenstra, J. K.
Parallel algorithms.

KIND 1986-1 Kindervater, G. A. P.; Lenstra, J. K.
An introduction to parallelism in combinatorial optimization.

KIND 1986-2 Kindervater, G. A. P.; Lenstra, J. K.
Parallel computing in combinatorial optimization.

KNOP 1984-1 Knop, J. V.; Szymanski, K.; Trinajstic, N.; Krivka, P.
Computer generation of all 1-factors for a class of graphs with all vertices of degree two or three.

LEEU 1985-1 Leeuwen, J. van
Parallel computers and algorithms.

MALS 1984-1 Malstrom, E. M.; Moraniec, D.
Complex lot sizing analyses with the pocket calculator.

MARK 1986-1 Markland, R. E.; Vickery, S. K.
The efficient computer implementation of a large-scale integer goal programming model.

MARTD 1983-1 Martin, E.
Lösung von Transportproblemen mit dem Programmsystem OPSI.

MINOA 1986-1 Minoux, M.; Bartnik, G.
Graphes, algorithmes, logiciels.

POWE 1985-1 Powell, S.
Software.

PURD 1985-1 Purdom, P. W.; Brown, C. A.
The analysis of algorithms.

PUTU 1986-1 Puturidze, Z. S.
Computational aspects of a modified algorithm for constructing a sequence of plans for the solution of discrete optimization problems (Russian).

RIBE 1987-1 Ribeiro, C. C.
Parallel computer models and combinatorial algorithms.

ROSH 1986-1 Roshchin, V. A.
An algorithm for solving a covering problem by means of parallel computations (Russian).

SIGA 1986-1 Sigal, I. K.
Computer implementation of a combined branch and bound algorithm for the travelling salesman problem.

SIGA 1987-1 Sigal, I. K.
An approximate algorithm for the large-scale travelling salesman problem and its computer implementation (Russian).

TINHA 1984-1 Tinhofer, G.; Schreck, H.
Linear time tree codes.

TOML 1985-1 Tomlin, J. A.; Welch, J. S.
Integration of a primal simplex network algorithm with a large-scale mathematical programming system.

VACH 1983-1 Vachajitpan, P.
A computer program for solving job sequencing problems.

VINO 1985-1 Vinokur, A. B.
One method for storage and retrieval of information on the shortest paths of graphs with distinguished characteristics (Russian).

WASCA 1987-1 Wäscher, G.; Chamoni, P.
MICROLAY: an interactive computer program for factory layout planning on microcomputers.

WILLD 1985-1 Willis, R. J.
An algorithm for constructing project network diagrams on a ordinary line printer.

AGAR	1985-1	Agarwal, S.; Mittal, A. K.; Sharma, P. A decomposition algorithm for linear relaxation of the weighted r-covering problem.
AGEE	1983-1	Ageev, A. A. On the complexity of the minimization problems for polynomials of Boolean functions (Russian).
AKOP	1984-1	Akopyan, G. C. Inclusion of a subset of edges of a hypergraph in some terminal c-covering (Russian).
ALEKE	1984-1	Alekseev, O. G.; Gregorjev, V. F. Some algorithms of the set-covering problem and their experimental estimation with computer (Russian).
ALONB	1985-1	Alon, N.; Tarsi, M. Covering multigraphs by simple circuits.
AMARC	1985-1	Amar, D.; Fournier, I.; Germa, A.; Häggkvist, R. Covering of vertices of a simple graph with given connectivity and stability number.
ARCH	1986-1	Archdeacon, D. Coverings of graphs by cycles.
ASRA	1984-1	Asratyan, A. S. Investigation of imbedded c-combinations and c-coverings (Russian).
BABAC	1987-1	Babaev, R. D. Generation of test problems on covering and partitioning (Russian).
BALAM	1986-1	Balas, E.; Shu, M.-N. On the set covering polytope: I. All the facets with coefficients in (0,1,2) (revised version).
BALLA	1985-1	Ball, M. O.; Roberts, A. A graph partitioning approach to airline crew scheduling.
BAR-	1985-1	Bar-Yehuda, R.; Even, S. A local-ratio theorem for approximating the weighted vertex cover problem.
BARAI	1986-1	Barany, I.; Edmonds, J.; Wolsey, L. A. Packing and covering a tree by subtrees.
BARAJ	1987-1	Barany, I.; Füredi, Z. Covering all secants of a square.
BAREA	1986-1	Barefoot, C. A.; Jones, K. F.; Lundgren, J. R.; Maybee, J. S. Biclique covers of regular digraphs.
BEAS	1987-1	Beasley, J. E. An algorithm for set covering problems.
BELAA	1984-1	Belardo, S.; Harrald, J.; Wallace, W. A.; Ward, J. E. A partial covering approach to siting response resources for major maritime oil spills.
BERTA	1986-1	Bertolazzi, P.; Sassano, A. A class of polynomially solvable set-covering problems.
BIENE	1987-1	Bienstock, D.; Monma, C. L. On the complexity of covering vertices by faces in a planar graph.
BOND	1985-1	Bondarenko, V. A. Nonpolynomial lower bounds for the complexity of optimization problems of clique and vertex covering in a class of direct algorithms (Russian).
BORZA	1984-1	Borzacchini, L.; Pulito, C. Coverings enumeration for graphical partially ordered sets.
BROI	1986-1	Broin, M. W.; Lowe, T. J. A dynamic programming algorithm for covering problems with (greedy) totally balanced constraint matrices.
CAENA	1985-1	Caen, D. de; Gregory, D. A.; Pullman, N. J. Clique coverings of complements of paths and cycles.
CARSA	1985-1	Carstensen, P. J.; Murty, K. G.; Perin, C. Intermediate feasibility in 0-1 integer linear systems.
CAVA	1986-1	Cavalier, T. M.; Chandra, M. J. A heuristic algorithm for assigning crews among bases in an airlift operation.
CERD	1985-1	Cerdeira, J. O.; Luz, C. J. Covering nodes with a fixed number of trees.
CHANT	1985-1	Chang, G. J.; Nemhauser, G. L. Covering, packing and generalized perfection.
CHANX	1984-1	Chang, L.; Poe, A. T. Surjections and coverings.
CHAUA	1987-1	Chaudhry, S. S.; Moon, I. D.; Mac Cormick, S. T. Conditional covering: greedy heuristics and computational results.
CHOP	1987-1	Chopra, S. On ternary problems.
CONF	1984-1	Conforti, M.; Corneil, D. G.; Mahjoub, A. R. K index i - covers 1: complexity and polytopes.
CONF	1986-1	Conforti, M.; Corneil, D. G.; Mahjoub, A. R. K(i)-covers I: complexity and polytopes.
CONFA	1984-1	Conforti, M.; Corneil, D. G.; Majoub, A. R. K index i - covers 2: K index i - perfect graphs.
COPPB	1985-1	Coppersmith, D.; Vishkin, U. Solving NP-hard problems in 'almost trees': vertex cover.
CORNJ	1986-1	Cornuejols, G.; Sassano, A. On the 0,1 facets of the set covering polytope.
CRAI	1987-1	Crainic, T. G.; Rousseau, J.-M. The column generation principle and the airline crew scheduling problem.
CRAMA	1987-1	Crama, Y.; Hammer, P. L.; Ibaraki, T. Packing, covering and partitioning problems with strongly unimodular constraint matrices.
CURRB	1985-1	Current, J. R.; Revelle, C. S.; Cohon, J. L. The maximum covering/shortest path problem: a multiobjective network design and routing formulation.
DAHB	1984-1	Dahbura, A. T. An efficient minimum vertex cover set algorithm for a broad class of undirected graphs.

DANH 1983-1 Danhof, K. J.; Varol, Y. L.
Generalized set covering problem.

EISED 1987-1 Eiselt, H. A.; Laporte, G.
Combinatorial optimization problems with soft and hard requirements.

EMELD 1987-1 Emelichev, V. A.; Perepelitsa, V. A.; Shungarov, K. D.
Asymptotic approach to a multicriteria problem of the covering of a graph by stars (Russian).

EMMO 1985-1 Emmons, H.
Work-force scheduling with cyclic requirements and constraints on days off, weekends off, and work stretch.

FALK 1987-1 Falkner, J. C.; Ryan, D. M.
A bus crew scheduling system using a set partitioning model.

FISHI 1981-1 Fisher, M. L.; Wolsey, L. A.
On the greedy heuristic for covering and packing problems.

FLEIA 1986-1 Fleischner, H.
Proof of the strong 2-cover conjecture for planar graphs.

FRAIA 1985-2 Fraisse, P.
Cycle covering in bridgeless graphs.

FRAND 1984-1 Frank, A.; Sebö, A.; Tardos, E.
Covering directed and odd cuts.

GAMBB 1987-1 Gamble, A. B.; Pulleyblank, W. R.
Forest covers and a polyhedral intersection theorem.

GLAN 1984-1 Glanert, W.
A timetable approach to the assignment of pilots to rotations.

GONS 1987-1 Gonsalvez, D. J.; Hall, N. G.; Rhee, W. T.; Siferd, S. P.
Heuristic solutions and confidence intervals for the multicovering problem.

GORD 1981-1 Gordeev, E. N.
New estimates in the covering problem (Russian).

GRANA 1984-1 Granot, D.; Granot, F.
Minimal covers, minimal sets and canonical facets of the posynomial knapsack polytope.

GUANB 1985-1 Guan, M.-G.; Fleischner, H.
On the minimum weighted cycle covering problem for planar graphs.

GUSFB 1986-1 Gusfield, D.; Pitt, L.
Equivalent approximation algorithms for node cover.

GYARA 1985-1 Gyarfas, A.; Lehel, J.
Covering and coloring problems for relatives of intervals.

HALLB 1986-1 Hall, N. G.; Hochbaum, D. S.
A fast approximation algorithm for the multicovering problem.

HALLD 1986-1 Hall, N. G.; Vohra, R. V.
Absolute bounds on optimal costs for a class of set covering problems.

HAMM 1984-1 Hammer, G.
Greedy solutions for general set covering problems.

HEYD 1985-1 Heydemann, M. C.
Minimum number of circuits covering the vertices of a strong digraph.

HOCHA 1983-1 Hochbaum, D. S.; Maass, W.
Fast approximation algorithms for the robot placement problem.

HOCHA 1985-1 Hochbaum, D. S.; Maass, W.
Approximation schemes for covering and packing problems in image processing and VLSI.

HOCHA 1987-1 Hochbaum, D. S.; Maass, W.
Fast approximation algorithms for a nonconvex covering problem.

HOFFA 1985-1 Hoffman, A. J.; Kolen, A. W. J.; Sakarovitch, M.
Totally-balanced and greedy matrices.

HOLS 1984-1 Holst, O.; Sorensen, B.
Combined scheduling and maitainance planning for an aircraft fleet.

HONG 1987-1 Hong, D. W.
An efficient algorithm for determining the intersection of all minimal coverings in a bipartite graph, and a related algorithm for finding all minimal coverings (Chinese).

JAEGA 1985-1 Jaeger, F.; Khelladi, A.; Mollard, M.
On shortest cocycle covers of graphs.

JOHN 1985-1 John, C. G.; Kochenberger, G. A.
Using surrogate constraints in a Lagrangian relaxation approach to set covering problems.

JONEA 1985-1 Jones, K. F.; Lundgren, J. R.; Maybee, J. S.
Clique covers of digraphs II.

KESE 1987-1 Keselman, D. Y.
Cycle coverage of the graph ribs (Russian).

KHANA 1987-1 Khan, M. R.; Lewis, D. A.
A network model for nursing staff scheduling.

KIM D 1985-1 Kim, S.-I.; Park, Y. S.
Models maximizing covering reliability.

KIM F 1986-1 Kim, T. U.; Lowe, T. J.; Ward, J. E.; Francis, R. L.
A minimal length covering subtree of a tree.

KOCH 1985-1 Kochkarov, A. M.
Asymptotically exact algorithms for solving a multicriterial problem of covering a graph by paths (Russian).

KOCHA 1985-1 Kochkarov, A. M.; Perepelitsa, V. A.
A multicriterial covering problem of a graph by paths of large and small length (Russian).

KOLE 1985-1 Kolen, A. W. J.
The round-trip p-center and covering problem on a tree.

KOLEB 1984-1 Kolen, A. W. J.; Tamir, A.
Covering problems.

KORS 1985-1 Korshunov, A. D.
Discrete extremal problems on covering.

KOVAH 1980-1 Kovalev, S. M.
On the determination of minimal coverings in hypergraphs (Russian).

LAUR 1986-1 Laurent, M.
A generalization of antiwebs to independence systems and their relation to set covering polytopes.

LEFK 1985-1 Lefkovitch, L. P.
Entropy and set covering.

LEMO 1985-1 Lemos, M.
On Seymour's question about packing and covering with matroid circuits.

LEVCA 1985-1 Levcopoulos, C.
On covering regions with minimum number of rectangles.

LI H 1984-1 Li, X.-M.; Slyke, R. van
Primal-dual heuristics for multiple set covering.

LIKH 1985-1 Likholat, N. A.
On an algebraic algorithm for solving a problem of the covering of a graph by a growing directed tree (Russian).

MAGA 1986-1 Maga, F.; Vizvari, B.
The relaxation of a special polynomial zero-one programming problem to a set covering problem (Hungarian).

MINO 1986-2 Minoux, M.
Optimal traffic assignment in a SS/TDMA frame: a new approach by set covering and column generation.

MINO 1987-1 Minoux, M.
A class of combinatorial problems with polynomially solvable large scale set covering/ partitioning relaxations.

MONIA 1984-1 Monien, B.; Speckenmeyer, E.
Ramsey numbers and an approximation algorithm for the vertex cover problem.

MONIA 1985-1 Monien, B.; Speckenmeyer, E.
Ramsey numbers and an approximation algorithm for the vertex cover problem.

NIGM 1985-1 Nigmatullin, R. G.; Shtivelman, V. K.
Unique and multiple coverings of graphs by edges (Russian).

NTAF 1984-1 Ntafos, S.; Gonzalez, T.
On the computational complexity of path cover problems.

NYU 1982-1 Nyu, V.
Covering of a set of words by circuits (Russian).

NYU A 1983-1 Nyu, V.; Evdokimov, A. A.
Coverings of graphs by paths (Russian).

OLEJ 1984-1 Olejnik, F.
On total matching numbers and total covering numbers for k-uniform hypergraphs.

ORDM 1987-1 Ordman, E. T.
Dining philosophers and graph covering problems.

PAIX 1985-1 Paixao, J. P.; Almeida, M. T.; Coelho, J. D.
Problems in graph theory and their relation with set covering and partitioning problems (Portuguese).

PELEA 1985-1 Peled, U. N.; Simeone, B.
Polynominal-time algorithms for regular set-covering and threshold synthesis.

PETEB 1987-1 Peter-Orth, C.
On solving computationally difficult set covering problems.

PRIS 1987-1 Prisakar, K. F.; Prisakar, A. V.
On the minimal covering of the vertices of a graph by d-convex subsets (Russian).

PYBE 1986-1 Pyber, L.
Clique covering of graphs.

ROBEC 1985-1 Roberts, F. S.
Applications of edge coverings by cliques

RODI 1986-1 Roditty, Y.
The packing and covering of the complete graph. I: The forests of order five.

ROSH 1986-1 Roshchin, V. A.
An algorithm for solving a covering problem by means of parallel computations (Russian).

ROSHA 1984-1 Roshchin, V. A.; Sergienko, I. V.
An approach to the solution of the set-covering problem.

ROUSA 1984-1 Rousseau, J.-M.
Crew scheduling methods in the operation and planning of transportation systems.

SASS 1985-1 Sassano, A.
On the facial structure of the set covering polytope.

SCHRE 1983-1 Schrijver, A.
Packing and covering of crossing families of cuts.

SCOTB 1985-1 Scott, J. L.; Vignaux, G. A.
A Benders' decomposition algorithm to help airline schedules planning.

SPER 1985-1 Speranza, M. G.
A remark on the Karp algorithm for the set covering problem (Italian).

STANF 1983-1 Stanton, R. G.; Stinson, D. R.
Perfect pair-coverings with block sizes two, three, and four.

TINHA 1986-1 Tinhofer, G.; Schreck, H.
The bounded subset sum problem is almost everywhere randomly decidable in O(n).

TODO 1984-1 Todorov, D. T.
A method of constructing coverings.

TOME 1984-2 Tomescu, I.
Colorings and irreducible coverings by cliques of graphs and hypergraphs.

TOWN 1986-1 Townsend, W.
An application of the assignment model to bus crew rostering.

TROT 1985-1 Trotter, L. E. Jr.
Discrete packing and covering.

VASK 1986-1 Vasko, F. J.; Wilson, G. R.
Hybrid heuristics for minimum cardinality set covering problems.

VASKA 1987-1 Vasko, F. J.; Wolf, F. E.; Stott, K. L.
Optimal selection of ingot sizes via set covering.

VERC 1984-1 Vercellis, C.
A probabilistic analysis of the set covering problem.

WALT 1987-1 Walther, H.
On a conjecture of W. Klotz concerning clique-decomposition of graphs.

WERR 1987-2 Werra, D. de
Partitions into odd chains.

WEST 1985-1 West, D. B.
Parameters of partial orders and graphs: packing, covering and representation.

WOLS 1981-1 Wolsey, L. A.
An analysis of the greedy algorithm for the submodular (set) covering problem.

YOUN 1985-1 Young, M. H.; Muroga, S.
Symmetric minimal covering problem and minimal PLA's with symmetric variables.

ZIMMC 1986-3 Zimmermann, U.
Sharing problems.

ALEKD 1987-1 Alekseev, O. G.
Complex application of discrete optimization methods (Russian).

AUST 1985-1 Austin, L. M.; Ghandforoush, P.
A surrogate cutting plane algorithm for all-integer programming.

AUSTA 1986-1 Austin, L. M.; Ruparel, B. C.
The mixed cutting plane algorithm for all-integer programming.

BARAC 1986-1 Barahona, F.; Grötschel, M.; Jünger, M.; Reinelt, G.
An application of combinatorial optimization to statistical physics and circuit layout design.

BARAG 1986-1 Barahona, F.; Manjoub, A. R.
On the cut polytope.

CHVA 1985-1 Chvatal, V.
Cutting planes in combnatorics.

COOKA 1986-1 Cook, W.
Cutting-plane proofs in polynomial space.

COOKB 1987-1 Cook, W.; Coullard, C. R.; Turan, G.
On the complexity of cutting-plane proofs.

COOKE 1986-1 Cook, W.; Kannan, R.; Schrijver, A.
Chvatal closures for mixed integer programming problems.

FAIGG 1987-1 Faigle, U.; Schrader, R.; Suletzki, R.
A cutting-plane algorithm for optimal graph partitioning.

FLEI 1985-1 Fleischmann, B.
A cutting plane procedure for the travelling salesman problem on road networks.

FLEI 1988-1 Fleischmann, B.
Cutting planes for the symmetric travelling salesman problem.

GHAN 1985-1 Ghandforoush, P.; Austin, L. M.
Solution approaches for highly primal- and dual-degenerate all-integer programming problems.

GROT 1984-1 Grötschel, M.
Developments in combinatorial optimization.

GROT 1986-2 Grötschel, M.
Polyedrische Kombinatorik und Schnittebenenverfahren.

GROTA 1985-1 Grötschel, M.; Holland, O.
Solving matching problems with linear programming.

GROTA 1987-1 Grötschel, M.; Holland, O.
A cutting plane algorithm for minimum perfect 2-matchings.

GROTB 1984-1 Grötschel, M.; Jünger, M.; Reinelt, G.
A cutting plane algorithm for the linear ordering problem.

GROTB 1985-1 Grötschel, M.; Jünger, M.; Reinelt, G.
Acyclic subdigraphs and linear orderings: polytopes, facets, and a cutting plane algorithm.

GROTB 1987-1 Grötschel, M.; Jünger, M.; Reinelt, G.
Calculating exact ground states of spin glasses: a polyhedral approach.

GROTE 1986-1 Grötschel, M.; Pulleyblank, W. R.
Clique tree inequalities and the symmetric travelling salesman problem.

GROTG 1987-1 Grötschel, M.; Wakabayashi, Y.
A cutting plane algorithm for a clustering problem.

HANN 1985-1 Hanna, M. E.; Austin, L. M.
An advanced start algorithm for all-integer programming.

JENK 1987-1 Jenkins, L.
Parametric-objective integer programming using knapsack facets and Gomory cutting planes.

JENKA 1987-1 Jenkins, L.; Peters, D.
A computational comparison of Gomory and knapsack cuts.

KHOK 1983-1 Khokhlyuk, V. I.
Cutting plane algorithms (Russian).

KIVI 1984-1 Kivistik, L.
A generalized rule for selection of a generating row for Gomory's all-integer algorithm (Russian).

KOLOA 1984-1 Kolokolov, A. A.; Tsepkova, E. V.
An upper estimate of the number of regular cutting planes for a family of problems on a cone (Russian).

KRAH 1986-1 Krahl, R.
Surrogate constraints und Schnittverfahren in der linearen 0-1-Optimierung.

LAPOC 1987-1 Laporte, G.; Mercure, H.; Nobert, Y.
Cutting planes based on bin packing solutions for the capacitated vehicle routing problem.

LOGA 1987-1 Loganathan, G. V.; Sherali, H. D.
A convergent interactive cutting-plane algorithm for multiobjective optimization.

MINO 1986-1 Minoux, M.
Mathematical Programming. Theory and Algorithms.

NEMHA 1984-1 Nemhauser, G. L.; Wolsey, L. A.
A recursive procedure for generating all mixed integer cuts.

ORLI 1985-1 Orlin, J. B.
A finitely converging cutting plane technique.

PADBA 1985-1 Padberg, M. W.; Grötschel, M.
Polyhedral computations.

PIEH 1987-1 Piehler, J.
Parallelschnitte in der ganzzahligen Optimierung.

PLASA 1987-1 Plastria, F.
Solving general continuous single facility location problems by cutting planes.

POCHA 1986-1 Pochet, Y.; Wolsey, L. A.
Lot-size models with backlogging: strong reformulations and cutting planes.

REIN 1985-1 Reinelt, G.
The linear ordering problem: algorithms and application.

RUDNB 1981-1 Rudneva, T. L.; Yaranov, M. G.
Program realization of the completely integer Gomory algorithm (Russian).

SAKAA 1984-1 Sakarovitch, M.
Optimisation combinatoire - methodes mathematiques et algorithmiques. Vol. 1: Programmation discrete.

SCHRE 1986-2 Schrijver, A.
Geometric methods in discrete optimization.

SCHRE 1986-5 Schrijver, A.
Theory of linear and integer programming.

SCHRE 1987-2 Schrijver, A.
Polyhedral combinatorics.

SEN A 1986-1 Sen, S.; Sherali, H. D.
Facet inequalities from simple disjunctions in cutting plane theory.

SHERA 1985-1 Sherali, H. D.
A disjunctive cutting plane algorithm for the extreme point mathematical programming problem.

SHERG 1985-1 Sherali, H. D.; Sen, S.
On generating cutting planes from combinatorial disjunctions.

SUNA 1985-1 Sunaga, T.; Hayakawa, A. J.; Maruki, Y.
A computational method for all-integer interval linear programming.

VENIA 1986-1 Veniaminov, S. S.
The effectiveness of the hybrid method for solving discrete optimization problems.

WAKA 1986-1 Wakabayashi, Y.
Aggregation of binary relations: algorithmic and polyhedral investigations.

ZABL 1984-1 Zablotskaya, O. A.
On the comparisons of completely regular cutting-off algorithms.

AGAR	1985-1	Agarwal, S.; Mittal, A. K.; Sharma, P. A decomposition algorithm for linear relaxation of the weighted r-covering problem.
AVIS	1985-1	Avis, D.; Devroye, L. An analysis of a decomposition heuristic for the assignment problem.
BAHLB	1987-1	Bahl, H. C.; Zionts, S. Multi-item scheduling by Benders' decomposition.
BARTD	1985-1	Barton, D. R.; Zippel, R. Polynomial decomposition algorithms.
BHATB	1984-1	Bhatt, S. N.; Leighton, F. T. A framework for solving VLSI graph layout problems.
BOROC	1984-1	Boronin, A. B.; Pershin, O. Y. Decomposition of a network into minimally connected parts.
BRUC	1984-3	Brucker, P. Über gewisse Prinzipien des Entwurfs effizienter Algorithmen.
BURK	1985-1	Burkard, R. E. Time-slot assignment for TDMA-systems.
BURK	1985-2	Burkard, R. E. Time-slot assignment for TDMA-systems.
BURKC	1985-1	Burkard, R. E.; Hamacher, H. W.; Tind, J. On general decomposition schemes in mathematical programming.
CARA	1987-1	Caramanis, M. Production system design: a discrete event dynamic system and generalized Benders' decomposition approach.
CHANAA	1984-1	Chang, R. C.; Lee, R. C. T. The average performance analysis of a closest-pair algorithm.
CHANAA	1986-1	Chang, R. C.; Lee, R. C. T. An O(N log N) minimal spanning tree algorithm for N points in the plane.
COURA	1985-1	Courtois, P.-J. On time and space decomposition of complex structures.
DEBA	1984-1	Debanne, J. G. Tailoring decomposition algorithms to efficiently solve large scale energy planning problems.
DEOR	1985-1	Deorditsa, Y. S.; Morozova, T. Y. Method of block decomposition for engineer networks (Russian).
DURA	1987-1	Duran, F. A large mixed integer production and distribution program.
EGAWB	1986-1	Egawa, Y.; Fukuda, T.; Nagoya, S.; Urabe, M. A decomposition of complete bipartite graphs into edge-disjoint subgraph with star components.
FAIGC	1986-1	Faigle, U.; Goecke, O.; Schrader, R. Church-Rosser decomposition in combinatorial structures.
FARRA	1983-1	Farrell, E. J. Decomposition of complete graphs and complete bipartite graphs into node disjoint paths.
FARRA	1984-1	Farrell, E. J. A note on cycle decompositions of complete bipartite graphs.
GLIC	1985-1	Glickman, T. S.; Sherali, H. D. Large-scale network distribution of pooled empty freight cars over time, with limited substitution and equitabie benefits.
GLOVC	1985-2	Glover, F.; Klingman, D. Layering strategies for creating exploitable structure in linear and integer programs.
GOLDM	1986-1	Goldengorin, B. I. Decomposition of the location problem.
HABIB	1985-1	Habib, M.; Maurer, M. C. 1-intersecting families.
HEAR	1984-1	Hearn, D. W. Practical and theoretical aspects of aggregation problems in transportation planning models.
HILTB	1986-1	Hilton, A. J. W.; Rodger, C. A. Hamiltonian decompositions of complete regular s-partite graphs.
HU A	1985-1	Hu, T. C.; Shing, M. T. A decomposition algorithm for circuit routing.
HUMM	1985-1	Hummeltenberg, W. Capital budgeting with Benders decomposition.
KRIA	1985-1	Kriauciukas, V. Free decomposition of graph search problems (Russian).
LAZI	1986-1	Lazimy, R. Interactive relaxation method for a broad class of integer and continuous nonlinear multiple criteria problems.
LOVEB	1981-1	Love, R. R. Jr. Traffic scheduling via Benders decomposition.
MAAM	1985-1	Maamoun, M. Decompositions of digraphs into paths and cycles.
MAGN	1986-1	Magnanti, T. L.; Mireault, P.; Wong, R. T. Tailoring Benders decomposition for uncapacitated network design.
MARTJ	1983-1	Martinov, M. K. Some decompositions of a complete graph into wheels (Bulgarian).
MARTL	1983-2	Martinova, M. K. On the decomposition of a complete graph into wheels (Bulgarian).
MILK	1984-1	Milkiewicz, F. Decomposition of production scheduling in production systems with switched over continuous technological processes.
MLYN	1985-1	Mlynarovic, V. Benders' decomposition and some problems of the production-transport type (Czechoslovakian).
MOHR	1986-1	Möhring, R. H. Algorithmic aspects of the substitution decomposition in optimization over relations, set systems and Boolean functions.

NERA	1984-1	Neralic, L.; Sirovitsa, L.; Skok, S.; Khunet, D. Solution of a production-transportation problem by Benders' decomposition method (Russian).
POTTA	1987-1	Potts, C. N.; Wassenhove, L. N. van Dynamic programming and decomposition approaches for the single machine total tardiness problem.
ROUH	1985-1	Rouhani, R.; Lasdon, L.; Lebow, W.; Warren, A. D. A generalized Benders decomposition approach to reactive source planning in power systems.
SCOTB	1985-1	Scott, J. L.; Vignaux, G. A. A Benders' decomposition algorithm to help airline schedules planning.
SHERB	1984-1	Sherali, H. D.; Adams, W. P. A decomposition algorithm for a discrete location-allocation problem.
SHI	1985-1	Shi, R.-H. Graph decomposition under a minimum degree restriction (Chinese).
SHINA	1986-1	Shing, M. T.; Hu, T. C. A decomposition algorithm for multi-terminal network flows.
SKILA	1985-1	Skilton, D. K. Chain decompositions of graphs.
TANGA	1984-1	Tang, C. Y.; Lee, R. C. T. Optimal speeding up of parallel algorithms based upon the divide-and-conquer strategy.
TOID	1985-1	Toida, S. A decomposition of a graph into dense subgraphs.
TOPK	1985-1	Topka, V. V. Decomposition in a problem of resource allocation on a network (Russian).
TRUE	1985-3	Truemper, K. A flexible decomposition approach for finding submatrices with certain inherited properties.
TRUSB	1983-1	Truszczynski, M. Decompositions of complete multipartite graphs into isomorphic stars.
TYSHA	1985-3	Tyshkevich, R. I.; Chernjak, A. A. Decomposition of graphs.
VELO	1985-1	Veloso, P. A. S. How to "divide-and-conquer": a logical basis.
VULI	1985-1	Vulis, V. I.; Sherman, Y. S. Decomposition method for solving large-scale problems of transportation type (Russian).
WALS	1984-1	Walsh, T. R. How evenly should one divide to conquer quickly?
WOJC	1984-1	Wojciechowski, J. Generation of trees of a graph with the use of decomposition.

BABA 1986-1 Babaev, D. A.; Aliev, A. A.
A method of surrogate constraints for solution of discrete programming problems.

BIXBF 1986-1 Bixby, R. E.; Wagner, D. K.
A note on detecting simple redundancies in linear systems.

BOROE 1986-1 Borosh, I.; Flahive, M.; Treybig, B.
Small solutions of linear Diophantine equations.

BRUA 1987-1 Brualdi, R. A.; Katz, M.
An extremal problem concerning matrices of 0's and 1's.

BUZY 1983-1 Buzytskij, P. L.; Freiman, G. A.
An effective formula for the number of solutions of a system of 0,1 linear equations.

CARSA 1985-1 Carstensen, P. J.; Murty, K. G.; Perin, C.
Intermediate feasibility in 0-1 integer linear systems.

CHOP 1987-1 Chopra, S.
On ternary problems.

DAVY 1986-1 Davydov, E. G.; Bolshakov, S. Y.
Group-theoretic methods for aggregation in network problems (Russian).

ELIM 1985-1 Elimam, A. A.; Elmaghraby, S. E.
On the reduction method for integer linear programs, II.

FISHA 1985-1 Fishburn, P. C.; Kochenberger, G. A.
Aggregating assignment constraints.

HAST 1985-1 Hastad, J.; Helfrich, B.; Lagarias, J. C.; Schnorr, C. P.
Polynomial time algorithms for finding integer relations among real numbers.

HASTA 1986-1 Hastad, J.; Just, B.; Lagarias, J. C.; Schnorr, C. P.
Polynomial time algorithms for finding integer relations among real numbers.

LAGA 1982-1 Lagarias, J. C.
The computational complexity of simultaneous diophantine approximation problems.

LEE M 1987-1 Lee, S.-M.; Luebbe, R. L.
A zero-one goal-programming algorithm using partitioning and constraint aggregation.

LIN 1986-1 Lin, G. D.
A note on aggregating constraints in integer programming.

LOVA 1985-1 Lovasz, L.
An algorithmic theory of numbers, graphs, and convexity.

MARCE 1987-1 Marcotorchino, F.
Block seriation problems: a unified approach. Reply to the problem of H. Garcia and J. M. Proth.

MILA 1985-1 Milanov, P. B.
The linear diophantine problem of Frobenius and discrete optimization.

PASCA 1986-1 Pascoletti, K.-H.
Diophantische Systeme und Lösungsmethoden zur Bestimmung aller Invarianten in Petri-Netzen.

PLATA 1984-1 Plateau, G.; Guerch, M. T.
Aggregation of equalities in integer programming.

ROSEB 1956-1 Rosenblatt, D.
On aggregation and consolidation in linear systems.

RUBI 1985-1 Rubin, D. S.
Polynomial algorithms for m*(m+1) integer programs and m*(m+k) diophantine systems.

SEEL 1984-2 Seeländer, J.
Über ein Kriterium zur Lösbarkeit reinganzzahliger linearer Optimierungsprobleme.

SHCHA 1983-1 Shcherbina, O. A.; Matveev, V. V.
On the efficiency of local algorithms for the solution of quasi-block problems of integer linear programming (Russian).

SMAR 1981-1 Smarandache, F. G.
General solution properties in whole numbers for linear equations.

SPRI 1986-1 Springer, J.
Algorithm 641. Exact solution of general integer systems of linear equations.

TRUE 1985-3 Truemper, K.
A flexible decomposition approach for finding submatrices with certain inherited properties.

TUNGB 1987-1 Tung, S. P.
Computational complexities of diophantine equations with parameters.

VEN 1985-1 Ven, V. L.
Partial aggregation in problems of integer linear programming with variable parameters (Russian).

VESE 1985-1 Veselov, S. I.
Aggregation of linear integer equations.

WAKA 1986-1 Wakabayashi, Y.
Aggregation of binary relations: algorithmic and polyhedral investigations.

BALAG 1985-2 Balas, E.
Disjunctive programming and a hierarchy of relaxations for discrete optimization problems.
BALAN 1986-1 Balas, E.; Tama, J. M.; Tind, J.
Sequential convexification in reverse convex and disjunctive programming.
CONC 1986-1 Concalves, A. S.
Symmetric duality for disjunctive programming with absolute value functionals.
JEROA 1985-1 Jeroslow, R. G.
A simplification for some disjunctive formulations.
PATK 1985-1 Patkar, V.; Stancu-Minasian, I. M.
Duality in disjunctive linear fractional programming.
SEN A 1986-1 Sen, S.; Sherali, H. D.
Facet inequalities from simple disjunctions in cutting plane theory.
SHERA 1985-1 Sherali, H. D.
A disjunctive cutting plane algorithm for the extreme point mathematical programming problem.

AHARA 1985-1 Aharoni, R.; Erdös, P.; Linial, N.
Dual integer linear programs and the relationship between their optima.

ALEKC 1985-1 Alekseev, A. O.; Alekseev, O. G.; Anisimov, V. G.; Anisimov, E. G.
The use of duality to increase the effectiveness of the branch and bound method when solving the knapsack problem.

ANEJ 1985-1 Aneja, Y. P.; Chandrasekaran, R.; Nair, K. P. K.
Classes of linear programs with integral optimal solutions.

BACHA 1986-1 Bachem, A.
Dualität und Polarität in diskreten Strukturen.

BARCA 1987-1 Barcia, P.
Constructive dual methods for discrete programming.

BORC 1985-1 Borchardt, M.
Einige Resultate zur Dualitätstheorie in der diskreten Optimierung.

BOROD 1986-1 Boros, E.
On the complexity of the surrogate dual of 0-1 programming.

BRUCD 1984-1 Brucker, P.; Papenjohann, W.; Zimmermann, U.
A dual optimality criterion for algebraic linear programs.

CHANL 1984-1 Chandrasekaran, R.; Shirali, S.
Total weak unimodularity: testing and applications.

COOKA 1985-1 Cook, W.
A note on matchings and separability.

COOKA 1986-2 Cook, W.
On box totally dual integral polyhedra.

COOKC 1986-1 Cook, W.; Fonlupt, J.; Schrijver, A.
An integer analogue of Caratheodory's theorem.

COOKF 1984-1 Cook, W.; Lovasz, L.; Schrijver, A.
A polynomial-time test for total dual integrality in fixed dimension.

CRAMC 1986-1 Crama, Y.; Hansen, P.; Williams, A. C.
Use of the dual in branch and bound methods for mixed 0-1 programming.

CRYS 1987-1 Crystal Llewellyn, D.
Min algebraic duality.

DAVYA 1985-1 Davydov, G. V.; Davydova, I. M.
A duality criterion for optimality in allocation problems.

DAVYA 1987-1 Davydov, G. V.; Davydova, I. M.
Dual algorithms in discrete optimization (Russian).

EDMO 1984-1 Edmonds, J.; Giles, R.
Total dual integrality of linear inequality systems.

EULEC 1985-1 Euler, R.; Mahjoub, A. R.
On a composition of independence systems by circiut-identification.

FUJIA 1984-1 Fujishige, S.
A system of linear inequalities with a submodular function on {0,+-1}vectors.

GERAC 1987-1 Gerards, A. M. H.; Sebö, A.
Total dual integrality implies local strong unimodularity.

GROF 1987-1 Gröflin, H.
On switching path polyhedra.

GROS 1985-1 Grosser, E.
Über eine Methode zur Lösung des Zuordnungsproblems.

GUIGC 1987-1 Guignard, M.; Spielberg, K.
On combining Lagrangean relaxation, Benders inequalities and Lagrangean dual ascent.

GUNN 1986-1 Gunn, E. A.; Kusiak, A.
The procurement problem: an integer programming problem well suited to a solution using duality.

HANSF 1987-1 Hansen, P.; Lu, S. H.
On the equivalence of paved-duality and standard linearization in 0-1 optimization.

HOLM 1987-1 Holm, S.
Adding activities to the dual instead of cuts to the primal problem.

JUEL 1987-1 Juel, H.; Love, R. F.
Duality in constrained location problems.

KABA 1985-1 Kabadi, S. N.; Chandrasekaran, R.
On totally dual integral systems.

KIVI 1985-2 Kivistik, L.
New sufficient finiteness conditions for dual cutoff algorithms (Russian).

KOHLA 1984-2 Köhler, J.
A note on the solution of integer linear programming problems using duality properties.

KOLO 1983-1 Kolokolov, A. A.
Lower bound for the number of iterations for a class of truncation algorithms (Russian).

KORA 1982-1 Korach, E.
Packing of t-cuts, and other aspects of dual integrality.

LEBEA 1983-1 Lebedev, S. S.; Shejnman, O. K.
Dual approach to integer programming.

LIBU 1984-1 Libura, M.
Integer programming duality and its application in sensitivity analysis and approximate algorithms (Polish).

LU 1987-1 Lu, S. H.; Simeone, B.
On the equivalence of roof-duality and paved-duality in quadratic 0-1 optimization.

LU A 1985-1 Lu, S. H.; Williams, A. C.
Roof duality for nonlinear 0-1 optimization.

LU A 1987-1 Lu, S. H.; Williams, A. C.
Roof duality and linear relaxation for quadratic and polynomial 0-1 optimization.

LU A 1987-2 Lu, S. H.; Williams, A. C.
Roof duality for polynomial 0-1 optimization.

LUPS 1984-1 Lupsa, L.
Duality theorems for integer programming problems.

MISHC 1984-1 Mishra, B. K.; Das, C.
Minimax and symmetric duality for nonlinear mixed-integer programming problem.

MISHC 1985-1 Mishra, B. K.; Das, C.
Maximin problem and a duality theorem for mixed-integer quadratic programming.

NEMH 1985-1 Nemhauser, G. L.
Duality for integer optimization.

PALE 1987-2 Palekar, U. S.; Karwan, M. H.; Zionts, S.
Duality-based bounds for the fixed charge transportation problem.

PATK 1985-1 Patkar, V.; Stancu-Minasian, I. M.
Duality in disjunctive linear fractional programming.

PAZ 1986-1 Paz, A.
A duality property for the set of all feasible solutions to an integer program.

QI 1986-1 Qi, L.-Q.
Directed submodular functions, ditroids and total dual integrality.

QI 1987-2 Qi, L.-Q.
Odd submodular functions, Dilworth functions and discrete convex functions.

SCHRE 1984-1 Schrijver, A.
Total dual integrality from directed graphs, crossing families, and sub- and supermodular functions.

SCHRE 1986-3 Schrijver, A.
Polyhedral combinatorics - some recent developments and results.

SCHRE 1986-5 Schrijver, A.
Theory of linear and integer programming.

SCHRE 1987-2 Schrijver, A.
Polyhedral combinatorics.

SEBO 1985-2 Sebö, A.
The Schrijver system of odd join polyhedra.

SEEL 1987-1 Seeländer, J.
On two duality concepts of discrete vector optimization (German).

SHEV 1984-1 Shevchenko, A. N.
Application of the dynamic programming method and duality theory to a multi-index transportation problem (Russian).

SINGB 1986-1 Singer, I.
A general duality approach to min-max results in combinatorial optimization. I.

TANGB 1987-1 Tang, H.; Ni, M.
An algorithm for solving surrogate dual problems in integer linear programming (Chinese).

TROTA 1987-1 Trotter, L. E. Jr.; West, D. B.
Two easy duality theorems for product partial orders.

VENTA 1987-1 Ventura, J. A.; Hearn, D. W.
A Lagrangian dual approach for networks with quadratic costs.

WILL 1985-1 Williams, A. C.
Quadratic 0-1 programming using the proof dual - with computational results.

ZIMMC 1986-1 Zimmermann, U.
Duality for balanced submodular flows.

ZIMMC 1986-3 Zimmermann, U.
Sharing problems.

ALEKD	1987-1	Alekseev, O. G. Complex application of discrete optimization methods (Russian).
BEAS	1985-2	Beasley, J. E. Algorithms for unconstrained two-dimensional guillotine cutting.
BEEK	1985-1	Beek, P. van; Bremer, A.; Putten, C. van Design and optimization of multi-echelon assembly networks: savings and potentials.
BRUC	1984-3	Brucker, P. Über gewisse Prinzipien des Entwurfs effizienter Algorithmen.
BYER	1984-1	Byers, T. H.; Waterman, M. S. Determining all optimal and near-optimal solutions when solving shortest path problems by dynamic programming.
CARRB	1986-1	Carraway, R. L.; Morin, T. L.; Moskowitz, H. Generalized dynamic programming for multi-criteria optimization.
CARRB	1987-1	Carraway, R. L.; Morin, T. L.; Moskowitz, H. The stochastic traveling salesman problem revisited: a generalized dynamic programming approach.
CEPU	1984-1	Cepulic, V. An algorithm for finding minimal intersection sets.
COOPA	1984-1	Cooper, M. W.; Farhangian, K. A dynamic programming algorithm for multiple-choice constraints.
DESRD	1986-1	Desrosiers, J.; Dumas, Y.; Soumis, F. A dynamic programming solution of the large-scale single-vehicle dial-a-ride problem with time windows.
DUDZ	1984-1	Dudzinski, K. A dynamic programming approach to solving the multiple choice knapsack problem.
FREY	1986-1	Frey, J.-J.; Yehia-Alcoutlabi, A. Comparaisons par paires: une interpretation et une generalisation de la methode des scores.
GLYA	1983-1	Glyakov, P. V. Algorithm for a partitioning problem and its investigation on an electronic computer (Russian).
GROP	1985-1	Groppen, V. O. Optimization of dynamic programming methods in the solution of extremal combinatorial problems (Russian).
GROP	1985-2	Groppen, V. O. Optimization of dynamic programming methods when solving extremal combinatorial problems.
GROP	1987-1	Groppen, V. O. Optimizing the realization of dynamic programming for a class of combinatorial problems.
GURA	1984-1	Gurari, E. M.; Sudborough, I. H. Improved dynamic programming algorithms for bandwidth minimization and the min-cut linear arrangement problem.
HANSA	1986-1	Hansen, J. V.; Giauque, W. C. Task allocation in distributed processing systems.
HARTA	1985-1	Hartley, R. Vector optimal routing by dynamic programming.
HENI	1986-1	Henig, M. I. Extensions of the dynamic programming method in the deterministic and stochastic assembly-line balancing problems.
IBAR	1987-2	Ibaraki, T. Enumerative Approaches to Combinatorial Optimization - Part 2.
IWAI	1985-1	Iwainsky, A. Layout optimization for a hierarchy of star networks.
JERO	1984-1	Jeromin, B.; Körner, F. A hybrid method for solving integer linear programming problems.
JERO	1986-1	Jeromin, B.; Körner, F. A hybrid method for solving integer linear programming problems.
JEROC	1987-1	Jeroslow, R. G.; Wang, J. Dynamic programming, integral polyhedra and Horn clause knowledge bases.
KLINB	1985-1	Klingman, D.; Armstrong, R. D.; Partow-Navid, P. Implementation and analysis of alternative algorithms for generalized shortest path problems.
KOVAC	1987-1	Kovalenko, A. G. Algorithms of intervals and their application to solve problems of discrete optimization of multistage processes (Russian).
LAGE	1985-1	Lageweg, B.; Lenstra, J. K.; Stougie, L.; Rinnooy Kan, A. H. G. Stochastic integer programming by dynamic programming.
LEE I	1986-1	Lee, I.-S. On sensitivity analysis for shortest-path dynamic programming models.
LEE K	1985-1	Lee, K. T. Knapsack problems solved by dynamic programming.
MARTF	1987-1	Martin, R. K.; Rardin, R. L.; Campbell, B. A. Polyhedral characterization of discrete dynamic programming.
MARU	1987-1	Marucheck, A. S.; Sulek, J. M. An algorithm for technology choice in local area network design.
MERE	1986-1	Merenzon, M. N. A dynamic programming algorithm for optimal scheduling of a nonhomogeneous multiprocessor system (English).
NGUY	1983-1	Nguyen Van Thieu On the problem of production location in an interbranch formulation (Vietnamese).
NGUYB	1986-1	Nguyen, S.; Pallottino, S. Assignment of passengers to an urban transport system: Determination of minimal hyperpaths (Italian).
PERLB	1984-1	Perl, Y. Optimum split trees.
PINO	1982-1	Pino Mejias, J. L. Solving scheduling problems by means of dynamic programming (Spanish).

POTTA 1987-1 Potts, C. N.; Wassenhove, L. N. van
Dynamic programming and decomposition approaches for the single machine total tardiness problem.

RICHF 1987-1 Richter, K.
Stability of the constant cost dynamic lot size model.

ROSEC 1986-1 Rosenblatt, M. J.
The dynamics of plant layout.

SAKSA 1985-1 Saksena, J. P.
Applications of decision networks.

SANCA 1985-1 Sancho, N. G. F.
Maximum reliability through a network with resource constraints.

SEEL 1984-1 Seeländer, J.
Einige Bemerkungen zur Lösung von rein ganzzahlingen linearen Optimierungsproblemen.

SHEV 1984-1 Shevchenko, A. N.
Application of the dynamic programming method and duality theory to a multi-index transportation problem (Russian).

STAND 1986-1 Stanfel, L. E.
A recursive Lagrangian method for clustering problems.

TSIL 1984-1 Tsiligirides, T.
Heuristic methods applied to orienteering.

VASS 1985-1 Vassilacopoulos, G.
Allocating doctors to shifts in an accident and emergency department.

WAJS 1984-1 Wajs, W.
Scheduling problem for metallurgical processes (Polish).

YANGA 1985-1 Yang, C. E.; Lin, Y. X.
Board welding problem (Chinese).

YEH 1986-1 Yeh, D. Y.
A dynamic programming approach to the complete set partitioning problem.

ARAOB 1984-1 Araoz, J.; Edmonds, J.; Griffin, V.
Lifting the facets of polyhedra.
BALAG 1985-3 Balas, E.
On the facial structure of scheduling polyhedra.
BALAL 1984-1 Balas, E.; Saltzman, M. J.
Some facets of the three-index assignment polytope.
BALIB 1984-1 Balinski, M. L.; Russakoff, A.
Faces of dual transportation polyhedra.
BARAD 1985-1 Barahona, F.; Grötschel, M.; Mahjoub, A. R.
Facets of the bipartite subgraph polytope.
BARAF 1985-1 Barahona, F.; Mahjoub, A. R.
Compositions in the acyclic subdigraph polytope.
BARAG 1986-1 Barahona, F.; Manjoub, A. R.
On the cut polytope.
BIXB 1987-1 Bixby, R. E.
Notes on combinatorial optimization.
BOLO 1986-1 Bolotashvili, G. G.
Facets of a permutation polyhedron (Russian).
BOLO 1986-2 Bolotashvili, G. G.
On the facets of a permutation polyhedron (Russian).
CONF 1984-1 Conforti, M.; Corneil, D. G.; Mahjoub, A. R.
K index i - covers 1: complexity and polytopes.
GERA 1985-1 Gerards, A. M. H.
Testing the odd bicycle wheel inequalities for the bipartite subgraph polytope.
GRANA 1984-1 Granot, D.; Granot, F.
Minimal covers, minimal sets and canonical facets of the posynomial knapsack polytope.
GROT 1986-2 Grötschel, M.
Polyedrische Kombinatorik und Schnittebenenverfahren.
GROTB 1985-1 Grötschel, M.; Jünger, M.; Reinelt, G.
Acyclic subdigraphs and linear orderings: polytopes, facets, and a cutting plane algorithm.
GROTB 1985-2 Grötschel, M.; Jünger, M.; Reinelt, G.
Facets of the linear ordering polytope.
GROTB 1985-3 Grötschel, M.; Jünger, M.; Reinelt, G.
On the acyclic subgraph polytope.
GROTB 1987-1 Grötschel, M.; Jünger, M.; Reinelt, G.
Calculating exact ground states of spin glasses: a polyhedral approach.
GROTG 1987-2 Grötschel, M.; Wakabayashi, Y.
Composition of facets of the clique partitioning polytope.
GROTG 1987-3 Grötschel, M.; Wakabayashi, Y.
Facets of the clique partitioning polytope.
JUNGA 1985-1 Jünger, M.
Polyhedral combinatorics and the acyclic subdigraph problem.
LEUN 1986-1 Leung, J. M. Y.; Magnanti, T. L.
Valid inequalities and facets of the capacitated plant location problem.
PADBC 1985-1 Padberg, M. W.; Vanroy, T. J.; Wolsey, L. A.
Valid linear inequalities for fixed charge problems.
PALU 1985-1 Palubetskis, G. S.
Subgraph polytopes (Russian).
POLJA 1987-1 Poljak, S.; Turzik, D.
On a facet of the balanced subgraph polytope.
REIN 1985-1 Reinelt, G.
The linear ordering problem: algorithms and application.
VANRA 1985-1 Vanroy, T. J.; Wolsey, L. A.
Valid inequalities and separation for uncapacitated fixed charge networks.

BOEN 1985-1 Boenchendorf, K.
Improving the Kennington/Unger penalty method for the fixed-charge transportation problem.
BOEN 1986-1 Boenchendorf, K.
Combining penalties for the fixed-charge transportation problem.
KHAN 1986-1 Khan, A. M.
Solid-waste disposal with intermediate transfer stations: an application of the fixed-charge location problem.
MOTE 1986-1 Mote, J.
The family constrained network problem.
PADBC 1985-1 Padberg, M. W.; Vanroy, T. J.; Wolsey, L. A.
Valid linear inequalities for fixed charge problems.
PALE 1987-1 Palekar, U. S.; Karwan, M. H.; Zionts, S.
A branch-and-bound method for the fixed charge transportation problem.
PALE 1987-2 Palekar, U. S.; Karwan, M. H.; Zionts, S.
Duality-based bounds for the fixed charge transportation problem.
SUHL 1985-1 Suhl, U. H.
Solving large scale mixed-integer programs with fixed charge variables.
VANRA 1985-1 Vanroy, T. J.; Wolsey, L. A.
Valid inequalities and separation for uncapacitated fixed charge networks.
WRIG 1986-1 Wright, D. D.; Haehling von Lanzenauer, C.
COAL: A new heuristic approach for solving the fixed charge problem - computational results.
WRIG 1986-2 Wright, D. D.; Haehling von Lanzenauer, C.
Solving the fixed charge problem with Lagrangian relaxation and cost allocation heuristics.

ABDE 1984-1 Abdel-Kader, I.
Path-number of digraphs and graphs.

ABDE 1986-1 Abdel-Kader, I.
Cheminements remarquables dans les graphes antisymetriques.

ACHAA 1983-1 Acharya, B. D.
A characterization of consistent marked graphs.

ACHAA 1983-2 Acharya, B. D.
On d-sequential graphs.

ACHAA 1984-1 Acharya, B. D.
Some further properties of consistent marked graphs.

ACHAA 1986-1 Acharya, B. D.
Connectedness and acyclicity of the transposed square of a graph.

ACHAB 1983-1 Acharya, B. D.; Acharya, B. R.
The contraction polynomial of a graph.

ADAM 1984-2 Adam, A.
On the laxity of rooted trees with vertices of degree three.

AFRA 1985-1 Afrati, F.; Papadimitriou, C. H.; Papageorgiou, G.
The complexity of cubical graphs.

AHARB 1985-1 Aharoni, R.; Hartman, I. B.-A.; Hoffman, A. J.
Path partitions and packs of acyclic digraphs.

AINO 1985-1 Ainouche, A.; Christofides, N.
Conditions for the existence of Hamiltonian circuits in graphs based on vertex degrees.

AINO 1987-1 Ainouche, A.; Christofides, N.
Semi-independence number of a graph and the existence of hamiltonian circuits.

AKIY 1984-1 Akiyama, J.; Ando, K.; Harary, F.
A graph and its complement with specified properties. VIII. Interval graphs.

AKIYA 1983-1 Akiyama, J.; Harary, F.
The regulation number of a graph.

AKIYB 1985-1 Akiyama, J.; Kano, M.
Path factors of a graph.

AKIYB 1985-2 Akiyama, J.; Kano, M.
1-factors of triangle graphs.

AKIYC 1985-1 Akiyama, J.; Ruiz, M.-J.
Intersection graphs and their applications.

AKKA 1985-1 Akka, D. G.
Graph equations for line graphs, entire graphs and semi-entire graphs.

AKL 1983-1 Akl, S. G.
A note on Euclidean matchings, triangulations and spanning trees.

AKSE 1984-1 Aksenov, V. A.
The degree of perfection of a graph (Russian).

ALAM 1983-1 Alameddine, A. F.
Centers of maximal planar graphs with two vertices of degree three.

ALAM 1987-1 Alameddine, A. F.
On a characterization of a class of outerplanar graphs.

ALAV 1986-1 Alavi, Y.; Chartrand, G.; Lesniak, L.; Oellermann, O. R.
Bipartite regulation numbers.

ALBEA 1987-1 Albertson, M. O.
Finding Hamiltonian cycles in ore graphs.

ALBEB 1983-1 Albertson, M. O.; Berman, D. M.
Cliques, colorings and locally perfect graphs.

ALBEC 1984-1 Albertson, M. O.; Collins, K. L.
Duality and perfection for edges in cliques.

ALDR 1984-1 Aldred, R. E. L.
Paths through m vertices in 3-connected cubic graphs.

ALDRA 1987-1 Aldred, R. E. L.; Holton, D. A.
Cycles through five edges in 3-connected cubic graphs.

ALDRB 1985-1 Aldred, R. E. L.; Holton, D. A.; Thomassen, C.
Cycles through four edges in 3-connected cubic graphs.

ALDRC 1987-1 Aldred, R. E. L.; Sheng, B.; Holton, D. A.
Primitive graphs.

ALEKF 1982-1 Alekseev, V. E.
The effect of local constraints on the complexity of determination of the graph independence number (Russian).

ALEKG 1981-1 Alekseev, V. E.; Markov, A. A.; Noskov, V. V.
Optimization of independent and support sets in hypergraphs (Russian).

ALEX 1982-1 Alexander, C.; Hopkins, G.; Staton, W.
Maximal bipartite subgraphs in triangle-free graphs.

ALLEA 1985-1 Allender, E. W.
On the number of cycles possible in digraphs with large girth.

ALLEB 1986-1 Alles, P.
Dimension of amalgamated graphs and trees.

ALON 1985-1 Alon, N.
Asynchronous threshold networks.

ALSP 1985-1 Alspach, B.; Durnberger, E.; Parsons, T. D.
Hamilton cycles in metacirculant graphs with prime cardinality blocks.

AMAH 1985-1 Amahashi, A.
On factors with all degrees odd.

AMARA 1983-1 Amar, D.
On the connectivity of some telecommunications networks.

AMARA 1986-1 Amar, D.
Partition of a bipartite Hamiltonian graph into two cycles.

AMARB 1985-1 Amar, D.; Fournier, I.; Germa, A.
Some conditions for digraphs to be Hamiltonian.

ANAS	1985-1	Anastasyan, V. G. One combinatorial line location problem (Russian).
ANDO	1987-1	Ando, K.; Enomoto, H.; Saito, A. Contractible edges in 3-connected graphs.
ANDOA	1985-1	Ando, K.; Mizuno, H. Biconnectedness of graphs.
ANDOB	1987-1	Ando, K.; Usami, Y. Critically (k,k)-connected graphs.
ANDR	1983-1	Andrasfai, B. Graph theory. Network flows. Matrices. (Hungarian).
ANDRA	1985-1	Andreae, T. On an extremal problem concerning the interval number of a graph.
ANDRA	1985-2	Andreae, T. On superperfect noncomparability graphs.
ANDRA	1985-3	Andreae, T. On the interval number of a triangulated graph.
ANDRA	1985-4	Andreae, T. On the unit interval number of a graph.
ANDRB	1982-1	Andreatta, A.; Liburdi, A.; Medagliani, G. On maximum genus of cubic graphs (Italian).
ANDRF	1987-1	Andrews, J. A.; Jacobson, M. S. On a generalization of chromatic number and two kinds of Ramsey numbers.
ANST	1985-1	Anstee, R. P. An algorithmic proof of Tutte's f-factor theorem.
APOS	1986-1	Apostolico, A.; Hambrusch, S. E. New cliques and independent set algorithms for circle graphs.
APOS	1987-1	Apostolico, A.; Hambrusch, S. E. Finding maximum cliques on circular-arc graphs.
ARAI	1986-1	Arai, M.; Yamashita, M.; Hirata, T.; Ibaraki, T.; Honda, N. Lower bounds on time complexity for some subgraph detection problems (Japanese).
ARANA	1983-1	Arany, I. Another method for finding pseudo-peripheral nodes.
ARANA	1985-1	Arany, I. An efficient algorithm for finding peripheral nodes.
ARANA	1986-1	Arany, I. An algorithm for getting a minimum cut-set of a graph.
ARCHB	1986-1	Archdeacon, D.; Richter, B. Circuits in 4-regular plane graphs.
ARKI	1985-1	Arkin, E. M.; Papadimitriou, C. H. On negative cycles in mixed graphs.
ARNB	1985-1	Arnborg, S. Efficient algorithms for combinatorial problems on graphs with bounded decomposability - a survey.
ARNBA	1984-1	Arnborg, S.; Corneil, D. G.; Proskurowski, A. Complexity of finding embeddings in a k-tree.
ARNBB	1984-1	Arnborg, S.; Proskurowski, A. Linear time algorithms for NP-hard problems on graphs embedded in k-trees.
ARNBB	1985-1	Arnborg, S.; Proskurowski, A. Characterization and recognition of partial k-trees.
ARNBB	1986-1	Arnborg, S.; Proskurowski, A. Algorithms on graphs with bounded decomposability.
AROC	1984-1	Arocha, J. L. Properties of the independent polynomial of a graph (Spanish).
ASAN	1985-1	Asano, T. An approach to the subgraph homeomorphism problem.
ASAN	1987-1	Asano, T. An application of duality to edge-deletion problems.
ASANB	1983-1	Asano, T.; Hirata, T. Edge-contraction problems.
ASH	1983-1	Ash, P. Two sufficient conditions for the existence of Hamilton cycles in bipartite graphs.
ASH	1985-1	Ash, P. The maximum number of diagonals of a cycle in a block.
ASH A	1984-1	Ash, P.; Jackson, B. Dominating cycles in bipartite graphs.
ASRA	1984-1	Asratyan, A. S. Investigation of imbedded c-combinations and c-coverings (Russian).
ASRAA	1985-1	Asratyan, A. S.; Khachatryan, N. K. Investigation of the Hamiltonian property of a graph using neighborhoods of vertices (Russian).
ATALB	1984-1	Atallah, M. J.; Vishkin, U. Finding Euler tours in parrallel.
AUSI	1986-1	Ausiello, G.; D'Atri, A.; Moscarini, M. Chordality properties on graphs and minimal conceptual connections in semantic data models.
AVRA	1984-1	Avram-Nitchi, R. An algorithm based on graph theory for solving the time-criterion assignment problem.
AYEL	1984-1	Ayel, J.; Favaron, O. Helms are gaceful.
AZAR	1987-1	Azarenok, A. S. Packing of graphs of width two (Russian).
BABAE	1986-1	Babai, L. A Las Vegas - NC algorithm for isomorphism of graphs with bounded multiplicity of eigenvalues (draft).
BABAF	1987-1	Babai, L.; Turan, G. The complexity of defining a relation on a finite graph.

BAGA 1983-1 Bagaev, G. N.
Enumeration of many-colored trees (Russian).

BAGAA 1983-1 Bagaev, G. N.; Dmitriev, E. F.
Enumeration of connected two-color graphs (Russian).

BAGAA 1984-1 Bagaev, G. N.; Dmitriev, E. F.
Enumeration of connected labeled bipartite graphs (Russian).

BALA 1985-1 Balaban, A. T.; Filip, P. A.; Balaban, T.-S.
Computer program for finding all possible cycles in graphs.

BALA 1986-1 Balaban, A. T.; Filip, P. A.; Balaban, T.-S.
Erratum: Computer program for finding all possible cycles in graphs.

BALAE 1983-1 Balakrishnan, R.; Paulraja, P.
Existence of graphs with edge-connectivity 3 and 1-Hamiltonian index 2.

BALAE 1983-2 Balakrishnan, R.; Paulraja, P.
Powers of chordal graphs.

BALAE 1984-1 Balakrishnan, R.; Paulraja, P.
Counterexample to a conjecture of Häggkvist and Thomassen on circuits containing specified edges.

BALAE 1984-2 Balakrishnan, R.; Paulraja, P.
Note on the existence of directed (k+1)-cycles in diconnected complete k-partite digraphs.

BALAE 1986-1 Balakrishnan, R.; Paulraja, P.
Chordal graphs and some of their derived graphs.

BALAE 1986-2 Balakrishnan, R.; Paulraja, P.
Self-clique graphs and diameters of iterated clique graphs.

BALAG 1985-1 Balas, E.
An $O(|V| + |E|)$ algorithm for finding an edge-maximal subgraph with a tr-formative coloring.

BALAG 1986-1 Balas, E.
A fast algorithm for finding an edge-maximal subgraph with TR-formative coloring.

BALAH 1984-1 Balas, E.; Chang, S. Y.
Finding a maximum clique in an arbitrary graph.

BALAI 1987-1 Balas, E.; Chvatal, V.; Nesetril, J.
On the maximum weight clique problem.

BALAP 1986-1 Balas, E.; Yu, C. S.
Finding a maximum clique in an arbitrary graph.

BALAP 1986-2 Balas, E.; Yu, C. S.
More on graphs with polynomially solvable maximum-weight clique problem.

BAND 1984-1 Bandelt, H. J.
Hereditary modular graphs.

BAND 1985-1 Bandelt, H. J.
Networks with Condorcet solutions.

BANG 1987-1 Bange, D. W.; Barkauskas, A. E.; Host, L. H.; Slater, P. J.
Efficient near-domination of grid graphs.

BANKA 1982-1 Banks, J.; Mitchem, J.
On the acyclic point-connectivity of the n-cube.

BARAC 1986-1 Barahona, F.; Grötschel, M.; Jünger, M.; Reinelt, G.
An application of combinatorial optimization to statistical physics and circuit layout design.

BARAD 1985-1 Barahona, F.; Grötschel, M.; Mahjoub, A. R.
Facets of the bipartite subgraph polytope.

BARAF 1985-1 Barahona, F.; Mahjoub, A. R.
Compositions in the acyclic subdigraph polytope.

BARAF 1987-1 Barahona, F.; Mahjoub, A. R.
Compositions of graphs and polyhedra II. Stable sets.

BARAH 1987-1 Barahona, F.; Pulleyblank, W. R.
Exact arborescences, matchings and cycles.

BARE 1985-1 Barefoot, C. A.
R-partial line graphs.

BARE 1987-1 Barefoot, C. A.
Hamiltonian connectivity of the Halin graphs.

BAROA 1987-1 Baronaigien, D. R. van; Ruskey, F.
A Hamilton path in the rotation lattice of binary trees.

BASA 1983-1 Basangova, E. O.; Erusalimskij, Y. M.
Mixed attainability on partially directed graphs (Russian).

BATA 1984-1 Batagelj, V.; Cvetkovic, D.; Gutman, I.
A formula involving the number of 1-factors in a graph.

BAUE 1985-1 Bauer, D.
A note on degree conditions for Hamiltonian cycles in line graphs.

BAUEA 1987-1 Bauernöppel, F.
Degree bounds and subhamiltonian cycles in planar graphs.

BAUM 1983-1 Bauman, E. V.; Muchnik, I. B.
Simplification of graph structure with the aid of Boolean matrices.

BAUMA 1985-1 Baumann, U.
On fixed points of partition preserving automorphisms of complete graphs with minimal colourings.

BAUMC 1987-1 Baumann, U.; Lesch, M.; Schmeichel, I.
Farb- und zerlegungstreue Automorphismen von Graphen.

BECKC 1985-1 Becker, R. I.; Perl, Y.
Finding the two-core of a tree.

BEHZ 1987-1 Behzad, M.
Total concepts in graph theory.

BEIS 1987-1 Beissinger, J. S.; Peled, U. N.
Enumeration of labelled threshold graphs and a theorem of Frobenius involving Eulerian polynomials.

BEKA 1985-1 Beka, J.
On the tripartite conjecture.

BEND 1985-1 Bender, E. A.; Richmond, L. B.; Wormald, N. C.
Almost all chordal graphs split.

BENDB 1985-1 Bender, E. A.; Wormald, N. C.
The number of loopless planar maps.
BENH 1986-1 Benhocine, A.
Pancyclism and Meyniel's conditions.
BENHA 1987-1 Benhocine, A.; Fouquet, J. L.
The Chvatal-Erdös condition and pancyclic line-graphs.
BENHB 1987-1 Benhocine, A.; Wojda, A. P.
The Geng-Hua Fan conditions for pancyclic or Hamilton-connected graphs.
BENT 1987-1 Bent, S. W.; Manber, U.
On non-intersecting Eulerian circuits.
BENZ 1983-1 Benzaken, C.; Boyd, S. C.; Hammer, P. L.; Simeone, B.
Adjoints of pure bidirected graphs.
BENZA 1987-1 Benzaken, C.; Crama, Y.; Hammer, P. L.; Maffray, F.
More characterisations of triangulated and cotriangulated graphs.
BENZB 1985-1 Benzaken, C.; Hammer, P. L.
Boolean techniques for matroidal decomposition of independence systems and applications to graphs.
BENZC 1985-1 Benzaken, C.; Hammer, P. L.; Werra, D. de
Split graphs of Dilworth number 2.
BERG 1983-1 Berge, C.
Graphes (3. edition).
BERG 1984-1 Berge, C.
A property of k-optimal path-partitions.
BERG 1984-3 Berge, C.
Minimax theorems for normal hypergraphs and balanced hypergraphs - a survey.
BERG 1984-4 Berge, C.
On the minimum number of arcs to reverse to make a graph strongly connected.
BERG 1985-1 Berge, C.
Graphs (second revised edition).
BERG 1987-3 Berge, C.
On two conjectures to generalize Vizing's theorem.
BERM 1985-1 Berman, A.
Convexity, graph theory and non-negative matrices.
BERMA 1987-1 Berman, K. A.
Balanced vertex weightings and arborescences.
BERMG 1984-1 Bermond, J. C.; Delorme, C.; Farhi, G.
Large graphs with given degree and diameter. II.
BERMH 1984-1 Bermond, J. C.; Fouquet, J. L.; Habib, M.; Peroche, B.
On linear k-arboricity.
BERMI 1985-1 Bermond, J. C.; Paoli, M.
Hamiltonian problems in powers of graphs.
BERNA 1985-1 Bern, M. W.; Lawler, E. L.; Wong, A. L.
Why certain subgraph computations require only linear time.
BERNA 1987-1 Bern, M. W.; Lawler, E. L.; Wong, A. L.
Linear-time computation of optimal subgraphs of decomposable graphs.
BERTB 1986-1 Bertossi, A. A.
Total domination in interval graphs.
BERTC 1987-1 Bertossi, A. A.; Benuccelli, M. A.
Some parallel algorithms on interval graphs.
BERTD 1985-2 Bertossi, A. A.; Bonuccelli, M. A.
Some parallel algorithms on interval graphs.
BERTD 1986-1 Bertossi, A. A.; Bonuccelli, M. A.
Hamiltonian circuits in interval graph generalizations.
BESS 1985-1 Bessonov, Y. E.
On solution of the problem of search for the maximal intersections of graphs on the basis of analysis of projections of subgraphs of the modular product (Russian).
BESSA 1981-1 Bessonov, Y. E.; Skorobogatov, V. A.
Recursive analysis of graphs (Russian).
BESSA 1982-1 Bessonov, Y. E.; Skorobogatov, V. A.
On a family of schemes of recursive stripping of graphs (Russian).
BEUT 1984-1 Beutelspacher, A.; Hering, P.-R.
Minimal graphs for which the chromatic number equals the maximal degree.
BHAT 1986-1 Bhat-Nayak, V. N.; Gokhale, S. K.
Validity of Hebbare's conjecture.
BHATD 1985-1 Bhattacharya, D.
The minimum order of n-connected n-regular graphs with specified diameters.
BIAL 1987-1 Bialostocki, A.; Roditty, Y.
A monotone path in an edge-ordered graph.
BIDA 1985-1 Bidamon, E.; Meyniel, H.
On the Shannon capacity of a directed graph.
BIEL 1983-1 Bielak, H.
On graphs with nonisomorphic 2-neighbourhoods.
BIEL 1987-1 Bielak, H.
Some remarks about medians in graphs.
BIELA 1983-1 Bielak, H.; Soczewinska, E.
Some remarks about digraphs with nonisomorphic 1- or 2-neighbourhoods.
BIENC 1985-1 Bienstock, D.
An algorithm for reliability analysis of planar graphs.
BIENC 1987-1 Bienstock, D.
On embedding graphs in trees.
BIEND 1986-1 Bienstock, D.; Brickell, E. F.; Monma, C. L.
Properties of k-connected networks.
BIENE 1987-1 Bienstock, D.; Monma, C. L.
On the complexity of covering vertices by faces in a planar graph.

BIENE	1987-2	Bienstock, D.; Monma, C. L. On the complexity of embedding planar graphs to minimize certain distance measures.
BIENE	1987-3	Bienstock, D.; Monma, C. L. Optimal enclosing regions in planar graphs.
BILLC	1986-1	Billionnet, A. On interval graphs and matrice profiles.
BILLD	1985-1	Billionnet, A.; Minoux, M. Maximizing a supermodular pseudoboolean function: a polynomial algorithm for supermodular cubic functions.
BIXB	1984-2	Bixby, R. E. Recent algorithms for two versions of graph realization and remarks on applications to linear programming.
BIXBF	1984-1	Bixby, R. E.; Wagner, D. K. An almost linear-time algorithm for graph realization.
BLID	1986-1	Blidia, M. A parity digraph has a kernel.
BLOO	1982-1	Bloom, G. S.; Hsu, D. F. On graceful digraphs and a problem in network addressing.
BLOO	1985-1	Bloom, G. S.; Hsu, D. F. On graceful directed graphs that are computational models of some algebraic systems.
BLOO	1985-2	Bloom, G. S.; Hsu, D. F. On graceful directed graphs.
BODE	1980-1	Bodendiek, R.; Schumacher, H. Zur Numerierung von Raupen.
BODEA	1984-1	Bodendiek, R.; Schumacher, H.; Wagner, K. Über 1-optimale Graphen.
BOES	1985-1	Boesch, F. T. Is reliable network synthesis a graph theory problem?
BOES	1985-2	Boesch, F. T. The cut frequency vector.
BOESA	1984-1	Boesch, F. T.; Suffel, C. L. An overview of discrete methods in the synthesis of reliable networks.
BOESB	1985-1	Boesch, F. T.; Tindell, R. Connectivity and symmetry in graphs.
BOESC	1986-1	Boesch, F. T.; Wang, J. F. Super line-connectivity properties of circulant graphs.
BOHM	1983-1	Böhme, T.; Sachs, H. Über räumliche Darstellungen endlicher Graphen.
BOKH	1987-1	Bokhari, S. H. Assigment Problems in Parallel and Distributed Computing.
BOLD	1986-1	Bold, C. Order-degree sequences.
BOLL	1985-1	Bollobas, B. Geodesics in oriented graphs.
BOLL	1985-2	Bollobas, B. Random graphs.
BOLL	1986-1	Bollobas, B. Extremal graph theory with emphasis on probabilistic methods.
BOLLA	1983-1	Bollobas, B.; Chung, F. R. K.; Graham, R. L. On complete bipartite subgraphs contained in spanning tree complements.
BOLLB	1985-1	Bollobas, B.; Fenner, T. I.; Frieze, A. M. An algorithm for finding Hamilton cycles in random graphs.
BOLLB	1987-1	Bollobas, B.; Fenner, T. I.; Frieze, A. M. An algorithm for finding Hamilton paths and cycles in random graphs.
BOLLC	1983-1	Bollobas, B.; Frieze, A. M. On matchings and Hamiltonian cycles in random graphs.
BOLLE	1984-1	Bollobas, B.; Hell, P. Sorting and graphs.
BOND	1985-1	Bondarenko, V. A. Nonpolynomial lower bounds for the complexity of optimization problems of clique and vertex covering in a class of direct algorithms (Russian).
BONDB	1985-1	Bondy, J. A. Kotzig's conjecture on generalized friendship graphs - a survey.
BONDC	1987-1	Bondy, J. A.; Fan, G. A sufficient condition for dominating cycles.
BONDD	1985-1	Bondy, J. A.; Jackson, B. Long paths between specified vertices of a block.
BONU	1985-1	Bonuccelli, M. A. Dominating sets and domatic number of circular arc graphs.
BORD	1985-1	Bordat, J. P. Parcours dans les graphes: un outil pour l'algorithmique des ensembles ordonnes.
BORI	1982-1	Borisov, K. Y. Convex extremal problems on graphs (Russian).
BOROC	1983-1	Boronin, A. B.; Pershin, O. Y. An algorithm for partition of a weighted graph into minimally connected subgraphs (Russian).
BOROC	1984-1	Boronin, A. B.; Pershin, O. Y. Decomposition of a network into minimally connected parts.
BOROF	1984-1	Borowiecki, M. Point partition numbers and generalized Nordhaus-Gaddum problems.
BOROF	1987-1	Borowiecki, M. Characterization of graphs having minimaximal kernels cardinality half their order.
BOROF	1987-2	Borowiecki, M. Partition numbers, connectivity and Hamiltonicity.

BOROG 1985-1 Borowiecki, M.; Mihok, P.
On (n,P)-partitionable graphs.
BORZ 1987-1 Borzacchini, L.
Graphical partially ordered sets.
BOSA 1984-1 Bosak, J.
Induced subgraphs.
BOUCA 1985-1 Bouchet, A.
Un algorithme polynomial pour reconnaitre les graphes d'alternance.
BOUCA 1986-1 Bouchet, A.
Characterizing and recognizing circle graphs.
BOUCA 1987-4 Bouchet, A.
Reducing prime graphs and recognizing circle graphs.
BOUCA 1987-6 Bouchet, A.
Unimodularity and circle graphs.
BOUCC 1985-1 Bouchitte, V.
Chordal bipartite graphs and crowns.
BOURA 1984-1 Bourjolly, J.-M.; Hammer, P. L.; Simeone, B.
Node-weighted graphs having the König-Egervary property.
BRANA 1985-1 Brandstädt, A.
A characterization of split permutation graphs.
BRANA 1985-2 Brandstädt, A.
Partitions of graphs into one or two independent sets and cliques.
BRANA 1986-1 Brandstädt, A.
Characterization of split permutation graphs.
BRANA 1987-1 Brandstädt, A.
The computational complexity of feedback vertex set, Hamiltonian circuit, dominating set, Steiner tree, and bandwidth on special perfect graphs.
BRANB 1985-1 Brandstädt, A.; Kratsch, D.
On the restriction of some NP-complete graph problems to permutation graphs.
BRANB 1987-1 Brandstädt, A.; Kratsch, D.
On domination problems for permutation and other graphs.
BREC 1985-1 Brecht, T. B.; Colbourn, C. J.
Lower bounds on two-terminal network reliability.
BREC 1986-1 Brecht, T. B.; Colbourn, C. J.
Improving bounds on network reliability: some examples.
BREN 1984-1 Brenner, J. L.; Lyndon, R. C.
Doubly Eulerian trails on rectangular grids.
BRID 1987-1 Bridgland, M. F.
Universal traversal sequences for paths and cycles.
BRIG 1984-1 Brigham, R. C.; Chinn, P. Z.; Dutton, R. D.
On the clique graph of the line graph.
BRIGA 1984-1 Brigham, R. C.; Dutton, R. D.
Upper bounds on the edge clique cover number of a graph.
BRODA 1982-1 Brodskij, G. M.; Soloveva, N. A.
The partially ordered set of subtrees of a graph (Russian).
BROEC 1987-1 Broersma, H. J.
Subgraph conditions for dominating circuits in graphs and pancyclicity of line graphs.
BROU 1985-1 Brouwer, A. E.; Mesner, D. M.
The connectivity of strongly regular graphs.
BROWC 1985-1 Brown, W. G.; Erdös, P.; Simonovits, M.
Algorithmic solution of extremal digraph problems.
BRUC 1981-1 Brucker, P.
The Chinese postman problem for mixed graphs.
BRUD 1985-1 Brudaru, O.
A parallel algorithm for a variant of the acyclic partition problem.
BRUD 1987-1 Brudaru, O.
Hamiltonian circuits with generalized cost.
BRUI 1984-1 Bruin, R. de
Some software for graph theoretical problems.
BU 1984-1 Bu, Y. H.
Generating trees obtained from a cycle basis (Chinese).
BUCKA 1984-1 Buckley, F.
Equalities involving certain graphical distributions.
BUEL 1983-1 Buell, K. E.; Lesniak, L.; Ringeisen, R. D.
More on mission networks.
BUI 1984-1 Bui, T. N.; Chaudhuri, S.; Leighton, F. T.; Sipser, M.
Graph bisection algorithms with good average case behavior.
BUI 1987-1 Bui, T. N.; Chaudhuri, S.; Leighton, F. T.; Sipser, M.
Graph bisection algorithms with good average case behavior.
BURL 1984-1 Burlet, M.; Fonlupt, J.
Polynomial algorithm to recognize a Meyniel graph.
BURL 1984-2 Burlet, M.; Fonlupt, J.
Polynomial algorithm to recognize a Meyniel graph.
BURLA 1984-1 Burlet, M.; Uhry, J.-P.
Parity graphs.
BURN 1984-1 Burns, D.
Biregular edge-symmetric graphs.
BURNA 1984-1 Burns, D.; Kapoor, S. F.; Ostrand, P. A.
On line-symmetric graphs.
BURRA 1985-1 Burr, S. A.; Nesetril, J.; Rödl, V.
On the use of senders in generalized Ramsey theory for graphs.
BUSN 1984-1 Busnyuk, N. N.
The number of 3-cycles in graphs with given degrees of vertices (Russian).

BUSNA	1985-1	Busnyuk, N. N.; Sarvanov, V. I. Graphs of partition of a rectangle. I (Russian).
BUZA	1984-1	Buzacott, J. A.; Chang, J. S. K. Cut-set intersections and node partitions.
CACC	1987-1	Caccetta, L.; Erdös, P.; Ordman, E. T.; Pullman, N. J. The difference between the clique numbers of a graph.
CACCA	1986-1	Caccetta, L.; Vijayan, K. On minimal graphs with prescribed adjacency property.
CAEN	1987-1	Caen, D. de; Gregory, D. A. On the decomposition of a directed graph into complete bipartite subgraphs.
CAHI	1987-1	Cahit, I. Cordial graphs: a weaker version of graceful and harmonious graphs.
CAI A	1983-1	Cai, M.-C. Edge-disjoint mappings in a bipartite graph (Chinese).
CAI A	1987-1	Cai, M.-C. The maximal size of a digraph with maximal local connectivity. I.
CAI C	1984-1	Cai, X.-T. On the panconnectivity of Ore graphs.
CAI C	1985-1	Cai, X.-T. A long cycle in a 2-connected graph with f>=n-4 (Chinese).
CALV	1983-1	Calvillo, G.; Neumann-Lara, V. Minimal graphs whose N-path graphs are connected.
CAPO	1984-1	Caponi, M.; Pieroni, P.; Saviozzi, G. A Benders dual decomposition approach to the optimum network orientation.
CARP	1984-1	Carpaneto, G. Relaxation techniques and applications to the optimal Hamiltonian tour problem (Italian).
CART	1984-1	Carter, M. A practical algorithm for finding the largest clique in a graph.
CERD	1985-1	Cerdeira, J. O.; Luz, C. J. Covering nodes with a fixed number of trees
CERD	1985-2	Cerdeira, J. O.; Luz, C. J. The 2-components of minimum cost of a graph.
CHAKE	1985-1	Chakroun, N.; Manoussakis, Y. Degrees, diameter and structure of oriented graphs.
CHANA	1987-1	Chan, O.; Chen, C. C.; Slater, P. J. On a characterization of comparability graphs.
CHANQ	1986-1	Chang, G. J. Labeling algorithms for domination problems in sun-free chordal graphs.
CHANQ	1986-3	Chang, G. J. Total domination in block graphs.
CHANT	1985-1	Chang, G. J.; Nemhauser, G. L. Covering, packing and generalized perfection.
CHANU	1986-1	Chang, G. J.; Wu, C.-S. Graphs whose cycles are of length r modulo k.
CHAO	1984-1	Chao, C.-Y.; Zhao, L.-C. Chromatic polynomials of connected graphs.
CHARB	1983-1	Chartrand, G.; Cooper, J. K. Jr.; Lesniak-Foster, L.; Wall, C. E. The traceability of a graph.
CHARC	1985-1	Chartrand, G.; Harary, F.; Oellermann, O. R. On the regulation number of a multigraph
CHARD	1987-1	Chartrand, G.; Holbert, K. S.; Oellermann, O. R.; Swart, H. C. F-degrees in graphs.
CHARE	1986-1	Chartrand, G.; Johnson, M.; Oellermann, O. R. Connected graphs containing a given connected graph as a unique greatest common subgraph.
CHARF	1984-1	Chartrand, G.; Kapoor, S. F.; Lesniak, L.; Schuster, S. Near 1-factors in graphs.
CHARG	1985-1	Chartrand, G.; Kapoor, S. F.; Lick, D. R.; Schuster, S. The partial complement of graphs.
CHARI	1986-1	Chartrand, G.; Oellermann, O. R.; Saba, F.; Zou, H.-B. The uniformity number of a graph.
CHARJ	1986-1	Chartrand, G.; Ruiz, S.; Wall, C. E. Smallest regular graphs without near 1-factors.
CHARK	1984-1	Chartrand, G.; Saba, F.; Wormald, N. C. Bridge and cycle degrees of vertices of graphs.
CHARL	1985-1	Chartrand, G.; Saba, F.; Zou, H.-B. Edge rotations and distance between graphs.
CHARL	1987-1	Chartrand, G.; Saba, F.; Zou, H.-B. Greatest common subgraphs of graphs.
CHARM	1985-1	Chartrand, G.; Wilson, R. J. The Petersen graph.
CHAUB	1987-1	Chaudhuri, P. Algorithms for some graph problems on a distributed computational model.
CHAUB	1987-2	Chaudhuri, P. Efficient algorithms for some graph problems on a tree-structured parallel computer.
CHEN	1987-1	Chen, B.-L.; Lih, K.-W. Hamiltonian uniform subset graphs.
CHENI	1985-1	Chen, R. Y. Long cycles in regular 2-connected bipartite graphs (Chinese).
CHENJ	1986-1	Chen, Z. B. A simple proof of a theorem about strongly connected digraphs (Chinese).
CHENK	1982-1	Chen, Z. M. Some results about the graphs with unique 1-factor (Chinese).

CHENL	1986-1	Chen, Z. Z. A generalization of the Bodendiek conjecture about graceful graphs (Chinese).
CHEP	1986-1	Chepoi, V. D. Some properties of d-convexity in triangular graphs (Russian).
CHERB	1984-1	Chernjak, A. A. Connectivity of graphs with prescribed degree set and order unigraphicity (Russian).
CHERB	1985-1	Chernjak, A. A. Switching-complete properties of graphs (Russian).
CHERC	1987-1	Chernjak, A. A.; Chernjak, Z. A.; Tyshkevich, R. I. On forcibly hereditary P-graphical seqeuences.
CHERD	1986-1	Chernjak, Z. A.; Chernjak, A. A. About recognizing (alpha, beta) classes of polar graphs.
CHET	1985-1	Chetwynd, A. G.; Hilton, A. J. W. Regular graphs of high degree are 1-factorizable.
CHET	1986-1	Chetwynd, A. G.; Hilton, A. J. W. Critical star multigraphs.
CHET	1987-1	Chetwynd, A. G.; Hilton, A. J. W. The edge-chromatic class of regular graphs of degree 4 and their complements.
CHIA	1985-1	Chia, G.-L.; Lim, C.-K. On supercompact graphs I: the nucleus.
CHIB	1985-1	Chiba, N.; Nishizeki, T. Arboricity and subgraph listing algorithms.
CHIBA	1985-1	Chiba, N.; Onoguchi, K.; Nishizeki, T. Drawing plane graphs nicely.
CHINC	1983-1	Chinn, P. Z.; Dewdney, A. K.; Greene, T. M. Bandwidth and peripheral vertices of trees.
CHOG	1984-1	Chogovadze, G. G.; Chachanedze, G. G.; Kachibaya, V. V.; Surguladze, G. G. A method for determining the minimal structure of functional dependencies (Russian).
CHRE	1984-1	Chretienne, P. Chemins extremaux d'un graphe doublement value.
CHRID	1984-1	Christofides, N.; Valls Verdejo, V. Finding all elementary circuits of zero cost in an optimal incremental graph (Spanish).
CHUNA	1985-1	Chung, F. R. K. On the cutwidth and the topological bandwidth of a tree.
CHUNB	1983-1	Chung, F. R. K.; Erdös, P.; Spencer, J. On the decomposition of graphs into complete bipartite subgraphs.
CHUNC	1984-1	Chung, F. R. K.; Garey, M. R. Diameter bounds for altered graphs.
CHUNE	1987-1	Chung, F. R. K.; Graham, R. L.; Saks, M. Dynamic search in graphs.
CHUNF	1985-1	Chung, F. R. K.; Graham, R. L.; Winkler, P. M. On the addressing problem for directed graphs.
CHVA	1984-5	Chvatal, V. Probabilistic methods in graph theory.
CHVA	1985-2	Chvatal, V. Hamiltonian cycles.
CHVAE	1986-1	Chvatalova, J.; Opatrny, J. The bandwidth problem and operations on graphs.
COCAA	1987-1	Coçan, M. Über Matrizen, die den Multigraphen zugeordnet werden und ihre Anwendung im Studium der Eigenschaften der Multigraphen.
COCH	1986-1	Cochand, M.; Werra, D. de Generalized neighbourhoods and a class of perfectly orderable graphs.
COCK	1986-1	Cockayne, E. J.; Gamble, B.; Shepherd, B. Domination parameters for the bishops graph.
COLB	1986-1	Colbourn, C. J. Exact algorithms for network reliability.
COLBA	1985-1	Colbourn, C. J.; Colbourn, M. J.; Rosa, A. Indecomposable 1-factorizations of the complete multigraph.
COLBB	1985-1	Colbourn, C. J.; Keil, J. M.; Stewart, L. K. Finding minimum dominating cycles in permutation graphs.
COLBD	1986-1	Colbourn, C. J.; Pulleyblank, W. R.; Rosa, A. Hybrid triple systems and cubic feedback sets.
COLBE	1987-1	Colbourn, C. J.; Slater, P. J.; Stewart, L. K. Locating dominating sets in series parallel networks.
COLBF	1985-1	Colbourn, C. J.; Stewart, L. K. Permutation graphs: connected domination and Steiner trees.
COLBF	1987-1	Colbourn, C. J.; Stewart, L. K. Dominating cycles in series-parallel graphs.
COOK	1984-1	Cook, R. J. Graph factorization and theorems of the Nordhaus-Gaddum class.
COPPA	1987-1	Coppersmith, D.; Raghavan, P.; Tompa, M. Parallel graph algorithms that are efficient on average.
CORNA	1984-1	Corneil, D. G.; Goldberg, M. A non-factorial algorithm for canonical numbering of a graph.
CORNC	1984-1	Corneil, D. G.; Perl, Y.; Stewart, L. Cographs: recognition, applications and algorithms.
CORND	1985-1	Corneil, D. G.; Perl, Y.; Stewart, L. K. A linear recognition algorithm for cographs.
CORNE	1987-1	Cornuejols, G. General factors of graphs.
CORNF	1985-1	Cornuejols, G.; Cunningham, W. H. Compositions for perfect graphs.

COULB 1987-1 Coullard, C. R.; Pulleyblank, W. R.
On cycle cones and polyhedra.
COZZ 1984-1 Cozzens, M. B.; Leibowitz, R.
Threshold dimension of graphs.
COZZB 1984-1 Cozzens, M. B.; Wang, D. I.
Closed neighborhood containment graphs.
CRAM 1985-1 Crama, Y.; Hammer, P. L.
The complexity of recognizing partition-quadratic graphs.
CSIM 1984-1 Csima, J.; Ralston, T. A.
Crossing-free Hamiltonian paths in Euclidean spaces.
CUNNA 1985-2 Cunningham, W. H.
Optimal attack and reinforcement of a network.
CURR 1985-1 Curran, S. J.; Witte, D.
Hamilton paths in Cartesian products of directed cycles.
CVET 1983-1 Cvetkovic, D.; Petric, M. V.
Connectedness of the noncomplete extended p-sum of graphs.
DAHL 1987-1 Dahlhaus, E.; Duchet, P.
On strongly chordal graphs.
DAHLA 1987-1 Dahlhaus, E.; Karpinski, M.
A fast parallel algorithm for computing all maximal cliques in a graph and the related problems.
DAHLA 1987-2 Dahlhaus, E.; Karpinski, M.
Fast parallel computation of perfect and strongly perfect elimination schemes.
DAI 1986-1 Dai, H. T.
A graceful labelling of the complete binary trees (Chinese).
DAI A 1985-1 Dai, X.-S.; Xue, Y.-F.
Some Hamiltonian properties of connected graph (Chinese).
DAI A 1985-2 Dai, X.-S.; Xue, Y.-F.
Some properties of nth order minimally 2-connected graphs (Chinese).
DARB 1982-1 Darbinyan, S. K.
Cycles of arbitrary length in digraphs with large semi degrees (Russian).
DARB 1985-1 Darbinyan, S. K.
Pancyclicity of digraphs with large semidegrees (Russian).
DARB 1985-2 Darbinyan, S. K.
Pancyclicity of digraphs with the Meyniel condition (Russian).
DARB 1986-1 Darbinyan, S. K.
A sufficient condition for the Hamilton property of digraphs with large semidegrees (Russian).
DARB 1986-2 Darbinyan, S. K.
On the pancyclicity of digraphs with large semidegrees (Russian).
DAS B 1983-1 Das, P.; Gangopadhyay, T.
Some results on alpha-valuations and average numberings.
DAS C 1984-1 Das, P.; Rao, S. B.
Center graphs of chordal graphs.
DAS D 1986-1 Das, S. K.; Deo, N.
Stirling graphs and their properties.
DAVI 1985-1 Davitt, R. M.; Fink, J. F.; Jacobson, M. S.
On isopart parameters of complete bipartite graphs and n-cubes.
DAWE 1983-1 Dawes, R. W.
Constructions of minimally k-connected graphs for k=1,2 and 3.
DAWE 1986-1 Dawes, R. W.
Minimally 3 connected graphs.
DAWS 1985-1 Dawson, J. E.
Algorithms to find directed packings.
DELF 1983-1 Delfrate, M. G.
Oriented 2-graphs and their generalization to dimension n (Italian).
DELF 1983-2 Delfrate, M. G.
The oriented 2-graphs and a generalization to the n-dimensional case (Italian).
DELO 1985-1 Delorme, C.
Grands graphes de degre et diametre donnes.
DELOA 1984-1 Delorme, C.; Farhi, G.
Large graphs with given degree and diameter. I.
DEMAA 1979-1 DeMaria, F.; Gionfriddo, M.
Some parameters associated with graphs and hypergraphs (Italian).
DESI 1986-1 DeSimone, C.; Lucertini, M.; Pallottino, S.; Simeone, B.
Fair dissections of spiders, worms and caterpillars.
DEWD 1985-1 Dewdney, A. K.; Walsh, T. R.
A methodology for average-time testing of exact and approximate graph algorithms.
DICK 1984-1 Dick, W. E.
Discrete characterization of planarity. I: The classical viewpoint.
DINGC 1985-1 Ding, S. K.
The structure of a maximum critical 2-edge-connected graph (Chinese).
DINY 1985-1 Diny, D. A.; Harary, F.; Mac Morris, F. R.
A survey of results on a graph and its complement with specified hereditary properties.
DJID 1984-1 Djidjev, H. N.
On some properties of nonplanar graphs.
DJID 1985-1 Djidjev, H. N.
A linear algorithm for partitioning graphs of fixed genus.
DJID 1985-2 Djidjev, H. N.
A separator theorem for graphs of fixed genus.
DJID 1985-3 Djidjev, H. N.
Efficient parallel algorithms of graphs.
DMIT 1982-1 Dmitriev, I. G.
Minimal degree of uniquely colored graphs with minimal number of edges (Russian).

DONA 1987-1 Donald, J.; Elwin, J.; Hager, R.; Salamon, P.
Handle bases and bounds on the number of subgraphs.
DONGA 1985-1 Dong, J. Q.
Graceful labelings of a special class of lobster trees (Chinese).
DOOB 1986-1 Doob, M.
Pseudocyclic graphs.
DORIA 1985-1 Döring, S.; Iwainsky, A.
Graph reduction for the optimization of node-disjoint paths in networks.
DU 1986-1 Du, D. Z.
An optimization problem on graphs.
DUCHA 1987-1 Duchet, P.; Hamidoune, Y. O.; Meyniel, H.
Sur les quasi-noyaux d'un graphe.
DUCHC 1986-1 Duchet, P.; Vergnas, M. Las; Meyniel, H.
Connected cutsets of a graph and triangle bases of the cycle space.
DUKE 1982-1 Duke, R. A.; Erdös, P.
Subgraphs in which each pair of edges lies in a short common cycle.
DYERA 1985-1 Dyer, M. E.; Foulds, L. R.; Frieze, A. M.
Analysis of heuristics for finding a maximum weight planar subgraph.
DYERB 1985-2 Dyer, M. E.; Frieze, A. M.
On the complexity of partitioning graphs into connected subgraphs.
DZHI 1985-1 Dzhidzhev, K. N.
A linear algorithm for partitioning graphs of fixed genus.
EADEA 1984-1 Eades, P.; Pullman, N. J.; Robinson, P. J.
On clique partition numbers of regular graphs.
EBEN 1984-1 Ebenegger, C.; Hammer, P. L.; Werra, D. de
Pseudo-boolean functions and stability of graphs.
EBERA 1987-1 Ebert, J.
A versatile data structure for edge-oriented graph algorithms.
EDELA 1987-1 Edelsbrunner, H.; Rote, G.; Welzl, E.
Testing the necklace condition for shortest tours and optimal factors in the plane.
EGAW 1987-1 Egawa, Y.
Cycles in k-connected graphs whose deletion results in a (k-2)-connected graph.
EGAWA 1986-1 Egawa, Y.; Enomoto, H.; Saito, A.
Contractible edges in triangle-free graphs.
EGAWA 1986-2 Egawa, Y.; Enomoto, H.; Saito, A.
On component factors.
EGAWB 1986-1 Egawa, Y.; Fukuda, T.; Nagoya, S.; Urabe, M.
A decomposition of complete bipartite graphs into edge-disjoint subgraph with star components.
EGGL 1984-1 Eggleton, R. B.; Skilton, D. K.
Chain decompositions of graphs, 1: Abstract graphs.
EGGL 1984-2 Eggleton, R. B.; Skilton, D. K.
Chain decompositions of graphs, 2: Surface embeddings.
EGGL 1984-3 Eggleton, R. B.; Skilton, D. K.
Double tracings of graphs.
EGGL 1984-4 Eggleton, R. B.; Skilton, D. K.
Graphs with Eulerian chains.
EGIA 1983-1 Egiazaryan, E. V.
The external stability number of graphs (Russian).
EGYH 1985-1 Egyhazy, C. J.
An algorithm for generating the minimum length fundamental cycle set in a graph.
EL-M 1987-1 El-Mallah, E. S.; Colbourn, C. J.
On two dual classes of planar graphs.
ELLIB 1984-1 Ellingham, M. N.; Holton, D. A.; Little, C. H. C.
Cycles through ten vertices in 3-connected cubic graphs.
ENNS 1984-1 Enns, T. C.
An efficient algorithm determining when a fabric hangs together.
ENOM 1986-1 Enomoto, H.
Toughness and the existence of k-factors.II.
ENTR 1985-1 Entringer, R. C.
A short proof of Rubin's block theorem.
ENTRA 1983-1 Entringer, R. C.; Jackson, D. E.
Degree sums of paths in extremal trees.
ERA 1984-1 Era, H.
A note on semiregular factorizations of regular graphs.
ERA 1985-1 Era, H.
Semiregular factorizations of regular graphs.
ERDOC 1985-1 Erdös, P.; West, D. B.
A note on the interval number of a graph.
ERDOD 1985-1 Erdös, P. L.; Györi, E.
Any four independent edges of a 4-connected graph are contained in a circuit.
EREM 1987-1 Eremin, G. S.
Partition of an arbitrary set of vertices of an acyclic digraph into convex subsets (Russian).
EVER 1986-1 Everett, H.
Finding Hamiltonian paths in non-rectangular grid graphs.
EVERA 1985-1 Everett, M. G.; Seidman, S. B.
The hull number of a graph.
EVST 1979-1 Evstigneev, V. A.
A local algorithm for isolating blocks in a graph (Russian).
EVST 1981-1 Evstigneev, V. A.
Local numerical algorithms and determination of vertices with the greatest degree in a non-directed graph (Russian).
EVST 1983-1 Evstigneev, V. A.
On properties of local algorithms on graphs (Russian).

EVST 1985-1 Evstigneev, V. A.
Application of graph theory to computer programming (Russian).

EXOO 1984-1 Exoo, G.; Harary, F.
Smallest cubic graphs with prescribed properties.

FAIG 1987-1 Faigle, U.
The clique polytope of a graph.

FAIGF 1984-1 Faigle, U.; Schrader, R.
Comparability graphs and order invariants.

FAIGF 1985-3 Faigle, U.; Schrader, R.
Setup minimization techniques for comparability graphs.

FALC 1987-1 Falconer, K. J.
Cut-set sums and tree processes.

FAN 1984-1 Fan, G.-H.
New sufficient conditions for cycles in graphs.

FAN 1985-1 Fan, G.-H.
Longest cycles in regular graphs.

FANGA 1984-1 Fang, W.-W.
An algorithm for the directed T-immersion of a graph.

FARBA 1986-1 Farber, M.; Jamison, R. E.
Convexity in graphs and hypergraphs.

FARBB 1985-1 Farber, M.; Keil, J. M.
Domination in permutation graphs.

FARL 1984-1 Farley, A. M.; Proskurowski, A.
Computing the maximum order of an open irredundant set in a tree.

FARR 1985-1 Farr, G.; Mac Diarmid, C.
The complexity of counting homeomorphs.

FARRA 1983-1 Farrell, E. J.
Decomposition of complete graphs and complete bipartite graphs into node disjoint paths.

FARRA 1983-2 Farrell, E. J.
On the number of 2-factors in the short ladder.

FARRA 1983-3 Farrell, E. J.
The subgraph polynomial and its relation to some other graph polynomials.

FARRA 1984-1 Farrell, E. J.
A note on cycle decompositions of complete bipartite graphs.

FARRA 1984-2 Farrell, E. J.
Decompositions of graphs with cyclomatic number 2 into node disjoint paths.

FARRA 1984-3 Farrell, E. J.
On coefficients of path polynomials.

FARRA 1985-1 Farrell, E. J.
On coefficients of circuit polynomials and characteristic polynomials.

FARRA 1986-1 Farrell, E. J.
A survey of the unifying effects of F-polynomials in combinatorics and graph theory.

FARRB 1983-1 Farrell, E. J.; Grell, J. C.
Some analytical properties of the circuit polynomial of a graph.

FARRB 1983-2 Farrell, E. J.; Grell, J. C.
The circuit polynomial and its relation to other graph polynomials.

FARRB 1985-1 Farrell, E. J.; Grell, J. C.
Some further constructions of cocircuit and cospectral graphs.

FARRC 1987-1 Farrell, E. J.; Guo, J. M.
Circuit characterizations of unions of graphs.

FARRD 1985-1 Farrell, E. J.; Gutman, I.
A note on the circuit polynomial and its relation to the mu-polynomial.

FARRE 1987-2 Farrell, E. J.; Wahid, S. A.
Some general classes of comatching graphs.

FAUD 1987-1 Faudree, R. J.; Gould, R. J.; Schelp, R. H.
Neighborhood conditions and edge disjoint Hamiltonian cycles.

FAUDB 1985-1 Faudree, R. J.; Rousseau, C. C.; Schelp, R. H.
Edge-disjoint Hamiltonian cycles.

FAVA 1985-1 Favaron, O.
On a conjecture of Fink and Jacobson concerning k-domination and k-dependence.

FAVAA 1985-1 Favaron, O.; Kouider, M.
Path partitions and cycle partitions of Eulerian graphs of maximum degree 4.

FAVAB 1985-1 Favaron, O.; Kouider, M.; Maheo, M.
Edge-vulnerability and mean distance.

FAVAC 1985-1 Favaron, O.; Lonc, Z.; Truszczynski, M.
Decompositions of graphs into graphs with three edges.

FAVAD 1986-1 Favaron, O.; Ordaz, O.
A sufficient condition for oriented graphs to be Hamiltonian.

FEDO 1985-1 Fedorova, N. P.
On homogeneous Hamiltonian graphs contractible to complete ones (Russian).

FELL 1987-1 Fellows, M.; Hickling, F.; Syslo, M. M.
A topological parametrization and hard graph problems.

FENG 1983-1 Feng, C. J.
A necessary and sufficient condition for a class of graceful graphs (Chinese).

FENGA 1985-1 Feng, T.; Shi, R.-H.
Smallest regular graphs with girth pair (4,2t+1).

FENN 1984-1 Fenner, T. I.; Frieze, A. M.
Hamiltonian cycles in random regular graphs.

FENN 1984-2 Fenner, T. I.; Frieze, A. M.
On the complexity of finding sets of edges in graphs and digraphs.

FENT 1985-1 Fenton, N. E.
The structural complexity of flow graphs.

FEOF	1987-1	Feofiloff, P.; Younger, D. H. Directed cut transversal packing for source-sink connected graphs.
FIED	1985-1	Fiedler, M. Signed bigraphs of monotone matrices.
FINK	1983-1	Fink, J. F. Randomly bitraceable graphs.
FINKA	1985-1	Fink, J. F.; Jacobson, M. S. N-domination in graphs.
FINKA	1985-2	Fink, J. F.; Jacobson, M. S. On n-domination, n-dependence and forbidden subgraphs.
FINKB	1985-1	Fink, J. F.; Jacobson, M. S.; Kinch, L. F.; Roberts, J. A. On graphs having domination number half their order.
FIOL	1985-1	Fiol, M. A.; Alegre, I.; Yebra, J. L. A.; Fabrega, J. Digraphs with walks of equal length between vertices.
FIOLA	1983-1	Fiol, M. A.; Yebra, J. L. A.; Fabrega, J. Sequence graphs and interconnection networks.
FISH	1985-1	Fishburn, P. C. Interval graphs and interval orders.
FISH	1985-2	Fishburn, P. C. Interval orders and interval graphs. A study of partially ordered sets.
FLAM	1985-1	Flament, C. Comparability graphs with constraint, partial semi-orders and interval orders.
FLAN	1985-1	Flandrin, E. Cycles hamiltoniens et dominants dans le carre d'un graphe connexe.
FLEIA	1983-1	Fleischner, H. A way of enumerating Eulerian trails in a connected Eulerian graph.
FLEIA	1984-1	Fleischner, H. Cycle decompositions, 2-coverings, removable cycles and the four-color-disease.
FLEIB	1985-1	Fleischner, H.; Jackson, B. Removable cycles in planar graphs.
FLEIB	1987-1	Fleischner, H.; Jackson, B. Compatible path-cycle-decompositions on plane graphs.
FOULD	1984-1	Foulds, L. R.; Robinson, R. W. Enumeration of phylogenetic trees without points of degree two.
FOUQA	1985-1	Fouquet, J. L.; Thuillier, H. Cycles through given vertices in planar 3-connected cubic graphs.
FOUR	1985-1	Fournier, I. Longest cycles in 2-connected graphs of independence number alpha.
FOURA	1985-1	Fournier, I.; Fraisse, P. On a conjecture of Bondy.
FOURB	1984-1	Fournier, J.-C.; Vergnas, M. Las A class of bichromatic hypergraphs.
FRAE	1985-1	Fraenkel, A. S.; Hammer, P. L. Pseudo-Boolean functions and their graphs.
FRAIA	1985-1	Fraisse, P. A new sufficient condition for Hamiltonian graphs.
FRAIB	1986-1	Fraisse, P.; Hell, P.; Kirkpatrick, D. G. A note on f-factors in directed and undirected multigraphs.
FRAIC	1985-1	Fraisse, P.; Ordaz, O. Chvatal-Erdos condition for (tau, tau)-factors in digraphs using given arcs.
FRAID	1987-1	Fraisse, P.; Thomassen, C. Hamiltonian dicycles avoiding prescribed arcs in tournaments.
FRANC	1985-1	Frank, A. Edge-disjoint paths in planar graphs.
FRANC	1986-1	Frank, A. On connectivity properties of Eulerian digraphs.
FRANC	1987-1	Frank, A. Graph connectivity and network flows.
FRAND	1984-1	Frank, A.; Sebö, A.; Tardos, E. Covering directed and odd cuts.
FRANG	1983-1	Franke, H.-G. Analyse der mittleren Rechenzeit für den transitiven Abschluß azyklischer Graphen.
FRAY	1984-1	Fraysseix, H. de A characterization of circle graphs.
FRAYA	1985-1	Fraysseix, H. de; Rosenstiehl, P. A characterization of planar graphs by Tremaux orders.
FREDA	1987-1	Frederickson, G. N.; Hambrusch, S. E. Planar linear arrangements of outerplanar graphs.
FRIEC	1984-2	Frieze, A. M. Limit distribution for the existence of Hamiltonian cycles in random bibartite graphs.
FRIEC	1986-2	Frieze, A. M. On large matchings and cycles in sparse random graphs.
FRIEC	1987-2	Frieze, A. M. Parallel algorithms for finding hamilton cycles in random graphs.
FRUCA	1985-1	Frucht, R. W.; Salinas, L. C. Graceful numbering of snakes with constraints on the first label.
FU	1986-1	Fu, Y.-T.; Anderson, P. W. Application of statistical mechanics to NP-complete problems in combinatorial optimization.
FUKUA	1985-1	Fukui, M.; Kamae, T. Information polynomials of graphs.
GABB	1986-1	Gabbani, D.; Magazine, M. An interactive heuristic approach for multiobjective integer-programming problems.

GABO 1985-1 Gabor, C. P.; Hsu, W.-L.; Supowit, K. J.
Recognizing circle graphs in polynomial time.
GABOB 1984-1 Gabow, H. N.; Galil, Z.; Spencer, T.
Efficient implementation of graph algorithms using contraction.
GAIN 1987-1 Gainanov, D. N.
On the connectedness of the graphs of some classes of independence systems (Russian).
GALEA 1986-1 Galeana-Sanchez, H.
A theorem about a conjecture of H. Meyniel on kernel-perfect graphs.
GALEB 1986-1 Galeana-Sanchez, H.; Neumann-Lara, V.
On kernel-perfect critical digraphs.
GALIA 1987-1 Galil, Z.; Hoffmann, C. M.; Luks, E. M.; Schnorr, C. P.; Weber, A.
An O(n**3 log n) deterministic and an O(n**3) Las Vegas isomorphism test for trivalent graphs.
GAMBB 1987-1 Gamble, A. B.; Pulleyblank, W. R.
Forest covers and a polyhedral intersection theorem.
GAO 1987-1 Gao, J.-Z.
Quasi-1-Hamilton connectedness of tree graphs (Chinese).
GARC 1987-1 Garcia-Aviles, A.
Writing cut-set and loop equations by inspection.
GARFA 1984-1 Garfinkel, R. S.; Sundararaghavan, P. S.
Hamiltonian cycles in striped graphs: the two-stripe problem.
GARGA 1985-1 Gargano, M.; Quintas, L. V.
A digraph generalization of balanced signed graphs.
GAZI 1986-1 Gazit, H.
An optimal randomized parallel algorithm for finding connected components in a graph.
GAZIA 1987-1 Gazit, H.; Miller, G. L.
A parallel algorithm for finding a separator in planar graphs.
GERA 1985-1 Gerards, A. M. H.
Testing the odd bicycle wheel inequalities for the bipartite subgraph polytope.
GERAB 1986-2 Gerards, A. M. H.; Schrijver, A.
Signed graphs - regular matroids - grafts.
GERN 1986-1 Gernert, D.
Equations and inequalities related to the diameter of a graph.
GERN 1987-1 Gernert, D.
The cycle-length sequence of a graph and some of its applications.
GHOS 1984-1 Ghosh, R. K.; Bhattacharjee, G. P.
Parallel breadth-first search algorithms for trees and graphs.
GIBB 1985-1 Gibbons, A.
Algorithmic graph theory.
GIFF 1984-1 Giffin, J. W.; Foulds, L. R.
Efficient graph planarity updating tests.
GILB 1984-1 Gilbert, J. R.; Hutchinson, J. P.; Tarjan, R. E.
A separator theorem for graphs of bounded genus.
GODD 1986-1 Goddard, W.; Winter, P. A.
A characterization of stable graphs on a maximum (minimum) number of edges.
GODDA 1985-1 Goddyn, L.
A girth requirement for the double cycle cover conjecture.
GODS 1984-1 Godsil, C. D.
Real graph polynomials.
GODSA 1987-1 Godsil, C. D.; Nowakowski, R.; Nesetril, J.
The chromatic connectivity of graphs.
GOETB 1985-1 Goetze, B.; Staedler, G.
Algorithms on graphs with costly adjacency-information.
GOLB 1987-1 Golberg, A. I.; Gurvich, V. A.
On the maximal number of edges for a graph with n vertices in which a given subgraph with k vertices has no more than l edges (Russian).
GOLDB 1986-1 Goldberg, M. K.
Parallel algorithms for three graph problems.
GOLDC 1984-1 Goldberg, M. K.; Gardner, R.
On the minimal cut problem.
GOLI 1987-1 Golitschek, M. von
The cost-to-time ratio problem for large or infinite graphs.
GOLU 1985-1 Golumbic, M. C.
Interval graphs and related topics.
GOLUA 1985-1 Golumbic, M. C.; Jamison, R. E.
Edge and vertex intersection of paths in a tree.
GOLUA 1985-2 Golumbic, M. C.; Jamison, R. E.
The edge intersection graphs of paths in a tree.
GOLUB 1982-1 Golumbic, M. C.; Monma, C. L.
A generalization of interval graphs with tolerances.
GOLUB 1982-2 Golumbic, M. C.; Monma, C. L.
A generalization of interval graphs with tolerances.
GOLUC 1984-1 Golumbic, M. C.; Monma, C. L.; Trotter, W. T. Jr.
Tolerance graphs.
GORKA 1986-1 Gorkovoi, V. F.; Egorov, M. Y.
Decomposition of graphs (Russian).
GORS 1985-1 Gorshkov, A. F.
A method for finding extremal subgraphs (Russian).
GORS 1987-1 Gorshkov, A. F.
A method for finding extremal subgraphs in bipartite graphs.
GOUL 1983-1 Gould, R. J.
Traceability in the square of a tree.
GOULA 1983-1 Gould, R. J.; Jacobson, M. S.
A note on mixed Ramsey numbers: total chromatic number versus complete graphs.

GOULB 1984-1 Gould, R. J.; Lick, D. R.
Degree sets and graph factorizations.
GRAC 1982-1 Grace, T.
New proof techniques for gracefully labelling graphs.
GRAC 1983-1 Grace, T.
More graceful trees.
GRAND 1987-1 Grant, D. D.; Mac Avaney, K. L.
Antidirected cycles in planar graphs.
GRAOA 1984-1 Graovac, A.; Jashari, G.; Strunje, M.
On the distance spectrum of a cycle.
GREE 1985-1 Green, C. D.; Suraweera, F.
A path entropy function for rooted acyclic digraphs.
GREEC 1983-1 Greenberg, H. J.; Lundgren, J. R.; Maybee, J. S.
The inversion of 2-step graphs.
GRES 1984-1 Greshnev, S. N.
The polyhedron of a problem on an m-vertex subgraph of a complete graph.
GRIM 1985-1 Grimaldi, R. P.
Properties of integer-interval graphs.
GRIMA 1986-1 Grimmett, G. R.
An exact threshold theorem for random graphs and the node-packing problem.
GROF 1984-1 Gröflin, H.
Path-closed sets.
GROTB 1985-1 Grötschel, M.; Jünger, M.; Reinelt, G.
Acyclic subdigraphs and linear orderings: polytopes, facets, and a cutting plane algorithm.
GROTB 1985-3 Grötschel, M.; Jünger, M.; Reinelt, G.
On the acyclic subgraph polytope.
GROTC 1986-1 Grötschel, M.; Lovasz, L.; Schrijver, A.
Relaxations of vertex packing.
GROTG 1987-1 Grötschel, M.; Wakabayashi, Y.
A cutting plane algorithm for a clustering problem.
GROTG 1987-2 Grötschel, M.; Wakabayashi, Y.
Composition of facets of the clique partitioning polytope.
GROTG 1987-3 Grötschel, M.; Wakabayashi, Y.
Facets of the clique partitioning polytope.
GUAN 1983-1 Guan, M.-G.
Another improvement in the assignment method on graphs.
GUAN 1984-1 Guan, M.-G.
The maximum weighted cycle-packing problem and its relation to the Chinese postman problem.
GUANC 1985-1 Guan, M.-G.; Pulleyblank, W. R.
Eulerian orientations and circulations.
GUGL 1986-1 Guglielmann Grieco, M.; Zuchetti, R.; Maria, B.
Digraphs and bigraphs: relationship between strong connectedness and strong alpha-connectedness, alpha-hamiltonian digraphs.
GULD 1987-1 Guldan, F.
On a problem of linear arboricity.
GUNT 1985-1 Gunther, G.
Neighbour-connectivity in regular graphs.
GUO 1986-1 Guo, X. F.; Zhu, B. W.
A construction of critically 2-connected graphs (Chinese).
GUO A 1985-1 Guo, Z.-Y.
A theorem on cohesiveness of points of a graph (Chinese).
GUO A 1986-1 Guo, Z.-Y.
Minimally and critically 2-connected graph (Chinese).
GURE 1987-1 Gurevich, Y.; Shelah, S.
Expected computation time for Hamiltonian path problem.
GURG 1986-1 Gurgel, M. A.; Wakabayashi, Y.
On k-leaf-connected graphs.
GUSF 1987-1 Gusfield, D.
Optimal mixed graph augmentation.
GUST 1985-1 Gustafson, K.; Harary, F.
The curl of graphs and networks.
GUSTA 1985-1 Gustafson, K.; Hartman, R.
Graph theory and fluid dynamics.
GUTIA 1984-1 Gutin, G. M.
Cycles in strong n-partite tournaments (Russian).
GUTIA 1987-1 Gutin, G. M.
Finding of the largest contour in a complete bipartite graph (Russian).
GUTM 1983-1 Gutman, I.
Graphs with maximum and minimum independence numbers.
GUTM 1985-1 Gutman, I.
On some graphic polynomials whose zeros are real.
GUTM 1986-1 Gutman, I.
Some relations for graphic polynomials.
GUTMB 1984-1 Gutman, I.; Cvetkovic, D.
Finding tricyclic graphs with a maximal number of matchings - another example of computer aided research in graph theory.
GUTMD 1986-1 Gutman, I.; Zhang, F. J.
On a quasi-ordering of bipartite graphs.
GYARC 1987-1 Gyarfas, A.; Lehel, J.; Schelp, R. H.; Tuza, Z.
Ramsey numbers for local colorings.
GYORA 1985-1 Györi, E.; Rothschild, B. L.; Rucinski, A.
Every graph is contained in a sparsest possible balanced graph.

GYUL 1987-1 Gyulumyan, S. M.
Colour-critical graphs with an independent line cut (Russian).
HABI 1984-1 Habib, M.
Comparability invariants.
HAGE 1985-1 Hager, M.
Pendant tree-connectivity.
HAGE 1986-1 Hager, M.
Path-connectivity in graphs.
HAGG 1980-1 Haggard, G.
Excursions in graph theory.
HAGGA 1985-1 Häggkvist, R.
A lemma on cycle decompositions.
HAGGA 1985-2 Häggkvist, R.
A note on Hamilton cycles.
HAGGA 1985-3 Häggkvist, R.
Equicardinal disjoint cycles in sparse graphs.
HAGGB 1985-1 Häggkvist, R.; Jackson, B.
A note on maximal cycles in 2-connected graphs.
HAKI 1985-1 Hakimi, S. L.
Further results on a generalization of edge-coloring.
HALI 1981-1 Halin, R.
Graphentheorie 2.
HALL 1985-1 Hall, J. I.; Shult, E. E.
Locally cotriangular graphs.
HAMA 1985-2 Hamacher, H. W.
Min cost tensions.
HAMI 1984-1 Hamidoune, Y. O.; Vergnas, M. Las
Directed switching games on graphs and matroids (French).
HAMI 1986-1 Hamidoune, Y. O.; Vergnas, M. Las
Switching games on graphs and matroids.
HAMMA 1986-1 Hammer, P. L.; Maffray, F.
Completely separable graphs.
HAMMB 1985-1 Hammer, P. L.; Mahadev, N. V. R.
Bithreshold graphs.
HAMMB 1985-2 Hammer, P. L.; Mahadev, N. V. R.
Intershold graphs.
HAMMC 1985-1 Hammer, P. L.; Mahadev, N. V. R.; Werra, D. de
Stability in CAN-free graphs.
HAMMC 1985-2 Hammer, P. L.; Mahadev, N. V. R.; Werra, D. de
The struction of a graph: application to CN-free graphs.
HANSM 1985-1 Hanson, D.; Seyffarth, K.
Point saturation of graphs.
HARAB 1984-1 Harary, F.
General connectivity.
HARAB 1985-2 Harary, F.
Graph theoretic approaches to finite mathematical structures.
HARAB 1985-3 Harary, F.
Petersen's graph and its unique properties (Danish).
HARAC 1982-1 Harary, F.; Hoede, C.; Kadlecek, D.
Graph-valued functions related to step graphs
HARAD 1985-1 Harary, F.; Iri, M.
Generalized Ramsey theory for graphs. XVI. Reduced Ramsey numbers.
HARAE 1986-1 Harary, F.; Jones, K. F.
Degree conditional bipartition numbers in graphs.
HARAF 1984-1 Harary, F.; Karabed, R.
Regular digraphs containing a given digraph.
HARAG 1985-1 Harary, F.; Kolasinska, E.; Syslo, M. M.
Cycle basis interpolation theorems.
HARAH 1987-1 Harary, F.; Peled, U. N.
Hamiltonian threshold graphs.
HARAI 1983-1 Harary, F.; Plantholt, M.
The derived signed graph of a digraph.
HARAJ 1985-1 Harary, F.; Robinson, R. W.
The diameter of a graph and its complement.
HARAK 1984-1 Harary, F.; Schmidt, R.
Realization of graph regulation numbers.
HARAL 1987-1 Harary, F.; Schuster, S.; Vestergaard, P. D.
Interpolation theorems for the invariants of spanning trees of a given graph: edge-covering.
HARAM 1986-1 Harary, F.; Slater, P. J.
A linear algorithm for the cutting center of a tree.
HARAN 1985-1 Harary, F.; Tindell, R.
The minimal blocks of diameter two and three.
HARB 1985-1 Harborth, H.
Darstellungen von Graphen in der Ebene.
HARB 1985-2 Harborth, H.
Drawings of graphs and multiple crossings.
HARE 1985-1 Hare, E. O.; Hare, W. R.; Hedetniemi, S. T.
Another upper bound for the bandwidth of trees.
HAREA 1986-1 Hare, E. O.; Hedetniemi, S. T.; Hare, W. R.
Algorithms for computing the domination number of KxN complete grid graphs.
HAREB 1987-1 Hare, E. O.; Hedetniemi, S. T.; Laskar, R.; Peters, K.; Wimer, T. V.
Linear-time computability of combinatorial problems on generalized series-parallel graphs.

HAREC 1985-1 Hare, W. R.; Hare, E. O.; Hedetniemi, S. T.
Bandwidth of grid graphs.
HAREE 1984-1 Hare, W. R.; Laskar, R.; Peters, K.
Max-min parameters and disconnected colorings.
HARTB 1985-1 Hartman, A.; Rosa, A.
Cyclic one-factorization of the complete graph.
HARTC 1983-1 Hartman, J.; Katchalski, M.
On k-cycled refinements of certain graphs.
HARTE 1987-1 Hartvigsen, D.; Zemel, E.
Is every cycle basis fundamental?
HASSA 1984-1 Hässelbarth, W.
Die Verzweigtheit von Graphen.
HASSD 1986-1 Hassin, R.; Tamir, A.
Efficient algorithms for optimization and selection on series-parallel graphs.
HAVE 1984-1 Havel, I.
On Hamiltonian circuits and spanning trees of hypercubes.
HAYW 1985-1 Hayward, R. B.
Weakly triangulated graphs.
HAYW 1987-1 Hayward, R. B.
A lower bound for the optimal crossing-free Hamiltonian cycle problem.
HE 1985-1 He, W.-J.
On Jackson's conjecture of 2-connected and "almost regular" graphs (Chinese).
HEAT 1987-1 Heath, L. S.
Embedding outerplanar graphs in small books.
HEDE 1985-1 Hedetniemi, S.; Hedetniemi, S.; Laskar, R.
Domination in trees: models and algorithms.
HEDEA 1986-1 Hedetniemi, S.; Laskar, R.
A bipartite theory of graphs: I.
HEDEC 1985-1 Hedetniemi, S. M.; Proskurowski, A.; Syslo, M. M.
Interior graphs of maximal outerplane graphs.
HEDED 1983-1 Hedetniemi, S. T.
A max-min relationship between matchings and domination in graphs.
HEDEE 1985-1 Hedetniemi, S. T.; Laskar, R.; Pfaff, J.
Irredundance in graphs: a survey.
HEDEE 1986-1 Hedetniemi, S. T.; Laskar, R.; Pfaff, J.
A linear algorithm for finding a minimum dominating set in a cactus.
HEDM 1985-1 Hedman, B.
The maximum number of cliques in dense graphs.
HEDM 1986-1 Hedman, B.
A polynomial algorithm for constructing the clique graph of a line graph.
HEDM 1987-1 Hedman, B.
Another extremal problem for Turan graphs.
HEES 1985-1 Heesch, H.
E-Reduktion.
HEFN 1987-1 Hefner, K. A. S.; Lundgren, J. R.
Biclique covers and partitions of acyclic digraphs.
HEINA 1987-1 Heinrich, K.; Hell, P.
On the problem of bandsize.
HEINB 1985-1 Heinrich, K.; Nonay, G. M.
Path and cycle decompositions of complete multigraphs.
HEMM 1985-1 Hemminger, R. L.; Wilson, E. L.
Gamma-regularity: multigraphs and digraphs.
HEND 1986-1 Hendry, G. R. T.
The size of graphs uniquely Hamiltonian-connected from a vertex.
HENDA 1987-1 Hendy, M. D.; Penny, D.
Edge lengths of trees from sequence data.
HEXE 1985-2 Hexel, E.
On vertex sets in graphs with neighbourhood constraints - the component greedoid of a graph.
HEYDA 1984-1 Heydemann, M. C.; Sotteau, D.
Number of arcs and cycles in digraphs.
HEYDA 1985-1 Heydemann, M. C.; Sotteau, D.
About some cyclic properties in digraphs.
HIC 1984-1 Hic, P.
On partially directed p-graphs.
HIC 1985-1 Hic, P.
A construction of geodetic blocks.
HIC 1986-1 Hic, P.
Criticality concepts in geodetic graphs.
HIGA 1984-1 Higashiyama, Y.; Ariyoshi, H.
An algorithm for generation of s-t acyclic graphs.
HILB 1986-1 Hilbig, F.
Edge structures in nonhamiltonian graphs (German).
HILTA 1987-1 Hilton, A. J. W.; Johnson, P. D.
Graphs which are vertex-critical with respect to the edge-chromatic number.
HILTB 1986-1 Hilton, A. J. W.; Rodger, C. A.
Hamiltonian decompositions of complete regular s-partite graphs.
HIRA 1984-1 Hirata, T.; Kubota, K.; Saito, O.
A sufficient condition for a graph to be weakly k-linked.
HIRS 1985-1 Hirschberg, D. S.; Larmore, L. L.
The least weight subsequence problem.
HOAN 1987-1 Hoang, C. T.
Alternating orientation and alternating colouration of perfect graphs.

HOANB 1987-1 Hoang, C. T.; Reed, B.
A note on short cycles in digraphs.
HOCH 1985-1 Hochbaum, D. S.
Easy solutions for the K-center problem or the dominating set problem on random graphs.
HOCHB 1985-2 Hochbaum, D. S.; Shmoys, D. B.
An 0(||V||**2) algorithm for the planar 3-cut problem.
HOES 1986-1 Hoesel, C. van; Schrijver, A.
Edge-disjoint homotopic paths in a planar graph with one hole.
HOFFB 1986-1 Hoffman, A. J.; Lee, C. W.
On the cone of nonnegative circuits.
HOLR 1985-1 Holroyd, F. C.; Wingate, W. J. G.
Cycles in the complement of a tree or other graph.
HOLT 1982-1 Holton, D. A.
Cycles in 3-connected cubic graphs.
HOLT 1985-1 Holton, D. A.
Cycles in 3-connected cubic planar graphs.
HOLTA 1986-1 Holton, D. A.; Mac Kay, B. D.
Cycles in 3-connected cubic planar graphs. II.
HOLTB 1985-1 Holton, D. A.; Manvel, B.; Mac Kay, B. D.
Hamiltonian cycles in cubic 3-connected bipartite planar graphs.
HOLTC 1987-1 Holton, D. A.; Plummer, M. D.
On the cyclability of k-connected (k+1)-regular graphs.
HOLU 1985-1 Holub, P.
A remark on covering graphs.
HOPK 1985-1 Hopkins, G.; Staton, W.
Graphs with unique maximum independent sets.
HORA 1982-1 Horak, P.; Niepel, L.
A short proof of a linear arboricity theorem for cubic graphs.
HORT 1987-1 Horton, J. D.
A polynomial-time algorithm to find the shortest cycle basis of a graph.
HORTA 1985-1 Horton, J. D.; Roy, B. K.; Schellenberg, P. J.; Stinson, D. R.
On decomposing graphs into isomorphic uniform 2-factors.
HSU 1985-1 Hsü, L.-H.
On a strongly multiplicative graph function conjecture (Chinese).
HSU A 1984-1 Hsu, W.-L.
Efficient algorithms for the maximum weight clique problem on circular-arc graphs and circle graphs.
HSU A 1984-3 Hsu, W.-L.
On the existence of induced odd cycles in planar graphs (revised).
HSU A 1985-1 Hsu, W.-L.
Maximum weight clique algorithms for circular-arc graphs and circle graphs.
HSU A 1986-1 Hsu, W.-L.
Coloring planar perfect graphs by decomposition.
HUANC 1984-1 Huang, J.-H.; Cai, C.-T.
Numbers of Hamilton cycles on generalized Petersen graph P(n,2) (Chinese).
HUAND 1984-1 Huang, S.-H. S.; Wong, C. K.
Optimal binary split trees.
HUANE 1984-1 Huang, S.-M.
A note on graphs with equal edge-connectivity and minimum degree (Chinese).
HWANB 1987-1 Hwang, F. K.
The Hamiltonian property of linear functions.
IDZI 1987-1 Idzik, A.; Komar, J.; Malawski, M.
Edge-coloured complete graphs: connectedness of some subgraphs.
IHM 1982-1 Ihm, H. S.; Ntafos, S.
On finding legal paths.
ILEVA 1983-1 Ilev, V. P.; Fridman, G. S.
On a problem of approximation by three-component graphs (Russian).
IMAIA 1986-1 Imai, H.; Asano, T.
Efficient algorithms for geometric graph search problems.
IMAS 1985-1 Imase, M.; Soneoka, T.; Okada, K.
Connectivity of regular directed graphs with small diameters.
IOI 1986-1 Ioi, S.
A special "tree" problem.
IORD 1986-1 Iordache, M.; Dumitriu, L.; Mierlea, C.; Georgescu, M.
Generation without duplication of all trees in a connected graph.
IORDA 1982-1 Iordanskii, M. A.
Problems of graph arrangement. II. (Russian).
IVAND 1983-1 Ivanov, A. A.
Bounding the diameter of a distance-regular graph (Russian).
IVANE 1984-1 Ivanov, A. A.; Ivanov, A. V.; Faradzhev, I. A.
Distance-transitive graphs of valency 5, 6 and 7 (Russian).
IWAH 1983-1 Iwahori, N.
On some algorithm of the number of the branch points and their degrees starting from the distance matrix of a given tree.
IWAI 1985-2 Iwainsky, A.
Optimal trees - a short overview on problem formulations.
JACK 1985-1 Jackson, B.
Long cycles in bipartite graphs.
JACK 1986-1 Jackson, B.
Longest cycles in 3-connected cubic graphs.
JACK 1987-1 Jackson, B.
A Chvatal-Erdös condition for Hamilton Cycles in Digraphs.
JACK 1987-2 Jackson, B.
Compatible Euler tours for transition systems in Eulerian graphs.

JACK 1987-3 Jackson, B.
Small r-regular r-connected nonhamiltonian graphs.
JACKA 1985-1 Jackson, B.; Ordaz, O.
A Chvatal-Erdös condition for (1,1)-factors on digraphs.
JACKB 1984-2 Jackson, B.; Ringel, G.
The splitting number of complete bipartite graphs.
JACKC 1982-1 Jackson, D. E.; Entringer, R. C.
Minimum k-geodetically connected graphs.
JAEG 1984-2 Jaeger, F.
On some algebraic properties of graphs.
JAEG 1985-1 Jaeger, F.
A survey of the cycle double cover conjecture.
JAEG 1985-2 Jaeger, F.
Graphic description of binary spaces.
JAMI 1987-1 Jamison, R. E.
Alternating Whitney sums and matchings in trees, Part 1.
JAMIA 1984-1 Jamison, R. E.; Laskar, R.
Elimination orderings of chordal graphs.
JAMIB 1984-1 Jamison, R. E.; White, M. M.
Coronation numbers and circular systems in block graphs.
JASMA 1987-1 Jasmon, G. B.; Foong, K. W.
A method for evaluating all the minimal cuts of a graph.
JAUM 1986-1 Jaumard, B.; Minoux, M.
An efficient algorithm for the transitive closure and a linear worst-case complexity result for a class of sparse graphs.
JAYAA 1987-1 Jayaram, S. R.
Line domination in graphs.
JAYAA 1987-2 Jayaram, S. R.
The nomatic number of a graph.
JEUR 1985-1 Jeurissen, R. H.
The Petersen graph as a box of Pandora.
JEZN 1983-1 Jezny, S.; Trenkler, M.
Characterization of magic graphs.
JIN 1986-1 Jin, M. Y.; Diao, Z. Y.; Zhang, S. L.
An optimal-depth ordering algorithm on large scale graphs (Chinese).
JIRA 1987-1 Jirasek, J.
On a certain class of multidigraphs, for which reversal of no arc decreases the number of their cycles.
JOHNJ 1987-1 Johnson, E. L.; Mosterts, S.
On four problems in graph theory.
JOHNK 1986-1 Johnson, E. L.; Pieroni, P.
A linear programming approach to the optimum network orientation problem.
JONE 1985-1 Jones, K. F.
Maximum bipartite subgraphs and independence.
JORD 1982-1 Jordan, S. A.; Shier, D. R.
Algorithms for determining fundamental cycles in a graph.
JUNG 1984-1 Jung, H. A.
Longest circuits in 3-connected graphs.
JUNGA 1985-1 Jünger, M.
Polyhedral combinatorics and the acyclic subdigraph problem.
JUNGC 1987-1 Jungnickel, D.
Graphen, Netzwerke und Algorithmen.
KAJI 1981-1 Kajitani, Y.; Ueno, S.
Redundant elements of a fundamental tie matrix of a graph.
KALA 1985-1 Kalai, G.
Hyperconnectivity of graphs.
KALI 1983-1 Kalinin, V. B.
On graphs of intersections (Russian).
KANEC 1984-1 Kanetkar, S. V.; Rao, P. R.
Connected, locally 2-connected, K(1,3)-free graphs are panconnected.
KANED 1987-1 Kanevsky, A.; Ramachandran, V.
Improved algorithms for graph four-connectivity.
KANG 1984-1 Kang, T.; Lin, K. R.
Necessary and sufficient conditions for a planar graph G with chi(G)=3 (Chinese).
KANN 1985-1 Kannan, R.
Unraveling k-page graphs.
KANO 1984-1 Kano, M.
Graph factors with given properties.
KANO 1985-1 Kano, M.
(a,b)-factorizations of nearly bipartite graphs.
KANO 1986-1 Kano, M.
Factors of regular graphs.
KANO 1987-2 Kano, M.
Ranking the vertices of an r-partite paired comparison digraph.
KANOA 1985-1 Kano, M.; Sakamoto, A.
Ranking the vertices of a paired comparison digraph.
KANTA 1987-1 Kanter, I.; Sompolinsky, H.
Graph optimization problems and the Potts glass.
KARA 1985-1 Karapetyan, I. A.
Chord graphs (Russian).
KARM 1984-1 Karmakar, S. B.
An algorithm for finding a circuit of even length in a directed graph.
KARM 1986-1 Karmakar, S. B.
Determination of a circuit in a directed graph.

KARM 1987-1 Karmakar, S. B.
A heuristic method for the determination of a Hamiltonian circuit in a graph.

KARPH 1985-1 Karp, R. M.; Wigderson, A.
A fast parallel algorithm for the maximal independent set problem.

KARZ 1984-2 Karzanov, A. V.
On closed sets of a directed graph.

KARZ 1985-1 Karzanov, A. V.
Cuts that are minimal with respect to mean weight, and cycles of a digraph (Russian).

KARZA 1986-1 Karzanov, A. V.; Timofeev, E. A.
An effective algorithm for finding all minimal branch sections of a non-oriented graph (Russian).

KATA 1987-1 Katajainen, J.; Nevalainen, O.; Teuhola, J.
A linear expected-time algorithm for computing planar relative neigbourhood graphs.

KATAA 1985-1 Kataranchuk, S. G.
D-convex simple planar graphs (Russian).

KATAA 1986-1 Kataranchuk, S. G.
D-convex simple bipartite graphs (Russian).

KATEA 1983-1 Katerinis, P.
Some results on the existence of 2n-factors in terms of vertex-deleted subgraphs.

KATEA 1985-1 Katerinis, P.
A Chvatal-Erdös condition for an r-factor in a graph.

KATEA 1985-2 Katerinis, P.
Minimum degree of a graph and the existence of k-factors.

KATEB 1987-1 Katerinis, P.; Woodall, D. R.
Binding numbers of graphs and the existence of k-factors.

KAUFA 1984-1 Kaufmann, M.
Nearly tight bounds for the longest edge of a tree in a VLSI layout.

KEAT 1985-1 Keating, K.
Multiple-ply Hamiltonian graphs and digraphs.

KEIL 1985-1 Keil, J. M.
Finding Hamiltonian circuits in interval graphs.

KEIL 1986-1 Keil, J. M.
Total domination in interval graphs.

KELL 1985-1 Kelly, D.
Comparability graphs.

KELL 1986-1 Kelly, D.
Invariants of finite comparability graphs.

KELM 1983-1 Kelmans, A. K.
A criterion for the existence of two crossing chain-chords of given cycle (Russian).

KELM 1984-1 Kelmans, A. K.
A strengthening of the Kuratowski planarity criterion for 3-connected graphs.

KELM 1986-1 Kelmans, A. K.
On 3-connected graphs without essential 3-cuts or triangles.

KENN 1985-1 Kennedy, J. W.; Quintas, L. V.; Syslo, M. M.
The theorem on planar graphs.

KHAC 1986-1 Khachatryan, N. K.
The Hamilton property and systems of different representatives in families of subsets of vertices of a graph (Russian).

KHADA 1985-1 Khadzhiivanov, N. G.; Nenov, N. D.
Relations and graphs (Bulgarian).

KHAR 1984-1 Kharazishvili, A. B.
Independent families and universal colored graphs (Russian).

KHO 1985-1 Kho, P.-L.
Tree growing.

KIDW 1987-1 Kidwell, M. E.; Richter, R. B.
Trees and Euler tours in a planar graph and its relatives.

KIE 1986-1 Kie, L.-T.
On the legitimate edge deck problem.

KIKUA 1986-1 Kikusts, P.
Another algorithm determining the independence number of a graph.

KIRK 1985-1 Kirkpatrick, D. G.; Klawe, M.; Pippenger, N.
Some graph-colouring theorems with applications to generalized connection networks.

KIRO 1985-1 Kirousis, L. M.; Papadimitriou, C. H.
Interval graphs and searching.

KLEEA 1984-1 Klee, V.; Kleinschmidt, P.
An upper bound on graph diameters.

KLEIB 1986-1 Klein, P. N.; Reif, J. H.
An efficient parallel algorithm for planarity.

KLIM 1984-1 Klimov, A. N.
Structural functions of labeled graphs (Russian).

KNOP 1984-1 Knop, J. V.; Szymanski, K.; Trinajstic, N.; Krivka, P.
Computer generation of all 1-factors for a class of graphs with all vertices of degree two or three.

KOCH 1985-1 Kochkarov, A. M.
Asymptotically exact algorithms for solving a multicriterial problem of covering a graph by paths (Russian).

KOGA 1987-1 Koganov, L. M.
An oriented graph connected with an ordered pair of ordered partitions (Russian).

KOH 1985-1 Koh, K. M.; Poh, K. S.
On a construction of critical graphs which are not sensitive.

KOH A 1985-1 Koh, K. M.; Yap, K. Y.
Graceful numberings of cycles with a P(3)-chord.

KOLA 1987-1 Kolaitis, P. G.; Prömel, H. J.; Rothschild, B. L.
K(l+1)-free graphs: asymptotic structure and a 0-1 law.

KOMA	1984-1	Komarek, P. Some new good characterizations for directed graphs.
KOMA	1986-1	Komarek, P. On good characterizations for digraphs.
KOMAA	1986-1	Komarovskii, I. V. Alpha-weighted t-ary trees (Russian).
KORA	1982-1	Korach, E. Packing of t-cuts, and other aspects of dual integrality.
KORNA	1984-1	Kornienko, N. M. Combinatorial algorithms on a class of graphs (Russian).
KOROA	1983-1	Korotkin, A. A. The accuracy of a greedy algorithm in a problem of the choice of an optimal acyclic subgraph (Russian).
KORTD	1987-1	Korte, N. Intervallgraphen und MPQ-Bäume: Algorithmen und Datenstrukturen zur Manipulation von Intervallgraphen und der Lösung von Seriationsproblem.
KORTE	1985-1	Korte, N.; Möhring, R. H. Transitive orientation of graphs with side constraints.
KORTE	1987-1	Korte, N.; Möhring, R. H. A simple linear-time algorithm to recognize interval graphs.
KOSK	1983-1	Koska, A. On partition graphs.
KOST	1985-2	Kostochka, A. V. Maximum set of edges no two covered by a clique.
KOST	1985-3	Kostochka, A. V. On the number of graph edges not contained in common cliques.
KOTI	1984-1	Kotin, L. Connectivity and the design of optimally survivable systems.
KOTZ	1983-1	Kotzig, A. Regularly k-path connected graphs.
KOTZ	1984-1	Kotzig, A. Recent results and open problems in graceful graphs.
KOU	1984-1	Kou, L. T. Fixed length sources and sinks of a graph.
KOUB	1985-1	Koubek, V.; Krsnakova, J. Parallel algorithms for connected components in a graph.
KOVAA	1985-1	Kovacs, P. The minimal trivalent graphs with given smallest odd cycle.
KOVAE	1984-1	Kovalev, M. M.; Kotov, V. M. Analysis of algorithms for constructing a Hamiltonian cycle of maximal weight (Russian).
KOZE	1985-1	Kozen, D.; Vazirani, U. V.; Vazirani, V. V. NC algorithms for comparability graphs, interval graphs, and testing for unique perfect matching.
KOZY	1985-1	Kozyrev, V. P.; Yushmanov, S. V. Graph theory (algorithmic, algebraic and metric problems) (Russian).
KRAT	1986-1	Kratochvil, J.; Goljan, M.; Kucera, P. String graphs.
KRATA	1986-1	Kratochvil, J.; Zeps, D. On the minimum number of Hamiltonian cycles in triangulations.
KRATB	1987-1	Kratsch, D. Finding the minimum bandwidth of an interval graph.
KRAU	1985-1	Krause-Solberg, H. Über die Konstruktion regulärer Pläne gegebener Weite.
KRIA	1981-1	Kriauciukas, V. A tree-matching algorithm (Russian).
KRIA	1985-1	Kriauciukas, V. Free decomposition of graph search problems (Russian).
KRIS	1986-1	Krishnamoorthy, M. S.; Murthy, K. On the total dominating set problem.
KRIZ	1986-1	Kriz, I. A hypergraph-less construction of highly chromatic graphs without short cycles.
KULI	1983-1	Kulikowski, J. L. On some formal properties of tri-partite graphs.
KUMAA	1984-1	Kumar, K. Efficient algorithms for grouping component-processor families.
KWAS	1985-1	Kwasnik, M. Primitivity of products of directed graphs.
KYS	1982-1	Kys, P. A note on characterization of diameter stable graphs.
KYS	1985-1	Kys, P. Remarks on domatic number.
KYS A	1983-1	Kys, P.; Wawruch, A. Point-maximal subgraphs with a given diameter.
KYS A	1985-1	Kys, P.; Wawruch, A. Maximal sizes of planar digraphs without transititve edges.
LACH	1982-1	Lachlan, A. H. Finite homogeneous simple digraphs.
LAI	1987-1	Lai, P.-Y.; Goldschmidt, Y. Y. Application of statistical mechanics to combinatorial optimization problems: the chromatic number problem and q-partitioning of a graph.
LAKS	1986-1	Lakshmipathy, N.; Winklmann, K. "Global" graph problems tend to be intractable.
LAN	1985-1	Lan, J.-L.; Chen, W.-K. On f-factors of a graph.

LANG 1985-1 Lang, Z.
Untersuchung spezieller Fälle des Layout-Problems mit Hilfe von Graphenmethoden.

LAPOD 1983-1 Laporte, G.; Nobert, Y.
Finding the shortest cycle through k specified nodes.

LASK 1985-1 Laskar, R.; Peters, K.
Vertex and edge domination parameters.

LAUM 1985-1 Laumond, J.-P.
Enumeration of articulation pairs of a planar graph.

LAURA 1984-1 Lauritzen, S. L.; Speed, T. P.; Vijayan, K.
Decomposable graphs and hypergraphs.

LAWLD 1985-1 Lawler, E. L.; Slater, P. J.
A linear time algorithm for finding an optimal dominating subforest of a tree.

LEE L 1985-1 Lee, S.-M.
K-graceful labelling of Mongolian tents and related graphs.

LEHM 1985-1 Lehmer, D. H.
The chromatic polynomial of a graph.

LENG 1985-1 Lengauer, T.
Efficient solution of biconnectivity problems on hierarchically defined graphs.

LEPI 1985-1 Lepin, V. V.
Minimax numberings of weighted trees (Russian).

LEPS 1986-1 Lepschy, A.; Romanin-Jacur, G.; Viaro, U.
An extension to 'An algorithm for finding a circuit of even length in a directed graph'.

LEPS 1986-2 Lepschy, A.; Romanin-Jacur, G.; Viaro, U.
An extension to: "An algorithm for finding a circuit of even length in a directed graph" by S. Karmakar.

LESCA 1983-1 Lesch, M.; Schmeichel, I.
Über farbtreue Automorphismen schlichter Graphen mit Kantenfärbungen.

LESK 1984-1 Lesk, M.
Diametre de graphes et qualite de service d'un reseau de donnees.

LEUNA 1984-1 Leung, J. Y.-T.
Fast algorithms for generating all maximal independent sets of interval, circular-arc and chordal graphs.

LEVC 1985-1 Levchenko, V. I.
A generalization of the notion of homogeneity of a graph (Russian).

LI B 1984-1 Li, H.; Lai, Z. K.
Some theorems on self-centered graphs (Chinese).

LI B 1985-1 Li, H.; Lai, Z. K.
Self-centered tree graphs (Chinese).

LI C 1981-1 Li, Q.; Tao, M. Q.; Shen, Y. Q.
The bandwidth of the discrete tori C(m)*C(n) (Chinese).

LI D 1984-1 Li, W.-X.
Graphs decomposable into paths of length at least three.

LI G 1984-1 Li, W.-Z.
On lower bounds of Ramsey numbers (Chinese).

LIH 1981-1 Lih, K.-W.
Connected planar graphs are Sperner posets.

LIKH 1985-1 Likholat, N. A.
On an algebraic algorithm for solving a problem of the covering of a graph by a growing directed tree (Russian).

LIN A 1986-1 Lin, G. N.; Li, X. L.
On strongly harmonious labelings of graphs (Chinese).

LIN C 1986-1 Lin, Y. C.
The adjunction algorithm for planarity of graphs (Chinese).

LIN E 1985-1 Lin, Y. X.; Zhang, F. J.
Hamilton circuits in minimum tree graphs and generation of all minimum trees (Chinese).

LINI 1984-1 Linial, N.
Graph decompositions without isolates.

LINIA 1986-1 Linial, N.; Lovasz, L.; Wigderson, A.
A physical interpretation of graph connectivity and its algorithmic applications.

LIPA 1986-1 Lipatov, E. P.
Graph theory and its applications (Russian).

LIPM 1985-1 Lipman, M. J.
Hamiltonian cycles and paths in vertex-transitive graphs with abelian and nilpotent groups.

LIPS 1983-1 Lipski, W. Jr.
The consecutive retrieval property, interval graphs and related topics - a survey.

LITT 1985-1 Little, C. H. C.; Holton, D. A.
No graph has a maximal 3-ring of bonds.

LIU 1986-1 Liu, B. L.
Errata to: "Several classes of self-centered graphs with cyclomatic number 3" (Chinese).

LIU C 1984-3 Liu, G.-Z.
A theorem on the 1-factors of hypergraphs without cycles (Chinese).

LIU C 1984-4 Liu, G.-Z.
A theorem on the 1-factors of 2-trees (Chinese).

LIU C 1984-5 Liu, G.-Z.
On independent circuits of 3-circuit graphs (Chinese).

LIU C 1985-2 Liu, G.-Z.
A property of minimally 2-connected graphs (Chinese).

LIU C 1985-3 Liu, G.-Z.
A theorem on the 1-factors of (m,n)-trees.

LIU C 1985-4 Liu, G.-Z.
An algorithm for lexicographically generating ordered rooted trees (Chinese).

LIU C 1987-1 Liu, G.-Z.
The connectivities of adjacent tree graphs.

LIU E 1986-1 Liu, J.-H.
Sequences of path lengths and their applications.
LIU F 1984-1 Liu, J.-Q.; Tian, S.-L.
The binding number of the cartesian product of n circuits (Chinese).
LIU G 1983-1 Liu, J.-Z.
Lexicographic generation of rooted trees and trees.
LIU H 1984-1 Liu, Y.-P.
Enumeration of rooted separable planar maps.
LIU H 1985-1 Liu, Y.-P.
Enumeration equation for inseparable planar map node decompositions (Chinese).
LIU H 1985-2 Liu, Y.-P.
Some enumeration problems for node decompositions (Chinese).
LIU H 1986-1 Liu, Y.-P.
An equation for the enumeration of the node decompositions of nonseparable Eulerian planar maps (Chinese).
LIU H 1986-2 Liu, Y.-P.
Enumeration of node decompositions of nonseparable outerplanar maps (Chinese).
LIU H 1986-4 Liu, Y.-P.
On the linearity of testing planarity of graphs.
LIU H 1987-1 Liu, Y.-P.
Boolean planarity characterization of graphs.
LIU H 1987-3 Liu, Y.-P.
Enumeration of rooted outerplanar maps with vertex partition.
LIU I 1987-1 Liu, Z. H.
A new sufficient condition for the existence of a Hamiltonian cycle (Chinese).
LIU J 1984-1 Liu, Z. S.
Two generalizations of Rosa's theorem on graceful graphs (Chinese).
LIU J 1985-1 Liu, Z. S.
A class of graceful graphs and two classes of graceful graphs derived from it (Chinese).
LO 1985-1 Lo, S.-P.
On edge-graceful labellings of graphs.
LOCK 1985-1 Locke, S. C.
A basis for the cycle space of a 3-connected graph.
LOCK 1985-2 Locke, S. C.
A generalization of Dirac's theorem.
LONC 1985-1 Lonc, Z.; Truszczynski, M.
Decomposition of graphs into graphs with bounded maximum degrees.
LOUK 1983-1 Loukakis, E.
Two algorithms for generating and ranking the family of solutions of a finite irreflexive relation.
LOUK 1984-1 Loukakis, E.
Two algorithms for determining a minimum independent dominating set.
LOVA 1984-1 Lovasz, L.
Normal hypergraphs and the weak perfect graph conjecture.
LOVA 1985-1 Lovasz, L.
An algorithmic theory of numbers, graphs, and convexity.
LOVA 1985-2 Lovasz, L.
Computing ears and branchings in parallel.
LOVA 1986-1 Lovasz, L.
An algorithmic theory of numbers, graphs and convexity.
LOVA 1986-2 Lovasz, L.
Connectivity algorithms using rubber-bands.
LOVA 1986-3 Lovasz, L.
Theory of numbers, graphs and convexity.
LOVAB 1987-1 Lovasz, L.; Saks, M.; Schrijver, A.
Orthogonal representations and connectivity of graphs.
LOVAC 1983-1 Lovasz, L.; Simonovits, M.
On the number of complete subgraphs of a graph. II.
LU B 1980-1 Lu, S. X.
Necessary and sufficient conditions for planar graphs to be Hamiltonian (Chinese).
LUBI 1985-1 Lubiw, A.
Orderings and some combinatorial optimization problems with geometric applications.
LUBY 1985-1 Luby, M.
A simple parallel algorithm for the maximal independent set problem.
LUBY 1986-1 Luby, M.
A simple parallel algorithm for the maximal independent set problem.
LUND 1983-1 Lundgren, J. R.; Maybee, J. S.
A characterization of upper bound graphs.
LUND 1985-1 Lundgren, J. R.; Maybee, J. S.
Block graphs and their matrix duals.
LUND 1985-2 Lundgren, J. R.; Maybee, J. S.
Food webs with interval competition graph.
LUO 1984-1 Luo, Y.-X.
The binding number of a product graph - proof of a conjecture of V. C. Kane et al. (Chinese).
MA 1984-1 Ma, H.-K.
A counterexample to the conjecture of Aksionov and Mel'nikov on non-3-colorable planar graphs.
MA B 1984-1 Ma, K. J.; Feng, C. J.
About the Bodendiek's conjecture of graceful graph (Chinese).
MA B 1984-2 Ma, K. J.; Feng, C. J.
On gracefulness of gear graphs (Chinese).
MA D 1986-1 Ma, S.-H.; Wallis, W. D.
Clique numbers of threshold graphs.
MA F 1986-1 Ma, X. D.
Some classes of graceful graphs (Chinese).

MAAM	1985-1	Maamoun, M. Decompositions of digraphs into paths and cycles.
MAC	1984-1	Mac Avaney, K. L. Composite graphs with stability index one.
MAC B	1985-1	Mac Call, J. On the basis number of some complete bipartite graphs.
MAC C	1983-1	Mac Carthy, P. J. Pseudographs with degrees in prescribed intervals.
MAC E	1985-1	Mac Cuaig, W. D.; Rosenfeld, M. Parity of cycles containing specified edges.
MAC F	1986-1	Mac Curley, R.; Schneider, F. B. Derivation of a distributed algorithm for finding paths in directed networks.
MAC J	1983-1	Mac Kee, T. A. Validity up to complementation in graph theory.
MAC J	1984-1	Mac Kee, T. A. Balance and duality in signed graphs.
MAC J	1984-2	Mac Kee, T. A. Recharacterizing Eulerian: intimations of new duality.
MAC J	1984-3	Mac Kee, T. A. Relative expressiveness of the edge/adjacency language for graph theory.
MAC J	1985-1	Mac Kee, T. A. The logic of graph-theoretic duality.
MAC J	1987-1	Mac Kee, T. A. A local analogy between directed and signed graphs.
MAC J	1987-2	Mac Kee, T. A. Dual properties within graph theory.
MAC J	1987-4	Mac Kee, T. A. Multiterminal duality and three-terminal series-parallelness.
MAC K	1983-1	Mac Morris, F. R.; Shier, D. R. Representing chordal graphs on K(1,n).
MADE	1984-1	Mader, W. On k-critically n-connected graphs.
MADE	1985-1	Mader, W. Degree and local connectivity in digraphs.
MADE	1985-2	Mader, W. Disjunkte Fragmente in kritisch n-fach zusammenhängenden Graphen.
MADE	1985-3	Mader, W. Minimal n-fach zusammenhängende Digraphen.
MADE	1985-4	Mader, W. Paths in graphs, reducing the edge connectivity only by two.
MADE	1986-1	Mader, W. Kritisch n-fach kantenzusammenhängende Graphen.
MAFFB	1986-1	Maffray, F. On kernels in i-triangulated graphs.
MAHA	1986-1	Mahadev, N. V. R.; Peled, U. N. Strict 2-threshold graphs.
MAI A	1984-1	Mai, J.-H. The bandwidth of the product of two gamma-type condensed graphs.
MAI B	1984-1	Mai, J.-H.; Luo, H.-P. Some theorems for the bandwidth of a graph (Chinese)
MAI B	1984-2	Mai, J.-H.; Luo, H.-P. Some theorems on the bandwidth of a graph (Chinese).
MAI C	1984-1	Mai, J.-H.; Wang, S. H.; Luo, H.-P. The number of edges of graphs with girth exceeding n+1 (Chinese).
MAJC	1984-1	Majcher, Z. On some regularities of graphs. II.
MAJC	1985-1	Majcher, Z. Matrices representable by directed graphs.
MAJS	1983-1	Majstrenko, V. E. A method for the determination of the smallest dominating set of vertices for a graph of a certain form (Russian).
MAJS	1983-2	Majstrenko, V. E. The graph in problems of the description of recognition objects (Russian).
MALE	1983-1	Malenchenko, S. I. An algorithm for enumeration of cliques of a graph (Russian).
MANAA	1986-1	Manaster, A. B.; Remmel, J. B.; Schmerl, J. H. Planarity and minimal path algorithms.
MANV	1985-1	Manvel, B. Extremely greedy coloring algorithms.
MAO	1986-1	Mao, J. Z. An extremal problem for minimally k-edge-connected graphs (Chinese).
MAON	1986-1	Maon, Y.; Schieber, B.; Vishkin, U. Parallel ear decomposition search (EDS) and st-numbering in graphs.
MAON	1986-2	Maon, Y.; Schieber, B.; Vishkin, U. Parallel ear decomposition search (EDS) and st-numbering in graphs.
MARCD	1985-1	Marchioro, P.; Morgana, A.; Petreschi, R.; Simeone, B. Adamant digraphs.
MARCF	1986-2	Marcotte, O. On the chromatic index of multigraphs and a conjecture of Seymour. I.
MARCJ	1983-2	Marcu, D. On some vector spaces associated to a digraph.
MARCJ	1983-3	Marcu, D. Sur les bases d'un digraphe.

MARCJ 1984-4 Marcu, D.
On the strictly quasistrong connected digraphs without bridges.
MARCJ 1985-1 Marcu, D.
A new upper bound for the domination number of a graph.
MARCJ 1985-2 Marcu, D.
K-graphs in regular graphs.
MARCJ 1985-3 Marcu, D.
Note on the regular digraphs.
MARCJ 1985-5 Marcu, D.
Note on the 2-connectivity of a graph.
MARCJ 1985-6 Marcu, D.
On finding the elementary paths of a digraph.
MARCJ 1985-7 Marcu, D.
On finding the Eulerian paths and circuits of a digraph.
MARCJ 1985-8 Marcu, D.
On the paths of a digraph without circuits.
MARCJ 1985-9 Marcu, D.
Some results on the bipartite graphs.
MARCJ 1986-1 Marcu, D.
A note on Bermond's conjecture.
MARCJ 1986-3 Marcu, D.
An upper bound on the domination number of a graph.
MARCJ 1986-4 Marcu, D.
Another upper bound for the domination number of a graph.
MARCJ 1986-5 Marcu, D.
Note on the number of Hamiltonian paths in tournaments.
MARCJ 1986-7 Marcu, D.
Paths, connectivity and matchings in graphs (French).
MARCJ 1987-1 Marcu, D.
Note on the domination number of a graph.
MARCJ 1987-2 Marcu, D.
Note on the square of a graph.
MARCL 1985-1 Marczyk, A.; Skupien, Z.
General approach to line graphs of graphs.
MARI 1979-1 Maria, F. de; Gionfriddo, M.
On some parameters associated with graphs and hypergraphs (Italian).
MARIB 1982-1 Marino, M. C.
Some bounds for the parameter triangle in a graph (Italian).
MARKA 1985-1 Markosyan, A. S.
On imperfect critical graphs with incomplete critical component (Russian).
MARTJ 1983-1 Martinov, M. K.
Some decompositions of a complete graph into wheels (Bulgarian).
MARTK 1980-1 Martinov, N.
Uncontractable 4-connected graphs (Bulgarian).
MARTK 1986-1 Martinov, N.
Locally tree-like graphs and their generalizations (Russian).
MARTL 1983-2 Martinova, M. K.
On the decomposition of a complete graph into wheels (Bulgarian).
MARTL 1984-1 Martinova, M. K.
Cyclic and acyclic invariants for a class of outerplanar graphs (Bulgarian).
MARTL 1984-2 Martinova, M. K.
On the cyclicity of a complete n-partite graph (Bulgarian).
MARUA 1987-1 Marusic, D.
Hamiltonian cycles in vertex symmetric graphs of order 2p**2.
MASU 1987-1 Masuda, S.; Nakajima, K.
An optimal algorithm for generating all maximal independent sets of an interval graph.
MASU 1987-2 Masuda, S.; Nakajima, K.
Optimal algorithms for interval containment graphs.
MASUC 1987-1 Masuzawa, T.; Hagihara, K.; Tokura, N.
An optimal time algorithm for the k-vertex-connectivity unweighted augmentation problem for rooted directed trees.
MASUD 1986-1 Masuzawa, T.; Hagihara, K.; Tokura, N.; Wada, K.
Optimal-time algorithm for the k-node-connectivity augmentation problem for ternary trees.
MASUE 1984-1 Masuzawa, T.; Hagihara, K.; Wada, K.; Tokura, N.
The k-node connectivity augmentation problem for directed binary trees.
MATI 1987-1 Matitskas, I.-K. L.; Palubetskis, G. S.
Graph cutting and 0-1 quadratic programming (Russian).
MATSA 1986-1 Matsumoto, K.; Nishizeki, T.; Saito, N.
Planar multicommodity flows, maximum matchings and negative cycles.
MATU 1985-1 Matula, D. W.
Concurrent flow and concurrent connectivity in graphs.
MATU 1987-1 Matula, D. W.
Determining edge connectivity in O(nm).
MAYB 1983-1 Maybee, J. S.; Maybee, S. J.
An algorithm for identifying Morishima and anti-Morishima matrices and balanced digraphs.
MEHL 1984-1 Mehlhorn, K.
Data structures and algorithms 2: Graph algorithms and NP-completeness.
MEHLD 1986-1 Mehlhorn, K.; Schmidt, B. H.
On BF-orderable sets.
MELN 1984-1 Melnikov, L. S.; Ponomarev, V. V.
Solution of a problem of the Turan type (Russian).
MELNB 1985-1 Melnikov, O. I.; Kozhich, P. P.
Algorithms for recognizing the polarity of a graph with bounded parameters (Russian).

METE 1984-1 Metelskij, N. N.
Methods of local optimization for the problem of permutating bipartite graphs.
MEYN 1985-1 Meyniel, H.
A note about a theorem by Maamoun on decompositions of digraphs into elementary directed paths or cycles.
MICH 1983-1 Michalak, D.
On line and middle graphs with point-coarseness number one.
MICH 1983-2 Michalak, D.
On line graphs with coarseness number 1.
MICH 1985-1 Michalak, D.
On the binding number of a graph.
MICHA 1983-1 Michalski, M.
Spectra and direct derivations of graphs.
MIDD 1987-1 Middendorf, M.
Symmetric matroids and connectivity properties of graphs.
MIHO 1985-1 Mihok, P.
Additive hereditary properties and uniquely partitionable graphs.
MILIA 1984-1 Militaru, V.; Nicula, A.
Base transformations through the technique of the structural numbers.
MILLB 1987-1 Miller, G.; Ramachandran, V.
A new graph triconnectivity algorithm and its parallelization.
MILLC 1986-1 Miller, G. L.
Finding small simple cycle separators for 2-connected planar graphs.
MILN 1984-1 Milner, E. C.
Lectures on the marriage theorem of Aharoni, Nash-Williams and Shelah.
MINI 1987-1 Minieka, E.
The delivery man problem on a tree network.
MINOA 1986-1 Minoux, M.; Bartnik, G.
Graphes, algorithmes, logiciels.
MISHA 1986-1 Mishin, Y. S.
A method to describe graph on the plane.
MISHB 1984-1 Mishra, B.
An efficient algorithm to find all bidirectional edges of an undirected graph.
MITCA 1985-1 Mitchem, J.; Schmeichel, E.
Pancyclic and bipancyclic graphs - a survey.
MITCB 1986-1 Mitchison, G.; Durbin, R.
Optimal numberings of an N x N array.
MITE 1984-1 Mitev, J. G.
Numerical method for combinatorial optimization problem (Bulgarian).
MOHAA 1984-2 Mohar, B.
Search for minimal non-Hamiltonian simple 3-polytopes.
MOHAA 1984-3 Mohar, B.
Simplicial schemes and some combinatorial applications.
MOHAA 1986-2 Mohar, B.
On the cycle space of graphs.
MOHAA 1987-1 Mohar, B.
Generalization of Maclane's characterization of planar graphs.
MOHR 1985-1 Möhring, R. H.
Algorithmic aspects of comparability graphs and interval graphs.
MOITA 1985-1 Moitra, A.; Iyengar, S. S.
A maximally parallel balancing algorithm for obtaining complete balanced binary trees.
MONI 1985-2 Monien, B.
The complexity of embedding graphs into binary trees.
MONI 1986-1 Monien, B.
The bandwidth minimization problem for caterpillars with hair length 3 is NP-complete.
MONMA 1986-1 Monma, C. L.; Munson, B. S.; Pulleyblank, W. R.
Minimum-weight two-connected spanning networks.
MONMC 1986-1 Monma, C. L.; Reed, B.; Trotter, W. T. Jr.
A generalization of threshold graphs with tolerances.
MONMC 1987-1 Monma, C. L.; Reed, B.; Trotter, W. T. Jr.
Threshold tolerance graphs.
MONMH 1985-1 Monma, C. L.; Wei, V. K.
Intersection graphs of paths in a tree.
MONMH 1986-1 Monma, C. L.; Wei, V. K.
Intersection graphs of paths in a tree.
MONT 1984-1 Montejano, L.; Neumann-Lara, V.
A variation of Menger's theorem for long paths.
MORA 1985-1 Moran, G.
On planar Eulerian graphs and permutations.
MORA 1987-1 Moran, G.
A remark on embedded bipartite graphs.
MUKHC 1985-1 Mukherjee, P.; Mukherjee, T.
Studies on interrelations of matrices associated with a graph.
MUKHC 1986-1 Mukherjee, P.; Mukherjee, T.
An algorithmic approach for determination of iterated line graphs.
MULLA 1987-1 Müller, H.; Brandstädt, A.
The NP-completeness of STEINER TREE and DOMINATING SET for chordal bipartite graphs.
MURA 1986-1 Muradyan, D. O.
A polynomial algorithm for finding the bandwidth of interval graphs (Russian).
MURO 1987-1 Murota, K.
Homotopy base of an acyclic graph - a combinatorial analysis of commutative diagrams by means of preordered matroid.

MUROA	1987-1	Murota, K.; Fujishige, S. Finding a homotopy base for directed paths in an acyclic graph.
MYNHA	1987-1	Mynhardt, C. M.; Cockayne, E. J. K-minimal domination numbers of cycles.
NADDA	1984-1	Naddef, D. J.; Pulleyblank, W. R. Hamiltonicity in (0-1)-polyhedra.
NAJI	1985-1	Naji, W. Reconnaissance des graphes de cordes.
NAKAF	1985-1	Nakayama, H.; Nishizeki, T.; Saito, N. Lower bounds for combinatorial problems on graphs.
NAOR	1987-1	Naor, J.; Naor, M.; Schaffer, A. Fast parallel algorithms for chordal graphs.
NARA	1984-1	Nara, C. The maximal number of edges of a self-placeable graph.
NARUB	1983-1	Narushima, H.; Kusaka, Y. A move problem of weighted digraphs.
NASH	1985-1	Nash-Williams, C. St. J. A. Connected detachments of graphs and generalized Euler trails.
NASH	1985-2	Nash-Williams, C. St. J. A. Detachments of graphs and generalised Euler trails.
NEBE	1984-1	Nebesky, L. Edge-disjoint 1-factors in powers of connected graphs.
NEBE	1984-2	Nebesky, L. On quasistars in n-cubes.
NEBE	1984-3	Nebesky, L. Upper embeddable factorizations of graphs.
NEBE	1985-1	Nebesky, L. On locally quasiconnected graphs and their upper embeddability.
NEBEA	1985-1	Nebesky, L.; Wisztova, E. Two edge-disjoint Hamiltonian cycles of powers of a graph.
NEGA	1987-1	Negami, S. Polynomial invariants of graphs.
NEGAA	1986-1	Negami, S.; Xu, G.-H. Locally geodesic cycles in 2-self-centered graphs.
NEKR	1983-1	Nekrasov, V. P. Planar graphs having an external access (Russian).
NENO	1986-1	Nenov, N. D. On the existence of a minimal t-graph with nine vertices (Bulgarian).
NENO	1986-2	Nenov, N. D. Up to isomorphism there exists only one minimal t-graph with nine vertices (Bulgarian).
NESE	1986-1	Nesetril, J. For graphs there are only 4 types of hereditary Ramsey classes.
NESEA	1985-1	Nesetril, J.; Poljak, S. On the complexity of the subgraph problem.
NESEA	1986-1	Nesetril, J.; Poljak, S. A remark on max-cut problem with an application to digital-analogue convertors.
NEUF	1985-1	Neufeld, E. M. Optimal expansion of existing networks.
NEUM	1985-1	Neumaier, A. Characterization of a class of distance regular graphs.
NI A	1986-1	Ni, Z. Z. Bond graph construction from planar circuits (Chinese).
NIEM	1983-1	Nieminen, J. On linear and semilinear pointsets in graphs.
NIEM	1984-1	Nieminen, J. A characterization of median graphs.
NIEM	1984-2	Nieminen, J. Centrality, convexity and intersections in graphs.
NIEM	1984-3	Nieminen, J. New binary operations on graphs.
NIEM	1984-4	Nieminen, J. On convexies and related congruences on graphs.
NIEM	1987-1	Nieminen, J. Distance center and centroid of a median graph.
NIEP	1983-1	Niepel, L.; Safarikova, D. On a generalization of Menger's theorem.
NIKO	1983-1	Nikogosyan, Z. G. The maximum cycle of a graph (Russian).
NIKO	1984-1	Nikogosyan, Z. G. A sufficient condition for a graph to be Hamiltonian (Russian).
NIKO	1985-1	Nikogosyan, Z. G. A sufficient condition for the Hamilton property of a graph (Russian).
NIKO	1985-2	Nikogosyan, Z. G. On longest cycles in graphs (Russian).
NOMU	1987-1	Nomura, K. An inequality between intersection numbers of a distance-regular graph.
NOSO	1985-1	Nosov, V. A.; Sachkov, V. N.; Tarakanov, V. E. Combinatorial analysis (nonnegative matrices, algorithmic problems).
NOVO	1986-1	Novozhilova, L. M. Ordered sorting, with a successive decrease in the size of the problem, of enumeration problems on graphs (Russian).
ODLY	1987-1	Odlyzko, A. M.; Wilf, H. S. Bandwidths and profiles of trees.

OELL	1987-1	Oellermann, O. R. On the l-connectivity of a graph.
OELLA	1984-1	Oellermann, O. R.; Swart, H. C. Graphs for which all strong orientations are Eulerian.
OKAM	1984-2	Okamura, H. Path and edge-connectivity in graphs.
OKAM	1984-3	Okamura, H. Paths and edge-connectivity in graphs.
OKAM	1985-1	Okamura, H. Paths and edge-connectivity in graphs. II.
OKAM	1987-1	Okamura, H. Paths and edge-connectivity in graphs III. Six-terminal k paths.
OLAH	1981-1	Olah, G. Path-matrices of trees (Hungarian).
OLARA	1983-1	Olaru, E.; Antohe, S. Decompositions with respect to the graph-join operation.
OLARA	1984-1	Olaru, E.; Antohe, S. Algebraic aspects of S-partitions of graphs (German).
OLARA	1985-1	Olaru, E.; Antohe, S. Some results concerning graph decomposition.
ORZE	1984-1	Orzech, G. Elementary properties of vector space graphs.
OUBI	1984-1	Oubina, L.; Zucchello, R. A generalization of outerplanar graphs.
OUDH	1985-1	Oudheusden, D. L. van; Khan, M. L. R. Note on ILP flow formulation for subgraph selection.
OZAW	1987-1	Ozawa, T.; Kajitani, Y.; Ueno, S. The principal partition of vertex-weighted graphs and its applications.
PAIX	1985-1	Paixao, J. P.; Almeida, M. T.; Coelho, J. D. Problems in graph theory and their relation with set covering and partitioning problems (Portuguese).
PALK	1983-1	Palka, Z. A new algorithm calculating the distance between binary arborescences.
PALU	1985-1	Palubetskis, G. S. Subgraph polytopes (Russian).
PAN	1985-1	Pan, Z. T.; Zhao, S. C. Graph-theoretic proof of an inequality (Chinese).
PAOL	1984-1	Paoli, M. Powers of connected graphs and hamiltonicity.
PAOL	1986-1	Paoli, M. Some Hamiltonian properties of the powers of a graph.
PAOLA	1984-1	Paoli, M.; Trotter, W. T. Jr. The maximum number of edges in a strongly connected oriented graph without long directed paths.
PAPAD	1986-1	Papadimitriou, C. H.; Yannakakis, M. A note on succinct representations of graphs.
PATI	1984-1	Patil, H. P. A note on the edge-arboricity of maximal k-degenerate graphs.
PAUL	1987-1	Paulraja, P. Counterexample to a conjecture on Hamilton cycles.
PAUL	1987-2	Paulraja, P. On graphs admitting spanning Eulerian subgraphs.
PAWA	1987-1	Pawagi, S. Maximum weight independent set in trees.
PEEL	1985-1	Peele, R. Finite partition sublattice representations for some geometric lattices.
PEEM	1984-1	Peemöller, J. Hamiltonsche Graphen - notwendige Bedingungen und Algorithmen.
PEEM	1985-1	Peemöller, J. Necessary conditions for Hamiltonian split graphs.
PELE	1987-1	Peled, U. N. The threshold measure of a graph.
PELEB	1987-1	Peled, U. N.; Srinivasan, M. K. The polytope of degree sequences.
PERLB	1984-1	Perl, Y. Optimum split trees.
PERM	1984-1	Perminov, E. A. Imbeddings of graphs in rigid graphs (Russian).
PERO	1984-1	Peroche, B. Products and path numbers I: the cartesian sum.
PEROA	1986-1	Peroche, B.; Sadi, B. Recouvrement et partition en chaines des aretes d'un graphe cubique.
PETEA	1984-1	Peteanu, V.; Tigan, S. On some discrete fractional max-min problems. Application to max-min problems in graphs.
PETED	1986-1	Peters, K.; Laskar, R.; Hedetniemi, S. Maximal/minimal connectivity in graphs.
PETEE	1986-1	Peters, K.; Laskar, R.; Hedetniemi, S. T. On the strength of a graph.
PETKA	1984-1	Petkovsek, M. Comparability line graphs.
PETR	1986-1	Petrov, K. On strongly regular graphs (Bulgarian).
PETRC	1984-1	Petrovic, V. Antidirected Hamiltonian circuits in tournaments.

PFAF 1984-1 Pfaff, J.; Laskar, R.; Hedetniemi, S. T.
Linear algorithms for independent domination and total domination in series-parallel graphs.

PIAZ 1984-1 Piazza, B. L.
On Hamiltonian cycles in cycle permutation graphs.

PIAZA 1985-1 Piazza, B. L.; Stueckle, S.
On the cut frequency vector of permutation graphs.

PICAB 1982-1 Pica, G.; Pisanski, T.; Shawe-Taylor, J.; Ventre, A. G. S.
The permutation girth of a graph.

PISA 1983-1 Pisanski, T.; Shawe-Taylor, J.; Mohar, B.
1-factorization of the composition of regular graphs.

PLES 1985-1 Plesnik, J.
Remarks on diameters of orientations of graphs.

PLES 1986-2 Plesnik, J.
On minimal graphs of diameter 2 with every edge in a 3-cycle.

PLON 1984-1 Plonka, J.
Graphs of parital partitions.

PLUM 1985-1 Plummer, M. D.
Some recent results on cycle traversability in graphs.

PLUM 1986-1 Plummer, M. D.
Matching extension in bipartite graphs.

POLA 1985-1 Polansky, O. E.
Die Entwicklung des mü-Polynoms eines schlichten Graphen, entsprechend seiner Zerlegung an einem Knoten.

POLI 1986-1 Politof, T.; Satyanarayana, A.
Network reliability and inner-four-cycle-free graphs.

POLJA 1986-1 Poljak, S.; Turzik, D.
A polynomial time heuristic for certain subgraph optimization problems with guaranteed worst case bound.

POLJA 1987-1 Poljak, S.; Turzik, D.
On a facet of the balanced subgraph polytope.

POLJB 1987-1 Poljak, S.; Tuza, Z.
Maximum bipartite subgraphs of Kneser graphs.

PONO 1982-1 Ponomarenko, T. Y.
The telephone center of a tree (Russian).

POPE 1983-1 Popescu, E.
Algorithme matriciel pour la determination du graphe minimal qui conserve la relation d'ordre donne d'un graphe d'ordre sans circuits.

POSH 1985-1 Poshtaru, A. I.
Medians of planar graphs (Russian).

PRAB 1984-1 Prabhu, G. M.; Deo, N.
The ellipsoid algorithm and the graph isomorphism problem.

PREIB 1986-1 Preissmann, M.; Werra, D. de; Mahadev, N. V. R.
A note on superbrittle graphs.

PRET 1985-1 Pretzel, O.
On graphs that can be oriented as diagrams of ordered sets.

PRET 1986-1 Pretzel, O.
On reorienting graphs by pushing down maximal vertices.

PRET 1986-2 Pretzel, O.
Orientations and reorientations of graphs.

PRIS 1987-1 Prisakar, K. F.; Prisakar, A. V.
On the minimal covering of the vertices of a graph by d-convex subsets (Russian).

PROC 1983-1 Procesi Ciampi, R.
A criterion for acyclic orientations.

PROC 1983-2 Procesi Ciampi, R.
A minimality property for acyclic orientations.

PROC 1984-1 Procesi Ciampi, R.
Acyclic orientations and domination of a graph.

PROC 1985-1 Procesi Ciampi, R.
Spanning trees and acyclic orientations.

PROM 1986-1 Prömel, H. J.
Almost bipartite-making graphs.

PROM 1986-3 Prömel, H. J.
Counting unlabeled structures.

PROMA 1986-1 Prömel, H. J.; Voigt, B.
Ramsey theorems for finite graphs.

PROV 1986-1 Provan, J. S.
The complexity of reliability computations in planar and acyclic graphs.

PUEB 1987-1 Puebla, J. G.
Spatial structures of network flows: A graph theoretical approach.

PULLA 1987-1 Pullman, N. J.; Stanford, M.
The biclique numbers of regular graphs.

PUNN 1987-1 Punnim, N.; Pabhapote, N.
On graceful graphs: cycles with a P*(k)-chord, k>=4.

PYBE 1987-1 Pyber, L.
On one-factorizations of the complete graph.

QI A 1987-1 Qi, N.
On separation and adjacency problems for perfectly matchable subgraph polytopes of a graph.

QUINA 1984-1 Quinn, M. J.; Deo, N.
Parallel graph algorithms.

RABI 1985-1 Rabinowitz, J. H.; Proulx, V. K.
An asymptotic approach to the channel assignment problem.

RADO 1983-1 Radosavljevic, Z.; Simic, S.; Syslo, M. M.; Topp, J.
A note on generalized line graphs.

RALL 1984-1 Rall, D. F.; Slater, P. J.
On location-domination numbers for certain classes of graphs.
RAMO 1986-1 Ramos, R. E.
On 1-factorization of line graphs.
RAMR 1983-1 Ramras, M.
Maximal independent sets in graphs with vertex m-covers.
RAO 1985-1 Rao, I. H. N.; Narasimham, M. S.
A note on graceful and supergraceful numbering of a class of trees.
RAO C 1987-1 Rao, S.
Finding near optimal separators in planar graphs.
RAO D 1985-1 Rao, S. B.
On regular and strongly-regular self-complementary graphs.
RAYC 1985-1 Raychaudhuri, A.; Roberts, F. S.
Generalized competition graphs and their applications.
READ 1987-1 Read, R. C.
A new method for drawing a planar graph given the cyclic order of the edges at each vertex.
READ 1987-2 Read, R. C.
Connectivity and chromatic uniqueness.
REID 1985-1 Reid, K. B.
Two complementary circuits in two-connected tournaments.
REIDA 1983-1 Reid, K. B.; Wormald, N. C.
Embedding oriented n-trees in tournaments.
REIF 1987-1 Reif, J. H.
A topological approach to dynamic graph connectivity.
REIN 1985-1 Reinelt, G.
The linear ordering problem: algorithms and application.
RICH 1986-1 Richards, D.
Finding short cycles in planar graphs using separators.
RICHA 1985-1 Richards, D.; Liestman, A. L.
Finding cycles of a given length.
RICHB 1987-1 Richey, M. B.; Parker, R. G.
On computational issues related to series-parallel solvability.
RICHC 1985-1 Richey, M. B.; Parker, R. G.; Rardin, R. L.
On finding spanning Eulerian subgraphs.
RICHD 1985-1 Richmond, L. B.; Robinson, R. W.; Wormald, N. C.
On Hamilton cycles in 3-connected cubic maps.
RICHE 1984-1 Richter, B.; Shank, H.
The cycle space of an embedded graph.
RIHA 1983-1 Riha, S.; Sekanina, M.
Squares of triangular cacti.
RING 1985-1 Ringel, G.
250 Jahre Graphentheorie.
RINS 1987-1 Rinsma, I.
Embedding a tree in a maximal outerplanar graph.
RIVE 1986-1 Rivera Campo, E.
Interchanging of edges in graceful graphs (Spanish).
ROBEA 1985-1 Robert, Y.; Tchuente, M.
Connection-graph and iteration-graph of monotone boolean functions.
ROBED 1984-1 Robertson, N.
Minimal cyclic-4-connected graphs.
ROBEE 1984-1 Robertson, N.; Seymour, P. D.
Graph minors. III. Planar tree-width.
ROBEE 1984-2 Robertson, N.; Seymour, P. D.
Graph width and well-quasi-ordering: a survey.
ROBEE 1985-1 Robertson, N.; Seymour, P. D.
Disjoint paths - a survey.
ROBEE 1985-2 Robertson, N.; Seymour, P. D.
Graph minors - a survey.
ROBEE 1985-3 Robertson, N.; Seymour, P. D.
Graph minors IV. Tree-width and well-quasi-ordering.
ROBEE 1985-4 Robertson, N.; Seymour, P. D.
Graph minors VII: Disjoint paths on a surface.
ROBEE 1985-5 Robertson, N.; Seymour, P. D.
Graph minors VIII. A Kuratowski theorem for general surfaces.
ROBEE 1986-1 Robertson, N.; Seymour, P. D.
Graph minors. II. Algorithmic aspects of tree-width.
ROBEE 1986-2 Robertson, N.; Seymour, P. D.
Graph minors. V. Excluding a planar graph.
ROBEE 1986-3 Robertson, N.; Seymour, P. D.
Graph minors. VI. Disjoint paths across a disc.
ROBI 1985-1 Robinson, D. F.; Janjic, I.
The constructability of floorplans with given outerplanar adjacency graph and room areas.
ROBS 1986-1 Robson, J. M.
Algorithms for maximum independent sets.
ROSEA 1987-1 Rosenberg, I. G.
Characterization of diagrams by 0-1 inequalities.
ROSEB 1956-1 Rosenblatt, D.
On aggregation and consolidation in linear systems.
ROSEF 1986-1 Rosenstiehl, P.; Tarjan, R. E.
Rectilinear planar layouts and bipolar orientations of planar graphs.
ROTK 1986-1 Rotko, V. F.
Efficient algorithms for synthesis of graphs with a prescribed set of fundamental sections or cycles.

ROUB 1985-1 Roubens, M.
Families of nested interval graphs (French).
ROUBA 1983-1 Roubens, M.; Vincke, P.
Linear fuzzy graphs.
ROUCA 1987-1 Roucairol, C.; Hansen, P.
Probleme de la bipartition minimale d'un graphe.
ROUN 1987-1 Rounault, A.
Superconcentrateurs.
RYJA 1987-1 Ryjacek, Z.; Zelinka, B.
Locally disconnected graphs with large numbers of edges.
RYU 1983-1 Ryu, H. D.
On invariants of graphs. IV. (Korean).
RYU 1984-1 Ryu, H. D.
On invariants of graphs. V. (Korean).
RYU 1985-1 Ryu, H. D.
On invariants of graphs, VI (Korean).
RYU A 1985-1 Ryu, H. D.; Bak, T. B.
On dual problem of Hadwiger's conjecture (Korean).
RYU B 1982-1 Ryu, H. D.; Chong, C. J.
On a graph invariant and its application (Korean).
RYU C 1984-1 Ryu, H. D.; Li, C. G.
Criteria for the efficiency of algorithms in graph theory and its application (Korean).
RYU D 1984-1 Ryu, H. D.; Ri, C. K.
Criteria for the efficiency of algorithms in graph theory and its application (Korean).
SACH 1983-1 Sachs, H.
On a spatial analogue of Kuratowski's theorem on planar graphs - an open problem.
SACHA 1982-1 Sachs, H.; Stiebitz, M.
Simple eigenvalues of transitive graphs.
SAIT 1985-1 Saito, A.; Tian, S.-L.
The binding number of line graphs and total graphs.
SAKAA 1984-2 Sakarovitch, M.
Optimisation combinatoire - methodes mathematiques et algorithmiques. Vol. 2: Graphes et programmation lineaire.
SAKS 1986-1 Saks, M.
Some sequences associated with combintorial structures.
SAMO 1984-1 Samodivkin, V. D.
Hamiltonian line graphs (Russian).
SAMP 1983-1 Sampathkumar, E.
A generalization of Menger's theorem for trees.
SAMP 1984-1 Sampathkumar, E.
Connectivity of a graph - a generalization.
SAMP 1984-2 Sampathkumar, E.
Convex sets in a graph.
SAMP 1984-3 Sampathkumar, E.
Point signed and line signed graphs.
SAMPA 1982-1 Sampathkumar, E.; Bhave, V. N.
Arbitrarily traceable graphs.
SAMPB 1985-1 Sampathkumar, E.; Neeralagi, P. S.
The neighbourhood number of a graph.
SAMPB 1986-1 Sampathkumar, E.; Neeralagi, P. S.
The line neighbourhood number of a graph.
SAMPC 1982-1 Sampathkumar, E.; Walikar, H. B.
On a conjecture of Cockayne and Hedetniemi.
SANC 1987-1 Sanchez-Flores, A.
A counterexample to a generalization of Richardson's theorem.
SARV 1984-1 Sarvanov, V. I.
Sufficient conditions for traceability of planar 3-connected graphs (Russian).
SARW 1983-1 Sarwate, D. G.
A note on strongly regular Hamiltonian equivalence of K(2n) and K(2n-1,2n-1) and its generalization.
SAST 1987-1 Sastry, P. S.; Jayakumar, N.; Veni Madhavan, C. E.
Efficient algorithms for domination and Hamilton circuit problems on permutation graphs.
SATY 1987-1 Satyanarayana, A.; Tindell, R.
Chromatic polynomials and network reliability.
SAVA 1984-1 Savage, C.
A systolic design for connectivity problems.
SCHA 1986-1 Schaar, G.
Remarks on k-path traceable graphs.
SCHE 1985-1 Scheinerman, E. R.
Characterization and computational complexity questions for representation classes of graphs.
SCHE 1985-2 Scheinerman, E. R.
Characterizing intersection classes of graphs.
SCHE 1985-3 Scheinerman, E. R.
Irrepresentability by multiple intersection, or why the interval number is unbounded.
SCHM 1982-1 Schmerl, J. H.
The effective version of Brooks' theorem.
SCHMA 1984-1 Schmidt-Steup, M.
Hamiltonsche und Hypohamiltonsche unendliche Graphen.
SCHOB 1986-1 Schoone, A. A.; Bodlaender, H. L.; Leeuwen, J. van
Improved diameter bounds for altered graphs.
SCHRE 1983-1 Schrijver, A.
Packing and covering of crossing families of cuts.
SCHRE 1986-1 Schrijver, A.
Distances and cuts in planar graphs.

SCHRE	1986-2	Schrijver, A. Geometric methods in discrete optimization.
SCHRE	1987-1	Schrijver, A. Disjoint circuits of prescribed homotopies in a graph on a surface.
SCHUB	1985-1	Schulz, R. Über das Feedback-Vertex-Set-Problem.
SCOTA	1987-1	Scott, D. D. The competition-common enemy graph of a digraph.
SEBO	1986-2	Sebö, A. Finding the t-join structure of graphs.
SEBO	1987-2	Sebö, A. A quick proof of Seymour's theorem on t-joins.
SEDL	1985-1	Sedlacek, J. Finite graphs with distinct neighbourhoods.
SEDL	1985-2	Sedlacek, J. Über eine spezielle Klasse von asymmetrischen Graphen.
SEES	1985-1	Seese, D. Baum-partite Graphen und die Komplexität von Algorithmen.
SEES	1985-2	Seese, D. Tree-partite graphs and the complexity of algorithms.
SEID	1983-1	Seidman, S. B. Internal cohesion of LS sets in graphs.
SEID	1983-2	Seidman, S. B. LS sets as cohesive subsets of graphs and hypergraphs.
SEID	1983-3	Seidman, S. B. Network structure and minimum degree.
SEKA	1985-1	Sekanina, M. Factors in powers of graphs.
SEN	1983-1	Sen, M.; Roy, A. B. A Boolean approach to graph labelling. II.
SERD	1986-1	Serdyukov, A. I. A polynomial algorithm with an estimate of accuracy of (6+root2)/8 for a problem on the maximal Hamiltonian cycle on the Euclidean plane (Russian).
SEYMA	1984-1	Seymour, P. D.; Robertson, N. Some new results on the well-quasi ordering of graphs.
SEYMB	1987-1	Seymour, P. D.; Thomassen, C. Characterization of even directed graphs.
SHAMA	1985-1	Shamir, E.; Upfal, E. A fast parallel construction of disjoint paths in networks.
SHAN	1983-1	Shan, K. Y.; Zhuang, S. Q. On graceful labelings of some particular lobsters (Chinese).
SHAO	1985-1	Shao, J.-Y. The connectivity of the interchange graph of class U(R,S) of (0,1) matrices.
SHAW	1982-1	Shawe-Taylor, J.; Pisanski, T. Cycle permutation graphs with large girth.
SHEE	1987-1	Shee, S.-C. On harmonious and related graphs.
SHERI	1985-1	Sherlekar, D. D.; Pawagi, S.; Ramakrishnan, I. V. O(1) parallel time incremental graph algorithms.
SHI	1985-1	Shi, R.-H. Graph decomposition under a minimum degree restriction (Chinese).
SHI	1985-2	Shi, R.-H. Graph decomposition with constraint on the minimum degree.
SHI	1985-3	Shi, R.-H. Smallest regular graphs with girth pair (4,5).
SHI	1985-4	Shi, R.-H. The binding number of a graph and its circuits.
SHI	1985-5	Shi, R.-H. The binding number of a graph and its triangle.
SHI	1987-1	Shi, R.-H. The binding number of a graph and its pancyclism.
SHI A	1986-1	Shi, R.-H.; Sun, R.-G.; Nian, K. Hamiltonian property in the square of a connected graph.
SHI C	1981-1	Shi, Y.-B. On USP graphs (Chinese).
SHI C	1985-1	Shi, Y.-B. On UPC graphs (Chinese).
SHI C	1986-1	Shi, Y.-B. A counting theorem for Hamilton cyclic number of a graph (Chinese).
SHI C	1986-2	Shi, Y.-B. Further results on uniquely pancyclic graphs, II (Chinese).
SHI C	1986-3	Shi, Y.-B. Further results on uniquely pancyclic graphs, I (Chinese).
SHIH	1984-1	Shih, C.-H. Cographic subspaces of cocycle spaces for line graphs of complete graphs.
SHIM	1984-1	Shimizu, M.; Nakamura, G. Expressions for the number of 2-3 trees with n leaves.
SHLY	1986-1	Shlyk, V. A. Polyhedral paths connecting two vertices of a graph (Russian).
SHU	1986-1	Shu, J.-J. The minimum degree of a critically n-line-connected graph (Chinese).
SIME	1985-1	Simeone, B. Consistency of quadratic boolean equations and the König-Egervary property for graphs.

SIMOB 1987-1 Simonson, S.
A variation on the min cut linear arrangement problem.

SINGE 1987-1 Singh, H. G.; Pargas, R. P.
A parallel implementation for the domination number of a grid graph.

SIRA 1983-1 Siran, J.
Edges and Kuratowski subgraphs of non-planar graphs.

SIRA 1984-1 Siran, J.
Additivity of the crossing number of graphs with connectivity 2.

SKILA 1984-1 Skilton, D. K.
Eulerian chains and segment reversals.

SKILA 1986-1 Skilton, D. K.
Digraphs with Eulerian chains.

SKORA 1981-1 Skorobogatov, V. A.; Khvorostov, P. V.
Analysis of the metric properties of graphs (Russian).

SKORA 1984-1 Skorobogatov, V. A.; Khvorostov, P. V.
Methods and algorithms for the analysis of symmetries of graphs (Russian).

SKOW 1983-1 Skowronska, M.
Characterizations of middle graphs and properties of Halin graphs (Polish).

SKOW 1983-2 Skowronska, M.
Hamiltonian properties of Halin-like graphs.

SKOW 1985-1 Skowronska, M.
The pancyclicity of Halin graphs and their exterior contractions.

SKRI 1984-1 Skrien, D.
Interval line graphs.

SKRIA 1985-1 Skrien, D.; Gimbel, J.
Homogeneously representable interval graphs.

SKUP 1984-1 Skupien, Z.
Homogeneously traceable and Hamiltonian connected graphs.

SKUP 1985-1 Skupien, Z.
An improvement of Jung's condition for hamiltonicity.

SKUP 1985-2 Skupien, Z.
N-closures of large nonhamiltonian n-vertex graphs.

SKUP 1986-1 Skupien, Z.
On counting maximum path-factors of a tree.

SLAT 1982-1 Slater, P. J.
On k-graceful graphs.

SMAD 1986-1 Smadici, C.
A note on interval graphs.

SMITB 1983-1 Smith, D. H.
Reliable networks and (k,2k-2)-connected graphs.

SMITB 1985-1 Smith, D. H.
Connectivity preserving transformations of graphs and the reliability of mobile communication networks.

SMYT 1987-1 Smyth, W. F.
Sharp bounds on the diameter of a graph.

SOBI 1981-1 Sobik, F.; Sommerfeld, E.
Untersuchungen zur Klassifikation strukturierter Objekte auf der Grundlage des Enthaltenseins bestimmter Untergraphen.

SOLT 1983-1 Soltan, V. P.
D-convexity in graphs.

SOLTA 1984-1 Soltan, V. P.; Chepoi, V. D.
D-convex sets in triangulated graphs (Russian).

SOLTA 1984-2 Soltan, V. P.; Chepoi, V. D.
The number of d-convex sets in a graph (Russian).

SOLTA 1985-1 Soltan, V. P.; Chepoi, V. D.
D-convexity and Steiner functions on a graph (Russian).

SOLTB 1986-1 Soltes, L.
Orientations of graphs minimizing the radius or the diameter.

SOMM 1985-1 Sommerfeld, E.
Zur Stabilität der Korona zweier Graphen.

SONE 1987-1 Soneoka, T.; Nakada, H.; Imase, M.; Peyrat, C.
Sufficient conditions for maximally connected dense graphs.

SONGB 1985-1 Song, Z.-M.
The arc constrained optimum arborescence problem (Chinese).

SONGB 1987-1 Song, Z.-M.
Number of arcs and cycles in digraphs (Chinese).

SONN 1985-1 Sonntag, M.
Some Hamiltonian properties of products of digraphs.

SONN 1986-1 Sonntag, M.
Pancyclicity and strong path-connectedness of the Cartesian sum of digraphs.

SONN 1987-1 Sonntag, M.
Some results on Hamiltonian properties of products of digraphs I. t-Hamiltonian properties and strong path-connectedness.

SONN 1987-2 Sonntag, M.
Some results on Hamiltonian properties of products of digraphs II. Pancyclicity properties.

SOTT 1986-1 Sotteau, D.; Wojda, A. P.
Digraphs without directed path of length two or three.

SPAT 1985-1 Späth, H.
Heuristically determining cliques of given cardinality and with minimal cost within weighted complete graphs.

SPEC 1983-1 Speckenmeyer, E.
Untersuchungen zum Feedback Vertex Problem in ungerichteten Graphen (Dissertation).

SPIN 1985-1 Spinrad, J.
On comparability and permutation graphs.

SPINA 1987-1 Spinrad, J.; Brandstädt, A.; Stewart, L.
Bipartite permutation graphs.

SRID 1983-1 Sridharan, M. R.; George, O. T.
On indecomposable graphs.

SRIDB 1987-1 Sridharan, S.
Biregular subgraphs of biregular graphs.

SRIN 1983-1 Srinivasan, N.
Block-cut-point incidence pattern of separable geodetic graphs of diameter three and four.

STANC 1983-1 Standley, H. M.
Minimal area graph embedding into a directed grid is NP-hard.

STEI 1985-1 Steiner, G.
A compact labeling scheme for series-parallel graphs.

STIE 1985-2 Stiebitz, M.
Subgraphs of colour-critical graphs.

STIE 1987-1 Stiebitz, M.
K*(5) is the only double-critical 5-chromatic graph.

STRA 1982-1 Straight, H. J.
Extremal problems concerning the cochromatic number of a graph.

STRAA 1985-1 Strassner, K.
Decomposing graphs and hypergraphs (German).

STUE 1984-1 Stueckle, S.; Ringeisen, R. D.
Generalized Petersen graphs which are cycle permutation graphs.

SU 1984-1 Su, J. J.
Minimally k-edge connected graphs of low order with maximum number of edges (Chinese).

SU 1985-1 Su, J. J.
Upper bounds for the maximum degree and for the number of vertices having degree greater than four in a critically 2-edge-connected graph (Chinese).

SU 1986-1 Su, J. J.
The number of vertices with low degree in critically h-connected graphs (Chinese).

SU A 1986-1 Su, X. Y.
A note on clique graphs (Chinese).

SUCI 1985-1 Suciu, D.
Labelled graphs and their associated values (Romanian).

SUN 1985-1 Sun, H.-C.
On the cycle decomposition of complete graphs (Chinese).

SUN 1985-2 Sun, H.-C.
Two sufficient conditions for simple graphs to be of class one (Chinese).

SUN A 1986-1 Sun, H.-Q.
The connectivity of line graphs (Chinese).

SYSL 1984-1 Syslo, M. M.
Series-parallel graphs and depth-first search trees.

SYSL 1985-1 Syslo, M. M.
A graph-theoretic approach to the jump-number problem.

SYSL 1985-4 Syslo, M. M.
Triangulated edge intersection graphs of paths in a tree.

SYSL 1987-2 Syslo, M. M.
Minimizing the jump number for partially-ordered sets: a graph-theoretic approach, II.

SYSL 1987-3 Syslo, M. M.
On some generalizations of outerplanar graphs: results and open problems.

SYSLA 1983-1 Syslo, M. M.; Topp, J.
A characterization of generalized line graphs whose complements are also generalized line graphs.

SYSLA 1985-1 Syslo, M. M.; Topp, J.
On the planarity and outerplanarity of generalized line, middle and total graphs.

SZWAC 1987-1 Szwarcfiter, J. L.
A note on the computation of the k-closure of a graph.

SZWAD 1985-1 Szwarcfiter, J. L.; Persiano, R. C. M.; Oliveira, A. A. F.
Orientations with single source and sink.

TAMAA 1987-1 Tamassia, R.
On embedding a graph in the grid with the minimum number of bends.

TAN 1986-1 Tan, E. L.
Some classes of cycle graphs.

TANGD 1986-1 Tang, Q. F.
An algorithm for constructing minimal trees with designated roots (Chinese).

TAO A 1982-1 Tao, K. M.; Hsia, P.
On the structure of directed graphs with applications. I. A rapprochement with systems theory.

TAO B 1983-1 Tao, M. Q.; Shen, Y. Q.
A conjecture of Erdös - the planar case (Chinese).

TARAB 1986-1 Tarasova, L. G.
Clutters of the set of paths between two vertices of a graph (Russian).

TARJ 1985-1 Tarjan, R. E.
Decomposition by clique separators.

TARJA 1985-1 Tarjan, R. E.; Vishkin, U.
An efficient parallel biconnectivity algorithm.

TARJB 1985-1 Tarjan, R. E.; Yannakakis, M.
Simple linear-time algorithms to test chordality of graphs, test acyclicity of hypergraphs, and selectively reduce acyclic hypergraphs (Addendum).

TASH 1985-1 Tashchuk, D. A.
Properties of special type subgraphs in an oriented graph (Russian).

TASHA 1984-1 Tashkinov, V. A.
Three-regular parts of four-regular graphs.

TAY 1984-1 Tay, T.-S.
Rigidity of multi-graphs II.

TAZA 1984-1 Tazawa, S.
A short proof of Turan theorem.

TAZA 1985-1 Tazawa, S.
Decomposition of a complete multi-partite graph into isomorphic claws.

TAZZ 1983-1 Tazzi, G. C.; Aragno Marauta, E.
The maximum genus of certain graphs (Italian).

TEIC 1983-1 Teichert, H.-M.
Hamiltonian properties of the lexicographic product of undirected graphs.

TEIC 1984-1 Teichert, H.-M.
Zusammenhangs- und Durchlaufbarkeitseigenschaften der Disjunktion ungerichteter Graphen.

TERW 1985-1 Terwilliger, P.
Distance-regular graphs with Girth 3 or 4: I.

THOM 1984-1 Thomas, R.
Series-parallel graphs and well- and better-quasi-orderings.

THOMA 1984-1 Thomassen, C.
A refinement of Kuratowski's theorem.

THOMA 1984-2 Thomassen, C.
Graph theory - an introduction (Danish).

THOMA 1984-3 Thomassen, C.
Plane representations of graphs.

THOMA 1985-1 Thomassen, C.
Even cycles in directed graphs.

THOMA 1985-2 Thomassen, C.
Hamilton circuits in regular tournaments.

THOMA 1985-3 Thomassen, C.
The 2-linkage problem for acyclic digraphs.

THOMA 1986-2 Thomassen, C.
Interval representations of planar graphs.

THOMA 1986-3 Thomassen, C.
Sign-nonnegative matrices and even cycles in directed graphs.

THOMA 1987-1 Thomassen, C.
A characterization of locally finite vertex-transitive graphs.

THOMA 1987-2 Thomassen, C.
Counterexamples to Adam's conjecture on arc reversals in directed graphs.

THOMA 1987-3 Thomassen, C.
On digraphs with no two disjoint directed cycles.

THOMD 1985-1 Thompson, G. L.; Singhal, S.
A successful algorithm for the undirected Hamiltonian path problem.

THUL 1987-1 Thulasiraman, K.; Comeau, M. A.
Maximum-weight markings in marked graphs: algorithms and interpretations based on the simplex method.

TIAN 1986-1 Tian, F.
The progress on some problems in graph theory (Chinese).

TIANA 1986-1 Tian, F.; Shi, R.-H.
A new class of pancyclic graphs (Chinese).

TIANB 1985-1 Tian, S.-L.; Wang, J. F.; Liu, J.-Q.
The binding number of two product graphs.

TING 1985-1 Ting, S.-K.
The structure of a maximum critical 2-edge-connected graph (Chinese).

TINHA 1984-1 Tinhofer, G.; Schreck, H.
Linear time tree codes.

TOFT 1985-1 Toft, B.
Some problems and results related to subgraphs of colour critical graphs.

TOID 1985-1 Toida, S.
A decomposition of a graph into dense subgraphs.

TOME 1984-1 Tomescu, I.
A Hamiltonian connectivity property of regular graphs with forbidden subgraphs.

TOME 1985-1 Tomescu, I.
Problems in combinatorics and graph theory.

TOMO 1984-1 Tomova, E.
Decomposition of complete bipartite graphs into factors with given diameters and radii.

TOPA 1984-1 Topala, O. I.
Star property on graphs (Russian).

TOSI 1986-1 Tosic, R.
Vertex-to-vertex search in a graph.

TRUSB 1983-1 Truszczynski, M.
Decompositions of complete multipartite graphs into isomorphic stars.

TRUSB 1983-2 Truszczynski, M.
On a conjecture of Mohar and Pisanski.

TRUSB 1984-1 Truszczynski, M.
Graceful unicyclic graphs.

TRUSB 1985-1 Truszczynski, M.
Centers and centroids of unicyclic graphs.

TRUSB 1985-3 Truszczynski, M.
Note on decomposition of LambdaK(m,n) (LambdaK(m,n)*) into paths.

TSUC 1983-1 Tsuchiya, M.
Another proof of Bela Bollobas' theorem on edge disjoint cycles.

TUCK 1984-1 Tucker, A.
Applied Combinatorics (second edition).

TUTT 1985-1 Tutte, W. T.
Graph theory as I have known it (Part III) (Chinese).

TUTZ 1985-1 Tutzauer, F.
Toward a theory of disintegration in communication networks.
TYSH 1985-1 Tyshkevich, R. I.
Polar graphs and the decomposition.
TYSHA 1985-1 Tyshkevich, R. I.; Chernjak, A. A.
Algorithms of a canonical decomposition of a graph and polarity recognition (Russian).
TYSHA 1985-2 Tyshkevich, R. I.; Chernjak, A. A.
Box-threshold graphs: The structure and the enumeration.
TYSHA 1985-3 Tyshkevich, R. I.; Chernjak, A. A.
Decomposition of graphs.
URRU 1984-1 Urrutia, J.
Intersection graphs, (B*1)-orientable graphs and proper circular arc graphs.
USAM 1984-1 Usami, Y.
Minimum number of edges in P(4)-connected graphs.
USAN 1979-1 Usanov, V. V.
Representation and recognition of triangulated graphs (Russian).
USHI 1985-1 Ushio, K.
Factorization of complete bipartite graphs into Hamilton paths and Hamilton cycles.
USHI 1986-1 Ushio, K.
K(2,2)-factorization of complete multipartite graphs.
VAGI 1983-1 Vagin, V. N.; Georgieva, N. T.; Pospelov, D. A.
On segmentation of graphs.
VAINB 1986-1 Vainshtein, Y. B.
An algorithm for complete planarization of graphs of schemes (Russian).
VANNB 1984-1 Vannelli, A.; Vidyasagar, M.
Three approximate solution techniques for the optimal k-decomposition problem.
VARM 1986-1 Varma, B. N.
On pancyclic line graphs.
VASH 1982-1 Vashchenko, V. P.
An algorithm for graph coloring on the basis of fixed homomorphisms (Russian).
VASI 1985-1 Vasilev, N. S.
On the properties of a flow polyhedron and their applications in the numerical solution of extremal problems on graphs.
VENI 1984-1 Veni Madhavan, C. E.
Approximation algorithm for maximum independent set in planar triangle-free graphs.
VERG 1986-1 Vergnas, M. Las
Hamilton paths in tournaments and a problem of McMullen on projective transformations in R**d.
VEST 1987-1 Vestergaard, P. D.
On graphs with prescribed spanning subgraphs.
VICH 1984-1 Viches, S. A.
An asymptotically optimal algorithm for enumerating intersections of edges of a bipartite graph (Russian).
VINI 1983-1 Vinichenko, S. M.
Calculation of Hamiltonian cycles and paths in complete n-colored graphs (Russian).
VINN 1985-1 Vinnichenko, N. G.
On graphs with the given nondensity and length of a maximal cycle (Russian).
VISH 1984-1 Vishkin, U.
An optimal parallel connectivity algorithm.
VOHRA 1987-1 Vohra, R. V.; Golden, B. L.
Finding the most vital arcs in a network
VOIT 1984-1 Voitishin, Y. V.
Application of network fronts for solution of certain problems on graphs (Russian).
VOLGB 1986-1 Volgenant, A.; Jonker, R.; Kindervater, G. A. P.
A note on finding a shortest complete cycle in an undirected graph.
VOLO 1984-1 Volodina, V. V.; Cherkasskii, B. V.
On the realization of an algorithm for recognizing the planarity and packing of a graph in O(v) operations (Russian).
VOLOC 1982-1 Voloshin, V. I.
Properties of triangulated graphs (Russian).
VOLOD 1984-1 Voloshina, A. A.; Feinberg, V. Z.
Planarity of hypergraphs (Russian).
VORN 1981-1 Vornberger, O.
Alternating cycle covers and paths.
VORO 1983-1 Vorobev, V. A.
Relative addressing of elements of a cyclic graph (Russian).
VOSS 1983-1 Voss, H.-J.
Unabhängige Mengen in k-chromatischen Graphen.
VOSS 1985-1 Voss, H.-J.
Bridges of circuits through specified edges.
VOSS 1985-2 Voss, H.-J.
Chords of circuits in graphs with given minimum degree.
VOSS 1985-4 Voss, H.-J.
Symmetric subgraphs and the reconstruction conjecture.
VU 1984-1 Vu Dinh Hoa
Hamiltonkreise in ungerichteten und schlichten Graphen.
VU 1986-1 Vu Dinh Hoa
Ein Struktursatz für 2-fach zusammenhängende Graphen mit grosser Minimalvalenz.
WAGNA 1984-2 Wagner, K.
The complexity of problems concerning graphs with regularities.
WAKA 1986-1 Wakabayashi, Y.
Aggregation of binary relations: algorithmic and polyhedral investigations.

WALKA 1987-1 Walker, K.
Fully-coloured Hamiltonian cycles in edge-colorings of K(n) (n odd) when there are no fully-coloured C(4)'s.
WALLB 1985-1 Wallis, W. D.
The clique partition number of the complement of a cycle.
WALLC 1985-1 Wallis, W. D.; Wang, Z.-J.
On one-factorizations of Cartesian products.
WALT 1984-1 Walther, H.
Ten applications of graph theory.
WALT 1987-1 Walther, H.
On a conjecture of W. Klotz concerning clique-decomposition of graphs.
WALTA 1987-1 Walther, H.; Nägler, G.
Graphen, Algorithmen, Programme.
WANG 1985-1 Wang, C.-C.; Lloyd, E. L.; Soffa, M. L.
Feedback vertex sets and cyclically reducible graphs.
WANGA 1985-1 Wang, C.-D.
On the sum of two parameters concerning independence and irredundance in graphs (Chinese).
WANGB 1986-1 Wang, H.
The maximum number of edges of a critical 3-edge connected graph on n vertices (Chinese).
WANGC 1987-1 Wang, J. F.; Tian, S.-L.; Liu, J.-Q.
On binding numbers of graphs (Chinese).
WANGE 1986-1 Wang, J. S.
A general algorithm for enumerating some subgraphs of a simple graph (Chinese).
WANGF 1982-1 Wang, J.-Z.
A necessary and sufficient condition for a Hamiltonian tournament (Chinese).
WANGH 1986-1 Wang, N. S.
On the specific properties of the characteristic polynomial of a tree (Chinese).
WANGI 1986-1 Wang, S. H.; Tao, M. Q.
The structure of k-connected digraphs and mixed (p,q)-connected graphs (Chinese).
WANGN 1983-1 Wang, Z.-Y.
Additive enumeration problems over trees.
WANGN 1983-2 Wang, Z.-Y.
On the endpoint number of free trees.
WANGN 1983-3 Wang, Z.-Y.
On the problem of the external path length of trees (Chinese).
WANGN 1983-4 Wang, Z.-Y.
Some combinatorial problems on ordered trees.
WANGN 1984-1 Wang, Z.-Y.
Some combinatorial problems on t-ary trees.
WANGN 1984-2 Wang, Z.-Y.
Some combinatorial problems on t-ary trees.
WANGN 1985-1 Wang, Z.-Y.
On the degree path length of trees.
WANGN 1986-1 Wang, Z.-Y.
An enumeration problem on ordered trees.
WATA 1987-1 Watanabe, T.; Nakamura, A.
Edge-connectivity augmentation problems.
WATAA 1984-1 Watanabe, T.; Nakamura, A.; Takahashi, M.
Problems of graph augmentation by the addition of sides (Japanese).
WEGE 1985-1 Wegener, I.
The critical complexity of all (monotone) Boolean functions and monotone graph properties.
WEGE 1985-2 Wegener, I.
The critical complexity of all (monotone) Boolean functions and monotone graph properties.
WERNA 1986-1 Werner, R.
Kreise im Strukturgraphen eines Netzplanes.
WERR 1984-2 Werra, D. de
Variation on a theorem of König.
WESS 1985-1 Wessel, W.
Kreissehnengraphen - ein neues Hilfsmittel zur Darstellung von Graphen.
WEST 1985-1 West, D. B.
Parameters of partial orders and graphs: packing, covering and representation.
WHITB 1984-1 White, K.; Faber, M.; Pulleyblank, W. R.
Steiner trees, connected domination and strongly chordal graphs.
WHITG 1984-2 Whitesides, S. H.
A method for solving certain graph recognition and optimization problems, with applications to perfect graphs.
WIEG 1987-1 Wiegers, M.
Recognizing outerplanar graphs in linear time.
WILF 1986-2 Wilf, H. S.
The number of maximal independent sets in a tree.
WILFA 1987-1 Wilf, H. S.; Yoshimura, N. A.
Ranking rooted trees, and a graceful application.
WILLC 1984-1 Williamson, S. G.
Depth-first search and Kuratowski subgraphs.
WILSB 1985-1 Wilson, R. J.
Introduction to graph theory (third edition).
WIME 1987-1 Wimer, T. V.
Linear algorithms for the dominating cycle problems in series-parallel graphs, partial k-trees, and Halin graphs.
WIMEA 1985-1 Wimer, T. V.; Hedetniemi, S. T.; Laskar, R.
A methodology for constructing linear graph algorithms.
WINT 1986-1 Winter, P.
An algorithm for the enumeration of spanning trees.

WINTA 1984-1 Winter, P. A.; Swart, H. C.
On alpha, beta destructible graphs.

WINTA 1984-2 Winter, P. A.; Swart, H. C.
On stable graphs.

WISZ 1985-1 Wisztova, E.
A Hamiltonian cycle and a 1-factor in the fourth power of a graph.

WOJC 1985-1 Wojciechowski, J.
An approximate formula for counting trees in a graph.

WORM 1985-1 Wormald, N. C.
Counting labelled chordal graphs.

WU A 1985-1 Wu, W.-J.
On the planar imbedding of linear graphs. I.

WU A 1986-1 Wu, W.-J.
On the planar imbedding of linear graphs (continued).

WU B 1983-1 Wu, W.-M.
Robbins' theorem for fuzzy mixed multigraphs (Chinese).

WU C 1984-1 Wu, X.-M.
Recent developments on graph theory (Chinese).

XIE 1983-1 Xie, L.-T.
Methods of algebraic topology applied to the theory of graphs and matroids (Chinese).

XIE 1984-1 Xie, L.-T.
Connected partition problems on graphs and n-dimensional complexes (Chinese).

XIE A 1984-1 Xie, L.-T.; Liu, G.-Z.
A survey of the problem of graceful trees (Chinese).

XIE B 1982-1 Xie, L.-T.; Zheng, H.-D.; Liu, J.-Z.
On the valence sequence of an ordered tree.

XU A 1982-1 Xu, C.-D.
Diameter-stable graphs and line-persistence (Chinese).

XU B 1984-1 Xu, J.-M.
On the proof of Simon and Murty conjecture (Chinese).

XU B 1984-2 Xu, J.-M.
Proof of the mixed form of Menger's theorem in graph theory (Chinese).

XU B 1986-1 Xu, J.-M.
A note on the algebraic connectivity of graphs (Chinese).

XU C 1986-1 Xu, J.-M.; Tian, F.
On the lower bound of the maximum size of a critically h-edge-connected and a critically h-connected graph (Chinese).

XU E 1986-1 Xu, O.
A class of graceful graphs (Chinese).

XU F 1983-2 Xu, S.-J.
Complete-graph-basis and graphs with more edges (Chinese).

XU F 1987-1 Xu, S.-J.
Some parameters of graph and its complement.

XUE A 1986-1 Xue, Y.-F.; Dai, X.-S.
On maximum degree and the number of vertices of degree >4 in minimally 2-(edge) connected graphs (Chinese).

YANGB 1982-1 Yang, S. H.
Isomorphic factorizations of complete tripartite graphs (Chinese).

YANGC 1986-1 Yang, S. M.; Zhang, Z. F.
The enumeration of the circles of a lattice plane graph (Chinese).

YANGE 1985-1 Yang, W.-L.
On graphs which contain all trees with n edges I. A lower bound of S(T(n)) (Chinese).

YANGE 1985-2 Yang, W.-L.
On graphs which contain all trees with n edges II - the exact values of s(T(n)) for n<or=16 (Chinese).

YANN 1985-1 Yannakakis, M.
A polynomial algorithm for the min-cut linear arrangement of trees.

YANN 1986-1 Yannakakis, M.
Four pages are necessary and sufficient for planar graphs.

YANNA 1987-1 Yannakakis, M.; Gavril, F.
The maximum k-colorable subgraph problem for chordal graphs.

YAP 1983-1 Yap, H. P.
A class of chromatic index critical graphs.

YAP 1984-1 Yap, H. P.
Computational complexity of graph properties.

YEH A 1984-1 Yeh, D. Y.; Lee, D. T.
Graph algorithms on a tree-structured parallel computer.

YU 1984-1 Yu, J.-L.
Remarks on the critically n-connected graphs (Chinese).

YU A 1984-1 Yu, J.-L.; Ma, Y.
A remark on critically n-connected graphs (Chinese).

YU B 1984-1 Yu, Z.-G.
A generalization of Bondy-Chvatal's theorem on the k-closure of a bipartite graph (Chinese).

ZAGA 1983-1 Zagaglia Salvi, N.
On the chromatic index of permutation graphs (Italian).

ZAKS 1985-1 Zaks, J.
How does a complete graph split into bipartite graphs and how are neighborly cubes arranged?

ZAMF 1986-1 Zamfirescu, T.; Zucco, A.
On the least Hamiltonian traceable graphs (French).

ZASL 1987-1 Zaslavsky, T.
Balanced decompositions of a signed graph.

ZELI 1983-1 Zelinka, B.
Double covers and logics of graphs II.

ZELI 1983-2 Zelinka, B.
Double covers and logics of graphs.
ZELI 1984-1 Zelinka, B.
On k-ply domatic numbers of graphs.
ZELI 1984-2 Zelinka, B.
Semidomatic numbers of directed graphs.
ZELI 1985-1 Zelinka, B.
Bichromaticity and domatic number of a bipartite graph.
ZELI 1985-2 Zelinka, B.
Comparison of various distances between isomorphism classes of graphs.
ZELI 1985-3 Zelinka, B.
Cyclicity of a complete 3-equipartite graph.
ZELI 1985-4 Zelinka, B.
Domatic numbers of uniform hypergraphs.
ZELI 1985-5 Zelinka, B.
Graphs with prescribed neighbourhood graphs.
ZELI 1985-6 Zelinka, B.
Odd graphs.
ZELI 1986-1 Zelinka, B.
Connected domatic number of a graph.
ZELI 1986-2 Zelinka, B.
Disconnected neighbourhood graphs.
ZELI 1986-3 Zelinka, B.
Domatic number and linear arboricity of cacti.
ZELI 1986-4 Zelinka, B.
Edge neighbourhood graphs.
ZELI 1986-5 Zelinka, B.
Neighbourhood tournaments.
ZELI 1987-1 Zelinka, B.
Distances between directed graphs.
ZELI 1987-2 Zelinka, B.
Natural bijections between directed and undirected self-complementary graphs.
ZELI 1987-3 Zelinka, B.
On a problem of P. Vestergaard concerning circuits in graphs.
ZELI 1987-4 Zelinka, B.
The distance between a graph and its complement.
ZHAN 1986-1 Zhan, S.-M.
Hamiltonian connectedness of line graphs.
ZHANB 1985-1 Zhang, C.-Q.
Circuits and Hamilton circuits in domination orientable tournaments.
ZHANC 1985-1 Zhang, F. J.; Chen, R. S.
On two kinds of 2-edge-connected graphs (Chinese).
ZHANE 1986-1 Zhang, F. J.; Chen, Z. B.
Connectivity of (adjacency) tree graphs.
ZHANE 1987-1 Zhang, F. J.; Chen, Z. B.
Lower bounds for the maximum number of end-vertices of spanning trees (Chinese).
ZHANF 1983-1 Zhang, F. J.; Guo, X. F.
H-transformations of Hamiltonian cycles (Chinese).
ZHANF 1985-1 Zhang, F. J.; Guo, X. F.
Connectedness of generalized directed tree graphs and an interpolation theorem for the numbers of terminal vertices of directed trees (Chinese).
ZHANF 1985-2 Zhang, F. J.; Guo, X. F.
Some properties of minimally strongly connected digraphs (Chinese).
ZHANF 1986-1 Zhang, F. J.; Guo, X. F.
Hamilton cycles in Euler tour graph.
ZHANF 1986-2 Zhang, F. J.; Guo, X. F.
Hamiltonian cycles in perfect matching graphs (Chinese).
ZHANF 1987-1 Zhang, F. J.; Guo, X. F.
Hamilton cycles in directed Euler tour graphs.
ZHANG 1983-1 Zhang, F. J.; Lin, G. N.
A graphical method for generating all Euler tours of an Eulerian digraph (Chinese).
ZHANI 1985-1 Zhang, F. J.; Lin, Y. X.
Two classes of general polynomials for graphs (Chinese).
ZHANR 1986-1 Zhang, Z. F.; Chen, Y.
Operations on graphs and the chromatic number of a graph (Chinese).
ZHANS 1983-1 Zhang, Z. F.; Yao, B.; Xie, J. G.
On the bandwidth of graphs (Chinese).
ZHANT 1985-1 Zhang, Z. F.; Zhang, J. X.
On some sufficient conditions for graphs to be of first kind (Chinese).
ZHAO 1985-2 Zhao, C. K.
A note on bipartite graphs (Chinese).
ZHAOB 1987-1 Zhao, G.-F.
Graphs isomorphic to subgraphs of their iterated line graphs.
ZHAOC 1986-1 Zhao, L.-C.; Liu, C. F.; Wang, H.
A note on Hamiltonian line graphs (Chinese).
ZHAOD 1984-1 Zhao, S. L.
The theory of pleasant stringing trees and some kinds of pleasant trees (Chinese).
ZHAOD 1986-1 Zhao, S. L.
Double isomorphic graceful trees are graceful stringing trees (Chinese).
ZHEN 1986-1 Zheng, M. L.
The Prim algorithm and the uniqueness of minimal trees of graphs (Chinese).
ZHOU 1984-1 Zhou, B.
A note on the Erdös-Sos conjecture.

ZHOUA 1985-1 Zhou, M. K.
Some properties about distance degree regular graphs.
ZHOUB 1984-1 Zhou, Z. H.
Minimal and maximal line digraphs (Chinese).
ZHU A 1985-2 Zhu, B. W.
The circumference of minimal 2-edge-connected graphs (Chinese).
ZHU B 1985-1 Zhu, B. W.; Liu, Z. S.; Chen, Z. Q.
A theorem on the edge-cohesiveness of a vertex of a graph (Chinese).
ZHU C 1987-1 Zhu, B. W.; Su, J. J.
Some problems in the research on the connectedness of graphs (Chinese).
ZHU D 1985-1 Zhu, D. X.; Cai, G. Q.
On an exhaustive circuit matrix of an oriented graph (Chinese).
ZHU F 1983-1 Zhu, R. Y.
A remark on Jackson's theorem on regular graphs (Chinese).
ZHU F 1983-2 Zhu, R. Y.
On maximal circuits in 2-connected graphs (Chinese).
ZHU G 1986-1 Zhu, S. Y.
A strong graceful path (Chinese).
ZHU H 1985-1 Zhu, W. S.
Proof of Bodendiek's conjecture on graceful graphs (Chinese).
ZHU I 1983-1 Zhu, X. D.
Pansystems graph theory and (l,d)-stable graphs.
ZHU J 1986-1 Zhu, X. F.
On arc-Hamiltonian tournaments (Chinese).
ZHU K 1983-1 Zhu, Y.-J.
A survey of the research on the Hamilton problem in graph theory (Chinese).
ZHU K 1986-1 Zhu, Y.-J.
Constructive characteristics appearing in some problems of graph theory and algorithms (Chinese).
ZHU L 1985-1 Zhu, Y.-J.; Li, H.
Progress on the Hamilton problem in graph theory (Chinese).
ZHU M 1985-1 Zhu, Y.-J.; Liu, Z. H.; Yu, Z.-G.
An improvement of Jackson's result on Hamilton cycles in 2-connected regular graphs.
ZHU M 1986-1 Zhu, Y.-J.; Liu, Z. H.; Yu, Z.-G.
2-connected k-regular graphs (k greater equal 6) on at most 3k+3 vertices are Hamiltonian.
ZHU M 1986-2 Zhu, Y.-J.; Liu, Z. H.; Yu, Z.-G.
2-connected k-regular graphs (k greater equal 6) on at most 3k+3 vertices are Hamiltonian. II.
ZHU N 1984-1 Zhu, Y.-J.; Tian, T.; Chen, C.-P.; Zhang, C.-Q.
Arc-pancyclic property of tournaments under some degree conditions.
ZIKA 1985-1 Zikan, K.; Schmeichel, E.
A path counting problem in digraphs.
ZNOI 1983-1 Znoiko, D. V.
Vertex-transitive graphs with a given degree and girth (Russian).
ZORN 1985-1 Zörnig, P.
Strukturuntersuchungen an 2xn-Entartungsgraphen.
ZUCC 1984-1 Zucco, A.
On hyperhomogeneously traceable oriented graphs.

ARAO 1984-1 Araoz, J.
Packing problems in semigroup programming.
CHOPA 1987-1 Chopra, S.; Johnson, E. L.
Dual row modules and polyhedra of blocking group problems.
CUNIA 1986-1 Cuninghame-Green, R. A.; Burkard, R. E.
Saddlepoints in group and semigroup minimization.
DAVY 1986-1 Davydov, E. G.; Bolshakov, S. Y.
Group-theoretic methods for aggregation in network problems (Russian).
GAST 1986-1 Gastou, G.; Johnson, E. L.
Binary group and Chinese postman polyhedra.
LUCE 1984-1 Lucertini, M.
Lagrangian and group-theoretic relaxation in integer programming (Italian).
MINO 1986-1 Minoux, M.
Mathematical Programming. Theory and Algorithms.
PIEH 1985-1 Piehler, J.
Über die irreduziblen Punkte des Eckenpolyeders.

AIEL	1985-1	Aiello, A.; Burattini, E.; Massarotti, A.; Ventriglia, F. Heuristic evaluation techniques for bin packing approximation algorithms.
ALI	1984-1	Ali, A. I. A greedy algorithm for the transshipment along a single road problem.
ALI A	1986-1	Ali, A. I.; Kennington, J. L. The asymmetric m-travelling salesmen problem: a duality based branch-and-bound algorithm.
ALIEA	1986-1	Aliev, A. A.; Akhmedov, F. B.; Babaev, D. A. Application package KNAPSACK-1 for approximate solution of the generalized problem of knapsack (Russian).
ALTI	1987-1	Altinkemer, K.; Gavish, B. Heuristics for unequal weight delivery problems with a fixed error guarantee.
AVIS	1985-1	Avis, D.; Devroye, L. An analysis of a decomposition heuristic for the assignment problem.
BAR-	1985-1	Bar-Yehuda, R.; Even, S. A local-ratio theorem for approximating the weighted vertex cover problem.
BART	1984-1	Bartholdi, J. J. III; Platzman, L. K. Heuristics based on spacesfilling curves for combinatorial problems in the plane.
BEIN	1986-1	Bein, W. W.; Brucker, P. Greedy concepts for network flow problems.
BENA	1985-1	Benavent, E.; Campos, V.; Corberan, A.; Mota, E. Analysis of heuristics for the rural postman problem (Spanish).
BIGI	1985-1	Bigildeeva, T. B. Methods and algorithms of approximate solution of a quadratic assignment problem of large dimensions (a survey) (Russian).
BIXBD	1986-1	Bixby, R. E.; Fourer, F. Finding embedded network rows in linear programs I: extraction heuristics.
BOUCD	1986-1	Bouchitte, V.; Habib, M. Some NP-completeness properties about linear extensions.
BRUC	1984-3	Brucker, P. Über gewisse Prinzipien des Entwurfs effizienter Algorithmen.
BURR	1985-1	Burr, S. A.; Falk, J. E.; Karr, A. F. Integer prim-read solutions to a class of target defense problems.
CAMEF	1985-1	Cameron, J.; Thomas, G. An approximate graph partitioning algorithm and its computational complexity.
CAVA	1986-1	Cavalier, T. M.; Chandra, M. J. A heuristic algorithm for assigning crews among bases in an airlift operation.
CHAOA	1986-1	Chao, M.-T.; Franco, J. Probabilistic analysis of two heuristics for the 3-satisfiability problem.
CHAUA	1987-1	Chaudhry, S. S.; Moon, I. D.; Mac Cormick, S. T. Conditional covering: greedy heuristics and computational results.
COFFB	1984-1	Coffman, E. G. Jr.; Garey, M. R.; Johnson, D. S. Approximation algorithms for bin-packing - an updated survey.
COFFE	1984-1	Coffman, E. G. Jr.; Langston, M. A. A performance guarantee for the greedy set-partitioning algorithm.
CROI	1986-1	Croitoru, C.; Moscovici, J. A "greedy" O—V—**3-algorithm for the maximum cardinality matching problem.
DEMBA	1981-1	Dembo, R. S.; Klincewicz, J. G. An approximate second-order algorithm for network flow problems with convex separable costs.
DEMP	1983-1	Dempe, S. Zwei greedy-Algorithmen für ein gemischt-ganzzahliges Optimierungsproblem.
DEMP	1985-1	Dempe, S. A simple but NP-hard problem of mixed-discrete programming and its solution by approximate algorithms.
DERI	1985-4	Derigs, U. Using confidence limits for the global optimum in combinatorial optimization.
DONAA	1986-1	Donath, G.; Hofstedt, K. Über eine spezielle lineare 0-1-Optimierungsaufgabe.
DUBO	1983-1	Dubois, D. A fuzzy, heuristic, interactive approach to the optimal network problem.
DYERA	1985-1	Dyer, M. E.; Foulds, L. R.; Frieze, A. M. Analysis of heuristics for finding a maximum weight planar subgraph.
DYERC	1984-1	Dyer, M. E.; Frieze, A. M.; Mac Diarmid, C. Partitioning heuristics for two geometric maximization problems.
EISED	1987-1	Eiselt, H. A.; Laporte, G. Combinatorial optimization problems with soft and hard requirements.
EMELB	1985-1	Emelichev, V. A.; Ovchinnikov, V. G. Greedy algorithms for solution some problems of discrete optimization.
EREN	1984-2	Erenguc, S. S.; Tufekci, S. A transportation type aggregate production model with bounds on inventory and backordering.
EVAN	1987-1	Evans, J. R. Structural analysis of local search heuristics in combinatorial optimization.
FABI	1985-1	Fabian, C. Economic pre-optimizing algorithms for integer programming.
FABI	1985-2	Fabian, C. Greedy algorithm for a type of integer programming problems.
FAIGF	1985-1	Faigle, U.; Schrader, R. A setup heuristic for interval orders.
FISC	1987-1	Fischetti, M. A new linear storage, polynomial time approximation scheme for the subset-sum problem.
FISHI	1981-1	Fisher, M. L.; Wolsey, L. A. On the greedy heuristic for covering and packing problems.
FONLB	1986-1	Fonlupt, J.; Zemirline, A. Polynomial recognition algorithm of perfect 3-chromatic K(4)/(e)-free graphs.

FOX 1985-1 Fox, G. E.; Scudder, G. D.
A heuristic with tie breaking for certain 0-1 integer programming models.
FREV 1986-1 Freville, A.; Plateau, G.
Heuristics and reduction methods for multiple constraints 0-1 linear programming problems.
GLAS 1986-1 Glassey, C. R.; Mizrach, M.
A decision support system for assigning classes to rooms.
GLOV 1986-1 Glover, F.
Future paths for integer programming and links to artificial intelligence.
GOLDB 1984-1 Goldberg, M. K.
An approximate algorithm for the edge-coloring problem.
GOLDF 1984-1 Golden, B. L.; Assad, A. A.
A decision-theoretic framework for comparing heuristics.
GOLDK 1985-1 Golden, B. L.; Stewart, W. R. Jr.
Empirical analysis of heuristics.
GRIGC 1986-1 Grigoriadis, M. D.; Kalantari, B.; Lai, C. Y.
On the existence of weak greedy matching heuristics.
GRIGC 1987-1 Grigoriadis, M. D.; Kalantari, B.; Lai, C. Y.
Weak greedy heuristics for perfect matching.
GUIG 1984-2 Guignard, M.
Dual and primal heuristics for network location problems.
GULY 1983-1 Gulyanitskii, L. F.; Khodzinskii, A. N.
Two algorithms for solution of optimization problems on permutations (Russian).
GUSFB 1986-1 Gusfield, D.; Pitt, L.
Equivalent approximation algorithms for node cover.
HACK 1986-1 Hackbarth, K.-D.
A heuristic procedure for calculating telecommunication transmission networks in consideration of network reliability.
HAIM 1985-1 Haimovich, M.; Rinnooy Kan, A. H. G.
Bounds and heuristics for capacitated routing problems.
HALLB 1986-1 Hall, N. G.; Hochbaum, D. S.
A fast approximation algorithm for the multicovering problem.
HAMM 1984-1 Hammer, G.
Greedy solutions for general set covering problems.
HART 1987-1 Hart, J. P.; Shogan, A. W.
Semi-greedy heuristics: an empirical study.
HASH 1987-1 Hashizume, S.; Fukushima, M.; Katoh, N.; Ibaraki, T.
Approximation algorithms for combinatorial fractional programming problems.
HEIN 1984-1 Heinrich, H.
Ein heuristisches Lösungsverfahren zur Stundenplanung und Raumbelegung an Hochschulen.
HOCHA 1983-1 Hochbaum, D. S.; Maass, W.
Fast approximation algorithms for the robot placement problem.
HOCHA 1985-1 Hochbaum, D. S.; Maass, W.
Approximation schemes for covering and packing problems in image processing and VLSI.
HOCHA 1987-1 Hochbaum, D. S.; Maass, W.
Fast approximation algorithms for a nonconvex covering problem.
HOCHB 1986-3 Hochbaum, D. S.; Shmoys, D. B.
A unified approach to approximation algorithms for bottleneck problems.
HOCHB 1986-4 Hochbaum, D. S.; Shmoys, D. B.
Best possible heuristics for the bottleneck wandering salesperson and bottleneck vehicle routing problems.
HOFS 1985-1 Hofstedt, K.
Zur Kopplung eines exakten Verfahrens mit einem heuristischen Verfahren für die Lösung ganzzahliger linearer Optimierungsprobleme.
HOSK 1984-1 Hoskins, W. D.; Lambert, T. D.; Stanton, R. G.
An improved algorithm for determining optimal sums of disjoint sets.
ILEVA 1983-1 Ilev, V. P.; Fridman, G. S.
On a problem of approximation by three-component graphs (Russian).
IMAI 1986-1 Imai, H.
Worst-case analysis for planar matching and tour heuristics with bucketing techniques and spacefilling curves.
JENKB 1986-1 Jenkyns, T. A.
The greedy algorithm is a shady lady.
JOHNG 1985-2 Johnson, D. S.; Papadimitriou, C. H.
Performance guarantees for heuristics.
JOHNH 1985-1 Johnson, D. S.; Papadimitriou, C. H.; Yannakakis, M.
How easy is local search?
KARMB 1982-1 Karmarkar, N.; Karp, R. M.
An efficient approximation scheme for the one-dimensional bin-packing problem.
KARPF 1985-1 Karp, R. M.; Steele, J. M.
Probabilistic analysis of heurictics.
KASA 1984-1 Kasai, T.; Yamada, S.; Ohno, T.
An approximation method for planar assignment problem.
KIER 1986-1 Kierstead, H. A.
NP-completeness results concerning greedy and super greedy linear extensions.
KORB 1983-1 Korbut, A. A.; Finkelstein, Y. Y.
Approximate methods of discrete programming.
KOROA 1983-1 Korotkin, A. A.
The accuracy of a greedy algorithm in a problem of the choice of an optimal acyclic subgraph (Russian).
KOVAE 1985-1 Kovalev, M. M.; Kotov, V. M.
Serien von Greedy-Algorithmen (Russian).
KRAR 1986-1 Krarup, J.; Pruzan, P. M.
Assessment of approximate algorithms: the error measure's crucial role.

KUCE 1986-1 Kucera, L.
Greedy coloring is a bad probabilistic algorithm.

KUMAB 1986-1 Kumar, K. R.; Kusiak, A.; Vannelli, A.
Grouping of parts and components in flexible manufacturing systems.

KUZY 1984-1 Kuzyurin, N. N.
The complexity of approximate algorithms for the problem of integer programming.

LANGC 1987-1 Langston, M. A.
A study of composite heuristic algorithms.

LIBU 1984-1 Libura, M.
Integer programming duality and its application in sensitivity analysis and approximate algorithms (Polish).

LIU I 1983-1 Liu, Z. H.
Approximate algorithms for combinatorial optimization problems. I (Chinese).

LIU I 1983-2 Liu, Z. H.
Approximate algorithms for combinatorial optimization problems. II (Chinese).

LIU I 1983-3 Liu, Z. H.
NP-complete problems and some approximate algorithms II (Chinese).

LIU I 1983-4 Liu, Z. H.
NP-complete problems and some approximate algorithms I (Chinese).

MANG 1985-1 Mangoubi, R. S.; Mathaisel, D. F. X.
Optimizing gate assignments at airport terminals.

MANV 1985-1 Manvel, B.
Extremely greedy coloring algorithms.

MARCB 1985-1 Marchetti-Spaccamela, A.; Romano, G.
On different approximation criteria for subset product problems.

MARTB 1985-2 Martello, S.; Toth, P.
Approximation schemes for the subset-sum problem: survey and experimental analysis.

MARTB 1986-1 Martello, S.; Toth, P.
A heuristic approach to the bus driver scheduling problem.

MELAA 1985-1 Melachrinoudis, E.; Cullinane, T. P.
A heuristic approach to the single facility maximin location problem.

MILA 1986-1 Milanov, P. B.
On the worst behaviour of the greedy procedure.

MONIA 1984-1 Monien, B.; Speckenmeyer, E.
Ramsey numbers and an approximation algorithm for the vertex cover problem.

MONIA 1985-1 Monien, B.; Speckenmeyer, E.
Ramsey numbers and an approximation algorithm for the vertex cover problem.

ONG A 1984-1 Ong, H. L.; Magazine, M.; Wee, T. S.
Probabilistic analysis of bin packing heuristics.

ORLI 1980-1 Orlin, J. B.
Fast approximation schemes for combinatorial optimization problems (extended abstract).

PIRK 1987-1 Pirkul, H.
A heuristic solution procedure for the multiconstraint zero-one knapsack problem.

PLAI 1984-1 Plaisted, D. A.
Heuristic matching for graphs satisfying the triangle inequality.

POLJA 1986-1 Poljak, S.; Turzik, D.
A polynomial time heuristic for certain subgraph optimization problems with guaranteed worst case bound.

PSAR 1984-1 Psaraftis, H. N.
On the practical importance of asymptotic optimality in certain heuristic algorithms.

RAGHA 1986-1 Raghavan, P.
Probabilistic construction of deterministic algorithms: approximating packing integer programs.

REEV 1985-1 Reeves, C. R.
An improved heuristic for the quadratic assignment problem.

RIBEA 1985-1 Ribeiro, C. C.; Minoux, M.
A heuristic approach to hard constrained shortest path problems.

RINN 1984-1 Rinnooy Kan, A. H. G.
An introduction to approximation algorithms.

RINN 1986-1 Rinnooy Kan, A. H. G.
An introduction to the analysis of approximation algorithms.

RIVAA 1986-1 Rival, I.; Zaguia, N.
Constructing greedy linear extensions by interchanging chains.

RONEA 1984-1 Ronen, D.; Perl, Y.
Heuristics for finding a maximum number of disjoint bounded paths.

SAKAA 1984-1 Sakarovitch, M.
Optimisation combinatoire - methodes mathematiques et algorithmiques. Vol. 1: Programmation discrete.

SEMEB 1984-1 Semenova, N. V.
Solution of one generalized integer programming problem(Russian).

SERGA 1982-1 Sergienko, I. V.
Some questions of the development and use of approximation methods of solution of discrete optimization problems (Russian).

SERGB 1984-1 Sergienko, I. V.; Artemenko, V. I.
Search for approximate solutions of integer linear problems by the methods of linear and parametric programming (Russian).

SERGF 1985-1 Sergienko, I. V.; Lebedeva, T. T.; Konzeratskaya, L. N.
Some parametric problems of integer convex programming and their approximate solution (Russian).

SERGG 1980-1 Sergienko, I. V.; Lebedeva, T. T.; Roshchin, V. A.
Approximate methods for solving discrete problems of optimization.

SGURC 1981-1 Sgurev, V. S.; Vasilev, V. S.
Approximate algorithm for integer programming (Russian).

SGURC 1985-1 Sgurev, V. S.; Vasilev, V. S.
An approximate algorithm solving mixed integer programming problems (Bulgarian).

SHEN 1982-1 Shen, L.
Effective use of heuristic algorithms in network design.
SHERF 1986-1 Sherali, H. D.; Rajgopal, P.
A flexible, polynomial-time, construction and improvement heuristic for the quadratic assignment problem.
SIME 1986-1 Simeone, B.
An asymptotically exact polynomial algorithm for equipartition problems.
SPAT 1985-1 Späth, H.
Heuristically determining cliques of given cardinality and with minimal cost within weighted complete graphs.
STOIA 1986-1 Stoica, M.; Hartulari, C.
Modalities and principles to design heuristic algorithms.
STRZ 1985-1 Strzelec, K. R.; Gadecki, H.
Exact or approximate algorithms? A combinatorial approach to integer programming.
SYSL 1986-2 Syslo, M. M.
On some new types of greedy chains and greedy linear extensions of partially ordered sets.
SZKA 1987-1 Szkatula, K.; Libura, M.
On probabilistic properties of greedy-like algorithms for the binary knapsack problem.
TAO 1987-1 Tao, G.-Y.; Zhou, P.-D.; Wang, Q.
The heuristic shortest-path algorithm for transportation (Chinese).
TINH 1984-2 Tinhofer, G.
Naive greedy matching - a probabilistic analysis.
TRAU 1987-1 Trauwaert, E.; Decrock, A.
A heuristic algorithm for allocation of components in manufacturing processes.
TSAR 1985-1 Tsarevskij, N. A.
An approximational decomposition algorithm for a combinatorial problem of developing production capacities of an industry with block production constraints II.
TURA 1985-1 Turan, G.
On the greedy algorithm for an edge-partitioning problem.
VAID 1984-1 Vaidya, P. M.
A fast approximation algorithm for minimum spanning trees in k-dimensional space.
VANN 1984-1 Vannelli, A.
Approximating a class of graph decomposition problems by linear transportation problems.
VANNB 1984-1 Vannelli, A.; Vidyasagar, M.
Three approximate solution techniques for the optimal k-decomposition problem.
VASIA 1983-1 Vasilev, V. S.
An approximate algorithm, solving a class of nonlinear discrete programming problems (Bulgarian).
VASK 1986-1 Vasko, F. J.; Wilson, G. R.
Hybrid heuristics for minimum cardinality set covering problems.
VENI 1984-1 Veni Madhavan, C. E.
Approximation algorithm for maximum independent set in planar triangle-free graphs.
VIZV 1983-1 Vizvari, B.
The heuristic methods of discrete programming. I. The method of neighbourhood.
VIZV 1987-1 Vizvari, B.
On the greedy solution in integer linear programming.
VOHR 1985-1 Vohra, R. V.
On the greedy heuristic for the vertex packing problem.
VOLK 1987-2 Volkovich, O. V.
Approximation of the admissible domain in a problem of integer programming (Russian).
WAJS 1984-1 Wajs, W.
Scheduling problem for metallurgical processes (Polish).
WARB 1985-1 Warburton, A.
Worst case analysis of greedy and related heuristics for some min-max combinatorial optimization problems.
WERR 1987-1 Werra, D. de
Design and operation of flexible manufacturing systems: the kingdom of heuristic methods.
WOLS 1981-1 Wolsey, L. A.
An analysis of the greedy algorithm for the submodular (set) covering problem.
WRIG 1986-1 Wright, D. D.; Haehling von Lanzenauer, C.
COAL: A new heuristic approach for solving the fixed charge problem - computational results.
WU D 1986-1 Wu, Y. F.; Widmayer, P.; Wong, C. K.
A faster approximation algorithm for the Steiner problem in graphs.

BALI 1984-2 Balinski, M. L.
The Hirsch conjecture for dual transportation polyhedra.

BURKA 1985-1 Burkard, R. E.; Euler, R.; Grommes, R.
On the facial structure of polytopes related to the planar 3-dimensional assignment problem.

COOKA 1986-2 Cook, W.
On box totally dual integral polyhedra.

CORNG 1985-1 Cornuejols, G.; Fonlupt, J.; Naddef, D. J.
The traveling salesman problem on a graph and some related integer polyhedra.

FINKH 1984-1 Finkelstein, Y. Y.; Shmelev, A. A.
Über die Abschätzung der Koeffizienten von Seiten gewisser ganzzahliger Polyeder (Russisch).

FINKH 1986-1 Finkelstein, Y. Y.; Shmelev, A. A.
On an estimate of coefficients of the faces of some integer polyhedra (Russian).

FUJIA 1987-2 Fujishige, S.
Optimization over the polyhedron determined by a submodular function on a co-intersection family.

GROT 1985-1 Grötschel, M.
Polyhedral Combinatorics.

GROT 1986-1 Grötschel, M.
Fortschritte in der polyedrischen Kombinatorik.

GROT 1986-2 Grötschel, M.
Polyedrische Kombinatorik und Schnittebenenverfahren.

GROTC 1984-2 Grötschel, M.; Lovasz, L.; Schrijver, A.
Geometric methods in combinatorial optimization.

GROTD 1985-1 Grötschel, M.; Padberg, M. W.
Polyhedral theory.

HOFFD 1985-1 Hoffman, K.; Padberg, M. W.
LP-based combinatorial problem solving.

HOFFD 1986-1 Hoffman, K.; Padberg, M. W.
LP-based combinatorial problem solving.

JEROC 1987-1 Jeroslow, R. G.; Wang, J.
Dynamic programming, integral polyhedra and Horn clause knowledge bases.

JUNGA 1985-1 Jünger, M.
Polyhedral combinatorics and the acyclic subdigraph problem.

NADD 1984-1 Naddef, D. J.
Pancyclic properties of the graph of some 0-1-polyhedra.

NADDA 1984-1 Naddef, D. J.; Pulleyblank, W. R.
Hamiltonicity in (0-1)-polyhedra.

PALU 1985-1 Palubetskis, G. S.
Subgraph polytopes (Russian).

SCAR 1985-1 Scarf, H. E.
Integral polyhedra in three-space.

SCHRE 1986-2 Schrijver, A.
Geometric methods in discrete optimization.

SCHRE 1986-3 Schrijver, A.
Polyhedral combinatorics - some recent developments and results.

SCHRE 1986-4 Schrijver, A.
Polyhedral proof methods in combinatorial optimization.

VESEA 1981-1 Veselov, S. I.; Shevchenko, V. N.
Faces and extreme points of discrete programming problems (Russian).

WERR 1984-1 Werra, D. de
A decomposition property of polyhedra

WERR 1986-2 Werra, D. de
Variations on the integral decomposition property.

ALIE	1983-1	Aliev, A. A. A method of solution of the integer linear programming problem (Russian).
ANEJ	1985-1	Aneja, Y. P.; Chandrasekaran, R.; Nair, K. P. K. Classes of linear programs with integral optimal solutions.
ARTE	1986-1	Artemenko, V. I.; Chernyakova, A. M. On the determination of the parameters of algorithms for solving integer linear programming problems (Russian).
AUSTA	1986-1	Austin, L. M.; Ruparel, B. C. The mixed cutting plane algorithm for all-integer programming.
AYUP	1984-1	Ayupov, R. S. A combinatorial continuous method of solving block problems of discrete programming.
BABAE	1985-1	Babai, L. On Lovasz' lattice reduction and the nearest lattice point problem.
BABAE	1986-2	Babai, L. On Lovasz' lattice reduction and the nearest lattice point problem.
BATU	1984-1	Batukhtin, V. D.; Maiboroda, L. A.; Tyulyukin, V. A. Optimization of funcitons of an integer-valued argument (Russian).
BEAL	1985-1	Beale, E. M. L. Integer Programming.
BUZY	1981-1	Buzytskij, P. L.; Freiman, G. A. Integer programming and number theory.
BYRD	1987-1	Byrd, R. H.; Goldman, A. J.; Heller, M. Recognizing unbounded integer programs.
CARRB	1986-1	Carraway, R. L.; Morin, T. L.; Moskowitz, H. Generalized dynamic programming for multi-criteria optimization.
CHALA	1986-1	Chalmet, L. G.; Lemonidis, L.; Elzinga, D. J. An algorithm for the bi-criterion integer programming problem.
CHANJ	1984-1	Chandrasekaran, R. Integer programming problems for which a simple rounding-type algorithm works.
CONL	1984-1	Conley, W. Satisfying the customer.
COOKD	1986-1	Cook, W.; Gerards, A. M. H.; Schrijver, A.; Tardos, E. Sensitivity theorems in integer linear programming.
DEL	1985-1	Del Ojo Mesa, J.; Fernandez Garcia, F. R.; Pelegrin, B. A global otpimization algorithm for integer programming (Spanish).
FABI	1985-1	Fabian, C. Economic pre-optimizing algorithms for integer programming.
FEIR	1986-1	Feiring, B. R. A parametric programming procedure for solving pure integer programming problems.
GHAN	1985-1	Ghandforoush, P.; Austin, L. M. Solution approaches for highly primal- and dual-degenerate all-integer programming problems.
GLOV	1986-1	Glover, F. Future paths for integer programming and links to artificial intelligence.
IKURB	1985-1	Ikura, Y.; Nemhauser, G. L. Simplex pivots on the set packing polytope.
JERO	1984-1	Jeromin, B.; Körner, F. A hybrid method for solving integer linear programming problems.
JERO	1986-1	Jeromin, B.; Körner, F. A hybrid method for solving integer linear programming problems.
JEROB	1984-1	Jeroslow, R. G.; Lowe, J. K. Modelling with integer variables.
KABE	1985-1	Kabe, D. G. On discrete linear and quadratic programming problems.
KABE	1986-2	Kabe, D. G. On solving linear and quadratic integer programming problems.
KABE	1986-3	Kabe, D. G. On solving two-variables programming problems.
KANN	1987-1	Kannan, R. Algorithmic geometry of numbers.
KANNA	1986-1	Kannan, R.; Lovasz, L. Covering minima and lattice point free convex bodies.
KHOK	1982-1	Khokhlyuk, V. I. Algorithms of integer optimization (Russian).
KHOK	1987-1	Khokhlyuk, V. I. Parallel algorithms for integer optimization (Russian).
KIVI	1984-1	Kivistik, L. A generalized rule for selection of a generating row for Gomory's all-integer algorithm (Russian).
KIVI	1985-1	Kivistik, L. Acceleration of Gomory's all-integer algorithm (Russian).
KOLO	1983-1	Kolokolov, A. A. Lower bound for the number of iterations for a class of truncation algorithms (Russian).
LEBE	1986-1	Lebedev, S. S.; Medvedovskaya, I. T. A generalized method of marks for a problem of integer linear programming.
LEDE	1980-1	Ledermann, W.; Vajda, S. Handbook of applicable mathematics. Vol. 1: Algebra.
LUPS	1986-1	Lupsa, L. An integer theorem of Farkas lemma type.
MINA	1985-1	Minaev, Y. N. On a formally heuristic algorithm for solving integer optimization problems (linear and quadratic models) (Russian).
NAKE	1985-1	Nakechbandi, M.; Trehel, M. Systolic array for integer and linear programming.

NGUE 1986-1 Nguen Ngoc Tu
Algorithm for maximizing a linear function on the set of integral points of a convex polyhedron.
NOVI 1983-1 Novikov, N. A.
An algorithm for solving an all-integer problem of linear programming (Russian).
NYGR 1987-1 Nygreen, B.
A possible way to reduce degeneracy in integer programming computations.
OGUZ 1985-1 Oguz, O.
An integral transformation for integer programming problems.
PAVL 1985-1 Pavlov, A. A.
Analysis of complexity estimates of exact algorithms for solving the general linear integer programming problem.
PAVLA 1985-1 Pavlov, A. A.; Gershgorin, A. E.
Reducing a general linear integer programming problem to a knapsack problem.
RUBI 1985-1 Rubin, D. S.
Polynomial algorithms for m*(m+1) integer programs and m*(m+k) diophantine systems.
RUDNA 1980-1 Rudneva, T. L.
A modification of Balas' algorithm for a class of integer programming problems. (Russian).
RUDNB 1981-1 Rudneva, T. L.; Yaranov, M. G.
Program realization of the completely integer Gomory algorithm (Russian).
SCAR 1985-1 Scarf, H. E.
Integral polyhedra in three-space.
SCHNB 1987-1 Schnorr, C. P.
Geometry of numbers and integer programming (summary).
SCHO 1982-1 Schoch, M.
Zur Lösung einer speziellen linearen diskreten Optimierungsaufgabe mit Zusatzforderungen.
SCHRE 1986-5 Schrijver, A.
Theory of linear and integer programming.
SEEL 1984-1 Seeländer, J.
Einige Bemerkungen zur Lösung von rein ganzzahlingen linearen Optimierungsproblemen.
SEEL 1984-2 Seeländer, J.
Über ein Kriterium zur Lösbarkeit reinganzzahliger linearer Optimierungsprobleme.
SERGA 1985-1 Sergienko, I. V.
Mathematical models and methods for solving problems of discrete optimization.
SERGD 1983-1 Sergienko, I. V.; Filonenko, N. V.
Investigation of the stability of the vector of the right-hand sides of a system of constraints for an integer linear programming problem (Russian).
SERGG 1980-1 Sergienko, I. V.; Lebedeva, T. T.; Roshchin, V. A.
Approximate methods for solving discrete problems of optimization.
SHCH 1986-3 Shcherbina, O. A.
On local algorithms with variable neighbourhoods and indicator information (Russian).
SHCHA 1983-1 Shcherbina, O. A.; Matveev, V. V.
On the efficiency of local algorithms for the solution of quasi-block problems of integer linear programming (Russian).
SHEVA 1979-1 Shevchenko, V. N.
Convex polyhedral cones, comparison systems and regular truncations in integer programming (Russian).
SITN 1984-1 Sitnikova, O. D.
Calculation of integer linear programming problems by the method of consecutive investigation of the level planes using filter-limitation (Russian).
SKAL 1983-1 Skaletskii, V. V.
An algorithm for solution of a problem of integer programming (Russian).
SOLO 1983-1 Sologashvili, V. G.; Dularidze, N. G.
An algorithm for solution of a problem of integer linear programming (Russian).
STRZ 1985-1 Strzelec, K. R.; Gadecki, H.
Exact or approximate algorithms? A combinatorial approach to integer programming.
SUNA 1985-1 Sunaga, T.; Hayakawa, A. J.; Maruki, Y.
A computational method for all-integer interval linear programming.
SURO 1986-1 Surovtsov, L. K.; Vilkov, V. B.
An approach to the solution of linear problems of integer programming (Russian).
VASIA 1985-1 Vasilev, V. S.
Problems of discrete programming and algorithms for their solution (Bulgarian).
VOLK 1986-1 Volkovich, O. V.
Determination of an admissible solution for the problem of integer linear programming (Russian).
VOLK 1987-1 Volkovich, O. V.
An algorithm for finding a feasible solution of the integral linear programming (Russian).
VOLK 1987-2 Volkovich, O. V.
Approximation of the admissible domain in a problem of integer programming (Russian).
WEID 1986-1 Weidner, H. G.
Ein Periodizitätssatz der ganzzahligen Optimierung.
WHITE 1985-1 White, P. M.; Caplin, A. S.; Heyden, L. van der
Scarf's procedure for integer programming and a dual simplex algorithm.
WILLA 1985-1 Williams, H. P.
Model building in linear and integer programming.
XIA 1985-1 Xia, D.-L.
A new method of integer linear programming - branch direction search method.
ZHENA 1984-1 Zheng, Q.
Global optimization of integer and mixed programming (Chinese).

AITT	1985-1	Aittoniemi, L.; Oehlandt, K. A note on the Martello-Toth algorithm for the one-dimensional knapsack problems.
ALEKC	1985-1	Alekseev, A. O.; Alekseev, O. G.; Anisimov, V. G.; Anisimov, E. G. The use of duality to increase the effectiveness of the branch and bound method when solving the knapsack problem.
ALIEA	1986-1	Aliev, A. A.; Akhmedov, F. B.; Babaev, D. A. Application package KNAPSACK-1 for approximate solution of the generalized problem of knapsack (Russian).
ANDOC	1987-1	Andonov, R. A. Finding facets of the knapsack polytope which cut off a given point.
ATKIA	1987-1	Atkinson, M. D.; Negro, A.; Santoro, N. Sums of lexicographically ordered sets.
BABAG	1984-1	Babayev, D. A.; Babayev, R. D.; Jysifov, M. M. General multiple-choice knapsack problem.
BALAJ	1987-1	Balas, E.; Nauss, R.; Zemel, E. Comment on:'Some computational results on real 0-1 knapsack problems'.
BRIC	1983-1	Brickell, E. F. Are most low density knapsacks solvable in polynomial time.
BRIC	1984-1	Brickell, E. F. Solving low density knapsacks.
BRUC	1984-2	Brucker, P. An o(n) algorithm for quadratic knapsack problems.
CAMEE	1984-1	Camerini, P. M.; Vercellis, C. The matroidal knapsack: a class of (often) well-solvable problems.
DEMP	1984-1	Dempe, S. Gradient algorithms for solving a generalized knapsack problem (Russian).
DUDZ	1984-1	Dudzinski, K. A dynamic programming approach to solving the multiple choice knapsack problem.
DUDZA	1984-1	Dudzinski, K.; Walukiewicz, S. A fast algorithm for the linear multiple-choice knapsack problem.
DUDZA	1984-2	Dudzinski, K.; Walukiewicz, S. An algorithm for multiple choice knapsack problem.
DUDZA	1985-1	Dudzinski, K.; Walukiewicz, S. On the multiperiod binary knapsack problem.
DUDZA	1985-2	Dudzinski, K.; Walukiewicz, S. On the multiperiod binary knapsack problem.
DUDZA	1986-1	Dudzinski, K.; Walukiewicz, S. A multiperiod binary knapsack problem (Polish).
DYERD	1984-1	Dyer, M. E.; Kayal, N.; Walker, J. A branch and bound algorithm for solving the multiple-choice knapsack problem.
EISEA	1986-1	Eiselt, H. A. Continuous maximin knapsack problems with GLB constraints.
FEIR	1987-1	Feiring, B. R. A knapsack procedure for solving integer programming problems.
FISC	1987-1	Fischetti, M. A new linear storage, polynomial time approximation scheme for the subset-sum problem.
FISCA	1987-1	Fischetti, M.; Toth, P. Bounding and dominating procedure for combinatorial optimization problems (Italian).
FURU	1984-1	Furukawa, M. The 1-3-dimensional knapsack problem in production planning: formulation and heuristic methods (Japanese).
GALLA	1987-1	Gallo, G.; Simeone, B. On the supermodular knapsack problem.
GAVI	1985-1	Gavish, B.; Pirkul, H. Efficient algorithms for solving multiconstraint zero-one knapsack problems to optimality.
GRANA	1984-1	Granot, D.; Granot, F. Minimal covers, minimal sets and canonical facets of the posynomial knapsack polytope.
GREEA	1985-1	Greenberg, H. An algorithm for the periodic solutions in the knapsack problem.
GREEA	1986-1	Greenberg, H. On equivalent knapsack problems.
HAYA	1985-1	Hayashi, Y. Constuction of the F-, P- and K-trees of a knapsack problem and their computational experiments.
HAYA	1986-1	Hayashi, Y. A solution method for a knapsack problem and its variant.
ISHIA	1984-1	Ishii, H.; Nishida, T. Stochastic linear knapsack problem : probability maximization model.
IVANB	1984-1	Ivanchev, M. D.; Janev, N. I.; Mitev, J. G. A knapsack approach to a given class of distribution problems.
JENK	1987-1	Jenkins, L. Parametric-objective integer programming using knapsack facets and Gomory cutting planes.
KARNA	1984-1	Karnin, E. D. A parallel algorithm for the knapsack problem.
KAUF	1985-1	Kaufman, L.; Tubeeckx, S.; Plastria, F. The zero-one knapsack problem with equality constraint.
KORN	1986-1	Körner, F. A new bound for the quadratic knap sack problem and its use in a branch and bound algorithm.
LEE K	1985-1	Lee, K. T. Knapsack problems solved by dynamic programming.
LEE N	1986-1	Lee, Y.-D. Integer programming applications and their relationship (Korean).
LENGA	1986-1	Lengyel, I. A practical generalization of the knapsack problem for cutting stock (Hungarian).

LIEP 1983-1 Lieponite-Yukhimenko, B. I.
Effectiveness of preference functions in integer linear programming algorithms (Russian).

MA A 1986-1 Ma, J.-L.
An optimum solution for a knapsack problem (Chinese).

MA C 1985-1 Ma, S.-H.
Solvability of a class of knapsack problems (Chinese).

MACH 1987-1 Machida, H.
A remark on the complexity of the knapsack problem.

MARCC 1987-1 Marchetti-Spaccamela, A.; Vercellis, C.
Efficient on-line algorithms for the knapsack problem.

MARTB 1985-1 Martello, S.; Toth, P.
Algorithm 632. A program for the 0-1 multiple knapsack problem.

MARTB 1987-1 Martello, S.; Toth, P.
Algorithms for knapsack problems.

MATHA 1986-1 Mathur, K.; Salkin, H. M.; Bohanty, B. B.
A note on a general non-linear knapsack problem.

MATHB 1986-1 Mathur, K.; Salkin, H. M.; Morito, S.
An efficient algorithm for the general multiple-choice knapsack problem (GMKP).

MEAN 1984-1 Meanti, M.; Rinnooy Kan, A. H. G.; Stougie, L.; Vercellis, C.
A probabilistic analysis of the multi-knapsack value function.

MEAN 1987-1 Meanti, M.; Rinnooy Kan, A. H. G.; Stougie, L.; Vercellis, C.
A probabilistic analysis of the multi-knapsack value function.

MELC 1986-1 Melching, C. S.; Liebman, J. S.
Solving binary knapsack problems with precedence constraints.

MILA 1984-1 Milanov, P. B.
A polynomial time algorithm for a large subclass of the integer knapsack problem.

MILAA 1987-1 Milanov, P. B.; Brimkov, V. E.
On the knapsack problem.

MINOC 1984-1 Minoux, M.; Ribeiro, C. C.
A transformation of hard (equality constrained) knapsack problems into constrained shortest path problems.

MURP 1986-1 Murphy, R. A.
Some computational results on real 0-1 knapsack problems.

NURI 1983-1 Nuriev, U. G.
A hybrid method for solving the multidimensional knapsack problem (Russian).

NURI 1986-1 Nuriev, U. G.
On the solution of the knapsack problem with guaranteed estimates (Russian).

ORLI 1985-4 Orlin, J. B.
Some very easy knapsack/partition problems.

PAVLA 1985-1 Pavlov, A. A.; Gershgorin, A. E.
Reducing a general linear integer programming problem to a knapsack problem.

PETEC 1987-1 Peters, J. G.; Rudolph, L.
Parallel approximation schemes for subset sum and knapsack problems.

PIEH 1985-1 Piehler, J.
Über die irreduziblen Punkte des Eckenpolyeders.

PIRK 1987-1 Pirkul, H.
A heuristic solution procedure for the multiconstraint zero-one knapsack problem.

PIRKA 1986-1 Pirkul, H.; Narasimhan, S.
Efficient algorithms for the multiconstraint general knapsack problem.

PLAT 1985-1 Plateau, G.; Elkihel, M.
A hybrid method for the 0-1 knapsack problem.

ROUS 1985-1 Rousseau, C. K.; Rousseau, C. C.; Liu, A. L.; Klamkin, M. S.
Greedy algorithm and coinage systems.

SCHRF 1984-1 Schröter, N.
Average-case analysis of an approximative algorithm for the knapsack problem.

SHIO 1987-1 Shiode, S.; Ishii, H.; Nishida, T.
A stochastic linear knapsack problem.

STOU 1988-1 Stougie, L.
The stochastic knapsack problem.

SZKA 1987-1 Szkatula, K.; Libura, M.
On probabilistic properties of greedy-like algorithms for the binary knapsack problem.

TIGA 1984-1 Tigan, S.
On some nonlinear knapsack problems.

VIZV 1984-1 Vizvari, B.
On the connection of the Frobenius problem and the knapsack problem.

VOLGD 1986-1 Volgenant, A.; Zoon, J. A.
An improved heuristic for multidimensional 0-1 knapsack problems (Dutch).

VOLOB 1986-1 Voloshin, A. F.; Mashchenko, S. O.
A subexponential algorithm for solving the knapsack problem (Russian).

WALU 1987-1 Walukiewicz, S.; Dudzinski, K.
Exact methods for the knapsack problem and its generalizations.

YANA 1987-1 Yanasse, H. H.; Soma, N. Y.
A new enumeration scheme for the knapsack problem.

YANGA 1985-1 Yang, C. E.; Lin, Y. X.
Board welding problem (Chinese).

ZEME 1986-1 Zemel, E.
On the computational complexity of facets of the knapsack problem.

ZHANN 1986-1 Zhang, L. A.; Geng, S. Y.
The complexity of the 0/1 multi-knapsack problem.

ZHANN 1987-1 Zhang, L. A.; Geng, S. Y.
Computation of multiknapsack problems (Chinese).

ZIMMC 1986-3 Zimmermann, U.
Sharing problems.

ABAI	1987-1	Abaiyldanov, K. N.; Astakhov, N. D. Problem of determining the optimal order of priority in setting up production centers.
ABDEA	1985-1	Abdel-Malek, L. L. Optimum positioning of moving service facility.
ABDUA	1985-1	Abdullaev, A. K.; Kiselev, V. G. A method for solving the Benders integer subproblem in some problems of plant location (Russian).
AIKE	1985-1	Aikens, C. H. Facility location models for distribution planning.
AKIN	1985-1	Akinc, U. Multi-activity facility design and location problems.
ANDRD	1986-1	Andreatta, G.; Mason, F. Review of recent results about the k-centrum of a tree.
ANDRE	1985-1	Andreatta, G.; Romanin-Jacur, G. An integer quadratic allocation problem.
ANTHA	1988-1	Anthonisse, K.; Lageweg, B. Aircraft-stand allocation at Schiphol airport: a decision support system.
ANTHA	1988-2	Anthonisse, K.; Lageweg, B. Aircraft-stand allocation at Schiphol airport: an optimization procedure.
ARAU	1984-1	Araujo, E. O.; Franca, P. M.; Soares, S.; Tavares, H. M. F. Optimal location of telephone exchanges.
ASTA	1983-1	Astakhov, N. D. On the dynamical problem of positioning factories with warehouses of production inventory.
ASTA	1986-1	Astakhov, N. D. The production-transportation problem of optimal plant location with startup priorities.
ASTAA	1985-1	Astakhov, N. D.; Kaliev, E. Z. The problem of optimal distribution of points for processing of an agricultural product and its solution by the approximation-combinatorial method (Russian).
AVRA	1981-1	Avram-Nitchi, R. An application of graph theory to the time-criterion allocation problem (Romanian).
AYKI	1987-1	Aykin, T.; Babu, A. J. G. Constrained large-region multifacility location problems.
AYKI	1987-2	Aykin, T.; Babu, A. J. G. Multifacility location problems on a sphere.
BACH	1988-1	Bach, L. Regionale Standortplanung für medizinisch-technische Grossgeräte.
BACHF	1985-1	Bachmann, P. Ein Beitrag zur optimalen Standortfindung von Teilen im Arbeitsbereich eines Montageroboters.
BAKEA	1986-1	Baker, B. M. A partial dual algorithm for the capacitated warehouse location problem.
BAKH	1981-1	Bakhtin, A. E. A modification of the algorithm for the solution of location problems (Russian).
BALAC	1987-1	Balakrishnan, A.; Ward, J. E.; Wong, R. T. Integrated facility location and vehicle routing models: recent work and future prospects.
BAND	1985-1	Bandelt, H. J. Networks with Condorcet solutions.
BANDA	1986-1	Bandelt, H. J.; Labbe, M. How bad can a voting locating be.
BART	1984-1	Bartholdi, J. J. III; Platzman, L. K. Heuristics based on spacesfilling curves for combinatorial problems in the plane.
BATT	1987-1	Batta, R. Comment on "The dynamics of plant layout" by M. J. Rosenblatt.
BAXT	1984-1	Baxter, J. Depot location: a technique for the avoidance of local optima.
BEAS	1985-1	Beasley, J. E. A note on solving large p-median problems.
BECKF	1986-1	Beckmann, M. J.; Thisse, J.-F. The location of production activities.
BERMC	1987-1	Berman, O.; Larson, R. C.; Parkan, C. The stochastic queue p-median problem.
BERMD	1984-1	Berman, O.; LeBlanc, B. Location-relocation of mobile facilities on a stochastic network.
BERME	1986-1	Berman, O.; Mandowsky, R. R. Location-allocation on congested networks.
BERMF	1985-1	Berman, O.; Rahnama, M. R. Optimal location-relocation decisions on stochastic networks.
BEUL	1988-1	Beulens, A. J. M.; Kolen, A. W. J.; Niekamp, E. G. A decision support system for a location-allocation-routing problem.
BIEN	1983-1	Bienert, K. Modell- und computergestützte Planung logistischer Systeme.
BIRG	1985-1	Birge, J. R.; Malyshko, V. V. Methods for a network design problem in solar power systems.
BOEN	1984-1	Boenchendorf, K. Identifying redundant facilities in the capacitated facility location problem.
BORIA	1985-1	Borisova, L. N. The telephone center and the plant centroid (Russian).
BRAN	1986-1	Brandeau, M. L.; Chiu, S. S.; Batta, R. Locating the 2-median of a tree network with continuous link demands.
BUCKA	1987-1	Buckley, F. Facility location problems.
BUHL	1988-1	Buhl, H. U. Zielsetzungen bei der Modellierung von Standortproblemen.
CALA	1987-1	Calamai, P. H.; Conn, A. R. A projected Newton method for l(p) norm location problems.

CAME 1986-1 Camerini, P. M.; Colorni, A.; Maffioli, F.
Some experience in applying a stochastic method to location problems.
CAPT 1985-1 Captivo, M. E.; Branco, I. M.; Coelho, J. D.
An exact algorithm for the problem of the p-median (Portuguese).
CAPT 1986-1 Captivo, M. E.; Branco, I. M.; Coelho, J. D.
Errata: "An exact algorithm for the problem of the p-median" (Portuguese).
CAVAA 1985-1 Cavalier, T. M.; Sherali, H. D.
Sequential location-allocation problems on chains and trees with probabilistic link demands.
CAVAA 1986-1 Cavalier, T. M.; Sherali, H. D.
Euclidean distance location-allocation problems with uniform demands over convex polygons.
CAVAA 1986-2 Cavalier, T. M.; Sherali, H. D.
Network location problems with continuous link demands: p-medians on a chain and 2-medians on a tree.
CERNA 1987-1 Cerny, J.; Guldan, F.
Location of polygon vertices on circles and its application in transport studies.
CHAR 1985-1 Charalambous, C.
Acceleration of the HAP approach for the multifacility location problem.
CHAU 1985-1 Chaudhry, S. S.; Mac Cormick, S. T.; Moon, I. D.
Locating independent facilities with maximum weight: greedy heuristics.
CHENF 1984-1 Chen, R.
Location problems with costs being sums of powers of Euclidean distances.
CHENG 1987-1 Chen, R.; Handler, G. Y.
Relaxation method for the solution of the minimax location-allocation problem in Euclidean space.
CHIU 1987-1 Chiu, S. S.
The minisum location problem on an undirected network with continuous link demands.
CHIUA 1985-1 Chiu, S. S.; Berman, O.; Larson, R. C.
Locating a mobile server queueing facility on a tree network.
CHIUB 1985-1 Chiu, S. S.; Larson, R. C.
Locating an n-server facility in a stochastic environment.
CHUN 1986-1 Chung, C.-H.
Recent applications of the maximal covering location planning (M.C.L.P.) model.
CHUR 1986-1 Church, R. L.; Weaver, J. R.
Theoretical links between median and coverage location problems.
COOP 1987-1 Cooper, M. W.
Modelling nonlinearities in multicriteria locational decisions.
CURRC 1987-1 Current, J. R.; Schilling, D. A.
Elimination of source A and B errors in p-median location problems.
CZUC 1986-1 Czuchra, W.
A graphical method to solve a maximin allocation problem.
DADU 1986-1 Daduna, J. R.
Solving a p-center transportation problem.
DAHLB 1986-1 Dählmann, A; Schütte, H.
Optimale Standorte und Abstimmungsparadoxa auf Graphen.
DARZ 1987-1 Darzentas, J.
On fuzzy location models.
DARZA 1986-1 Darzentas, J.; Bairaktaris, P.
On the set partitioning type formulation for the discrete location problem.
DAS A 1985-1 Das, C.; Verma, A.
A heuristic method of finding the optimal location and size of facilities.
DAVYA 1985-1 Davydov, G. V.; Davydova, I. M.
A duality criterion for optimality in allocation problems.
DAVYB 1985-1 Davydova, I. M.
Enumeration schemes in location problems (Russian).
DAX 1986-1 Dax, A.
A note on optimality conditions for the euclidean multifacility location problem.
DAX 1986-2 Dax, A.
An efficient algorithm for solving the rectilinear multifacility location problem.
DEAK 1984-1 Deak, I.; Hoffer, J.; Mayer, J.; Nemeth, A.; Potecz, B.; Prekopa, A.; Strazicky, B.
Optimal, short-distance plant scheduling for electrical energy systems based on thermal plants, taking conditions of network locations into consideration (Hungarian).
DEAR 1985-1 Dearing, P. M.
Location problems.
DHARA 1986-1 Dhar, U. R.
Spherical location under restricted distance.
DHARB 1981-1 Dhar, U. R.; Rao, J. R.
Large region location problems under distance constraints.
DOBSA 1987-1 Dobson, G.; Karmarkar, U. S.
Competitive location on a network.
DOMSA 1985-1 Domschke, W.; Drexl, A.
ADD- and DROP-procedures for capacitated plant location models with different capacities.
DOMSA 1985-2 Domschke, W.; Drexl, A.
Location and layout planning. An international bibliography.
DOSS 1984-1 Doss, H.
Bayesian estimation in the symmetric location problem.
DOWL 1986-1 Dowling, P. D.; Love, R. F.
Bounding methods for facilities location algorithms.
DOWL 1987-1 Dowling, P. D.; Love, R. F.
An evaluation of the dual as a lower bound in facilities location problems.
DREZ 1984-1 Drezner, Z.
The planar two-center and two-median problems.
DREZ 1985-1 Drezner, Z.
A solution to the Weber location problem on the sphere.

DREZ	1985-2	Drezner, Z. O (N log N) algorithm for the rectilinear round-trip location problem.
DREZ	1985-3	Drezner, Z. Sensitivity analysis of the optimal location of a facility.
DREZ	1986-1	Drezner, Z. Location of regional facilities.
DREZ	1986-2	Drezner, Z. The p-cover problem.
DREZ	1987-1	Drezner, Z. A heuristic procedure for the layout of a large number of facilities.
DREZ	1987-2	Drezner, Z. Heuristic solution methods for two location problems with unreliable facilities.
DREZ	1987-3	Drezner, Z. On the rectangular p-center problem.
DREZB	1985-1	Drezner, Z.; Gavish, B. Epsilon-approximations for multi-dimensional weighted location problems.
DREZD	1987-1	Drezner, Z.; Shelah, S. On the complexity of the Elzinga-Hearn algorithm for the 1-center problem.
DREZE	1985-1	Drezner, Z.; Steiner, G.; Wesolowsky, G. O. One-facility location with rectilinear tour distances.
DREZF	1985-1	Drezner, Z.; Wesolowsky, G. O. Layout of facilities with some fixed points.
DREZF	1985-2	Drezner, Z.; Wesolowsky, G. O. Location of multiple obnoxious facilities.
DURI	1985-1	Durier, R.; Michelot, C. Geometrical properties of the Fermat-Weber problem.
DYER	1986-1	Dyer, M. E. On a multidimensional search technique and its application to the Euclidean one-centre problem.
DYERB	1985-1	Dyer, M. E.; Frieze, A. M. A simple heuristic for the p-center problem.
EDAH	1984-1	Edahiro, M.; Kokubo, I.; Asano, T. A new-point location algorithm and its practical efficiency - comparison with existing algorithms.
EISEA	1985-1	Eiselt, H. A. A location problem with minimax objective.
EISEB	1986-1	Eiselt, H. A.; Charlesworth, G. A note on p-center problems in the plane.
ERKU	1985-1	Erkut, E.; Francis, R. L.; Lowe, T. J. Locating two facilities on a tree subject to a distance constraint.
ERKU	1987-1	Erkut, E.; Francis, R. L.; Lowe, T. J. A multimedian problem with interdistance constraints.
ERLE	1985-1	Erlenkotter, D.; Leonardi, G. Facility location with spatially interactive travel behavior.
FORTA	1986-1	Fortenberry, J. C.; Mitra, A. A multiple criteria approach to the location-allocation problem.
FORTB	1985-1	Fortenberry, J. C.; Snyder, C. A.; Cox, J. F. A multiple criteria decision support model for facilities layout.
FOUL	1985-1	Foulds, L. R. Enumeration of graph theoretic solutions for facilities layout.
FOULA	1985-1	Foulds, L. R.; Gibbons, P. B.; Giffin, J. W. Facilities layout adjacency determination: An experimental comparison of three graph theoretic heuristics.
FOULB	1985-1	Foulds, L. R.; Giffin, J. W. A graph-theoretic heuristic for minimizing total transport cost in facilities layout.
FOULB	1985-2	Foulds, L. R.; Giffin, J. W. Facilities layout via new graph theoretic methods.
FOULC	1986-1	Foulds, L. R.; Giffin, J. W.; Evans, J. D. Towards a new decision support system for facilities layout planners.
FRIC	1985-1	Frick, H. On the solution of the facility location problem when the dominance method fails.
FUJIB	1986-1	Fujishige, S.; Katoh, N.; Ichimori, T. The fair resource allocation problem with submodular constraints and its solution algorithm.
GARCA	1985-1	Garcia, R. An optimization method for multiplexer locations.
GARR	1986-1	Garrett, S. E. A microcomputer implementation of a facility location model.
GELD	1987-1	Gelders, L. F.; Pintelon, L.; Wassenhove, L. N. van A location-allocation problem in a large Belgian brewery.
GERT	1986-1	Gerth, C. Duality properties of the vector-valued location problem (German).
GHANA	1986-1	Ghandforoush, P.; Greber, B. J. Solving allocation and scheduling problems inherent in forest resource management using mixed-integer programming.
GIFF	1987-1	Giffin, J. W.; Foulds, L. R. Facilities layout generalized model solved by n-boundary shortest path heuristics.
GIMA	1983-1	Gimadi, E. K. An effective algorithm for the solution of the location problem with service domains connected with respect to an acyclic network (Russian).
GOLDJ	1986-1	Golden, B. L.; Skiscim, C. C. Using simulated annealing to solve routing and location problems.
GOLDM	1984-1	Goldengorin, B. I. A correcting algorithm for solving allocation type problems.
GOLDM	1986-1	Goldengorin, B. I. Decomposition of the location problem.

GONS 1987-1 Gonsalvez, D. J.; Hall, N. G.; Rhee, W. T.; Siferd, S. P.
Heuristic solutions and confidence intervals for the multicovering problem.
GORU 1985-1 Gorunovich, S. A.
An algorithm for finding p-medians of a graph (Russian).
GRAN 1987-1 Granot, D.
The role of cost allocation in locational models.
GROB 1987-1 Grobelny, J.
On one possible 'fuzzy' approach to facilities layout problems.
GROB 1987-2 Grobelny, J.
The fuzzy approach to facilities layout problems.
GROG 1985-1 Gröger, M.
Location problems in graphs.
GUIG 1984-1 Guignard, M.
A Lagrangean dual ascent algorithm for simple plant location problems.
GUIG 1984-2 Guignard, M.
Dual and primal heuristics for network location problems.
GUIGA 1983-1 Guignard, M.; Kim, S.
A strong Lagrangean relaxation for capacitated plant location problems.
GUIGB 1984-1 Guignard, M.; Opaswongkarn, K.
Langrangean dual ascent algorithms for the capacitated plant location problem.
HAKI 1986-1 Hakimi, S. L.
P-median theorems for competitive locations.
HALLE 1985-1 Hall, R. W.
Heuristics for selecting facility locations.
HAND 1984-1 Handler, G. Y.
The absolute m-center problem on a directed network.
HAND 1985-1 Handler, G. Y.
Medi-centers of a tree.
HANJ 1985-1 Hanjoul, P.; Peeters, D.
Comparison of two dual-based procedures for solving the p-median problem.
HANSE 1987-1 Hansen, P.; Labbe, M.; Peeters, D.; Thisse, J.-F.
Single facility location on networks.
HANSH 1985-1 Hansen, P.; Peeters, D.; Richard, D.; Thisse, J.-F.
The minisum and minimax locations problems revisited.
HANSI 1985-1 Hansen, P.; Peeters, D.; Thisse, J.-F.
Facility location analysis.
HANSK 1985-1 Hansen, P.; Thisse, J.-F.; Wendell, R. E.
Location by competitive and voting processes.
HANSK 1986-1 Hansen, P.; Thisse, J.-F.; Wendell, R. E.
Equivalence of solutions to network location problems.
HOCH 1984-1 Hochbaum, D. S.
When are NP-hard location problems easy?
HOCH 1985-1 Hochbaum, D. S.
Easy solutions for the K-center problem or the dominating set problem on random graphs.
HOCHB 1985-1 Hochbaum, D. S.; Shmoys, D. B.
A best possible heuristic for the k-center problem.
HODD 1984-1 Hodder, J. E.
Financial market approaches to facility location under uncertainty.
HODDA 1986-1 Hodder, J. E.; Dincer, M. C.
A multifactor model for international plant location and financing under uncertainty.
HODDB 1985-1 Hodder, J. E.; Jucker, J. V.
A simple plant-location model for quantity-setting firms subject to price uncertainty.
HODG 1986-1 Hodgson, M. J.
A hierarchical location-allocation model with allocations based on facility size.
HODGA 1987-1 Hodgson, M. J.; Wong, R. T.; Honsaker, J.
The P-centroid problem on an inclined plane.
HOLZ 1986-1 Holzman, R.
An axiomatic approach to location on networks.
HOOK 1986-1 Hooker, J.
Solving nonlinear single-facility network location problems.
HOSA 1986-1 Hosage, C. M.; Goodchild, M. F.
Discrete space location-allocation solutions from genetic algorithms.
HWAN 1985-1 Hwang, C.-L.
Manufacturing plant location analysis by multiple attribute decision making: part II - multi-plant strategy and plant relocation.
ICHI 1985-1 Ichimori, T.
A shortest path approach to a multifacility minimax location problem with rectilinear distances.
ICHIA 1985-1 Ichimori, T.; Nishida, T.
Note on a rectilinear distance round-trip location problem.
IRI A 1984-1 Iri, M.; Murota, K.; Ohya, T.
A fast Voronoi-diagram algorithm with applications to geographical optimization problems.
IYER 1987-1 Iyer, A. V.; Ratliff, H. D.
Location issues in guaranteed time distribution systems.
JASI 1984-1 Jasinska, E.; Woytych, E.
Location of depots in a sugar-beet distribution system (case study).
JOHA 1986-1 Johansson, B.; Leonardi, G.
Public facility location: a multiregional and multiauthority decision context.
JUEL 1985-1 Juel, H.; Love, R. F.
The facility location problem for hyper-rectilinear distances.
JUEL 1986-1 Juel, H.; Love, R. F.
A geometrical interpretation of the existing facility solution condition for the Weber problem.
JUEL 1987-1 Juel, H.; Love, R. F.
Duality in constrained location problems.

KANTB 1983-1 Kantorovich, E. G.
Investigation of a model of the location of service in an urban system (Russian).

KARK 1986-1 Karkazis, J.; Karagiorgis, P.
A method to locate the maximum circle(s) inscribed in a polygon.

KATH 1987-1 Kathawala, Y.; Gholamnezhad, H.
New approach to facility locations decisions.

KATO 1985-1 Katoh, N.; Ibaraki, T.; Mine, H.
An algorithm for the equipollent resource allocation problem.

KHACA 1986-1 Khachaturov, V. R.
Models and methods for solving multiextremal allocation problems by the use of properties of supermodular functions defined on Boolean lattices (Russian).

KHAN 1986-1 Khan, A. M.
Solid-waste disposal with intermediate transfer stations: an application of the fixed-charge location problem.

KIM G 1986-1 Kim, Y.
Computational study on the simple plant location problem: variations of the Benders decomposition method.

KINC 1987-1 Kincaid, R. K.
Finding two-centers of graphs that are almost trees.

KINCA 1987-1 Kincaid, R. K.; Lowe, T. J.; Morin, T. L.
The location of central structures in trees.

KINCB 1987-1 Kincaid, R. K.; Maimon, O.
The variance measure for location problems on 3-cacti.

KISE 1985-1 Kiseleva, E. M.; Shor, N. Z.
Calculation algorithm for multiproduct problem of optimal division with constraints (Russian).

KLIN 1986-1 Klincewicz, J. G.; Luss, H.
A Lagrangian relaxation heuristic for capacitated facility location with single-source constraints.

KLIN 1987-1 Klincewicz, J. G.; Luss, H.
A dual-based algorithm for multiproduct uncapacitated facility location.

KLINA 1986-1 Klincewicz, J. G.; Luss, H.; Rosenberg, E.
Optimal and heuristic algorithms for multiproduct uncapacitated facility location.

KLINF 1984-1 Klingman, D.; Ross, G. T.; Schneider, R. F.
Optimization-based facility location planning: an application and some insights.

KOHL 1983-1 Kohlberg, E.
Equilibrium store locations when consumers minimize travel time plus waiting time.

KOLE 1985-1 Kolen, A. W. J.
The round-trip p-center and covering problem on a tree.

KOLE 1986-1 Kolen, A. W. J.
Tree network and planar rectilinear location theory.

KORK 1987-1 Körkel, M.
On Karkazis' bounding procedure for the simple plant location problem.

KORK 1987-2 Körkel, M.
On the exact solution of large-scale simple plant location problems.

KRAR 1986-1 Krarup, J.; Pruzan, P. M.
Assessment of approximate algorithms: the error measure's crucial role.

KUBY 1987-1 Kuby, M. J.
Programming models for facility dispersion: the p-dispersion and maxisum-dispersion problems.

KULS 1987-1 Kulshrestha, D. K.
A mini-max location problem with demand points arbitrarily distributed in a compact connected space.

KUMAD 1987-1 Kumara, S. R. T.; Kashyap, R. L.; Moodie, C. L.
Expert system for industrial facilities layout planning and analysis.

KUSIB 1987-1 Kusiak, A.; Heragu, S. S.
The facility layout problem.

KWAK 1985-1 Kwak, N. K.; Schniederjans, M. J.
A goal programming model as an aid in facility location analysis.

KWAK 1985-2 Kwak, N. K.; Schniederjans, M. J.
A goal programming model for selecting a facility location site.

LABB 1985-1 Labbe, M.
Essays in network location theory.

LABB 1985-2 Labbe, M.
Outcomes of voting and planning in single facility location problems.

LABBA 1986-1 Labbe, M.; Laporte, G.
Maximizing user convenience and postal service efficiency in post box location.

LALO 1986-1 Lalov, P. I.; Vasileva, T. V.
Extremal problems of the triangular metric (Bulgarian).

LAPOE 1986-1 Laporte, G.; Nobert, Y.; Arpin, D.
An exact algorithm for solving a capacitated location-routing problem.

LAUN 1985-1 Laundy, R. S.
A tree search algorithm for the multi-commodity location problem.

LEE H 1986-1 Lee, D. T.; Wu, Y. F.
Geometric complexity of some location problems.

LEE N 1986-1 Lee, Y.-D.
Integer programming applications and their relationship (Korean).

LEUN 1986-1 Leung, J. M. Y.; Magnanti, T. L.
Valid inequalities and facets of the capacitated plant location problem.

LIN D 1984-1 Lin, Y. X.
The discrete location problems (Chinese).

LORI 1984-1 Loridan, P.
A dual approach to the generalized Weber problem under locational uncertainty.

LOUV 1986-1 Louveaux, F.
Discrete stochastic location models.

LOUVA 1985-1 Louveaux, F.; Thisse, J.-F.
Production and location on a network under demand uncertainty.

LOVEA 1985-1 Love, R. F.; Truscott, W. G.; Walker, J.
Terminal location problem: a case study supporting the status quo.

LU C 1987-1 Lü, W. J.
A new convergent algorithm for location problems of the type min(sum||A(i)x-b(i)||) (Chinese).

MAC I 1983-1 Mac Kay, B. D.; Tollis, I. G.
A resource location problem on graphs.

MAI 1984-1 Mai, C.-C.; Shieh, Y.-N.
Transport rates structure, optimum location, and theory of production: reexamination.

MAIM 1986-1 Maimon, O.
The variance equity measure in locational decision theory.

MAKR 1985-1 Makris, A.; Tziafetas, G.; Koutsopoulos, K.
Simultaneous location allocation of facilities and servers.

MARTI 1985-1 Martinec, I.
Methods for a class of discrete location problems.

MARTI 1987-1 Martinec, I.
An algorithm for the epsilon-optimal solution of a discrete location problem (Czech).

MATSB 1985-1 Matsutomi, T.; Ishii, H.; Nishida, T.
Stochastic location problem of emergency service facility.

MATSB 1987-1 Matsutomi, T.; Ishii, H.; Nishida, T.
Chance constrained round trip location problem.

MATTA 1986-1 Mattsson, L.-G.
Residential location and school planning in a tightening urban economy.

MAUR 1981-1 Maurer, H.
The post-office problem and related questions.

MAZE 1985-1 Maze, T. H.; Khasnabis, S.
Bus garage location planning with dynamic vehicle assignments: a methodology.

MEGI 1986-1 Megiddo, N.
Dynamic location problems.

MEGIA 1986-1 Megiddo, N.; Zemel, E.
An O(n log n) randomizing algorithm for the weighted Euclidean 1-center problem.

MEHRA 1985-1 Mehrez, A.; Sinuany-Stern, Z.; Stulman, A.
A single facility location problem with a weighted maximin-minimax rectilinear distance.

MEHRC 1984-1 Mehrez, A.; Stulman, A.
The facility location problem when the underlying distribution is either dominated or dominating.

MELA 1985-1 Melachrinoudis, E.
Determining an optimum location for an undesirable facility in a workroom environment.

MELAA 1982-1 Melachrinoudis, E.; Cullinane, T. P.
A maximin approach to the location of an undesirable facility in a nonconvex region.

MELAA 1985-1 Melachrinoudis, E.; Cullinane, T. P.
A heuristic approach to the single facility maximin location problem.

MELAA 1986-1 Melachrinoudis, E.; Cullinane, T. P.
Locating an obnoxious facility within a polygonal region.

MELAA 1986-2 Melachrinoudis, E.; Cullinane, T. P.
Locating an undesirable facility with a minimax criterion.

MESA 1985-1 Mesa Lopez-Colmenar, J. A.
Stochastic dynamic location with the minimax criterion (Spanish).

MICHB 1987-1 Michelot, C.
Localization in multifacility location theory

MICHB 1987-2 Michelot, C.
Properties of a class of location-allocation problems.

MIRC 1987-1 Mirchandani, P. B.
Generalized hierarchical facility locations.

MIRCA 1985-1 Mirchandani, P. B.; Oudjit, A.; Wong, R. T.
Multidimensional extensions and a nested dual approach for the m-median problem.

MIRCB 1986-1 Mirchandani, P. B.; Reilly, J. M.
"Spatial nodes" in discrete location problems.

MONMD 1986-1 Monma, C. L.; Schrijver, A.; Todd, M. J.; Wei, V. K.
A class of convex programs for resource allocation problems on directed acyclic graphs.

MONTA 1985-1 Monterosso, C. D. B.; Wright, C. L.; Lacerda, M. C. S.; Ofugi, N.
Grain storage in developing areas: Location and size of facilities.

MORE 1985-1 Moreno, J. A.
A correction to the definition of local center.

MOSL 1987-1 Mosler, K. C.
Continuous location of transportation networks.

MUNO 1984-1 Munoz Perez, J.
Minimax location when destinations are positioned randomly (Spanish).

MYUN 1987-1 Myung, Y.-S.; Tcha, D.-W.
The stochastic uncapacitated facility location problem: a new formulation and extensions to the capacitated problem.

NARU 1986-1 Narula, S. C.
Minisum hierarchical location-allocation problems on a network: a survey.

NARUA 1985-1 Narula, S. C.; Ogbu, U. I.
Lagrangean relaxation and decomposition in a uncapacitated 2-hierarchal location-allocation problem.

NGUY 1983-1 Nguyen Van Thieu
On the problem of production location in an interbranch formulation (Vietnamese).

NIEM 1987-1 Nieminen, J.
Distance center and centroid of a median graph.

NKAN 1986-1 Nkansah, P. T.; David, H. T.
Network median problems with continuously distributed demand.

NUDE 1983-1 Nudelman, G. A.
An algorithm for solution of a class of location problems (Russian).

O'KE 1986-1 O'Kelly, M. E.
The location of interacting hub facilities.

O'KE 1987-1 O'Kelly, M. E.
A quadratic integer program for the location of interacting hub facilities.

ODON 1987-1 Odoni, A. R.
Stochastic facility location problems.

OHSA 1987-1 Ohsawa, Y.; Suzuki, A.
Location-allocation problem of multi-person facility.

OPAS 1984-1 Opaswongkarn, K.; Guignard, M.
A multi-commodity location problem.

PANY 1985-1 Panyukov, A. V.; Shtraus, V. A.
Nonatomic problems of location of extended objects (Russian).

PAPAF 1986-1 Papathanassiu, B.; Tsouros, C.
Determining the optimal place and capacity of spare parts warehouses in the construction industry.

PARKA 1986-1 Parkhomenko, N. V.; Korneiko, V. G.
An attempt at solving, in an automatic subject-indexed control system, an allocation problem in a discrete formulation (Russian).

PELEE 1983-1 Pelegrin, B.
The expected value criterion in an industrial location model (Spanish).

PELEE 1985-1 Pelegrin, B.
Minimax location: the problem of a center (Spanish).

PELEF 1985-1 Pelegrin, B.; Michelot, C.; Plastria, F.
On the uniqueness of optimal solutions in continuous location theory.

PERL 1987-1 Perl, J.
The multidepot routing allocation problem.

PERLA 1985-1 Perl, J.; Daskin, M. S.
A warehouse location-routing problem.

PIRK 1987-2 Pirkul, H.
Efficient algorithms for the capacitated concentrator location problem.

PLASA 1984-1 Plastria, F.
Localization in single facility location.

PLASA 1987-1 Plastria, F.
Solving general continuous single facility location problems by cutting planes.

PLES 1987-1 Plesnik, J.
A heuristic for the p-center problem in graphs.

PLES 1987-2 Plesnik, J.
On the interchange heuristic for locating centers and medians in a graph.

POCHA 1986-1 Pochet, Y.; Wolsey, L. A.
Lot-size models with backlogging: strong reformulations and cutting planes.

POHL 1986-1 Pöhler, K.
An application of duality for the solution of special vector optimization problems (German).

PRAK 1984-1 Prakash, J.; Sinha, S. B.
A goal programming model for express bus-stop location.

PRIC 1986-1 Price, W. L.; Turcotte, M.
Locating a blood bank.

RANDB 1986-1 Randow, R. von
On an optimal location problem in a triangle.

RAO A 1983-1 Rao, J. R.; Rao, S. N.
Facility location in stochastic environment.

RAO B 1985-1 Rao, J. R.; Varma, N. H.
Stochastic multi-facility minisum location problem involving Euclidean distances.

RAVIB 1985-1 Ravindranath, K.; Vrat, P.; Singh, N.
Bicriteria single facility rectilinear location problems in the presence of a single forbidden region.

RO 1984-1 Ro, H.-B.; Tcha, D.-W.
A branch and bound algorithm for the two-level uncapacitated facility location problem with some side constraints.

ROSED 1987-1 Rosenblatt, M. J.; Lee, H. L.
A robustness approach to facilities design.

SCHRC 1988-1 Schreuder, J. A. M.
Location of Rotterdam fire stations.

SCOT 1985-1 Scott, C. H.; Murtagh, B. A.; Sirri, E.
Solution of constrained minmax location with Euclidean distances via conjugate duality.

SEMEA 1984-1 Semenov, A. S.; Semenov, A. V.
On the Steiner-Weber problem in a rectangular metric (Russian).

SENG 1987-1 Sengoku, M.; Shinoda, S.; Chen, W.-K.
A fault diagnosis in a nonlinear location network.

SENG 1987-2 Sengoku, M.; Shinoda, S.; Chen, W.-K.
A fault diagnosis in a nonlinear location network.

SERGE 1984-1 Sergienko, I. V.; Kaspshitskaya, M. F.
Stability of a discrete allocation problem (Russian).

SFOR 1986-1 Sforza, A.; Sepe, A.
An algorithm for the problem of the absolute centre of a network (Italian).

SHERB 1984-1 Sherali, H. D.; Adams, W. P.
A decomposition algorithm for a discrete location-allocation problem.

SHERE 1985-1 Sherali, H. D.; Nordai, F. L.
A capacitated, balanced, 2-median location-allocation problem on a tree network with a continuum of link demands.

SHERE 1985-2 Sherali, H. D.; Nordai, F. L.
Capacitated, balanced location-allocation problems on a chain graph with a continuum of link demands.

SHIO 1985-1 Shiode, S.; Ishii, H.; Nishida, T.
A chance constrained minimax facility location problem.

SIMC 1987-1 Simchi-Levy, D.; Berman, O.
A heuristic algorithm for the travelling salesman location problem on networks.
SIMC 1987-2 Simchi-Levy, D.; Berman, O.
Heuristics and bounds for the travelling salesman location problem on the plane.
SING 1985-1 Singer, D.; Elek, J.
Algorithm for determining the near optimal centrum locations in large graph structures.
SINGF 1984-1 Singh, N.; Vrat, P.
A location-allocation model for a two-echelon repair-inventory system.
SLAT 1984-1 Slater, P. J.
Some definitions of central structures.
SLAT 1985-1 Slater, P. J.
Minimum average distance partitions of the vertices on a line network.
SLUS 1985-1 Slusarek, M.
A note on the dynamic storage allocation problem.
SPAT 1986-1 Späth, H.
Heuristic solution methods for the m-mean median location problem.
SPIL 1988-1 Spilker, I.
Aircraft-stand allocation at Schiphol airport: problem description.
SRIDA 1986-1 Sridharan, R.
A heuristic Lagrangean algorithm for the capacitated plant location problem - a comment.
STANB 1985-1 Stanczak, W.; Kacprzyk, J.
On an extension of Weber problem with Euclidean distance.
STEP 1983-1 Stepanov, G. D.
An approach to the solution of a problem of the multimedian on a tree (Russian).
STIN 1983-1 Stinson, J. P.; Khumawala, B. M.; Steinberg, E.
An improved algorithm for the multi-activity, multi-facility location problem.
STIR 1984-1 Stiranka, A. I.
A problem of distribution of productive capacities (Russian).
STOK 1988-1 Stokx, C. F. M.
A strategic model for the solution of the location-allocation problem of a major oil company.
SZID 1984-1 Szidarovszky, F.; Carrera, J.; Usunoff, E.
Optimum location of measuring points in the quality analysis of groundwater (Hungarian).
TAMI 1985-1 Tamir, A.
A finite algorithm for the continuous p-center location problem on a graph.
TAMI 1987-1 Tamir, A.
On the solution value of the continuous p-center location problem on a graph.
TAMI 1987-2 Tamir, A.
Totally balanced and totally unimodular matrices defined by center location problems.
TARAA 1985-2 Tarasov, O.; Richter, P.
Placement and routing for the optimization of star networks in graphs.
TCHAA 1987-1 Tcha, D.-W.; Jung, S.-H.; Myung, Y.-S.
A comparative study on optimization algorithms for capacitated facility location (Korean).
TCHAB 1984-1 Tcha, D.-W.; Lee, B.-I.
A branch-and-bound algorithm for the multi-level uncapacitated facility location problem.
TCHAC 1985-1 Tcha, D.-W.; Yoon, M.-G.
A dual-based heuristic for the simple facility location problem with stochastic demand.
THIS 1984-1 Thisse, J.-F.; Ward, J. E.; Wendell, R. E.
Some properties of location problems with block and round norms.
TIEN 1983-1 Tien, J. M.; El-Tell, K.; Simons, G. R.
Improved formulations to the hierarchical health facility location allocation problem.
TIES 1987-1 Tiesis, V.; Saltenis, V.; Juska, F.
A problem of the allocation of a discrete resource (Russian).
TILA 1988-1 Tilanus, C. B.
Transportation planning: recent developments in the Netherlands.
TOBI 1985-1 Tobin, R. L.; Friesz, T. L.
A new look at spacially competitive facility location models.
TOBI 1986-1 Tobin, R. L.; Friesz, T. L.
Spatial competition facility location models: definition, formulation, and solution approach.
TOPK 1985-1 Topka, V. V.
Decomposition in a problem of resource allocation on a network (Russian).
TRUBA 1986-1 Trubin, V. A.; Sharifov, F. A.
An efficient method for solving a class of allocation problems (Russian).
TRYF 1986-1 Tryfos, P.
An integer programming approach to the apparel sizing problem.
VANR 1986-1 Vanroy, T. J.
A cross decomposition algorithm for capacitated facility location.
VENT 1985-1 Venta, E. R.; Hurter, A. P. Jr.
Production-location problems with demand considerations.
VIJA 1985-1 Vijay, J.
An algorithm for the p-center problem in the plane.
VOHRB 1986-1 Vohra, R. V.; Hall, N. G.
Bounds and heuristics for the maximal covering location problem: a preliminary report.
VOLOA 1987-1 Voloshin, A. F.
A method for localization of the optimum region in problems of mathematical programming (Russian).
WARD 1985-1 Ward, J. E.; Wendell, R. E.
Using block norms for location modeling.
WARDA 1987-1 Ward, J. E.; Wong, R. T.; Lemke, P.; Oudjit, A.
Properties of the tree k-median linear programming relaxation.
WATS 1984-1 Watson-Gandy, C. D. T.
The multi-facility min-max Weber problem.
WATS 1985-1 Watson-Gandy, C. D. T.
The solution of distance constrained mini-sum location problems.

WEAV 1985-1 Weaver, J. R.; Church, R. L.
A median location model with nonclosest facility service.
WEAV 1986-1 Weaver, J. R.; Church, R. L.
A location model based on multiple metrics and multiple facility assignment.
WEAV 1987-1 Weaver, J. R.; Church, R. L.
The formal and computational relationship of the supporting median problem to the p-median problem.
WHIT 1985-1 Whitaker, R. A.
Some add-drop and drop-add interchange heuristics for non-linear warehouse location.
WILH 1985-1 Wilhelm, M. R.; Evans, G. W.
The quadratic assignment problem: applications and solution techniques.
WILHB 1986-1 Wilhelm, M. R.; Ward, T. L.; Lichtefeld, R. A.
Computer-aided facility location.
WONGA 1985-1 Wong, R. T.
Location and network design.
WONGA 1986-1 Wong, R. T.
Recent advances in location and network design models for transportation planning.
WRIG 1986-1 Wright, D. D.; Haehling von Lanzenauer, C.
COAL: A new heuristic approach for solving the fixed charge problem - computational results.
WRIG 1986-2 Wright, D. D.; Haehling von Lanzenauer, C.
Solving the fixed charge problem with Lagrangian relaxation and cost allocation heuristics.
XUE 1987-1 Xue, G. L.
A quickly converging algorithm for optimal location problems of the form min(sum c(i)||x-a(i)||) on a closed convex set (Chinese).
YOON 1985-1 Yoon, K.; Hwang, C.-L.
Manufacturing plant location analysis by multiple attribute decision making: part I - single-plant strategy.
YOU 1985-1 You, Z. Y.; Zhu, S. Q.
Convex combined iteration and non-tip point iteration for solving constrained optimal location problems (Chinese).
ZAMB 1985-1 Zambitskij, D. K.
Study of the stability of the optimal solution of Weber's problem in a metric space (Russian).
ZAMBA 1985-1 Zambitskij, D. K.; Sibirskij, V. K.
Nonlinear Weber problem on graphs (Russian).
ZEME 1984-1 Zemel, E.
Probabilistic analysis of geometric location problems.
ZEME 1985-1 Zemel, E.
Probabilistic analysis of geometric location problems.
ZHANP 1985-1 Zhang, T.-C.
A brief survey of plane location problems (Chinese).
ZORK 1983-1 Zorkin, V. N.; Stepanov, G. D.
Optimal arrangement of additional centers on a graph of simple structure (Russian).

AHAR 1985-1 Aharoni, R.
Fractional matchings and covers in infinite hypergraphs.

AHAR 1985-2 Aharoni, R.
Matchings in n-partite n-graphs.

AJTA 1984-1 Ajtai, M.; Komlos, J.; Tusnady, G.
On optimal matchings.

AKL 1983-1 Akl, S. G.
A note on Euclidean matchings, triangulations and spanning trees.

ANST 1987-1 Anstee, R. P.
A polynomial algorithm for b-matchings: an alternative approach.

ARCHA 1987-1 Archdeacon, D.; Dinitz, J. H.
Factorizations and orthogonal matchings.

ATALA 1987-1 Atallah, M. J.; Hambrusch, S. E.
On bipartite matchings of minimum density.

BABI 1986-1 Babic, D.; Graovac, A.; Mohar, B.
Recursion for the matching polynomials of fasciagraphs and rotagraphs.

BABIA 1986-1 Babic, D.; Graovac, A.; Mohar, B.; Pisanski, T.
Recursion for the matching polynomials of fasciagraphs and rotagraphs.

BALAK 1987-1 Balas, E.; Pulleyblank, W. R.
The perfectly matchable subgraph polytope of an arbitrary graph.

BALLB 1986-1 Ball, M. O.; Taverna, R.
Sensitivity analysis for the matching problem and its use in solving matching problems with a single side constraint.

BAND 1983-1 Bandelt, H. J.
Matchings in squares of odd cycles.

BARAH 1987-1 Barahona, F.; Pulleyblank, W. R.
Exact arborescences, matchings and cycles.

BARTA 1986-1 Bartholdi, J. J. III; Trick, M. A.
Stable matching with preferences derived from a psychological model.

BERG 1983-1 Berge, C.
Graphes (3. edition).

BERG 1985-1 Berge, C.
Graphs (second revised edition).

BLAI 1986-1 Blair, C.
The lattice structure of the set of stable matchings with multiple partners.

BOLLC 1983-1 Bollobas, B.; Frieze, A. M.
On matchings and Hamiltonian cycles in random graphs.

BOUCA 1987-1 Bouchet, A.
Delta-matroids and matchings.

BREZ 1987-1 Brezovec, C.; Cornuejols, G.; Glover, F.
A matroid algorithm and its application to the efficient solution of two optimization problems on graphs.

CARR 1986-1 Carraresi, P.; Sodini, C.
An efficient algorithm for the bipartite matching problem.

CARSA 1985-1 Carstensen, P. J.; Murty, K. G.; Perin, C.
Intermediate feasibility in 0-1 integer linear systems.

CHARH 1987-1 Chartrand, G.; Kapoor, S. F.; Oellermann, O. R.; Ruiz, S.
On maximum matchings in cubic graphs with a bounded number of bridge-covering paths.

CHEG 1987-1 Chegireddy, C. R.; Hamacher, H. W.
Algorithms for finding K best perfect matchings.

CHOU 1984-1 Choudum, S. A.
A necessary condition for disjoint perfect matchings in graphs.

CHRO 1986-1 Chrobak, M.; Poljak, S.
On common edges in optimal solutions to traveling salesman and other optimization problems.

COOKA 1985-1 Cook, W.
A note on matchings and separability.

COOKG 1987-1 Cook, W.; Pulleyblank, W. R.
Linear systems for constrained matching problems.

CORNH 1986-1 Cornuejols, G.; Hartvigsen, D.
An extension of matching theory.

CROI 1986-1 Croitoru, C.; Moscovici, J.
A "greedy" O|V|**3-algorithm for the maximum cardinality matching problem.

CUNNC 1986-1 Cunningham, W. H.; Green-Krotki, J.
Dominants and submissives of matching polyhedra.

DAHLA 1986-1 Dahlhaus, E.; Karpinski, M.
Perfect matching for regular graphs is AC**0-hard for the general matching problem.

DAHLA 1986-2 Dahlhaus, E.; Karpinski, M.
The matching problem for strongly chordal graphs is in NC.

DAHLA 1987-3 Dahlhaus, E.; Karpinski, M.
On the parallel complexity of matching for chordal and path graphs.

DEMA 1987-1 Demange, G.; Gale, D.; Sotomayor, M.
A further note on the stable matching problem.

DERI 1981-1 Derigs, U.
F-factors, perfect matchings and related concepts.

DERI 1984-3 Derigs, U.
Programming in networks and graphs - on the combinatorial background and near-equivalence of network flow and matching algorithms.

DERI 1984-4 Derigs, U.
The cardinality matching problem - a state of the art tutorial.

DERI 1985-2 Derigs, U.
Postoptimal analysis for matching problems.

DERI 1986-1 Derigs, U.
A short note on matching algorithms.

DERIA 1984-1 Derigs, U.; Goecke, O.; Schrader, R.
Bisimplicial edges, Gaussian elimination and matchings in bipartite graphs.

DERIC 1987-1 Derigs, U.; Hilbrand, C.
Optimal matchings with GUP-constraints.

DERID 1986-3 Derigs, U.; Metz, A.
On the use of optimal fractional matchings for solving the (integer) matching problem.

DING 1987-1 Ding, G.-L.
On the maximum 2-1 matching.

FARL 1984-2 Farley, A. M.; Proskurowski, A.
Extremal graphs with no disconnecting matching.

FARRA 1986-2 Farrell, E. J.
On the occurrences of Fibonacci sequences in the counting of matchings in linear polygonal chains.

FARRA 1987-1 Farrell, E. J.
Counting matchings in graphs.

FARRA 1987-2 Farrell, E. J.
Matchings in hexagonal cacti.

FARRE 1984-1 Farrell, E. J.; Wahid, S. A.
Matchings in heptagonal and octagonal chains.

FARRE 1984-2 Farrell, E. J.; Wahid, S. A.
Matchings in nonagonal and decagonal chains.

FARRE 1985-1 Farrell, E. J.; Wahid, S. A.
Matchings in long benzene chains.

FARRE 1986-1 Farrell, E. J.; Wahid, S. A.
Matching polynomials: a matrix approach and its applications.

FARRE 1987-1 Farrell, E. J.; Wahid, S. A.
On the reconstruction of the matching polynomial and the reconstruction conjecture.

FARRE 1987-2 Farrell, E. J.; Wahid, S. A.
Some general classes of comatching graphs.

FLAC 1987-1 Flach, P.; Volkmann, L.
Abschätzungen gesättigter Matchings nach unten.

FRIEC 1986-1 Frieze, A. M.
Maximum matchings in a class of random graphs.

FRIEC 1986-2 Frieze, A. M.
On large matchings and cycles in sparse random graphs.

GABOA 1985-1 Gabow, H. N.
A scaling algorithm for weighted matching on general graphs.

GABOB 1984-1 Gabow, H. N.; Galil, Z.; Spencer, T.
Efficient implementation of graph algorithms using contraction.

GALE 1985-1 Gale, D.; Sotomayor, M.
Some remarks on the stable matching problem.

GALI 1986-1 Galil, Z.
Efficient algorithms for finding maximum matching in graphs.

GALI 1986-2 Galil, Z.
Sequential and parallel algorithms for finding maximum matchings in graphs.

GALIB 1982-1 Galil, Z.; Micali, S.; Gabow, H.
Priority queues with variable priority and an O(EVlogV) algorithm for finding a maximal weighted matching in general graphs.

GALIB 1986-1 Galil, Z.; Micali, S.; Gabow, H.
An 0(EVlogV) algorithm for finding a maximal weighted matching in general graphs.

GIBB 1985-1 Gibbons, A.
Algorithmic graph theory.

GLOVB 1987-1 Glover, F.; Harary, F.
A simple criterion for a graph to have a perfect matching.

GRAO 1985-1 Graovac, A.; Babic, D.
Comment on "Matchings in long benzene chains" by E. J. Farrell and S. A. Wahid.

GREGA 1986-1 Gregory, D. A.; Mac Guinness, S.; Wallis, W. D.
Clique partitions of the cocktail party graphs.

GRIGB 1986-1 Grigoriadis, M. D.; Kalantari, B.
A lower bound to the complexity of Euclidean and rectilinear matching algorithms.

GRIGC 1986-1 Grigoriadis, M. D.; Kalantari, B.; Lai, C. Y.
On the existence of weak greedy matching heuristics.

GRIGC 1987-1 Grigoriadis, M. D.; Kalantari, B.; Lai, C. Y.
Weak greedy heuristics for perfect matching.

GRIGD 1987-1 Grigoriev, D. Y.; Karpinski, M.
The matching problem for bipartite graphs with polynomially bounded permanents is in NC.

GROTA 1985-1 Grötschel, M.; Holland, O.
Solving matching problems with linear programming.

GROTA 1987-1 Grötschel, M.; Holland, O.
A cutting plane algorithm for minimum perfect 2-matchings.

GROTF 1986-1 Grötschel, M.; Truemper, K.
Decomposition and optimization over cycles in binary matroids.

GUIG 1983-1 Guignard, M.
A Lagrangean dual ascent method based on (separable) relaxation strengthened by valid inequalities (illustrated for the matching problem).

GUTI 1983-1 Gutierrez Diez, J. M.
A variant of the Edmonds Algorithm for maximum matching (Spanish).

GUTMA 1987-1 Gutman, I.; Cioslowski, J.
Bounds for the number of perfect matchings in hexagonal systems.

GUTMB 1984-1 Gutman, I.; Cvetkovic, D.
Finding tricyclic graphs with a maximal number of matchings - another example of computer aided research in graph theory.

GUTMC 1983-1 Gutman, I.; Farrell, E. J.; Wahid, S. A.
On the matching polynomials of graphs containing benzenoid chains.

HAMAC 1986-1 Hamacher, H. W.; Queyranne, M.
K best solutions to combinatorial optimization problems.

HARTE 1987-2 Hartvigsen, D.; Zemel, E.
On the complexity of lifted inequalities for the knapsack problem.

HEDED 1983-1 Hedetniemi, S. T.
A max-min relationship between matchings and domination in graphs.

HOLL 1983-1 Holland, O.
Optimierung kapazitierter und nicht kapazitierter b-Matchings unter Verwendung der linearen Programmierung.

HOU 1987-1 Hou, M.
Enumeration of 2-matchings of complete graphs (Chinese).

IMAI 1986-1 Imai, H.
Worst-case analysis for planar matching and tour heuristics with bucketing techniques and spacefilling curves.

IRVIB 1987-1 Irving, R. W.; Leather, P.; Gusfield, D.
An efficient algorithm for the "optimal" stable marriage.

ISRA 1986-1 Israeli, A.; Itai, A.
A fast and simple randomized parallel algorithm for maximal matching.

ISRAA 1986-1 Israeli, A.; Shiloach, Y.
An improved parallel algorithm for maximal matching.

JAMI 1984-1 Jamison, R. E.
Whitney sums and matchings in trees (part I).

JAMI 1987-1 Jamison, R. E.
Alternating Whitney sums and matchings in trees, Part 1.

JOHNA 1985-1 John, P.; Rempel, J.
Counting perfect matchings in hexagonal systems.

JOHNB 1985-1 John, P.; Sachs, H.
Calculating the number of perfect matchings and Pauling's bond orders in hexagonal systems whose inner dual is a tree.

KAKH 1985-1 Kakhichko, A. A.
On the problem of the existence of perfect matchings of a graph (Russian).

KARG 1987-1 Karger, P. B.
Untersuchungen zur Matching-Relaxation für Set-Partitioning-Probleme.

KARLA 1986-1 Karloff, H. J.
A Las Vegas RNC algorithm for maximum matching.

KARPG 1985-1 Karp, R. M.; Upfal, E.; Wigderson, A.
Constructing a perfect matching is in random NC.

KARPG 1986-1 Karp, R. M.; Upfal, E.; Wigderson, A.
Constructing a perfect matching is in random NC.

KARPI 1987-1 Karpinski, M.; Lingas, A.
Subtree isomorphism and bipartite perfect matchings are mutually nc-reducible.

KARZ 1987-2 Karzanov, A. V.
On maximal matchings of a prescribed weight in complete and complete bipartite graphs.

KIM E 1987-1 Kim, T.; Chwa, K.-Y.
An O(nlognloglogn) parallel maximum matching algorithm for bipartite graphs.

KORO 1987-1 Korolev, A. G.
A cubic algorithm for the construction of matching with minimal weight using linear memory (Russian).

KOST 1985-1 Kostochka, A. V.
Criterion for existence of perfect matchings in hexagonal systems (Russian).

KOZE 1985-1 Kozen, D.; Vazirani, U. V.; Vazirani, V. V.
NC algorithms for comparability graphs, interval graphs, and testing for unique perfect matching.

KRUSA 1985-1 Kruskal, C. P.; West, D. B.
Compatible matchings in bipartite graphs.

KUNG 1987-1 Kung, J. P. S.
Matchings and Radon transforms in lattices. II. Concordant sets.

LAI B 1985-1 Lai, T.-H.; Sprague, A.
On the complexity of a family of generalized matching problems.

LECLA 1986-2 Leclerc, M.
Slices of the matching polytopes.

LIU C 1982-1 Liu, G.-Z.
Pseudo-matchings in graphs (Chinese).

LIU C 1985-1 Liu, G.-Z.
A dual algorithm for optimum matching.

LIU C 1985-6 Liu, G.-Z.
The perfect matching in quasi-forest complexes (Chinese).

LOIZ 1984-1 Loizou, G.
Certain NP-complete matching problems.

LOVA 1987-1 Lovasz, L.
Matching structure and the matching lattice.

LOVAA 1986-1 Lovasz, L.; Plummer, M. D.
Matching theory.

LUTT 1986-1 Lutton, J.-L.; Bonomi, E.
Simulated annealing algorithm for the minimum weighted perfect euclidean matching problem.

MARCJ 1986-7 Marcu, D.
Paths, connectivity and matchings in graphs (French).

MATSA 1986-1 Matsumoto, K.; Nishizeki, T.; Saito, N.
Planar multicommodity flows, maximum matchings and negative cycles.

MENG 1984-1 Meng, J. X.
Some classes of hypomatchable graphs (Chinese).

MICA 1980-1 Micali, S.; Vazirani, V. V.
An O(|E|.root|v|) algorithm for finding maximum matching in general graphs.

MONTB 1987-1 Montreuil, B.; Ratliff, H. D.; Goetschalckx, M.
Matching based interactive facility layout.
MUHL 1984-1 Mühlbacher, J. R.; Steinparz, F. X.; Tinhofer, G.
On certain classes of fractional matchings.
MULM 1987-1 Mulmuley, K.; Vazirani, U. V.; Vazirani, V. V.
Matching is as easy as matrix inversion.
OLEJ 1984-1 Olejnik, F.
On total matching numbers and total covering numbers for k-uniform hypergraphs.
PETRA 1985-1 Petrova, G. L.
Perfect matchings in n-dimensional parallelepipeds (Russian).
PLAI 1984-1 Plaisted, D. A.
Heuristic matching for graphs satisfying the triangle inequality.
PLUM 1986-1 Plummer, M. D.
Matching extension in bipartite graphs.
POLAA 1983-1 Polat, N.
Couplages dans les systemes combinatoires flous. I.
POLAA 1983-2 Polat, N.
Couplages dans les systemes combinatoires flous. II.
POLJ 1986-1 Poljak, S.; Chrobak, M.
On common edges in optimal solutions to traveling salesman and other optimization problems.
PULL 1987-1 Pulleyblank, W. R.
Fractional matchings and the Edmonds-Gallai theorem.
QI A 1987-1 Qi, N.
On separation and adjacency problems for perfectly matchable subgraph polytopes of a graph.
QUINA 1984-1 Quinn, M. J.; Deo, N.
Parallel graph algorithms.
REIC 1984-1 Reichmeider, P. F.
The equivalence of some combinatorial matching theorems.
REND 1986-1 Rendl, F.
On the Euclidean assignment problem.
ROCK 1984-1 Rockafellar, R. T.
Network flows and monotropic optimization.
ROOS 1984-1 Roos, C.
Network programming with umbrellas.
ROTH 1985-1 Roth, A. E.
The college admissions problem is not equivalent to the marriage problem.
RUCI 1982-1 Rucinski, A.
Matchings and k-factors in a random graph.
SACH 1986-1 Sachs, H.
On the number of perfect matchings in bipartite plane graphs.
SAKS 1986-1 Saks, M.
Some sequences associated with combintorial structures.
SCHRE 1986-2 Schrijver, A.
Geometric methods in discrete optimization.
SCHRE 1986-3 Schrijver, A.
Polyhedral combinatorics - some recent developments and results.
SCHRE 1987-2 Schrijver, A.
Polyhedral combinatorics.
SCUT 1985-1 Scutella, M. G.; Scevola, G.
An algorithm for the maximum cardinality matching problem in convex bipartite graphs.
STEF 1985-1 Steffens, K.
Maximal tight sets and the Edmonds-Gallai decomposition for matchings.
STEIC 1982-1 Steinparz, F. X.
On the existence of F-factors.
TINH 1984-1 Tinhofer, G.
A probabilistic analysis of some greedy cardinality matching algorithms.
TINH 1984-2 Tinhofer, G.
Naive greedy matching - a probabilistic analysis.
TOCZ 1984-1 Toczylowski, E.
A perfect matching algorithm for sparse bipartite graphs.
TONG 1984-1 Tong, P.; Lawler, E. L.; Vazirani, V. V.
Solving the weighted parity problem for gammoids by reduction to graphic matching.
TRIC 1987-1 Trick, M. A.
Matching problems and generalized networks. Chapter IV of: Networks with additional structured constraints.
VENK 1985-1 Venkatesan, S. M.
Approximation algorithms for planar matching.
VENK 1987-1 Venkatesan, S. M.
Approximation algorithms for weighted matching.
WEBE 1986-1 Weber, M.; Liebling, T. M.
Euclidean matching problems and the Metropolis algorithm.
WEI 1987-1 Wei, S.-B.
Some classes of hypomatchable graphs (Chinese).
WEI A 1986-1 Wei, S.-B.; Ye, D.-B.
Some classes of secondary randomly matchable graphs (Chinese).
WESTA 1985-1 West, D. B.; Banerjee, P.
Partial matching in degree-restricted bipartite graphs.
ZASL 1984-1 Zaslavsky, T.
Generalized matchings and generalized Hermite polynomials.
ZHAND 1985-1 Zhang, F. J.; Chen, R. S.; Guo, X. F.
Perfect matchings in hexagonal systems.
ZHANF 1986-2 Zhang, F. J.; Guo, X. F.
Hamiltonian cycles in perfect matching graphs (Chinese).

ZHANH 1987-1 Zhang, F. J.; Lin, G. N.; Zheng, M. L.
An algorithm for generating all perfect matchings of a graph (Chinese).

ZHANJ 1985-1 Zhang, F. J.; Liu, Y.-T.; Guo, X. F.
The enumeration of perfect matchings of treelike polyhexes (Chinese).

ZHANL 1987-1 Zhang, F. J.; Zhou, M. K.
The characteristic and matching polynomials of a class of graphs (Chinese).

ZHOUA 1986-1 Zhou, M. K.
Matching polynomials of polyphenylenes.

ZHOUA 1986-2 Zhou, M. K.
Numbers of perfect matchings in three classes of graphs.

ZHUR 1984-1 Zhuravlev, Y. I.; Platonenko, I. M.
On an economical product of Boolean equations (Russian).

ACKE 1984-1 Acketa, D. M.
A construction of non-simple matroids on at most 8 elements.
ACKE 1984-2 Acketa, D. M.
On the essential chains and squares.
ACKE 1985-1 Acketa, D. M.
Some results on "small" matroids.
ACKE 1986-1 Acketa, D. M.
A construction of all the non-isomorphic non-simple matroids on 9 elements.
ACKE 1986-2 Acketa, D. M.
On the number of graphic representations of graphic matroids.
AL-S 1986-1 Al-Salti, B.
Minimization of a concave function on a polymatroid (Russian).
ALON 1986-1 Alon, N.
The number of polytopes, configurations and real matroids.
ASANC 1984-1 Asano, T.; Nishizeki, T.; Seymour, P. D.
A note on nongraphic matroids.
BACHA 1984-1 Bachem, A.
Optimization and geometry in discrete structures.
BACHA 1986-1 Bachem, A.
Dualität und Polarität in diskreten Strukturen.
BACHB 1984-1 Bachem, A.; Euler, R.
Recent trends in combinatorial optimization.
BACHD 1984-1 Bachem, A.; Kern, W.
Adjoints of oriented matroids.
BACHD 1985-1 Bachem, A.; Kern, W.
On sticky matroids.
BACHD 1986-1 Bachem, A.; Kern, W.
Adjoints of oriented matroids.
BACHD 1986-2 Bachem, A.; Kern, W.
Extension equivalence of oriented matroids.
BACHE 1985-1 Bachem, A.; Wanka, A.
On intersection properties of (oriented) matroids (extended abstract).
BACHE 1985-2 Bachem, A.; Wanka, A.
Separation theorems for oriented matroids.
BACHE 1986-1 Bachem, A.; Wanka, A.
Euclidean intersection properties.
BACHE 1986-2 Bachem, A.; Wanka, A.
Matroids without adjoint.
BACHF 1984-1 Bachmann, P.
Ein Beitrag zur Maximierung submodularer und isotoner Funktionen über Unabhängigkeitssystemen.
BARAA 1984-1 Barahona, F.; Cunningham, W. H.
A submodular network simplex method.
BARAB 1986-1 Barahona, F.; Grötschel, M.
On the cycle polytope of a binary matroid.
BATTA 1982-1 Batten, L. M.
A problem in transversal theory.
BECKE 1985-1 Beckert, R.; Girlich, E.
Kombinatorische Typen von Polymatroiden im E(3,+).
BENZB 1985-1 Benzaken, C.; Hammer, P. L.
Boolean techniques for matroidal decomposition of independence systems and applications to graphs.
BERMA 1984-1 Berman, K. A.
A generalization to bases common to r binary matroids of the weighted matrix-tree theorem in the case when the weights are Boolean.
BERT 1986-1 Bertani, L.
On near dimensions in Steiner systems with parallelism and matroids.
BIENA 1987-1 Bienia, W.; Cordovil, R.
An axiomatic of non-radon partitions of oriented matroids.
BILL 1984-1 Billera, L. J.; Munson, B. S.
Oriented matroids and triangulations of convex polytopes.
BILL 1984-2 Billera, L. J.; Munson, B. S.
Polarity and inner products in oriented matroids.
BILL 1984-3 Billera, L. J.; Munson, B. S.
Triangulations of oriented matroids and convex polytopes.
BIXB 1987-1 Bixby, R. E.
Notes on combinatorial optimization.
BIXBA 1986-1 Bixby, R. E.; Coullard, C. R.
On chins of 3-connected matroids.
BIXBA 1987-1 Bixby, R. E.; Coullard, C. R.
Finding a small 3-connected minor maintaining a fixed minor and a fixed element.
BIXBB 1987-1 Bixby, R. E.; Cunningham, W. H.
Short cocircuits in binary matroids.
BIXBC 1985-1 Bixby, R. E.; Cunningham, W. H.; Topkis, D. M.
The partial order of a polymatroid extreme point.
BIXBE 1986-1 Bixby, R. E.; Rajan, A.
A short proof of a decomposition theorem for max-flow min-cut matroids.
BIXBE 1987-1 Bixby, R. E.; Rajan, A.
A short proof of the Truemper-Tseng theorem on max-flow min-cut matroids.
BJOR 1985-1 Björner, A.
On matroids, groups and exchange matroids.
BJORA 1985-1 Björner, A.; Korte, B.; Lovasz, L.
Homotopy properties of greedoids.
BJORB 1987-1 Björner, A.; Lovasz, L.
Pseudomodular lattices and continuous matroids.

BLAN 1986-1 Bland, R. G.; Dietrich, B. L.
Abstract duality.
BOKO 1985-1 Bokowski, J.; Sturmfels, B.
Coordinatization of oriented matroids.
BOKO 1986-1 Bokowski, J.; Sturmfels, B.
On the coordinatization of oriented matroids.
BOKO 1986-2 Bokowski, J.; Sturmfels, B.
Reell realisierbare orientierte Matroide.
BOND 1982-1 Bondarenko, V. A.
Neighbouring vertices of polyhedra generated by matroids (Russian).
BOUCA 1987-1 Bouchet, A.
Delta-matroids and matchings.
BOUCA 1987-2 Bouchet, A.
Greedy algorithm and symmetric matroids.
BOUCA 1987-3 Bouchet, A.
Maps and delta-matroids.
BOUCA 1987-5 Bouchet, A.
Representability of delta-matroids.
BREZ 1986-1 Brezovec, C.; Cornuejols, G.; Glover, F.
Two algorithms for weighted matroid intersection.
BREZ 1987-1 Brezovec, C.; Cornuejols, G.; Glover, F.
A matroid algorithm and its application to the efficient solution of two optimization problems on graphs.
BRUC 1984-1 Brucker, P.
A greedy algorithm for solving network flow problems in trees.
BRYA 1984-1 Bryant, V. W.
Undirected strict gammoids.
BRYA 1984-2 Bryant, V. W.
Undirected strict grammoids.
BRYL 1985-1 Brylawski, T.
Coordinatizing the Dilworth truncation.
BRYL 1986-1 Brylawski, T.
Blocking sets and the Möbius function.
BRYL 1986-2 Brylawski, T.
Iterated parallel union of matroids.
CAI A 1985-1 Cai, M.-C.
Progress in network flow theory and matroid theory (Chinese).
CAMEB 1986-1 Camerini, P. M.; Hamacher, H. W.
Intersections of two matroids: (condensed) border graphs and ranking.
CAMEC 1985-1 Camerini, P. M.; Maffioli, F.
Some applications of unlabelled partition systems.
CAMEE 1984-1 Camerini, P. M.; Vercellis, C.
The matroidal knapsack: a class of (often) well-solvable problems.
CECC 1986-1 Ceccherini, P. V.; Sappa, A.
F-binomial coefficients and related combinatorial topics: perfect matroid designs, posets of full binomial type and F-geodetic graphs.
CHAB 1984-1 Chabukiani, T. A.
On a combinatorial geometry of three-dimensional space (Russian).
CHANK 1987-1 Chandrasekaran, R.; Kabadi, S. N.
Pseudomatroids.
CHANN 1985-1 Chandru, V.; Coullard, C. R.; Wagner, D. K.
On the complexity of recognizing a class of generalized networks.
CHANQ 1985-1 Chang, G. J.
Greedoid polyhedra for MPP-greedoids.
CHANQ 1986-2 Chang, G. J.
MPP-greedoids.
COLB 1986-2 Colbourn, C. J.
The reliability polynomial.
COLBC 1987-1 Colbourn, C. J.; Pulleyblank, W. R.
Matroid Steiner problems, the Tutte polynomial and network reliability.
CONFB 1985-1 Conforti, M.; Laurent, M.
A characterization of matroidal systems of inequalities.
CONFB 1985-2 Conforti, M.; Laurent, M.
On the facial structure of independence systems polyhedra.
CONFB 1985-3 Conforti, M.; Laurent, M.
On the geometric structure of independence systems.
CONFB 1986-1 Conforti, M.; Laurent, M.
On the geometric structure of independence systems.
CONFC 1987-1 Conforti, M.; Rao, M. R.
Some new matroids on graphs: cut sets and the max cut problem.
CORD 1985-1 Cordovil, R.
On simplicial matroids and Sperner's lemma.
CORD 1987-1 Cordovil, R.
Polarity and point extensions in oriented matroids.
CORDA 1986-1 Cordovil, R.; Duchet, P.
Separation par une droite dans les matroides orientes de rang 3.
CORDB 1987-1 Cordovil, R.; Fukuda, K.
Oriented matroids and combinatorial manifolds.
CORDC 1987-1 Cordovil, R.; Silva, I. P.
Determining a matroid polytope by non-Radon partitions.
COUL 1986-1 Coullard, C. R.
Counterexamples to conjectures on 4-connected matroids.

COULA 1987-1 Coullard, C. R.; DelGreco, J. G.; Wagner, D. K.
Representations of bicircular matroids.

CRAM 1986-1 Crama, Y.; Hammer, P. L.
Some results on the recognition of bimatroidal independence systems.

CRAM 1987-1 Crama, Y.; Hammer, P. L.
Optimization over a bimatroidal independence system.

CRAP 1984-1 Crapo, H.
Selectors: a theory of formal languages, semimodular lattices, and branching and shelling processes.

CRAP 1985-1 Crapo, H.
The combinatorial theory of structures.

CRYSA 1985-1 Crystal Llewellyn, D.; Trotter, L. E. Jr.
On abstract integral dependence.

CUNNA 1984-1 Cunningham, W. H.
Testing membership in matroid polyhedra.

CUNNA 1985-1 Cunningham, W. H.
On submodular function minimization.

CUNNA 1986-1 Cunningham, W. H.
Improved bounds for matroid partition and intersection algorithms.

CUNNB 1985-1 Cunningham, W. H.; Frank, A.
A primal-dual algorithm for submodular flows.

DAMA 1985-1 Damaschke, P.; Stern, M.
A characterization of generalized matroid lattices.

DAWS 1983-1 Dawson, J. E.
Balanced sets in an independence structure induced by a submodular function.

DAWS 1984-1 Dawson, J. E.
A collection of sets related to the Tutte polynomial of a matroid.

DAWS 1984-2 Dawson, J. E.
Solubility of the matroid sum equations.

DAWS 1985-2 Dawson, J. E.
Decomposition of binary matroids.

DEMP 1985-1 Dempe, S.
A simple but NP-hard problem of mixed-discrete programming and its solution by approximate algorithms.

DEZA 1984-1 Deza, M.
Perfect matroid designs.

DEZAA 1986-1 Deza, M.; Fukuda, K.
On bouquets of matroids and orientations.

DEZAB 1987-1 Deza, M.; Laurent, M.
Bouquets of matroids, d-injection geometries and diagrams.

DIET 1987-1 Dietrich, B. L.
A circuit set characterization of antimatroids.

DIMO 1985-1 Dimow, E. T.; Kovalev, M. M.; Milanov, P. B.
Qualitätsabschätzungen von Greedy-Algorithmen (Russian).

DINGB 1987-1 Ding, L.-I.; Yue, M.-Y.
On a generalization of the Rado-Hall theorem to greedoids.

DRES 1986-1 Dress, A.
Chirotopes and oriented matroids.

DRES 1986-2 Dress, A.
Duality theory for finite and infinite matroids with coefficients.

DRESA 1986-1 Dress, A.; Havel, T. F.
Some combinatorial properties of discriminants in metric vector spaces.

DRESC 1987-1 Dress, A.; Lovasz, L.
On some combinatorial properties of algebraic matroids.

DRESD 1987-1 Dress, A.; Wenzel, W.
Endliche Matroide mit Koeffizienten.

EDEL 1984-1 Edelman, P. H.
The acyclic sets of an oriented matroid.

EMELE 1985-1 Emelicheva, E. V.; Kovalev, M. M.
The method of partial orderings on matroid structures (Russian).

EMELE 1986-1 Emelicheva, E. V.; Kovalev, M. M.
Treelike polymatroids and extremal problems on permutations (Russian).

EULE 1984-1 Euler, R.
Independence systems and perfect k-matroid-intersections.

EULEA 1986-1 Euler, R.; Burkard, R. E.; Grommes, R.
On latin squares and the facial structure of related polytopes.

EULEB 1987-1 Euler, R.; Jünger, M.; Reinelt, G.
Generalizations of cliques, odd cycles and anticycles and their relation to independence system polyhedra.

EULEC 1985-1 Euler, R.; Mahjoub, A. R.
On a composition of independence systems by circiut-identification.

FAIG 1977-1 Faigle, U.
Geometries on partially ordered sets.

FAIG 1984-1 Faigle, U.
Matroids on ordered sets and the greedy algorithm.

FAIG 1984-2 Faigle, U.
A vector exchange property of submodular systems.

FAIG 1985-1 Faigle, U.
On ordered languages and the optimization of linear functions by greedy algorithms.

FAIG 1985-2 Faigle, U.
Orthogonal sets, matroids, and theorems of the alternative.

FAIG 1986-1 Faigle, U.
Exchange properties of combinatorial closure spaces.

FAIGC 1984-1 Faigle, U.; Goecke, O.; Schrader, R.
On simplicial decomposition in combinatorial structures.

FAIGF 1987-2 Faigle, U.; Schrader, R.
Setup optimization problems with matroid structure.

FEDE 1986-3 Federgruen, A.; Groenevelt, H.
The greedy procedure for resource allocation problems: necessary and sufficient conditions for optimality.

FENN 1985-1 Fenner, T. I.; Frieze, A. M.
An algorithm for finding a matroid basis which maximizes the products of the weights of the elements.

FENT 1984-1 Fenton, N. E.
Matroid representations - an algebraic treatment.

FENTA 1987-1 Fenton, W. E.
Completeness in oriented matroids.

FRANC 1984-1 Frank, A.
Finding feasible vectors of Edmonds-Giles polyhedra.

FRANC 1984-2 Frank, A.
Generalized polymatroids.

FRANC 1984-3 Frank, A.
Submodular flows.

FRANE 1984-1 Frank, A.; Tardos, E.
An algorithm for the unbounded matroid intersection polyhedron.

FRANE 1984-2 Frank, A.; Tardos, E.
Matroids from crossing families.

FRANE 1985-2 Frank, A.; Tardos, E.
Generalized polymatroids and submodular flows.

FREDC 1987-1 Frederickson, G. N.; Srinivas, M. A.
Algorithms and data structures for an expanded family of matroid intersection problems.

FREDC 1987-2 Frederickson, G. N.; Srinivas, M. A.
On-line updating of solutions to a class of matroid intersection problems.

FUJIA 1984-1 Fujishige, S.
A system of linear inequalities with a submodular function on $\{0,+-1\}$vectors.

FUJIA 1984-2 Fujishige, S.
Combinatorial optimization problems described by submodular functions.

FUJIA 1984-3 Fujishige, S.
Submodular systems and related topics.

FUJIA 1987-1 Fujishige, S.
An out-of-kilter method for submodular flows.

FUJIA 1987-2 Fujishige, S.
Optimization over the polyhedron determined by a submodular function on a co-intersection family.

FUJIB 1986-1 Fujishige, S.; Katoh, N.; Ichimori, T.
The fair resource allocation problem with submodular constraints and its solution algorithm.

FUJID 1986-1 Fujishige, S.; Röck, H.; Zimmermann, U.
A strongly polynomial algorithm for minimum cost submodular flow problems.

FUKU 1987-1 Fukuda, K.; Tamura, A.
Local deformation and orientation transformation in oriented matroids.

GABOD 1985-1 Gabow, H. N.; Stallmann, M.
Efficient algorithms for graphic matroid intersection and parity.

GABOD 1986-1 Gabow, H. N.; Stallmann, M.
An augmenting path algorithm for linear matroid parity.

GABOE 1984-1 Gabow, H. N.; Tarjan, R. E.
Efficient algorithms for a family of matroid intersection problems.

GAIN 1987-1 Gainanov, D. N.
On the connectedness of the graphs of some classes of independence systems (Russian).

GAINA 1987-1 Gainanov, D. N.; Gusak, I. Y.
On the number of bases of a vector matroid that is generated by a positive basis (Russian).

GALLA 1987-1 Gallo, G.; Simeone, B.
On the supermodular knapsack problem.

GAST 1986-1 Gastou, G.; Johnson, E. L.
Binary group and Chinese postman polyhedra.

GELF 1987-1 Gelfand, I. M.; Goresky, R. M.; Mac Pherson, R. D.; Serganova, V. V.
Combinatorial geometries, convex polyhedra, and Schubert cells.

GELFA 1987-1 Gelfand, I. M.; Serganova, V. V.
On the general definition of a matroid and a greedoid.

GERAB 1986-2 Gerards, A. M. H.; Schrijver, A.
Signed graphs - regular matroids - grafts.

GIRL 1985-1 Girlich, E.; Kovalev, M. M.
Klassifizierung von Polymatroiden.

GIRLA 1985-1 Girlich, E.; Schlechtweg, M.
Struktur von Polymatroiden mit submodularen Funktionen des Problems der optimalen Standardisierung.

GLEB 1983-1 Glebov, N. I.
Minimization of a convex separable function on an intersection of polymatroids (Russian).

GLEB 1983-2 Glebov, N. I.
On the problem of minimizing a convex separable function on the intersection of polymatroids.

GLOVH 1986-1 Glover, F.; Novick, B.
The 2-quasi-greedy algorithm for cardinality constrained matroid bases.

GOEC 1986-1 Goecke, O.
A greedy algorithm for hereditary set systems and a generalization of the Rado-Edmonds characterization of matroids.

GOEC 1986-2 Goecke, O.
Eliminationsprozesse in der Kombinatorischen Optimierung - ein Beitrag zur Greedoidtheorie.

GOECA 1987-1 Goecke, O.; Korte, B.; Lovasz, L.
Examples and algorithmic properties of greedoids.
GOECB 1986-1 Goecke, O.; Schrader, R.
Minor characterization of undirected branching greedoids - a short proof.
GOET 1986-1 Goetschel, R. H. Jr.
Linear objective functions on certain classes of greedoids.
GORDA 1984-1 Gordon, G.
Matroids over F(p) which are rational excluded minors.
GORG 1982-1 Gorgos, I. M.
Internally stable sets and the method of alternating sequences (Russian).
GREED 1983-1 Greene, C.
On the interpretation of Whitney numbers through arrangements of hyperplanes, zonotopes, non-Radon partitions, and orientations of graphs.
GRIS 1983-1 Grishukhin, V. P.
Submodular problems and algorithms for solving them.
GROTC 1984-1 Grötschel, M.; Lovasz, L.; Schrijver, A.
Corrigendum to our paper "The ellipsoid method and its consequences in combinatorial optimization".
GROTF 1986-1 Grötschel, M.; Truemper, K.
Decomposition and optimization over cycles in binary matroids.
GROTF 1986-2 Grötschel, M.; Truemper, K.
Master polytopes for cycles of binary matroids.
GUANA 1985-1 Guan, M.-G.; Chen, Q.-H.
A simple algorithm for finding the second order-constrained base of matroid.
GUANA 1985-2 Guan, M.-G.; Chen, Q.-H.
Order of minimal bases of restricted degree in a matroid (Chinese).
HALI 1981-1 Halin, R.
Graphentheorie 2.
HAMA 1985-1 Hamacher, H. W.
A time expanded matroid algorithm for finding optimal dynamic matroid intersections.
HAMA 1986-1 Hamacher, H. W.
Maximal dynamic polymatroid flows and applications.
HAMAC 1986-1 Hamacher, H. W.; Queyranne, M.
K best solutions to combinatorial optimization problems.
HAMI 1984-1 Hamidoune, Y. O.; Vergnas, M. Las
Directed switching games on graphs and matroids (French).
HAMI 1986-1 Hamidoune, Y. O.; Vergnas, M. Las
Switching games on graphs and matroids.
HASSD 1987-1 Hassin, R.; Tamir, A.
Maximizing classes of two parameter objectives over matroids.
HELLB 1985-1 Hell, P.; Speer, E. R.
Matroids with weighted bases and Feynman integrals.
HEXE 1985-1 Hexel, E.
On a class of greedoids (German).
HEXE 1985-2 Hexel, E.
On vertex sets in graphs with neighbourhood constraints - the component greedoid of a graph.
HIGG 1984-1 Higgs, D.
Topological spaces which are also matroidal.
HOFF 1985-1 Hoffman, A. J.
On greedy algorithms that succeed.
HOFFA 1985-1 Hoffman, A. J.; Kolen, A. W. J.; Sakarovitch, M.
Totally-balanced and greedy matrices.
HOPP 1987-1 Hoppe, V.
Polymatroide im E(3,+).
ILEV 1984-2 Ilev, V. P.
On the basis graphs of polymatroids (Russian).
ILEV 1985-1 Ilev, V. P.
Structure of intersection of the partitioning matroids (Russian).
IMAI 1985-1 Imai, H.
On combinatorial structures of line drawings of polyhedra.
INOU 1984-1 Inoue, M.; Onozawa, A.
On quasimatroids (Japanese).
IRI 1984-1 Iri, M.
Structural theory for the combinatorial systems characterized by submodular flows.
IRI B 1975-1 Iri, M.; Tomizawa, N.
A unifying approach to fundamental problems in network theory by means of matroids.
ISAC 1984-1 Isachenko, A. N.
Polynomial reducibility of matroid oracles (Russian).
IWAM 1987-1 Iwamura, K.
Lexicographically optimal base of a submodular system with respect to a weight vector.
JAEG 1985-2 Jaeger, F.
Graphic description of binary spaces.
JAMIB 1984-1 Jamison, R. E.; White, M. M.
Coronation numbers and circular systems in block graphs.
JAMS 1987-1 Jamshy, U.; Raspaud, A.; Tarsi, M.
Short circuit covers for regular matroids with a nowhere zero 5-flow.
JOHNJ 1987-1 Johnson, E. L.; Mosterts, S.
On four problems in graph theory..
JUNGC 1986-1 Jungnickel, D.
Transversaltheorie: ein Überblick.
JUNGC 1987-1 Jungnickel, D.
Graphen, Netzwerke und Algorithmen.
KAHN 1984-2 Kahn, J.
A geometric approach to forbidden minors for GF(3).

KAHN 1985-1 Kahn, J.
A problem of P. Seymour on nonbinary matroids.

KAHNA 1987-1 Kahn, J.; Seymour, P. D.
On forbidden minors for GF(3).

KAJI 1984-1 Kajitani, Y.; Ueno, S.
A matroid generalization of theorems of Lewin and Gallai.

KANE 1986-1 Kaneko, H.; Fushimi, M.
A polymatroid associated with convex games.

KARPG 1985-2 Karp, R. M.; Upfal, E.; Wigderson, A.
The complexity of parallel computation on matroids.

KASC 1985-1 Kaschube, P. A.
Determining when a graphic matroid is transversal in linear time.

KELLA 1984-1 Kelly, D. G.; Oxley, J. G.
On random representable matroids.

KELMA 1986-1 Kelmans, A. K.; Polesskii, V. P.
Extremal sets and covering and packing problems in matroids (Russian).

KERN 1985-2 Kern, W.
Lattice theoretic duality in combinatorial geometries and oriented matroids (German).

KERN 1985-3 Kern, W.
On finite locally projective planar spaces.

KERN 1985-4 Kern, W.
On modular embeddings of geometric lattices (extended abstract).

KERN 1985-5 Kern, W.
Verbandstheoretische Dualität in kombinatorischen Geometrien und orientierten Matroiden.

KHACA 1986-1 Khachaturov, V. R.
Models and methods for solving multiextremal allocation problems by the use of properties of supermodular functions defined on Boolean lattices (Russian).

KINDB 1986-1 Kindler, J.
A Mazur-Orlicz type theorem for submodular set functions.

KLAF 1985-1 Klafszky, E.; Terlaky, T.
Remarks on the feasibility problem of oriented matroids.

KLAF 1986-1 Klafszky, E.; Terlaky, T.
A new approach to the feasibility problem of oriented matroids (Hungarian).

KORTA 1984-1 Korte, B.
Kombinatorische Optimierung und algorithmische Prinzipien.

KORTA 1986-1 Korte, B.
Kombinatorische Optimierung und algorithmische Prinzipien.

KORTB 1984-1 Korte, B.; Lovasz, L.
Greedoids - a structural framework for the greedy algorithm.

KORTB 1985-1 Korte, B.; Lovasz, L.
A note on selectors and greedoids.

KORTB 1985-2 Korte, B.; Lovasz, L.
Basis graphs of greedoids and two-connectivity.

KORTB 1985-3 Korte, B.; Lovasz, L.
On submodularity in greedoids and a counterexample.

KORTB 1985-4 Korte, B.; Lovasz, L.
Polyhedral results for antimatroids.

KORTB 1985-5 Korte, B.; Lovasz, L.
Polymatroid greedoids.

KORTB 1985-6 Korte, B.; Lovasz, L.
Posets, matroids and greedoids.

KORTB 1985-7 Korte, B.; Lovasz, L.
Relations between subclasses of greedoids.

KORTB 1986-1 Korte, B.; Lovasz, L.
Homomorphisms and Ramsey properties of antimatroids.

KORTB 1986-2 Korte, B.; Lovasz, L.
Non-interval greedoids and the transposition property.

KORTB 1986-3 Korte, B.; Lovasz, L.
The intersection of matroids and antimatroids.

KOVAD 1985-1 Kovalev, M. M.
Cones of symmetric and flow submodular functions (Russian).

KOVAD 1987-1 Kovalev, M. M.
Matroids in discrete optimization (Russian).

KOVAF 1984-1 Kovalev, M. M.; Milanov, P. B.
Monotonic functions of many-valued logic and supermatroids.

KOVAG 1984-1 Kovalev, M. M.; Pisaruk, N. N.
Generalized supermatroids (Russian).

KOVAG 1984-2 Kovalev, M. M.; Pisaruk, N. N.
Independent flows and polymatroids (Russian).

KUNG 1986-1 Kung, J. P. S.
A source book in matroid theory.

KUNG 1986-2 Kung, J. P. S.
Growth rates and critical exponents of classes of binary combinatorial geometries.

LAUR 1985-1 Laurent, M.
Upper bounds for the cardinality of s-distance codes.

LAUR 1986-1 Laurent, M.
A generalization of antiwebs to independence systems and their relation to set covering polytopes.

LAWL 1985-1 Lawler, E. L.
Submodular functions and polymatroid optimization.

LAWLC 1986-1 Lawler, E. L.; Martel, C. U.
Polymatroidal flows with lower bounds.

LEAH 1985-1 Leahu, V.
Über p-konforme Unabhängigkeitssysteme.

LECLA	1987-1	Leclerc, M. Eine Min-Max Beziehung für das Exakte Matroid Problem.
LEE J	1986-1	Lee, J. The incidence structure of subspaces with well-scaled frames.
LEE J	1987-1	Lee, J. Turan's triangle theorem and binary matroids.
LEMO	1985-1	Lemos, M. On Seymour's question about packing and covering with matroid circuits.
LINDA	1983-1	Lindström, B. Some examples of non-regular simplicial matroids.
LINDA	1983-2	Lindström, B. The non-Pappus matroid is algebraic.
LINDA	1984-1	Lindström, B. A simple nonalgebraic matroid of rank three.
LINDA	1984-2	Lindström, B. On algebraic representations of matroids.
LINDA	1985-1	Lindström, B. A Desarguesian theorem for algebraic combinatorial geometries.
LINDA	1985-2	Lindström, B. On the algebraic characteristic set for a class of matroids.
LINDA	1986-1	Lindström, B. A non-linear algebraic matroid with infinite characteristic set.
LINDA	1986-2	Lindström, B. On harmonic conjugates in full algebraic combinatorial geometries.
LINDA	1986-3	Lindström, B. The non-Pappus matroid is algebraic over any finite field.
LINDA	1987-1	Lindström, B. A class of nonalgebraic matroids of rank 3.
LINDA	1987-2	Lindström, B. A reduction of algebraic representations of matroids.
LINDA	1987-3	Lindström, B. An elementary proof in matroid theory using Tutte's coordinatization theorem.
LIU C	1984-2	Liu, G.-Z. A result on Welsh's conjecture (Chinese).
LIU C	1984-6	Liu, G.-Z. The connectivity of matroid base graphs (Chinese).
LIU C	1985-5	Liu, G.-Z. Matroid complexes - geometrical representations of matroids.
LIU C	1985-7	Liu, G.-Z. Verification of the Welsh conjecture for simple binary matroids (Chinese).
LIU C	1985-8	Liu, G.-Z. Welsh's conjecture is true for simple binary matroids.
LIU I	1987-2	Liu, Z. H. Matroid theory and its applications (Chinese).
LOEB	1986-1	Loebl, M.; Poljak, S. On matroids induced by packing subgraphs.
LOVA	1986-3	Lovasz, L. Theory of numbers, graphs and convexity.
LOVEC	1987-1	Lovetskij, S. E.; Melamed, I. I. Static flows in networks.
LOVI	1985-1	Lovinescu, C. Conditions for the existence of flux in a matroid.
LOVI	1986-1	Lovinescu, C. Flow in matroids with parallel elements.
MAC G	1986-1	Mac Diarmid, C. On the greedy algorithm with random costs.
MAHO	1985-1	Mahoney, C. On the unimodality of the independent set numbers of a class of matroids.
MARCH	1984-1	Marcotte, O.; Trotter, L. E. Jr. An application of matroid polyhedral theory to unit-execution time, tree-precedence job scheduling.
MARCJ	1983-1	Marcu, D. Note on the standard matrix representation of a matroid.
MARCJ	1984-1	Marcu, D. Note on the circuits of a perfect matroid design.
MARCJ	1986-2	Marcu, D. A note on the independence number of an identically self-dual perfect matroid design.
MARCJ	1986-6	Marcu, D. On the closed sets of a matroid (Romanian).
MART	1986-1	Martel, C. U. Lower bounds on parallel algorithms for finding the first maximal independent set.
MAZZ	1983-1	Mazzocca, F. On a characterization of Dilworth truncations of combinatorial geometries.
MAZZ	1984-1	Mazzocca, F. Extensions of combinatorial geometries by the addition of a unique line.
MIDD	1987-1	Middendorf, M. Symmetric matroids and connectivity properties of graphs.
MIHO	1987-1	Mihok, P. On graphs matroidal with respect to additive hereditary properties.
MILA	1987-1	Milanov, P. B. On the treshold independence systems.
MILAB	1983-1	Milanov, P. B.; Girlich, E. Zum Kegel der monoton wachsenden submodularen Funktionen.

MILL 1987-1 Miller, D. A.
Oriented matroids from smooth manifolds.

MORI 1985-1 Morikawa, T.
Chemical matroids, I: independence and dependence in chemical reaction systems.

MORR 1986-1 Morris, W. D.
Oriented matroids and the linear complementarity problem.

MORT 1985-1 Morton, G.; Randow, R. von; Ringwald, R.
A greedy algorithm for solving a class of convex programming problems and its connection with polymatroid theory.

MOSH 1986-1 Moshchenskii, A. V.
Semimatroids and Gale diagrams (Russian).

MUCH 1987-1 Muchnik, I. B.; Shvartser, L. V.
Submodular set functions and monotone systems in aggregation problems. I (Russian).

MUCH 1987-2 Muchnik, I. B.; Shvartser, L. V.
Submodular set functions and monotone systems in aggregation problems. II (Russian).

MUMM 1986-1 Mummy, M. S.
Full transversal matroids, strict gammoids, and the matroid components problem.

MUMM 1986-2 Mummy, M. S.
Strict matching matroids and matroid algorithms.

MUMM 1986-3 Mummy, M. S.
The generalized column incidence graph and a matroid base-listing algorithm.

MURO 1985-1 Murota, K.
Combinatorial canonical form of layered mixed matrices and block-triangularization of large-scale systems of linear/nonlinear equations.

MURO 1986-1 Murota, K.
Bimatroid network and its application to the structural solvability of systems of equations.

MURO 1986-2 Murota, K.
Combinatorial dynamical system theory.

MURO 1987-1 Murota, K.
Homotopy base of an acyclic graph a combinatorial analysis of commutative diagrams by means of preordered matroid.

MURO 1987-2 Murota, K.
Refined study on structural controllability of descriptor systems by means of matroids.

MURO 1987-3 Murota, K.
Systems analysis by graphs and matroids. Structural solvability and controllability.

MUROB 1985-1 Murota, K.; Iri, M.
Structural solvability of systems of equations - a mathematical formulation for distinguishing accurate and inaccurate numbers in structural analysis of systems.

NADD 1984-1 Naddef, D. J.
Pancyclic properties of the graph of some 0-1-polyhedra.

NAKAB 1986-1 Nakamura, M.
A decomposition of poly-linking systems.

NAKAB 1986-2 Nakamura, M.
Structural theorems for submodular functions, polymatroids and polymatroid intersections.

NAKAC 1981-1 Nakamura, M.; Iri, M.
A structural theory for submodular functions, polymatroids and polymatroid intersections.

NESEB 1985-1 Nesetril, J.; Poljak, S.; Turzik, D.
Special amalgams and Ramsey matroids.

ONOZ 1984-1 Onozawa, A.; Inoue, M.
Some problems on principal partition of 2 matroids (Japanese).

ORLIB 1987-1 Orlin, J. B.; Vandevate, J. H.
Solving the linear matroid parity problem as a sequence of matroid intersection problems.

OXLE 1984-1 Oxley, J. G.
On singleton 1-rounded sets of matroids.

OXLE 1984-2 Oxley, J. G.
On the intersections of circuits and cocircuits in matroids.

OXLE 1985-1 Oxley, J. G.
Proof of a conjecture of Kahn for non-binary matroids.

OXLE 1986-1 Oxley, J. G.
On matroids with no 5-point-line-minor.

OXLE 1986-2 Oxley, J. G.
On ternary transversal matroids.

OXLE 1986-3 Oxley, J. G.
On the matroids representable over GF(4).

OXLE 1987-1 Oxley, J. G.
A characterization of the ternary matroids with no M(K(4))-minor.

OXLE 1987-2 Oxley, J. G.
On nonbinary 3-connected matroids.

OXLE 1987-3 Oxley, J. G.
The binary matroids with no 4-wheel minor.

OXLEA 1987-1 Oxley, J. G.; Reid, T. J.
The smallest rounded sets of binary matroids.

PAZZ 1984-1 Pazzoli, L.
Hyperplanes and circuits of matroids on partially ordered sets (Italian).

PERF 1986-1 Perfilova, V. N.
On the number of transversal matroids.

PEZZ 1984-1 Pezzoli, L.
Hyperplanes and circuits of matroids on partially ordered sets (Italian).

POLJA 1984-1 Poljak, S.; Turzik, D.
Amalgamation of matroids along hypergraphs.

POLJA 1984-2 Poljak, S.; Turzik, D.
Amalgamation over uniform matroids.

POTY 1983-1 Potyagailo, Z. D.
Construction of a matroid of states of a dynamical system and algorithms for its investigation (Russian).
POTY 1984-1 Potyagailo, Z. D.
Matroid theory methods in the study of information systems dynamics (Russian).
PRODA 1984-1 Prodon, A.
A note on the separation problem for the matching matroid.
PRUS 1984-1 Pruss, V.
A characterization of matroidal families of multigraphs.
PUJA 1983-1 Pujari, A. K.; Das, P. K.
Monotonicity of polymatroids.
QI 1986-1 Qi, L.-Q.
Directed submodular functions, ditroids and total dual integrality.
QI 1987-1 Qi, L.-Q.
A rank characterization of 2-matroid intersections.
QI 1987-2 Qi, L.-Q.
Odd submodular functions, Dilworth functions and discrete convex functions.
RAMA 1986-2 Ramamurthy, K. G.; Parthasarathy, T.
Matroidal games.
RECS 1984-1 Recski, A.
Some problems of self-dual matroids.
RECS 1984-2 Recski, A.
Statics and electric network theory: a unifying role of matroids.
RECS 1985-1 Recski, A.
Some open problems of matroid theory, suggested by its applications.
RECS 1987-2 Recski, A.
Elementary strong maps of graphic matroids.
RIIV 1984-1 Riives, K.
A classification problem for polymatroids (Russian).
RIIV 1986-1 Riives, K.
A classification of three-dimensional polymatroids (Russian).
ROUS 1985-1 Rousseau, C. K.; Rousseau, C. C.; Liu, A. L.; Klamkin, M. S.
Greedy algorithm and coinage systems.
SCHL 1983-1 Schlechtweg, M.; Hoppe, V.
Polymatroide mit baumartigen Facettensystemen. I.
SCHL 1983-2 Schlechtweg, M.; Hoppe, V.
Polymatroide mit baumartigen Facettensystemen. II.
SCHL 1983-3 Schlechtweg, M.; Hoppe, V.
Polymatroide mit baumartigen Facettensystemen. III.
SCHMC 1985-1 Schmidt, W.
A characterization of undirected branching greedoids.
SCHMC 1985-2 Schmidt, W.
A min-max theorem for greedoids.
SCHMC 1985-3 Schmidt, W.
Greedoids and searches in directed graphs.
SCHRE 1984-1 Schrijver, A.
Total dual integrality from directed graphs, crossing families, and sub- and supermodular functions.
SCHRE 1985-1 Schrijver, A.
Supermodular colourings.
SEYM 1985-1 Seymour, P. D.
Applications of the regular matroid decomposition.
SEYM 1985-2 Seymour, P. D.
Minors of 3-connected matroids.
SEYM 1986-1 Seymour, P. D.
Adjacency in binary matroids.
SEYM 1986-2 Seymour, P. D.
Triples in matroid circuits.
SHAM 1985-1 Shameeva, O. V.
Algebraic representability of matroids (Russian).
SHAM 1985-2 Shameeva, O. V.
On the algebraic representability of matroids (Russian).
SOCZ 1985-1 Soczewinska, E.; Szamkolowicz, L.
Chordal characterization of binary matroids.
STAL 1984-1 Stallmann, M.; Gabow, H. N.
An augmenting path algorithm for the parity problem on linear matroids.
STUR 1986-1 Sturmfels, B.
Oriented matroids and combinatorial convex geometry.
STURA 1985-1 Sturmfels, B.; Bokowski, J.
Programmsystem zur Realisierung orientierter Matroide.
SUSH 1985-1 Sushkov, Y. A.
Matroids and hypergraphs.
TARD 1985-2 Tardos, E.
Generalized matroids and supermodular colourings.
TARD 1986-1 Tardos, E.
A strongly polynomial algorithm to solve combinatorial linear programs.
TARDA 1986-1 Tardos, E.; Tovey, C. A.; Trick, M. A.
Layered augmenting path algorithms.
TARS 1985-1 Tarsi, M.
Nowhere zero flow and circuit covering in regular matroids.
TERL 1985-1 Terlaky, T.
A finite criss-cross method for oriented matroids (Hungarian).
TERL 1987-1 Terlaky, T.
A finite crisscross method for oriented matroids.

TINH 1984-1 Tinhofer, G.
A probabilistic analysis of some greedy cardinality matching algorithms.

TODD 1984-1 Todd, M. J.
Complementarity in oriented matroids.

TODD 1985-1 Todd, M. J.
Linear and quadratic programming in oriented matroids.

TOPKA 1984-1 Topkis, D. M.
Adjacency on polymatroids.

TRUE 1984-1 Truemper, K.
Elements of a decomposition theory for matroids.

TRUE 1984-2 Truemper, K.
Partial matroid representations.

TRUE 1985-1 Truemper, K.
A decomposition theory for matroids. I. General results.

TRUE 1985-2 Truemper, K.
A decomposition theory for matroids. II. Minimal violation matroids.

TRUE 1985-3 Truemper, K.
A flexible decomposition approach for finding submatrices with certain inherited properties.

TRUE 1986-1 Truemper, K.
A decomposition theory for matroids. III. Decomposition conditions.

TRUE 1987-1 Truemper, K.
A decomposition theory for matroids V. Testing of matrix total unimodularity.

TRUE 1987-2 Truemper, K.
Max-flow min-cut matroids: polynomial testing and polynomial algorithms for maximum flow and shortest routes.

TRUSB 1985-2 Truszczynski, M.
Jump number problem: the role of matroids.

TSEN 1986-1 Tseng, F. T.; Truemper, K.
A decomposition of the matroids with the max-flow min-cut property.

TUMA 1985-1 Tuma, J.
Dilworth truncations and modular cuts.

TUYT 1986-1 Tuyttens, D.; Teghem, J. Jr.
Theorie des matroides et optimisation combinatoire.

VADE 1986-1 Vaderlind, P.
Clutters and atomistic lattices.

VADE 1986-2 Vaderlind, P.
Clutters and semimatroids.

VAND 1987-1 Vandevate, J. H.
Structural properties of matroid matchings.

VERG 1984-1 Vergnas, M. Las
Oriented matroids as signed geometries real in corank 2.

VERG 1984-2 Vergnas, M. Las
The Tutte polynomial of a morphism of matroids II. Activities of orientations.

VERG 1986-1 Vergnas, M. Las
Hamilton paths in tournaments and a problem of McMullen on projective transformations in R**d.

VINC 1983-1 Vince, A.
Generalization of the cycle matroid of a graph.

WAGN 1985-1 Wagner, D. K.
Connectivity in bicircular matroids.

WAGN 1985-2 Wagner, D. K.
On theorems of Whitney and Tutte.

WAGN 1986-1 Wagner, D. K.
Equivalent factor matroids of graphs.

WANGM 1984-1 Wang, Z.-M.
A proof of constructive type for a theorem by Welsh (Chinese).

WANGM 1987-1 Wang, Z.-M.
A finite conformal-elimination free algorithm over oriented matroid programming.

WANGM 1987-2 Wang, Z.-M.
An extended version of the cardinality matroid intersection algorithm (Chinese).

WANK 1986-1 Wanka, A.
Matroiderweiterungen zur Existenz endlicher LP-Algorithmen, von Hahn-Banach Sätzen und Polarität in orientierten Matroiden.

WARB 1985-1 Warburton, A.
Worst case analysis of greedy and related heuristics for some min-max combinatorial optimization problems.

WENZ 1986-1 Wenzel, W.
Über die Struktur der Tutte-Gruppe von Matroiden.

WHITC 1986-1 White, N.
Theory of matroids.

WHITC 1987-1 White, N.
Combinatorial Geometries.

WHITD 1986-1 White, N. L.
A pruning theorem for linear count matroids.

WHITH 1984-1 Whittle, G. P.
On the critical exponent of transversal matroids.

WHITH 1985-1 Whittle, G. P.
An elementary proof that every matroid is an intersection of principal transversal matroids.

WHITH 1987-1 Whittle, G. P.
Modularity in tangential k-blocks.

WHITH 1987-2 Whittle, G. P.
Some aspects of the critical problem for matroids.

WOLS 1981-1 Wolsey, L. A.
An analysis of the greedy algorithm for the submodular (set) covering problem.

XIE 1983-1 Xie, L.-T.
Methods of algebraic topology applied to the theory of graphs and matroids (Chinese).
XIE 1987-1 Xie, L.-T.
A supplement to the paper: "Method of algebraic topology applied to the theory of graphs and matroids" (Chinese).
YAN 1985-1 Yan, J.-Y.
Greedoid polyhedra.
ZASL 1987-2 Zaslavsky, T.
The biased graphs whose matroids are binary.
ZHAO 1985-1 Zhao, C. K.
A conjecture on matroids (Chinese).
ZIMMC 1984-1 Zimmermann, U.
Minimization of combined objective functions on integral submodular flows.
ZIMMC 1984-2 Zimmermann, U.
Submodulare Flüsse: Verfahren zur Minimierung linearer Zielfunktionen.
ZIMMC 1985-1 Zimmermann, U.
Augmenting circuit methods for submodular flow problems.
ZIMMC 1986-1 Zimmermann, U.
Duality for balanced submodular flows.
ZIMMC 1986-3 Zimmermann, U.
Sharing problems.

ABDUA 1985-1 Abdullaev, A. K.; Kiselev, V. G.
A method for solving the Benders integer subproblem in some problems of plant location (Russian).

ADAMC 1985-1 Adams, W. P.; Sherali, H. D.
Mixed-integer bilinear programming problems.

AKINA 1986-1 Akinc, U.; Roodman, G. M.
A new approach to aggregate production planning.

ARTI 1984-1 Artina, S.; Walker, J.
The use of mixed-integer linear programming in designing hydraulic networks (Italian).

BABAC 1983-1 Babaev, R. D.
A problem for the choice of an optimal sequence of construction of objects (Russian).

BANK 1987-1 Bank, B.; Mandel, R.
The existence of optimal points for (mixed) integer optimization problems.

BANK 1987-2 Bank, B.; Mandel, R.
Mixed-integer solutions of quasiconvex polynomial inequalities.

BANK 1987-4 Bank, B.; Mandel, R.
Quantitative stability of (mixed) integer linear optimization problems.

BECHA 1984-1 Bechtold, S. E.; Janaro, R. E.; Sumners, D. L.
Maximization of labor productivity through optimal rest-break schedules.

BENDC 1985-1 Benders, J. F.; Beulens, A. J. M.; Nunen, J. A. E. E. van
On using the linear programming relaxation of assignment type mixed integer problems.

BLAIA 1984-1 Blair, C. E.; Jeroslow, R. G.
Constructive characterizations of the value-function of a mixed-integer program, I.

BLAIA 1985-1 Blair, C. E.; Jeroslow, R. G.
Constructive characterizations of the value-function of a mixed-integer program, II.

BRUV 1985-1 Bruvold, N. T.; Evans, J. R.
Flexible mixed-integer programming formulations for production scheduling problems.

BUJA 1985-1 Bujan, M. F.
Mathematical model for planning in an industrial branch.

CAPO 1984-1 Caponi, M.; Pieroni, P.; Saviozzi, G.
A Benders dual decomposition approach to the optimum network orientation.

COOKE 1986-1 Cook, W.; Kannan, R.; Schrijver, A.
Chvatal closures for mixed integer programming problems.

CRAMC 1986-1 Crama, Y.; Hansen, P.; Williams, A. C.
Use of the dual in branch and bound methods for mixed 0-1 programming.

DEAK 1983-1 Deak, I.; Hoffer, J.; Mayer, J.; Nemeth, A.; Potecz, B.; Prekopa, A.; Strazicky, B.
A large scale, mixed variable mathematical programming model for the short-range optimal schedule of electric power systems with thermal power plants taking network constraints into account (Hungarian).

DEMP 1983-1 Dempe, S.
Zwei greedy-Algorithmen für ein gemischt-ganzzahliges Optimierungsproblem.

DEMP 1985-1 Dempe, S.
A simple but NP-hard problem of mixed-discrete programming and its solution by approximate algorithms.

DUCR 1985-1 Ducrocq, J.-C.
Optimal energy management in a complex system with mixed-integer programming.

DURA 1987-1 Duran, F.
A large mixed integer production and distribution program.

DURAA 1986-1 Duran, M. A.; Grossmann, I. E.
An outer-approximation algorithm for class of mixed-integer nonlinear programs.

EPPEA 1986-1 Eppen, G. D.; Martin, R. K.; Schrage, L.
A mixed-integer programming model for evaluating distribution configurations in the steel industry.

ERIC 1987-1 Erickson, R. E.; Monma, C. L.; Veinott, A. F. Jr.
Send-and-split method for minimum-concave-cost network flows.

GHANA 1986-1 Ghandforoush, P.; Greber, B. J.
Solving allocation and scheduling problems inherent in forest resource management using mixed-integer programming.

GLOVE 1986-1 Glover, F.; Klingman, D.; Phillips, N.; Ross, G. T.
Integrating modeling, algorithm design, and computational implementation to solve a large-scale nonlinear mixed integer programming problem.

GOIC 1985-1 Goicoechea, A.; Soland, R. M.; Sharp, F. M.
A transportation model for inland waterways navigation via mathematical programming.

GREEE 1986-1 Greene, T. J.; Sadowski, R. P.
A mixed integer program for loading and scheduling multiple flexible manufacturing cells.

GUIGC 1985-1 Guignard, M.; Spielberg, K.
Separable-Lagrangean relaxation for weakly linked mixed integer programming problems.

GUIGC 1987-1 Guignard, M.; Spielberg, K.
On combining Lagrangean relaxation, Benders inequalities and Lagrangean dual ascent.

HAN 1982-1 Han, F.
An integer programming algorithm for the plan of a freight car classification system (Chinese).

HANS 1984-1 Hanscom, M. A.; Nguyen, V. H.; Strodiot, J.-J.
A reduced subgradient algorithm for network flow problems with convex nondifferentiable costs.

HANSC 1987-1 Hansen, P.; Huge, J.
Implicit treatment of zero or range constraints in a model for minimun cost foundry alloys.

HANSG 1987-1 Hansen, P.; Minoux, M.; Labbe, M.
Extension de la programmation lineaire generalisee au cas des programmes mixtes.

HIRSA 1984-1 Hirshfeld, D. S.
An overview of mathematical programming for strategic planning.

HOFS 1985-1 Hofstedt, K.
Zur Kopplung eines exakten Verfahrens mit einem heuristischen Verfahren für die Lösung ganzzahliger linearer Optimierungsprobleme.

JEROA 1984-1 Jeroslow, R. G.
On monotone chaining procedures of the CF type.

JEROA 1985-2 Jeroslow, R. G.
Alternative formulations of mixed integer programs.
JEROA 1985-3 Jeroslow, R. G.
An extension of mixed-integer programming models and techniques to some database and artificial intelligence settings.
JEROA 1985-4 Jeroslow, R. G.
Computation-oriented reductions of predicate to propositional logic.
JEROA 1986-1 Jeroslow, R. G.
Ten lectures on mixed integer model formulation for logic-based decision support.
JEROA 1987-1 Jeroslow, R. G.
Representability in mixed integer programming, I: characterization results.
JEROB 1984-1 Jeroslow, R. G.; Lowe, J. K.
Modelling with integer variables.
JEROB 1985-1 Jeroslow, R. G.; Lowe, J. K.
Experimental results on the new techniques for integer programming formulations.
KHANC 1983-1 Khan, Z. A.; Ahsan, M. J.; Khan, E. A.
On mixed integer concave quadratic programs.
KIM C 1985-1 Kim, S.-H.
Sufficient conditions for zero duality gap of Lagrangean relaxation.
LAZI 1985-1 Lazimy, R.
Improved algorithm for mixed-integer quadratic programs and a computational study.
LE 1984-1 Le, W.-L.; Liu, Z. Y.; Wang, Z.-Y.
The algorithm for mixed programming with upper bounded variables and its application to energy planning (Chinese).
LEISA 1983-1 Leiserson, C. E.; Saxe, J. B.
A mixed-integer linear programming problem which is efficiently solvable.
MARTE 1985-1 Martin, R. K.
Generating alternative mixed - 0/1 linear programming models using variable redefintion (revised version).
MARTG 1985-1 Martin, R. K.; Schrage, L.
Subset coefficient reduction cuts for 0/1 mixed-integer programming.
MARTH 1985-1 Martin, R. K.; Sweeney, D. J.; Doherty, M. E.
The reduced cost branch and bound algorithm for mixed integer programming.
MASH 1983-1 Mashchenko, S. O.
Sequential algorithm to solve problems of mixed linear programming (Russian).
MEHRB 1985-1 Mehrez, A.; Stern, H. I.
Optimal refueling strategies for a mixed-vehicle fleet.
MELAB 1986-1 Melachrinoudis, E.; Rumpf, D.; Venegas, R.
Spray operation planning aided by a mixed integer programming model.
MIKHA 1986-1 Mikhalevich, M. V.
A stochastic algorithm for solving nonlinear mixed integer problems (Russian).
MISH 1983-1 Misha, M. K.; Das, C.
Maximin problem and symmetric duality for mixed-integer quadratic programs.
MISHC 1984-1 Mishra, B. K.; Das, C.
Minimax and symmetric duality for nonlinear mixed-integer programming problem.
MISHC 1985-1 Mishra, B. K.; Das, C.
Maximin problem and a duality theorem for mixed-integer quadratic programming.
MISHD 1985-1 Mishra, M. S.; Nanda, S.
On a pair of nonlinear mixed integer programming problems.
MULLB 1987-1 Müller, P.
Untersuchungen zu speziellen linearen gemischt-ganzzahligen 0-1-Optimierungsaufgaben.
MUST 1985-1 Mustafi, C. K.; Xavier, M. J.
Mixed-integer linear-programming formulation of a multi-attribute threshold model of choice.
NAND 1987-1 Nanda, S.; Mishra, M. S.; Acharya, D.
A note on a pair of nonlinear mixed integer programming problems.
NEMHA 1984-1 Nemhauser, G. L.; Wolsey, L. A.
A recursive procedure for generating all mixed integer cuts.
NUNE 1983-1 Nunen, J. A. E. E. van; Beulens, A. J. M.; Benders, J. F.
On using the linear programming relaxation of assignment type mixed integer problems.
NUNE 1984-1 Nunen, J. A. E. E. van; Beulens, A. J. M.; Benders, J. F.
On solving assignment type mixed integer linear programming problems within decision support systems.
OHTA 1985-1 Ohtake, Y.; Nishida, N.
A branch-and-bound algorithm for 0-1 parametric mixed integer programming.
OLEY 1983-1 Oley, L. A.; Sjoquist, R. J.
Automatic reformulation of mixed and pure integer models to reduce solution time in APEX IV.
ONUR 1987-1 Onur, L.; Fabrycky, W. J.
An input/output control system for the dynamic job shop.
POPP 1985-1 Popp, W.
Zur Analyse strategischer Aspekte mit gemischt-ganzzahliger Programmierung.
SANNA 1985-1 Sannomiya, N.; Okamoto, K.
A method for decomposing mixed-integer linear programs with staircase structure.
SELE 1986-1 Selen, W. J.; Hott, D. D.
A mixed-integer goal-programming formulation of the standard flow-shop scheduling problem.
SERGG 1980-1 Sergienko, I. V.; Lebedeva, T. T.; Roshchin, V. A.
Approximate methods for solving discrete problems of optimization.
SGURC 1985-1 Sgurev, V. S.; Vasilev, V. S.
An approximate algorithm solving mixed integer programming problems (Bulgarian).
SHAT 1983-1 Shatalina, L. S.
The compiling of a guaranteed spare-parts supply schedule formulated as a mixed linear programming problem.
SMIT 1987-1 Smith-Daniels, D. E.; Smith-Daniels, V. L.
Optimal project scheduling with materials ordering.

SMITA 1986-1 Smith-Daniels, V. L.; Smith-Daniels, D. E.
A mixed integer programming model for lot sizing and sequencing packaging lines in the process industries.

SRIK 1986-1 Srikar, B. N.; Ghosh, S.
A MILP model for the n-job, M-stage flowshop with sequence dependent set-up times.

SUHL 1985-1 Suhl, U. H.
Solving large scale mixed-integer programs with fixed charge variables.

SWEE 1984-1 Sweeney, D. J.
A practical evaluation of the potential of mixed integer programming as a strategic planning tool.

TANGC 1986-1 Tang, H.-W.; Sun, Z.-X.; Gao, G.-Q.
An algorithm for a class of nonlinear mixed integer programming problems, and its convergence (Chinese).

TAYI 1985-1 Tayi, G. K.
Sensitivity analysis of mixed integer programs: an application to environmental policy making.

TCHA 1985-1 Tcha, D.-W.
On the selection of primary paths for a communication network.

TSAR 1985-1 Tsarevskij, N. A.
An approximational decomposition algorithm for a combinatorial problem of developing production capacities of an industry with block production constraints II.

VANRA 1986-1 Vanroy, T. J.; Wolsey, L. A.
Valid inequalities for mixed 0-1 programs.

VANRA 1987-1 Vanroy, T. J.; Wolsey, L. A.
Solving mixed integer programming problems using automatic reformulation.

VOLKB 1986-1 Volkovich, S. V.
A parallel algorithm for the sequential analysis of variants for a problem of the block-diagonal structure of mixed linear programming (Russian).

VOLKC 1987-1 Volkovich, S. V.; Voloshin, A. F.
A successive approach to the solution of problems of linear partially integer programming of special structure (Russian).

WILL 1986-1 Williams, A. C.
Marginal values in mixed integer linear programming.

ZHENA 1984-1 Zheng, Q.
Global optimization of integer and mixed programming (Chinese).

ZINC 1984-1 Zinchenko, A. B.
Properties and conditions for solvability of a special problem of optimal geometric allocation (Russian).

ALI D 1985-1 Ali, I.; Charnes, A.; Song, T.
Karmarkar's projective algorithm: a null space variant for multi-commodity generalized networks.

ANEJA 1985-1 Aneja, Y. P.; Nair, K. P. K.
Network flows with age dependent decay rates.

ARON 1986-1 Aronson, J. E.
The multiperiod assignment problem: a multicommodity network flow model and specialized branch and bound algorithm.

BARA 1987-2 Barahona, F.
Planar multicommodity flows, max cut and the Chinese postman problem.

BATY 1983-1 Batycki, T.; Kasprzak, A.
Flow assignment algorithms for computer communication networks (Polish).

BIELB 1984-1 Bielli, M.; Cini, M.; Calicchio, G.; Magliozzi, L.; Oddone, M.
Network models in planning the new generation of submarine communication systems.

CALI 1985-1 Calicchio, G.
Introduction to the methodology and techniques for the solution of multicommodity flow problems (Italian).

DESRA 1988-1 Desrochers, M.; Kolen, A. W. J.; Lucena, A.
A 2-commodity flow formulation for the vehicle routing problem.

FANG 1985-1 Fang, S. C.; Peterson, E. L.
An economic equilibrium model on a multicommodity network.

FINKF 1984-1 Finke, G.; Claus, A.; Gunn, E. A.
A two-commodity network flow approach to the traveling salesman problem.

FLORA 1986-1 Florian, M.
Nonlinear cost network models in transportation analysis.

FRANC 1985-1 Frank, A.
Edge-disjoint paths in planar graphs.

GAUL 1984-1 Gaul, W.
Optimal routes in compound transportation systems.

GOME 1987-1 Gomes, L. F. A. M.
On the problem of assigning traffic flows to a multicriteria, multicommodity, free-flow network.

HACK 1986-1 Hackbarth, K.-D.
A heuristic procedure for calculating telecommunication transmission networks in consideration of network reliability.

HASSB 1987-2 Hassin, R.
Solution bases of multiterminal cut problems.

HEMP 1985-1 Hempel, L.; Herrmann, U.; Hölzer, W.
Einige spezielle Probleme im Zusammenhang mit Multiströmen.

HURK 1986-1 Hurkens, C. A. J.; Schrijver, A.; Tardos, E.
On fractional multicommodity flows and distance functions.

JORNA 1986-1 Jörnsten, K. O.; Värbrand, P.
An augmented Lagrangian approach to multi-commodity flow problems with conversions.

KAPO 1986-1 Kapoor, S. F.; Vaidya, P. M.
Fast algorithms for convex quadratic programming and multicommodity flows.

KARZ 1984-1 Karzanov, A. V.
A generalized MFMC-property and multicommodity cut problems.

KARZ 1985-3 Karzanov, A. V.
Metrics and undirected cuts.

KARZ 1985-4 Karzanov, A. V.
Multiproduct flow problems with integer optimal solutions (Russian).

KARZ 1985-5 Karzanov, A. V.
On multicommodity flow problems with integer-valued optimal solutions.

KARZ 1987-1 Karzanov, A. V.
Half-integral five-terminus flows.

KIM B 1984-1 Kim, K. V.; Cherkasskii, B. V.
A program for solving a multicommodity transportation problem (Russian).

KOPY 1984-1 Kopylov, G. N.; Levin, A. Y.
On the realizability of a stochastic multiproduct flow.

LAMA 1988-1 Lamatsch, A.
Wagenumlaufplanung im ÖPNV bei begrenzten Betriebshofkapazitäten - ein ganzzahliges Mehrgüterflussmodell mit Lösungsansätzen.

LINS 1985-1 Lins, S.
An uncrossing algorithm fo the multicommodity flow problem in plane graphs.

LOMO 1985-2 Lomonosov, M. V.
Combinatorial approaches to multiflow problems.

MARTN 1985-1 Martins, E. Q. V.; Rosa, M. S.
An algorithm for the maximal multicommodity funnel-node flow in an undirected network.

MATS 1983-1 Matsui, K.; Takada, J.
SLP method for a nonlinear multicommodity flow assignment problem.

MATSA 1984-1 Matsumoto, K.; Nishizeki, T.; Saito, N.
Algorithms for plane multicommodity flows.

MATSA 1985-1 Matsumoto, K.; Nishizeki, T.; Saito, N.
An efficient algorithm for finding multicommodity flows in planar networks.

MATSA 1986-1 Matsumoto, K.; Nishizeki, T.; Saito, N.
Planar multicommodity flows, maximum matchings and negative cycles.

MATUA 1987-1 Matula, D. W.; Shahrokhi, F.
Sparsest cuts and bottlenecks in graphs.

MOIS 1984-1 Moiseenko, G. E.
A multicommodity network flow problem with commodity conversion and dependent flows.

OKAM 1984-1 Okamura, H.
Multicommodity flows in graphs, II.

PILI 1984-1 Pilipchuk, L. A.
An algorithm for solving a two-commodity transport problem with supplementary matrix of constraints by a direct method (Russian).

PILI 1984-2 Pilipchuk, L. A.
On dual methods for solving a two-commodity transportation network problem (Russian).
PILI 1986-1 Pilipchuk, L. A.
Algorithms for solving an inhomogeneous distributive network problem (Russian).
SAVI 1986-1 Saviozzi, G.
Advanced start for the multicommodity network flow problem.
SCHMB 1986-1 Schmidt, G.
A fertilizer transportation problem.
SCHRA 1986-1 Schreier, H.
A variant for checking a solution criterion for the balanced multicommodity transportation problem (German).
SHAH 1987-1 Shahrokhi, F.
A primal dual approximation algorithm for a multicommodity flow problem.
SHAHA 1987-1 Shahrokhi, F.; Matula, D. W.
The maximum concurrent flow problem.
SHEVB 1986-1 Shevchik, V. I.
A problem of multicommodity flow (Russian).
SOKO 1983-1 Sokolov, A. V.
A heuristic algorithm for the distribution of multicommodity flow (Russian).
SUZU 1985-1 Suzuki, H.; Nishizeki, T.; Saito, N.
Multicommodity flows in planar undirected graphs and shortest paths.
TARD 1986-1 Tardos, E.
A strongly polynomial algorithm to solve combinatorial linear programs.
ZHAOA 1982-1 Zhao, F.-Z.
The potential method for solving the multicommodity capacity transportation problem (Chinese).

AGAR 1985-1 Agarwal, S.; Mittal, A. K.; Sharma, P.
A decomposition algorithm for linear relaxation of the weighted r-covering problem.

AHLF 1985-1 Ahlfield, D. P.; Dembo, R. S.; Mulvey, J. M.; Zenios, S. A.
Nonlinear programming on generalized networks.

AHMA 1986-1 Ahmad, S. H.
GERT models in an educational institution.

AHUJB 1987-1 Ahuja, R. K.; Orlin, J. B.
A fast and simple algorithm for the maximum flow problem.

AJYA 1984-1 Ajyashi, M.
An optimality criterion in a dynamic problem on minimal cost flow with additional constraints (Russian).

AKUL 1983-1 Akulich, F. E.
A problem on maximal flow on bottlenecks (Russian).

ALI B 1986-1 Ali, A. I.; Kennington, J. L.; Shetty, B.
The linear equal flow problem.

ALI C 1986-1 Ali, A. I.; Padman, R.; Thiagarajan, H.
Dual simplex-based reoptimization procedures for network problems.

ALI C 1986-2 Ali, A. I.; Padman, R.; Thiagarajan, H.
Dual simplex-based reoptimization procedures for network problems.

ALI C 1986-3 Ali, A. I.; Padman, R.; Thiagarajan, H.
The dual algorithm for network problems: a computational study.

ALI D 1984-1 Ali, I.; Charnes, A.; Song, T.
Design and implementation of data structures for generalized networks.

ALI E 1985-1 Ali, I.; Kennington, J. L.; Shetty, B.
The equal flow problem.

ANDR 1983-1 Andrasfai, B.
Graph theory. Network flows. Matrices. (Hungarian).

ANKL 1986-1 Anklesaria, K. P.; Drezner, Z.
A multivariate approach to estimating the completion time for PERT networks.

ANST 1986-1 Anstee, R. P.
Invariant sets of arcs in network flow problems.

ARON 1985-1 Aronson, J. E.
A survey of dynamic network flows.

ARONB 1984-1 Aronson, J. E.; Chen, B. D.
A computational study of empirical decision horizons in infinite horizon, multiperiod network flow problems.

ARONB 1985-1 Aronson, J. E.; Chen, B. D.
A forward network simplex algorithm for solving multiperiod network flow problems.

ARONB 1985-2 Aronson, J. E.; Chen, B. D.
A primary/secondary memory implementation of a forward network simplex algorithm for multiperiod network flow problems.

ARONB 1986-1 Aronson, J. E.; Chen, B. D.
A forward network simplex algorithm for solving multiperiod network flow problems.

BAKE 1985-1 Bakenrot, V. Y.
The efficiency of planning algorithms in multiprocessor networks.

BAKEA 1984-1 Baker, B. M.
A network-flow algorithm for project selection.

BAKEF 1981-1 Baker, T. E.
A branch and bound network algorithm for interactive process scheduling.

BAKO 1980-1 Bako, A.
Network realization with flow method.

BALAD 1987-1 Balakrishnan, N.; Wong, R. T.
A network model for the rotating workforce scheduling problem.

BARA 1987-1 Barahona, F.
On via minimization.

BARA 1987-2 Barahona, F.
Planar multicommodity flows, max cut and the Chinese postman problem.

BARAA 1984-1 Barahona, F.; Cunningham, W. H.
A submodular network simplex method.

BARRB 1986-1 Barros, O.; Weintraub, A.
Spatial market equilibrium problems as network models.

BARTE 1986-1 Bartusch, M.
Optimierung von Netzplänen mit Anordungsbeziehungen bei knappen Betriebsmitteln (Nachdruck der Dissertation, RWTH Aachen, 1983).

BARTF 1986-1 Bartusch, M.; Möhring, R. H.; Radermacher, F. J.
Scheduling project networks with resource constraints and time windows.

BATUA 1983-1 Baturina, L. N.; Lepeshinskij, N. A.
Algorithms for the optimization of parameters ensuring a given vitality level of a network (Russian).

BEIN 1986-1 Bein, W. W.; Brucker, P.
Greedy concepts for network flow problems.

BEINA 1985-1 Bein, W. W.; Brucker, P.; Tamir, A.
Minimum cost flow algorithms for series-parallel networks.

BELL 1984-1 Belling-Seib, K.
Methods for solving nonlinear network flow problems.

BELL 1985-1 Belling-Seib, K.
A hybrid algorithm for solving convex separable network flow problems.

BELL 1987-1 Belling-Seib, K.
Network flow problems with one side constraint: a comparison of three solution methods.

BERG 1983-1 Berge, C.
Graphes (3. edition).

BERG 1985-1 Berge, C.
Graphs (second revised edition).

BERTF 1982-1 Bertsekas, D. P.
Optimal routing and flow control methods for communication networks.

BERTF 1985-1 Bertsekas, D. P.
A unified framework for primal-dual methods in minimum cost network flow problems.

BERTG 1987-1 Bertsekas, D. P.; Eckstein, J.
Distributed asynchronous relaxation methods for linear network flow problems.

BERTH 1987-1 Bertsekas, D. P.; El-Baz, D.
Distributed asynchronous relaxation methods for convex network flow problems.

BERTI 1987-1 Bertsekas, D. P.; Hossein, P. A.; Tseng, P.
Relaxation methods for network flow problems with convex arc costs.

BERTJ 1985-1 Bertsekas, D. P.; Tseng, P.
Relaxation methods for minimum cost network flow problems.

BIENC 1985-2 Bienstock, D.
Efficient reliability computation of some highly structured networks.

BIENC 1986-1 Bienstock, D.
Some lattice-theoretic tools for network reliability analysis.

BILLD 1985-1 Billionnet, A.; Minoux, M.
Maximizing a supermodular pseudoboolean function: a polynomial algorithm for supermodular cubic functions.

BIXBD 1986-1 Bixby, R. E.; Fourer, F.
Finding embedded network rows in linear programs I: extraction heuristics.

BIXBE 1986-1 Bixby, R. E.; Rajan, A.
A short proof of a decomposition theorem for max-flow min-cut matroids.

BLAZB 1984-1 Blazewicz, J.; Bovet, D. P.; Gambosi, G.
Deadlock-resistant flow control procedures for store-and-forward networks.

BOES 1982-1 Boesch, F. T.
Introduction to basic network problems.

BOKH 1987-1 Bokhari, S. H.
Assigment Problems in Parallel and Distributed Computing.

BOROC 1984-2 Boronin, A. B.; Pershin, O. Y.
Decomposition of a network into minimally connected parts.

BROO 1981-1 Brooks, R. E.
Using generalized networks to forecast natural gas distribution and allocation during periods of shortage.

BROWA 1984-1 Brown, G. G.; Mac Bride, R. D.
Solving generalized networks.

BROWB 1985-1 Brown, G. G.; Mac Bride, R. D.; Wood, R. K.
Extracting embedded generalized networks from linear programming problems.

BRUC 1984-1 Brucker, P.
A greedy algorithm for solving network flow problems in trees.

BRUC 1985-1 Brucker, P.
Parametric programming and circulation problems with one additional linear constraint.

BRUCC 1985-1 Brucker, P.; Papenjohann, W.
An out-of-kilter method for the algebraic circulation problem.

BRUG 1984-1 Brugha, C. M.
Finding embedded networks in linear programmes.

BURK 1985-1 Burkard, R. E.
Time-slot assignment for TDMA-systems.

BURK 1985-2 Burkard, R. E.
Time-slot assignment for TDMA-systems

CAI A 1985-1 Cai, M.-C.
Progress in network flow theory and matroid theory (Chinese).

CARE 1987-1 Carey, M.
Optimal time-varying flows on congested networks.

CARS 1984-1 Carstensen, P. J.
Parametric cost shortest chain problem.

CHANB 1987-1 Chanas, S.
Fuzzy optimization in networks.

CHAND 1984-1 Chanas, S.; Kolodziejczyk, W.
Real-valued flows in a network with fuzzy arc capacities.

CHANE 1985-2 Chanas, S.; Kolodziejczyk, W.; Machaj, A.
The min-cost flow problem - a fuzzy approach.

CHANN 1985-1 Chandru, V.; Coullard, C. R.; Wagner, D. K.
On the complexity of recognizing a class of generalized networks.

CHANR 1985-1 Chang, G. J.; Edmonds, J.
The poset scheduling problem.

CHANW 1985-1 Chang, H.-J.
A minimax-arc-path problem.

CHANY 1986-1 Chang, M. D.; Chen, C.-H. J.; Engquist, M.
An improved primal simplex variant for pure processing networks.

CHANZ 1986-1 Chang, M. D.; Engquist, M.
On the number of quasi-trees in an optimal generalized network basis.

CHAU 1985-1 Chaudhry, S. S.; Mac Cormick, S. T.; Moon, I. D.
Locating independent facilities with maximum weight: greedy heuristics.

CHAUC 1986-1 Chaudhuri, P.; Ghosh, R. K.
Parallel algorithms for analyzing activity networks.

CHENA 1986-1 Chen, C.-H. J.; Engquist, M.
A primal simplex approach to pure processing networks.

CHOIA 1985-1 Choi, W.; Francis, R. L.
Some optimization models of building evacuation problems with flow-dependent arc capacities.

CHOIB 1984-1 Choi, W.; Francis, R. L.; Hamacher, H. W.; Tufekci, S.
Network models of building evacuation problems with flow-dependent exit capacities.

CHOIC 1985-1 Choi, W.; Hamacher, H. W.; Tufekci, S.
Modeling of building evacuation problems by network flows with side constraints.
CHRID 1986-1 Christofides, N.; Valls Verdejo, V.
Finding all optimal solutions to the network flow problem.
CHUNG 1982-1 Chung, M.-J.; Makedon, F.; Sudborough, I. H.; Turner, J.
Polynomial time algorithms for the MIN CUT problem on degree restricted trees.
CHUNG 1985-1 Chung, M.-J.; Makedon, F.; Sudborough, I. H.; Turner, J.
Polynomial time algorithms for the min-cut problem on degree restricted trees.
CIUR 1979-1 Ciurea, E.
On dynamic flows (French).
CIUR 1982-1 Ciurea, E.
Two remarks on minimal flows in networks (Romanian).
CIUR 1985-1 Ciurea, E.
Deux extensions sur les flots dynamiques.
CIUR 1985-2 Ciurea, E.
Les flots dynamiques maximaux speciaux.
CIUR 1985-3 Ciurea, E.
Two classes of maximal dynamic flows.
CIUR 1985-4 Ciurea, E.
Un algorithme pour les flots dynamiques maximals avec stationnements.
CIUR 1986-1 Ciurea, E.
Le probleme de l'approvisionnement et de la demande pour les flots dynamiques.
CIUR 1987-1 Ciurea, E.
Les flots dynamique avec les temps de parcours nonstationnaires.
CIUR 1987-2 Ciurea, E.
Two simple versions of dynamic flow algorithms.
CIURA 1983-1 Ciurea, E.; Radescu, N.
Minimal dynamical flows (Romanian).
CIURA 1984-1 Ciurea, E.; Radescu, N.
Classes of minimal dynamical flows (Romanian).
COLB 1986-2 Colbourn, C. J.
The reliability polynomial.
CRUM 1981-1 Crum, R. L.; Nye, D. J.
A network model of insurance company cash flow management.
CUNNB 1985-1 Cunningham, W. H.; Frank, A.
A primal-dual algorithm for submodular flows.
DADU 1987-1 Daduna, J. R.
An efficient algorithm for solving bottleneck transportation problems with euclidean distances.
DANU 1984-1 Danut, M.
Note on the flow and tension in a network.
DAVY 1986-1 Davydov, E. G.; Bolshakov, S. Y.
Group-theoretic methods for aggregation in network problems (Russian).
DEMB 1986-1 Dembo, R. S.
The performance of NLPNET, a large-scale nonlinear network optimizer.
DEMB 1987-1 Dembo, R. S.
A primal truncated Newton algorithm with application to large-scale nonlinear network optimization.
DEMBA 1981-1 Dembo, R. S.; Klincewicz, J. G.
An approximate second-order algorithm for network flow problems with convex separable costs.
DERI 1984-3 Derigs, U.
Programming in networks and graphs - on the combinatorial background and near-equivalence of network flow and matching algorithms.
DEWE 1984-1 Dewess, G.
Zur Anwendung der Filtermethode auf das kombinatorische Fundamentalproblem der Ablaufplanung.
DINE 1984-1 Dinescu, C.; Savulescu, B.
On transportation of a maximum flow under some conditions on arcs.
DODI 1984-1 Dodin, B.
Determining the k most critical paths in PERT networks.
DODI 1985-1 Dodin, B.
Bounding the project completion time distribution in PERT networks.
DODIA 1985-1 Dodin, B.; Elmaghraby, S. E.
Approximating the criticality indices of the activities in PERT networks.
DODIB 1985-1 Dodin, B.; Sirvanci, M.
The extreme value distribution and stochastic activity networks.
DRUD 1986-1 Drud, A.; Zenios, S. A.; Mulvey, J. M.
Balancing large social accounting matrices with nonlinear network programming.
DUBO 1983-1 Dubois, D.
A fuzzy, heuristic, interactive approach to the optimal network problem.
EL-M 1985-2 El-Mallah, E. S.; Colbourn, C. J.
Reliability of delta-Y reducible networks.
ENGQ 1985-1 Engquist, M.; Chang, M. D.
New labeling procedures for the basis graph in generalized networks.
ENGQA 1986-1 Engquist, M.; Chen, C.-H. J.
Computational comparison of two solution procedures for allocation/processing networks.
FANE 1985-1 Fanelli, S.; Lombari, T.
An application of the technique of Q-GERT networks to the evaluation of logistic parameters of electronic systems (Italian).
FEDE 1986-1 Federgruen, A.; Groenevelt, H.
Optimal flows in networks with multiple sources and sinks, with applications to oil and gas lease investment programs.
FINC 1987-1 Fincke, U.; Hamacher, H. W.
On the expected value of stochastic linear programs and (dynamic) network flow problems.
FINKC 1987-1 Finke, G.
Network flow based branch and bound method for asymmetric traveling salesman problems.

FINKD 1987-1 Finke, G.; Ahmed, S.
Network flow methods for the asymmetric traveling salesman problem.

FINKG 1984-1 Finke, U.; Hamacher, H. W.
Some bounds for stochastic network flow problems.

FISHB 1985-1 Fisher, D. L.; Saisi, D.; Goldstein, W. M.
Stochastic PERT networks: OP diagrams, critical paths and the project completion time.

FISHJ 1987-1 Fishman, G. S.
Maximum flow and critical cutset as descriptors of multi-stage systems with randomly capacitated components.

FLORA 1984-1 Florian, M.
An introduction to network models used in transportation planning.

FLORA 1986-1 Florian, M.
Nonlinear cost network models in transportation analysis.

FOUL 1984-1 Foulds, L. R.
Combinatorial optimization for undergraduates.

FOUL 1985-3 Foulds, L. R.
Traffic network arc elimination by branch and bound enumeration.

FRANA 1984-1 Francis, R. L.
A negative exponential solution to an evacuation problem.

FRANC 1984-3 Frank, A.
Submodular flows.

FRANC 1987-1 Frank, A.
Graph connectivity and network flows.

FRIEC 1984-1 Frieze, A. M.
Algebraic flows.

FUJIA 1986-1 Fujishige, S.
A capacity-rounding algorithm for the minimum-cost circulation problem: a dual framework of the Tardos algorithm.

FUJIA 1987-1 Fujishige, S.
An out-of-kilter method for submodular flows.

FUJIC 1986-1 Fujishige, S.; Nakayama, A.; Cui, W.-T.
On the equivalence of the maximum balanced flow problem and the weighted minimax flow problem.

FUJID 1986-1 Fujishige, S.; Röck, H.; Zimmermann, U.
A strongly polynomial algorithm for minimum cost submodular flow problems.

FUNG 1985-1 Fung, K. T.
An entropy approach to solving some network reliability problems.

GABOA 1985-2 Gabow, H. N.
Scaling algorithms for network problems.

GALIC 1986-1 Galil, Z.; Tardos, E.
An O(n**2(m+nlogn)logn) min-cost flow algorithm.

GAMBA 1986-1 Gamble, A. B.; Conn, A. R.; Pulleyblank, W. R.
A network penalty method.

GAZD 1983-1 Gazdik, I.
Fuzzy-network planning - FNET.

GEY 1985-1 Gey, H.; Lehne, D.
Netzplantechnik: Grundlagen, Methoden, praxisnahe Einführung, Anwendungsbeispiele, Netzplanprogramm in MBASIC.

GHIU 1985-1 Ghius, V.
The types of flows in fan-shaped graphs (French).

GIBB 1985-1 Gibbons, A.
Algorithmic graph theory.

GLIC 1985-1 Glickman, T. S.; Sherali, H. D.
Large-scale network distribution of pooled empty freight cars over time, with limited substitution and equitable benefits.

GLOVC 1981-1 Glover, F.; Klingman, D.
The simplex SON algorithm for LP/embedded network problems.

GLOVC 1985-1 Glover, F.; Klingman, D.
Basis exchange characterization for the simplex SON algorithm for LP/embedded networks.

GOLDA 1986-1 Goldberg, A. V.; Tarjan, R. E.
A new approach to the maximum flow problem.

GOLDA 1987-1 Goldberg, A. V.; Tarjan, R. E.
Solving minimum-cost flow problems by successive approximation.

GRABB 1985-1 Grabowski, J.; Skubalska, E.
Optimization of the transportation network structure with respect to the flow costs (Polish).

GRANB 1986-1 Granot, F.; Hassin, R.
Multi-terminal maximum flows in node-capacitated networks.

GRANC 1985-1 Granot, F.; Veinott, A. F. Jr.
Substitutes, complements and ripples in network flows.

GREEB 1986-1 Greenberg, H. J.
Diagnosing infeasibility in min-cost network flow problems. I: Dual infeasibility.

GRIG 1986-1 Grigoriadis, M. D.
An efficient implementation of the network simplex method.

GRIGA 1984-1 Grigoriadis, M. D.; Hsu, T.
Numerical methods for basic solutions of generalized flow networks.

GRIMA 1985-2 Grimmett, G. R.
Random flows: network flows and electrical flows through random media.

GROTB 1987-2 Grötschel, M.; Jünger, M.; Reinelt, G.
Via minimization with pin preassignments and layer preference.

GUANC 1985-1 Guan, M.-G.; Pulleyblank, W. R.
Eulerian orientations and circulations.

GUPTH 1985-1 Gupta, S. K.
Linear programming and network models.

GURV 1984-1 Gurvich, V. A.
Kirchhoff's second law for optimal transportation flows.
GUSF 1987-3 Gusfield, D.
Very simple methods for all pairs network flow analysis.
GUSFA 1987-1 Gusfield, D.; Martel, C.; Fernandez-Baca, D.
Fast algorithms for bipartite network flow.
GVIS 1984-1 Gvishiani, A. D.; Gurvich, V. A.
Balanced flow in a multipole network.
HAGS 1984-1 Hagstrom, J. N.
Note on independence of arcs in antiparallel for network flow problems.
HAMA 1981-1 Hamacher, H. W.
Optimal (s,t)-cuts.
HAMA 1986-1 Hamacher, H. W.
Maximal dynamic polymatroid flows and applications.
HAMAA 1986-1 Hamacher, H. W.; Foulds, L. R.
Algorithms for flows with parametric capacities.
HAMAB 1984-1 Hamacher, H. W.; Picard, J.-C.; Queyranne, M.
Ranking the cuts and cut-sets of a network.
HAMAC 1986-1 Hamacher, H. W.; Queyranne, M.
K best solutions to combinatorial optimization problems.
HAMAD 1984-1 Hamacher, H. W.; Tufekci, S.
Algebraic flows and time-cost tradeoff problems.
HAMAD 1987-1 Hamacher, H. W.; Tufekci, S.
On the use of lexiographic min cost flows in evacuation modeling.
HAREF 1985-1 Harel, D.
A linear algorithm for finding dominators in flow graphs and related problems.
HASSB 1987-1 Hassin, R.
A computing scheme for network problems with random edge lengths.
HASSC 1985-1 Hassin, R.; Johnson, D. B.
An 0(n(log n)**2) algorithm for maximum flow in undirected planar networks.
HAVEA 1984-1 Haverly, C. A.
The number of simplex iterations for a general LP code on network problems.
HEARA 1984-1 Hearn, D. W.; Lawphongpanich, S.; Ventura, J. A.
Restricted simplicial decomposition: computation and extensions.
HEARA 1987-1 Hearn, D. W.; Lawphongpanich, S.; Ventura, J. A.
Optimization algorithms for congested network models.
HERN 1985-1 Hernandez, M. G.; Alvarez, A.; Ruiz J.; Blanco, R.
Synthesis of a transportation graph with a minimum number of edges, given an upper bound for the flow (Spanish).
HERNA 1985-1 Hernandez, M. G.; Canizares, L. C.; Gonzalez, I. Y.; Cedeno, R. V.
Synthesis of a transportation graph with a minimum number of intermediate nodes (Spanish).
HERNB 1985-1 Hernandez, M. G.; Cedeno, R. V.; Gonzalez, I. Y.
Synthesis of a transportation network with linear restrictions on the flow capacities (Spanish).
HERNC 1986-1 Hernandez, M. G.; Cornell, A. M.; Gamboa, D.
Synthesis of transport-type networks with restrictions in the output-edges (Spanish).
HERND 1985-1 Hernandez, M. G.; Gonzalez, I. Y.; Canizares, L. C.; Cedeno, R. V.
Analysis of the existence of a linear flow in a transportation graph (Spanish).
HOHM 1986-1 Hohmann, C.; Kern, W.
Optimization and optimality test for the max-cut problem.
HOUSA 1982-1 Houska, R. V.; Brockus, C. G.
Stochastic optimal control applied to deterministic network problems.
IKURA 1986-1 Ikura, Y.; Gross, G.; Hall, G. S.
PGandE's state-of-the-art scheduling tool for hydro systems.
ILEV 1984-1 Ilev, V. P.
Basis graphs of polymatroids (Russian).
ILUM 1987-1 Ilumoka, A. A.
A tolerance analysis approach to network scheduling for engineering project management.
IOI 1986-1 Ioi, S.
A special "tree" problem.
ITAI 1985-1 Itai, A.; Rodeh, M.
Scheduling transmissions in a network.
ITAK 1984-1 Itakura, H.; Nishikawa, Y.
Fuzzy network technique for technological forecasting.
IVAN 1985-1 Ivanchev, D. T.
Finding the foremost arcs in a network with parametric arc capacities.
IVANA 1984-1 Ivanchev, D. T.; Ruhe, G.
Flows in generalized networks with parametric arc capacities.
IVANA 1986-1 Ivanchev, D. T.; Ruhe, G.
Finding the k bicriterial foremost edges in planar flow networks.
JACO 1985-1 Jacobs, K.
A measure theoretical max-flow-problem (Italian).
JAEG 1984-1 Jaeger, F.
On circular flows in graphs.
JAEG 1985-3 Jaeger, F.
On five-edge colorings of cubic graphs and nowhere-zero flow.
JANIB 1985-1 Janiga, L.; Koubek, V.
A note on finding minimum cuts in directed planar networks by parallel computations.
JASM 1985-1 Jasmon, G. B.
Cutset analysis of networks using basic minimal paths and network decomposition.
JOHND 1987-1 Johnson, D. B.
Parallel algorithms for minimum cuts and maximum flows in planar networks.
JOHNE 1982-1 Johnson, D. B.; Venkatesan, S. M.
Parallel algorithms for minimum cuts and maximum flows in planar networks (preliminary version).

JOHNL 1984-1 Johnson, R.
Network reliability and acyclic orientations.

JUNGC 1986-1 Jungnickel, D.
Transversaltheorie: ein Überblick.

JUNGC 1987-1 Jungnickel, D.
Graphen, Netzwerke und Algorithmen.

KALB 1984-1 Kalb, G.; Röck, H.; Schmidt, G.
Netzflussmodelle zur präferenzorientierten dynamischen Einsatz- und Ablaufplanung.

KAMB 1984-1 Kamburowski, J.
Criticality indices in the PERT-type network (Polish).

KAMB 1984-2 Kamburowski, J.
Flow distributions through a stochastic acyclic network.

KAMB 1985-2 Kamburowski, J.
An upper approximation of the expected completion time of a project described by the PERT-type network (Polish).

KAMB 1985-3 Kamburowski, J.
An upper bound on the expected completion time of PERT networks.

KAMB 1985-4 Kamburowski, J.
Two-point approximations for activity times in PERT network.

KAMB 1986-1 Kamburowski, J.
On the computational complexity of the shortest route and maximum flow problems in stochastic networks.

KAMB 1987-1 Kamburowski, J.
An overview of the computational complexity of the PERT shortest route and maximum flow problems in stochastic networks.

KARP 1982-1 Karp, R. M.
The computational complexity of network problems.

KARZ 1985-2 Karzanov, A. V.
Families of cuts with the MFMC-property.

KELLB 1985-1 Kelly, F. P.
Some probabilistic aspects of network flow.

KHOD 1985-1 Khodzinskii, A. N.
Sequential algorithm for the solution of problems of combinatorial optimization on permutations.

KIME 1985-1 Kimemia, J.; Gershwin, S. B.
Flow optimization in flexible manufacturing systems.

KLEI 1986-1 Klein Haneveld, W. K.
Robustness against dependence in PERT: an application of duality and distributions with known marginalS.

KLEIA 1984-1 Klein Haneveld, W. K.; Schweigman, C.
Robustness against dependence in PERT: an application of duality and distributions with known marginals.

KLEID 1982-1 Kleitman, D. J.
Some practical network problems.

KLINC 1984-1 Klingman, D.; Mead, M.; Phillips, N. V.
Network optimization models for military manpower planning.

KLINE 1987-1 Klingman, D.; Phillips, N. V.; Steiger, D.; Wirth, R.; Padman, R.; Krishnan, R.
An optimization based integrated short-term refined petroleum product planning system.

KOBA 1984-1 Kobayashi, T.
The lexico-bounded flow algorithm for solving the minimum cost project scheduling problem with an additional linear constraint.

KOEN 1983-1 Koene, J.
Minimal cost flow in processing networks, a primal approach.

KOIT 1984-1 Koita, Y.
The dynamic form of the generalized problem of minimal cost flow (Russian).

KOUR 1984-1 Kourtz, P.
A network approach to least-cost daily transfers of forest fire control resources.

KOVAG 1984-2 Kovalev, M. M.; Pisaruk, N. N.
Independent flows and polymatroids (Russian).

KOVAG 1985-2 Kovalev, M. M.; Pisaruk, N. N.
Independent flows with a discrete-concave expenditure function.

KOVAG 1986-1 Kovalev, M. M.; Pisaruk, N. N.
Generalized flows with discrete convex cost function (Russian).

KRES 1984-1 Kress, M.
The chance constrained critical path with location-scale distributions.

KULKA 1984-1 Kulkarni, V. G.
A compact hash function for paths in PERT networks.

KULKB 1985-1 Kulkarni, V. G.; Adlakha, V. G.
Maximum flow in planar networks with exponentially distributed arc capacities.

KULKB 1986-1 Kulkarni, V. G.; Adlakha, V. G.
Markov and Markov-regenerative PERT networks.

KUZMA 1984-1 Kuzmin, S. L.
A problem of maximum flow in networks with relaying.

KUZMA 1985-1 Kuzmin, S. L.
A problem on a flow with limitations on the number of transits.

KUZMA 1985-2 Kuzmin, S. L.
On the complexity of flow problems with path length constraints.

LAPOC 1985-1 Laporte, G.; Mercure, H.; Nobert, Y.
Finding the shortest Hamiltonian circuit through n-clusters: a Lagrangean relaxation approach.

LAPOG 1987-1 Laporte, G.; Nobert, Y.; Nga, T. K.
An optimal flow circulation algorithm for the asymmetrical multiple travelling salesman problem.

LAWLC 1986-1 Lawler, E. L.; Martel, C. U.
Polymatroidal flows with lower bounds.

LEDE 1985-1 Ledermann, W.; Vajda, S.
Handbook of applicable mathematics. Vol. 5 : Combinatorics and geometry. Part A.
LI 1985-1 Li, C.-K.
On the extremal solutions for capacitated network problems.
LI J 1987-1 Li, Z.-J.
A matrix representation for the overall planning network and relative calculations (Chinese).
LIMN 1986-1 Limnios, N.; Ziani, R.
An algorithm for reducing cut sets in fault-tree analysis.
LIU A 1986-1 Liu, C. C.
An improved lower bound on the max flow of postoutage network configurations.
LOMO 1985-1 Lomonosov, M. V.
A multiflow problem without the max-flow min-cut property (Abstract).
LOZO 1984-1 Lozovanu, D. D.
A generalized problem of network synthesis and some of its properties (Russian).
LUCEA 1984-1 Lucertini, M.; Paletta, G.
A decomposition method for network design with multicommodity non-simultaneous flows.
LUH 1985-1 Luh, J. Y. S.; Lin, C. S.
Scheduling parallel operations in automation for minimum execution time based on PERT.
LUO A 1985-1 Luo, Z.-H.
Fundamental characteristics of distributions of random flow in PERT (Chinese).
MAC A 1985-1 Mac Bride, R. D.
Solving embedded generalized network problems.
MAIE 1986-1 Maier, G.
Postoptimal analysis of minimal cost network flow problems.
MAIE 1987-1 Maier, G.
Determining all nondegenerate bounds in sensitivity analysis of minimal cost network flow problems.
MALYA 1986-1 Malyshko, V. V.; Harary, F.
A problem of the minimum blocking flow in a network (Russian).
MANA 1986-1 Manabe, R.
Scheduling buses by the network flow method on a personal computer.
MARCJ 1984-2 Marcu, D.
Note on the flow and tension in a network.
MARCK 1984-1 Marcus, R.
An application of artificial intelligence to operations research.
MARTM 1987-1 Martins, E. Q. V.
On a particular quadratic network problem.
MASO 1985-1 Mason, L. G.
Equilibrium flows, routing patterns and algorithms for store-and-forward networks.
MATUA 1987-1 Matula, D. W.; Shahrokhi, F.
Sparsest cuts and bottlenecks in graphs.
MEYE 1985-1 Meyer, R. R.
Network optimization.
MINO 1984-1 Minoux, M.
A polynomial algorithm for minimum quadratic cost flow problems.
MINO 1986-3 Minoux, M.
Solving integer minimum cost flows with separable convex cost objective polynomially.
MINO 1987-2 Minoux, M.
Network synthesis and dynamic network optimization.
MINOB 1987-1 Minoux, M.; Pinson, E.
Lower bounds to the graph partitioning problem through generalized linear programming and network flows.
MIRCC 1987-1 Mirchandani, P. B.; Soroush, H.
Routes and flows in stochastic networks.
MONH 1987-1 Monhor, D.
An approach to PERT: application of Dirichlet distribution.
MONM 1986-1 Monma, C. L.; Erickson, R. E.; Veinott, A. F. Jr.
Send-and-split method for minimum-concave-cost network flows.
MROZ 1984-1 Mrozek, M.
A note on minimum-dummy-activities Pert networks.
MULVA 1985-1 Mulvey, J. M.; Zenios, S. A.
Integrated risk/cost planning models for the U.S. air traffic system.
MULVA 1985-2 Mulvey, J. M.; Zenios, S. A.
Solving large scale generalized networks.
NAGL 1985-1 Nägler, G.
Lokal optimale Lösungen der kombinierten Strom- und Spannungsoptimierung auf Netzwerken im Anwendungsfall der Ablaufplanung.
NAGL 1986-1 Nägler, G.
Coupling between network flow and network tension and its application in project planning.
NAKAA 1984-1 Nakamori, M.; Mizutani, A.; Kakegawa, M.; Shimizu, K.
Solution of the minimum flow problem in selective networks and application to the design of arrangements and distributions of integrated circuits (Japanese).
NAKAE 1986-1 Nakayama, A.
A polynomial algorithm for the maximum balanced flow problem with a constant balancing rate function.
NAKAE 1987-1 Nakayama, A.
A polynomial-time dual simplex algorithm for the minimum cost flow problem.
NARUB 1983-1 Narushima, H.; Kusaka, Y.
A move problem of weighted digraphs.
NEUMC 1984-1 Neumann, K.
Recent developments in stochastic activity networks.
NEUMC 1985-1 Neumann, K.
EOR project networks.

NEUMC 1985-2 Neumann, K.
GERT networks with tree stucture: properties, temporal analysis, cost minimization and scheduling.

NGUYA 1984-1 Nguyen, S.
Estimating origin-destination matrices from observed flows.

OHUC 1984-1 Ohuchi, A.; Kaji, I.
Lagrangian dual coordinatewise maximization algorithm for network transportation problems with quadratic costs.

ONAG 1982-1 Onaga, K.
Time-optimal transmission of time-variant particle flow streams through time-invariant networks.

ORLI 1985-2 Orlin, J. B.
Genuinely polynomial simplex and nonsimplex algorithms for the minimum cost flow problem.

ORLI 1985-3 Orlin, J. B.
On the simplex algorithm for networks and generalized networks.

ORLI 1986-1 Orlin, J. B.
A dual version of Tardos's algorithm for linear programming.

ORLIA 1985-1 Orlin, J. B.; Rothblum, U. G.
Computing optimal scalings by parametric network algorithms.

OUDH 1985-1 Oudheusden, D. L. van; Khan, M. L. R.
Note on ILP flow formulation for subgraph selection.

OZAW 1987-1 Ozawa, T.; Kajitani, Y.; Ueno, S.
The principal partition of vertex-weighted graphs and its applications.

PAND 1983-1 Pandit, S. N. N.; Reddy, B. K.
A maximum flow problem with intermediate node requirements.

PASC 1987-1 Pasche, C.
Flot a cout convexe lineaire par morceaux.

PETE 1984-1 Peteanu, V.
Algebraical structures in the network theory.

PETK 1986-1 Petkov, M. G.
The union of two transportation flows - additional investigations (Bulgarian).

PETK 1986-2 Petkov, M. G.
Transportation flow passing through a bottleneck (Bulgarian).

POIA 1986-1 Poiaga, L.
Resolution by reduction of a network, by means of its assimilation to a semi-Markov process (Italian).

PUEB 1987-1 Puebla, J. G.
Spatial structures of network flows: A graph theoretical approach.

PULA 1984-1 Pulat, P. S.; Elmaghraby, S. E.
On maximizing flow in generalized flow networks.

RADEA 1982-1 Radescu, N.
An algorithm for determining maximum flow through a transportation network (Romanian).

RADEA 1982-2 Radescu, N.
Determination of a flow compatible with capacity intervals by means of epsilon-minimal paths (Romanian).

RADEA 1982-3 Radescu, N.
Determination of minimal flow through a network with accumulated capacities (Romanian).

ROCK 1984-1 Rockafellar, R. T.
Network flows and monotropic optimization.

ROOS 1984-1 Roos, C.
Network programming with umbrellas.

ROSEG 1986-1 Rosenthal, R. E.
Representing inverses in pure network flow optimization.

ROTEA 1986-1 Rote, G.; Rendl, F.
Minimizing the density of terminal assignments in layout design.

ROUNA 1986-1 Roundy, R.
A 98% effective lot sizing rule for a multi-product, multi-stage production/inventory system.

RUGE 1984-1 Rüger, W.
Berechnung der Zusammenhangswahrscheinlichkeit von Netzgraphen.

RUHE 1984-1 Ruhe, G.
Ein Verfahren zum Auffinden eingebetteter Netzwerkstrukturen in linearen Programmen.

RUHE 1985-1 Ruhe, G.
Characterization of all optimal solutions and parametric maximal flows in networks.

RUHEA 1985-1 Ruhe, G.; Ivanchev, D. T.
The solution of generalized network flow problems with parametric capacities and stochastic demands.

RUHEB 1986-1 Ruhe, G.; Sieber, N.
Zur Komplexität zweier vektorieller Netzwerkflussprobleme.

RUMC 1984-1 Rumchev, V. G.; Mateev, L. N.
Solution of a dynamic transportation problem in network formulation by the method of a time expanded network (Russian).

SAKAA 1984-1 Sakarovitch, M.
Optimisation combinatoire - methodes mathematiques et algorithmiques. Vol. 1: Programmation discrete.

SAKSA 1985-1 Saksena, J. P.
Applications of decision networks.

SAWI 1987-1 Sawik, T. J.
A critical path algorithm for scheduling a batch-type production on nonidentical machines.

SCHN 1986-1 Schneider, M. H.
A complementary pivoting algorithm for linear network problems.

SCHRA 1986-1 Schreier, H.
A variant for checking a solution criterion for the balanced multicommodity transportation problem (German).

SCHW 1983-1 Schwarze, J.
Netzplantheorie.

SEBO 1986-1 Sebö, A.
Application of network flow methods in the scheduling of terminal sessions (Hungarian).

SGUR 1983-1 Sgurev, V. S.
A flow on a net with linear capacity constraints (Bulgarian).
SGUR 1983-2 Sgurev, V. S.
A min cost network flow with linear capacity constraints (Bulgarian).
SGUR 1983-3 Sgurev, V. S.
Capacity and maximum flow in a net with linear capacity constraints (Bulgarian).
SGUR 1985-1 Sgurev, V. S.
Flows on graphs with multiplicative capacity constraints.
SGURA 1986-1 Sgurev, V. S.; Dzhelatova, M. N.
Maximal linear flow and capacity of cutting set.
SGURB 1985-1 Sgurev, V. S.; Nikolova, M.
A linear network flow (Russian).
SHAKA 1984-1 Shakhbazyan, K. V.; Tushkina, T. A.
On the maximum number of flows in a pipeline mode (Russian).
SHAKB 1987-1 Shakhbazyan, K. V.; Tushkina, T. A.; Lebedinskii, M. M.
A spanning tree algorithm for the construction of flows in unconditional pipeline computers (Russian).
SHERC 1984-1 Sherali, H. D.; Baines, S. G.
Advanced basis or block pivoting methods for network structured linear programs.
SHET 1986-1 Shetty, B.
The linear equal flow problem.
SHIE 1986-1 Shier, D. R.
Algebraic aspects of computing network reliability.
SHIEB 1985-1 Shier, D. R.; Whited, D. E.
Algorithms for generating minimal cutsets by inversion.
SHINA 1986-1 Shing, M. T.; Hu, T. C.
A decomposition algorithm for multi-terminal network flows.
SHIR 1987-1 Shirakawa, I.; Shinoda, S.; Sengoku, M.; Chen, W.-K.
A problem of diagnosis of mixed networks.
SIEB 1984-1 Sieber, N.
Einige Ergebnisse der Forschungsrichtung "Optimierung auf Netzwerken und Netzplantechnik".
SIEB 1985-1 Sieber, N.
Entwicklungsrichtungen in der Netzwerkoptimierung.
SIED 1982-1 Siedersleben, J.
Optimierung von deterministischen und stochastischen Entscheidungsnetz- plänen (Dissertation).
SIED 1986-1 Siedersleben, J.
Common aspects of several network flow algorithms.
SINGC 1986-1 Singh, C.; Asgarpoor, S.
Reliability evaluation of flow networks using delta-star transformations.
SINGG 1986-1 Singh, S.
Improved methods for storing and updating information in the out-of-kilter algorithm.
SMIT 1987-1 Smith-Daniels, D. E.; Smith-Daniels, V. L.
Optimal project scheduling with materials ordering.
SOTSA 1982-1 Sotskov, Y. N.; Chernjak, Z. A.
Determination of parameters of a network graph (Russian).
SRINA 1985-1 Srinivasan, V.; Kim, Y. H.
A network optimization approach to efficient cash management in India.
TAMAA 1987-1 Tamassia, R.
On embedding a graph in the grid with the minimum number of bends.
TARD 1985-1 Tardos, E.
A strongly polynomial minimum cost circulation algorithm.
TARDA 1986-1 Tardos, E.; Tovey, C. A.; Trick, M. A.
Layered augmenting path algorithms.
TARJ 1983-1 Tarjan, R. E.
Data structures and network algorithms.
TARJ 1986-1 Tarjan, R. E.
Algorithms for maximum network flow.
TAVA 1987-1 Tavares, L. V.
Optimal resource profiles for program scheduling.
TOML 1985-1 Tomlin, J. A.; Welch, J. S.
Integration of a primal simplex network algorithm with a large-scale mathematical programming system.
TRIC 1987-1 Trick, M. A.
Matching problems and generalized networks. Chapter IV of: Networks with additional structured constraints.
TRUE 1987-2 Truemper, K.
Max-flow min-cut matroids: polynomial testing and polynomial algorithms for maximum flow and shortest routes.
TUNG 1986-1 Tung, P. C. T.
A bicriterion Pareto path algorithm.
VALK 1985-1 Valk, M.
Generalized flows and stresses on graphs (Russian).
VANRA 1985-1 Vanroy, T. J.; Wolsey, L. A.
Valid inequalities and separation for uncapacitated fixed charge networks.
VASI 1985-1 Vasilev, N. S.
On the properties of a flow polyhedron and their applications in the numerical solution of extremal problems on graphs.
VASI 1986-1 Vasilev, N. S.
A model of a communication network development.
VENTA 1987-1 Ventura, J. A.; Hearn, D. W.
A Lagrangian dual approach for networks with quadratic costs.
VILA 1984-1 Vilar, J.
Minimal cuts up to third order in a planar graph.

VOIT 1984-1 Voitishin, Y. V.
Application of network fronts for solution of certain problems on graphs (Russian).

WALK 1985-1 Walk, M.
Generalized flows and tensions on graphs (Russian).

WALLA 1987-1 Wallace, S. W.
A piecewise linear upper bound on the network recourse problem.

WALT 1984-1 Walther, H.
Ten applications of graph theory.

WALTA 1987-1 Walther, H.; Nägler, G.
Graphen, Algorithmen, Programme.

WANGG 1985-1 Wang, K.-Y.; Zhan, J.; Wang, Z.-T.
Models and algorithms for integrated energy transportation system planning (Chinese).

WANGK 1986-1 Wang, S.-M.; Sun, Q.-H.; Wang, X.-F.
A study of flows in complete stochastic networks - theoretical study of a new method for evaluating the reliability of interconnected electrical networks (Chinese).

WEISA 1984-2 Weiss, G.
Stochastic bounds on distributions of optimal value functions with applications to PERT, network flows and reliability.

WEISA 1986-1 Weiss, G.
Stochastic bounds on distributions of optimal value functions with applications to PERT.

WHITI 1985-1 Whittle, P.
Scheduling and characterization problems for stochastic networks. With discussion.

WILF 1986-1 Wilf, H. S.
Algorithms and complexity.

WILLD 1985-1 Willis, R. J.
An algorithm for constructing project network diagrams on a ordinary line printer.

WILLD 1985-2 Willis, R. J.
Critical path analysis and resource constrained project scheduling - theory and practice.

WOLL 1985-1 Wollmer, R. D.
Critical path planning under uncertainty.

WU 1984-1 Wu, S.-Z.; Wu, X.-M.
Generalized principle of optimality in pansystems network analysis.

YANGF 1987-1 Yang, Y. M.
A comment on the discussion of the PERT time distribution (Chinese).

YAO A 1982-1 Yao, F.
Maximum flows in networks.

YUNU 1984-1 Yunussa, K.
Dynamic form of generalized problem on minimal cost flow (Russian).

ZACH 1984-1 Zachary, W. W.
Modeling social network processes using constrained flow representations.

ZAWA 1987-1 Zawack, D. J.; Thompson, G. L.
A dynamic space-time network flow model for city traffic congestion.

ZENI 1985-1 Zenios, S. A.; Mulvey, J. M.
A distributed algorithm for convex network optimization problems.

ZENI 1986-1 Zenios, S. A.; Mulvey, J. M.
Nonlinear network programming on vector supercomputers.

ZENI 1986-2 Zenios, S. A.; Mulvey, J. M.
Relaxation techniques for strictly convex network problems.

ZIMM 1987-1 Zimmermann, H.-J.
Methoden und Modelle des Operations Research. Für Ingenieure, Ökonomen und Informatiker.

ZIMMC 1984-1 Zimmermann, U.
Minimization of combined objective functions on integral submodular flows.

ZIMMC 1984-2 Zimmermann, U.
Submodulare Flüsse: Verfahren zur Minimierung linearer Zielfunktionen.

ZIMMC 1985-1 Zimmermann, U.
Augmenting circuit methods for submodular flow problems.

ZIMMC 1986-1 Zimmermann, U.
Duality for balanced submodular flows.

ZYKA 1987-1 Zyka, O.
Nowhere-zero 30-flow on bidirected graphs.

ADAMC 1985-1 Adams, W. P.; Sherali, H. D.
Mixed-integer bilinear programming problems.
ADAMC 1986-1 Adams, W. P.; Sherali, H. D.
A tight linearization and an algorithm for zero-one quadratic programming problems.
AGEE 1984-1 Ageev, A. A.
On the minimization of quadratic polynomials of Boolean functions (Russian).
AGRAA 1985-1 Agrawal, S. C.; Chand, M.
An alternative method on integer solutions to complementary programming problems with linear fractional objective function by a branch and bound technique.
AHLF 1985-1 Ahlfield, D. P.; Dembo, R. S.; Mulvey, J. M.; Zenios, S. A.
Nonlinear programming on generalized networks.
AL-S 1986-1 Al-Salti, B.
Minimization of a concave function on a polymatroid (Russian).
ALLE 1987-1 Allen, J. D.; Helgason, R. V.; Kennington, J. L.
The frequency assignment problem: a solution via nonlinear programming.
ANDRE 1985-1 Andreatta, G.; Romanin-Jacur, G.
An integer quadratic allocation problem.
ASSAB 1985-1 Assad, A. A.; Xu, W.-X.
On lower bounds for a class of quadratic 0, 1 programs.
ATKI 1984-1 Atkins, D. R.; Granot, D.; Raghavendra, B. G.
Application of mathematical programming to the plywood design and manufacturing problem.
AVDO 1986-1 Avdonin, L. N.; Kodalaev, V. I.
A combinatorial approach to multicriteria optimization (Russian).
BABAG 1984-1 Babayev, D. A.; Babayev, R. D.; Jysifov, M. M.
General multiple-choice knapsack problem.
BANK 1987-3 Bank, B.; Mandel, R.
Nonlinear parametric integer programming.
BARA 1986-1 Barahona, F.
A solvable case of quadratic 0-1 programming.
BARAE 1987-1 Barahona, F.; Jünger, M.; Reinelt, G.
Experiments in quadratic 0-1 programming.
BEAN 1984-1 Bean, J. C.
A Langrangian algorithm for the multiple choice integer program.
BECHA 1984-1 Bechtold, S. E.; Janaro, R. E.; Sumners, D. L.
Maximization of labor productivity through optimal rest-break schedules.
BELAA 1984-1 Belardo, S.; Harrald, J.; Wallace, W. A.; Ward, J. E.
A partial covering approach to siting response resources for major maritime oil spills.
BELL 1984-1 Belling-Seib, K.
Methods for solving nonlinear network flow problems.
BELL 1985-1 Belling-Seib, K.
A hybrid algorithm for solving convex separable network flow problems.
BERTH 1987-1 Bertsekas, D. P.; El-Baz, D.
Distributed asynchronous relaxation methods for convex network flow problems.
BERTI 1987-1 Bertsekas, D. P.; Hossein, P. A.; Tseng, P.
Relaxation methods for network flow problems with convex arc costs.
BIGI 1985-1 Bigildeeva, T. B.
Methods and algorithms of approximate solution of a quadratic assignment problem of large dimensions (a survey) (Russian).
BONO 1986-1 Bonomi, E.; Lutton, J.-L.
The asymptotic behaviour of quadratic sum assignment problems: a statistical mechanics approach.
BOOK 1985-1 Bookhold, I.
Untersuchungen zum quadratischen Zuordnungsproblem.
BORC 1987-1 Borchardt, M.
Zur Anwendung eines Dualitätskonzeptes für quadratische Optimierungsprobleme mit stetigen und Booleschen Variablen. I.
BORC 1987-2 Borchardt, M.
Zur Lösung von nichtlinearen Booleschen Optimierungsproblemen.
BORI 1982-1 Borisov, K. Y.
Convex extremal problems on graphs (Russian).
BOURA 1984-1 Bourjolly, J.-M.; Hammer, P. L.; Simeone, B.
Node-weighted graphs having the König-Egervary property.
BRUC 1984-2 Brucker, P.
An o(n) algorithm for quadratic knapsack problems.
BRUN 1983-1 Brunn, B.
Ein Verfahren zur Lösung von ganzzahligen nichtlinearen Optimierungsproblemen.
BURKD 1984-1 Burkard, R. E.; Krarup, J.; Pruzan, P. M.
Some relationships between multicriteria and parametric discrete optimization problems with bottleneck criteria.
BYST 1987-1 Bystrov, V. A.; Shcherbina, O. A.
Solution of block problem of discrete optimization (Russian).
CABO 1986-1 Cabot, A. V.; Erenguc, S. S.
A branch and bound algorithm for solving a class of nonlinear integer programming problems.
CARRC 1983-1 Carrera, J.; Szidarovszky, F.
A multiobjective branch and bound algorithm for optimal observation network design.
CHAKA 1986-1 Chakraborty, T. K.
A preemptive single sampling attribute plan of given strength.
CHALA 1986-1 Chalmet, L. G.; Lemonidis, L.; Elzinga, D. J.
An algorithm for the bi-criterion integer programming problem.
CHANH 1987-1 Chandra, S.; Saxena, P. K.
Cost/completion-date tradeoffs in quadratic fractional transportation problem.
COOP 1987-1 Cooper, M. W.
Modelling nonlinearities in multicriteria locational decisions.

COOPA 1984-1 Cooper, M. W.; Farhangian, K.
A dynamic programming algorithm for multiple-choice constraints.

COOPA 1985-1 Cooper, M. W.; Farhangian, K.
Multicriteria optimization for nonlinear integer-variable problems.

CRAM 1985-1 Crama, Y.; Hammer, P. L.
The complexity of recognizing partition-quadratic graphs.

CURRB 1985-1 Current, J. R.; Revelle, C. S.; Cohon, J. L.
The maximum covering/shortest path problem: a multiobjective network design and routing formulation.

CURRB 1987-1 Current, J. R.; Revelle, C. S.; Cohon, J. L.
The median shortest path problem: A multiobjective approach to analyse cost vs. accessibility in the design of transportation networks.

DEMB 1986-1 Dembo, R. S.
The performance of NLPNET, a large-scale nonlinear network optimizer.

DEMB 1987-1 Dembo, R. S.
A primal truncated Newton algorithm with application to large-scale nonlinear network optimization.

DIAK 1987-1 Diakonu, M. M.
An algorithm for solving a bicriteria problem of integer linear programming (Russian).

DO 1987-1 Do, Z. C.
Transformations of the noninteger maximum of a concave function into the integer maximum (Russian).

DRUD 1986-1 Drud, A.; Zenios, S. A.; Mulvey, J. M.
Balancing large social accounting matrices with nonlinear network programming.

DURAA 1986-1 Duran, M. A.; Grossmann, I. E.
An outer-approximation algorithm for class of mixed-integer nonlinear programs.

DYERC 1986-1 Dyer, M. E.; Frieze, A. M.; Mac Diarmid, C.
On linear programs with random costs.

DYERD 1984-1 Dyer, M. E.; Kayal, N.; Walker, J.
A branch and bound algorithm for solving the multiple-choice knapsack problem.

DYERE 1987-1 Dyer, M. E:; Walker, J.
An algorithm for a separable integer programming problem with cumulatively bounded variables.

EMELC 1985-1 Emelichev, V. A.; Perepelitsa, V. A.
On multicriterial assignment problems.

EMELC 1986-1 Emelichev, V. A.; Perepelitsa, V. A.
On the estimation of the complexity of multicriterial transportation problems (Russian).

EMELD 1987-1 Emelichev, V. A.; Perepelitsa, V. A.; Shungarov, K. D.
Asymptotic approach to a multicriteria problem of the covering of a graph by stars (Russian).

ERIC 1987-1 Erickson, R. E.; Monma, C. L.; Veinott, A. F. Jr.
Send-and-split method for minimum-concave-cost network flows.

FEDE 1986-3 Federgruen, A.; Groenevelt, H.
The greedy procedure for resource allocation problems: necessary and sufficient conditions for optimality.

FERN 1982-1 Fernandez Garcia, F. R.; Pino Mejias, J. L.
Study of multiple objectives in sequencing problems on one machine (Spanish).

FINKE 1987-1 Finke, G.; Burkard, R. E.; Rendl, F.
Quadratic assignment problems.

FLORA 1986-1 Florian, M.
Nonlinear cost network models in transportation analysis.

FORTA 1986-1 Fortenberry, J. C.; Mitra, A.
A multiple criteria approach to the location-allocation problem.

FORTB 1985-1 Fortenberry, J. C.; Snyder, C. A.; Cox, J. F.
A multiple criteria decision support model for facilities layout.

FRENB 1985-1 Frenk, J. B. G.; Houweninge, M. van; Rinnooy Kan, A. H. G.
Asymptotic properties of the quadratic assignment problem.

FREY 1986-1 Frey, J.-J.; Yehia-Alcoutlabi, A.
Comparaisons par paires: une interpretation et une generalisation de la methode des scores.

GABB 1986-1 Gabbani, D.; Magazine, M.
An interactive heuristic approach for multiobjective integer-programming problems.

GILBB 1985-1 Gilbert, K. C.; Holmes, D. D.; Rosenthal, R. E.
A multiobjective discrete optimization model for land allocation.

GLOVE 1986-1 Glover, F.; Klingman, D.; Phillips, N.; Ross, G. T.
Integrating modeling, algorithm design, and computational implementation to solve a large-scale nonlinear mixed integer programming problem.

GOME 1987-1 Gomes, L. F. A. M.
On the problem of assigning traffic flows to a multicriteria, multicommodity, free-flow network.

GONZ 1985-1 Gonzalez, J. J.; Reeves, G. R.; Franz, L. S.
An interactive procedure for solving multiple-objective integer linear programming problems.

GRIB 1985-1 Gribik, P. R.; Kortanek, K. O.
Extremal methods of operations research.

GULY 1983-1 Gulyanitskii, L. F.; Khodzinskii, A. N.
Two algorithms for solution of optimization problems on permutations (Russian).

GUPTE 1985-1 Gupta, O. K.; Ravindran, A.
Branch and bound experiments in convex nonlinear integer programming.

HABE 1984-1 Habenicht, W.
Interaktive Lösungsverfahren für diskrete Vektoroptimierungsprobleme unter besonderer Berücksichtigung von Wegeproblemen in Graphen.

HACK 1986-1 Hackbarth, K.-D.
A heuristic procedure for calculating telecommunication transmission networks in consideration of network reliability.

HAMAD 1987-1 Hamacher, H. W.; Tufekci, S.
On the use of lexiographic min cost flows in evacuation modeling.

HAMMD 1987-2 Hammer, P. L.; Simeone, B.
Quadratic functions of binary variables.

HANS 1984-1 Hanscom, M. A.; Nguyen, V. H.; Strodiot, J.-J.
A reduced subgradient algorithm for network flow problems with convex nondifferentiable costs.

HASH 1987-1 Hashizume, S.; Fukushima, M.; Katoh, N.; Ibaraki, T.
Approximation algorithms for combinatorial fractional programming problems.

HENI 1986-2 Henig, M. I.
The shortest path problem with two objective functions.

HOFFC 1985-2 Hoffman, A. J.; Wolfe, P.
Minimizing a unimodal function of two integer variables.

HOOK 1986-1 Hooker, J.
Solving nonlinear single-facility network location problems.

HWANA 1984-1 Hwang, C.-L.; Lee, H. B.; Tillmann, F. A.; Lie, C. H.
Nonlinear integer goal programming applied to optimal system reliability.

IGNI 1985-1 Ignizio, J. P.
Integer goal programming via goal aggregation.

IGNIA 1984-1 Ignizio, J. P.; Thomas, L. C.
An enhanced conversion scheme for lexicographic, multiobjective integer programs.

JABL 1984-1 Jablonsky, J.
Integer vector optimization problems and their solutions (Czechoslovakian).

JHA 1986-1 Jha, N. K.
Optimizing the number of tools and cutting parameters in multi-tool turning for multiple objectives through geometric programming.

JORG 1987-1 Jorgensen, C.; Powell, S.
Solving 0-1 minimax problems.

KABE 1985-1 Kabe, D. G.
On discrete linear and quadratic programming problems.

KABE 1986-2 Kabe, D. G.
On solving linear and quadratic integer programming problems.

KABE 1986-3 Kabe, D. G.
On solving two-variables programming problems.

KANA 1984-1 Kanaeva, N. N.; Shcherbina, O. A.
On the efficiency of a modified local algorithm for solving discrete optimization problems.

KANA 1986-1 Kanaeva, N. N.; Shcherbina, O. A.
On a modified local algorithm for problems of discrete programming.

KARWA 1985-1 Karwan, M. H.; Zionts, S.; Villarreal, B.; Ramesh, R.
An improved interactive multicriteria integer programming algorithm.

KHACC 1983-1 Khachiyan, L. G.
Polynomial solvability of convex diophantine programming problems with a fixed number of variables.

KHANB 1984-1 Khan, Z. A.; Ahsan, M. J.
On integer solutions to the concave quadratic programs using branch and bound technique.

KHANC 1983-1 Khan, Z. A.; Ahsan, M. J.; Khan, E. A.
On mixed integer concave quadratic programs.

KHANC 1986-1 Khan, Z. A.; Ahsan, M. J.; Khan, E. A.
A branch and bound method for integer solutions to the concave quadratic programs.

KHOD 1985-1 Khodzinskii, A. N.
Sequential algorithm for the solution of problems of combinatorial optimization on permutations.

KHOD 1985-2 Khodzinskii, A. N.
Solution of the quadratic assignment problem by the mean value method (Russian).

KHRI 1985-1 Khristov, G.; Karamiteva, Z.; Stoyanov, T.
Numerical method and program for the solution of a linear fractional integer optimization problem (Bulgarian).

KIM 1985-1 Kim, J.-G.
Lagrangean relaxation approach for interactive multiobjective integer programming problem.

KLIND 1984-1 Klingman, D.; Phillips, N. V.
Topological and computational aspects of preemptive multicriteria military personnel assignment problems.

KOCH 1985-1 Kochkarov, A. M.
Asymptotically exact algorithms for solving a multicriterial problem of covering a graph by paths (Russian).

KOCHA 1985-1 Kochkarov, A. M.; Perepelitsa, V. A.
A multicriterial covering problem of a graph by paths of large and small length (Russian).

KOHLA 1984-1 Köhler, J.
A method for the solution of a special nonlinear integer programming problem.

KORH 1984-1 Korhonen, P.; Wallenius, J.; Zionts, S.
Solving the discrete multiple criteria problem using convex cones.

KORN 1985-1 Körner, F.
Ein effektiver Branch and Bound-Algorithmus für Boolesche quadratische Optimierungsprobleme.

KORN 1985-2 Körner, F.
Integer quadratic optimization.

KORN 1985-4 Körner, F.
Nichtlineare, ganzzahlige Optimierung.

KORN 1986-1 Körner, F.
A new bound for the quadratic knap sack problem and its use in a branch and bound algorithm.

KOVAD 1985-2 Kovalev, M. M.
Gradient methods of maximization of convex functions on discrete structures.

KOVAG 1985-1 Kovalev, M. M.; Pisaruk, N. N.
Gradient methods in convex integer programming.

KOVAG 1985-2 Kovalev, M. M.; Pisaruk, N. N.
Independent flows with a discrete-concave expenditure function.

KOVAG 1985-3 Kovalev, M. M.; Pisaruk, N. N.
Ordnungskonvexität (Russian).

KOVAG 1986-1 Kovalev, M. M.; Pisaruk, N. N.
Generalized flows with discrete convex cost function (Russian).

KREJ 1983-1 Krejnes, M. G.; Novikova, N. M.
An algorithm for solving a class of discrete multicriterion problems.

KWAK 1985-1 Kwak, N. K.; Schniederjans, M. J.
A goal programming model as an aid in facility location analysis.

KWAK 1985-3 Kwak, N. K.; Schniederjans, M. J.
Goal programming solutions to transportation problems with variable supply and demand requirements.

LAWR 1984-1 Lawrence, K. D.; Reeves, G. R.; Lawrence, S. M.
A multiple objective shift allocation model.

LAZI 1985-1 Lazimy, R.
Improved algorithm for mixed-integer quadratic programs and a computational study.

LAZI 1986-1 Lazimy, R.
Interactive relaxation method for a broad class of integer and continuous nonlinear multiple criteria problems.

LEE M 1987-1 Lee, S.-M.; Luebbe, R. L.
A zero-one goal-programming algorithm using partitioning and constraint aggregation.

LIEP 1983-1 Lieponite-Yukhimenko, B. I.
Effectiveness of preference functions in integer linear programming algorithms (Russian).

LIGG 1985-1 Liggett, R. S.
Optimal spatial arrangement as a quadratic assignment problem.

LOGA 1987-1 Loganathan, G. V.; Sherali, H. D.
A convergent interactive cutting-plane algorithm for multiobjective optimization.

LOZO 1984-2 Lozovanu, D. D.
Some properties of a network nonlinear transportation problem and the problem of synthesis of a network in the form of an outgoing tree with a source (Russian).

LU 1987-1 Lu, S. H.; Simeone, B.
On the equivalence of roof-duality and paved-duality in quadratic 0-1 optimization.

LU A 1985-1 Lu, S. H.; Williams, A. C.
Roof duality for nonlinear 0-1 optimization.

LU A 1987-1 Lu, S. H.; Williams, A. C.
Roof duality and linear relaxation for quadratic and polynomial 0-1 optimization.

LU A 1987-2 Lu, S. H.; Williams, A. C.
Roof duality for polynomial 0-1 optimization.

LUPS 1985-1 Lupsa, L.
On quasiconvex integer programming problems.

LYSK 1985-1 Lyska, W.
Ein kombinatorischer Algorithmus zur Lösung der linearen 0-1 Vektoroptimierungsaufgabe.

LYSK 1985-2 Lyska, W.
Über ein Ersatzproblem zur ganzzahligen linearen Vektoroptimierung.

MAHE 1985-1 Mahey, P.
Subgradient techniques and combinatorial optimization.

MARCG 1986-1 Marcotte, O.; Soland, R. M.
An interactive branch-and-bound algorithm for multiple criteria optimization.

MARK 1986-1 Markland, R. E.; Vickery, S. K.
The efficient computer implementation of a large-scale integer goal programming model.

MARTM 1987-1 Martins, E. Q. V.
On a particular quadratic network problem.

MATHA 1986-1 Mathur, K.; Salkin, H. M.; Bohanty, B. B.
A note on a general non-linear knapsack problem.

MATHC 1985-1 Mathur, K.; Salkin, H. M.; Nishimura, K.; Morito, S.
Applications of integer programming in radio frequency management.

MATI 1987-1 Matitskas, I.-K. L.; Palubetskis, G. S.
Graph cutting and 0-1 quadratic programming (Russian).

MATS 1983-1 Matsui, K.; Takada, J.
SLP method for a nonlinear multicommodity flow assignment problem.

MAWE 1986-1 Mawengkang, H.; Murtagh, B. A.
Solving nonlinear integer programs with large-scale optimization software.

MIKHA 1986-1 Mikhalevich, M. V.
A stochastic algorithm for solving nonlinear mixed integer problems (Russian).

MINA 1985-1 Minaev, Y. N.
On a formally heuristic algorithm for solving integer optimization problems (linear and quadratic models) (Russian).

MINO 1984-1 Minoux, M.
A polynomial algorithm for minimum quadratic cost flow problems.

MINO 1986-3 Minoux, M.
Solving integer minimum cost flows with separable convex cost objective polynomially.

MISH 1983-1 Misha, M. K.; Das, C.
Maximin problem and symmetric duality for mixed-integer quadratic programs.

MISHC 1984-1 Mishra, B. K.; Das, C.
Minimax and symmetric duality for nonlinear mixed-integer programming problem.

MISHC 1985-1 Mishra, B. K.; Das, C.
Maximin problem and a duality theorem for mixed-integer quadratic programming.

MISHD 1985-1 Mishra, M. S.; Nanda, S.
On a pair of nonlinear mixed integer programming problems.

MOTS 1985-1 Motskus, J.; Teshis, V.; Shaltyanis, V.; Yushka, F.
On a problem allocation of resources (Russian).

MULVA 1985-1 Mulvey, J. M.; Zenios, S. A.
Integrated risk/cost planning models for the U.S. air traffic system.

MURTA 1985-1 Murty, K. G.; Kabadi, S. N.
Some NP-complete problems in quadratic and nonlinear programming.

MUSA 1984-1 Musa, A. A.; Saxena, U.
Scheduling nurses using goal-programming techniques.

NAND	1987-1	Nanda, S.; Mishra, M. S.; Acharya, D. A note on a pair of nonlinear mixed integer programming problems.
NAUM	1986-1	Naumovich, N. A. Dichotomous realization of gradient algorithms for solving integer programming problems (Russian).
NEHS	1985-1	Nehse, R. Diskrete Vektoroptimierung - Möglichkeiten der numerischen Bearbeitung.
O'KE	1987-1	O'Kelly, M. E. A quadratic integer program for the location of interacting hub facilities.
OGRY	1984-1	Ogryczak, W. Branch-and-bound procedures for multicriteria integer programming (Polish).
OHUC	1984-1	Ohuchi, A.; Kaji, I. Lagrangian dual coordinatewise maximization algorithm for network transportation problems with quadratic costs.
PAL	1984-1	Pal, B. B.; Basu, M. Optimal solution of a transportation problem by goal programming technique.
PASC	1987-1	Pasche, C. Flot a cout convexe lineaire par morceaux.
PELI	1984-1	Pelikan, J. Container transportation problem (Czechoslovakian).
PELI	1985-1	Pelikan, J. Container transportation problem.
PHIL	1987-1	Phillips, N. V. A weigthing function for pre-emptive multicriteria assignment problems.
PICO	1984-1	Picone, C. J.; Wilhelm, W. E. A perturbation scheme to improve Hillier's solution to the facilities layout problem.
PICO	1986-1	Picone, C. J.; Wilhelm, W. E. "A perturbation scheme to improve Hillier's solution to the facilities layout problem": a clarification.
POTE	1980-1	Potemkina, A. V. Solution of a problem of integer convex programming (Russian).
PRAK	1984-1	Prakash, J.; Sinha, S. B. A goal programming model for express bus-stop location.
QUEY	1986-1	Queyranne, M. Performance ratio of polynomial heuristics for triangle inequality quadratic assignment problems.
RAME	1986-1	Ramesh, R.; Karwan, M. H.; Zionts, S. A class of practical interactive branch and bound algorithms for multicriteria integer programming.
RASM	1986-1	Rasmussen, L. M. Zero-one programming with multiple criteria.
RAVIA	1986-1	Ravindran, A.; Shin, W. S.; Arthur, J. L.; Moskowitz, H. Nonlinear integer goal programming models for acceptance sampling.
REDD	1984-1	Reddy, K. P.; Mittal, A. K.; Gupta, S. K. Bivalent quadratic programming problem - a computational study.
REEV	1985-1	Reeves, C. R. An improved heuristic for the quadratic assignment problem.
REND	1985-1	Rendl, F. Ranking scalar products to improve bounds for the quadratic assignment problem.
REND	1986-2	Rendl, F. Quadratic assignment problems on series-parallel digraphs.
ROSE	1987-1	Rosenberg, E. A nonlinear programming heuristic for computing optimal link capacities in a multi-hour alternate routing communications network.
ROUC	1987-1	Roucairol, C. A parallel branch and bound algorithm for the quadratic assignment problem.
SAAT	1987-1	Saatcioglu, Ö. A multi-attribute assignment goal-programming model with incentives.
SCHNA	1987-1	Schniederjans, M. J.; Kim, G. C. A goal programming model to optimize departmental preference in course assignments.
SEEL	1986-1	Seeländer, J. Comments on special scalar auxiliary problems in discrete optimization (German).
SELE	1986-1	Selen, W. J.; Hott, D. D. A mixed-integer goal-programming formulation of the standard flow-shop scheduling problem.
SEMEB	1984-1	Semenova, N. V. Solution of one generalized integer programming problem(Russian).
SEN A	1985-1	Sen, S.; Sherali, H. D. A branch and bound algorithm for extreme point mathematical programming problems.
SERA	1984-1	Serafini, P. Dual relaxation and branch-and-bound techniques for multiobjective optimization.
SERG	1987-1	Sergeev, S. I. A new lower bound for the quadratic assignment problem (Russian).
SERGF	1985-1	Sergienko, I. V.; Lebedeva, T. T.; Konzeratskaya, L. N. Some parametric problems of integer convex programming and their approximate solution (Russian).
SERGI	1986-1	Sergienko, I. V.; Sokirko, S. V. Solution of a nonlinear integer programming problem (Russian).
SHERD	1986-1	Sherali, H. D.; Myers, D. C. The design of branch and bound algorithms for a class of nonlinear integer programs.
SHERF	1986-1	Sherali, H. D.; Rajgopal, P. A flexible, polynomial-time, construction and improvement heuristic for the quadratic assignment problem.
SHOR	1985-1	Shor, N. Z.; Davydov, A. S. On the method of obtaining estimates in quadratic extremal problems with Boolean variables.
SIME	1985-1	Simeone, B. Consistency of quadratic boolean equations and the König-Egervary property for graphs.
SIME	1986-1	Simeone, B. An asymptotically exact polynomial algorithm for equipartition problems.

SINC 1986-1 Sinclair, M.
An exact penalty function approach for nonlinear integer programming problems.
SKOR 1987-1 Skorin-Kapov, J.; Granot, F.
Non-linear integer programming: sensitivity analysis for branch and bound.
SLOW 1984-1 Slowinski, R.
Modelling and solving multicriteria project scheduling problems.
SLOW 1984-2 Slowinski, R.
Modelling and solving multicriteria project scheduling problems.
SMOR 1983-1 Smorodinskii, S. S.; Tanaevskii, V. G.
A means of scalarization of vector estimates for the solution of discrete problems of multicriteria choice (Russian).
SOLOA 1984-1 Solomon, D. I.
Generalization of the linear and fractional-linear transportation problem (Russian).
STET 1985-1 Stetsenko, S. I.
Numerical experiments in the solution of quadratic problems with Boolean variables (Russian).
STEU 1986-1 Steuer, R. E.; Wood, E. F.
On the 0-1 implementation of the Tchebycheff solution approach: a water quality illustration.
SZIL 1983-1 Szilagyine-Szemkeö, J.
A multiple objective scheduling problem (Hungarian).
TANAA 1984-1 Tanaka, K.; Maruyama, Y.
The multiobjective optimization problem of set function.
TANGC 1986-1 Tang, H.-W.; Sun, Z.-X.; Gao, G.-Q.
An algorithm for a class of nonlinear mixed integer programming problems, and its convergence (Chinese).
TAVA 1986-1 Tavares, L. V.
Multicriteria scheduling of a railway renewal program.
TAYL 1985-1 Taylor, B. W. III; Moore, L. J.; Clayton, E. R.; Davis, K. R.; Rakes, T. R.
An integer nonlinear goal programming model for the deployment of state highway patrol units.
TEGH 1986-1 Teghem, J. Jr.; Kunsch, P. L.
A survey of techniques for finding efficient solutions to multi-objective integer linear programming.
TEGH 1986-2 Teghem, J. Jr.; Kunsch, P. L.
Interactive methods for multiobjective integer linear programming.
TERZ 1985-1 Terzi, D. G.
Generators of vectors in problems of discrete optimization (Russian).
THOMB 1986-1 Thompson, G. L.; Kaku, B. K.
An exact algorithm for the general quadratic assignment problem.
TIGA 1984-1 Tigan, S.
On some nonlinear knapsack problems.
TODD 1985-1 Todd, M. J.
Linear and quadratic programming in oriented matroids.
TUNG 1986-1 Tung, P. C. T.
A bicriterion Pareto path algorithm.
TYUL 1986-1 Tyulyukin, V. A.
On a numerical method for the minimization of functions of an integer argument (Russian).
VASIA 1983-1 Vasilev, V. S.
An approximate algorithm, solving a class of nonlinear discrete programming problems (Bulgarian).
VENIA 1986-1 Veniaminov, S. S.
The effectiveness of the hybrid method for solving discrete optimization problems.
VICK 1985-1 Vickery, S. K.; Markland, R. E.
Integer goal programming for multistage lot sizing: experimentation and implementation.
VOLKA 1987-1 Volkovich, O. V.; Roshchin, V. A.; Sergienko, I. V.
Models and methods for solution of integer quadratic programming problems (Russian).
WARB 1987-1 Warburton, A.
Approximation of Pareto optima in multiple-objective, shortest path problems.
WILH 1985-1 Wilhelm, M. R.; Evans, G. W.
The quadratic assignment problem: applications and solution techniques.
WILHA 1987-1 Wilhelm, M. R.; Ward, T. L.
Solving quadratic assignment problems by "simulated annealing".
WILL 1985-1 Williams, A. C.
Quadratic 0-1 programming using the proof dual - with computational results.
XIE 1982-1 Xie, L.-T.
On a kind of integer programming problems on n-complexes (Chinese).
XU H 1987-1 Xu, W.-X.
A new bounding technique for the quadratic assignment problem (Chinese).
YAN A 1982-1 Yan, T. Y.; Tan, H. H.
Quadratic integer programming for large scale banned matrices.
YARL 1983-1 Yarlagadda, R.
Unit integer quadratic binary programming.
ZENI 1985-1 Zenios, S. A.; Mulvey, J. M.
A distributed algorithm for convex network optimization problems.
ZENI 1986-1 Zenios, S. A.; Mulvey, J. M.
Nonlinear network programming on vector supercomputers.
ZENI 1986-2 Zenios, S. A.; Mulvey, J. M.
Relaxation techniques for strictly convex network problems.

AIEL	1985-1	Aiello, A.; Burattini, E.; Massarotti, A.; Ventriglia, F. Heuristic evaluation techniques for bin packing approximation algorithms.
ARAO	1984-1	Araoz, J. Packing problems in semigroup programming.
ASSM	1984-1	Assmann, S. F.; Johnson, D. S.; Kleitman, D. J.; Leung, J. Y.-T. On a dual version of the one-dimensional bin packing problem.
AZAR	1987-1	Azarenok, A. S. Packing of graphs of width two (Russian).
BAKEB	1985-1	Baker, B. S. A new proof for the first-fit decreasing bin-packing algorithm.
BALAF	1987-1	Balakrishnan, R.; Paulraja, P.; Basha, A. R. Packing half-complete graphs with trees.
BARAI	1986-1	Barany, I.; Edmonds, J.; Wolsey, L. A. Packing and covering a tree by subtrees.
BEINB	1985-1	Beineke, L. W. Packings of complete bipartite graphs.
BERK	1987-1	Berkey, J. O.; Wang, P. Y. Two-dimensional finite bin-packing algorithms.
BOUR	1986-1	Bourjolly, J.-M. Integral and fractional node packings, and pseudo-Boolean programming.
BRUNB	1985-1	Bruno, J. L.; Downey, P. J. Probabilistic bounds for dual bin-packing.
BUTZ	1984-1	Butz, L.; Hammer, P. L.; Hausmann, D. Reduction methods for the vertex packing problem.
CARPC	1985-1	Carpenter, H.; Dowsland, W. B. Practical considerations of the pallet-loading problem.
CHANT	1985-1	Chang, G. J.; Nemhauser, G. L. Covering, packing and generalized perfection.
CHAUD	1987-1	Chauny, F.; Loulou, R.; Sadones, S.; Soumis, F. A two-phase heuristic for strip packing: algorithm and probabilistic analysis.
CHAZ	1983-1	Chazelle, B. The bottom-left bin-packing heuristic: an efficient implementation.
CHOP	1987-1	Chopra, S. On ternary problems.
COFFB	1984-1	Coffman, E. G. Jr.; Garey, M. R.; Johnson, D. S. Approximation algorithms for bin-packing - an updated survey.
COFFB	1987-1	Coffman, E. G. Jr.; Garey, M. R.; Johnson, D. S. Bin packing with divisible item sizes.
COFFD	1984-1	Coffman, E. G. Jr.; Gilbert, E. N. Dynamic, first-fit packings in two or more dimensions.
COFFF	1986-1	Coffman, E. G. Jr.; Lueker, G. S.; Rinnooy Kan, A. H. G. An introduction to the probabilistic analysis of sequencing and packing heuristics.
CORN	1985-1	Corneil, D. G. The complexity of generalized clique packing.
COUR	1986-1	Courcoubetis, C.; Weber, R. R. Necessary and sufficient conditions for stability of a bin-packing system.
CRAMA	1987-1	Crama, Y.; Hammer, P. L.; Ibaraki, T. Packing, covering and partitioning problems with strongly unimodular constraint matrices.
CSIR	1986-1	Csirik, J. Bin packing as a random walk: a note on Knödel's paper.
CSIRB	1986-1	Csirik, J.; Galambos, G. An O(n) bin-packing algorithm for uniformly distributed data.
CSIRB	1986-2	Csirik, J.; Galambos, G. Nonlinear sequences for bin-packing (Hungarian).
CSIRC	1986-1	Csirik, J.; Galambos, G.; Frenk, J. B. G.; Frieze, A. M.; Rinnooy Kan, A. H. G. A probabilistic analysis of the next fit decreasing bin packing heuristic.
CSIRD	1986-1	Csirik, J.; Mate, E. The probabilistic behaviour of the NFD bin packing algorithm.
DAWS	1985-1	Dawson, J. E. Algorithms to find directed packings.
DOWS	1984-1	Dowsland, K. A. The three-dimensional pallet chart: an analysis of the factors affecting the set of feasible layouts for a class of two-dimensional packing problems.
DOWS	1985-1	Dowsland, K. A. A graph-theoretic approach to a pallet loading problem.
DOWS	1985-2	Dowsland, K. A. Determining an upper bound for a class of rectangular packing problems.
DOWS	1987-1	Dowsland, K. A. A combined data-base and algorithmic approach to the pallet-loading problem.
DOWS	1987-2	Dowsland, K. A. An exact algorithm for the pallet loading problem.
DOWSA	1985-1	Dowsland, W. B. Two and three dimensional packing problems and solution methods.
DREZC	1984-1	Drezner, Z.; Nof, S. Y. On optimizing bin picking and insertion plans for assembly robots.
FEOF	1987-1	Feofiloff, P.; Younger, D. H. Directed cut transversal packing for source-sink connected graphs.
FISHI	1981-1	Fisher, M. L.; Wolsey, L. A. On the greedy heuristic for covering and packing problems.
FOX	1986-1	Fox, G. E.; Scudder, G. D. A simple strategy for solving a class of 0-1 integer programming models.
FRENA	1987-1	Frenk, J. B. G.; Galambos, G. Hybrid next-fit algorithm for the two-dimensional rectangle bin-packing problem.

FRIEA 1983-1 Friesen, D. K.; Langston, M. A.
Variable sized bin packing.
FRIEA 1986-1 Friesen, D. K.; Langston, M. A.
Variable sized bin packing.
FRIEA 1987-1 Friesen, D. K.; Langston, M. A.
Bin packing: on optimizing the number of pieces packed.
GALA 1985-1 Galambos, G.
A new heuristic for the classical bin-packing problem.
GALA 1985-2 Galambos, G.
Algorithms for different bin packing problems.
GALA 1986-1 Galambos, G.
Parametric lower bound for on-line bin-packing.
GALAA 1985-1 Galambos, G.; Csirik, J.
Next-fit type algorithms for solving different bin-packing problems.
GIMAA 1984-1 Gimadi, E. K.; Zalyubovskij, V. V.
An asymptotically exact approach to the solution of one-dimensional problem of packing in containers (Russian).
GRIMA 1986-1 Grimmett, G. R.
An exact threshold theorem for random graphs and the node-packing problem.
GRIMB 1985-1 Grimmett, G. R.; Pulleyblank, W. R.
Random near-regular graphs and the node packing problem.
GROTC 1986-1 Grötschel, M.; Lovasz, L.; Schrijver, A.
Relaxations of vertex packing.
GUAN 1984-1 Guan, M.-G.
The maximum weighted cycle-packing problem and its relation to the Chinese postman problem.
HELL 1986-1 Hell, P.; Kirkpatrick, D. G.
Packings by complete bipartite graphs.
HOBB 1984-1 Hobbs, A. M.
Packing sequences.
HOBBA 1987-1 Hobbs, A. M.; Bourgeois, B. A.; Kasiraj, J.
Packing trees in complete graphs.
HOCHA 1985-1 Hochbaum, D. S.; Maass, W.
Approximation schemes for covering and packing problems in image processing and VLSI.
HOCHB 1986-1 Hochbaum, D. S.; Shmoys, D. B.
A packing problem you can almost solve by sitting on your suitcase.
HOFR 1984-1 Hofri, M.
A probabilistic analysis of the next-fit bin packing algorithm.
HOSK 1984-1 Hoskins, W. D.; Lambert, T. D.; Stanton, R. G.
An improved algorithm for determining optimal sums of disjoint sets.
HUANF 1986-1 Huang, W.-Q.; Li, Q.-H.; Yu, X.-D.
A quasi-physical method for solving the three-dimensional packing problem.
IKURB 1985-1 Ikura, Y.; Nemhauser, G. L.
Simplex pivots on the set packing polytope.
ISER 1987-1 Isermann, H.
Ein Planungssystem zur Optimierung der Palettenbeladung mit kongruenten rechteckigen Versandgebinden.
IVANC 1984-1 Ivani, A. M.
Estimation of efficiency of algorithms for packing into containers (Russian).
IVANF 1984-1 Ivanyi, A.
Estimation of the efficiency of bin packing algorithms (Russian).
IVANF 1984-2 Ivanyi, A.
Performance bounds for simple bin packing algorithms.
IVANF 1985-1 Ivanyi, A.
Tight worst-case bounds for bin packing algorithms.
JOHNF 1985-1 Johnson, D. S.; Garey, M. R.
A 71/60 theorem for bin packing.
KARMA 1982-1 Karmarkar, N.
Probabilistic analysis of some bin-packing problems.
KARMB 1982-1 Karmarkar, N.; Karp, R. M.
An efficient approximation scheme for the one-dimensional bin-packing problem.
KARZ 1984-1 Karzanov, A. V.
A generalized MFMC-property and multicommodity cut problems.
KARZ 1986-1 Karzanov, A. V.
An algorithm for the maximal packing of odd-pole cuts and its applications (Russian).
KORNB 1986-1 Kornienko, N. M.; Lepin, V. V.
Optimal packings of a weighted cycle in a line (Russian).
KORNB 1986-2 Kornienko, N. M.; Lepin, V. V.
Optimal packings of a weighted cycle into a line (Russian).
KRAUA 1987-1 Krause, K.; Larmore, L. L.; Volper, D. J.
Packing items from a triangular distribution.
LAPOC 1987-1 Laporte, G.; Mercure, H.; Nobert, Y.
Cutting planes based on bin packing solutions for the capacitated vehicle routing problem.
LEE 1985-1 Lee, C. C.; Lee, D. T.
A simple on-line packing algorithm.
LEE N 1986-1 Lee, Y.-D.
Integer programming applications and their relationship (Korean).
LEMO 1985-1 Lemos, M.
On Seymour's question about packing and covering with matroid circuits.
LOUL 1984-1 Loulou, R.
Probabilistic behaviour of optimal bin-packing solutions.
LOVA 1985-4 Lovasz, L.
Vertex packing algorithms.

MART	1985-1	Martel, C. U. A linear time bin-packing algorithm.
MARTB	1984-1	Martello, S.; Toth, P. Lower bounds and reduction procedures for the bin packing problem.
MURG	1987-1	Murgolo, F. D. An efficient approximation scheme for variable-sized bin packing.
NOWI	1987-1	Nowicki, E.; Smutnicki, C. On lower bounds on the minimum maximum lateness on one machine subject to release date.
ONG A	1984-1	Ong, H. L.; Magazine, M.; Wee, T. S. Probabilistic analysis of bin packing heuristics.
PROD	1985-1	Prodinger, H. Einige Bemerkungen zu einer Arbeit von W. Knödel über das mittlere Verhalten von on-line-Packungsalgorithmen.
RAGHA	1986-1	Raghavan, P. Probabilistic construction of deterministic algorithms: approximating packing integer programs.
RAMAB	1987-1	Ramanathan, A.; Colbourn, C. J. Bounds for all-terminal reliability by arc-packing.
RHEE	1985-1	Rhee, W.-S. T. Convergence of optimal stochastic bin packing.
RHEE	1987-1	Rhee, W.-S. T. Probabilistic analysis of the next fit decreasing algorithm for bin-packing.
ROBE	1985-1	Robert, P. Etude de la version temps reel d'un algorithme de bin-packing asymptotiquement optimal.
RODI	1986-1	Roditty, Y. The packing and covering of the complete graph. I: The forests of order five.
SCHRE	1983-1	Schrijver, A. Packing and covering of crossing families of cuts.
SHORA	1984-1	Shor, P. W. The average-case analysis of some on-line algorithms for bin packing.
SHORA	1986-1	Shor, P. W. The average-case analysis of some on-line algorithms for bin packing.
TROT	1985-1	Trotter, L. E. Jr. Discrete packing and covering.
VAINA	1984-1	Vainshtein, A. D. The problem of packing of rectangles into a strip - a survey (Russian).
VERC	1986-1	Vercellis, C. A probabilistic analysis of the set packing problem.
VOHR	1985-1	Vohra, R. V. On the greedy heuristic for the vertex packing problem.
WERR	1987-2	Werra, D. de Partitions into odd chains.
WEST	1985-1	West, D. B. Parameters of partial orders and graphs: packing, covering and representation.
ZHANV	1987-1	Zhang, Z. Z.; Masuyama, S.; Ibaraki, T.; Mine, H. Graph packing over a rooted tree.

BALLB 1986-1 Ball, M. O.; Taverna, R.
Sensitivity analysis for the matching problem and its use in solving matching problems with a single side constraint.

BANK 1987-3 Bank, B.; Mandel, R.
Nonlinear parametric integer programming.

BIALA 1985-1 Bialy, H.
Stabilitätsbetrachtungen zur linearen Transportoptimierung.

BLAIA 1986-1 Blair, C. E.; Jeroslow, R. G.
Computational complexity of some problems in parametric discrete programming. I.

BOYAA 1983-1 Boyarchuk, D. A.
Investigation of the admissible domain of an integer programming problem with a parameter in one of the constraints (Russian).

BOYAA 1985-1 Boyarchuk, D. A.
An algorithm for solving a parametric integer linear programming problem (Russian).

BRUC 1985-1 Brucker, P.
Parametric programming and circulation problems with one additional linear constraint.

BURKD 1984-1 Burkard, R. E.; Krarup, J.; Pruzan, P. M.
Some relationships between multicriteria and parametric discrete optimization problems with bottleneck criteria.

CHANE 1984-1 Chanas, S.; Kolodziejczyk, W.; Machaj, A.
A fuzzy approach to the transportation problem.

COOKD 1986-1 Cook, W.; Gerards, A. M. H.; Schrijver, A.; Tardos, E.
Sensitivity theorems in integer linear programming.

DERI 1985-2 Derigs, U.
Postoptimal analysis for matching problems.

DREZ 1985-3 Drezner, Z.
Sensitivity analysis of the optimal location of a facility.

FEIR 1986-1 Feiring, B. R.
A parametric programming procedure for solving pure integer programming problems.

GLAZ 1987-2 Glazebrook, K. D.
Sensitivity analysis for stochastic scheduling problems.

GOLDM 1982-1 Goldengorin, B. I.
Stability of solutions in problems with one component of connection of local minima (Russian).

HAMAA 1986-1 Hamacher, H. W.; Foulds, L. R.
Algorithms for flows with parametric capacities.

IVANA 1984-1 Ivanchev, D. T.; Ruhe, G.
Flows in generalized networks with parametric arc capacities.

JENK 1987-1 Jenkins, L.
Parametric-objective integer programming using knapsack facets and Gomory cutting planes.

JENK 1987-2 Jenkins, L.
Using parametric integer programming to plan the mix of an air transport fleet.

KASP 1985-1 Kaspshitskaya, M. F.
Stability of the solution of a certain combinatorial optimization problem (Russian).

KOZEA 1986-1 Kozeratskaya, L. N.
Stability regions for some integer programming problems (Russian).

LEE I 1986-1 Lee, I.-S.
On sensitivity analysis for shortest-path dynamic programming models.

LIBU 1984-1 Libura, M.
Integer programming duality and its application in sensitivity analysis and approximate algorithms (Polish).

MAIE 1987-1 Maier, G.
Determining all nondegenerate bounds in sensitivity analysis of minimal cost network flow problems.

MIKH 1986-1 Mikhailyuk, V. A.
Certain aspects of postoptimal analysis of the minimization problem for the Boolean functions (Russian).

OHTA 1985-1 Ohtake, Y.; Nishida, N.
A branch-and-bound algorithm for 0-1 parametric mixed integer programming.

PETEA 1984-1 Peteanu, V.; Tigan, S.
On some discrete fractional max-min problems. Application to max-min problems in graphs.

RUHE 1985-1 Ruhe, G.
Characterization of all optimal solutions and parametric maximal flows in networks.

RUHEA 1985-1 Ruhe, G.; Ivanchev, D. T.
The solution of generalized network flow problems with parametric capacities and stochastic demands.

SCHR 1985-1 Schrage, L.; Wolsey, L. A.
Sensitivity analysis for branch and bound integer programming.

SCHU 1986-1 Schubert, I. S.; Zimmermann, U.
One-parametric bottleneck transportation problems.

SERGB 1984-1 Sergienko, I. V.; Artemenko, V. I.
Search for approximate solutions of integer linear problems by the methods of linear and parametric programming (Russian).

SERGD 1983-1 Sergienko, I. V.; Filonenko, N. V.
Investigation of the stability of the vector of the right-hand sides of a system of constraints for an integer linear programming problem (Russian).

SERGF 1985-1 Sergienko, I. V.; Lebedeva, T. T.; Konzeratskaya, L. N.
Some parametric problems of integer convex programming and their approximate solution (Russian).

SKOR 1987-1 Skorin-Kapov, J.; Granot, F.
Non-linear integer programming: sensitivity analysis for branch and bound.

TARJ 1986-2 Tarjan, R. E.
Corrigendum to: Sensitivity analysis of minimum spanning trees and shortest path trees.

TAYI 1985-1 Tayi, G. K.
Sensitivity analysis of mixed integer programs: an application to environmental policy making.

THOME 1986-1 Thompson, G. L.; Zawack, D. J.
A problem expanding parametric programming method for solving the job shop scheduling problem.

VEN 1985-1 Ven, V. L.
Partial aggregation in problems of integer linear programming with variable parameters (Russian).

ADAM 1984-1 Adam, A.
On certain partitions of finite directed graphs and of finite automata.

AGARA 1985-1 Agarwal, Y. K.
Set partitioning approach to vehicle routing.

AHARB 1985-1 Aharoni, R.; Hartman, I. B.-A.; Hoffman, A. J.
Path partitions and packs of acyclic digraphs.

ANTO 1983-1 Antonelli, S.; Conte, G.; Delli Guanti, G.
An optimal partitioning algorithm for a new class of graphs (Italian).

BABAC 1987-1 Babaev, R. D.
Generation of test problems on covering and partitioning (Russian).

BALLA 1985-1 Ball, M. O.; Roberts, A.
A graph partitioning approach to airline crew scheduling.

BARN 1985-1 Barnes, E. R.
Partitioning the nodes of a graph.

BART 1984-1 Bartholdi, J. J. III; Platzman, L. K.
Heuristics based on spacesfilling curves for combinatorial problems in the plane.

BAUMB 1985-1 Baumann, U.; Lesch, M.
Klassen von Zerlegungen zu Kantenmenge vollständigen Graphen.

BECKD 1984-1 Becker, R. I.; Schach, S. R.
A bottom-up algorithm for weight- and height-bounded minimal partition of trees.

BECKD 1984-2 Becker, R. I.; Schach, S. R.
A bottom-up algorithm for weight- and height-bounded minimal partition of trees.

BERG 1985-2 Berge, C.
Path-partitions in directed graphs and posets.

BIENB 1986-1 Bienia, W.; Meyniel, H.
Partition of digraphs into paths or circuits.

BOROC 1983-1 Boronin, A. B.; Pershin, O. Y.
An algorithm for partition of a weighted graph into minimally connected subgraphs (Russian).

BOROG 1985-1 Borowiecki, M.; Mihok, P.
On (n,P)-partitionable graphs.

BRANA 1985-2 Brandstädt, A.
Partitions of graphs into one or two independent sets and cliques.

BROW 1987-1 Brown, G. G.; Graves, G. W.; Ronen, D.
Scheduling ocean transportation of crude oil.

BUSNA 1985-1 Busnyuk, N. N.; Sarvanov, V. I.
Graphs of partition of a rectangle. I (Russian).

BUZA 1984-1 Buzacott, J. A.; Chang, J. S. K.
Cut-set intersections and node partitions.

CAEN 1985-1 Caen, D. de; Gregory, D. A.
Partitions of the edge-set of a multigraph by complete subgraphs.

CAEN 1987-1 Caen, D. de; Gregory, D. A.
On the decomposition of a directed graph into complete bipartite subgraphs.

CAMEF 1985-1 Cameron, J.; Thomas, G.
An approximate graph partitioning algorithm and its computational complexity.

CAMEG 1986-1 Cameron, K.
On k-optimum dipath partitions and partial k-colourings of acyclic digraphs.

CAMPA 1985-1 Campello, R. E.
Updating a hybrid algorithm for set partitioning problems.

CAMPB 1987-1 Campello, R. E.; Maculan, N. F.
Lagrangean relaxation for a lower bound to a set partitioning problem with side constraints: properties and algorithms.

CARO 1986-1 Caro, Y.
Chromatic decomposition of a graph.

CHRO 1986-1 Chrobak, M.; Poljak, S.
On common edges in optimal solutions to traveling salesman and other optimization problems.

CHUNB 1983-1 Chung, F. R. K.; Erdös, P.; Spencer, J.
On the decomposition of graphs into complete bipartite subgraphs.

CIOBA 1985-1 Ciobanu, G.; Galupa, A.; Fabian, C.; Hartulari, C.
Reduction of the large size graphs with a view of their scheduling.

COFFE 1984-1 Coffman, E. G. Jr.; Langston, M. A.
A performance guarantee for the greedy set-partitioning algorithm.

COLBC 1987-1 Colbourn, C. J.; Pulleyblank, W. R.
Matroid Steiner problems, the Tutte polynomial and network reliability.

CRAMA 1987-1 Crama, Y.; Hammer, P. L.; Ibaraki, T.
Packing, covering and partitioning problems with strongly unimodular constraint matrices.

DARBA 1985-1 Darby-Dowman, K.; Mitra, G.
An extension of set partitioning with application to scheduling problems.

DARZA 1986-1 Darzentas, J.; Bairaktaris, P.
On the set partitioning type formulation for the discrete location problem.

DERI 1985-3 Derigs, U.
Some basic exchange properties in combinatorial optimization and their application to constructing the K-best solutions.

DIEH 1985-1 Diehr, G.
Evaluation of a branch and bound algorithm for clustering.

DYERB 1985-2 Dyer, M. E.; Frieze, A. M.
On the complexity of partitioning graphs into connected subgraphs.

DYERC 1984-1 Dyer, M. E.; Frieze, A. M.; Mac Diarmid, C.
Partitioning heuristics for two geometric maximization problems.

DZHI 1985-1 Dzhidzhev, K. N.
A linear algorithm for partitioning graphs of fixed genus.

EADEA 1984-1 Eades, P.; Pullman, N. J.; Robinson, P. J.
On clique partition numbers of regular graphs.

EDEN 1986-1 Edenbrandt, A.
Quotient tree partitioning of undirected graphs.
EDMOA 1985-1 Edmonds, J.; Lubiw, A.
Bipartition systems and how to partition polygons.
EL-D 1986-1 El-Dessouki, O.; Huen, W.; Evens, M.
Optimal partitioning of sequential jobs on a network computer.
ELST 1984-1 Elster, K.-H.; Harant, J.; Hutschenreuther, H.
Ein Partitionierungsalgorithmus zur Distanzberechnung auf Verkehrs- netzen.
FABE 1987-1 Faber, Z.; Vannelli, A.
Partitioning a graph with weighted nodes and edges.
FAIGG 1987-1 Faigle, U.; Schrader, R.; Suletzki, R.
A cutting-plane algorithm for optimal graph partitioning.
FAVAA 1985-1 Favaron, O.; Kouider, M.
Path partitions and cycle partitions of Eulerian graphs of maximum degree 4.
FAVAC 1985-1 Favaron, O.; Lonc, Z.; Truszczynski, M.
Decompositions of graphs into graphs with three edges.
FISH 1987-1 Fishburn, P. C.
Decomposing weighted digraphs into sums of chains.
FISHD 1986-1 Fisher, M. L.
A dual algorithm for large sale set partitioning.
FLEJ 1985-1 Flejshman, S. B.
A pseudo-polynomial algorithm for partitioning graphs of special form.
FUJI 1984-1 Fujii, T.; Wakabayashi, S.; Kikuno, T.
Partition of graph into forest in LSI layout design (Japanese).
GLYA 1983-1 Glyakov, P. V.
Algorithm for a partitioning problem and its investigation on an electronic computer (Russian).
GLYAA 1987-1 Glyakov, P. V.; Kovalev, M. M.; Kotov, V. M.
A series of gradient algorithms for a problem of the partitioning of a tree (Russian).
GONZB 1982-1 Gonzalez, T. F.
On the computational complexity of clustering and related problems.
GORKA 1986-1 Gorkovoi, V. F.; Egorov, M. Y.
Decomposition of graphs (Russian).
GREGA 1986-1 Gregory, D. A.; Mac Guinness, S.; Wallis, W. D.
Clique partitions of the cocktail party graphs.
HANSD 1987-2 Hansen, P.; Jaumard, B.
Minimum sum of diameters clustering.
HEDG 1983-1 Hedge, M.; Sridharan, M. R.
Partitions of some classes of digraphs.
HILT 1984-1 Hilton, A. J. W.
Hamiltonian decompositions of complete graphs.
KARG 1987-1 Karger, P. B.
Untersuchungen zur Matching-Relaxation für Set-Partitioning-Probleme.
KLAS 1985-1 Klastorin, T. D.
The p-median problem for cluster analysis: a comparative test using the mixture model approach.
KRATB 1987-2 Kratsch, D.
On the partition of graphs into cliques or independent sets.
KUMAA 1984-1 Kumar, K.
Efficient algorithms for grouping component-processor families.
KUMAB 1986-1 Kumar, K. R.; Kusiak, A.; Vannelli, A.
Grouping of parts and components in flexible manufacturing systems.
LENT 1985-1 Lentsevichyus, R.
Clustering of the graph with vertex labels using a rooted tree (Russian).
LEON 1985-1 Leonas, V.; Simniskis, S.
Automatic partitioning of large graphs and systems of linear algebraic equations of network systems into blocks (Russian).
LINI 1984-1 Linial, N.
Graph decompositions without isolates.
LIU D 1984-1 Liu, J.; Wang, X.-M.
A kind of efficient algorithm on the optimal k-partitioning of the graphs.
LIU H 1986-1 Liu, Y.-P.
An equation for the enumeration of the node decompositions of nonseparable Eulerian planar maps (Chinese).
LONC 1985-1 Lonc, Z.; Truszczynski, M.
Decomposition of graphs into graphs with bounded maximum degrees.
MA D 1986-2 Ma, S.-H.; Wallis, W. D.
Clique partitions of triangulated graphs.
MAND 1985-1 Mandrescu, E.
On the alpha-partitionability of X-join of graphs.
MARC 1987-1 Marchetti-Spaccamela, A.; Pelaggi, A.
Worst case analysis of two heuristics for the set partitioning problem.
MINO 1987-1 Minoux, M.
A class of combinatorial problems with polynomially solvable large scale set covering/ partitioning relaxations.
MINOB 1987-1 Minoux, M.; Pinson, E.
Lower bounds to the graph partitioning problem through generalized linear programming and network flows.
MULV 1984-1 Mulvey, J. M.; Beck, M. P.
Solving capacitated clustering problems.
NODA 1985-1 Noda, R.
Partitioning strongly regular graphs.
NOUH 1987-1 Nouh, A.
A sequential aggregation algorithm for the set partitioning problem.

NOWIB	1984-1	Nowicki, T.; Stanczak, W. An algorithmic approach to partitioning a set of elements due to their similarity.
ORDM	1987-1	Ordman, E. T. Dining philosophers and graph covering problems.
ORLI	1985-4	Orlin, J. B. Some very easy knapsack/partition problems.
OZAW	1987-1	Ozawa, T.; Kajitani, Y.; Ueno, S. The principal partition of vertex-weighted graphs and its applications.
PAIX	1985-1	Paixao, J. P.; Almeida, M. T.; Coelho, J. D. Problems in graph theory and their relation with set covering and partitioning problems (Portuguese).
PERO	1985-1	Peroche, B. Complexite de l'arboricite lineaire d'une graph II.
POLJ	1986-1	Poljak, S.; Chrobak, M. On common edges in optimal solutions to traveling salesman and other optimization problems.
ROSHA	1986-1	Roshchin, V. A.; Sergienko, I. V. An algorithm for the exact solution of the set partitioning problem (Russian).
SHI	1985-2	Shi, R.-H. Graph decomposition with constraint on the minimum degree.
SINH	1983-1	Sinha, S. An algorithm for network decompositions.
STANA	1984-1	Stanczak, W. An efficient algorithm for partitioning a network into minimally interconnected subnetworks.
STAND	1985-1	Stanfel, L. E. Partitioning problems in cellular manufacturing.
STAND	1986-1	Stanfel, L. E. A recursive Lagrangian method for clustering problems.
SYSL	1987-1	Syslo, M. M. A solvable case of the set-partitioning problem.
THOMA	1986-1	Thomassen, C. Decomposition of graphs into chains.
TRUSB	1986-1	Truszczynski, M. Decomposing graphs into forests of stars.
TRUSB	1986-2	Truszczynski, M. Decompositions of graphs into forests of equal size with bounded maximum degrees.
TURA	1985-1	Turan, G. On the greedy algorithm for an edge-partitioning problem.
TYSHA	1985-1	Tyshkevich, R. I.; Chernjak, A. A. Algorithms of a canonical decomposition of a graph and polarity recognition (Russian).
VAGI	1983-1	Vagin, V. N.; Georgieva, N. T.; Pospelov, D. A. On segmentation of graphs.
VANN	1984-1	Vannelli, A. Approximating a class of graph decomposition problems by linear transportation problems.
VANNB	1984-1	Vannelli, A.; Vidyasagar, M. Three approximate solution techniques for the optimal k-decomposition problem.
VOIT	1986-1	Voitishin, Y. V. A network approach to the solution of problems of set partitioning (Russian).
WAH	1984-1	Wah, B. W.; Chen, K. L. A partitioning approach to the design of selection networks.
WALUA	1987-1	Walukiewicz, S.; Lucertini, M. Constraint generation for the graph partitioning problem.
YAKU	1983-1	Yaku, T. Partitions of graphs into chains of the minimum block number.
YEH	1986-1	Yeh, D. Y. A dynamic programming approach to the complete set partitioning problem.
ZAKS	1985-1	Zaks, J. How does a complete graph split into bipartite graphs and how are neighborly cubes arranged?
ZHANM	1983-1	Zhang, G. Q. NP-completeness and restricted partitions (Chinese).
ZHU A	1985-1	Zhu, B. W. A partition of the set of vertices of a connected graph (Chinese).

AGAR	1985-1	Agarwal, S.; Mittal, A. K.; Sharma, P. A decomposition algorithm for linear relaxation of the weighted r-covering problem.
AGGAA	1985-1	Aggarwal, V. A Lagrangean-relaxation method for the constrained assignment problem.
ALI A	1986-1	Ali, A. I.; Kennington, J. L. The asymmetric m-travelling salesmen problem: a duality based branch-and-bound algorithm.
ARAN	1985-1	Arani, T.; Karwan, M. H.; Lotfi, V. A Lagrangian relaxation approach for solving an exam scheduling problem.
BALAG	1985-2	Balas, E. Disjunctive programming and a hierarchy of relaxations for discrete optimization problems.
BEAN	1984-1	Bean, J. C. A Langrangian algorithm for the multiple choice integer program.
BEAS	1985-4	Beasley, J. E. An exact two-dimensional non-guillotine cutting tree search procedure.
BERTI	1987-1	Bertsekas, D. P.; Hossein, P. A.; Tseng, P. Relaxation methods for network flow problems with convex arc costs.
BERTJ	1985-1	Bertsekas, D. P.; Tseng, P. Relaxation methods for minimum cost network flow problems.
BORC	1985-2	Borchardt, M. Straf- und Multiplikatormethoden für Optimierungsprobleme mit diskreten und stetigen Variablen. I.
CAMPB	1987-1	Campello, R. E.; Maculan, N. F. Lagrangean relaxation for a lower bound to a set partitioning problem with side constraints: properties and algorithms.
CARP	1984-1	Carpaneto, G. Relaxation techniques and applications to the optimal Hamiltonian tour problem (Italian).
CHANO	1987-1	Chandru, V.; Trick, M. A. On the complexity of Lagrange multiplier search.
DESRG	1984-1	Desrosiers, J.; Soumis, F.; Sauve, M. Lagrangian relaxation for routing with time windows.
DORH	1988-1	Dorhout, B. Solution of a tinned iron purchasing problem using Lagrangean relaxation.
FEO	1986-1	Feo, T. A.; Hochbaum, D. S. Lagrangian relaxation for testing infeasibility in VLSI routing.
FISHD	1985-1	Fisher, M. L. An applications oriented guide to Lagrangian relaxation.
FISHF	1984-1	Fisher, M. L.; Jaikumar, R. The local delivery problem: algorithms and applications.
FISHG	1986-1	Fisher, M. L.; Jaikumar, R.; Wassenhove, L. N. van A multiplier adjustment method for the generalized assignment problem.
FREV	1986-1	Freville, A.; Plateau, G. Heuristics and reduction methods for multiple constraints 0-1 linear programming problems.
GLOVC	1985-2	Glover, F.; Klingman, D. Layering strategies for creating exploitable structure in linear and integer programs.
GUIG	1983-1	Guignard, M. A Lagrangean dual ascent method based on (separable) relaxation strengthened by valid inequalities (illustrated for the matching problem).
GUIG	1984-1	Guignard, M. A Lagrangean dual ascent algorithm for simple plant location problems.
GUIGA	1983-1	Guignard, M.; Kim, S. A strong Lagrangean relaxation for capacitated plant location problems.
GUIGA	1986-1	Guignard, M.; Kim, S. Lagrangean decomposition: a model yielding better bounds (French).
GUIGA	1987-1	Guignard, M.; Kim, S. Lagrangean decomposition for integer programming: theory and applications.
GUIGB	1984-1	Guignard, M.; Opaswongkarn, K. Langrangean dual ascent algorithms for the capacitated plant location problem.
GUIGC	1985-1	Guignard, M.; Spielberg, K. Separable-Lagrangean relaxation for weakly linked mixed integer programming problems.
GUIGC	1987-1	Guignard, M.; Spielberg, K. On combining Lagrangean relaxation, Benders inequalities and Lagrangean dual ascent.
HARK	1982-1	Harker, P. T.; Friesz, T. L. A simultaneous freight network equilibrium model.
JOHN	1985-1	John, C. G.; Kochenberger, G. A. Using surrogate constraints in a Lagrangian relaxation approach to set covering problems.
JORN	1986-1	Jörnsten, K. O.; Näsberg, M. A new Lagrangian relaxation approach to the genralized assignment problem.
JORNA	1986-1	Jörnsten, K. O.; Värbrand, P. An augmented Lagrangian approach to multi-commodity flow problems with conversions.
KARW	1984-1	Karwan, M. H.; Rardin, R. L.; Sarin, S. A new surrogate dual multiplier search procedure.
KARW	1985-1	Karwan, M. H.; Rardin, R. L.; Sarin, S. Surrogate duality in a branch-and-bound procedure for integer programming.
KIM	1985-1	Kim, J.-G. Lagrangean relaxation approach for interactive multiobjective integer programming problem.
KIM C	1985-1	Kim, S.-H. Sufficient conditions for zero duality gap of Lagrangean relaxation.
KLIN	1986-1	Klincewicz, J. G.; Luss, H. A Lagrangian relaxation heuristic for capacitated facility location with single-source constraints.
LAPOC	1985-1	Laporte, G.; Mercure, H.; Nobert, Y. Finding the shortest Hamiltonian circuit through n-clusters: a Lagrangean relaxation approach.
LUCE	1984-1	Lucertini, M. Lagrangian and group-theoretic relaxation in integer programming (Italian).

MAHE 1985-1 Mahey, P.
Subgradient techniques and combinatorial optimization.

MINA 1985-1 Minaev, Y. N.
On a formally heuristic algorithm for solving integer optimization problems (linear and quadratic models) (Russian).

MINO 1986-1 Minoux, M.
Mathematical Programming. Theory and Algorithms.

NARUA 1985-1 Narula, S. C.; Ogbu, U. I.
Lagrangean relaxation and decomposition in a uncapacitated 2-hierarchal location-allocation problem.

NEEB 1986-1 Neebe, A. W.; Rao, M. R.
Sequencing capacity expansion projects in continuous time.

POTT 1985-1 Potts, C. N.
A Lagrangean based branch and bound algorithm for single machine sequencing with precedence constraints to minimize total weighted completion time.

RIBEA 1986-1 Ribeiro, C. C.; Minoux, M.
Solving hard constrained shortest path problems by Lagrangean relaxation and branch-and-bound algorithms.

SRIDA 1986-1 Sridharan, R.
A heuristic Lagrangean algorithm for the capacitated plant location problem - a comment.

STAND 1986-1 Stanfel, L. E.
A recursive Lagrangian method for clustering problems.

THIZ 1985-1 Thizy, J. M.; Wassenhove, L. N. van
Lagrangean relaxation for the multi-item capacitated lot-sizing problem: a heuristic implementation.

WRIG 1986-2 Wright, D. D.; Haehling von Lanzenauer, C.
Solving the fixed charge problem with Lagrangian relaxation and cost allocation heuristics.

AGARA 1985-1 Agarwal, Y. K.
Set partitioning approach to vehicle routing.
AGARA 1985-2 Agarwal, Y. K.
Vehicle routing with limited fleet and common carrier option.
AICA 1986-1 Aicardi, M.; Casalino, G.; Davoli, F.; Minciardi, R.; Zoppoli, R.
A decentralized closed-loop solution to the routing problem in networks.
AICA 1986-2 Aicardi, M.; Casalino, G.; Davoli, F.; Minciardi, R.; Zoppoli, R.
Dynamic routing policies in communication networks: the finite buffer case.
ANTH 1987-1 Anthonisse, J. M.; Lenstra, J. K.; Savelsbergh, M. W. P.
Functional description of CAR, an interactive system for computer aided routing.
ASSAA 1987-1 Assad, A. A.; Pearn, W. L.; Golden, B. L.
The capacitated Chinese postman problem: lower bounds and solvable cases.
BAKEB 1986-1 Baker, B. S.
A provably good algorithm for the two module routing problem.
BAKEC 1986-1 Baker, E. K.; Schaffer, J. R.
Solution improvement heuristics for the vehicle routing and scheduling problem with time window constraints.
BALAC 1987-1 Balakrishnan, A.; Ward, J. E.; Wong, R. T.
Integrated facility location and vehicle routing models: recent work and future prospects.
BARA 1987-2 Barahona, F.
Planar multicommodity flows, max cut and the Chinese postman problem.
BART 1984-1 Bartholdi, J. J. III; Platzman, L. K.
Heuristics based on spacesfilling curves for combinatorial problems in the plane.
BARTC 1984-1 Bartolini, G.; Casalino, G.; Davoli, F.; Minciardi, R.; Zoppoli, R.
A team theoretical approach to routing and multiple access in data communication networks.
BATY 1983-1 Batycki, T.; Kasprzak, A.
Flow assignment algorithms for computer communication networks (Polish).
BECH 1986-1 Bechara, J. J. B.; Galvao, R. D.
The use of interactive computing for vehicle routing.
BECKB 1986-1 Becker, M.; Mehlhorn, K.
Algorithms for routing in planar graphs.
BELA 1985-1 Belardo, S.; Duchessi, P.; Seagle, J. P.
Microcomputer graphics in support of vehicle fleet routing.
BENA 1985-1 Benavent, E.; Campos, V.; Corberan, A.; Mota, E.
Analysis of heuristics for the rural postman problem (Spanish).
BERND 1984-1 Bernussou, J.; LeGall, F.; Garcia, J. M.
Optimisation et acheminement dynamique dans les reseaux telephoniques.
BERTF 1982-1 Bertsekas, D. P.
Optimal routing and flow control methods for communication networks.
BEUL 1988-1 Beulens, A. J. M.; Kolen, A. W. J.; Niekamp, E. G.
A decision support system for a location-allocation-routing problem.
BHAS 1986-1 Bhaskar, J.; Sahni, S.
Via assignment in single row routing.
BIRK 1985-1 Birkhoff, I. J. M.
A direct routing algorithm for the bit-reversal permutation on a shuffle-exchange network.
BLAID 1986-1 Blair, J. R. S.; Lloyd, E. L.
Generalized river routing - algorithms and performance bounds.
BOTT 1986-1 Bott, K.; Ballou, R. H.
Research perspectives in vehicle routing and scheduling.
BREB 1985-1 Brebner, G.
Relating routing graphs and 2-dimensional grids.
BROD 1984-1 Broder, A.; Dolev, D.; Fischer, M.; Simons, B.
Efficient fault-tolerant routings in networks.
BUXE 1985-1 Buxey, G.
Designing routes to minimize fleet operating costs.
CAMP 1987-1 Campbell, B. A.; Rardin, R. L.
Steiner tree problems on series-parallel block graphs I: polynomial recognition and solution.
CHANAB 1987-1 Chang, S. G.
A heuristic for non-bifurcated routing in communication networks.
CHANAD 1987-1 Chang, Y.; Hambrusch, S. E.; Simon, J.
On the computational complexity of continuous routing.
CHANP 1986-1 Chang, F.; Wu, L.
An optimal adaptive routing algorithm.
CHAP 1985-1 Chapleau, L.; Ferland, J. A.; Rousseau, J.-M.
Clustering for routing in dense area.
CHEND 1987-1 Chen, M. S.; Meditch, J. S.
A distributed adaptive routing algorithm.
CHENM 1985-1 Cheng, D.-Q.
The solution to the Chinese postman problem by branch and bound (Chinese).
CHIE 1987-1 Chien, T. W.; Balakrishnan, A.; Wong, R. T.
An integrated inventory allocation and vehicle routing problem.
CHINB 1987-1 Chin, W.-P.; Ntafos, S.
Optimum zoo-keeper routes.
CHRI 1985-1 Christofides, N.
Vehicle routing.
CHRI 1985-2 Christofides, N.
Vehicle routing.
CHRIB 1984-1 Christofides, N.; Benavent, E.; Campos, V.; Corberan, A.; Mota, E.
An optimal method for the mixed postman problem.
CHRIC 1986-1 Christofides, N.; Campos, V.; Corberan, A.; Mota, E.
An algorithm for the rural postman problem on a directed graph.
COLE 1984-1 Cole, R.; Siegel, A.
River routing every which way, but loose.

CROW	1985-1	Crowley, J. A.; Scott, P.; Heeney, B. International truck routing problems with ferry crossing time window and legal constraints in driving hours.
CURRB	1985-1	Current, J. R.; Revelle, C. S.; Cohon, J. L. The maximum covering/shortest path problem: a multiobjective network design and routing formulation.
DAGA	1984-1	Daganzo, C. F. The distance traveled to visit n points with a maximum of c stops per vehicle: an analytic model and an application.
DESRB	1987-1	Desrochers, M.; Lenstra, J. K.; Savelsbergh, M. W. P. A classification scheme for vehicle routing and scheduling problems.
DESRC	1987-1	Desrochers, M.; Lenstra, J. K.; Savelsbergh, M. W. P.; Soumis, F. Vehicle routing with time windows: optimization and approximation.
DESRD	1986-1	Desrosiers, J.; Dumas, Y.; Soumis, F. A dynamic programming solution of the large-scale single-vehicle dial-a-ride problem with time windows.
DESRE	1984-1	Desrosiers, J.; Soumis, F.; Desrochers, M. Routing with time windows by column generation.
DESRF	1986-1	Desrosiers, J.; Soumis, F.; Desrochers, M.; Sauve, M. Methods for routing with time windows.
DESRF	1986-2	Desrosiers, J.; Soumis, F.; Desrochers, M.; Sauve, M. Vehicle routing and scheduling with time windows.
DESRG	1984-1	Desrosiers, J.; Soumis, F.; Sauve, M. Lagrangian relaxation for routing with time windows.
DOLE	1987-1	Dolev, D.; Halpern, J. Y.; Simons, B.; Strong, H. R. A new look at fault-tolerant network routing.
DREZA	1986-1	Drezner, Z.; Barak, A. Efficient algorithms for routing information in a multicomputer system.
DROR	1987-1	Dror, M.; Ball, M. O. Inventory/routing: reduction from an annual to a short-period problem.
DRORA	1986-1	Dror, M.; Ball, M. O.; Golden, B. L. A computational comparison of algorithms for the inventory routing problem.
DRORB	1986-1	Dror, M.; Levy, L. A vehicle routing improvement algorithm comparison of a greedy and a matching implementation for inventory routing.
DRORD	1986-1	Dror, M.; Trudeau, P. Stochastic vehicle routing with modified savings algorithm.
EBER	1988-1	Eberhard, U.; Vaterrodt, H.-J. Mehr-Depot-Tourenplanung.
EGGL	1984-4	Eggleton, R. B.; Skilton, D. K. Graphs with Eulerian chains.
ESFA	1985-1	Esfahanian, A.-H.; Hakimi, S. L. Fault-tolerant routing in de Bruijn communication networks.
EVANB	1985-1	Evans, S. R.; Norback, J. P. The impact of a decision-support system for vehicle routing in a foodservice supply situation.
FAND	1985-1	Fandel, G.; Reese, J. Organisation und Planung des Fahrzeugeinsatzes in einem Chemieunternehmen.
FEDEA	1984-1	Federgruen, A.; Zipkin, P. A combined vehicle routing and inventory allocation problem.
FEO	1986-1	Feo, T. A.; Hochbaum, D. S. Lagrangian relaxation for testing infeasibility in VLSI routing.
FILI	1983-1	Filipiak, J. Joint optimization of routing and congestion control in data communication networks.
FISHF	1984-1	Fisher, M. L.; Jaikumar, R. The local delivery problem: algorithms and applications.
FLIK	1985-1	Flikop, Z. Routing optimization in packet switching communication networks.
FORL	1986-1	Forler, C. Modified transport optimization under restrictions.
FREDB	1986-1	Frederickson, G. N.; Janardan, R. Space-efficient message routing in c-decomposable networks.
FREDB	1987-1	Frederickson, G. N.; Janardan, R. Designing networks with compact routing tables.
FREDB	1987-2	Frederickson, G. N.; Janardan, R. Efficient message routing in planar networks.
FRICA	1987-1	Fricker, J.; Nafakh, J. Finding optimal routes in doubly weighted graphs using the multiplane technique.
FROL	1987-1	Froleyks, B.; Korte, B.; Prömel, H. J. Routing in VLSI-layout.
GAO A	1985-1	Gao, S.-D. An algorithm for two-layer channel routing.
GAO B	1986-1	Gao, S.-D.; Hambrusch, S. E. Two-layer channel routing with vertical unit-length overlap.
GAST	1986-1	Gastou, G.; Johnson, E. L. Binary group and Chinese postman polyhedra.
GHEY	1984-1	Gheysens, F.; Golden, B. L.; Assad, A. A. A comparison of techniques for solving the fleet size and mix vehicle routing problem.
GHEY	1986-1	Gheysens, F.; Golden, B. L.; Assad, A. A. A new heuristic for determining fleet size and composition.
GIBB	1985-1	Gibbons, A. Algorithmic graph theory.
GIRA	1985-1	Girard, A. Nonhierarchical sequential routing: An approximate solution to the optimal routing problem.

GIRAA 1984-1 Girard, A.; Cote, Y.
Sequential routing optimization for circuit switched networks.

GOLDE 1984-1 Golden, B. L.
Introduction to and recent advances in vehicle routing methods.

GOLDF 1986-1 Golden, B. L.; Assad, A. A.
Perspectives on vehicle routing: exciting new developments.

GOLDG 1984-1 Golden, B. L.; Assad, A. A.; Dahl, R.
Analysis of a large scale vehicle routing problem with an inventory component.

GOLDH 1986-1 Golden, B. L.; Bodin, L. D.; Goodwin, T.
Microcomputer-based vehicle routing and scheduling software.

GOLDI 1987-1 Golden, B. L.; Levy, L.; Vohra, R. V.
The orienteering problem.

GOLDJ 1986-1 Golden, B. L.; Skiscim, C. C.
Using simulated annealing to solve routing and location problems.

GOLDL 1987-1 Golden, B. L.; Wasil, E. A.
Computerized vehicle routing in the soft drink industry.

GOLU 1987-1 Golumbic, M. C.
A general method for avoiding cycling in a network.

GONZC 1987-1 Gonzalez, T. F.; Lee, S.-L.
A 1.6 approximation algorithm for routing multiterminal nets.

GREG 1985-1 Grega, W.
A decomposition algorithm for the determination of optimal bus frequencies.

GUAN 1984-1 Guan, M.-G.
The maximum weighted cycle-packing problem and its relation to the Chinese postman problem.

HAIM 1985-1 Haimovich, M.; Rinnooy Kan, A. H. G.
Bounds and heuristics for capacitated routing problems.

HAIN 1985-1 Haines, C. L.
An algorithm for carrier routing in a flexible material-handling system.

HAJE 1984-1 Hajek, B.; Ogier, R. G.
Optimal dynamic routing in communication networks with continuous traffic.

HALLE 1987-1 Hall, R. W.
Comparison of strategies for routing shipments through transportation terminals.

HALLE 1987-2 Hall, R. W.
Direct versus terminal freight routing on a network with concave costs.

HARAA 1984-1 Harao, M.; Lomtong, P.
Set-up and routing algorithms for a cellular interconnection network.

HART 1987-1 Hart, J. P.; Shogan, A. W.
Semi-greedy heuristics: an empirical study.

HARTA 1985-1 Hartley, R.
Vector optimal routing by dynamic programming.

HELLC 1984-1 Hellmann, A.
Theorie und Praxis von Routing-Problemen: Lösungsverfahren zur Tourenplanung im Eindepot- und Mehrdepot-Fall.

HOCHB 1986-4 Hochbaum, D. S.; Shmoys, D. B.
Best possible heuristics for the bottleneck wandering salesperson and bottleneck vehicle routing problems.

HSIE 1984-1 Hsieh, W. N.
Routing strategies in computer networks.

HU A 1985-1 Hu, T. C.; Shing, M. T.
A decomposition algorithm for circuit routing.

JAFF 1985-1 Jaffe, J. M.
Distributed multi-destination routing: the constraints of local information.

JAIK 1987-1 Jaikumar, R.; Solomon, M. M.
The tug fleet size problem for barge line operations: a polynomial algorithm.

JAIL 1987-2 Jaillet, P.
Stochastic routing problems.

JASIA 1984-1 Jasinski, L. J.
An exact method for solving the vehicle routing problem (Polish).

JEWO 1987-1 Jeworrek, T.; Schulz, G.
Zur Lösung des Briefträgerproblems auf gemischten Graphen.

JOHNJ 1987-1 Johnson, E. L.; Mosterts, S.
On four problems in graph theory.

KARPA 1987-1 Karp, R. M.; Leighton, F. T.; Rivest, R. L.; Thompson, C. D.; Vazirani, U. V.; Vazirani, V. V.
Global wire routing in two-dimensional arrays.

KATE 1985-1 Katehakis, M. N.; Levine, A.
A dynamic routing problem - numerical procedures for light traffic conditions.

KAUFB 1985-1 Kaufmann, M.; Mehlhorn, K.
Routing through a generalized switchbox.

KAUFB 1986-1 Kaufmann, M.; Mehlhorn, K.
On local routing of two-terminal nets.

KIME 1985-1 Kimemia, J.; Gershwin, S. B.
Flow optimization in flexible manufacturing systems.

KLAU 1983-1 Klauz, S. A.; Kulkova, L. I.
On a method of routing in a network of common signaling channels.

KOEB 1985-1 Koebe, M.
Graphentheoretische Modelle zur Behandlung des Kanaltrassierungsproblems.

KOEB 1986-1 Koebe, M.
Über die Anzahl der Spuren für die Lösung des Kanaltrassierungsproblems.

KOLEA 1987-1 Kolen, A. W. J.; Rinnooy Kan, A. H. G.; Trienekens, H. W. J. M.
Vehicle routing with time windows.

KOLEC 1988-1 Kolenbrander, W. G.
Tourenplanung mit einem Personal Computer.

KORT 1985-1 Kortanek, K. O.
Network design and dynamic routing under queueing demand.

KRAMB 1982-1 Kramer, M. R.; Leeuwen, J. van
Wire-routing is NP-complete.

KREU 1985-1 Kreuter, P.
Kreuztrassierung - ein algebraischer Ansatz zur Trassierungstheorie mikroelektronischer Schaltkreise.

KULK 1985-1 Kulkarni, R. V.; Bhave, P. R.
Integer programming formulations of vehicle routing problems.

KUMAC 1985-1 Kumar, P.; Miller, F. G.
Shortest route network heuristic for the dynamic vehicle routing problem.

LAMA 1988-1 Lamatsch, A.
Wagenumlaufplanung im ÖPNV bei begrenzten Betriebshofkapazitäten - ein ganzzahliges Mehrgüterflussmodell mit Lösungsansätzen.

LANGB 1986-1 Langen, D.; Jandl, H.
T O U R - Transport Optimization Under Restrictions.

LAPA 1980-1 LaPaugh, A. S.
A polynomial time algorithm for optimal routing around a rectangle.

LAPO 1986-1 Laporte, G.
Generalized subtour elimination constraints and connectivity constraints.

LAPOC 1987-1 Laporte, G.; Mercure, H.; Nobert, Y.
Cutting planes based on bin packing solutions for the capacitated vehicle routing problem.

LAPOD 1984-1 Laporte, G.; Nobert, Y.
Comb inequalities for the vehicle routing problem.

LAPOD 1987-1 Laporte, G.; Nobert, Y.
Exact algorithms for the vehicle routing problem.

LAPOE 1984-1 Laporte, G.; Nobert, Y.; Arpin, D.
Optimal solutions to capacitated multidepot vehicle routing problems.

LAPOE 1986-1 Laporte, G.; Nobert, Y.; Arpin, D.
An exact algorithm for solving a capacitated location-routing problem.

LAPOF 1985-1 Laporte, G.; Nobert, Y.; Desrochers, M.
Optimal routing under capacity and distance restrictions.

LAPOH 1987-1 Laporte, G.; Nobert, Y.; Taillefer, S.
A branch-and-bound algorithm for the asymmetrical distance-constrained vehicle routing problem.

LEIS 1985-1 Leiserson, C. E.; Maley, F. M.
Algorithms for routing and testing routability of planar VLSI layouts.

LIEB 1986-1 Liebl, F.
Der simulative Test verschiedener Routing-Algorithmen in Datennetzen.

LIEB 1987-1 Liebl, F.
Routing Algorithmen in Datennetzen - eine vergleichende Simulationsstudie.

LLOYA 1984-1 Lloyd, E. L.; Ravi, S. S.
One-layer routing without component constraints.

LODI 1986-1 Lodi, E.; Pagli, L.
Some new results on a restricted channel routing problem.

MARCA 1984-1 Marchetti-Spaccamela, A.; Rinnooy Kan, A. H. G.; Stougie, L.
Hierarchical vehicle routing problems.

MASO 1985-1 Mason, L. G.
Equilibrium flows, routing patterns and algorithms for store-and-forward networks.

MATH 1986-1 Mathews, B. P.; Waters, C. D. J.
Computerized routeing for community nurses - a pilot study.

MAXE 1987-1 Maxemchuk, N. F.
Routing in the Manhattan street network.

MEHL 1986-1 Mehlhorn, K.
Über Verdrahtungsalgorithmen.

MEHLA 1987-1 Mehlhorn, K.; Kaufmann, M.
On local routing of 2-terminal nets.

MEHLB 1986-1 Mehlhorn, K.; Preparata, F. P.
Routing through a rectangle.

MEHLC 1986-1 Mehlhorn, K.; Preparata, F. P.; Sarrafzadeh, M.
Channel routing in knock-knee mode: simplified algorithms and proofs.

MIRCC 1987-1 Mirchandani, P. B.; Soroush, H.
Routes and flows in stochastic networks.

MIRCD 1986-1 Mirchandani, P. B.; Veatch, M. H.
"Hot job" routing through a stochastic job-shop network.

MIRZ 1987-1 Mirzaian, A.
River routing in VLSI.

MOLE 1983-1 Mole, R. H.; Johnson, D. G.; Wells, K.
Combinatorial analysis for route first-cluster second vehicle routing.

MUKHB 1986-1 Mukhamadiev, E. M.; Aralbaev, T. Z.; Nazimov, A. B.
An algorithm for calculating routes in a problem with cyclic servicing of the subscribers (Russian).

MURPA 1987-1 Murphy, R. A.; Mueller, T. E.
Routing full truckload movements.

NARE 1985-1 Narendra, K. S.; Wheeler, R. M. Jr.
Routing in communication networks - a case study of learning in large scale systems.

NELS 1985-1 Nelson, M. D.; Nygard, K. E.; Griffin, J. H.; Shreve, W. E.
Implementation techniques for the vehicle routing problem.

NEUMC 1987-1 Neumann, K.
The postman problem in mixed graphs (German).

NEWE 1986-1 Newell, G. F.
Design of multiple-vehicle delivery tours - III. Valuable goods.

NEWEA 1986-1 Newell, G. F.; Daganzo, C. F.
Design of multiple-vehicle delivery tours - I. A ring-radial network.

NEWEA 1986-2 Newell, G. F.; Daganzo, C. F.
Design of multiple-vehicle delivery tours - II. Other metrics.

NOGO 1985-1 Nogova, E. G.
Construction of an optimal system of isolated routes for passenger transport (Russian).
ODON 1985-1 Odoni, A. R.
The "third generation" of vehicle routing and scheduling problems: two illustrative examples.
ORDA 1986-1 Orda, A.; Rom, R.
Duplicate routing in distributed networks.
PAIB 1986-1 Paiboon, C.-P.; Pojana, P.-P.; Rosenthal, R. E.
Optimal ship routing and personnel assignment for naval recruitment in Thailand.
PEAR 1987-1 Pearn, W. L.; Assad, A. A.; Golden, B. L.
Transforming arc routing into node routing problems.
PELEC 1987-1 Peleg, D.; Simons, B.
On fault tolerant routings in general networks.
PELED 1987-2 Peleg, D.; Upfal, E.
The generalized packet routing problem.
PERL 1987-1 Perl, J.
The multidepot routing allocation problem.
PERLA 1985-1 Perl, J.; Daskin, M. S.
A warehouse location-routing problem.
PREP 1985-1 Preparata, F. P.; Sarrafzadeh, M.
Channel routing of nets bounded degree.
PSAR 1986-1 Psaraftis, H. N.
Scheduling large-scale advance-request dial-a-ride systems.
RADA 1986-1 Radaramanan, R.; Choi, L. I.
A branch and bound algorithm for the travelling salesman and the transportation routing problems.
RAGHB 1985-1 Raghavan, P.; Thompson, C. D.
Provably good routing in graphs: regular arrays.
RAJA 1987-1 Rajasekaran, S.; Tsantilas, T.
An optimal randomized routing algorithm for the mesh and a class of efficient mesh-like routing networks.
RICH 1984-1 Richards, D.
Complexity of single-layer routing.
ROSBA 1985-1 Rosberg, Z.; Towsley, D.
Customer routing to parallel servers with different rates.
RUBL 1983-1 Rublev, V. S.
A generalized assignment problem and the scheduling of a transportation route (Russian).
RUSSA 1986-1 Russell, R. A.; Morrel, R. B.
Routing special-education school buses.
SALH 1987-1 Salhi, S.; Rand, G. K.
Improvements to vehicle routing heuristics.
SANT 1985-1 Santoro, N.; Khatib, R.
Labelling and implicit routing in networks.
SARR 1985-1 Sarrafzadeh, M.; Preparata, F. P.
Compact channel routing of multiterminal nets.
SAVE 1985-1 Savelsbergh, M. W. P.
Vehicle routing and computer graphics.
SAVE 1986-1 Savelsbergh, M. W. P.
Local search in routing problems with time windows.
SAVE 1988-1 Savelsbergh, M. W. P.
Local search for constrained routing problems.
SEBA 1986-1 Sebastian, H.-J.
Optimization and optimal control of a general transportation, exchange and inventory process II: routing problems with time interval restrictions (German).
SEBO 1985-1 Sebö, A.
On the Chinese postman problem: algorithms, structure and applications.
SEXT 1985-2 Sexton, T. R.; Bodin, L. D.
Optimizing single vehicle many-to-many operations with desired delivery times: II. Routing.
SHAMA 1985-1 Shamir, E.; Upfal, E.
A fast parallel construction of disjoint paths in networks.
SHIN 1987-1 Shin, K. G.
Performance analysis of distributed routing strategies free of ping-pong-type looping.
SINCA 1987-1 Sinclair, M.; Dyk, E. van
Combined routeing and scheduling for the transportation of containerized cargo.
SKIT 1985-1 Skitt, R. A.; Levary, R. R.
Vehicle routing via column generation.
SKLA 1984-1 Sklar, M. G.; Armstrong, R. D.; Davis, K. R.
A routing problem for a plant nursery.
SKUB 1984-1 Skubalska, E.
Application of the branch-and-bound method to the optimal vehicle routing (Czechoslovakian).
SKUB 1984-2 Skubalska, E.
Vehicle routing problems, mathematical models (Polish).
SOLOB 1987-1 Solomon, M. M.
Algorithms for the vehicle routing and scheduling problem with time window constraints.
SONGA 1987-1 Song, S. H.; Park, S. D.
An exact algorithm for the asymmetrical vehicle routing problem (Korean).
SOUM 1984-1 Soumis, F.; Desrosiers, J.; Desrochers, M.
Optimal urban bus routing with scheduling flexibilities.
STAS 1987-1 Stassinopoulos, G. I.
Optimal dynamic routing in multidestination networks.
STASA 1985-1 Stassinopoulos, G. I.; Konstantopoulos, P.
Optimal congestion control in single destination networks.
SUZU 1987-1 Suzuki, H.; Nishizeki, T.; Saito, N.
Algorithms for routing around a rectangle (extended abstract).

TADE 1983-1 Tadeusz, U.
Schnelle Routing-Algorithmen in Rechnernetzen.
TILA 1988-1 Tilanus, C. B.
Transportation planning: recent developments in the Netherlands.
TSIT 1986-1 Tsitsiklis, J. N.; Bertsekas, D. P.
Distributed asynchronous optimal routing in data networks.
ULUS 1985-1 Ulusoy, G.
The fleet size and mix problem for capacitated arc routing.
UYTT 1988-1 Uyttenhove, H. J. J.
Transportation planning - as easy as 1-2-3.
VASYA 1982-1 Vasyanin, V. A.; Savenkov, A. I.
A suboptimal algorithm for solving a transportation routing problem.
VENE 1985-1 Venetsanopoulos, A. N.; Waung, W.
Adaptive bifurcation routing algorithms for computer-communication networks.
WADA 1987-1 Wada, K.; Kawaguchi, K.; Luo, Y.-P.
Fault-tolerant network routings for (k+1)-node connected and (k+1)-edge connected graphs.
WATE 1984-1 Waters, C. D. J.
Interactive vehicle routeing.
WATE 1984-2 Waters, C. D. J.
Interactive vehicle routing. An approach to complex problems.
WETZ 1985-1 Wetzel, R.
Ein Branch and Bound Algorithmus zur Lösung eines Problems der Fahrplangestaltung.
WETZ 1986-1 Wetzel, R.
A branch and bound concept for solving a vehicle schedule designing problem.
WHITI 1984-1 Whittle, P.
Optimal routing in Jackson networks.
WONG 1984-1 Wong, K. F.; Beasley, J. E.
Vehicle routing using fixed delivery areas.
XIA A 1985-1 Xia, W.-M.; Huang, B.; Cheng, S.-Z.
An optimization design of vehicle routes in urban public transportation systems (Chinese).
YANOA 1987-1 Yano, C. A.; Chan, T. J.; Richter, L. K.; Cutler, T.; Murty, K. G.; Mac Gettigari, D.
Vehicle routing at quality stores.
ZAW 1986-1 Zaw Win
On the windy postman problem on Eulerian graphs.
ZAW 1987-1 Zaw Win
On the windy postman problem.

ABDU 1983-1 Abdullaev, A. A.; Rzaev, T. G.
Optimization of operative planning and scheduling in integrated ACS for enterprises (Russian).

ABEL 1986-1 Abelis, E.; Pavlovskii, Y. A.; Spilbergs, A.; Yubase, V.
The use of interactive computers for solving scheduling problems (Russian).

ADAMA 1987-1 Adam, N. R.; Bertrand, J. W. M.; Surkis, J.
Priority assignment procedures in multi-level assembly job shops.

ADAMB 1986-1 Adams, J.; Balas, E.
The shifting bottleneck procedure for job shop scheduling.

ADIR 1987-1 Adiri, I.; Aizikowitz, N.; Lev, V.
Reversible-shop scheduling.

ADIRA 1984-1 Adiri, I.; Amit, N.
Openshop and flowshop scheduling to minimize sum of completion times.

ADIRB 1987-1 Adiri, I.; Yehudai, Z.
Scheduling on machines with variable service rates.

AFEN 1985-1 Afentakis, P.
Simultaneous lot sizing and sequencing for multistage production systems.

AFEN 1987-1 Afentakis, P.
A parallel heuristic algorithm for lot sizing in multistage production systems.

AFENA 1986-1 Afentakis, P.; Gavish, B.
Optimal lot-sizing algorithms for complex product structures.

AGGA 1987-1 Aggarwal, S. P.; Goel, V. P.
Deterministic lot size model with nth power time proportional demand.

AGGAD 1984-1 Aggarwal, V. B.; Bagga, P. C.; Kalra, K. R.
Optimal algorithms for job-blocks in 2xn sequencing problem with transportation times – a computational comparison.

AGRA 1985-1 Agrawal, P. K.
The related activity concept in assembly line balancing.

AGRAB 1984-1 Agrawala, A. K.; Coffman, E. G. Jr.; Garey, M. R.; Tripathi, S. K.
A stochastic optimization algorithm minimizing expected flow times on uniform processors.

AIRA 1984-1 Airapetyan, L. R.; Eksuzyan, S. A.
A necessary and sufficient condition for existence of a correct schedule for a two-processor system (Russian).

AKEL 1986-1 Akella, R.; Kumar, P. R.
Optimal scheduling of a flexible manufacturing system: a stochastic control problem for a system with jump Markov disturbances.

ALBE 1987-1 Alberton, I. B.
Polynomial algorithms for problems in the theory of scheduling on "narrow" networks.

ALVAA 1984-1 Alvarez-Valdes, R.; Tamarit, J. M.
Lower bounds for the problem of sequencing with resource restrictions (Spanish).

AMAN 1987-1 Aman, A.; Balakrishnan, A.; Chandru, V.
Maintenance of optimal schedules.

AMAR 1986-1 Amar, A. D.; Gupta, J. N. D.
Simulated versus real life data in testing the efficiency of scheduling algorithms.

ANGE 1986-1 Anger, F. D.; Lee, C.-Y.; Martin-Wega, L. A.
Single machine scheduling with tight windows.

ARAN 1985-1 Arani, T.; Karwan, M. H.; Lotfi, V.
A Lagrangian relaxation approach for solving an exam scheduling problem.

ARKIA 1987-1 Arkin, E. M.; Silverberg, E. B.
Scheduling jobs with fixed start and end times.

ASHI 1985-1 Ashimov, A. A.; Burkov, V. N.; Dzhaparov, B. A.; Kondratev, V. V.
Mechanisms of operative coordinated control of industrial systems. I. Coordinated scheduling of continuous production (Russian).

ASSA 1985-1 Assad, A. A.
Nested optimal policies for set functions with applications to scheduling.

AUCA 1987-1 Aucamp, D. C.
A lot-sizing policy for production planning with applications in MRP.

AUCAA 1986-1 Aucamp, D. C.; Kuzdrall, P. J.
Lot sizes for one-time-only sales.

AXSA 1985-1 Axsäter, S.
Performance bounds for lot sizing heuristics.

AXSA 1986-1 Axsäter, S.
Design and evaluation of a lot sizing heuristic.

AXSA 1986-2 Axsäter, S.
Evaluation of lot-sizing techniques.

AXSA 1987-1 Axsäter, S.
An extension of the extended basic period approach for economic lot scheduling problems.

AXSAA 1987-1 Axsäter, S.; Nuttle, H. L. W.
Combining items for lot sizing in multi-level assembly systems.

BABU 1985-1 Babushkin, A. I.; Belov, I. S.
Optimization of the sequence of an assembly using resources of one kind (Russian).

BABU 1986-1 Babushkin, A. I.; Belov, I. S.
Optimal combinations of priority scheduling rules.

BAGC 1987-1 Bagchi, U.; Chang, Y.-L.; Sullivan, R. S.
Minimizing absolute and squared deviations of completion times with different earliness and tardiness penalties and a common due date.

BAGG 1982-1 Bagga, P. C.
Single machine scheduling with penalty cost as an m-degree monomial function of completion times.

BAGGA 1983-1 Bagga, P. C.; Khurana, K.
Minimizing waiting cost of jobs in renting situations.

BAGGA 1983-2 Bagga, P. C.; Khurana, K.
The single machine problem with quadratic penalty function.

BAGGA 1984-1 Bagga, P. C.; Khurana, K.
Transportation of job-blocks with separated setup times and removal times.

BAGGA 1985-1 Bagga, P. C.; Khurana, K.
Special cases with set up considerations for n*3 flowshop problem.

BAHL 1984-1 Bahl, H. C.; Ritzman, L. P.
A cyclical scheduling heuristic for lot sizing with capacity constraints.

BAHLA 1987-1 Bahl, H. C.; Ritzman, L. P.; Gupta, J. N. D.
Determining lot sizes and resource requirements: A review.

BAHLB 1987-1 Bahl, H. C.; Zionts, S.
Multi-item scheduling by Benders' decomposition.

BAKEC 1986-1 Baker, E. K.; Schaffer, J. R.
Solution improvement heuristics for the vehicle routing and scheduling problem with time window constraints.

BAKED 1984-1 Baker, K. R.
Sequencing rules and due-date assignments in a job shop.

BAKEE 1984-1 Baker, K. R.; Kanet, J. J.
Improved decision rules in a combined system for minimizing job tardiness.

BAKEF 1981-1 Baker, T. E.
A branch and bound network algorithm for interactive process scheduling.

BALAD 1987-1 Balakrishnan, N.; Wong, R. T.
A network model for the rotating workforce scheduling problem.

BALAG 1985-3 Balas, E.
On the facial structure of scheduling polyhedra.

BAPT 1984-1 Baptiste, P.; Favrel, J.
Resolution de problemes d'ordonnancement par les treillis de Galois et les graphes d'intervalle.

BARAK 1984-1 Barany, I.; Vanroy, T. J.; Wolsey, L. A.
Strong formulations for multi-item capacitated lot sizing.

BARAK 1984-2 Barany, I.; Vanroy, T. J.; Wolsey, L. A.
Uncapacitated lot-sizing: the convex hull of solutions.

BARK 1985-1 Barker, J. R.; Mac Mahon, G. B.
Scheduling the general job-shop.

BARRA 1986-1 Barron, E. N.; Jensen, R.
On the Kleinrock-Nilsson problem of optimal scheduling algorithms for time-shared systems.

BARTF 1985-1 Bartusch, M.; Möhring, R. H.; Radermacher, F. J.
Some remarks on the m-machine unit time scheduling problem (extended abstract).

BARTF 1986-1 Bartusch, M.; Möhring, R. H.; Radermacher, F. J.
Scheduling project networks with resource constraints and time windows.

BAYB 1986-1 Baybars, I.
A survey of exact algorithms for the simple assembly line balancing problem.

BAYB 1986-2 Baybars, I.
An efficient heuristic method for the simple assembly line balancing problem.

BEANA 1985-1 Bean, J. C.; Hall, D. H.
Accuracy of the modified due date rule.

BEANB 1987-1 Bean, J. C.; Smith, R. L.; Yano, C. A.
Forecast horizons for the discounted dynamic lot-size problem allowing speculative motive.

BECHA 1984-1 Bechtold, S. E.; Janaro, R. E.; Sumners, D. L.
Maximization of labor productivity through optimal rest-break schedules.

BEEK 1985-1 Beek, P. van; Bremer, A.; Putten, C. van
Design and optimization of multi-echelon assembly networks: savings and potentials.

BEEKA 1983-1 Beek, P. van; Schuylenburg, P. van
An application of aggregation and decomposition to the N-product, 1-machine problem.

BELE 1980-1 Belenkij, A. S.; Ilenkova, S. D.; Levner, E. V.; Rezer, S. M.; Timkovskii, V. G.; Tsurkan, V. M.
Application of methods of scheduling theory to the optimization of the loading of containers (Russian).

BELEA 1985-1 Beletskii, S. A.; Levental, V. T.
An effective approximate algorithm for a general problem of scheduling theory (Russian).

BELY 1984-1 Belyalov, T. K.; Khabibullin, R. F.
On the problem of choice of optimal sizes of lots (Russian).

BENTA 1986-1 Benton, W. C.
Evaluating a modified heuristic for the multiple-vehicle scheduling problem.

BENV 1986-1 Benveniste, R.
An integrated plant design and scheduling problem in the food industry.

BERR 1986-1 Berrada, M.; Stecke, K. E.
A branch and bound approach for machine load balancing in flexible manufacturing systems.

BERTD 1985-1 Bertossi, A. A.; Bonuccelli, M. A.
A polynomial feasibility test for preemptive periodic scheduling of unrelated processors.

BERTE 1985-1 Bertrand, J. W. M.
Multiproduct optimal batch sizes with in-process inventories and multi work centers.

BIAN 1982-1 Bianco, L.; Ricciardelli, S.
On scheduling with ready times to minimize total weighted completion time.

BILLA 1986-1 Billington, P. J.
The capacitated multi-item dynamic lot-sizing problem.

BILLB 1986-1 Billington, P. J.; Mac Clain, J. O.; Thomas, L. J.
Heuristics for multilevel lot-sizing with a bottleneck.

BITRA 1984-1 Bitran, G. R.; Magnanti, T. L.; Yanasse, H. H.
Approximation methods for the uncapacitated dynamic lot size problem.

BITRB 1986-1 Bitran, G. R.; Matsuo, H.
Approximation formulations for the single-product capacitated lot size problem.

BITRB 1986-2 Bitran, G. R.; Matsuo, H.
The multi-item capacitated lot size problem: error bounds of Manne's formulations.

BLAC 1985-1 Blackburn, J. D.; Millen, R. A.
An evaluation of heuristic performance in multi-stage lot-sizing systems.

BLANC 1982-1 Blankenship, G. L.; Menaldi, J. L.
Optimal stochastic scheduling of power generation systems with scheduling delays and large cost differentials.

BLAZ 1984-1 Blazewicz, J.
Scheduling preemptible tasks on parallel processors with information loss.
BLAZ 1987-1 Blazewicz, J.
Selected topics in scheduling theory.
BLAZA 1986-1 Blazewicz, J.; Barcelo, J.; Kubiak, W.; Röck, H.
Scheduling tasks on two processors with deadlines and additional resources.
BLAZD 1986-1 Blazewicz, J.; Cellary, W.; Slowinski, R.; Weglarz, J.
Scheduling under resource constraints - Deterministic models.
BLAZE 1986-1 Blazewicz, J.; Drabowski, M.; Weglarz, J.
Scheduling multiprocessor tasks to minimize schedule length.
BLAZF 1983-1 Blazewicz, J.; Ecker, K.
A polynomial in time algorithm for task scheduling under fixed resource constraints.
BLAZG 1987-1 Blazewicz, J.; Finke, G.
Minimizing mean weighted execution time loss on identical and uniform processors.
BLAZH 1984-1 Blazewicz, J.; Kubiak, W.; Röck, H.
Complexity of minimizing mean flow time under resource constraits.
BLAZI 1987-1 Blazewicz, J.; Kubiak, W.; Röck, H.; Szwarcfiter, J. L.
Minimizing mean flow-time with parallel processors and resource constraints.
BLAZJ 1984-1 Blazewicz, J.; Weglarz, J.
The complexity of scheduling under resource constraints (Polish).
BOCT 1985-1 Boctor, F. F.
Single machine lot scheduling: a comparison of some solution procedures.
BONDA 1981-1 Bondarenko, V. A.; Kuksa, A. I.
A solution method for the problem of optimally planning independent jobs in time under constraints on stored and unstored resources (Russian).
BORO 1986-1 Borodich, S. A.; Strusevich, V. A.
Schedules for servicing a partially ordered set of customers for a class of determinate systems (Russian).
BOROA 1985-1 Borodich, S. A.; Tuzikov, A. V.
On the complexity of constructing time-optimal schedules for some two-stage service systems with non-fixed routes (Russian).
BOTT 1986-1 Bott, K.; Ballou, R. H.
Research perspectives in vehicle routing and scheduling.
BOUC 1987-1 Boucher, T. O.
Choice of assembly line design under task learning.
BOXM 1985-1 Boxma, O. J.
A probabilistic analysis of the LPT scheduling rule.
BOXMA 1986-1 Boxma, O. J.; Forst, F. G.
Minimizing the expected weighted number of tardy jobs in stochastic flow shops.
BRAS 1986-1 Bräsel, H.
Zu einigen Eigenschaften des open shop Problems.
BRAS 1987-1 Bräsel, H.
Rank minimal machine scheduling problems (German).
BROW 1987-1 Brown, G. G.; Graves, G. W.; Ronen, D.
Scheduling ocean transportation of crude oil.
BRUC 1988-1 Brucker, P.
Scheduling periodically recurring events and processes.
BRUCA 1987-1 Brucker, P.; Hamacher, H. W.
K-optimal solution sets for some polynomially solvable scheduling problems.
BRUCB 1986-1 Brucker, P.; Hurink, J.
A railway scheduling problem.
BRUNA 1985-1 Bruno, J. L.
On scheduling tasks with exponential service times and in-tree precedence constraints.
BUNN 1985-1 Bunnag, P.; Smith, S. B.
A multifactor priority rule for jobshop scheduling using computer search.
BURK 1986-2 Burkard, R. E.
Optimal schedules of periodically recurring events.
BURNB 1987-1 Burns, L. D.; Daganzo, C. F.
Assembly line job sequencing principles.
BURNC 1985-1 Burns, R. N.; Carter, M. W.
Work force size and single shift schedules with variable demands.
BURND 1987-1 Burns, R. N.; Koop, G. J.
A modular approach to optimal multiple-shift manpower scheduling.
BUZAA 1985-1 Buzacott, J. A.; Shanthikumar, J. G.
On approximate queueing models of dynamic job shops.
CALD 1985-1 Calderbank, A. R.; Coffman, E. G. Jr.; Flatto, L.
Sequencing problems in two-server systems.
CAMEC 1985-1 Camerini, P. M.; Maffioli, F.
Some applications of unlabelled partition systems.
CARL 1987-1 Carlier, J.
The m machines problem.
CARLA 1985-1 Carlier, J.; Chretienne, P.; Girault, C.
Modelling scheduling problems with timed Petri nets.
CARRA 1987-1 Carraway, R. L.; Chambers, R. J.; Morin, T. L.; Moskowitz, H.
Scheduling with multiple performance measures: generalized dynamic programming for nonlinear multicriteria cost functions.
CERN 1985-1 Cerny, J.; Cerna, A.; Pesko, S.; Vasek, K.
Optimal scheduling of transport.
CHAKB 1986-1 Chakravarthy, S.
A single-machine scheduling problem with random processing times.
CHAKB 1987-1 Chakravarthy, S.
On two cyclic exponential machines sharing a common tool magazine.

CHAKD 1987-1 Chakravarty, A. K.; Shtub, A.
Capacity, cost and scheduling analysis for a multiproduct flexible manufacturing cell.

CHAL 1985-1 Chalmet, L. G.; Bodt, M. de; Wassenhove, L. N. van
The effect of engineering changes and demand uncertainty on MRP lot sizing: a case study.

CHAN 1987-1 Chan, L. K.; Falkenberg, J.; Rosenbloom, E. S.
Implementation problems of a nurse preference mathematical programming approach to scheduling.

CHANF 1986-1 Chand, S.; Morton, T. E.
Minimal forcast horizon procedures for dynamic lot size models.

CHANG 1984-1 Chand, S.; Schneeberger, H.
A note on Smith's schedule to minimize the weighted sum of completion time with maximum allowable tardiness, (revised).

CHANG 1986-1 Chand, S.; Schneeberger, H.
A note on the single-machine scheduling problem with minimum weighted completion time and maximum allowable tardiness.

CHANR 1985-1 Chang, G. J.; Edmonds, J.
The poset scheduling problem.

CHARA 1984-1 Charnetski, J. R.
Scheduling operating room surgical procedures with early and late completion penalty costs.

CHENN 1987-1 Cheng, S.-C.; Stankovic, J. A.; Ramaritham, K.
Dynamic scheduling of groups of tasks with precedence constraints in distributed hard real-time systems.

CHENO 1985-1 Cheng, T. C. E.
Analysis of job flow-time in a job shop.

CHENO 1986-1 Cheng, T. C. E.
A note on minimizing the average deviation of job completion times about a common due-date.

CHENO 1986-2 Cheng, T. C. E.
A note on the common due-date assignment problem.

CHENO 1986-3 Cheng, T. C. E.
On optimal common due-date determination.

CHENO 1986-4 Cheng, T. C. E.
Optimal due-date assignment for a single machine sequencing problem with random processing times.

CHENO 1987-1 Cheng, T. C. E.
An algorithm for the CON due-date determination and sequencing problem.

CHENO 1987-2 Cheng, T. C. E.
Minimizing the average deviation of job completion times about a common due-date: an extension.

CHENO 1987-3 Cheng, T. C. E.
Optimal due-date determination and sequencing with random processing times.

CHENO 1987-4 Cheng, T. C. E.
Optimal TWK-power due-date determination and sequencing.

CHERE 1985-1 Chernyak, V. V.; Shcherbina, I. N.
On a stochastic model of volume-scheduling planning (Russian).

CHINA 1984-1 Chin, F.; Wong, M.
An optimal schedule for printing and binding.

CHOI 1987-1 Choi, H.-A.; Hakimi, S. L.
Scheduling file transfers for trees and odd cycles.

CHOO 1987-1 Choobineh, F.; Khattab, M. M.
Heuristic models for resource-constrained project scheduling.

CHRIA 1987-1 Christofides, N.; Alvarez-Valdes, R.; Tamarit, J. M.
Project scheduling with resource contraints: a branch and bound algorithm.

CIKA 1987-1 Cikanek, D. G.; Slater, P. J.
The k-processing center of a graph.

CIOBA 1985-1 Ciobanu, G.; Galupa, A.; Fabian, C.; Hartulari, C.
Reduction of the large size graphs with a view of their scheduling.

CLEG 1987-1 Clegg, W. H.
Optimum machine scheduling.

COFF 1987-1 Coffman, E. G. Jr.; Flatto, L.; Garey, M. R.; Weber, R. R.
Minimizing expected makespans on uniform processor systems.

COFFA 1985-1 Coffman, E. G. Jr.; Flatto, L.; Lueker, G. S.
Expected makespans for largest-first multiprocessor scheduling.

COFFC 1985-1 Coffman, E. G. Jr.; Garey, M. R.; Johnson, D. S.; LaPaugh, A. S.
Scheduling file transfers.

COFFD 1985-1 Coffman, E. G. Jr.; Gilbert, E. N.
On the expected relative performance of list scheduling.

COFFF 1986-1 Coffman, E. G. Jr.; Lueker, G. S.; Rinnooy Kan, A. H. G.
An introduction to the probabilistic analysis of sequencing and packing heuristics.

COLBC 1985-1 Colbourn, C. J.; Pulleyblank, W. R.
Minimizing setups in ordered sets of fixed width.

COLEA 1986-1 Cole, R.; Vishkin, U.
Approximate and exact parallel scheduling with applications to list, tree, and graph problems.

CONT 1984-1 Conti, A.; Rossetto, S.
Scheduling work loads on one maschine in the presence of an objective function of mixed type (Italian).

CSIRA 1987-1 Csirik, J.; Frenk, J. B. G.; Galambos, G.; Rinnooy Kan, A. H. G.
Probabilistic analysis of algorithms for dual bin packing problems.

CUNI 1984-1 Cuninghame-Green, R. A.; Borawitz, W. F.
Scheduling by non-commutative algebra.

CUNN 1986-1 Cunningham, P.; Browne, J.
A LISP-based heuristic scheduler for automatic insertion in electronics assembly.

CURRE 1986-1 Curry, G. L.; Deuermeyer, B. L.
Optimal scheduling of multiple feedstock batch biogas production systems.

DADU 1988-1 Daduna, J. R.
Verfahren zur Lösung von Vehicle Scheduling Problemen im öffentlichen Personennahverkehr.

DANIA	1985-1	Danilchenko, A. M.; Levchenko, S. N.; Panishev, A. V. An approximate algorithm for the three-machine problem.
DARBA	1985-1	Darby-Dowman, K.; Mitra, G. An extension of set partitioning with application to scheduling problems.
DAUL	1987-1	Dauler, J. C.; Nuttle, H. L. W. Addendum to "A comparison of heuristics for a school bus scheduling problem".
DAUS	1985-1	Dauscha, W.; Modrow, H. D.; Neumann, A. On cyclic sequence types for constructing cyclic schedules.
DAYA	1986-1	Dayan, I.; Harel, D. Fair termination with cruel schedulers.
DEAK	1984-1	Deak, I.; Hoffer, J.; Mayer, J.; Nemeth, A.; Potecz, B.; Prekopa, A.; Strazicky, B. Optimal, short-distance plant scheduling for electrical energy systems based on thermal plants, taking conditions of network locations into consideration (Hungarian).
DESRF	1986-2	Desrosiers, J.; Soumis, F.; Desrochers, M.; Sauve, M. Vehicle routing and scheduling with time windows.
DICA	1986-1	Dicasali, R. L. Job shops can use repetitive manufacturing methods to facilitate just-in-time production.
DIXO	1987-1	Dixon, P. S. Comments on 'lot sizing for material requirements planning systems'.
DOBS	1987-1	Dobson, G. The economic lot-scheduling problem: Achieving feasibility using time-varying lot sizes.
DOBSA	1986-1	Dobson, G.; Karmarkar, U. S. Large scale shop scheduling: formulations and decomposition.
DOBSA	1986-2	Dobson, G.; Karmarkar, U. S. Simultaneous resource scheduling to minimize flow times.
DOBSB	1985-1	Dobson, G.; Karmarkar, U. S.; Rummel, J. L. Single machine sequencing with lot sizing.
DOBSB	1985-2	Dobson, G.; Karmarkar, U. S.; Rummel, J. L. Single-product multi-machine flow minimization with batching.
DOBSB	1987-1	Dobson, G.; Karmarkar, U. S.; Rummel, J. L. Batching to minimize flow times on one machine.
DODI	1984-2	Dodin, B. Scheduling N products on a single facility with backordering and fixed terminal inventory.
DODI	1985-2	Dodin, B. Scheduling N products on a single facility with allowed backordering.
DOKE	1985-1	Dokev, N. B. An algorithm for solving a class of scheduling tasks (Bulgarian).
DOLEA	1985-1	Dolev, D.; Upfal, E.; Warmuth, M. K. Scheduling trees in parallel.
DOLEB	1984-1	Dolev, D.; Warmuth, M. K. Scheduling precedence graphs of bounded height.
DOLEB	1985-1	Dolev, D.; Warmuth, M. K. Profile scheduling of opposing forests and level orders.
DOLEB	1985-2	Dolev, D.; Warmuth, M. K. Scheduling flat graphs.
DORO	1984-1	Dorofeeva, M. P.; Ovsyankin, B. P.; Smirnov, A. V. Multiprocessor schedules with due dates (Russian).
DOYA	1981-1	Doyar, A. J.; Peltek, Y. A. An algorithm to solve scheduling problems in project tasks (Russian).
DREX	1988-1	Drexl, A. A stochastic assignment approach to resource-constrained multi-project scheduling.
DRID	1986-1	Dridi, N.; Portmann, M. C.; Proth, J. M. A decomposition method for production scheduling problems (French).
DRIDA	1984-1	Dridi, N.; Proth, J. M.; Sedillot, S. Ordonnancement de taches constraintes dans le cas d'un serveur.
DRORC	1987-1	Dror, M.; Stern, H. I.; Lenstra, J. K. Parallel machine scheduling: Processing rates dependent on number of jobs in operation.
EAST	1986-1	Easton, F.; Faaland, B.; Klastorin, T. D.; Schmitt, T. An improved network based algorithm for the assembly line balancing problem.
EHNE	1984-1	Ehnert, G.; Harant, J.; Walther, H. Ein exakter in Polynomialzeit arbeitender Algorithmus zur Fahrzeugminimierung in einem Kraftverkehrsbetrieb.
EHNE	1985-1	Ehnert, G.; Harant, J.; Walther, H. Ein in Polynomialzeit arbeitender Algorithmus zur Minimierung einer skalarisierten Zielfunktion in einem Kraftverkehrsbetrieb mit einem Depot, einem Fahrzeugtyp und Aufträgen mit unterschiedlichen Anforderungen an die zu transportierenden Mengen.
EL-D	1986-1	El-Dessouki, O.; Huen, W.; Evens, M. Optimal partitioning of sequential jobs on a network computer.
ELIZ	1984-1	Elizarov, Y. D.; Stolov, E. L. An analogue of methods of a deformable polyhedron in problems of scheduling theory (Russian).
ELSA	1986-1	Elsayed, E. A.; Nasr, N. Z. Heuristics for resource constrained scheduling.
ELVE	1985-1	Elvers, D. A.; Treleven, M. D. Job-shop vs. hybrid flow-shop routing in a dual resource constrained system.
EMMO	1987-1	Emmons, H. Scheduling to a common due date on parallel uniform processors.
EPPE	1985-1	Eppen, G. D.; Martin, R. K. Solving multi-item capacitated lot sizing problems using variable redefinition.
EREN	1987-1	Erenguc, S. S.; Tufekci, S. A branch and bound algorithm for a single-item multi-source dynamic lot sizing problem with capacity constraints.
ERSC	1985-1	Erschler, J.; Fontan, G.; Merce, C. Un nouveau concept de dominance pour l'ordonnancement de travaux sur une machine.

ETSC 1985-1 Etschmaier, M. M.; Mathaisel, D. F. X.
Airline scheduling: an overview.
EVAN 1987-1 Evans, J. R.
Structural analysis of local search heuristics in combinatorial optimization.
FAAL 1987-1 Faaland, B.; Schmitt, T.
Scheduling tasks with due dates in a fabrication/assembly process.
FAHR 1987-1 Fahrion, R.
A parallelization approach for sequencing problems.
FAIGB 1985-1 Faigle, U.; Gierz, G.; Schrader, R.
Algorithmic approaches to setup minimization.
FALK 1987-1 Falkner, J. C.; Ryan, D. M.
A bus crew scheduling system using a set partitioning model.
FARD 1986-1 Fardanesh, B.; Villaseca, F. E.
Two-step optimal thermal generation scheduling.
FEDE 1986-2 Federgruen, A.; Groenevelt, H.
Preemptive scheduling of uniform machines by ordinary network flow techniques.
FEJG 1983-1 Fejgin, L. I.
Scheduling problems with fuzzy activity durations.
FERL 1986-1 Ferland, J. A.
The single depot vehicle scheduling problem.
FERN 1982-1 Fernandez Garcia, F. R.; Pino Mejias, J. L.
Study of multiple objectives in sequencing problems on one machine (Spanish).
FERNA 1987-1 Fernandez, E. B.; Ngo, T. A.
Scheduling of task graphs into transputer networks.
FISHE 1986-1 Fisher, M. L.; Huang, J.-G.; Tang, B.-X.
Scheduling bulk-pickup-delivery vehicles in Shanghai.
FISHH 1981-1 Fisher, M. L.; Krieger, A. M.
Analysis of a linearization heuristic for single-machine scheduling to maximize profit.
FLOY 1986-1 Floyd, S.; Karp, R. M.
FFD bin-packing for item sizes with distributions on (0, 1/2).
FOLE 1984-1 Foley, R. D.; Suresh, S.
Minimizimg the expected flowtime in stochastic flow shops.
FOLE 1984-2 Foley, R. D.; Suresh, S.
Stochastically minimizing the makespan in flow shops.
FOLE 1986-1 Foley, R. D.; Suresh, S.
Scheduling n overlapping jobs and two stochastic jobs in a flow shop.
FORS 1984-1 Forst, F. G.
A review of the static, stochastic job sequencing literature.
FOUL 1984-1 Foulds, L. R.
Combinatorial optimization for undergraduates.
FRAI 1984-1 Frair, L.
Airport noise modelling and aircraft scheduling so as to minimize community annoyance.
FREDA 1987-1 Frederickson, G. N.; Hambrusch, S. E.
Planar linear arrangements of outerplanar graphs.
FREE 1986-1 Freeman, D. S.
Design of automated assembly systems: issues and procedures.
FREEA 1986-1 Freeman, D. S.; Swain, R. W.
A heuristic technique for balancing automated assembly lines.
FRENC 1987-1 Frenk, J. B. G.; Rinnooy Kan, A. H. G.
The asymptotic optimality of the LPT rule.
FREND 1984-1 Frenk, J. B. G.; Rinnooy Kan, A. H. G.; Stougie, L.
A hierarchical scheduling problem with a well-solvable second stage.
FRIE 1987-1 Friesen, D. K.
Tighter bounds for LPT scheduling on uniform processors.
FRIED 1986-1 Frieze, A. M.; Baybars, I.
Expected behaviour of line-balancing heuristics.
FROS 1985-1 Frostig, E.; Adiri, I.
Stochastic flowshop no-wait scheduling.
FROS 1985-2 Frostig, E.; Adiri, I.
Three-machine flowshop stochastic scheduling to minimize distribution of schedule length.
FRY 1987-1 Fry, T. D.; Armstrong, R. D.; Blackstone, J. H.
Minimizing weighted absolute deviation in single machine scheduling.
FRY A 1987-1 Fry, T. D.; Leong, G. K.
A bi-criterion approach to minimizing inventory costs on a single machine when early shipments are forbidden.
FUCHA 1985-1 Fuchs, K.; Kafura, D.
Memory-constrained task scheduling on a network of dual processors.
FUKUB 1987-1 Fukumori, K.; Sano, H.; Hasegawa, T.; Sakai, T.
Fundamental algorithm for train scheduling based on artificial intelligence.
GALV 1985-1 Galvin, B.
Machine scheduling models with application to the loading of bulk carriers in tidal ports.
GAZM 1985-1 Gazmuri, P. G.
Probabilistic analysis of a machine scheduling problem.
GAZM 1986-1 Gazmuri, P. G.
Probabilistic analysis of algorithms for some combinatorial optimization problems.
GE 1984-1 Ge, X. K.; Zhou, G. R.; Jiang, J. R.; Guo, R.
Heuristic algorithms for vehicle routing of a fleet (Chinese).
GERS 1985-1 Gershwin, S. B.; Akella, R.; Choong, Y. F.
Short-term production scheduling of an automated manufacturing facility.
GERSA 1984-1 Gershwin, S. B.; Akella, R.; Choong, Y. F.; Mitter, S. K.
Short term production scheduling of an automated manufacturing facility.

GHANA 1986-1 Ghandforoush, P.; Greber, B. J.
Solving allocation and scheduling problems inherent in forest resource management using mixed-integer programming.
GILBA 1987-1 Gilbert, K. C.; Hofstra, R. B.
An algorithm for a class of three-dimensional assignment problems arising in scheduling applications.
GLAZ 1985-1 Glazebrook, K. D.
Semi-Markov models for single-machine stochastic scheduling problems.
GLAZ 1987-1 Glazebrook, K. D.
Evaluating the effects of machine breakdowns in stochastic scheduling problems.
GLAZ 1987-2 Glazebrook, K. D.
Sensitivity analysis for stochastic scheduling problems.
GLAZA 1987-1 Glazebrook, K. D.; Fay, N. A.
On the scheduling of alternative stochastic jobs on a single machine.
GLOVG 1986-1 Glover, F.; Mac Millan, C.
The general employee scheduling problem: an integration of MS and AI.
GOETA 1985-1 Goetz, J.; Makuchowski, M.
Heuristic methods for the general problem of resources allocation and operation scheduling (Polish).
GOLDH 1986-1 Golden, B. L.; Bodin, L. D.; Goodwin, T.
Microcomputer-based vehicle routing and scheduling software.
GOLE 1985-1 Golenko, D. I.
Approximated statistical methods in optimization problems of scheduling theory.
GOLE 1985-2 Golenko, D. I.
Statistical methods of optimizing varied-route problems in the scheduling theory.
GONS 1987-1 Gonsalvez, D. J.; Hall, N. G.; Rhee, W. T.; Siferd, S. P.
Heuristic solutions and confidence intervals for the multicovering problem.
GONZA 1986-1 Gonzalez, R. L.; Rofman, E.
Computational methods in scheduling optimization.
GOODA 1986-1 Goodwin, J. S.; Weeks, J. K.
Evaluating scheduling policies in a multi-level assembly system.
GORDB 1985-1 Gordon, V. S.; Yanova, O. V.
Uniform parallel machine scheduling under precedence constraints and due-dates (Russian).
GOYAA 1985-1 Goyal, S. K.; Demirbag-Büyükkurt, M.
A note on "Determination of coordinated maintenance scheduling frequencies for a group of machines".
GOYAB 1986-1 Goyal, S. K.; Szendrovits, A. Z.
A constant lot size model with equal and unequal sized batch shipments between production stages.
GRAB 1984-1 Grabowski, J.; Janiak, A.
On job-shop scheduling with resource constraints.
GRAB 1987-1 Grabowski, J.; Janiak, A.
Job-shop scheduling with resource-time models of operations.
GRABA 1986-1 Grabowski, J.; Nowicki, E.; Zdrzalka, S.
A block approach for single-machine scheduling with release dates and due dates.
GRABC 1983-1 Grabowski, J.; Smutnicki, C.
The flow-shop problems. Properties, algorithms and applications. I. (Polish).
GRABC 1984-1 Grabowski, J.; Smutnicki, C.
Minimizing maximum cost in flow-shop problems. I: basic properties and algorithms (Polish).
GRABC 1984-2 Grabowski, J.; Smutnicki, C.
Minimizing maximum cost in flow-shop problems. II: Lower bounds, computational results, applications (Polish).
GRABC 1984-3 Grabowski, J.; Smutnicki, C.
The flow-shop problems, Properties, algorithms and applications. II (Polish).
GRABC 1986-1 Grabowski, J.; Smutnicki, C.
Scheduling problems with minimax goal function (Polish).
GRABD 1985-1 Grabowski, J.; Zdrzalka, S.
Problems on sequencing jobs on an assembly line (Polish).
GRAH 1986-1 Graham, D.; Nuttle, H. L. W.
A comparison of heuristics for a school bus scheduling problem.
GRANE 1986-1 Grant, T. J.
Lessons for O. R. from A. I.: a scheduling case study.
GRAV 1986-1 Graves, S. C.
A tactical planning model for a job shop.
GREEE 1986-1 Greene, T. J.; Sadowski, R. P.
A mixed integer program for loading and scheduling multiple flexible manufacturing cells.
GUIN 1984-1 Guinet, A.
Transports industriels routiers, un probleme d'affectation avec reemploi sous contraintes.
GUMI 1986-1 Guminskij, A. P.; Martynov, V. V.
Construction and implementation of a scheduling algorithm in a calculus based on universal semantic code.
GUNTA 1987-1 Gunther, H. O.
Planning lot sizes and capacity requirements in a single stage production system.
GUPTB 1984-1 Gupta, J. N. D.
Optimal schedules for single facility with two job classes.
GUPTB 1986-1 Gupta, J. N. D.
Flowshop schedules with sequence dependent setup times.
GUPTC 1986-1 Gupta, J. N. D.; Darrow, W. P.
The two-machine sequence dependent flowshop scheduling problem.
GUPTD 1987-1 Gupta, J. N. D.; Shanthikumar, J. G.; Szwarc, W.
Generating improved dominance conditions for the flowshop problem.
GUPTF 1984-1 Gupta, R. G.; Srivastava, V. S. P.
On synthesis of scheduling algorithms.
GUPTG 1984-1 Gupta, S.; Sen, T.
Minimizing the range of lateness on a single machine.

GUSF 1984-1 Gusfield, D.
Bounds for naive multiple machine scheduling with release times and deadlines.

HABIC 1987-1 Habib, M.; Möhring, R. H.
On some complexity properties of N-free posets and posets with bounded decomposition diameter.

HABID 1987-1 Habib, M.; Möhring, R. H.; Steiner, G.
Computing the bump number of ordered sets is easy.

HALLA 1986-1 Hall, N. G.
Production problems with deadline penalties.

HALLA 1986-2 Hall, N. G.
Scheduling problems with generalized due dates.

HALLC 1986-1 Hall, N. G.; Rhee, W. T.
Average and worst-case analysis of heuristics for the maximum tardiness problem.

HALLE 1985-2 Hall, R. W.
Vehicle scheduling at a transportation terminal with random delay en route.

HAN A 1987-1 Han, J.-Y.; Xu, B.-S.
Advances in the theory of sequencing and scheduling (Chinese).

HARA 1985-1 Harant, J.
Neue Ergebnisse in der Dienstplanoptimierung.

HELM 1986-1 Helmbold, D.; Mayr, E.
Two processor scheduling is in NC.

HELM 1987-1 Helmbold, D.; Mayr, E.
Applications of parallel scheduling to perfect graphs.

HELM 1987-2 Helmbold, D.; Mayr, E.
Two processor scheduling is in NC .

HENI 1986-1 Henig, M. I.
Extensions of the dynamic programming method in the deterministic and stochastic assembly-line balancing problems.

HOCHB 1985-3 Hochbaum, D. S.; Shmoys, D. B.
Using dual approximation algorithms for scheduling problems: theoretical and practical results.

HOCHB 1986-1 Hochbaum, D. S.; Shmoys, D. B.
A packing problem you can almost solve by sitting on your suitcase.

HOCHB 1986-2 Hochbaum, D. S.; Shmoys, D. B.
A polynomial approximation scheme for machine scheduling on uniform processors: using the dual approximation approach.

HOCHB 1987-1 Hochbaum, D. S.; Shmoys, D. B.
Using dual approximation algorithms for scheduling problems: theoretical and practical results.

HODGB 1986-1 Hodgson, T. J.; Nuttle, H. L. W.
A note on linear programming and the single machine lot size scheduling problem.

HODS 1985-1 Hodson, A.; Muhlemann, A. P.; Price, D. H. R.
A microcomputer based solution to a practical scheduling problem.

HOFFE 1985-1 Hoffmann, T. R.; Scudder, G. D.
An evaluation of value-based dispatching rules in a flow shop.

HU 1987-1 Hu, T. C.; Kuo, Y. S.; Ruskey, F.
Some optimum algorithms for scheduling problems with changeover costs.

HUANA 1984-1 Huang, H. C.; Ong, H. L.
Expected performance of some lot-sizing heuristics in a rolling-schedule environment.

IGNA 1984-1 Ignatev, V. M.
Some algorithms for inactive schedules (Russian).

IGNI 1984-1 Ignizio, J. P.
A generalized goal programming approach to the minimal interference, multicriteria N|1 scheduling problem.

IKRA 1986-1 Ikram, M.
A note on minimization of lateness cost function and determination of optimal due-date in a 2-machine scheduling problem.

IKUR 1986-1 Ikura, Y.; Gimple, M.
Efficient scheduling algorithms for a single batch processing machine.

IKURA 1986-1 Ikura, Y.; Gross, G.; Hall, G. S.
PGandE's state-of-the-art scheduling tool for hydro systems.

ISHI 1987-1 Ishii, H.; Masuda, T.; Nishida, T.
Two machine mixed shop scheduling problem with controllable machine speeds.

ISHIA 1986-1 Ishii, H.; Nishida, T.
Two machine open shop scheduling problem with controllable machine speeds.

JAINA 1986-1 Jain, M.; Sharma, G. C.
A seminumerical iterative method for solving a machine interference problem.

JAMR 1985-1 Jamroz, L.
Suboptimal algorithm for job-shop problem with intermediate constraints (Polish).

JANI 1986-1 Janiak, A.
On a single machine sequencing to minimize the maximum job cost subject to resource and precedence constraints.

JANI 1986-2 Janiak, A.
One-machine scheduling problems with resource constraints.

JANI 1986-3 Janiak, A.
Time-optimal control in a single machine problem with resource constraints.

JANI 1987-1 Janiak, A.
One-machine scheduling with allocation of continuously-divisible resource and with no precedence constraints.

JANIA 1984-1 Janiak, A.; Grabowski, J.
Job-shop problem with resource constraints.

JAW 1986-1 Jaw, J.-J.; Odoni, A. R.; Psaraftis, H. N.; Wilson, N. H. M.
A heuristic algorithm for the multi-vehicle advance request dial-a-ride problem with time windows.

JOHNC 1984-1 John, T. C.; Sadowski, R. P.
Compromise solutions in single stage production scheduling for minimizing total flow time and maximum penalty.

JURK 1984-1 Jurkowska, T.; Komorowska, E.; Mazbic-Kulma, B.; Partyka, A.; Pogorzelec, A.
Computer method of bus routes determining for the inter-urban transportation.
KALIA 1987-1 Kalinin, V. N.; Sokolov, B. V.
Dynamic model and optimal scheduling algorithm for a set of jobs with preemption constraints.
KALR 1982-1 Kalra, K. R.; Bagga, P. C.
Single machine sequencing problem with polynomial production cost functions: modifications in the branch-and-bound technique of Townsend.
KALRA 1985-1 Kalra, K. R.; Bagga, P. C.; Singh, S. P.
Some remarks on the paper "Minimizing the average deviation of job completion times about a common due-date" by John J. Kanet.
KAMB 1986-2 Kamburowski, J.
Upper bounds on the expected project duration under incomplete probabilistic information (Polish).
KAMP 1987-1 Kämpke, T.
Necessary optimality conditions for priority policies in stochastic weighted flowtime scheduling problems.
KAMP 1987-2 Kämpke, T.
On the optimality of static priority policies in stochastic scheduling on parallel machines.
KANEA 1986-1 Kanet, J. J.
Tactically delayed versus non-delay scheduling: An experimental investigation.
KANEB 1987-1 Kanet, J. J.; Adelsberger, H. H.
Expert systems in production scheduling.
KANOB 1983-1 Kanopa, R.
On the analysis of the heuristic solution for a problem concerning the composition of a time-table of works in network planning (Russian).
KANTC 1983-1 Kantsedal, S. A.
A decomposition approach to the solution of large-scale scheduling problems.
KARL 1983-1 Karla, K. R.; Khurana, K.
Single machine scheduling to minimize waiting cost with secondary criterion.
KARMC 1987-1 Karmarkar, U. S.
Lot-sizing and sequencing delays.
KARMD 1985-1 Karmarkar, U. S.; Kekre, S.; Kekre, S.
Lotsizing in multi-item multi-machine job shops.
KARMD 1987-1 Karmarkar, U. S.; Kekre, S.; Kekre, S.
The dynamic lot-sizing problem with startup and reservation costs.
KARME 1985-1 Karmarkar, U. S.; Kekre, S.; Kekre, S.; Freeman, S.
Lot-sizing and lead-time performance in a manufacturing cell.
KARN 1986-1 Karni, R.
A study of lot sizing algorithms.
KAWA 1986-1 Kawaguchi, T.; Kyan, S.
Worst case bound of an LRF schedule for the mean weighted flow-time problem.
KAWAB 1984-1 Kawanago, T.; Yamashiro, M.
A study on the single machine production-line system with multiple arrival sources. II: On the development of simulation programs (Japanese).
KHAB 1984-1 Khabibullin, R. F.; Khalimov, F. R.
The problem of optimal operation-wise scheduling (Russian).
KHAB 1984-2 Khabibullin, R. F.; Khalimov, F. R.
The problem of optimization of a manager's workload in the execution of a set of of independent jobs (Russian).
KHANA 1987-1 Khan, M. R.; Lewis, D. A.
A network model for nursing staff scheduling.
KHUR 1984-1 Khurana, K.; Bagga, P. C.
Minimizing the makespan in a 2-machine flowshop with time lags and setup conditions.
KHUR 1985-1 Khurana, K.; Bagga, P. C.
Scheduling of job-block with deadline in n*2 flowshop problem with separated set up times.
KIRA 1984-1 Kiran, A. S.
On ordered flow shop-mean flowtime scheduling problem.
KIRA 1986-1 Kiran, A. S.
Complexity of FMS loading and scheduling problems.
KOBA 1984-1 Kobayashi, T.
The lexico-bounded flow algorithm for solving the minimum cost project scheduling problem with an additional linear constraint.
KOEL 1984-1 Koelling, C. P.; Bailey, J. E.
A multiple criteria decision aid for personnel scheduling.
KOOP 1986-1 Koop, G. J.
Cyclic scheduling of offweekends.
KRAMA 1985-1 Krämer, M.
A processor-sharing model with several types of jobs and preemptive priorities.
KU 1986-1 Ku, P.-S.; Niu, S.-C.
On Johnson's two-machine flow shop with random processing times.
KUBA 1987-1 Kubale, M.
The complexity of scheduling independent two-processor tasks on dedicated processors.
KUMAA 1984-1 Kumar, K.
Efficient algorithms for grouping component-processor families.
KUND 1985-1 Kunde, M.; Steppat, H.
First fit decreasing scheduling on uniform multiprocessors.
KUND 1987-1 Kunde, M.; Steppat, H.
On the worst-case ratio of a compound multiprocessor scheduling algorithm.
KURT 1985-1 Kurtulus, I.
Multiproject scheduling: Analysis of scheduling strategies under unequal delay penalties.
KURTA 1985-1 Kurtulus, I.; Narula, S. C.
Multi-project scheduling: analysis of project performance.
KUSI 1986-1 Kusiak, A.
Efficient implementation of Johnson's scheduling algorithm.

KUSI 1987-1 Kusiak, A.
Designing expert systems for scheduling automated manufacturing.

LABE 1984-1 Labetoulle, J.; Lawler, E. L.; Lenstra, J. K.; Rinnooy Kan, A. H. G.
Preemptive scheduling of uniform machines subject to release dates.

LAFO 1985-1 Laforge, R. L.
MRP lot sizing with multiple purchase discounts.

LAI A 1983-1 Lai, T.-H.; Sahni, S.
Preemptive scheduling of uniform processors with memory.

LANGA 1986-1 Lange, E. G.; Batyrkanov, M. S.
Sufficient conditions for the existence of a nonstoppage schedule for the work of a transportation-loading-unloading complex (Russian).

LANGB 1986-1 Langen, D.; Jandl, H.
T O U R - Transport Optimization Under Restrictions.

LANGC 1987-2 Langston, M. A.
Interstage transportation planning in the deterministic flow-shop environment.

LAPO 1987-1 Laporte, G.
Solving a family of permutation problems on 0-1 matrices.

LARS 1985-1 Larson, R. E.; Dessouky, M. I.; Devor, R. E.
A forward-backward procedure for the single machine problem to minimize maximum lateness.

LASS 1985-1 Lasserre, J. B.; Bes, C.; Roubellat, F.
The stochastic discrete dynamic lot size problem: an open-loop solution.

LAWLA 1981-1 Lawler, E. L.; Lenstra, J. K.
Machine scheduling with precedence constraints.

LAWLB 1981-1 Lawler, E. L.; Lenstra, J. K.; Rinnooy Kan, A. H. G.
Recent developments in deterministic sequencing and scheduling: a survey.

LAZA 1984-1 Lazarev, A. A.
Algorithms in scheduling theory that are based on necessary conditions for optimality (Russian).

LAZA 1984-2 Lazarev, A. A.
Investigation of problems of scheduling theory by means of transformations (Russian).

LEBEC 1985-1 Lebedinskaya, N. B.
An algorithm for constructing an optimal structural schedule.

LEBED 1987-1 Lebedinskaya, N. B.; Safonova, T. E.
The stability of an optimal structural schedule (Russian).

LEE A 1985-1 Lee, C.-Y.
Worst case performance of heuristics for the joint replenishment problem.

LEE A 1986-1 Lee, C.-Y.
The dynamic lot size model with quantity discount.

LEE B 1987-1 Lee, C.-Y.; Danusaputro, S.
Economic lot scheduling problem: a resolution of two-product problem.

LEE C 1985-1 Lee, C.-Y.; Denardo, E. V.
Economic lot scheduling problem: test of feasibility and heuristic method.

LEE C 1986-1 Lee, C.-Y.; Denardo, E. V.
Rolling planning horizons: error bounds for the dynamic lot size model.

LEE D 1987-1 Lee, C.-Y.; Hwang, J.-J.; Chow, Y.-C.; Anger, F. D.
Multiprocessor scheduling with interprocessor communication delays.

LEE E 1986-1 Lee, C.-Y.; Massey, J. D.
Multiprocessor scheduling: an extension of multifit algorithm.

LEE E 1986-2 Lee, C.-Y.; Massey, J. D.
Multiprocessor scheduling: combining LPT and MULTIFIT.

LEFE 1985-1 Lefevre, C.; Laloyaux, P.
Note sur les problemes d'ordonnancement Pert aleatoires.

LENSA 1984-1 Lenstra, J. K.; Rinnooy Kan, A. H. G.
Two open problems in precedence constrained scheduling.

LENSA 1985-1 Lenstra, J. K.; Rinnooy Kan, A. H. G.
Sequencing and scheduling.

LENSD 1987-1 Lenstra, J. K.; Shmoys, D. B.; Tardos, E.
Approximation algorithms for scheduling unrelated parallel machines.

LENSD 1988-1 Lenstra, J. K.; Shmoys, D. B.; Tardos, E.
Approximation algorithms for scheduling unrelated parallel machines.

LEV 1984-1 Lev, V.; Adiri, I.
V-shop scheduling.

LEVIA 1987-1 Levin, V. I.
The m-machine problem with parts-sequencing constraints.

LICA 1986-1 Licata, R.; Johnston, R. E.
A scheduling procedure to minimize wastages in cable manufacturing processes.

LIES 1986-1 Liestman, A. L.; Campbell, R. H.
A fault-tolerant scheduling problem.

LIN D 1986-1 Lin, Y. X.
Dynamic programming method for m x n flow-shop scheduling problems (Chinese).

LIU B 1985-1 Liu, C. Y.; Bulfin, R. L.
On the complexity of preemptive open-shop scheduling problems.

LIU B 1987-1 Liu, C. Y.; Bulfin, R. L.
Scheduling ordered open shops.

LLOY 1986-1 Lloyd, E. L.
Two processor scheduling with limited preemption.

LLOY 1987-1 Lloyd, E. L.
Scheduling with semaphore constraints.

LOTF 1986-1 Lotfi, V.; Sarin, S.
A graph coloring algorithm for large scale scheduling problems.

LOVEB 1981-1 Love, R. R. Jr.
Traffic scheduling via Benders decomposition.

LU D 1985-1 Lu, Z. D.
A job scheduling algorithm for duplex processor systems (Chinese).

LUTH 1985-1 Lüthi, H.-J.; Polymeris, A.
Scheduling to minimize maximum workload.
LYUL 1986-1 Lyulyakov, A. V.
Algorithms for the construction of dense schedules for pipeline computers (Russian).
MA A 1987-1 Ma, J.-L.
A class of scheduling problems for multiprocessors (Chinese).
MAC D 1985-1 Mac Clain, J. O.; Trigeiro, W. W.
Cyclic assembly schedules.
MAC H 1984-1 Mac Hugh, J. A. M.
Hu's precedence tree scheduling algorithm: a simple proof.
MAES 1986-1 Maes, J.; Wassenhove, L. N. van
A simple heuristic for the multi-item single level capacitated lot sizing problem.
MAES 1986-2 Maes, J.; Wassenhove, L. N. van
Multi item single level capacitated dynamic lotsizing heuristics: A computational comparison (Part I: Static case).
MAES 1986-3 Maes, J.; Wassenhove, L. N. van
Multi item single level capacitated dynamic lotsizing heuristics: A computational comparison (Part II: Rolling horizon).
MAGG 1981-1 Maggu, P. L.
The two-machine maximum flow time problem with break-down intervals of machines.
MAGGA 1983-1 Maggu, P. L.; Mohd, H. N.; Prakash, O.
On an extension of n*3 flow-shop scheduling problem involving 'job-weight' and 'job-transportation' time.
MAGGA 1983-2 Maggu, P. L.; Mohd, H. N.; Prakash, O.
Weighted-job flow-shop scheduling problem involving transportation time for 'n-job, 3-machine' problem.
MAGGB 1984-1 Maggu, P. L.; Prakash, O.
Principle of dominance in flow-shop scheduling models.
MAGGC 1984-1 Maggu, P. L.; Prakash, O.; Singh, T.
A note on transportation time in dual three-stage production scheduling problem with set up time included.
MAGGC 1984-2 Maggu, P. L.; Prakash, O.; Singh, T.
On dual optimal three-stage production scheduling problem with set-up times included.
MAGGD 1985-1 Maggu, P. L.; Sharma, B. K.
'n-job, 2-machine' sequencing problem with stochastic service time involving equivalent-jobs for job-blocks.
MAGGD 1986-1 Maggu, P. L.; Sharma, B. K.
On equivalent-job for job-block in a 'n-job, 2-machine' flow-shop scheduling with random processing times.
MAGGD 1986-2 Maggu, P. L.; Sharma, B. K.
On 2-machine flow shop with additional times and equivalent jobs.
MAGGE 1984-1 Maggu, P. L.; Singh, T.; Sharma, V. K.
A short note on first generalized dual scheduling problem.
MAGGE 1984-2 Maggu, P. L.; Singh, T.; Sharma, V. K.
A short note on second generalised dual scheduling problem.
MAGGE 1984-3 Maggu, P. L.; Singh, T.; Sharma, V. K.
A short note on third generalised dual scheduling problem.
MAGGF 1984-1 Maggu, P. L.; Yadav, S. K.; Singh, T. P.; Dev, A.
Flow-shop scheduling problem involving job-weight and transportation time.
MALO 1983-1 Malova, S.; Milashin, S.
A problem of scheduling theory for two identical parallel devices (Russian).
MALS 1984-1 Malstrom, E. M.; Moraniec, D.
Complex lot sizing analyses with the pocket calculator.
MANA 1986-1 Manabe, R.
Scheduling buses by the network flow method on a personal computer.
MANN 1985-1 Manning, J. R.
How not to schedule a stitching room.
MARCH 1984-1 Marcotte, O.; Trotter, L. E. Jr.
An application of matroid polyhedral theory to unit-execution time, tree-precedence job scheduling.
MARE 1985-1 Marecki, F.
Asymmetrical assembly line balancing with the aid of a multi-stage programming method (Polish).
MARE 1985-2 Marecki, F.
Generalized assembly line balancing problem.
MART 1983-1 Martel, C. U.
Scheduling multiple processors with memory constraits.
MART 1985-2 Martel, C. U.
Preemptive scheduling to minimize maximum completion time on uniform processors with memory constraints.
MARTB 1986-1 Martello, S.; Toth, P.
A heuristic approach to the bus driver scheduling problem.
MASUA 1985-1 Masuda, T.; Ishii, H.
Two-machine scheduling problem for jobs with generalized due dates.
MASUB 1985-1 Masuda, T.; Ishii, H.; Nishida, T.
The mixed shop scheduling problem.
MASUB 1985-2 Masuda, T.; Ishii, H.; Nishida, T.
Two-machine scheduling problem for jobs with generalized due dates.
MATT 1984-1 Matta, K. F.
Job sequencing: economic approach provides best order.
MAXW 1986-1 Maxwell, W. L.; Singh, H.
Scheduling cyclic production on several identical machines.
MAZE 1985-1 Maze, T. H.; Khasnabis, S.
Bus garage location planning with dynamic vehicle assignments: a methodology.

MAZZA 1987-1 Mazzola, J. B.; Mac Coy, W. F.; Wagner, H. M.
Algorithms and heuristics for variable-yield lot sizing.
MEHRA 1986-1 Mehrez, A.; Sinuany-Stern, Z.; Stulman, A.
An enhancement of the Drezner-Wesolowsky algorithm for single-facility location with maximin of rectilinear distance.
MELNA 1984-1 Melnikov, O. I.
On Bellman-Johnson's (2*n)-problem for which all schedules are optimal (Russian).
MELNC 1985-1 Melnyk, S. A.; Piper, C. J.
Leadtime errors in MRP: the lot-sizing effect.
MERE 1986-1 Merenzon, M. N.
A dynamic programming algorithm for optimal scheduling of a nonhomogeneous multiprocessor system (English).
MILLA 1987-1 Miller, D. M.
An interactive, computer-aided ship scheduling system.
MIRCD 1986-1 Mirchandani, P. B.; Veatch, M. H.
"Hot job" routing through a stochastic job-shop network.
MITC 1987-1 Mitchell, J. S. B.
98%-effective lot-sizing for one-warehouse, multi-retailer inventory systems with backlogging.
MIYA 1987-1 Miyazaki, S.
A scheduling method based on local optimum for minimizing total tardiness.
MIYAA 1984-1 Miyazaki, S.; Nishiyama, N.
Scheduling to minimize total penalties due to earliness and tardiness.
MOHRA 1985-1 Möhring, R. H.; Radermacher, F. J.
Generalized results on the polynomiality of certain weighted sum scheduling problems.
MOHRA 1985-2 Möhring, R. H.; Radermacher, F. J.
Introduction to stochastic scheduling problems.
MOHRB 1984-1 Möhring, R. H.; Radermacher, F. J.; Weiss, G.
Stochastic scheduling problems I - general strategies.
MOHRB 1985-1 Möhring, R. H.; Radermacher, F. J.; Weiss, G.
Stochastic scheduling problems II - set strategies.
MOIL 1986-1 Moily, J. P.
Optimal and heuristic procedures for component lot-splitting in multi-stage manufacturing systems.
MOIT 1986-1 Moitra, A.
Scheduling of hard real-time systems.
MONA 1983-1 Monakhov, O. G.
One method of decentralized job allocation in computer systems with a programmable structure.
MONMB 1987-1 Monma, C. L.; Potts, C. N.
On the complexity of scheduling with batch set-up times.
MONMG 1987-1 Monma, C. L.; Sidney, J. B.
Optimal sequencing via modular decomposition: characterization of sequencing functions.
MOSI 1984-1 Mosier, C. T.; Elvers, D. A.; Kelly, D.
Analysis of group technology scheduling heuristics.
MOSK 1985-1 Moskalyuk, L. V.
The discrete maximum principle in the solution of scheduling problems (Russian).
MURD 1985-1 Murdeshwar, T. M.; Sathe, Y. S.
Some aspects of lot size models with two levels of storage.
MUSA 1984-1 Musa, A. A.; Saxena, U.
Scheduling nurses using goal-programming techniques.
NABE 1982-1 Nabeshima, I.
Notes on the analytical results in flow-shop scheduling. XVII. Special single criterion sequences. I.
NABE 1982-2 Nabeshima, I.
Notes on the analytical results in flow-shop scheduling. XVIII. Special single criterion sequences. II.
NABE 1983-1 Nabeshima, I.
Notes on the analytical results in flow-shop scheduling. XIX. Generalization of job-base results in part 13.
NABE 1983-2 Nabeshima, I.
Notes on the analytical results in flow-shop scheduling. XX. Another generalization of job-base results in part 13.
NABE 1983-3 Nabeshima, I.
Notes on the analytical results in flow-shop scheduling. XXI. Machine-base counterpart of job-base results for r=2 in part 20.
NABE 1984-1 Nabeshima, I.
Notes on the analytical results in flow shop scheduling. XXIII. Weaker sufficient optimality conditions for w(2) by machine-base transformation method.
NABE 1984-2 Nabeshima, I.
Notes on the analytical results in flow shop scheduling. XXIV. Job-base counterpart of the results in part XXII.
NABE 1984-3 Nabeshima, I.
Notes on the analytical results in flow-shop scheduling. XXII. Weaker sufficient optimality conditions for a specified sequence w(2) in parts 2 and 8.
NABE 1984-4 Nabeshima, I.
On sufficient optimality conditions with convex parameter in three-machine flow-shop makespan problem: a supplement to: "Notes on the analytical results in flow shop scheduling. XI".
NABE 1985-1 Nabeshima, I.
General formulation and applications of a machine-base transformation method for the NP-complete permutation scheduling problems.
NABE 1985-2 Nabeshima, I.
Notes on the analytical results in flow shop scheduling. XXV. Results by job-base transformation method for conditions in part XXIV.
NABE 1985-3 Nabeshima, I.
Notes on the analytical results in flow shop scheduling. XXVI. Generalization of job-base results in part XXIV.

NABE	1985-4	Nabeshima, I. Solvable cases in the m-machine flow-shop problem with separated sequence-independent setup times.
NABE	1986-1	Nabeshima, I. Another solvable case in the three-machine flow-shop makespan scheduling problem with separated sequence-independent setup times.
NABE	1986-2	Nabeshima, I. Job-base conditions in the three-machine flow-shop makespan scheduling problem with separated sequence-independent setup times.
NABE	1986-3	Nabeshima, I. Solvable cases extended by transformation method in the m-machine flow-shop problem with separated sequence-independent setup times. I, II.
NABE	1987-1	Nabeshima, I. Solvable cases in the m-machine flow shop mean completion time scheduling problem with separated sequence-independent setup times.
NABEA	1984-1	Nabeshima, I.; Maruyama, S. Two- and three-machine flow-shop makespan scheduling problems with additional times separated from processing times.
NABEA	1985-1	Nabeshima, I.; Maruyama, S. Correction to: Two- and three-machine flow-shop makespan scheduling problems with additional times separated from processing times.
NAID	1986-1	Naidu, M. M.; Singh, N. Lot sizing for material requirements planning systems - an incremental cost approach.
NAID	1987-1	Naidu, M. M.; Singh, N. Further investigations on the performance of incremental cost approach for lot sizing for material requirements planning systems.
NARAA	1984-1	Narasimhan, R.; Melnyk, S. A. Assessing the transient impact of lot sizing rules following MRP implementation.
NEEB	1986-1	Neebe, A. W.; Rao, M. R. Sequencing capacity expansion projects in continuous time.
NELSA	1986-1	Nelson, R. T.; Sarin, R. K.; Daniels, R. L. Scheduling with multiple performance measures: The one-machine case.
NEUMC	1985-3	Neumann, K. Stochastic single-machine scheduling with GERT-like precedence constraints.
NEVI	1985-1	Nevison, C. H. A cost adjustment heuristic for dynamic lot-sizing with uncertain demand timing.
NI	1985-1	Ni, L. M.; Hwang, K. Optimal load balancing in a multiple processor system with many job classes.
NORB	1986-1	Norbis, M. I.; Smith, J. M. Two level heuristic for the resource constrained scheduling problem.
NOWIA	1985-1	Nowicki, E.; Zdrzalka, S. Two-machine flow-shop scheduling problem with controllable job processing times (Polish).
NOWIA	1986-1	Nowicki, E.; Zdrzalka, S. A note on minimizing maximum lateness in a one-machine sequencing problem with release dates.
O'GRA	1985-1	O'Grady, P. J.; Harrison, C. A general search sequencing rule for job shop sequencing.
ODON	1985-1	Odoni, A. R. The "third generation" of vehicle routing and scheduling problems: two illustrative examples.
OLHA	1984-1	Olhager, J.; Rapp, B. Lot sizing and turnover rate.
OLHA	1985-1	Olhager, J.; Rapp, B. Balancing capacity and lot sizes.
OLIF	1985-1	Oliff, M. D.; Burch, E. E. Multiproduct production scheduling at Owens-Corning Fiberglas.
OLIN	1986-1	Olinger, W. Interactive algorithm for scheduling under renewable resource constraints (Polish).
ONDA	1984-1	Ondas, J. Algorithms for scheduling inhomogeneous multiprocessor computers.
ORLI	1985-5	Orlin, J. B. The complexity of dynamic/periodic languages and optimization problems.
OW	1985-1	Ow, P. S. Focused scheduling in proportionate flowshops.
OW A	1985-1	Ow, P. S.; Morton, T. E. An investigation of beam search for scheduling.
OYAM	1987-1	Oyama, T.; Han-I, S.-J. Application of discrete optimization techniques to train scheduling problems.
PAES	1988-1	Paessens, H. Fahrzeugeinsatzprobleme in Städtereinigungsunternehmen.
PAL	1984-1	Pal, B. B.; Basu, M. Optimal solution of a transportation problem by goal programming technique.
PALEA	1984-1	Palem, K. V.; Fussel, D. S. The complexity of scheduling systems of non-identical processors.
PALEB	1987-1	Paletta, G. A heuristic algorithm for a pseudo-cyclic delivery problem under window constraints.
PANI	1987-1	Panishev, A. V. The dichotomy method in the three-machine problem.
PAPAA	1987-1	Papadimitriou, C. H.; Tsitsiklis, J. N. On stochastic scheduling with in-tree precedence constraints.
PARA	1984-1	Paramonov, A. V.; Popov, Y. V.; Stępanenko, I. D.; Frenkin, B. R. The use of a distributive problem for determining the optimal schedules for processing stocks of sugar beets (Russian).
PARK	1984-1	Park, K. S.; Yun, D. K. A stepwise partial enumeration algorithm for the economic lot scheduling problem.

PARK 1985-1 Park, K. S.; Yun, D. K.
Optimal scheduling of periodic activities.
PERU 1986-1 Perusch, M.
Simulated annealing applied to a single machine scheduling problem with sequence dependent set up times and due dates.
PINE 1984-1 Pinedo, M.
A note on the flow time and the number of tardy jobs in stochastic open shops.
PINE 1984-2 Pinedo, M.
Optimal policies in stochastic shop scheduling.
PINE 1985-1 Pinedo, M.
A note on stochastic shop models in which jobs have the same processing requirements on each machine.
PINEA 1985-1 Pinedo, M.; Schechner, Z.
Inequalities and bounds for the scheduling of stochastic jobs on parallel machines.
PINEB 1985-1 Pinedo, M.; Weiss, G.
Scheduling jobs with exponentially distributed processing times and intree precedence constraints on two parallel machines.
PINEC 1986-1 Pinedo, M.; Wie, S.-H.
Inequalities for stochastic flow shops and job shops.
PINO 1982-1 Pino Mejias, J. L.
Solving scheduling problems by means of dynamic programming (Spanish).
POCH 1986-1 Pochet, Y.
Valid inequalities and separation for capacitated economic lot sizing.
POCHA 1986-1 Pochet, Y.; Wolsey, L. A.
Lot-size models with backlogging: strong reformulations and cutting planes.
POGU 1986-1 Pögün, G.; Satir, A.
Alternative bus scheduling policies for an exclusive bus lane.
POGUA 1986-1 Poguntke, W.
Order-theoretic aspects of scheduling.
PORT 1986-1 Porteus, E. L.
Optimal lot sizing, process quality improvement and setup cost reduction.
POSN 1985-1 Posner, M. E.
A sequencing problem with release dates and clustered jobs.
POSN 1985-2 Posner, M. E.
Minimizing weighted completion times with deadlines.
POTT 1985-1 Potts, C. N.
A Lagrangean based branch and bound algorithm for single machine sequencing with precedence constraints to minimize total weighted completion time.
POTT 1985-2 Potts, C. N.
Analysis of a linear programming heuristic for scheduling unrelated parallel machines.
POTT 1985-3 Potts, C. N.
Analysis of heuristics for two-machine flow-shop sequencing subject to release dates.
POTTA 1985-1 Potts, C. N.; Wassenhove, L. N. van
A branch and bound algorithm for the total weighted tardiness problem.
POTTA 1987-1 Potts, C. N.; Wassenhove, L. N. van
Dynamic programming and decomposition approaches for the single machine total tardiness problem.
PRAKA 1983-1 Prakash, S.
On the machine-assignment problem with objectives to minimize duration of completing work and processing cost.
PRAS 1985-1 Prasad, V. R.
Scheduling on a single processor subject to interruptions in processing.
QUAD 1987-1 Quaddus, M. A.
A generalized model of optimal due-date assignment by linear programming.
QUEY 1985-1 Queyranne, M.
Bounds for assembly line balancing heuristics.
RACH 1987-1 Rachamadugu, R. M. V.
A note on the weighted tardiness problem.
RADE 1986-1 Radermacher, F. J.
Schedule-induced posets.
RADE 1986-2 Radermacher, F. J.
Scheduling of project networks.
RAGH 1986-1 Raghavachari, M.
A V-shape property of optimal schedule of jobs about a common due date.
RAIS 1987-1 Raiszadeh, F. M. E.; Dileepan, P.; Sen, T.
A single machine bicriterion scheduling problem and an optimizing branch-and-bound procedure.
RAYW 1987-1 Rayward-Smith, V. J.
The complexity of preemptive scheduling given interprocessor communication delays.
RAYW 1987-2 Rayward-Smith, V. J.
UET scheduling with unit interprocessor communication delays.
RAZ 1987-1 Raz, T.; Bricker, D.
Sequencing of imperfect inspection operations subject to constraints on the quality of accepted and rejected units.
REIT 1984-1 Reiter, S.
How a network of processors can schedule its work.
RICHF 1985-1 Richter, K.
Complexity of the robot sequencing problem.
RICHF 1985-2 Richter, K.
Stability of the constant cost dynamic lot size model.
RICHF 1985-3 Richter, K.
The robot sequencing problem: polynomial algorithm and complexity.
RICHF 1986-1 Richter, K.
The two-criterial dynamic lot size problem.

RICHF	1987-1	Richter, K. Stability of the constant cost dynamic lot size model.
RINNA	1986-1	Rinnooy Kan, A. H. G.; Frenk, J. B. G. The rate of convergence to optimality of the LPT rule.
RIST	1984-1	Ristroph, J. H.; Ashley, C. E. Measurement of the effectiveness of alternative priority rules in network scheduling.
RITC	1986-1	Ritchie, E. Lot-sizing for non-stationary demand.
RIVA	1985-1	Rival, I. Some order theoretical ideas about scheduling.
RONE	1986-1	Ronen, D. Short term scheduling of vessels for shipping bulk or semi-bulk commodities originating in a single area.
ROSB	1985-1	Rosberg, Z. Process scheduling in a computer system.
ROSEE	1987-1	Rosenbloom, E. S.; Goertzen, N. F. Cyclic nurse scheduling.
ROUNA	1985-1	Roundy, R. 98%-effective integer-ratio lot-sizing for one-warehouse multi-retailer systems.
ROUNA	1986-1	Roundy, R. A 98% effective lot sizing rule for a multi-product, multi-stage production/inventory system.
ROWI	1986-1	Rowicki, A. Preemptive scheduling for two-processor systems.
RUSS	1986-1	Russell, R. A. A comparison of heuristics for scheduling projects with cash flows and resource restrictions.
RUSSB	1985-1	Russell, R. S.; Taylor, B. W. III An evaluation of scheduling policies in a dual resource constrained assembly shop.
RUSSB	1985-2	Russell, R. S.; Taylor, B. W. III An evaluation of sequencing rules for an assembly shop.
RZEM	1986-1	Rzemykowski, Z. On the method of maximal sets of activities (Polish).
SAFO	1985-1	Safonova, T. E. Stability of compatible schedules.
SAFO	1987-1	Safonova, T. E. Reconstruction of a consistent schedule in connection with change of the volume of a job.
SAKSB	1986-1	Saksena, J. P.; Ramachandran, A. An optimal fleet utilisation.
SALT	1987-1	Saltzman, M. J.; Baybars, I. A two-process implicit enumeration algorithm for the simple assembly line balancing problem.
SANN	1985-1	Sannikova, A. K.; Shafranskij, Y. M. NP-completeness of some optimal sequencing problems with a given grouping of elements.
SANTB	1985-1	Santos, C.; Magazine, M. Batching in single operation manufacturing systems.
SAWI	1985-1	Sawik, T. J. Scheduling lots of dependent unit-time operations on identical machines to minimize schedule length.
SAWI	1987-1	Sawik, T. J. A critical path algorithm for scheduling a batch-type production on nonidentical machines.
SAWI	1987-2	Sawik, T. J. Multilevel scheduling of multistage production with limited in-process inventory.
SCHMB	1984-1	Schmidt, G. Scheduling on semi-identical processors.
SCUD	1985-1	Scudder, G. D.; Hoffmann, T. R. Composite cost-based rules for priority scheduling in a randomly routed job shop.
SCUL	1987-1	Sculli, D.; Mok, K. C.; Cheung, S. H. Scheduling vehicles for refuse collection.
SEIDA	1987-1	Seidmann, A. On-line scheduling of a robotic manufacturing cell with stochastic sequence-dependent processing rates.
SEKI	1987-1	Sekiguchi, Y. A decomposition theory based on a dominance relation and composite jobs.
SELE	1986-1	Selen, W. J.; Hott, D. D. A mixed-integer goal-programming formulation of the standard flow-shop scheduling problem.
SELE	1986-2	Selen, W. J.; Hott, D. D. A new formulation and solution of the flowshop scheduling problem with no in-process waiting.
SEPE	1986-1	Sepehri, M.; Silver, E. A.; New, C. A heuristic for multiple lot sizing for an order under variable yield.
SERAA	1985-1	Serafini, P.; Ukovich, W. An approach towards solving periodic scheduling problems.
SERAB	1984-1	Serazzi, G. A bottleneck-driven scheduler.
SERV	1983-1	Servakh, V. V. On the Akers-Friedman problem (Russian).
SEVA	1982-1	Sevastyanov, S. V. Algorithms with estimates for the Johnson and Akers-Fridman problems in the case of three benches (Russian).
SEVA	1984-1	Sevastyanov, S. V. Efficient construction of near-optimal schedules for the cases of arbitrary and alternative routes of parts.
SEXT	1985-1	Sexton, T. R.; Bodin, L. D. Optimizing single vehicle many-to-many operations with desired delivery times: I. Scheduling.
SGURD	1984-1	Sgurev, V. S.; Vasilev, V. S.; Dokev, N. B.; Genova, K. B. An algorithm for the solution of operative scheduling problems in an averaging mode (Bulgarian).

SHAK	1985-1	Shakhbazyan, K. V. Algorithm of shifts for optimal structural schedules.
SHAK	1985-2	Shakhbazyan, K. V. Solution of two sequencing problems.
SHAK	1985-3	Shakhbazyan, K. V. Structured solutions in scheduling problems.
SHANA	1984-1	Shanker, K.; Kumar, P. Coordinate repair scheduling for a group of goods-producting equipment.
SHANB	1985-1	Shanthikumar, J. G.; Wu, Y. B. Decomposition approaches in permutation scheduling problems with application to the M-machine flow shop scheduling problems.
SHARA	1984-1	Sharma, R. K. A note on three machines flowshop sequencing problem (in special case).
SHARB	1987-1	Sharma, S.; Rani, P. On weighted-jobs scheduling problem involving machine break-down time interval.
SHER	1981-1	Sher, S. Y. On a problem of composing a schedule (Russian).
SIDN	1986-1	Sidney, J. B.; Steiner, G. Optimal sequencing by modular decomposition: polynomial algorithms.
SILV	1985-1	Silver, E. A.; Miltenburg, J. Errata: two modifications of the Silver-Meal lot sizing heuristic.
SILVA	1986-1	Silverman, F. N.; Carter, J. C. A cost-based methodology for stochastic line balancing with intermittent line stoppages.
SIMOA	1980-1	Simons, B. A fast algorithm for multiprocessor scheduling.
SINCA	1987-1	Sinclair, M.; Dyk, E. van Combined routeing and scheduling for the transportation of containerized cargo.
SINGA	1987-1	Singer, D.; Moritz, P. Expert system for scheduling cleaning operations for municipal canal systems.
SINGD	1987-1	Singh, H.; Foster, J. B. Production scheduling with sequence dependent setup costs.
SINGH	1985-1	Singh, T. P. On 2*n flow-shop problem involving job-block, transportation-time, arbitrary time and break-down machine times.
SINU	1986-1	Sinuany-Stern, Z.; Teomi, Y. Multi-objective scheduling plans for security guards.
SIVA	1987-1	Sivazlian, B. D. Optimum scheduling of a new maintenance program under stochastic degradation.
SLOW	1984-1	Slowinski, R. Modelling and solving multicriteria project scheduling problems.
SLOW	1984-2	Slowinski, R. Modelling and solving multicriteria project scheduling problems.
SMIR	1985-1	Smirnov, A. N. Mathematical model of three-dimensional scheduling of technologically linked activities.
SMITA	1986-1	Smith-Daniels, V. L.; Smith-Daniels, D. E. A mixed integer programming model for lot sizing and sequencing packaging lines in the process industries.
SOLOB	1987-1	Solomon, M. M. Algorithms for the vehicle routing and scheduling problem with time window constraints.
SOTS	1985-1	Sotskov, Y. N. The network approach for the solution of problems of scheduling theory (Russian).
SPES	1986-1	Spesivtsev, A. V. The error of certain approximate list scheduling algorithms (English).
SPIN	1985-2	Spinrad, J. Worst-case analysis of a scheduling algorithm.
SRIDC	1987-1	Sridharan, V.; Berry, W. L.; Udayabhanu, V. Freezing the master production schedule under rolling planning horizons.
SRIK	1986-1	Srikar, B. N.; Ghosh, S. A MILP model for the n-job, M-stage flowshop with sequence dependent set-up times.
SRIS	1986-1	Sriskandarajah, C.; Ladet, P. Some no-wait shops scheduling problems: complexity aspects.
STANE	1985-1	Stanley, J. D. Bottleneck structures and planning horizons.
STAV	1982-1	Stavre, P. Applications of mathematics in engineering and to the controversial lot-size problem (Romanian).
STEI	1985-1	Steiner, G. A compact labeling scheme for series-parallel graphs.
STEI	1985-2	Steiner, G. An algorithm for minimizing setups in precedence constrained scheduling.
STEIA	1987-1	Steiner, G.; Stewart, L. K. A linear time algorithm to find the jump number of 2-dimensional bipartite partial orders.
STRU	1985-1	Strusevich, V. A. The three-machine problem: sufficient condition for reducibility in a convex form (Russian).
STRU	1986-1	Strusevich, V. A. On the possiblity for constructing time-optimal schedules for a multi-stage system with non-fixed routes of stage passage (Russian).
SUMI	1986-1	Sumichrast, R. T. A new cutting-stock heuristic for scheduling production.
SUMI	1987-1	Sumichrast, R. T. Scheduling parallel processors to minimize setup time.
SUNGA	1986-1	Sung, C. S.; Chang, S. H. A capacity-constrained single-facility multi-product production planning model.

SUNGB 1986-1 Sung, C. S.; Park, M. H.
An approach for the optimal production rate of a single product system with dynamic demands.
SURE 1985-1 Suresh, S.; Foley, R. D.; Dickey, S. E.
On Pinedo's conjecture for scheduling in a stochastic flow shop.
SUSS 1984-1 Sussams, J.
The future for computerized vehicle-load planning systems.
SZIL 1983-1 Szilagyine-Szemkeö, J.
A multiple objective scheduling problem (Hungarian).
SZWA 1986-1 Szwarc, W.
The flow shop problem with time lags and separated setup times.
SZWAA 1987-1 Szwarc, W.; Gupta, J. N. D.
A flow-shop problem with sequence-dependent additive setup times.
SZWAC 1987-2 Szwarcfiter, J. L.
Job shop scheduling with unit time operations under resource constraints and release dates.
SZYM 1986-1 Szymanowski, W.
COLOS: a collection optimization system as an application of the vehicle scheduling problem.
TADA 1986-1 Tada, M.; Ishii, H.; Nishida, T.
A generalized version of 1-machine mean flow time minimization problem.
TALB 1986-1 Talbot, F. B.; Patterson, J. H.; Gehrlein, W. V.
A comparative evaluation of heuristic line balancing techniques.
TANA 1984-1 Tanaev, V. S.; Gordon, V. S.; Shafranskij, Y. M.
Scheduling theory, one-machine systems (Russian).
TANE 1985-1 Taneja, H. R.
On a new simple heuristic algorithm for scheduling n weighted jobs in a flow shop of two machines.
TANG 1986-1 Tang, C.; Denardo, E. V.
Models arising from a flexible manufacturing machine. Part 1: minimization of the number of tool switches.
TANG 1986-2 Tang, C.; Denardo, E. V.
Models arising from a flexible manufacturing machine. Part 2: minimization of the number of switching instants.
TAVA 1986-1 Tavares, L. V.
Multicriteria scheduling of a railway renewal program.
TEGZ 1984-1 Tegze, M.; Vlach, M.
Parallel flow shop problems.
TEGZ 1985-1 Tegze, M.; Vlach, M.
Matrix given scheduling problems.
TEGZ 1987-1 Tegze, M.; Vlach, M.
Improved bounds for the range of lateness on a single machine.
TEMP 1985-1 Templeton, W. D.; Daellenbach, H. G.
Interactive simulation for production scheduling.
TEOD 1985-1 Teodorovic, D.
Model for designing meteorologically most reliable airline schedules.
TERS 1986-1 Tersine, R. J.; Fisher, W. W.; Morris, J. S.
Varying lot sizes as an alternative to undertime and layoffs in aggregate scheduling.
THIZ 1985-1 Thizy, J. M.; Wassenhove, L. N. van
Lagrangean relaxation for the multi-item capacitated lot-sizing problem: a heuristic implementation.
THOME 1986-1 Thompson, G. L.; Zawack, D. J.
A problem expanding parametric programming method for solving the job shop scheduling problem.
TIMK 1985-1 Timkovskii, V. G.
On the complexity of scheduling an arbitrary system.
TIMK 1986-1 Timkovskii, V. G.
Approximate solution of the scheduling problem for a cyclic system (Russian).
TIMM 1988-1 Timmer, G. T.
Scheduling the construction of Dutch roads.
TOCZ 1986-1 Toczylowski, E.
On aggregation in a two-dimensional cutting stock scheduling problem.
TOCZ 1986-2 Toczylowski, E.
On aggregation of items in the single-stage lot size scheduling problem.
TOCZ 1987-1 Toczylowski, E.
Two-phase algorithm for lot size scheduling in single-stage production systems with parallel facilities.
TOCZA 1986-1 Toczylowski, E.; Nowosad, K.; Jagdev, H.; Hindi, K. S.
Aggregate lot-size scheduling for a class of two-stage batch manufacturing systems.
TOKA 1986-1 Tokarev, D. D.; Dishliev, A. B.
On a problem of scheduling theory (Bulgarian).
TOVE 1986-1 Tovey, C. A.
Rescheduling to minimize makespan on a changing number of identical processors.
TRIG 1987-1 Trigeiro, W. W.
A dual-cost heuristic for the capacitated lot sizing problem.
TRZA 1984-1 Trzaskalik, T.
Optimal multiperiod planning for manufacturing plant - general problems.
TUSH 1985-1 Tushkina, T. A.
Computation of a lower bound of the lengths of schedules.
TUZI 1984-1 Tuzikov, A. V.
A bicriterion problem of the scheduling theory taking the variation of servicing time into account.
TUZI 1984-2 Tuzikov, A. V.
A two-criterion scheduling problem allowing for variation in job execution time (Russian).
TUZI 1984-3 Tuzikov, A. V.
On a bicriterion problem of scheduling theory with regard to the change of service length (Russian).
TUZI 1984-4 Tuzikov, A. V.
On bicriterion scheduling problem with variable executing time of jobs (Russian).
UVAR 1985-1 Uvarov, S. I.
Estimate of the quality of a method for all construction of a schedule for carrying out a set of operations in a sequence specified by a graph (Russian).

UVAR 1985-2 Uvarov, S. I.
Investigation of a method for constructing a schedule of jobs with graph-specified precedence (Russian).
VACH 1983-1 Vachajitpan, P.
A computer program for solving job sequencing problems.
VANI 1987-1 Vani, V.; Raghavachari, M.
Deterministic and random single machine sequencing with variance minimization.
VARA 1985-1 Varakin, A. S.; Danilchenko, A. M.; Panishev, A. V.
Optimization of schedules in a multiprocessor system.
VAZI 1985-1 Vazirani, U. V.; Vazirani, V. V.
The 2-processor scheduling problem is in R-NC.
VELD 1988-1 Velde, S. L. van de
Minimizing makespan on unrelated parallel machines.
VEPS 1987-1 Vepsalainen, A. P. J.; Morton, T. E.
Priority rules for job shops with weighted tardiness costs.
VICK 1985-1 Vickery, S. K.; Markland, R. E.
Integer goal programming for multistage lot sizing: experimentation and implementation.
VIZI 1985-1 Vizing, V. G.
Schedules ensuring the most uniform machine loading.
VOLY 1982-1 Volynskij, E. I.; Nilova, L. I.; Sejts, A. I.; Filonenko, V. L.
Application of random search to solve a problem of time scheduling (Russian).
WAJS 1984-1 Wajs, W.
Scheduling problem for metallurgical processes (Polish).
WAJS 1986-1 Wajs, W.
Polynomial algorithm for dynamic sequencing problem (Polish).
WATE 1987-1 Waters, C. D. J.
A solution procedure for the vehicle-scheduling problem based on iterative route improvement.
WEBEA 1986-1 Weber, R. R.; Varaiya, P.; Walrand, J.
Scheduling jobs with stochastically ordered processing times on parallel machines to minimize expected flowtime.
WEGL 1985-1 Weglarz, J.
Scheduling independent activities to meet deadlines with continuous constrained resources.
WEIS 1987-1 Weiss, E. N.
An optimization based heuristic for scheduling parallel project networks with constrained renewable resources.
WEISA 1984-1 Weiss, G.
Scheduling spares with exponential lifetimes in a two-component parallel system.
WERN 1986-2 Werner, F.
On the structure of special 2-machine problems.
WERN 1987-1 Werner, F.
On some neighborhood graphs for the development of suitable iteration methods to approximately solve a special permutation problem (German).
WERN 1987-2 Werner, F.
On the quasi-convexity of a special job shop scheduling problem.
WERR 1984-1 Werra, D. de
A decomposition property of polyhedra.
WERR 1986-2 Werra, D. de
Variations on the integral decomposition property.
WETZ 1985-1 Wetzel, R.
Ein Branch and Bound Algorithmus zur Lösung eines Problems der Fahrplangestaltung.
WETZ 1986-1 Wetzel, R.
A branch and bound concept for solving a vehicle schedule designing problem.
WHITA 1985-1 White, C.
Production scheduling with applications in the printing industry.
WIE 1986-1 Wie, S.-H.; Pinedo, M.
On minimizing the expected makespan and flow time in stochastic flow shops with blocking.
WIJN 1988-1 Wijngaard, P.
A heuristic for scheduling problems especially for scheduling farm operations.
WILHC 1985-1 Wilhelm, W. E.; Sarin, S.
A structure for sequencing robot activities in machine tending applications.
WILLD 1985-2 Willis, R. J.
Critical path analysis and resource constrained project scheduling - theory and practice.
WILS 1984-1 Wilson, H. G.; Ezzat, M. O.
Scheduling of material handling vehicles in a jobshop environment.
WITT 1985-1 Wittrock, R. J.
Scheduling algorithms for flexible flow lines.
WOER 1988-1 Woerlee, A. P.
Scheduling techniques with limited intermediate storage.
WOOL 1986-1 Woollam, C. R.
Flowshop with no idle machine time allowed.
XU 1984-1 Xu, B.-S.
A family of sequencing problems with earliness and tardiness penalties.
YAKO 1981-1 Yakovlev, S. V.
Solution of scheduling theory problems by directed random search (Russian).
YAMA 1985-1 Yamamoto, M.; Nof, S. Y.
Scheduling/rescheduling in the manufacturing operating system environment.
YAMAA 1984-1 Yamashiro, M.; Kawanago, T.
A study on the single machine production-line system with multiple arrival sources. I: On the analysis of the system with two different types of arrival sources (Japanese).
YANE 1985-1 Yanev, N. I.
Algorithms for three polynomially solvable cases of the lot-size problem.
YANG 1987-1 Yang, C. E.
A class of scheduling problems and their solution (Chinese).

YANGD 1987-1 Yang, T.-Y.; Ignizio, J. P.
An algorithm for the scheduling of army battalion training exercises.
YANO 1985-1 Yano, C. A.; Carlson, R. C.
An analysis of scheduling policies in multiechelon production systems.
YAO 1987-1 Yao, D. D.; Kim, S. C.
Reducing the congestion in a class of job shops.
YE 1983-1 Ye, M.-D.
On the passing through a rod problem and multi-machine scheduling. II (Chinese).
YUE 1985-1 Yue, M.-Y.; Han, J.-Y.; Chen, Y.-M.
Elimination conditions and lower bounds for the permutation flow-shop sequencing problem.
ZAITA 1986-1 Zaitsev, I. D.; Vainer, V. G.
An optimal scheduling algorithm taking account of the injury factor (Russian).
ZALO 1987-1 Zaloom, V. A.; Chu, H.-W.; Li, D.
An algorithm for sequencing jobs to minimize total tardiness.
ZDRZ 1985-1 Zdrzalka, S.
Single machine scheduling on an assembly line (Polish).
ZHANO 1982-1 Zhang, S. K.
Generalization of sequencing problems of the optimum service for model P-J.
ZHU E 1985-1 Zhu, F.
Notes on the elimination method in the m x n flow-shop scheduling problem (Chinese).
ZIND 1986-1 Zinder, Y. A.
An efficient algorithm for a deterministic problem of the scheduling theory with parallel devices (Russian).
ZINDA 1985-1 Zinder, Y. A.; Shkurba, V. V.
Effective iterative algorithms of scheduling theory (Russian).
ZOLL 1987-1 Zoller, K.; Robrade, A.
Dynamische Bestellmengen- und Losgrössenplanung. Verfahrensübersicht und Vergleich.

ACHAC 1987-1 Acharya, B. D.; Sampathkumar, E.
Graphoidal covers and graphoidal covering number of a graph.

AGGAB 1985-1 Aggarwal, V.; Suresh, R.
A shortest-route approach for optimum aggregation of data with an application to income data.

AKMA 1987-1 Akman, V.
Unobstructed shortest paths in polyhedral environments.

ANCO 1986-1 Ancona, M.; Floriani, L. de; Deogun, J. S.
Path problems in structured graphs.

ANDE 1985-1 Anderson, R.
A parallel algorithm for the maximal path problem.

ANDE 1987-1 Anderson, R.
A parallel algorithm for the maximal path problem.

ANDEA 1987-1 Anderson, R.; Mayr, E.
Parallelism and the maximal path problem.

ANDRC 1987-1 Andreatta, G.
Shortest path models in stochastic networks.

ANDRG 1987-1 Androsov, S. M.; Gluzman, B. Y.; Oslon, V. A.
Program realization of shortest path algorithms in transportation MIS.

ANIS 1986-1 Anisimov, A. V.
Local algorithm for the one-to-all shortest-path problem (Russian).

ASANA 1985-1 Asano, T.; Asano, T.; Guibas, L.; Hershberger, J.; Imai, H.
Visibility-polygon search and euclidean shortest paths.

ASAND 1985-1 Asano, T.; Sato, S.
Long path enumeration algorithms for timing verification on large digital systems.

BAKO 1980-1 Bako, A.
Network realization with flow method.

BERG 1983-1 Berge, C.
Graphes (3. edition).

BERG 1985-1 Berge, C.
Graphs (second revised edition).

BERMB 1987-1 Berman, K. A.; Paul, J. L.
Long monochromatic paths in vertex 3-colored planar graphs.

BERNC 1984-1 Bernstein, R.
Shortes paths in undirected graphs with negative edges.

BERTB 1986-1 Bertossi, A. A.
Total domination in interval graphs.

BIXB 1987-1 Bixby, R. E.
Notes on combinatorial optimization.

BOVE 1986-1 Bovet, J.
Une amelioration de la methode de Dijkstra pour la recherche d'un plus court chemin dans un reseau.

BRON 1984-1 Bronshtein, M. L.
An algorithm for finding the shortest path for a network with alternate arc lengths (Russian).

BYER 1984-1 Byers, T. H.; Waterman, M. S.
Determining all optimal and near-optimal solutions when solving shortest path problems by dynamic programming.

CAPO 1984-1 Caponi, M.; Pieroni, P.; Saviozzi, G.
A Benders dual decomposition approach to the optimum network orientation.

CHAKC 1985-1 Chakravarty, A. K.
Multiproduct purchase scheduling with limited budget and/or group discounts.

CHANC 1985-1 Chanas, S.; Kamburowski, J.
The fuzzy shortest route problem.

CHANV 1985-1 Chang, H. C.
A minimax-arc-path problem.

CHERA 1986-1 Cherkassky, B.
An algorithm for finding shortest paths in a maze.

COLB 1986-2 Colbourn, C. J.
The reliability polynomial.

CORLA 1985-1 Corley, H. W.; Moon, I. D.
Shortest paths in networks with vector weights.

CURRA 1986-1 Current, J. R.; Min, H.-K.
Multiobjective design of transportation networks: Taxonomy and annotation.

CURRB 1985-1 Current, J. R.; Revelle, C. S.; Cohon, J. L.
The maximum covering/shortest path problem: a multiobjective network design and routing formulation.

CURRB 1987-1 Current, J. R.; Revelle, C. S.; Cohon, J. L.
The median shortest path problem: A multiobjective approach to analyse cost vs. accessibility in the design of transportation networks.

DEPA 1987-1 DePalma, A.; Hansen, P.; Labbe, M.
Commuters' paths with penalties for early or late arrival time.

DERI 1985-1 Derigs, U.
An efficient Dijkstra-like labeling method for computing shortest odd/even paths.

DESR 1986-1 Desrochers, M.
An algorithm for the shortest path problem with resource constraints.

DESR 1987-1 Desrochers, M.
A note on the partitioning shortest path algorithm.

DIVO 1987-1 Divoky, J. J.
Improvements for the Thresh X2 shortest path algorithm.

DONG 1984-1 Dong, C. F.; Liu, W. G.
The shortest path problem on a kind of variable weighted networks (Chinese).

DORI 1985-1 Döring, S.
Optimization of node-disjoint paths in networks.

DORIA 1985-1 Döring, S.; Iwainsky, A.
Graph reduction for the optimization of node-disjoint paths in networks.

EASA 1985-1 Easa, S. M.
Shortest route algorithm with movement prohibitions.
EIGE 1985-1 Eiger, A.; Mirchandani, P. B.; Soroush, H.
Path preferences and optimal paths in probabilistic networks.
ELST 1984-1 Elster, K.-H.; Harant, J.; Hutschenreuther, H.
Ein Partitionierungsalgorithmus zur Distanzberechnung auf Verkehrs- netzen.
ENOMA 1984-1 Enomoto, H.; Saito, A.
Disjoint shortest paths in graphs.
ENTRB 1982-1 Entringer, R. C.; Mac Kendrick, S.
Longest paths in locally Hamiltonian graphs.
EVEN 1985-1 Even, S.; Gazit, H.
Updating distances in dynamic graphs.
FAUDA 1987-1 Faudree, R. J.; Ordman, E. T.; Schelp, R. H.; Jacobson, M. S.; Tuza, Z.
Menger's theorem and short paths.
FELD 1985-1 Feldman, P.
Fault tolerance of minimal path routings in a network.
FILI 1984-1 Filipiak, J.
A dynamic flow assignment problem for a telephone trunk network.
FOUL 1985-2 Foulds, L. R.
One-way street system design.
FRED 1985-1 Frederickson, G. N.
A single source shortest path algorithm for a planar distributed network.
FRED 1987-1 Frederickson, G. N.
A new approach to all pairs shortest paths in planar graphs.
FRED 1987-2 Frederickson, G. N.
Fast algorithms for shortest paths in planar graphs, with applications.
FREDD 1984-1 Fredman, M. L.; Tarjan, R. E.
Fibonacci heaps and their uses in improved network optimization algorithms.
FRIEE 1985-1 Frieze, A. M.; Grimmett, G. R.
The shortest-path problem for graphs with random arc-lengths.
FUCH 1985-1 Fuchs, A.
1-N Verfahren zur Bestimmung von k-kürzesten Wegen.
GAFN 1987-1 Gafni, E. M.; Bertsekas, D. P.
Asymptotic optimality of shortest path routing algorithms.
GALL 1984-1 Gallo, G.; Pallottino, S.
Shortest path methods in transportation models.
GALL 1986-1 Gallo, G.; Pallottino, S.
Shortest path methods: A unifying approach.
GARA 1987-1 Garashchuk, M. M.
On the testing of paths in schemes of functional elements (Russian).
GHOS 1986-1 Ghosh, R. K.; Bhattacharjee, G. P.
Parallel algorithm for shortest paths.
GLOVD 1985-1 Glover, F.; Klingman, D.; Phillips, N.
A new polynomially bounded shortest path algorithm.
GLOVF 1985-1 Glover, F.; Klingman, D.; Phillips, N. V.; Schneider, R. F.
New polynomial shortest path algorithms and their computational attributes.
GOLDO 1986-1 Goldman, A. J.; Tiwari, P.
Allowable processing orders in the accelerated cascade algorithm.
GUIB 1987-1 Guibas, L.; Hershberger, J.; Leven, D.; Sharir, M.; Tarjan, R. E.
Linear-time algorithms for visibility and shortest path problems inside triangulated simple polygons.
GUSH 1987-1 Gushchinskii, N. N.; Levin, G. M.
Two-level minimization of a composite function and its application to a problem of the optimization of a path in a graph (Russian).
HABE 1984-1 Habenicht, W.
Interaktive Lösungsverfahren für diskrete Vektoroptimierungsprobleme unter besonderer Berücksichtigung von Wegeproblemen in Graphen.
HACK 1984-1 Hackbarth, K.-D.
Implementation of algorithms for telecommunication network planning.
HALLE 1986-1 Hall, R. W.
The fastest path through a network with random time-dependent travel times.
HANSB 1984-1 Hansen, P.
Shortest paths in signed graphs.
HASSE 1985-1 Hassin, R.; Zemel, E.
On shortest paths in graphs with random weights.
HENI 1986-2 Henig, M. I.
The shortest path problem with two objective functions.
HIC 1985-1 Hic, P.
A construction of geodetic blocks.
HOUS 1984-1 Houska, R. V.
Using stochastic optimal control to find shortest paths in networks.
HSIE 1984-1 Hsieh, W. N.
Routing strategies in computer networks.
HU B 1987-1 Hu, Z.-Q.
On the maximal path in connected graphs (Chinese).
ICHI 1985-1 Ichimori, T.
A shortest path approach to a multifacility minimax location problem with rectilinear distances.
IVAN 1984-1 Ivanchev, D. T.
Minimization of the processing time for metal cutting machines considered as a shortest path problem in a network (Bulgarian).
JASM 1985-1 Jasmon, G. B.
Cutset analysis of networks using basic minimal paths and network decomposition.
JUNG 1986-1 Jung, H. A.
Longest paths joining given vertices in a graph.

JUNGC 1987-1 Jungnickel, D.
Graphen, Netzwerke und Algorithmen.

KAMB 1985-1 Kamburowski, J.
A note on the stochastic shortest route problem.

KAMB 1986-1 Kamburowski, J.
On the computational complexity of the shortest route and maximum flow problems in stochastic networks.

KAMB 1987-1 Kamburowski, J.
An overview of the computational complexity of the PERT shortest route and maximum flow problems in stochastic networks.

KARPD 1983-1 Karp, R. M.; Pearl, J.
Searching for an optimal path in a tree with random costs.

KLINB 1985-1 Klingman, D.; Armstrong, R. D.; Partow-Navid, P.
Implementation and analysis of alternative algorithms for generalized shortest path problems.

KLING 1986-1 Klingman, D.; Schneider, R. F.
Microcomputer-based algorithms for large scale shortest path problems.

KULKA 1984-1 Kulkarni, V. G.
A compact hash function for paths in PERT networks.

KUMAC 1985-1 Kumar, P.; Miller, F. G.
Shortest route network heuristic for the dynamic vehicle routing problem.

LAKH 1984-1 Lakhani, G. D.
An improved distribution algorithm for shortest paths problem.

LAPAA 1984-1 LaPaugh, A. S.; Papadimitriou, C. H.
The even-path problem for graphs and digraphs.

LAU 1986-1 Lau, H. T.
A note on the shortest path through a given link in a network.

LEDE 1985-1 Ledermann, W.; Vajda, S.
Handbook of applicable mathematics. Vol. 5 : Combinatorics and geometry. Part A.

LEE G 1984-1 Lee, D. T.; Preparata, F. P.
Euclidean shortest paths in the presence of rectilinear barriers.

LEE I 1986-1 Lee, I.-S.
On sensitivity analysis for shortest-path dynamic programming models.

LI A 1986-1 Li, C.-L.; Mac Cormick, S. T.; David, S.-L.
The complexity of finding two disjoint paths with min-max objective function.

LIPS 1984-1 Lipski, W. Jr.
An O(n log n) Manhattan path algorithm.

LUBYA 1985-1 Luby, M.; Ragde, P.
A bidirectional shortest-path algorithm with good average-case behavior.

LUCEA 1984-2 Lucertini, M.; Paletta, G.
Network design with non-simultaneous flows.

MAHR 1981-1 Mahr, B.
A bird's-eye view to path problems.

MANAA 1986-1 Manaster, A. B.; Remmel, J. B.; Schmerl, J. H.
Planarity and minimal path algorithms.

MARTM 1984-1 Martins, E. Q. V.
An algorithm for ranking paths that may contain cycles.

MAY 1985-1 May, M.
On longest paths in digraphs (German).

MAY 1985-2 May, M.
Über längste Wege in Digraphen.

MEHLD 1986-1 Mehlhorn, K.; Schmidt, B. H.
On BF-orderable sets.

MINO 1986-1 Minoux, M.
Mathematical Programming. Theory and Algorithms.

MINOC 1984-1 Minoux, M.; Ribeiro, C. C.
A transformation of hard (equality constrained) knapsack problems into constrained shortest path problems.

MIRCC 1985-1 Mirchandani, P. B.; Soroush, H.
Optimal paths in probabilistic networks: a case with temporary preferences.

MOFF 1985-1 Moffat, A.; Takaoka, T.
An all pairs shortest path algorithm with expected running time O(n**2.logn).

MOFF 1987-1 Moffat, A.; Takaoka, T.
An all pairs shortest path algorithm with expected time O(n**2 log n).

MOHA 1986-1 Mohanty, H.; Bhattacharjee, G. P.
A distributed algorithm for edge-disjoint path problem.

MONI 1985-1 Monien, B.
How to find long paths efficiently.

NGUYB 1986-1 Nguyen, S.; Pallottino, S.
Assignment of passengers to an urban transport system: Determination of minimal hyperpaths (Italian).

NOSH 1985-1 Noshita, K.
A theorem on the expected complexity of Dijkstra's shortest path algorithm.

NOVIA 1982-1 Novitskaya, I. I.; Safronenko, V. A.
Determination of the shortest way satisfying the given requirements on a sequential or-graph (Russian).

O'RO 1985-1 O'Rourke, J.; Suri, S.; Booth, H.
Shortest paths on polyhedral surfaces.

ORDI 1987-1 Ordin, V. V.
The method of Markov iterations for finding paths in a network of general form (Russian).

PAPA 1985-1 Papadimitriou, C. H.
An algorithm for shortest-path motion in three dimensions.

PAPA 1986-1 Papadimitriou, C. H.
Shortest-path motion.

PAUM 1986-1 Paumelle, I.
Un indicateur informatique d'itineraire pour les reseaux d'autobus.

PEDE 1987-1 Pederzoli, G.; Sancho, N. G. F.
A shortest path routing problem with resource allocation.

PELED 1987-1 Peleg, D.; Upfal, E.
Constructing disjoint paths on expander graphs.

PERE 1985-1 Perez Prados, A.
Obtaining extremal paths in an associated graph (Spanish).

PERK 1984-1 Perko, A.
Storing paths in kth shortest path problems.

PHIL 1986-1 Phillips, N. V.
A new variant of the partitioning shortest path algorithm.

POCHA 1986-1 Pochet, Y.; Wolsey, L. A.
Lot-size models with backlogging: strong reformulations and cutting planes.

PUTU 1984-1 Puturidze, Z. S.
Problem on the wiring of the electrical net in technical products (Russian).

QIN 1985-1 Qin, Y.
On generalized Floyd algorithm.

QIN 1985-2 Qin, Y.
On Jar-metric principle - a unified approach to solve optimum paths problems on multistage directed graphs.

QUINA 1984-1 Quinn, M. J.; Deo, N.
Parallel graph algorithms.

RANC 1983-1 Ranchich, M. V.
Determination of shortest paths of a graph (Russian).

REUT 1986-1 Reutenauer, C.
Free sets of paths in a graph (French).

RIBE 1983-1 Ribeiro, C. C.
Algorithmes de recherche de plus courts chemins avec contraintes: Etude theorique, implementation et parallelisation.

RIBEA 1985-1 Ribeiro, C. C.; Minoux, M.
A heuristic approach to hard constrained shortest path problems.

RIBEA 1986-1 Ribeiro, C. C.; Minoux, M.
Solving hard constrained shortest path problems by Lagrangean relaxation and branch-and-bound algorithms.

ROBEB 1986-1 Robert, Y.; Trystram, D.
An orthogonal systolic array for the algebraic shortest-path problem (French).

ROHN 1985-1 Rohnert, H.
A dynamization of the all pairs least cost path problem.

ROHN 1986-1 Rohnert, H.
Shortest paths in the plane with convex polygonal obstacles.

RONEA 1984-1 Ronen, D.; Perl, Y.
Heuristics for finding a maximum number of disjoint bounded paths.

ROOS 1984-1 Roos, C.
Network programming with umbrellas.

ROTE 1985-1 Rote, G.
A systolic array algorithm for the algebraic path problem (shortest paths; matrix inversion).

RUGE 1984-2 Rüger, W.
Ermittlung der maximalen Anzahl kantendisjunkter Wege in einem Netz.

SANCA 1985-1 Sancho, N. G. F.
Maximum reliability through a network with resource constraints.

SCHEA 1987-1 Scheithauer, G.
On the determination of a restricted shortest path (German).

SEDGA 1984-1 Sedgewick, R.; Vitter, J. S.
Shortest paths in Euclidean graphs.

SEDGA 1986-1 Sedgewick, R.; Vitter, J. S.
Shortest paths in Euclidean graphs.

SEDU 1985-1 Sedukhin, S. G.
Computational structures of algorithms for finding paths in a graph (Russian).

SHAR 1986-1 Sharir, M.; Schorr, A.
On shortest paths in polyhedral spaces.

SHIEB 1985-1 Shier, D. R.; Whited, D. E.
Algorithms for generating minimal cutsets by inversion.

SHINB 1985-1 Shinoda, S.; Tsukiyama, S.; Shirakawa, I.; Omura, K.
On a second shortest k-tuple of edge-disjoint paths.

SKILA 1985-1 Skilton, D. K.
Chain decompositions of graphs.

SOKOB 1983-1 Sokolova, T. E.
Determining the group of shortest paths in a graph (Russian).

SONGB 1986-1 Song, Z.-M.
Long paths and cycles in oriented graphs (Chinese).

SONGB 1986-2 Song, Z.-M.
Longest paths and cycles in bipartite oriented graphs (Chinese).

TAKA 1984-1 Takaoka, T.; Ko, K.
An 0((N**2)LOGN) expected time algorithm for the all pairs shortes path problem over a certain class of graphs.

TAO 1987-1 Tao, G.-Y.; Zhou, P.-D.; Wang, Q.
The heuristic shortest-path algorithm for transportation (Chinese).

TARAA 1985-1 Tarasov, O.; Richter, P.
Efficient algorithms for the determination of optimal paths in non-negatively weighted digraphs.

TARJ 1985-2 Tarjan, R. E.
Shortest path algorithms.

TARJ	1986-2	Tarjan, R. E. Corrigendum to: Sensitivity analysis of minimum spanning trees and shortest path trees.
TCHA	1985-1	Tcha, D.-W. On the selection of primary paths for a communication network.
TOPK	1984-1	Topka, V. V. An exact algorithm for the uniform distribution of resources in a network (Russian).
TRUE	1987-2	Truemper, K. Max-flow min-cut matroids: polynomial testing and polynomial algorithms for maximum flow and shortest routes.
TUNG	1986-2	Tung, P. C. T. Finding a path with minimal cost-to-time ratio in a network.
TUNGA	1987-1	Tung, P. C. T.; Chan, G. H.; Chew, K. L. Finding a minimal-ratio elementary path in a network.
VAINA	1985-1	Vainshtein, A. D. A vector problem of the shortest path in a uniform norm (Russian).
VIEG	1985-1	Viegas, J.; Hansen, P. Finding shortest paths in the plane in the presence of barriers to travel (for any l(p)-norm).
VINO	1985-1	Vinokur, A. B. One method for storage and retrieval of information on the shortest paths of graphs with distinguished characteristics (Russian).
WARB	1987-1	Warburton, A. Approximation of Pareto optima in multiple-objective, shortest path problems.
WIDMA	1985-1	Widmayer, P.; Wu, Y. F.; Schlag, M. D. F.; Wong, C. K. Rectilinear shortest paths and minimum spanning trees in the presence of rectilinear obstacles.
XU D	1987-1	Xu, M. L. A simple method for finding shortest paths (Chinese).
ZIMM	1987-1	Zimmermann, H.-J. Methoden und Modelle des Operations Research. Für Ingenieure, Ökonomen und Informatiker.

AHUJA 1987-1 Ahuja, R. K.; Murty, V. V. S.
Exact und heuristic algorithms for the optimum communication spanning tree problem.

AKL 1983-1 Akl, S. G.
A note on Euclidean matchings, triangulations and spanning trees.

AKL 1986-1 Akl, S. G.
An adaptive and cost-optimal parallel algorithm for minimum spanning trees.

ALTI 1985-1 Altinkemer, K.; Gavish, B.
Heuristics with constant error guarantees for the topological design of local access tree networks.

AUSI 1986-1 Ausiello, G.; D'Atri, A.; Moscarini, M.
Chordality properties on graphs and minimal conceptual connections in semantic data models.

AWER 1987-1 Awerbuch, B.
Optimal distributed algorithms for minimum weight spanning tree, counting, leader election, and related problems.

BABAD 1981-1 Babaev, T. A.
The problem of the shortest tree as a linear programming problem (Russian).

BALL 1987-1 Ball, M. O.; Liu, W. G.; Pulleyblank, W. R.
Two terminal Steiner tree polyhedra.

BAR-A 1987-1 Bar-Yehuda, R.; Kutten, S.; Wolfstahl, Y.; Zaks, S.
Making distributed spanning tree algorthms fault-resilient.

BARO 1985-1 Baron, G.; Prodinger, H.; Tichy, R. F.; Boesch, F. T.; Wang, J. F.
The number of spanning trees in the square of a cycle.

BERG 1985-1 Berge, C.
Graphs (second revised edition).

BERMA 1986-1 Berman, K. A.
Bicycles and spanning trees.

BERN 1986-1 Bern, M. W.
Two probabilistic results on rectilinear Steiner trees.

BIXB 1987-1 Bixby, R. E.
Notes on combinatorial optimization.

BOKH 1987-1 Bokhari, S. H.
Assigment Problems in Parallel and Distributed Computing.

BOLLA 1983-1 Bollobas, B.; Chung, F. R. K.; Graham, R. L.
On complete bipartite subgraphs contained in spanning tree complements.

BRANA 1987-1 Brandstädt, A.
The computational complexity of feedback vertex set, Hamiltonian circuit, dominating set, Steiner tree, and bandwidth on special perfect graphs.

BREZ 1987-1 Brezovec, C.; Cornuejols, G.; Glover, F.
A matroid algorithm and its application to the efficient solution of two optimization problems on graphs.

CAI A 1984-1 Cai, M.-C.
A solution of Chartrand's problem on spanning trees.

CAI B 1985-1 Cai, W.-Q.
The nth minimal spanning trees with degree constraints having two nodes (Chinese).

CAMEA 1985-1 Camerini, P. M.; Galbiati, G.; Maffioli, F.
The complexity of weighted multi-constrained spanning tree problems.

CAMED 1986-1 Camerini, P. M.; Maffioli, F.; Martello, S.; Toth, P.
Most and least uniform spanning trees.

CAMP 1986-1 Campbell, B. A.; Rardin, R. L.
Polynomial time solution of Steiner tree problems on special planar graphs.

CAO 1985-1 Cao, H.-G.
Storage and search of data with minimal spanning trees (Chinese).

CHANAA 1983-1 Chang, R. C.; Lee, R. C. T.
An O(|V|) parallel algorithm to construct minimal spanning trees based upon Voronoi diagrams.

CHANAA 1986-1 Chang, R. C.; Lee, R. C. T.
An O(N log N) minimal spanning tree algorithm for N points in the plane.

CHENB 1986-1 Chen, J. H.
Spanning trees with the maximum number of end vertices (Chinese).

CHENE 1983-1 Chen, Q.-H.
Finding r-dimensional minimum spanning trees with a fixed incident number at an assigned simplex D(S) (Chinese).

CHER 1987-1 Cherkasskii, B. V.
New algorithms of spanning trees generation (Russian).

CHES 1984-1 Cheston, G. A.
Generating a spanning tree with a specified node as a centroid point for biconnected graphs.

CHIN 1985-1 Chin, F.; Ting, H. F.
An almost linear time and O(nlogn + e) messages distributed algorithm for minisum-weight spanning trees.

CHRO 1986-1 Chrobak, M.; Poljak, S.
On common edges in optimal solutions to traveling salesman and other optimization problems.

CHUND 1985-1 Chung, F. R. K.; Graham, R. L.
A new bound for Euclidean Steiner minimal trees.

CLAR 1985-1 Clark, L.; Entringer, R. C.
Extremal spanning trees of cubic graphs.

COCKA 1986-1 Cockayne, E. J.; Hewgill, D. E.
Exact computation of Steiner minimal trees in the plane.

COCKB 1983-1 Cockayne, E. J.; Peters, D. B.
MFST82: a program for minimum full Steiner trees in the plane.

COLB 1986-2 Colbourn, C. J.
The reliability polynomial.

COLBC 1987-1 Colbourn, C. J.; Pulleyblank, W. R.
Matroid Steiner problems, the Tutte polynomial and network reliability.

COLBF 1985-1 Colbourn, C. J.; Stewart, L. K.
Permutation graphs: connected domination and Steiner trees.

COPP 1986-1 Coppersmith, D.; Klawe, M.; Pippenger, N.
Alphabetic minimax trees of degree at most t.

CORL 1985-1 Corley, H. W.
Efficient spanning trees.

DINGA 1984-1 Ding, J. Y.; Song, G. D.; Pan, Z. T.
The Steiner problem for triangle-trapezoid zigzag lines (Chinese).

DRESB 1987-1 Dress, A.; Krüger, M.
Parsimonious phylogenetic trees in metric spaces and simulated annealing.

DU A 1987-1 Du, D. Z.; Hwang, F. K.
Steiner minimal trees for bar waves.

DU B 1985-1 Du, D. Z.; Hwang, F. K.; Chao, S. C.
Steiner minimal tree for points on a circle.

DU C 1987-1 Du, D. Z.; Hwang, F. K.; Song, G. D.; Ting, G. Y.
Steiner minimal trees on sets of four points.

DU D 1987-1 Du, D. Z.; Hwang, F. K.; Weng, J. F.
Steiner minimal trees for regular polygons.

DUNN 1987-1 Dunne, P. E.
A result on k-valent graphs and its application to a graph embedding problem.

EL-M 1984-1 El-Mallah, E. S.; Colbourn, C. J.
Optimum communication spanning trees in chordal graphs.

EL-M 1985-1 El-Mallah, E. S.; Colbourn, C. J.
Optimum communication spanning trees in series-parallel networks.

ERNV 1984-1 Ernvall, J.
On the construction of spanning paths by Gray-code in compression of files.

ERZI 1983-1 Erzin, A. I.; Mordvinova, T. B.
A problem of the construction of an optimal tree (Russian).

FARRB 1986-1 Farrell, E. J.; Grell, J. C.
An application of circuit polynomials to the counting of spanning trees in graphs.

FRED 1985-2 Frederickson, G. N.
Data structures for on-line updating of minimum spanning trees, with applications.

FREDD 1984-1 Fredman, M. L.; Tarjan, R. E.
Fibonacci heaps and their uses in improved network optimization algorithms.

GABOB 1984-1 Gabow, H. N.; Galil, Z.; Spencer, T.
Efficient implementation of graph algorithms using contraction.

GABOC 1986-1 Gabow, H. N.; Galil, Z.; Spencer, T.; Tarjan, R. E.
Efficient algorithms for finding minimum spanning trees in undirected and directed graphs.

GAVR 1987-1 Gavril, F.
Generating the maximum spanning trees of a weighted graph.

GEOR 1987-1 Georgakopoulos, G.; Papadimitriou, C. H.
The 1-Steiner tree problem.

GIBB 1985-1 Gibbons, A.
Algorithmic graph theory.

GOMM 1984-1 Gommers, J. H. M.; Volgenant, A.
A Langrangean appraoch to the degree-constrained minimum spanning tree problem.

GRAHA 1984-1 Graham, R. L.; Foulds, L. R.
It is unlikely that minimal phylogenies for a realistic biological study can be constructed in reasonable computational time.

GRAHB 1985-1 Graham, R. L.; Hell, P.
On the history of the minimum spanning tree problem.

HARAI 1987-1 Harary, F.; Schuster, S.; Vestergaard, P. D.
Interpolation theorems for the invariants of spanning trees of a given graph: edge-covering.

HAVE 1984-1 Havel, I.
On Hamiltonian circuits and spanning trees of hypercubes.

HAYM 1984-1 Haymond, R. E.; Jarvis, J. P.; Shier, D. R.
Algorithm 613. Minimum spanning tree for moderate integer weights.

HEDEE 1985-1 Hedetniemi, S. T.; Laskar, R.; Pfaff, J.
Irredundance in graphs: a survey.

HWANB 1986-1 Hwang, F. K.
A linear time algorithm for full Steiner trees.

HWANC 1986-1 Hwang, F. K.; Weng, J. F.
Hexagonal coordinate systems and Steiner minimal trees.

IWAI 1985-3 Iwainsky, A.
Some notes on the Steiner tree problem in graphs.

JAIN 1987-1 Jain, A.
Probabilistic analysis of LP relaxation bounds for the Steiner problem in networks.

JAYA 1984-1 Jayakumar, R.; Thulasiraman, K.; Swamy, M. N. S.
Complexity of computation of a spanning tree enumeration algorithm.

JIANA 1986-1 Jiang, J.-W.; Tang, P.-S.
An algorithm for rectilinear Steiner trees (Chinese).

KANO 1987-1 Kano, M.
Maximum and k-th maximal spanning trees of a weighted graph.

KAWAA 1981-1 Kawamoto, T.; Kajitani, Y.; Shinoda, S.
Eccentricity of tree set.

KOML 1984-1 Komlos, J.
Linear verification for spanning trees.

KOML 1985-1 Komlos, J.
Linear verification for spanning trees.

KORAA 1987-1 Korach, E.; Moran, S.; Zaks, S.
The optimality of distributive constructions of minimum weight and degree restricted spanning trees in a complete network of processors.

KOTIA 1985-1 Kotin, L.; Stark, W. R.
A parallel minimum spanning tree algorithm in LISP.

KOU A 1987-1 Kou, L. T.; Makki, K.
An even faster approximation algorithm for the Steiner tree problem in graphs.

LAVA 1984-1 Lavallee, I.
An efficient parallel algorithm for computing a minimum spanning tree.

LAVA 1985-1 Lavallee, I.
Un algorithme parallele efficace pour construire un arbre de poids minimal dans un graph.

LAVAA 1986-1 Lavallee, I.; Roucairol, C.
A fully distributed (minimal) spanning tree algorithm.

LECL 1987-1 Leclerc, B.
Arbres minimums communs et compatibilite de donnees de types varies.

LENG 1986-1 Lengauer, T.
Efficient algorithms for finding minimum spanning forests of hierarchically defined graphs.

LENG 1987-1 Lengauer, T.
Efficient algorithms for finding minimum spanning forests of hierarchically defined graphs.

LEWI 1986-1 Lewinter, M.
Spanning trees of product graphs.

LEWI 1987-1 Lewinter, M.
Interpolation theorem for the number of degree-preserving vertices of spanning trees.

LI I 1985-1 Li, Y.-S.; Li, S.-H.; Liu, J.-J.
A new algorithm for determining the minimum spanning tree.

LIN D 1985-1 Lin, Y. X.
A simpler proof of interpolation theorem for spanning trees.

LIN D 1985-2 Lin, Y. X.
All minimum-weight spanning trees in a graph (Chinese).

LIU C 1984-1 Liu, G.-Z.
A lower bound on solutions of Chartrand's problem.

MA C 1985-2 Ma, S.-H.
The Steiner tree problem on graphs and a heuristic algorithm for it (Chinese).

MACU 1987-1 Maculan, N. F.
The Steiner problem in graphs.

MAI D 1984-1 Mai, S.-W.; Evans, D. J.
A parallel algorithm for the enumeration of the spanning trees of a graph.

MALY 1985-1 Malyshko, V. V.
An effective approach to some practical capacitated tree problems.

MARCJ 1984-3 Marcu, D.
Note on the spanning trees and cotrees of a connected graph.

MARCJ 1985-4 Marcu, D.
Note on the spanning trees of a connected graph.

MELN 1984-1 Melnikov, L. S.; Ponomarev, V. V.
Solution of a problem of the Turan type (Russian).

MILI 1983-1 Militaru, V.
A procedure of gradual generating trees in a linear graph.

MULLA 1987-1 Müller, H.; Brandstädt, A.
The NP-completeness of STEINER TREE and DOMINATING SET for chordal bipartite graphs.

NIKO 1982-1 Nikogosyan, Z. G.
N-spanning and hypo-n-spanning graphs (Russian).

PAPAB 1984-1 Papadimitriou, C. H.; Vazirani, U. V.
On two geometric problems related to the traveling salesman problem.

PARD 1987-1 Pardalos, P. M.; Rosen, J. B.
Constrained global optimization: Algorithms and applications.

PAWAA 1986-1 Pawagi, S.; Ramakrishnan, I. V.
An O(log n) algorithm for parallel update of minimum spanning trees.

PESA 1987-1 Pesamosca, G.
On the solution of the generalized Steiner problem by the subgradient method.

POLJ 1986-1 Poljak, S.; Chrobak, M.
On common edges in optimal solutions to traveling salesman and other optimization problems.

PROC 1985-1 Procesi Ciampi, R.
Spanning trees and acyclic orientations.

PRODB 1985-1 Prodon, A.; Liebling, T. M.; Gröflin, H.
Steiner's problem on two-trees.

PROV 1987-1 Provan, J. S.
An approximation scheme for finding Steiner trees with obstacles.

PUEL 1985-1 Puel, L.
Unavoidable sets of trees and an application to generalizing Kruskal's theorem (French).

QUINA 1984-1 Quinn, M. J.; Deo, N.
Parallel graph algorithms.

RICHB 1985-1 Richey, M. B.; Parker, R. G.
On multiple Steiner subgraph problems.

RODR 1984-1 Rodriguez, R.; Kakes, A.; Casas, O.
Suboptimal partial trees.

ROSK 1985-1 Roskind, J.; Tarjan, R. E.
A note on finding minimum-cost edge-disjoint spanning trees.

RUDO 1984-1 Rudowski, M. H.
Hypothesis on the length of minimal Steiner tree in multidimensional spaces with a rectilinear metric.

SANK 1985-1 Sankaran, V.; Krishnamoorthy, V.
On the diameters of spanning trees.

SAVA 1984-1 Savage, C.
A systolic design for connectivity problems.

SAVEA 1985-1 Savelsbergh, M. W. P.; Volgenant, T.
Edge exchanges in the degree-constrained minimum spanning tree problem.

SCHEA 1985-1 Scheithauer, G.
On recursive computation of minimum spanning trees for special partial and subgraphs.

SCHEA 1985-2 Scheithauer, G.
On recursive computation of minimum spanning trees for special partial graphs.
SCHEA 1985-3 Scheithauer, G.
Rekursive Berechnung von Minimalgerüsten für spezielle Subgraphen.
SCHI 1985-1 Schiemangk, C.
Convergence of the thermodynamically motivated simulation and experiments for solving the Steiner-Tree-Problem.
SEGE 1985-1 Segev, A.
The node-weighted Steiner tree problem.
SHAKB 1987-1 Shakhbazyan, K. V.; Tushkina, T. A.; Lebedinskii, M. M.
A spanning tree algorithm for the construction of flows in unconditional pipeline computers (Russian).
SMITC 1985-1 Smith, J. M.
Generalized Steiner network problems in engineering design.
SOLOB 1986-1 Solomon, M. M.
The minimum spanning tree problem with time window constraints.
SONG 1986-1 Song, G. D.; Ding, J. Y.
A note on the four-point complete Steiner tree theorem (Chinese).
STEE 1987-1 Steele, J. M.
On Frieze's zeta(3) limit for lengths of minimal spanning trees.
TARJ 1986-2 Tarjan, R. E.
Corrigendum to: Sensitivity analysis of minimum spanning trees and shortest path trees.
TRIEA 1984-1 Trietsch, D.
An improved algorithm for Steiner trees.
TRIEA 1985-1 Trietsch, D.
Augmenting Euclidean networks - the Steiner case.
TRIEA 1985-2 Trietsch, D.
Connecting Euclidean networks–the Steiner case.
TRUB 1985-1 Trubin, V. A.
On one subclass of the Steiner problems on the plane with rectangular metric (Russian).
VAID 1984-1 Vaidya, P. M.
A fast approximation algorithm for minimum spanning trees in k-dimensional space.
VOHRC 1984-1 Vohra, R. V.; Washington, L.
Counting spanning trees in the graphs of Kleitman and Golden and a generalization.
VOLG 1986-1 Volgenant, A.; Duin, C. W.
Reduction tests for the Steiner problem in graphs.
VOSSA 1987-1 Voss, S.; Mehr, K. D.
On a generalized Steiner problem with 2-edge connectivity.
WALD 1982-1 Wald, J. A.; Colbourn, C. J.
Steiner trees in outerplanar graphs.
WALDA 1984-1 Wald, J. A.; Sorenson, P. G.
Resolving the query inference problem using Steiner trees.
WALTA 1987-1 Walther, H.; Nägler, G.
Graphen, Algorithmen, Programme.
WANGD 1984-1 Wang, J. F.; Yang, C. S.
On the number of spanning trees of circulant graphs.
WANGJ 1985-1 Wang, S.-M.
A multiple source algorithm for suboptimum Steiner trees in graphs.
WENG 1985-1 Weng, J. F.
Generalized Steiner problem and hexagonal coordinate system (Chinese).
WENG 1985-2 Weng, J. F.
Steiner minimal trees on vertices of regular polygons (Chinese).
WHITB 1984-1 White, K.; Faber, M.; Pulleyblank, W. R.
Steiner trees, connected domination and strongly chordal graphs.
WHITJ 1985-1 Whitty, R. W.
Spanning trees in program flowgraphs.
WIDM 1987-1 Widmayer, P.
On approximation algorithms for Steiner's problem in graphs.
WIDMA 1985-1 Widmayer, P.; Wu, Y. F.; Schlag, M. D. F.; Wong, C. K.
Rectilinear shortest paths and minimum spanning trees in the presence of rectilinear obstacles.
WINT 1985-1 Winter, P.
Generalized Steiner problems in outerplanar networks.
WINT 1987-1 Winter, P.
Steiner problem in Halin networks.
WOJC 1984-1 Wojciechowski, J.
Generation of trees of a graph with the use of decomposition.
WU D 1986-1 Wu, Y. F.; Widmayer, P.; Wong, C. K.
A faster approximation algorithm for the Steiner problem in graphs.
XUE B 1987-1 Xue, Y.-F.; Li, Y.
A minimum spanning tree algorithm for solving the transportation problem (Chinese).
ZAMA 1986-1 Zamanskij, L. Y.
Properties of the number of spanning trees and of partial connected graphs (Russian).
ZENG 1983-1 Zeng, F. K.
Finding the tree-set of linear graphs by the DIDORTG-method (Chinese).
ZHANK 1986-1 Zhang, F. J.; Zheng, M. L.
Some notes on an algorithm for minimal spanning trees of undirected graphs, and their uniqueness (Chinese).

AART 1987-1 Aarts, E. H. L.; Laarhoven, P. J. M. van
A pedestrian review of the theory and application of the simulated annealing algorithm.
ABDU 1983-1 Abdullaev, A. A.; Rzaev, T. G.
Optimization of operative planning and scheduling in integrated ACS for enterprises (Russian).
AGRAB 1984-1 Agrawala, A. K.; Coffman, E. G. Jr.; Garey, M. R.; Tripathi, S. K.
A stochastic optimization algorithm minimizing expected flow times on uniform processors.
AHMA 1986-1 Ahmad, S. H.
GERT models in an educational institution.
AHME 1982-1 Ahmed, M.
Modeles probabilistes pour choisir un itineraire dans un reseau de trafic.
AKEL 1986-1 Akella, R.; Kumar, P. R.
Optimal scheduling of a flexible manufacturing system: a stochastic control problem for a system with jump Markov disturbances.
ANDRC 1987-1 Andreatta, G.
Shortest path models in stochastic networks.
APOL 1984-1 Apolloni, B.; Pezzella, F.
Confidence intervals in the solution of stochastic integer linear programming problems.
BABAB 1987-1 Babaev, D. A.; Nasibov, E. N.
Dialog approach to the solution of fuzzy integer linear programming problems (Russian).
BERE 1985-1 Berezinski, M.; Krus, L.
Aids for multiobjective decision making on development of transportation network under uncertainty.
BERG 1984-5 Berge, C.
Stochastic graphs and strongly perfect graphs: a survey.
BERMC 1987-1 Berman, O.; Larson, R. C.; Parkan, C.
The stochastic queue p-median problem.
BERMD 1984-1 Berman, O.; LeBlanc, B.
Location-relocation of mobile facilities on a stochastic network.
BERMF 1985-1 Berman, O.; Rahnama, M. R.
Optimal location-relocation decisions on stochastic networks.
BERN 1986-1 Bern, M. W.
Two probabilistic results on rectilinear Steiner trees.
BIENC 1985-2 Bienstock, D.
Efficient reliability computation of some highly structured networks.
BIENC 1986-1 Bienstock, D.
Some lattice-theoretic tools for network reliability analysis.
BLANC 1982-1 Blankenship, G. L.; Menaldi, J. L.
Optimal stochastic scheduling of power generation systems with scheduling delays and large cost differentials.
BOENA 1987-1 Boender, C. G. E.; Rinnooy Kan, A. H. G.; Vercellis, C.
Stochastic optimization methods.
BOLL 1985-2 Bollobas, B.
Random graphs.
BOLL 1986-1 Bollobas, B.
Extremal graph theory with emphasis on probabilistic methods.
BOLLB 1985-1 Bollobas, B.; Fenner, T. I.; Frieze, A. M.
An algorithm for finding Hamilton cycles in random graphs.
BOLLC 1983-1 Bollobas, B.; Frieze, A. M.
On matchings and Hamiltonian cycles in random graphs.
BOXMA 1986-1 Boxma, O. J.; Forst, F. G.
Minimizing the expected weighted number of tardy jobs in stochastic flow shops.
BREC 1986-1 Brecht, T. B.; Colbourn, C. J.
Improving bounds on network reliability: some examples.
CAME 1986-1 Camerini, P. M.; Colorni, A.; Maffioli, F.
Some experience in applying a stochastic method to location problems.
CARRB 1987-1 Carraway, R. L.; Morin, T. L.; Moskowitz, H.
The stochastic traveling salesman problem revisited: a generalized dynamic programming approach.
CAVAA 1985-1 Cavalier, T. M.; Sherali, H. D.
Sequential location-allocation problems on chains and trees with probabilistic link demands.
CHANB 1987-1 Chanas, S.
Fuzzy optimization in networks.
CHANC 1985-1 Chanas, S.; Kamburowski, J.
The fuzzy shortest route problem.
CHAND 1984-1 Chanas, S.; Kolodziejczyk, W.
Real-valued flows in a network with fuzzy arc capacities.
CHANE 1984-1 Chanas, S.; Kolodziejczyk, W.; Machaj, A.
A fuzzy approach to the transportation problem.
CHANE 1985-1 Chanas, S.; Kolodziejczyk, W.; Machaj, A.
The fuzzy transportation problem with fuzzy constraints.
CHENO 1986-4 Cheng, T. C. E.
Optimal due-date assignment for a single machine sequencing problem with random processing times.
CHERE 1985-1 Chernyak, V. V.; Shcherbina, I. N.
On a stochastic model of volume-scheduling planning (Russian).
CHIUB 1985-1 Chiu, S. S.; Larson, R. C.
Locating an n-server facility in a stochastic environment.
CHVA 1984-5 Chvatal, V.
Probabilistic methods in graph theory.
COFF 1987-1 Coffman, E. G. Jr.; Flatto, L.; Garey, M. R.; Weber, R. R.
Minimizing expected makespans on uniform processor systems.
COLB 1986-1 Colbourn, C. J.
Exact algorithms for network reliability.
DALA 1986-1 Dalang, R. C.; Trotter, L. E. Jr.; Werra, D. de
On randomized stopping points and perfect graphs.

DARZ	1987-1	Darzentas, J. On fuzzy location models.
DELG	1987-1	Delgado, M.; Verdegay, J. L.; Vila, M. A. Fuzzy transportation problems: a general analysis.
DHAR	1987-1	Dhar, D.; Barma, M.; Chakrabarti, B. K.; Taraphder, A. The travelling salesman problem on a randomly diluted lattice.
DODI	1985-1	Dodin, B. Bounding the project completion time distribution in PERT networks.
DODIB	1985-1	Dodin, B.; Sirvanci, M. The extreme value distribution and stochastic activity networks.
DRORD	1986-1	Dror, M.; Trudeau, P. Stochastic vehicle routing with modified savings algorithm.
DUBO	1983-1	Dubois, D. A fuzzy, heuristic, interactive approach to the optimal network problem.
DUPA	1984-1	Dupac, V.; Herkenrath, U. On integer stochastic approximation.
DURR	1984-1	Durrett, R. Oriented percolation in two dimensions.
DYERC	1986-1	Dyer, M. E.; Frieze, A. M.; Mac Diarmid, C. On linear programs with random costs.
EIGE	1985-1	Eiger, A.; Mirchandani, P. B.; Soroush, H. Path preferences and optimal paths in probabilistic networks.
FABIA	1984-1	Fabian, C.; Stoica, M. Fuzzy integer programming.
FAIGF	1986-4	Faigle, U.; Schrader, R. On the convergence of stationary distributions in simulated annealing algorithms.
FANE	1985-1	Fanelli, S.; Lombari, T. An application of the technique of Q-GERT networks to the evaluation of logistic parameters of electronic systems (Italian).
FEJG	1983-1	Fejgin, L. I. Scheduling problems with fuzzy activity durations.
FENN	1984-1	Fenner, T. I.; Frieze, A. M. Hamiltonian cycles in random regular graphs.
FINC	1987-1	Fincke, U.; Hamacher, H. W. On the expected value of stochastic linear programs and (dynamic) network flow problems.
FINKG	1984-1	Finke, U.; Hamacher, H. W. Some bounds for stochastic network flow problems.
FISHB	1985-1	Fisher, D. L.; Saisi, D.; Goldstein, W. M. Stochastic PERT networks: OP diagrams, critical paths and the project completion time.
FISHJ	1987-1	Fishman, G. S. Maximum flow and critical cutset as descriptors of multi-stage systems with randomly capacitated components.
FOLE	1984-1	Foley, R. D.; Suresh, S. Minimizimg the expected flowtime in stochastic flow shops.
FOLE	1984-2	Foley, R. D.; Suresh, S. Stochastically minimizing the makespan in flow shops.
FOLE	1986-1	Foley, R. D.; Suresh, S. Scheduling n overlapping jobs and two stochastic jobs in a flow shop.
FORS	1984-1	Forst, F. G. A review of the static, stochastic job sequencing literature.
FRANB	1986-1	Franco, J. On the probabilistic performance of algorithms for the satisfiability problem.
FRIEC	1984-2	Frieze, A. M. Limit distribution for the existence of Hamiltonian cycles in random bibartite graphs.
FRIEC	1986-1	Frieze, A. M. Maximum matchings in a class of random graphs.
FRIEC	1986-2	Frieze, A. M. On large matchings and cycles in sparse random graphs.
FRIEC	1987-1	Frieze, A. M. On the exact solution of random travelling salesman problems with medium size integer coefficients.
FRIEC	1987-2	Frieze, A. M. Parallel algorithms for finding hamilton cycles in random graphs.
FRIEE	1985-1	Frieze, A. M.; Grimmett, G. R. The shortest-path problem for graphs with random arc-lengths.
FROS	1985-1	Frostig, E.; Adiri, I. Stochastic flowshop no-wait scheduling.
FROS	1985-2	Frostig, E.; Adiri, I. Three-machine flowshop stochastic scheduling to minimize distribution of schedule length.
FUNG	1985-1	Fung, K. T. An entropy approach to solving some network reliability problems.
GAZD	1983-1	Gazdik, I. Fuzzy-network planning - FNET.
GELFB	1987-1	Gelfand, S. B.; Mitter, S. K. Simulated annealing.
GLAZ	1985-1	Glazebrook, K. D. Semi-Markov models for single-machine stochastic scheduling problems.
GLAZ	1987-1	Glazebrook, K. D. Evaluating the effects of machine breakdowns in stochastic scheduling problems.
GLAZ	1987-2	Glazebrook, K. D. Sensitivity analysis for stochastic scheduling problems.
GLAZA	1987-1	Glazebrook, K. D.; Fay, N. A. On the scheduling of alternative stochastic jobs on a single machine.

GRIMA 1985-1 Grimmett, G. R.
Large deviations in subadditive processes and first-passage percolation.

GRIMA 1985-2 Grimmett, G. R.
Random flows: network flows and electrical flows through random media.

GRIMA 1986-1 Grimmett, G. R.
An exact threshold theorem for random graphs and the node-packing problem.

GRIMB 1985-1 Grimmett, G. R.; Pulleyblank, W. R.
Random near-regular graphs and the node packing problem.

GROB 1987-1 Grobelny, J.
On one possible 'fuzzy' approach to facilities layout problems.

HALLE 1985-2 Hall, R. W.
Vehicle scheduling at a transportation terminal with random delay en route.

HALLE 1986-1 Hall, R. W.
The fastest path through a network with random time-dependent travel times.

HASSB 1987-1 Hassin, R.
A computing scheme for network problems with random edge lengths.

HASSE 1985-1 Hassin, R.; Zemel, E.
On shortest paths in graphs with random weights.

HENI 1986-1 Henig, M. I.
Extensions of the dynamic programming method in the deterministic and stochastic assembly-line balancing problems.

HODD 1984-1 Hodder, J. E.
Financial market approaches to facility location under uncertainty.

HODDA 1986-1 Hodder, J. E.; Dincer, M. C.
A multifactor model for international plant location and financing under uncertainty.

HOUS 1984-1 Houska, R. V.
Using stochastic optimal control to find shortest paths in networks.

HOUSA 1982-1 Houska, R. V.; Brockus, C. G.
Stochastic optimal control applied to deterministic network problems.

ISHIA 1984-1 Ishii, H.; Nishida, T.
Stochastic linear knapsack problem : probability maximization model.

ITAK 1984-1 Itakura, H.; Nishikawa, Y.
Fuzzy network technique for technological forecasting.

JAIL 1985-1 Jaillet, P.
Analysis of the probabilistic traveling salesman problem in the plane.

JAIL 1987-1 Jaillet, P.
On some probabilistic combinatorial optimization problems defined on graphs.

JAIL 1987-2 Jaillet, P.
Stochastic routing problems.

JOHNL 1984-1 Johnson, R.
Network reliability and acyclic orientations.

KAMB 1984-2 Kamburowski, J.
Flow distributions through a stochastic acyclic network.

KAMB 1985-1 Kamburowski, J.
A note on the stochastic shortest route problem.

KAMB 1985-3 Kamburowski, J.
An upper bound on the expected completion time of PERT networks.

KAMB 1985-4 Kamburowski, J.
Two-point approximations for activity times in PERT network.

KAMB 1986-1 Kamburowski, J.
On the computational complexity of the shortest route and maximum flow problems in stochastic networks.

KAMB 1987-1 Kamburowski, J.
An overview of the computational complexity of the PERT shortest route and maximum flow problems in stochastic networks.

KAMP 1987-2 Kämpke, T.
On the optimality of static priority policies in stochastic scheduling on parallel machines.

KARPD 1983-1 Karp, R. M.; Pearl, J.
Searching for an optimal path in a tree with random costs.

KELLB 1985-1 Kelly, F. P.
Some probabilistic aspects of network flow.

KESEA 1984-1 Keselman, V. O.
The statistical approach to the solution of a class of transportation problems with fuzzy constraints (Russian).

KHOD 1985-1 Khodzinskii, A. N.
Sequential algorithm for the solution of problems of combinatorial optimization on permutations.

KIM D 1985-1 Kim, S.-I.; Park, Y. S.
Models maximizing covering reliability.

KLEI 1986-1 Klein Haneveld, W. K.
Robustness against dependence in PERT: an application of duality and distributions with known marginalS.

KOPY 1984-1 Kopylov, G. N.; Levin, A. Y.
On the realizability of a stochastic multiproduct flow.

KRES 1984-1 Kress, M.
The chance constrained critical path with location-scale distributions.

KU 1986-1 Ku, P.-S.; Niu, S.-C.
On Johnson's two-machine flow shop with random processing times.

KUCE 1986-1 Kucera, L.
Greedy coloring is a bad probabilistic algorithm.

KULKB 1985-1 Kulkarni, V. G.; Adlakha, V. G.
Maximum flow in planar networks with exponentially distributed arc capacities.

KULKB 1986-1 Kulkarni, V. G.; Adlakha, V. G.
Markov and Markov-regenerative PERT networks.

LAAR 1987-1 Laarhoven, P. J. M. van; Aarts, E. H. L.
Simulated annealing: theory and applications.

LAGE 1985-1 Lageweg, B.; Lenstra, J. K.; Stougie, L.; Rinnooy Kan, A. H. G.
Stochastic integer programming by dynamic programming.

LASS 1985-1 Lasserre, J. B.; Bes, C.; Roubellat, F.
The stochastic discrete dynamic lot size problem: an open-loop solution.

LEFE 1985-1 Lefevre, C.; Laloyaux, P.
Note sur les problemes d'ordonnancement Pert aleatoires.

LESH 1984-1 Leshchenko, L. L.
On a stochastic transportation problem (Russian).

LEVA 1984-1 Levary, R. R.
An experimental sequential solution procedure to stochastic linear programming problems with 0-1 variables.

LIU I 1983-1 Liu, Z. H.
Approximate algorithms for combinatorial optimization problems. I (Chinese).

LIU I 1983-2 Liu, Z. H.
Approximate algorithms for combinatorial optimization problems. II (Chinese).

LOUV 1986-1 Louveaux, F.
Discrete stochastic location models.

LOUVA 1985-1 Louveaux, F.; Thisse, J.-F.
Production and location on a network under demand uncertainty.

MAC G 1984-1 Mac Diarmid, C.
Colouring random graphs.

MAFF 1986-1 Maffioli, F.
Randomized algorithms in combinatorial optimization: a survey.

MAFFA 1985-1 Maffioli, F.; Speranza, M. G.; Vercellis, C.
Randomized algorithms.

MAGGD 1986-1 Maggu, P. L.; Sharma, B. K.
On equivalent-job for job-block in a 'n-job, 2-machine' flow-shop scheduling with random processing times.

MATSB 1985-1 Matsutomi, T.; Ishii, H.; Nishida, T.
Stochastic location problem of emergency service facility.

MATSB 1987-1 Matsutomi, T.; Ishii, H.; Nishida, T.
Chance constrained round trip location problem.

MIKHA 1986-1 Mikhalevich, M. V.
A stochastic algorithm for solving nonlinear mixed integer problems (Russian).

MIRCC 1985-1 Mirchandani, P. B.; Soroush, H.
Optimal paths in probabilistic networks: a case with temporary preferences.

MIRCC 1987-1 Mirchandani, P. B.; Soroush, H.
Routes and flows in stochastic networks.

MOHRA 1985-2 Möhring, R. H.; Radermacher, F. J.
Introduction to stochastic scheduling problems.

MOHRB 1984-1 Möhring, R. H.; Radermacher, F. J.; Weiss, G.
Stochastic scheduling problems I - general strategies.

MOHRB 1985-1 Möhring, R. H.; Radermacher, F. J.; Weiss, G.
Stochastic scheduling problems II - set strategies.

MYUN 1987-1 Myung, Y.-S.; Tcha, D.-W.
The stochastic uncapacitated facility location problem: a new formulation and extensions to the capacitated problem.

NAKA 1985-1 Nakai, T.
Optimal assignment for a random sequence with an unknown number of jobs.

NAKAD 1984-1 Nakano, H.; Nakanishi, Y.
Simulated annealing method for combinatorial optimization problems (Japanese).

NEUMC 1984-1 Neumann, K.
Recent developments in stochastic activity networks.

NEUMC 1985-1 Neumann, K.
EOR project networks.

NEUMC 1985-2 Neumann, K.
GERT networks with tree stucture: properties, temporal analysis, cost minimization and scheduling.

NEUMC 1985-3 Neumann, K.
Stochastic single-machine scheduling with GERT-like precedence constraints.

NEVI 1985-1 Nevison, C. H.
A cost adjustment heuristic for dynamic lot-sizing with uncertain demand timing.

ODON 1987-1 Odoni, A. R.
Stochastic facility location problems.

PAPAA 1987-1 Papadimitriou, C. H.; Tsitsiklis, J. N.
On stochastic scheduling with in-tree precedence constraints.

PELT 1987-1 Peltsverger, B. V.; Khavronin, O. V.
A decompositional approach to the problem of synthesizing a transport network with a fuzzy use period (Russian).

PINE 1984-1 Pinedo, M.
A note on the flow time and the number of tardy jobs in stochastic open shops.

PINE 1984-2 Pinedo, M.
Optimal policies in stochastic shop scheduling.

PINE 1985-1 Pinedo, M.
A note on stochastic shop models in which jobs have the same processing requirements on each machine.

PINEC 1986-1 Pinedo, M.; Wie, S.-H.
Inequalities for stochastic flow shops and job shops.

POIA 1986-1 Poiaga, L.
Resolution by reduction of a network, by means of its assimilation to a semi-Markov process (Italian).

POLAA 1983-1 Polat, N.
Couplages dans les systemes combinatoires flous. I.

POLAA	1983-2	Polat, N. Couplages dans les systemes combinatoires flous. II.
QI	1985-1	Qi, L.-Q. Forest iteration method for stochastic transportation problem.
QI	1987-3	Qi, L.-Q. The A-forest iteration method for the stochastic generalized transportation problem.
RAGHA	1986-1	Raghavan, P. Probabilistic construction of deterministic algorithms: approximating packing integer programs.
RAO A	1983-1	Rao, J. R.; Rao, S. N. Facility location in stochastic environment.
RAO B	1985-1	Rao, J. R.; Varma, N. H. Stochastic multi-facility minisum location problem involving Euclidean distances.
RHEE	1985-1	Rhee, W.-S. T. Convergence of optimal stochastic bin packing.
RINN	1986-2	Rinnooy Kan, A. H. G. Stochastic integer programming: the distribution problem.
ROSS	1987-1	Rossi, F. A.; Gavioli, I. Aspects of heuristic methods in the "Probabilistic Travelling Salesman Problem".
ROSSA	1986-1	Rossier, Y.; Troyon, M.; Liebling, T. M. Probabilistic exchange algorithms and Euclidean traveling salesman problems.
ROUBA	1983-1	Roubens, M.; Vincke, P. Linear fuzzy graphs.
RUCI	1982-1	Rucinski, A. Matchings and k-factors in a random graph.
RUGE	1984-1	Rüger, W. Berechnung der Zusammenhangswahrscheinlichkeit von Netzgraphen.
RUHEA	1985-1	Ruhe, G.; Ivanchev, D. T. The solution of generalized network flow problems with parametric capacities and stochastic demands.
SAKA	1984-1	Sakaguchi, M. A sequential stochastic assignment problem associated with a non-homogeneous Markov process.
SAKA	1984-2	Sakaguchi, M. A sequential stochastic assignment problem with an unknown number of jobs.
SANCA	1985-1	Sancho, N. G. F. Maximum reliability through a network with resource constraints.
SEIDA	1987-1	Seidmann, A. On-line scheduling of a robotic manufacturing cell with stochastic sequence-dependent processing rates.
SERGE	1983-1	Sergienko, I. V.; Kaspshitskaya, M. F. Stability of algorithms of the method of descent vector in a class of combinatorial optimization problems (Russian).
SERGH	1987-1	Sergienko, I. V.; Shilo, V. P. Polynomial asymptotically-optimal random search algorithm for a multidimensional problem of Boolean linear programming (Russian).
SERGH	1987-2	Sergienko, I. V.; Shilo, V. P. Probabilistic approach to evaluate heuristics for problems of the Boolean linear programming (Russian).
SHIO	1985-1	Shiode, S.; Ishii, H.; Nishida, T. A chance constrained minimax facility location problem.
SHIO	1987-1	Shiode, S.; Ishii, H.; Nishida, T. A stochastic linear knapsack problem.
SIED	1982-1	Siedersleben, J. Optimierung von deterministischen und stochastischen Entscheidungsnetz- plänen (Dissertation).
SILVA	1986-1	Silverman, F. N.; Carter, J. C. A cost-based methodology for stochastic line balancing with intermittent line stoppages.
SIMO	1981-1	Simon, I. Probabilistic algorithms and complexity (Portuguese).
SIVA	1987-1	Sivazlian, B. D. Optimum scheduling of a new maintenance program under stochastic degradation.
SPEN	1985-1	Spencer, J. Probabilistic methods.
STEE	1986-1	Steele, J. M. Probabilistic algorithm for the directed traveling salesman problem.
STOU	1987-1	Stougie, L. Design and analysis of algorithms for stochastic integer programming.
STOU	1988-1	Stougie, L. The stochastic knapsack problem.
SURE	1985-1	Suresh, S.; Foley, R. D.; Dickey, S. E. On Pinedo's conjecture for scheduling in a stochastic flow shop.
SZKA	1987-1	Szkatula, K.; Libura, M. On probabilistic properties of greedy-like algorithms for the binary knapsack problem.
TCHAC	1985-1	Tcha, D.-W.; Yoon, M.-G. A dual-based heuristic for the simple facility location problem with stochastic demand.
TIGAA	1985-1	Tigan, S.; Stancu-Minasian, I. M. The stochastic bottleneck transportation problem.
TINH	1984-3	Tinhofer, G. Probabilistische Ansätze in der diskreten Optimierung.
TINHA	1986-1	Tinhofer, G.; Schreck, H. The bounded subset sum problem is almost everywhere randomly decidable in O(n).
VANI	1987-1	Vani, V.; Raghavachari, M. Deterministic and random single machine sequencing with variance minimization.
WANGK	1986-1	Wang, S.-M.; Sun, Q.-H.; Wang, X.-F. A study of flows in complete stochastic networks - theoretical study of a new method for evaluating the reliability of interconnected electrical networks (Chinese).

WEBEA 1986-1 Weber, R. R.; Varaiya, P.; Walrand, J.
Scheduling jobs with stochastically ordered processing times on parallel machines to minimize expected flowtime.
WHITI 1985-1 Whittle, P.
Scheduling and characterization problems for stochastic networks. With discussion.
WIE 1986-1 Wie, S.-H.; Pinedo, M.
On minimizing the expected makespan and flow time in stochastic flow shops with blocking.
WILHA 1987-1 Wilhelm, M. R.; Ward, T. L.
Solving quadratic assignment problems by "simulated annealing".
WOLL 1985-1 Wollmer, R. D.
Critical path planning under uncertainty.
WU 1984-1 Wu, S.-Z.; Wu, X.-M.
Generalized principle of optimality in pansystems network analysis.
WU B 1983-1 Wu, W.-M.
Robbins' theorem for fuzzy mixed multigraphs (Chinese).
YAKO 1987-1 Yakovlev, S. V.
Probabilistic methods for optimization on discrete sets (Russian).
ZHANQ 1984-1 Zhang, Y.; Zhang, Y. C.
A limit theorem for N(0n)/n in first-passage percolation.
ZIMM 1987-1 Zimmermann, H.-J.
Methoden und Modelle des Operations Research. Für Ingenieure, Ökonomen und Informatiker.
ZIMMA 1984-1 Zimmermann, H.-J.; Pollatschek, M. A.
Fuzzy 0-1 linear programs.

AART 1987-1 Aarts, E. H. L.; Laarhoven, P. J. M. van
A pedestrian review of the theory and application of the simulated annealing algorithm.
AIKE 1985-1 Aikens, C. H.
Facility location models for distribution planning.
ALEKG 1981-1 Alekseev, V. E.; Markov, A. A.; Noskov, V. V.
Optimization of independent and support sets in hypergraphs (Russian).
ANDR 1983-1 Andrasfai, B.
Graph theory. Network flows. Matrices. (Hungarian).
ANDRD 1986-1 Andreatta, G.; Mason, F.
Review of recent results about the k-centrum of a tree.
ARNB 1985-1 Arnborg, S.
Efficient algorithms for combinatorial problems on graphs with bounded decomposability - a survey.
ARON 1985-1 Aronson, J. E.
A survey of dynamic network flows.
BACHA 1984-1 Bachem, A.
Optimization and geometry in discrete structures.
BACHA 1986-1 Bachem, A.
Dualität und Polarität in diskreten Strukturen.
BACHB 1984-1 Bachem, A.; Euler, R.
Recent trends in combinatorial optimization.
BAHLA 1987-1 Bahl, H. C.; Ritzman, L. P.; Gupta, J. N. D.
Determining lot sizes and resource requirements: A review.
BALAC 1987-1 Balakrishnan, A.; Ward, J. E.; Wong, R. T.
Integrated facility location and vehicle routing models: recent work and future prospects.
BALAO 1985-1 Balas, E.; Toth, P.
Branch and bound methods.
BAYB 1986-1 Baybars, I.
A survey of exact algorithms for the simple assembly line balancing problem.
BEHZ 1987-1 Behzad, M.
Total concepts in graph theory.
BERG 1984-3 Berge, C.
Minimax theorems for normal hypergraphs and balanced hypergraphs - a survey.
BERG 1984-5 Berge, C.
Stochastic graphs and strongly perfect graphs: a survey.
BIGI 1985-1 Bigildeeva, T. B.
Methods and algorithms of approximate solution of a quadratic assignment problem of large dimensions (a survey) (Russian).
BOES 1982-1 Boesch, F. T.
Introduction to basic network problems.
BONDB 1985-1 Bondy, J. A.
Kotzig's conjecture on generalized friendship graphs - a survey.
BURK 1986-1 Burkard, R. E.
Assignment problems: recent solution methods and applications.
CACCA 1987-1 Caccetta, L.; Vijayan, K.
Applications of graph theory.
CAI A 1985-1 Cai, M.-C.
Progress in network flow theory and matroid theory (Chinese).
CARTA 1986-1 Carter, M. W.
A survey of practical applications of examination timetabling algorithms.
CHRI 1985-1 Christofides, N.
Vehicle routing.
CLAU 1986-1 Clausen, J.; Krarup, J.
Combinatorial optimization: challenges and trends.
COFFB 1984-1 Coffman, E. G. Jr.; Garey, M. R.; Johnson, D. S.
Approximation algorithms for bin-packing - an updated survey.
COFFF 1986-1 Coffman, E. G. Jr.; Lueker, G. S.; Rinnooy Kan, A. H. G.
An introduction to the probabilistic analysis of sequencing and packing heuristics.
CURRA 1986-1 Current, J. R.; Min, H.-K.
Multiobjective design of transportation networks: Taxonomy and annotation.
DEAR 1985-1 Dearing, P. M.
Location problems.
DERI 1984-4 Derigs, U.
The cardinality matching problem - a state of the art tutorial.
DICK 1984-1 Dick, W. E.
Discrete characterization of planarity. I: The classical viewpoint.
DINY 1985-1 Diny, D. A.; Harary, F.; Mac Morris, F. R.
A survey of results on a graph and its complement with specified hereditary properties.
DOMSA 1985-2 Domschke, W.; Drexl, A.
Location and layout planning. An international bibliography.
DURR 1984-1 Durrett, R.
Oriented percolation in two dimensions.
EMEL 1981-2 Emelichev, V. A.; Kravtsov, M. K.
Transportation polyhedra (Russian).
EMEL 1983-1 Emelichev, V. A.; Kravtsov, M. K.
Combinatorial problems on transport polyhedra - a survey (Russian).
ETSC 1985-1 Etschmaier, M. M.; Mathaisel, D. F. X.
Airline scheduling: an overview.
FARRA 1986-1 Farrell, E. J.
A survey of the unifying effects of F-polynomials in combinatorics and graph theory.
FINKE 1987-1 Finke, G.; Burkard, R. E.; Rendl, F.
Quadratic assignment problems.
FISHD 1985-1 Fisher, M. L.
An applications oriented guide to Lagrangian relaxation.

FLORA	1984-1	Florian, M. An introduction to network models used in transportation planning.
FORS	1984-1	Forst, F. G. A review of the static, stochastic job sequencing literature.
FRENE	1986-1	Frenkel, K. A. Complexity and parallel processing: an interview with Richard M. Karp.
FRIEB	1985-1	Friesz, T. L. Transportation network equilibrium, design and aggregation: Key developments and research opportunities.
FUJIA	1984-2	Fujishige, S. Combinatorial optimization problems described by submodular functions.
GALL	1984-1	Gallo, G.; Pallottino, S. Shortest path methods in transportation models.
GARF	1985-1	Garfinkel, R. S. Motivation and modeling.
GERAA	1987-1	Gerards, A. M. H.; Kolen, A. W. J. Polyhedral combinatorics in combinatorial optimization.
GERN	1985-1	Gernert, D. A survey of partial proofs for Read's conjecture and some recent results.
GILM	1985-1	Gilmore, P. C.; Lawler, E. L.; Shmoys, D. B. Well-solved special cases.
GION	1987-1	Gionfriddo, M. A short survey on some generalized colourings of graphs.
GLOV	1986-1	Glover, F. Future paths for integer programming and links to artificial intelligence.
GOLDE	1984-1	Golden, B. L. Introduction to and recent advances in vehicle routing methods.
GOLDF	1986-1	Golden, B. L.; Assad, A. A. Perspectives on vehicle routing: exciting new developments.
GOLU	1985-1	Golumbic, M. C. Interval graphs and related topics.
GRAHB	1985-1	Graham, R. L.; Hell, P. On the history of the minimum spanning tree problem.
GRIMA	1985-2	Grimmett, G. R. Random flows: network flows and electrical flows through random media.
GROT	1984-1	Grötschel, M. Developments in combinatorial optimization.
GROT	1985-1	Grötschel, M. Polyhedral Combinatorics.
GROT	1986-1	Grötschel, M. Fortschritte in der polyedrischen Kombinatorik.
HANSE	1987-1	Hansen, P.; Labbe, M.; Peeters, D.; Thisse, J.-F. Single facility location on networks.
HARAB	1985-3	Harary, F. Petersen's graph and its unique properties (Danish).
HELD	1984-1	Held, M.; Hoffman, A. J.; Johnson, E. L.; Wolfe, P. Aspects of the traveling salesman problem.
HILT	1987-1	Hilton, A. J. W. Recent progress on edge-coloring graphs.
HIRSA	1984-1	Hirshfeld, D. S. An overview of mathematical programming for strategic planning.
HOFFC	1985-1	Hoffman, A. J.; Wolfe, P. History.
HOFFD	1985-1	Hoffman, K.; Padberg, M. W. LP-based combinatorial problem solving.
HOFFD	1986-1	Hoffman, K.; Padberg, M. W. LP-based combinatorial problem solving.
HSU A	1984-4	Hsu, W.-L. The perfect graph conjecture on special graphs - a survey.
IORDA	1982-1	Iordanskii, M. A. Problems of graph arrangement. II. (Russian).
IWAI	1985-2	Iwainsky, A. Optimal trees - a short overview on problem formulations.
JAEG	1985-1	Jaeger, F. A survey of the cycle double cover conjecture.
JOHNG	1985-1	Johnson, D. S.; Papadimitriou, C. H. Computational complexity.
JUNGB	1986-1	Junginger, W. Timetabling in Germany - a survey.
JUNGC	1986-1	Jungnickel, D. Transversaltheorie: ein Überblick.
KARP	1986-1	Karp, R. M. Combinatorics, complexity and randomness.
KARPB	1985-1	Karp, R. M.; Lenstra, J. K.; Mac Diarmid, C.; Rinnooy Kan, A. H. G. Probabilistic analysis.
KHOK	1982-1	Khokhlyuk, V. I. Algorithms of integer optimization (Russian).
KHOK	1983-1	Khokhlyuk, V. I. Cutting plane algorithms (Russian).
KHOK	1987-1	Khokhlyuk, V. I. Parallel algorithms for integer optimization (Russian).
KIDW	1987-1	Kidwell, M. E.; Richter, R. B. Trees and Euler tours in a planar graph and its relatives.

KIND 1985-2 Kindervater, G. A. P.; Lenstra, J. K.
Parallel algorithms.

KIND 1986-1 Kindervater, G. A. P.; Lenstra, J. K.
An introduction to parallelism in combinatorial optimization.

KIND 1986-2 Kindervater, G. A. P.; Lenstra, J. K.
Parallel computing in combinatorial optimization.

KLEE 1983-1 Klee, V.
Combinatorial optimization: what is the state of the art? (Chinese).

KORTA 1984-1 Korte, B.
Kombinatorische Optimierung und algorithmische Prinzipien.

KORTA 1985-1 Korte, B.
Combinatorial optimization (Part I) (Chinese).

KORTA 1985-2 Korte, B.
Was ist kombinatorische Optimierung?

KORTA 1986-1 Korte, B.
Kombinatorische Optimierung und algorithmische Prinzipien.

LAPOD 1987-1 Laporte, G.; Nobert, Y.
Exact algorithms for the vehicle routing problem.

LAWL 1985-1 Lawler, E. L.
Submodular functions and polymatroid optimization.

LAWLA 1981-1 Lawler, E. L.; Lenstra, J. K.
Machine scheduling with precedence constraints.

LAWLB 1981-1 Lawler, E. L.; Lenstra, J. K.; Rinnooy Kan, A. H. G.
Recent developments in deterministic sequencing and scheduling: a survey.

LENSA 1985-1 Lenstra, J. K.; Rinnooy Kan, A. H. G.
Sequencing and scheduling.

LENSB 1982-1 Lenstra, J. K.; Rinnooy Kan, A. H. G.; Emde Boas, P. van
An appraisal of computational complexity for operations research.

LENSC 1985-1 Lenstra, J. K.; Rinnooy Kan, A. H. G.; Guan, M.-G.
On travelling salesman problem (Chinese).

LIBU 1984-1 Libura, M.
Integer programming duality and its application in sensitivity analysis and approximate algorithms (Polish).

LIPS 1983-1 Lipski, W. Jr.
The consecutive retrieval property, interval graphs and related topics - a survey.

LIU I 1983-1 Liu, Z. H.
Approximate algorithms for combinatorial optimization problems. I (Chinese).

LIU I 1983-2 Liu, Z. H.
Approximate algorithms for combinatorial optimization problems. II (Chinese).

LIU I 1987-2 Liu, Z. H.
Matroid theory and its applications (Chinese).

LOVA 1985-1 Lovasz, L.
An algorithmic theory of numbers, graphs, and convexity.

LOVA 1986-2 Lovasz, L.
Connectivity algorithms using rubber-bands.

MACU 1987-1 Maculan, N. F.
The Steiner problem in graphs.

MAFF 1986-1 Maffioli, F.
Randomized algorithms in combinatorial optimization: a survey.

MAFFA 1985-1 Maffioli, F.; Speranza, M. G.; Vercellis, C.
Randomized algorithms.

MAHR 1981-1 Mahr, B.
A bird's-eye view to path problems.

MARTB 1987-1 Martello, S.; Toth, P.
Algorithms for knapsack problems.

MEIN 1984-1 Meinel, C.
Effiziente Algorithem.

MEYE 1985-1 Meyer, R. R.
Network optimization.

MILA 1985-1 Milanov, P. B.
The linear diophantine problem of Frobenius and discrete optimization.

MINK 1980-1 Minker, J.; Minker, R. G.
Optimization of Boolean expressions - historical developments.

MINO 1987-2 Minoux, M.
Network synthesis and dynamic network optimization.

MITCA 1985-1 Mitchem, J.; Schmeichel, E.
Pancyclic and bipancyclic graphs - a survey.

MURT 1985-1 Murtagh, F.
A survey of algorithms for contiguity-constrained clustering and relating problems.

NARU 1986-1 Narula, S. C.
Minisum hierarchical location-allocation problems on a network: a survey.

NEMH 1985-1 Nemhauser, G. L.
Duality for integer optimization.

NEUMC 1984-1 Neumann, K.
Recent developments in stochastic activity networks.

NOSO 1985-1 Nosov, V. A.; Sachkov, V. N.; Tarakanov, V. E.
Combinatorial analysis (nonnegative matrices, algorithmic problems).

OLARA 1985-1 Olaru, E.; Antohe, S.
Some results concerning graph decomposition.

ORDM 1987-1 Ordman, E. T.
Dining philosophers and graph covering problems.

PAIX 1985-1 Paixao, J. P.; Almeida, M. T.; Coelho, J. D.
Problems in graph theory and their relation with set covering and partitioning problems (Portuguese).

PAPA 1985-2 Papadimitriou, C. H.
Computational complexity.

PAPA 1986-1 Papadimitriou, C. H.
Shortest-path motion.

POWE 1985-1 Powell, S.
Software.

RING 1985-1 Ringel, G.
250 Jahre Graphentheorie.

RINN 1984-1 Rinnooy Kan, A. H. G.
An introduction to approximation algorithms.

ROBEE 1984-2 Robertson, N.; Seymour, P. D.
Graph width and well-quasi-ordering: a survey.

ROBEE 1985-1 Robertson, N.; Seymour, P. D.
Disjoint paths - a survey.

ROBEE 1985-2 Robertson, N.; Seymour, P. D.
Graph minors - a survey.

ROCK 1984-1 Rockafellar, R. T.
Network flows and monotropic optimization.

ROSEB 1956-1 Rosenblatt, D.
On aggregation and consolidation in linear systems.

ROUSA 1984-1 Rousseau, J.-M.
Crew scheduling methods in the operation and planning of transportation systems.

SAKSA 1985-1 Saksena, J. P.
Applications of decision networks.

SCHRE 1986-2 Schrijver, A.
Geometric methods in discrete optimization.

SCHRE 1986-3 Schrijver, A.
Polyhedral combinatorics - some recent developments and results.

SCHRE 1986-4 Schrijver, A.
Polyhedral proof methods in combinatorial optimization.

SEBO 1985-1 Sebö, A.
On the Chinese postman problem: algorithms, structure and applications.

SEKA 1985-1 Sekanina, M.
Factors in powers of graphs.

SPEN 1985-1 Spencer, J.
Probabilistic methods.

SWEE 1984-1 Sweeney, D. J.
A practical evaluation of the potential of mixed integer programming as a strategic planning tool.

TARD 1986-1 Tardos, E.
A strongly polynomial algorithm to solve combinatorial linear programs.

TARJ 1985-2 Tarjan, R. E.
Shortest path algorithms.

TEGH 1986-1 Teghem, J. Jr.; Kunsch, P. L.
A survey of techniques for finding efficient solutions to multi-objective integer linear programming.

THOMA 1984-2 Thomassen, C.
Graph theory - an introduction (Danish).

TINH 1984-3 Tinhofer, G.
Probabilistische Ansätze in der diskreten Optimierung.

TROT 1985-1 Trotter, L. E. Jr.
Discrete packing and covering.

TUTT 1985-1 Tutte, W. T.
Graph theory as I have known it (Part III) (Chinese).

TUYT 1986-1 Tuyttens, D.; Teghem, J. Jr.
Theorie des matroides et optimisation combinatoire.

VAINA 1984-1 Vainshtein, A. D.
The problem of packing of rectangles into a strip - a survey (Russian).

VIZV 1986-1 Vizvari, B.
Some recent Hungarian results in discrete programming (Hungarian).

WERR 1985-1 Werra, D. de
An introduction to timetabling.

WILH 1985-1 Wilhelm, M. R.; Evans, G. W.
The quadratic assignment problem: applications and solution techniques.

WILLA 1985-1 Williams, H. P.
Model building in linear and integer programming.

WILLD 1985-2 Willis, R. J.
Critical path analysis and resource constrained project scheduling - theory and practice.

WONGA 1984-1 Wong, R. T.
Introduction and recent advances in network design: models and algorithms.

WONGA 1985-1 Wong, R. T.
Location and network design.

WONGA 1986-1 Wong, R. T.
Recent advances in location and network design models for transportation planning.

WU C 1984-1 Wu, X.-M.
Recent developments on graph theory (Chinese).

XIE A 1984-1 Xie, L.-T.; Liu, G.-Z.
A survey of the problem of graceful trees (Chinese).

ZHU C 1987-1 Zhu, B. W.; Su, J. J.
Some problems in the research on the connectedness of graphs (Chinese).

ZHU K 1983-1 Zhu, Y.-J.
A survey of the research on the Hamilton problem in graph theory (Chinese).

ABDU 1983-1 Abdullaev, A. A.; Rzaev, T. G.
Optimization of operative planning and scheduling in integrated ACS for enterprises (Russian).

ALEKD 1987-1 Alekseev, O. G.
Complex application of discrete optimization methods (Russian).

ANDR 1983-1 Andrasfai, B.
Graph theory. Network flows. Matrices. (Hungarian).

AVEN 1985-1 Aven, O. I.; Lovetskij, S. E.; Moiseenko, G. E.
Optimization of transportation flows (Russian).

BERG 1983-1 Berge, C.
Graphes (3. edition).

BERG 1985-1 Berge, C.
Graphs (second revised edition).

BLAZD 1986-1 Blazewicz, J.; Cellary, W.; Slowinski, R.; Weglarz, J.
Scheduling under resource constraints - Deterministic models.

BOKH 1987-1 Bokhari, S. H.
Assigment Problems in Parallel and Distributed Computing.

BOLL 1985-2 Bollobas, B.
Random graphs.

BOLL 1986-1 Bollobas, B.
Extremal graph theory with emphasis on probabilistic methods.

DAVYB 1985-1 Davydova, I. M.
Enumeration schemes in location problems (Russian).

DUTA 1983-1 Duta, L. D.; Fabian, C.
Mathematical methods in optimal cutting (Romanian).

EVST 1985-1 Evstigneev, V. A.
Application of graph theory to computer programming (Russian).

FOUL 1984-1 Foulds, L. R.
Combinatorial optimization for undergraduates.

FREN 1986-1 French, S.; Hartley, R.; Thomas, L. C.; White, D. J.
Operational Research Techniques.

GEY 1985-1 Gey, H.; Lehne, D.
Netzplantechnik: Grundlagen, Methoden, praxisnahe Einführung, Anwendungsbeispiele, Netzplanprogramm in MBASIC.

GIBB 1985-1 Gibbons, A.
Algorithmic graph theory.

GRIB 1985-1 Gribik, P. R.; Kortanek, K. O.
Extremal methods of operations research.

GROP 1984-1 Groppen, V. O.
Models and algorithms of combinatorial programming (Russian).

GUPTH 1985-1 Gupta, S. K.
Linear programming and network models.

HAGG 1980-1 Haggard, G.
Excursions in graph theory.

HALI 1981-1 Halin, R.
Graphentheorie 2.

HELLC 1984-1 Hellmann, A.
Theorie und Praxis von Routing-Problemen: Lösungsverfahren zur Tourenplanung im Eindepot- und Mehrdepot-Fall.

JUNGC 1987-1 Jungnickel, D.
Graphen, Netzwerke und Algorithmen.

KIM A 1982-1 Kim, K. H.
Boolean matrix theory and applications.

KOVAD 1987-1 Kovalev, M. M.
Matroids in discrete optimization (Russian).

KUNG 1986-1 Kung, J. P. S.
A source book in matroid theory.

KWAK 1987-1 Kwak, N. K.; Schniederjans, M. J.
Introduction to Mathematical Programming.

LAAR 1987-1 Laarhoven, P. J. M. van; Aarts, E. H. L.
Simulated annealing: theory and applications.

LEDE 1980-1 Ledermann, W.; Vajda, S.
Handbook of applicable mathematics. Vol. 1: Algebra.

LIPA 1986-1 Lipatov, E. P.
Graph theory and its applications (Russian).

LOVAA 1986-1 Lovasz, L.; Plummer, M. D.
Matching theory.

MEHL 1984-1 Mehlhorn, K.
Data structures and algorithms 2: Graph algorithms and NP-completeness.

MIKHC 1986-1 Mikhalevich, V. S.; Trubin, V. A.; Shor, N. Z.
Optimization problems of production-transportation planning. Models, Methods, Algorithms (Russian).

MINO 1986-1 Minoux, M.
Mathematical Programming. Theory and Algorithms.

MINOA 1986-1 Minoux, M.; Bartnik, G.
Graphes, algorithmes, logiciels.

MOSL 1987-1 Mosler, K. C.
Continuous location of transportation networks.

MURO 1987-3 Murota, K.
Systems analysis by graphs and matroids. Structural solvability and controllability.

SAKAA 1984-1 Sakarovitch, M.
Optimisation combinatoire - methodes mathematiques et algorithmiques. Vol. 1: Programmation discrete.

SAKAA 1984-2 Sakarovitch, M.
Optimisation combinatoire - methodes mathematiques et algorithmiques. Vol. 2: Graphes et programmation lineaire.
SCHRE 1986-5 Schrijver, A.
Theory of linear and integer programming.
SERGA 1985-1 Sergienko, I. V.
Mathematical models and methods for solving problems of discrete optimization.
SERGG 1980-1 Sergienko, I. V.; Lebedeva, T. T.; Roshchin, V. A.
Approximate methods for solving discrete problems of optimization.
TANA 1984-1 Tanaev, V. S.; Gordon, V. S.; Shafranskij, Y. M.
Scheduling theory, one-machine systems (Russian).
TARA 1985-1 Tarakanov, V. E.
Combinatorial problems and (0,1)-matrices (Russian).
TOME 1985-1 Tomescu, I.
Problems in combinatorics and graph theory.
TUCK 1984-1 Tucker, A.
Applied Combinatorics (second edition).
WALT 1984-1 Walther, H.
Ten applications of graph theory.
WALTA 1987-1 Walther, H.; Nägler, G.
Graphen, Algorithmen, Programme.
WHITC 1986-1 White, N.
Theory of matroids.
WHITC 1987-1 White, N.
Combinatorial Geometries.
WILF 1986-1 Wilf, H. S.
Algorithms and complexity.
WILSB 1985-1 Wilson, R. J.
Introduction to graph theory (third edition).
ZIMM 1987-1 Zimmermann, H.-J.
Methoden und Modelle des Operations Research. Für Ingenieure, Ökonomen und Informatiker.

ARAN 1985-1 Arani, T.; Karwan, M. H.; Lotfi, V.
A Lagrangian relaxation approach for solving an exam scheduling problem.
BENDC 1985-1 Benders, J. F.; Beulens, A. J. M.; Nunen, J. A. E. E. van
On using the linear programming relaxation of assignment type mixed integer problems.
CARTA 1986-1 Carter, M. W.
A survey of practical applications of examination timetabling algorithms.
CEDE 1984-1 Ceder, A.; Stern, H. I.
Optimal transit timetables for a fixed vehicle fleet.
CROU 1984-1 Croucher, J. S.
A goal programming model for timetable scheduling.
EGLE 1987-1 Eglese, R. W.; Rand, G. K.
Conference seminar timetabling.
FERLA 1985-1 Ferland, J. A.; Roy, S.
Timetabling problem for university as assignment of activities to resources.
FISHC 1983-1 Fisher, J. G.; Shier, D. R.
A heuristic procedure for large-scale examination scheduling problems.
GLAN 1984-1 Glanert, W.
A timetable approach to the assignment of pilots to rotations.
GLAS 1986-1 Glassey, C. R.; Mizrach, M.
A decision support system for assigning classes to rooms.
GOLE 1985-3 Golenko, D. I.
Using metric spaces in problems of optimal calendar planning.
HEIN 1984-1 Heinrich, H.
Ein heuristisches Lösungsverfahren zur Stundenplanung und Raumbelegung an Hochschulen.
JUNGB 1985-1 Junginger, W.
Stundenpläne aus dem Computer - Ein Bericht zur momentanen Lage in Deutschland.
JUNGB 1985-2 Junginger, W.
Stundenpläne aus dem Computer - ein Bericht zur momentanen Lage in Deutschland.
JUNGB 1986-1 Junginger, W.
Timetabling in Germany - a survey.
LAPOA 1984-1 Laporte, G.; Desroches, S.
Examination timetabling by computer.
LAPOA 1986-1 Laporte, G.; Desroches, S.
The problem of assigning students to course sections in a large engineering school.
NUNE 1983-1 Nunen, J. A. E. E. van; Beulens, A. J. M.; Benders, J. F.
On using the linear programming relaxation of assignment type mixed integer problems.
RUDN 1984-1 Rudnev, Y. P.
A method for solution of a calendar planning problem with discrete time (Russian).
SABI 1986-1 Sabin, G. C. W.; Winter, G. K.
The impact of automated timetabling on universities - a case study.
SCHRC 1983-1 Schreuder, J. A. M.
The challenge of program-timetables: a new concept.
SCHRD 1984-1 Schreuder, J. A. M.; Velde, J. A. van der
Timetables in Dutch high schools.
SEBO 1986-1 Sebö, A.
Application of network flow methods in the scheduling of terminal sessions (Hungarian).
SEBO 1987-1 Sebö, A.
A particular timetable problem: terminal scheduling.
TRIP 1984-1 Tripathy, A.
School timetabling - a case in large binary integer linear programming.
WERR 1985-1 Werra, D. de
An introduction to timetabling.
WERR 1985-2 Werra, D. de
Graphs, hypergraphs and timetabling.
WERR 1985-3 Werra, D. de
Some uses of hypergraphs in timetabling.
WERR 1986-1 Werra, D. de
Some models of graphs for scheduling sports competitions.
WERR 1986-2 Werra, D. de
Variations on the integral decomposition property.
WERRA 1986-1 Werra, D. de; Pasche, C.; Petter, A.
Time-tabling problems: should they be canonical?

ABDA 1987-1 Abdallaoui, G.
Maintainability of a grade structure as a transportation problem.

ABUO 1986-1 Abuov, K. A.; Cherenin, V. P.
Synthesis of one-product two-stage production-transport networks.

ACHA 1983-1 Achary, K. K.; Tikekar, V. G.
On trade-off solution pairs in a special type of transportation problem.

AHUJ 1986-1 Ahuja, R. K.
Algorithms for the minimax transportation problem.

AKINA 1986-1 Akinc, U.; Roodman, G. M.
A new approach to aggregate production planning.

ALEKB 1984-1 Alekseev, A. O.
The transportation problem with respect to the time criterion in case of limited amount of transport facilities (Russian).

ALEKB 1985-2 Alekseev, A. O.
Multivariant transport problem by the time criterion.

ALI 1984-1 Ali, A. I.
A greedy algorithm for the transshipment along a single road problem.

ALI 1986-1 Ali, A. I.
An efficient algorithm for the transshipment along a single road problem.

AROR 1986-1 Arora, S. R.
Multi-index transportation problem with planar constraints.

ASTA 1983-1 Astakhov, N. D.
On the dynamical problem of positioning factories with warehouses of production inventory.

AVEN 1985-1 Aven, O. I.; Lovetskij, S. E.; Moiseenko, G. E.
Optimization of transportation flows (Russian).

AVRA 1984-2 Avram-Nitchi, R.
An operatorial model of transportation type problems.

BALAA 1981-1 Balachandran, V.; Srinivasan, V.; Thompson, G. L.
Applications of the operator theory of parametric programming for the transportation and generalized transportation problems.

BALI 1984-2 Balinski, M. L.
The Hirsch conjecture for dual transportation polyhedra.

BALIB 1984-1 Balinski, M. L.; Russakoff, A.
Faces of dual transportation polyhedra.

BANDB 1985-1 Bandopadhyay, L.
The capacitated solid transportation problem (planar constraints) with upper and lower bounds on rim conditions.

BARNA 1985-1 Barnes, E. R.; Hoffman, A. J.
On transportation problems with upper bounds on leading rectangles.

BARR 1981-1 Barr, R. S.; Turner, J.
Microdata file merging through large-scale network technology.

BATI 1987-1 Batishchev, D. I.; Kogan, D. I.
Transportation problems with linear estimates of participants.

BATIA 1983-1 Batishchev, D. I.; Kogan, D. I.; Shakhriev, K.
Transportation problem with dichotomous preferences (Russian).

BATIA 1985-1 Batishchev, D. I.; Kogan, D. I.; Shakhriev, K.
Solution of a multicriterial transport problem in an interactive regime (Russian).

BERE 1985-1 Berezinski, M.; Krus, L.
Aids for multiobjective decision making on development of transportation network under uncertainty.

BIALA 1985-1 Bialy, H.
Stabilitätsbetrachtungen zur linearen Transportoptimierung.

BIALA 1987-1 Bialy, H.
Duale Transportoptimierung.

BLYU 1984-1 Blyumin, S. L.; Kozlov, P. A.; Milovidov, S. P.
A dynamic transportation problem with delays (Russian).

BOEN 1985-1 Boenchendorf, K.
Improving the Kennington/Unger penalty method for the fixed-charge transportation problem.

BOEN 1986-1 Boenchendorf, K.
Combining penalties for the fixed-charge transportation problem.

BOEN 1987-1 Boenchendorf, K.
An optimality criterion for degenerated transportation problems.

BORZB 1983-1 Borzenkov, A. V.; Gabasov, R.
An algorithm for solving a special minimax problem (Russian).

BRAM 1984-1 Brameller, A.
Sparsity in transportation problems.

BULU 1985-1 Bulut, H.
Decomposition properties of the transportation problem (Turkish).

CHANE 1984-1 Chanas, S.; Kolodziejczyk, W.; Machaj, A.
A fuzzy approach to the transportation problem.

CHANE 1985-1 Chanas, S.; Kolodziejczyk, W.; Machaj, A.
The fuzzy transportation problem with fuzzy constraints.

CHANH 1987-1 Chandra, S.; Saxena, P. K.
Cost/completion-date tradeoffs in quadratic fractional transportation problem.

CHANI 1987-1 Chandra, S.; Seth, K.; Saxena, P. K.
Time minimizing transportation problem with impurities.

CURRA 1986-1 Current, J. R.; Min, H.-K.
Multiobjective design of transportation networks: Taxonomy and annotation.

CURRD 1986-1 Currin, D. C.
Transportation problems with inadmissible routes.

DADU 1986-1 Daduna, J. R.
Solving a p-center transportation problem.

DAS 1987-1 Das, C.; Patel, G.
A variant of the solid transportation problem with mixed constraints.

DELG 1987-1 Delgado, M.; Verdegay, J. L.; Vila, M. A.
Fuzzy transportation problems: a general analysis.

DOMS 1981-1 Domschke, W.
Logistik: Transport. Grundlagen, lineare Transport- und Umladeprobleme.

DONGB 1986-1 Dong, Z. F.
An investigation of degenerate cases of the transportation problem (Chinese).

DUBU 1986-1 Dubuc, S.; Tanguay, M.
Deplacement de materiel continu unidimensionnel a moindre cout.

EISEC 1984-1 Eiselt, H. A.; Gerchak, Y.
Solution structures and sensitivity of special assignment problems.

EMEL 1981-1 Emelichev, V. A.; Kravtsov, M. K.
Transport polytopes (Russian).

EMEL 1981-2 Emelichev, V. A.; Kravtsov, M. K.
Transportation polyhedra (Russian).

EMEL 1983-1 Emelichev, V. A.; Kravtsov, M. K.
Combinatorial problems on transport polyhedra - a survey (Russian).

EMEL 1984-1 Emelichev, V. A.; Kravtsov, M. K.
Transportpolyeder mit maximaler Anzahl von Seiten aller Dimensionen.

EMELA 1985-1 Emelichev, V. A.; Kravtsov, M. K.; Krachkovskii, A. P.
Kombinatorische und Extremalprobleme auf Transportpolyedern.

EMELA 1985-2 Emelichev, V. A.; Kravtsov, M. K.; Krachkovskii, A. P.
Transportation polyhedra with a maximum number of k-faces (Russian).

EMELC 1986-1 Emelichev, V. A.; Perepelitsa, V. A.
On the estimation of the complexity of multicriterial transportation problems (Russian).

EREN 1984-2 Erenguc, S. S.; Tufekci, S.
A transportation type aggregate production model with bounds on inventory and backordering.

EVAN 1984-1 Evans, J. R.
The factored transportation problem.

EVAN 1985-1 Evans, J. R.
On equivalent transportation models for production planning problems.

FINKC 1987-1 Finke, G.
Network flow based branch and bound method for asymmetric traveling salesman problems.

FLOR 1986-1 Florescu, G.
Aspects of modelling transportation through the dynamic transportation problem.

FOTI 1985-1 Foti, G.; Tadei, R.
New transportation models for urban planning: fundamental theory and use (Italian).

FRID 1983-1 Fridman, V. V.
A transportation problem with respect to time for networks of a special type (Russian).

GAMB 1987-1 Gambarov, L. A.
On a method of decomposition for a multiproduct transportation problem with intermediate nodes (Russian).

GARG 1985-1 Garg, R.; Prakash, S.
Time minimizing transshipment problem.

GOYA 1984-1 Goyal, S. K.
Improving VAM for unbalanced transportation problems.

GRIB 1985-1 Gribik, P. R.; Kortanek, K. O.
Extremal methods of operations research.

GUPTA 1987-1 Gupta, A. K.; Sharma, J. K.
An iterative technique for time-cost trade-off in solid transportation problems.

GURV 1984-1 Gurvich, V. A.
Kirchhoff's second law for optimal transportation flows.

HARR 1986-1 Harris, B.
A continuous dual method for the Hitchcock problem.

HIRC 1985-1 Hirche, J.
Optimalität der Nordwesteckenlösung beim Transportproblem mit monoton kreuzdifferenten Kosten.

HUNG 1986-1 Hung, M. S.; Rom, W. O.; Waren, A. D.
Degeneracy in transportation problems.

IKURB 1983-1 Ikura, Y.; Nemhauser, G. L.
A polynomial-time dual simplex algorithm for the transportation problem: condensed version.

IKURB 1986-1 Ikura, Y.; Nemhauser, G. L.
Computational experience with a polynomial-time dual simplex algorithm for the transportation problem.

ILIC 1981-1 Ilichev, A. P.; Shevchenko, V. N.
Extreme points of polyhedra of multi-index transportation problems (Russian).

INTR 1985-1 Intrator, J.
A note on "The assignment problem under categorized jobs".

INTR 1985-2 Intrator, J.
Experimental comparisons of codes for long transportation problems.

INTRA 1984-1 Intrator, J.; Ehrlich, G.
Reusing augmenting paths in long transportation problems.

INTRA 1984-2 Intrator, J.; Ehrlich, G.
Tree modification in the transportation problem: a new method.

INTRB 1987-1 Intrator, J.; Peleg, D.
An efficient algorithm for the long transportation problem.

INTRC 1986-1 Intrator, J.; Weiss, J.
Improving techniques for transportation problems.

INTRC 1987-1 Intrator, J.; Weiss, J.
Exhaustive iterations - an improving approach for long transportation problems.

IPEK 1983-1 Ipek, M.; Dökmeci, V.
Optimum design of transportation networks.

ISER 1984-1 Isermann, H.
Linear bottleneck transportation problems.

ISTO 1984-1 Istomin, L. A.; Golovnya, S. G.
An approach to the solution of large-scale transportation problems (Russian).

JAIL 1985-1 Jaillet, P.
Analysis of the probabilistic traveling salesman problem in the plane.

KABE 1986-1 Kabe, D. G.
On a linear costs standard transportation problem optimality test.

KALT 1987-1 Kaltinska, R. N.
A method for solving a class of quadratic transportation problems (Russian).

KESEA 1984-1 Keselman, V. O.
The statistical approach to the solution of a class of transportation problems with fuzzy constraints (Russian).

KHAND 1983-1 Khanna, S.
Multi-index time transportation problem.

KIKU 1985-1 Kikuchi, S.
Empty freight car dispatching model under freight car pool concept.

KLEI 1986-1 Klein Haneveld, W. K.
Robustness against dependence in PERT: an application of duality and distributions with known marginalS.

KLEIC 1987-1 Kleinschmidt, P.; Lee, C. W.; Schannath, H.
Transportation problems which can be solved by the use of Hirsch-paths for the dual problems.

KORSA 1987-1 Korsnikov, A. D.
Planar three-index transportation problems with dominating index.

KORSB 1987-1 Korsnikov, A. D.; Burkard, R. E.
On the dimension of polytopes of planar three-index transportation problems.

KRAV 1984-1 Kravtsov, M. K.
On an upper bound for the radius of a transportation polyhedron (Russian).

KRAV 1985-1 Kravtsov, M. K.
A proof of the maximal diameter conjecture for the transportation polyhedron.

KRAV 1986-1 Kravtsov, M. K.
On the structure of the set of admissible solutions of a transportation problem with limited traffic capacities (Russian).

KRAVA 1984-1 Kravtsov, M. K.; Emelichev, V. A.
Facets, diameter and radius of a transportation polytope (Russian).

KRAVA 1984-2 Kravtsov, M. K.; Emelichev, V. A.
On transportation polytopes with a maximum number of facets of all dimensions (Russian).

KRAVB 1987-1 Kravtsov, M. K.; Galaktionova, E. S.
The number of lattice points in integer transportation polyhedra (Russian).

KUMAB 1986-1 Kumar, K. R.; Kusiak, A.; Vannelli, A.
Grouping of parts and components in flexible manufacturing systems.

KWAK 1985-3 Kwak, N. K.; Schniederjans, M. J.
Goal programming solutions to transportation problems with variable supply and demand requirements.

LEE N 1986-1 Lee, Y.-D.
Integer programming applications and their relationship (Korean).

LESH 1984-1 Leshchenko, L. L.
On a stochastic transportation problem (Russian).

LEWIA 1986-1 Lewis, A.
Extreme points of infinite transportation problems.

LIN B 1986-1 Lin, T. F.
A system of linear equations related to the transportation problem with application to probability theory.

LOZO 1984-2 Lozovanu, D. D.
Some properties of a network nonlinear transportation problem and the problem of synthesis of a network in the form of an outgoing tree with a source (Russian).

MARTD 1983-1 Martin, E.
Lösung von Transportproblemen mit dem Programmsystem OPSI.

MATV 1986-1 Matveenko, V. D.
A combinatorial approach to the problem of solvability of the three-index transportation problem.

MAZU 1984-1 Mazurov, V. D.
Committee solution of an unbalanced transportation problem (Russian).

MOTE 1986-1 Mote, J.
The family constrained network problem.

NERA 1984-1 Neralic, L.; Sirovitsa, L.; Skok, S.; Khunet, D.
Solution of a production-transportation problem by Benders' decomposition method (Russian).

NYKO 1986-1 Nykowski, I.
Bicriterion linear transportation problem (Polish).

OHUC 1984-1 Ohuchi, A.; Kaji, I.
Lagrangian dual coordinatewise maximization algorithm for network transportation problems with quadratic costs.

OSMA 1986-1 Osman, M. S. A.; Ellaimony, E. E. M.
An algorithm for solving large scale transhipment problems.

PAIB 1986-1 Paiboon, C.-P.; Pojana, P.-P.; Rosenthal, R. E.
Optimal ship routing and personnel assignment for naval recruitment in Thailand.

PALE 1987-1 Palekar, U. S.; Karwan, M. H.; Zionts, S.
A branch-and-bound method for the fixed charge transportation problem.

PALE 1987-2 Palekar, U. S.; Karwan, M. H.; Zionts, S.
Duality-based bounds for the fixed charge transportation problem.

PELI 1984-1 Pelikan, J.
Container transportation problem (Czechoslovakian).

PELI 1985-1 Pelikan, J.
Container transportation problem.

PLAS 1985-1 Plasil, J.
Optimal organizing of flows in transportation network.

POSNA 1985-1 Posner, M. E.; Szwarc, W.
Transshipment along a single road.
POTA 1984-1 Potapova, L. I.
On a procedure for solving a transportation problem with respect to a time criterion (Russian).
PURU 1984-1 Purushotam, S.; Prakash, S.
A transportation problem with minimization of duration and total cost of transportation as high and low priority objectives respectively.
QI 1985-1 Qi, L.-Q.
Forest iteration method for stochastic transportation problem.
QI 1987-3 Qi, L.-Q.
The A-forest iteration method for the stochastic generalized transportation problem.
RADA 1986-1 Radaramanan, R.; Choi, L. I.
A branch and bound algorithm for the travelling salesman and the transportation routing problems.
RINGA 1987-1 Ringuest, J. L.; Rinks, D. B.
Interactive solutions for the linear multiobjective transportation problem.
ROOS 1984-1 Roos, C.
Network programming with umbrellas.
RUMC 1984-1 Rumchev, V. G.; Mateev, L. N.
Solution of a dynamic transportation problem in network formulation by the method of a time expanded network (Russian).
RYAN 1986-1 Ryang, S. Y.
A solution method of the fractional n-acyclic transportation problem (Korean).
RYAN 1987-1 Ryang, S. Y.
A solution method of the fractional n-index transportation problem with pass capacity (Korean).
SAND 1986-1 Sandi, C.
On a nonbasic dual method for the transportation problem.
SAXE 1985-1 Saxena, P. K.
Shipping cost/completion-date tradeoffs in fractional transportation problems with penalty cost.
SCHRB 1985-1 Schreier, H.; Tzscharschuch, D.
Ein Verfahren zur Lösung eines mehrstufigen Transportproblems mit nach oben beschränkten Variablen und rechentechnische Realisierung.
SCHU 1986-1 Schubert, I. S.; Zimmermann, U.
One-parametric bottleneck transportation problems.
SCHUA 1987-1 Schulz, G.
Ein duales Verfahren zur Lösung des Transportproblems mit zusätzlichen linearen Restriktionen.
SCOB 1984-1 Scobey, P.; Kabe, D. G.
Direct solutions to some constrained transportation problems.
SERGJ 1985-1 Sergienko, T. I.
Method for solving a special integer programming problem.
SHEV 1984-1 Shevchenko, A. N.
Application of the dynamic programming method and duality theory to a multi-index transportation problem (Russian).
SOKOC 1982-1 Sokolovskij, V. Z.; Zadorozhnyj, M. V.
Experiences with random search in problems of the traveling salesman (Russian).
SOLOA 1984-1 Solomon, D. I.
Generalization of the linear and fractional-linear transportation problem (Russian).
STAN 1985-1 Stancu-Minasian, I. M.; Tigan, S.
The minimum-risk approach to the bottleneck transportation problem.
SZWAB 1985-1 Szwarc, W.; Posner, M. E.
The caterer problem.
TAYI 1986-1 Tayi, G. K.
Bicriteria transportation problem: an alternate approach.
THOMC 1986-1 Thompson, G. L.; Sethi, A. P.
Solution of constrained generalized transportation problems using the pivot and probe algorithm.
TIGAA 1985-1 Tigan, S.; Stancu-Minasian, I. M.
The stochastic bottleneck transportation problem.
TSYG 1985-1 Tsygankov, I. V.
Solving the transportation problem by using the method of invariants (Russian).
VANN 1984-1 Vannelli, A.
Approximating a class of graph decomposition problems by linear transportation problems.
VASY 1983-1 Vasyanin, V. A.
On a problem of discrete optimization in processes of control over transportations in multiflow transportation networks (Russian).
VLAC 1986-1 Vlach, M.
Conditions for the existence of solutions of the three-dimensional planar transportation problem.
VULI 1985-1 Vulis, V. I.; Sherman, Y. S.
Decomposition method for solving large-scale problems of transportation type (Russian).
WALL 1986-1 Wallace, S.
Decomposing the requirement space of a transportation problem into polyhedral cones.
WALT 1984-1 Walther, H.
Ten applications of graph theory.
WALTA 1987-1 Walther, H.; Nägler, G.
Graphen, Algorithmen, Programme.
WANGL 1985-1 Wang, Y. C.
An improvement on an algorithm for a transport problem with a feasibility constraint (Chinese).
XUE B 1987-1 Xue, Y.-F.; Li, Y.
A minimum spanning tree algorithm for solving the transportation problem (Chinese).
YON 1986-1 Yon, R. S.
A solution method of the fractional n-index transportation problem.
YON 1987-1 Yon, R. S.
A solution method of the fractional n-index transportation problem with pass capacity (Korean).

ZAITB 1983-1 Zaitsev, R. V.; Sibirko, A. N.; Kostenko, E. I.
Application of sequential optimization algorithms for the solution of a problem of transportation type (Russian).

ZHAOA 1982-1 Zhao, F.-Z.
The potential method for solving the multicommodity capacity transportation problem (Chinese).

ADRA 1983-1 Adrabinski, A.; Syslo, M. M.
Computational experiments with some approximation algorithms for the travelling salesman problem.

ALEKB 1985-1 Alekseev, A. O.
A multivariant travelling-salesman problem.

ALI A 1986-1 Ali, A. I.; Kennington, J. L.
The asymmetric m-travelling salesmen problem: a duality based branch-and-bound algorithm.

ALTI 1987-1 Altinkemer, K.; Gavish, B.
Heuristics for unequal weight delivery problems with a fixed error guarantee.

ALVA 1985-1 Alvarez-Valdes, R.; Corberan, A.; Tamarit, J. M.
Polyhedral combinatorics and the traveling salesman problem. Application to the case of 103 Spanish cities (Spanish).

ARTH 1985-1 Arthur, J. L.; Frendewey, J. O.
A computational study of tour construction procedures for the traveling salesman problem.

ARTH 1985-2 Arthur, J. L.; Frendewey, J. O.
Generating traveling salesman problems with known optimal tours.

BALAG 1987-1 Balas, E.
The prize collecting traveling salesman problem.

BALAO 1985-1 Balas, E.; Toth, P.
Branch and bound methods.

BARAB 1985-1 Barahona, F.; Grötschel, M.
The travelling salesman problem for graphs not contractible to k(5)-e.

BASK 1986-1 Baskaran, G.; Fu, Y.-T.; Anderson, P. W.
On the statistical mechanics of the traveling salesman problem.

BIXB 1987-1 Bixby, R. E.
Notes on combinatorial optimization.

BLANB 1987-1 Bland, R. G.; Shallcross, D. F.
Large traveling salesman problems arising from experiments in x-ray crystallography: a preliminary report on computation.

BOND 1983-1 Bondarenko, V. A.
Nonpolynomial lower bounds for the complexity of the traveling salesman problem in a class of algorithms.

BONO 1984-1 Bonomi, E.; Lutton, J.-L.
The n-city travelling salesman problem: statistical mechanics and the metropolis algorithm.

BOYD 1986-1 Boyd, S. C.
The subtour polytope of the travelling salesman problem.

BOYDA 1984-1 Boyd, S. C.; Pulleyblank, W. R.
Optimizing over the subtour polytope of the travelling salesman problem.

BOYDB 1987-1 Boyd, S. C.; Pulleyblank, W. R.; Cornuejols, G.
Travel - an interactive travelling salesman problem package for the IBM-personal computer.

CARRB 1987-1 Carraway, R. L.; Morin, T. L.; Moskowitz, H.
The stochastic traveling salesman problem revisited: a generalized dynamic programming approach.

CHAK 1986-1 Chakrabarti, B. K.
Directed travelling salesman problem.

CHANJ 1986-1 Chandrasekaran, R.
Recognition of Gilmore-Gomory traveling salesman problem.

CHRI 1985-2 Christofides, N.
Vehicle routing.

CHRO 1986-1 Chrobak, M.; Poljak, S.
On common edges in optimal solutions to traveling salesman and other optimization problems.

CHVA 1985-2 Chvatal, V.
Hamiltonian cycles.

CORNG 1985-1 Cornuejols, G.; Fonlupt, J.; Naddef, D. J.
The traveling salesman problem on a graph and some related integer polyhedra.

CORNI 1985-1 Cornuejols, G.; Naddef, D. J.; Pulleyblank, W. R.
The traveling salesman problem in graphs with 3-edge cutsets.

DEIN 1985-1 Deineko, V. G.
A polynomial algorithm for the search for the optimum in exponential neighborhoods for traveling salesman problems (Russian).

DHAR 1987-1 Dhar, D.; Barma, M.; Chakrabarti, B. K.; Taraphder, A.
The travelling salesman problem on a randomly diluted lattice.

DILL 1987-1 Dillencourt, M. B.
Traveling salesman cycles are not always subgraphs of Delaunay triangulations or of minimum weight triangulations.

ERNVA 1985-1 Ernvall, J.; Katajainen, J.; Penttonen, M.
NP-completeness of the Hamming salesman problem.

EVANA 1984-1 Evans, J. R.; Hall, R. A.
Probabilistic analysis of assignment ranking: the traveling salesman problem.

FEIR 1987-2 Feiring, B. R.
Some properties of the solution space of the N-city traveling-salesman problem.

FINKC 1984-1 Finke, G.
Bulk shipment formulations for the traveling salesman problem.

FINKD 1987-1 Finke, G.; Ahmed, S.
Network flow methods for the asymmetric traveling salesman problem.

FINKF 1984-1 Finke, G.; Claus, A.; Gunn, E. A.
A two-commodity network flow approach to the traveling salesman problem.

FISC 1986-1 Fischetti, M.
An improved bound for the asymmetric traveling salesman problem.

FISCA 1987-1 Fischetti, M.; Toth, P.
Bounding and dominating procedure for combinatorial optimization problems (Italian).

FLEI 1985-1 Fleischmann, B.
A cutting plane procedure for the travelling salesman problem on road networks.

FLEI 1988-1 Fleischmann, B.
Cutting planes for the symmetric travelling salesman problem.

FONL 1985-1 Fonlupt, J.; Naddef, D. J.
The traveling salesman problem in graphs with some excluded minors.

FOUL 1984-1 Foulds, L. R.
Combinatorial optimization for undergraduates.

FRIEC 1987-1 Frieze, A. M.
On the exact solution of random travelling salesman problems with medium size integer coefficients.

FRIS 1983-1 Fris, P.; Volgenant, A.
Solving the traveling salesman problem in Basic.

FRUH 1987-1 Fruhwirth, B.
Untersuchung über genetisch motivierte Heuristiken für das Euklidische Rundreiseproblem.

GANI 1987-1 Gani, A. Z.
The application of the interaction theory for solving the traveling salesman problem.

GARF 1985-1 Garfinkel, R. S.
Motivation and modeling.

GAVIA 1986-1 Gavish, B.; Srikanth, K.
An optimal solution method for large-scale multiple traveling salesmen problems.

GERAD 1983-1 Gerasimov, V. A.
On a modification of the traveling salesman problem (Russian).

GILM 1985-1 Gilmore, P. C.; Lawler, E. L.; Shmoys, D. B.
Well-solved special cases.

GLUS 1985-1 Glushkova, V. V.
An algorithm for solving the traveling salesman problem with moving objects (Russian).

GOLDI 1985-1 Golden, B. L.; Levy, L.; Vohra, R. V.
Some heuristics for the generalized traveling salesman problem.

GOLDK 1985-1 Golden, B. L.; Stewart, W. R. Jr.
Empirical analysis of heuristics.

GROTD 1985-1 Grötschel, M.; Padberg, M. W.
Polyhedral theory.

GROTE 1986-1 Grötschel, M.; Pulleyblank, W. R.
Clique tree inequalities and the symmetric travelling salesman problem.

GROTE 1987-1 Grötschel, M.; Pulleyblank, W. R.
Corrigenda: clique tree inequalities and the symmetric travelling salesman problem.

GULYA 1986-1 Gulyanitskii, L. F.; Sapko, A. V.
On an approach to the solution of nonsymmetric travelling salesman problems (Russian).

HELD 1984-1 Held, M.; Hoffman, A. J.; Johnson, E. L.; Wolfe, P.
Aspects of the traveling salesman problem.

HENR 1987-1 Henriques, N.; Safayeni, F.; Fuller, D.
Human heuristics for the travelling salesman problem.

HOFFC 1985-1 Hoffman, A. J.; Wolfe, P.
History.

HOFFD 1985-1 Hoffman, K.; Padberg, M. W.
LP-based combinatorial problem solving.

HOFFD 1986-1 Hoffman, K.; Padberg, M. W.
LP-based combinatorial problem solving.

IMAI 1986-1 Imai, H.
Worst-case analysis for planar matching and tour heuristics with bucketing techniques and spacefilling curves.

JERO 1985-1 Jeromin, B.; Körner, F.
On the refinement of bounds of heuristic algorithms for the traveling salesman problem.

JERO 1986-2 Jeromin, B.; Körner, F.
Sharp bounds for Karp's "patching"-algorithm for the approximate solution of the traveling salesman problem.

JOHNG 1985-1 Johnson, D. S.; Papadimitriou, C. H.
Computational complexity.

JOHNG 1985-2 Johnson, D. S.; Papadimitriou, C. H.
Performance guarantees for heuristics.

JONG 1985-1 Jongens, K.; Volgenant, T.
The symmetric clustered traveling salesman problem.

JONK 1986-1 Jonker, R.
Traveling salesman and assignment algorithms: design and implementation.

JONKA 1984-1 Jonker, R.; Volgenant, T.
Nonoptimal edges for the symmetric traveling salesman problem.

JONKA 1986-2 Jonker, R.; Volgenant, T.
Transforming asymmetric into symmetric traveling salesman problems: Erratum.

JUNGC 1987-1 Jungnickel, D.
Graphen, Netzwerke und Algorithmen.

KALAA 1985-1 Kalantari, B.; Hill, A. V.; Arora, S. R.
An algorithm for the traveling salesman problem with pickup and delivery customers.

KAME 1986-1 Kamenarov, S. N.
Two special cases for the traveling salesman problem with premiums (Bulgarian).

KARPF 1985-1 Karp, R. M.; Steele, J. M.
Probabilistic analysis of heurictics.

KASPA 1986-1 Kaspshitskaya, M. F.; Glushkova, V. V.
On the stability to solve travelling salesman problem with moving objects (Russian).

KERN 1986-1 Kern, W.
A probabilistic analysis of the switching algorithm for the euclidean TSP.

KHOD 1985-1 Khodzinskii, A. N.
Sequential algorithm for the solution of problems of combinatorial optimization on permutations.

KIND 1986-3 Kindervater, G. A. P.; Lenstra, J. K.
The parallel complexity of TSP heuristics.

KIRKA 1985-1 Kirkpatrick, S.; Toulouse, G.
Configuration space analysis of travelling salesman problems.

KOLI 1987-1 Kolinek, M.
A polynomial-time linear decision tree for the traveling salesman problem and other NP-complete problems.

KORN 1985-3 Körner, F.
Linear transformations associated with the traveling salesman problem.

KORN 1986-2 Körner, F.
On the relationship of approximation algorithms for the minimum and the maximum travelling salesman problem.

KORN 1987-1 Körner, F.
On the degree of asymmetry in the travelling salesman problem.

KOTO 1985-1 Kotov, V. M.; Cherenkova, E. P.
A parametric algorithm for the solution of the travelling salesman problem in the plane (Russian).

KOVAB 1985-1 Kovacs, W. J.; Goodin, D. T.
A statistical approach to the traveling salesman problem.

KUBO 1986-1 Kubo, M.; Kasugai, H.
Traveling purchaser problem (Japanese).

KUBO 1987-1 Kubo, M.; Kasugai, H.
Errata: "Traveling purchaser problem" (Bull. Sci. Engrg. Res. Lab. Waseda Univ., No. 116 (1986), 7-12).

LAPO 1986-1 Laporte, G.
Generalized subtour elimination constraints and connectivity constraints.

LAPO 1987-1 Laporte, G.
Solving a family of permutation problems on 0-1 matrices.

LAPOB 1987-1 Laporte, G.; Martello, S.
The selective travelling salesman problem.

LAPOC 1985-1 Laporte, G.; Mercure, H.; Nobert, Y.
Finding the shortest Hamiltonian circuit through n-clusters: a Lagrangean relaxation approach.

LAPOC 1987-2 Laporte, G.; Mercure, H.; Nobert, Y.
Generalized travelling salesman problem through n sets of nodes: the asymmetrical case.

LAPOG 1987-1 Laporte, G.; Nobert, Y.; Nga, T. K.
An optimal flow circulation algorithm for the asymmetrical multiple travelling salesman problem.

LENSC 1985-1 Lenstra, J. K.; Rinnooy Kan, A. H. G.; Guan, M.-G.
On travelling salesman problem (Chinese).

LITK 1984-1 Litke, J. D.
An improved solution to the traveling salesman problem with thousands of nodes.

MAGI 1986-1 Magirou, V. F.
The efficient drilling of printed circuit boards.

MIKA 1984-1 Mikami, K.; Machida, H.
On approximate solutions of the rectilinear travelling salesman problem (Japanese).

MIKHB 1987-1 Mikhalevich, M. V.; Sapko, A. V.
The efficiency of an algorithm for solving the traveling salesman problem (Russian).

MORAA 1984-1 Moran, S.
On the length of optimal TSP circuits in sets of bounded diameter.

NADD 1987-1 Naddef, D. J.
The symmetric traveling salesman polytope I: valid inequalities derived from the graphical relaxation.

ONG 1984-1 Ong, H. L.
Probabilistic analysis of some travelling salesman heuristics.

PADBA 1985-1 Padberg, M. W.; Grötschel, M.
Polyhedral computations.

PADBB 1987-1 Padberg, M. W.; Rinaldi, G.
Optimization of a 532-city symmetric traveling salesman problem by branch and cut.

PAPAB 1984-1 Papadimitriou, C. H.; Vazirani, U. V.
On two geometric problems related to the traveling salesman problem.

PAPAC 1985-1 Papadimitriou, C. H.; Wolfe, D.
The complexity of facets resolved.

PAPAE 1986-1 Papakostantinou, N.
The complexity of the travelling repairman problem.

PILC 1987-1 Pilcher, M. G.; Rardin, R. L.
A random cut generator for symmetric traveling salesman problems with known optimal solutions.

PLAN 1987-1 Plante, R. D.; Lowe, T. J.; Chandrasekaran, R.
The product matrix traveling salesman problem: An application and solution heuristic.

PLATB 1984-1 Platzman, L. K.; Bartholdi, J. J. III
Spacefilling curves and the planar travelling salesman problem (revised).

PLES 1986-1 Plesnik, J.
Bad examples of the metric traveling salesman problem for the 2-change heuristic.

POLJ 1986-1 Poljak, S.; Chrobak, M.
On common edges in optimal solutions to traveling salesman and other optimization problems.

QUINA 1984-1 Quinn, M. J.; Deo, N.
Parallel graph algorithms.

RADA 1986-1 Radaramanan, R.; Choi, L. I.
A branch and bound algorithm for the travelling salesman and the transportation routing problems.

RAND 1986-1 Randelman, R. E.; Grest, G. S.
N-city travelling salesman problem: optimization by simulated annealings.

REES 1987-1 Rees, S.; Ball, R. C.
Criteria for an optimum simulated annealing schedule for problems of the traveling salesman type.

REND 1988-1 Rendl, F.
Combinatorial improvements of the 1-tree bound for the traveling salesman problem.

ROOS 1984-1 Roos, C.
Network programming with umbrellas.

ROSS 1987-1 Rossi, F. A.; Gavioli, I.
Aspects of heuristic methods in the "Probabilistic Travelling Salesman Problem".

ROSSA 1986-1 Rossier, Y.; Troyon, M.; Liebling, T. M.
Probabilistic exchange algorithms and Euclidean traveling salesman problems.

SANCA 1986-1 Sancho, N. G. F.
Optimal and near-optimal solutions to the travelling salesman problem.
SERD 1984-1 Serdyukov, A. I.
An algorithm with an estimate for the travelling salesman problem of the maximum (Russian).
SIGA 1985-1 Sigal, I. K.
Composite algorithms for solving the traveling salesman problem (Russian).
SIGA 1986-1 Sigal, I. K.
Computer implementation of a combined branch and bound algorithm for the travelling salesman problem.
SIGA 1987-1 Sigal, I. K.
An approximate algorithm for the large-scale travelling salesman problem and its computer implementation (Russian).
SILL 1986-1 Sillaber, A. H.
A continuous linear optimization model for the exact solution of travelling-salesman-problems in connexion with expansion planning of ring networks.
SIMC 1987-1 Simchi-Levy, D.; Berman, O.
A heuristic algorithm for the travelling salesman location problem on networks.
SIMC 1987-2 Simchi-Levy, D.; Berman, O.
Heuristics and bounds for the travelling salesman location problem on the plane.
SOLI 1984-1 Solis, R.
Evaluation of heuristic algorithms for the TSP: a new statistical approach.
STEE 1986-1 Steele, J. M.
Probabilistic algorithm for the directed traveling salesman problem.
STEWA 1985-1 Stewart, W. R. Jr.
An accelerated branch exchange heuristic for the traveling salesman problem.
STEWA 1985-2 Stewart, W. R. Jr.
An improved assignment lower bound for the Euclidean traveling salesman problem.
SUND 1984-1 Sundararaghavan, P. S.
The unit capacity pickup and delivery problem on a one-way loop.
TELE 1986-1 Telemtaev, M. M.
An algorithm for the travelling salesman problem (Russian).
VOLGA 1985-1 Volgenant, A.; Jonker, R.
Improving Christofides' lower bound for the traveling salesman problem.
VOLGA 1986-1 Volgenant, A.; Jonker, R.
A transformation of a generalized traveling salesman problem.
VOLGA 1986-2 Volgenant, A.; Jonker, R.
An improved transformation for the symmetric multiple traveling salesman problem.
VOLGC 1987-1 Volgenant, A.; Sluis, H. J. van der; Jonker, R.
Better assignment lower bounds for the Euclidean traveling salesman problem.
VOLGE 1987-1 Volgenant, T.; Jonker, R.
On some generalizations of the travelling-salesman problem.
WALA 1984-1 Walas, R. A.; Askin, R. G.
An algorithm for NC turret punch press tool location and hit sequencing.
WALT 1984-1 Walther, H.
Ten applications of graph theory.
WERN 1986-1 Werner, F.
On the method of interchanging networks for permutation problems.
ZIYA 1984-1 Ziyatdinov, R. K.
Estimate of the accuracy of certain heuristic algorithms for the solution of the travelling salesman problem (Russian)

AKSE 1984-1 Aksenov, V. A.
The degree of perfection of a graph (Russian).
ALBEB 1983-1 Albertson, M. O.; Berman, D. M.
Cliques, colorings and locally perfect graphs.
ANSTA 1984-1 Anstee, R. P.; Farber, M.
Characterizations of totally balanced matrices.
BAAS 1988-1 Baas, S. M.
Berge's strong perfect graph conjecture for 4-chromatic graphs.
BASAA 1987-1 Basavayya, D.; Ravindra, G.
Strongly perfect line graphs of (0,1)-graphs.
BERG 1983-1 Berge, C.
Graphes (3. edition).
BERG 1984-2 Berge, C.
Diperfect graphs.
BERG 1984-5 Berge, C.
Stochastic graphs and strongly perfect graphs: a survey.
BERG 1985-1 Berge, C.
Graphs (second revised edition).
BERG 1986-1 Berge, C.
New classes of perfect graphs.
BERGA 1984-1 Berge, C.; Duchet, P.
Strongly perfect graphs.
BERGA 1986-1 Berge, C.; Duchet, P.
Solvability of perfect graphs.
BIXB 1984-1 Bixby, R. E.
A composition for perfect graphs.
BLANA 1984-1 Bland, R. G.; Huang, H. C.; Trotter, L. E. Jr.
Graphical properties related to minimal imperfection.
BOUCA 1987-6 Bouchet, A.
Unimodularity and circle graphs.
BRANA 1987-1 Brandstädt, A.
The computational complexity of feedback vertex set, Hamiltonian circuit, dominating set, Steiner tree, and bandwidth on special perfect graphs.
BROI 1986-1 Broin, M. W.; Lowe, T. J.
A dynamic programming algorithm for covering problems with (greedy) totally balanced constraint matrices.
BUCK 1985-1 Buckingham, M. A.; Golumbic, M. C.
Recent results on the strong perfect graph conjecture.
BURL 1984-1 Burlet, M.; Fonlupt, J.
Polynomial algorithm to recognize a Meyniel graph.
BURL 1984-2 Burlet, M.; Fonlupt, J.
Polynomial algorithm to recognize a Meyniel graph.
BURLA 1984-1 Burlet, M.; Uhry, J.-P.
Parity graphs.
CAMEH 1986-1 Cameron, K.; Edmonds, J.; Lovasz, L.
A note on perfect graphs.
CHANL 1984-1 Chandrasekaran, R.; Shirali, S.
Total weak unimodularity: testing and applications.
CHANM 1984-1 Chandrasekaran, R.; Tamir, A.
On the integrality of an extreme solution to pluperfect graph and balanced systems.
CHVA 1984-1 Chvatal, V.
A semi-strong perfect graph conjecture.
CHVA 1984-2 Chvatal, V.
An equivalent version of the strong perfect graph conjecture.
CHVA 1984-3 Chvatal, V.
Notes on perfect graphs.
CHVA 1984-4 Chvatal, V.
Perfectly ordered graphs.
CHVA 1985-3 Chvatal, V.
Star-cutsets and perfect graphs.
CHVA 1986-1 Chvatal, V.
On the P(4)-structure of perfect graphs III: partner decompositions.
CHVA 1987-1 Chvatal, V.
On the P(4)-structure of perfect graphs. III. Partner decompositions.
CHVA 1987-2 Chvatal, V.
Perfect graphs.
CHVA 1987-3 Chvatal, V.
Recent results on perfect graphs.
CHVAA 1984-1 Chvatal, V.; Graham, R. L.; Perold, A. F.; Whitesides, S. H.
Combinatorial designs related to the perfect graph conjecture.
CHVAB 1985-1 Chvatal, V.; Hoang, C. T.
On the P(4)-structure of perfect graphs. I. Even decompositions.
CHVAC 1986-1 Chvatal, V.; Lenhart, W. J.; Sbihi, N.
2-colourings that decompose perfect graphs.
CHVAD 1987-1 Chvatal, V.; Sbihi, N.
Bull-free Berge graphs are perfect.
CONFA 1984-1 Conforti, M.; Corneil, D. G.; Majoub, A. R.
K index i - covers 2: K index i - perfect graphs.
CONFC 1987-2 Conforti, M.; Rao, M. R.
Structural properties and recognition of restricted and strongly unimodular matrices.
COOKC 1986-1 Cook, W.; Fonlupt, J.; Schrijver, A.
An integer analogue of Caratheodory's theorem.

CORNB 1983-1 Corneil, D. G.; Kirkpatrick, D. G.
Families of recursively defined perfect graphs.
CORNF 1985-1 Cornuejols, G.; Cunningham, W. H.
Compositions for perfect graphs.
CRAMA 1986-1 Crama, Y.; Hammer, P. L.; Ibaraki, T.
Strong unimodularity for matrices and hypergraphs.
CRAMA 1987-1 Crama, Y.; Hammer, P. L.; Ibaraki, T.
Packing, covering and partitioning problems with strongly unimodular constraint matrices.
DALA 1986-1 Dalang, R. C.; Trotter, L. E. Jr.; Werra, D. de
On randomized stopping points and perfect graphs.
DUCH 1984-1 Duchet, P.
Classical perfect graphs.
DUCHB 1986-1 Duchet, P.; Olariu, S.
Graphes parfaitement ordonnables generalises.
FONLA 1984-1 Fonlupt, J.; Raco, M.
Orientation of matrices.
FOURB 1984-1 Fournier, J.-C.; Vergnas, M. Las
A class of bichromatic hypergraphs.
GERAB 1986-1 Gerards, A. M. H.; Schrijver, A.
Matrices with the Edmonds-Johnson property.
GERAC 1987-1 Gerards, A. M. H.; Sebö, A.
Total dual integrality implies local strong unimodularity.
GILE 1984-1 Giles, R.; Trotter, L. E. Jr.; Tucker, A.
The strong perfect graph theorem for a class of partitionable graphs.
GOLU 1984-1 Golumbic, M. C.
Algorithmic aspects of perfect graphs.
GOLUB 1982-1 Golumbic, M. C.; Monma, C. L.
A generalization of interval graphs with tolerances.
GOLUB 1982-2 Golumbic, M. C.; Monma, C. L.
A generalization of interval graphs with tolerances.
GOLUC 1984-1 Golumbic, M. C.; Monma, C. L.; Trotter, W. T. Jr.
Tolerance graphs.
GRIN 1984-1 Grinstead, C.
The perfect graph conjecture for toroidal graphs.
GROTC 1984-3 Grötschel, M.; Lovasz, L.; Schrijver, A.
Polynomial algorithms for perfect graphs.
GROTC 1986-1 Grötschel, M.; Lovasz, L.; Schrijver, A.
Relaxations of vertex packing.
GYAR 1985-2 Gyarfas, A.
Problems from the world surrounding perfect graphs.
HAYW 1985-1 Hayward, R. B.
Weakly triangulated graphs.
HAYW 1986-1 Hayward, R. B.
Murky graphs.
HOAN 1985-1 Hoang, C. T.
On the P(4)-structure of perfect graphs. II. Odd decompositions.
HOAN 1987-1 Hoang, C. T.
Alternating orientation and alternating colouration of perfect graphs.
HOAN 1987-2 Hoang, C. T.
On a conjecture of Meyniel.
HOANA 1986-1 Hoang, C. T.; Maffray, F.
Weakly triangulated graphs are strict quasi-parity graphs.
HOFFA 1985-1 Hoffman, A. J.; Kolen, A. W. J.; Sakarovitch, M.
Totally-balanced and greedy matrices.
HSU A 1984-2 Hsu, W.-L.
Finding maximum independent sets in planar perfect graphs.
HSU A 1984-4 Hsu, W.-L.
The perfect graph conjecture on special graphs - a survey.
HSU A 1987-1 Hsu, W.-L.
Decomposition of perfect graphs.
HSU A 1987-2 Hsu, W.-L.
Recognizing planar perfect graphs.
HSU B 1984-1 Hsu, W.-L.; Nemhauser, G. L.
Algorithms for maximum weight cliques, minimum weighted clique covers and minimum colorings of claw-free perfect graphs.
KAHN 1984-1 Kahn, J.
A family of perfect graphs associated with directed graphs.
LEHE 1986-1 Lehel, J.; Tuza, Z.
Neighborhood perfect graphs.
LOEB 1986-2 Loebl, M.; Poljak, S.
Remark on restricted and strongly unimodular matrices and a class between them.
LOVA 1984-1 Lovasz, L.
Normal hypergraphs and the weak perfect graph conjecture.
MARKB 1984-1 Markosyan, S. E.; Karapetyan, I. A.
Critically imperfect graphs (Russian).
MEYN 1984-1 Meyniel, H.
The graphs whose odd cycles have at least two chords.
MEYN 1986-1 Meyniel, H.
A new property of critical imperfect graphs and some consequences.
MEYN 1987-1 Meyniel, H.
A new property of critical imperfect graphs and some consequences.
OLAR 1986-1 Olariu, S.
No antitwins in minimal imperfect graphs.

OLARB 1984-1 Olaru, E.; Sachs, H.
Contributions to a characterization of the structure of perfect graphs.
PADB 1984-1 Padberg, M. W.
A characterization of perfect matrices.
PREI 1985-1 Preissmann, M.
A class of strongly perfect graphs.
PREIA 1985-1 Preissmann, M.; Werra, D. de
A note on strong perfectness of graphs.
RAVI 1984-1 Ravindra, G.
Meyniel's graphs are strongly perfect.
RAVI 1984-2 Ravindra, G.
Strongly perfect line graphs and total graphs.
REED 1985-1 Reed, B.
A note on the semi-strong perfect graph conjecture.
REED 1987-1 Reed, B.
A note on even pairs.
REED 1987-2 Reed, B.
A semi-strong perfect graph theorem.
SBIH 1984-1 Sbihi, N.; Uhry, J.-P.
A class of h-perfect graphs.
SCHRE 1986-5 Schrijver, A.
Theory of linear and integer programming.
SCHRE 1987-2 Schrijver, A.
Polyhedral combinatorics.
TAMI 1987-2 Tamir, A.
Totally balanced and totally unimodular matrices defined by center location problems.
TUCK 1984-2 Tucker, A.
The validity of the perfect graph conjecture for K(4)-free graphs.
TUCK 1987-1 Tucker, A.
A reduction procedure for coloring perfect K(4)-free graphs.
TUCK 1987-2 Tucker, A.
Coloring Perfect (K(4)-e)-free graphs.
WERR 1984-1 Werra, D. de
A decomposition property of polyhedra.
WERR 1986-2 Werra, D. de
Variations on the integral decomposition property.
WHITG 1984-1 Whitesides, S. H.
A classification of certain graphs with minimal imperfection properties.
WHITG 1984-2 Whitesides, S. H.
A method for solving certain graph recognition and optimization problems, with applications to perfect graphs.
YANN 1980-1 Yannakakis, M.
On a class of totally unimodular matrices.
YANN 1985-2 Yannakakis, M.
On a class of totally unimodular matrices.

ADAMC 1986-1 Adams, W. P.; Sherali, H. D.
A tight linearization and an algorithm for zero-one quadratic programming problems.
AGEE 1984-1 Ageev, A. A.
On the minimization of quadratic polynomials of Boolean functions (Russian).
ARZO 1984-1 Arzola, J.
The task of selection of proposals, formulation, properties, solution method (Spanish).
ASSAB 1985-1 Assad, A. A.; Xu, W.-X.
On lower bounds for a class of quadratic 0, 1 programs.
BABAA 1986-1 Babaev, D. A.; Babaev, R. D.
Generating 0-1 integer programming test problems.
BABAC 1983-1 Babaev, R. D.
A problem for the choice of an optimal sequence of construction of objects (Russian).
BABAC 1985-1 Babaev, R. D.
The construction of test problems in integer-value programming with binary unknowns.
BABAH 1985-1 Babayev, R. D.
On test integer programming problems with binary unknowns.
BALAJ 1987-1 Balas, E.; Nauss, R.; Zemel, E.
Comment on:'Some computational results on real 0-1 knapsack problems'.
BARA 1986-1 Barahona, F.
A solvable case of quadratic 0-1 programming.
BARAE 1987-1 Barahona, F.; Jünger, M.; Reinelt, G.
Experiments in quadratic 0-1 programming.
BARCA 1985-1 Barcia, P.
The bound improving sequence algorithm.
BEREA 1984-1 Bereznev, V. A.; Palmova, I. E.; Tretyakov, A. A.
A method for solving a class of problems of linear integer programming with Boolean variables (Russian).
BORC 1987-1 Borchardt, M.
Zur Anwendung eines Dualitätskonzeptes für quadratische Optimierungsprobleme mit stetigen und Booleschen Variablen. I.
BOROD 1986-1 Boros, E.
On the complexity of the surrogate dual of 0-1 programming.
BOURA 1984-1 Bourjolly, J.-M.; Hammer, P. L.; Simeone, B.
Node-weighted graphs having the König-Egervary property.
CONL 1984-1 Conley, W.
Satisfying the customer.
CRAMC 1986-1 Crama, Y.; Hansen, P.; Williams, A. C.
Use of the dual in branch and bound methods for mixed 0-1 programming.
DILLA 1988-1 Dillmann, R.
Hierarchisierung von Restriktionen und der Balas-Algorithmus.
DONAA 1986-1 Donath, G.; Hofstedt, K.
Über eine spezielle lineare 0-1-Optimierungsaufgabe.
EL-B 1985-1 El-Bakry, M.; Nassar, S.
The solution of the 0-1 programming problem by parallel processing.
EL-BA 1985-1 El-Bakry, M.; Wahsh, S.; Nassar, S.
A new pulse width modulation technique based on 0-1 programming.
EVAN 1987-1 Evans, J. R.
Structural analysis of local search heuristics in combinatorial optimization.
FINKH 1986-1 Finkelstein, Y. Y.; Shmelev, A. A.
On an estimate of coefficients of the faces of some integer polyhedra (Russian).
FISHA 1985-1 Fishburn, P. C.; Kochenberger, G. A.
Aggregating assignment constraints.
FOX 1985-1 Fox, G. E.; Scudder, G. D.
A heuristic with tie breaking for certain 0-1 integer programming models.
FOX 1986-1 Fox, G. E.; Scudder, G. D.
A simple strategy for solving a class of 0-1 integer programming models.
FREV 1986-1 Freville, A.; Plateau, G.
Heuristics and reduction methods for multiple constraints 0-1 linear programming problems.
GAVI 1985-2 Gavish, B.; Pirkul, H.
Zero-one integer programs with few constraints - lower bounding theory.
GAVI 1985-3 Gavish, B.; Pirkul, H.
Zero-one integer programs with few constraints - efficient branch and bound algorithms.
GRANA 1984-1 Granot, D.; Granot, F.
Minimal covers, minimal sets and canonical facets of the posynomial knapsack polytope.
HAMMD 1987-1 Hammer, P. L.; Simeone, B.
Order relations of variables in 0-1 programming.
HAMMD 1987-2 Hammer, P. L.; Simeone, B.
Quadratic functions of binary variables.
HANSF 1987-1 Hansen, P.; Lu, S. H.
On the equivalence of paved-duality and standard linearization in 0-1 optimization.
HARN 1984-1 Harnett, R. M.
A single-branch implicit enumeration algorithm for zero-one programs with geometrical constraints.
HOFFD 1985-1 Hoffman, K.; Padberg, M. W.
LP-based combinatorial problem solving.
HOFFD 1986-1 Hoffman, K.; Padberg, M. W.
LP-based combinatorial problem solving.
ICHIB 1984-1 Ichino, M.; Sklansky, J.
Optimum feature selection by zero-one integer programming.
JOHNI 1985-1 Johnson, E. J.; Kostreva, M. M.; Suhl, U. H.
Solving 0-1 integer programming problems arising from large scale planning models.
JORG 1987-1 Jorgensen, C.; Powell, S.
Solving 0-1 minimax problems.

KALAB 1987-1 Kalantari, B.; Rosen, J. B.
Penalty formulation for zero-one nonlinear programmming.
KOVA 1984-1 Kovacs, L. B.; Boros, E.; Inotay, F.
A two-stage mathematical model and interactive program system for planning networks of sewer systems and waste water treatment plants - with application to the Lake Balaton area (Hungarian).
KRAH 1986-1 Krahl, R.
Surrogate constraints und Schnittverfahren in der linearen 0-1-Optimierung.
LEE M 1987-1 Lee, S.-M.; Luebbe, R. L.
A zero-one goal-programming algorithm using partitioning and constraint aggregation.
LEVA 1984-1 Levary, R. R.
An experimental sequential solution procedure to stochastic linear programming problems with 0-1 variables.
LIEP 1986-1 Lieponite-Yukhimenko, B. I.
Sequential design algorithm for solving linear programming problems with Boolean variables (Russian).
LU 1987-1 Lu, S. H.; Simeone, B.
On the equivalence of roof-duality and paved-duality in quadratic 0-1 optimization.
LU A 1985-1 Lu, S. H.; Williams, A. C.
Roof duality for nonlinear 0-1 optimization.
LU A 1987-1 Lu, S. H.; Williams, A. C.
Roof duality and linear relaxation for quadratic and polynomial 0-1 optimization.
LU A 1987-2 Lu, S. H.; Williams, A. C.
Roof duality for polynomial 0-1 optimization.
LYSK 1985-1 Lyska, W.
Ein kombinatorischer Algorithmus zur Lösung der linearen 0-1 Vektoroptimierungsaufgabe.
MAGA 1986-1 Maga, F.; Vizvari, B.
The relaxation of a special polynomial zero-one programming problem to a set covering problem (Hungarian).
MAMEA 1986-1 Mamedov, K. S.; Shikhaliev, N. S.
Application of the subgradient descent method for approximation solution of a Boolean programming problem with an estimate of the deviation from the optimum (Russian).
MARTE 1985-1 Martin, R. K.
Generating alternative mixed - 0/1 linear programming models using variable redefintion (revised version).
MARTG 1985-1 Martin, R. K.; Schrage, L.
Subset coefficient reduction cuts for 0/1 mixed-integer programming.
MATI 1987-1 Matitskas, I.-K. L.; Palubetskis, G. S.
Graph cutting and 0-1 quadratic programming (Russian).
MATVA 1984-1 Matveev, V. V.; Shcherbina, O. A.
On one class of local algorithm for solving quasi-unit problems of integer linear programming (Russian).
MEHR 1985-1 Mehrez, A.
A resource allocation problem with the aid of an integer zero-one programming model in the plastic industry.
MULLB 1987-1 Müller, P.
Untersuchungen zu speziellen linearen gemischt-ganzzahligen 0-1-Optimierungsaufgaben.
OHTA 1985-1 Ohtake, Y.; Nishida, N.
A branch-and-bound algorithm for 0-1 parametric mixed integer programming.
ONUR 1987-1 Onur, L.; Fabrycky, W. J.
An input/output control system for the dynamic job shop.
PADBC 1985-1 Padberg, M. W.; Vanroy, T. J.; Wolsey, L. A.
Valid linear inequalities for fixed charge problems.
PUTU 1983-1 Puturidze, Z. S.
A modified algorithm for constructing a sequence of schemes for the solution of discrete optimization problems (Russian).
RASM 1986-1 Rasmussen, L. M.
Zero-one programming with multiple criteria.
REDD 1984-1 Reddy, K. P.; Mittal, A. K.; Gupta, S. K.
Bivalent quadratic programming problem - a computational study.
SCHNA 1987-1 Schniederjans, M. J.; Kim, G. C.
A goal programming model to optimize departmental preference in course assignments.
SERGH 1987-1 Sergienko, I. V.; Shilo, V. P.
Polynomial asymptotically-optimal random search algorithm for a multidimensional problem of Boolean linear programming (Russian).
SERGH 1987-2 Sergienko, I. V.; Shilo, V. P.
Probabilistic approach to evaluate heuristics for problems of the Boolean linear programming (Russian).
SGURE 1982-1 Sgurev, V. S.; Vasilev, V. S.; Dzhelatova, M. N.
An experimental investigation of algorithms solving linear problems with binary variables (Bulgarian).
SGURF 1982-1 Sgurev, V. S.; Vasilev, V. S.; Ivanov, S. S.; Dzhelatova, M. N.
Algorithmic maintenance of a packet of application programmes solving optimization problems with binary variables (Bulgarian).
SHCH 1985-1 Shcherbina, O. A.
Local algorithms for block-tree-like problems of discrete programming.
SHCH 1986-2 Shcherbina, O. A.
A unimodular problem of integer programming.
SHOR 1985-1 Shor, N. Z.; Davydov, A. S.
On the method of obtaining estimates in quadratic extremal problems with Boolean variables.
SIME 1985-1 Simeone, B.
Consistency of quadratic boolean equations and the König-Egervary property for graphs.
SMIT 1987-1 Smith-Daniels, D. E.; Smith-Daniels, V. L.
Optimal project scheduling with materials ordering.
STET 1985-1 Stetsenko, S. I.
Numerical experiments in the solution of quadratic problems with Boolean variables (Russian).

TCHA 1985-1 Tcha, D.-W.
On the selection of primary paths for a communication network.
TRIP 1984-1 Tripathy, A.
School timetabling - a case in large binary integer linear programming.
VANRA 1986-1 Vanroy, T. J.; Wolsey, L. A.
Valid inequalities for mixed 0-1 programs.
WILL 1985-1 Williams, A. C.
Quadratic 0-1 programming using the proof dual - with computational results.
ZABLA 1983-1 Zablotskaya, O. A.; Kolokolov, A. A.
Completely regular truncations in Boolean programming (Russian).
ZIMMA 1984-1 Zimmermann, H.-J.; Pollatschek, M. A.
Fuzzy 0-1 linear programs.

```
PPPPPPPPPPP     AAAAAAAAAA    RRRRRRRRRRR    TTTTTTTTTTTT              3333333333
PPPPPPPPPPPP   AAAAAAAAAAAA   RRRRRRRRRRRR   TTTTTTTTTTTT             333333333333
PP        PP   AA        AA   RR        RR        TT                  33        33
PP        PP   AA        AA   RR        RR        TT                            33
PP        PP   AA        AA   RR        RR        TT                            33
PPPPPPPPPPPP   AAAAAAAAAAAA   RRRRRRRRRRRR        TT                          3333
PPPPPPPPPPP    AAAAAAAAAAAA   RRRRRRRRRRR         TT                          3333
PP             AA        AA   RR     RR           TT                            33
PP             AA        AA   RR      RR          TT                            33
PP             AA        AA   RR       RR         TT                  33        33
PP             AA        AA   RR        RR        TT                  333333333333
PP             AA        AA   RR         RR       TT                   3333333333
```

LIST OF AUTHORS

Vol. 292: I. Tchijov, L. Tomaszewicz (Eds.), Input-Output Modeling. Proceedings, 1985. VI, 195 pages. 1987.

Vol. 293: D. Batten, J. Casti, B. Johansson (Eds.), Economic Evolution and Structural Adjustment. Proceedings, 1985. VI, 382 pages. 1987.

Vol. 294: J. Jahn, W. Krabs (Eds.), Recent Advances and Historical Development of Vector Optimization. VII, 405 pages. 1987.

Vol. 295: H. Meister, The Purification Problem for Constrained Games with Incomplete Information. X, 127 pages. 1987.

Vol. 296: A. Börsch-Supan, Econometric Analysis of Discrete Choice. VIII, 211 pages. 1987.

Vol. 297: V. Fedorov, H. Läuter (Eds.), Model-Oriented Data Analysis. Proceedings, 1987. VI, 239 pages. 1988.

Vol. 298: S.H. Chew, Q. Zheng, Integral Global Optimization. VII, 179 pages. 1988.

Vol. 299: K. Marti, Descent Directions and Efficient Solutions in Discretely Distributed Stochastic Programs. XIV, 178 pages. 1988.

Vol. 300: U. Derigs, Programming in Networks and Graphs. XI, 315 pages. 1988.

Vol. 301: J. Kacprzyk, M. Roubens (Eds.), Non-Conventional Preference Relations in Decision Making. VII, 155 pages. 1988.

Vol. 302: H.A. Eiselt, G. Pederzoli (Eds.), Advances in Optimization and Control. Proceedings, 1986. VIII, 372 pages. 1988.

Vol. 303: F.X. Diebold, Empirical Modeling of Exchange Rate Dynamics. VII, 143 pages. 1988.

Vol. 304: A. Kurzhanski, K. Neumann, D. Pallaschke (Eds.), Optimization, Parallel Processing and Applications. Proceedings, 1987. VI, 292 pages. 1988.

Vol. 305: G.-J.C.Th. van Schijndel, Dynamic Firm and Investor Behaviour under Progressive Personal Taxation. X, 215 pages. 1988.

Vol. 306: Ch. Klein, A Static Microeconomic Model of Pure Competition. VIII, 139 pages. 1988.

Vol. 307: T.K. Dijkstra (Ed.), On Model Uncertainty and its Statistical Implications. VII, 138 pages. 1988.

Vol. 308: J.R. Daduna, A. Wren (Eds.), Computer-Aided Transit Scheduling. VIII, 339 pages. 1988.

Vol. 309: G. Ricci, K. Velupillai (Eds.), Growth Cycles and Multisectoral Economics: the Goodwin Tradition. III, 126 pages. 1988.

Vol. 310: J. Kacprzyk, M. Fedrizzi (Eds.), Combining Fuzzy Imprecision with Probabilistic Uncertainty in Decision Making. IX, 399 pages. 1988.

Vol. 311: R. Färe, Fundamentals of Production Theory. IX, 163 pages. 1988.

Vol. 312: J. Krishnakumar, Estimation of Simultaneous Equation Models with Error Components Structure. X, 357 pages. 1988.

Vol. 313: W. Jammernegg, Sequential Binary Investment Decisions. VI, 156 pages. 1988.

Vol. 314: R. Tietz, W. Albers, R. Selten (Eds.), Bounded Rational Behavior in Experimental Games and Markets. VI, 368 pages. 1988.

Vol. 315: I. Orishimo, G.J.D. Hewings, P. Nijkamp (Eds.), Information Technology: Social and Spatial Perspectives. Proceedings, 1986. VI, 268 pages. 1988.

Vol. 316: R.L. Basmann, D.J. Slottje, K. Hayes, J.D. Johnson, D.J. Molina, The Generalized Fechner-Thurstone Direct Utility Function and Some of its Uses. VIII, 159 pages. 1988.

Vol. 317: L. Bianco, A. La Bella (Eds.), Freight Transport Planning and Logistics. Proceedings, 1987. X, 568 pages. 1988.

Vol. 318: T. Doup, Simplicial Algorithms on the Simplotope. VIII, 262 pages. 1988.

Vol. 319: D.T. Luc, Theory of Vector Optimization. VIII, 173 pages. 1989.

Vol. 320: D. van der Wijst, Financial Structure in Small Business. VII, 181 pages. 1989.

Vol. 321: M. Di Matteo, R.M. Goodwin, A. Vercelli (Eds.), Technological and Social Factors in Long Term Fluctuations. Proceedings. IX, 442 pages. 1989.

Vol. 322: T. Kollintzas (Ed.), The Rational Expectations Equilibrium Inventory Model. XI, 269 pages. 1989.

Vol. 323: M.B.M. de Koster, Capacity Oriented Analysis and Design of Production Systems. XII, 245 pages. 1989.

Vol. 324: I.M. Bomze, B.M. Pötscher, Game Theoretical Foundations of Evolutionary Stability. VI, 145 pages. 1989.

Vol. 325: P. Ferri, E. Greenberg, The Labor Market and Business Cycle Theories. X, 183 pages. 1989.

Vol. 326: Ch. Sauer, Alternative Theories of Output, Unemployment, and Inflation in Germany: 1960–1985. XIII, 206 pages. 1989.

Vol. 327: M. Tawada, Production Structure and International Trade. V, 132 pages. 1989.

Vol. 328: W. Güth, B. Kalkofen, Unique Solutions for Strategic Games. VII, 200 pages. 1989.

Vol. 329: G. Tillmann, Equity, Incentives, and Taxation. VI, 132 pages. 1989.

Vol. 330: P.M. Kort, Optimal Dynamic Investment Policies of a Value Maximizing Firm. VII, 185 pages. 1989.

Vol. 331: A. Lewandowski, A.P. Wierzbicki (Eds.), Aspiration Based Decision Support Systems. X, 400 pages. 1989.

Vol. 332: T.R. Gulledge, Jr., L.A. Litteral (Eds.), Cost Analysis Applications of Economics and Operations Research. Proceedings. VII, 422 pages. 1989.

Vol. 333: N. Dellaert, Production to Order. VII, 158 pages. 1989.

Vol. 334: H.-W. Lorenz, Nonlinear Dynamical Economics and Chaotic Motion. XI, 248 pages. 1989.

Vol. 335: A. G. Lockett, G. Islei (Eds.), Improving Decision Making in Organisations. Proceedings. IX, 606 pages. 1989.

Vol. 336: T. Puu, Nonlinear Economic Dynamics. VII, 119 pages. 1989.

Vol. 337: A. Lewandowski, I. Stanchev (Eds.), Methodology and Software for Interactive Decision Support. VIII, 309 pages. 1989.

Vol. 338: J.K. Ho, R.P. Sundarraj, DECOMP: an Implementation of Dantzig-Wolfe Decomposition for Linear Programming. VI, 206 pages. 1989.

Vol. 339: J. Terceiro Lomba, Estimation of Dynamic Econometric Models with Errors in Variables. VIII, 116 pages. 1990.

Vol. 340: T. Vasko, R. Ayres, L. Fontvieille (Eds.), Life Cycles and Long Waves. XIV, 293 pages. 1990.

Vol. 341: G.R. Uhlich, Descriptive Theories of Bargaining. IX, 165 pages. 1990.

Vol. 342: K. Okuguchi, F. Szidarovszky, The Theory of Oligopoly with Multi-Product Firms. V, 167 pages. 1990.

Vol. 343: C. Chiarella, The Elements of a Nonlinear Theory of Economic Dynamics. IX, 149 pages. 1990.

Vol. 344: K. Neumann, Stochastic Project Networks. XI, 237 pages. 1990.

Vol. 345: A. Cambini, E. Castagnoli, L. Martein, P. Mazzoleni, S. Schaible (Eds.), Generalized Convexity and Fractional Programming with Economic Applications. Proceedings, 1988. VII, 361 pages. 1990.

Vol. 346: R. von Randow (Ed.), Integer Programming and Related Areas. A Classified Bibliography 1984–1987. XIII, 514 pages. 1990.